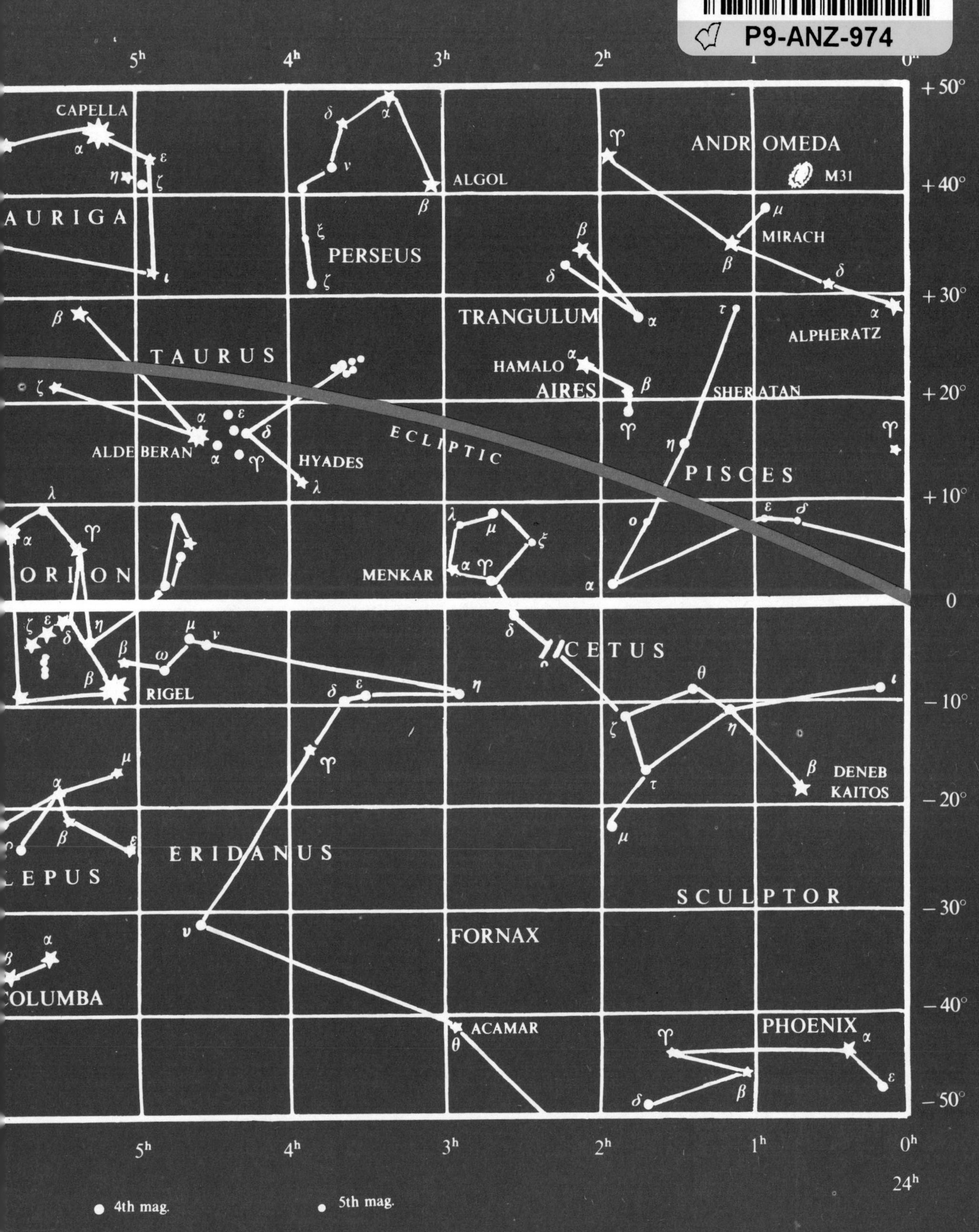

HART
12 HOURS
50°

ESSENTIALS OF ASTRONOMY

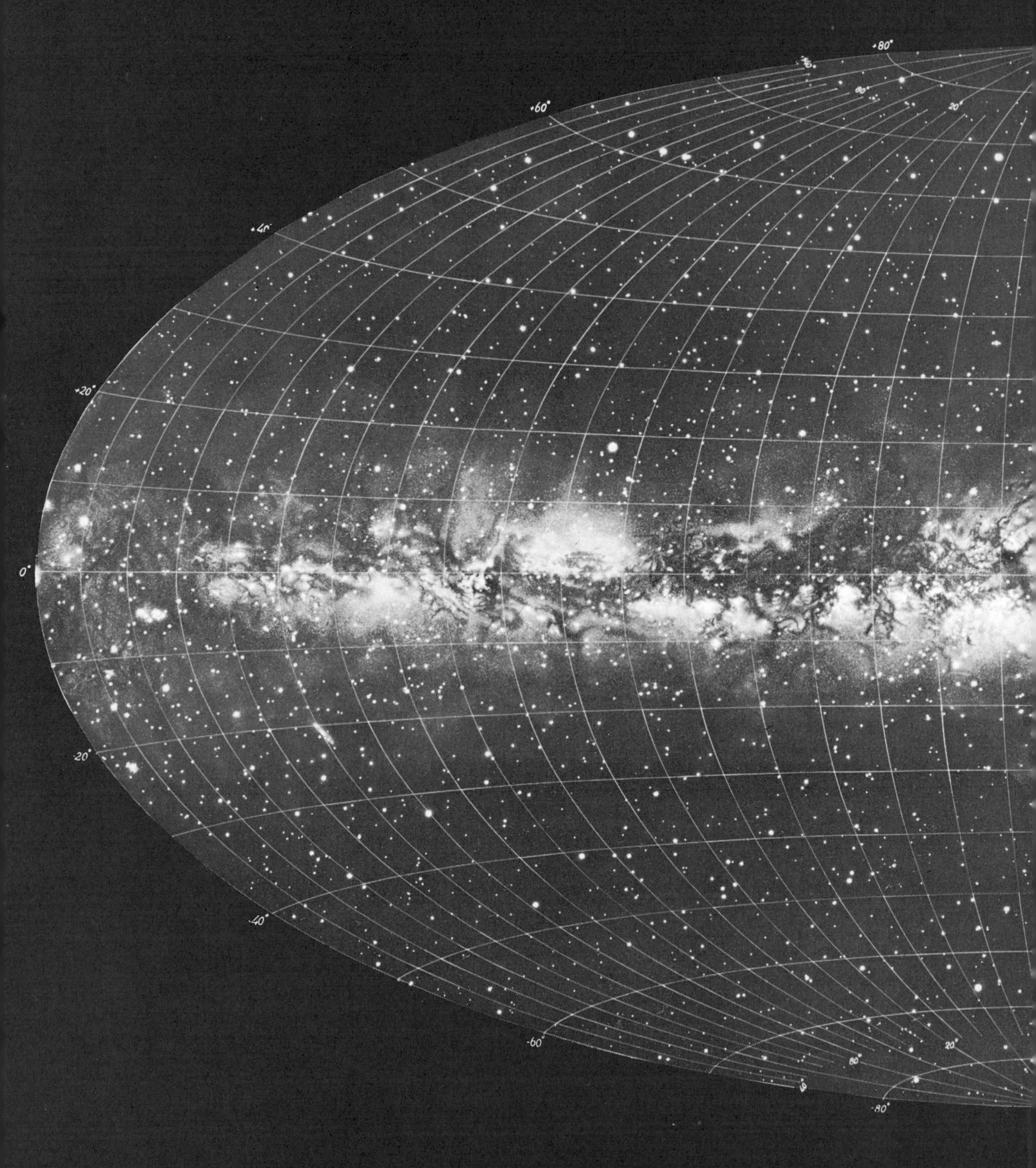

ESSENTIALS OF

LLOYD MOTZ *and* ANNETA DUVEEN, T.O.F.

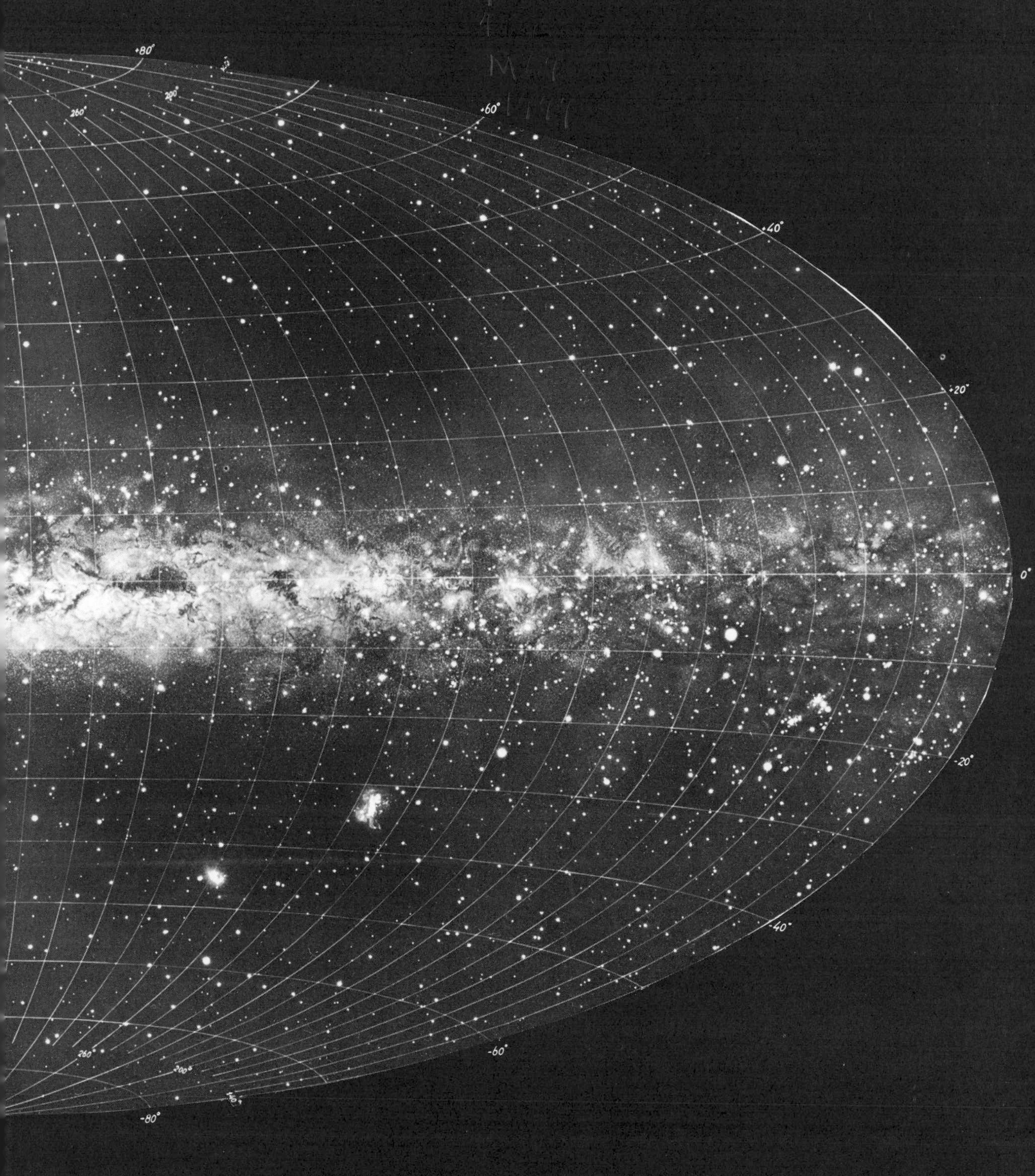

ASTRONOMY

SECOND EDITION

COLUMBIA UNIVERSITY PRESS
New York

Illustrations by Anneta Duveen, coauthor

Columbia University Press, New York and Guildford, Surrey.

First Edition, 1966
Second Edition, 1977

Library of Congress Cataloging in Publication Data

Motz, Lloyd, 1910–
Essentials of astronomy.

Bibliography: p.
Includes indexes.
1. Astronomy. I. Duveen, Anneta, joint author.
II. Title.
QB43.2.M67 1976 523 76-19068
ISBN 0-231-04009-1

Printed in the United States of America
10 9 8 7 6 5 4 3 2

Tables 2.1, 4.1, 4.2, 6.1, 12.2, 12.3, 12.4, 13.1, 13.2, 15.1, 15.2, 18.1, 21.1, 21.2, 23.1, 24.2, 24.3, 24.4, 24.5, 24.6, 25.1, 26.1, 26.3, 27.1, 27.2, 27.3, 27.4, 27.5, 27.6, 27.7, 29.1, 30.2, and 30.3 are excerpted by permission from C. W. Allen, Astrophysical Quantities, *Second Edition (London: The Athlone Press, University of London; New York: Oxford University Press, Inc., 1963).*

Frontispiece: A reproduction of the Milky Way by Martin and Tatjana Kesküla. Courtesy of the Lund Observatory.

Cover: The bending of light in a gravitational field.

PREFACE TO THE FIRST EDITION

This book is the expanded version of a two-semester introductory course in astronomy given at Columbia University over a period of years. It departs from traditional first-year astronomy texts in its emphasis on physical principles, a knowledge of which is indispensable to the study of modern astronomy. Today the college student is not content with an astronomy course that is primarily descriptive in nature; he wants to know not only the physical laws that lie behind the observational data but also how astronomers use these laws as guides to further discoveries. The understanding of stellar systems that we now have has resulted from this cooperation between physics and astronomy. For this reason physical laws are introduced and developed in as simple a manner as possible wherever they are required for the student's grasp of the astronomical material or wherever knowledge of these laws illuminates the text.

But physics has also gained from astronomy. The magnificent development of classical mechanics and its elegant presentations in the eighteenth and nineteenth centuries stemmed from the investigations into celestial mechanics during those centuries; and certainly our comprehension of gravity and of the whole field of potential theory owes much to astronomical research. Similarly, during the present century the inquiries of astronomers have resulted in substantial contributions to the penetration of atomic processes, the laws of radiation, and the physics of space-time. In recent years astronomical research into stellar interiors has supplemented the work of nuclear physicists and has helped us in our study of the synthesis of heavy nuclei. Indeed, the stars are physical laboratories that we can never hope to reproduce here on earth, particularly as far as enormously high temperatures and highly condensed states of matter are concerned.

It is therefore necessary for students to learn the laws of physics right along with the descriptive material; such concepts as velocity, acceleration, force, mass, gravity, angular momentum, and energy must be assimilated, and we have introduced these topics and developed them as part of the text. In the same way we have introduced relativity theory, quantum theory, atomic structure, and nuclear structure as an integral part of the astronomical material.

We have presented these topics without the use of advanced mathematics, emphasizing the physical concepts at each point so that a student who has had no more than intermediate algebra can follow the material step by step. Whatever laws of trigonometry are required are presented in the text. For the student who has a background in calculus, the formulas as developed here can easily be re-expressed in differential form by replacing the small differences (indicated by Δ in the equations), wherever they appear, with differentials. We feel that the use of differences rather than differentials in physical formulas is advantageous from three points of view:

1. It enables the student who has a background only in algebra to follow the arithmetic that is involved;
2. It more nearly conforms to the actual physical processes involved, since actual calculations always deal with differences rather than differentials; and
3. It is possible to put even complicated differential equations, such as the stellar interior equations, the relativistic field equations, and the equations of model universes, into simple algebraic forms.

It is possible to derive most of the results of modern physics and astronomy by such algebraic means. Moreover, a differential involves passing to a mathematical limit, which is a very sophisticated concept, whereas a difference is a small quantity that has meaning even to the student with modest mathematical training.

For pedagogical reasons we repeat and redefine concepts after they have been first introduced and come into the text again. Throughout the text there are careful references to the original definition whenever a topic reappears. That this way of presenting a course in astronomy is pedagogically sound is borne out by the success it has enjoyed at Columbia University.

Our text is divided into five parts; four deal with the subject material of astronomy and the fifth (the Appendixes), among other things, deals with optical and radio astronomical instruments. Part I treats the solar system and includes two chapters on special and general relativity. Since the sun is properly a member of the solar system, a small chapter (dealing with those of its properties that are relevant to solar system astronomy) is introduced here as well. Part II deals with stellar properties and the structure and evolution of stars. In this connection we have discussed the sun throughout and introduced into the text all of its stellar characteristics. One chapter in this part is devoted to the solar atmosphere. This section of the book also introduces those phases of modern physics, such as thermodynamics, quantum theory, atomic structure, and nuclear forces, that are essential to modern astronomy. Part III is concerned with stellar systems, with special emphasis on statistical astronomy and the structure of galaxies. Part IV is devoted to cosmology and the model universes.

Because radio astronomy has enjoyed such an enormous growth in the last decade, we have placed it on an equal footing with optical astronomy, and have woven the threads of its discoveries into the fabric of the text wherever it has made pertinent contributions.

We are grateful to a number of people who have aided us in various ways. Dr. Emilia P. Belserene of Rutherfurd Observatory, Columbia University, and Hunter College has been especially helpful in her penetrating and constructive criticisms, which grew out of her reading of the final manuscript. We are most grateful to Dr. Allen Hynek of the Dearborn Observatory, Northwestern University, for his meticulous reading of the entire text during its preparation and his many valuable recommendations. Thanks are also due Dr. Raymond T. Grenchick of Louisiana State University, who offered numerous suggestions that helped us in revising the final manuscript. Finally, special appreciation should be expressed to Carol Goslin and Charles Goslin for their assurance and technical suggestions during the preparation of the art, and to Irene Wilson, Arthur Hopkins, Robert Makla, and Benjamin Shain for their encouragement.

Lloyd Motz
Anneta Duveen, T.O.F.

PREFACE TO THE SECOND EDITION

When the first edition of this book was completed, such things as lunar landings, planetary missions, large radio telescopes, orbiting observatories, and the application of high-speed computers to astrophysical calculations were still being planned. The introduction of these scientific tools opened extensive new areas of research, and great quantities of observational data and numerous theoretical calculations were published. To keep the additions to *Essentials of Astronomy* to a reasonable length, we have usually limited ourselves to material that is broadly accepted and firmly established.

The areas that we have up-dated include lunar and planetary properties; the sun; the evolution of stars onto and away from the main sequence; the interstellar medium and interstellar molecules; variable stars; neutron stars, pulsars, X-ray sources, and black holes; quasars; the cosmic background radiation; and cosmology.

Since the new material is fairly extensive, most of it is gathered into a separate division at the end of the book, Part 5, with section headings that correspond to those in the main text. Sections in the main text for which there is new material in Part 5 are marked with an asterisk. To retain the continuity and overall structure of the book, we have made few revisions in the main body of the text, introducing them only where we found a contradiction between old and new data and where we felt additional clarification was required.

For assistance and cooperation with the new edition, we wish to thank especially Nicholas Panagokos, Margaret Ware, Les Sauer, Miles Waggoner, Dr. Thomas Gold, Robert Tilley, John Moore, and Maria Caliandro.

Lloyd Motz
Anneta Duveen, T.O.F.

CONTENTS

PART ONE THE SOLAR SYSTEM

PART TWO STELLAR PROPERTIES AND THE STRUCTURE OF STARS

PART ONE
THE SOLAR SYSTEM

I

COORDINATE SYSTEMS IN ASTRONOMY

Astronomy, unlike physics and chemistry, is an observational rather than an experimental science, since it deals with objects at such great distances as to be beyond the reach of experimentation. Because the astronomer must obtain his information by observing celestial objects in all parts of the sky, he must introduce an accurate method of cataloguing their positions in order to keep track of them. He does this by superimposing on the sky what the mathematicians call a **coordinate system**, or a **frame of reference.**

A coordinate system is a geometrical lattice or framework, whether superimposed by the astronomer on the sky, by the mathematician on abstract surfaces, or by the physicist on his laboratory to study a series of events taking place in his apparatus. Whatever the application, all coordinate systems have certain fundamental features in common.

1.1 The One-Dimensional Coordinate System

Since the purpose of the coordinate system is to enable us to locate objects or events with respect to some arbitrary but convenient reference point (the **origin**) and a system of lines (the **coordinate axes**), we first consider the concept of the origin of a coordinate system.

Although, in general, the origin alone does not define a coordinate system, there exists a very simple case in which it does. Consider the straight line passing through the point O, and assume that the entire universe of objects consists of points on this line. This situation defines a linear universe, and any point on this line (say O) may be chosen as the origin (starting point). The location of any point in this linear universe can be determined by measuring its distance from the origin (O). *Since the position of any object can be located by means of a single number, this universe is called a one-dimensional manifold or a one-dimensional domain.* If the distances on one side of O are taken as positive, and the distances on the other as negative, the line is covered completely by a one-dimensional coordinate system, consisting of all the real numbers (Figure 1.1).

1.2 Two-Dimensional Coordinate Systems in a Plane

To locate objects that lie outside a given line—for example, all the points in a **plane**—the concept of the coordinate system must be extended from one to two dimensions. The reason is that after choosing any line (Figure 1.2) on a plane, there is still one other line on the plane that can be drawn perpendicular to the given line through any point (P) on the given line. We see that only one such perpendicular line can be drawn (the dotted line represents the nonperpendicular lines). In other words, only two independent (that is, mutually perpendicular) directions can be drawn through any point on a plane. For this reason *the coordinate system on a plane must be two-dimensional—that is, must involve two sets of numbers.*

To introduce a coordinate system on a plane, we first designate some point O as the origin. To locate an object on the plane we must have more information than just the distance from the object to O. The mere statement that an object is 5 inches

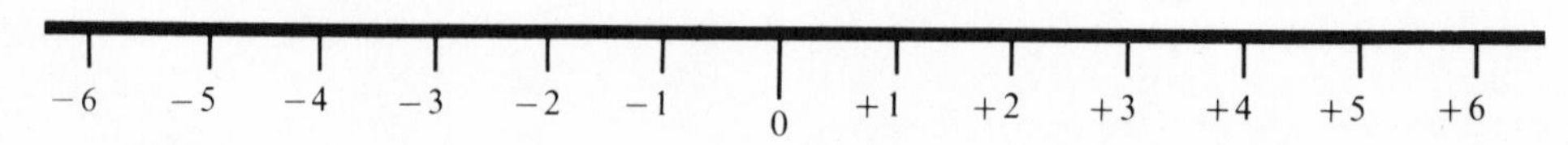

FIG. I–I *One-dimensional coordinate system.*

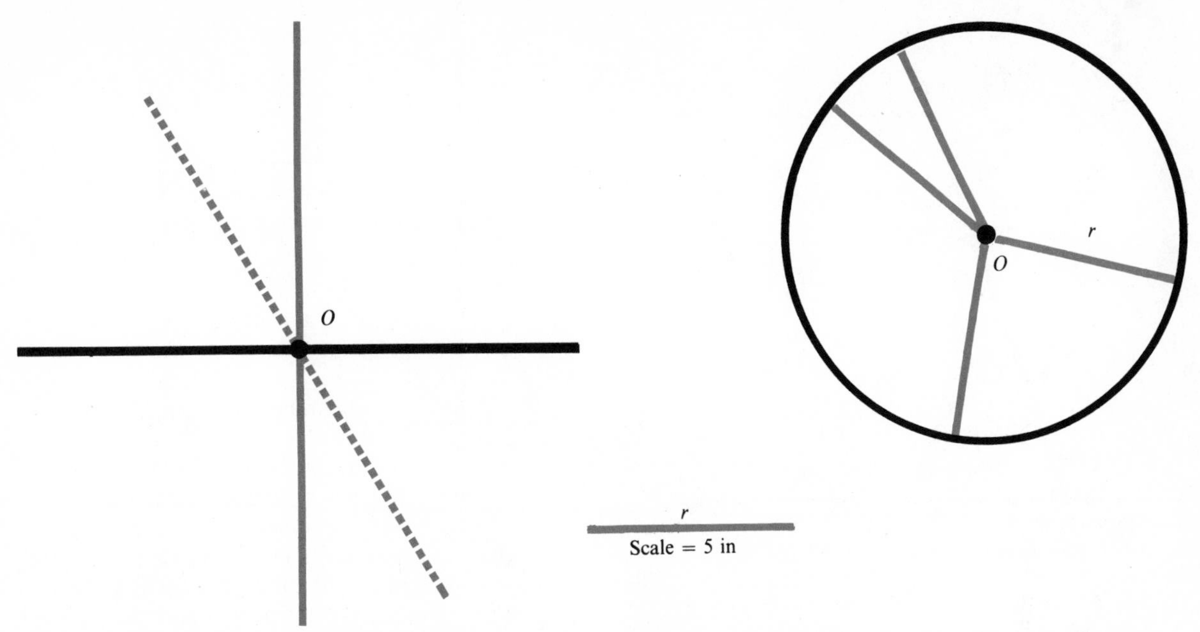

FIG. 1–2 *Two-dimensional coordinate system showing that there are only two independent directions on a plane.*

FIG. 1–3 *The position of a point whose distance from the origin O is r may lie anywhere on the circumference of a circle.*

from point O on the plane refers to an infinitude of points, all of which lie on the circumference of a circle having a radius of 5 inches, and O as its center (Figure 1.3).

1.3 Polar Coordinates in a Plane

Besides the distance from the point O we must specify another number in order to locate the given point on the circumference. One simple method is first to choose some fixed direction in the plane and to pass a line through O (line OX in Figure 1.4) in this direction. We take this as the reference line in our coordinate system. The object P is now located with respect to the origin O and the line OX by specifying two numbers:

1. the distance from the point O to P, and
2. the amount through which we must turn from the direction OX in order to face the direction OP.

This amount of turning (rotation), performed in a counterclockwise direction, is defined as the angle formed by the two lines OX and OP. *These two numbers, a distance and an angle, are called the polar coordinates of P relative to O and OX.*

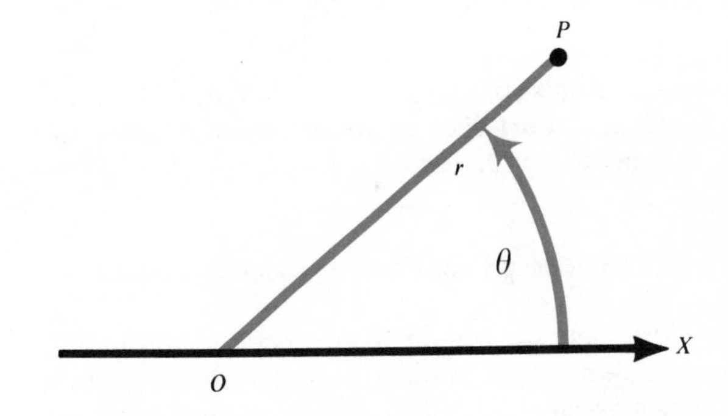

FIG. 1–4 *Polar coordinate system on a plane, showing the two coordinates r and θ of a point.*

1.4 Cartesian Coordinate System in a Plane

Another method of introducing a coordinate system in a plane is to construct a network composed of two sets of intersecting parallel lines which can be arranged to cut at some desired angle (Figure 1.5). If we choose any two lines as reference lines (OX and OY), the position of any object, such as the point P, is specified by giving the distances of this point from the lines OX and OY, as measured along the two lines (parallel to OX and OY, respectively) that intersect at P. *These two distances are*

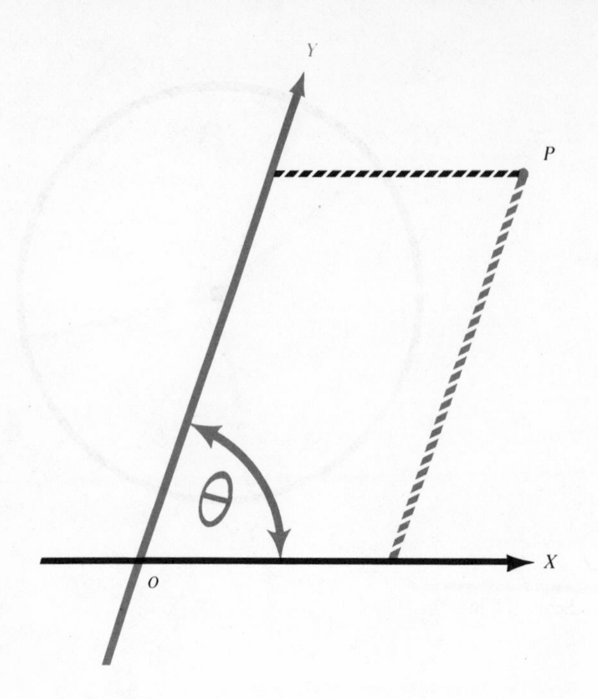

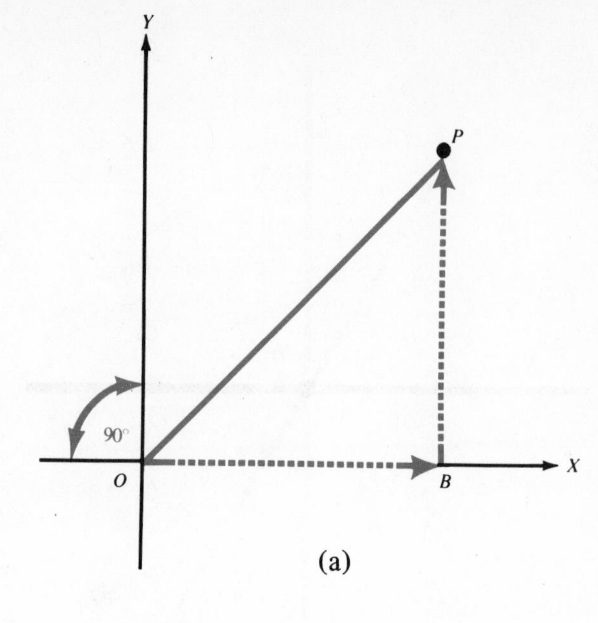

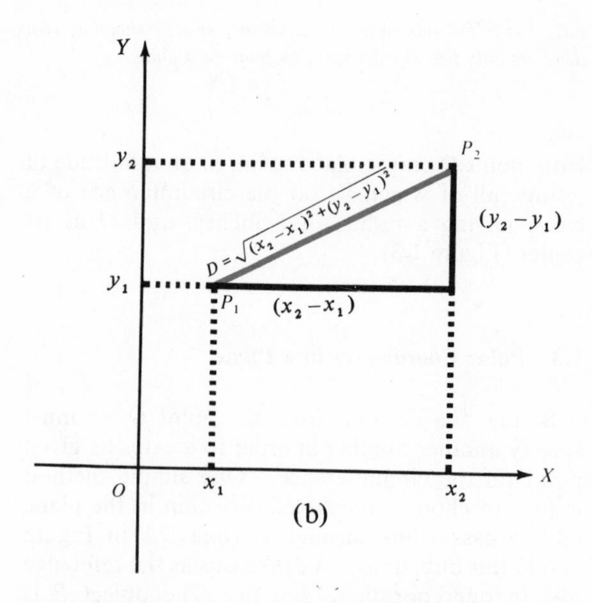

FIG. 1–5 *An oblique Cartesian coordinate system in which the reference lines are not orthogonal (perpendicular) to each other.*

FIG. 1–6 *(a) A rectangular Cartesian coordinate system. (b) Diagram illustrating the formula for the distance between two points in a Cartesian coordinate system.*

the coordinates of the point. A system of this sort is called a **Cartesian coordinate system,** after the mathematician Descartes.

1.5 Curvilinear Coordinate Systems in a Plane

Although straight lines are used in Figure 1.5, other families of parallel curves such as ellipses, circles, parabolas, and so on may be used to obtain what mathematicians call a curvilinear coordinate system. A polar coordinate system is an example.

1.6 Rectangular Coordinate System

Although in setting up the type of coordinate system shown in Figure 1.5 we may assign any value we wish to the angle θ, we obtain the simplest and most convenient system when the intersecting lines are mutually perpendicular. This is called a **rectangular coordinate system.** In a rectangular coordinate system the distance between any two points can be expressed by means of a very simple formula involving only the differences between the coordinates of the two points, as shown in Figure 1.6.

We see that the east-west direction of OX and the north-south direction of OY are independent of each other. No amount of advancing along OX results in any motion along the direction of OY, but a displacement along any arbitrary direction OP may be described as the sum of the displacements OB in the eastward direction and BP in the northward direction, as shown in Figure 1.6(a).

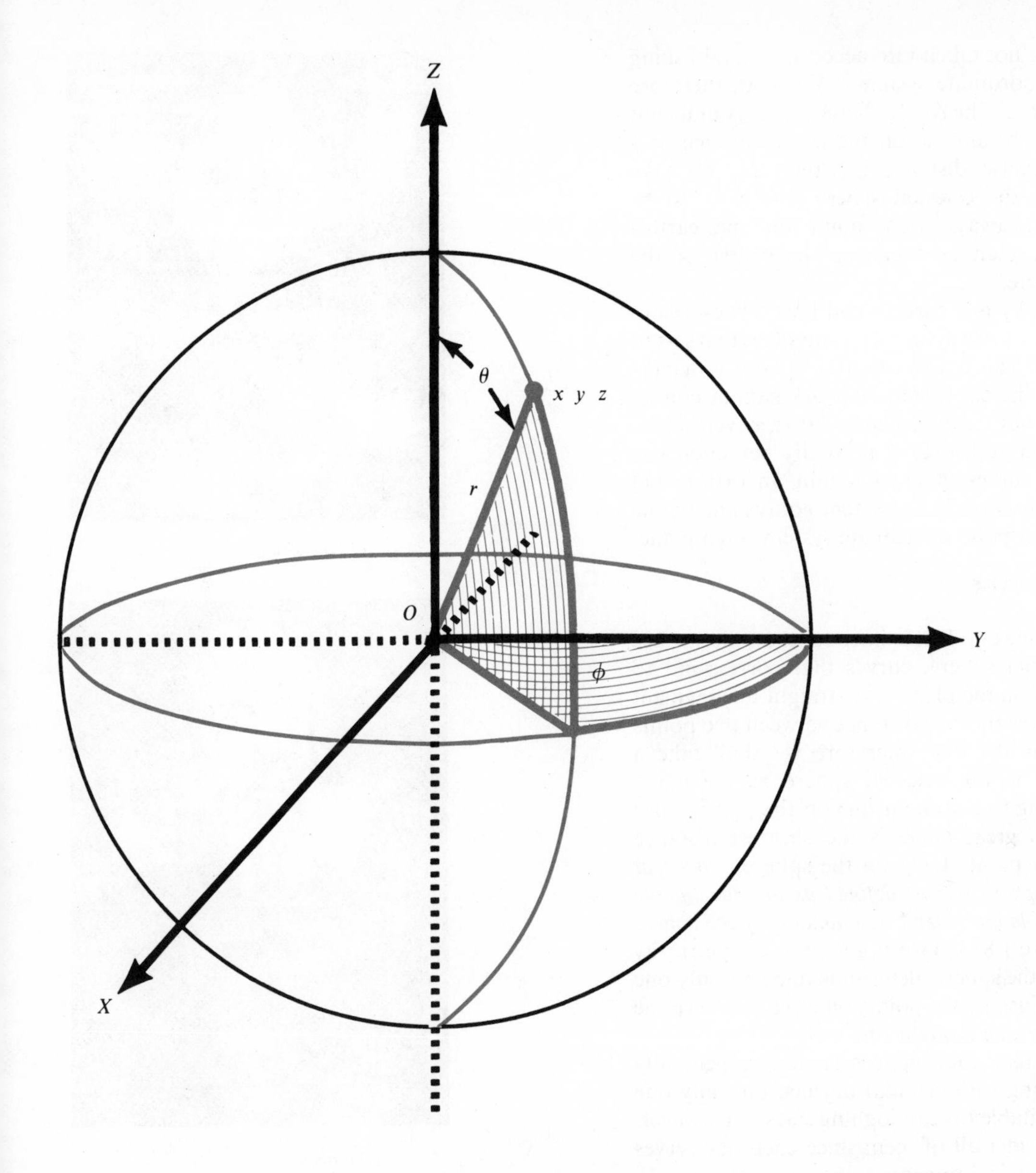

FIG. 1–7 *Two types of three-dimensional coordinate systems superimposed: Cartesian coordinate system, x, y, z; spherical coordinate system, r, θ, ϕ.*

1.7 Three-Dimensional Coordinate System

If we leave the plane and consider all of space, we see that two numbers no longer suffice to locate an object. At any moment of time three numbers or coordinates are required to locate each object in space. *Therefore, space is said to be three-dimensional.* We may now proceed as above and introduce three sets of mutually perpendicular lines to obtain a three-dimensional coordinate system, as shown in Figure 1.7. We may also use the three numbers r, θ, ϕ, the so-called spherical coordinates, where r is the distance of P from the origin and θ and ϕ are the two angles that give the position of P on the sphere. The angles θ and ϕ are called the co-latitude and longitude of P.

1.8 Coordinate Systems on the Sky

Since the main consideration in cataloguing celestial objects is their positions on the sky, their distances are not taken into account in establishing a celestial coordinate system. We shall therefore assume that all heavenly bodies are equidistant from the earth, and lie on the inside surface of a sphere. Because distance is irrelevant, we take this surface, the **celestial sphere** (the sky), to be infinitely far away. Any point on the earth's surface may then be taken as the center of the celestial sphere.

Since the sky is a surface and hence two-dimensional, a coordinate system on it involves two sets of numbers. Because the celestial sphere is a curved surface, the coordinate network cannot consist of straight lines, as on a plane. It is, nevertheless, possible to lay off a set of mutually perpendicular intersecting curves, and to obtain an orthogonal (right-angled) coordinate system equivalent to the rectangular or polar coordinate systems on a plane.

Tycho Brahe.

1.9 Great Circles

What we must do first is introduce, on the surface of the celestial sphere, curves that correspond to straight lines on the plane. A straight line segment on a plane is the shortest distance between two points connected by the line; therefore we shall take a **great circle** on the celestial sphere as the curve corresponding to a straight line on the plane, since the arc of a great circle is the shortest distance between two points lying on the sphere. *A great circle on a sphere is now defined as one having the largest possible diameter* (*the diameter of the sphere itself*) (Figure 1.8). Two points, not diametrically opposite on the sphere, determine one and only one great circle, just as two points on a plane determine one and only one straight line.

Four different coordinate systems are generally used in dealing with celestial objects, but only one of these is suitable for cataloguing stars. However, we shall consider all of them since each one serves some astronomical purpose.

Galileo.

Johannes Kepler.

William Herschel.

Isaac Newton.

Galileo's drawings of the satellites of Jupiter.

Copernicus, by Arthur Szyk.

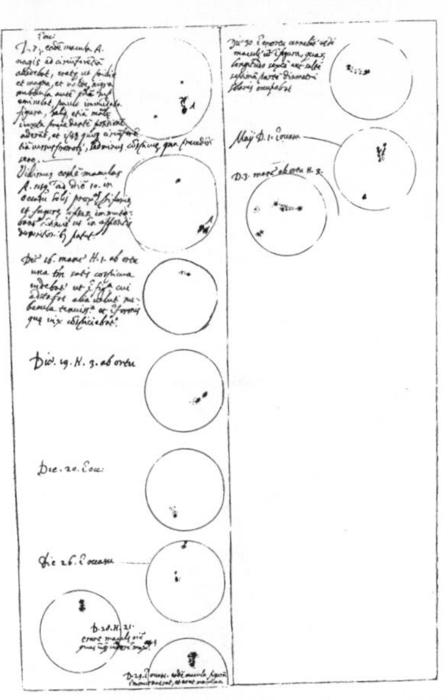

Galileo's drawings of sunspots.

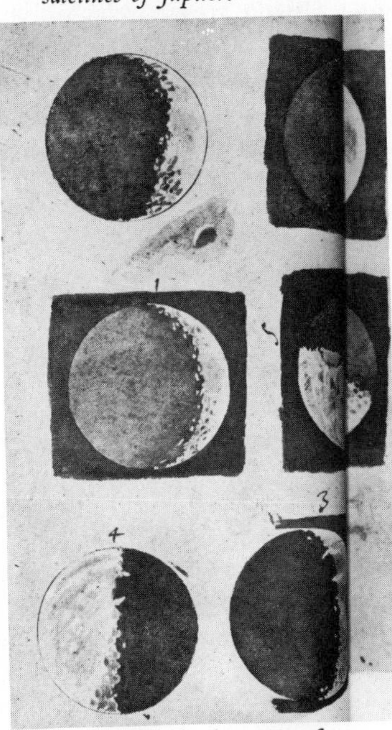

Galileo's drawings of mountains on the moon.

Einstein.

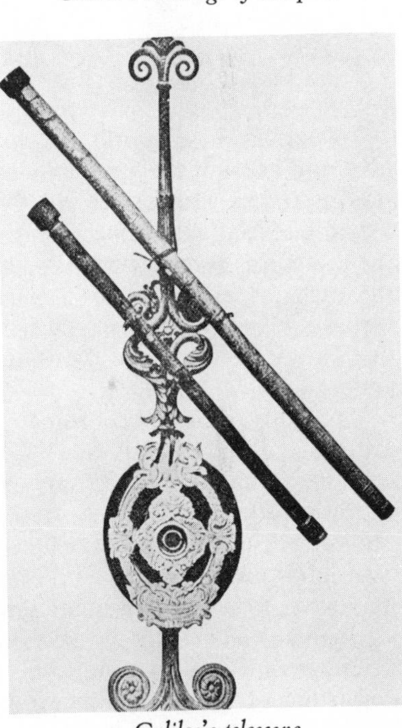

Galileo's telescope.
From the telescope by Louis Bell.

Halley's comet in 1066, as depicted on the Bayeux tapestry.

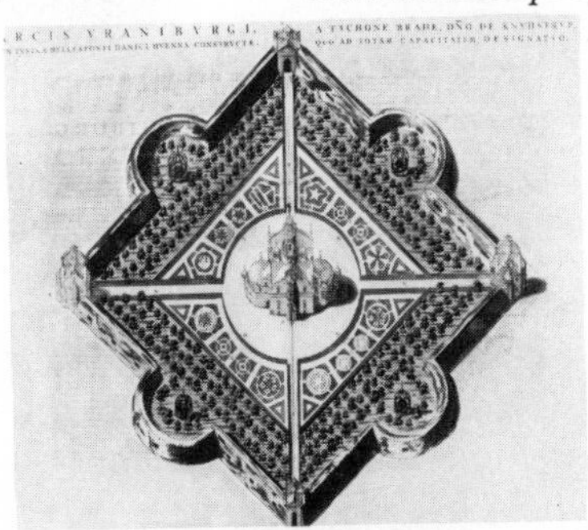

Tycho Brahe's observatory (about 1584).

Illustrations on pages 8 and 9 courtesy of Yerkes Observatory.

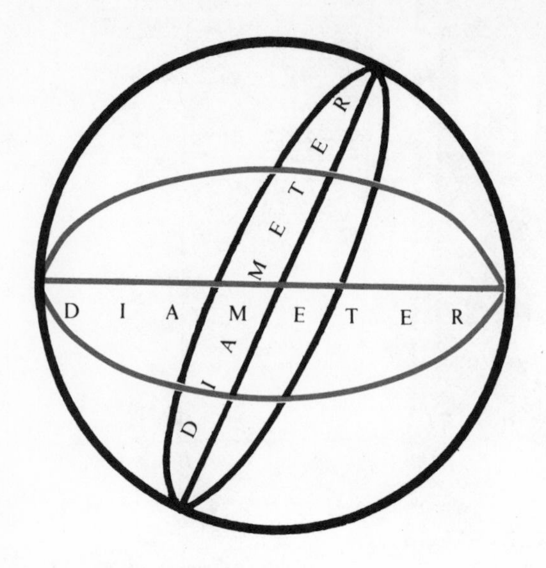

FIG. 1–8 *The diameters of great circles are equal to the diameter of the sphere on the surface of which they are drawn.*

To discuss these coordinate systems we have to take into account the apparent **diurnal** (daily) rotation of the sky, and since we are concerned with locating objects with respect to an observer fixed on the earth, it is convenient for the time being to take the earth as fixed and to assume that the sky is turning. This, of course, is a fiction, but a useful one in discussing basic astronomical coordinate systems.

To set up a celestial coordinate system, we must introduce fundamental great circles in the sky corresponding to the OX and OY lines of the Cartesian coordinate system on the plane, or to the radiating lines of the polar coordinate system. Since coordinate systems on the celestial sphere are analogous to polar coordinate systems on a plane, we shall again consider the polar coordinate system in a plane. This is an orthogonal coordinate system consisting of lines radiating from a point O (the origin) and a family of concentric circles having O as their center (the circles are thus everywhere perpendicular to the radiating lines). One of the radiating lines OX is chosen as a fundamental line of reference.

To establish such a system on the celestial sphere, we must therefore choose some point O on the sphere and a system of great circles passing through this point. These correspond to the system of radiating lines on the plane. Although the choice of this point is arbitrary, it is, in general, dictated by the use to which the coordinate system is put.

1.10 The Horizon Coordinate System

The horizon coordinate system is related to the direction of gravity at a point on the earth's surface. The observer at this point (Q) determines the origin of this coordinate system by constructing the plumb line (direction of gravity) from the center of the earth, C (assuming the earth is spherical), through Q, then upward until it pierces the celestial sphere at the point directly overhead for this observer. This point, Z, which is called the **zenith** of the observer, is chosen as the origin for a system of great circles that correspond to the radiating lines of the polar coordinate system on a plane. *The point Z applies to one observer only; it must be pictured as being attached to this observer, not to the celestial sphere.* As the observer moves about, the point Z moves about with him, taking the coordinate system with it, as shown in Figure 1.9.

If we now consider all the great circles on the celestial sphere that intersect at Z, we see that they also intersect at another point, N (the **nadir**), diametrically opposite Z on the celestial sphere. The nadir is the point at which the plumb line pierces the celestial sphere when extended downward.

1.11 Correspondence between the Horizon Coordinate System on the Sky and the Polar Coordinate System on the Plane

In Figure 1.10 we see how the radiating lines of the polar coordinate system on the plane can be made to correspond to the family of great circles on the sphere. We bring the plane and the sphere in contact so that the point Z on the sphere and the point O on the plane coincide. If we now draw straight lines from the nadir (N) until they pierce the plane, we see that each of these lines relates one point on the sphere to a single point on the plane. We note in particular that the great circles connecting the points Z and N on the sphere correspond to the straight lines radiating from the point O to infinity on the plane. We may note also that corresponding to the fundamental reference line OX on the plane there is a fundamental reference great circle ZP on the sphere, which may be arbitrarily chosen just as OX was.

Just as we see from the diagram (Figure 1.10) that the radiating lines of the polar coordinate system on the plane correspond to the great circles passing through Z on the celestial sphere, we see also that corresponding to the concentric circles in the polar coordinate system on the plane there is a

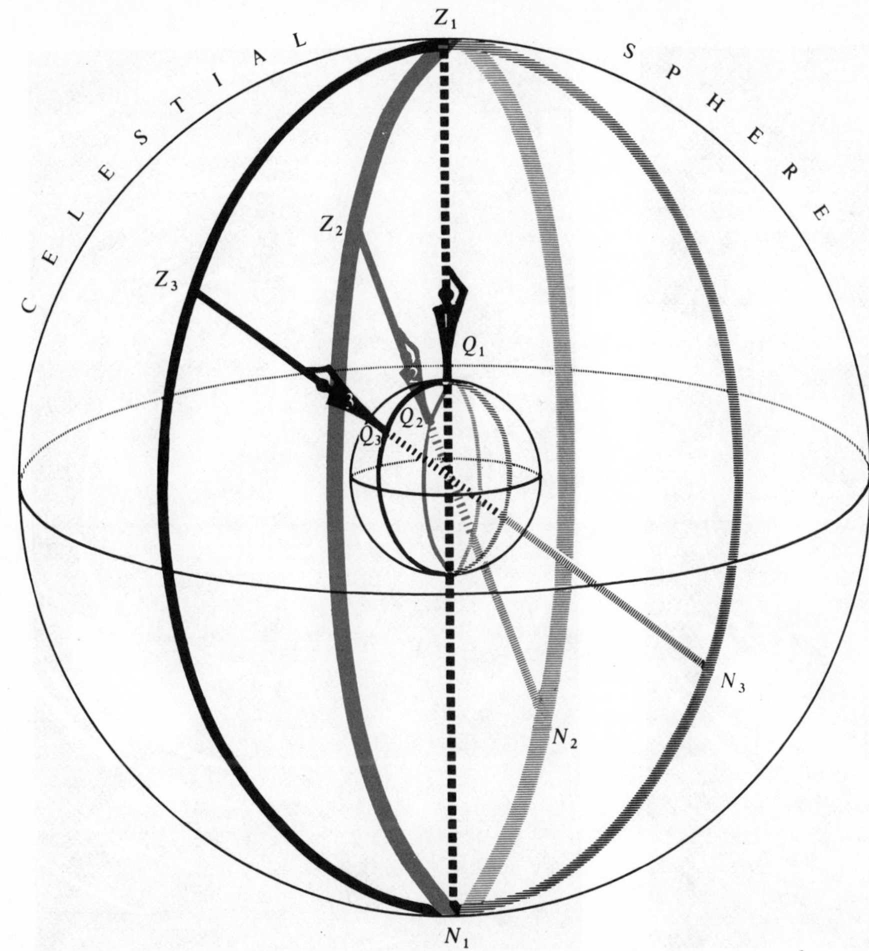

Fig. 1–9 *The observer's zenith on the celestial sphere changes as the observer moves from point to point on the earth.*

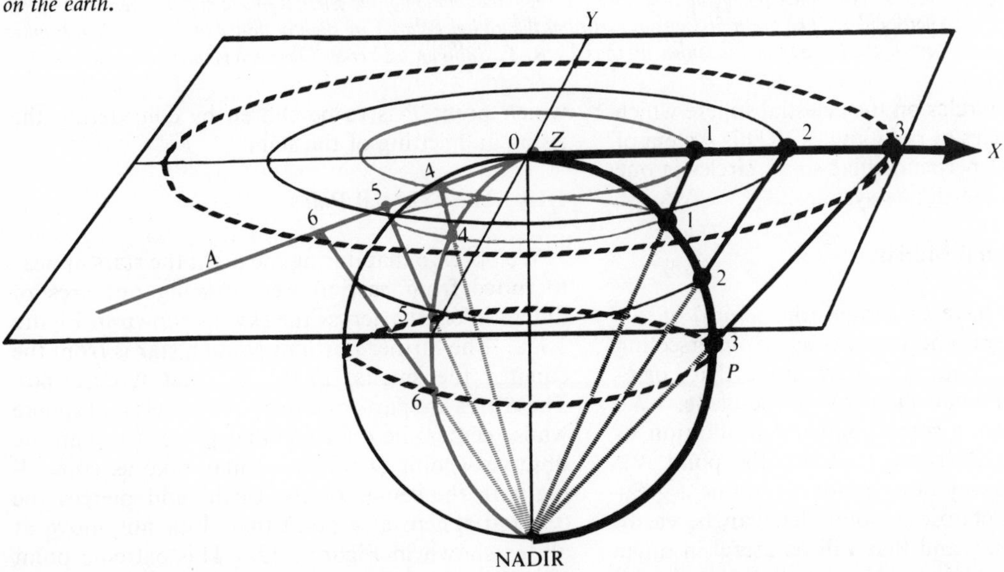

Fig. 1–10 *The correspondence between a polar coordinate system on a plane and on the celestial sphere.*

FIG. 1–11 *These photographs of star trails show the rotation of the celestial sphere and the precession of the north pole. The star trails (note in particular the trail of Polaris) are slightly shifted in the two photographs, one taken in 1925 and the other in 1935 by R. F. Sullivan. (Yerkes Observatory photographs)*

system of small circles on the celestial sphere which cut the great circles at right angles. This system of great circles and perpendicular small circles is one of our celestial coordinate systems.

1.12 The Celestial Meridian

Now that we have the origin (the zenith) of our coordinate system, and the two sets of intersecting curves, we must choose a great circle that corresponds to the reference line OX on the plane. We do this by choosing some point, P, in addition to Z on the celestial sphere (but not the point N). Although Z and any other point determine a great circle, we shall choose a point that can be easily located on the sky and that will be useful in other ways. The great circle ZP so determined is the observer's **celestial meridian**. It is easy to establish which point P is to be chosen by considering the rising and setting of the stars.

1.13 The Celestial Poles

We observe that during the night the stars appear to move from east to west, tracing out arcs of different lengths across the sky, as shown in Figure 1.11. The farther north or south a star is from the equator the smaller is the arc that it describes. These arcs are parts of circles on the celestial sphere whose centers lie on a line that passes through the observer (point Q which we may take as coinciding with the center of the earth) and pierces the celestial sphere at a point that does not move at all, as shown in Figure 1.12. This extreme point is called the **north celestial pole** (N.C.P.). The diametrically opposite point in the southern sky is

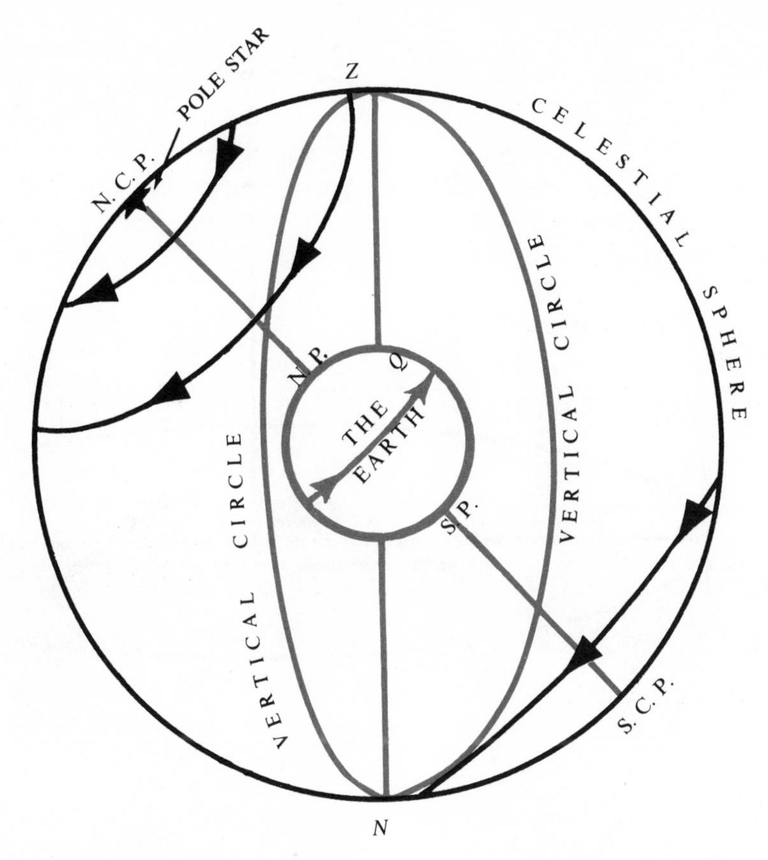

FIG. 1–12 *Diagram showing north and south celestial poles as determined by the rotation of the earth; the zenith of an observer and his vertical circles are also shown.*

the **south celestial pole** (**S.C.P.**), and the line from the N.C.P. to the S.C.P. coincides with the earth's axis of rotation. It is a line of points that have no motion as the sky turns. A star, Polaris, is now conveniently close to the north celestial pole so that by visual methods this point can be determined with sufficient accuracy for many purposes.

1.14 The Vertical Circles

The N.C.P. is the second point, P, used to determine the fundamental great circle of the **horizon coordinate system.** This fundamental great circle, determined by the two points Z and N.C.P., is called the celestial meridian of the observer. It passes through the four points Z, N.C.P., N, and S.C.P. *The celestial meridian is a **vertical circle**, the term given to any great circle passing through the zenith and nadir of the observer.* Vertical circles comprise the first set of coordinate lines of the horizon coordinate system and correspond to the radial lines of the polar coordinate system, the celestial meridian corresponding to OX. Note that all observers having the same terrestrial longitude on the earth have the same celestial meridian, although their respective zenith points differ.

1.15 Almucantars and the Horizon Great Circle

The small circles in the horizon coordinate system that correspond to the circles of the polar coordinate system on a plane are those circles on the celestial sphere that have their centers on the straight line ZQN (zenith, observer, nadir) and are perpendicular to the set of vertical circles. They are called **almucantars.** The almucantar having the diameter of the celestial sphere lies midway between the zenith and the nadir. This great circle, lying in the plane tangent to the earth at Q (the position of the observer), is called the **horizon.** *The horizon circle changes as the observer changes position.* (See Figure 1.13.)

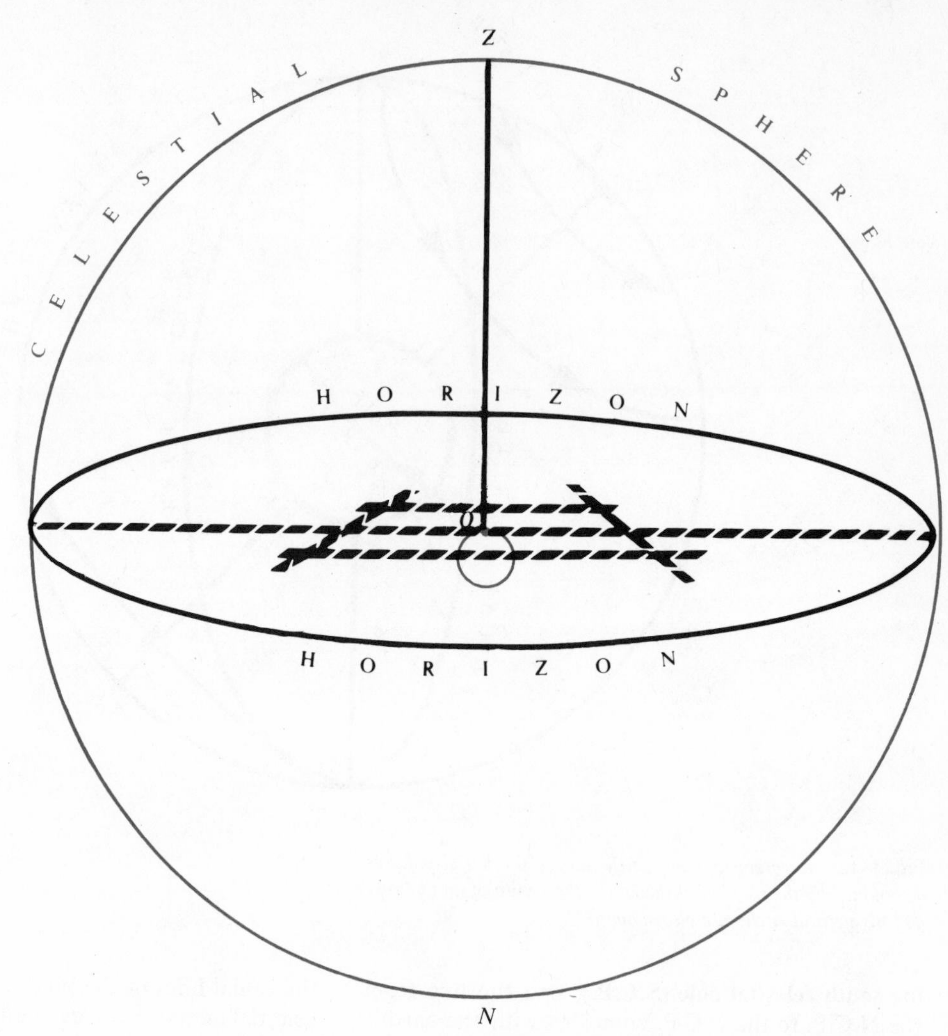

FIG. 1–13 *The zenith and the plane of the horizon of an observer.*

The two fundamental reference circles of the horizon coordinate system are:

1. the celestial meridian, and
2. the horizon.

We have introduced the horizon as another reference circle because it will be convenient to measure positions relative to it rather than relative to the zenith, although in complete analogy with the polar coordinate system on the plane we should work with the celestial meridian and the zenith.

1.16 The Cardinal Points of the Compass

The celestial meridian and the horizon intersect in two points 90° from the zenith. These are the **north** and **south points** of the observer. If the observer faces the south point and turns to his right along the horizon through an angle of 90° he faces the **west point**; if he turns 90° further to the right he faces the **north point**; and 90° still further to the right he faces the **east point**. These four points on the horizon define the **cardinal** points of the compass. When an object is on the celestial meridian, it is either north or south of the observer. Note also that all objects fixed on the celestial sphere spend as much time rising (moving from the eastern horizon to the celestial meridian) as they do setting (moving from the celestial meridian to the western horizon). The vertical circle passing through the zenith, east point, nadir, and west point is called the **prime vertical.** (See Figure 1.14.)

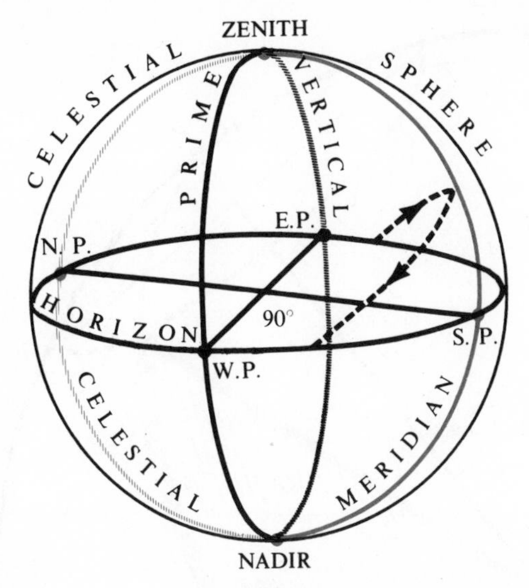

FIG. I–14 *Cardinal points for an observer. The dotted line is the path across the celestial sphere of an object rising and setting.*

1.17 The Azimuth

Since the celestial sphere is infinite, the two coordinates of any point on the celestial sphere will be expressed in angular measure. To find the coordinates of such a point (for example, some star) we first pass a vertical circle from our zenith through this point and note where this circle cuts the horizon, let us say at point M, as shown in Figure 1.15. Now face the south point and turn **westwardly**, measuring the arc SM along the horizon. This westwardly measured arc is called the **azimuth** of the star. It is the length of the arc (in angular measure) from the intersection (south point) of the celestial meridian and the horizon to the intersection M of the horizon and the vertical circle passing through P (object on the celestial sphere). The azimuth can also be defined as the angle SQM in the horizon plane, or the angle between the plane of the celestial meridian and the plane of the vertical circle passing through the star. It ranges from 0° to 360°.

1.18 The Altitude

The **altitude** is the second coordinate and is measured along the vertical circle from the point M on the horizon to the point P (celestial object) to be located. The altitude (the angle MQP) is directed from the horizon towards the zenith, and it ranges from 0° to 90°.

The azimuth and the altitude of the horizon coordinate system are generally measured in degrees, minutes, and seconds. A **degree** (°) is the 360th part of one complete rotation and subtends an arc along the circumference of a circle which in length is the 360th part of the complete circumference. A degree is subdivided into 60 minutes (′), and a minute is subdivided into 60 seconds (″). In general, an angle is written in the form

15°	12′	18″
fifteen degrees	twelve minutes	eighteen seconds

Three examples of the location of celestial objects are as follows:

1. Azimuth 15° Altitude 60°.
2. Azimuth 110° Altitude 40°.
3. Azimuth 225° Altitude 10°.

1.19 The Zenith Distance

The zenith distance of a point on the celestial sphere is the arc of the vertical circle (in angular measure) extending from the point to the zenith. It is the complement of the altitude of the point (i.e., the sum of the zenith distance and the altitude is 90°). In strict analogy with the polar coordinate system on the plane we should use the zenith distance and the azimuth as the coordinates in the horizon system. The following examples refer to the above star locations:

1. Zenith distance 30°.
2. Zenith distance 50°.
3. Zenith distance 80°.

1.20 Limitations of the Horizon Coordinate System

Because the horizon coordinate system is attached to the observer and not to a fixed point on the celestial sphere, the coordinates of any celestial object in this system change from moment to moment and from observer to observer. Therefore, this coordinate system cannot be used to catalogue stars; instead we must introduce an origin and reference circles *attached to the sky* and not to the observer.

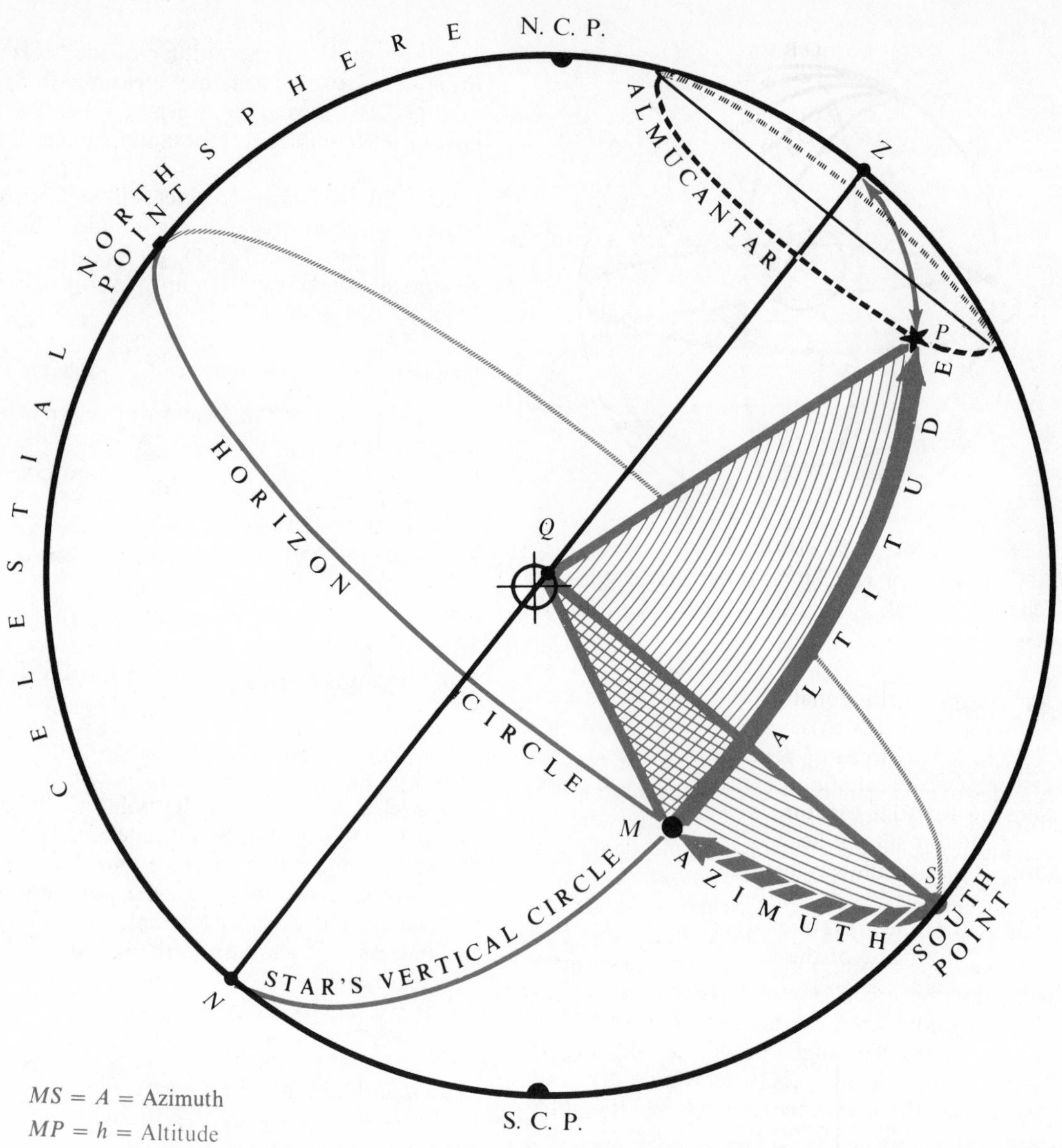

FIG. 1–15 *The horizon coordinate system for an observer, Q. The azimuth (A) of the point P is the arc SM of the horizon (measured westwardly in degrees from the south point); the altitude (h) of the point P is the arc MP of the vertical circle through P (measured in degrees from the horizon).*

1.21 The Equatorial Coordinate System

We can obtain a set of great circles fixed in the sky instead of the vertical circles if we start not with the zenith, but with some fixed point on the celestial sphere. Since the N.C.P. is such a point to a very good approximation (it is the same for all observers, but it has a very slight westward motion which we shall neglect for the time being), all great circles passing through it are fixed on the celestial sphere. We therefore use these half-circles (they correspond to the vertical circles in the horizon coordinate system) in the **equatorial coordinate system**. Note that all these circles also pass through the S.C.P. They are called **hour circles**. (See Figure 1.16.)

1.22 Parallels of Declination and the Celestial Equator

A line from the N.C.P. to the S.C.P. defines the axis of rotation of the celestial sphere, and a set of

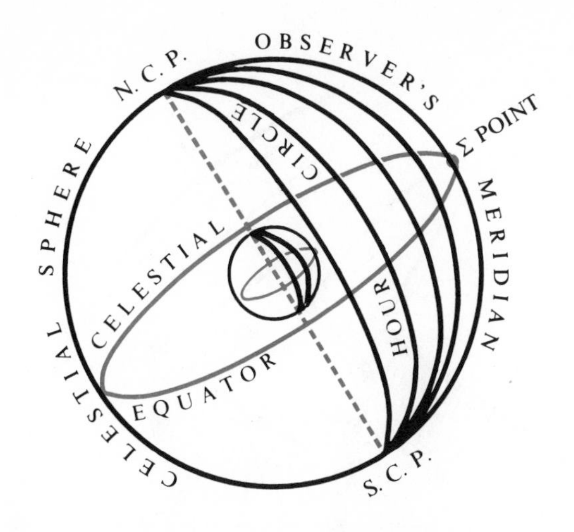

FIG. 1–16 *The hour circles and celestial equator drawn on the celestial sphere. The hour circles correspond to the circles of longitude on the earth.*

parallel circles orthogonal to the hour circles have their centers on this axis. Among these **parallels of declination** or **circles of declination** is one that is a great circle of the celestial sphere. This great circle, called the **celestial equator**, is concentric with the earth's equator since it lies in the plane of the latter and its center is midway between the two celestial poles. In other words, the celestial equator may be pictured as the great circle that defines the intersection of the plane of the earth's equator and the celestial sphere.

Celestial		*Terrestrial*
Hour circles	*concentric with and correspond to*	Circles of longitude
Celestial equator		Terrestrial equator
Parallels of declination		Parallels of latitude

The equatorial coordinate system has one of the prerequisites of a system designed for cataloguing stars. Its fundamental circles are the same for all observers since they are defined with respect to the celestial poles, which are the same for all observers.

1.23 The Sigma Point

Note that the hour circle that passes through the observer's zenith at any moment coincides with the celestial meridian of the observer. *At each instant a different hour circle is in this position, so that any one hour circle continuously changes its position relative to the celestial meridian.*

Consider now a coordinate system obtained by taking as the two reference circles:

1. the celestial equator,
2. the observer's celestial meridian.

These two circles intersect on the sky in a point called the **sigma point** (Σ), as shown in Figure 1.16. Note that the Σ point is attached to the observer, and is not a fixed point on the sky. It is different for observers having different longitudes.

1.24 The Hour Angle

To locate a *fixed celestial point* (P) with respect to the new fundamental circles (hour circles and circles parallel to the celestial equator) we first note that this point always lies on the same hour circle. As time goes on, this hour circle intersects the celestial equator at different consecutive points (Figure 1.17). Let P' be the point of intersection of these two circles at the **moment** that we wish to locate the point P. We measure from Σ **westward** along the celestial equator to P' (it is at the foot of the hour circle passing through P) sweeping out the angle $\sphericalangle \Sigma QP'$. This angle is called the **hour angle** (**H.A.**) of the point P and is measured either in degrees or in **15°** **units**, each of which is called an hour.

$$15° = 1\ \textbf{hour},$$

$$15' = 1\ \textbf{minute},$$

$$15'' = 1\ \textbf{second}.$$

When the hour circle passing through a celestial point coincides with the observer's meridian, its hour angle is 0°. Note that the hour angle of any point continuously increases as the point moves westward relative to the meridian, ranging from 0° to 360°, or from 0 hours to 24 hours.

Knowledge of the hour angle is important for locating celestial objects with a telescope, since telescopes are generally mounted equatorially, that is, with freedom to rotate about an axis parallel to the earth's axis of rotation, and about another axis lying in the plane of the equator.

1.25 Declination

After determining the hour angle of P' on the celestial equator, we can locate the point P by moving **northward** from P' (if the star is above the celestial equator) along the hour circle to point P. The angle $\sphericalangle P'QP$ thus described is the **declination**

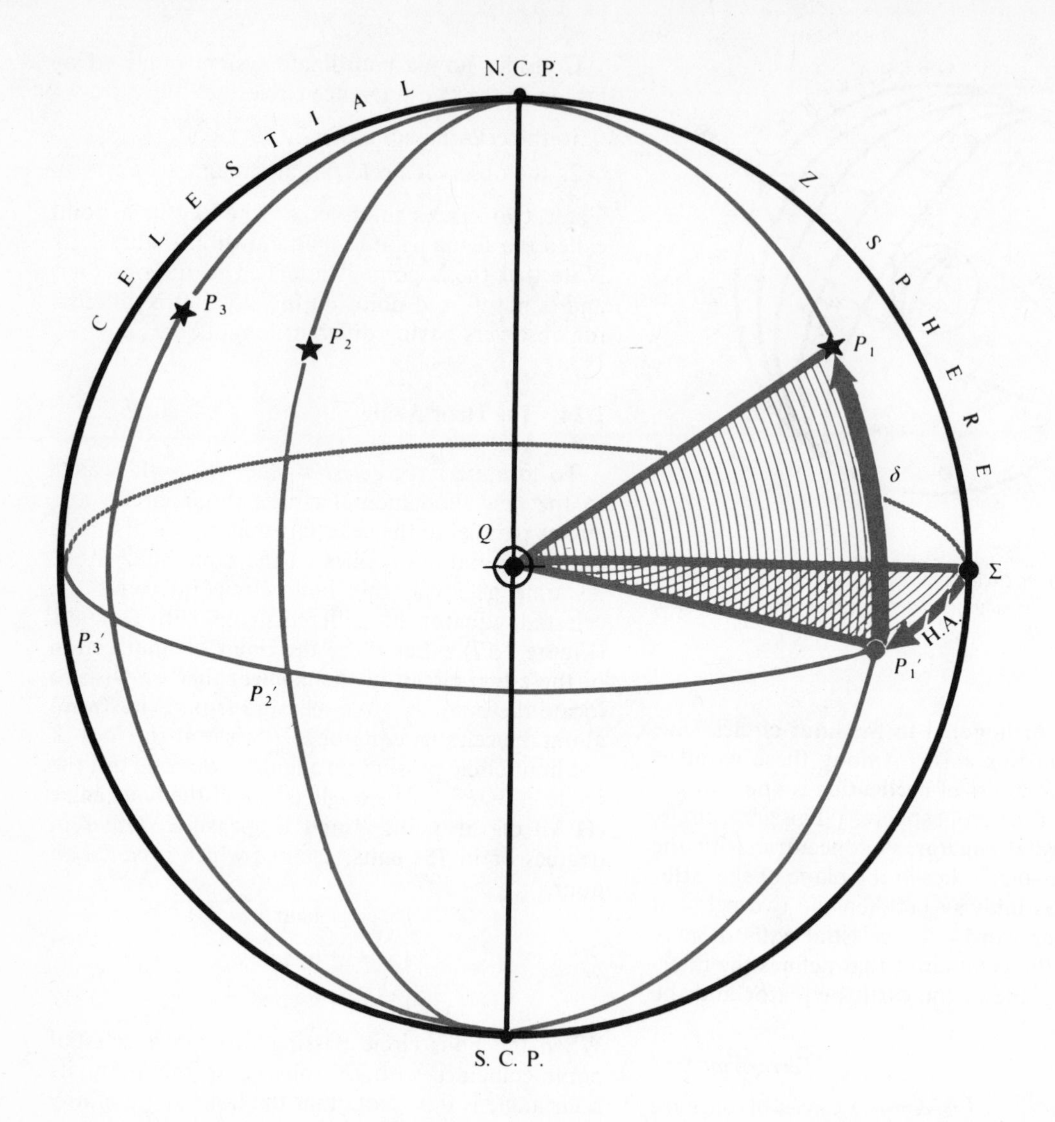

FIG. 1–17 *The hour angle and declination of a point, P, on the celestial sphere. As time goes on, the hour angle increases. The arc* $\Sigma P_1'$ *is the hour angle of the point* P_1. *The arc* $P_1'P_1$ *is the declination,* δ, *of* P_1.

(δ) of P. The declination is measured in degrees. All points on the celestial equator have 0° declination; points above the equator have positive declinations, ranging from 0° to +90° at the north celestial pole; points below the equator have negative declinations, ranging from 0° to −90° at the south celestial pole.

The declination of a star can be used as a cataloguing coordinate since it is measured from a fixed point on the celestial sphere (the N.C.P.). Actually it is measured from a fixed circle, the celestial equator (the equator is only approximately fixed, but like the N.C.P. it changes so slightly that we neglect it here), but that is equivalent to measuring from the north celestial pole since the two angles so obtained are complementary to each other. The declination of a fixed celestial object is constant, not changing from moment to moment, since it is the arc of the object's hour circle lying between the celestial equator and the parallel of declination passing through the object, as shown in Figure 1.18 (P_1, P_2, P_3). A fixed point on the sky on a parallel of declination 10° north of the equator always has a declination of +10°.

Although the declination is a good coordinate for cataloguing stars, the hour angle is not, since,

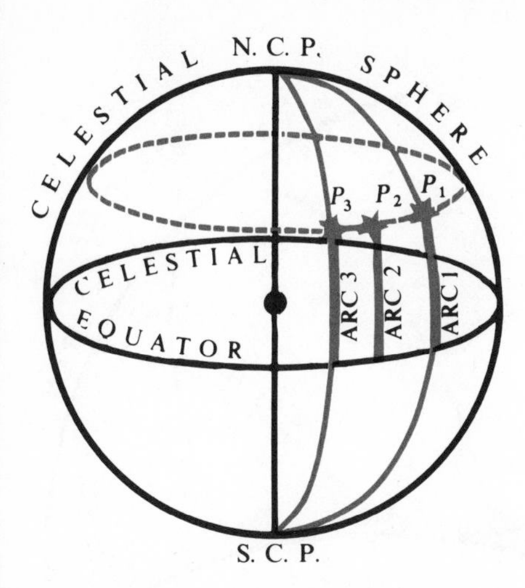

FIG. 1–18 *As the celestial sphere turns, the declination of the fixed point P remains constant, as shown by the three equal arcs.*

as we have seen, it changes continuously. The reason is that it is measured with respect to a point Σ attached to the observer and not with respect to a point fixed on the celestial sphere. The hour angle must therefore be replaced by a coordinate that does not change, and to do this we must start not from the sigma point, but from a point that is the same for all observers, and hence fixed on the celestial sphere.

1.26 Using the Apparent Motion of the Sun to Establish a Fixed Point on the Celestial Sphere

We shall impose two conditions on this point:

1. it must lie on the equator, and
2. it must be a point that can be located easily by observing some celestial body.

It would be very convenient if there were some bright object in the sky easily identifiable by everyone and always in one fixed position on the equator. Unfortunately this is not the case, for even if there were some bright star on the equator it would, in time, move away from it. However, we have what is the next best thing, an easily observed object that periodically moves across the equator, and therefore defines a point on it. This object is the sun.

Casual observation of the sun reveals two different apparent motions of that body across the sky:

1. diurnal; which is the daily rising and setting from east to west;
2. motion in declination, which is the change of the position of the sun relative to the celestial equator.

The sun does not rise and set at the same points on the horizon throughout the year. In summer in the northern hemisphere it rises and sets considerably to the north of the east and west points, respectively, and during this time of the year it attains its highest altitudes (Figure 1.19). The arc of its diurnal path across the sky is longer from rising to setting than from setting to rising, so that the daylight period is proportionately longer than the darkness period. The opposite is true when the sun is at its minimum declination in the winter; nights then are longer and days shorter in the northern hemisphere.

Whereas the declination of the sun is positive in summertime (it lies between the N.C.P. and the celestial equator), it is negative in wintertime (lying between the S.C.P. and the celestial equator). This implies that twice during the year its declination is 0°, since it must pass through zero declination in going from positive to negative, and vice versa. It follows, therefore, that the sun is directly on the celestial equator at those two instants. Because the equator and the horizons of all observers intersect at the observers' east and west points, the sun rises exactly in the east point and sets exactly in the west point for all observers at those times. The length of the day then equals the length of the night. *We shall take one of these points, namely, the point on the equator occupied by the sun when it is passing from negative to positive declination (from below to above the equator), as the reference point to replace the Σ point.*

1.27 The Diurnal Rising and Setting of the Stars and the Apparent Eastward Motion of the Sun

Although we have just indicated how we might locate the reference point we are seeking by following the sun, it is more convenient to define this point in terms of the intersection of the equator and one other great circle fixed on the celestial sphere. This other great circle is described by the apparent annual motion of the sun on the celestial sphere. The sun, in addition to rising and setting, changes its position from day to day with respect to the fixed stars, as we have already noted in discussing its motion in declination. *This motion can be*

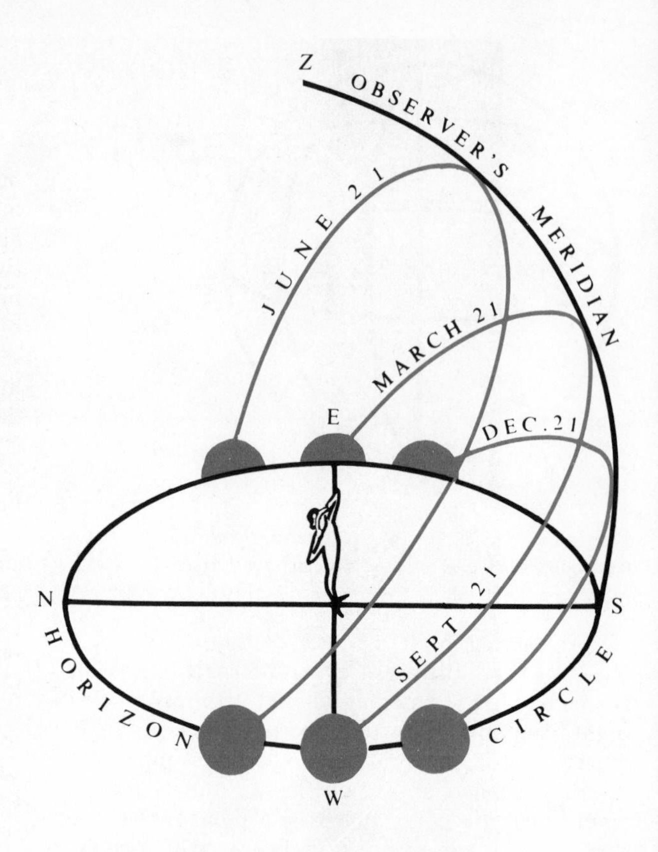

FIG. 1–19 *The noontime altitude of the sun changes from month to month as seen by a fixed observer. (Here a northern-hemisphere observer is shown.)*

detected by observing that the stars rise about 4 minutes earlier each night.

Imagine a star and the sun occupying the same point on the celestial sphere at 12 o'clock noon on some day (the sun will then be on the observer's meridian). When the sun is on the meridian again the following day (12 o'clock noon), the star will be about 1° to the west of the meridian. This means that the star must have risen about 4 minutes earlier than the sun, or, put differently, that the sun has apparently moved eastwardly by about 1° with respect to the star.

1.28 The Ecliptic and the Equinoxes

This apparent eastward motion of the sun describes a great circle, the ***ecliptic****, on the celestial sphere* as shown in Figure 1.20. We take one of the points of intersection (*the equinoxes*) of this circle with the equator as our fixed reference point to replace the Σ point. The ecliptic passes through a band of 12 constellations in the sky called the **zodiac** (see Figure 1.20). The equinox that the sun passes in going from negative to positive declination (from south to north of the equator) is called the **vernal equinox**, or the **first of Aries** (♈). This passage occurs about March 21. The equinox that the sun passes in going from positive to negative declination (from north to south of the equator) is called the **autumnal equinox**. This passage occurs about September 23. As examples we have for the years 1961, 1962, and 1963 the following passage dates:

♈ { 1961 March 20—20^h32^m G.T.
1962 March 21— 2^h30^m G.T.
1963 March 21— 8^h20^m G.T.

♎ { 1961 September 23— 6^h43^m G.T.
1962 September 23—12^h35^m G.T.
1963 September 23—18^h24^m G.T.

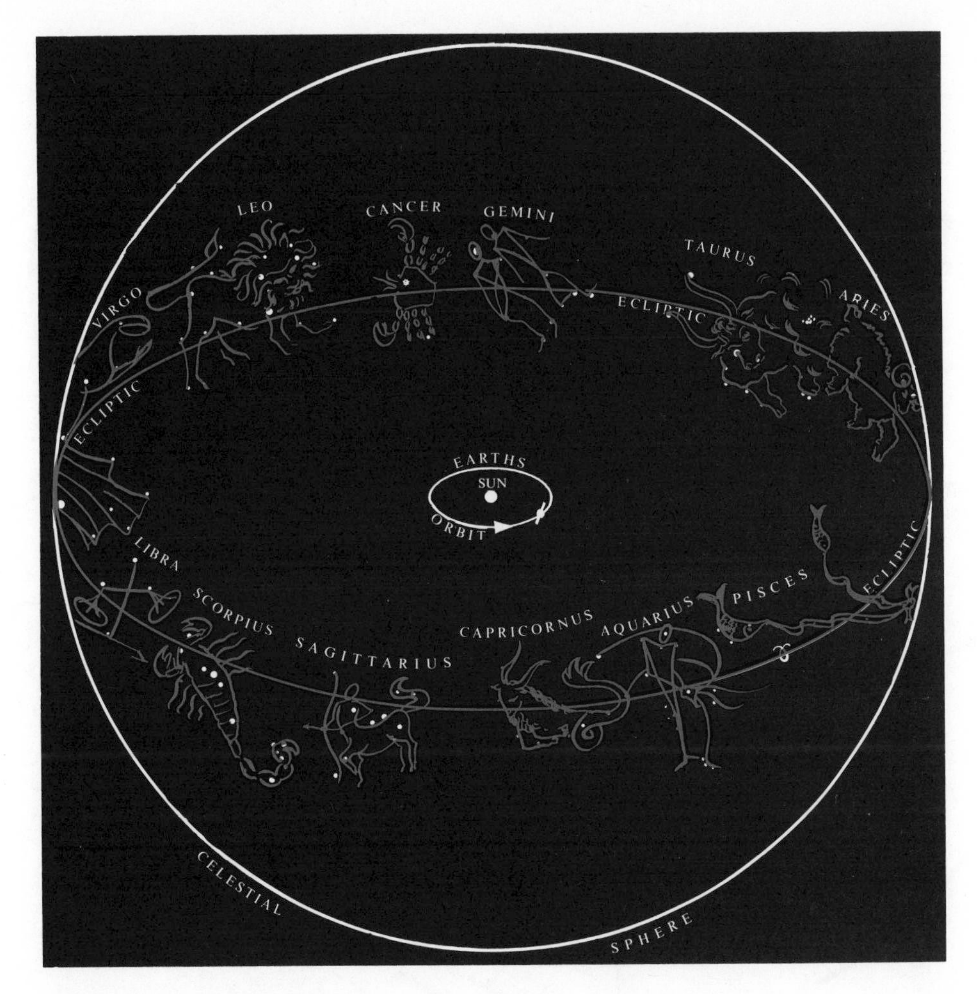

FIG. 1–20 *The belt of the Zodiac and its relationship to the ecliptic.*

1.29 The Solstices

The term **solstice** refers to the position of the sun at minimum and maximum declinations. When the sun's declination is a maximum ($+23\frac{1}{2}°$), it is at the **summer solstice**, and is directly overhead (at the observer's zenith at noon) for an observer on the *Tropic of Cancer*. When the sun's declination is a minimum ($-23\frac{1}{2}°$), the sun is directly overhead for an observer on the *Tropic of Capricorn*, and it is at the **winter solstice**. The sun is at the summer solstice about June 22, and at the winter solstice about December 22.

1.30 The Right Ascension, α

We now take the vernal equinox ♈ as the fixed reference point on the celestial equator to supersede the Σ point. Since the vernal equinox is fixed, coordinates referred to it are also fixed. The position of the point P on the celestial sphere is now defined by its declination and a new coordinate, the **right ascension (R.A. or α)**, which we shall define below.

We again consider the hour circle through the point P, on the celestial sphere, and note where it intersects the equator at P' just as before. The right ascension of point P is now defined as the arc of the celestial equator extending **eastward** from the vernal equinox ♈ to the point P'. It is the angle $P'Q$♈ subtended by the arc $\widehat{P'♈}$, as shown in Figure 1.21.

Right ascension is measured in hours, minutes, and seconds. If the right ascension of a star is known, its hour angle at any moment for any particular observer can be found. The reason is

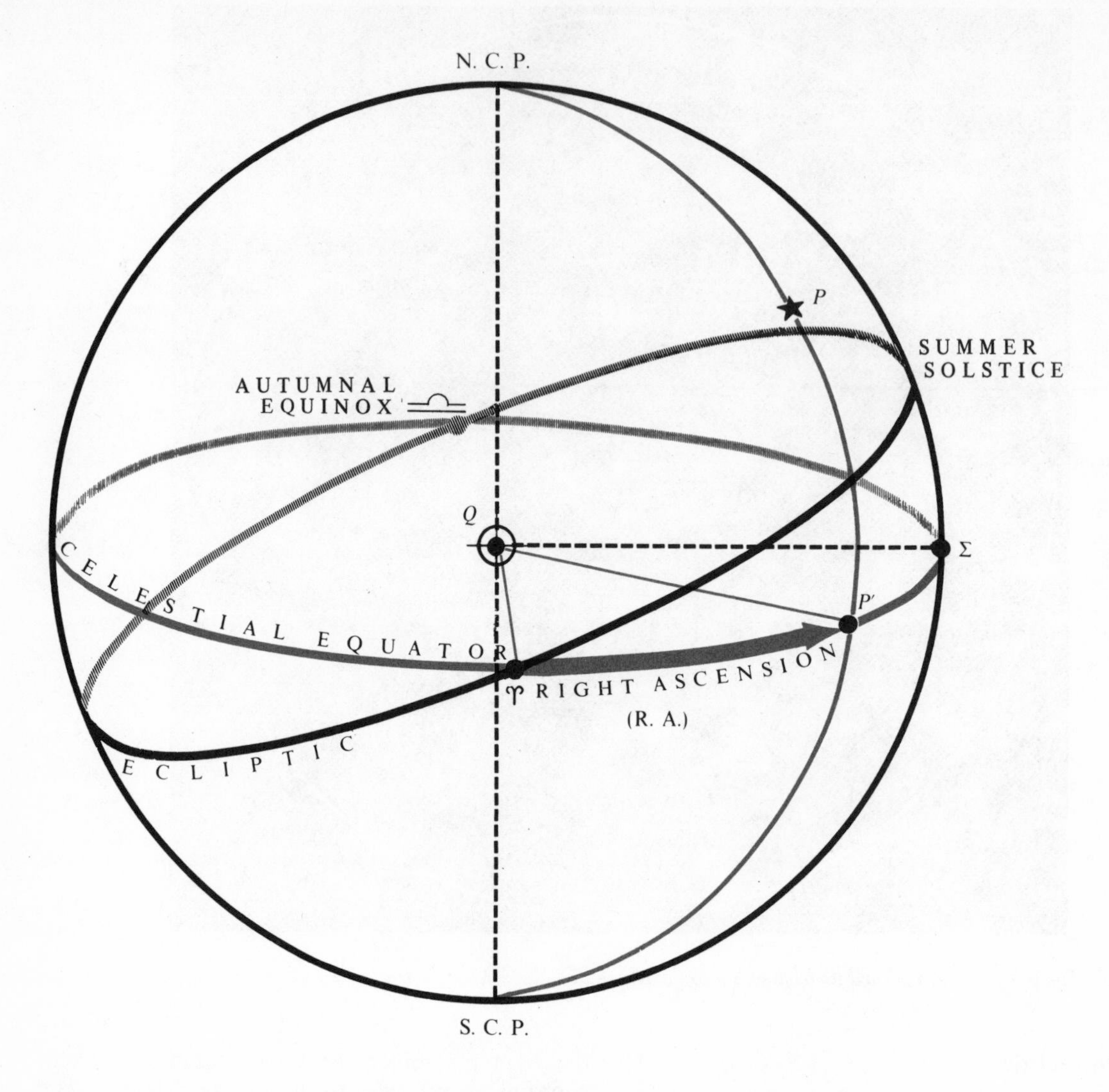

FIG. 1–21 *The right ascension of the point P is the angle ♈ QP′ measured eastwardly from ♈ along the celestial equator.*

that the hour angle of the star at that moment is equal to the difference between the known hour angle of ♈ for the observer at that moment (the angle ΣQ♈) and the right ascension of the star as defined above. The positions of stellar objects in star charts and catalogues are given in terms of declination and right ascension.

	R.A.		δ	
	h	m	°	′
1. Sirius	6	40.8	−16	35
2. Rigel	5	9.7	+8	19
3. Vega	18	33.6	+38	41

1.31 Locating the Vernal Equinox

We have just noted that the hour angle of a stellar object can be found at any moment if the hour angle of the vernal equinox is known at that moment. This would offer no problem if the point ♈ were marked by some visible object and if it were always above the horizon. But since this is not the case (it rises and sets, passing the observer's meridian once a day), we must devise some way to locate this point, or (what amounts to the same thing) to find its hour angle at any moment. Because ♈ is a

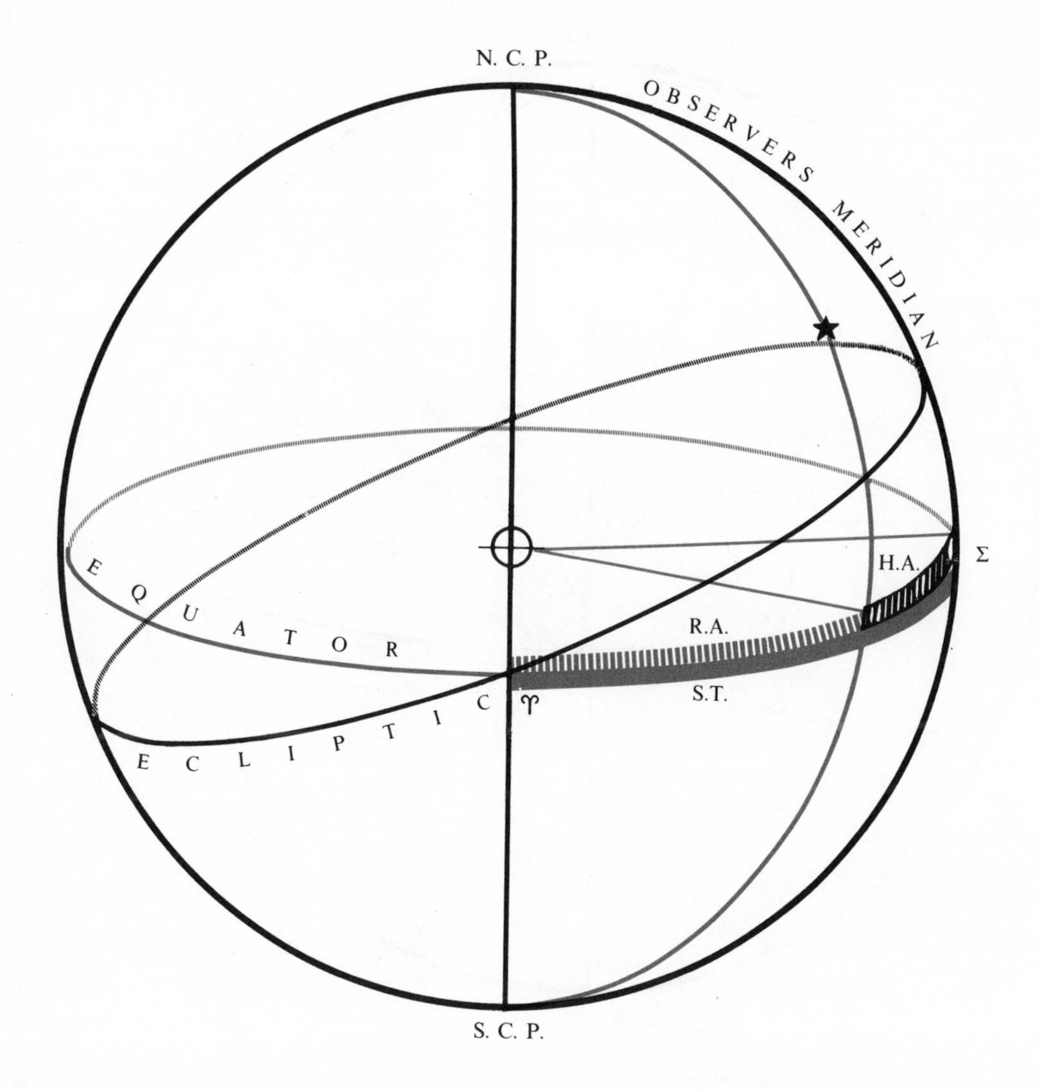

Fig. 1–22 *The relationship between the hour angle, H.A., the right ascension, R.A., and sidereal time, S.T.*

point attached to the celestial sphere (neglecting a small westward motion arising from precession—see Section 5.17) we can use a mechanical device for this purpose. This is essentially a clock mechanism synchronized with the rotation of the celestial sphere rather than with the diurnal rising and setting of the sun.

1.32 The Sidereal Clock and Sidereal Time

We consider a clock with twenty-four subdivisions on its face, so constructed that the hour hand makes one complete revolution for each rotation of the celestial sphere. If we now set this clock to read zero (0^h) when ♈ is on the observer's meridian, the hour angle of the vernal equinox is given by the reading on this clock. Since the time measured by this clock is different from our ordinary solar time (see Section 2.3) but *corresponds to the time as measured by the rising and setting of the stars, we shall call this time* ***sidereal time*** *(**S.T.**)*. We can now say that the *hour angle of an object in the sky for a given observer at any given time is equal to that observer's sidereal time minus the right ascension of the object*, as shown in Figure 1.22.

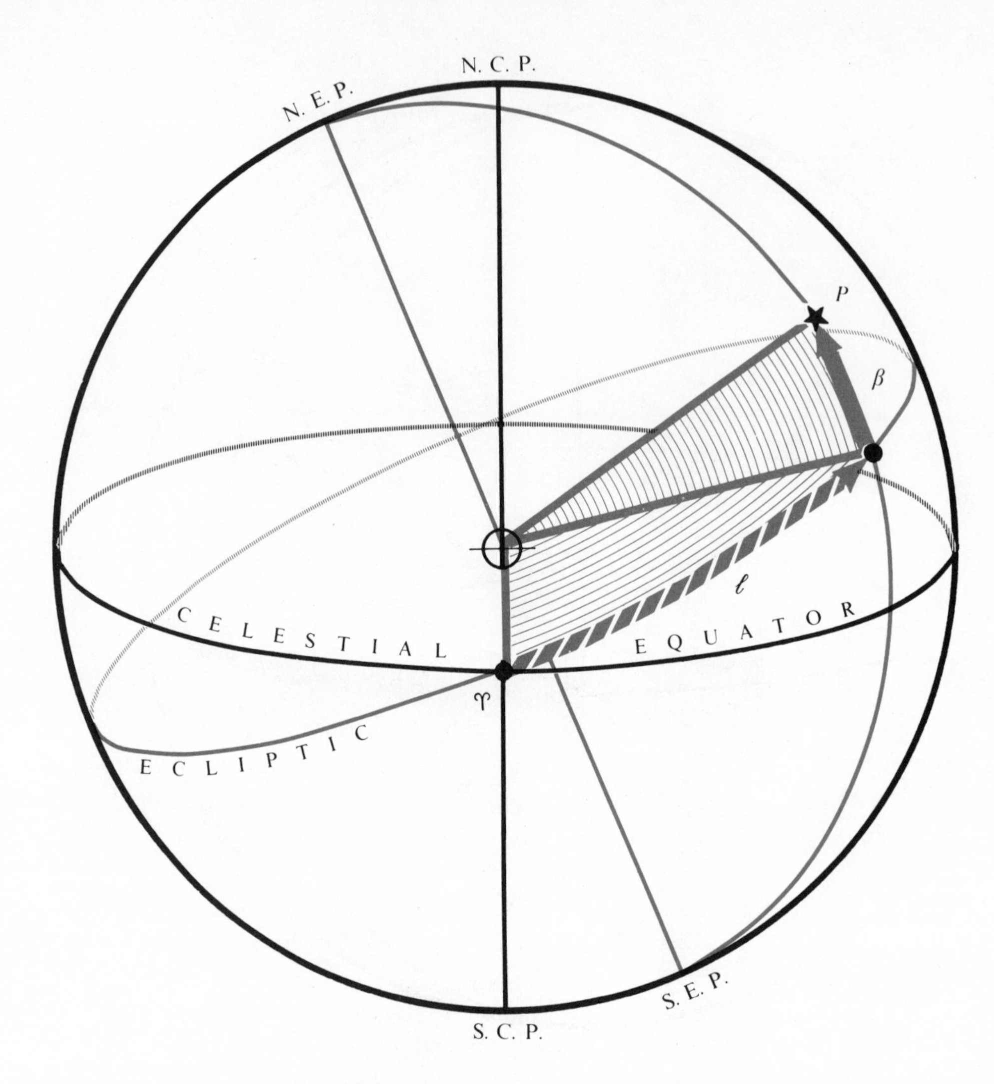

FIG. 1–23 *Celestial longitude and celestial latitude. Celestial longitude is the arc ℓ measured from the vernal equinox eastwardly along the ecliptic. The celestial latitude, β, is measured along an hour circle from the ecliptic.*

$$\boxed{\text{H.A.} = \text{S.T.} - \text{R.A.}} \tag{1.1}$$

It should be noted that the sidereal time at any moment differs from observer to observer.

It is fairly easy to find the sidereal time at almost any moment during the night if the right ascensions of enough stars are known. If a star is on the observer's meridian, the observer's sidereal time at that moment is equal to the right ascension of that star, since the hour angle of a star is zero when it is on the observer's meridian. As we shall see, this method is used to determine the error of an observer's sidereal clock.

From the positions of stellar objects given in sky charts and catalogues in terms of their right ascension and declination, together with a knowledge of the sidereal time, we can locate these objects in the eyepiece of an equatorially mounted telescope without any difficulty.

1.33 The Ecliptic Coordinate System

If we use the ecliptic instead of the equator as the fundamental great circle, we obtain the **ecliptic co-**

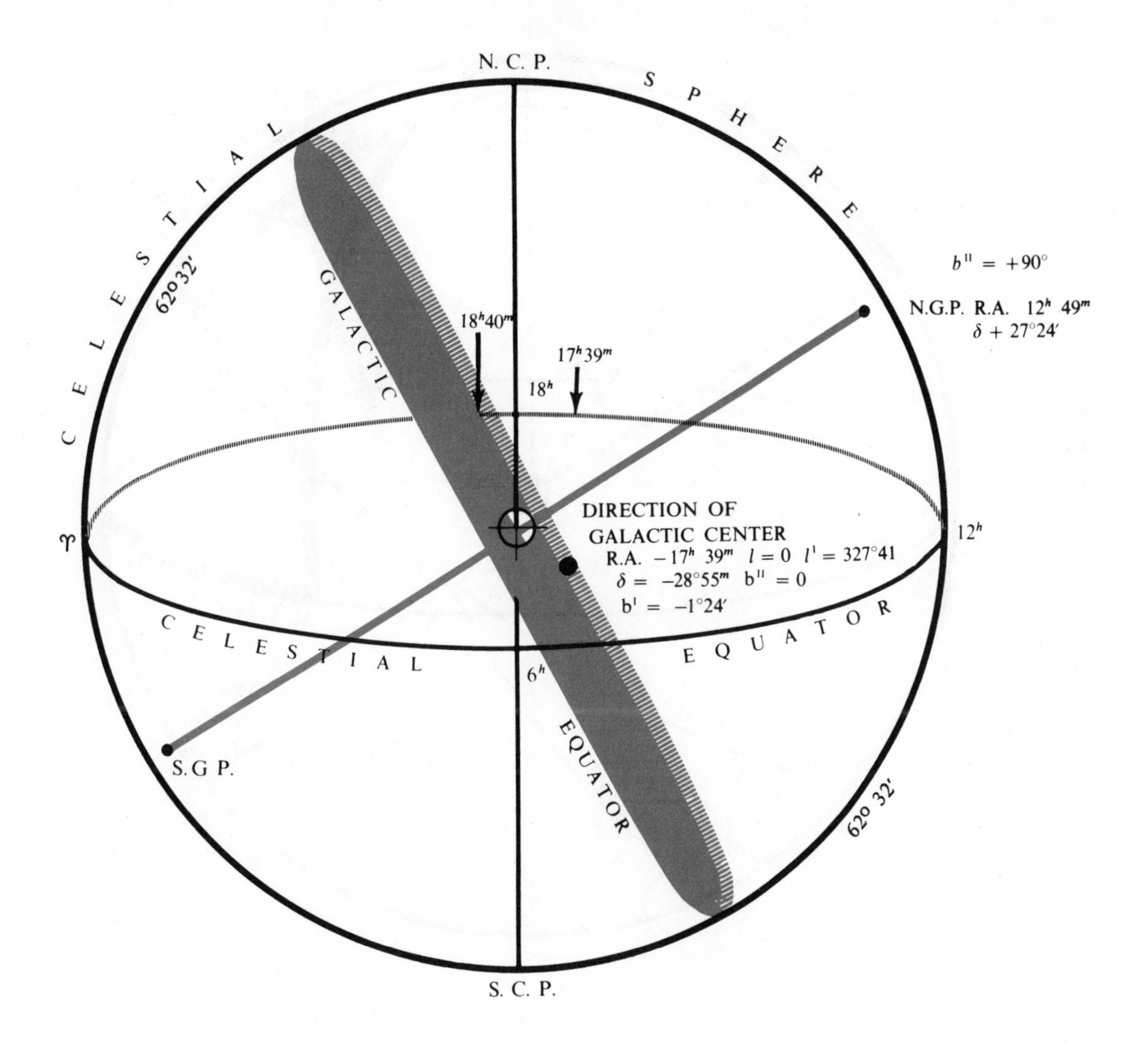

FIG. 1–24 *The plane of the galaxy and the galactic coordinate system shown with respect to the celestial equator. SGP = South Galactic Pole; NGP = North Galactic Pole.*

ordinate system, in which the coordinates are *celestial longitude* and *celestial latitude*. As shown in Figure 1.23, the celestial longitude of an object is defined as the angle ♈ QP' measured **eastward** along the ecliptic from the vernal equinox to the point where the great circle passing through the pole of the ecliptic and through the object meets the ecliptic. The celestial latitude is defined as the angle $P'QP$ measured along this great circle from the ecliptic to the point P. The position of the north pole of the ecliptic in equatorial coordinates is given as R.A. 18^h and declination $+66\frac{1}{2}°$, since it lies on the hour circle that passes through the position of the sun when it is farthest south of the equator, that is, the winter solstice.

1.34 The Galactic Coordinate System

As a final example we consider the **galactic coordinate system**, which is of great use in the study of stellar dynamics and galactic structure. The sun and all the neighboring stars are part of a vast flattened stellar system, which appears to the observer as a narrow band of stars in the sky—

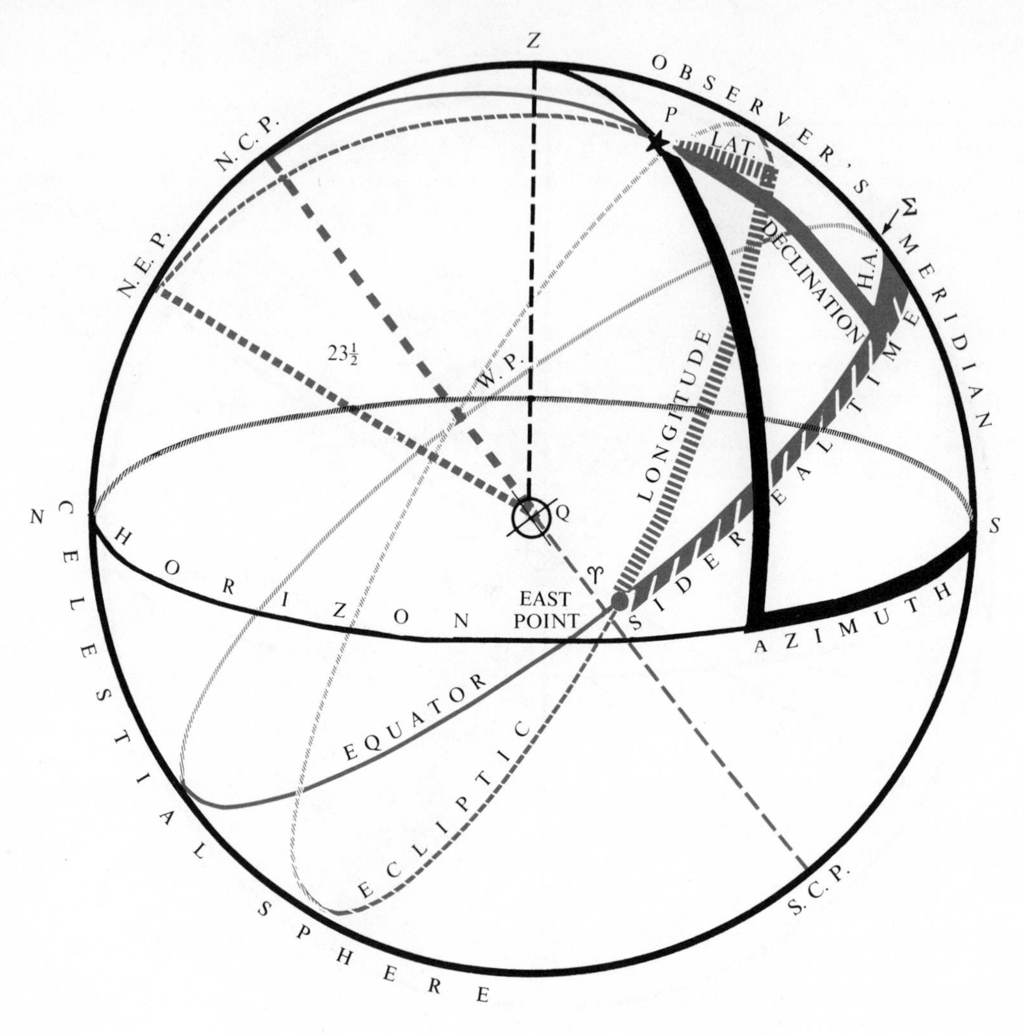

FIG. 1–25 *Composite diagram of the various coordinate systems.*

the **Milky Way**. If we take the intersection of the plane of the Milky Way with the celestial sphere, we obtain the fundamental great circle (the galactic equator) of the galactic coordinate system, as shown in Figure 1.24.

Galactic latitude is measured from the galactic equator, and *galactic longitude* is measured along this circle starting from the point at which it intersects the celestial equator (R.A. 18^h30^m). The pole of this coordinate system (the north galactic pole) is at R.A. 12^h40^m and declination $+28°$ in equatorial coordinates. This defines the galactic equator so that it cuts the celestial equator at an angle of 62° 36′. Figure 1.25 shows the relationship between the various coordinate systems.

In 1958 the International Astronomical Union introduced the I.A.U. galactic coordinate system, l^{II} and b^{II}, which is defined in the following way: The direction of I.A.U. zero galactic longitude and I.A.U. zero galactic latitude, $l^{II} = 0$ and $b^{II} = 0$, coincides with the direction of the center of the Galaxy, whose R.A. $= 17^h39^m\!.3$ and $\delta = -28° 55'$. In the old galactic coordinate system this direction has the longitude $l^I = 327° 41'$ and galactic latitude $b^I = -1° 24'$. The galactic north pole ($b^{II} = +90°$) has the R.A. $= 12^h49^m$ and $\delta = +27° 24'$.

TABLE 1.1
SUMMARY OF COORDINATE SYSTEMS

	Fundamental circles	Reference points	Coordinates
Horizon system	**Vertical circles** (celestial meridian)	**Zenith**	**Azimuth** (measured along horizon from south point)
	Almucantars (horizon)	**South point**	**Altitude** or **zenith distance** (measured along meridian from horizon or from zenith)
Equatorial system	**Hour circles** (celestial meridian)	**N.C.P., north celestial pole**	**H.A., hour angle** (measured westward along equator from Σ point)
	Circles of declination (equator)	**Σ, sigma point**	δ, **declination** (measured from equator along hour circle)
		N.C.P.	**R.A.** or α, **right ascension** (measured along equator from ♈)
		♈, **vernal equinox**	δ, **declination** (as above)
Ecliptic system	**Circles of longitude**	**N.E.P., north ecliptic pole**	**Celestial longitude, λ** (measured eastward along the ecliptic from the vernal equinox)
	Ecliptic circle (determined by the plane of the earth's orbit about the sun)	♈, **vernal equinox**	**Celestial latitude, β** (measured perpendicular to the ecliptic)
Galactic system	**Circles of longitude**	**N.G.P., north galactic pole**	**I.A.U. galactic longitude and galactic latitude** (zero galactic longitude $l^{II} = 0$ and zero galactic latitude $b^{II} = 0$ coincide with the direction of the galactic center)
	Circles of declination (galactic equator, determined by the plane of the Galaxy)	**Zero point** on galactic equator	

EXERCISES

1. (a) Using a Cartesian coordinate system on a plane, locate the following points (where the first number in the parenthesis is the x coordinate and the second number is the y coordinate): (0, 5), (4, 0), (4, 3), (6, 2), (−4, −3), (−7, 3), (8, −4). (b) Express each point given in (a) in terms of polar coordinates θ, r, where θ is the angle measured with respect to the x-axis and r is the distance of the point from the origin.

2. Give the altitude of the following points if your latitude is 40°N: (a) the North Celestial Pole; (b) the Σ point; (c) the vernal equinox at sidereal zero time; (d) the summer and winter solstices at solar noon on the longest day of the year in the Northern Hemisphere; (e) the summer and winter solstices at noon on the shortest day of the year in the Northern Hemisphere.

3. What is the azimuth of any object when its altitude is a maximum?

4. What is the right ascension of the full moon that occurs on the first day of spring? On the first day of fall? On the first day of summer? On the first day of winter?

5. Give the zenith distances of the following points if your latitude is 40°N: (a) the North Celestial Pole; (b) the Σ point; (c) the vernal equinox when its altitude is a maximum; (d) the summer solstice when its altitude is a maximum; (e) the winter solstice when its altitude is a maximum.

6. Give the maximum and minimum altitudes of the sun at noon for the entire year if your latitude is 40°N.

7. If you are on the equator, where will you find the North Celestial Pole and the South Celestial Pole?

8. What is the sidereal time for you when the following stars are on your celestial meridian: Rigel, Arcturus, Betelgeuse, Sirius, α Centauri, and Aldebaran?

9. What is the right ascension of a star on the celestial equator if it is rising at sidereal zero time? If it is setting at sidereal zero time?

10. What is the azimuth of the vernal equinox when it is just rising? When it is just setting?

11. Give the declination and right ascension of the following points on the celestial sphere: (a) the vernal equinox; (b) the autumnal equinox; (c) the North Celestial Pole; (d) the solstices.

12. When a star whose right ascension is 23^h is on your meridian one hour before solar noon, what is the date? What is the approximate date when it is on your meridian one hour after solar noon?

13. At what times of the day on June 21 will the stars with the following right ascensions cross your meridian: 0^h, 6^h, 12^h, 18^h?

14. What angle does the plane of the celestial equator make with your horizon if you are: (a) on the Arctic Circle; (b) at the North Pole; (c) at latitude 40°N; (d) at the Tropic of Cancer?

15. (a) What is the maximum angle during the day that the plane of the ecliptic can make with the plane of the horizon of an observer whose latitude is λ? Where is the vernal equinox when this happens? (b) Substitute "minimum" for "maximum" in (a) and answer the two questions.

16. What is the smallest declination that a star can have if it is always to be above the horizon of an observer whose latitude is λ?

17. (a) Where would you have to be on the earth's surface for the plane of the ecliptic to make an angle of 90° with your horizon plane twice every 24 hours? (b) Where would you have to be on the earth's surface for the plane of the ecliptic always to make the same angle with your horizon plane? What is this angle?

18. (a) If the right ascension of a star is 6 hours, what is its celestial longitude? (b) If the declination of this star is δ, what is its celestial latitude?

2

TIME

2.1 The Measurement of Time

The positions of objects on the celestial sphere can be specified at any instant by using one of the various coordinate systems outlined in the previous chapter, but to follow these objects as they change their positions on the sky we need an additional entity. This entity is a unit of **time**.

The astronomer does not attempt to define the concept of time, but rather to describe a time-measuring system and to introduce a unit of time. A little thought shows at once that to introduce a unit of time (*an interval between two nonsimultaneous events*) one must have recourse to some type of *periodic motion*. Thus, if we consider the crude water clock, the hourglass, the pendulum clock, the modern watch, or the earth, we see that their time-telling ability lies in their periodic motions.

Although the astronomer makes use of the rotation of the earth to establish his time scale, he has found it necessary to introduce not one but three different time scales based on this fundamental periodic motion. These are **sidereal time**, introduced in Chapter 1, the **apparent solar time**, and the **mean solar time**. The need for these different scales arises because the astronomer must concern himself not only with the time as given by the rising and setting of the stars (sidereal time), but also with the time as determined by the rising and setting of the sun (solar time).

2.2 Sidereal Time and the Sidereal Day

Sidereal time, as we have already noted, is measured by the hour angle of the vernal equinox. The sidereal day for a particular observer begins when the vernal equinox is on his celestial meridian. In other words, *the local sidereal time is* 0^h *when the hour angle of* ♈ *is* 0^h; and the sidereal day is defined as the interval between two consecutive passages of ♈ across the observer's meridian. The sidereal day is divided into 24 sidereal hours, and the sidereal hour into sidereal minutes and seconds.

The sidereal clock, introduced in Section 1.32, marks off sidereal intervals. Since the sidereal clock suffers from the imperfections of all mechanical instruments, it cannot keep time with perfect accuracy. Instead of continuously correcting the clock itself, however, the error of the clock is constantly determined by making frequent observations of the local sidereal times of transit (passage across the meridian) of stars whose right ascensions are known. *The difference between the reading on the sidereal clock at the time of transit of a star and the known right ascension of this star is the clock error.*

As we shall see later, when we discuss the precession of the equinox, the first of Aries is not a fixed point on the sky; it has a very slow westward motion along the ecliptic. As a result, *the length of the sidereal day is, on the average,* $\frac{1}{120}$ *second shorter than the interval between two consecutive transits of a given star across the observer's meridian.* Since the vernal equinox moves irregularly in its westward precession, the actual length of the sidereal day is not constant but shows a slight variation, which is quite difficult to detect.

2.3 Local Solar Time

Because the sidereal clock is not in step with the sun, we cannot regulate our daily affairs according to sidereal time. For this reason we must introduce a solar time scale and a solar clock geared to the rising and setting of the sun. *We define the local apparent solar time for an observer at any moment as the hour angle of the sun.* Thus, when the center of the sun is on the celestial meridian of the observer, the local apparent solar time for that observer is zero hours (noon). A sun dial can give us the local apparent solar time but a mechanical timepiece cannot because, as we have seen, the sun, besides rising and setting, has an apparent irregular eastward motion among the stars. This motion varies from day to day, and is faster in January than in June. This means that the length of the solar day (the interval between two consecutive transits of the sun across the meridian

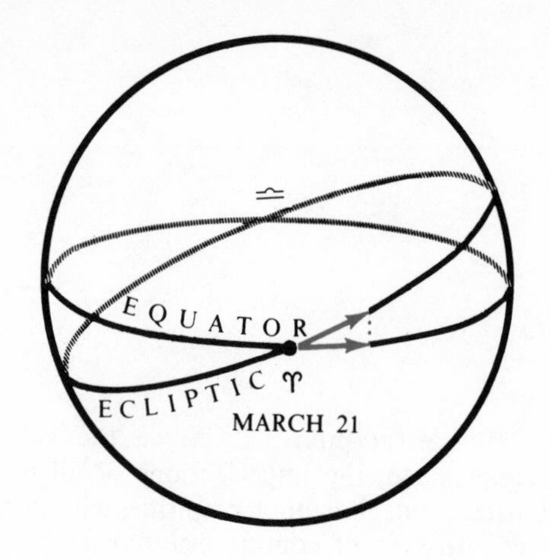

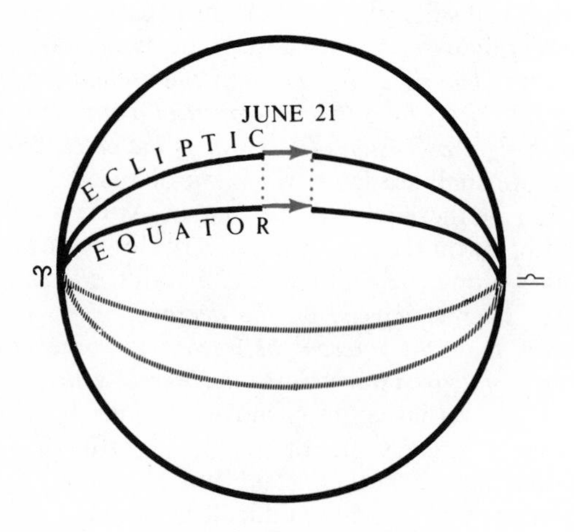

FIG. 2–1 *The different rates at which the sun advances along the equator for two different times of the year.*

of the observer) varies from season to season; on December 23 it is 51 seconds longer than it is in September.

2.4 Variations in the Apparent Solar Day

There are two reasons for this variation in the length of the apparent solar day from month to month as given by the rising and setting of the true sun.

First, the true sun moves along the ecliptic and not along the equator, the circle along which time is measured. Even if the motion of the sun along the ecliptic were regular, its projected motion along the equator would not always be the same because at the equinoxes the angle between the ecliptic and the equator is greater than at the solstices (see Figure 2.1).

Second, the sun appears to move faster during the northern winter months, because the earth is then closer to the sun and is moving faster in its orbit.

2.5 The Mean Sun

These irregularities of the true sun can be eliminated from our time scales by introducing *a fictitious sun, the* ***mean sun****, which is assumed to move with uniform speed along the equator.* Its rate of motion is the average of the true sun's angular motion during the year along the ecliptic. During the year the mean sun runs on the average as much behind the real sun as ahead of it.

2.6 Mean Solar Time: The Equation of Time

The mean sun is used to define the mean solar time, which we use in regulating our daily activities. *Mean solar time is defined as the hour angle of the mean sun.* The interval between two consecutive transits of the mean sun is called the **mean solar day.**

TABLE 2.1
THE EQUATION OF TIME
(Apparent Time Minus Mean Time)

Date		Minutes
January	1	− 3.2
	16	− 9.6
February	1	−13.5
	16	−14.2
March	1	−12.6
	16	− 9.0
April	1	− 4.2
	16	0.0
May	1	+ 2.8
	16	+ 3.7
June	1	+ 2.4
	16	− 0.4
July	1	− 3.5
	16	− 5.9
August	1	− 6.3
	16	− 4.4
September	1	− 0.2
	16	+ 4.8
October	1	+10.0
	16	+14.2
November	1	+16.3
	16	+15.3
December	1	+11.3
	16	+ 4.8

Every mean solar day is of equal length and is equal to the length of the average true solar day. The difference between the apparent solar time and the mean solar time is called the **equation of time**. During the course of the year this quantity varies between -14.2 minutes and $+16.3$ minutes.

2.7 Civil Time

For most practical purposes (both in astronomy and in civil affairs) it is convenient to define the civil day as starting at midnight. Thus, a clock that keeps mean solar time accurately tells us the civil time if it reads 0^h at mean midnight. *This means simply that civil time is defined as mean solar time plus twelve hours.* From the definition of civil time we see that it varies from observer to observer because it is defined in terms of the observer's meridian. Since, however, it is impossible for everybody to carry a watch that adjusts itself to the variation in longitude as he travels about, it is convenient to define the mean time of a locality in terms of a *standard meridian*, or several such standard meridians passing through the region in question.

2.8 Time Zones

In terms of these standard meridians the continental United States (excluding Alaska) lies in four *standard time zones*: Eastern, Central, Mountain, and Pacific. The standard times defined by these zones are five, six, seven, and eight hours, respectively, behind Greenwich time. Although the boundaries between time zones are quite irregular, the time in any one zone is uniform, and always differs by precisely one hour from that of each neighboring zone, as shown in Figure 2.2.

The prime meridian of Greenwich, England, is taken as the standard for terrestrial longitude, and Greenwich time is now adopted throughout the world. Thus, Central European Standard Time is one hour ahead of Greenwich, whereas Standard Atlantic Time is four hours behind. From the definition of terrestrial longitude it follows that the local time of any observer is equal to the Greenwich local time plus or minus the longitude of the observer.

2.9 The 180th Meridian

Since Greenwich is chosen as the international standard, it is convenient to choose the 180th meridian from Greenwich as the place where the new day begins. To understand this idea of the beginning of the day, imagine traveling westwardly at such a speed that the mean sun is always on your meridian. You then require no watch to tell you the time, since for you it is always 12 o'clock noon. This does not mean that time stands still for you, since the days pass one by one, just as they do when you are standing still. At some moment in your westward journey you go from 12 o'clock noon today into 12 o'clock noon tomorrow. That happens when you pass the 180th degree of longitude (the International Date Line). Thus ships that cross this line in going from east to west must skip one day on their calendar (Sunday would immediately become Monday at the same standard time), and ships passing in the opposite direction must count the same day twice (Sunday would become Saturday).

2.10 Conversion of Time

It is often necessary to convert one type of time at a given place into some other time (e.g., local civil time into sidereal time, or vice versa). Methods for doing this are outlined in great detail in the *Nautical Almanac*, published by the Naval Observatory at Washington, D.C. The simplest procedure is first to obtain the desired time at the given instant at Greenwich and then to convert this to the proper time at the given locality by adding or subtracting the appropriate longitude. The reason it is convenient to do this is that the *American Ephemeris* and *Nautical Almanac* give the equation of time as well as the sidereal time at Greenwich at the beginning of each Greenwich civil day. It is therefore easy, by using this kind of information, to convert from one kind of time to another at Greenwich. The equation of time is zero four times during the year—on or about April 16, June 14, September 21, and December 25.

2.11 Approximating Sidereal Time

An approximate value for the sidereal time at any instant of solar time can be found if one counts the number of days that have passed since the sun was at the autumnal equinox, and then adds four minutes for each such day to the apparent solar time. We do this because the solar and sidereal clocks coincide on or about September 22, when the sun is at the autumnal equinox and hence 180° or twelve hours

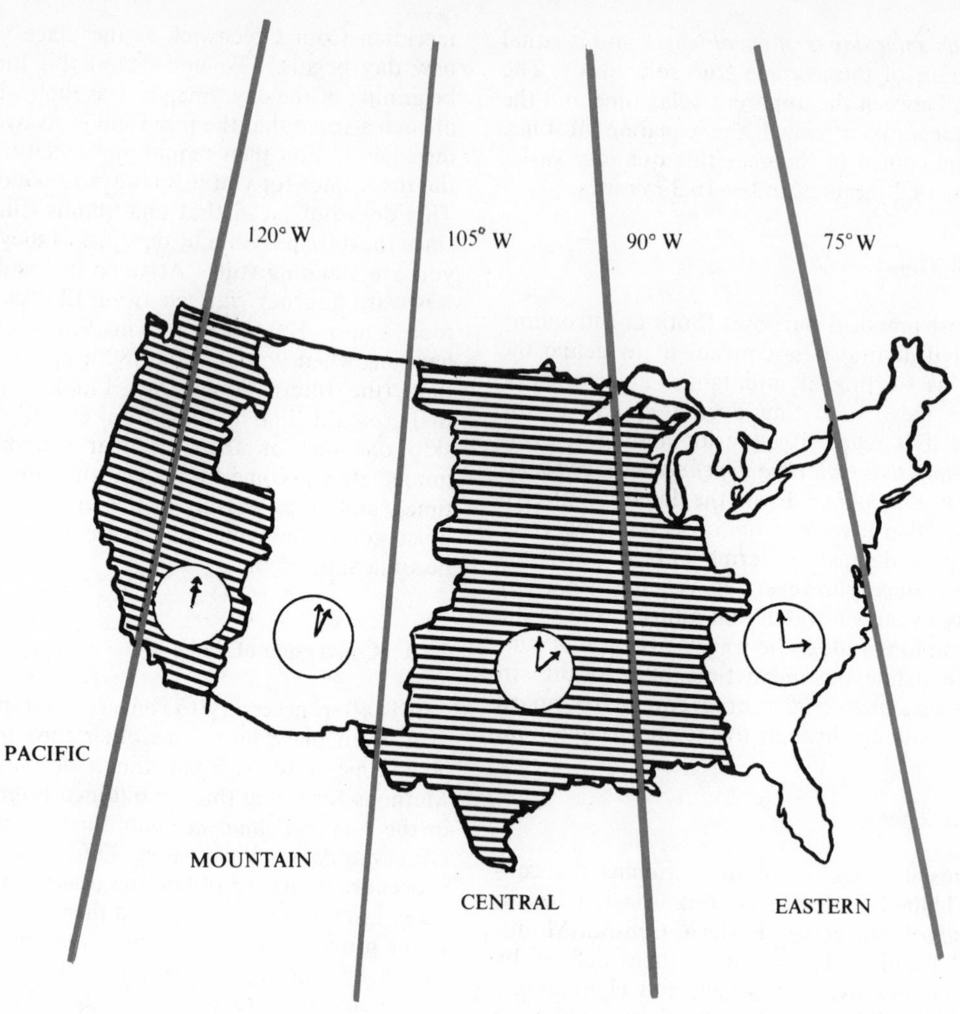

FIG. 2–2 *Time zones for continental United States.*

away from the vernal equinox. Hence, when it is midnight (0^h solar time) for an observer on or about September 22, the vernal equinox is on the observer's meridian and it is then also 0^h sidereal time. From then on the sidereal clock runs about four minutes ahead of the solar clock each day (or about two hours every month), giving us the following approximate table of Greenwich Sidereal Times at midnight Greenwich Solar Time for the various quarter-year periods:

midnight September 22: 0^h,
midnight December 22: 6^h,
midnight March 23: 12^h,
midnight June 23: 18^h.

2.12 Relationship between Solar and Sidereal Time Intervals

Since the tropical year contains one more sidereal day than solar days (the earth rotates $366\frac{1}{4}$ times per year on its axis), it follows that the number of mean solar seconds contained in any time interval is smaller than the number of sidereal seconds in that interval. *To go from solar seconds to sidereal seconds in such an interval, we multiply by* 1.002 738, *which is the ratio* 366.2422/365.2422. We should note that one sidereal day consists of 24 sidereal hours, just as one solar day consists of 24 solar hours. However, one solar day contains 24.065 712 sidereal hours, whereas one sidereal day contains 23.934 467 solar hours.

EXERCISES

1. If a star rises tonight at 8 P.M., at approximately what time will it rise 30 days from tonight?

2. What are the approximate sidereal times when it is noon solar time on: (a) the first day of spring; (b) the first day of summer; (c) April 21; (d) October 21?

3. How much does a sidereal clock gain on a solar clock in 5 hours?

4. If an observer's longitude is 75° west of Greenwich, what is his sidereal time at the moment that Greenwich's sidereal time is 2^h?

5. (a) Assume that the sun moves eastwardly along the ecliptic by an average of 1° per day on March 21. What are its celestial longitude and its right ascension 10 days after it has passed the vernal equinox? (b) With the same assumption as in (a), what would the right ascension and celestial longitude of the sun be 10 days after it has passed the summer solstice?

3

THE EARTH

Although radio telescopes have extended the observable part of our universe to 6 billion light years, or 36×10^{21} miles, the astronomer still finds it necessary to study the earth in order to interpret the vast domain uncovered by his telescopes. Before it is possible to interpret the observations of the planets, the sun, and all the other stars, one must have an accurate picture of the structure of the earth and know just how it is moving with respect to the surrounding stars. In a sense, the earth must still be looked upon as part of the astronomer's apparatus and as the structure to which he must attach the coordinate systems that he uses to study the other heavenly bodies. With the rapid strides now being made in space technology, it will soon be possible to detach ourselves from the earth and set up observatories on other celestial bodies. We shall then have no need for earthbound coordinate systems. Until then, our astronomical studies depend on continued investigation of the earth.

3.1 Fundamental Quantities in Physics and Astronomy

The only way we can arrive at an understanding of the nature of the earth, or of any other astronomical body, is by means of extremely accurate measurements. Here the astronomer operates in a domain common to other scientists, particularly to the physicists. Indeed, insofar as the astronomer deals with motions and structures of bodies (planets, stars, etc.) he must make use of the laws of physics, and work with the measurable quantities with which the physicist also deals.

Although in physics and astronomy we have to make many different kinds of measurements of a great variety of quantities, all of them can be reduced to a few fundamental operations involving the measurement of distances and time intervals. There are, as we shall see, two types of measurable quantities in physics:

1. **fundamental quantities**, which are introduced without definition and receive their meaning from the physical operations performed during their measurement; and
2. **derived quantities**, which are defined in terms of the fundamental quantities.

3.2 Measuring the Distance between Two Points: The Unit of Length

To measure the size and shape of the earth, we must introduce the first of the fundamental quantities that are basic to physics. This measurable quantity is *the distance between two points*. We shall not attempt to define distance, but we shall outline the set of operations that must be performed to measure this quantity.

We start by introducing a *unit of distance*. All of us have a sense of distance which has grown with us from the moment of our birth; even an infant can estimate fairly accurately how far he must extend his hand to reach the bottle that feeds him. However, this built-in sense of distance, although quite satisfactory for daily activity, must be replaced by a *precise unit and a set of precise operations* if we are to carry out scientific measurements.

We take as our unit a very accurately constructed straight edge of arbitrary length which we call, say, a foot, or a meter, or a yard. If we now wish to determine the distance between any two points A and B we proceed as follows: we place one end of our unit length in coincidence with one of the points (say the left end in coincidence with point A, as shown in Figure 3.1) and then see how many times the unit must be laid off in this manner before point B is reached. This number of times is called the **distance** between the two points. Since, in general, the distance will not be equal to an integral number of units, the unit itself must be subdivided

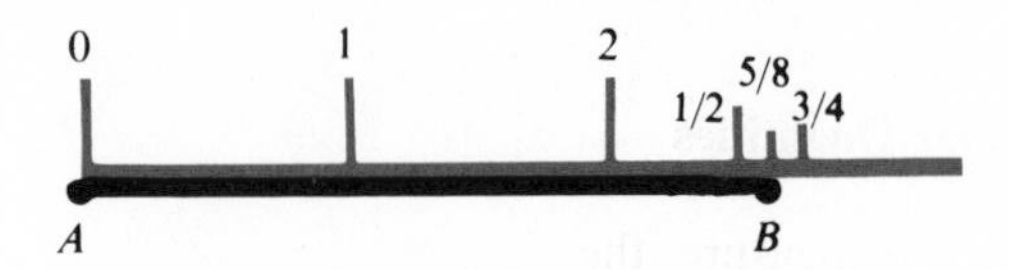

FIG. 3–1 *Measurement of distance AB using a unit length.*

as often as may be needed to achieve the desired accuracy in the measurement. In physics the centimeter (the hundredth part of the meter) is commonly used as a unit of length.

3.3 Units of Length in Astronomy

The size of the unit used in expressing distances depends on the magnitude of these distances. Thus, for distances of the same order of magnitude as the height of a man, inches, feet, or centimeters (cm) are most suitable. For distances between cities, the mile or kilometer ($\frac{5}{8}$ mile) is most convenient. In astronomy, various units are used. *In considering the solar system, we generally use the* ***astronomical unit*** (*the mean distance between the earth and the sun*), which is approximately 1.49×10^{13} cm or 93 000 000 miles, whereas, *for stellar distances it is more convenient to use either the* ***light year*** (*the distance that light travels in one year*), about 9.6×10^{17} cm (about 6 000 000 000 000 miles), *or the* ***parsec***, about 3.26 light years.

3.4 Limitations in Measuring Distances

The above procedure cannot be applied to the measurements of extremely small (atomic) or extremely large (astronomical) distances. Atomic and astronomical dimensions must be measured indirectly by methods involving (at times complex) geometrical concepts and the laws of physics and of astronomy. However, we must understand that the usual geometrical interpretation of distance may not apply to these measurements. Experience has shown that when the distance-measuring operation differs radically from that used in introducing the unit, unusual properties are obtained. Even in these extreme instances, however, the direct method must be introduced at some stage of the measurement process. Thus, all measurements of astronomical distances must ultimately depend on the precise determination of some fundamental length on the surface of the earth (such, for example, as the distance between two points on the arc of a great circle on the earth's surface).

3.5 Scalar Quantities

When we measure the distance between two points, we obtain a number that tells us nothing at all about the direction from one point to the other. Such a number is called a **scalar** quantity. We may generalize this idea by stating that *a scalar is any measurable quantity that is completely specified by a single number*. Thus, the price of wheat, the amount of money a person has in the bank, the temperature of the atmosphere, and the speed of a body are scalars. We note that *two scalars referring to the measurements of similar quantities may be added like ordinary numbers*. Thus, if a man first walks a distance of 5 miles, then a distance of 6 miles, he has walked altogether a distance of 11 miles, as shown in Figure 3.2.

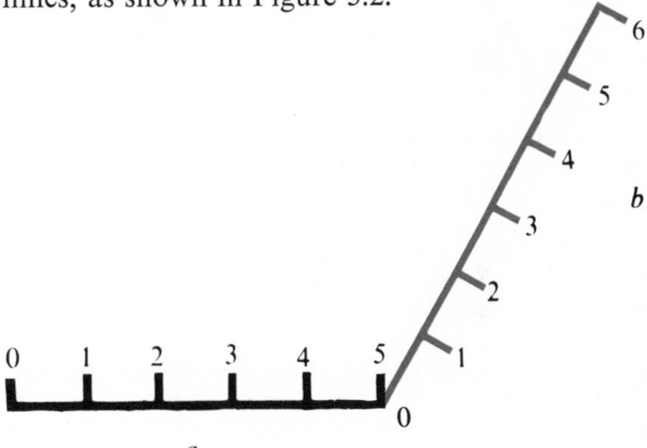

FIG. 3–2 *The total distance walked by an observer is the sum of the distinct distances regardless of the directions of these distances.*

3.6 Vector Quantities

If we now go back again to our two points, we see that the knowledge of the scalar distance between them does not give us a complete picture, since there is also a *direction associated* with the two points with respect to some frame of reference, as shown in Figure 3.3. We shall very often be concerned with quantities that can be completely specified only if *in addition to a magnitude a direction is also given*: *such quantities are called* **vector** *quantities*.

If we consider the displacement of an object from the point A to the point B in Figure 3.3 we see that this is a vector quantity; the *direction* from A to B is as important in defining the displacement as is the *distance* from A to B. *Vector quantities cannot be added like numbers.* This can be seen in Figure 3.3 where we consider the displacement from A to C, which is the sum of the displacements from A to B and from B to C. If the distance from A to B is three units of length, and the distance from B to C is four units, then the total distance traveled by the object in moving from A to B and then from B to C is seven units of length, but the length of the

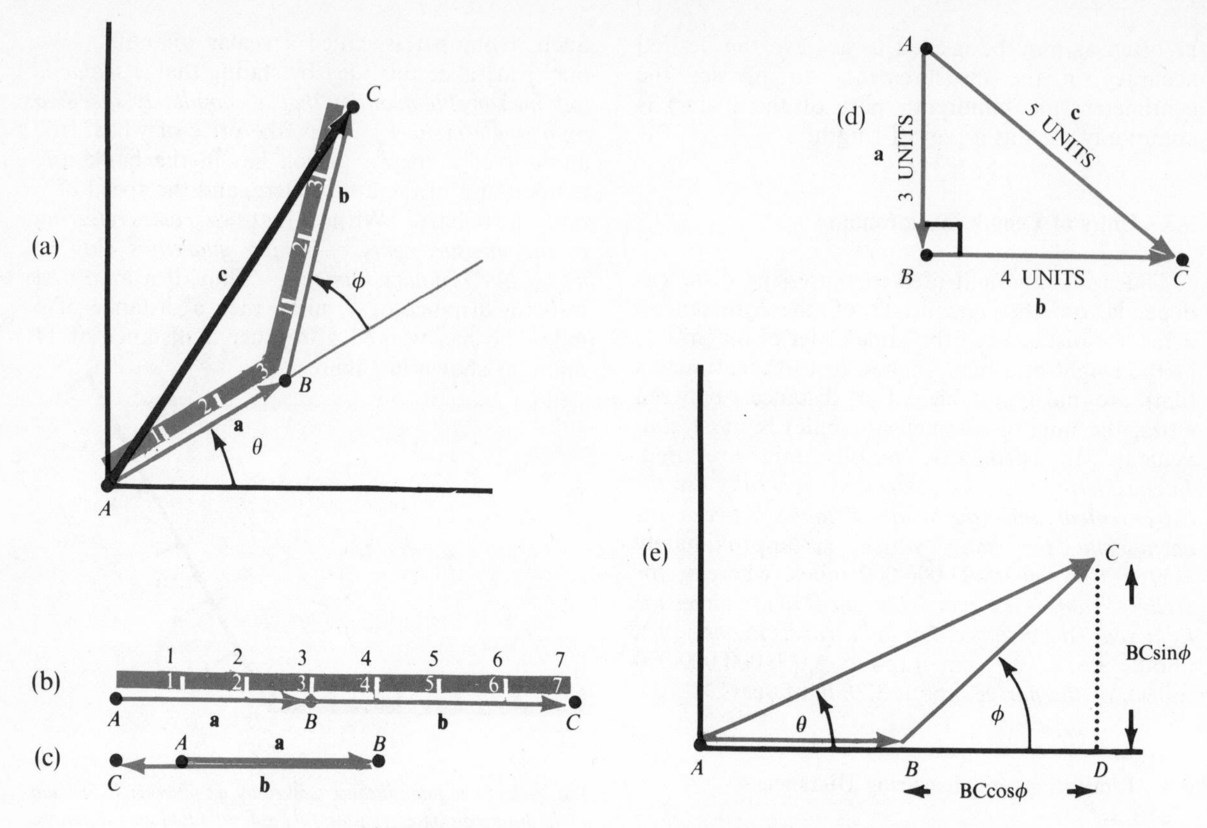

FIG. 3–3 *(a) The displacement from A to C is the sum of the displacements* **a** *and* **b***; its length depends on the angle ϕ between these two vectors. (b), (c), and (d) The length of the resultant displacement AC depends on the angle between the two vectors; it varies from a + b to a − b as the angle varies from 0° to 180°. (e) The derivation of the formula for the length and the direction of the resultant vector from A to C.*

displacement AC may vary anywhere from seven units to one unit depending on the angle between the two displacements AB and BC.

3.7 The Addition of Vector Quantities

If the second displacement above is in the same direction as the first, the length of AC is seven units, and if the two displacements are opposite to each other, the length of AC is one unit. A very simple situation arises if the two displacements are at right angles, in which case AC forms the hypotenuse of a right triangle and its length is given (by the theorem of Pythagoras) as the square root of the sum of the squares of the lengths of AB and BC, which in this case is five units.

In general, the length of the displacement AC is obtained from the lengths of the two displacements AB and BC, making an angle ϕ with each other by applying the formula (the bar over the figure signifies length)

$$\overline{AC} = \sqrt{\overline{AB}^2 + \overline{BC}^2 + 2(\overline{AB})(\overline{BC}) \cos \phi}. \quad (3.1)$$

The direction θ of the resultant displacement AC with respect to the displacement AB is given by

$$\tan \theta = \frac{BC \sin \phi}{AB + BC \cos \phi}. \quad (3.2)$$

For the definitions of sine, cosine, and tangent of an angle ϕ see Section 3.10.

3.8 The Measurement of Angles: The Degree

We have seen that to add two vectors we must take into account the angles between them. This then leads us to the consideration of angles, which also must be measured to determine the size and shape of the earth. An angle is a measurement of turning about an axis; and we must now introduce a unit of turning, or of **angular measure**. Distance and angle are related to the two different kinds of

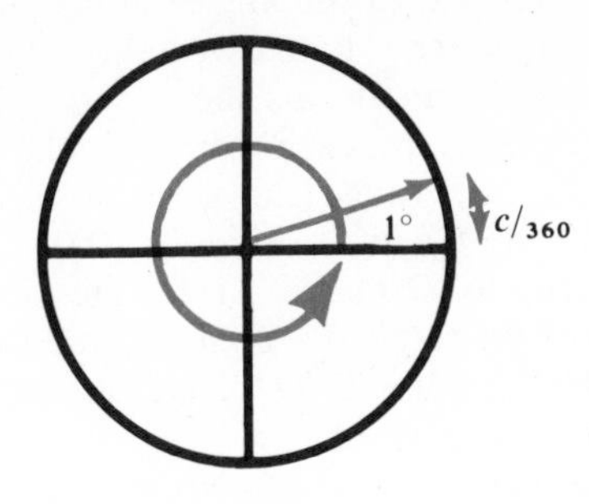

c = Circumference

FIG. 3–4 *The definition of the degree.*

motion that are possible in our universe. Whereas distance measures the displacement from one point to another (translational motion), angle measures the rotation from one direction to another.

We shall define the unit of angle or turning as a certain fraction of one complete rotation, where a complete rotation means the amount through which we must turn in order to come back to the direction we were facing initially, as shown in Figure 3.4. The most common unit employed is the degree (the 360th part of a complete turn), which we have already used in the general discussion of celestial coordinates. *An angle is a vector quantity* whose direction is given by the direction of the axis about which the rotation associated with the angle takes place. (See the right-hand screw diagram in Figure 3.5, where the angle is described in a counterclockwise direction.)

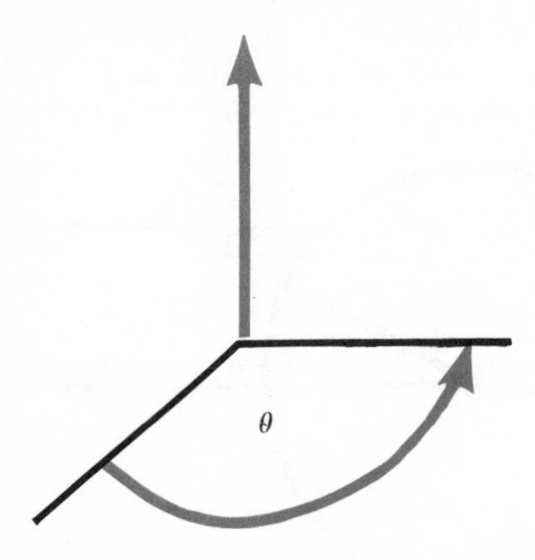

FIG. 3–5 *An angle is a vector quantity whose direction is perpendicular to the plane of the angle θ.*

3.9 The Radian

Another important unit of angle that has extensive application in physics and astronomy is obtained by considering a point moving along the circumference of a circle. If, while standing at the center of the circle, we keep facing the point as it moves along the circumference through a distance equal in length (measured along the circumference) to the radius of this circle, we turn through an angle called one **radian**. If we turn through two radians the amount of circumference covered is twice the radius, through three radians the amount covered is three times the radius, etc., as shown in Figure 3.6. We see from this that the radian is a very

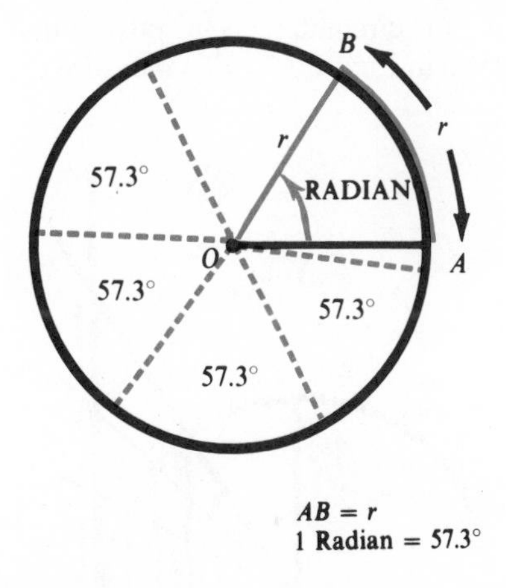

FIG. 3–6 *The definition of the radian.*

useful unit of angle. The length of arc through which a point has moved along the circumference of a circle can be found at once by multiplying the radius of the circle by the angle (expressed in radians) subtended by the arc, as shown in Figure 3.6. We may express this algebraically as follows:

$$s = r\theta, \tag{3.3}$$

where $s =$ the length of arc,
$r =$ the radius of the circle,
$\theta =$ the subtended angle expressed in radians.

To find the relationship between the radian and the degree, all we need note is that the number of times the radius is contained in the circumference of a circle is 2π, where π is very nearly equal to 3.141 59. Since every time we lay off one radius along the circumference we turn through one radian and the complete circumference corresponds to a turning through 360°, it follows that 2π radians equal 360°. In other words

$$1 \text{ radian} = \frac{360°}{2\pi} = 57.295\,78°$$

or

$$1 \text{ radian} = 206\,265''$$

(radian measure in seconds of arc).

3.10 Sines, Cosines, and Tangents

Although the most common way of measuring an angle is by means of a unit of rotation, we can also measure angles by using a unit of length. To see this we consider a circle having a radius of unit length (1 cm, 1 ft, or whatever other unit may be used). We lay off radius OA and allow a point to move along the circumference to positions B, C, D, etc. as shown in Figure 3.7. To measure the angles

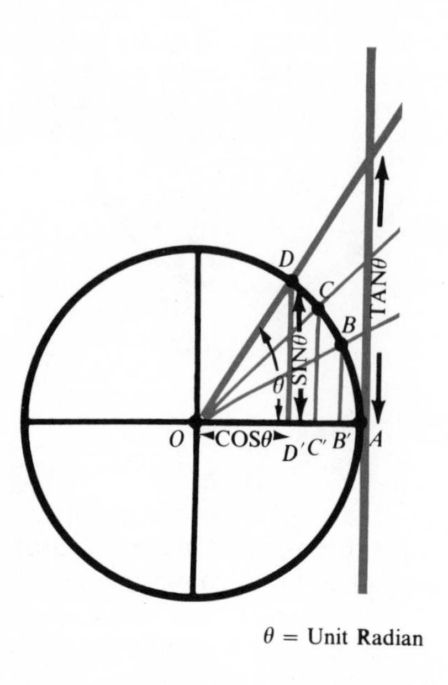

FIG. 3–7 *The circular definition of the sine, cosine, and tangent.*

θ_B, θ_C, θ_D, etc. (subtended respectively by the arcs AB, AC, AD, etc.), all we need note is that if we drop perpendiculars from the points B, C, D, etc. to the radius OA, the lengths of these perpendiculars increase with the angles. It should therefore be possible to get an idea of how big the angles are by measuring these perpendicular distances.

This, indeed, turns out to be the case, for although these lengths (the lines BB', CC', DD', etc.) are not proportional to the angles, they are nevertheless related to the angles in a unique and well-known manner. Extensive tables have been prepared by mathematicians showing how large an angle corresponds to each perpendicular length. These lengths are called the **sines** of the angles. Thus $BB' = \sin\theta_B$, $CC' = \sin\theta_C$, $DD' = \sin\theta_D$, etc. The lengths OB', OC', OD', etc. are called the **cosines** of the respective angles θ, and the ratios BB'/OB', CC'/OC', DD'/OD', etc. are called the **tangents** of the angles.

Note that these definitions apply only if the circle has a unit radius. If the radius is different from unity, then each perpendicular distance must be divided by the radius to give the sines, and, similarly, the distances OB', OC', OD', etc. must be divided by the radius to obtain the cosines. From the above definitions we see at once that $\sin^2\theta + \cos^2\theta = 1$ and that $\tan\theta = \sin\theta/\cos\theta$.

3.11 The Size and Shape of the Earth

Now that we have introduced a unit of length and a unit of angle, we are in a position to discuss the shape and size of the earth. Although to a very good approximation we may consider the earth as being spherical, it is really somewhat flatter at the poles than at the equator (as a consequence of its rotation about an axis). Because of the flattening it is almost an oblate spheroid, although this too is an approximation, since the data gathered by artificial satellites during and after the I.G.Y. (International Geophysical Year) indicate that the amounts of flattening at the two poles are not equal.

If we now assume the earth to be spherical, we can determine its size to a fairly good approximation by making simultaneous observations of the sun from two different points on the earth's surface. In Figure 3.8 two observers A and B are stationed along the same meridian and separated by an accurately measured arc of length AB. At the very

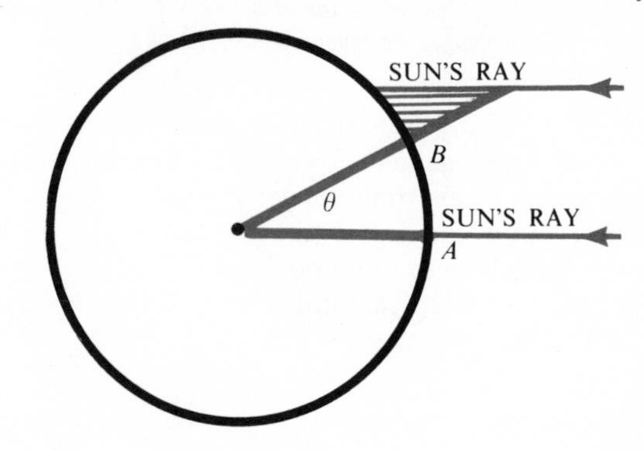

FIG. 3–8 *Eratosthenes' method of measuring the circumference of the earth.*

moment that the observer at A finds the sun exactly at his zenith (observer A must have a latitude such that the sun passes across his zenith at least once a year), the observer at B measures the zenith distance of the sun and finds, let us say, the angle θ. It follows that the arc AB is contained as many times in the circumference of the earth as the angle θ is contained in 360°. This method was first applied to the determination of the earth's radius by the Greek Eratosthenes, in about the year 250 B.C. Observations were made under his direction at two stations in Egypt, one at Alexandria, and the other at Syene, where the sun was at the zenith.

The most accurate procedure for determining the size and shape of the earth today is a variation of this method, which consists essentially in the measurement of a number of arcs of a great circle (for example, a meridian) at widely separated latitudes. Careful measurements of this sort indicate that the length of arc along a given meridian corresponding to a single astronomical degree increases with increasing latitude, as shown in Figure 3.9. *This means simply that the earth must be flatter the closer one approaches the poles* (as

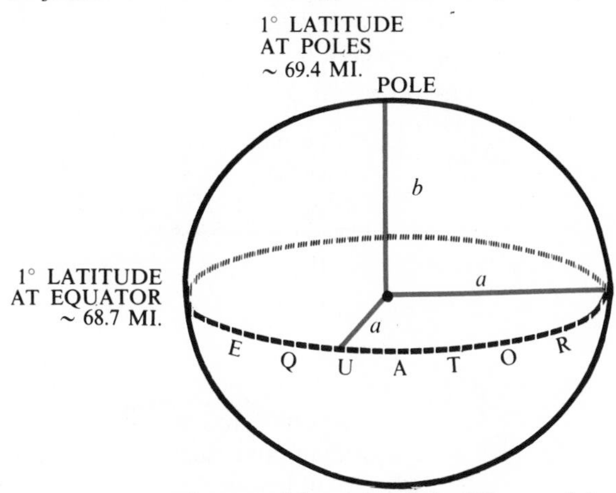

FIG. 3–9 *The oblateness of the earth and the difference of the length of the degree at the equator and at the poles.*

we have already noted). Accurate measurements show that one degree of latitude at the poles of the earth is about 0.7 mile longer than at the equator. The length of the degree (that is, the distance along a meridian between two points differing in latitude by 1°) varies from 68.7 miles at the equator to 69.4 miles at the poles.

3.12 Areas

If we add two lengths, or subtract them, we are still left with a length, so that no new quantities can be obtained in this way. However, we do obtain an important *new quantity by multiplying two lengths together*. This product is a measure of the area enclosed in the parallelogram formed from the two lengths, as shown in Figure 3.10. *An element of area is a vector quantity because it can be tilted in various directions*, and can be represented by a directed line (similar to a displacement) whose length is taken equal to the numerical value of the area, $(AB)(AC) \sin \theta$, and whose direction is taken perpendicular to the plane of the area.

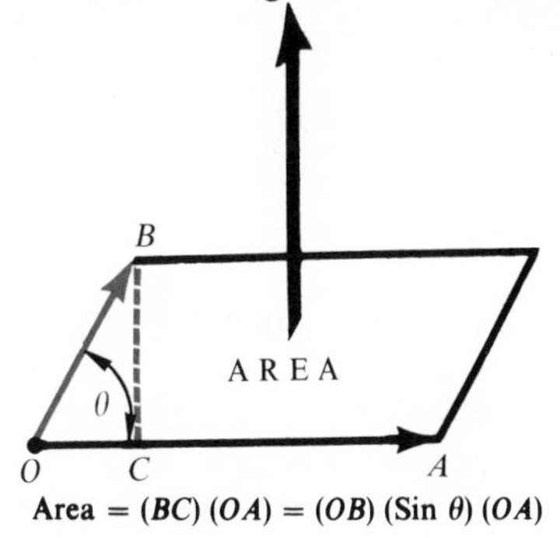

FIG. 3–10 *The vectorial definition of an area bounded by two vectors.*

3.13 Volumes

We can obtain another important physical quantity by multiplying three lengths together. This product defines a volume, and it is easy to see that *volumes are scalar quantities, because there can be no direction associated with them.* Thus, volumes can be added or subtracted like any other numbers, but this is not true of areas. Since areas are vectors, they must be added like vectors.

3.14 Mean Radius of the Earth: Its Surface Area and Its Volume

If the earth were a perfect sphere, one could obtain its surface area and its volume from the respective formulae $4\pi R_\oplus^2$ and $\frac{4}{3}\pi R_\oplus^3$, where $R_\oplus$ is the radius of the earth. Since the earth is an oblate spheroid, however, these formulae do not apply. We can, nevertheless, use them to get a very good approximation if, instead of $R_\oplus$, we use the mean radius of the earth. For an oblate spheroid, the mean semidiameter is equal to $(2a + b)/3$ (as shown in Figure 3.9), where a is the equatorial radius, and b is the polar radius, the two differing by about $13\frac{1}{2}$ miles (21.8 km).

This formula for the mean value has to be used because two of the three mutually perpendicular axes of symmetry that can be drawn through the center of the earth are equatorial diameters, and the third is the polar diameter or the axis of rotation. Since the mean radius of the earth is found to be 3 981.89 miles (6 371.03 km), its surface area is 196 950 000 mi^2 (5.1×10^{18} cm^2), and its volume is 1.083×10^{27} cm^3.

3.15 Time as a Fundamental Physical Quantity

Having introduced the concepts of length and angle, we were able to determine the shape and size of the earth. However, there are other physical properties of bodies which cannot be represented only by spatial and angular measurements. One of these is **mass**, that is, the amount of material contained in a body. To measure the mass of the earth we must introduce additional fundamental quantities, as well as certain derived quantities and certain relationships among them.

We could, of course, take the mass itself as a fundamental quantity and outline a procedure for measuring the masses of bodies in terms of a unit of mass, arbitrarily introduced. However, we do not have as intuitive a grasp of the concept of mass as we do of the concept of length. Most people have incorrect notions of the meaning of the mass of a body, and hardly any idea at all as to how one determines the mass of a body by comparing it with some standard unit. It is preferable, therefore, to introduce other fundamental quantities which we can grasp intuitively, and then derive the mass in terms of these intuitive fundamental quantities. The first of these additional fundamental quantities we shall introduce is **time**.

In the previous chapter we saw that the astronomer finds it necessary to introduce a time scale in order to keep a record of the motions of the heavenly bodies; but *in the formulation of the laws of physics time must be introduced as a fundamental quantity on the same footing as distance.* Just as in the case of distance, we shall not attempt to define this concept, but simply introduce a *unit of time*, and describe a set of operations that will enable us to express any interval of time in terms of this unit.

The unit of time, the **second**, that appears in the laws of physics is the same as that used by astronomers. This interval is approximated very closely by the beat of a pendulum constructed to definite specifications, and set oscillating at a specific point on the earth's surface. In recent years highly accurate atomic clocks, such as the **maser** clock, have been developed; these employ the oscillations of molecules. Such clocks are accurate to one part in ten billion, which means that they will not gain or lose more than one second every three hundred years. *The length of any interval of time can now be defined as the number of beats the pendulum or atomic clock executes during this time interval.*

This method of measuring intervals of time can be applied directly only where the time intervals to be measured are not too long or too short. When we speak of an event as occurring over a period of millions of years, or of some atomic process as having taken place in a hundred-millionth of a second, obviously we are not referring to the sort of operation described above. Where very long or very short time intervals are involved, the results are not obtained by direct measurements, but are derived by the applications of the laws of physics.

3.16 Velocity of a Particle

With the two fundamental quantities, length and time, it is now possible to obtain two derived quantities that play important roles in astronomy. We consider an object (such as a planet) that is in motion, and follow it as it changes its position along a straight line from the point A to the point B as shown in Figure 3.11. If we measure the distance,

(a)

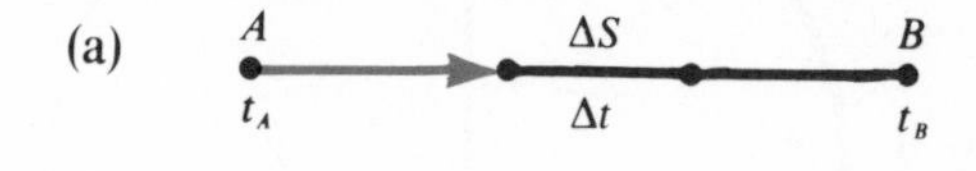

a

(b) A B

FIG. 3–11 *(a) The definition of the average speed of a particle moving along a straight line from A to B. (b) Both the magnitude and the direction of the velocity of a particle change as it moves along an arbitrary curve.*

s, from A to B, and measure the time interval during which the particle was moving over this distance, we may define the average velocity of the body as

$$\bar{v} = \text{average velocity} = \frac{s}{t} = \frac{\text{distance}}{\text{time}}. \quad (3.4)$$

The reason we speak of this as the average velocity is that the motion of the body may have undergone some changes during the time interval, t, and the bare ratio of s to t cannot take these changes into account. If we wish to find the instantaneous velocity of a body, we must choose s and t so small that during such a small interval of time and over such a small distance the velocity undergoes practically no measurable change. This is the basic idea behind Newton's development of the **calculus**. If, then, we use the symbol Δ to signify a small quantity, we may define the instantaneous velocity as

$$v = \frac{\Delta s}{\Delta t}.$$

Following Newton's notation, this is often written as $\dot{s}$ if Δt is allowed to become infinitesimal.

Since the motion of a body is always in some direction, *velocity is a vector quantity*, so that, in general, velocities cannot be added like ordinary numbers, but must be added vectorially (see Section 3.7). Thus, if a boat is moving in an eastwardly direction at a speed of 4 miles per hour and a man is walking on the deck of the boat, at right angles to its motion, at a speed of 3 miles per hour relative to the boat, his motion (velocity) relative to an observer on land is 5 miles per hour in a southeasterly direction, as shown in Figure 3.3 (d).

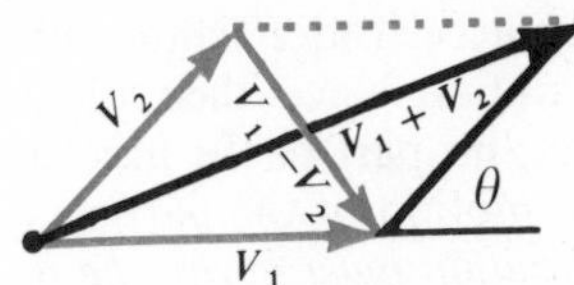

FIG. 3–12 *The vectorial sum and difference of two velocities.*

In the case of the particle in Figure 3.11(a) we assumed that it moved from A to B along the straight line connecting these two points, but we may also consider the more general case of motion from A to B along the arc of a curve, as shown in Figure 3.11(b). In this case the average speed $\overline{v}$ is defined in terms of the total distance measured along the curve from A to B. At each instant during the motion of the particle the direction of the velocity is given by the direction of the tangent drawn to the curve at the point occupied by the particle at that instant. We note that over an extended time the direction of the velocity of a body, as well as its magnitude, may change. The magnitude of the velocity of a body is always expressed in units of distance per time, such as centimeters per second, miles per hour, light years per century.

3.17 Acceleration of a Particle

We have noted that the speed and the direction of motion of a particle may change during a finite interval of time, and we shall see that this leads to the introduction of a second important derived quantity involving space and time. We consider a particle in motion at two different instants of time, t_1 and t_2, and suppose that its corresponding velocities at these two moments are $\mathbf{v}_1$ and $\mathbf{v}_2$. If we now subtract $\mathbf{v}_1$ from $\mathbf{v}_2$ (this must be done vectorially as in Figure 3.13), and divide this difference by the interval of time between t_1 and t_2, we obtain the expression:

$$\mathbf{a} = \frac{\mathbf{v}_2 - \mathbf{v}_1}{t_2 - t_1}. \qquad (3.5)$$

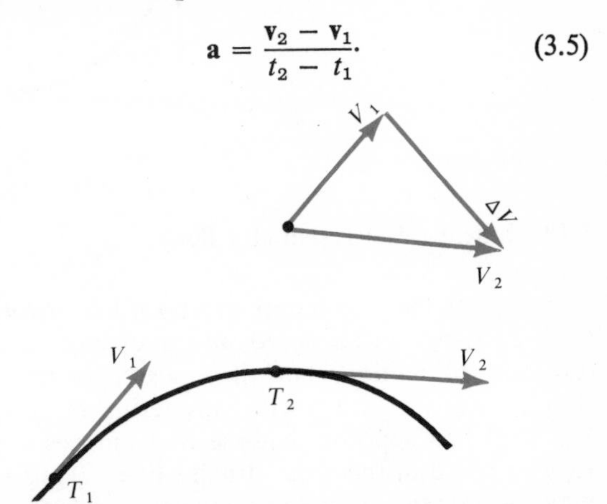

FIG. 3–13 *Vectorial subtraction of the velocity at time t_2 from the velocity at time t_1.*

We may define this as the average acceleration suffered by the particle during this time interval. If we take the time interval as very short, we obtain the instantaneous acceleration of the particle; that is, we have

$$\mathbf{a}_{\text{instantaneous}} = \frac{\Delta \mathbf{v}}{\Delta t};$$

if Δt is infinitesimal this is written as $\dot{\mathbf{v}}$ or $\ddot{\mathbf{s}}$.

We see that the acceleration of a particle is simply the rate at which its velocity is changing, and is therefore expressed as a velocity per time. Since velocity itself is distance per time, it is clear that acceleration can be expressed as a distance per time per time, or distance per time squared. We shall illustrate this later with a few examples.

Since acceleration is essentially a velocity divided by a time, and velocity is a vector quantity whereas time is a scalar, it follows that acceleration is a vector quantity. In other words, the acceleration of a body must have a definite direction at each instant of time.

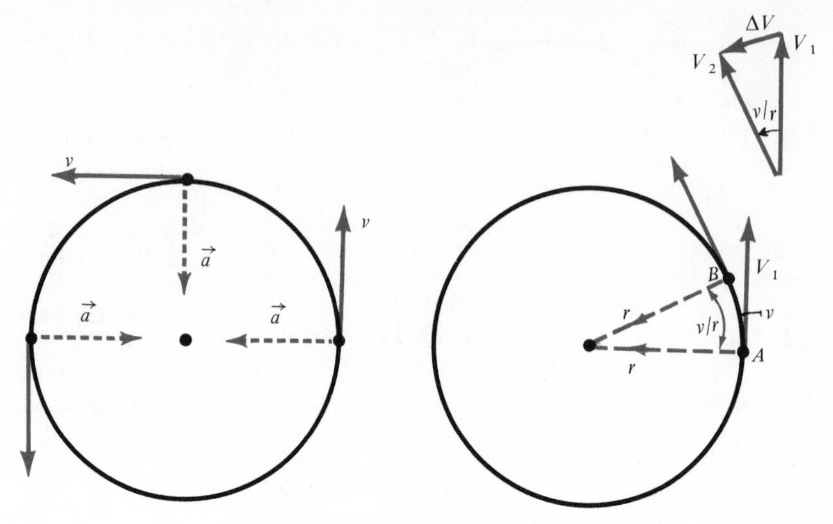

FIG. 3-14 *The acceleration of a particle moving with constant speed in a circle.*

3.18 Linear Acceleration of a Body

In general, the acceleration of a body may involve both a change in its speed and a change in its direction of motion, but there are two special simple cases that are worth considering in detail. We shall first suppose that the body moves with varying speed in the same straight line. Thus we may consider an automobile moving with increasing speed along some fixed direction. If, as we watch it, its speed increases from 15 miles an hour to 18 miles an hour after 2 seconds, and then increases to 21 miles an hour after another 2 seconds, and so on, we say that the automobile is moving with uniform acceleration. If we apply our formula for the acceleration, we see that the difference between the speed of the automobile at the beginning and the end of each 2-second interval is 3 miles per hour, so that the acceleration is 3 miles per hour per 2 seconds, or $1\frac{1}{2}$ miles per hour per second.

$$a = \frac{v_2 - v_1}{t_2 - t_1} = \frac{18 \text{ mph} - 15 \text{ mph}}{2 \text{ sec}} = \frac{3 \text{ mph}}{2 \text{ sec}} = \frac{\Delta v}{\Delta t},$$

$$a = \frac{v_3 - v_2}{t_3 - t_2} = \frac{21 \text{ mph} - 18 \text{ mph}}{2 \text{ sec}} = \frac{3 \text{ mph}}{2 \text{ sec}} = \frac{\Delta v}{\Delta t},$$

where the symbol Δ again stands for a small difference.

We may change this to miles per minute per second or miles per second per second by dividing first by 60, then by 60 again. Thus the acceleration is $\frac{1}{40}$ mile per minute per second, or $\frac{1}{2400}$ mile per second per second or *per second squared.* The acceleration here is in the direction of the body's motion.

3.19 Acceleration for Circular Motion

As another special case of acceleration we consider the situation in which a body's speed remains constant but the direction of its motion changes. In this case the acceleration cannot be in the direction of the motion since then the speed would change; in fact, no part of the acceleration can lie parallel to the motion of the particle. *This means that the acceleration must always be at right angles to the velocity*, as shown in Figure 3.14, so that the particle must be moving along the arc of a circle with constant speed.

As an example, we may consider an automobile moving southwardly at a speed of 10 miles per hour and then turning eastwardly without changing speed. The automobile suffers an acceleration even though its speed is constant, and the effect of the acceleration is merely to change the direction of its motion.

It is easy to see how to determine the acceleration of a body if the acceleration is always parallel to the direction of the body's motion, but the problem is a bit more difficult if the acceleration is at right angles. Since motion accelerated at right angles is very important in the study of the motion of the earth and planets around the sun, we shall write down the mathematical expression for it.

The simplest example of this kind of accelerated motion is that of a particle moving with constant

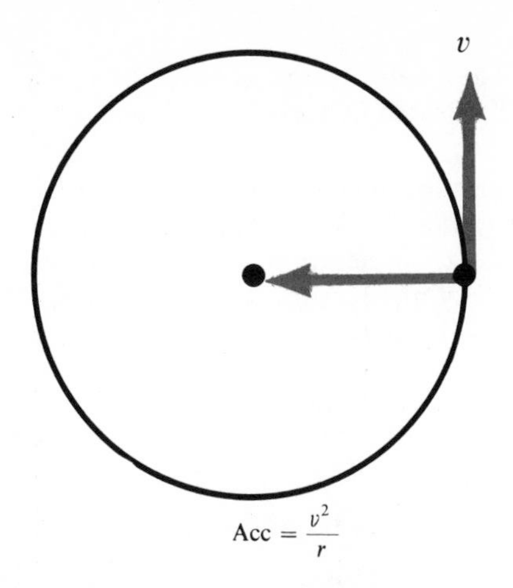

FIG. 3–15 *The diagrammatic representation of the acceleration of a particle moving with constant speed in a circle in terms of the speed of the particle and the radius of the circle.*

speed v in a circular orbit of radius r, as shown in Figure 3.15. We can obtain the formula for the acceleration of a body in this case if we note the following things:

1. this formula can involve only the speed v and the radius r, since these are the only physical quantities associated with the motion, and
2. the acceleration must always be expressed as a length divided by the square of a time.

Since velocity itself is a length divided by a time, we see that the only combination of v and r that gives a length divided by the square of a time is

$$\boxed{\frac{v^2}{r}} \tag{3.6}$$

This proves to be exactly the formula that we want for the acceleration of a body moving with constant speed in a circle. The acceleration in this case is always directed from the particle along a radius towards the circle's center.

We have based this formula only on general dimensional arguments, and it remains to be proved. To obtain a rigorous proof (which we shall not give here) one must apply the laws of geometry to Figure 3.15.

Since $v = 2\pi r/P$, where P is the period of the particle (the time it takes to go around once), we may also write the above formula for the circular acceleration as

$$\boxed{a_{\text{circular}} = \frac{4\pi^2 r}{P^2}}. \tag{3.7}$$

3.20 The Acceleration of a Freely Falling Body on the Earth's Surface

As an important example of accelerated motion we consider a body falling freely on the surface of the earth. Obviously we are dealing here with the first case discussed above since the body falls along a straight line (towards the center of the earth) and only the speed changes. An important thing to note about this **acceleration of gravity** is that *it is the same for all bodies (neglecting air resistance) falling at the same point on the earth's surface.*

This remarkable fact was first discovered by Galileo. It marked the beginning of modern dynamics because it directly contradicted the Aristotelian idea, held at that time, that heavy bodies fall more rapidly than lighter ones. By allowing bodies to slide down inclined planes of differing slopes, Galileo was able to make measurements on the acceleration of gravity and to arrive at a fairly good idea of its value.

Today, accurate determinations can be made by measuring the period of a pendulum and then using a simple formula relating this period to the length of the pendulum and the acceleration of gravity. *We find that this value changes slightly from point to point on the earth's surface.* This indicates that the earth is not a perfect sphere and that its internal matter is not distributed around its center in uniform shells. As we move away from the equator, the acceleration of gravity increases slightly towards the poles. The average value of this acceleration, which we designate with the letter g, is 980.665 cm per sec per sec (or cm/sec^2). Since this is very nearly 980 cm/sec^2, we shall take this value as sufficiently accurate for our purposes. We may also express it as 32.2 ft/sec^2.

This means that if a body is allowed to fall freely from rest near the earth's surface, its speed at the end of the first second is 980 cm/sec, at the end of the second second is twice this value, at the end of the third second is three times this value, etc.

3.21 The Speed of a Freely Falling Body

If the falling body already has a speed v_0 when we first observe it, then its speed at the end of the first second is v_0 + 980 cm/sec, at the end of the second second v_0 + twice 980 cm/sec, etc. We thus have the following simple prescription for finding the speed in cm/sec of a freely falling body after a given interval of time: to the speed it has at the moment we observe it add 980 as many times as there are seconds in the interval of time. We may express this by means of the following formula:

$$v_f = v_i + 980t \equiv v_i + gt, \qquad (3.8)$$

where v_f = final velocity (at the end of time interval t), and

v_i = initial velocity (at the start of time interval t).

This equation applies to a body falling towards the earth. If a body is rising (it is considered to be in free fall even then), we can still find its speed at any moment by subtracting gt from its initial speed v_i.

3.22 Distance Traveled by a Freely Falling Body

We can find the distance through which a freely falling body moves by using the fact that the distance is equal to the average velocity times the time. Since in the case of a freely falling body the speed increases regularly (by 980 cm/sec every second), the average speed is found by taking one-half the sum of the initial and final speeds. In other words, the average speed is equal to one-half the sum of v_i (initial speed) and $v_i + gt$ (final speed), or

$$\bar{v} = \tfrac{1}{2}(v_i + v_i + gt) = v_i + \tfrac{1}{2}gt.$$

To obtain the distance through which a body falls during the time t, we must multiply this average speed ($\bar{v}$) by the time t, which gives us the following formula:

$$\boxed{d = \bar{v}t = v_i t + \tfrac{1}{2}gt^2}\,. \qquad (3.9)$$

This formula applies to any object moving with uniform acceleration if g is replaced by the appropriate acceleration.

This leads to a simple procedure for measuring the acceleration of gravity. All we need note is that if we allow a body to fall from rest, the initial speed v_i is zero—which means that the first term of the formula ($v_i t$) drops out. If we allow the body to fall for one second, then $t = 1$, and we have the formula

$$d = \tfrac{1}{2}g,$$

or

$$g = 2d. \qquad (3.10)$$

In other words, the acceleration of gravity is twice the distance through which a body falls from rest during its first second of motion.

3.23 The Velocity of a Freely Falling Body Expressed in Terms of Distance

Another formula which we shall find useful and which applies to a body moving with uniform (that is, constant) acceleration relates the velocity v of a body at any moment to its initial velocity v_i, to the distance d through which it has moved in the time t, and to its acceleration a. Since the distance d the body has moved in the time t is

$$d = v_i t + \tfrac{1}{2}at^2,$$

and the velocity at the end of this time is

$$v = v_i + at,$$

we may solve for t from the second equation:

$$t = \frac{v - v_i}{a}$$

and substitute it in the first to obtain

$$d = v_i\left(\frac{v - v_i}{a}\right) + \frac{1}{2}a\left(\frac{v - v_i}{a}\right)^2,$$

or

$$d = \frac{v^2}{2a} - \frac{v_i^2}{2a},$$

so that

$$\boxed{v^2 = v_i^2 + 2ad}\,. \qquad (3.11)$$

Thus, if a body falls freely from rest (initial velocity $v_i = 0$) through a height h, its speed is

$$v = \sqrt{2gh},$$

where g is the acceleration of gravity.

3.24 The Unit of Force: The Dyne

Velocity and acceleration are the only new derived quantities that we can obtain by combining length and time, but if we wish to continue our pursuit of the concept of mass, and to understand the dynamics of the solar system and the stars, we must introduce another fundamental quantity at this point. This we shall take to be force, rather than mass, which is customarily taken as a fundamental quantity in physics. Once we have introduced force in this way, we shall see that the mass of a body is obtained as a derived quantity.

Since we are subjected to all kinds of forces throughout our lives, we soon develop a force sense which enables us to estimate the magnitude of forces almost subconsciously, so that we automatically flex our muscles just the right amount if we wish to open a door, or pick up a chair, or hit a tennis ball. But this way of estimating force is not good enough for scientific work; we must introduce the unit of force by means of an accurate instrument which we can properly calibrate. *A device of this sort is a helical spring, which has the property that the amount by which it stretches under the action of a force is proportional to the force.*

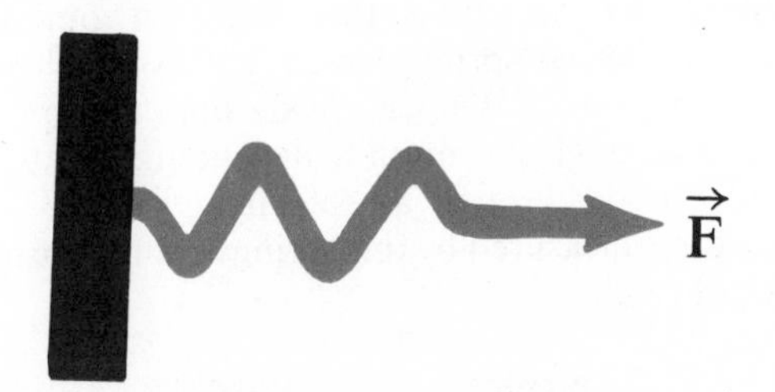

FIG. 3–16 *A force, as represented by the stretch of a spring.*

We can now define the unit of force, which we shall call the **dyne**, as that force which stretches a particular kind of spring a specific amount. Any force that stretches this same spring twice this amount has a value of two dynes, etc. Another unit of force that is in common use is the **pound**, which is defined as approximately 445 000 dynes. We could, of course, have defined the pound in terms of the stretch of an appropriate type of spring, and then introduced the dyne by means of the approximate conversion factor 445 000.

Since a force can be exerted in different directions, it is a vector quantity, and the resultant of the forces acting in different directions must be obtained by vector addition. If more than two forces are acting on a body, they can be combined vectorially in pairs until the final resultant of all forces is obtained, as shown in Figure 3.17.

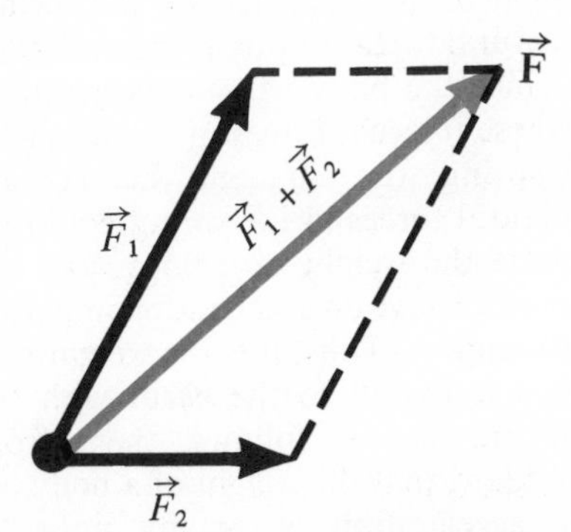

FIG. 3–17 *The vectorial sum of two forces.*

3.25 Types of Forces in Nature

Although we are constantly under the action of forces in our universe, only four fundamental kinds of forces have been discovered, and all other known forces can be reduced to these. The forces resulting from the contact of bodies, such as frictional forces, collisional forces, pressures, etc. are all due to molecular interactions, and therefore can be explained in terms of the electrical and magnetic forces between the electrical particles in an atom. Besides the electrical and magnetic forces there is the force of gravity, which is exerted over a large distance by one body on another. Finally, we have the very short-range forces exerted by nuclear particles on each other, the so-called nuclear forces. Astronomy is concerned with all of these types of forces.

3.26 The Effect of a Force on a Body

If a force acts on a body it causes a change in its shape, such as bending, twisting, or stretching, or a change in its state of motion. *A body continues to change its shape until the deforming force is just balanced by the restoring forces in the body itself.* If the restoring forces (the structural forces in the body) are always *smaller* than the force acting on the body, the body ultimately is torn away from its mooring, or torn apart, and the effect of the applied force manifests itself in a *change in the state of motion of the body.*

3.27 Newton's First Law of Motion

If a force that is not balanced by any other force acts on a particle, the particle undergoes a change in its state of motion, that is, it *suffers an acceleration.* This is the content of Newton's first law of motion, which can be stated as follows:

> **A body remains at rest or continues to move in the same straight line with constant speed unless it is acted upon by a force.**

In other words, a force is that which changes the state of motion of a body—that is, imparts an acceleration to it.

3.28 Inertia

If two different particles or bodies are acted upon by the same force, we find, in general, that they do not react in the same way; their resulting accelerations are not the same. Thus there are associated with different bodies different amounts of some entity which determine how readily these bodies change their state of motion under the action of the given force. This is sometimes referred to as the **inertia** of the body, and we shall see how this leads, by means of Newton's second law, to the concept of the mass of a body.

Let us consider a particle that is acted on by a series of different forces. If in each case we carefully measure the force and also the acceleration which the force imparts to the body, and then divide the force by the respective acceleration, we find that this ratio is always the same for a given body. This is the content of Newton's second law.

3.29 Newton's Second Law of Motion: The Definition of Inertial Mass

We may state the second law mathematically as follows: **if the forces $\mathbf{F}_1, \mathbf{F}_2, \ldots, \mathbf{F}_n$ are applied to a particle, and the accelerations imparted to the particle by these forces in succession are $\mathbf{a}_1, \mathbf{a}_2, \ldots \mathbf{a}_n$, respectively, then we must have**

$$\frac{F_1}{a_1} = \frac{F_2}{a_2} = \cdots = \frac{F_n}{a_n} = \textbf{constant.}$$

We call this constant the **Mass** of the body, and label it with the letter m.

We can introduce the unit of mass by applying a unit of force, that is, a force of one dyne, to particle after particle until we find just the proper one that suffers *an acceleration of one centimeter per second per second* (*unit acceleration*) *under the action of this unit force.* We shall say that this particle has a **unit mass** and call it **one gram**. In other words, one gram is that amount of mass which acquires an acceleration of one cm per sec per sec when a force of one dyne acts on it. If a body acquires an acceleration of $\frac{1}{2}$ cm/sec^2 under the action of one dyne, its mass is 2 grams, and so on.

Newton's second law is often stated as follows:

The force applied to a body is in the direction of the acceleration imparted to the body, and is equal to the mass of the body times this acceleration.

The algebraic formula is

$$\boxed{F = ma}\,. \qquad (3.12)$$

We shall discuss the third law of motion later. At this point we want to consider the application of Newton's second law to a freely falling body.

3.30 The Weight of a Body: Its Relationship to the Mass of the Body

If we apply Newton's second law of motion to a freely falling body, we see that such a body must be acted on by a force since it is constantly being accelerated. We can measure this force easily enough by attaching this body to our force-defining helical spring, which itself is attached to a fixed support. We see, then, that the spring is stretched a given amount, depending on the particular body we use, as shown in Figure 3.18(a). The force measured by this spring is called the body's **weight** (w).

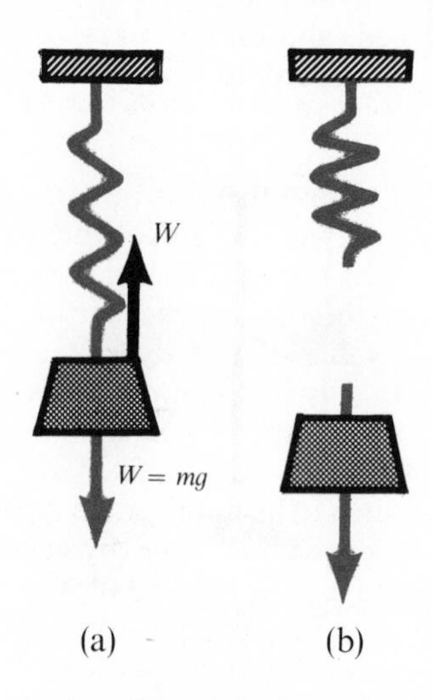

FIG. 3–18 *(a) The weight of a body as determined by the stretch of a spring. Here the force of the spring just balances the weight of the body. (b) The weight of the body imparts an acceleration to it when the body is falling freely.*

Since the weight of the body acts on the spring, the internal structural forces of the spring are called into play, and since the body has no acceleration, its weight and these internal forces must be exactly equal. If the spring were suddenly cut (Figure 3.18b), the structural forces in the spring could no longer compensate the weight of a body, and the body would therefore have an acceleration imparted to it by its own weight. Thus, it is the weight of a body that causes it to fall to the earth with the acceleration of gravity. It follows, then, from Newton's second law, that the weight of a body, its mass, and the acceleration of gravity must be related by the formula

$$w = m \quad g \qquad (3.13)$$

(weight) = (mass)(acceleration).

This relationship is extremely important since with its aid we can measure the mass of any object just by weighing the object. We note that the weight of a body is a force, and therefore cannot be equal to the body's mass, but we see that the weight is proportional to the mass, since the value of g is the same for all bodies at a given point on the earth's surface. This means that if one body weighs twice or three times as much as another it must have a mass twice or three times that of the other body. We can thus obtain the mass of any body by comparing its weight with the weight of the unit of mass. This can be done by means of a simple balance, on one side of which we place the body whose mass is to be determined, and on the other as many units of mass as are needed to balance the weight, as shown in Figure 3.19.

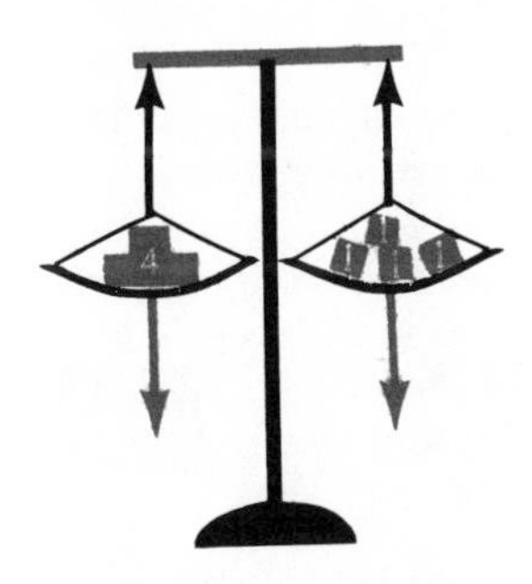

FIG. 3–19 *The mass of a body is equal to the number of unit masses necessary to balance it in a gravitational field.*

If we know the mass of a body, we can easily determine its weight by using the formula we have just written down. As an example, if a body has a mass of 100 grams we obtain for its weight (by multiplying by g) 98 000 dynes.

3.31 Newton's Law of Gravity: Its Dependence on Distance

The following question naturally arises: what is the nature and origin of this force (the weight of the body)? This question was first answered by Newton in his famous law of gravity. Newton realized that the force arises somehow or other from the earth, and that it has its origin in the center of the earth since bodies fall along a plumb line. His great contribution was to show that this force (gravity) that acts between any two bodies is universal, and that its magnitude depends on the masses of the bodies involved. To see how we may arrive at Newton's law of gravity we consider two particles (masses m_1 and m_2, each one concentrated at a point) separated by a distance r and exerting a gravitational force on each other, as shown in Figure 3.20.

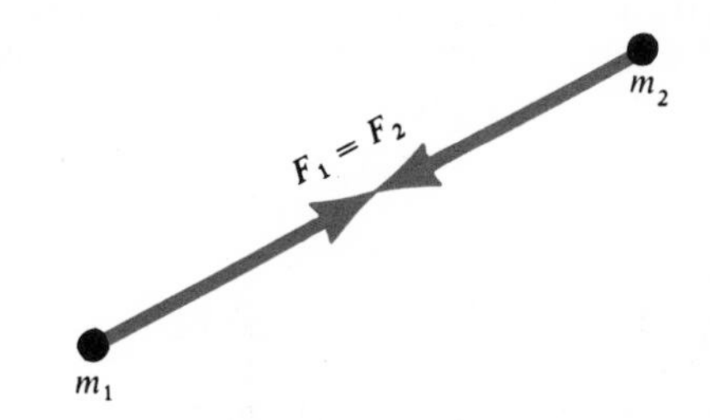

FIG. 3–20 *The gravitational attraction between two particles. The gravitational force on particle 1 is equal and opposite to that on particle 2, and both forces are along the line connecting the two particles.*

We first consider how the distance enters into Newton's formula. Since the force exerted by particle 1 is always the same no matter where particle 2 is, as long as the distance between the two is the same, we can picture the gravitational force as radiating outward from the particle in all directions, as shown in Figure 3.21. It follows,

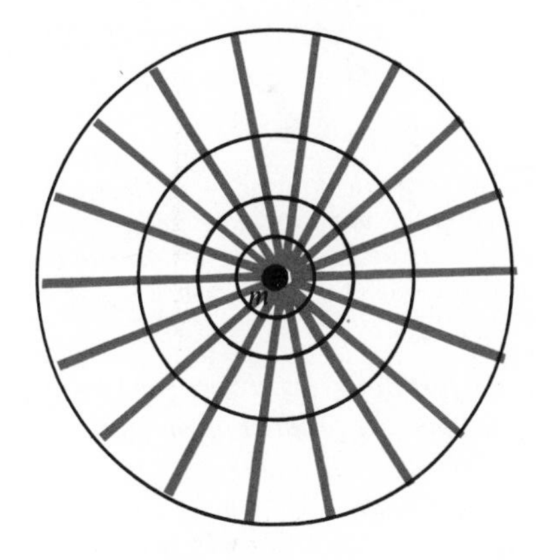

FIG. 3–21 *A cross section showing the gravitational lines of force thinning out with distance from the source of these lines of force.*

then, that as particle 2 gets further and further away from particle 1 the force gets weaker and weaker, since the lines of force as shown in the diagram thin out. The amount of thinning out depends on the square of the distance, since the end points of the lines of force lie on the surfaces of spheres whose areas increase as the square of the distance. *This means that the gravitational force diminishes as the square of the distance.* The formula for this force must therefore contain the square of this distance in the denominator. That is:

$$F \propto \frac{1}{r^2},$$

where $\propto$ means *proportional to*, or *varies as.* Newton himself actually described this dependence on distance of the gravitational force by analyzing Kepler's third law (harmonic law) of planetary motion.

3.32 Dependence of Gravitational Force on Mass

If we place another particle of the same mass together with particle 1, we have twice as many lines of force from these two particles as before, so that the gravitational force on particle 2 is twice as great as before. The same happens if we double the mass of particle 2 leaving the mass of particle 1 unchanged, as shown in Figure 3.22. *Thus, it is clear that the force of gravity between two mass points must depend upon the product of the masses, since the force increases by the same factor by which we increase either mass.* This also follows from the fact that the weight of a body is proportional to its mass.

FIG. 3–22 *A schematic drawing of the lines of force in the attraction between 2 units of mass and 3 units of mass. Here we arbitrarily assign one line of force to each unit of mass. There are 6 lines of force, which is just the product of 2 and 3.*

If we combine the distance factor and the mass factor, we have for the gravitational force the formula:

$$F \propto \frac{m_1 m_2}{r^2}.$$

3.33 The Universal Gravitational Constant

However, we can see at once that this formula as it stands cannot be correct or complete, because it gives us gravitational forces that are much too large, and also because it does not have the right length, time, and mass ingredients. Force must be expressible as a mass times an acceleration, and this cannot be done with the formula as it stands. Some unknown factor is missing. This factor is called **Newton's universal gravitational constant**, and is represented by G (capital, not to be confused with the acceleration of gravity, g). *It has the properties of the cube of a length divided by the product of a mass and a time squared,* so that in terms of our fundamental units it is expressed as cm^3 per gm per sec^2. In terms of these units the value of G is

$$\boxed{G = 6.67 \times 10^{-8} \text{ dynes } \frac{\text{cm}^2}{\text{gm}^2}}\,.$$

The formula for the gravitational force is thus

$$\boxed{F = G\,\frac{m_1 m_2}{r^2}}\,. \qquad (3.14)$$

We should observe that the gravitational force between any two particles is always directed along the line connecting the two particles, and moreover, the force of particle 1 on particle 2 is exactly equal to the force of particle 2 on particle 1. This last statement is a consequence of Newton's third law of motion, which we shall discuss later.

3.34 Gravitational Force between Extended Bodies

We can apply this law of force in the way we have stated it only if the two mass particles are no bigger than points. If we are dealing with two bodies of arbitrary shape and size, we cannot use the formula because there is no way of defining a distance between the two bodies. What we must do in that case is imagine each body divided into a large number of point-size particles, and then take all the mutual forces between these and *add them vectorially* to give us the final gravitational force, as shown in Figure 3.23.

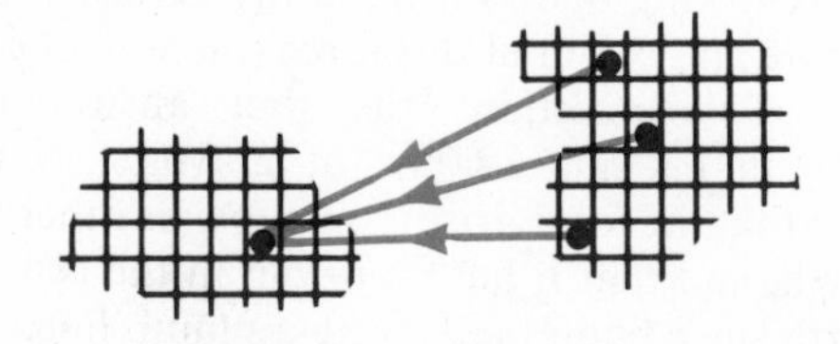

FIG. 3–23 *The gravitational attraction of one irregular body on another.*

It should be noted that in the case of an irregularly shaped body there is no such thing as the center of gravity. In other words, for purposes of computing the gravitational force between two such bodies there are no two points in the bodies at which we may picture the two masses as being concentrated. This is obvious if we just note that by rotating the two bodies about any two points the gravitational force between them changes. If there were such a thing as the center of gravity, the rotation of either body about its center of gravity would cause no change in the gravitational force between it and the other body.

3.35 Gravitational Force between Two Regular Spheres

Although in general there is no simple mathematical formula for dealing with the gravitational force between extended bodies, mathematicians have discovered ways of treating gravitation between bodies with regular geometric shapes in which the mass is distributed in a symmetrical way. Thus, in the case of two spheres in the interior of which the mass is distributed uniformly in shells, the gravitational force can be treated as though we were dealing with two point masses concentrated at the centers of the spheres. *This is very important since the sun and planets can be treated as spheres of this sort. We may, therefore, compute the gravitational force between such spheres by taking as the distance r the separation of their centers.*

3.36 The Mass of the Earth

Now that we have introduced the concept of the mass, we may consider the mass of the earth. To do this we apply Newton's second law of motion and his gravitational law to a freely falling body on the surface of the earth. We have seen that a body falls to the surface of the earth because of its weight, mg. On the other hand, we also know that the weight of a body is nothing more than the gravitational force of the earth exerted on the body. We can now compute this force by using Newton's law of gravity, since we may treat the earth as though all its mass were concentrated at its center. In other words, the weight of a body on the surface of the earth is equal to the gravitational force exerted on that body by a particle situated at the center of the earth and having the mass of the earth. Observe that the body pulls upon the earth with a force exactly equal to that which the earth exerts upon it.

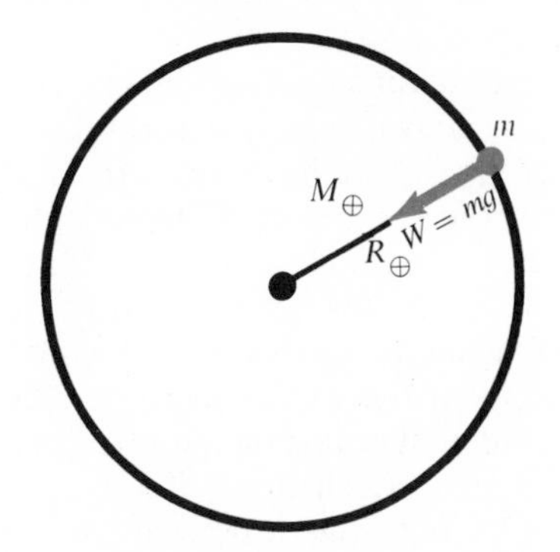

FIG. 3-24 *Calculating the mass of the earth by equating the weight of a body on the earth's surface to the gravitational pull exerted by the earth on this body.*

If we let $M_\oplus$ equal the mass of the earth, and $R_\oplus$ the radius of the earth, then we see from Newton's law of gravitation that the weight of a particle of mass m at the surface of the earth is equal to

$$w = G\frac{mM_\oplus}{R_\oplus^2}.$$

But we have also observed that $w = mg$, so that we must have

$$mg = G\frac{mM_\oplus}{R_\oplus^2}.$$

Since the small m's on both sides of the equation are the same, we may cancel them and obtain an expression for the acceleration of gravity on the surface of the earth:

$$\boxed{g = G\frac{M_\oplus}{R_\oplus^2}}. \qquad (3.15)$$

We see at once that this is the same for all particles on the surface of the earth since it depends only on the universal constant G and the parameters $M_\oplus$ and $R_\oplus$, which are independent of the falling particle.

If we put into this equation the known values for G, g (acceleration of gravity), and $R_\oplus$, we are left with one unknown, the mass of the earth. We then find that

$$\boxed{M_\oplus = 5.98 \times 10^{27} \text{ gm}}.$$

If we now divide $M_{\oplus}$ by the volume of the earth, we find that the mean density of the earth is very nearly equal to 5.52 gm/cm^3. It is a mark of Newton's great genius that he conjectured that the mean density of the earth is about five or six times that of water.

A very important conclusion about the shape of the earth can be drawn from the formula for the acceleration of gravity we have just written down. We note that if the earth were perfectly spherical, bodies would have the same acceleration (fall at the same speed) at all points on the earth's surface because all points on the earth's surface would then be equidistant from the earth's center. The fact that bodies at different points on the earth's surface cannot have the same acceleration of gravity means that the earth is not spherical; and, indeed, by measuring the acceleration of gravity, as we have already indicated, we can ascertain the flattening at the poles.

If a pendulum of length L is allowed to swing freely through a small angle, then the square of its period (the time required for one full swing) is given by

$$\boxed{p^2 = 4\pi^2\left(\frac{L}{g}\right)}. \tag{3.16}$$

Thus the pendulum beats more rapidly (the period is smaller) where g is larger, and since g is larger where the earth is flattest, because the earth's surface there is closest to the earth's center, it follows then that a pendulum beats faster at the poles than at the equator, as shown in Figure 3.25.

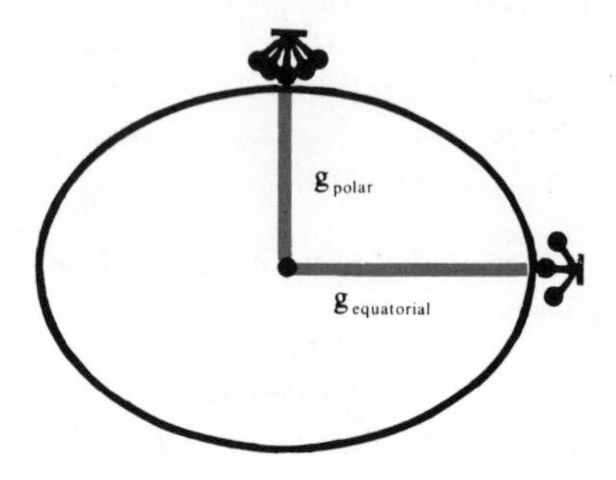

FIG. 3–25 *The period of a pendulum is longer at the equator than it is at the poles.*

EXERCISES

For all exercises in this chapter use $G = 6.7 \times 10^{-8}$ and $g = 32$ ft/sec^2 or 980 cm/sec^2.

1. Two co-planar vectors, one of length 5 units and the other of length 8 units, make an angle of 30° with each other. (a) Construct their resultant graphically. (b) Calculate the resultant and compute the direction it makes with the larger vector.

2. As seen from the sun, Venus is observed to change its direction by 10°. Through how large an arc has it moved? (Hint: use the value of Venus' distance from the sun given in the tables of planets, p. 227.)

3. From the data given in the planetary tables, calculate the mean speeds in their orbits of the planets (a) Venus, (b) Earth, (c) Jupiter, and (d) Neptune (assume their orbits to be circular).

4. If a body is thrown vertically upward from the ground with a speed of 161 ft per sec, how long will it continue to rise? How high will it rise? How long will it fall? (Neglect air resistance.)

5. Assume that (a) Venus, (b) Earth, (c) Mars, and (d) Saturn move in circular orbits around the sun. Calculate their accelerations towards the sun. (Use only the formulae in 3.19.)

6. If each planet in problem 5 were brought to rest in its orbit and allowed to fall towards the sun, how far would it fall in the first second?

7. A body falls from rest to the earth's surface through a height of 100 ft. With what speed does it strike the earth if air resistance is neglected?

8. (a) A body subjected to a force of 1 000 dynes acquires an acceleration of 50 cm/sec^2. What is the mass of the body? (b) If this body were placed on a scale in an elevator that has an upward vertical acceleration of 32.2 ft/sec^2, what weight would the scale register?

9. If a body in a circular orbit (radius 100 cm) has a force of 10^5 dynes acting on it and its constant speed is 50 cm per sec, what is its mass?

10. Use Newton's law of gravity to calculate the pull of the sun on (a) Mercury, (b) Ceres, (c) Jupiter, and (d) Neptune (use the data for the masses and distances of these bodies given on p. 227).

11. If the period of a pendulum changes by 0.1 per cent as we go from one point to another on the earth's surface, by what amount does the acceleration of gravity change?

12. (a) Show that if two bodies have velocities $\mathbf{v}_1$ and $\mathbf{v}_2$ respectively, which make an angle θ with each other, their relative speed (the speed of one with respect to the other) is

$$v_{\text{relative}} = \sqrt{v_1^2 + v_2^2 - 2v_1v_2\cos\theta}.$$

(b) Show that the angle ϕ, which the relative velocity makes with the vector $\mathbf{v}_2$, is given by

$$\tan\phi = \frac{v_1 \sin\theta}{v_2 - v_1\cos\theta}.$$

(Hint: the relative velocity is also the vectorial difference between the velocities $\mathbf{v}_1$ and $\mathbf{v}_2$.)

13. Given a sphere of matter of radius r and of constant density ρ at all points in its interior, write down a formula for the acceleration of gravity in terms of ρ that holds at all points in its interior. What is the value of this acceleration at the center of the sphere?

14. If space were infinite and filled with a uniform homogeneous distribution of gas and dust of constant density, what would the acceleration of gravity be at each point of space? Prove your statement.

15. Derive the formula for the circular acceleration on a body moving in a circle with constant speed.

In the following problems, express the results in terms of the letter g for the acceleration of gravity and not in terms of its numerical value.

16. Assume a body is thrown upwardly with an initial speed v_0 and at an angle θ with the horizontal. (a) What is its horizontal velocity after a time t and what is its vertical velocity then if we neglect air resistance and suppose that the body always remains very close to the surface of the earth? (b) What is the total velocity of the body after the time t and in what direction is it moving then? (Neglect air resistance.)

17. (a) In problem 16, if X is the distance the body has moved in the horizontal direction and Y the distance it has moved vertically in the time t, derive the formula for X and Y in terms of t, θ, and g. (b) Obtain the expression for the total distance of the body from its starting point.

18. (a) Eliminate t between the two expressions for X and Y in problem 17, and obtain an equation for Y in terms of X. (b) If you plotted Y as a function of X, what kind of curve would you get? (c) Why is the path that you derived for the body in (b) not part of an ellipse?

19. (a) Obtain an expression for the highest point above the ground reached by the body in problem 16. (b) How fast and in what direction is it moving at the highest point? (c) Obtain an expression for the range of the body. (d) Show that when the angle θ is 45°, the range is a maximum.

4

THE STRUCTURE OF THE EARTH

4.1 The Earth's Atmosphere

For purposes of studying the structure of the earth, we may consider it as being divided into a series of concentric shells. We begin by studying the outermost shell, or gaseous envelope, which we call the **atmosphere**. The structure of the atmosphere is exceedingly complex, and to analyze it in detail would involve us in advanced problems of hydrodynamics, so that all we shall do here is outline some of the essential features that are of interest to astronomers.

The upper limit of this shell is qualified in the following words by R. Jastrow and L. Kyle: "In any case the definition of the atmosphere must be extended considerably beyond 5 000 km, if the atmosphere is defined in a broader sense as the region *surrounding the earth in which the presence of our planet appreciably perturbs the interplanetary medium.* The proper significance of this definition has only become apparent with the discovery of the Van Allen zones. It appears that solar particles, as well as protons and electrons produced by the beta decay of cosmic ray neutrons, may be injected into the region around the earth and trapped by the geomagnetic field for appreciable periods, creating conditions quite different from those in the surrounding interplanetary space. The flights of U.S. and U.S.S.R. space rockets and satellites have shown that the population of geomagnetically trapped particles may extend out as far as 100 000 km in periods of unusual solar activity. This great distance marks the outer limit of the atmosphere in the sense of the present definition. Since the particle populations in this farthest region of the atmosphere are determined by the geomagnetic field, Thomas Gold and J. F. Clark have suggested that it be designated as the *magnetosphere*."[1]

[1] In H. H. Kölle, ed., *Handbook of Astronautical Engineering*, Chapter 2, Section 2.122, pages 2–4, "Atmospheric Regions." New York: McGraw-Hill Book Co., Inc., 1961.

4.2 Zones of the Atmosphere

Because of varying thermal and electrically charged atomic conditions at different heights, it is convenient to divide the atmosphere into zones. The shell closest to the earth is called the **troposphere**, and extends to a height of 8 or 9 miles (about 13 km). Starting at the upper layers of this zone and extending into the next shell, the **stratosphere**, we find an abundance of ozone (the triatomic oxygen molecule) which readily absorbs ultraviolet radiation. If this ozone were not present, life as we know it on the earth would not be possible, since intense ultraviolet radiation is destructive of most present forms of life.

By the time we have reached the stratosphere, the temperature has dropped to about −55°C, and remains constant up to a height of about 15½ miles (25 km), at which point it begins to rise, reaching a value of about −8°C. This last value marks the beginning of another zone known as the **mesosphere**.

The temperature does not increase steadily in the mesosphere, but falls off again and drops to a minimum value of about −80°C at a height of 85–90 km. The next zone, which begins at this point, is called the **thermosphere** and is one of rapidly increasing temperature which probably reaches a plateau at a height of 450 miles (700 km). The present temperature evaluations for this maximum level range from about 1 220°C (daytime) through about 750°C (nighttime). This temperature, however, is what is known as the kinetic temperature, and it does not mean that an object would get very hot at these heights. This high temperature means the molecules of the atmosphere are moving about very rapidly, but so few of them would hit a body every second that the body would remain very cold.

The **exosphere** extends from a height of about 450 to 3 000 miles (700 to 5 000 km) where the extremely low density that characterizes this zone ranges from about 10^8 particles/cm^3 to 10^2 particles/cm^3. The outermost layer has been described above, and is called the **magnetosphere**.

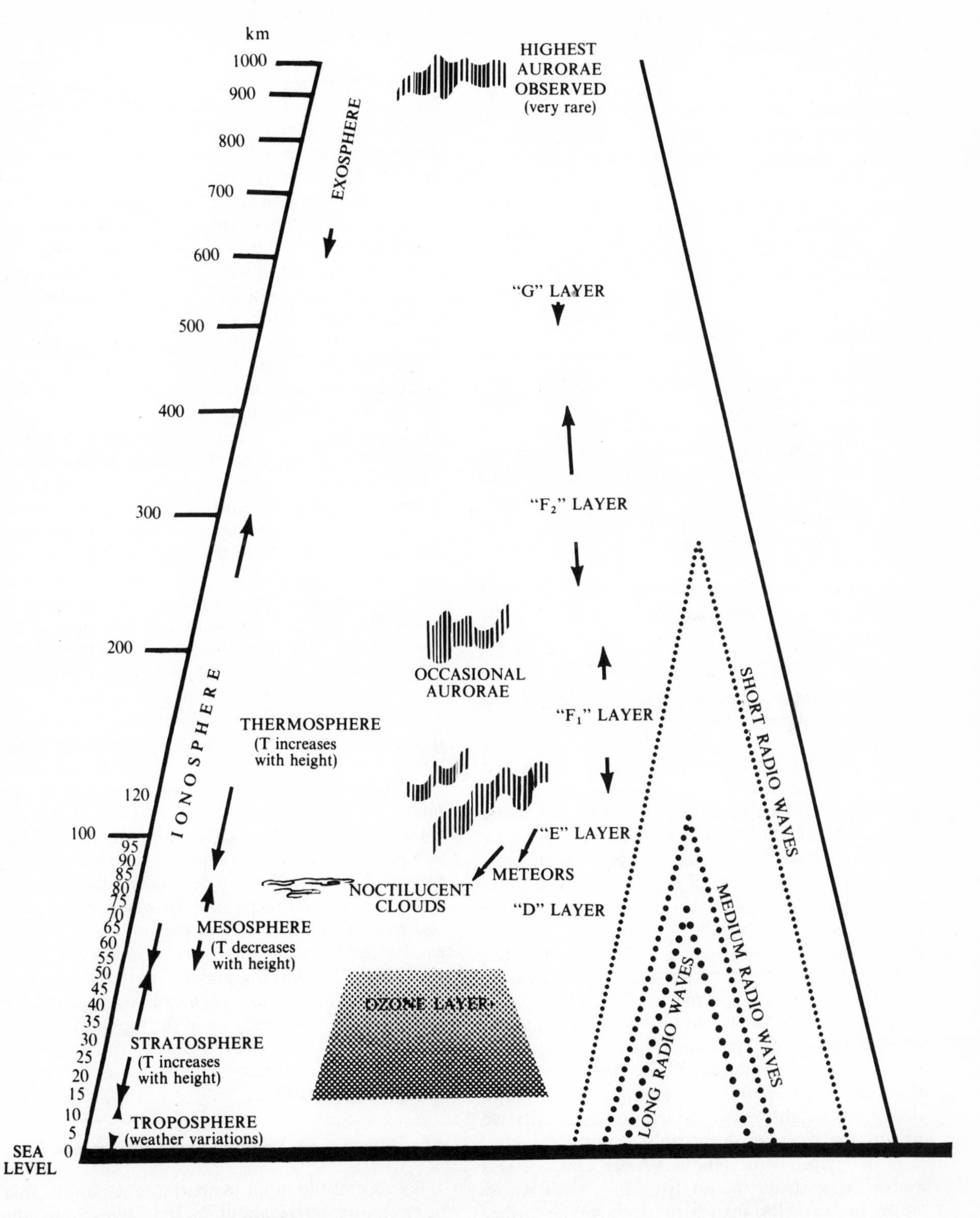

FIG. 4–1 *Various layers of the earth's atmosphere.*

In the troposphere, stratosphere, and mesosphere the atoms and molecules are all neutral, but above this layer ionization sets in and free electrons are present. The importance of free electrons in the atmosphere is that they reflect radio waves and therefore have an important bearing on radio communications. As we go above the mesosphere the kinetic temperature begins to increase and does so steadily, reaching a value of about 1 200°C at about 260 miles (420 km). This is just about where the number of free electrons is greatest and where the **ionosphere** lies. We can send radio waves around the world by bouncing them off this ionized layer back to the earth.

4.3 Density of the Atmosphere

The **density** (that is, the mass contained in a unit volume) of the air is greatest at the surface of the earth where it is equal to 1.2×10^{-3} gm/cm^3. It falls off quite rapidly as we move away from the earth, and at a height of about 125 miles (200 km) its value is only about 10^{-13} gm/cm^3. If we consider a column of atmosphere having a cross-sectional area of one square inch and extending all the way to the top of the atmosphere, *we can determine how much air it contains because the weight of this column is just the atmospheric pressure.* This turns out to be 14.7 lb/in.^{2}

4.4 Atmospheric Pressure and Weight of the Earth's Atmosphere

The atmospheric pressure is generally expressed in terms of the height of a column of mercury of unit cross-sectional area that weighs the same amount as the unit cross-sectional atmospheric column. Thus if we say that the atmospheric pressure at sea level is 76 cm of mercury, we mean that a column of mercury 76 cm high covering an area of one square inch weighs 14.7 lb. We can also express the pressure in dynes per square centimeter, and in these units the standard atmospheric pressure is equal to 1.013×10^6 dynes/cm^2.

If we multiply the weight of a column of atmosphere covering one square inch by the total number of square inches on the earth's surface, we can find the total weight of the earth's atmosphere. If we divide this weight by the acceleration of gravity we obtain the mass of the earth's atmosphere, viz. 5×10^{21} gm. Most (about 97 per cent) of this weight arises from the air lying in a shell whose upper surface is less than 20 miles above the surface of the earth.

4.5 Pressure at Any Point in the Earth's Atmosphere

If we know the density and the temperature at any point of the earth's atmosphere, we can determine the pressure at that point. Here we must stop a moment to define pressure precisely, since we will have occasion to use it later. Whether we are dealing with the pressure in a gas, or that exerted by a liquid, or by a solid resting on a surface, the general definition is the same. It is clear that pressure is related to force, but we can see that it is not the same as force, since *pressure is a measurement of how a force is distributed over an area.*

To see what the difference between pressure and force is, we may consider two men of equal weight walking over a layer of soft snow, one wearing ordinary shoes and the other one on snowshoes. Since both men weigh the same amount, the force exerted on the snow is the same in both cases, and yet we see that one man sinks into the snow while the other does not. Here it is the pressure that is important. In the case of the man on snowshoes, the force is distributed over a larger area, so that the amount of force per unit area is smaller. From this example we see that, to obtain the pressure exerted, *we must divide the total force by the area over which the force acts.* This is precisely what we did when we spoke of the atmospheric pressure at sea level. We took the weight of a column of air having a unit of cross-sectional area; in other words, we considered the weight per unit area. *We therefore define pressure as force per unit area.*

The pressure in the atmosphere (for that matter, in any gas) is the result of the motions of the molecules. *As these molecules move about and strike against the walls of their container, they push against these walls and thus exert a pressure.* Later we shall write down an exact formula for the pressure at any point inside a gas, and we shall see that this formula depends upon the density and the temperature of the gas at this point. However, all we need to know at this time is that the pressure increases as the temperature and density increase. From this we can see that the pressure in the atmosphere diminishes rapidly as we move away from the surface of the earth because of the rapid drop of the density. The atmospheric density diminishes by about one-half for every 4 miles distance from the surface of the earth.

4.6 Atmospheric Pressure and Air Currents

Because of the rapid temperature variations that occur as we move about in the atmosphere, the pressure does not always change smoothly from point to point but may undergo rapid variations. These variations give rise to localized air currents of a complex nature. Superimposed upon these local variations we have the over-all jet streams at various heights in the atmosphere. *At a given point on the earth's surface the atmospheric pressure changes from moment to moment because of temperature and density variations.* The figure we have given for atmospheric pressure is the *mean value about which fluctuations occur.* We shall see later that atmospheric pressure variations also arise because of the tides raised in the atmosphere by the sun and the moon.

4.7 Chemistry of the Earth's Atmosphere

The atmosphere is a mixture of various chemical compounds, the most abundant of which, N_2 (the diatomic nitrogen molecule), constitutes 78 per cent of the volume. Oxygen is present to the extent of 21 per cent by volume and the next most abundant element is argon, whose concentration is 0.934 per cent. The amount of carbon dioxide by volume is quite small; it is of the order of 0.034 per cent. Although water vapor is present, we cannot give its abundance since it occurs in variable quantities ranging from 0 to 2 per cent. It is remarkable to find such very small abundance of hydrogen, 5×10^{-5} per cent, in the earth's atmosphere since we know that this element constitutes more than 95 per cent of all matter in the universe. We shall see, however, that there is good reason for its absence from our atmosphere, and it is fortunate that this is so since hydrogen has a great affinity for oxygen, carbon, and nitrogen. If any large quantities of hydrogen had remained in our atmosphere, there would now be very little free oxygen and nitrogen, and instead of having an atmosphere that is pleasant and congenial to life, our earth would be surrounded by clouds of ammonia and methane.

4.8 Astronomical Aspects of the Atmosphere

Knowledge of the physical and chemical nature of the atmosphere is of particular importance to the astronomer, since the information that he wants to wring from the light coming from celestial bodies is distorted to some extent as the light passes through the atmosphere. Only by knowing the properties of the atmosphere can we allow for this distortion. The first point to be noted is that the atmosphere absorbs radiation, and therefore reduces the brightness of stellar objects. It has been estimated that under average clear-sky conditions about 25 per cent of the solar radiation is either scattered back or absorbed by the earth's atmosphere. This, of course, is also true for light coming from stars, so that in measuring stellar brightness astronomers must take into account this important factor.

The atmosphere does not absorb all the light equally. It tends to transmit the red colors more readily than the blue colors, so that celestial objects appear somewhat redder than they really are. This phenomenon of selective scattering of light makes the sky look blue and gives the sun its blood-red appearance during sunset or sunrise. During these times, as shown in Figure 4.2, the light passes through a much greater thickness of atmosphere than when the sun is high in the sky, so that much more of the blue light is scattered and most of the red light comes through.

FIG. 4–2 *The scattering of sunlight by the earth's atmosphere at noon and at sunset.*

4.9 The Atmosphere and Refraction of Light

Another important effect that the atmosphere has on astronomical observation is its influence on the apparent positions of stars. If a star is not directly overhead but at some intermediate altitude, light enters the atmosphere obliquely and refraction occurs, as shown in Figure 4.3. This tends to make a star appear higher up in the sky than it really is. For accurate observations, atmospheric refraction effects must be taken into account, and astronomers have prepared accurate refraction tables for the different altitudes of stars.

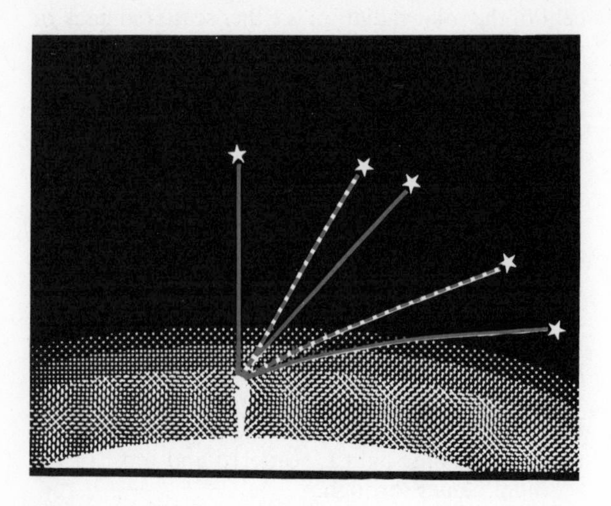

FIG. 4–3 *Refraction of light by the earth's atmosphere. The solid lines indicate the true positions, and the dotted lines the apparent positions of the stars.*

TABLE 4.1
ATMOSPHERIC REFRACTION

Apparent altitude	True zenith distance	Apparent altitude	True zenith distance
0°	90° 35′ 22″	10°	80° 5′ 19″
1°	89° 24′ 44″	15°	75° 3′ 35″
2°	88° 18′ 27″	20°	70° 2′ 39″
3°	87° 14′ 27″	25°	65° 2′ 5″
4°	86° 11′ 47″	30°	60° 1′ 41″
5°	85° 9′ 53″	40°	50° 1′ 10″
6°	84° 8′ 29″	50°	40° 0′ 49″
7°	83° 7′ 24″	60°	30° 0′ 34″
8°	82° 6′ 34″	70°	20° 0′ 21″
9°	81° 5′ 53″	80°	10° 0′ 10″
10°	80° 5′ 19″	90°	0° 0′ 0″

Note: The difference between the complement of the angle in column 1 and the angle in column 2 is the *amount of refraction* at the given apparent altitude (column 1) or the true zenith distance (column 2).

4.10 The Hydrosphere and Biosphere

As we leave the atmosphere and move in towards the center of the earth, we pass a very thin layer, the **hydrosphere**, which is mostly liquid. It is a discontinuous shell of fresh and salt water, composed of the totality of all the oceans, lakes, and rivers. The discontinuities in this hydrospheric shell occur where the underlying crust crops out in the form of dry land. Distributed over the surface of this crust and throughout the hydrosphere is the **biosphere**, which is the totality of all organic material both living and dead, and is the concern of the biologist.

The atmosphere, the hydrosphere, and the biosphere, together with the thin exposed surface of the crust which we speak of as the dry land, constitute less than 0.03 per cent of the total mass of the earth. The total amount of dry exposed land constitutes slightly more than 25 per cent of the earth's surface, and we may note that for each square centimeter of the earth's surface there are 278 kilograms of sea water. The hydrosphere has a mean depth of about $2\frac{1}{2}$ miles (4 km), and its greatest depth is about equal to the highest peak of dry land (Mt. Everest).

4.11 Surface Variations of the Earth

The generally regular features of the earth's surface above and below the hydrosphere do not remain the same at all times, but undergo slow variations, although occasional violent disturbances such as volcanoes and earthquakes may bring sudden and drastic alterations. The slow variations are the result of various weathering processes, the most important of which is the action of water. This occurs as a constant slow attrition in the case of rivers and streams, and also in the form of sudden and rapid erosion after heavy rainfall.

As an example of the action of rivers we may note that the Mississippi River annually carries away about 515 million tons of topsoil. It has been estimated that at the present time erosion in the United States is reducing the surface by one meter per 30 000 years. If a process of this sort had been going on continuously since the Paleozoic era (500 million years ago), a layer about 17 km thick would have been removed from the continental shelf of North America. Since we can assume that these erosive processes have been going on over the entire world, it is clear that unless some process were compensating for this erosion, the entire surface of the earth would be completely smooth and covered by water at the present time.

4.12 Isostasy

However, we know from geological considerations that as rapidly as erosion acts to smooth out the earth's surface, there are compensating forces that keep wrinkling it. This constant balance between eroding and wrinkling processes is called **isostasy**.

Although the earth may appear to us as being rather rough because of the irregularities in the formation of mountain peaks and valleys, it is, relatively speaking, very smooth and would appear much more regular than a billiard ball if it were reduced to that size. Nevertheless, these slight irregularities make it possible for life to exist on dry land since a smooth earth would be uniformly covered by water to a depth of a mile (1.6 km). Even as it is, if it were not for the great quantity of water that is frozen on Greenland and the Antarctic Continent, only the high peaks would be above water.

4.13 The Crust of the Earth

When we leave the hydrosphere and penetrate more deeply, we enter the crust of the earth, which extends down to an average depth of about 40 km. (The crust under the oceans is much thinner but denser than elsewhere.) This crust, which is composed of the normal silicate rocks, has a mean density of about 3 gm/cm^3, and, taken altogether, represents only 0.7 per cent of the mass of all the earth. The rock itself is mostly igneous, with sedimentary and metamorphic rock being present in only comparatively insignificant amounts. Although silicon and oxygen are the principal chemical elements found in the crust, most of the metals are present in fairly large quantities.

4.14 Outward Flow of Heat from the Earth's Crust

Of considerable importance is the presence of uranium and thorium, although the concentrations of these elements are rather small. These radioactive substances, which have a long half-life (billions of years), have some effect on the thermal conditions on the earth's surface, since they give rise to a heat flow from the interior to the surface. From the temperature readings in mines and various borings made in the rock structure, we find that the temperature increases with depth. Although this increase varies from place to place, it has a mean value of 25°C for each kilometer we penetrate into the earth. From a knowledge of the thermal conductivity of the rocks and this temperature gradient, we can compute the outward heat flow. We find the total over the earth's entire surface is about 6×10^{12} cal/sec. This, however, is 25 000 times smaller than that which reaches the earth's surface from the sun, so that the internal heat has very little influence on surface temperature conditions.

If we compute the total amount of heat lost by the earth over geologic times, we find that this is too large a value to be accounted for by processes other than radioactive decay. Thus, if we were to assume that this heat came from the original molten state of the earth (assuming the earth was in a molten state), it would follow that the earth could not now be more than 30 million years old, since for an earth older than this the layers of cooled rock above the molten core would be much too thick to allow very much heat to flow through. A careful analysis of the heat flow indicates that the radioactivity cannot extend much below the thickness of the crust itself, for if layers of the crust much beyond a depth of 30 miles were radioactive, the outward heat flow would be greater than that observed. We can say nothing about the heat from radioactive materials lying at depths greater than a few hundred miles, since that heat must pass through such a thickness of rock that it has not yet had time to reach the surface.

4.15 The Mantle and the Deep Interior of the Earth

Below the rock crust we have a layer which is referred to as the **mantle**; this extends to a depth of 3 000 km, with the density increasing slowly from 3.5 gm/cm^3 to 6 gm/cm^3. This shell is composed of rocklike material in a more or less plastic state in which flow occurs because of the enormous pressure exerted on it by the overlying layers.

To determine the structure of the deep interior of the earth, one must have recourse to an analysis of earthquake waves. Various models of the earth have been constructed based on earthquake-wave analysis. Extensive work in this field has been done by K. E. Bullen. To understand how one carries out this analysis in practice, we must note that there are two types of waves which originate from a point of disturbance (earthquake), and that the propagation of these waves depends on the elastic constants (the elasticity and rigidity) of the rock structure.

4.16 P- and S-Waves

One of the two types of elastic waves which we have referred to is a **compressional** or **P-wave**, in which the wave advances parallel to the vibrations in the wave itself. These waves are also sometimes referred to as **longitudinal waves**. The other type of wave is a **distortional** or **S-wave**, in which the displacement is at right angles to the direction of propagation of the wave. This type of wave is also known as a **transverse wave**. Figure 4.4 shows the two types of waves.

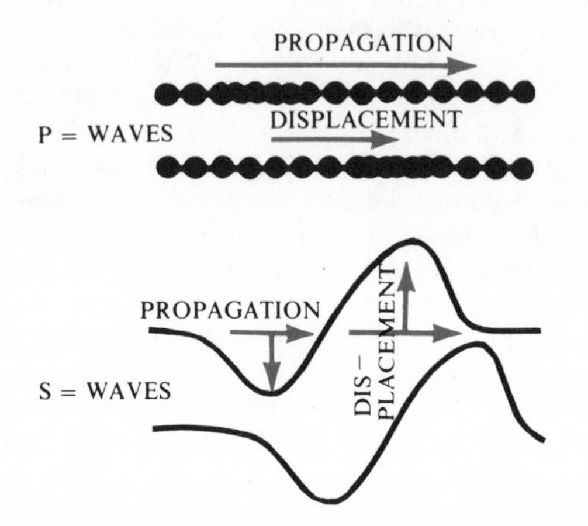

FIG. 4–4 *P (primary) longitudinal or compressional waves, and S (secondary) transverse waves.*

Since the velocities of these two waves are different, it is possible to separate one kind from the other and hence to use them as an analytical tool to study the internal structure of the earth. *We may also note that transverse waves are only transmitted through solids and are totally reflected from a solid-liquid interface. The P-waves, on the other hand, can be propagated through both solids and liquids.* One other feature about P- and S-waves must be understood. Whenever they pass from a medium of one density into a medium of another density they are partly reflected and partly refracted, thus suffering a change in their direction of propagation. It is possible, then, by analyzing P- and S-waves emanating from an earthquake source and reaching different seismic stations on the earth's surface, not only to determine sudden variations in the density of the deep interior of the earth, but also to discover at what point a liquid core sets in.

4.17 The Density of the Interior of the Earth from S- and P-Waves

Directions of propagation of waves are shown in Figure 4.5. It can be seen from Bullen's analysis that at a depth of 3 000 km the density increases suddenly to a value of about 9 gm/cm^3, and it is at this depth that we must picture the core of the

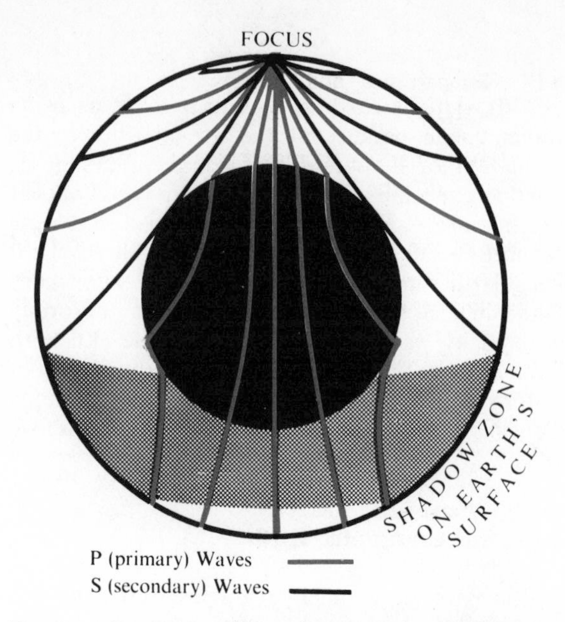

FIG. 4–5 *Propagation of P- and S-waves through the earth.*

earth as starting. Since we find that only P-waves are transmitted through this core, we must conclude that it is liquid rather than solid. The diameter of this core can be determined from the size of its shadow on the earth's surface cast by S-waves (as shown in the diagram, Figure 4.5). The P-waves do not go right through, however, and Bullen has shown that they suffer additional refraction at about 5 000 km. He has concluded that an inner core sets in at about this depth, and that starting at this interface the density increases from about 11.5 gm/cm^3 to 16 gm/cm^3. The inner structure of the earth is shown in Figure 4.6.

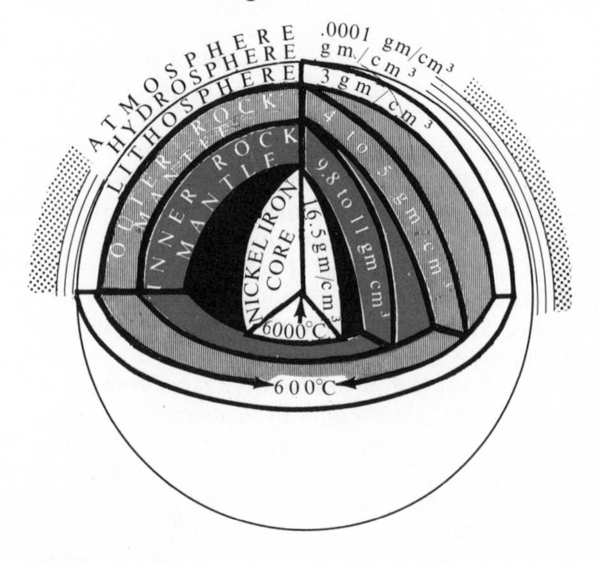

FIG. 4–6 *Schematic drawing of the earth's interior.*

4.18 Temperature and Pressure in the Earth's Interior

Just as the density increases as we go towards the center, so do the temperature and the pressure. From various considerations of the radioactive content of the rocks, and of the assumption that the earth is slowly contracting, the minimum central temperature cannot be less than 1 600°C and is probably closer to 5 000°C. It is also possible by analyzing the velocities of S- and P-waves to determine the pressure at various points in the earth's interior, and we find that at the center it is 3.64×10^{12} dynes/cm^2. Taken altogether, the earth behaves like a very rigid and highly elastic ball, its over-all elasticity and rigidity greatly exceeding that of steel.

4.19 Chemistry of the Earth's Interior

As far as the chemical nature of the interior of the earth goes, we must conclude that it is not rock-like all the way down to the center, for if it were, the mean density would be considerably smaller than 5.5 gm/cm^3, and it would be impossible to account for the very dense inner core. The only common substances that are heavier than the silicates, and can therefore contribute to a higher density, are iron and its oxides. We may, however, exclude the molten oxides, because they cannot exist together with the silicates since the two are mutually soluble. We are therefore left only with molten iron as a most likely substance for the interior. However, it is quite possible that some nickel is also present, since we find that these two metals are present in many meteorites.

4.20 Electromagnetic Properties of the Earth

Before ending the discussion of the structure of the earth we should mention its electromagnetic properties. Since it is known that there is a steady flow of electrical current from the interior of the earth out to the surface, it follows that there must be a charge distribution on the surface. In addition to this electrostatic charge, the earth has a magnetic field surrounding it. *This magnetic field is similar to one that would be produced by a magnetic dipole (bar magnet) at the center of the earth with its axis inclined 11.4° to the axis of rotation of the earth.* If this dipole were extended, it would cut the earth's surface at the geomagnetic north pole, which is not coincident with the geographic North Pole. Its latitude is 78.6°N. The magnetic field strength at the poles is 0.63 gauss, and at the equator it drops to about 0.31 gauss. Actually the magnetic field is extremely complex and is only approximately that of a dipole. We still do not have an acceptable theory as to its origin. The theory most widely accepted is that of Elsasser, which represents the phenomenon as being due to electrical currents within the core and therefore related to the rotation of the earth itself.

4.21 The Van Allen Belts

The magnetic field of the earth is responsible for the particle radiation belts such as the Van Allen zones (first discovered by Van Allen from early U.S. satellite data) surrounding the earth (Figure 4.7). As we have already noted, the magnetic field emanating from the earth creates a magnetic trap which captures the charged particles ejected into it by the sun and by cosmic rays. Vernov and Chudakov have shown by analyzing the data obtained from artificial satellites that the abundances of these charged particles in the particle radiation belts around the earth are so great that these belts completely nullify the earth's magnetic field at these distances. For a detailed discussion of how the earth's magnetic fields capture the charged particles see the discussion of magnetic mirrors in Chapter 32 (Section 32.13). The details of the Van Allen belts are shown in Figure 4.7.

TABLE 4.2

PROPERTIES OF THE EARTH

Property	Symbol	Value
Spheriod		
Equatorial radius		= 6 378 km
Polar radius		= 6 357 km
Mean radius		= 6 371 km
Ellipticity		= 1/298
Surface area		= 5.1×10^{18} cm^2
Volume		= 1.08×10^{27} cm^3
Earth's mass	$M_\oplus$	= 5.98×10^{27} gm
Earth's mean density	$\bar{\rho}_\oplus$	= 5.52 gm/cm^3
Angular velocity of earth's rotation		= 7.29×10^{-5} radians per sec
Angular momentum		= 5.86×10^{40} cm^2 gm per sec
Lengthening of day		= 0.0016 sec per century
Energy lost by tidal friction:		
spring tide		= 2.2×10^{19} ergs/sec
mean tide		= 1.1×10^{19} ergs/sec
Mean velocity of earth in its orbit		= 29.8 km/sec
Velocity of escape from earth's surface		= 11.2 km/sec
Standard surface acceleration of gravity, g, of the earth	$g_\oplus$	= 980.665 cm/sec^2
Age of earth		= 4.5×10^9 years

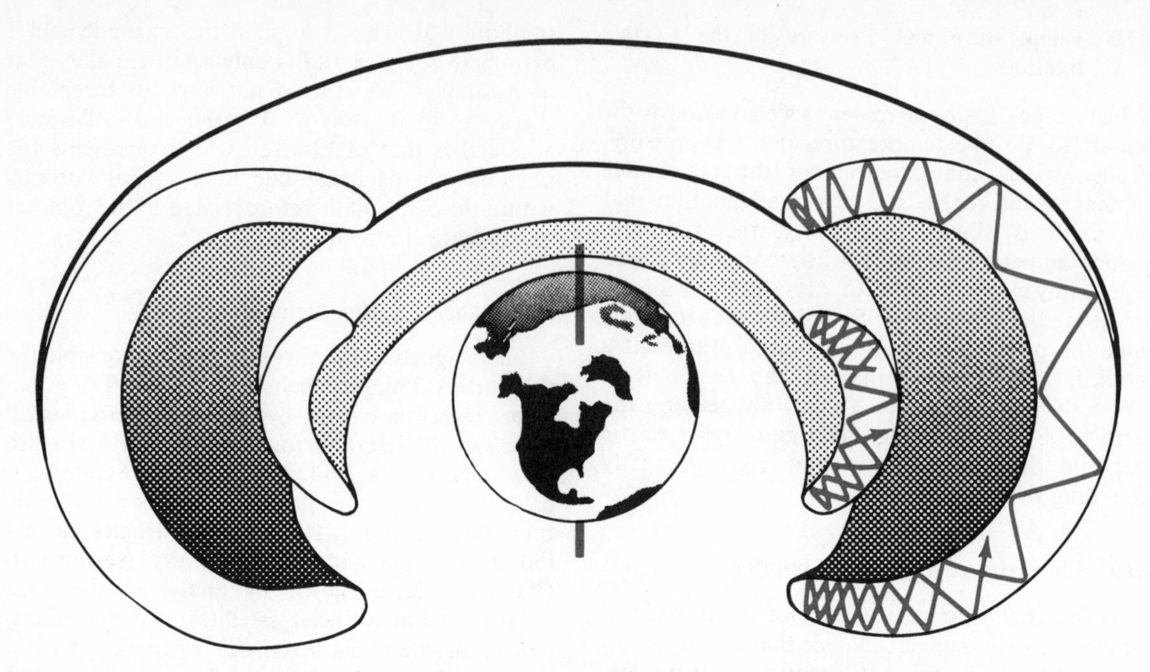

FIG. 4–7 *Van Allen belts. Charged particles in the earth's magnetic field oscillate back and forth.*

4.22 Plate Tectonics

Starting from the observation of **Alfred L. Wegener**, in the early part of this century, that the various continents look as though they fit together like pieces of a jigsaw puzzle, geologists have collected overwhelming evidence that the continents are drifting (**plate tectonics or continental drift**). The crust and mantle of the earth, down to a depth of about 100 mi, consists of about a dozen major plates that float and drift about very slowly on an underlying plastic layer. Some hundreds of million years ago, all of these plates were bound to one another and the continents on top of the plates formed a single unbroken land mass. Since then the plates have drifted apart, giving us the present distribution of the continents. When two plates slide against each other or plow into each other (as they do from time to time), geological faults and mountain ranges may be formed. When plates pull apart from each other, deep ocean ridges are formed by the new molten rock that surges up from the deep interior.

EXERCISES

1. Would an object weigh more or less on the surface of the earth if there were no atmosphere? Why?

2. If the atmospheric pressure is expressed as 76 cm of mercury, what is its value in dynes per sq cm if g is taken equal to 980 cm/sec^2? What is its value when expressed as the height of a column of water?

3. What is the effect of the atmosphere on the apparent disk of the sun when it is near the horizon?

4. Where is the atmospheric pressure greater, at the poles or at the equator? Why?

5. If there were a layer of rock 100 miles thick surrounding the earth at its surface, what would the pressure in dynes/cm^2 of this rock be on the layer of material on which it is resting? Assume that the rock density in the surface layer is 3 gm/cc.

5

THE MOTION OF THE EARTH

5.1 The Apparent Motions of Stars and the Motion of the Earth

When we introduced the celestial coordinate systems, we saw that the various heavenly bodies appear to have a fairly complex motion across the sky. This is true not only of the sun and the moon (which, in addition to rising and setting, drift eastwardly each day) but also of the stars and planets. The apparent motions of the planets show both an eastward and westward drift among the stars, which is superimposed upon their diurnal rising and setting. Although at first sight the stars seem only to rise and set, we find after careful observation that they, too, have an apparent, slow, complex motion.

We can explain the apparent stellar motions quite simply if we discard the fiction that the earth is stationary and examine the behavior of stellar bodies with the knowledge that the earth is really moving. We shall see that three important components of this terrestrial motion are primarily responsible for the complexities of the apparent motions of heavenly bodies. In addition to these well-defined motions of the earth, there are minor irregularities in its motion which, although small, must be well understood before a complete picture of stellar motions can be obtained.

That the earth is moving came to man's consciousness only late in the development of civilization. There is evidence that as long ago as 300 B.C. Aristarchus of Samos taught that the earth moves. Only in recent times, however, has this doctrine been completely accepted. As late as the seventeenth century Galileo was forced to suffer the indignities of house imprisonment in the latter days of his life for insisting that the earth moves.

In order properly to discuss the motion of any object it is necessary to introduce a frame of reference, for no meaning can be attached to motion in an absolute sense. For the time being, to analyze the motion of the earth, we shall use the sun as our frame of reference. Ultimately, we shall be concerned with how the sun itself is moving with respect to the stars and we shall then have to use another frame of reference.

5.2 Rotation of the Earth: The Foucault Pendulum

Since the motion of the earth relative to the sun is quite involved, we shall separate it into its various components in order to understand it fully. The largest part of the apparent stellar motions is their diurnal rising and setting, and we can explain this at once in terms of a *spinning earth*. It is difficult for people who have no background in science to understand how one can live on a spinning globe and still not be aware of it. But we must understand that the angular speed of rotation is very small and therefore impossible to detect by means of physiological reactions. However, there are many devices of a mechanical nature that can be used to measure this rotation; moreover, it has numerous observable effects on the structure of the earth, on the flow of rivers, on wind patterns, on the flight of projectiles, and so on.

The simplest way to detect the earth's rotation, without making observations on stellar bodies, is to observe the behavior of a pendulum. In this experiment, first performed by Foucault in 1851, a massive ball having a diameter of about one foot is suspended from a very high ceiling by means of a long wire, and set oscillating. Foucault used the dome of the Pantheon in Paris, and suspended the ball by means of a wire 200 ft long. If a pin is set into the bottom of the ball so that it can trace out a pattern in a thin layer of sand over which the oscillations are taking place, *one immediately observes (in the northern hemisphere) that the plane of oscillation of the pendulum rotates slowly to the right (that is, in the direction from east to west, which is opposite to that of the rotation of the earth). In the southern hemisphere, the plane of*

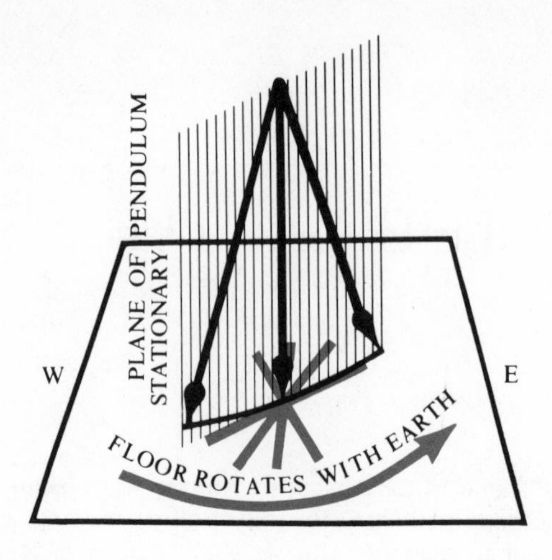

FIG. 5–1 *Schematic diagram of Foucault pendulum showing the ground rotating under the swinging pendulum while the plane of the swinging pendulum remains fixed in space.*

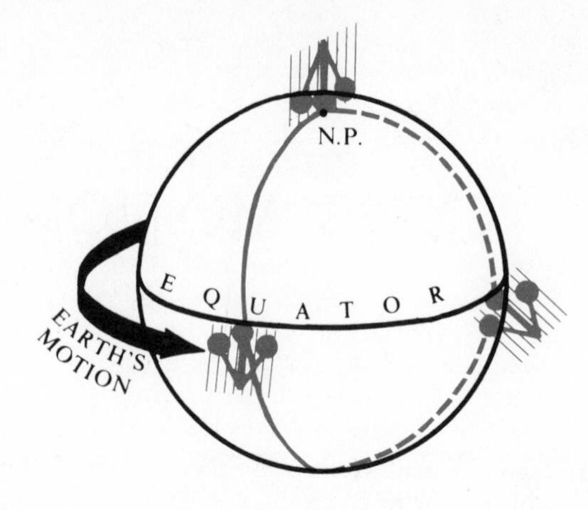

FIG. 5–2 *Foucault pendulum at the North Pole and at the equator.*

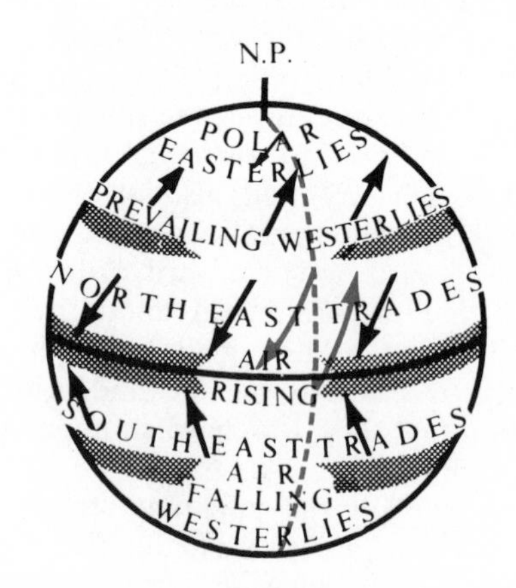

FIG. 5–3 *The prevailing winds are determined by the rotation of the earth and the temperature differences between the equatorial and polar regions.*

oscillation turns slowly to the left. This can be understood only in terms of a rotating earth since, according to Newton's second law of motion (Section 3.29), the bob of a pendulum always swings in one direction unless a force acts on it at right angles to the plane of oscillation, as shown in Figure 5.1. Since, however, this is not the case in this experiment, we must conclude that the floor is turning under the pendulum.

This effect is most pronounced if the pendulum is at either the North or the South Pole. The plane of oscillation of the pendulum then makes one complete rotation in 24 hours. However, if the pendulum is set swinging at the equator, no effect is observed. We can understand this if we note that in its swinging the bob of the pendulum always moves as much to the north as it does to the south of the equator, and hence the tendencies for the plane to turn both to the right and to the left offset each other. The time that it takes the plane to make a complete rotation at the equator can thus be considered as infinitely long, as shown in Figure 5.2.

If the pendulum is set up at any point between the North Pole and the equator, again a slow rightward rotation of the plane of oscillation occurs, but the period of rotation of the plane of the pendulum is then more than 24 hours. Foucault found in his experiment that the period of rotation was 32 hours. In New York City a similar pendulum set up at the U.N. rotates every 36 hours. One can show that the period of rotation of the Foucault pendulum is $24/\sin \lambda$ hours where λ is the latitude.

5.3 Effects of the Earth's Rotation on Projectiles and Air Masses

In analyzing the motion of projectiles on the surface of the earth, we must take the earth's rotation into account, since the trajectory of the projectile is affected just as the plane of oscillation of a pendulum is. The reason is clear if we keep in mind that the closer a point on the earth's surface is to the equator, the faster it moves because of the earth's rotation. If a projectile starts at a given point on the earth's surface in the northern hemisphere, it already has an eastwardly motion because of the earth's rotation. If it is propelled initially northward, its eastwardly motion (arising from the earth's rotation) is greater than that of points on the earth's surface directly below its path. *Hence to observers stationed at these points it appears that the projectile is deviated to the east (to the right as viewed in the direction of its flight).*

The same effect is present in large air masses moving to the north or to the south. As warm air masses rise above the equator and drift to the north and south temperate zones, they are deflected eastwardly and thus give rise to the prevailing westerly winds in the two hemispheres. For air masses drifting towards the equator from the north or south polar regions the deviation is to the west and one gets the north- and south-easterly trade winds. (See Figure 5.3.) The motions of these air masses give rise to cyclones wherever there are low- or high-pressure regions since the air masses tend to rush in or out of such areas, as shown in Figure 5.4.

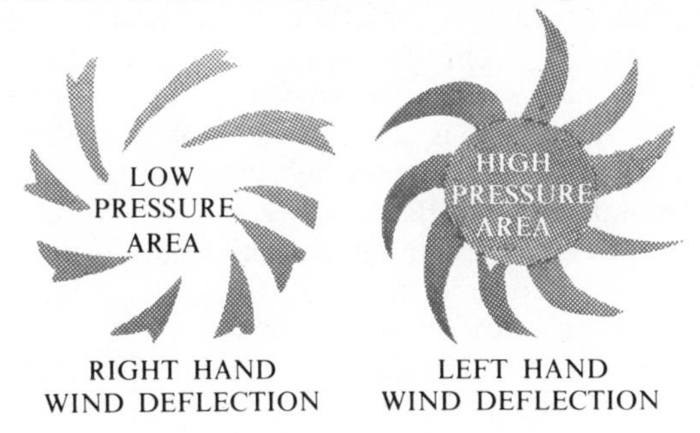

FIG. 5–4 *The different rotations of cyclones in regions of low and high atmospheric pressure.*

The rotation of the earth also affects freely falling bodies. If an object is dropped from a tall building, it does not fall along a plumb line to strike the earth directly under the point from which it starts, but is deflected eastwardly. The reason is that, when the object is at the top of the building, it has an eastwardly motion because of the earth's rotation which must carry it through a greater arc in the same time that the foot of the building is carried through a corresponding—but smaller—arc on the earth's surface. In other words, the object at the top of the building has a greater eastward speed than the foot of the building has, and hence is deviated to the east during the time of its fall, as shown in Figure 5.5.

FIG. 5–5 *The eastwardly deflection of a falling body resulting from the earth's rotation.*

5.4 Effect of Rotation on the Structure of the Earth

Just as the rotation of the earth affects the motions of bodies in various ways, it also affects the structure of the earth. It is, indeed, the rotation of the earth that is responsible for the earth's deviation from sphericity. We may understand this if we consider the forces that act on a body stationed at some point between the North Pole and the equator on a perfectly spherical earth. In order to simplify the analysis as much as possible, we suppose that our earth is covered by perfectly smooth ice on which a particle, free to move in any direction, is situated, as shown in Figure 5.6. As

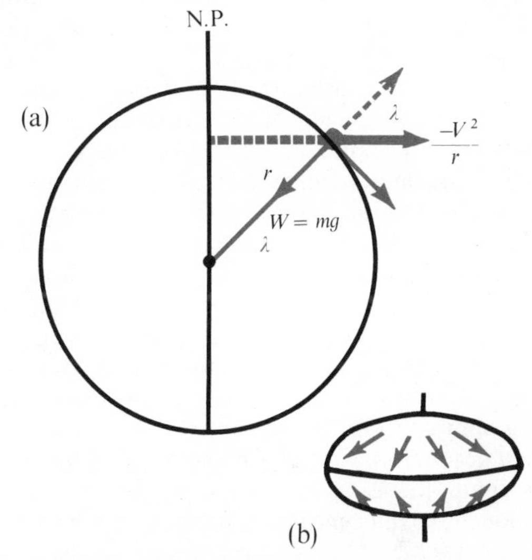

FIG. 5–6 (*a*) *Centrifugal and gravitational forces acting on a particle that is free to move on the surface of a perfectly spherical earth. The horizontal and vertical components of the centrifugal force are shown.* (*b*) *The effect of the uncompensated (horizontal) component of the centrifugal force is to distort the shape of the earth.*

we can see from the figure, where we picture a perfectly spherical earth, this particle has two forces acting on it: (1) its weight, which has the effect of keeping it pressed to the ground, and (2) a centrifugal force, which acts at right angles to the axis of rotation, and gets smaller as we move towards the North Pole.

We can now break up this centrifugal force into a vertical component and a horizontal component as shown in Figure 5.6. The vertical component is found by multiplying the centrifugal force, mv^2/r, by the cosine of the latitude of the point, and the horizontal component is found by multiplying mv^2/r by the sine of the latitude. Here r is the perpendicular distance of the body from the earth's axis of rotation, v is the speed of the body resulting from the earth's rotation, and m is the mass or the body. We see that the vertical component, which is very small, has the effect of reducing the weight of the body somewhat. In fact, what really happens is that the acceleration of a freely falling body at this point on the earth's surface is not g, but $g - (v^2/r) \cos \lambda$, where λ is the latitude of the point. This effect is very small.

The horizontal component, $(mv^2/r) \sin \lambda$ *(which acts towards the equator), of the centrifugal force is uncompensated and, hence, imparts an acceleration to the body on the ice.* This body, therefore, moves towards the equator. This is true for all points on the spherical earth's surface, but we do not observe bodies moving this way because the earth is not spherical now. The tendency for all objects to move toward the equator on a spinning sphere caused the earth to suffer a distortion and a bulge appeared at the equator (see Figure 5.6).

This horizontal component of the centrifugal force is also responsible for the flow of rivers such as the Mississippi, which run from north to south. Because of the bulge of the equator we would really be climbing a mountain about $13\frac{1}{2}$ miles high if we walk from the North Pole to the equator and if the earth were not spinning but still had its present shape. Hence, from this point of view, the Mississippi appears to flow uphill from its source in Minnesota to its mouth in the Gulf of Mexico. This uphill flow is made possible by the horizontal component of the centrifugal force, which we have just discussed. We may also account for the slow motion of glaciers by calling upon this force.

5.5 Variations in the Rotation of the Earth

In our discussion of time we saw that sidereal and solar time are based on the rotation of the earth, and for many years it was thought that the earth was, indeed, a good timepiece. However, within recent years careful *analysis of the apparent motions of various celestial bodies has shown us that there are small departures from uniform motion.* These variations in the period of rotation of the earth have been detected by comparing the theory giving us the positions of such bodies as the moon and sun and certain planets with the actual observations of these positions. In addition, it is now possible to compare the unit of time as defined by the rotation of the earth with the unit defined by highly precise atomic clocks that are now available. From this type of analysis we have found that there are three kinds of variations in the period of rotation:

1. a slow, steady increase in the length of the day,
2. irregular fluctuations in the period of rotation resulting in increases at some times and decreases at others,
3. seasonal variations.

5.6 Evidence from Eclipses of the Increase in the Day

The evidence for the slow increase in the length of the day goes back to observations made by Halley in 1695. By comparing his observations of the actual positions of the moon with the positions the moon would have had if they had been derived theoretically from very early observations of eclipses, he was able to show that the moon was ahead of where it should have been had the rotation of the earth and the motion of the moon been uniform since these early eclipses.

In the eighteenth and nineteenth centuries it was felt that a proper application of Newton's law of gravity would account for the apparent acceleration of the moon's motion, since it was thought that the observed discrepancy in the times of the eclipses was due to the motion of the moon. However, all attempts to account for the discrepancy in this way failed, and it was then realized that the apparent speeding up of the moon was really due to an annual slowing down in the rotation of the earth. We shall see later that this is due to the frictional action of the tides on the earth.

The slowing down of the earth is very small since it results in a lengthening of the day by 0.0016 *second per century.* However, in spite of the smallness of the effect, the actual amount by which the earth, taken as a timepiece, lags behind a uniform clock is fairly large over a period of a few thousand years. The reason is that the slowing down is a slowing down in the speed of rotation and hence is an acceleration. The amount, therefore, by which the earth as a clock lags behind a uniform clock is determined by the square of the time that has elapsed and not just by the time itself.

As an example of how large this discrepancy may be let us suppose that an eclipse occurred at noon just 4 000 years ago and that theory shows that another eclipse is to occur today. If the time for this predicted eclipse is based on the assumption that the earth is a perfect clock and does not slow down, we will be 8 hours late in looking for the eclipse. In other words, the earth as a timepiece lags behind a perfect clock about 8 hours in 4 000 years.

5.7 Fluctuations in the Period of the Earth's Rotation

Just as we have found that the moon's true position is different from that derived from eclipse observations, we have also found that the motion of the moon appears to fluctuate. Again, since these fluctuations cannot be accounted for by gravitational theory *we must conclude that the result is due to small unpredictable variations in the earth's rotation.* As far as the seasonal variations in the earth's rotation go, the evidence is based on the performance of clocks at various observatories. By means of either quartz crystal clocks, or the more recent atomic clocks, we have found that the earth rotates more slowly in the spring, and more rapidly in the autumn, and that the total variation in the length of the day throughout the year amounts to about 0.0025 seconds ($2\frac{1}{2}$ milliseconds). This fluctuation is observed from year to year but the amount is variable.

Both the irregular and seasonal variations in the rotations of the earth are probably due to redistribution of air masses in the atmosphere, the waters in the ocean, and the material in the mantle and core of the earth. We may also note that the spring and fall seasons are marked by different distributions of organic matter (foliage on trees, for instance) on the earth's surface.

Because of the variations in the period of rotation of the earth, it is desirable to introduce a unit of time based on the motion of the earth around the sun, rather than on its rotational motion.

5.8 The Orbital Motion of the Earth

When we introduced the equatorial coordinate system, we found it necessary to discuss the apparent motion of the sun in the sky. We saw that the sun appears to move to the east by about 1° per day, tracing out a great circle (the ecliptic) on the celestial sphere. Up until the time that Copernicus introduced his heliocentric concept of the solar system, this apparent motion of the sun was believed to be real. Not until after the death of Copernicus, and even then rather slowly, did people generally accept the earth's motion around the sun as the direct explanation of this apparent solar motion.

Even a man of such great observational skill as Tycho Brahe was not receptive to the Copernican theory, because, on the one hand, it was contrary to his strong theological convictions that man's dignity was dependent on the earth's being at the center of the heavenly bodies, and, on the other hand, he found no observational evidence for the earth's moving. He reasoned that if the earth were, indeed, moving around the sun, the stars ought to appear to shift their positions semiannually. Since he could detect no such apparent shift (stellar parallax), he argued that this effect is not present. We shall see that this effect is, indeed, present, but it is so small and requires such accurate observations that not until about 300 years after the death of Copernicus were the parallaxes of stars first measured.

5.9 The Earth's Motion and the Changing Sky

Before we consider the evidence supporting the earth's motion we may note that it is easy, in terms of such a motion, to account for the observed variations in the sun's position in the sky, and also for the appearance of the sky from season to season. In Figure 5.7 we have indicated the earth in its orbit around the sun, and we see that an observer standing on the earth and facing the sun sees it projected on the sky in the direction of his line of sight. Thus, when the earth is at the point marked June 21, the sun appears to be in the constellation of Gemini. The stars in the neighborhood of Cancer are not visible, however, because of the sunlight. In the nighttime the observer sees the night sky diametrically opposite to the sun in the direction of the constellation of Sagittarius. At some later date the earth has a different position, and the sun therefore appears to change its position on the sky, so that it is in the constellation of Leo, the stars of which are now invisible. The stars in the night sky, diametrically opposite, are now those in the constellation of Aquarius, and so on, as shown in the figure.

Let us consider what happens during the period of one day. Figure 5.8 shows the earth as seen from the N.C.P. with the sun on the observer's meridian. If we consider the observer 23^h56^m later (the end of one sidereal day) we note that he no longer finds the sun on his meridian, but rather to the east of it. The reason is that the earth has advanced in its orbit by about 1°, so that the line from the sun to the earth falls on the circle of longitude that is 1° to the east of the observer. Four minutes later the observer will see the sun on his meridian because the earth will have rotated 1° to the east.

We may not conclude from these considerations that the earth is moving around the sun, for it is possible to account for these various effects by

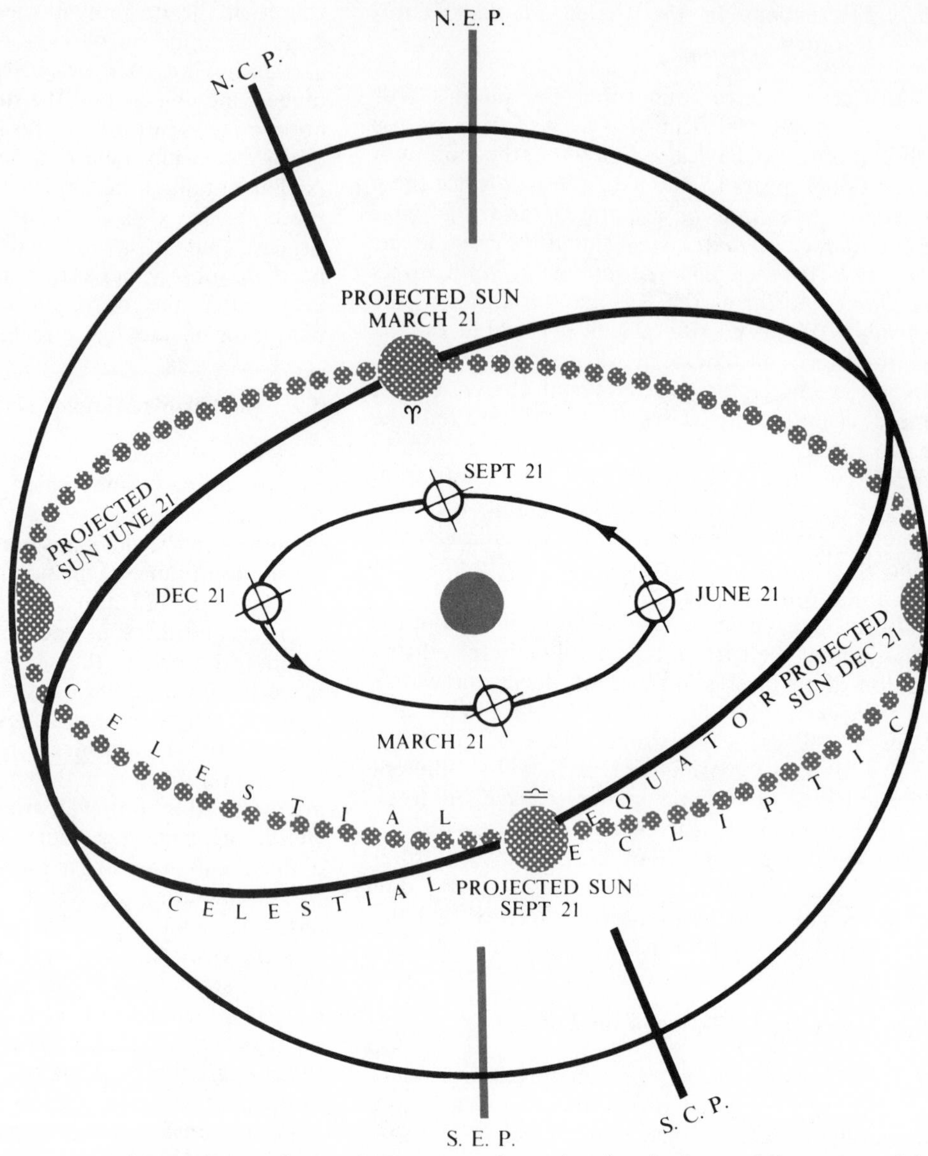

FIG. 5–7 *Schematic diagram showing the sun projected onto the celestial sphere at different times of the year. The earth's axis and its tilt to the ecliptic are also shown.*

assuming that the sun is really moving. As long as we limit ourselves to observation within the solar system, we are unable to obtain any evidence that requires a moving earth rather than a moving sun. Only when we make observations of the stars do we obtain conclusive evidence for the earth's motion. For if we find the stars shifting their positions or behaving in other ways which can be accounted for only by a moving earth, we may conclude that the earth is, indeed, moving.

5.10 Observations of Stars that Prove the Earth is Moving: Aberration of Light

We shall see that three different kinds of observations can be made on the stars to prove that the earth is moving:

1. the aberration of light;
2. the parallactic shift of stars; and
3. the Doppler effect.

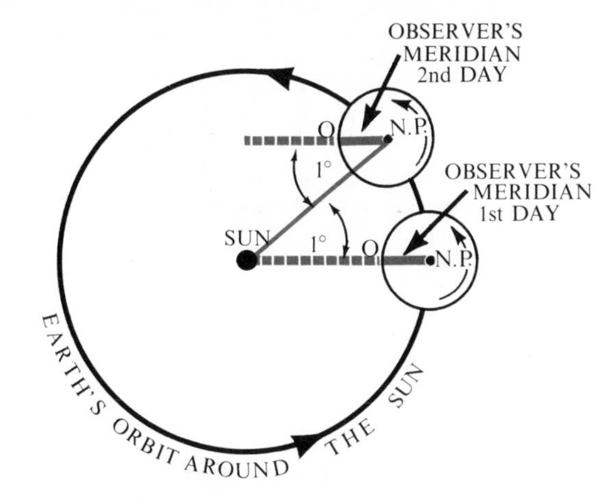

FIG. 5–8 *The earth's motion around the sun leads to a difference between the length of the solar and sidereal days. After a 23-hour 56-minute lapse, the direction from the observer to the sun differs from the direction to the observer's meridian by about 1°. This difference is exactly equal to the arc through which the earth has moved around the sun during this period.*

Once the Copernican theory was accepted, astronomers began to look for evidence for the earth's motion in the apparent behavior of stars. Not until 1725 was the first of the three effects we have mentioned observed. In that year Bradley (who had been looking for the parallactic effect) announced that the apparent positions of stars in the sky are different from their true positions. This effect, which is due not to a change in the position of the earth, but rather to the *motion of the earth, is called the* **aberration of light**. It really results from the fact that the direction from which light appears to come to us, when it comes from a very distant source, is affected by our motion. The same thing, by the way, occurs if we watch falling raindrops or snowflakes. When no wind is present and we sit in a stationary bus, the snowflakes or raindrops make vertical streaks on the window. But as soon as the bus starts moving, the drops or flakes appear to be coming from in front of us and we find diagonal streaks on the window, as shown in Figure 5.9.

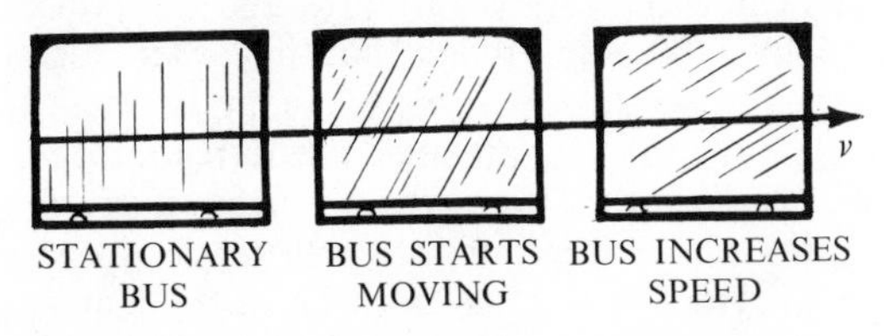

FIG. 5–9 *The apparent direction of rainfall changes as the speed of the moving observer changes.*

We can understand the nature of this effect by means of a simple mechanical analogy. If an observer inside a boxcar at rest on a railroad track watches a bullet moving at right angles to the car with a speed c, he notes that it enters the first side wall at A, and leaves the second side wall at B, as shown in Figure 5.10. If he can make no observations of what is happening outside the car, he will conclude that the direction of the bullet is that of the line connecting A and B, and in this case he will be right.

If, however, the car is moving at a speed v to the right, the bullet, after entering at A (we assume that the wall of the car is paper thin, and therefore has no effect on the flight of the bullet), leaves the car not at the point B (directly across from A) but rather at some point D to the left of B. Of course, the bullet, while inside the car, still moves in the same direction it had before entering the car, but the motion of the car brings the point D to the position previously occupied by point B at the moment the bullet was entering the car at point A, as shown in Figure 5.10. To the observer in the car, it now appears that the direction of the flight of the bullet is given by the direction of a line from A to D; we might call this the aberration of the flight of a bullet. It is clear from the diagram that this is simply the result of the combination of the velocity of the bullet and the velocity of the train.

5.11 Aberration Formula

The observer can relate the angle of aberration θ (that is, the apparent deviation of the direction of motion of the bullet) to the speed of the train, and the speed of the bullet. It is clear that the distance DB is just exactly the distance through which the train has moved during the time that the bullet has traveled the distance AB. This means that the ratio of DB to AB is equal to the ratio of v to c:

$$\frac{DB}{AB} = \frac{v}{c}.$$

On the other hand, from trigonometric considerations we know that the ratio DB/AB is equal to the tangent of the angle θ. *We thus see that the tangent of the angle of aberration is given by the ratio of the speed of the observer to the speed of the bullet provided the two velocities are at right angles to each other*:

$$\boxed{\tan\theta = \frac{v}{c}}\,. \tag{5.1}$$

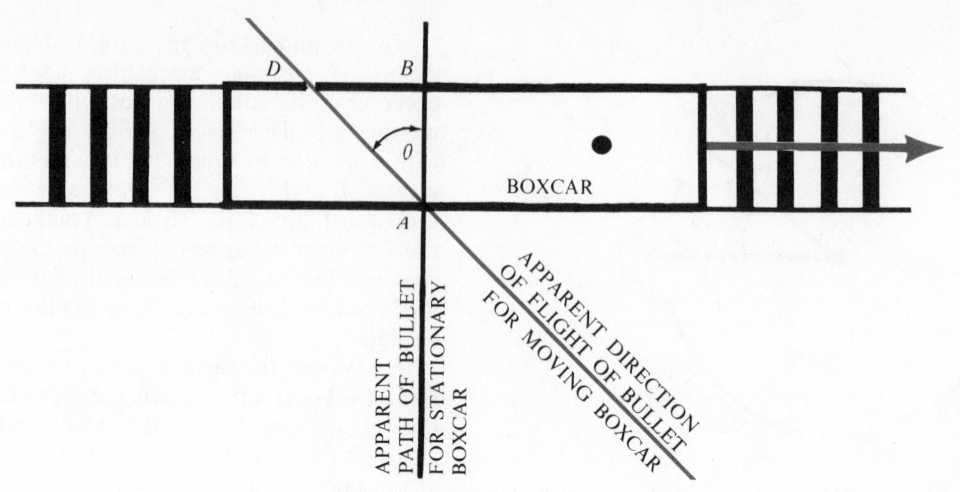

FIG. 5-10 *The aberration of the flight of a bullet as seen by an observer in a moving train.*

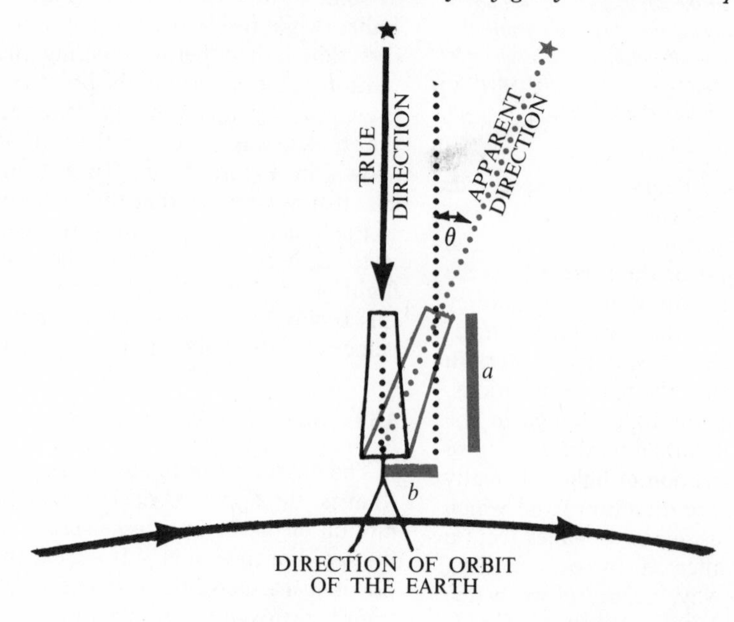

FIG. 5-11 *The aberration of starlight as seen from the moving earth. The observer's telescope must be tilted (as is the telescope tube in the diagram) in the direction of the earth's motion. Line a indicates the distance the light travels through the telescope tube, and line b the distance the eyepiece travels while the light travels through the tube.*

We may note that if the observer measures this angle of aberration, and knows the speed of the bullet, he can determine his own speed.

5.12 The Aberration of Light from Stars and the Orbital Speed of the Earth

This same sort of analysis can be applied to light from a distant star, as shown in Figure 5.11. In this case, the train is the earth and the bullet is the light. We see, then, that in order for an observer to see a star in his telescope he must tilt the telescope through an angle θ with respect to the true direction of the star, whose tangent is given by the ratio of the speed of the earth to the speed of light, provided the actual position of the star is at right angles to the direction of the earth's motion.

This effect could not be observed if the earth

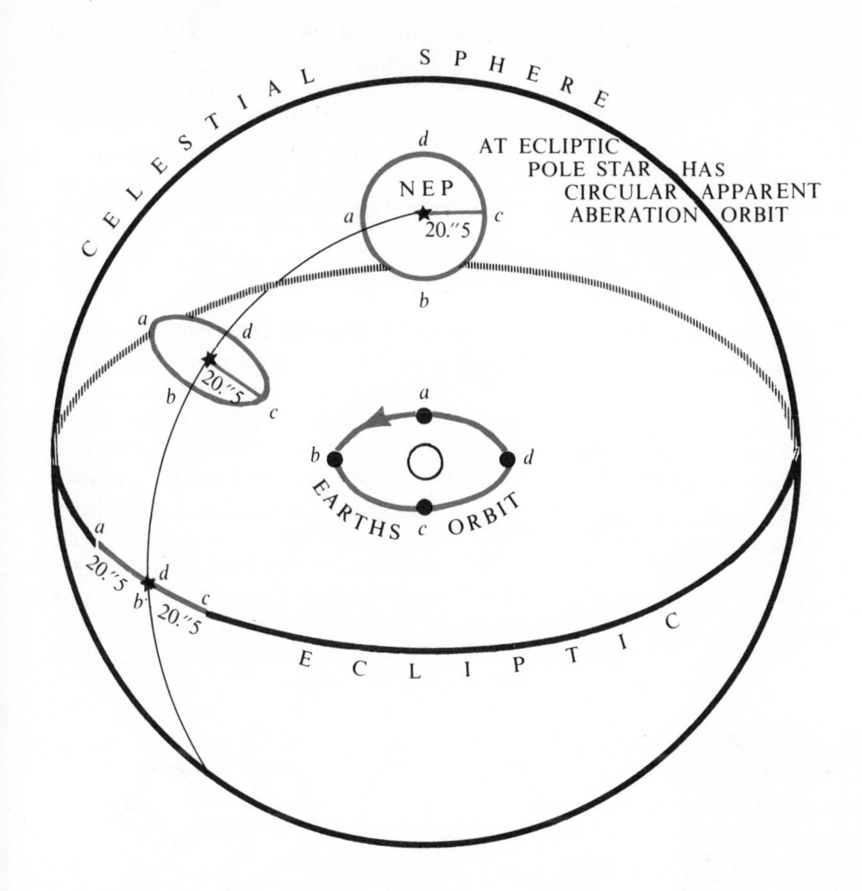

FIG. 5–12 *The apparent aberration orbits of stars at different positions relative to the ecliptic. The length of the major axis of the apparent orbit in each case is 41″ of arc.*

were moving in the same direction, for then there would be no way of our knowing that we had tilted our telescopes. However, the earth reverses its direction of motion six months later, so that the telescope must then be tilted in the opposite direction. *Note that the telescope must always be tilted in the direction of the motion of the earth.*

Bradley, in 1727–1728, first measured this effect, and today we find with very accurate measurements that the angle of tilt, which is called the **angle of aberration**, is equal to 20″.49. Since we can easily look up the tangent of this angle, and since we know the speed of light, we can use our simple formula to find the speed of the earth. *The value we obtain in this way for the average speed of the earth is about* 29.80 *km/sec* ($18\frac{1}{2}$ *mi/sec*).

From this value of the speed of the earth we can compute the circumference of the earth's orbit, if we assume this orbit to be a circle—which is a good approximation. There are about 3.15×10^7 seconds in a year, and this means that the earth's orbit is equal to this number multiplied by 18.5 mi/sec, which gives 5.83×10^8 miles or 9.33×10^8 kilometers. *From this we find that the mean-distance between the earth and the sun (the astronomical unit, A.U.) is* 93×10^6 *miles or* 1.49×10^8 *kilometers.*

Note that the aberration of light is a maximum if the direction of the star is at right angles to the earth's motion, and vanishes for stars whose directions from us are parallel to the earth's motion. For a star in an intermediate position the aberration is obtained by multiplying the maximum value by the sine of the angle between the earth's motion and the direction to the star. This means the aberration has a constant value for stars at the poles of the ecliptic, with the result that these stars appear to move in small annual nearly circular orbits whose diameters are 41″ of arc.

The stars lying in the plane of the ecliptic have aberrations ranging from a maximum value to zero. Any star in the plane of the ecliptic at right angles to the earth's motion appears to oscillate back and forth along a straight line through an angle of 41″. For stars lying between the poles and the plane of the ecliptic, the orbits of the apparent motions resulting from aberration are ellipses, as shown in Figure 5.12. Note that the aberration is independent of the distance of a star.

5.13 Parallax of Stars

When Tycho Brahe rejected the Copernican system because he was unable with his naked-eye instruments to detect the parallactic shift of stars, he was under the impression that the stars were much closer than they actually are. Although he was able to make measurements with an accuracy of 1 minute of arc, this was hardly good enough to allow him to observe the shifting of the nearby stars arising from the earth's motion. This effect is so small, even for the closest stars, that it was not until 1838 that such measurements were made. See Figure 5.13.

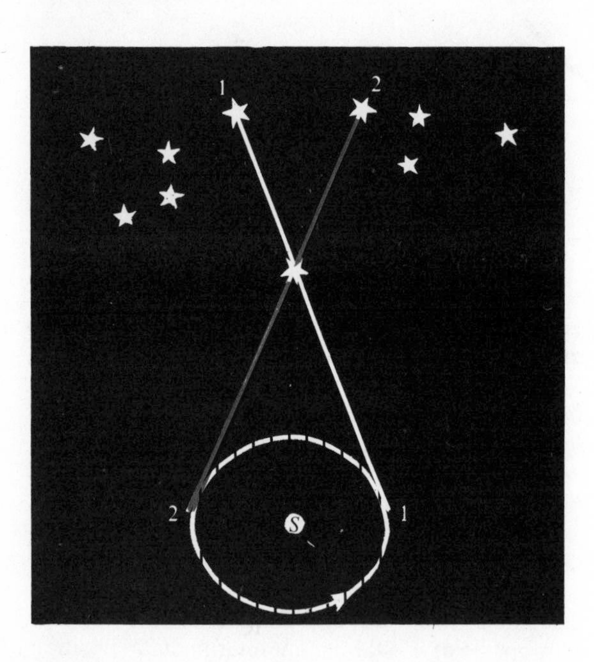

FIG. 5–13 *The parallactic displacement of a star resulting from the annual motion of the earth.*

In 1837, Bessel, using the most advanced instruments available at that time, began a series of observations on 61 Cygni which enabled him to announce in December 1838 that he had measured its heliocentric parallax, and that the star was at a distance of about 600 000 A.U. Following upon this, Henderson, in South Africa in 1839, measured the parallax of alpha Centauri; and Struve, in Russia in 1840, determined the parallax of Vega. These were the first direct observations of the effect of the change in the earth's position in space on the apparent position of a star, and they present conclusive evidence that the earth is moving.

5.14 The Doppler Effect

To understand the **Doppler effect** we first give a brief description of the properties of light (for a detailed discussion of this subject see Chapter 16). Although there are certain difficulties in representing light entirely as a wave phenomenon, many of its behavior patterns can be explained by the wave picture. The Doppler effect is one of these. According to the wave picture, a monochromatic beam of light can be represented by a simple sinusoidal curve, as shown in Figure 5.14. What is important about this is the concept of the wavelength, which is simply the distance between two successive crests of the wave.

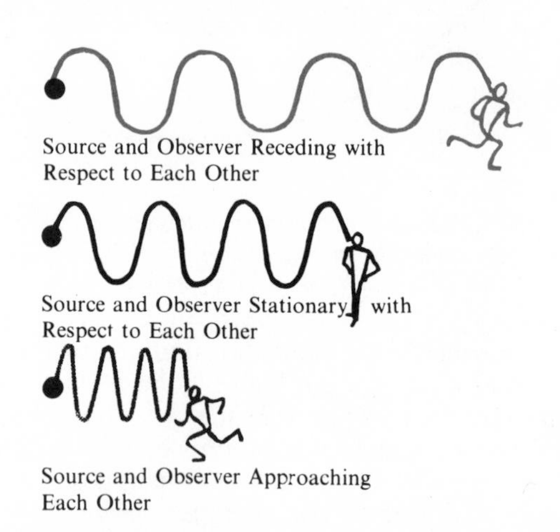

FIG. 5–14 *The change in the wavelength of light (the Doppler effect) as measured by an observer receding from the source of light and an observer approaching the source of light, compared with a stationary observer.*

The wavelength determines the color of the light as we see it. The longer wavelengths correspond to red light, and the shorter wavelengths to blue (violet) light. If a source emits light of different wavelengths, one can use a spectroscope to separate out these various wavelengths (the distinct colors); we call this a spectrum of the source.

Let us now consider the light (from a star) of one particular wavelength, as measured by an observer

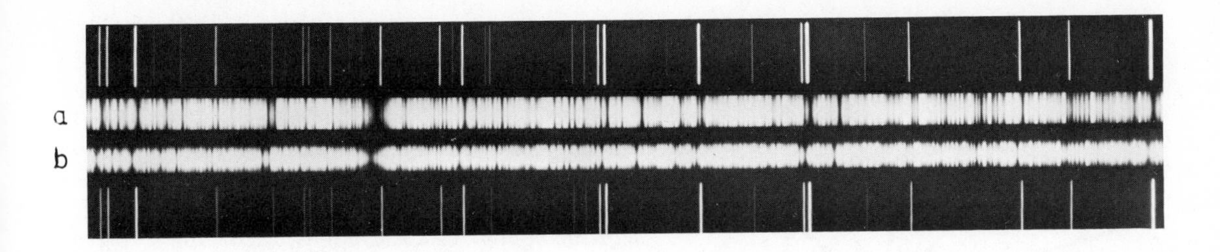

FIG. 5-15 *Two spectra of constant-velocity star Arcturus taken 6 months apart to show orbital velocity of the earth by means of the Doppler shift. (Photograph from the Mt. Wilson and Palomar Observatories)*

who is moving neither away from nor towards the star, and compare this wavelength with that measured by an observer who is moving with respect to the star. If this second observer is moving towards the star, his eye receives more waves per second than it would if he were not moving, so that the waves appear to be crowded together—that is, the wavelength appears to be smaller. If the observer is moving away from the star, he receives fewer waves per second and therefore assigns a longer wavelength to these waves, because the wave train appears to be stretched out. This is called the Doppler effect, and it can tell us whether we are moving away from or towards a star. *If we are moving towards a star, all the wavelengths appear smaller to us than they do to a fixed observer, and the light appears bluer. If we are moving away from the star, the wavelengths as measured by us are longer, and the light appears redder than to a fixed observer.* Since the earth is moving around the sun, it is always approaching some stars and receding from others, so that the light from these various stars should exhibit Doppler effects which change periodically as the direction of the earth's motion changes. This is, indeed, found to be the case. (See Figure 5.15.)

5.15 The Shape of the Earth's Orbit

It is a fairly easy matter, by making observations on the variations of the size of the sun from week to week, to determine the shape of the earth's orbit. Although we have spoken of the earth's orbit as circular, actual measurements prove it is not. We may first note that the earth moves more rapidly in its orbit from the autumnal equinox through the fall and winter months to the vernal equinox than it does during the spring and summer months from the vernal to the autumnal equinox. During the period of rapid motion the earth is closer to the sun than it is during the time of its slower motion.

Precise comparisons of the distances of the earth from the sun at various times of the year can be made by comparing the variations in the apparent size of the sun. On the average, the sun appears to subtend an angle of $\frac{1}{2}°$ at the eye, but the actual apparent size changes as we get closer or farther away. We can now obtain a graphical representation of the earth's orbit by laying off lines radiating from a point, *S*, which represents the position of the sun, as shown in Figure 5.16. Let each line give the direction from the earth to the sun at the given moment of observation, and let the length of that line be inversely proportional to the angle subtended at that moment at the earth by the diameter of the sun. Each line then represents a measurement of the direction of the sun from the earth, and a measurement of the apparent size of the sun at that moment. *If all the end points of the lines are connected, we obtain the earth's orbit properly scaled down. If the distance corresponding to any one of these lines is known, then all of the other distances can be found. From this we find that the earth's orbit is an ellipse, with the sun at one focus of the ellipse.*

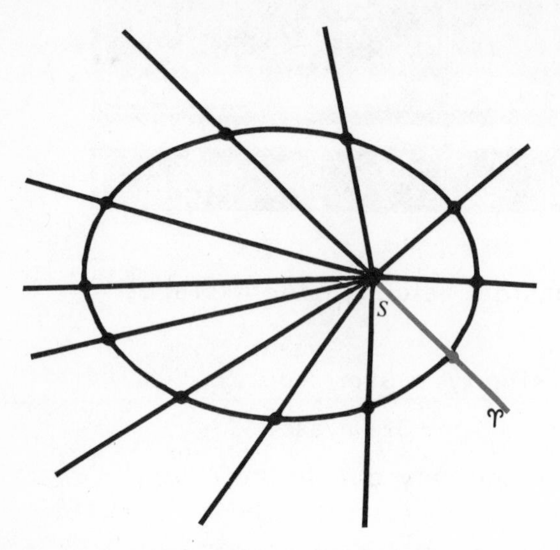

FIG. 5–16 *The shape of the earth's orbit determined graphically by using the variations in the apparent size of the sun's disk from time to time during the year.*

5.16 Precession of the Equinoxes

We come now to two additional components of the earth's motion which are quite small as compared with the rotation and revolution of the earth. These are referred to as the **precession of the equinoxes** and **nutation**. When we introduced the equatorial coordinate system, we defined the right ascension of a star in terms of the vernal equinox, which we assumed to be a fixed point on the celestial equator. Although for day-to-day observations of stars this approximation is satisfactory, it is necessary to take into account the continual displacement of the equinoxes along the ecliptic if we want to keep accurate records of stellar positions from year to year.

In the year 125 B.C. Hipparchus discovered, as a result of careful observations of the positions of bright stars over a period of years, that although the ecliptic appears to be a fixed circle in the sky, the equator appears to change its position in such a way that the vernal equinox moves backward (westwardly, or opposite to the direction of the sun's apparent motion) along the ecliptic. That is, he discovered that the year of the seasons (the year as measured by the interval of time between two successive coincidences of the sun with the vernal equinox) is shorter by twenty minutes than that given by the interval of time between two successive coincidences of the sun with the same point in a given constellation (the sidereal year).

What this means is that the vernal equinox appears to move towards the sun along the ecliptic, and to meet it twenty minutes before the sidereal year ends. Since twenty minutes is about the one twenty-six-thousandth part of the year, it follows that the vernal equinox revolves once every twenty-six thousand years. This annual precession of the equinoxes corresponds, according to precise measurements made by Newcomb in 1925, to 50″.2619 per year.

5.17 Planetary and Luni-Solar Precession of the Equinoxes

Although it appears that the ecliptic itself is fixed, and that the observed precession is due entirely to the shifting of the celestial equator, this is not so. Actually, there are two components of the precession, as shown in Figure 5.17, *one of which arises from the motion of the ecliptic and is called the* ***planetary precession****, and the other of which arises from the motion of the equator and is called the* ***luni-solar precession****.* The sum of these two contributions gives rise to the general precession. In the case of the luni-solar precession, which is about 40 times larger than the planetary precession, the north celestial pole revolves around the pole of the ecliptic, as shown in the diagram.

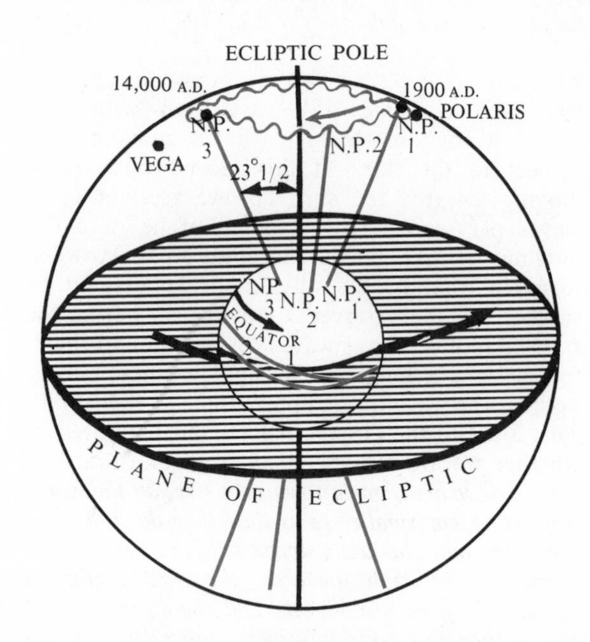

FIG. 5–17 *The precession of the equinoxes and nutation. The axis of rotation in positions 1 and 2 corresponds to the positions 1 and 2 of the vernal equinox on the ecliptic.*

The planetary precession is opposite to that of the luni-solar precession and therefore tends to reduce the latter slightly.

Because of the luni-solar precession the appearance of the sky changes appreciably every few thousand years, so that different stars lie near the north celestial pole at different times. This is illustrated in Figure 5.17, where we see that in the year 14 000 A.D. Vega will lie close to the celestial pole. In Section 5.28 we shall explain how the precession of the equinox arises after we have introduced the concept of angular momentum and the torque acting on a spinning object.

5.18 Nutation

Superimposed on the progressive motion of the mean pole, there is a rapid periodic motion which is known as **nutation,** which is also shown in Figure 5.17. This periodic motion of the celestial pole arises primarily because the plane of the moon's orbital motion around the earth is not coincident with the ecliptic, so that periodically the moon tends to displace the axis. The maximum effect is 9″.23, and a complete oscillation occurs in about 19 years. Although the moon's contribution to the nutation is the largest effect, there is a small periodic contribution from the sun also. This arises because the sun's distance from the earth and its position relative to the equatorial bulge change.

5.19 The Concept of Momentum

To understand the way in which precession arises, we must begin by introducing a new physical concept which we can derive from previous considerations. Since we have already introduced the concepts of velocity and mass, we may now combine these two by multiplying them together (giving the quantity $m\mathbf{v}$) to obtain the concept of the momentum of a body. *As we can see, the momentum of a body is a vector quantity since velocity is a vector.*

When Newton introduced his second law of motion, he did so in terms of the momentum, rather than the acceleration, and stated that *the force acting on a body is proportional to the time rate of change of the momentum of the body.* If, during the action of the force, the mass of the body does not change, then this statement of the second law is equivalent to the usual statement, but if the mass of the body does undergo a change, then this statement of the second law is more general and tells us more than the usual statement does.

5.20 Newton's Third Law of Motion

We can now obtain an important property of the momentum of a body if we consider the third of Newton's laws of motion, which we have not yet discussed. This law states:

> **For every force that is exerted on a given body, there is always an opposite and equal force exerted by this given body.**

This law means, essentially, that forces always appear in pairs, and that the two components in any pair are always opposite and equal, so that all forces in the universe cancel out, leaving no net force. Thus, if we consider the earth pulling upon a body and causing it to fall towards the center of the earth, then this body pulls on the earth's center with a force that is equal to the weight of the body itself. Thus, since the sun exerts a gravitational pull on the earth, the earth exerts an exactly equal and opposite pull on the sun. Again, if we consider a baseball player exerting a push on the ball that he is throwing, it follows that the ball exerts a push of an exactly equal amount against the pitcher.

At first sight this appears very strange to the student just beginning physics, because he wonders why, if the forces in a given pair are always equal, one body moves and the other does not appear to do so. Thus, he wonders why it is the ball that moves and not the pitcher, and why the earth and not the sun, or why the falling object and not the earth. Actually, this is a misunderstanding of what is taking place, because both objects do, indeed, move, if they are free to do so, but by different amounts (different accelerations). We must remember that the acceleration produced by a force depends not only on the force that is acting, but also on the mass of the body on which it acts.

In the case of the earth and the sun, the force of the earth on the sun is exactly equal to the force of the sun on the earth, but since the mass of the sun is 332 000 times larger than the mass of the earth, the acceleration imparted to the sun is negligible and hardly observable. In the case of the pitcher and the baseball, the effect of the ball on the pitcher cannot be observed because he is anchored to the ground, but if he were standing on a perfectly smooth surface, he would start moving backwards as soon as the ball acquired a forward motion.

5.21 Conservation of Momentum

Let us now consider two particles, one of mass m_1 and the other of mass m_2, moving towards each other along a straight line, just before they collide. If the velocity of m_1 is v_1, and that of m_2 is v_2, then the total momentum of the two particles before collision is $m_1v_1 - m_2v_2$. The minus sign appears because momentum is a vector quantity and in this case the two momenta are in opposite directions. Here, m_1v_1 represents the momentum of the first particle before the collision, and $-m_2v_2$ the momentum of the second particle before the collision.

We shall suppose that the collision lasts for a very short interval of time, Δt (the Greek symbol delta will always be used in this text to represent a small amount of a physical quantity). During this time of collision the two bodies exert forces on each other and, according to Newton's third law of motion, the force exerted by body 1 on body 2 is exactly equal to that exerted by body 2 on body 1. We shall further suppose that after the collision the two bodies separate and retrace their paths, but that the speed of body 1 is v_1' and that of body 2 is v_2'. Thus the momentum of body 1 after the collision is $-m_1v_1'$, and that of body 2 is m_2v_2'. As a result of the collision the momentum of body 1 suffers a change (this change is given by the difference between the final momentum and the initial momentum as shown in Figure 5.18), which is equal to

$$\underset{\text{final momentum}}{-m_1v_1'} - \underset{\text{initial momentum}}{m_1v_1}.$$

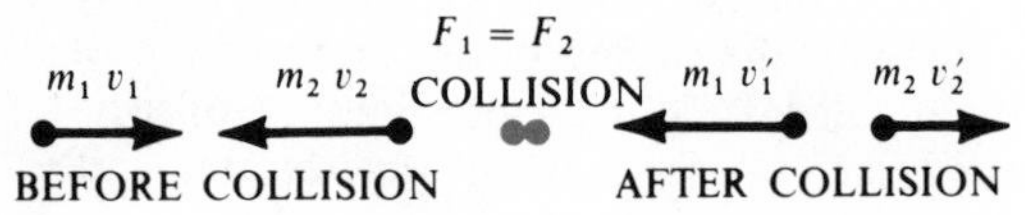

$F_1 = F_2$

$m_1 v_1$ $m_2 v_2$ COLLISION $m_1 v_1'$ $m_2 v_2'$

BEFORE COLLISION AFTER COLLISION

FIG. 5–18 *The motion of two bodies before, during, and after collision. The arrows indicate the momenta of the particles.*

Hence, by Newton's second law, the force acting on body 1 is this difference divided by Δt,

$$\frac{-m_1v_1' - m_1v_1}{\Delta t}.$$

This is the time rate of change of the momentum of body 1, during the collision.

In the same way, it is easy to see that the time rate of change of the momentum of body 2 during the collision is

$$\frac{m_2v_2' - (-m_2v_2)}{\Delta t}.$$

Since during the collision the two bodies experience equal but opposite forces, we must have

$$\frac{-m_1v_1' - m_1v_1}{\Delta t} = -\left(\frac{m_2v_2' - (-m_2v_2)}{\Delta t}\right),$$

where the minus sign on the right-hand side is introduced because the forces are opposite in direction.

If in this equation we cancel Δt on both sides and transpose all primed quantities to one side and all unprimed quantities to the other, we obtain

$$m_2v_2' - m_1v_1' = m_1v_1 - m_2v_2.$$

But we see that on the left-hand side of this equation we have the total momentum of the system after collision and, on the right-hand side, the total momentum before collision. This simple equation, therefore, is an expression of a very profound principle, namely, that *if there are no external forces acting on a system (the two particles taken as a single system have no external forces acting on them), then the total momentum of the system remains constant no matter how the component particles of a system interact with each other.* This principle of the conservation of momentum is a consequence of Newton's third law of motion.

5.22 The Conservation of Momentum and the Center of Mass of a System of Particles

Although we have discussed the principle of conservation of momentum for two bodies only, it holds no matter how many bodies may be present in a system. This, then, leads us to another important concept, the nature of which we can understand by considering a system composed of a number of different particles (for example, the fragments of an exploding bomb) which has no external forces acting on it. According to what we have just said, the momentum of this entire system always remains constant regardless of how the individual particles of the system are moving. If we take the momenta of all particles and add them vectorially, we obtain a single resultant momentum which is given by the total mass of the system of bodies times a velocity.

In other words, the system behaves as though all its mass were concentrated in a single point, the **center of mass** of the system, which moves with a velocity **V**. We may now state the **principle of conservation of momentum** as follows: *if no external forces are acting on a system of bodies whose total mass is* M, *the center of mass of the system moves with uniform velocity* **V** *in a straight line (constant*

$$m = m_1 + m_2 + m_3 + \cdots$$

$$m\mathbf{v}_c = m_1\mathbf{v}_1 + m_2\mathbf{v}_2 + m_3\mathbf{v}_3 + \cdots$$

$$X_c = \frac{m_1x_1 + m_2x_2 + m_3x_3 + \cdots\cdots\cdots}{M}$$

$$Y_c = \frac{m_1y_1 + m_2y_2 + m_3y_3 + \cdots\cdots\cdots}{M}$$

$$Z_c = \frac{m_1\,z_1 + m_2\,z_2 + m_3\,z_3 + \cdots\cdots\cdots}{M}$$

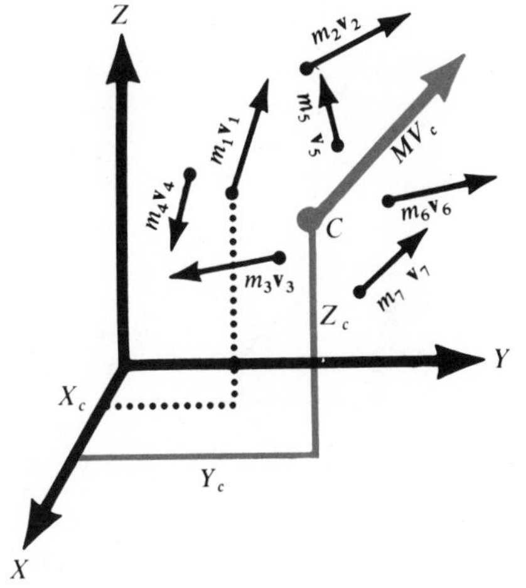

FIG. 5–19 *The motion and the momentum of the center of mass of a system of particles in a three-dimensional rectangular coordinate system.*

velocity), as shown in Figure 5.19, *such that* MV *is the total momentum of the system.*

It often happens in textbooks that the term "center of gravity" is introduced to refer to the center of mass. This is an error since, in general, there is no such thing as the center of gravity of a body. The center of gravity implies that one can picture a gravitational force on a body as behaving as though it were acting on a single point in the body. This is true under very limited conditions—for example, in the case of a uniform sphere or in a region where the gravitational force is a constant. The reason that it is permissible on the surface of the earth to speak of the center of gravity of a small body and to identify it with its center of mass is that the earth's gravitational force, per unit mass, is practically constant and parallel for all points of the body, as shown in Figure 5.20.

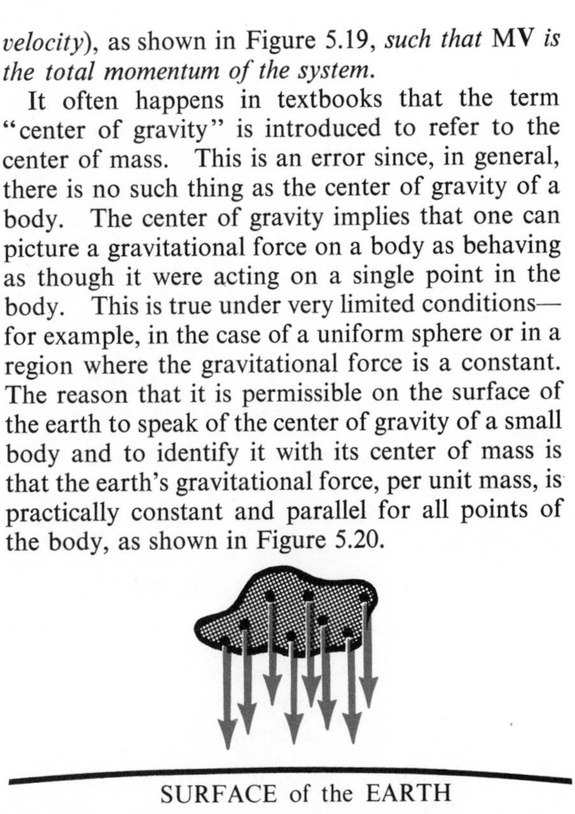

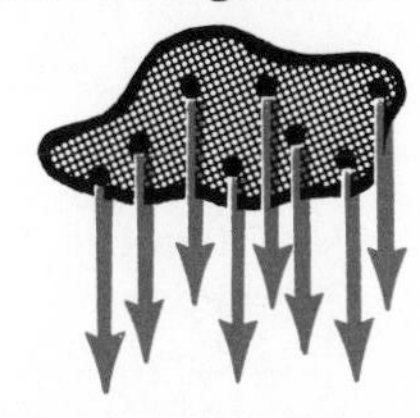

SURFACE of the EARTH

FIG. 5–20 *Near the surface of the earth the gravitational forces acting on the constituent particles of a small body are nearly parallel.*

5.23 Angular Momentum

We have just noted that if a body or a system of bodies has no force acting on it, its momentum is conserved, but now we shall see that even when a force is acting on a system, it is still possible to introduce a quantity that is conserved under certain other conditions. We consider a particle of mass m moving with constant speed v in a circular orbit. It is clear that the momentum of this particle is not conserved, since its momentum is changing continuously as a result of the change of direction of its velocity, as shown in Figure 5.21. This continual

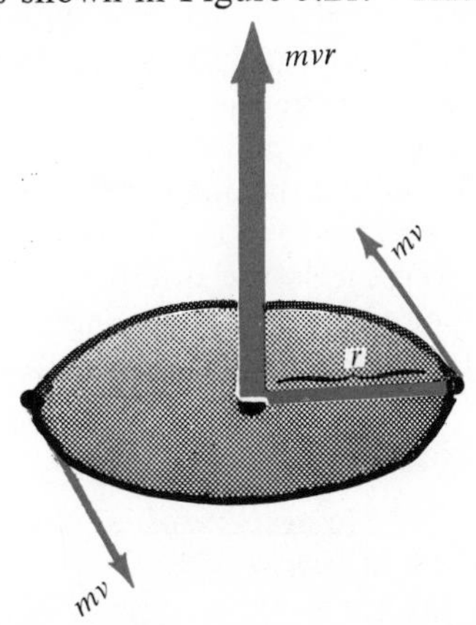

FIG. 5–21 *The angular momentum of a particle moving in a circle. The direction of the angular momentum vector is perpendicular to the plane of the circle and its magnitude is equal to mvr.*

change in the momentum is due to the centripetal force on the particle acting towards the center of the orbit. However, even though the momentum is not conserved, we can see that there is a vector quantity associated with this motion which is conserved. To discover this vector quantity we draw an arrow at the center of the orbit and at right angles to the plane of the orbit. If we let the size of this arrow represent the quantity mvr, where r is the radius of the circular orbit, then this vector remains unchanged as long as the particle moves with constant speed in its orbit. This vector is called the **angular momentum** of the particle.

We can now speak of the angular momentum of any moving particle with respect to any point, and this holds even if the particle is moving in a straight line as shown in Figure 5.22. In the general case

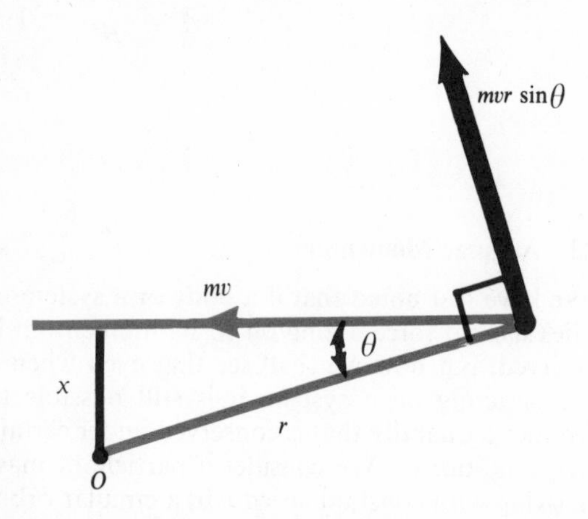

FIG. 5–22 *The angular momentum of a particle moving in a straight line with constant velocity, v, relative to an observer, o.*

we define this angular momentum with respect to a point as the momentum mv of the particle multiplied by the perpendicular distance, x, of the point from the line of motion (extended if necessary). Thus in Figure 5.22 the angular momentum of the particle m with respect to an observer, O, is mvx, or $mvr \sin \theta$, where r is the distance of the particle from the point, and θ is the angle between the velocity of the particle and the line from the point to the particle.

5.24 Angular Momentum of a Rotating Body: Moment of Inertia

Just as we can speak of the angular momentum of a single particle with respect to a point, so we can consider the angular momentum of a system of particles. We obtain this quantity by taking the angular momenta of all the component parts of the system separately, and adding them all together vectorially. In the case of a body that is spinning about an axis, such as is shown in Figure 5.23, we

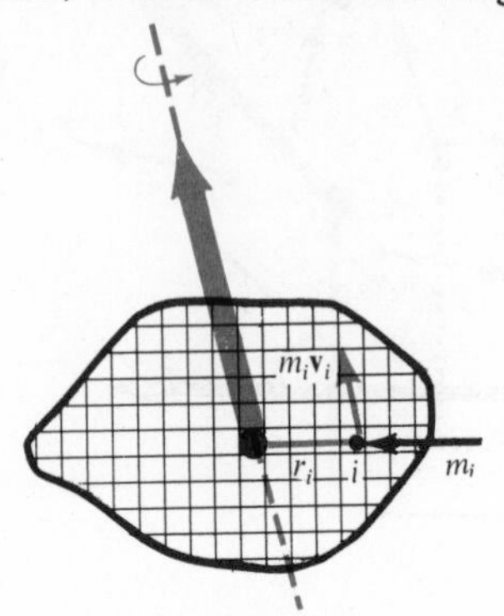

FIG. 5–23 *The angular momentum of a rigidly rotating body is the vectorial sum of the angular momenta of all its constituent particles, such as the particle i in the figure which is moving in a circle of radius r_i.*

may introduce the angular momentum of the body with respect to this axis in terms of the angular velocity ω of the body about the axis and another quantity called the **moment of inertia.**

We picture the body as being divided into a large number of small pieces, the ith one of which has a mass m_i. If r_i is its perpendicular distance from the axis of rotation, and v_i is its circular velocity about this axis, then its angular momentum about this axis is mv_ir_i—by definition. The direction of the angular momentum, of course, is parallel to the axis of rotation.

Since $v_i = r_i\omega$ (note that ω, the angular velocity, is the same for all particles in a body), we have

$$m_iv_ir_i = m_ir_i^2\omega.$$

This expression for the angular momentum applies to all the particles in the body, so that the total angular momentum about the axis of rotation is (if there are n particles)

$$m_1r_1^2\omega + m_2r_2^2\omega + \cdots + m_nr_n^2\omega$$

or

$$\left(\sum_{i=1}^{n} m_ir_1^2\right)\omega, \tag{5.2}$$

where $\sum_{i=1}^{n}$ means the sum over all the particles. Since the quantity $\sum m_ir_i^2$ is called the moment of inertia, I, of the body with respect to the axis of rotation, we may write for a spinning body

$$\text{angular momentum} = I\omega. \tag{5.3}$$

Note that the moment of inertia I plays the same role for a spinning body that the mass m does for a body in translational motion.

5.25 Changing the Angular Momentum of a System: Torque

Let us go back to Figure 5.21 and consider how we might change the angular momentum of the particle. We can do this by changing the direction of the axis about which the particle is revolving, or the speed of the particle, or the radius of its orbit. In any case, a force must be applied whose direction does not pass through the center of the orbit—otherwise it can neither change the speed of the particle nor affect the tilt of the axis of rotation.

To change the speed of the particle or to tilt the axis of rotation the force must not pass through the center of the orbit, but must have an orientation similar to that in Figure 5.24. In this case we say the force exerts a **torque** *on the system. This torque is given by the product of the force and the perpendicular distance* (x) *of the direction of the force from the center of rotation. This is the quantity fx as shown in the figure, or $fr \sin \theta$, where θ is the angle between the direction of the force and the line from the center of the orbit to the particle. Torque is a vector at right angles to the plane containing r and f.*

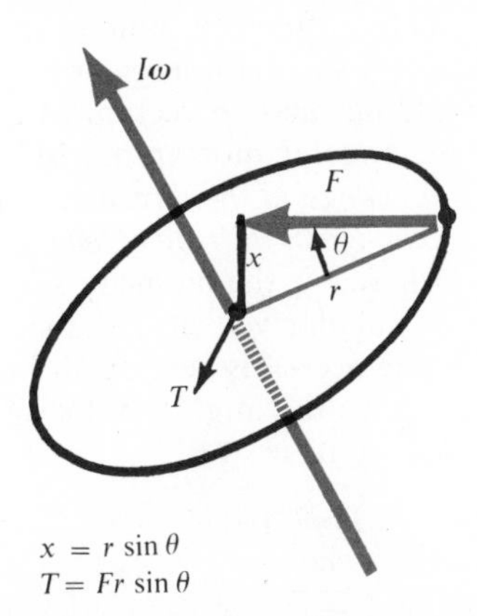

FIG. 5–24 *A force acting on a body gives rise to a torque, T, acting on the orbit of the body if the force does not pass through the center of the orbit.*

5.26 Conservation of Angular Momentum

We see, then, that if no torque acts on a system, its angular momentum remains constant. This is the content of the *principle of conservation of angular momentum.* We may express it somewhat differently by stating that *the torque acting on a system is equal to the time rate of change of the angular momentum of the system.*

We may illustrate the application of this principle, first by considering again a particle moving with constant speed along a straight line, as shown in Figure 5.25. Since in a unit interval of time this

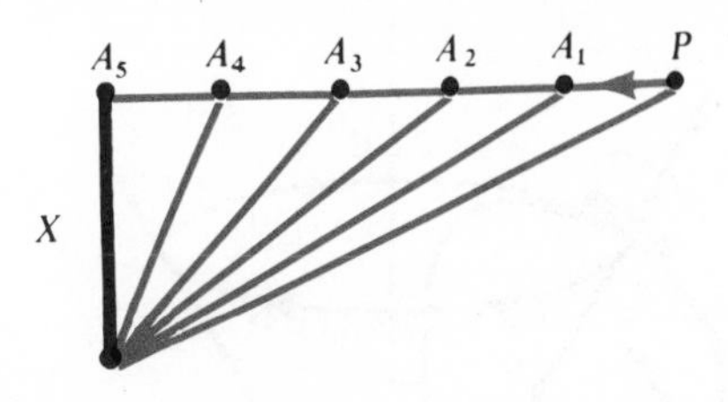

FIG. 5–25 *The area swept out in unit time by the line from the observer to a particle moving with constant speed is a constant. This area is proportional to the angular momentum of the moving particle relative to the observer.*

particle always moves through a distance v, the positions of the particle moving with constant speed along its path of motion at the end of successive unit intervals of time are $A_1, A_2, A_3, \ldots$ as shown in Figure 5.25, where the separation between two successive points is always v. If we now draw lines from the point P to $A_1, A_2, A_3, \ldots$, we obtain a set of triangles, whose areas are all equal. This follows since the area of each triangle is equal to $\frac{1}{2}vx$, where v is the base of the triangle, and x is the altitude (note that the altitude for each triangle is the same).

As we have already seen, the quantity $\frac{1}{2}vx$ is half the angular momentum of the particle divided by its mass. It follows, therefore, that the principle of conservation of angular momentum is equivalent to stating, in this case, that *the areas of triangles that are swept out by the line P to the particle in each unit of time are equal.* We shall see that this is but a special case of a more general law which holds for the motion of a particle in any kind of orbit around a central point. Note that the angular momentum of a body moving with uniform speed in a straight line (constant velocity) is not the same for all observers. The angular momentum of this body is zero relative to any observer standing on its line of motion, and it increases relative to observers farther and farther away from this line.

As another illustration of the principle of conservation of angular momentum we may consider a top that is spinning about an axis which is tilted with respect to the vertical, as shown in Figure 5.26. We observe that there is a torque acting on the top since the weight of the top, which acts at its center of mass, tends to rotate the top about the point of contact between the top and the

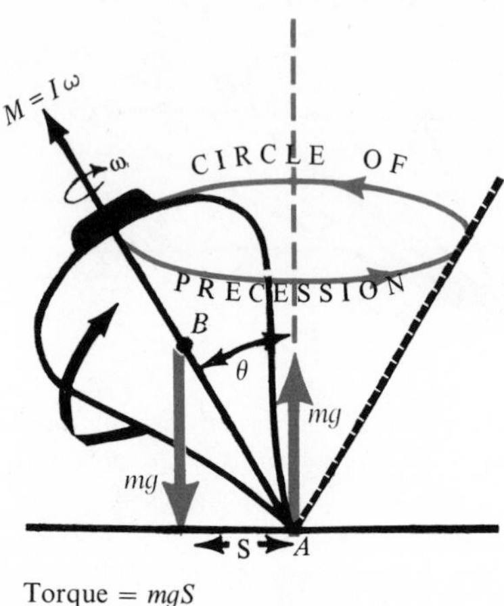

Torque = mgS

FIG. 5–26 *The torque acting on a top spinning on the surface of the earth. This torque causes the top to precess as shown.*

ground. At this contact point the ground exerts an upward force against the top which is exactly equal to the weight of the top. These two parallel but opposite forces constitute the torque on the top. If the top were falling freely there would be no torque exerted on it. The torque—acting on the top—imparts a rate of change to its angular momentum. Since the torque tends to induce a rotation about an axis which is at right angles to the axis of spin, the torque can have no effect upon the rate of spin, but only on the direction of spin.

5.27 Precession Induced by a Torque in a Spinning Top

Thus, there will be induced a continual change in the direction of the axis of the spinning top, so that *it precesses about the vertical direction in the same sense as it spins, as shown by the arrows in the diagram. The axis of the spinning top maintains the same angle with the vertical at all times.* The rate at which the top precesses (that is, the angular velocity of precession) depends directly on the torque that is applied, and inversely on the horizontal component of the angular momentum of the top itself. Thus, the faster the top is spinning, the more slowly it precesses. This is why we see a top precessing more and more rapidly as its spinning slows down. The bigger the torque is, the faster it precesses. From this it follows that the larger the angle is that the top makes with the vertical, the faster the precession is, since the torque is then also greater. If the axis of the top is vertical then the torque vanishes, and there is no precession.

Note that in the case of the top, the torque action is equal to the product of the weight, W, of the top and the distance AB of the center of mass of the top from the point of contact multiplied by the sine of the angle that the axis makes with the vertical:

$$\text{torque} = (AB)W \sin \theta. \tag{5.4}$$

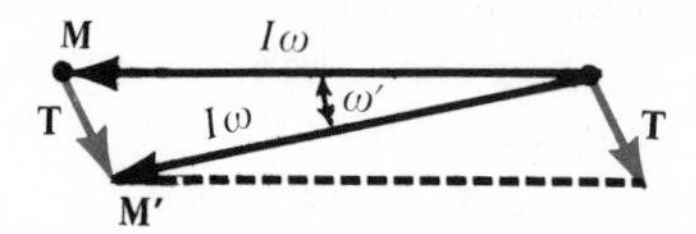

FIG. 5–27 *A vector diagram showing the relationship between the torque acting on a spinning body and its rate of precession, ω′.*

Figure 5.27 is a vector diagram showing how the torque **T** acts to change the angular momentum of the top if the angular momentum lies in a horizontal plane. The vector **M** represents the angular momentum of the top at any moment and the vector **T**, which is at right angles and pointing out of the plane of paper, is the change in the angular momentum of the top produced by the torque in a unit of time. Thus, after this unit time interval which we take as very small, the angular momentum is obtained by adding the two vectors **M** and **T** to give the resultant angular momentum **M′**. What we have here is *a change in the direction of the axis of the top, but no change in the rate of spin of the top.* Since $M = I\omega$, where I is the moment of inertia of the top and ω its angular velocity, and since from the figure we see (since ω' is very small here) that $T = M\omega'$, where ω' is the angular velocity of precession of the top, we have

$$T = (I\omega)\omega',$$

or

$$\boxed{\omega' = \frac{T}{I\omega}}\,. \tag{5.5}$$

This formula holds only if the angle θ is 90°. Otherwise, $\sin \theta$ must appear in the denominator multiplying $I\omega$.

5.28 Precession of the Earth's Axis

We can now apply the same analysis to the spinning earth. The torque in this case is produced

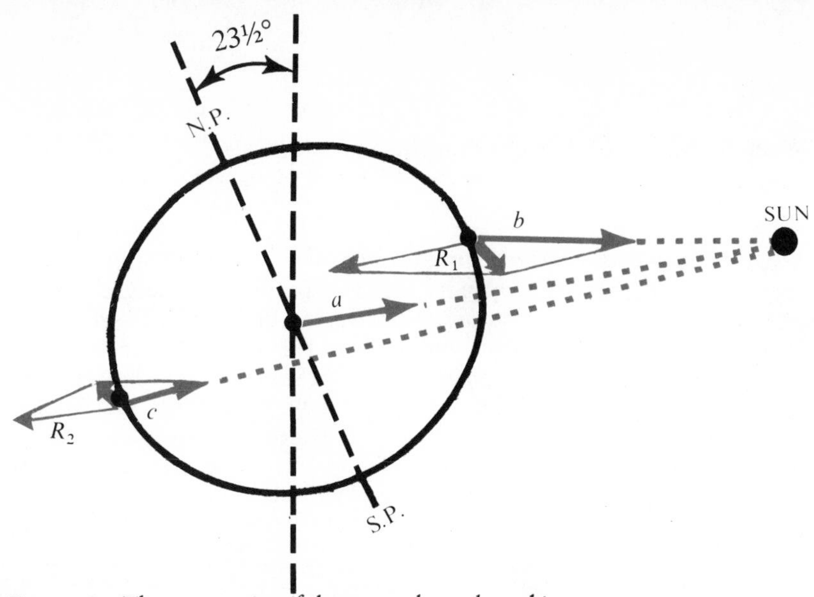

FIG. 5–28 *The torque action of the sun on the earth resulting from the bulge at the earth's equator. The two resultant forces shown at diametrically opposite points on the earth's equator give rise to the torque.*

by the sun and the moon, but it is simpler if we explain only the action of the sun.

If the earth were perfectly spherical and the matter inside it were distributed in uniform spherical layers, the sun could exert no torque, since its gravitational pull would act entirely along the line from the center of the earth to the center of the sun. But we have already noted that the earth bulges at the equator. Because of this bulge a torque arises. In Figure 5.28 we have shown a highly exaggerated diagram of the forces that are involved (note that the distances of the sun and the size of the earth are completely out of scale). Since the torque tends to produce a change in direction of the axis of the earth, all we need do to analyze this is to consider how a typical unit mass on the surface of the earth tends to move relative to the center of the earth as a result of the sun's gravitational pull. In other words, we shall consider a unit mass at the bulge facing the sun, a unit mass at the center of the earth, and a unit mass at the bulge away from the sun, and analyze the forces acting on them.

The earth as a whole is accelerated towards the sun as a result of the gravitational action at the center of the earth. This acceleration is shown by the vector **a** in the diagram. The acceleration of the bulge facing the sun is shown by the vector **b**, which is larger than the vector **a** because this part of the earth is closer to the sun than the center of the earth is. The acceleration of the bulge farthest from the sun is shown by the arrow **c** which is smaller than **a** since this bulge is farther from the sun than the center of the earth is. Note that **b** and **c** are not parallel to each other, nor to **a**, since they point to the sun.

If we now want to see how the torque arises from these three accelerations, we must find what the accelerations of the bulges are with respect to the center of the earth. We do this by subtracting from the accelerations **b** and **c** the acceleration **a**, as shown in Figure 5.28. When we carry out the vectorial subtraction, we obtain the resultant vectors $\mathbf{R}_1$ and $\mathbf{R}_2$ and we see that they tend to rotate the earth's axis perpendicular to the plane of the ecliptic. In other words, the action of the torque of the sun on the earth is to induce a rate of change in the angular momentum of the earth at right angles to the axis of spin of the earth and pointing into the plane of the paper. This is just opposite to what we had in the case of the top, so that the earth's precession in this case is opposite in direction to that of its spin. Note that the torque action of the sun varies from season to season, vanishing on March 21 and September 23 when the sun is on the equator. It is at a maximum on June 22 and December 22 when the sun is at its greatest positive and negative declination respectively. The moon's action also varies, but here the variation is more rapid, occurring monthly, and it is this that gives rise to the nutations.

5.29 The Seasons: Solar Radiation Striking the Earth

Because of the motion of the earth around the sun and the tilt of its axis, we have the seasonal variations of the climate. Our climate in its overall aspects is determined primarily by the radiation that comes to us from the sun. Since the northern and southern hemispheres do not share this radiation equally during different times of the year, it follows that there are climatic changes from month to month at any given region on the earth's surface.

To understand the broad seasonal variations of the climate, we shall begin by considering the solar radiation falling on the earth in a given time interval. Figure 5.29(a) shows rays of light from the sun when it is to the right and to the left of the earth. In this diagram the plane of the ecliptic is perpendicular to the plane of the paper, and the axis of the ecliptic is shown by the directed line *PE*. The earth is to be pictured in its orbit as moving out of the paper towards the reader, and spinning in such a way that its right-hand side is moving into the paper, and its left-hand side out of the paper.

We first consider rays coming in from the right and neglect the effect of the earth's atmosphere. The northernmost ray, which just grazes the earth, is given as r_N and enters the eye of a person standing on the Arctic Circle when the sun is just rising for him. He sees the sun for just an instant on the particular day represented by this drawing. For all observers north of the Arctic Circle the sun is completely invisible for that day. The southernmost ray is indicated by r_S and strikes an observer on the Antarctic Circle at sunset. Such an observer sees the sun above the horizon during the complete 24-hour period on this day, and for any observer between the Antarctic Circle and the South Pole the sun is continuously visible for more than 24 hours. To an observer standing at point *A* on the Tropic of Capricorn at the moment shown, the sun is directly overhead. A direct ray, r_0, is shown striking this observer.

All the radiation that strikes the earth's surface at a given time is contained in a cylinder bounded by the two rays r_N and r_S (it has the same cross section as the earth). Since light travels at 186 000 mi/sec (299 000 km/sec), we may thus say that the amount of solar radiation striking the earth per second is the amount contained in a cylinder 186 000 miles long and having a diameter of 8 000 miles. This radiation determines the climate.

5.30 Proportion of Total Solar Radiation Received by Northern and Southern Hemispheres

To return to our diagram, we see that the ray which strikes the equator separates the radiation in the cylinder into two parts: everything above this ray strikes the northern hemisphere in one second, and everything below this ray strikes the southern hemisphere in one second. It is clear then from this that it must be winter in the north and summer in the south, since the northern hemisphere receives a considerably smaller fraction of the total radiation than the southern hemisphere does.

The axis of the ecliptic which is denoted by the line *PE* divides the figure into a daylight and a nighttime zone, and we see at once that the daylight zone *NOE*, as shown in Figure 5.29(a), for people living in the northern hemisphere is much smaller than the nighttime zone *NOE'*; in the southern hemisphere just the reverse is the case. From the figure it is also clear that, if this situation prevailed all the time, people living above the Arctic Circle would be in perpetual darkness and those living below the Antarctic Circle would be in perpetual daylight.

However, the situation shown with the rays coming from the right applies for only one instant of time—that is, at some moment during December 22, when the declination of the sun is exactly $-23°27'8''.26$ (obliquity of ecliptic). Since the earth is in motion, the way in which the radiation is distributed over the two hemispheres changes continuously. As the earth moves away from this position, we must picture the cylinder of radiation as rotating around it, parallel to the plane of the ecliptic until finally it strikes the earth from the left. This occurs at some moment during June 22, when the declination of the sun is exactly $+23°27'8''.26$. On that day the direct rays strike an observer on the Tropic of Cancer, so that the northern hemisphere receives the major share of radiation and the conditions are the reverse of what they are on December 22. On June 22 summer begins in the northern hemisphere and winter begins in the southern hemisphere.

5.31 Radiation Received by Northern and Southern Hemispheres at Equinoxes

The situations on March 21 and September 23 are shown in Figure 5.29(b). We see that the direct rays strike the equator on those two days so that the northern and southern hemispheres share the

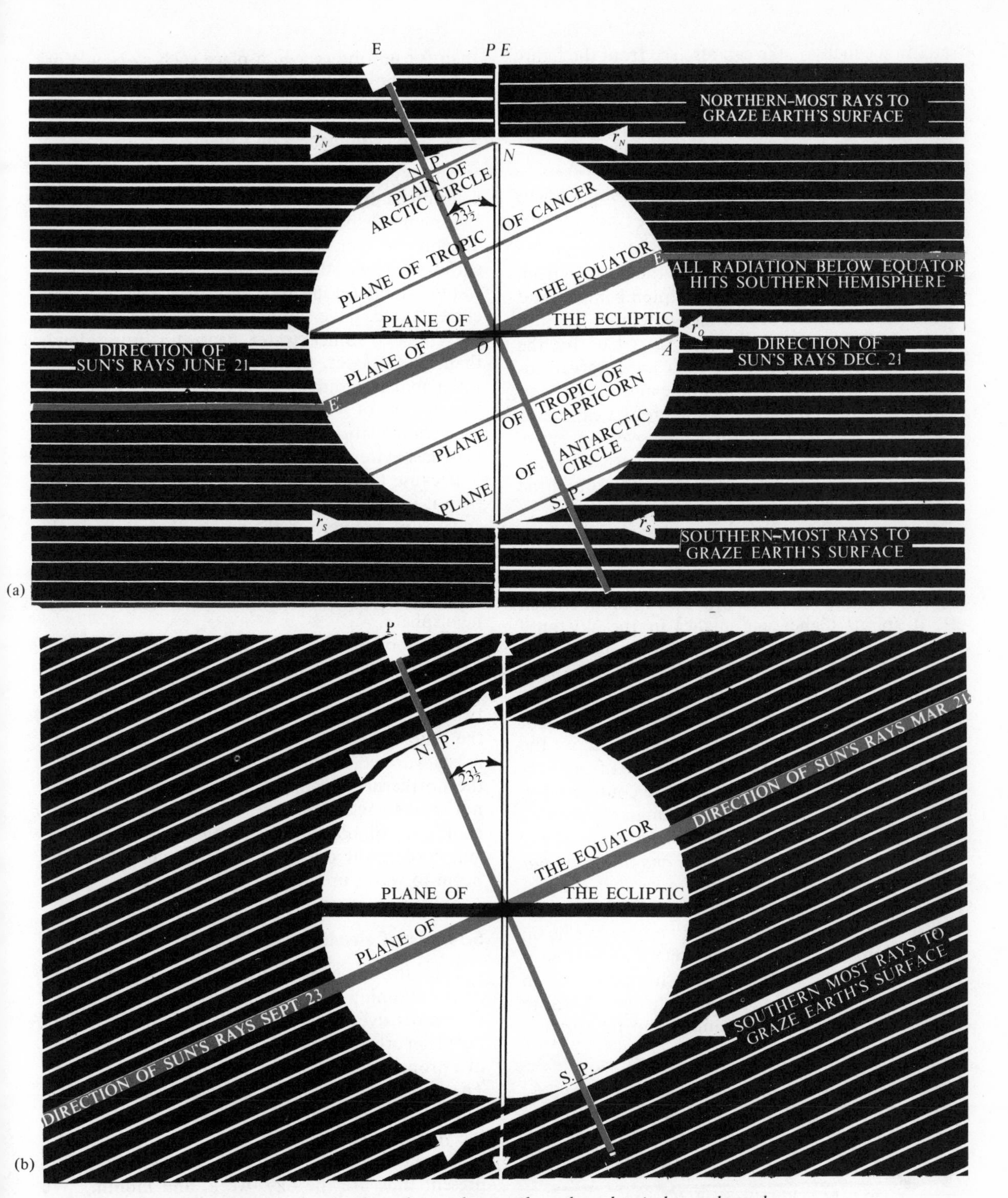

FIG. 5–29 *Schematic diagram showing how the northern and southern hemispheres share the light and heat from the sun at different times of the year. This variation in the distribution of the radiation from the sun over the two hemispheres from month to month accounts for the seasons.* (*a*) *Conditions on June 21 and December 21.* (*b*) *Conditions on March 21 and September 23. Here the sun is to be pictured as lying on a line at right angles to the page and all the rays of light are to be considered as being perpendicular to the page.*

radiation equally. We can also see from the figure that *on these two dates*, which mark the beginning of the spring and fall seasons in the respective hemispheres, *the days and nights are equal in length for all points on the surface of the earth.*

5.32 The Appearance of the Sky from Various Points on the Earth's Surface

The diagram shows us that for an observer at the North Pole the sun remains above the horizon from about March 21 until about September 23 (and below the horizon from about September 23 until about March 21); just the reverse is true for the South Pole. The zenith for such an observer coincides with the north celestial pole and his horizon with the celestial equator. He therefore can never see more than half the total sky over an entire 24-hour period. Thus for him the fixed stars neither rise nor set, but move in circles around the sky parallel to the horizon.

He does not experience the usual day of 24 hours, but instead finds that the entire year is divided into six months of daylight and six months of nighttime. After the sun first becomes visible on about March 21, it spirals higher and higher in the sky until about June 22 and then it spirals down again. *As an observer moves away from the North Pole, more and more of the sky becomes visible, although as long as he is above the equator, there are always stars that do not rise or set.* As long as he remains above the Arctic Circle, there is at least one period of 24 hours during which the sun remains visible continuously, and one 24-hour period when it remains invisible.

For an observer on the equator the entire sky is visible during each 24-hour period and all the stellar objects move across the sky in circles that are perpendicular to the observer's horizon. Both the north and south celestial poles for such an observer lie on his horizon. Because the sun sets and rises in a direction perpendicular to the horizon for an observer on the equator, the twilight period before sunrise and after sunset is much shorter for him than for an observer in a place such as New York City, as shown in Figure 5.30.

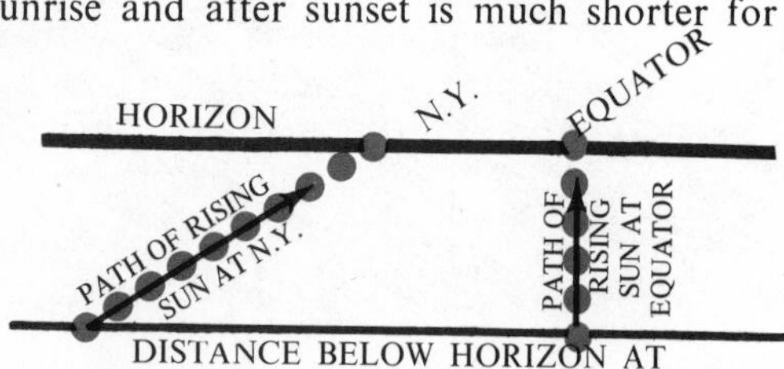

FIG. 5–30 *The duration of twilight at the equator and at New York City. The difference is due to the direction of the sun's rising or setting relative to the horizon as seen from these two latitudes.*

5.33 Other Influences on Climate

Although the over-all features of climate are determined by the radiation that comes from the sun, many elements besides direct radiation greatly affect the temperature at a given point on the earth's surface. Thus, the prevailing winds, bodies of water, mountainous terrain, valleys, warm ocean currents, etc. have important tempering effects on the climate. Since water heats up more slowly than land masses, the presence of large bodies of water makes summers cooler and winters warmer. The actions of winds and water currents introduce such complications that accurate weather prediction becomes extremely difficult.

Because the earth is closer to the sun in January than it is in July, the winters in the northern hemisphere are warmer, on the whole, than those in the southern hemisphere. For the same reason, it follows that the summers in the northern hemisphere are cooler than those in the southern hemisphere.

Even though the sun is highest in the sky on June 22 and lowest in the sky on December 22 the hottest and coldest days do not occur, in general, on these two dates. The reason is that there is a lag between the time the earth receives radiation and the time it has warmed up to its fullest. Thus, on June 22 the northern hemisphere is still warming up, and not until sometime in early August does it begin to cool off. Similarly, the northern hemisphere is still cooling off on December 22 and it does not begin to warm up until early February.

5.34 The Calendar

The astronomer is concerned not only with the daily reckoning of time, but also with the keeping of records over long periods. This is true also of people in other walks of life, so that the introduction of a proper calendar is a matter of great importance. The essential element in the development of the calendar is the introduction of a scheme for dividing the year into appropriate units.

The natural units that have been used up to the present time are the day, week, and month. We start by considering the number of days in a year. Not until comparatively recently was this number accurately known. Before the year 2781 B.C. it was generally thought by the more advanced civilizations that the year was 360 days long. However, at about that time the Egyptians introduced a 365-day year on the basis of the heliacal (simultaneous with sunrise) rising of the stars. They were able to correlate the periodic flooding of the Nile, which comes annually, with the great regularity in the appearance of the star Sirius.

5.35 Sidereal and Tropical Year

Still later, as a result of more accurate observations of the heliacal rising of Sirius, and the flooding of the Nile, the Egyptians added another quarter-day to the year. This $365\frac{1}{4}$-day year was accepted as substantially correct until quite recently. Today we differentiate between two year lengths: the sidereal year and the tropical year. *The **sidereal year**, which from a purely mechanical point of view may be spoken of as the true year, contains 365.256 36 solar days, and is the interval of time required by the sun to make one complete circle in the sky, returning to its initial position relative to the fixed stars. The **tropical year** (the basis for our calendar), which is 365.242 2 solar days long, is the interval of time between two successive coincidences of the sun with the vernal equinox.* The Egyptians were among the first, if not the first, people to recognize that the tropical and sidereal year are of different lengths. They observed that there is a slight displacement in time (about 20 minutes) between the heliacal rising of Sirius and the flooding of the Nile.

5.36 The Lunar Calendar

It was quite natural in the early chronology of the ancients to divide the year into months and to use lunar calendars, since the lunar period is one of the most striking in nature. However, since the lunar period and the tropical year are not commensurate, as we shall see, calendars based entirely on the lunar period (these were the earliest calendars in use) are quite unsatisfactory for civil purposes: for the most part they are out of step with the seasons. Because of this, lunar calendars were replaced by luni-solar calendars such as the one used in ancient Rome and by the Hebrews. In this type of calendar an additional month was intercalated whenever the discrepancy between the calendar and the seasons became large enough to warrant such a correction.

5.37 The Julian Calendar

The first modern calendar, the Julian Calendar, containing twelve months of varying length, was introduced in 45 B.C. by Julius Caesar, who was greatly concerned with the practical aspects of administering a military state. Up to that time the maintaining of the calendar was in the hands of the priesthood and subject to arbitrary manipulation for the benefit of those in control. Julius Caesar was too hardheaded a politician to allow theological considerations to stand in the way of a properly run state, and so, on the advice of the Greek astronomer Sosigenes, he abandoned the idea of a luni-solar calendar and introduced the Julian Calendar, which except for minor modifications is still in use.

In this calendar, which is divided into twelve months and in which the year begins in January, three out of four years consist of 365 days and the fourth year consists of 366 days (leap year). In the original Julian Calendar, the odd months were assigned 31 days, and the even months 30 days, with the exception of February which was given 29. Later, at the insistence of Augustus, the eighth month was given 31 days and named August in his honor, so that February was left with 28 days except for leap year, when it has 29 days.

5.38 The Gregorian Calendar

Since in the Julian Calendar the civil year is 11 minutes and 14 seconds longer than the seasonal year, this calendar is out of phase with (falls behind) the seasons by one day in 128 years. By 1583, the Julian Calendar was about 10 days behind the seasons. This discrepancy was so serious that Pope Gregory XIII revised the calendar and introduced what is now called the Gregorian Calendar. October 5 of the year 1582 became October 15 by an edict of the Pope. This calendar was accepted at once by all of the Catholic countries, and later by the Protestant European countries. The essential difference between the Gregorian and the Julian calendars is in the number of leap years that are introduced every four centuries. According to the Julian Calendar every year divisible by four is a leap year; in the Gregorian Calendar this rule applies except for century years. For a century year to be a leap year in the Gregorian Calendar, it must be divisible by 400, *so that only one out of every four century years is a leap year.*

EXERCISES

1. (a) By what amount does the earth's rotation reduce the weight of a unit mass at a latitude of 45°? (b) How large is the horizontal component of the acceleration imparted to a body by the earth's rotation at a latitude of 30°?

2. The text states that the earth lags behind a perfect clock by about 8 hours in 4 000 years. How is this figure obtained?

3. If a freely falling body near the surface of the earth has a mass of 10 000 gm, what force does it exert on the center of the earth? Does this impart an acceleration to the earth?

4. A baseball having a mass of 120 gm is thrown by a pitcher and leaves his hand at a speed of 220 cm/sec. If the pitcher exerted a uniform force on the ball through a distance of one meter before it left his hand, with what force did the ball push against the pitcher?

5. Calculate the momentum of the earth in its orbit. Do the same for Mars and Uranus. Use the data in the planetary table (p. 227).

6. If a man moving at a speed of 10 ft/sec jumps into a stationary cart whose mass is 10 times that of the man, what is the final speed of the cart (with the man at rest on the cart) if air resistance and frictional forces are neglected?

7. Calculate the orbital angular momenta of (a) Venus, (b) Earth, (c) Mars, (d) Jupiter, and (e) Saturn.

8. Two bodies moving in opposite directions collide head-on and stick together. If the mass of one body is 150 gm and its speed is 60 ft/sec while the mass of the second body is 450 gm and its speed is 10 ft/sec, at what speed and in what direction will the two bodies move off after the collision?

9. If two ball bearings, each of mass 50 gm, collide head-on, what force does each exert on the other during the collision if the collision lasts for 0.01 sec and if each ball has a speed of 90 cm per sec before and after the collision?

10. What is the angular momentum of a hoop of radius 30 cm if it is spinning with an angular speed of 360° per sec and if its mass distribution is a uniform 10 gm per cm along its circumference?

11. If a torque of 1 000 dynes/cm were applied to the hoop in exercise 10 at right angles to its rotation, at what rate would it precess (i.e. what would its angular velocity of precession be)?

12. Assume the moment of inertia I of the earth to be $(2/5)MR^2$, where M is its mass and R is its radius. Use the known values of its angular velocity of spin, ω, and its known angular velocity of precession, ω', to compute the torque I exerted on it.

13. (a) On March 21, how does the amount of sunlight striking a sq cm of the earth's surface per unit time at latitude 60° compare with the amount per unit time at latitude 45° and at the equator? Neglect the effect of the atmosphere. (b) Consider the same exercise as in (a) for June 21 and for December 21.

14. (a) Show that if an object is dropped vertically downward from the top of a building of height h, the object will strike the ground at a distance

$$\frac{2^{3/2}\pi h^{3/2}\cos\lambda}{g^{1/2}P}$$

to the east of the building, where λ is the latitude of the building, g is the acceleration of gravity, and P is the period of rotation of the earth. (b) Apply this formula to a building 100 ft high.

15. (a) By definition the center of mass of a group of particles is the point whose perpendicular distance from any plane is the mass-weighted average of the perpendicular distances of all the particles from that plane. Use the definition to show that if X_c, Y_c, Z_c are the cartesian coordinates of the center of mass of n particles in a rectangular coordinate system, then

$$X_c = \frac{m_1x_1 + m_2x_2 + \cdots + m_nx_n}{m_1 + m_2 + \cdots + m_n}$$

$$Y_c = \frac{m_1y_1 + m_2y_2 + \cdots + m_ny_n}{m_1 + m_2 + \cdots + m_n}$$

$$Z_c = \frac{m_1z_1 + m_2z_2 + \cdots + m_nz_n}{m_1 + m_2 + \cdots + m_n},$$

where m_i is the mass of the ith particle and x_i, y_i, z_i are its coordinates. (b) Use this formula to find the distance from the sun of the center of mass of Mercury, Venus, Earth, and Mars when they are all lined up on the same side of the sun. Assume that all these bodies lie in the plane of the ecliptic.

16. Calculate the maximum angles of aberration of light for observers on Mercury, Venus, and Mars from the known speeds of these planets in their orbits.

6

THE MOON*

Lying no more than a rocket's throw from us is the earth's only natural satellite, the moon. For thousands of years a subject of philosophers' and scientists' speculation and of poets' praise, the moon is now the object of intensive scientific probing and study in preparation for exploration by man. As such, it represents man's first extraterrestrial frontier in space. But, unlike the geographical frontiers that faced the early explorers on earth, this one is so well known that the first party to land on it will be well prepared for whatever difficulties may present themselves.

6.1 The Motion of the Moon

The most striking aspect of the moon to the person who studies it seriously for the first time is *the continual change in its appearance*, *or its* **phases.** We can understand these changes in terms of the position of the moon with respect to the earth and the sun. Since the moon itself emits no visible radiation, we can see it only because of the sunlight it reflects to us; thus its appearance at any moment depends upon how much of the solar radiation striking it is reflected back to earth. Because the moon is continually moving about the earth, *this amount of radiation varies from day to day.*

To analyze this we first consider how the moon appears to move as seen from the earth. Even the most casual observer knows that the moon does not always rise at the same place or time from day to day. *Careful observations show that the moon moves eastwardly in the sky from day to day the way the sun does*, *except much more rapidly*, *moving by about* 13° (*instead of* 1°) *to the east in each* 24-*hour period.* Since one degree corresponds to a 4-minute time lag, this means that the moon lags in its rising by about 51 minutes each day. This is only an average effect and varies from season to season. During the period of the autumnal equinox at full moon, the lag may be as little as 20 minutes for a few successive nights, so that during that time one gets the impression of being subjected to a great deal of moonlight. During the vernal equinox just the opposite is true, and the lag is greater than 52 minutes.

6.2 Elongation and Phases

Because of this eastward motion, the angle that the line from the earth to the moon makes with the line from the earth to the sun (the **elongation** of the moon) varies continuously from 0° to 360°. When the elongation is 0°, the moon and the sun are in the same part of the sky (the moon is said to be in **conjunction**) and we speak of the **new** phase of the moon, as shown in Figure 6.1. When the elongation is less than 90°, some of the radiation striking the moon is reflected back to the earth, and **crescent** phase occurs. When the elongation is 90° (the moon is said to be in **quadrature**) about half of the surface of the moon that receives sunlight is visible from the earth. This phase is spoken of as the **first quarter.** For an elongation greater than 90° we have the **gibbous** phase. When the elongation is 180°, the moon and the sun are in opposite parts of the sky (the moon is said to be in **opposition)** and light from the entire lunar surface that is irradiated by the sun is reflected back to earth. This phase of the moon is referred to as the **full** phase, and the moon is said to be in the **second quarter.** As the elongation continues to increase, the previous phases all recur, but in reverse order, as shown in the diagram.

It should be noted that the crescent phase after the new moon is visible immediately after sunset, whereas the crescent phase prior to the new moon is visible just before sunrise. In the early crescent phase, the entire face of the moon can be faintly seen. This phenomenon is due to **earthshine**—the light reflected from the earth onto the moon and back again, as shown in Figure 6.2. Note that the full moon generally rises in the early evening just about when the sun is setting. Figure 6.3 shows some actual photographs of the moon in various phases.

* NOTE An asterisk to a chapter or section title indicates that new information on the topic will be found in Part Five.

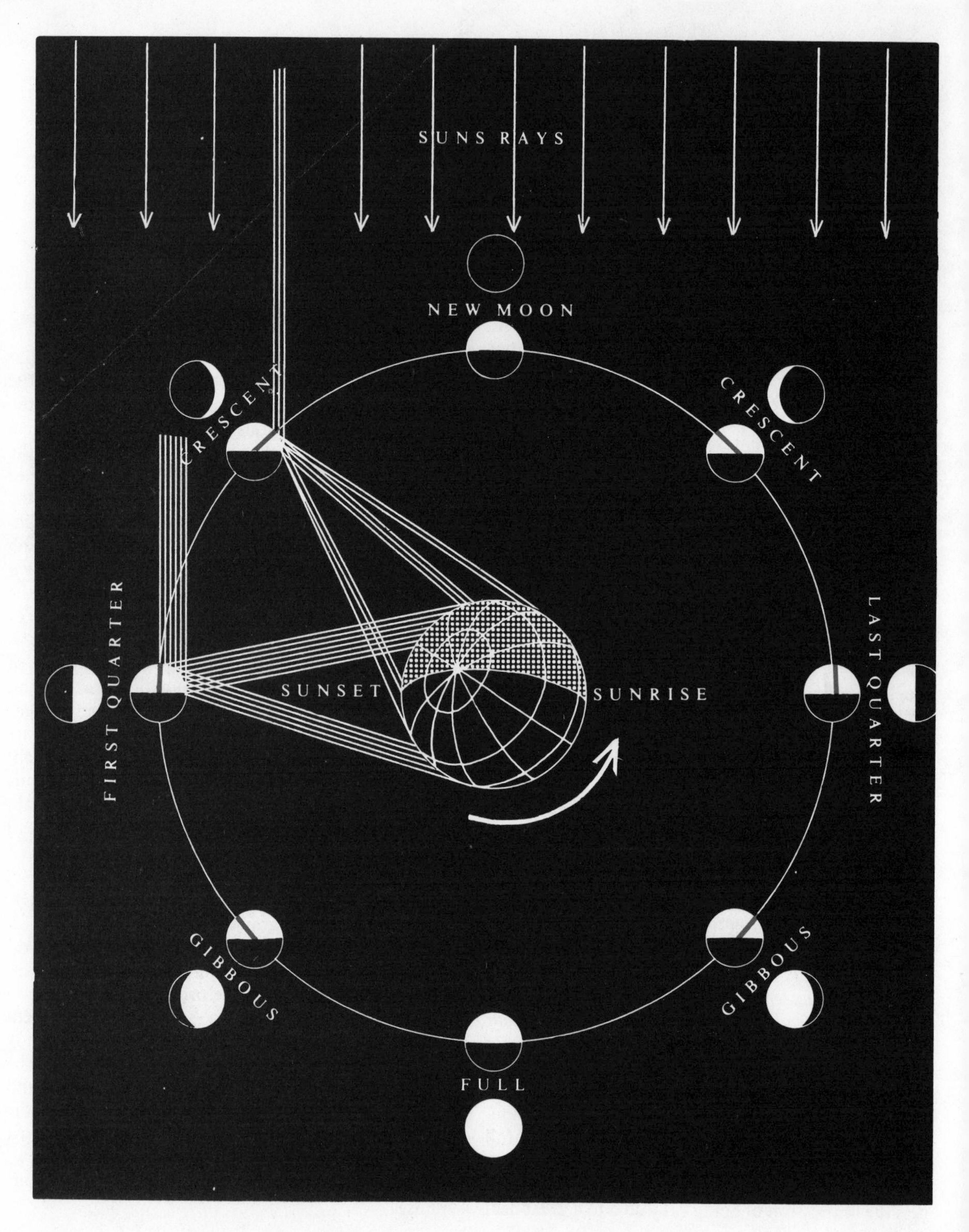

Fig. 6–1 *The phases of the moon are determined by its elongation as seen from the earth. The observer on the earth can see only the part of the moon that is illuminated by the sun's rays and that lies between the red line and the earth. This is the portion of the moon's surface which reflects the sun's rays to the earth.*

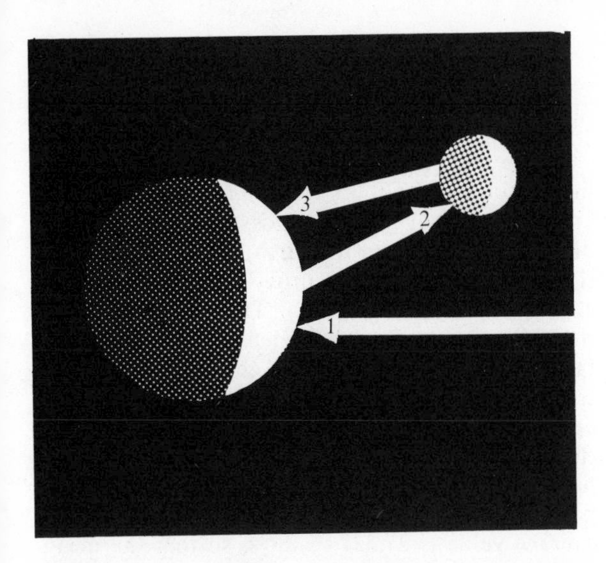

Fig. 6–2 *The earthshine is due to the sunlight (arrow 1), reflected by the earth onto the moon (arrow 2), and then back to the earth again (arrow 3).*

Fig. 6–3 (*a*) *Photographs of the moon, crescent phase. The orientation is inverted, as seen through a telescope. (I) Age three days, (II) age five days.*

FIG. 6–3 (*b*) *The moon, first quarter, age eight days.*

FIG. 6–3(*c*) *The moon, gibbous phase, age 17 days. (Fig. 6.3a–c, 100-inch photographs from the Mount Wilson and Palomar Observatories)*

6.3 Lunar Sidereal Period

As we have noted, the moon moves eastwardly in the sky by about 13° per day (disregarding its apparent diurnal motion across the sky, which is due to the rotation of the earth). This means that the moon completes a journey around the sky in slightly more than 27 days. This period is called the **sidereal** month, and is defined as the time it takes the moon to make one complete revolution from a given point on the sky (some fixed star) back again to this same point as seen from the center of the earth. This period varies by an amount that never exceeds 7 hours, and that arises because of the perturbing action of other planets and the sun. Its average value is 27.321 66 days, so that the mean daily motion of the moon is 13°11′.

6.4 Lunar Synodic Period

If we compare the lunar sidereal period with the interval of time between two identical successive phases, the ***synodic*** *month, we find that they are not equal.* The synodic period (which is what one generally refers to as **the month**) is approximately 2 days longer than the sidereal month. Its average value is 29.530 59 days, but there is a variation during the year of about 13 hours because the moon's orbit is elliptical.

The reason for the difference in length between the sidereal and synodic months is that the earth and moon together are moving about the sun, so that when the moon has gone all the way around the earth, the sun is no longer in the same relative position that it occupied initially. This is shown in Figure 6.4, where in the first position we note that the moon, earth, and sun are lined up (not necessarily in a straight line) so that the moon is new. If the earth were not moving about the sun, the moon would again be lined up in the same relative position with respect to the earth and the sun (the same elongation—that is, 0°) after $27\frac{1}{3}$ days (the sidereal period). However, during these $27\frac{1}{3}$ days the earth has moved around the sun (to position 2 in the diagram) through an angle of about $27\frac{1}{3}$ degrees, so that the moon, earth, and sun are no longer lined up. In fact, for the moon to be lined up with the earth and sun again, it must revolve another $27\frac{1}{3}$ degrees in its orbit around the earth, and this takes about two more days, as shown in the diagram.

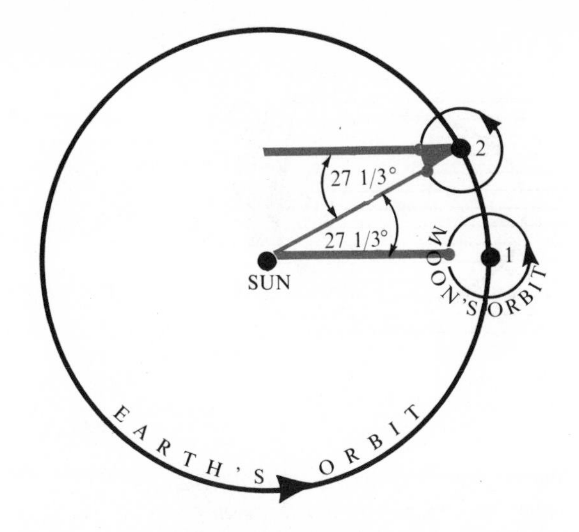

FIG. 6–4 *The synodic period of the moon is about 2 days longer than the sidereal period because during one sidereal period the earth–moon system has advanced by about 27° in its orbit around the sun.*

6.5 Relation between Synodic and Sidereal Periods

There is a simple relationship between the synodic and sidereal periods, which is important for the study not only of the moon's motion but also of planetary motion. In order to derive this formula, we picture the moon and the sun as racing around the sky at different speeds like two automobiles on a racetrack. By the synodic period we then mean the time it takes the moon (the moon corresponds to the faster-moving automobile) to overtake the sun, starting from the moment the two are together in the sky. Since each day the sun moves about 1° and the moon about 13° to the east we see that the moon gets ahead of the sun by about 12° per day. To catch up with the sun, it must get ahead by 360°; hence the synodic period is close to 30 days.

We may obtain the precise algebraic formula for the relationship between the synodic and sidereal periods by noting that *the reciprocal of its period represents the average rate at which an object is moving around an orbit.* Thus, if the sidereal period of the moon in days is P, $1/P$ is a measure of the average rate of the moon's apparent motion, since it represents the average angle through which the moon moves in its eastward motion in 1 day. In the same way, if E is the sidereal period of the sun, then $1/E$ is the average rate at which the sun moves to the east, that is, the average apparent angular motion of the sun to the east each day. Finally, if S is the synodic period of the moon, then $1/S$ is the average rate at which the moon is surging ahead of the sun. In other words, $1/S$ must equal the difference between the other two rates, so that we have

$$\boxed{\frac{1}{S} = \frac{1}{P} - \frac{1}{E}}. \tag{6.1}$$

From this it follows that

$$P = \frac{ES}{S+E}$$

and

$$S = \frac{EP}{E-P}.$$

Lunar phases recur on the same day of the month in a cycle that is very nearly equal to 19 years. The reason is that 19 is the smallest number of years that is a multiple of the synodic month. There are very nearly 235 synodic months in 19 Julian Years. This is known as the Metonic Cycle, named after Meton, who discovered it in about 433 B.C.

6.6 The Moon's Apparent Orbit on the Celestial Sphere: The Line of Nodes

Since the moon moves to the east by about 13° per day, it traces out an apparent path (great circle) among the stars on the celestial sphere. Although at first sight this apparent orbit seems to coincide with the ecliptic, careful observations of the right ascension and declination of the moon from day to day show quite clearly that it does not. *In fact the plane of the moon's orbit makes an angle of about 5°9′ with the plane of the ecliptic.*

The line along which these two planes intersect is called the **line of nodes**, and it pierces the celestial sphere in two diametrically opposite points called **lunar nodes**, as shown in Figure 6.5. These are the points on the celestial sphere in which the ecliptic intersects the great circle that is the projection of the moon's orbit on the celestial sphere. The node A, with which the moon coincides when it is crossing the ecliptic from **south** to **north**, is called the **ascending node**; the other is called the **descending node**.

If the earth and the moon were the only bodies present in space (and if both were spherical), the moon would always come back to the same point

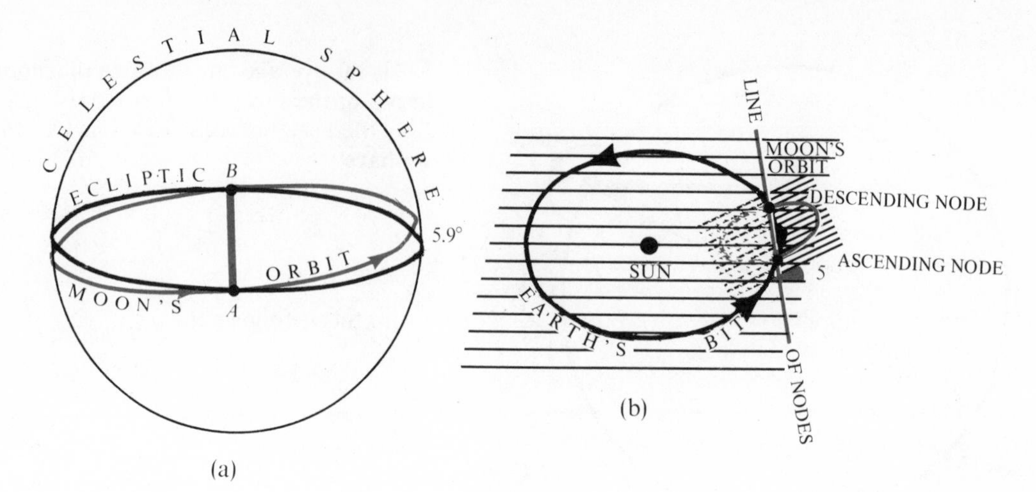

FIG. 6–5 *(a) The nodes are the intersections, A and B, of the ecliptic and the projection of the moon's orbit on the celestial sphere. (b) A spatial diagram: the nodes are shown as two points where the moon's orbit pierces the plane of the ecliptic. The line connecting these two points is the line of nodes. The line of nodes precesses westwardly, making one complete rotation every 19 years.*

in its orbit after one complete revolution. However, this is not the case since a third body, the sun, is present. The effect of the sun's presence is to introduce perturbations in the moon's motion which prevent it from returning to its initial position. *As a result of these perturbations the nodes do not remain fixed on the celestial sphere, but regress* ***westwardly*** *on the ecliptic, similar to—but much more rapidly than—the precession of the equinoxes.* The nodes make one complete revolution in about 19 years, so that the average period between two successive coincidences of the moon with the same node is 27.212 2 days. This is called the **nodical month**. During this 19-year period the inclination of the moon's apparent orbit with the ecliptic varies from 4°59′ to 5°18′, with an average value of 5°9′.

Because of the regression of the nodes, the angle that the plane of the moon's orbit makes with the plane of the celestial equator varies continuously from a maximum value of 28°35′, *when the ascending node coincides with the vernal equinox, to a minimum value of* 18°35′ *when the descending node coincides with the vernal equinox.* The explanation of this is left to the student as an exercise.

6.7 The Moon's Orbit around the Earth

As seen from the earth the moon revolves around it in a closed orbit just the way the earth revolves around the sun and for exactly the same reason. In fact it was by studying the moon's orbit that Newton first checked his law of gravity. Although the actual orbit of the moon around the earth (neglecting the perturbing effects of the sun and other planets) is an ellipse with the earth at the focus, it is sufficient to take the orbit as circular in order to use it to test Newton's law.

If we take a circular orbit, the radial acceleration of the moon is v^2/r, where v is its speed and r is the radius of its orbit. Since v is equal to $(2\pi r)/p$, where p is the sidereal period of the moon, we see that the acceleration of the moon towards the earth (on substituting this expression for v into the above formula for the acceleration) is equal to $(4\pi^2 r)/p^2$. On the other hand, it follows from Newton's law of gravity that the acceleration of the moon towards the earth resulting from the gravitational pull of the earth should be 3 600 times smaller than the acceleration of gravity on the surface of the earth, since the distance of the moon from the earth is 60 times the earth's radius (the acceleration of gravity according to Newton's law falls off inversely as the square of the distance). This value for the acceleration of the moon towards the earth must equal the value obtained from the formula $(4\pi^2 r)/p^2$ if Newton's law of gravity is correct, since it is the gravitational pull of the earth that keeps the moon in its orbit. This is, indeed, the case, since they both give the result that the moon falls a distance of 0.053 4 inches towards the earth each second [see Equation (3.10)] which is one-half the acceleration of the moon towards the earth.

6.8 Rotation of the Moon

Not only does the moon revolve around the earth, but it also rotates about an axis just the way the earth does, but much more slowly. At first sight this seems very puzzling to the layman, who knows that we can see only one side of the moon and concludes from this that the moon cannot be spinning. *However, it is precisely because we can see only one face of the moon that we know the moon must be spinning.* We can best explain this by referring to Figure 6.6, which shows the moon's orbit around the earth projected on a plane which is perpendicular to the moon's axis of rotation (the plane of the page). If the moon were not rotating about an axis (taken perpendicular to the plane of the paper) and a stake AB were driven through the moon, each new position of the stake would be parallel to each previous position. In position 1, an observer on the earth would then see point A of the stake, and 14 days later he would see the point B on the other side of the moon (Figure 6.6).

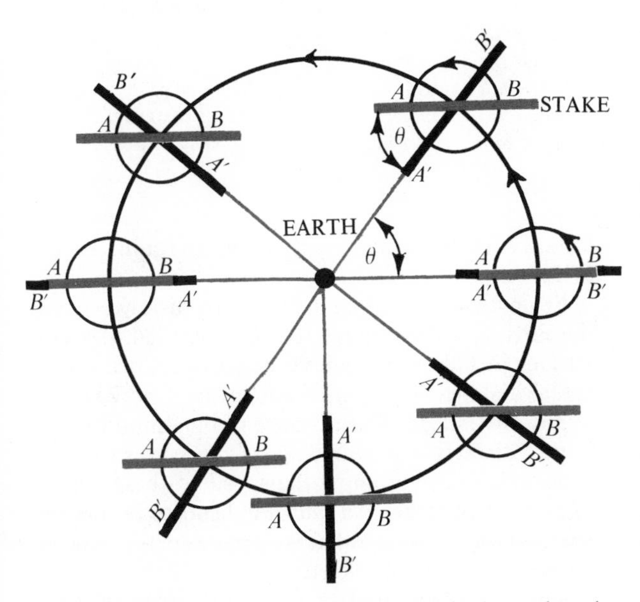

FIG. 6–6 *If the moon were* not *rotating, a stake driven through it (as shown by line AB) would always have the same direction in space. Hence, an observer on the earth would see side A at one time, and 14 days later side B. However, since we can see only side A from the earth, we conclude that the moon rotates through the angle θ in the same time it revolves through this angle around the earth, as shown by line A′B′.*

Since we never see its other side, the moon must turn continuously in such a way that the line AB always points to the earth. This means that when the moon has revolved through an angle (θ) in going from position 1 to position 2, it must at the same time rotate in the same sense about its own axis through the same angle θ. *In other words, it follows from the fact that the same face is visible at all times, that the period of rotation of the moon about its own axis must be equal to the sidereal period of its revolution around the earth.* However, even though the moon rotates once every time it goes around the earth, the rotation and the revolution are not exactly in step at all times because the moon's true orbit is an ellipse—so that it moves at different speeds at different times, whereas its rotation is always constant.

6.9 The Geocentric Parallax of a Celestial Body

Before we can determine the physical characteristics of any celestial body, we must first obtain its distance, and this is usually done by measuring what is called the parallax of the body. We can understand the nature of the parallax by noting that an object viewed against a distant background appears to shift when the position from which it is viewed is altered. Thus, a pencil held up at arm's length, viewed first with one eye and then with the other, appears to jump back and forth. Since the angle through which the pencil appears to jump depends on the distance of the pencil from the eyes and the separation between the two eyes, it is clear that we should be able to determine the distance of the pencil from the eye (as shown in Figure 6.7) by measuring the angle through which it appears to jump, if we know the separation between the two eyes.

This is essentially the procedure used to determine the distance of celestial bodies. What is involved here is the measurement of an angle (parallax) subtended at the celestial body by a known distance. In the case of bodies such as the sun, moon, and planets in our own solar system, the known distance is generally taken as the radius of the earth, so that one measures the apparent displacement of the object whose distance is to be determined when viewed (against the celestial sphere) first from a given point on the earth's surface, and then from a point equivalent to the center of the earth. *The angle thus found is called the geocentric parallax of the object*—even though one does not really view the object from the center of the earth, but rather

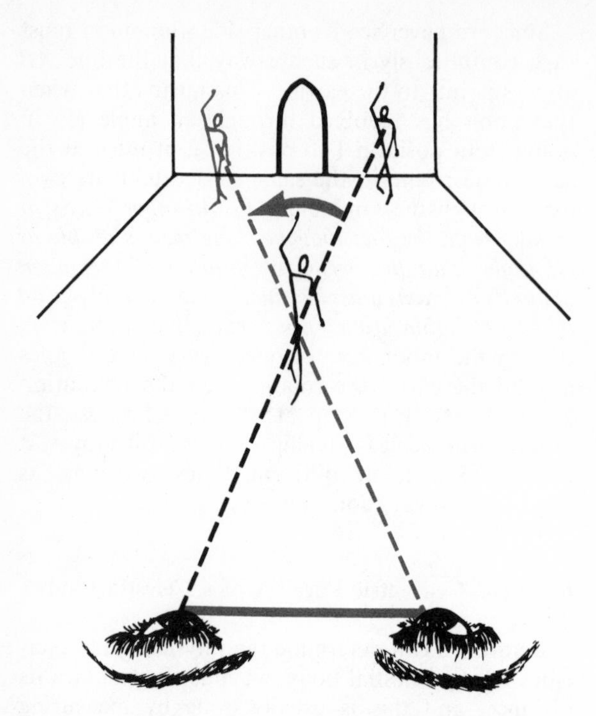

FIG. 6–7 *Viewed against a distant background, the position of a person appears to shift when looked at first with one eye and then the other. If we know the distance between the two eyes, we can determine the distance of the object by measuring the angle through which the object appears to shift.*

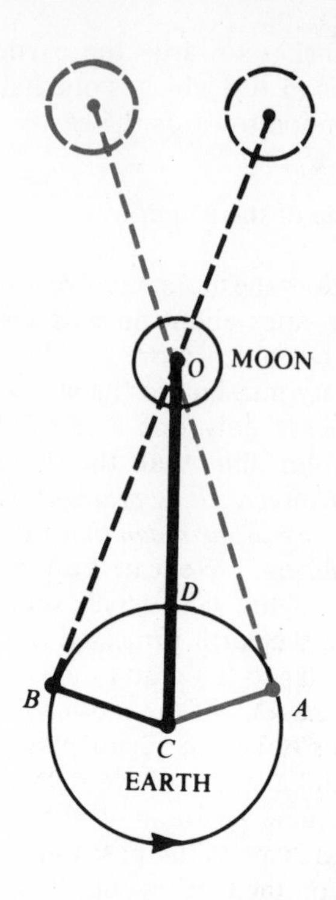

FIG. 6–8 *The distance of the moon as determined by the method of geocentric parallax. Two observers at A and B sight the same point, O, on the moon at the moment it is rising for observer A and setting for observer B. Since the arc AB can be measured, the angle BCA can be determined. Hence the two acute angles in the right triangle AOC can be found, and since CA, the radius of the earth, is known, the distance OC of the moon's center from the earth's center can be calculated trigonometrically.*

from two widely separated points on the earth's surface.

6.10 The Geocentric Parallax of the Moon: Its Distance

We illustrate this procedure in Figure 6.8 by considering two observers on the earth's surface, one at *A*, who sees a point on the moon (*O*) just rising, and the other at *B*, who at the same moment sees it just setting. Thus, the two rays *OA* and *OB* are tangent to the earth at *A* and *B*, respectively. Radii drawn from the center of the earth to these two points form two congruent triangles *OAC* and *OBC*.

Since the length of arc *ADB* along the surface of the earth can be measured, the angle *ACB* can be found because it is the same fraction of 360° as the length *ADB* is of the circumference of the earth. We now take half of this angle and divide its cosine into the earth's radius, and obtain the distance *CO* to the moon. Actually, the moon's equatorial horizontal parallax at mean distance (the moon's distance varies from day to day) is found to be 57′2″.7.

From this we find that the mean distance from the earth to the moon is 238 000 miles (384.4×10^3 km), or just about 60 times the earth's radius. The actual distance at any moment varies from 252 710 to 221 463 miles. The distance of the moon can be found directly by bouncing radar waves off its surface. Since we know the speed of these waves (the speed of light), we find the distance by taking one-half the product of this speed and the time it takes to receive the reflected waves.

Since we know the size and shape of the moon's orbit and the time it takes the moon to make one complete revolution, we can determine the mean speed of the moon in its orbit. This turns out to be 2 287 miles (3 659 km) per hour. The angular velocity across the celestial sphere is 33 minutes of arc per hour. In other words, the moon appears

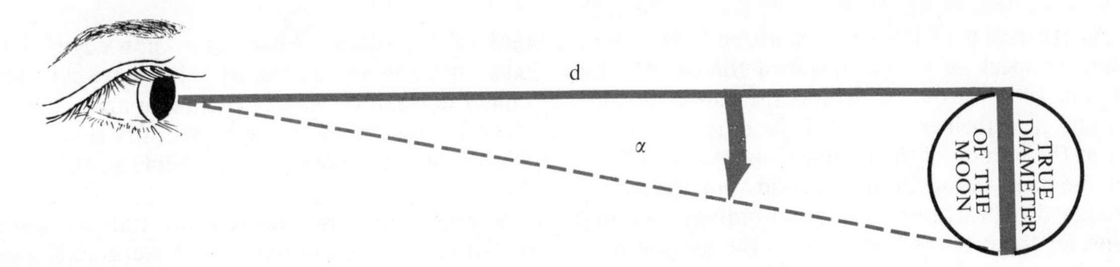

FIG. 6–9 *The true diameter of the moon can be calculated from a knowledge of its distance, d, from the earth, and the measured value of the angle α which the moon's diameter subtends at the eye (that is, its angular or apparent diameter).*

to move across the sky in one hour by an amount just about equal to its apparent diameter.

6.11 Size and Shape of the Moon

From our knowledge of the distance of the moon and its apparent diameter at any given moment we can determine the true linear dimensions of the moon. If α is the angle, in radians, subtended by the diameter of the moon at the eye of the observer (Figure 6.9), and d is the distance of the moon, then

$$\alpha\, d = \text{the true diameter of the moon.} \qquad (6.2)$$

Since the mean angle subtended by the moon is about $\frac{1}{2}°$ or 31′5″, we can compute its true diameter, which we find to be 2 160 miles (3 476 km) or 0.273 times the earth's diameter. On squaring 0.273 we find that the moon's surface area is 1/14 of the earth's surface area, and on cubing it, we find that the moon's volume is 1/49 of the earth's volume. Although the moon is very close to being a perfect sphere, it is slightly flattened at its poles because of its rotation. However, since the rotation is very slow, the flattening is not perceptible.

6.12 The Motion of the Center of Mass of the Earth-Moon System

To determine the mass of the moon, we must observe its gravitational influence on the orbit of some other body. The simplest approach is to see how the moon influences the earth's motion around the sun. If the moon were not present the center of the earth would move in an elliptical orbit around the sun, but since this is not the case, the earth and moon must be treated as a single system revolving around the sun. As a result, *it is not the earth's center that describes the elliptical orbit around the sun, but rather the common center of mass of the earth and the moon.* This is illustrated in Figure 6.10 where the earth and moon are shown in a series of different positions, and the center of mass is shown in its various positions along the elliptical orbit.

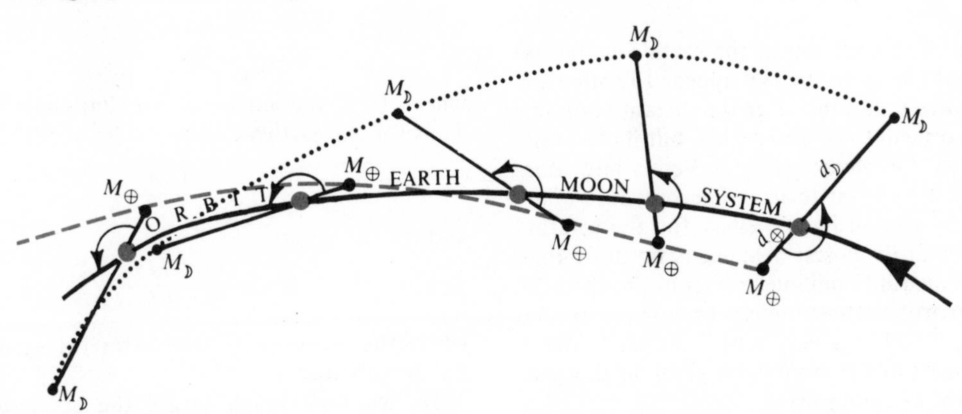

FIG. 6–10 *The motion of the earth-moon system around the sun, showing the path of the center of mass of the system (large red dots), the orbit of the moon's center (dotted line), and the orbit of the earth's center around the sun (dashed line). The earth's center and the moon's center revolve in almost circular orbits around the center of mass. The radii $d_\oplus$ and $d_☾$ are in the inverse ratio of the masses of the two bodies. The scales here are greatly distorted for clarity. The moon's orbit departs from the orbit of the center of mass by only 1/400 of the earth–sun distance.*

As the center of mass moves around the sun in what we speak of as the "orbit of the earth," the centers of both the earth and the moon revolve around it in their own orbits. In other words, as far as the motion of the earth-moon system around the sun goes, we may, to a very good approximation, disregard the earth and the moon individually and consider the orbit as being due to the motion of a single particle having a mass equal to the sum of the masses of the earth and the moon and concentrated at the center of mass.

Since the position of the center of mass on the line connecting the centers of the earth and the moon is determined by the ratio of the mass of the earth to the mass of the moon, we can find this ratio when we find the position of this point. The distance of this point from the center of the earth is as many times smaller than its distance from the center of the moon, as the mass of the earth is larger than the mass of the moon. In other words, the earth's center revolves around the center of mass in an orbit that is identical in shape with the moon's orbit around it but that is as many times smaller than the moon's orbit as the earth's mass is greater than the moon's mass, or

$$\frac{M_\oplus}{M_☾} = \frac{d_☾}{d_\oplus}, \tag{6.3}$$

where $M_\oplus$ and $M_☾$ are the masses of the earth and moon respectively, and $d_☾$ and $d_\oplus$ are the distances of the center of mass from the centers of the moon and the earth respectively.

6.13 The Mass of the Moon*

Because of this motion of the earth's center, all the celestial bodies in the sky appear to shift back and forth every month. For the distant stars this effect is too small to be observable, but it can easily be measured for planets such as Venus and Mars when they are closest to us. The best results are obtained by measuring the effect for the asteroid Eros, which at its closest distance is no more than 14 million miles (22 million km) from the earth.

Once an artificial satellite is set revolving around the moon, it will be a very simple matter to determine the mass of the moon with great accuracy by studying the satellite's orbit.

From observations of the monthly displacements of the nearby planets we find that the center of mass of the earth-moon system is 81 times closer to the earth's center than it is to the moon's. This means that the mass of the moon is 81 times smaller than that of the earth. From our knowledge of the mass of the moon, and of its volume, we can determine the mean density of the moon and we find it to be 3.3 gm/cm^3, which is very nearly equal to the density of the ordinary rock we find in the earth's crust.

In Figure 6.10, used above to illustrate the motion of the earth-moon system, we have greatly exaggerated the distances of the centers of the earth and the moon from the center of mass. Actually both of these bodies lie much closer to the orbit around the sun than it is possible to show in the diagram. Because of this, and because the speed of the center of mass around the sun is about 30 times as large as that of the moon around the earth, *the moon's orbit around the sun varies only minutely from the orbit of the center of mass around the sun.* As seen from the sun, the moon's orbit is always **concave** towards it although this orbit cuts the earth's orbit around the sun about 24 times a year. As we shall see when we discuss Kepler's third law of planetary motion, it is also possible to measure the mass of the moon by very careful observation of its sidereal period and its mean distance from the earth.

6.14 The Acceleration of Gravity on the Moon

Because the mass and the radius of the moon are different from those of the earth, the acceleration of a freely falling body on the surface of the moon is different from that on the surface of the earth. We have seen from Newton's law of gravitation that on the surface of a sphere of mass M and radius R, the acceleration of gravity is

$$\frac{GM}{R^2},$$

where G is the universal gravitational constant. Thus for the earth we have

$$g_\oplus = \frac{GM_\oplus}{R_\oplus^2},$$

and for the moon

$$g_☾ = \frac{GM_☾}{R_☾^2},$$

where the subscript $\oplus$ stands for the earth and ☾ for the moon.

To see how much larger the acceleration of gravity is on the earth than it is on the moon, we divide the first of these equations by the second and obtain

$$\frac{g_\oplus}{g_☾} = \frac{M_\oplus/M_☾}{(R_\oplus/R_☾)^2},$$

the constant G cancelling out. Since the numerator in this equation is equal to 81 and the denominator is equal to about 14, we find that the acceleration of gravity on the earth is about 6 times that on the moon. This means that a person who weighs 180 pounds on the earth will weigh only 30 pounds on the moon. A body allowed to fall freely from rest on the moon will have to fall 6 seconds before it acquires the same speed that a body has after falling 1 second from rest on the earth.

Once we land on the moon, walking will be something of a problem since we will have a great deal of trouble keeping our feet on the ground, and anything like a baseball game will be impossible since a 400-foot home run on the earth will become a half-mile home run on the moon.

6.15 Physical Characteristics of the Moon: The Absence of Atmosphere

The most striking thing about the moon, when we observe it through a telescope, is its rugged terrain, scarred as it is by a large number of craters and vast mountain ranges. Comparatively speaking, the peaks and crater indentations are much larger than anything of the sort we have on earth, so that we must conclude that there are no erosive forces at work tending to smooth down the moon's surface the way weathering does on the earth. From this we can be fairly certain that the moon has no atmosphere—a conclusion further supported by the fact that the shadows cast by peaks are pitch black with no indication at all of the diffusion of light that occurs in the presence of an atmosphere.

Additional evidence that the moon has no atmosphere can be obtained from the observation of occultation of stars by the moon. When a star passes behind the moon, it does so very suddenly and there is a sharp disappearance. This would not be the case if a lunar atmosphere were present, since then refraction effects would give rise to a more or less gradual disappearance of the star. Similarly, we find that there is no evidence of refraction during an eclipse of the sun. A lunar atmosphere would distort the rim of the sun during an eclipse. Still other observations can be made, such as spectroscopic analysis and the sharply defined appearance of the cusps in the crescent phase. All of these taken together indicate that if there is any permanent atmosphere on the moon, its amount cannot be greater than 1/100 000 000 of that on the earth.

Recently some evidence has been adduced that radioactive processes in the solid rocks of the lunar surface give rise to helium and argon atoms. These constitute components of a temporary atmosphere which at any moment has a density of 1/1 000 000 000 000 of the density of air at sea level here on earth.

6.16 The Motions of Molecules in an Atmosphere

To understand why the moon has no atmosphere we must consider two things: the nature of an atmosphere, and the concept of the speed of escape. An atmosphere is a gas composed of molecules that are free to move in any direction. In general, however, these molecules cannot move unhindered through any great distance because collisions occur among them. Thus, if we follow a single molecule, we discover that it has a very zig-zag path because its motion in any one direction is constantly being disrupted by collisions.

For any given pressure and temperature in the atmosphere a molecule, on the average, moves a definite distance before colliding with another molecule. This distance, as shown in Figure 6.11,

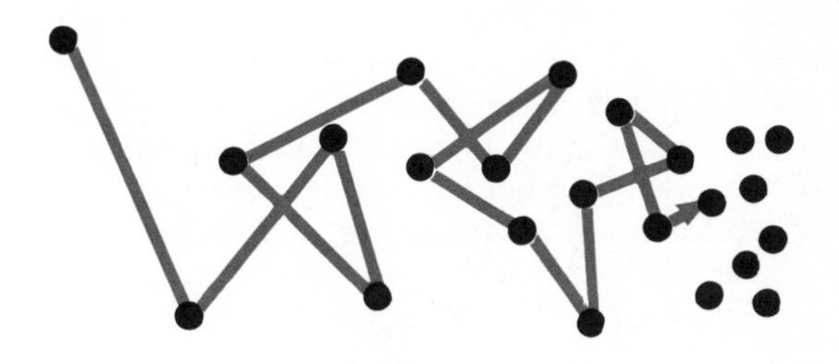

FIG. 6–11 *The mean free path of a molecule in a gas. This path gets smaller as the density of the gas increases.*

is called *the mean free path* of the molecule (as the density of the atmosphere increases the mean free path decreases). As a result of these constant collisions suffered by a molecule its speed is always changing, but it lies close to a certain average value which depends on the temperature of the atmosphere and the mass of the molecule. Thus, the average speed of hydrogen molecules in a gas at 0° centigrade is of the order of 2 to 3 kilometers per second, and that of oxygen molecules is one-fourth as large.

One might suppose that because of the random collisions a molecule suffers, it can never get very far away from where it happens to be at any particular moment. But this is not the case, and one can show that if l is the mean free path, then after n collisions have occurred, a molecule will, on the average, be at a distance equal to $\sqrt{n}\, l$ from its starting point. This means, in principle, that any particular molecule has a definite chance of reaching the outer regions of our atmosphere and therefore escaping entirely from the earth if it is moving fast enough. We see then that whether a body like the earth can retain an atmosphere or not depends upon how fast the molecules in such an atmosphere are moving when they reach the outskirts of the atmosphere. We see that *if molecules in an atmosphere surrounding a body move faster than a certain speed, called the speed of escape, the body in question cannot retain this atmosphere.*

6.17 The Speed of Escape

We observe that if a body is thrown vertically upward, it immediately begins to lose speed because of the downward acceleration of gravity. However, if we throw it up fast enough, we can make it go as high as we wish. One may then ask whether it is possible to throw up a body so fast that it continues moving forever. *If the acceleration of gravity remained constant at all distances from the earth's surface, then no matter how fast we threw the body, it would never escape.* But we have seen from Newton's law of gravity that the acceleration of a freely falling body diminishes with the square of its distance from the center of the earth. We can see, then, that if the body were projected outwardly with just the right speed, it would move so fast that the acceleration of gravity would fall off too quickly ever to reduce the speed of the body to zero. *A body leaving the earth with this projected speed is said to be moving with the speed of escape.* How are we to calculate this speed? For this we must introduce another derived quantity from physics.

6.18 The Concepts of Work and Energy

There exist in nature two physical entities that are intimately related to each other, but that exhibit very different characteristics. One of these is **matter**, which possesses very definite tactile properties and which can be localized in space; its primary characteristic is that of inertial mass. The other entity is **energy**, which manifests itself as a change in the behavior of a body when **work** (which we shall now define) is done on the body.

We have all had the experience of lifting, pushing, and throwing objects, and we know that we exert forces whenever we do these things. However, more than a force is involved in these actions, because we must exert the force through a certain distance in order to perform any of the operations which we have mentioned, so that both a force and a length are involved. Thus, we may push an object 10 feet across the floor, or lift it 3 feet above the ground, or push it a distance of 2 feet in the process of throwing it. *We shall now say that we do work on a body whenever we apply a force through a given distance, and that in virtue of this work the body acquires energy. When we do work on a body, either its state of motion or its distance above the ground, or both, change, provided no frictional forces are present.* (When we push an object across the floor at constant speed, we do work against the friction of the floor, and neither the state of motion of the body nor its height above the ground changes while we are pushing it.)

We may say then, if no friction is present, that the mechanical energy acquired or lost by a body is a measure of the change in its state of motion or in its position relative to the ground, as a result of the work done on it, or by it (note that a body can do work as well as have work done on it).

6.19 The Definition of Kinetic and Potential Energy

If only the motion of the body changes, we say that the work done changes only the kinetic energy of the body, so that the kinetic energy is energy of motion. If only the height of the body above the ground changes, we say that the work done changes only the potential energy of the body, so that potential energy is energy of position relative to the ground. At any moment during the motion of a body, it has both kinetic energy and potential energy. We shall now write down the formulae for these two types of mechanical energy.

To find an algebraic expression for kinetic energy we must first write down a precise formula for work itself. This is defined as the force applied times the distance through which the force acts; but only that component of the force which is parallel to the displacement is effective in doing work. To make this clear we consider Figure 6.12 where a force F is applied to a body, which as a result is moved through a distance d. If the force makes an angle θ with the displacement of the body, the work is not simply the force times the distance but, rather,

$$W = Fd \cos\theta. \quad (6.4)$$

The cosine of the angle θ takes care of choosing only that component of the force which is parallel to the displacement.

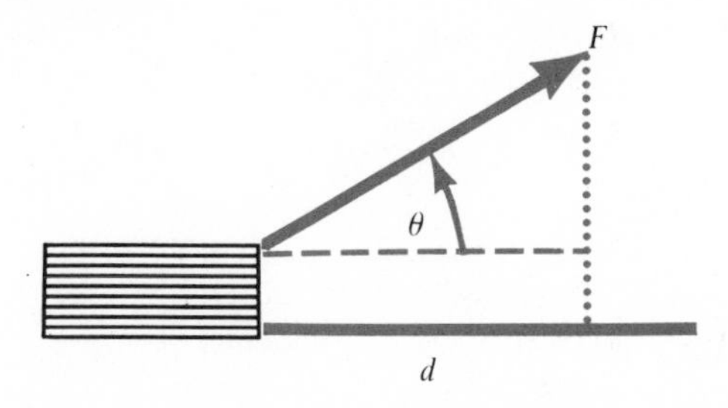

FIG. 6–12 *The work done by a force acting through a distance d at an angle θ with respect to this distance is equal to $Fd \cos\theta$.*

6.20 Formulae for Kinetic and Potential Energy

With this definition of work in mind, we can now derive an expression for the kinetic energy of a body set in motion from rest, and for the potential energy of a body lifted a certain height above the ground. If we throw a body horizontally (we neglect the frictional resistance of the air) and thus set it in motion, we do work on it because we are pushing it against its inertial resistance through a given distance, as shown in Figure 6.13 (the distance

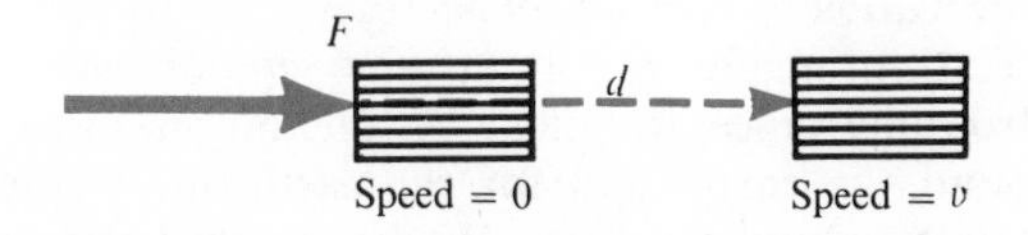

FIG. 6–13 *A constant force, F, does work on a body by pushing it from a state of rest, speed = 0, through a distance, d, to a final speed, v.*

through which we push it is the distance it moves while it is still in contact with our hand). At the moment it leaves our hand, we are no longer pushing it, so that the force we exert acts only through the distance d. But the force we exert is itself equal to the inertial resistance of the body, which is just the mass of the body times the acceleration being imparted to it. Hence the work (note that the angle θ is zero) we do when we throw a body horizontally is

$$W = dma. \quad (6.5)$$

If the body is uniformly accelerated through the distance d, it acquires a speed v at the end of this distance which is equal to the square root of twice the acceleration times the distance (see Equation 3.11). Thus

$$v^2 = 2ad, \quad (6.6)$$

where v is the speed acquired by the body. If we solve for the product ad in this equation, and substitute this in our expression for the work done, we obtain the kinetic energy (K.E.) of the body in terms of its mass and its final speed. We thus have

$$K.E. = \tfrac{1}{2}mv^2. \quad (6.7)$$

To find the potential energy (P.E.) of a body after it has been lifted through a height h, we note that *the force which is exerted is just the weight of the body, mg.* Hence by definition the work (the force and displacement are in the same direction so that the angle θ is again zero) we do in lifting the body is mgh, so that we have (as shown in Figure 6.14)

$$P.E. = mgh. \quad (6.8)$$

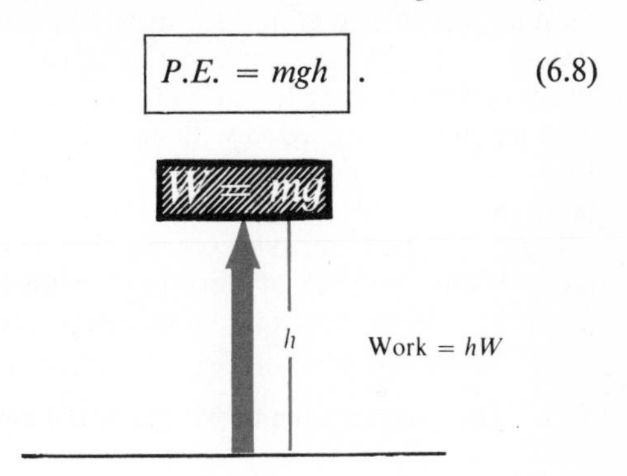

FIG. 6–14 *A body acquires potential energy when work is done on it to lift it a height h above the ground.*

6.21 Units of Energy

Before we go further, we must introduce the unit of energy, and we do this by starting from the concept of work. To obtain a unit of work we apply a unit of force through a unit distance, so that the product of two units is itself a unit. Thus, *if we apply a force of one dyne through a distance of one centimeter, we do a unit of work.* We call this unit the **erg** (this is also the unit of energy). The erg is a very tiny unit and is the work done when 1/980 gm is lifted a distance of 1·cm.

We may also relate the erg to another unit of energy, the **gram-calorie**. By definition the gram-calorie is the amount of heat required to raise the

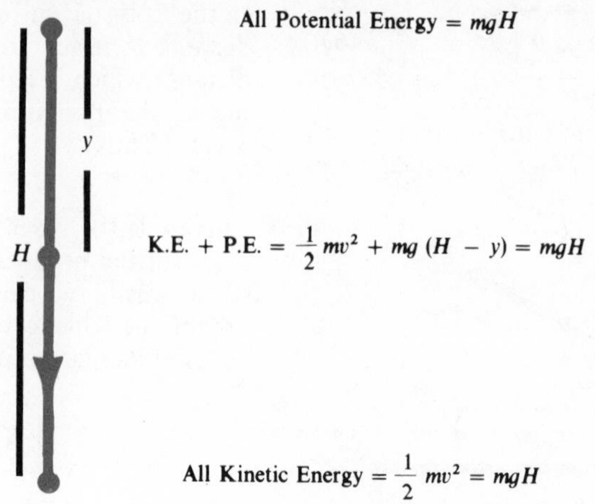

All Potential Energy = mgH

y

H

K.E. + P.E. = $\frac{1}{2}mv^2 + mg(H - y) = mgH$

All Kinetic Energy = $\frac{1}{2}mv^2 = mgH$

FIG. 6–15 *The principle of conservation of energy as illustrated by a body falling freely in a vacuum. At each stage the sum of its kinetic energy and potential energy is equal to its total energy mgH.*

temperature of one gram of water one degree centigrade. One calorie equals 4.185×10^7 ergs. In order to relate the erg to our daily experience, which is more closely associated with electrical appliances than with mechanical energy, we may note that one **joule** of electrical energy, which is the amount of energy delivered in one second by a one-watt current (one ampere of current flowing across a potential difference of one volt), equals 10^7 ergs.

6.22 The Conservation of Mechanical Energy

If we supply a body with mechanical energy by doing work on it, we naturally ask what finally happens to this energy. Two situations arise, depending upon whether the body moves in a vacuum or is subject to frictional resistance. We defer the discussion of the second case to a later chapter.

If a body is moving above the ground, it possesses both kinetic and potential energy at each point in its trajectory. We may therefore say that its total mechanical energy is given by the expression

$$\text{total mechanical energy} = \text{K.E.} + \text{P.E.}$$

$$\boxed{E = \tfrac{1}{2}mv^2 + mgh}, \qquad (6.9)$$

where v is its speed and h is its height above the ground at any moment. It is easy to show (left as an exercise for the student), if we neglect the friction of air (or if we suppose the body is moving in a vacuum), that this sum remains constant throughout the entire history of the motion of the body even though each term of the sum changes constantly. Thus, as the body rises, its P.E. increases and at the same time its speed, v, decreases, and hence the K.E. decreases. The decrease in the K.E. is just compensated for by the increase in the P.E. Just the reverse happens when the body falls as shown in Figure 6.15. *This is called the principle of the conservation of mechanical energy.*

This principle is of extreme importance in astronomy since it holds for the moon moving around the earth, and for the earth and planets moving around the sun. We have noted that the earth is sometimes closer to the sun than at other times. As it gets closer to the sun its P.E., relative to the sun, diminishes, but its K.E. increases by just the right amount to keep the total amount of M.E. of the earth relative to the sun constant. If this were not the case, the earth would either lose or gain energy; as a consequence it would either fall into the sun or recede farther and farther away from the sun, a most unhappy result for mankind in either case. Because conservation of mechanical energy does not hold for an artificial satellite moving through regions of the atmosphere (however rare the atmosphere may be) the satellite ultimately falls back to the earth. The friction of the satellite against the molecules of air robs it of its mechanical energy.

6.23 The Speed of a Freely Falling Body

We shall now make use of the principle of the conservation of mechanical energy to derive the formula for the speed of escape of a body from the surface of a sphere. To do this we first consider the trajectory of a body on the earth's surface as shown in Figure 6.16. We see that the body at any

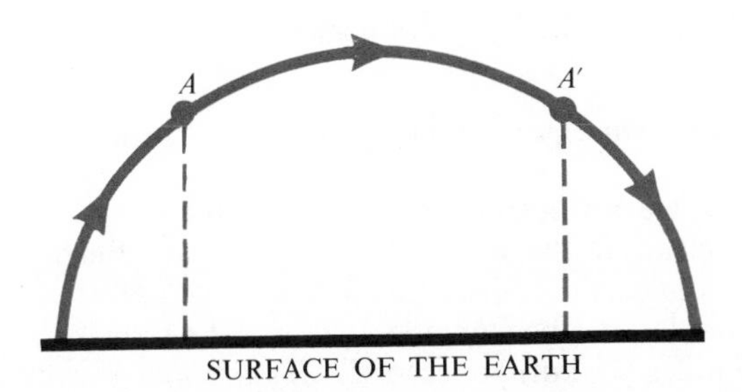

FIG. 6–16 *Since the potential energy of a particle at A and A′ in its trajectory is the same, because the height above the ground is the same in those two positions, the kinetic energy at A and A′ must be the same, by the principle of conservation of energy.*

point A on the ascending branch of its trajectory has the same P.E. as at the corresponding point A' on the descending branch, since A and A' are at equal heights above the ground. But from this it follows that the K.E. at A and A' must also be the same since the total energy (the conservation of mechanical energy) at all points of the trajectory (neglecting air resistance) is the same. In other words, if we disregard air resistance, *the speed of a projectile at a given point above the ground as it is moving upwardly is the same as its speed at the same height above the ground when it is moving downwardly.* In particular, the speed with which a body leaves the ground when it is thrown vertically upward to a given height must be exactly equal to the speed which it has when it hits the ground on the return trip (again neglecting air resistance).

If, then, we wish to know how rapidly we must throw a body if it is to reach a certain height, we can calculate this by computing the speed with which it hits the earth when it is allowed to fall from rest through the same height, as shown in Figure 6.17. But the speed it has when it reaches the earth is related to its kinetic energy at the moment it hits the earth, and this, in turn, by the principle of conservation of energy, must equal the potential energy it had when it was at the given height from which it was allowed to fall. In other words, we may com-

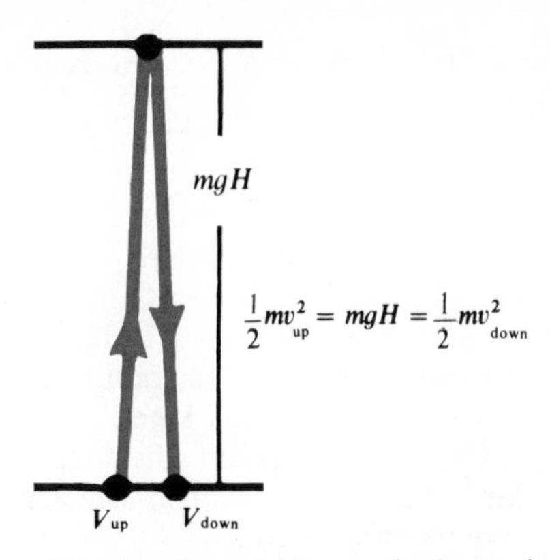

FIG. 6–17 *The initial speed (the speed when it leaves the ground) which a particle must have to reach the highest point in its ascent is exactly equal to the speed which a particle falling from this highest point has when it reaches the ground.*

pute the speed with which a body must leave the earth to reach a given height by computing its P.E. at the given height. This is what we shall do to compute the speed of escape.

6.24 The Speed of Escape

In the case of the speed of escape, the body must leave the surface of the earth with just the right speed to go infinitely far off. Hence, the speed of escape is the same as the speed with which a body starting from rest and falling towards the earth from an infinitely great distance strikes the earth. Our approach then is first to compute the work we have to do to move a body from the surface of the earth to a point infinitely far away, and then, since this gives us the potential energy of the body at that point, to equate this to the kinetic energy the body would have if it fell to the earth from an infinite distance. (Actually the distance need not be infinite but only very large.)

Since work is force times distance, and the force in this case is the gravitational force, we can write down an expression for the work done as we move the body away from the earth by moving the body along a straight line from the center of the earth. However, the problem is complicated because the gravitational force changes continuously, so that the total work done must be computed by breaking this work up into a large number of small contributions and then summing them all. There is a well-defined mathematical procedure for doing this which leads to the following expression for the work done in taking a body from the surface of the earth to infinity:

$$W = G\frac{M_\oplus m}{R_\oplus}, \tag{6.10}$$

where G is the gravitational constant, m is the mass of the body on which the work is done, $M_\oplus$ is the mass of the earth, and $R_\oplus$ is the radius of the earth.

This, then, must equal the kinetic energy of the body coming in from an infinite distance (in practice a very large but finite distance) when it hits the earth. We therefore have

$$\frac{1}{2}mv_{esc}^2 = G\frac{M_\oplus m}{R_\oplus},$$

where v_{esc} is the speed of escape. If we solve this equation we obtain for the speed of escape

$$v_{esc} = \sqrt{2G\frac{M_\oplus}{R_\oplus}}. \tag{6.11}$$

This equation is not quite correct because we have assumed that while the body is escaping from the earth, the earth itself does not move. Actually when an object is thrown away from the earth, the earth itself is given a backward push so that it too is set moving. The true picture is given by considering the system (earth and body) as seen from the center of mass of the earth and the body. We then find that both the body and the earth move away from the center of mass. Since the mass of the earth is vast in comparison with the mass of the body, the earth's motion is practically zero, and so may be neglected. However, if we were to picture another sphere almost as massive as the earth escaping from the earth, as shown in Figure 6.18,

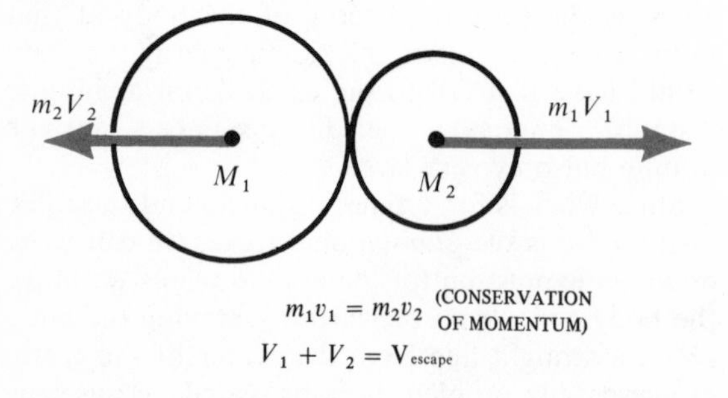

FIG. 6–18 *The speed of escape of one sphere from another depends on the masses of the two spheres and the distance between their centers when their surfaces are touching.*

we would have to replace $M_\oplus$ in the formula for the speed of escape by $M_\oplus + m$ because the formula must be symmetrical in both masses. In the same way, if the radius of the escaping sphere were r, we would have to replace $R_\oplus$ in the formula by $R_\oplus + r$. Thus, the general formula for the speed of escape is given by

$$v_{esc} = \sqrt{2G\frac{M_\oplus + m}{R_\oplus + r}}. \tag{6.12}$$

For our purpose it is sufficient to use the simpler formula (6.11), in which the motion of the earth and the size of the body are neglected.

6.25 The Speed of Escape from the Earth

If we now introduce the value of the gravitational constant G, the mass of the earth $M_\oplus$, and the radius of the earth $R_\oplus$ into this formula, we find that the speed of escape from the surface of the earth is about 7 miles (11 km) per second. This is well above the speeds of such molecules as oxygen, nitrogen, and water vapor, so that our earth can maintain an atmosphere. However, the molecules of hydrogen and helium will have speeds in excess of this, and that is why we find so little hydrogen in our atmosphere, even though hydrogen is the most abundant element in the universe.

It is fortunate for us that the speed of escape from the earth is not much larger than 7 miles (11 km) per second, for if it were, the earth would have retained a good deal of hydrogen in its atmosphere. This would have seriously upset the balance of the chemical compositions that are necessary for maintaining life. Thus, larger quantities of hydrogen would have eliminated most of the free oxygen, nitrogen, and carbon, giving rise to large quantities of water, ammonia vapor, and methane.

6.26 The Speed of Escape from the Moon

We can find the speed of escape from the surface of the moon by simply introducing the mass and the radius of the moon into the same formula that applies to the earth and dividing one formula by the other. We then obtain

$$\frac{v_{\text{escape from the earth}}}{v_{\text{escape from the moon}}} = \sqrt{\frac{M_\oplus/M_☾}{R_\oplus/R_☾}}. \tag{6.13}$$

Since the mass of the earth is 81 times that of the moon, and the radius is approximately 4 times that of the moon, this ratio is about 4.5, so that the speed of escape from the moon is less than one-fourth of that from the earth, or about 1.5 miles (2.38 km) per sec. From this it is clear that the moon cannot retain an atmosphere, since this speed is below that which molecules in such an atmosphere would have.

It should be noted that although the average speed of molecules in an atmosphere may be less than the speed of escape, the atmosphere may still escape, as has been pointed out by Jeans, since there are always many molecules moving at speeds four or five times the average and these escape first. Thus, if the mean molecular velocity is one-third the velocity of escape, half the atmosphere will disappear in a matter of weeks; and if the average speed is one-quarter the speed of escape, several thousand years will elapse before one-half the atmosphere is gone.

The speed of escape diminishes as one gets further and further away from the center of the attracting body, so that molecules high above the surface of the earth and the moon have an easier time escaping than those closer. Since the moon has no atmosphere, there can be no water on its surface, as the water would quickly evaporate to form a water-vapor atmosphere and this atmosphere would ultimately escape.

6.27 The Surface Features of the Moon

The most abundant topographical features on the moon are the craters, which vary in diameter from a few hundred feet to over 100 miles. The smallest craters are extremely numerous; at least 30 000 have been counted with ordinary-sized telescopes. The large craters surrounded by ring-shaped walls that rise thousands of feet (some to 20 000 ft) above the base of the crater, have been named for famous astronomers and philosophers.

Besides the craters there are vast mountain chains, named after terrestrial mountain chains such as the Apennines, the Alps, etc. Two other interesting features are the rills and rays, the latter appearing as white streaks running out radiantly from large craters. They form arcs of great circles which extend for hundreds of miles. The smooth regions of the moon's surface are referred to as maria (the Latin word for seas), although these are dry regions with no water whatsoever. See the accompanying photographs.

By measuring the lengths of shadows cast by the peaks on the moon's surface, we can determine the heights of these peaks, and although it is true that relatively speaking they are much higher than peaks on the earth, it is wrong to assume that if we were standing on the moon, we would be surrounded by large sharp spires and steep cliffs. Actually, the slopes up these peaks are rather small, seldom exceeding a grade of 10°.

FIG. 6–19 *A portion of the moon at last quarter from Ptolemaeus to Tycho. This central region shows the great variety of craters and mountains on the moon's surface. (100-inch photograph from the Mount Wilson and Palomar Observatories)*

Landing on the moon presents some problems, since there is evidence that the maria, which seem to be the most promising landing sites, are covered by a fine dust. This dust may have originated in the continual bombardment of the surface by micrometeorites. T. Gold has suggested that there is a continual smoothing out of the dust, owing to electrical charging of the dust particles. Such charged dust particles would repel one another, with the result that the slopes of the peaks would be swept clean and deep layers of dust would accumulate in the marias.

FIG. 6–20 *The moon, full phase, age 14 days. (100-inch photograph from the Mount Wilson and Palomar Observatories)*

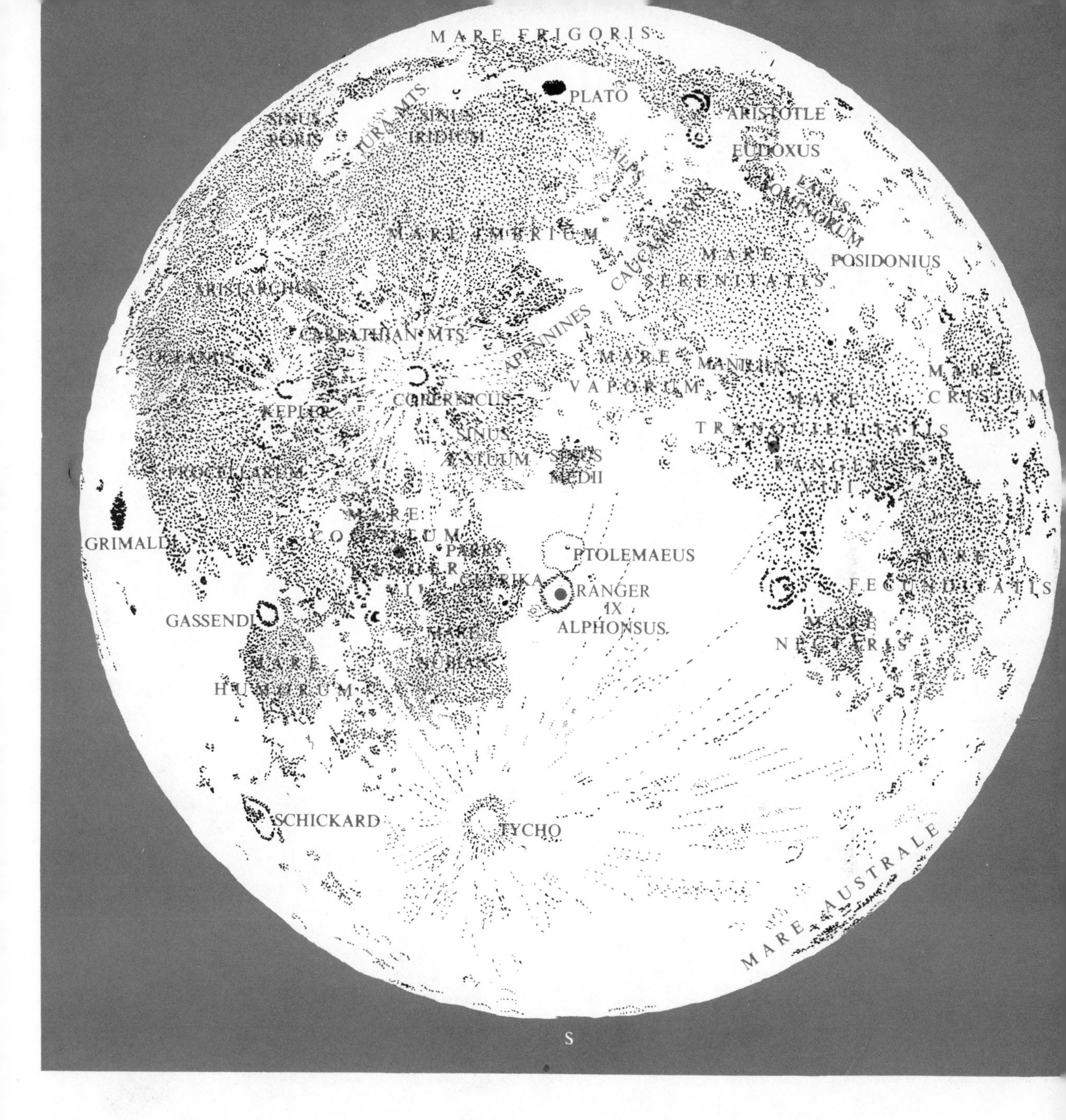

FIG. 6–21 *Map of the side of the moon facing the earth, showing the locations of the various natural features.*

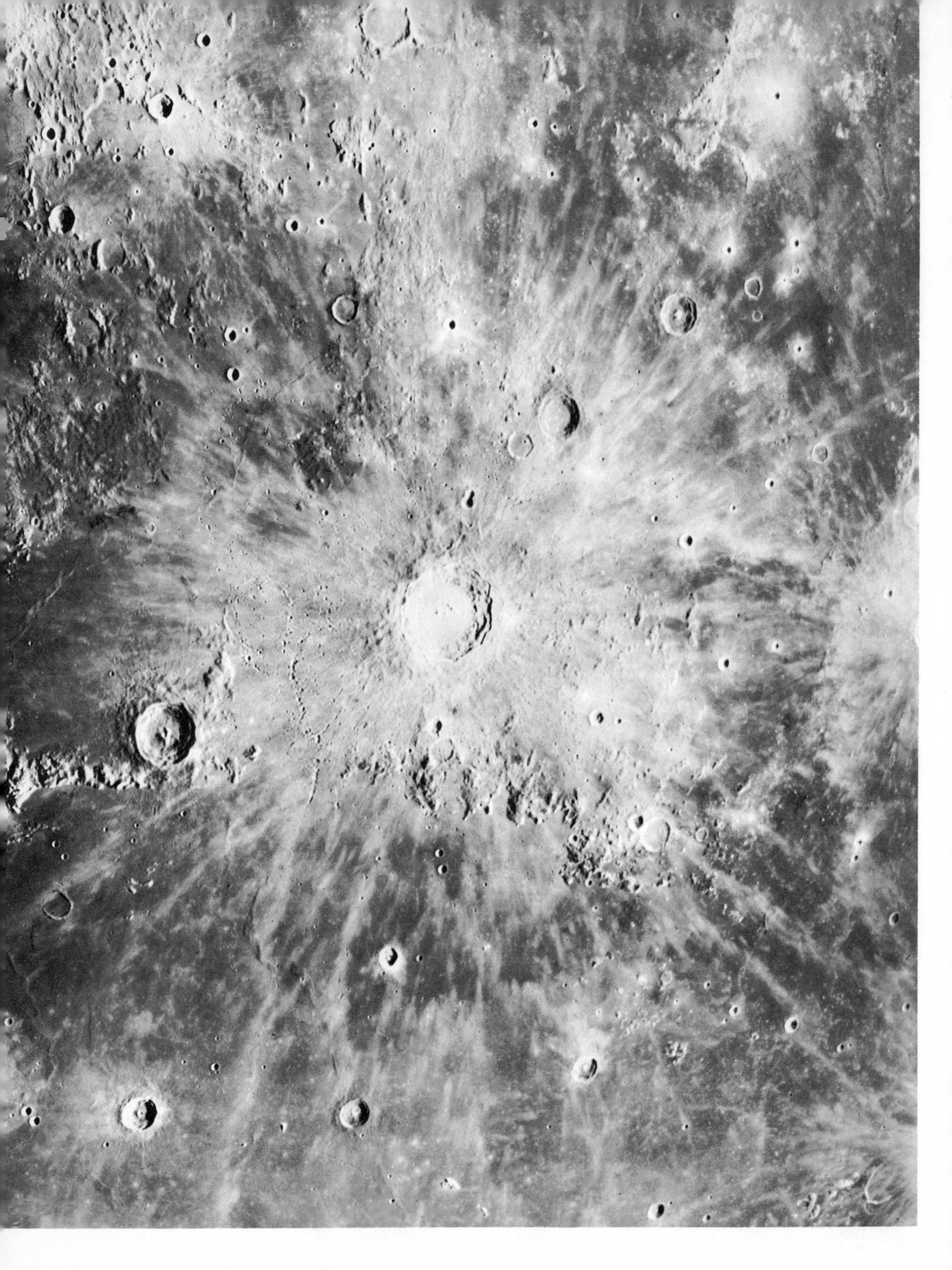

FIG. 6–22 *The moon, region of Copernicus, with the sun almost directly overhead.* (*100-inch photograph from the Mount Wilson and Palomar Observatories*)

At present many new projects are being planned to use artificial satellites instead of further manned landings to obtain additional detailed information about the surface of the moon.

FIG. 6–23 *The moon, region of Clavius. (200-inch photograph from the Mount Wilson and Palomar Observatories)*

6.28 Lunar Libration

Although the moon, on the average, keeps the same face towards the earth, we can actually see a little more than one-half of the moon's surface because of librations. These librations are continual eastwardly and westwardly oscillations relative to the earth, arising from various causes, but in no case exceeding 8° in either direction. Lunar libration may be divided into (1) **geometrical librations,** which arise because the rotation of the moon does not keep exactly in step with its revolution from day to day around the earth, as mentioned previously, and (2) **physical librations,** which are due to gravitational effects of other bodies on the moon.

If we take all of the librations into account, we find that we can see up to 59 per cent of the moon's surface—41 per cent being always visible and 18 per cent alternately visible and invisible. Although we have not been able to see directly the 41 per cent of the moon that is always invisible, we now know something about it from the photographs taken by manned and unmanned lunar expeditions. One fact revealed by these photographs is that the hidden surface is not nearly as pitted with craters and mountain ranges as are the visible portions of the moon.

6.29 Albedo of the Moon

The light that we get from the moon is reflected sunlight, so that its spectrum is identical with that of sunlight. However, at full moon, the ratio of the amount of light we get directly from the sun to the amount reflected to us from the moon ranges from 375 000 to 630 000, with an average value of 465 000. This means that even if the entire stellar hemisphere were covered with full moons, it would still send us only one-fifth the light that the sun does. From this figure we can determine the **albedo,** or the reflecting power, of the moon's surface. This is about 0.073, which means that only about 7 per cent of the sunlight striking the moon is reflected, the rest of it being absorbed. This is about what we would expect if the moon were covered with dark-colored rock.

6.30 The Origin of Craters

There has been a great deal of speculation and controversy about the origin of lunar craters, and two theories have been developed to account for them. One of these theories attributes these craters to volcanic activity, and recent observations of one such volcano by the Russian astronomer Kozyrev have strengthened the belief that at least some of the craters were formed by active volcanos. This theory is further supported by formations on the moon which, as Kuiper has pointed out, look like extinct volcanos. However, except for the one observation in 1958 by Kozyrev, no volcanic activity has ever been observed, so that volcanic craters must all have been formed in the distant past. Since there is probably radioactivity in the lunar

FIG. 6–24 *The moon, region of Copernicus, with the sun at an angle. The intense darkness in the shadow can be clearly seen. (200-inch photograph from the Mount Wilson and Palomar Observatories)*

formations, the moon must certainly have been a much hotter body a few billion years ago, and hence a source of volcanic activity.

The theory that seems to account for most of the craters is that of meteoric bombardment (see the accompanying photographs). Since the moon has no atmosphere, the number of direct meteoric collisions it suffers is in the billions every year, and the craters so formed remain to tell their story since no atmospheric weathering takes place there.

6.31 Lunar Climate

Because there is no atmosphere on the moon, the temperature there fluctuates very widely during the lunar month. During the fourteen days of sunlight, the surface of the moon facing the sun reaches a temperature of about 130°C, but a few hours after the sun has set there is an extremely sharp drop, and during the height of the sunless period the temperature falls to almost −200°C. These rapid changes in the temperature probably result in some exfoliation of the rock structure, so that in this sense some weathering does occur on the moon's surface.

6.32 The Influence of the Moon on the Earth: Tidal Action

The most important effect that the moon has on the earth is through its tidal action. Tides are known to rise and fall (flood tide and ebb tide) twice during each 24-hour, 50-minute interval, and since this is also the interval between the two succes-

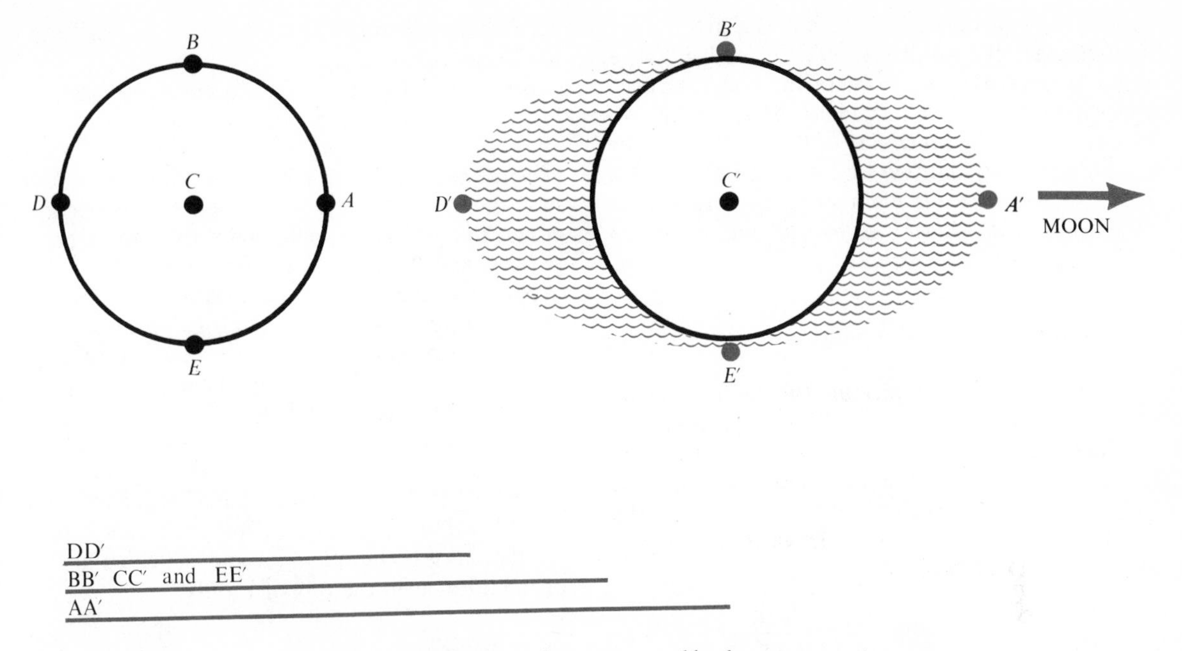

FIG. 6–25 *The tidal action of the moon is shown by the acceleration imparted by the moon to 4 unit masses of water at points A, B, D, E, and to a unit mass at the center of the earth, C. Since A is closer to the moon than C, and D is farther from the moon than C, A has a greater acceleration than C, and D a smaller one than C.*

sive transits of the moon across a meridian, we know the tides are related to the moon. Actually, however, both the moon and the sun are responsible for the total tidal action, but the contribution of the sun is less than one-half that of the moon. At first sight it appears that the sun ought to be much more effective than the moon in raising tides because the actual gravitational pull of the sun on the earth is about 180 times as large as that of the moon. But this conclusion is due to a misunderstanding of how the tides actually arise. The tide-raising effect is not due to the total force acting on the earth, but to the difference between the gravitational action on the earth as a whole (as though all its mass were concentrated at the center) and the gravitational attraction on the water on the surface.

6.33 The Explanation of the Tides *

To understand what happens we shall first suppose the earth's surface to be completely surrounded by water, and we shall represent this surface by a sphere as shown in Figure 6.25. Since we are interested only in the pull of the moon on portions of the earth relative to its center, we have shown just four particles of water, A, B, D, and E, and the center of the earth C. We can now understand the tidal action by considering the acceleration of these five points towards the moon, which we picture as being off to the right along the line DCA. The various accelerations are represented by arrows shown pointing to the moon.

Let us now suppose that the earth were to fall freely towards the moon. What would be the relative positions of these five points after the earth had fallen a certain distance? This is illustrated in Figure 6.25, where the center of the earth is shown as having fallen through the distance CC'. The point A, however, during the same interval will have fallen through a greater distance AA', since it is closer to the moon than C, and hence suffers a larger acceleration than the latter. On the other hand the point D will have fallen through a shorter distance DD' than C, because it is farther from the moon than the center of the earth, and hence experiences a smaller acceleration than the latter. Since B and E are practically at the same distance from the moon as C is, they will have fallen through the same distance as C, except that they will have moved along the lines BB' and EE', which converge towards the line from the center of the earth to the moon. Thus, the points B' and E' will be closer to the center of the earth than B and E are. If we now connect the points A', B', D', and E' by a smooth curve, we again obtain the water surface, and we see that it is ellipsoidal in shape, with two high tides, one on either side of the earth along the earth-moon line, and two low tides at right angles. *Thus we see that the tides arise because of the differential gravitational action of the moon on different parts of the earth.*

6.34 Formula for Tidal Action

It is easy to obtain a formula for the tide-raising acceleration at the point A by noting that this acceleration is the difference between the acceleration imparted to A by the moon and that imparted to C by the moon. If d is the distance from A to the moon, and r is the radius of the earth, then these two accelerations are

$$G\frac{M_{☾}}{d^2} \quad \text{and} \quad G\frac{M_{☾}}{(d+r)^2}$$

respectively. We obtain the differential acceleration by subtracting the second of these expressions from the first. If we note that r is very much smaller than d, so that we may simply disregard it where it just adds on to d, we obtain, when we carry out the subtraction, the following expression for the differential acceleration,

$$f = 2G\frac{M_{☾}r}{d^3}.$$

We see from this that the tide-raising acceleration is inversely proportional to the cube of the distance. This is why the sun's tidal action is less than one-half of the moon's. Because of the cube of the distance factor, the sun's much greater distance more than compensates for its larger mass.

6.35 Magnitude of the Tides

If we compare this tide-raising acceleration with the acceleration of gravity g, here on the earth, we have on dividing the above expression for f by the expression 3.15 for g

$$\frac{f}{g} = \frac{2M}{D^3}, \qquad (6.14)$$

where M is the mass of the moon (in earth mass units) and D is the distance of the moon (in units of the earth's radius). From this we find that the maximum value of f is about 1/9 000 000 of that of the acceleration of gravity here on earth. This acceleration gives rise to a tide of only a few feet in the deep oceans.

Since the sun is also effective in raising tides, the maximum tides occur when the moon is new or full. We then have the spring tides. At the first and last quarters of the moon, we have the neap tides, which have the smallest range. The highest tides occur at either full or new moon, with the moon at its perigee, and the earth at its perihelion.

Although the range in the tides is only a few feet in the deep oceans, they may range up to 50 or 60 feet in narrow water channels connecting large bodies of water. The actual heights of tides at various locations depend upon the shape of the water basin and the shore line. Since a body of water oscillates with its own frequency once it is disturbed, it is possible to have extremely high tides because of resonance between the natural oscillation of bodies of water and the ebb and flow of the tides, as in the Bay of Fundy.

6.36 Tidal Friction and the Earth's Rotation *

We have already noted that the length of the day is increasing by 1.6 second every 100 000 years, as evidenced by the discrepancy between the observations of positions of the moon and the positions as predicted from ancient eclipse records. We shall show that this slowing down of the earth's rotation is due to the tidal friction. In the deep oceans where the tides rise and fall slowly, there is very little frictional action between the water and the underlying ocean bed. But in the shallow regions along the shore lines there is a continual rubbing of the water against the land. This causes dissipation into heat of the kinetic energy of the rotation of the earth. Although the effect on the rotation is very small, the energy is dissipated at the rate of 1.6×10^{19} ergs per second, or 2 100 000 000 hp. Most of the tidal friction occurs in the Bering Sea because of the shallow water and strong currents there.

This tidal friction affects not only the earth's rotation but also the moon's, and we can best understand this by considering the diagram in Figure 6.26, in which the tides are shown. If the earth were not rotating, the high tide would be right under the moon, but because of the rotation of the earth, the tidal bulge is dragged around to the position given by the red area somewhat in advance of the moon. By pulling upon these two bulges of water, the moon exerts a frictional torque on the earth, causing it to

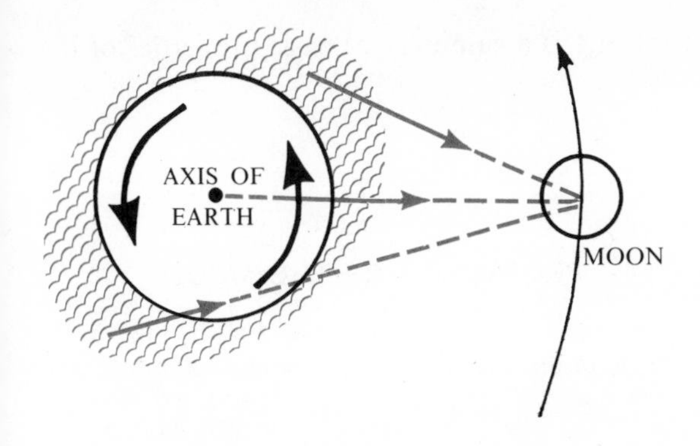

FIG. 6–26 *The frictional effect of the tides on the earth is shown. The torque action of the moon on the bulges of the water tends to slow down the earth's rotation. In turn these bulges act on the moon to speed it up in its orbit and thus to increase its orbital angular momentum relative to the earth.*

rotate more slowly, and hence to lose angular momentum and kinetic energy of rotation. But, as we have seen, a closed system experiencing no external torque (in this case the earth and the moon together) cannot lose angular momentum. *It follows, therefore, that the moon must gain the angular momentum lost by the spinning earth.* It also gains some of the kinetic energy of rotation that the earth loses (the part that is not dissipated into heat by the friction of the oceans).

The reason the moon gains this angular momentum is that the tidal bulges pull on the moon in the direction of its motion, and thus tend to speed it up. The effect is to increase the size of the moon's orbit and hence the length of the month.

As long as the earth keeps on rotating faster than the moon revolves (that is, as long as the day remains shorter than the month), this transfer of angular momentum from the rotation of the earth to the revolution of the moon goes on, and it will continue in this way until finally the month and the day become equal in length. This will not happen for billions of years, and when it does occur, the month will probably be twice its present length, and one day will then be equal in length to 60 current 24-hour days. The moon will then neither rise nor set, but will always be visible from one half of the earth and invisible from the other half.

Using the same reasoning we conclude that the day and the month must have been much shorter billions of years ago. It has been estimated by Jeffreys that about 4 billion years ago the day was somewhat more than 4 hours long and the moon was no more than 9 000 miles (14 000 km) from the earth. Since then tidal friction has brought about the present configuration.

6.37 The Ultimate Approach of the Moon to the Earth: The Roche Limit

Even after the day has become equal in length to the month, the solar tidal action will cause the earth to slow down still more. The solar tides will thus result in a transfer of angular momentum from the earth's rotation to its orbital motion around the sun. This means that the earth's orbit will become larger and the year correspondingly longer. Since, as a result, the day will then become longer than the month, the entire earth-moon system will lose angular momentum and the moon will gradually approach the earth. This will go on until the moon is so close to the earth that the tidal action of the earth on it will tear it to pieces. The limiting distance from the earth's center—approximately 10 000 miles—at which this will occur is called the **Roche limit.**

To calculate this limiting distance we picture the moon as composed of two equal liquid spheres, each one half the size of the moon as shown in Figure 6.27. These two spheres will remain together (that is, the moon will not be split in half) as long as their mutual gravitational attraction is

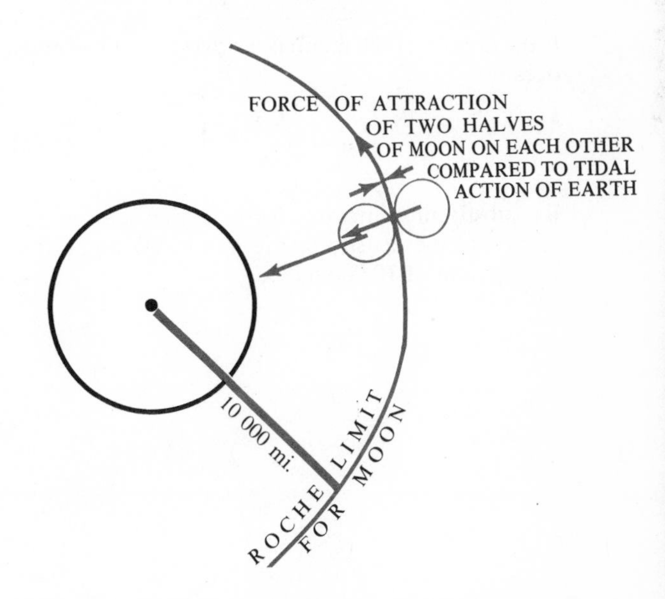

FIG. 6–27 *The tidal action of the earth on the moon. The earth exerts a greater gravitational force on that half of the moon which is nearer to it. When the moon is closer than the Roche limit, the gravitational attraction between these two halves is smaller than the earth's tidal action.*

greater than the destructive tidal action of the earth. If r is the radius of the moon, the mutual attraction of the two spheres is

$$\frac{G \dfrac{M_{☾}}{2} \dfrac{M_{☾}}{2}}{r^2},$$

where $M_{☾}$ is the mass of the moon. The tidal action of the earth tending to tear the moon apart is given approximately by

$$\frac{GM_{\oplus}M_{☾}}{2}\left[\left(\frac{1}{d-r}\right)^2 - \left(\frac{1}{d+r}\right)^2\right].$$

This is equal (on neglecting r^2 as compared with d^2 in the denominator after combining the two fractions into a single fraction) to

$$\frac{2GM_{\oplus}M_{☾}r}{d^3}.$$

Thus, the tidal disruptive force of the earth will equal the gravitational force tending to keep the moon together when the limiting distance d is so small that these two expressions are equal. We thus obtain this limiting distance d (the Roche limit) on equating these two expressions

$$2G\frac{M_{\oplus}M_{☾}r}{d^3} = \frac{GM_{☾}^2}{4r^2},$$

or

$$d^3 = \frac{8M_{\oplus}r^3}{M_{☾}}.$$

If the density of the moon is written as $\rho_{☾}$, we finally obtain

$$\boxed{d = \sqrt[3]{\frac{6M_{\oplus}}{\pi\rho_{☾}}}}. \qquad (6.15)$$

By substituting into this formula the mass of the earth and the density of the moon, we obtain the Roche limit of 10 000 miles.

Table 6.1 is a summary of basic properties of the moon.

TABLE 6.1

PROPERTIES OF THE MOON

Mean distance from the earth	= 384 404 km
range of distance	= 365 400 km through 406 700 km
Eccentricity of orbit	= 0.055
Inclination of orbit to ecliptic	= 5° 8′ 43″
Sidereal period	= 27.3217 days
Synodical month	= 29.5306 days
Tropical month	= 27.3216 days
Anomalistic month	= 27.5546 days
Nodical month	= 27.2122 days
Period of moon's node (equinox to same equinox)	= 18.6 tropical years
Period of rotation of moon's perigee	= 8.85 years
Moon's radius	= 1 737.9 km
Moon's mass	$M_{☾}$ = 7.349 × 10^{25} gm = (1/81.33) $M_{\oplus}$
Moon's semidiameter:	
(geocentric)	a = 15′ 32″.6
(topocentric, zenith)	a = 15′ 48″.3
Moon's volume	= 2.199 × 10^{25} cm^3
Moon's mean density	= 3.34 gm/cm^3
Surface acceleration of gravity	= 162.0 cm/sec^2
Velocity of escape from moon's surface	= 2.38 km/sec
Inclination of lunar equator:	
to ecliptic	= 1° 32′ 40″
to orbit	= 6° 41′

FIG. 6–28 *A nearly full moon, as photographed by Apollo 16 crewmen. This view is looking generally westerly toward the large circular Sea of Crises on the horizon. Most of the lunar area in this picture is on the far side of the moon. (From the National Aeronautics and Space Administration)*

FIG. 6–29 *A telescope-made photograph showing the Hadley-Apennine landing area for the Apollo 15 mission. The region in which the lunar module landed is near the exact center of the picture, near the most readily visible portion of the Hadley Rille (that part which appears about the size of an eyelash). The rille is approximately 70 miles long, and it is about 1 200 feet deep at the Apollo 15 site. The landing site was about 465 miles north of the lunar equator, on the edge of the Sea of Rains (more specifically, on the Marsh of Decay). The Sea of Serenity's western edge can be seen in the right-hand portion of the photograph. The three largest craters on the left side of the picture are Aristillus, Archimedes, and Autolycus. (From the National Aeronautics and Space Administration)*

FIG. 6–30 *A nearly vertical view of the Hadley-Apennine area, as photographed by Apollo 10 in lunar orbit. Hadley Rille meanders through the center of the photograph. The landing site of Apollo 15 was on the east side of the rille's "chicken beak" in the center of the picture (north is at the top). The smooth, dark area is the Marsh of Decay. (From the National Aeronautics and Space Administration)*

FIG. 6–31 *Crater Kepler and vicinity. Astronaut Richard Gordon took this photograph while in lunar orbit during the Apollo 12 mission.*

FIG. 6–32 *The large crater Gockenius, as photographed by Apollo 8 crewmen. An unusual feature of this crater is the prominent rille that crosses the crater rim. This crater is approximately 40 miles in diameter. (From the National Aeronautics and Space Administration)*

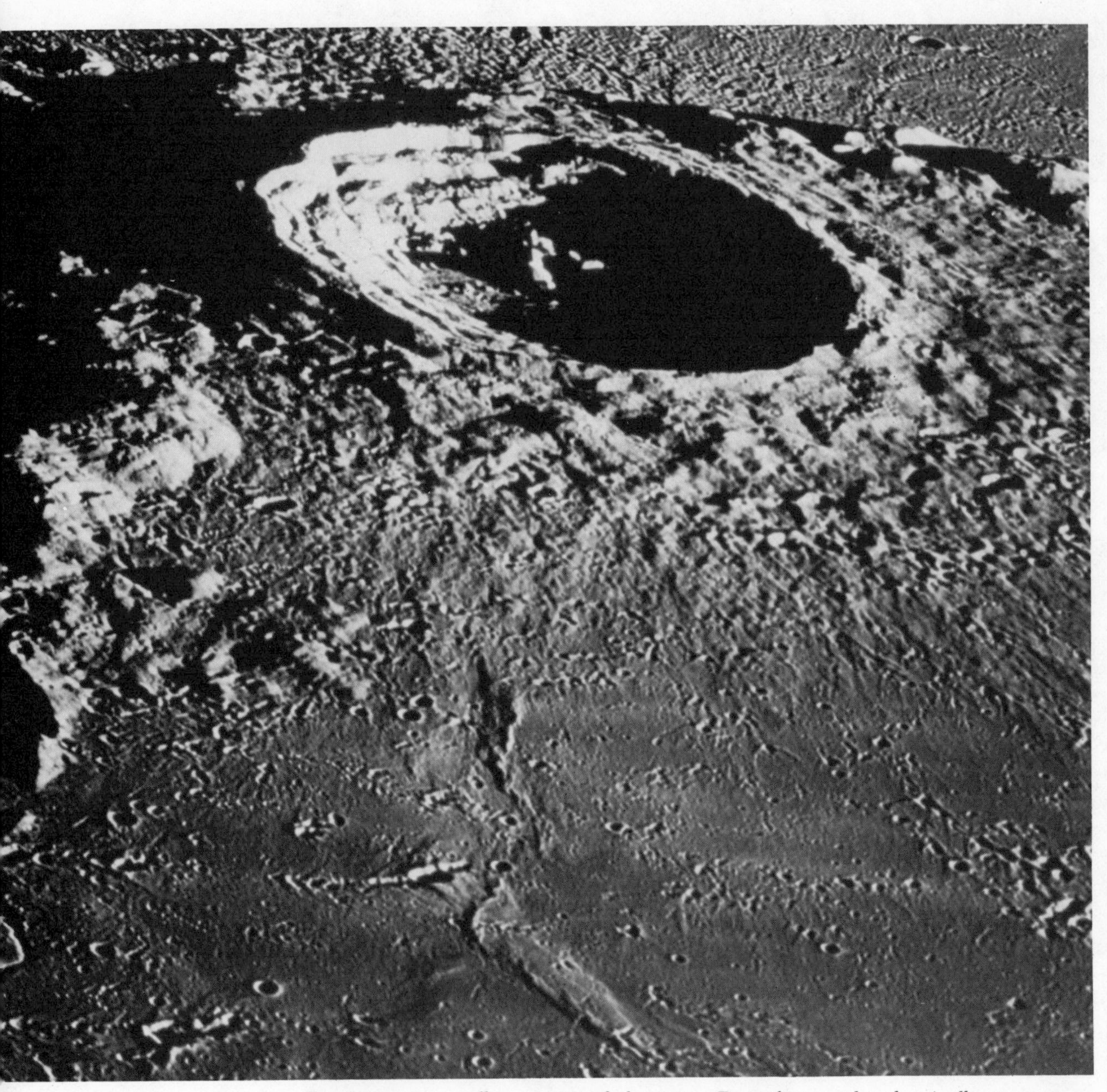

FIG. 6–33 *An excellent picture of the crater Eratosthenes, taken by Apollo 12 crewmen. Note the ejecta blanket surrounding the crater.* (*From the National Aeronautics and Space Administration*)

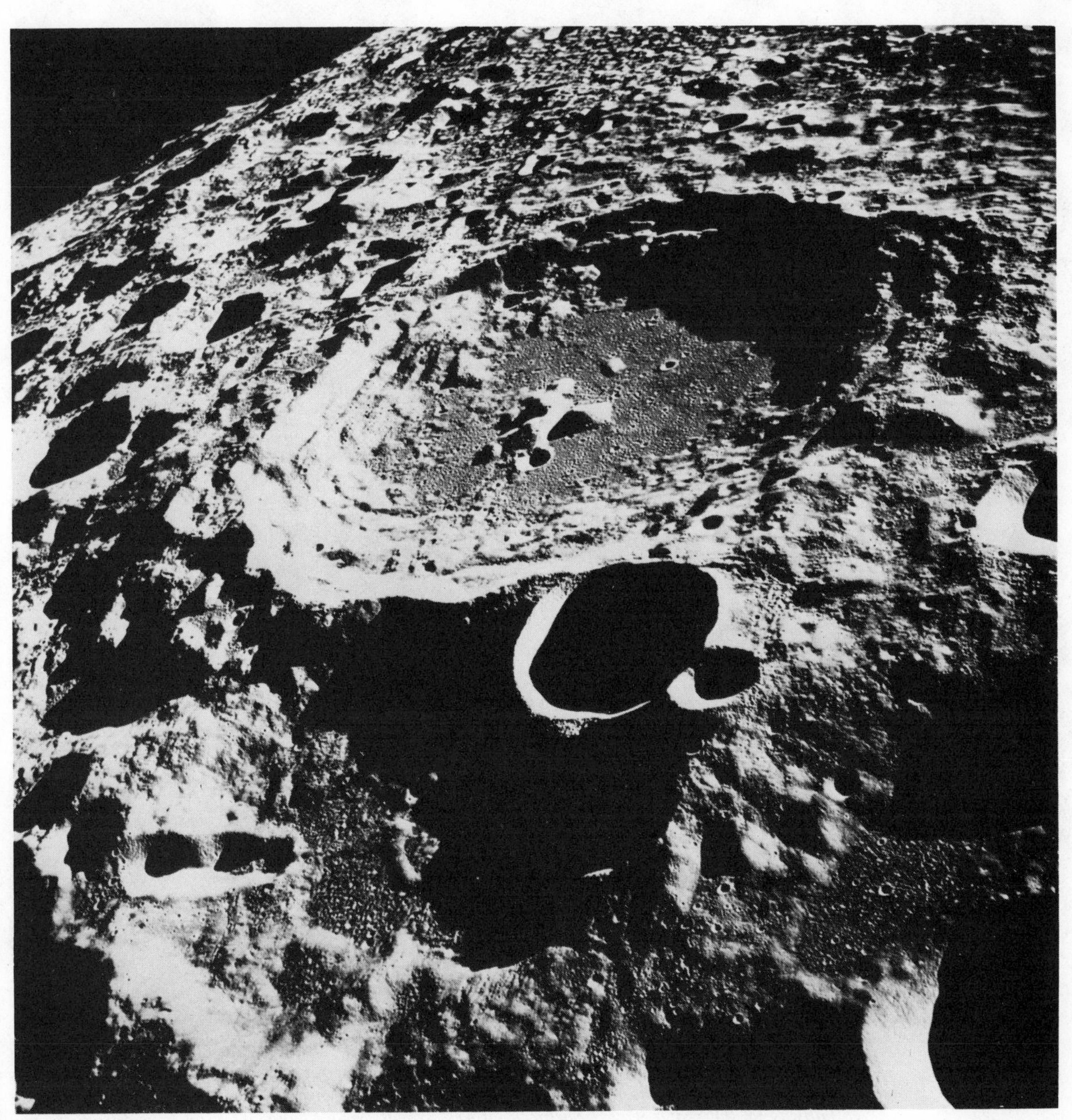

FIG. 6–34 *Oblique view of the lunar far side, photographed by Apollo 11. The large central crater is International Astronomical Union no. 308 and has a diameter of about 50 miles. (From the National Aeronautics and Space Administration)*

FIG. 6–35 *Vertical photograph by Apollo 8 of the western wall of a large farside crater (north is at the top of the picture). The western rim of a crater about 400 km (250 miles) in diameter is near the left side of the photograph. The smooth surface near the center of the photo is about halfway up the crater wall. Surface features strongly suggest that material flowed from the rim near the lower left corner and ponded to produce the high-level, smooth surface. Branching rilles are conspicuous on this surface and the smooth floor of the crater at right. (From the National Aeronautics and Space Administration)*

FIG. 6–36 *Oblique view of lunar far side, photographed by Apollo 13. The large mare area is Mare Moscoviense. (From the National Aeronautics and Space Administration)*

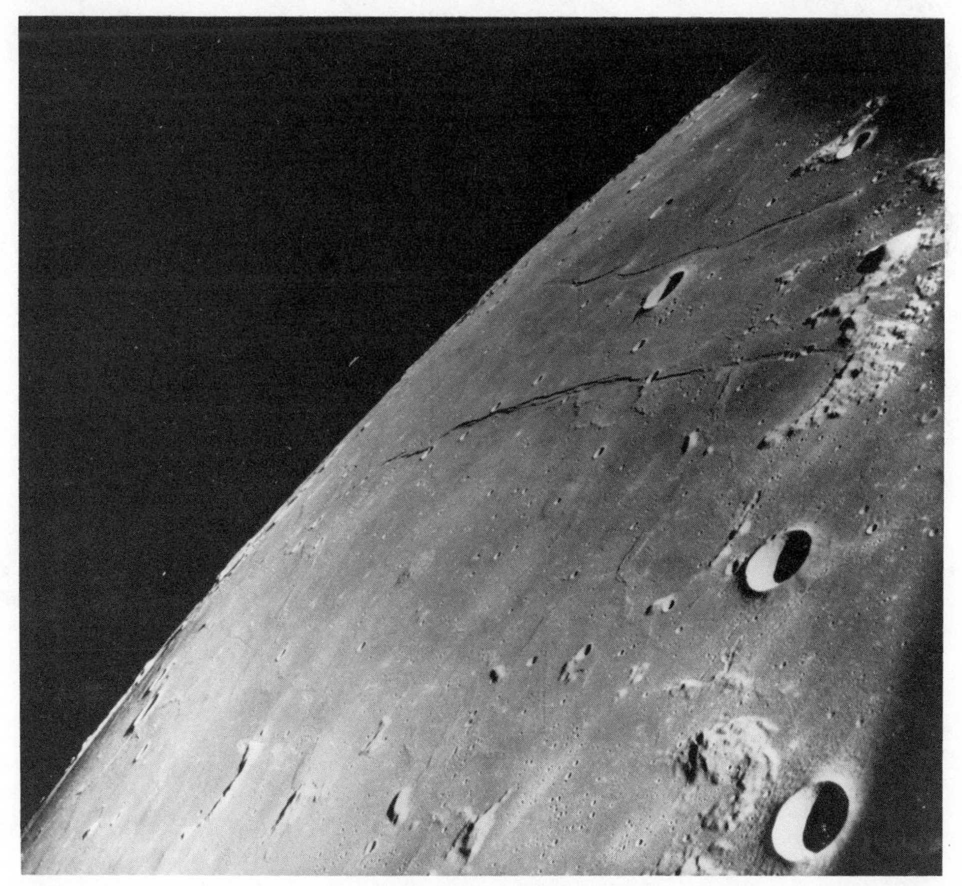

FIG. 6–37 *Oblique view of the Sea of Tranquility, photographed by Apollo 8. The lower linear feature is the Cauchy Scarp. The upper linear feature is the Cauchy Rille. The prominent crater Cauchy lies between the rille and the scarp.*

FIG. 6–38 *Photograph taken during extravehicular activity by Apollo 15 crewmen. Hadley Delta and St. George Crater are in the background. (From the National Aeronautics and Space Administration)*

FIG. 6–39 *A large boulder with multiple cracks, photographed by Apollo 17 crewmen in the Taurus-Littrow mountainous area of the moon, southeast of the Sea of Serenity. (From the National Aeronautics and Space Administration)*

FIG. 6–40 *A close-up view of the much-publicized "orange" soil that Apollo 17 crewmen found at Shorty Crater during extravehicular activity. (From the National Aeronautics and Space Administration)*

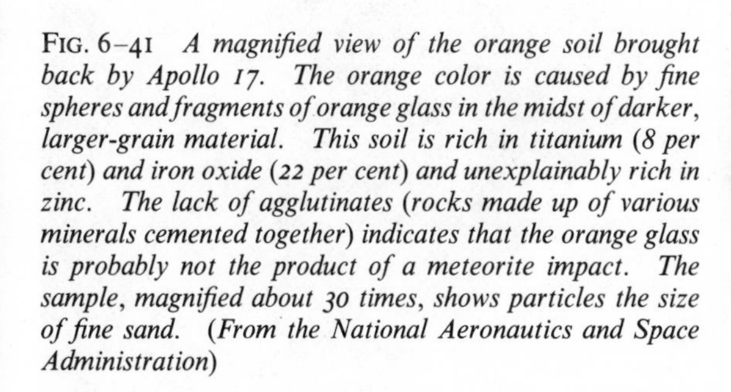

FIG. 6–41 *A magnified view of the orange soil brought back by Apollo 17. The orange color is caused by fine spheres and fragments of orange glass in the midst of darker, larger-grain material. This soil is rich in titanium (8 per cent) and iron oxide (22 per cent) and unexplainably rich in zinc. The lack of agglutinates (rocks made up of various minerals cemented together) indicates that the orange glass is probably not the product of a meteorite impact. The sample, magnified about 30 times, shows particles the size of fine sand. (From the National Aeronautics and Space Administration)*

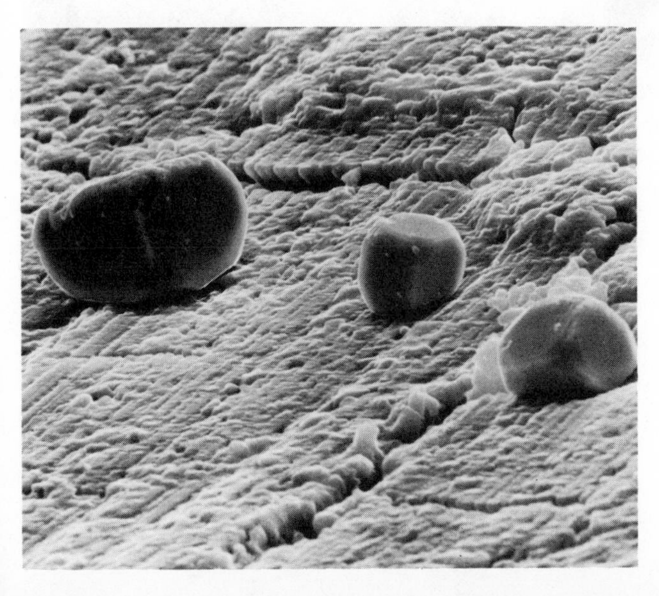

FIG. 6–42 *A scanning electron micrograph of iron crystals growing on a pyroxene crystal (calcium-magnesium-iron silicate) from the Apollo 15 Hadley-Apennine landing site. The largest iron crystal is about 8 microns across, and the smallest is about 4 microns across. The well-developed crystal facets indicate that these crystals were formed from a hot vapor as the rock was cooling. (From the National Aeronautics and Space Administration)*

FIG. 6–43 *Crescent earthrise with the far side of the moon in the foreground. This photograph was taken during the Apollo 17 mission. (From the National Aeronautics and Space Administration)*

EXERCISES

1. Since the moon moves to the east in its orbit around the earth by about 13° per day, how many miles does it move in one day around the earth if its distance from the earth is taken as 240 000 miles?

2. (a) Compute the gravitational force of the moon on the earth. (b) How does this compare with the gravitational pull of the sun on the earth?

3. How large an angle does the diameter of the earth subtend at the moon? How large would the earth appear to a lunar observer as compared to the apparent size of the full moon as seen from the earth?

4. (a) If a stone is dropped from a height of 30 ft to the surface of the moon, how long will it fall before striking the surface? (b) What speed will it have on striking the surface? (c) If the stone has a mass of 10^5 gm and is brought to rest in $\frac{1}{2}$ sec, with what force does it strike the surface of the moon? (d) Compare the answers with what would happen on the earth under the same condition, neglecting air resistance.

5. (a) How much work is done when a force of 1 000 dynes is exerted through a distance of 1 000 cm? (b) If this work were done on a body with a mass of 200 gm, what speed would the body have when all the work had been done on it? (c) How high above the ground could this amount of work lift the body in (b)?

6. A freely falling body having a mass of 150 gm is 200 cm above the ground and has a speed of 500 cm per sec. What is its total energy?

7. (a) Calculate the average kinetic energy of the moon in its orbit around the earth. (b) Calculate the average kinetic energies of the planets in their orbits around the sun.

8. (a) Show that if I is the moment of inertia of a spinning body and ω is its angular velocity, the kinetic energy of rotation of the body is $\frac{1}{2}I\omega^2$. (Hint: use the definition of I given in Section 5.24.) (b) Use this formula to calculate the kinetic energy of rotation of the earth if the moment of inertia of the earth is taken as $\frac{2}{5}MR^2$, where M is its mass and R is its radius.

9. Suppose Venus and the earth were just touching (neglecting the gravitational distortion of one body on the other). What speed would Venus have to have to escape from the earth?

10. Show that the period of the moon (assuming that its mass is negligible with respect to the earth's mass) is equal to that of a pendulum of length d in a gravitational field of acceleration g/d^2, where g is the acceleration of gravity on the surface of the earth and d is the mean distance of the earth's center from the moon's center. (Hint: assume the moon's orbit to be circular.)

11. If the moon were one-half its present mean distance from the earth, how much larger would its tide-raising acceleration be than it is now?

12. Compute the orbital angular momentum of the earth and moon together around their common center of mass and compare this with the rotational angular momentum of the earth.

13. Compute the angular speed of rotation of the moon in radians per sec.

14. How large is the tide-raising acceleration of the earth on the moon?

15. Assume a body on the surface of the moon is thrown up at an angle of 45° with the horizontal and with a speed of 5 000 cm per sec. (a) How high will it rise above the surface of the moon? (b) How long will it take to reach this height? (c) How far will it move horizontally before striking the surface of the moon again?

16. Describe the phases of the earth as they would appear to you if you were on the moon.

7

ECLIPSES OF THE SUN AND THE MOON

7.1 Shadows Cast by the Moon on the Earth: Types of Solar Eclipses

An eclipse of a celestial body occurs whenever it becomes totally or partially darkened, either because

1. *the light that ordinarily reaches us from it is blocked out by a third body, or*
2. *it passes into the shadow of the earth itself or the shadow of some opaque object like a planet (eclipses of Jovian satellites).*

Solar eclipses fall into the first category and **lunar eclipses** into the second. To understand how and with what frequency these two eclipses occur, we must consider the shadows in space cast by the moon and the earth. When the shadow of the moon strikes the earth, observers in the shadow see a total eclipse of the sun, and when the moon passes into the shadow of the earth, all observers on the side of the earth facing the moon see an eclipse of the moon.

In Figure 7.1, the sun, moon, and a portion of the earth's surface are shown, in order to illustrate an eclipse of the sun (the drawings are, of course, not to scale). An observer stationed at point A (within the **umbral region**) is in the total shadow of the moon, and sees the sun totally eclipsed. *The size of this shadow depends on how close to the earth the moon is.* If the earth is exactly at the apex of the shadow cone, the shadow is visible only over a small area at any one moment. If the earth is farther away from the moon than the length of the shadow, then the eclipse is **annular** with a dark central disk and a ring of light around it. If the earth is closer to the moon than the length of the shadow cone, the total eclipse is visible over a larger area. These situations, shown in the figure by the dotted curves, all occur since the distance of the moon from the earth ranges from about 220 000 to 250 000 miles (350 000 to 400 000 km). As the moon moves between the earth and the sun, the shadow traces a path across the face of the earth.

If an observer is outside the total shadow, but anywhere in the region from P to P' (the so-called **penumbral region**), the solar eclipse is partial. In the region from A to P the observer sees only the upper portions of the sun, and an observer in the region A to P' sees only the lower portions.

7.2 The Earth's Shadow and Lunar Eclipses

In Figure 7.2 the earth's shadow is shown with the moon passing through it to illustrate an eclipse of the moon. In this case we have a lunar eclipse; the moon is darkened because the light from the sun cannot reach it. All observers on the earth's surface facing the moon can see this lunar eclipse. The lunar eclipse is total if the entire disk of the moon is in the main shadow cone ABC, and it is partial if only a part of the moon passes through this cone. If the moon is in the penumbral region, it is visible but is fainter than it ordinarily is because it receives light from only part of the sun.

7.3 Eclipses and Phases of the Moon

Solar eclipses occur only when the moon is at or near its new phase, and lunar eclipses occur only when the moon is at or near its full phase. Since each of these phases occurs once a month, it may appear that solar and lunar eclipses should occur once a month. But they do not, because the plane of the moon's orbit does not coincide with the plane of the earth's orbit (the plane of the ecliptic) but cuts it at an angle of 5°. This means that during most of the full- or new-moon phases, the moon lies at such a large distance **above** or **below** the plane of the ecliptic that either its shadow does not hit the earth or the earth's shadow does not hit it, as shown in Figure 7.3.

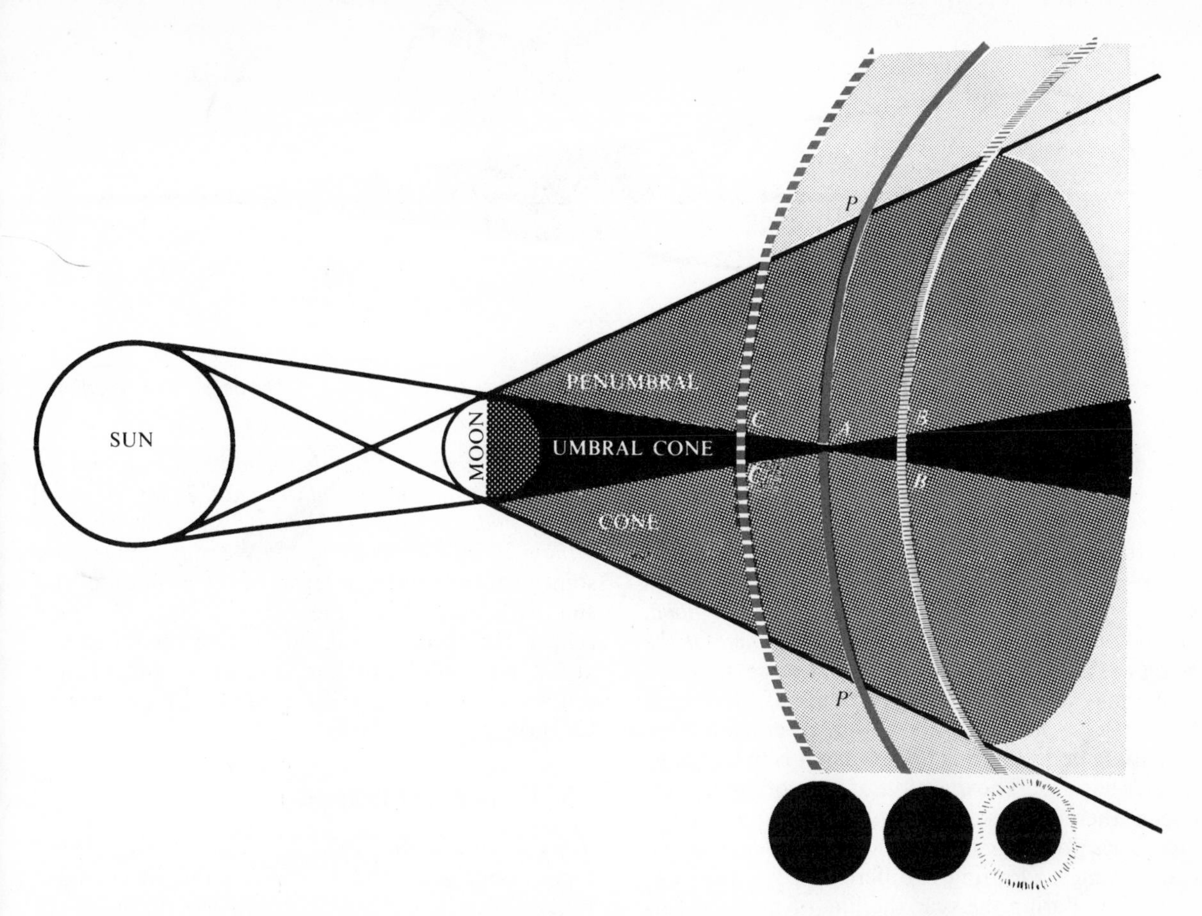

FIG. 7–1 *The intersection of the surface of the earth with the cone of the moon's shadow for varying distances between the earth and the moon. The appearance of the eclipse at the three distances is shown.*

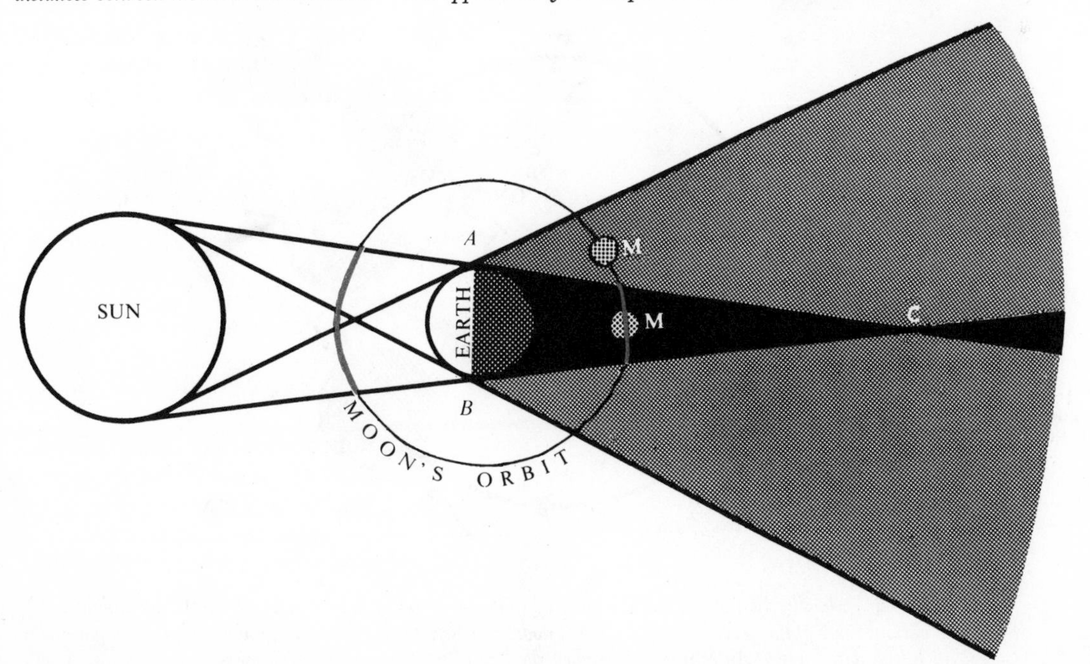

FIG. 7–2 *The moon passing through the earth's shadow. When this happens a lunar eclipse occurs. (The thin red arcs show the two ecliptic limits.)*

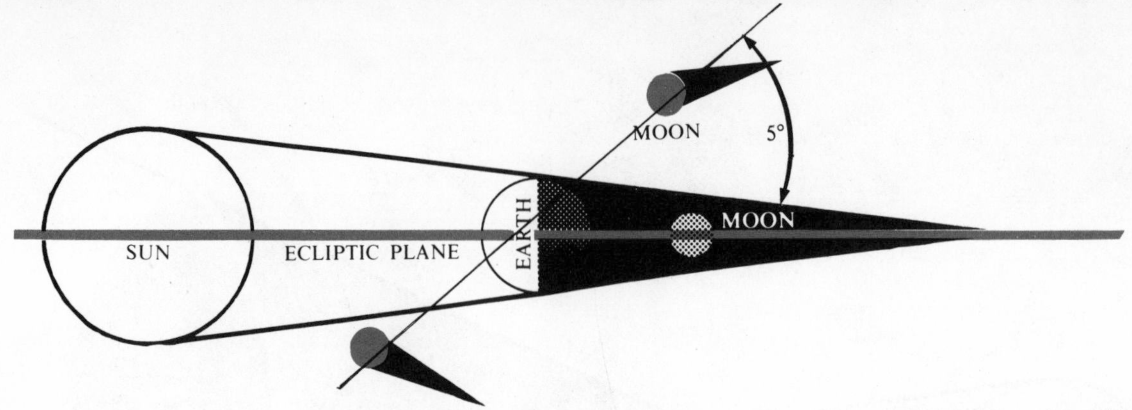

FIG. 7–3 *The shadow of the moon either above or below the plane of the ecliptic.*

7.4 Eclipses and the Lunar Nodes

To determine how often eclipses occur, we must therefore consider how often the moon lies close to the plane of the ecliptic during its new or full phases. *For it is only when we have a full or new moon, and when the moon is practically in the plane of the ecliptic, that either a lunar or solar eclipse occurs.*

We may also discuss this in terms of the nodes, for we have seen that the moon is either at the ascending or descending node when it is in the plane of the ecliptic. The question of how often eclipses occur, therefore, is equivalent to the question of *how often we have a full or new moon at the time of a node passage.* Putting it differently, we must ask how often during the year the line of nodes points to the sun, since this is a necessary condition for an eclipse to occur. *It is only when the moon is near or at its new or full phase, and when at the same time the line of nodes points nearly to the sun, that all conditions for an eclipse are met.* We may express this somewhat differently in terms of the position of the sun with respect to the lunar nodes. A solar eclipse can occur only if the sun and the moon are at (or near enough to) the same node, and a lunar eclipse can occur only if the sun and moon are at (or near) opposite nodes.

7.5 Frequency of Eclipses

We shall see that this can occur at most three times in one year, and to illustrate this we consider Figure 7.4, where the earth's orbit around the sun is shown as the large circle, and the moon's orbit

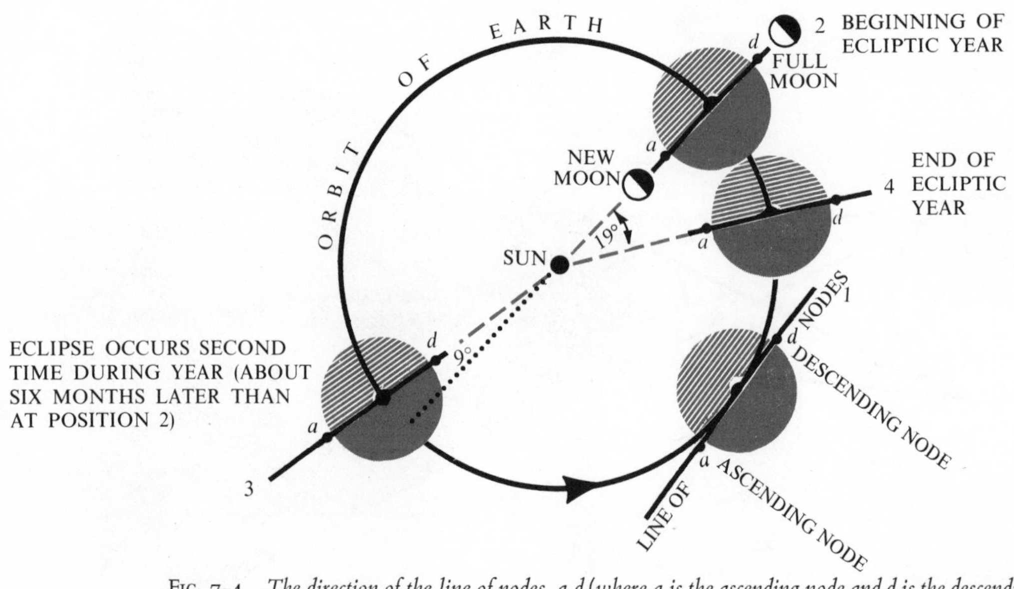

FIG. 7–4 *The direction of the line of nodes, a d (where a is the ascending node and d is the descending node), at various times of the year. Since the line of nodes precesses slowly westwardly, it doesn't remain parallel to its initial position 1. An eclipse occurs when the line of nodes passes through, or close to, the sun.*

around the earth is indicated by the partially dotted curve. In position 1 the line of nodes *a d* (where *a* is the ascending node, and *d* is the descending node) does not point to the sun. We can see that if the phase of the moon were new, the moon's shadow would fall beneath the earth, and if the moon were in its full phase the earth's shadow would fall below the moon; in either case there would be no eclipse. If the direction of the line of nodes in space were always the same, then, as the earth moved around the sun, this line would point to the sun just twice during one complete revolution. This is shown in positions 2 and 3.

Under these conditions lunar and solar eclipses could occur at most twice a year. However, the line of nodes does not remain in the same direction in space, because the nodes move westwardly around the ecliptic every 19 years, so that the line of nodes rotates westwardly by about 19° per year. In other words, the line of nodes will again be pointing to the sun about 19 days before the earth returns to position 1, as shown by position 4, in Figure 7.4. *Thus, eclipses may occur three times a year since the eclipse year, which is the time between two successive coincidences of the sun with the same node, is only* 346.62 *days.*

7.6 The Lunar Ecliptic Limit

Although we see from this that eclipses can occur three times during a 365¼-day year, we still do not know how many eclipses actually occur during this period since there may be an eclipse even if the moon is not exactly at one of the nodes, and even if the line of nodes does not point exactly to the sun. To illustrate this point, we shall consider lunar eclipses in detail. If the earth's shadow were a point, and the moon's disk were a point, a lunar eclipse could occur only at the three positions shown in the diagram, when the line of nodes points directly to the sun, and when the moon is right at the node. However, we must remember that the earth's shadow at the point where the moon passes through it is fairly large, and so too is the disk of the moon, so that these two disks have a chance of intersecting (resulting in an eclipse) even when their centers are not exactly at the node.

We can understand how frequently lunar eclipses occur by posing the following question: how often will two disks, of given apparent diameters (the moon's disk and the shadow disk of the earth), moving across the sky with different speeds along two great circles that intersect at an angle of about 5°, cut each other? In Figure 7.5, the earth's shadow is shown at various points along its great circle (the ecliptic) in the neighborhood of a node, and the moon's disk is shown along its orbit. For an eclipse to occur, the two disks must lie close enough to the node to cut each other. That means that they must lie somewhere between the points *A* and *A′*, since they just touch at these two points. The angular distance along the ecliptic from *A′* to *A* is therefore called the **lunar ecliptic limit**.

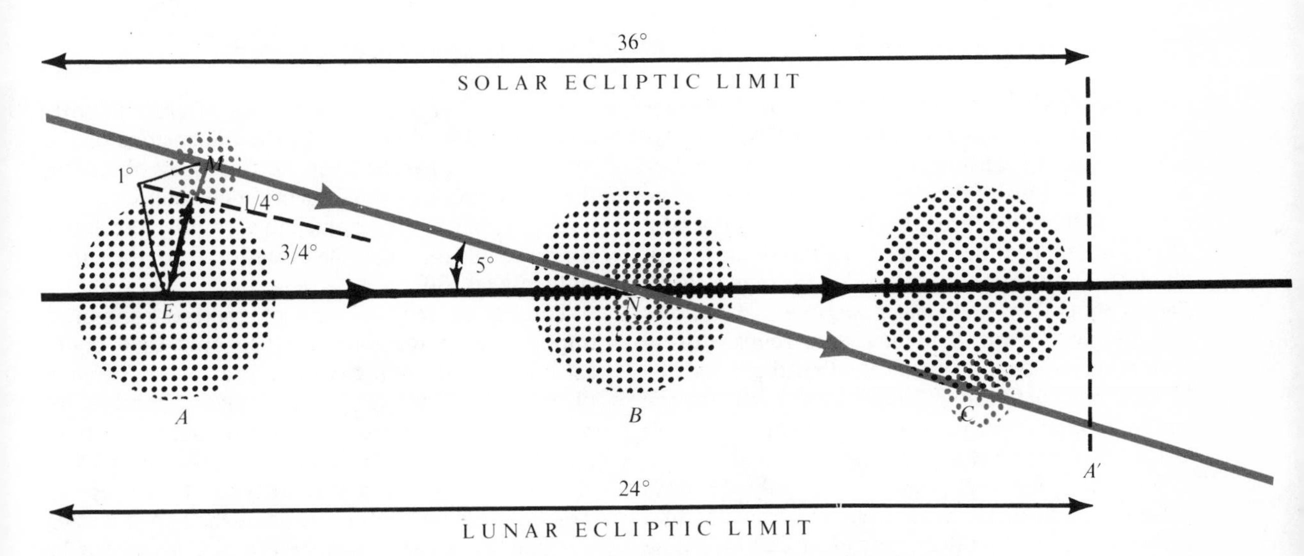

FIG. 7–5 *The ecliptic limit. A lunar eclipse occurs whenever the disk of the moon overtakes the disk of the earth's shadow in the region AA′. The same analysis applies to solar eclipses except that the solar ecliptic limit is about 1½ times larger. (See Figure 7–2.)*

We can find the numerical value for the lunar ecliptic limit since we know that the shadow disk of the earth is about three times the lunar disk, so that the arc *EM* from the center of the earth's shadow to the center of the lunar disk is just 1°. This means that the arc *EN* along the ecliptic is just 12°. *Hence, the lunar ecliptic limit is* 24°.

Since the earth's shadow disk travels across this ecliptic limit at a rate of 1° per day (the same as the apparent eastward motion of the sun) there is a 24-day period (12 days before the node passage, and 12 days after the node passage) during which a lunar eclipse can occur. *There are three such node passages during the year, so three lunar eclipses can occur during one calendar year.*

No more than three can occur because if one such eclipse occurs, let us say near the extreme position *A*, just when the moon's disk and the earth's shadow enter the ecliptic limit (this will be a very slight partial eclipse), there is not enough time during this node passage for another lunar eclipse to occur. The reason is that the shadow takes 24 days to pass over the ecliptic limit, whereas 29⅓ days (the lunar synodic period) must elapse before the moon's disk overtakes the shadow again. *Using the same reasoning, we can see that a node passage can occur without a lunar eclipse, so that no lunar eclipse need occur at all during the year.* Thus there may be a maximum of three lunar eclipses in one year, although none need occur.

7.7 Solar Ecliptic Limit

We can determine the frequency of solar eclipses in much the same way if we remember that the solar ecliptic limit is determined, not by the shadow disk of the earth, but by the diameter of the light cone in front of the earth through which the moon passes. This is about 1½ times larger than the shadow disk discussed in the last section, *so that the* ***solar ecliptic limit*** *is* 36° (18° *on each side of the node*). It follows that 36 days are available for a solar eclipse during each node passage. This means that there *must be at least one solar eclipse during each node passage and there may be as many as two.* Thus there are at least two solar eclipses during a sidereal year.

The third node passage occurs only 19 days before the end of the year, reckoned from the first node passage, so that there may be just enough time for either a solar eclipse or a lunar eclipse to occur at this third node passage. However, if one of them does occur, the other is definitely precluded.

There may be a maximum of seven eclipses, three of which may be lunar and four solar or two of which may be lunar and five solar. If two solar eclipses occur at the same node, a total lunar eclipse will always occur between them.

7.8 The Saros

Since eclipses were very important events to the ancients, Babylonian and Chaldean astronomers kept careful records of them, and discovered that similar eclipses recurred in cycles of 6 585.32 days (18 years 11⅓ days). This interval contains exactly 223 synodic months and almost exactly 19 eclipse years. Thus, if a total solar eclipse occurred today at noon, which means that both the sun and the moon were then at the same node (new moon), then after 18 years and 11⅓ days (the **saros interval**) the phase of the moon will again be new but the sun will not be exactly at the node because the saros period is about ½ day shorter than 19 eclipse years. The sun will therefore be about 28 minutes to the west of the node. Nevertheless, the sun and the moon will be close enough to the node for the eclipse to occur. However, the eclipse will not occur at our locality at noon, but at 8 hours longitude (120°) west of our meridian because of the third of a day in the saros period. About 71 solar eclipses of all types occur in a saros, and more than half of these are either total or annular.

7.9 The Importance of Eclipses *

Although lunar eclipses are not of great importance, they are of some use: the dimensions of the moon and its parallax can be found by observing the occultations of stars during such eclipses. It is also possible to determine how rapidly the lunar surface cools off when the radiation from the sun is suddenly cut off.

Total solar eclipses were of great astronomical importance in the past because only then could astronomers study the corona and the atmosphere of the sun. However, we no longer depend on an eclipse for this purpose since we now use the coronagraph, an instrument developed by Bernard Lyot. The coronagraph is a special type of telescope in which the bright central image of the sun is blotted out and the solar atmosphere and corona can be photographed.

However, one branch of physics in which the data obtained from the total solar eclipse are of great

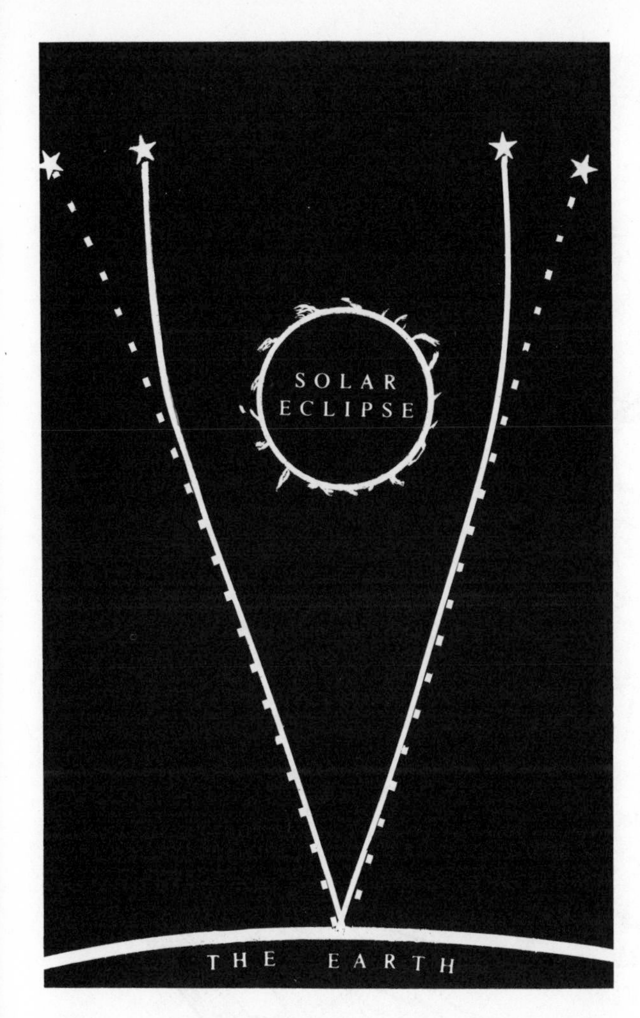

FIG. 7–6 *The bending of the rays by the gravitational field of the sun, shown during a total solar eclipse by the apparent displacement of the stars (dotted lines).*

importance is the general theory of relativity. One conclusion that follows from this theory is that light has gravitational mass and must therefore suffer a deflection when passing a massive body like the sun. In Figure 7.6 light from two stars on either side of the sun is shown reaching an observer on the earth. Since this light falls in toward the sun, according to this theory, the rays are bent so that to an observer on earth it appears that the two stars are separated by a greater angle than they really are. It is not possible to measure this effect under ordinary conditions since the star's light is lost in the intense glare of the sun.

When a total solar eclipse occurs, this effect is observable. One photographs the field of stars around the sun during a total solar eclipse, and then compares this photograph with a photograph of the same part of the sky taken with the same equipment six months later when the sun is no longer there. Measurements of this sort were first made in Africa during the total solar eclipse expedition of May 1919, under the direction of Sir Arthur Stanley Eddington. The results obtained were in amazingly good agreement with the value 1ʺ.75 of arc as predicted by Einstein. Since then many other measurements have been made during various eclipses. Although there are slight differences between theory and observations, it is now accepted that *light is bent in a gravitational field.*

The most recent analysis of all total solar-eclipse data from 1919 to 1952 was given by Mikhailov, of the U.S.S.R., in the 1959 George Darwin Lectures of the Royal Astronomical Society. From his analysis it is quite clear that light is deviated by the sun's gravitational field, but the numerical results are systematically larger than Einstein's value.

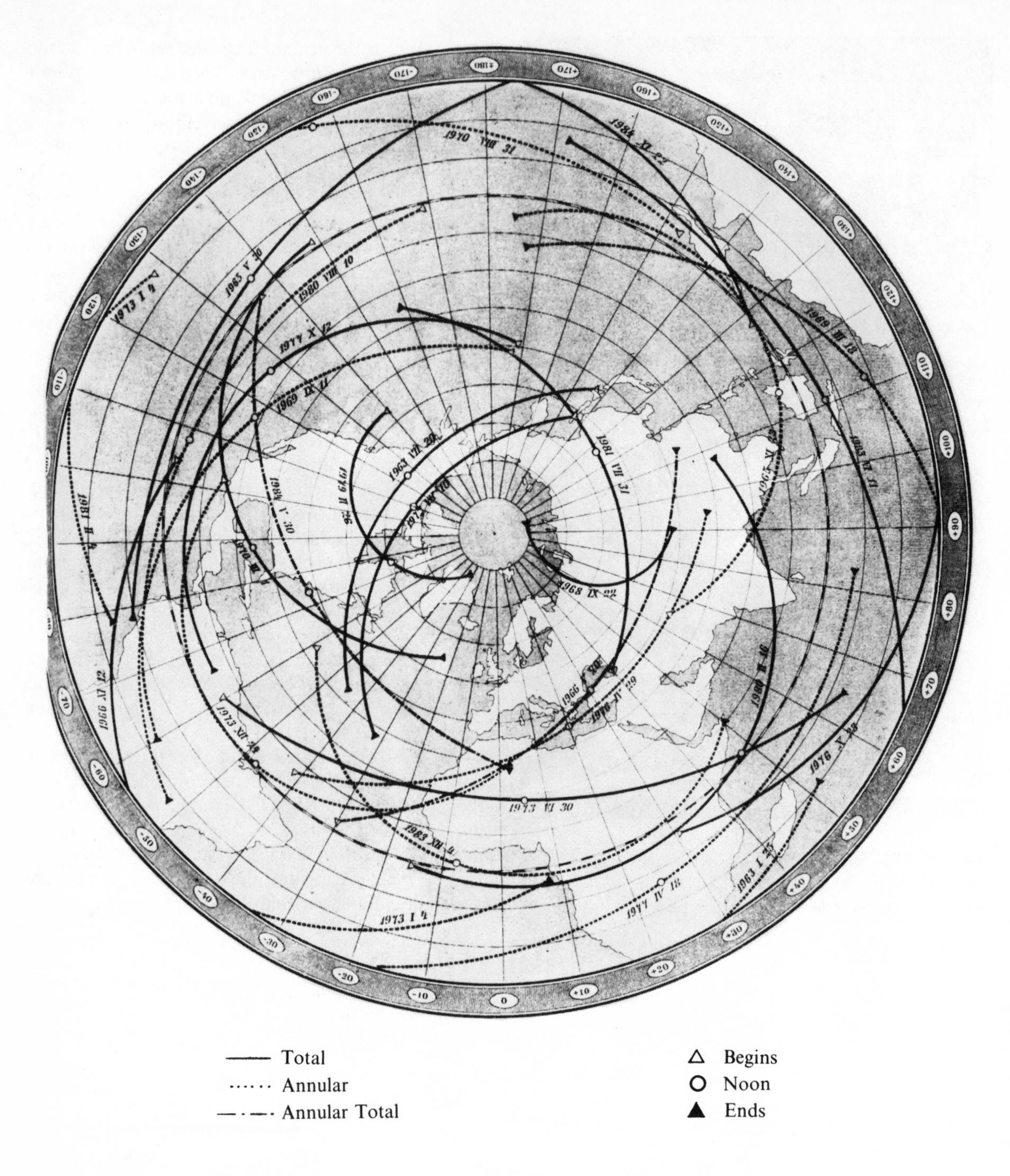

FIG. 7–7 *Diagram of tracks of all solar eclipses for the years 1963 to 1984* (Oppolozer). (*Yerkes Observatory photograph*)

Table 7.1

OCCURRENCE OF ECLIPSES

Solar Eclipses					Lunar Eclipses				
Year	Month	Day	G.M.T.		Year	Month	Day	G.M.T.	
			h	m				h	m
1963	1	25	13	42	1963	7	6	22	0
	7	20	20	43		12	30	11	7
1964	1	14	20	45	1964	6	25	1	7
	6	10	4	23		12	19	2	35
	7	9	11	31					
	12	4	1	19					
1965	5	30	21	14	1965	6	14	1	51
	6	23	4	11					
1966	5	20	9	43	1966	None			
	11	12	14	27					
1967	5	9	14	57	1967	4	24	12	7
	11	2	5	48		10	18	10	16
1968	3	28	22	48	1968	4	13	4	49
	9	22	11	9		10	6	11	41
1969	3	18	4	52	1969	None			
	9	11	19	56					
1970	3	7	17	43	1970	2	21	8	31
	8	31	22	3		8	17	3	25
1971	2	25	9	49	1971	2	10	7	42
	7	22	9	15		8	6	19	44
	8	20	22	54					
1972	1	16	10	53	1972	1	30	10	53
	7	10	19	40		7	26	7	18
1973	1	4	15	43	1973	12	10	1	48
	6	30	11	39					
	12	24	15	8					
1974	6	20	4	56	1974	6	4	22	14
	12	13	16	26		11	29	15	16
1975	5	11	7	6	1975	5	25	5	46
	11	3	13	5		11	18	22	24

EXERCISES

1. If the moon were twice as far away from the earth as it is now, what would the solar and lunar ecliptic limits be?

2. (a) How close to the earth would the moon have to be for a lunar eclipse to occur once a month? (b) For a solar eclipse to occur once a month?

3. (a) If the moon's orbit were tilted $2\frac{1}{2}°$ to the ecliptic, what would the solar and lunar ecliptic limits be? (b) How many solar and lunar eclipses could then occur?

4. (a) Is the moon totally dark during a total lunar eclipse? (b) Explain what you would expect the moon to look like during such an eclipse.

5. (a) If you were on the moon looking at the earth during a total solar eclipse, would the earth be in total eclipse? (b) What would it look like?

6. Explain how a solar eclipse enables one to analyze the structure of the solar atmosphere.

8

THE ORBITS OF THE PLANETS

8.1 The Apparent Motions of the Planets among the Stars

To the untrained observer all the objects in the sky, other than the sun and moon, appear pretty much the same. Only when one has looked at the sky over a period of weeks or months does he observe a few starlike objects that do not behave like the main body of stars. It is true that these few objects rise and set with the stars and look like the stars, but they exhibit definite differences in their behavior from night to night. Whereas the stars retain their same relative positions (the formations which we call the constellations) over very long periods, these objects—the planets—wander about among the stars in a rather complicated fashion. The early Greek astronomers were well aware of the difference between the behavior of these "wanderers" (the Greek word *planētēs* means "wandering") and the fixed stars.

Insofar as the planets change their positions relative to the fixed stars, they behave like the sun and the moon, but whereas the sun and the moon always appear to move eastwardly, each planet appears to move in both an eastwardly and westwardly direction at different times. The eastwardly motion is called **direct motion**, and the westwardly motion **retrograde motion**. In Figure 8.1 the direct and retrograde motions of the planet Venus are shown in relation to the ecliptic.

We shall see later that this apparent motion can be easily understood in terms of the actual revolution of the earth and the planet around the sun. When we look at a planet in the sky we really observe its relative motion with respect to the earth. This is a combination of the earth's motion and the planet's motion around the sun.

8.2 The Elongation of a Planet

We can best understand the behavior of the planets if we first introduce a few definitions. Since it is important to consider the position of the planet with respect to the sun, we define the **elongation** of the planet as *the angle formed by the line from the earth to the planet and the line from the earth to the sun.* We say the planet is:

in **conjunction** when the elongation is 0°,
in **quadrature** when the elongation is 90°, and
in **opposition** when the elongation is 180°.

8.3 Sidereal and Synodic Periods of Planets

Just as for the moon, we assign each planet a sidereal period and a synodic period. *The sidereal period of a planet is the time it takes the planet to make a complete revolution around the sun from a particular point on the celestial sphere back to that*

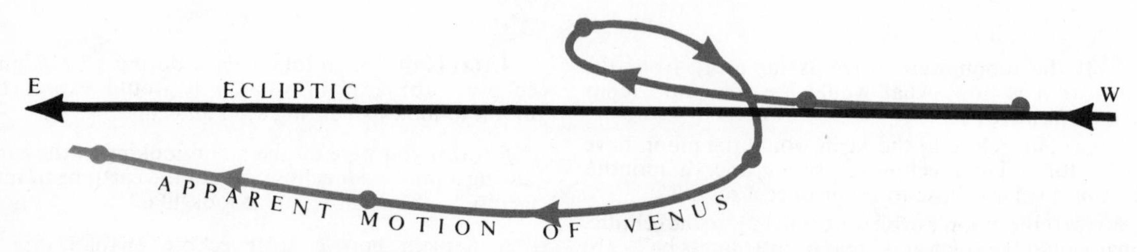

FIG. 8–1 *A schematic diagram of the apparent motion of the planet Venus.*

same point again as seen from the sun. The synodic period of a planet is the time it takes the planet to go from a given elongation back to the same elongation again. That is, it is the time it takes the planet to come back to the same relative position on the celestial sphere with respect to the sun as seen from the earth.

The algebraic relationship between the sidereal and synodic periods for a planet is the same as that for the moon. However, we must distinguish between the two planets, Venus and Mercury, which appear to move faster than the sun across the sky, and the remaining planets, which appear to move more slowly. The first two planets are referred to as the **inferior** planets, and for them the relationship is

$$\boxed{\frac{1}{S} = \frac{1}{P} - \frac{1}{E}}; \tag{8.1}$$

for the **superior** planets (that is, all the others) the relationship is

$$\boxed{\frac{1}{S} = \frac{1}{E} - \frac{1}{P}}, \tag{8.2}$$

where E is the sidereal period of the earth,
P is the sidereal period of the planet, and
S is the synodic period of the planet.

This relationship is important because the synodic period can be measured quite easily and with great accuracy, whereas the sidereal period cannot. We can, therefore, accurately determine the sidereal period by using this relationship.

8.4 The Apparent Motions of Mercury and Venus

Since the two inferior planets, Mercury and Venus, appear to behave quite differently from the superior planets, we shall consider their apparent motions separately. The apparent motions of these two planets are such that they always stay close to the sun in the sky; their maximum elongations are always less than 90°. *Thus, the greatest elongation of Mercury never exceeds* 28°, *and that of Venus never exceeds* 48°, as these planets oscillate from one side of the sun to the other. When Venus is to the east of the sun, as shown in position W of Figure 8.2, it is called an "evening star," and when it is to the west of the sun (position E) it is called a "morning star." At its maximum eastward elongation it can be seen as a bright object in the sky in the late afternoon or evening, and at its maximum westwardly elongation it is a bright object before sunrise.

FIG. 8–2 *Venus as a "morning star" and an "evening star."*

We can easily understand the apparent motion of these two planets if we consider the orbit of one of them—say, Venus—together with that of the earth. In Figure 8.3, the orbits of Venus, the earth, and a superior planet are shown as circles around the sun. If we consider Venus as seen from the earth (which we may assume to be fixed), we note that no matter where Venus may be in its orbit, its greatest elongation either to the east or to the west of the sun can never exceed the angle formed by the line from the earth to the sun, and the tangent lines from the earth to the orbit of Venus. If the direction of revolution of the planets around the sun is as shown by the

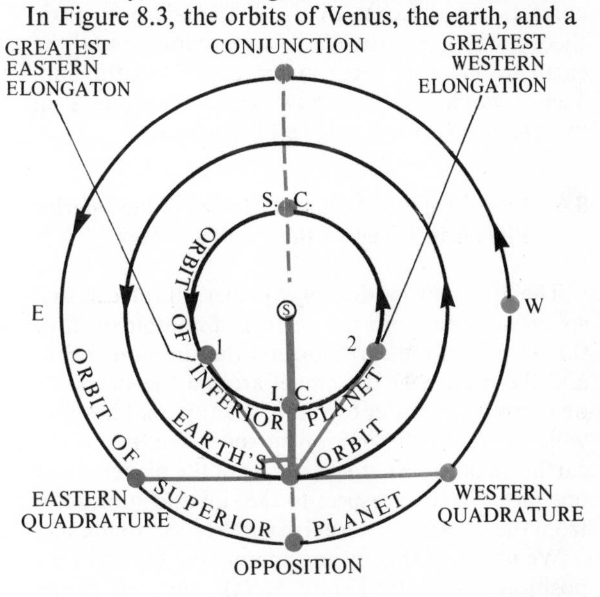

FIG. 8–3 *The elongations of an inferior and superior planet for various positions in their orbits (as seen from the earth).*

arrows, then position 1 represents the greatest eastward elongation, and position 2 the greatest westward elongation. We can see from this figure why it is impossible for Venus or Mercury ever to be in quadrature or opposition, and why the elongations of the superior planets can take on all values from 0° to 180°.

8.5 Phases of the Planets

It is interesting to note that when seen from the earth both Mercury and Venus (although this is difficult to observe in the case of Mercury) show the same phases as the moon does, since these planets, as they move around the sun, reflect the sunlight to the earth in much the same way as the moon. This important observation was first made by Galileo when he turned his home-built telescope on Venus. It convinced him that the Copernican theory is correct. In the case of the Ptolemaic geocentric theory, which pictures Venus as moving around the earth in an orbit between the earth and the sun, Venus would always show a crescent phase in disagreement with Galileo's observations.

8.6 Explaining the Apparent Motion of the Inferior Planets in Terms of the Earth's Motion

The apparent motion of a planet on the celestial sphere as seen from the earth is a combination of the motion of the earth around the sun in its orbit, and the motion of the planet around the sun in its own orbit. In order to obtain the apparent motion, which is really the motion in space relative to the earth, we take the actual velocity of the planet at any one moment with respect to the sun and subtract it from the velocity of the earth with respect to the sun.

We illustrate this by considering the earth, in the position shown in Figure 8.4(a), and the planet Venus in positions 1, 2, 3, 4. The four arrows (all taken to be equal) shown in these positions give the velocities of Venus in the four positions. The arrow attached to the earth in its position shows its velocity. All of these velocities are relative to the sun. Note that since Venus is moving faster around the sun than the earth is (we shall discuss this point in more detail when we consider Kepler's laws of planetary motion), the arrows assigned to Venus are longer than those assigned to the earth. To obtain the velocity of Venus relative to the earth in the four positions, we must subtract the earth's velocity vectorially from that of Venus. This is shown in Figure 8.4(b), where the relative velocity

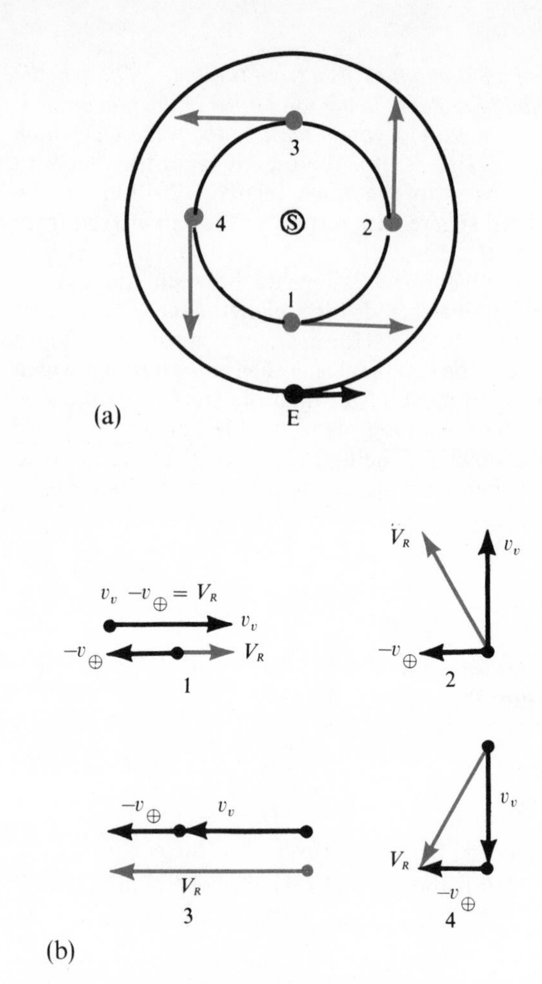

FIG. 8–4 *(a) The actual velocities of the planet Venus relative to the sun for various elongations. The velocities are not to scale. (b) The velocities relative to the earth (shown as red arrows) for the same elongations.*

is obtained as the resultant of Venus' velocity and the negative of the earth's velocity. We see from these constructions that in position 1, Venus, as seen on the celestial sphere from the earth, appears to be in retrograde motion, whereas in positions 2, 3, and 4 it appears to be in direct motion.

8.7 The Apparent Motions of the Superior Planets in Terms of the Earth's Motion

We can use the same sort of analysis to determine the relative motions of the superior planets. The projections of these relative motions on the celestial sphere give the planets their apparent epicyclical motions. We may describe this somewhat as follows: the apparent motion of the planet as seen from the earth (that is its geocentric motion) is the same as the true motion of a planet pictured as moving around the earth once a year in an orbit equal to the earth's orbit around the sun, while the earth itself is moving around the sun exactly as the planet actually moves around the sun. In Figure 8.5 we see that the apparent orbit of the planet is the projection (from the earth) on the celestial sphere of the various positions of the planet in its orbit around the sun.

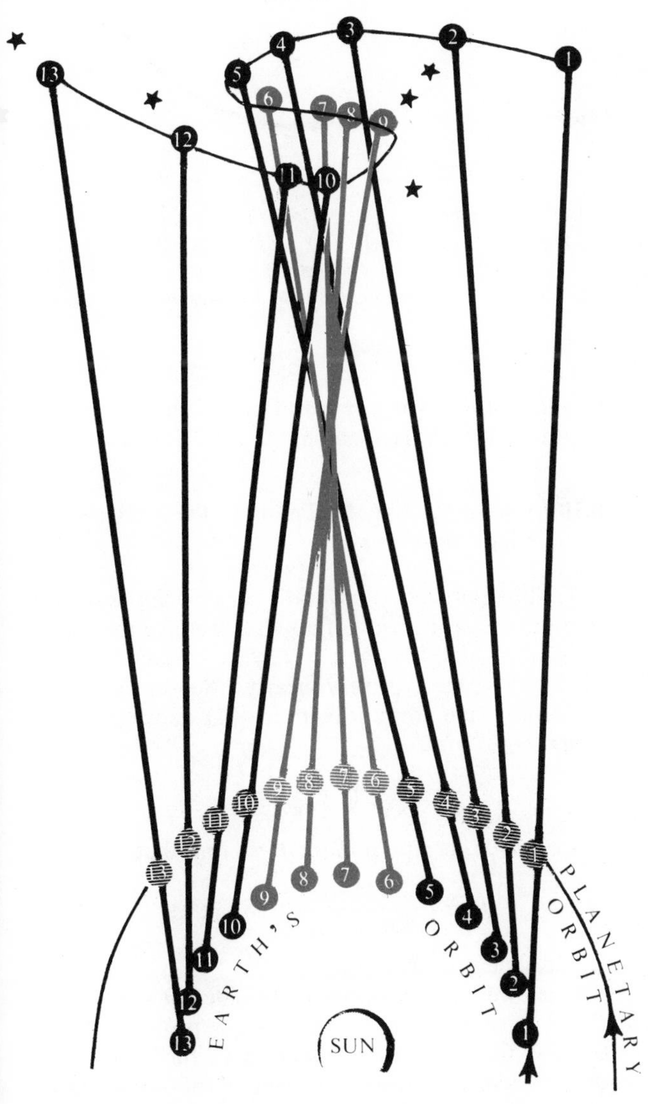

FIG. 8–5 *The apparent motion on the sky of a superior planet (e.g., Jupiter) as seen from the earth. The red lines indicate the apparent retrograde motion of the planet.*

8.8 The Ptolemaic and the Copernican Theory

In the Ptolemaic theory the apparent motion of the planet on the celestial sphere was taken as the planet's real motion. Copernicus' great contribution lay in pointing out that one can explain this observed motion by means of a **heliocentric theory** in which the planets revolve around the sun. Although Copernicus introduced the heliocentric theory of the solar system, he still found it necessary to introduce about 40 epicycles to account for the observations, but he considered this a great improvement since the Ptolemaic theory contained more than 240 such epicycles. The difficulty with the Copernican theory is that it pictures the planets as moving in circular orbits. Not until Kepler had analyzed the vast amount of observational data collected by Tycho Brahe was a true picture of the orbits of the planets obtained.

Since Kepler had no theory to guide him, he was forced to compute the orbits of the planets from Tycho Brahe's data by numerical and geometrical methods.

8.9 Kepler's Analysis of the Motions of Planets

We may briefly illustrate this procedure by considering the orbiting earth together with the planet whose orbit is being computed. To determine the planet's orbit, one must find its distances from the sun for many different orbital positions. Kepler approached the problem by noting first that if two different elongations of the planet are known at times when it is in the same position in its orbit, then its distance from the earth can be found in astronomical units (that is, in terms of the earth's distance from the sun) as shown in Figure 8.6(a). He then observed that if this distance is known for some position P of the planet, then one can compute the distance in A.U. of the planet from the sun in this position. For if, when the planet is in position P, EP is its distance from the earth, and ES is the astronomical unit, then SP, the planet's distance from the sun, is given by the cosine law (see Equation 3.1):

$$(SP)^2 = (SE)^2 + (PE)^2 - 2(SE)(EP)\cos\theta$$

or

$$\left(\frac{SP}{SE}\right)^2 = 1 + \left(\frac{PE}{SE}\right)^2 - 2\left(\frac{EP}{SE}\right)\cos\theta, \qquad (8.3)$$

where θ is the planet's elongation. The crux of this whole procedure is to find EP/SE first, and this can

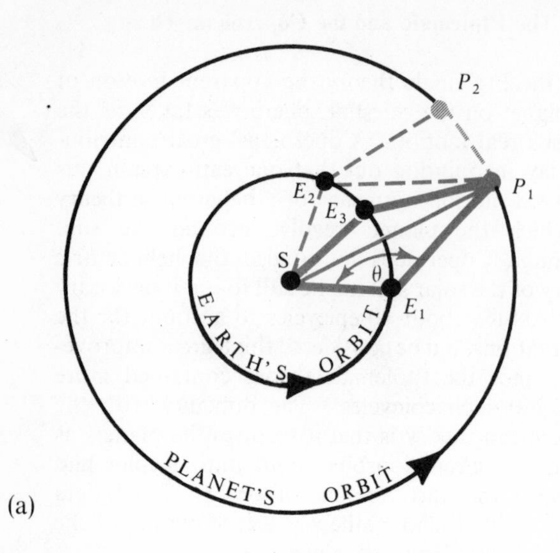

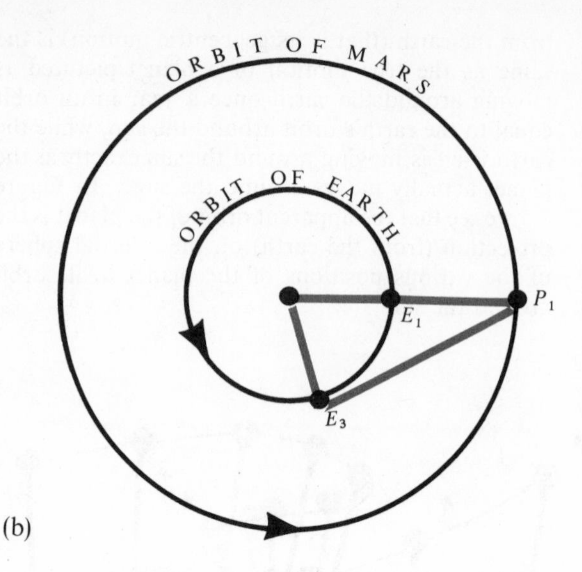

FIG. 8-6 *(a) The distance of a planet like Mars can be found by sighting it from two different points of the earth's orbit, E_1 and E_3, when the planet has returned to its position P_1 after 1 sidereal period. Note that the distance cannot be measured by sightings from E_1 and E_2 since the planet has moved to position P_2 by the time the earth has gone to position E_2. (b) The relative positions of the earth and Mars after 1 sidereal period of Mars.*

be done only if the angular change in the planet's apparent position can be found when the earth is at some orbital position other than E while the planet is again at the point P.

This is not as easy as it sounds, because both the earth and the planet are in motion. In Figure 8.6(a) let us again consider the earth and the planet at some moment and let the line E_1P_1 connect the two. If the planet did not move, it could be sighted from position 2 on the earth's orbit, and its distance could then be determined since the angle $E_1P_1E_2$ could be measured. However, while the earth moves from point 1 to point 2 in its orbit, the planet also moves from P_1 to P_2 in its own orbit, so that sighting it from position 2 of the earth in its orbit does not give us the angle that we seek.

But we can obtain such an angle if we keep in mind that after one sidereal period the planet will again be back at point P_1 in its orbit. At that time the earth will be at some point E_3 in its own orbit, so that when we sight the planet at that moment, we will have the angle $E_3P_1E_1$ that we want, as shown in Figure 8.6(b). We must, of course, know the true sidereal period of the planet, since without it we cannot know when the planet has returned to its original point P_1, nor will we know where E_3 is relative to E_1. However, we can easily find the sidereal period of the planet from its known synodic period with the aid of one of the algebraic relationships we have already written down.

8.10 Use of the Synodic Period in the Analysis of Planetary Orbits

To illustrate how the synodic period enables us to find the sidereal period of a planet we take the case of Mars, which is known to have a synodic period of 2.135 years (or very nearly 780 days). We therefore have from Equation (8.2) (since $E = 1$ year)

$$\frac{1}{1} - \frac{1}{P} = \frac{1}{2.135}.$$

If we solve this equation for P we find that

$$P = \frac{2.135}{2.135 - 1} = 1.881 \text{ years}$$

(or very nearly 687 days).

Once we have determined the sidereal period of the planet we then know that its true position in its orbit is the point of intersection of the two lines drawn from the positions E_1 and E_3 of the earth to the planet. In this way one point of the orbit of the planet can be determined, and if enough observations of this sort are available, many such points can be found.

The actual geometrical construction is shown in Figure 8.6(b), where the positions E_1 and E_3 of the earth in its orbit are given at the beginning and the end of one sidereal period of, say, the planet Mars.

Using this type of analysis Kepler found it impossible to fit circular orbits to Tycho Brahe's data. By working directly with Mars itself he was finally forced to discard the idea of circular orbits. He remarked that the error in Brahe's data for Mars was certainly smaller than the discrepancy that a circular orbit led to, for, as he went on to say, "The divine goodness has given to us in Tycho Brahe a very accurate observer."

Once we know the distance of a planet from the earth in miles or kilometers (which we can obtain by means of the solar parallax, see Section 14.1), we can easily find the planet's diameter, surface area, and volume, using the same procedure as we used for the moon (see Section 6.11). This is possible because we can measure the planet's apparent diameter by photographing its disk with a telescope. We shall discuss this material in later chapters devoted to the detailed characteristics of planets.

8.11 Kepler's Laws of Planetary Motion

When Kepler inherited the Tycho Brahe observational material in 1601, he was unable immediately to derive from it his famous laws of planetary motion. Only after years of the most careful and painstaking arithmetical computations could he see "the first glimmer of light" as to the true state of affairs. Being a man of great intellectual daring, he introduced many fanciful ideas in an attempt to account for the orbits of planets before his computations led him to his three laws of planetary motion. We shall first state the three laws, and then discuss them in detail.

THE FIRST LAW:

The orbit of each planet is an ellipse with the sun at one of its foci.

THE SECOND LAW (THE LAW OF AREAS):

The radius vector (the line from the sun to the planet) to a planet sweeps out equal areas in equal intervals of time as the planet revolves in its orbit. See Figure 8.7.

THE THIRD LAW (THE HARMONIC LAW):

The squares of the sidereal periods of the planets are proportional to the cubes of their mean distances from the sun.

Or, expressed differently,

The ratio of the square of the sidereal period of a planet to the cube of its mean distance to the sun is the same for all planets.

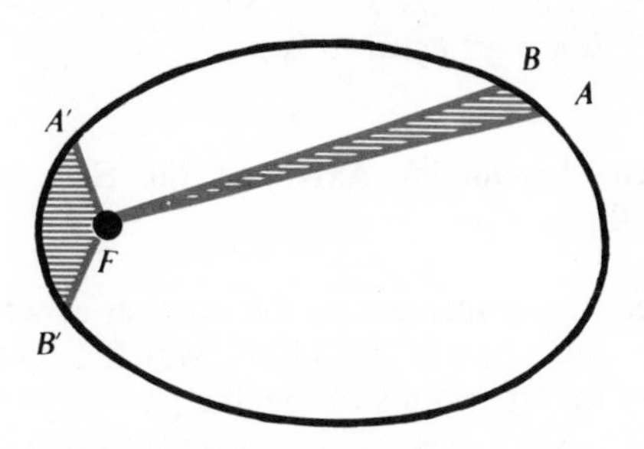

FIG. 8–7 *Schematic representation of the Law of Areas. If the planet takes the same time in going from A to B as in going from A′ to B′, the area of the ellipse in the sector AFB equals the area in the sector A′FB′.*

8.12 Defining an Ellipse

In order to discuss these laws in detail, we first consider briefly the properties of the ellipse and conic sections in general.

In Figure 8.8, let us consider two points F and F', and a third point P not on the line connecting F and F'. If we now add the distances FP and PF', we obtain a length which we shall designate as $2a$. The reason for this designation will become clear later. These three points and the magnitude $2a$

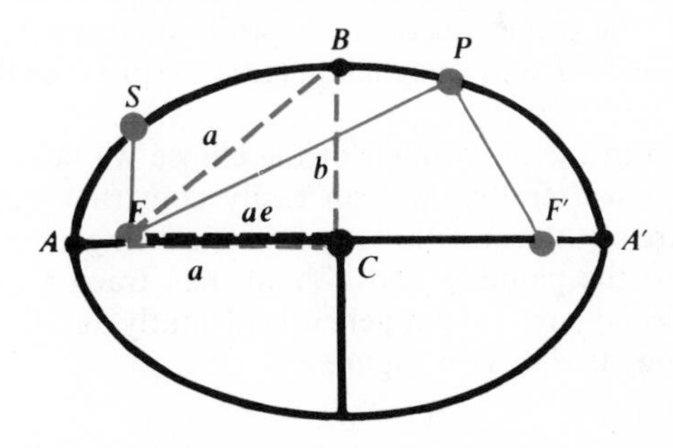

FIG. 8–8 *A schematic diagram of the ellipse showing the relationship of its various geometric properties to its semimajor axis and to its eccentricity.* $FP + PF' = 2a = AA'$; $b = a\sqrt{1 - e^2}$; $e = \dfrac{FC}{a}$.

determine an ellipse in the following way. Consider a fourth point P' and again take the sum of the two distances FP' and $P'F'$. We say that this point P' lies on the same ellipse as the point P if this last sum also equals $2a$. In this way we can test any point whatsoever, as to whether or not it lies on the given ellipse. We now say that *the ellipse is the class of all points such as P the sum of whose distances from F and F' is always equal to 2a.*

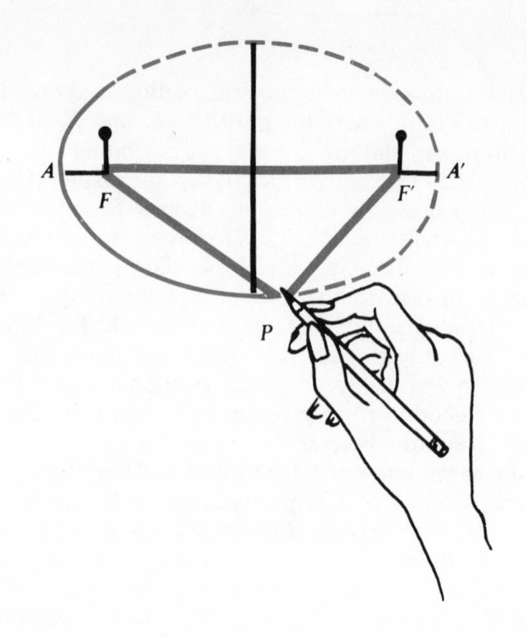

FIG. 8–9 *An ellipse can be constructed with a pencil, paper, and a piece of string.*

8.13 The Semimajor Axis and the Size of the Ellipse

To see the significance of the number a, we consider the two points A and A', which lie on the ellipse at the end points of the line passing through F and F'. Since these points are on the same ellipse as the point P, it follows from the definition of the ellipse that

$$FA + AF' = 2a,$$

and also

$$FA' + A'F' = 2a.$$

If we add these two equations we obtain

$$(FA + FA') + (AF' + A'F') = 4a.$$

But the distance in each parenthesis is just the total distance AA', so that it follows from this *that a itself is one-half the greatest diameter of the ellipse.* We shall therefore call a the **semimajor axis** of the ellipse. *The two points F and F' are called the foci of the ellipse and according to Kepler's first law the sun occupies one of these points.* If we place the sun at F, the planet's perihelion (position closest to the sun) is at A, and its aphelion (greatest distance from the sun) is at A'.

From the description of the ellipse we have just given we see that we can easily construct such a figure by attaching the two ends of a string of length $2a$ to the points F and F' and then tracing out a closed figure using a pencil held tautly against the string, as shown in Figure 8.9.

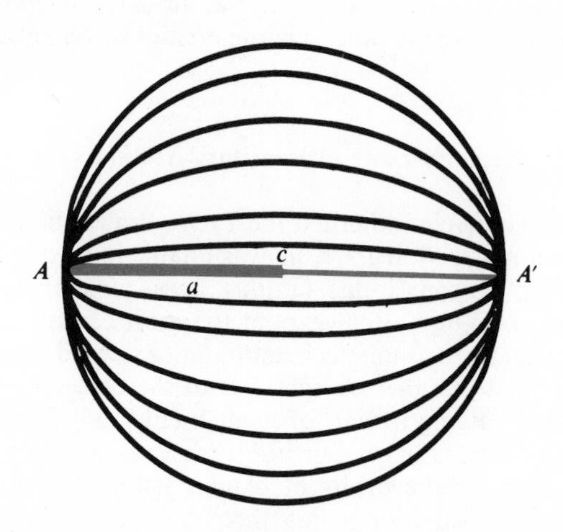

FIG. 8–10 *Associated with an ellipse of a given size (a given value of the semimajor axis, a) there are many different shapes ranging from the very flattest, $e \rightarrow 1$, to the very roundest, $e = 0$, a circle.*

8.14 The Eccentricity and the Shape of the Ellipse

The number a gives the size of the ellipse. But it is clear that for a given size a (semimajor axis) an infinite number of ellipses can be drawn, differing in shape all the way from the roundest figure, a circle, to the very thinnest figure, which is practically a straight line, as shown in Figure 8.10. In the case of the very thin ellipses the two foci move farther and farther apart, approaching the points A and A' (whereas for the very round ellipses the two foci move closer together and approach the center). To differentiate between two ellipses of the same size, we must therefore introduce some method of measuring the shape. *This is done by the number e, called the eccentricity of the ellipse (the word eccentricity signifies how far off center the two foci are; hence it is a measure of how flat the ellipse is—that is, how much it departs from circularity or roundness).*

We can see how we are to introduce this number if we note that the rounder the ellipse is, the smaller the distance FC is compared with AC (that is, the

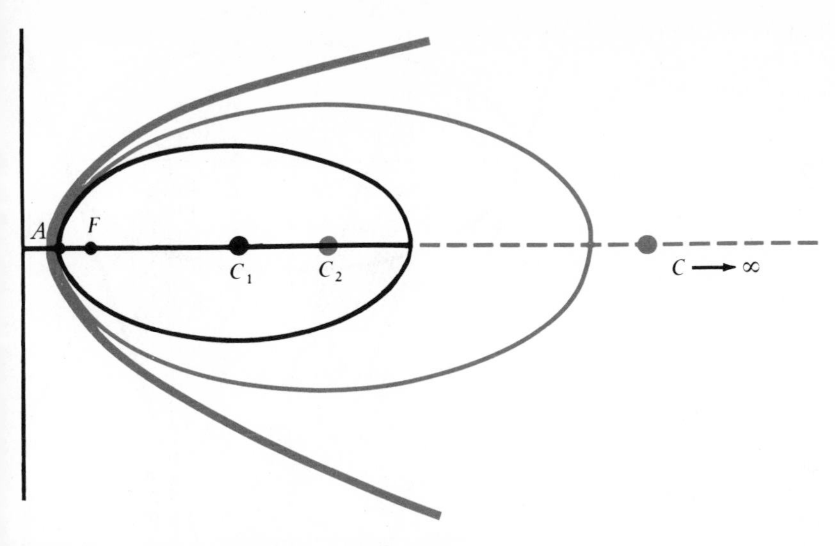

FIG. 8–11 *If the center of an ellipse is allowed to move further away from the focus, F, while the latter remains stationary at a fixed distance from A, the ellipse gets larger and larger and ultimately approaches a parabola.*

semimajor axis). We shall therefore define the eccentricity as

$$\boxed{e = \frac{FC}{a}}. \tag{8.4}$$

It is clear from this that the smaller e is, the rounder the ellipse is, and that the ellipse reduces to a circle when $e = 0$. We therefore have an ellipse only when e is bigger than 0.

8.15 The Relation of the Eccentricity to the Parabola and the Hyperbola

As e becomes larger and larger and approaches 1, the ellipse flattens out. We see what takes place as e gets closer to 1 if we consider the whole series of ellipses, all having the same focus F, and passing through the same point A, but with increasing semimajor axes. Of course, they all have different points A' and different centers. This is shown in Figure 8.11. Since the eccentricity in each case is AC/FC, this quantity gets closer and closer to 1 as the ellipses get bigger and bigger, because AC and FC both approach the same large number. Finally, if we consider the limiting case in which C moves off to infinity, AC becomes equal to FC so that the eccentricity equals 1. When this happens, the ellipse is no longer a closed figure, but is one in which the upper and lower branches meet at infinity, since these branches become more and more parallel to the axis. We call such an open figure a **parabola,** and we see from what we have just said that *for a parabola the eccentricity is* 1.

What happens when the eccentricity is greater than 1, and what kind of figure do we get in that case? We note that for the eccentricity to be larger than 1, A must lie between the points C and F. But, on the other hand, C must lie midway between F and F'. We thus see that these two things are possible if we have a figure with two branches as shown in Figure 8.12. For this figure, which is

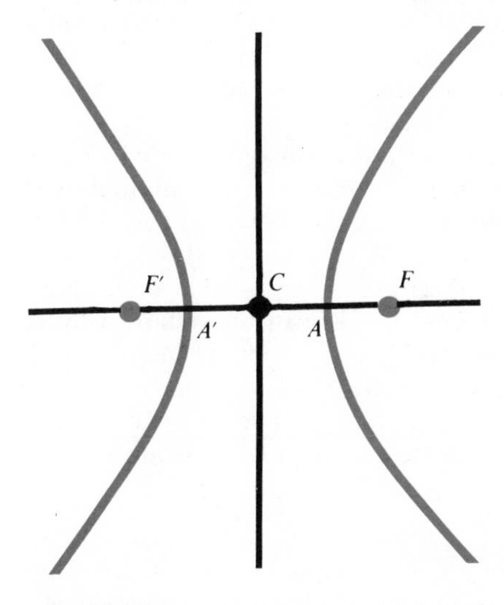

FIG. 8–12 *For the hyperbola the eccentricity is larger than 1. This means that AC must be smaller than AF and hence that A must lie between C and F.*

called a **hyperbola**, *the eccentricity is larger than* 1 since FC is larger than AC.

We may summarize these results as follows:

if $e = 0$, the conic is a circle,
if $0 < e < 1$, the conic is an ellipse,
if $e = 1$, the conic is a parabola,
if $e > 1$, the conic is a hyperbola.

8.16 Conic Sections

Since it is possible to pass continuously from a circle to a hyperbola by changing the eccentricity, it is clear that all of the figures we have been discussing must constitute a single family of curves. This is, indeed, the case, and this family is known as the **conic sections**. To illustrate conic sections, we consider a cone generated by spinning two intersecting lines about a line bisecting the angle of intersection as shown in Figure 8.13. In this way we obtain a right circular cone, with two domains. Each of the figures we have discussed above can be obtained by taking a section through this cone.

To obtain a circle we must cut the cone exactly at right angles to the axis of rotation. Any deviation from a right angle gives rise to an ellipse, as long as the section is not parallel to an element of the cone (that is, any line lying on the surface of the cone and passing through the vertex). *If the section is parallel to an element of the cone, we obtain a parabola*, which, as we can see, has branches which never meet, because the section is parallel to the element. If we now take a section at a still greater angle, we obtain two branches, since *such a section cuts both domains, top and bottom, of the cone: this figure is the hyperbola.*

We may note that for only one particular angle do we obtain a circle, and for only one particular angle do we get a parabola. It follows from this that if we take a random cut through a cone, the chance of getting either a circle or a parabola is zero. From this we can understand why there is very little likelihood of finding a circular orbit for a planet, for as we shall see, the conditions that determine a circular orbit are just as critical as those that determine a circular conic section.

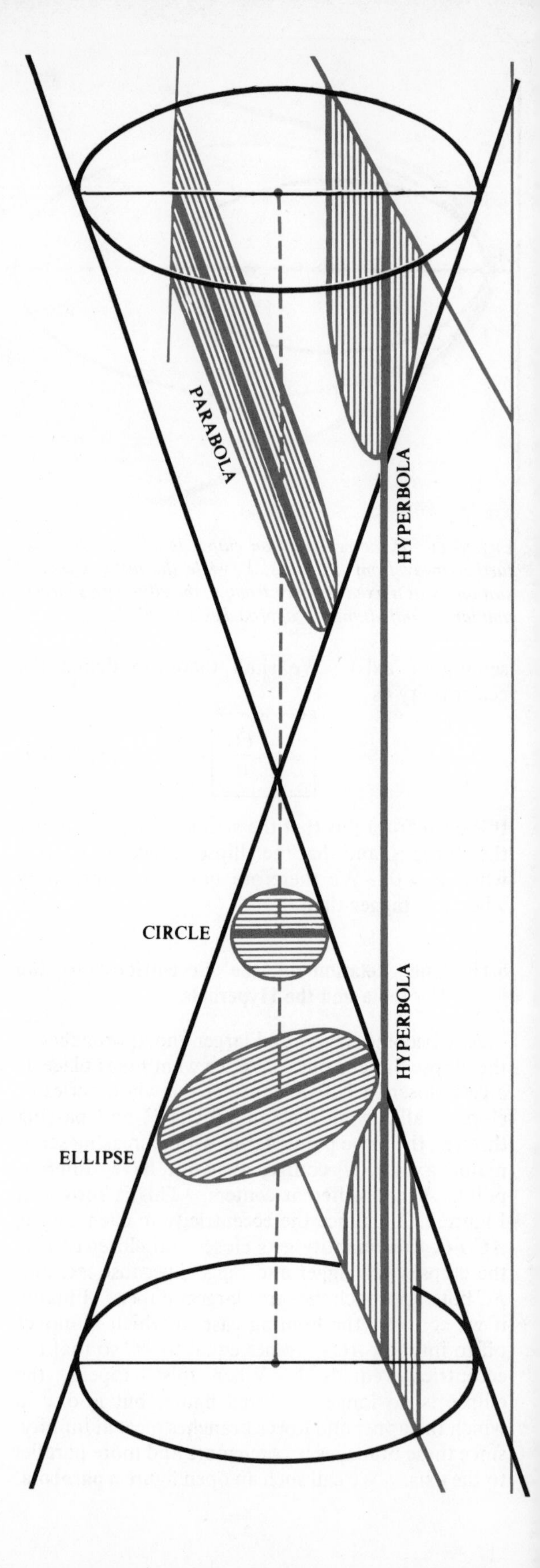

FIG. 8–13 *The circle, the ellipse, the parabola, and the hyperbola are shown as conic sections. To obtain a circle the section must be exactly perpendicular to the axis (dotted line) of the cone. To obtain a parabola the section must be exactly parallel to the edge of the cone. Any deviation from these two sections gives either an ellipse or a hyperbola.*

8.17 Additional Properties of the Ellipse

If we know the semimajor axis, a, of the ellipse, and its eccentricity, e, we can determine all its other geometrical properties. Thus, we can express the semiminor axis b (half the smallest diameter of the ellipse—that is, the line CB, see Figure 8.8) in terms of e and a. To do this, all we need note is

1. that the distance FB is just equal to a, since c lies midway between F and F', and
2. that this distance is the hypotenuse of the right triangle FCB.

But since FC is equal to ea (from the definition of the eccentricity), we can use the Pythagorean theorem to express b in terms of e and ea. We find from this that

$$b^2 = a^2 - a^2e^2$$

or,

$$b = a\sqrt{1 - e^2}. \tag{8.5}$$

We can also find the length p of the semiparameter of the ellipse in terms of e and a. Since the semiparameter is half the chord through the focus F perpendicular to the major axis of the ellipse, that is the line FS in the diagram, we can again use Pythagoras' theorem and the relationships $FS + SF' = 2a$ and $FF' = 2ae$ to show that

$$p = a(1 - e^2). \tag{8.6}$$

We can now write down an expression for the area of the ellipse in terms of a and e from the standard formula for the area, πab (obtained from analytical geometry which shows that the area of the ellipse is the geometric mean of the areas of a circle of radius a and a circle of radius b). If we express b in terms of a and e, we then find that the area is

$$A = \pi a^2\sqrt{1 - e^2}. \tag{8.7}$$

8.18 Perihelion and Aphelion Distance

Finally, both the perihelion and aphelion distances can also be expressed in terms of a and e just by making use of the definition of e. We have, in fact,

$$\text{perihelion distance} = AF = a - ae = a(1 - e) = \frac{a(1 - e^2)}{1 + e} = \frac{p}{1 + e}; \tag{8.8}$$

$$\text{aphelion distance} = FA' = a + ae = a(1 + e) = \frac{a(1 + e^2)}{1 - e} = \frac{p}{1 - e}. \tag{8.9}$$

8.19 The Mean Distance from Planet to Sun

If a planet moves around the sun in an ellipse, its distance from the sun varies continuously. If we try to get its mean distance by considering its distances at many different points in its orbit and then averaging these, we encounter difficulty. But there is, in fact, a very simple way of finding this average. We first take any pair of points which are situated in the orbit symmetrically on each side of the minor axis, and we find the average distance of these two points. The average distance for the entire orbit can then be found by averaging all such pairs of points. Thus if P and P' are two points of such a pair, as shown in Figure 8.14, the mean

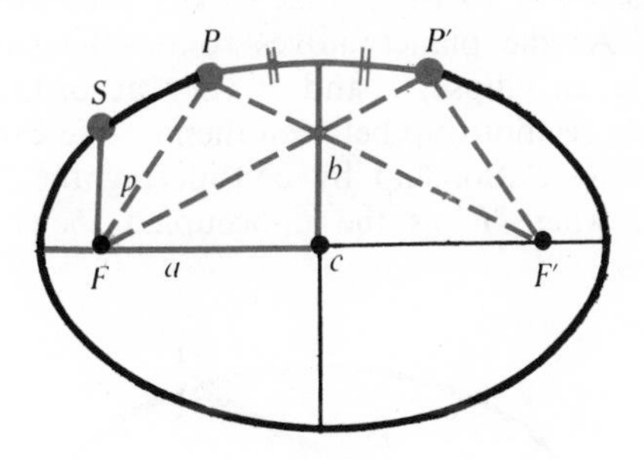

FIG. 8–14 *The mean distance from the focus, F, of the two points P and P′ situated symmetrically with respect to the semiminor axis, b, is a. This follows since PF = P′F′.*

distance for these two positions is $(FP + FP')/2$. But FP' is equal to PF', since P and P' are symmetrically situated with respect to the two foci. Hence the mean distance is given by $(FP + P'F)/2$. We see that this expression is just equal to a, since the numerator is equal to $2a$. From this it follows that the mean distance of a planet to the sun is equal to a (the semimajor axis) since all such pairs of points give this same average value.

8.20 The Equation of the Ellipse in Polar Coordinates

We must have one additional property of the ellipse in order to follow the motion of a planet. In Chapter 1 we discussed the general properties of coordinate systems, and we saw there that these are very useful devices for studying the motions of bodies. Since in the case of planetary motion our concern is with precisely this kind of problem, we must introduce a coordinate system. The most convenient system in this case is one with its origin at the sun since we are interested in the orbits of planets relative to the sun. However, because the sun is not at the center of any one of the elliptical orbits but at the focus, the algebraic expression that gives the orbit of the planet in this coordinate system takes on a simple form only when polar coordinates are used. We may note, by the way, that we need only consider a two-dimensional coordinate system, since, as we shall see later, the orbits of two bodies (relative to their center of mass) pulling upon each other gravitationally always lie in the same plane. (We assume that no other bodies that may affect these two gravitationally are present.)

In Figure 8.15 the position of the planet is given in terms of the distance r and the angle θ which the radius vector to the planet makes with the major axis. As the planet moves round the sun, describing an ellipse, r and θ vary according to a definite relationship between them. We can easily find this relationship by considering the triangle FPF', where F' is the unoccupied focus of the ellipse.

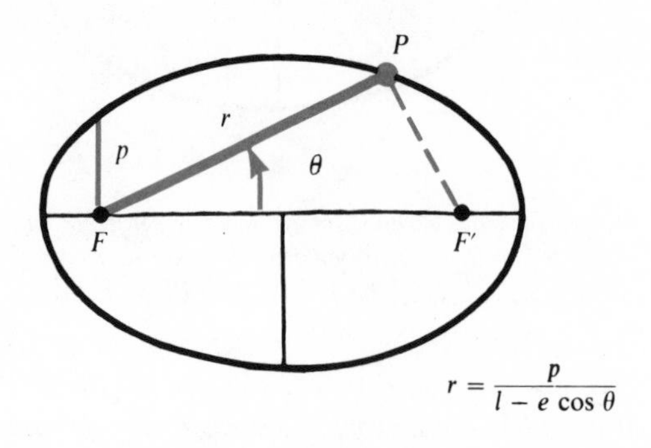

FIG. 8–15 *The polar coordinates of a planet, P, in an elliptical orbit.*

In the triangle we can express the distance PF', which (from the definition of an ellipse) is equal to $2a - r$, in terms of the two lengths r and FF' (which is equal to twice ea) and the cosine of θ by means of the formula (3.1) on p. 36. This leads to the equation

$$(F'P)^2 = (FP)^2 + (FF')^2 - 2(FP)(FF')\cos\theta$$

or

$$(2a - r)^2 = r^2 + 4e^2a^2 - 4rea\cos\theta,$$

$$r = \frac{a(1 - e^2)}{1 + e\cos(\pi - \theta)}$$

$$= \frac{p}{1 - e\cos\theta}. \qquad (8.10)$$

This is more than just the equation of an ellipse. It is, in fact, the polar-coordinate equation of any conic section, obtained by allowing e to take on the appropriate value for the given conic section.

Thus, for $e = 0$, we get the equation of a circle; for $e = 1$, the equation of a parabola; and for $e > 1$, the equation of a hyperbola.

8.21 Newton's Law of Gravity, and Kepler's First Law

When Newton tackled the problem of planetary motion using his law of gravity, he obtained precisely this equation for the motion of a planet around the sun. This means that *Newton's law of gravitation gives a much more general first law of motion than does Kepler's law, since it shows that any conic section is a permissible orbit for a celestial body.* Of course, since circles and ellipses are closed figures, planets can move only in such orbits. However, as we have seen, the circular orbit is excluded because of its zero probability.

Objects moving in either parabolas or hyperbolas (and we may exclude the former because of their zero probability) cannot remain attached to the sun but are momentarily part of the solar system. When we were discussing the fact that the moon has no atmosphere, we introduced the idea of the speed of escape of one body from another. We can now relate this concept to the various conic sections. We shall see that *if an object at any point in the solar system is moving with a speed less than the speed of escape from the sun at that point, it is moving in an elliptical orbit; if it is moving with just the speed of escape at that point, it is moving in a parabolic orbit; and if it is moving with a speed greater than the speed of escape at that point, it is moving in a hyperbolic orbit.*

8.22 The Law of Areas and Conservation of Angular Momentum

In Chapter 5 we introduced the concept of the angular momentum of a body with respect to some point of reference, and on p. 76 we illustrated this for a body moving in a straight line with constant speed. We saw there that as the body moves, a line drawn to it from the point of reference sweeps out triangles of equal areas in equal times—which means that the angular momentum of the body moving in a straight line with uniform speed is a constant. This is obviously true for such a body, but it is not so obvious for a planet. Nevertheless, **Kepler's** second law indicates that the angular momentum of a planet is conserved just as it is for a body moving with constant speed in a straight line. In other words, *Kepler's second law must be interpreted as a statement of the conservation of angular momentum for a planet moving around the sun.*

That this conservation principle applies to a planet is evident from the fact *that there is no torque acting on the planet-sun system, since the gravitational force passes through the sun*, and we saw in the discussion of angular momentum in Chapter 5 that only when a torque is exerted on a system can the angular momentum change. For a torque to act on the orbit of a planet, the force acting on the planet must not pass through the sun.

8.23 The Angular Momentum of a Planet and Its Areal Velocity

We shall now show that Kepler's second law is precisely equivalent to the principle of the conservation of angular momentum for a planet. Figure 8.16 shows a part of the planet's orbit around the sun; the arrow at A gives the velocity of the planet at that point. If we consider the motion of the planet from A to B (which is taken very close to A), we see that the section of the orbit AB can always be approximated by a straight line (this is always possible if B is taken close enough to A). If AB is the distance traveled in a unit time (we can define our unit of time as the time it takes our planet to go from A to B), then the area of the triangle FAB is the area swept out in a unit time, and is therefore the areal velocity of the planet. We may therefore restate Kepler's second law as follows:

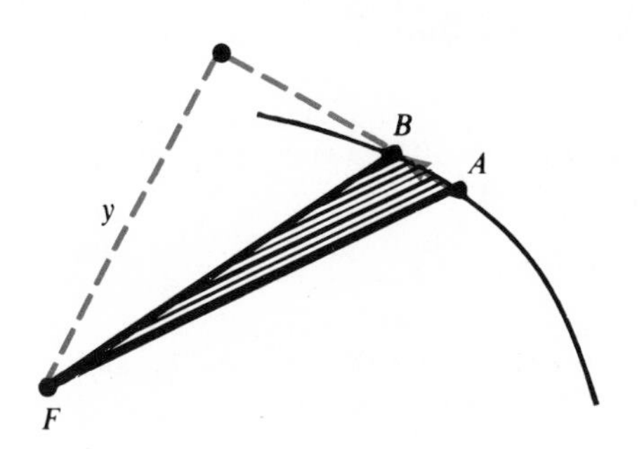

FIG. 8–16 *In this figure the area described by the radius vector from the sun to the planet (shaded area) in a unit time is just the triangle FAB. This is just equal to $\frac{1}{2}yv$, where y is the perpendicular distance from the sun to the direction of motion of the planet.*

The areal velocity of a planet in its orbit around the sun is constant.

To show that this is the same thing as saying that the angular momentum is a constant, all we need note is that the angular momentum of the planet relative to the sun at any point in its orbit A is given by its mass times its velocity (relative to the sun) times the perpendicular distance y from the sun to the velocity vector of the planet. In other words, the angular momentum J is given by

$$J = mvy. \tag{8.11}$$

But v (the distance AB) is the base of triangle AFB, and y is its altitude, so that J equals the mass of the planet times twice the area of the triangle swept out in unit time, or

$$J = 2m\mathscr{A}, \tag{8.12}$$

where $\mathscr{A}$ is the areal velocity of the planet. However, since Kepler's second law states that the areal velocity $\mathscr{A}$ is constant, J also is constant because twice the mass of the planet always remains constant. We see that we can now express the angular momentum of a planet in terms of the characteristics of its elliptical orbit.

8.24 Angular Momentum of a Planet in Terms of Its Sidereal Period

In the previous section we obtained an expression for the area of the planet's orbit in terms of its semimajor axis and the eccentricity, and we saw that this is equal to $\pi a^2\sqrt{1-e^2}$. Since, by definition, areal velocity is the area described in a given time divided by the time (during one sidereal period the complete area of the ellipse is described), we can get the areal velocity $\mathscr{A}$ of a planet by dividing the above expression for the area of its orbit by the planet's sidereal period, P.

$$\boxed{\mathscr{A} = \frac{\pi a^2\sqrt{1-e^2}}{P}}. \tag{8.13}$$

Thus, if we substitute this in the previous formula for J, we obtain for the angular momentum of the planet the constant

$$J = \frac{2\pi m a^2\sqrt{1-e^2}}{P}. \tag{8.14}$$

This is, indeed, a constant for all positions of the planet since all the quantities on the right-hand side depend only on the entire orbit and not on where the planet happens to be in its orbit.

8.25 The Speed of a Planet in Terms of Its Areal Velocity

We can also express the velocity of the planet at any point in its orbit in terms of the characteristics of the ellipse. We have noted (from equations 8.11 and 8.12) that the areal velocity is equal to one-half the velocity of the planet at any point times the perpendicular distance of the sun from the velocity vector at that point. We thus have

$$V = \frac{2\mathscr{A}}{y}$$

or

$$V = \frac{2\pi a^2\sqrt{1 - e^2}}{Py}, \tag{8.15}$$

using the expression we introduced above for the areal velocity.

There are four positions of a planet in its orbit for which the expressions for the velocity are very simple. These are the perihelion, the aphelion, and the two positions at the ends of the minor axis. Since $y = a(1 - e)$ for the planet at perihelion, we see that the speed of the planet in this position is just

$$V_{\text{perihelion}} = \frac{2\pi a^2\sqrt{1 - e^2}}{Pa(1 - e)}$$

$$= \frac{2\pi a}{P}\sqrt{\frac{1 + e}{1 - e}}. \tag{8.16}$$

Again since y for the aphelion is $a(1 + e)$, we have for the speed of the planet in this position

$$V_{\text{aphelion}} = \frac{2\pi a^2\sqrt{1 - e^2}}{Pa(1 + e)} = \frac{2\pi a}{P}\sqrt{\frac{1 - e}{1 + e}}. \tag{8.17}$$

We can also find the speed of the planet when it is at either end of the minor axis, for y is then just equal to the semiminor axis, b, since the planet is moving parallel to the major axis. But since $b = a\sqrt{1 - e^2}$, we then have

$$V_{\text{semiminor axis}} = \frac{2\pi a^2\sqrt{1 - e^2}}{Pa\sqrt{1 - e^2}} = \frac{2\pi a}{P}. \tag{8.18}$$

8.26 The Speed of a Planet at any Point in Its Orbit

From these three expressions we can obtain a general expression for the speed of a planet at any point in its orbit. We find such an expression by noting that as the planet moves around the sun, the angle θ which the radius vector makes with the major axis varies continuously from 0° in the aphelion position to 180° in the perihelion position. The expression for the speed of the planet must therefore vary with this angle in such a way that v takes on the three values given above for the planet when it is in the three special positions discussed. This is precisely the case if we write for the planet's speed at any point in its orbit the expression

$$V = \frac{2\pi a}{P}\sqrt{\frac{1 - 2e\cos\theta + e^2}{(1 - e^2)}}. \tag{8.19}$$

This follows since $\cos\theta$ equals 1 for the planet at aphelion; $\cos\theta$ equals e for the planet at the semiminor axis position; and $\cos\theta$ equals -1 for the perihelion position. These expressions show us that as the planet gets closer to the sun, its speed increases, and as it gets farther from the sun, its speed decreases.

We can also express the speed of the planet directly in terms of its distance from the sun by using the polar equation (8.10) for the ellipse to eliminate $\cos\theta$ from the above expression for the speed.

We have, in fact, from the equation for the ellipse (as the result of a simple algebraic rearrangement of terms)

$$2e\cos\theta = 2 - \frac{2p}{r} = 2 - \frac{2a(1 - e^2)}{r}.$$

If we now substitute this into the equation for the speed of the planet, we get

$$V = \frac{2\pi a}{P}\sqrt{\frac{1 - 2 + \left[\frac{2a(1 - e^2)}{r}\right] + e^2}{(1 - e^2)}}$$

$$= \frac{2\pi a}{P}\sqrt{\frac{2a}{r} - 1}$$

$$= \frac{2\pi a^{3/2}}{P}\sqrt{\frac{2}{r} - \frac{1}{a}}.$$

On squaring this expression, we obtain the very important formula

$$\boxed{V^2 = \frac{4\pi^2a^3}{P^2}\left(\frac{2}{r} - \frac{1}{a}\right)}. \tag{8.20}$$

If we could express the period of the planet in this formula in terms of the geometrical properties of the orbit and the physical properties (that is the masses) of the sun and the planet, we would then have an extremely useful formula. Since this is precisely what Kepler's third law enables us to do, we shall now consider that law in some detail.

8.27 Kepler's Third Law: The Constant in the Law

We have seen that Kepler's third law relates the square of the period of a planet to the cube of its mean distance from the sun by means of a constant that Kepler found to be the same for all planets. We may state this mathematically as follows:

If $P_1, P_2, \ldots, P_n$ are the periods of n planets and $a_1, a_2, \ldots, a_n$ are their respective mean distances from the sun, then

$$\frac{P_1^2}{a_1^3} = \frac{P_2^2}{a_2^3} = \cdots = \frac{P_n^2}{a_n^3} = \text{constant.} \quad (8.21)$$

We can easily determine the nature of the constant on the right-hand side by noting that it can refer only to some quantity that is related to the sun since only such a quantity is the same for all the planets. All we need do then to find this constant is determine what combination of solar characteristics has the dimensions of a time squared divided by a length cubed, since this is precisely the nature of the left-hand sides of these equations. Clearly since the relationship between period and distance of a planet must involve the gravitational force exerted on the planet by the sun, the constant must be some combination of the mass $M_\odot$ of the sun and the universal gravitational constant G.

The only combination of these two numbers that has the desired dimensional characteristics is

$$\frac{1}{GM}.$$

We may therefore express Kepler's third law for any planet as follows:

$$\frac{P^2}{a^3} = \frac{B}{GM}, \quad (8.22)$$

where B is a pure numerical factor that remains to be determined.

Kepler did not know the value of the constant on the right-hand side of Equation (8.21), and, as a matter of fact, the correct form of this constant and the correct value of the number B was not found until Newton's law of gravitation was applied to the motion of a planet. *Newton then discovered that the right-hand side of Equation* (8.21) *is not really a constant at all but only approximately so.* However, as we go from planet to planet this quantity varies so slightly that it is hardly to be wondered at that Kepler took it to be constant. In our discussion of Newton's derivation of Kepler's laws in the next chapter, we shall see exactly how Kepler's third law has to be corrected. However, for the time being we shall take it as it stands and assume the right-hand side is strictly constant as we go from planet to planet.

8.28 Kepler's Third Law Expressed in Astronomical Units

With this understood we can now express Kepler's third law in a very simple and instructive manner. To do this we note that if $P_\oplus$ is the period of the earth, and A is the astronomical unit (the earth's mean distance from the sun), we have then, from Kepler's third law, for any planet whose period is P and whose mean distance is a,

$$\frac{P^2}{a^3} = \frac{P_\oplus^2}{A^3}.$$

If we now agree to measure all periods in years and all mean distances in astronomical units, the right-hand side of this equation is unity, and we obtain

$$a^3 = P^2$$

or

$$\boxed{a = P^{2/3}}. \quad (8.23)$$

This equation is important because we can use it to determine the mean distance of any planet from the sun in astronomical units as soon as we know its sidereal period, which we can find with great accuracy from its synodic period. Kepler's third law thus enables us to construct the solar system accurately to scale, just from a knowledge of the periods of the planets. From this we see that knowledge of the A.U. is very important if we are to determine the actual scale of our solar system.

8.29 Finding the Value of B in Terms of the Speed of a Planet

We saw in Section 8.26, in discussing Kepler's second law, that if we could find the period of a planet in terms of the orbital characteristics, we would have an expression for the speed of the planet entirely in terms of these orbital characteristics. We can now in fact do so, since we have the expression $P^2 = Ba^3/GM$ for the square of the period. On substituting this into the expression (8.20) for the square of the speed, we have

$$V^2 = \frac{4\pi^2 GM}{B}\left(\frac{2}{r} - \frac{1}{a}\right). \quad (8.24)$$

From this expression we can find the value of the number B if we rearrange the terms somewhat. We first multiply each side by one-half of the mass of the planet, m, and transpose the term containing $1/r$ to the left-hand side, obtaining

$$\frac{1}{2}mV^2 + \left(-\frac{4\pi^2 GMm}{Br}\right) = -\frac{4\pi^2 GMm}{2Ba}.$$

We see that the first term of this expression on the left-hand side is just the kinetic energy of the planet in its orbit relative to the sun. It follows therefore that the second term must be an energy term, and since it varies with the distance from the sun, becoming larger as the planet gets farther from the sun (because of the minus sign) and smaller as the planet approaches the sun, it must be the potential energy of the planet with respect to the sun.

If we compare this term with the expression we obtained for the potential energy when we were discussing the speėd of escape, we see that this expression can be correct only if B is chosen equal to $4\pi^2$. We have thus evaluated the constant B by making use of energy considerations. Our equation now becomes

$$\frac{1}{2}mV^2 - \frac{GMm}{r} = -\frac{GMm}{2a}. \quad (8.25)$$

8.30 The Principle of Conservation of Energy Applied to a Planet

Since the two terms on the left-hand side are energy terms, the term on the right-hand side must also be an energy term. We note, however, that it always remains the same regardless of where the planet may be in its orbit. In other words, it is a constant, and hence must give the total value of the mechanized energy of the planet at all times. *The entire equation therefore expresses the very important principle of the conservation of mechanical energy applied to a planet.*

We note one more important point: *the total energy of a planet is given entirely by the size of the semimajor axis of its orbit.* The larger the mean distance of the planet from the sun (that is, the larger its orbit) the greater its total energy. Note that the total energy of a planet does not depend on the shape of its orbit, so that all objects having the same mean distance from the sun have the same energy regardless of the shape of their orbits. These considerations show us that *Kepler's third law is equivalent to the principle of conservation of energy, just as the second law is equivalent to the principle of conservation of angular momentum.*

8.31 The Motion of the Sun and Kepler's Third Law

The energy equation in the previous paragraph is not quite correct because it is based on the assumption that the sun stands still, and hence that it has no kinetic energy as the planet moves around it. However, the sun is also in motion (albeit very slowly) and one must replace the first term of the equation, which gives the kinetic energy of the planet with respect to a fixed sun, with a sum of two terms, one of which gives the kinetic energy of the planet and the other the kinetic energy of the sun, *both taken with respect to the center of mass of the planet and the sun.*

Since, with respect to the center of mass, the speed of the planet is greater than the speed of the sun in the same ratio that the mass of the sun is greater than the mass of the planet (see the discussion of mass and acceleration in Chapter 3 and also the discussion of Newton's third law) we must have

$$\frac{v_p}{v_\odot} = \frac{M_\odot}{m}, \quad (8.26)$$

where v_p and $v_\odot$ are the speeds of the planet and the sun respectively with respect to the center of mass. We also have, since V is the speed of the planet relative to the sun (the speed of two bodies relative to each other is equal to the sum of their speeds relative to their center of mass),

$$v_p + v_\odot = V.$$

From these two equations we deduce that

$$v_\odot = \frac{m}{M_\odot + m} V$$

and (8.27)

$$v_p = \frac{M_\odot}{M_\odot + m} V.$$

Hence the kinetic energy of the sun relative to the center of mass is

$$\frac{1}{2} M_\odot (v_\odot)^2 = \frac{1}{2} \frac{M_\odot m^2 V^2}{(M_\odot + m)^2}$$

and the kinetic energy of the planet relative to the center of mass is

$$\frac{1}{2}m(v_p)^2 = \frac{1}{2}\frac{mM_\odot^2V^2}{(M_\odot + m)^2}.$$

Since the sum of these two terms gives the total kinetic energy of the system, we must use it in place of the first term in our previous energy equation. If we do this, the equation for the conservation of energy becomes (note that the potential energy is the same as before)

$$\frac{1}{2}\frac{mM_\odot}{(m + M_\odot)}V^2 - \frac{GM_\odot m}{r} = -\frac{GM_\odot m}{2a}.$$

We now transpose the second term to the right-hand side and multiply the entire equation through by $2(m + M_\odot)/m$, and we obtain the correct equation for the square of the velocity:

$$V^2 = G(M_\odot + m)\left(\frac{2}{r} - \frac{1}{a}\right). \tag{8.28}$$

If we compare this with Equation (8.24) for the square of the velocity (derived on the assumption that the sun does not move) we see that it differs from it merely in that $M_\odot$ is replaced by $(M_\odot + m)$, the mass of the sun plus the mass of the planet (the constant B is, as noted, $4\pi^2$). If we now go back again to Kepler's third law as we originally stated it in Equation (8.22) we see that we must replace it by the more general equation

$$\boxed{\frac{P^2}{a^3} = \frac{4\pi^2}{G(M_\odot + m)}}, \tag{8.29}$$

since we must replace $M_\odot$ by $(M_\odot + m)$ and B by $4\pi^2$ just as we did in Equation (8.28). *This is the form of Kepler's third law which Newton obtained by using his law of gravitation. We see from this that the right-hand side is not really a constant, but changes very slightly as we go from planet to planet.* In the case of the most massive planet, Jupiter, it differs from the values for the other planets by only 0.1 per cent. We can see now why Kepler thought it was really a constant.

8.32 The Angular Momentum of a Planet in Terms of Orbital Characteristics

Now that we have a corrected expression for the period of the planet, we can substitute it into Equation (8.14) for the planet's angular momentum, and thus obtain

$$J = \frac{2\pi ma^2\sqrt{1 - e^2}}{\dfrac{2\pi a^{3/2}}{\sqrt{G(M_\odot + m)}}} = m\sqrt{G(M_\odot + m)a(1 - e^2)}$$

or

$$J = mp^{1/2}\sqrt{G(M_\odot + m)}, \tag{8.30}$$

where p is the semiparameter as defined by Equation (8.6).

We see from Equation (8.30) that for an orbit of given size (or, put differently, for all orbits having the same semimajor axis) the angular momentum depends only on the shape of the orbit. *The rounder the orbit, the greater the angular momentum, so that the angular momentum is greatest for a circular orbit. We also note that the angular momentum for orbits of the same shape increases if the size of the orbit increases.* Thus, whereas a determines the energy of the planet the eccentricity e (for a given a) determines the angular momentum.

8.33 Speeds of Planets Associated with Different Conic Sections

From the expression for the square of the velocity which we have just written down, we can determine how fast a particle must move if its orbit is to be a particular conic section. Since for a circular orbit the particle will always be at the same distance from the sun, r must be equal to a in our equation and we obtain

$$V_{\text{circle}}^2 = \frac{G(M_\odot + m)}{a}. \tag{8.31}$$

In this case the velocity is always a constant and always at right angles to the radius vector from the sun. Thus, if any particle is thrown into space in the neighborhood of the sun, *it can move in a circular orbit only if at the moment that it is thrown, it is moving exactly at right angles to the line from it to the sun and exactly at the speed given by the previous equation.* Any deviation from these conditions results in a noncircular orbit. Since the chance for achieving exactly these conditions is infinitesimal, we see that circular orbits are excluded.

As long as a is finite, the planet moves in an ellipse and its velocity is always smaller than the speed of escape from the solar system at any point in its orbit. However, if a becomes infinite, we are then dealing with a parabola, and the velocity of the particle at each point of its orbit is called its *parabolic velocity* at that point. In this case, of course, it is clear that the particle will leave the solar system never to return. Hence *the parabolic velocity at a given distance r from the sun is the same as the velocity of escape at that distance.* In fact if we place a equal to infinity we obtain for the parabolic velocity

$$V^2_{\text{parabolic}} = 2G\frac{(M_\odot + m)}{r},$$

so that if we take the square root of both sides we obtain

$$V_{\text{parabolic}} = \sqrt{2G\frac{(M_\odot + m)}{r}}, \qquad (8.32)$$

and we see that this is precisely the velocity of escape as given in Equation (6.12). Only if a particle has exactly this speed can it move in a parabola so that such orbits have zero probability.

Finally we may note that if a particle is moving in a parabolic orbit, its total energy must be zero, as can be seen from the equation of energy. This means that the kinetic energy at any point in a parabolic orbit is exactly equal to the numerical value of its potential energy (which is always negative) at that point. *If the total energy is negative, the planet's orbit is an ellipse, and if its total energy is positive its orbit is a hyperbola.*

In other words, whether a particle will move in a closed orbit (ellipse) around the sun, or in an open orbit (hyperbola), depends on whether its kinetic energy at any time is less than or greater than the numerical value of its potential energy. Only if these two are equal is the orbit a parabola. In the case of a circular orbit the kinetic energy is equal to one-half the numerical value of the potential energy. This is essentially the content of an important theorem on gravitational stability called the virial theorem, which we shall refer to later.

TABLE 8.1

PROPERTIES OF THE ELLIPSE AND THE MOTION OF ONE BODY ABOUT ANOTHER

a: semimajor axis: mean distance of planet from sun (1) determines size of ellipse, (2) determines total mechanical energy of the planet in its orbit (the larger a is, the larger the energy), and (3) determines sidereal period of the planet.

e: eccentricity $= FC/a$. For a given a, determines the angular momentum (the smaller e is, the larger the angular momentum).

b: semiminor axis $= a\sqrt{(1 - e^2)}$.

p: semiparameter $= a(1 - e^2)$.

A: area $= \pi ab = \pi a^2\sqrt{1 - e^2}$.

Perihelion distance $= a(1 - e)$.

Aphelion distance $= a(1 + e)$.

$\mathscr{A}$: areal velocity $= \dfrac{p^{1/2}}{2}\sqrt{G(M_\odot + m)}$; m = mass of planet.

V: velocity of planet relative to sun

$$= \sqrt{G(M_\odot + m)}\left(\frac{2}{r} - \frac{1}{a}\right)^{1/2};$$

r = distance of planet from sun.

J: angular momentum $= mp^{1/2}\sqrt{G(M_\odot + m)}$
$= 2m\mathscr{A}$.

E: total energy $= -\dfrac{GMm}{2a} = \dfrac{1}{2}mV^2 - \dfrac{GMm}{r}$.

$$r = \frac{p}{1 - e\cos\theta}.$$

θ = angular separation of planet from aphelion as seen from sun.

Parabolic velocity (orbit of zero energy)

$$= \sqrt{\frac{2G(M_\odot + m)}{r}}.$$

EXERCISES

1. (a) Using vectorial construction, show what the velocity of Mercury is, relative to the earth, at the two maximum elongations (east and west) and for opposition and conjunction. (Assume that speeds of both the earth and Mercury are constant and equal to their average speeds.) (b) Do the same thing for the planet Mars at opposition and conjunction and at the two quadratures.

2. If you were on a planet other than the earth, how would the synodic period of the earth as seen from that planet compare with the synodic period of that planet as seen from the earth?

3. (a) What would the synodic period of Jupiter be as seen from Saturn? (b) Of Mars as seen from Uranus?

4. If in 15 weeks the line from the sun to an asteroid sweeps out $\frac{1}{8}$ of the total area enclosed by the orbit of the asteroid, what is the period of the asteroid?

5. Show that the equation of an ellipse in a rectangular coordinate system whose origin is at the center of the ellipse and whose x-axis lies along the major axis of the ellipse is

$$\frac{x^2}{a^2} + \frac{y^2}{b^2} = 1.$$

6. Consider two concentric circles with radii a and b, and show that the area of the ellipse whose semimajor and semiminor axes are a and b respectively is equal to the geometric mean of the areas of the two circles.

7. Using the same two circles as in problem 6, draw a radius from the center intersecting the circles in two points. Show that if two mutually perpendicular lines are drawn through these two points on the circles, the point of intersection of these two lines lies on an ellipse whose semimajor axis is a and whose semiminor axis is b and that the semimajor axis of this ellipse is parallel to the line drawn from the point on the smaller circle.

8. Using the same two circles as in problem 7, show that if from a point on the larger circle a line is drawn tangent to the smaller circle, the cosine of the angle that this line makes with the radius passing through the point on the larger circle is equal to the eccentricity of the ellipse whose semimajor and minor axes are the radii of the two concentric circles.

9. If an artificial satellite with a mass of 2 000 kilograms is moving around the earth in such a way that the line from the center of the earth to it sweeps out an area of 19 200 square km per sec, what is the orbital angular momentum of the satellite?

10. Compute the areal velocities of (a) Mercury, (b) Venus, (c) Earth, and (d) Jupiter. Use data from tables and neglect the gravitational interaction between the planets.

11. Compute the aphelion and perihelion velocities of the planets in problem 10.

12. (a) Use Kepler's third law to determine the mean distance from the earth of an artificial satellite whose period is 3 hours. (b) If an asteroid is at a mean distance of 3 astronomical units from the sun, what is its sidereal period?

13. Compute the total orbital (around the sun) energies of the planets in problem 10 (neglect the gravitational interaction of the planets).

14. (a) Compare the orbital angular momentum of Jupiter with that of the earth. (b) How much larger could Mercury's orbital angular momentum become if the shape of its orbit could be altered with no change in the size of its orbit?

15. How much larger is the parabolic velocity than the circular velocity at a distance a from the sun?

16. What is the relationship between the kinetic energy and the potential energy of a particle in a circular orbit?

17. From the expression

$$r = \frac{a(1 - e^2)}{1 - e \cos \theta},$$

given in the text for the position of a planet, use the differential calculus to derive the expression for the velocity of a planet in its orbit. (Hint: use the fact that $r^2\omega$, where ω is the angular velocity of the planet, is equal to twice its areal velocity.)

18. Derive the expression for the radial velocity of the planet in problem 17.

19. Derive the expression for the radial acceleration of a planet at any point in its orbit from the expression for r.

9

CELESTIAL MECHANICS

In the previous chapter we discussed Kepler's laws of planetary motion and we saw that they are only special cases of more general principles. The most general formulation of these laws was first obtained by Newton, who applied his gravitational law of force and his three laws of motion to the analysis of the motion of a planet around the sun. Although Kepler had suspected that a planet moves as it does because of some force emanating from the sun, he had not attempted to obtain his planetary laws by analyzing this force. This, however, is what Newton did. He was able to do so because he had already expressed his second law of motion and his law of gravitation in mathematical form, and *he had also invented a new and powerful mathematical tool, the* **calculus**, *to handle problems in which the acceleration of a body varies continuously as the body moves.*

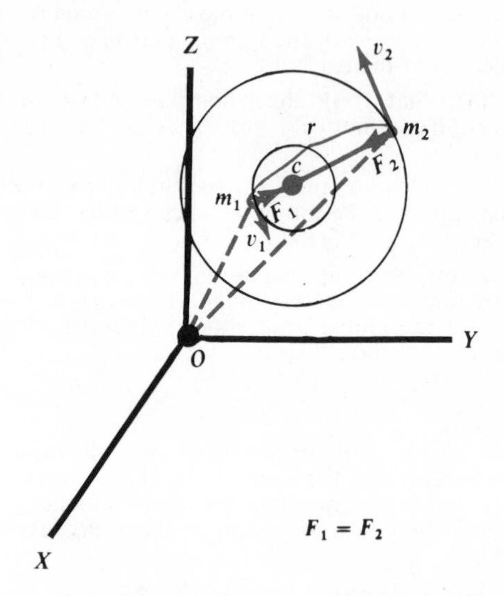

FIG. 9–1 *A schematic diagram of the gravitational action between 2 mass points, m_1 and m_2 (c is the center of mass), in a three-dimensional rectangular coordinate system (circular orbits shown).*

9.1 The Two-Body Problem

To understand the nature of the problem that confronted Newton, we shall consider briefly the two-body problem, which Newton was able to solve completely. In Figure 9.1, let m_1 and m_2 be the masses of two particles separated by a distance r. Newton asked himself the following question:

> How will two such particles move if they start out with a given initial velocity and if they interact gravitationally?

To solve the problem Newton applied to each particle the fundamental law of motion $\mathbf{F} = m\mathbf{a}$, where $\mathbf{F}$ is the gravitational force (a vector) acting on the particle and $\mathbf{a}$ is its acceleration (a vector). Thus, if we consider particle 1 we have, following Newton,

$$\mathbf{a}_1 = \frac{\mathbf{F}_1}{m_1},$$

where $\mathbf{a}_1$ is the acceleration of this particle at some instant, $\mathbf{F}_1$ is the force acting on it at that instant, and m_1 is its mass. In the same way, we have for particle 2

$$\mathbf{a}_2 = \frac{\mathbf{F}_2}{m_2}.$$

Since we know that $\mathbf{F}_1$ and $\mathbf{F}_2$ are equal and opposite ($\mathbf{F}_1$ is the gravitational force Gm_1m_2/r^2 of particle 2 on 1, and $\mathbf{F}_2$ is the negative of this), we can at each moment find the acceleration of each particle. *From this acceleration we can determine the velocity of each particle at a somewhat later time, since the velocity of a particle at any moment can be obtained from its velocity at a slightly earlier time by adding to this earlier velocity the acceleration of the particle at that earlier time.* In this way we can assign to each particle a velocity vector from moment to moment, and thus obtain a plot of its orbit, since these velocity vectors, as shown in Figure 9.1, are the tangents to the orbit.

However, obtaining an actual mathematical solution to this problem is not as easy as it sounds because the acceleration is continuously changing; the large change in the acceleration means that we cannot find the velocity of either particle at the end of any large interval of time from our knowledge of its velocity at the outset.

9.2 Newton's Method of Solving the Two-Body Problem

Newton overcame this difficulty by examining the behavior of a planet during a very short time interval. *He wrote down the equation* $\mathbf{a} = \mathbf{F}/m$ *for each particle in terms of very short time intervals.* If we allow these time intervals to become smaller and smaller, ultimately approaching zero, as Newton did, we obtain what we call **differential equations**. They are also called the **equations of motion** of the two particles. This is the type of equation which Newton had to solve, and when he did so, he obtained Kepler's laws in their most general form. Since the equation of motion for each particle is a vector equation, it really consists of three equations (a separate component equation for each of the three coordinate directions in space) so that there are six equations altogether for the two particles. But one can reduce these to two equations by using the principle of conservation of momentum of the center of mass of the two particles and by noting that the motion of the particles lies in one plane.

Although to obtain the general solution one must work with differential equations, there is one case for which we can solve the problem by using simple geometry and algebra. This is the case in which both particles move in circular orbits about their common center of mass. We shall therefore show how from Newton's law of gravitation we can obtain Kepler's laws for this case. Here we shall consider just the third law, since the first and second laws are obvious for circular orbits.

9.3 Kepler's Third Law Derived for Circular Orbits

In Figure 9.2, let M be the mass of the sun, which we shall picture as being a point concentrated at its center, and m the mass of the planet, also pictured as a point concentrated at its center, and let c be the center of mass of the system. The radius of the small circle, which is the orbit of the sun's center about the center of mass, is smaller than the radius of the planet's circular orbit about this point by the same factor by which m is smaller than M.

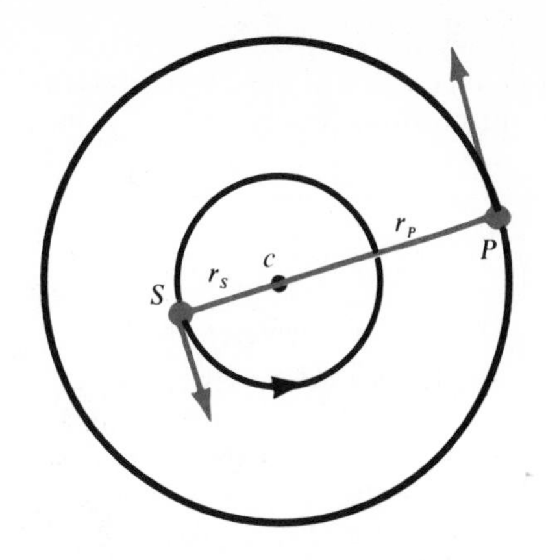

FIG. 9–2 *The center of the sun and the center of a planet moving in circular orbits about their center of mass, c. The scale is greatly exaggerated.*

Since the planet is moving about the point c in a circle, its centripetal acceleration is V^2/r, where r is the radius of its orbit and V is its constant velocity. This acceleration must be equal to the acceleration of gravity imparted to it by the sun, which is given by $GM_\odot/(r + R)^2$, where R is the radius of the sun's orbit around the center of mass. Note that the center of the planet and the sun must always lie on the line passing through the center of mass (the center of mass remains fixed), so that the distance of the sun from the planet is always $r + R$. We therefore have

$$\frac{V^2}{r} = G\frac{M_\odot}{(r + R)^2}.$$

From this we can find Kepler's third law by eliminating r on the left-hand side and placing $r + R = a$ (the mean distance of the planet from the sun) on the right-hand side.

We first note that the velocity V of the planet can be obtained by dividing the circumference of its orbit by its period, so that

$$V = \frac{2\pi r}{P}.$$

If we square this and substitute in our previous equation, we obtain

$$\frac{4\pi^2 r}{P^2} = G\frac{M_\odot}{a^2},$$

where we have placed $(r + R) = a$. Since the center of mass of the planet and the sun must divide the distance a into two segments whose ratio is equal to the ratio of the mass of the sun to the mass of the planet, we must have

$$r = \frac{M_\odot}{M_\odot + m}a \quad \text{and} \quad R = \frac{m}{M_\odot + m}a. \tag{9.1}$$

If we now use the first of these equations to eliminate r from our previous result, we finally obtain

$$\frac{4\pi^2}{P^2}\frac{M_\odot}{(M_\odot + m)}a = G\frac{M_\odot}{a^2}$$

or

$$P^2 = \frac{4\pi^2 a^3}{G(M_\odot + m)}, \quad (9.2)$$

which is just Kepler's third law.

Just as we can obtain Kepler's third law from Newton's law, so it is possible to reverse the procedure and show that Newton's inverse-square law of force is a necessary consequence of Kepler's third law of motion. We shall not do this here, but leave it as an exercise for the reader.

Just as we have derived Kepler's third law of planetary motion for circular orbits from Newton's law of gravity, we can show that if two bodies pull upon each other according to this law, the orbit of one body relative to the other (planet relative to sun) is a conic section. We can do this without having to solve the equations of motion if we use the principle of conservation of energy and conservation of momentum.

If M is the mass of one body and m that of the other, we have

Energy principle

$$\frac{1}{2}KmV^2 - \frac{GMm}{r} = E, \quad (9.3)$$

Angular momentum principle

$$mVr \sin\theta = J, \quad (9.4)$$

where G is the gravitational constant, E is the total energy of the two bodies (a constant), J is the angular momentum of the planet relative to the sun (also a constant), $K = M/(m + M)$, V is the velocity of the planet with respect to the sun, r is the distance of the planet from the sun at any moment, and θ is the angle between r and **V** as shown in Figure 9.3. It should be noted that Newton's law of gravity is not introduced explicitly into this derivation but is contained in the second term (the mutual potential energy) on the left-hand side of Equation (9.3).

If we now introduce (9.4) into (9.3) and thus eliminate V, we obtain a quadratic equation for r:

$$\frac{KJ^2}{(2mr^2 \sin^2\theta)} - \frac{GmM}{r} = E.$$

On rearranging terms in this equation we obtain (multiplying through by r^2 and dividing through by E)

$$r^2 + \frac{GmM}{E}r - \frac{KJ^2}{2mE\sin^2\theta} = 0. \quad (9.5)$$

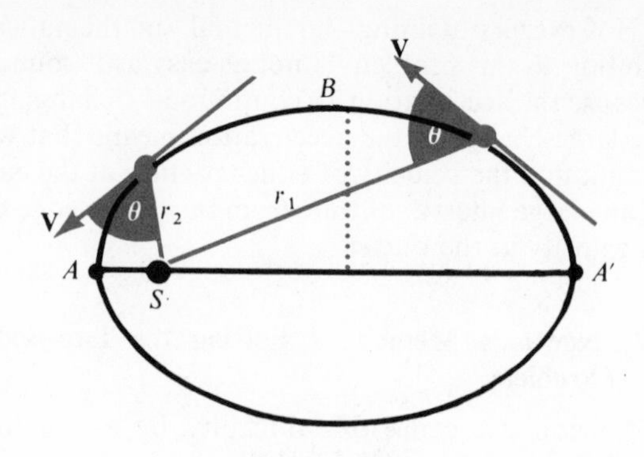

FIG. 9-3 *Schematic drawing showing that there are two positions of a planet, one to the right and one to the left of the semiparameter, for which the angle θ between the velocity vector and the radius vector is the same.*

We see from this equation that if E is less than zero (negative total energy) there are just two different *finite* values for r for each value of θ, which means, of course, that the orbit must be a closed figure since the velocity vector must turn in such a way as to give the same θ to the right and to the left of the semi-parameter of the orbit (see page 141). The two values of r are obtained as the two roots of the quadratic Equation (9.5):

$$r_1 = \frac{GmM}{2E}\left(-1 - \sqrt{1 + \frac{2EKJ^2}{G^2m^3M^2\sin^2\theta}}\right).$$

$$r_2 = \frac{GmM}{2E}\left(-1 + \sqrt{1 + \frac{2EKJ^2}{G^2m^3M^2\sin^2\theta}}\right).$$

For each value of θ there is a large (r_1) and a small (r_2) distance of the planet from the sun. The maximum r_1 (the aphelion distance) is given for θ equal to 90° ($\sin\theta = 1$) as shown in Figure 9.3, at the position A', and corresponding to this same value of θ we have the perihelion distance given by r_2 as shown in the figure at A. The perihelion and aphelion distances are thus given by placing $\sin\theta = 1$ in the above expressions for r_1 and r_2. If we add the two expressions above for r_1 and r_2 we obtain

$$r_1 + r_2 = -\frac{GmM}{E} \quad \text{or} \quad E = -\frac{GmM}{r_1 + r_2},$$

since the two square roots on the right-hand sides just cancel each other. This, by the way, is true for all values of θ, so that the average distance a of the planet from the sun is just half of this:

$$\text{average distance: } a = \frac{r_1 + r_2}{2} = -\frac{GmM}{2E}.$$

We thus obtain for the total energy of the two bodies

$$E = -\frac{GmM}{2a}.$$

The student should compare this with Equation (8.25) in Chapter 8.

If the planet is at a point in its orbit such that

$$\sin^2\theta = -\frac{2EKJ^2}{G^2m^3M^2} = \frac{KJ^2}{Gam^2M},$$

the square roots in the expressions for r_1 and r_2 vanish so that these two distances reduce to the same value, namely $-GmM/(2E)$, which is just equal to a, the semimajor axis. This clearly occurs when the planet is at the point B in the figure, since B is symmetrically situated with respect to A and A'. It is left as an exercise for the student to show that this expression for $\sin\theta$ leads to $\cos\theta = e$, where e is the eccentricity of the orbit.

We see from Equation (9.5) that for any given distance r from the sun there are two values of θ which are equal in magnitude but opposite in sign. Thus the orbit is symmetrical about the line from A to A' and hence is an ellipse.

We leave as an exercise for the student to show that when $E = 0$ the orbit is a parabola.

9.4 The *n*-Body Problem

Thus far we have limited ourselves to a discussion of two bodies moving in space under their mutual attractions, and we have seen that it is possible to obtain a complete solution for this simple case. Fortunately, because of the great mass of the sun as compared with that of any one of the planets, we may treat the motion of a planet around the sun by this method and obtain very accurate results. However, this is still only an approximation, even though a very good one.

If we wish to obtain results of the very highest accuracy, we must take into account that each planet interacts not only with the sun but with all the other planets. Of course, the disturbances (perturbations) in the motion of a particular planet around the sun resulting from the gravitational effects of all the other planets are very small because of the small masses of the planets; but these disturbances must, nevertheless, be taken into account. The most important perturbations are produced by the very massive, and also by the nearby, planets. In the case of the earth the greatest effect is due to Jupiter, because of its large mass, even though Mars and Venus both get much closer to the earth than Jupiter does.

Thus a complete solution to the problem of the motion of a planet involves the analysis of the interactions among many bodies and is referred to as the n-body problem. This problem is so complex that it has been impossible to obtain a general solution; in order to analyze all the motions one must proceed by numerical methods.

9.5 The Three-Body Problem

If, instead of considering any number (n) of interacting bodies, we limit ourselves just to three bodies—as, for example, the earth, the sun, and the moon—we still find that the problem is enormously complex, even though we have moved but one step from the two-body problem. Although the problem is quite definite and one can set down all the equations that are required, the mathematics is so involved that it has been impossible thus far to obtain a usable general solution.

To see why even the three-body problem is of great complexity, all we need note is that a third body introduced into a two-body system may have motions that can range from a simple, very nearly elliptical, orbit around either one of the bodies to a complex orbit involving all three bodies. Drastic variations in the orbit of any of the three bodies can occur as a result of slight changes in the order in which these bodies approach a particular point in space. We can see from this that a general solution, which must take into account all the variations that can occur, will have to be of enormous complexity, and this has, indeed, proved to be the case. About 20 years ago K. Sundman did succeed in writing down the general solution of the three-body problem, and it is of such complexity that one hardly knows how to begin to use it.

In the case of the three-body problem, each body has its own vector equation of motion so that for the three bodies there are altogether nine differential equations (three for each of the three coordinate directions) to be solved simultaneously. But three of these can be eliminated by using the principle of the conservation of momentum of the center of mass of the system, and the remaining equations can be reduced still further by using the principle of the conservation of angular momentum (law of areas) for the entire system. Finally, another simplification can be achieved by using the principle of the conservation of energy for the system so that fewer equations remain to be solved than one starts with.

But these are still mathematically forbidding. In the general n-body problem one starts with $3n$ differential equations, which can again be simplified by applying the three conservation principles, but the equations that are left are beyond the capabilities of modern mathematicians.

9.6 The Method of Successive Approximations

It is not necessary, however, to have a general mathematical solution of the three-body problem—or, for that matter, of the n-body problem—to obtain the orbit of any one of the bodies in the system to any desired degree of accuracy. This can be done numerically by a method of successive approximations, and today—with high-speed electronic computers available—this procedure is both practical and extremely useful.

The procedure used is the following: let us suppose that we wish to determine the position of some planet, let us say Mars, one month from today (assuming we know exactly where it is and how it is moving at this moment). We first divide the month into equal intervals, let us say days, and by using the law of gravity determine the velocity and position of Mars at the end of each such interval. In that way we obtain a step-by-step approximation to the position that Mars will occupy at the end of one month, and also an approximation to the velocity that it will have at that time. To see how this actually works, let us consider the sun, the earth, Mars, Jupiter, and Saturn, as shown in Figure 9.4, at the beginning of the month.

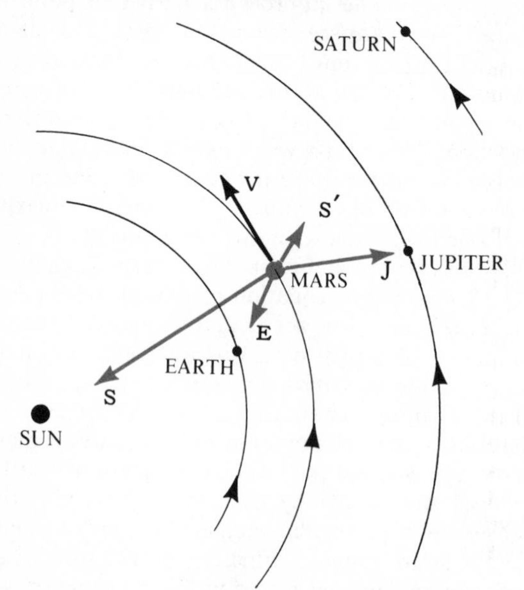

FIG. 9–4 *The acceleration imparted to the planet Mars by the sun, earth, Jupiter, and Saturn, shown schematically.*

The vector **v** represents the velocity of Mars at the beginning of the computation. If there were no forces acting on it, Mars would continue moving with constant speed in the direction given by this vector. Since, however, the sun, the earth, Jupiter, and Saturn are all pulling on Mars, each of these bodies imparts an acceleration to it (expressed in change in velocity per day) as shown by the four arrows marked **S**, **E**, **J**, and **S′**.

We now suppose that these accelerations do not affect the velocity of Mars during the course of any single day of our computation, but only at the end of that day (this, of course, is an approximation since we know that the accelerations are always present and have the effect of changing the velocity of the planet continuously). We may, therefore, suppose that the position of Mars at the end of the first day is what it would be if no other bodies were present, and Mars had just the velocity **v**. However, to find the position of Mars at the end of the second day, we must now (that is, at the end of the first day) alter its velocity **v** by adding to the velocity the four accelerations **S**, **E**, **J**, and **S′**.

This will give us a new velocity **v′** at the end of the first day, and using this new velocity, assuming again that no accelerations are acting during the second day, we can find the position and the velocity of Mars at the end of the second day, just as we did at the end of the first day. At the end of the second interval we must change the velocity **v′** by adding to it whatever accelerations are present. These accelerations again are due to the sun, earth, Jupiter, and Saturn, but of course they now do not equal the accelerations that we had at the end of the first day since the distances of Mars from the sun, the earth, Jupiter, and Saturn have changed because of Mars' own motion and the motion of the other bodies.

In this way we can trace the behavior of Mars from day to day by always adding to its velocity at the end of each day the appropriate accelerations. We can make the procedure somewhat more accurate by adding to the velocity at the end of each day not the accelerations at that moment but the average of the accelerations at the beginning and at the end of that day. The problem, however, is much more complicated than it sounds because not only are the three bodies pulling on Mars, but Mars is also pulling upon them in turn, thus changing their motions continuously. In addition the sun, the earth, Jupiter, and Saturn are all pulling upon one another.

We can improve the accuracy of the procedure described above still further by dividing the month into smaller time intervals. Indeed, there is no limit to how great an accuracy we may achieve in this way. But the larger the number of intervals we take, the greater is the amount of computation that has to be done.

9.7 The Method of Perturbations

Instead of using the step-by-step process that we outlined above, we can use the **method of perturbations**, as it is called, which makes use of the fact that the sun's gravitational force far exceeds that of the planets. What we do, therefore, is express the motion of any particular planet as a series of terms or successive approximations, the first and foremost of which arises from the action of the sun. In other words, we first assume that no other bodies except the sun and the planet are present, and obtain an exact two-body solution. This solution is an elliptical orbit, and since it is the first term in our series it is called the **first approximation**. This elliptical orbit gives us the approximate position of the planet at each instant of time in the interval during which we are studying its motion. To find the true orbit of the planet, we now correct the positions given by the elliptical orbit by introducing the higher approximations. These are nothing more than the corrections produced by the attractions of the other planets for each position of the given planet in its elliptical orbit.

When we carry out such calculations on various planets we find that their orbits are not closed figures at all, but rather complicated curves. The variations from ellipticity that occur in these orbits are referred to as the **perturbations** of the planets, and these may be either of a periodic or of an accumulative kind (the latter is called a secular perturbation). Thus it is found that the eccentricity of the orbit, the inclination of the plane of the orbit, and the size of the orbit change continuously, although very, very, slowly.

9.8 The Advance or the Precession of the Perihelion of a Planet

One of the most interesting of these perturbations is referred to as the precession or the advance of the perihelion of the planet. This can be best illustrated by supposing that the orbit of the planet is itself rotating in its own plane in the same direction in which the planet is revolving around the sun, so that the major axis (the **line of apsides**) is constantly changing its direction in space, as shown in Figure 9.5.

FIG. 9–5 *The precession of the perihelion of Mercury. The point precesses in the same direction as the planet revolves around the sun.*

This effect is most pronounced for planets such as Mercury whose orbits have large eccentricities. In the case of the planet Mercury very accurate observations of this advance of the perihelion have been made, and it has been known for a long time that this effect cannot be entirely accounted for by Newton's law of gravity. *After all the perturbations affecting Mercury are taken into account, there is still left over an advance of the perihelion at the rate of about* 42″.84 *per century. This effect is far too large to be accounted for by observational error.* We shall see in a later chapter that this discrepancy has been explained only by revision of our concepts of gravity.

9.9 The Launching of Artificial Satellites

Now that we have developed some of the ideas of celestial mechanics and the properties of planetary orbits, we are in a position to discuss the problems that arise in the launching of artificial earth satellites and space vehicles.

To treat such a problem accurately one must make use of the procedures we outlined above in our discussion of the n-body problem, but here we shall treat satellites and space vehicles only in terms of the two-body problem, and assume that the behavior of the satellite or vehicle is determined by some single massive body, such as the earth or the moon.

We shall first consider the problem of launching a satellite, neglecting the drag of the atmosphere. In doing this *we must keep in mind that, to a first approximation, the motion of the projectile must be considered with respect to the center of the earth and not to the surface of the earth.* We must remember that as soon as an object leaves a point on the earth's surface, its orbit is determined by the mass of the earth as though this mass were all concentrated at the center of the earth. It is clear from this that the point P on the surface of the earth from which the projectile is launched in a single stage must itself be a point of its orbit in space relative to the center of the earth, as shown in Figure 9.6. *From the figure we can see that no matter in what direction such a projectile is launched, and with whatever speed it leaves the earth, as long as this speed is below the speed of escape, the projectile must return to the surface of the earth in its attempt to come back to the point P again.*

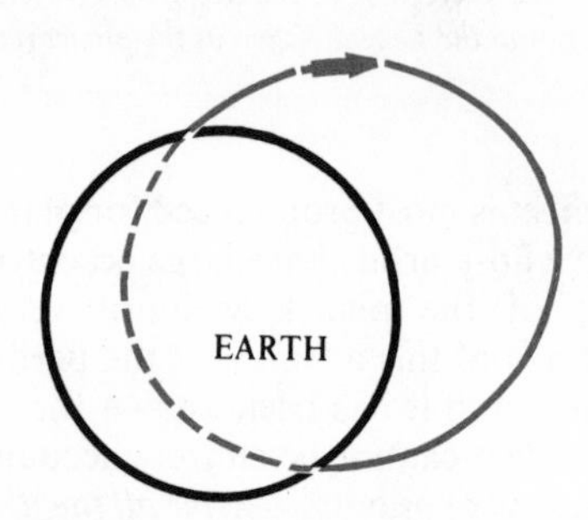

FIG. 9–6 *An object thrown away from the earth's surface in one single stage must move about the center of the earth in an ellipse; hence it must return to the earth's surface again and, therefore, cannot continue to orbit the earth. The point from which it is projected is itself a point on the ellipse and, hence, on the orbit.*

Because of this we cannot launch a satellite in a single stage unless we do so with a speed that is equal to or greater than the speed of escape, in which case it will leave the earth entirely. *Hence at least two stages must be used to launch an artificial satellite. The first stage serves the purpose of lifting the vehicle high enough above the earth's surface so that the first point in its orbit will be well above the earth's atmosphere. The second stage then gives it the necessary velocity for the required orbit.*

Of course, if we had a very high peak from which to launch a satellite, we could do it in a single stage. This is illustrated in Figure 9.7, where the arrows show the satellite being launched in different directions. Since the point at the top of the peak from which the satellite is launched is automatically a point in the orbit of the satellite, we should have to imagine the peak as being removed right after the launching if a collision is to be avoided. The first stage in the launching of a satellite can be pictured as playing the role of this peak.

9.10 Artificial Satellites and Circular Orbits

Now that we understand one of the reasons for multistage launching of satellites, we shall briefly

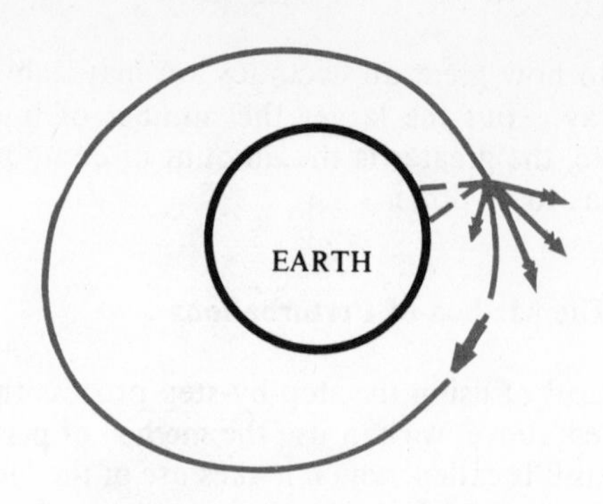

FIG. 9–7 *Launching a satellite from a high tower or peak above the earth's surface. In this case the satellite can be launched into a closed orbit in one single stage.*

consider some of the problems that are involved. Let us first suppose that we wish to have a satellite revolve around the earth in a circular orbit at a given height above the earth's surface. We know from previous discussions (see Section 8.33) that there is only one definite speed which a satellite can have, and one definite direction in which it can be launched if it is to move in a circular orbit. The speed is given by

$$\sqrt{G\frac{M_\oplus}{R_\oplus + d}},$$

where $M_\oplus$ is the mass of the earth, $R_\oplus$ is the radius of the earth, and d is the vertical distance of the satellite above the earth's surface, and it must be launched exactly parallel to the earth's surface. Any deviations from the proper speed or direction of launching will cause the satellite to move in an elliptical orbit.

If a satellite is to revolve around the earth in a circular orbit at a distance of 300 miles (480 km) above the earth's surface, it must move with a speed of 4.72 miles (7.66 km) per second, as can be immediately verified by putting the appropriate value for d in the above formula (note that if the above formula is used, and the value of G is taken as 6.6×10^{-8} units, then the mass of the earth must be expressed in grams and $R_\oplus$ and d in centimeters. The speed must then be expressed in cm/sec).

A simple way to compute the circular speed without having to put in numerical values for G and $M_\oplus$ is to note that the speed which a satellite must have to move in a circular orbit which just grazes the earth's surface at all points [in other words, an orbit whose radius is just R—that is, 4 000 miles (6 450 km)] is equal to

$$\sqrt{G\frac{M_\oplus}{R_\oplus}}.$$

If one introduces appropriate values for G, $M_{\oplus}$, and $R_{\oplus}$, one finds a speed of 4.89 miles (7.86 km) per second.

We can now compare this formula for the speed, $V_{R\oplus}$, in a circular orbit grazing the earth's surface with the formula previously written down for the speed, V_d, in a circular orbit at a distance d from the earth's surface. We find on taking the ratio of these two formulae that

$$\frac{V_d}{V_{R\oplus}} = \sqrt{\frac{R_{\oplus}}{R_{\oplus} + d}}$$

or

$$V_d = V_{R\oplus}\sqrt{\frac{R_{\oplus}}{R_{\oplus} + d}}$$

$$= 4.89\sqrt{\frac{R_{\oplus}}{R_{\oplus} + d}} \quad \text{mi/sec}, \qquad (9.6)$$

since the numerators $GM_{\oplus}$ under the square root sign are the same in both formulae and hence cancel out. From this we see that we can obtain the speed in a circular orbit at any distance d from the earth's surface by multiplying 4.89 miles (7.86 km) per second by the square root of $R_{\oplus}$ and dividing by the square root of the sum of the earth's radius and d.

9.11 The Period of an Artificial Satellite in a Circular Orbit

Once we know the radius of the circular orbit, as measured from the center of the earth, we can easily find the period of the satellite by using Kepler's third law. We know from this law that for a satellite moving around the earth at a mean distance $R_{\oplus} + d$ from the center of the earth the square of the period, P, is

$$P^2 = \frac{4\pi^2(R_{\oplus} + d)^3}{GM_{\oplus}}. \qquad (9.7)$$

We can easily find the period, P, with the aid of this formula by substituting the appropriate values of G, $M_{\oplus}$, and d, just as we did in calculating the speed. If we use the same units as we did in the formula for the speed (that is, centimeters and grams for distances and masses respectively),then the period is expressed in seconds.

Again we can simplify our calculation by noting that we can apply the same formula to the moon. We have in fact

$$P_{☾}^2 = \frac{4\pi^2(R_{\oplus} + d_{☾})^3}{G(M_{\oplus} + M_{☾})}.$$

Since the mass of the moon is 81 times smaller than that of the earth, we shall drop it from the denominator in this formula so that the denominator then becomes the same as that of the previous formula. If we now divide this formula (after having dropped the mass of the moon) by the previous one, we obtain

$$\frac{P^2}{P_{☾}^2} = \frac{(R_{\oplus} + d)^3}{(R_{\oplus} + d_{☾})^3}$$

or

$$P^2 = P_{☾}^2\left(\frac{R_{\oplus} + d}{R_{\oplus} + d_{☾}}\right)^3.$$

Since $R_{\oplus} + d_{☾}$, which is the mean distance of the moon's center from the earth's center, is just 60 times the radius of the earth (that is $R_{\oplus} + d_{☾} = 60R_{\oplus}$), we finally have

$$P^2 = P_{☾}^2\left(\frac{1}{60}\right)^3\left(\frac{R_{\oplus} + d}{R_{\oplus}}\right)^3$$

$$= \frac{P_{☾}^2}{216\,000}\left(1 + \frac{d}{R_{\oplus}}\right)^3,$$

$$P = \frac{27.322}{464.76}\sqrt{\left(1 + \frac{d}{R_{\oplus}}\right)^3}$$

$$= 0.058\,8\sqrt{\left(1 + \frac{d}{R_{\oplus}}\right)^3} \quad \text{days},$$

where the period is given in days since we have substituted 27.322 days for the period of the moon.

With the aid of this formula we find, for example, that a satellite moving in a circular orbit at a distance of 300 miles (483 km) above the earth's surface has a period of 1.59 hours. At a distance of 1 000 miles (1 610 km) above the earth's surface a satellite's period is about 2 hours, and its speed in its circular orbit is 4.37 miles (7.03 km) per second.

9.12 Artificial Satellite in Elliptical Orbit, Apogee and Perigee

We may note that even though a satellite is launched exactly parallel to the earth's surface at any given height above the earth's surface, it will move in a noncircular orbit if its speed is either greater or smaller than that for a circular orbit at that height. To see what is involved, we must keep in mind that if a satellite is in a circular orbit, it always moves at right angles to the line from it to the center of the earth. In an elliptical orbit, however, there are only two points at which the satellite moves at right angles to the line from itself to the center of the earth. These are the perigee and apogee (point closest to and farthest from the earth's center). If we keep this in mind, we see at once that if a satellite is launched exactly parallel to the earth's surface at a given height, but with a speed that is greater than the circular speed at that height (but less than the speed of escape), it must move in an elliptical orbit, and the point from which it is launched is then the perigee.

Since the speed of the satellite at the launching point is greater than the circular speed, the orbit must be larger than the circular orbit, because the kinetic energy and hence the total energy of the satellite is larger than its total energy would be in a circular orbit (see Section 8.30 for the relationship of the total energy of a planet to the size of its orbit), and hence the distance of the satellite from the center of the earth when it is diametrically opposite the launching point must be greater than the distance when it is launched. This is so because the major axis of the elliptical orbit (the distance from the launching point of the orbit to the opposite point of the orbit) is larger than the diameter of the circular orbit. As can be seen from Figure 9.8, this diametrically opposite point must then be the apogee of the satellite. If the satellite is launched with a speed less than the circular speed, then again the satellite moves in an elliptical orbit and again

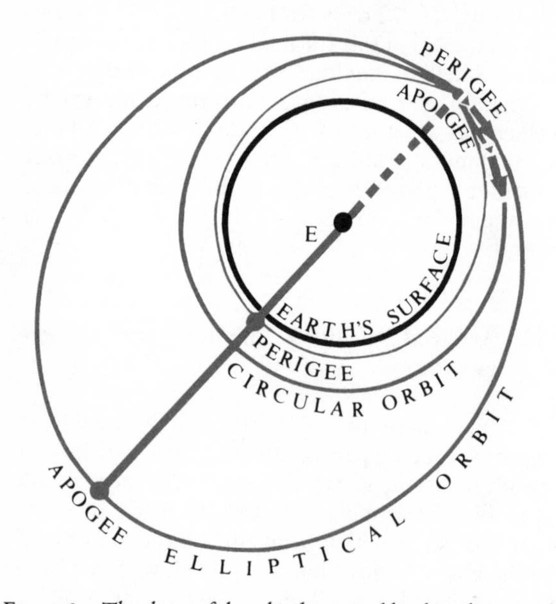

FIG. 9–8 *The shape of the orbit determined by the velocity with which the satellite is launched above the earth's surface. Only one velocity exactly perpendicular to the radius from the center of the earth gives a circular orbit. A greater perpendicular velocity has the perigee as its starting point, and a smaller perpendicular velocity has the apogee as its starting point.*

the launching point and the point diametrically opposite are the two points where the motion is exactly at right angles to the line from the center of the earth. These two points are therefore the apogee and perigee. However, the point from which the satellite is launched is now the apogee and the perigee is now the diametrically opposite point.

9.13 The Advantages of a Circular Orbit for an Artificial Satellite

It is, of course, advantageous to have a satellite move in as nearly circular an orbit as possible if the serious losses of energy resulting from atmospheric drag are to be avoided. Remember that the amount of energy required to put a satellite in orbit depends only on the size of the orbit and not on its shape. If we have a given amount of thrust available, the mean distance of the satellite from the center of the earth is the same whether the orbit is circular or elliptical. But, if the orbit is elliptical, the satellite at its apogee must be further from the earth than its mean distance, and at its perigee must be closer to the earth than its mean distance. This means that at its perigee it sweeps in too close to the earth and suffers a loss of energy because of atmospheric drag. As a result its orbit quickly becomes smaller and smaller, and the satellite spirals into the earth's atmosphere.

To illustrate this point, let us suppose that we wish to launch a satellite in an orbit whose mean distance from the earth's center is 4 400 miles (7 100 km), about 400 miles above the earth's surface. We know then that if the orbit has an eccentricity equal to e (see Sections 8.14 and 8.15), its distance from the earth's center at perigee is $4\,400(1 - e)$. If the eccentricity were equal to 0.1, the distance from the center of the earth at perigee would just be about 4 000 miles so that the satellite could not possibly continue in its orbit because it would practically be grazing the surface of the earth at its closest approach. Even if the eccentricity were 0.02, which is slightly greater than that of the earth's orbit, the satellite at perigee would come to within 300 miles of the earth's surface and hence would be subject to atmospheric drag.

We see from this discussion that launching a satellite that will stay in its orbit for a reasonably long time requires the greatest precision, since not only its speed but also the direction of its motion must be controlled with great accuracy at the moment of its launching. If the launching speed

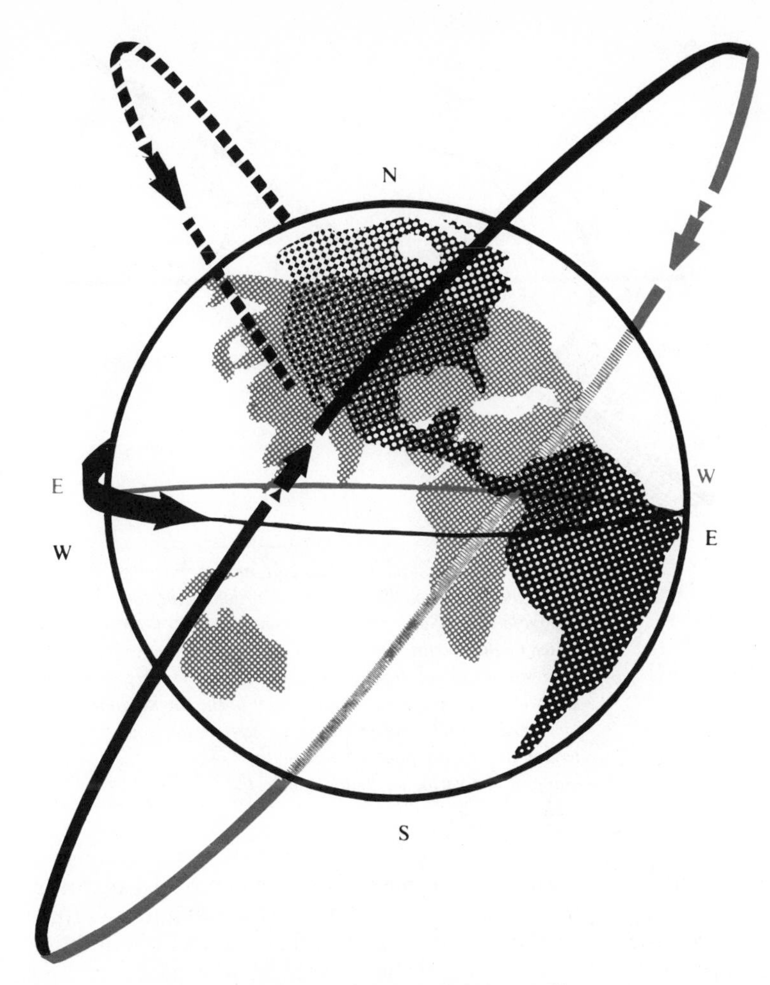

FIG. 9–9 *Changes in the apparent direction in which a satellite orbits the earth, as seen by an observer at different times of the day. The dotted line indicates the situation 12 hours after the orbit represented by the solid line.*

differs from the circular speed by as much as a few feet per second, the eccentricity of the orbit may be too large for the satellite to survive. The same kind of accuracy must be achieved in the direction of launching.

9.14 The Appearance of a Satellite

Once a satellite is put into orbit, it will continue to move in a fixed plane about the center of the earth (if we disregard the perturbations arising from bodies such as the moon and from the bulging of the earth at the equator) just the way the earth moves in a fixed plane about the center of the sun. However, since we are on a spinning earth, the satellite appears to us to be changing its direction constantly. The reason is that we are turning around under its orbit, so that a particular circle of longitude is always advancing eastwardly with respect to the satellite. The exact orientation of the observer with respect to the satellite at any moment depends on the angle which the plane of the satellite's orbit makes with the equatorial plane of the earth, and also on the relationship of the period of the satellite to the length of the sidereal day.

If a satellite is launched at some angle to the equator, it appears to move in different directions to two observers with the same latitudes but stationed on opposite sides of the earth. To one of these observers it appears to come over the southwestern horizon (assuming that it was launched in an eastwardly direction—the most suitable direction for a launching because it takes

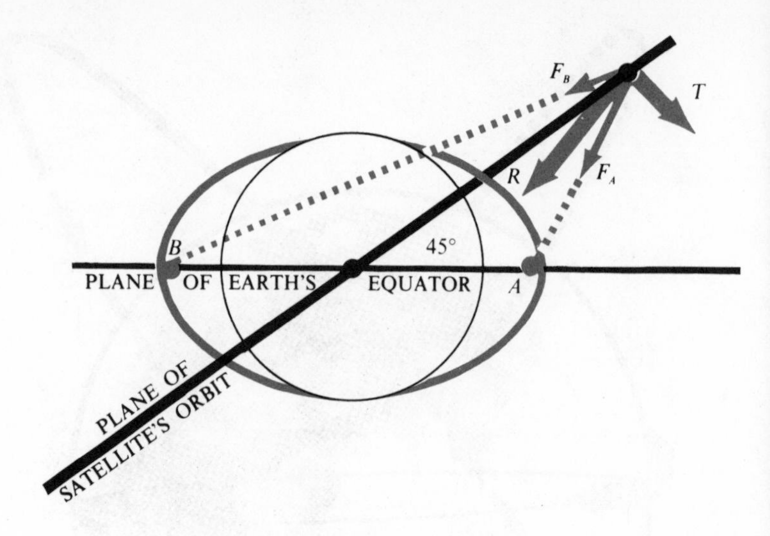

FIG. 9–10 *The gravitational pull of the bulge on each side of the earth on a satellite. (The distortion of the earth is greatly exaggerated.) Since the gravitational action of the bulge, A, closer to the satellite is greater than that of the distant bulge, B, the resultant, R, of these two forces has a component, T, which is at right angles to the plane of the orbit of the satellite and, hence, gives rise to a torque that always acts in the same direction, tending to pull the plane of the satellite's orbit into the equatorial plane of the earth, no matter where the satellite may be in its orbit.*

advantage of the slight additional boost given by the earth's rotation), and it appears to set in the northeastern horizon. To the observer on the opposite side of the earth the satellite appears to rise in the northwest and set in the southeast. Since the two observers interchange their positions relative to the orbit of the satellite every 12 hours, because of the earth's rotation, they see a continual variation in the direction of the motion of the satellite as it sweeps across their sky during each period of its revolution (Figure 9.9).

9.15 The Precession of the Orbit of a Satellite

Although we have assumed that the plane of the orbit of the satellite remains fixed in space, this is not really the case because of perturbations that arise from other bodies in the solar system, and from the fact that the earth bulges at the equator—the latter having the greater effect. To understand the effect, we consider a satellite the plane of whose orbit makes some angle (say, 45°) with the plane of the equator (as shown in Figure 9.10, in which the distortion of the earth is shown by means of a bulge at the equator). If we consider the earth's bulge that is closest to the satellite at some point in its orbit, and the bulge that is diametrically opposite, we see that these bulges exert a gravitational force on the satellite (an effect that is present in addition to the pull of the entire earth, which we must picture as emanating from the center of the earth), as indicated by the arrows.

We can break up each of these forces into two components: one lying in the plane of the orbit of the satellite, and the other perpendicular to this plane (as shown in Figure 9.10). The components of the two forces parallel to the plane of the orbit add up and simply have the same effect as though these forces were coming from the center of the earth, so that they cannot change the plane of the orbit. The two other components, however, are opposite to each other. Since the one arising from the distant bulge is smaller than the one arising from the bulge closest to the satellite, there is a net force on the satellite at right angles to the plane of its orbit. This, as we saw in our discussion of angular momentum in Chapter 5, gives rise to a torque on the orbit of the satellite and hence causes this orbit to precess in a westwardly direction. It does so because the torque acts in such a way as to try to bring the plane of the orbit of the satellite into the equatorial plane of the earth. Thus, if the orbit of the satellite crosses the plane of the equator at a point P at a given moment, it will cross at some point P' to the west of this at the end of one period.

9.16 Getting Away from the Earth

One may wonder why it is impossible at the present time to construct a vehicle that could leave the earth and move out into space at a comfortable, steady speed (at about 1 000 miles per hour) instead of by means of a single thrust which imparts an initial velocity sufficient to project it to the desired distance. Certainly it would be greatly to our advantage to travel out into space in this way, since everything could then be under careful control, and there would be no danger of high temperatures arising from the friction of the atmosphere. If one wants to travel from New York to Chicago, one does so at a steady speed, and one should, in principle, be able to do the same thing in going from the earth to the moon. The difficulty arises not because of a dynamical principle, but rather because of energy considerations.

The plane that travels at a steady speed from New York to Chicago must during the entire trip carry, on the average, one-half the total amount of fuel required for the entire trip. The same thing is true for the space traveler who wants to travel at a steady speed. Since the mass of fuel required to go from the earth to the moon is many times greater than that of the vehicle itself (each gram of payload requires about 5 grams of ordinary nonatomic fuels) refueling stations between the earth and the moon are necessary for a trip of this sort.

However, this problem can be solved, either by having refueling stations at various points in space, or by using atomic fuels for which the mass-to-fuel ratio is much more favorable, so that all the fuel that is needed can be carried along with the vehicle. Without such fuels, we must launch our vehicles by means of an initial thrust in which most of the fuel is used up at the very beginning of the launching.

9.17 The Speed of Escape from the Solar System

To get completely away from the earth a satellite must be launched at the speed of 7 miles (11 km) per second (speed of escape). But this does not mean that the satellite is then able to leave the solar system. The controlling factor here is the sun, and it is the speed of escape from the sun that we must reckon with if we want a satellite to leave the solar system. If a satellite were launched from the surface of the sun, it would have to have a speed of 375 miles per second to escape from the solar system, but at a distance of one astronomical unit from the sun the escape speed is about 26 miles per second (41 km/sec). Since the earth is already moving at a speed of $18\frac{1}{2}$ mi/sec (29.6 km/sec) with respect to the sun, we see that we can send a satellite into the distant regions of the solar system if we launch it from the earth at a speed of about 8 miles per second *above the speed of escape from the earth* in the same direction in which the earth is moving with respect to the sun. If the satellite is launched from the earth with a speed greater than 7 miles per second, but less than 44 mi/sec, in a direction opposite to the earth's motion, it will move in an orbit around the sun whose mean distance is smaller than 1 A.U.

By means of the method outlined in this chapter it is possible to determine fairly accurately what conditions must be imposed if a space vehicle is to have a desired orbit. But to achieve the accuracy that is required for launching interplanetary rockets, one has to determine the orbit by the methods outlined in the discussion of the n-body problem.

EXERCISES

1. Use vectorial construction to show the total acceleration imparted to Mars at some point in its orbit by the sun, Venus, the earth, and Jupiter. (Place Venus, the earth, and Jupiter at any position you wish in their orbits but not co-linear.)

2. Suppose an artificial satellite were launched horizontally (tangent to the earth's surface) with a speed sufficient to send it out to a maximum distance of 300 miles above the earth's surface. What would the eccentricity of the orbit be? What would its mean distance from the center of the earth be?

3. Suppose the earth were a perfectly smooth sphere of ice. What would be the maximum speed a skater could go and still stay in contact with the ice? (Neglect air resistance and the spin of the earth.)

4. Derive Newton's inverse square law of force from Kepler's third law of motion. (Assume circular orbits.)

5. What kind of third law of planetary motions would you have if the gravitational force varied inversely as the cube of the distance?

6. Would Kepler's second law still hold if the gravitational force varied as the inverse cube of the distance?

7. Consider a body approaching the sun from interstellar space. Is it possible for the sun to capture this body and force it to move in a closed orbit around the sun? (Neglect the action of the planets and other bodies in the solar system.) Give a reason for your answer.

8. If another star having the mass of the sun were to approach the sun and pass it at a closest distance of one-half an astronomical unit at a speed of 100 km/sec, what would the total angular momentum of the system be? What would the maximum gravitational attraction between the two bodies be?

9. Write down the equations of motion in a general three-dimensional Cartesian coordinate system for two mass points moving under their mutual gravitational attraction.

10. Reduce the equations in problem 9 to three equations by introducing relative coordinates and eliminating the motion of the center of mass of the two bodies.

11. Obtain Kepler's law of areas from the three equations in problem 10.

10

THE SPECIAL THEORY OF RELATIVITY

10.1 The Need for a New Gravitational Theory

When we discussed the n-body problem in the last chapter, we saw that the orbit of a planet would be a closed ellipse only if there were no gravitational perturbing effects from the other planets. If perturbations are taken into account, we find that there is an advance of the perihelion for each planet, which, except for a small residual effect, can be accounted for by Newtonian gravitational theory. This small residual effect, which in the case of Mercury amounts to 43″ of arc per century, cannot be explained in terms of the classical Newtonian gravitational theory and laws of motion.

We now know that to account for the discrepancy between the observed and computed positions (using Newtonian theory) of Mercury it is necessary to introduce entirely new concepts of space and time, and to replace Newton's law of gravitation by a new law which involves radical ideas about the relationship of geometry to the structure of the universe. These new ideas were developed by Einstein and constitute what we now call the theory of relativity.

The theory of relativity was developed in two parts:

1. *the special theory of relativity, which deals with the laws of physics as they are formulated by observers moving with uniform velocity with respect to each other; and*
2. *the general theory of relativity, which deals with observers in accelerated frames of reference or in gravitational fields.*

10.2 Einstein's Concept of Relativity

Although we must call upon the general theory of relativity, sometimes referred to as Einstein's theory of gravitation, to explain the anomalous behavior of Mercury, we shall have to start by considering the special theory because without the ideas developed there the general theory is difficult to understand. Before we discuss these theories we must have a clear understanding of Einstein's use of the word "relativity."

Everyone concedes that the way things appear in our universe depends upon the point of view of the observer. Thus the apparent brightness of a star, the apparent height of a tree, the apparent transverse motion of a car, all depend on the position of the observer relative to the object being observed. In that sense, of course, all of our knowledge is relative, so that the idea of relativity is nothing strange; it is an old concept in the world of physics and mathematics. However, Einstein did not introduce the concept of relativity in this sense, for his idea of relativity has to do with the very concepts of time and space themselves.

To comprehend the difference between Einstein's concept of relativity and the traditional concept, we need merely note that all fixed observers measuring the height of a tree accept the fact that *the tree possesses a real height* which is the same for all observers regardless of the visual angle that the tree subtends at the eye of each observer, as shown in Figure 10.1. However, according to Einstein's concept of relativity it is impossible to speak of the

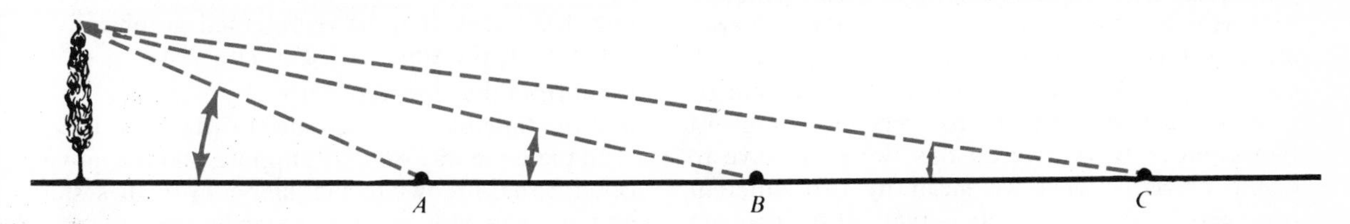

FIG. 10–1 *The apparent size of a tree depends on the distance of the observer relative to the tree.*

real dimensions of an object since these dimensions (and in fact all measurements of space and time) depend on how an observer is moving with respect to the object.

When we introduced Newton's law of motion and his law of gravitation, we implicitly accepted his ideas of absolute space and absolute time. According to Newton's picture of the behavior of particles subjected to forces, we must describe these particles as moving during time intervals that are the same for all observers and, at each instant, occupying positions in a three-dimensional spatial universe which is also the same for all observers. *In other words, according to Newton's picture, events in the universe unfold in a uniquely ordered way which is the same for all observers no matter where they are, or how they may be moving, so that space and time are absolute.*

This means that if any two observers in our universe arranged charts showing the same series of events in the order of their occurrence as they saw them, and on scales showing their correct separations as they measured them, they would both come up with identical charts. *The special theory of relativity DENIES this; and it is in this sense that we must understand the word "relativity" as it is used by Einstein, for, as we shall see, the special theory introduces the idea that both space and time are relative concepts, and that the ordering of events in our universe is not unique.* If an event A occurs one hour before event B, and these events are found to be 10 miles apart as determined by one observer, then according to Newtonian concepts of absolute space and time all observers should find event A occurring one hour before event B, and 10 miles distant from event B. However, according to the special theory of relativity, this is not so. We cannot, if we accept Einstein's theory, make any absolute statements about the distances or the time intervals between events.

10.3 Observers and Events

Why did Einstein find it necessary to deny the absolute concepts of space and time? To answer this question, we shall consider how different observers describe various events in the universe, and how these descriptions are to be compared with each other. We can best analyze the problem by considering two different observers recording just two events. Before we do this, we shall have to define precisely what we mean by two different observers. Of course, any two observers are different in the sense that they have different identities, but *we shall define two observers as being different, in our use of the term, only if they are moving with constant velocity relative to each other.* Such observers are referred to as **Galilean observers.** As long as *we assign uniform velocities to our observers, we shall be dealing with the special theory of relativity.* The general theory of relativity removes this restriction of constant velocity, and considers observers who may be moving with accelerated motions.

The two events which our two observers are to record are the explosions of two bombs at two different points and at two different instants of time (assuming that each explosion occurs at a different point and at a different instant of time). *We shall ask each observer to specify the position of each event, measured with respect to some frame of reference that he introduces, and the time of occurrence of each event, measured by a clock moving along with his frame of reference.* We shall assume throughout that all the measuring devices (rods, clocks, etc.) used by the two observers are identical in nature and exactly synchronized. Since we are concerned with observations involving distances and times, we shall have to introduce a frame of reference for each observer—that is, a coordinate system. We shall assume each of these coordinate systems to be of the rectangular Cartesian type.

10.4 The Use of a Coordinate System in Describing Events

Let one of our observers be attached to the earth, and let the other be moving with a speed v in a given direction with respect to the first observer. If X, Y, and Z are the three spatial coordinate axes of the fixed observer's rectangular coordinate system and T is his time axis, then this observer describes each event as he sees it (e.g., the explosion of one of the bombs) by assigning to it four numbers (x, y, z, t). The numbers x, y, and z give the spatial coordinates of the event as measured from the observer (the origin of this coordinate system) along the axes X, Y, and Z respectively. The number t, which gives the time of the event measured from some initial (zero) moment, may be represented graphically as a point on the time axis T. The distance of the event from the observer is given by $\sqrt{x^2 + y^2 + z^2}$, and the time of the occurrence of the event by t.

In the same way we shall suppose that the moving observer has his own rectangular coordinate system, and hence assigns to an event, as he sees it, a set of four numbers (x', y', z', t') where x', y', z' are the

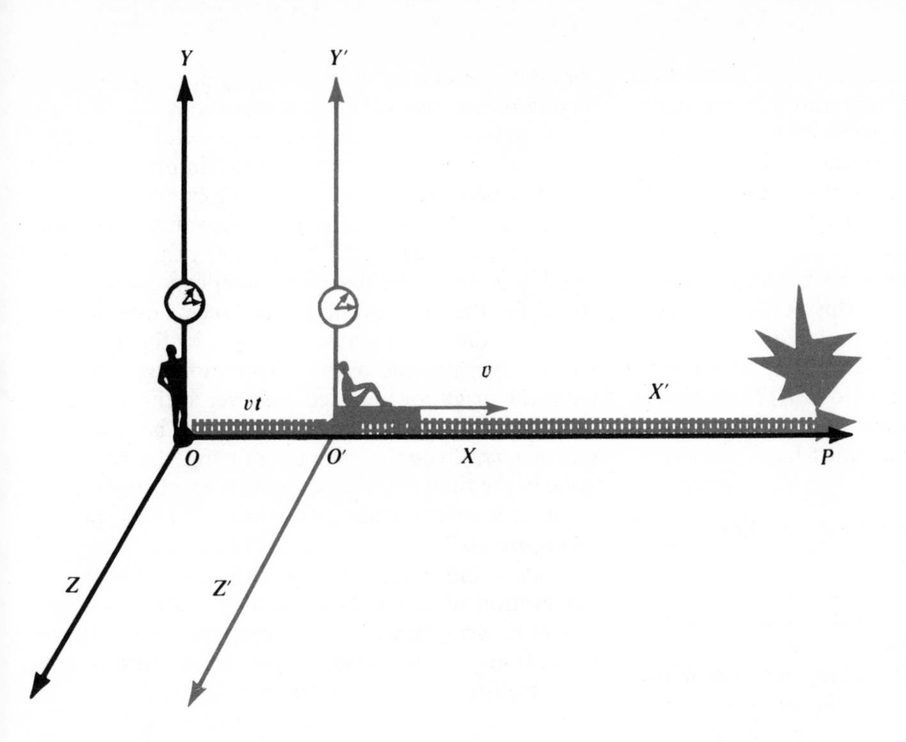

FIG. 10–2 *Two observers, one fixed at position O with respect to an event at P, and the other at O′, moving with a speed, v, with respect to the event at P.*

spatial coordinates of this event for the moving observer and t' is the time coordinate. The distance of this event from the moving observer (at the origin of his coordinate system) is $\sqrt{x'^2 + y'^2 + z'^2}$, and t' is the time the event occurred measured from some zero moment according to the moving observer's clock. (Note that we use unprimed quantities to represent the description of an event in the fixed system, and primed quantities for an event in the moving system.)

10.5 Transformations of Coordinates for Observers with Uniform Relative Velocity

To understand the nature of the problem that faced Einstein, we must ask ourselves how we are to pass from the description given by the first observer to the description given by the second observer. In other words, if (x, y, z, t) and (x', y', z', t') describe two events as seen by two different observers, how are we to know whether these descriptions refer to the same event occurring in our universe, or to two different events? It is clear that we can only determine this if we have some means of passing from the language of one observer to that of the other observer. This is what the mathematician calls a **transformation**. If we know how to pass from one description to the other, we can transform any set of measurements made by one observer in his system to the measurements made by the other observer, since all measurements, in the final analysis, can be reduced to sets of time and distance measurements. What a transformation of this sort enables us to do, then, is to relate each number of the set (x', y', z', t') to the set of numbers (x, y, z, t).

We can best see how to find such a transformation, or mathematical dictionary, by considering the very simplest case of two different observers. We suppose that the moving observer is situated on an open box car moving with constant speed v along a straight track, which we shall take to be in the X direction of both observers, whereas the fixed observer is stationed at some point on the track. There is to be no motion in the Y or Z directions, as shown in Figure 10.2. We suppose further that each observer has a clock with him, both of which are to read zero at the moment that the origin O' of the moving observer coincides with the origin O of the fixed observer, and that all times in both systems are to be measured from this zero instant. We now assume that at some instant t, as indicated by the clock attached to the fixed observer, a bomb explodes at a point down the road and at some later instant at a different point another bomb explodes. What is the relationship of the description of these events as given by the fixed observer to the description given by the moving observer?

The fixed observer notes that the first bomb explodes at a distance x from his origin, whereas the moving observer notes that this bomb explodes at the distance x' from his origin. How are x and x' to be related? That is, what are the transformation equations leading from (xt) to $(x't')$? In order to answer this we must understand the properties that transformation equations must have.

10.6 Properties of Transformation Equations: Coordinate Invariance

It is clear that a set of transformations which enable us to pass from the description of one observer to that given by the other observer must fulfill certain requirements. *On the one hand, these transformations must not violate the accepted truths about space and time, and on the other hand, they must leave the intrinsic features of an event unaltered when we use the equations to pass from one description to the other.* In other words, our mathematical dictionary must not inject into the description anything that is not really there, and must not tell us things that we know to be wrong about space and time. These two ideas together constitute what Einstein had in mind when he spoke of **invariance**. What this concept means essentially is that, in order to be valid, *transformations must leave the laws of nature unaltered, that is, invariant.* Or putting it differently, if our two observers are recording events (experiments) which ultimately enable them to formulate a law of nature, *the transformations must be such as to give the law the same mathematical form in both systems.*

10.7 The Galilean Transformations Based on Newtonian Concepts of Absolute Space and Time

If we wish to obtain the Galilean transformation equations (named after Galileo) from the fixed to the moving system in accordance with the Newtonian concepts of absolute space and time, we must impose the condition that these equations incorporate in them what Newton considered to be self-evident truths about space and time. Thus, since Newton believed that time intervals are the same for all observers, we must at once place t equal to t', as one of our transformation equations.

To go from x to x', we must use Newton's concept of absolute space, which means simply that the distance between the two exploding bombs must be the same for both observers. From this it follows that the x for one of the bombs (its distance along the X axis from the fixed observer) can differ from the x' for the same bomb (its distance along the X' axis from the moving observer) only by the distance which separates the origin of the moving observer from the origin of the fixed observer at the moment that this bomb explodes. But this is just the distance vt, through which the moving observer has gone in the time interval before this bomb exploded. In other words, x' must just equal $x - vt$, as shown in Figure 10.2.

Since y and z are the same as y' and z', because the motion occurs only along the X direction, we can now write down as the complete set of transformation equations that hold in a Newtonian universe for these two observers:

$$\begin{aligned} x' &= x - vt, \\ y' &= y, \\ z' &= z, \\ t' &= t. \end{aligned} \tag{10.1}$$

It is easy to see that these equations do, indeed, leave unaltered distances between the same events as measured by different observers and leave unaltered the time interval between events as measured by different observers. There is, therefore, incorporated in this set of equations a very definite concept of space and time, namely that space and time separately are absolute for all observers.

10.8 The Invariance of the Laws of Nature with Respect to These Equations

What about the laws of nature? Do these equations leave them invariant? It is again easy to see, if we limit ourselves to Newton's laws of motion, that these transformation equations do, indeed, leave them invariant. The reason is that these transformations have no effect upon the measured accelerations, so that the acceleration of an object measured by one observer is the same as that measured by the other observer. From this it follows that the law $F = ma$ is unaltered when we go from the fixed system to the moving system.

Since all of classical dynamics is built on this law, we see that these transformations do, indeed, have the desired property of leaving the laws of mechanics invariant. Why, then, was there a need for Einstein's theory of relativity? We shall see its necessity if we consider the laws that govern not mechanics, but another branch of physics, namely, **optics**. *We shall then see that the classical transformation equations that we wrote down above lead to contradictions with experimental evidence.*

10.9 The Speed of Light and the Galilean Transformations

We must first consider what the Newtonian transformations tell us about the speed of an object as viewed by the fixed and by the moving observer. If an object is moving along the x direction with constant uniform speed, U, as measured in the fixed system, we must have

$$U = \frac{x}{t}.$$

If its speed, as measured in the moving system, is U', we have

$$U' = \frac{x'}{t'} \quad \text{or} \quad \frac{x'}{t},$$

since t and t' are equal according to our transformation equations. We thus have from the first transformation equation, on dividing both sides through by t,

$$\frac{x'}{t} = \frac{x - vt}{t} = \frac{x}{t} - v,$$

or

$$U' = U - v. \tag{10.2}$$

In other words, *the speed of the object as measured by the moving observer is obtained by subtracting the speed of the moving observer from the speed of the object, as measured by the fixed observer.* This seems quite obvious to us for it simply tells us, for example, that if we are moving in a train at 60 mph, and an airplane is flying in the same direction at 100 mph, then we should observe it as passing us at a speed of 40 mph. However obvious this seems to us, it is nevertheless precisely this feature of the classical transformations that leads to a conflict with experimental evidence. To see why this is so we must compare the speed of a beam of light moving along the X and X' axes measured by the fixed observer with the speed of this beam measured by the moving observer.

In the second half of the nineteenth century James Clerk Maxwell, one of the greatest scientific geniuses of all time, showed by means of a series of mathematical investigations that light is an electromagnetic phenomenon, and that in a vacuum it is propagated with a speed that is the same in all directions, and is independent of the motion of the source of the light. Although Maxwell did not realize it at the time, a further consequence of his electromagnetic theory is that the speed of light in a vacuum is the same for all observers, and hence, according to Maxwell's theory, a *constant of nature*.

If we accept Maxwell's theory and his electromagnetic equations (which describe the propagation of light *in vacuo*), then these equations, as laws of nature, must by the principle of invariance have the same form in all Galilean systems (i.e., for all observers moving with uniform speed with respect to each other). This means that the speed of light, which is a constant, appearing in these equations must be the same for all observers and hence *independent of the motion of the observer.* But this at once throws us into conflict with the classical transformation equations. For let us suppose that a beam of light is advancing along the X direction, and passes both the fixed and the moving observer. What is the relationship between the speeds of the beam as measured by the two observers, as shown in Figure 10.3? According to the classical picture

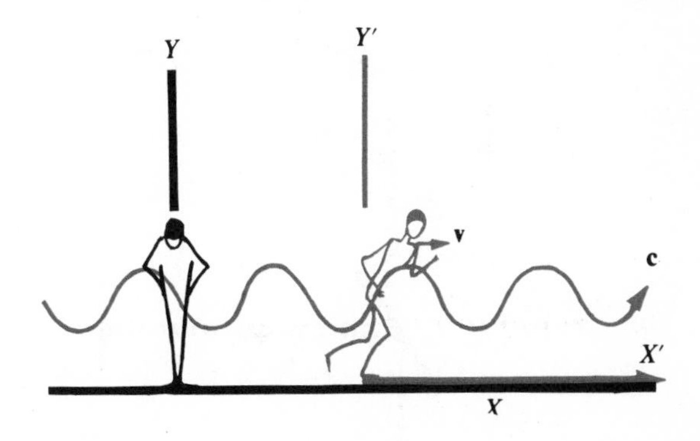

FIG. 10–3 *Two observers, one taken as fixed and the other moving with speed v with respect to the first, measure the speed of a ray of light passing them both.*

we must treat the beam just as we treated the moving particle, and if c is the speed of the beam as measured by the fixed observer, then according to our transformation equations the speed c' should be smaller than c by an amount equal to the speed of the moving observer. In other words we have

$$c' = c - v.$$

But we see that this is in contradiction of Maxwell's theory.

It follows, then, that if Maxwell's theory is a correct statement (that is, a law of nature), then the classical transformation equations cannot be correct and the Newtonian concepts of space and time which are built into these equations must be revised. On the other hand, if we accept the classical transformations and hence the Newtonian concepts of space and time, we must deny the validity of Maxwell's statement about the speed of light. Here then we have a precisely stated conflict between two theories which can be resolved only by experimental evidence.

10.10 The Michelson-Morley Experiment

In the year 1887 Michelson and Morley performed their famous experiment to determine what effect the motion of the earth has on the speed of light, as measured by an observer on the earth. They tried to measure the speed of the earth around the sun by comparing the speed of a beam of light moving parallel to the earth's direction of motion with the speed of a beam of light moving at right angles to the earth's motion (see Figure 10.4). If we accept the Galilean transformation equation as

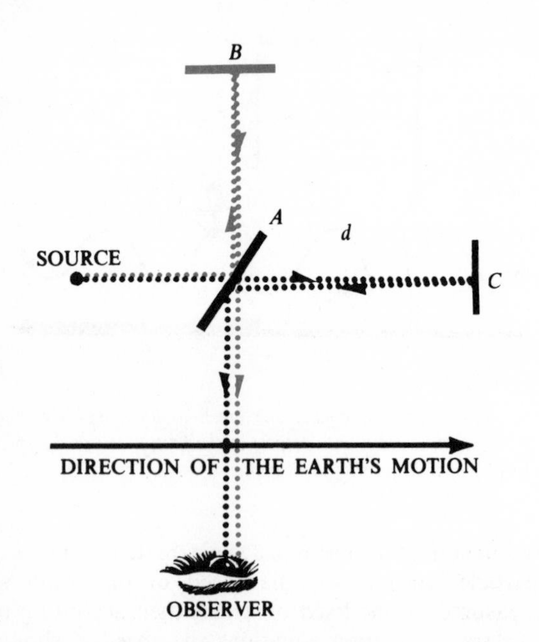

FIG. 10–4 *The arrangement of three mirrors, A, B, and C, in the Michelson–Morley interferometer experiment. The gray and black dotted lines show the splitting of the beam of light by the 45° mirror, A.*

correct, then, as we have seen, there should be a difference between the speeds of these two beams, and the earth's motion should be detectable. *However, the experiment failed to detect any difference in the speeds of these two beams, and hence proved conclusively that the speed of light measured in a vacuum is the same for all observers regardless of how they are moving.*

This result appeared so strange and contrary to common sense that all attempts to explain it—previous to Einstein—were based on the idea that the motion of an observer somehow or other has some physical effects on his measuring devices and hence distorts his measurements. It was Einstein, however, who first pointed out that we can obtain a clear understanding of this phenomenon only if we drastically revise our concepts of space and time.

10.11 Einstein's Concept of Simultaneity

We cannot go into a detailed analysis of Einstein's reasoning in this book but must refer the student to the many treatises on relativity for a complete picture of Einstein's procedure. However, we may note here that Einstein demonstrated by means of simple thought (gedanken) experiments that such concepts as the simultaneous occurrence of two events are not absolute concepts at all, but depend instead upon the motion of the observer. *Thus, if a fixed observer finds that two events separated in space occur simultaneously, a moving observer finds, in general, that they are not simultaneous.* This follows directly from the fact that the speed of light is the same for all observers. From an analysis of this kind Einstein finally concluded that one can accept Maxwell's result that the speed of light is the same for all observers only by rejecting the Newtonian concepts of absolute space and time, and hence rejecting the classical transformation equations.

As Einstein saw, the classical transformation equations must be replaced by a set of equations that leave the speed of light invariant when one goes from one moving observer to another. This means that both distance and time change when such a transformation is carried out, so that t' cannot be equal to t, and x' cannot be equal to $x - vt$.

What equations are we then to introduce instead of the Newtonian transformations? It is clear that whatever these equations are to be, two things can be said about them:

1. *they must contain c, the speed of light in a vacuum, in a very special way, and*

2. *they must differ from the classical transformation equations only by terms that are important when we are dealing with observers who are moving with speeds close to the speed of light.*

When the speed of the moving observer is small compared with the speed of light, the new transformations should reduce to the classical transformations.

10.12 The Einstein-Lorentz Transformations

To obtain the new transformation equations Einstein noted that when a beam of light is advancing along the X axis, the fixed observer describes its motion by means of the equation (the distance traveled by the beam of light is its speed, c, multiplied by the time, t)

$$x^2 = c^2t^2,$$

and the moving observer describes it by the equation

$$x'^2 = c^2t'^2,$$

since the speed of light in a vacuum, c, must be the same for both observers. In other words, the quantity $x^2 - c^2t^2$ for the fixed observer must equal the quantity $x'^2 - c^2t'^2$ for the moving observer. This condition, that is,

$$\boxed{x^2 - c^2t^2 = x'^2 - c^2t'^2}, \qquad (10.3)$$

is the basis upon which Einstein constructed his transformation equations.

Starting from Equation (10.3), which states that the speed of light (*in vacuo*) is the same for all observers, we now ask what the correct transformation equations are and how x' must be related to x, and t' to t. By means of algebra it is easy to show that Equation (10.3) is satisfied if the following transformation equations hold

$$x' = \frac{x - vt}{\sqrt{1 - \frac{v^2}{c^2}}}, \qquad x = \frac{x' + vt'}{\sqrt{1 - \frac{v^2}{c^2}}},$$

$$y' = y, \qquad \text{and} \qquad y = y', \qquad (10.4)$$

$$z' = z, \qquad z = z',$$

$$t' = \frac{t - \frac{v}{c^2}x}{\sqrt{1 - \frac{v^2}{c^2}}}. \qquad t = \frac{t' + \frac{v}{c^2}x'}{\sqrt{1 - \frac{v^2}{c^2}}}.$$

These equations, which are referred to as the Einstein-Lorentz transformations, are the basis of the special theory of relativity. We see that if v is very much smaller than c, these equations do, indeed, reduce to the classical Galilean transformations so that *the consequences of the special theory of relativity come into evidence only when the speed of the moving observer is sufficiently large compared with the speed of light; we then say this observer is moving with* ***relativistic speed.***

10.13 The Consequences of the Special Theory of Relativity: Contraction of Measuring Rods

Let us first consider the distance between two points fixed in the moving system, as measured by the fixed and the moving observer. If x'_1 is the position of the first point in the moving system as determined by the moving observer, and x'_2 is the position of the second point, then the distance between the two points as measured by the moving observer is $x'_2 - x'_1$.

To find this distance as measured by the fixed observer, we note that the fixed observer must look at these two points (x_2 and x_1 in his frame of reference) simultaneously—that is to say, at the time t as indicated by his clock. We therefore have, applying the Einstein transformations to the two points x'_1 and x'_2, the equations

$$x'_1 = \frac{x_1 - vt}{\sqrt{1 - \frac{v^2}{c^2}}},$$

$$x'_2 = \frac{x_2 - vt}{\sqrt{1 - \frac{v^2}{c^2}}}.$$

On subtracting the first equation from the second, we obtain

$$x'_2 - x'_1 = \frac{(x_2 - x_1)}{\sqrt{1 - \frac{v^2}{c^2}}}$$

or

$$x_2 - x_1 = (x'_2 - x'_1)\sqrt{1 - \frac{v^2}{c^2}}. \qquad (10.5)$$

Since the left-hand side is the distance between x_2 and x_1 as measured by the fixed observer, and $x'_2 - x'_1$ is the distance between these two points as measured by the moving observer, we can see that lengths as measured by an observer shrink by the factor $\sqrt{1 - (v^2/c^2)}$ if these lengths are moving with

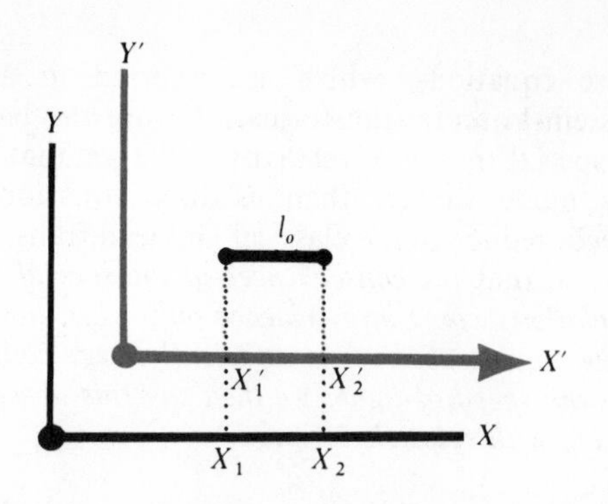

FIG. 10–5 *The measurement by a fixed observer of a moving length, l_0, (moving in the direction of its length). The fixed observer must sight the two end points x_1 and x_2 simultaneously in his coordinate system and compare the difference between these two coordinates with the difference between x_2' and x_1', which are the coordinates of the two end points in the moving system.*

respect to this observer with a speed v in the direction of their extension (Figure 10.5).

One can show that if the moving observer measures a distance fixed in the frame of reference of the fixed observer, the moving observer finds a shrinkage of this distance by exactly the same factor. In other words, there is complete reciprocity between the two observers, so that the only thing that matters in the measurement of a length is the relative speed of the length and the observer. *Whether it is the observer who is moving or the length that is moving makes no difference; the measured length is smaller by the same amount in both cases.*

10.14 The Contraction of Time for a Moving Observer

In the same manner we can show that if a fixed observer compares his clock with a clock attached to the moving observer he finds that the moving clock runs slower by the factor $\sqrt{1 - (v^2/c^2)}$. On the other hand, the moving observer on comparing his clock with the clock in the fixed system finds that the clock in the fixed system is slower by the factor $\sqrt{1 - (v^2/c^2)}$, where v is the relative speed of the two observers. In other words, it makes no difference which observer we consider as moving as far as these space and time effects are concerned—there is a shrinkage in either case.

10.15 The Transformation of the Velocity of a Body

Because the Einstein-Lorentz transformations lead to new concepts of space and time, it is clear that the velocity of a moving body as measured by a fixed observer is not related to the velocity of this body measured by the moving observer by means of the simple algebraic formula (10.2) that holds for the classical case. If U is the velocity of the body as measured in the fixed system (we assume the body to be moving along the X axis) and U' is the velocity as measured in the moving system, we have, on dividing the first of the Einstein transformation equations by the last,

$$\frac{x'}{t'} = \frac{x - vt}{t - \dfrac{v}{c^2}x} = \frac{\dfrac{x}{t} - v}{1 - \dfrac{v}{c^2}\dfrac{x}{t}},$$

since $\sqrt{1 - (v^2/c^2)}$ cancels out. Since $x'/t' = U'$, and $x/t = U$, we have

$$U' = \frac{U - v}{1 - \dfrac{v}{c^2}U}. \tag{10.6}$$

From this equation we see that adding velocities is a more complicated process in the theory of relativity than it is in the classical case. It is easy to see from this equation that no speed can exceed the speed of light, and that the speed of light must be the same for all observers. In fact, if $U = c$ (a beam of light moving along the positive X axis) we have from the above equation

$$U' = \frac{c - v}{1 - \dfrac{v}{c}} = \frac{c(c - v)}{c - v} = c,$$

so that the speed of the beam measured in the moving system is also c.

Just as we can use the Einstein-Lorentz transformations to compare rods and clocks in two coordinate systems moving with uniform speed with respect to each other, we can use these equations to find how the same wave of light appears to the two observers attached to these coordinate systems. When we do this we find that the observed frequency of the wave is different in the two systems. The relationship between the frequency measured by the fixed observer and the frequency measured by the moving observer is the **relativistic Doppler effect**, which differs from the classical Doppler effect (discussed in Section 5.14) by a slight relativistic correction. We also find that the direction of the wave in the fixed system is different from the direction in the moving system and that the relationship between these two directions is just the aberration of light. In other words, the Doppler effect and the aberration of light can be derived from the special theory of relativity.

10.16 The Space-Time Interval between Two Events

We can arrive at another consequence of the special theory of relativity if we consider the mass of a body as measured by our two observers. If an object is at rest relative to an observer, we assign to it an inertial mass in the usual way by measuring how much acceleration a known force imparts to it. We call this the **rest mass** of the particle, and designate it as m_0. *To the observer who is moving with a speed v relative to this body, however, the inertial mass is certainly not equal to the rest mass m_0.* The reason is that the concept of acceleration, involving as it does length and time measurements, is not an absolute entity, so that the inertial concept is not the same for all observers.

In order to see how the mass of a body changes when we go from one frame of reference to another, we again consider Einstein's fundamental hypothesis, namely, that the quantity $x^2 - c^2t^2$ (the space-time interval from the observer at the origin to an event) is the same for all observers. We shall, however, enlarge upon this somewhat to give it a more profound meaning. Let an observer measure the distance between two close events using a Cartesian coordinate system. If the spatial coordinates of one event are x_1, y_1, z_1 and those of the other event are x_2, y_2, z_2, then we know by the Pythagorean theorem that the distance (Δl) between the two events is given by

$$\begin{aligned}(\Delta l)^2 &= (x_2 - x_1)^2 + (y_2 - y_1)^2 + (z_2 - z_1)^2 \\ &= (\Delta x)^2 + (\Delta y)^2 + (\Delta z)^2,\end{aligned} \tag{10.7}$$

where the symbol Δ is used to indicate a small quantity, so that this notation tells us that we are dealing with neighboring events.

10.17 The Invariance of the Space-Time Interval

Let Δt be the small time interval between these two events. Now according to the Newtonian concept of absolute space and time, all observers must find the same Δl and the same Δt for the two events. However, this is not the case in the special theory of relativity, for, as Einstein pointed out, the quantity that must be the same for all observers is the following combination of Δl and Δt:

$$\begin{aligned}\Delta s^2 &= (\Delta l)^2 - c^2\,\Delta t^2 \\ &= \Delta x^2 + \Delta y^2 + \Delta z^2 - c^2\,\Delta t^2.\end{aligned} \tag{10.8}$$

We may look upon Δs as a new kind of interval, which we shall refer to as a space-time interval. *The statement that the space-time interval between any two events is an invariant (that is, the same for all Galilean observers) is the essence of the special theory of relativity.* This means that

$$\begin{aligned}\Delta s^2 &= \Delta x^2 + \Delta y^2 + \Delta z^2 - c^2\,\Delta t^2 \\ &= \Delta x'^2 + \Delta y'^2 + \Delta z'^2 - c^2\,\Delta t'^2 \\ &= \Delta s'^2.\end{aligned} \tag{10.9}$$

The space-time interval may be looked upon as a four-dimensional extension of the ordinary three-dimensional space interval Δl, as though the quantity $c\,\Delta t$ constituted a fourth spatial coordinate. However, the fact that $c^2\,\Delta t^2$ does not add on to $\Delta x^2 + \Delta y^2 + \Delta z^2$, as a real fourth spatial coordinate would according to Pythagoras' theorem, but, rather, must be subtracted from it, shows us there is something special in this temporal fourth dimension as compared with the three spatial dimensions. *In any case, the special theory of relativity introduces time on a kind of equal footing with space in our universe, and tells us that the only thing that has absolute significance to all observers is the space-time continuum of the universe, and not the space part separately nor the time part separately.*

10.18 World Lines of Events

To describe any event in the universe, therefore, we must always treat the spatial and temporal aspects of the event together, and always combine them in such a way that we obtain the space-time interval Δs as defined above. One can now imagine events portrayed graphically in a four-dimensional coordinate system in which there are three mutually perpendicular spatial axes and a fourth time axis at right angles to the other three. The history of any object in this extended coordinate system is represented by paths which have an absolute significance. *Such paths are called* **world lines.** For a complete discussion of this fascinating and important aspect of relativity, the student is referred to treatises on the special theory, such as Max Born's *Relativity Theory.*

In order to determine the relationship between the mass of a body as measured by a fixed observer and that measured by a moving observer, we use the fact that the space-time interval is an invariant for these two observers. We begin by considering the equation

$$\Delta s^2 = \Delta x^2 + \Delta y^2 + \Delta z^2 - c^2 \Delta t^2$$
$$= \left[\left(\frac{\Delta x}{\Delta t}\right)^2 + \left(\frac{\Delta y}{\Delta t}\right)^2 + \left(\frac{\Delta z}{\Delta t}\right)^2 - c^2\right] \Delta t^2 \qquad (10.10)$$

where we have factored out Δt^2 on the right-hand side.

Let v be the speed of a particle in the coordinate system X, Y, Z, T of the fixed observer, and let Δx, Δy, Δz, Δt refer to the coordinate intervals between neighboring points on the world line of this particle; then

$$\frac{\Delta x}{\Delta t} = v_x, \quad \frac{\Delta y}{\Delta t} = v_y \quad \text{and} \quad \frac{\Delta z}{\Delta t} = v_z, \qquad (10.11)$$

where v_x, v_y, v_z are the x, y, z components of the velocity of the particle. We therefore have

$$(\Delta s)^2 = (v_x^2 + v_y^2 + v_z^2 - c^2)\,\Delta t$$
$$= (v^2 - c^2)\,\Delta t^2$$
$$= -c^2\left(1 - \frac{v^2}{c^2}\right)\Delta t^2. \qquad (10.12)$$

10.19 The Variation of Mass of a Moving Particle

If we use this relationship for Δs^2, we may then write, using Equation (10.10),

$$-c^2\left(1 - \frac{v^2}{c^2}\right)\Delta t^2 = (v_x^2 + v_y^2 + v_z^2 - c^2)\,\Delta t^2,$$

or,

$$c^2 = -\left(\frac{v_x}{\sqrt{1 - \frac{v^2}{c^2}}}\right)^2 - \left(\frac{v_y}{\sqrt{1 - \frac{v^2}{c^2}}}\right)^2 - \left(\frac{v_z}{\sqrt{1 - \frac{v^2}{c^2}}}\right)^2 + \left(\frac{c}{\sqrt{1 - \frac{v^2}{c^2}}}\right)^2, \qquad (10.13)$$

where we have canceled out the Δt^2 on both sides and divided through by $-[1 - (v^2/c^2)]$. *Note that the quantity on the left-hand side, being just the speed of light, is the same for all observers and hence the quantity on the right-hand side is also the same for all observers.* This will still be true if we multiply through by the square of the rest mass, m_0^2, of the particle (that is, the mass of the particle as measured by an observer attached to the particle) since the rest mass of any particle is a definite number that is the same for all observers. We therefore have

$$(m_0c)^2 = -\left(\frac{m_0v_x}{\sqrt{1 - \frac{v^2}{c^2}}}\right)^2 - \left(\frac{m_0v_y}{\sqrt{1 - \frac{v^2}{c^2}}}\right)^2 - \left(\frac{m_0v_z}{\sqrt{1 - \frac{v^2}{c^2}}}\right)^2 + \left(\frac{m_0c}{\sqrt{1 - \frac{v^2}{c^2}}}\right)^2. \qquad (10.14)$$

We now consider $m_0v_x/\sqrt{1 - (v^2/c^2)}$, one of the three spatial terms that appear on the right-hand side of Equation (10.14). We know from our discussion of momentum in Chapter 5, Section 5.19, that from the classical point of view, m_0v_x is the x component of the momentum of a particle moving with a velocity v. This is not the case in relativity theory, since time and space are not absolute, so that the classical concept of momentum is not an absolute entity either. However, just as one must introduce a single four-dimensional space-time interval to replace the classical separate space and separate time intervals, so must we seek a four-dimensional space-time-momentum concept that is the same for all observers.

But we see that we have just such a quantity on the right-hand side of the above equation, for we have the sum of the squares of four terms equal to an invariant quantity $(m_0c)^2$, which is itself the square of a momentum. Therefore, each of the quantities on the right-hand side must be the square of a momentum, so that the x component of the momentum in relativity theory is no longer m_0v_x, but rather $m_0v_x/\sqrt{1 - (v^2/c^2)}$.

This then is the x component (the same thing applies to the y and z components) of the momentum of a body moving with a speed v as measured by a fixed observer. But we know that if the mass of this moving body, as measured by the fixed observer, is m, then by definition the x component of its momentum as measured by the fixed observer is mv_x. These two expressions for the x components of the momentum must therefore be equal, so that we obtain

$$mv_x = \frac{m_0v_x}{\sqrt{1 - \frac{v^2}{c^2}}}$$

or

$$m = \frac{m_0}{\sqrt{1 - \frac{v^2}{c^2}}}. \qquad (10.15)$$

This tells us that the mass of a moving body increases as its speed increases, and approaches infinity as the speed of the body approaches the speed of light. From this we see again that the speed of light is at an absolute maximum, which no material body can ever attain. We finally note that the increase in the mass of a body depends only on the relative motion of the body and the observer; it does not matter whether we think of the body as moving and the observer as being fixed, or vice versa.

10.20 The Relativistic Energy-Momentum Relationship

We know that the first three terms on the right-hand side of Equation (10.13) are the squares of the x, y, and z components of the momentum. What then is the significance of the fourth term on the right-hand side, or if we wish to speak of it in that way, the time component of the momentum? To see what this term means, we rewrite the equation as follows:

$$\left(\frac{m_0c}{\sqrt{1-\frac{v^2}{c^2}}}\right)^2 = (m_0c)^2 + \left(\frac{m_0v_x}{\sqrt{1-\frac{v^2}{c^2}}}\right)^2 + \left(\frac{m_0v_y}{\sqrt{1-\frac{v^2}{c^2}}}\right)^2 + \left(\frac{m_0v_z}{\sqrt{1-\frac{v^2}{c^2}}}\right)^2$$

$$= (m_0c)^2 + p_x^2 + p_y^2 + p_z^2$$

$$= m_0^2c^2 + p^2, \qquad (10.16)$$

where p_x, p_y, p_z are the x, y, z components of the momentum of the particle, and p^2 is the square of the total momentum. Since $m_0/\sqrt{1-(v^2/c^2)}$ is just the mass of the moving particle as measured by the fixed observer, we finally have

$$m^2c^2 = m_0^2c^2 + p^2$$

or, on multiplying through by c^2,

$$m^2c^4 = m_0^2c^4 + p^2c^2. \qquad (10.17)$$

We can now give a meaning to the term on the left-hand side of this equation, which we obtained from the fourth or time part of our original equation. It is simply the square of an energy, so that it must represent the square of the total energy of a freely moving particle (a particle with no forces acting on it). This leads to a very important and profound conclusion:

The square of the total energy, E^2, of a particle moving freely in space, as measured by a fixed observer, consists of two parts, one of which, c^2p^2, is related to the square of the old classical kinetic energy concept, and the other of which, $m_0^2c^4$, is entirely new and arises completely as a result of relativistic principles.

We may express this algebraically as follows:

$$\boxed{E^2_{\text{(total energy)}} = m_0^2c^4 + c^2p^2}. \qquad (10.18)$$

10.21 The Einstein Mass-Energy Relationship

We observe that even when a particle is not moving, in other words, *when the momentum p of the particle is equal to zero, it still possesses an amount of energy equal to m_0c^2.* This is the basis for the famous Einstein relationship

$$\boxed{E = mc^2}, \qquad (10.19)$$

which tells us that *matter and energy are equivalent.*

Equation (10.19) tells us that if we could, somehow or other, convert one gram of matter into energy, we would have available $(3 \times 10^{10})^2$ ergs of energy. Because measurable amounts of mass are converted into energy in nuclear reactions, vast amounts of energy are released by the hydrogen bomb.

10.22 The Relativistic Kinetic Energy

Since the total energy of a particle in motion is mc^2, and its energy when it is at rest in the system of the observer is m_0c^2, the relativistic kinetic energy of this particle (moving with speed v) is equal to $mc^2 - m_0c^2$. We thus have, using Equation (10.15),

$$K.E. = m_0c^2\left[\frac{1}{\sqrt{1-\frac{v^2}{c^2}}} - 1\right]. \qquad (10.20)$$

One can show algebraically that if v is very small as compared with c, this expression is very nearly equal to $\frac{1}{2}m_0v^2$; under these circumstances the classical expression for the kinetic energy, $\frac{1}{2}m_0v^2$, is approximately correct.

10.23 The Four-Vector in Relativity Theory

In classical physics the laws of nature are expressed as relationships among vector quantities such as displacement, velocity, acceleration, and force, and these vector quantities are the physical entities that we must measure to describe our universe. These vector quantities are the invariants or the absolutes in classical physics. Each of them, as we know, consists of three components in any observer's three-dimensional rectangular coordinate system, and although the individual components of any vector change from coordinate system to coordinate system, the length of that vector itself in classical physics does not change. This is what is meant by the principle of invariance in classical physics.

When we pass over into relativity theory, however, the situation is different; the vectors described above, consisting of three components in a three-dimensional rectangular coordinate system, are not invariant. Not only do the individual components of the vector change when we go from one coordinate system to another, but the length of the vector also changes. This means that according to relativity theory the laws of physics cannot be represented as relationships among three-dimensional vectors, since such relationships are not invariant when we transform from coordinate system to coordinate system because the vectors themselves are not invariant.

However, we can extend the idea of a vector to relativity theory by assigning to each of the vectors mentioned above a fourth component, to obtain what we call a **four-vector**. *In relativity theory the length of a four-vector is invariant.* As we go from one Galilean coordinate system (the fourth dimension is time) to another by means of the Einstein-Lorentz transformations, the four different components of any four-vector change, but the length itself of the four-vector remains unchanged.

If the laws of nature in relativity theory are expressed in terms of four-vectors, these laws meet the requirement of invariance in going from one coordinate system to another because the four-vectors themselves are invariant. Indeed, we use this as a criterion to check any suspected law. If the law can be expressed in terms of four-vectors, it is a correct law of nature (as long as the conditions of special relativity apply), but if not, we must discard it or seek to amend it.

10.24 Examples of Four-Vectors

We first consider the four-vector we obtain by assigning a fourth component to the vector $\boldsymbol{r}$, representing a classical displacement between two points. If the components of this displacement in a three-dimensional rectangular coordinate system are x, y, z we can assign to this displacement a fourth component (time) ict, where $i = \sqrt{-1}$. The quantity ict is the fourth component of our four-vector, which we may write as (x, y, z, ict) where c is the speed of light. The components x, y, z are called the **spatial components** and ict is the **temporal** or **time component** of the four-vector. The square of the length of the four-vector is obtained by taking the sum of the squares of the spatial components $x^2 + y^2 + z^2$ and adding to it the square of the time component $(ict)^2 = -c^2t^2$. When we do this we obtain just the square of the invariant space-time interval that we have already discussed.

To obtain another example of a four-vector we consider the three space components of the momentum p_x, p_y, p_z of a particle. If E is the total energy of this particle then we can show that iE/c is the fourth or temporal component of the momentum vector. We thus obtain the energy-momentum four-vector $(p_x, p_y, p_z, iE/c)$, and the square of the length of this vector is just $p_x^2 + p_y^2 + p_z^2 - E^2/c^2$. This quantity is an invariant since from Equation (10.18) we see that it is just equal to $-m_0^2c^2$, which is an invariant since m_0, the rest mass of the particle, and c, the speed of light, are invariants. In a similar way we can build other four-vectors in relativity theory which lead to important laws.

10.25 The Proper Time and the Aging of Fixed and Moving Observers *

If we change all the signs in Equation (10.9) we obtain

$$-\Delta s^2 = \left[c^2 - \left(\frac{\Delta x}{\Delta t}\right)^2 - \left(\frac{\Delta y}{\Delta t}\right)^2 - \left(\frac{\Delta z}{\Delta t}\right)^2\right]\Delta t^2$$

$$= [c^2 - v^2]\,\Delta t^2,$$

$$-\frac{\Delta s^2}{c^2} = \left[1 - \frac{v^2}{c^2}\right]\Delta t^2, \qquad (10.21)$$

where v represents the velocity of an observer moving with respect to some fixed observer, and where Δt represents the time interval measured by the clock in the frame of the fixed observer. The quantity $-\Delta s^2/c^2$, which we may write as $\Delta\tau^2$, is called the **square of the proper time of the moving observer**, since $\Delta\tau$ is precisely the time interval measured by a clock attached to the moving observer. We can demonstrate this by noting that for a clock moving along with an observer, the relative velocity v between the observer and the clock is zero and we have just $\Delta\tau = \Delta t'$.

From this we obtain a remarkable result if we consider two clocks that move away from each other and then meet again. Let us suppose that one clock remains here on earth and that the other clock moves off with a speed v to some distant point and then returns. How much time has elapsed for each clock? Or if we picture an observer associated with each clock, how much has the observer who remains here on earth aged as compared with the observer who goes out and returns with the moving clock? If the time T has elapsed as measured by the clock here on earth, and if T' is the time measured by the traveling clock, it is clear that we must have

$$T' = \sqrt{1 - \frac{v^2}{c^2}}\, T,$$

since T' is just the proper time of the moving clock and hence is governed by Equation (10.21). This shows us that the traveling observer ages more slowly than the observer remaining back on earth by the factor $\sqrt{1 - (v^2/c^2)}$. Thus if an observer moves off with a speed $\frac{3}{5}c$ (where c is the speed of light) to a distant point and returns after 20 years as measured by the clock here on earth, then the traveling observer will have aged by an amount $T' = 16$ years.

The faster the traveling observer moves, the more slowly he ages. We can look upon this in a different way to get the same result. If the observer moves out into space with a speed v, the distance he has to travel shrinks by the factor $\sqrt{1 - (v^2/c^2)}$ by the Einstein-Lorentz contraction of space, so that the time required to cover this distance is smaller by this factor. Therefore to this moving observer less time has elapsed on this trip than that measured by a fixed clock.

Of course to the moving observer it appears that the observer left back on the earth is moving away with a speed v, and that therefore the clock on the earth has slowed down by the factor $\sqrt{1 - (v^2/c^2)}$. As long as the two observers continue to recede from each other, they can never compare the effect of the journey on their two clocks or on their aging, and each will think that the other is aging more slowly. But if the traveling observer returns to the earth, he must have experienced a series of accelerations and decelerations which the earthbound observer has not and this makes the difference between the aging of the two observers. Because accelerations are involved in this problem a rigorous analysis cannot be given in the framework of the special theory of relativity (i.e., bý means of the Lorentz transformations) but must be treated by general relativity theory. The result obtained, however, is that given above. One can also obtain the above result by using the relativistic Doppler effect to analyze signals sent from one observer to the other (see problem 9).

EXERCISES

1. Show that if two equal swimmers were to start out simultaneously from the same point in a stream and one were to swim a given distance downstream and back, and the other were to swim the same distance across-stream and back, the second swimmer would return $1/\sqrt{1 - (v^2/c^2)}$ times sooner than the first swimmer, where v is the speed of the stream and c is the speed of both swimmers in still water.

2. What bearing does problem 1 have on the experimental basis of the special theory of relativity?

3. (a) As observed from the sun, how much smaller is Mercury's equatorial diameter than its polar diameter? (Assume that Mercury's speed in its orbit is always equal to its average speed.) (b) By how much does Mercury's mass increase, as measured from the sun, because of its motion?

4. When Mercury is in superior conjunction, what is its speed with respect to the earth if special relativity is taken into account? (Assume average speeds for the earth and for Mercury.)

5. (a) What is the space-time interval of an event that is one light year away from us and occurs in an hour from now? (b) Solve (a) when the event is one astronomical unit away from us.

6. Consider a beam of light coming from an infinite distance on your left and moving off to your right. What is its world line? What is the world line of a beam moving off to your left after coming from your right? (Hint: use a rectangular coordinate system with ct as your ordinate axis and the right-left line as the x axis.)

7. What is the relativistic kinetic energy of Mercury? Compare it with its nonrelativistic kinetic energy. Use the average speed of Mercury.

8. If one percent of the sun's mass were transformed into energy, how much energy would be released?

9. The correct Doppler effect as given by the special theory of relativity is as follows: If ν is the proper frequency of signals of any kind from a source that is at rest with respect to an observer, then the frequency ν' of these signals as detected by an observer moving relative to a source with a speed v is

$$\nu' = \nu \frac{\left(1 - \frac{v^2}{c^2}\right)^{1/2}}{1 \pm \frac{v}{c}}$$

where the + sign in the denominator holds when the source is receding from the observer, and the − sign when it is approaching the observer. (a) Use this to show that if a clock of any kind which beats N times per second when it is at rest moves out a distance L and back again with a constant speed v, the fixed observer will receive $N(2L/v)\sqrt{1 - (v^2/c^2)}$ beats from the clock. (b) Also show that an observer who moves back and forth once over the distance L with the moving clock will receive exactly N beats from a similar clock that is at rest.

10. (a) Show that if ν is the frequency of a wave of light and λ is the wave length, then the quantities $(h/\lambda)l$, $(h/\lambda)m$, $(h/\lambda)n$, $ih\nu/c$ are the components of a four-vector, where l, m, and n are the cosines of the angles that the direction of propagation of the wave makes with the x, y, z axis, respectively (the direction cosines), of a coordinate system, and h is some universal constant. (b) Show that the length of this four-vector is zero (a null vector).

11

THE GENERAL THEORY OF RELATIVITY

11.1 The Inadequacy of the Special Theory

Although the special theory of relativity, which we described briefly in the last chapter, had great impact on our space and time concepts when it was first announced by Einstein, still it did little to clear up the mystery of the motion of Mercury. The reason for this becomes clear if we note that the special theory deals only with systems that are moving with uniform speed with respect to each other (inertial frames of reference). A body in a gravitational field, however, is not moving with uniform speed, and hence cannot be treated in accordance with the principles developed in the special theory.

11.2 The Failure of Newton's Law of Gravity

Nevertheless, we can see with the aid of the special theory that Newton's law of gravitation cannot be correct, since it is not invariant when one applies to it the Einstein transformation equations. This law, as we recall, states that the gravitational force between two particles depends directly on the product of the masses of the two particles and varies inversely as the square of the distance between them. However, as we saw in the previous chapter, *neither the mass of each particle nor the distance between them is an absolute entity. These measured quantities depend upon the relative motion of the observer and the particles, so that one cannot speak in an absolute sense of a gravitational force between two particles.* From these considerations it was clear to Einstein that *the whole concept of the gravitational force would have to be revised*, if not entirely discarded, if one were to get a correct understanding of gravitation.

11.3 Einstein's Generalization of the Special Theory

But how is one to do this? From Einstein's point of view there was only one way to proceed and that was to enlarge the special theory of relativity (that is, to generalize it) to take into account observers situated in accelerated systems as well as those moving in inertial frames of reference. In other words, what he tried to do, and ultimately succeeded in doing, was to enlarge his concept of the *equality* of all observers in the eyes of nature, and to show *that all observers, regardless of how they are moving, must formulate the laws of nature in the same mathematical way.*

This extends the principle of invariance to include accelerated systems as well as those moving with uniform speed with respect to each other. This means essentially that as far as the laws of nature go, *there is no frame of reference that is to be preferred above any other for formulating these laws.* This may be expressed by saying *that no observer, regardless of how he is moving, should be able to determine his state of motion by any kind of experiment.* This is the assumption that Einstein made in order to extend his special theory of relativity to obtain the general theory of relativity.

Now offhand it appears to us that this generalization of the theory cannot be correct, since we know from direct experience that although we cannot tell when we are moving with uniform speed in a straight line, it is always possible for us to detect any departure from a uniform speed (that is, accelerated motion).

Not only does accelerated motion give rise to physical reactions in our own bodies which we can quickly detect, but we know that an object like a billiard ball, free to move on a flat surface, quickly responds to any acceleration that is imparted to the table on which it is moving or rests. How, then, could Einstein arrive at his generalization when we know, as we have stated above, that there is a definite difference between the behavior of bodies in inertial frames of reference and accelerated systems? He was able to do this by introducing a new principle, the famous **principle of equivalence**, which is based upon *the fact that a body's inertial mass and its gravitational mass are exactly equal.* We shall now see how this equality of inertial and gravitational mass leads to the principle of equivalence, which is basic to an understanding of the general theory of relativity.

11.4 The Equivalence of Inertial and Gravitational Forces

We first note that the equality of inertial and gravitational mass means that all bodies at the same point in a gravitational field (for example, at the same point above the earth's surface) experience the same gravitational acceleration, regardless of how heavy they are (regardless of their masses). Because of this, it was possible for Einstein to show that a person in an accelerated system cannot distinguish between the behavior of bodies resulting from this acceleration and their behavior if they were not in an accelerated system but rather at rest in a gravitational field. In other words, Einstein showed that *he could reproduce all the effects experienced by a body in a gravitational field by means of accelerations* (*inertial forces*). He made this point quite clear by introducing another of his famous thought (gedanken) experiments, namely, that of a man in an accelerated elevator.

To follow Einstein's reasoning we consider two observers, one on the surface of the earth and hence subject to all the effects of the gravitational field of the earth, and the other in an elevator, somewhere in empty space, where no gravitational fields are present. We now suppose that the elevator experiences a constant acceleration g exactly equal to the acceleration of gravity on the earth's surface, as shown in Figure 11.1. To make the conditions as nearly equal as possible for the two observers, we suppose further that the observer on the earth is also in a closed room so that he has no way of knowing that he is fixed on the earth. It is easy to see now that all the observations made on falling bodies by the man on the ground are exactly the same as those made by the man in the accelerated elevator. To begin with, we may note that each observer experiences a pressure against his feet, which exactly equals his weight, although in the case of the elevator the effect is not due to a gravitational pull but rather to the push of the floor that is constantly being accelerated upward.

11.5 Falling Bodies in an Accelerated System and a Gravitational Field

Let us now suppose that each of these observers holds, at the same height in front of him, two balls having different masses, and then releases them. We know by experience that the two balls on the earth (neglecting air resistance) hit the ground at exactly the same time because of the equality of the gravitational and inertial mass. But the same thing occurs for the man in the elevator, because the balls on leaving his hands are no longer accelerated, and hence appear to fall to the elevator floor, since the floor is constantly being accelerated towards the balls. Both balls in the elevator thus strike the floor at exactly the same time, since really the floor is coming up to meet them. In other words, as far as material bodies are concerned, they behave inside the elevator exactly as they would on the surface of the earth, so that conditions inside the elevator are equivalent to what they would be if the elevator were fixed (or moving with a uniform speed in a straight line) in the earth's gravitational field. *This means that we can create the equivalent of a gravitational field by means of acceleration.*

In the same way we can eliminate a gravitational field, or compensate for it, by introducing an accelerated system. Thus, if a man were perched in an enclosed stationary elevator above the earth's surface, he would experience all the effects of the earth's gravitational field, as long as the elevator were not accelerated. If, however, the elevator were suddenly cut free of its support, the man would discover that the gravitational field had vanished. Objects would no longer fall to the floor, but would stay suspended, and an object thrown across the elevator would move in a straight line. *In other words, the gravitational field would be compensated for by the downward acceleration of the elevator.*

11.6 The Principle of Equivalence

This analysis of the behavior of objects in accelerated systems led Einstein to the discovery of the principle of equivalence, which we may state as follows:

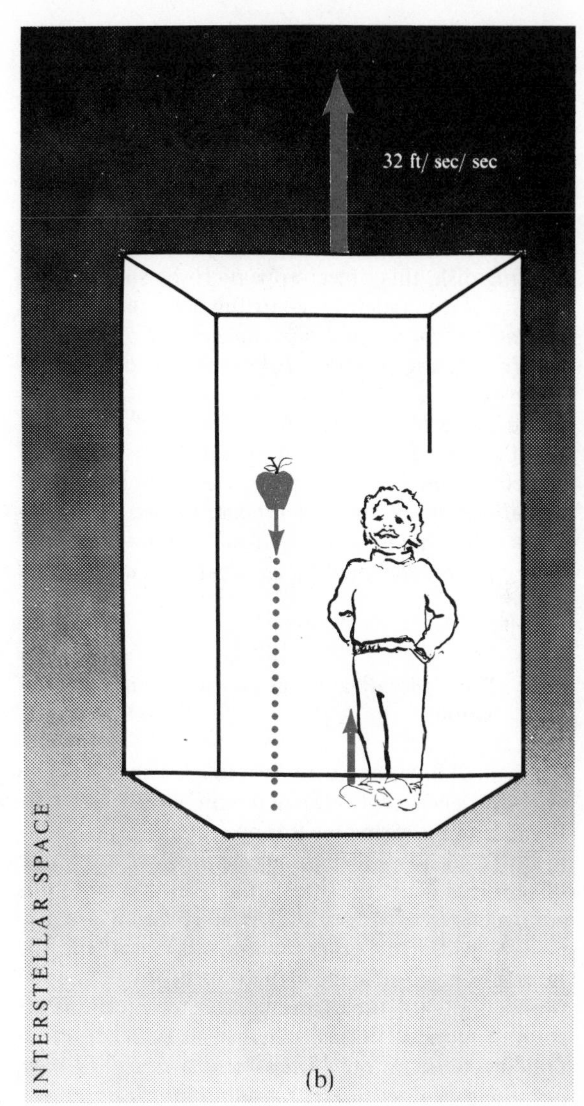

FIG. 11–1 *An apple falling on the earth as seen by a seventeenth century observer, and an apple falling in an accelerated elevator in empty space as seen by a twentieth century observer in the elevator. Both observers experience a push against their feet and both observers see the apple falling with accelerated motion.*

It is always possible to duplicate a gravitational field by an accelerated system (that is, to replace gravitational forces by inertial forces) to any desired degree of accuracy, if we choose a small enough region of space.

This last qualification about the size of the region of space to which the principle of equivalence can be applied is most important. Actually, if we consider the conditions in the accelerated elevator and for the man on earth, as shown in Figure 11.1, we note that the two metal balls in the elevator do not behave in exactly the same way as they would if they were falling freely on the surface of the earth. *In the elevator the two spheres move to the floor in exactly parallel lines, whereas on the earth the lines along which they move are not parallel, since they must converge to the center of the earth.* However, we can always take the region of space on the earth in which the metal balls are falling *so tiny that no available instruments can detect the convergence of the directions of motion of the two spheres.* It is in this sense that the principle of equivalence must be understood.

11.7 The Difference between Inertial and Gravitational Forces

We should note that in spite of the principle of equivalence there is a very important difference between inertial and gravitational forces. This, perhaps, we can best understand by considering an object on a rotating disk. To an observer on the rotating disk, this object appears to be subjected to definite forces (such as centrifugal forces). *These are inertial forces and one can get rid of them by simply stepping off the disk with the object.* In other words, all the inertial forces disappear as soon as we move to a nonspinning frame of reference. However, so far as gravitational forces are concerned, *no single frame of reference can be introduced that eliminates all the gravitational forces. Gravitational forces can be eliminated only in a small vicinity of a gravitational field by means of an accelerated system.*

11.8 The Generalization of Newton's First Law of Motion

We shall now see how Einstein used the principle of equivalence to generalize the theory of relativity. Because of this principle it is now no longer possible to distinguish between an accelerated system and a system that is at rest since an accelerated system is equivalent to a system that is at rest in a gravitational field. In other words, because of the principle of equivalence, there is no experimental procedure that enables an observer to determine his state of motion, that is, whether he is in an accelerated system or not. *The laws of nature must then, according to Einstein, be the same for all observers, regardless of how they are moving.*

This is the generalization of the principle of invariance, and it means, essentially, that if two observers, moving with respect to each other in any way whatsoever, formulate a law of nature, the law must have exactly the same mathematical form for each observer. If this were not the case, the formulation of a law would depend on the state of motion of the observer, and hence the law itself could be used to tell the observer whether or not he is accelerated.

How did this extension of the principle of invariance enable Einstein to introduce a new law of gravity? To understand this we must again consider Newton's first law of motion, which states that a body moves in a straight line with uniform velocity unless a force acts on it. We know now that according to the extended principle of invariance this cannot be a law because a body can deviate from straight-line motion for some observers even though no external force acts on it.

To see this we again consider the man in the accelerated elevator, and suppose that he throws an object across the elevator. Because the elevator is accelerated, it is clear that as far as the man in the elevator is concerned, the object does not move in a straight line. However, since no force acts on the object, this would be a violation of Newton's first law of motion if this were, indeed, a correct law. We are therefore forced to discard this law and hence must replace it with some other principle. But what is this principle to be? Clearly, it must involve some concept other than force, since the notion of a force is no longer tenable if, following Einstein, we adopt the extended principle of invariance.

It is precisely at this point that Einstein introduced a most revolutionary idea. For it occurred to him that he could still retain the essential features of Newton's first law if he simply stated the first part of it and dropped all references to forces. The law would then read as follows:

All bodies left to themselves must move in straight lines.

But how can this be correct if we know from experience that a body thrown horizontally on the surface of the earth moves in the arc of a parabola? This is certainly not a straight line, and hence can only arise, it seems, because of the action of a force.

11.9 The Generalized Concept of the Straight Line: The Geodesic

Einstein overcame this difficulty by pointing out that our concept of a straight line is the result of our having been brought up with the belief that Euclidean geometry correctly describes our world. Only if we rid ourselves of our preconceived geometrical ideas can we properly understand this new form of Newton's first law. According to this idea a body departs from what we ordinarily call a straight line not because of the action of a force (gravitational or inertial) but because the geometry of space-time, in which the particle is moving, is not the familiar geometry developed by Euclid. Hence the old idea of the straight line does not apply. In other words, *Einstein ascribed the gravitational field in a region of space-time to the departure of the geometry of that region from the familiar Euclidean geometry.* We shall see shortly

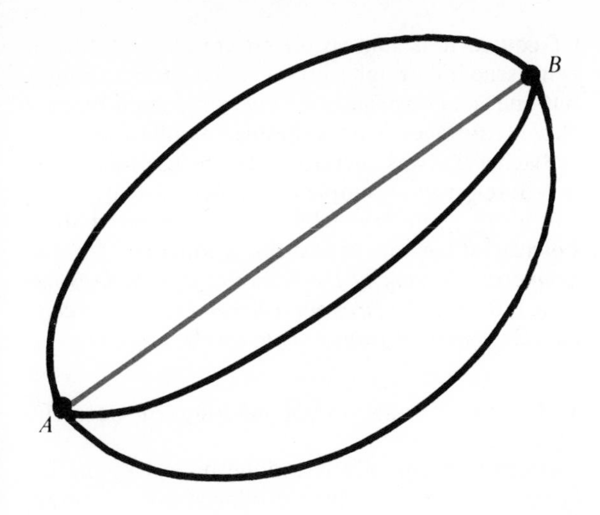

FIG. 11–2 *Between the points A and B an infinite number of paths can be drawn, but only one is the shortest distance, and this is what we call a straight line. In this figure four different paths are shown.*

just what, according to Einstein, causes the geometry in the neighborhood of any point to become non-Euclidean.

With this in mind we can still retain Newton's first law without having to refer to a force if we now say that all bodies left to themselves will always move along the equivalent of a straight line for the particular geometry involved. Since this equivalent of a straight line is a curve that changes as the geometry changes, we must have a universal criterion for recognizing it. The criterion is the following: among all the possible curves connecting two points *A* and *B* (as shown in Figure 11.2) in any geometry (for example, two points on the surface of a sphere), the equivalent of a straight line is the shortest curve. This is called a **geodesic** in relativity theory. In this way we generalize the first law and automatically take account of motion that deviates from Euclidean straight-line motion, by replacing the idea of a straight line by the concept of the geodesic or shortest distance between two points.

Since the shortest distance between two points is the ordinary straight line only in the case of Euclidean geometry, we see that a change in geometry leads to a curved shortest distance between two points and hence is equivalent to a field of force. This now makes unnecessary the concept of a gravitational force—which, as we have seen, has no absolute meaning—and it enables us to express the first law in terms of a physical concept, namely, the shortest distance between two points, which has the same mathematical form for all observers and hence has the desired property of a law of nature.

11.10 Non-Euclidean Geometries

Since we must describe a gravitational field in terms of a non-Euclidean geometry, we must say a few words about such geometries. As we know, a geometry is based on a collection of certain self-evident truths, the axioms of the geometry, which serve as the basis for the development of the theorems that constitute the complete body of the geometry. *If a particular axiom of the geometry is omitted the geometry is not complete*, so that in analyzing a geometry one must very carefully examine the axioms to see that a complete set is present. It is also necessary to see whether any of the axioms are really theorems that can be derived from the other axioms, and hence should not be included among the axioms.

11.11 The Parallel Axiom in Euclidean Geometry

Euclidean geometry has one axiom, the famous parallel axiom, which geometers in the mid-nineteenth century thought should be excluded from the body of axioms because they felt that it was really a theorem. This axiom can be stated as follows (see Figure 11.3):

Given a straight line and a point outside that line, there is one and only one line in the plane that can be drawn through the point parallel to the given line.

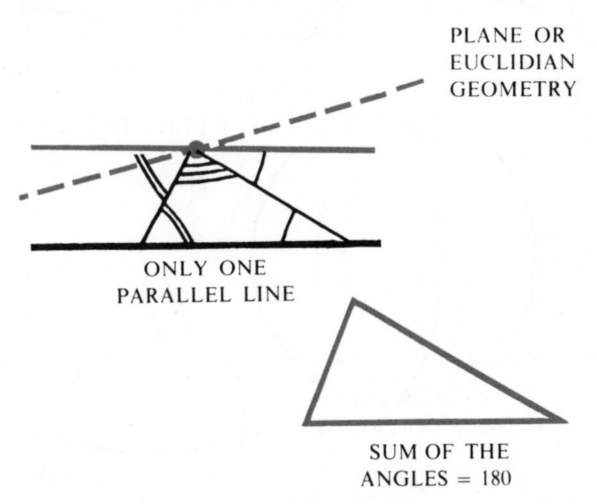

FIG. 11–3 *The Euclidean parallel axiom on a plane is equivalent to the statement that the sum of the angles of a triangle is equal to 180°.*

It is easy to show that this axiom is equivalent to the statement that the sum of the angles of a triangle must equal 180°, or to the statement that the circumference of a circle is equal to 2π times its radius. When geometers in the mid-nineteenth century tried to develop Euclidean geometry without calling upon this axiom, they found that they were unable to do so. In other words, this axiom is essential to Euclidean geometry.

11.12 The Parallel Axiom and Hyperbolic Geometry

It occurred to Lobachevsky, however, that one could develop a geometry by denying this axiom, and replacing it by the following axiom (see Figure 11.4):

Given a line and a point outside this line on a surface, it is possible to pass through the given point an infinitude of lines on the surface that do not cut the given line.

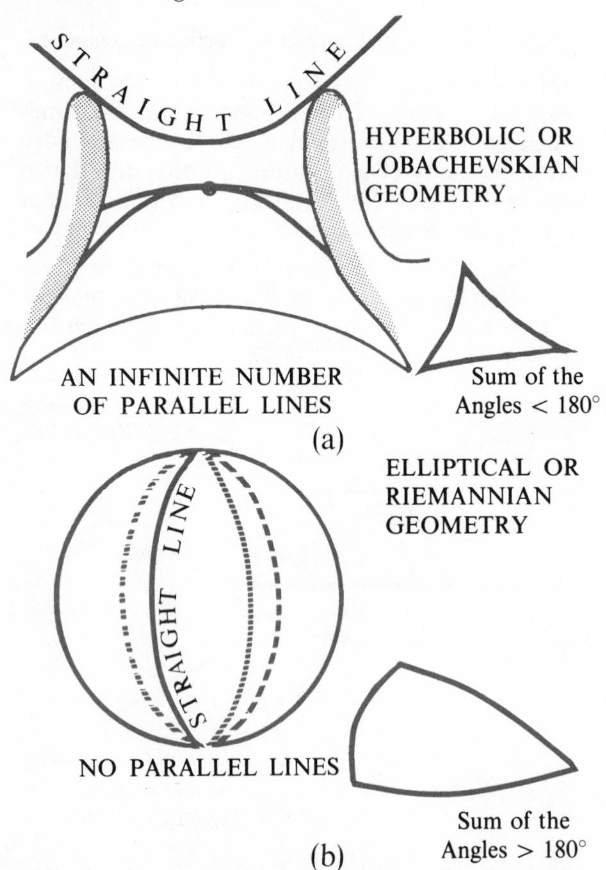

FIG. 11-4 *The geometries (a) on a saddle surface and (b) on a sphere are non-Euclidean.*

Of course it is impossible to conceive of this as being true for straight lines on a flat surface (plane), but one can construct such lines on certain types of curved surfaces (for example, saddle surfaces). Lobachevsky was therefore able to use this axiom to develop a complete, self-consistent, non-Euclidean geometry, which is now referred to as **Lobachevskian** or **hyperbolic** geometry. *In this geometry, the sum of the angles of a triangle is less than* 180°, *and the circumference of a circle is greater than* 2π *times the radius of the circle.*

11.13 The Parallel Axiom and Elliptical Geometry

After Lobachevsky had done his work, another type of geometry was introduced by a famous geometer, Riemann, who replaced the Euclidean parallel axiom by the following axiom:

Given a line and a point outside this line on a surface, there is no line on this surface that can be drawn through the given point parallel to the given line.

Again it is clear that this statement cannot hold for straight lines on a flat surface and applies only to certain curved surfaces—for example, to the sphere. On a sphere the equivalent of straight lines are great circles, and we know that all great circles must intersect, so that there can be no parallel straight lines on such a surface.

With this axiom, Riemann was able to develop another kind of complete, self-consistent, non-Euclidean geometry, the **Riemannian** or **elliptical** geometry, in which it can be shown that the sum of the angles of a triangle is greater than 180°, and that the circumference of a circle is less than 2π times its radius. Riemann worked out the most general case of this sort of geometry by developing it not for two or three dimensions, but for n dimensions (where n may be any number). This proved to be of great usefulness to Einstein, who concluded, as we have seen, that objects do not move in Euclidean straight lines when they are in a gravitational field or in a noninertial system (accelerated system) not because there are forces acting on them, but because the geometry of space-time along their world line is of a non-Euclidean nature.

11.14 The Gaussian Curvature and the Geometry of a Surface

Knowing that there are geometries other than the Euclidean, Einstein realized that he might be able to find the correct law of gravitation (which must, according to the general theory of invariance, be expressible in the same mathematical form for all observers) if he could find some way of formulating how a body like the sun, for example, causes the geometry of space-time in its neighborhood to differ from Euclidean geometry. This formulation of the deviation from Euclidean geometry would have to be expressed by means of mathematical formulae valid for all observers. The first thing Einstein did was to show that such a formulation does indeed exist.

In order to understand the nature of Einstein's problem, we consider a person living on the surface of a sphere (such as the earth), who does not realize that the surface is curved. Because of his small size as compared with that of the sphere, he has the impression that he is living on a flat surface, and he is, of course, prejudiced in favor of Euclidean geometry as the proper geometry to describe the surface. In other words, he really believes that he is living on a plane surface (the plane that is tangent to the point on which he stands, see Figure 11.5), a belief which arises because he can encompass only a small area surrounding his own neighborhood. If he moved any great distance over the surface, he would realize, if he were a good enough geometer, that his world is really curved and deviates from the fictitious plane he first thought it to be. To describe this surface accurately he would then have to substitute a non-Euclidean geometry for the plane geometry he had previously used.

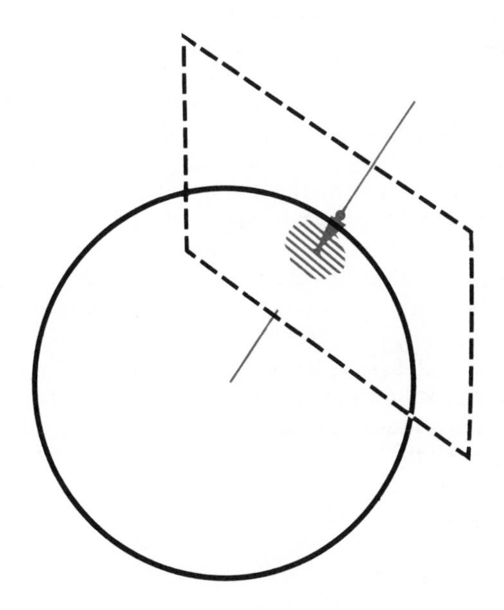

FIG. 11-5 *A person living on a very large sphere has the impression that the surface he is on is flat.*

Now the question that arises is this: is there any geometrical quantity that has an absolute meaning for all observers and that describes the way a geometrical curved surface differs from that of a plane? If there is such a quantity, then it can be used as the basis for constructing an invariant gravitational theory, *since the presence of a gravitational field must be interpreted as a departure from plane geometry.*

To see that such a mathematical quantity exists, and what its nature is, we again consider the man on his sphere and follow him as he moves farther and farther away from the point at which we constructed the imaginary tangent plane. We see that the spherical surface curves away from the plane more and more as we move farther and farther from this point. It is also clear that this departure of the sphere from the plane would be still greater if the radius of the sphere were smaller (that is, if the curvature of the sphere were larger). In other words, the curvature of the sphere tells us how drastically the geometry of its surface departs from Euclidean geometry. As it turns out, this applies not only to a sphere, but to all curved surfaces.

Karl Friedrich Gauss, the great nineteenth-century German mathematician, first recognized the importance of the curvature of a surface at a point and developed a mathematical procedure for using this curvature to describe the geometry of the surface. He introduced what is now called the **Gaussian curvature of a surface**, and showed that the value of this quantity at a point tells us whether the geometry of a surface at that point is hyperbolic, Euclidean, or elliptical.

Although the work of Gauss was very important for the description of properties of curved surfaces, it could not be used directly by Einstein since it deals with two-dimensional surfaces, and the theory of relativity is concerned with a universe of four dimensions. However, Riemann had extended Gauss' analysis to n-dimensional geometry, and Einstein was able to use Riemann's work directly. Just as Gauss found that there exists an invariant curvature (the Gaussian curvature) which tells us whether the geometry of a two-dimensional surface is hyperbolic, Euclidean, or elliptical (Gaussian curvature is negative for hyperbolic geometry, zero for Euclidean, and positive for elliptical geometry), so Riemann showed that there exists a quantity in the geometry of n dimensions (the Riemann-Christoffel tensor) that plays a role similar to that of the Gaussian curvature in two dimensions. Only if this quantity vanishes at all points is the geometry of any n-dimensional space Euclidean. In order to see what the nature of this quantity is, we shall have to give a brief description of *tensors*.

11.15 Vectors and n-Dimensional Coordinate Systems

We have seen that in order to describe velocity, force, and other directed quantities, it is necessary to introduce the concept of a vector, which we defined as an entity that possesses both magnitude and direction. In order to describe a vector completely we have to specify its length and give its direction with respect to some fixed reference line. There is, however, another way of doing this which is of great importance in geometry, and that is by making use of a coordinate system.

In Figure 11.6, let **A** represent a vector in the three-dimensional Cartesian coordinate system x, y, z. If we now project this vector first onto the x direction, and then onto the y and z directions in turn, we obtain the lengths A_x, A_y, and A_z, which we call the x, y, and z components of the vector in the given coordinate system. It is clear, of course, that *if we change our coordinate system by rotating it in some way, the three components change, but the vector itself remains unaltered or is invariant*. In other words, a vector in three-dimensional space is a quantity that has the same value in all coordinate systems (it is invariant to a change in coordinates) but the components of the vector differ as we go from one coordinate system to another.

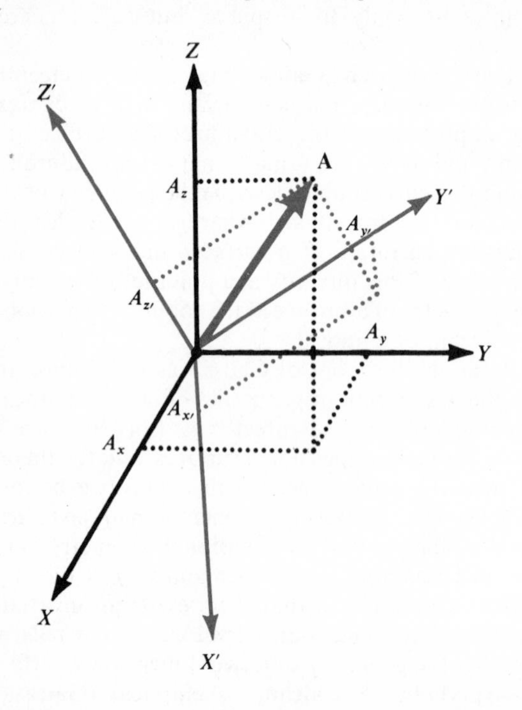

FIG. 11–6 *The coordinates of a vector,* **A**, *in two different coordinate systems are not the same, although the vector itself remains the same in both systems.*

There is a very definite mathematical way of finding the components of a vector in any coordinate system, if the components are all known in one particular coordinate system. It follows, then, that the vector is completely specified (i.e., known in all coordinate systems) if its components are given in any one coordinate system. Therefore, instead of speaking of the vector itself, we can just as well always refer to the quantities A_x, A_y, A_z.

We can extend this idea of a vector to a space of any number of dimensions by imagining the vector projected onto the additional axes of our enlarged coordinate system, thus obtaining other components in addition to the three we have already described. Of course, this is a mathematical fiction and has no physical reality, but a mathematician can indulge his fancy and speak of a vector in a space of any number of dimensions. Thus, if a mathematician were dealing with a space of n dimensions, he would introduce the n components $A_1, A_2, \ldots, A_n$ of the vector **A**, where the subscripts $1, 2, \ldots, n$ refer to the n-coordinate axes (instead of the letters x, y, z, etc.).

This extension of the idea of a vector to n-dimensional space became very important for geometry after Riemann had developed the n-dimensional geometry. It is also important in relativity theory since, as we know, this theory deals with a four-dimensional geometry. We made use of this idea in Section 10.23 where we introduced the concept of the four-vector. Here, we shall speak of the components A_1, A_2, A_3, A_4 when dealing with the four-vectors in relativity theory, where the subscripts 1, 2, 3, refer to the spatial coordinates x, y, z, and the subscript 4 refers to the time coordinate. We may use a shorthand notation by letting A_i stand for the components of the vector **A**, where it is understood that the subscript i takes on the values 1, 2, 3, 4, in turn.

11.16 Tensors

Now we shall see how the concept of a tensor is obtained from that of a vector. Consider a force

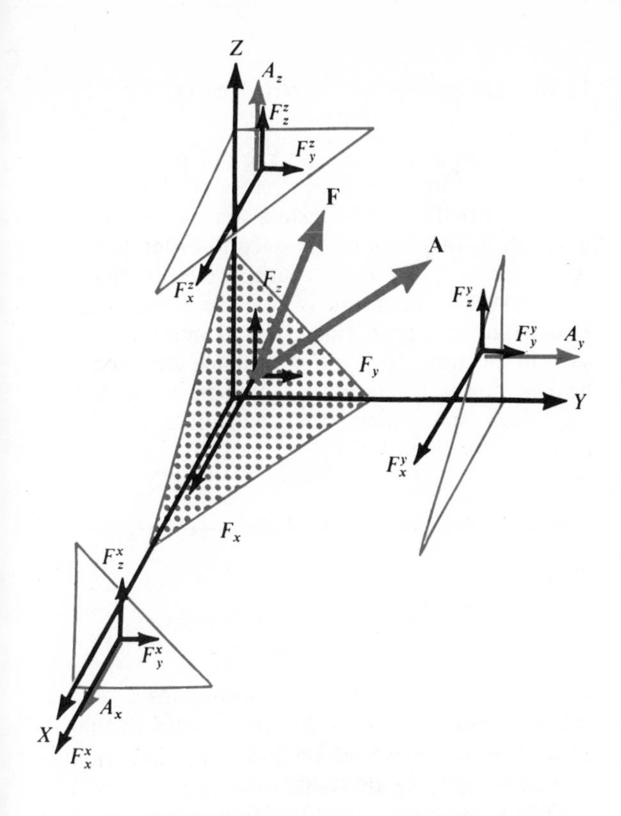

FIG. 11–7 *A tensor concept arises from the combination of two vectors; one,* **A**, *associated with a surface and the other,* **F**, *associated with a force acting on a surface. The components of the surface perpendicular to the 3 axes and the forces acting on these 3 components are shown.*

exerted against a surface, as shown in Figure 11.7. In this diagram we show three different surfaces with forces acting against them. If we consider any one of these three surfaces, we can identify it by specifying the line that is perpendicular to it. Thus, we label the surface that is perpendicular to the x axis the x surface, the surface perpendicular to the y axis the y surface, and the surface perpendicular to the z axis the z surface. We now identify the three forces by assigning to each force the letter identifying the surface to which it is applied. Thus, the force acting on the x surface is the $\mathbf{F}^x$ force, etc., as shown in the figure.

If we consider the $\mathbf{F}^x$ force and realize that it is a vector, we know that it has x, y, and z components, and hence that we may replace it by its three components F^x_x, F^x_y, F^x_z. In the same way we can specify the force against the y surface by means of its three components F^y_x, F^y_y, F^y_z, and the force against the z surface by means of its three components F^z_x, F^z_y, F^z_z. In other words, if we wish to specify in their entirety the forces acting on the three mutually perpendicular surfaces, we require nine quantities, which we may write in the form of an array:

$$\begin{matrix} F^x_x & F^x_y & F^x_z \\ F^y_x & F^y_y & F^y_z \\ F^z_x & F^z_y & F^z_z. \end{matrix}$$

We may look upon this array as a kind of enlarged or higher-order vector, but instead of calling it a vector we call it a **second-order tensor**. In fact according to this point of view we call a vector a **first-order tensor**, and a scalar a **zero-order tensor**. We see that in three-dimensional space a second-order tensor has nine components, and we may write any single component as F_{ij} where i and j both take on any one of the values 1, 2, and 3 corresponding to the three dimensions x, y, z, and where we have introduced both identifying letters as subscripts. We should note that a second-order tensor is obtained by combining two vectors in a definite way, as we have just seen in the example of forces acting on three mutually perpendicular surfaces. The two vectors involved in this case are the force and the vector representing the perpendicular surfaces.

If we extend the idea of a tensor to four dimensions, which is what we have to do in relativity theory, then we see that a tensor has sixteen components, since i and j can each take on values from 1 to 4.

11.17 The Invariance of Tensors in Coordinate Transformations

Now just as the components of a vector change when the coordinate system is changed, so do the components of a tensor, and there are definite mathematical relationships that enable one to obtain the components in one coordinate system if they are known in any other coordinate system. *Thus if all the components of a tensor are known in any one coordinate system, they can be found in any other coordinate system.*

There is another very important property of a tensor that must be introduced here: *regardless of how the components of a tensor change when we go from one coordinate system to another, the tensor itself remains unaltered.* This property gives tensors their great importance in relativity theory, because it means that *if a law can be expressed in tensor form, it is valid for all coordinate systems, and hence for all observers* (*that is, the law is invariant*). This is why Einstein made use of tensor mathematics. His purpose was to express the laws of nature in an invariant form, and this would automatically be insured if he wrote them in terms of tensors.

11.18 The Riemann-Christoffel Tensor and the Geometry of Space-Time

Now that we have some idea of the meaning of tensors, we shall consider just how Einstein used them to formulate an invariant law of gravity, that is, one that is valid for all observers. As we have seen, he first introduced the idea that what we call the gravitational field of force is really a departure of the geometry of space-time from Euclidean geometry. In order to express this mathematically he sought some method of representing how any geometry deviates from Euclidean geometry. In this he was guided by the work of Riemann who had proved that there is a fourth-order tensor, the famous **Riemann-Christoffel tensor**, having the following properties:

If this tensor vanishes everywhere (that is, each of its components is zero in any coordinate system), the geometry of space-time is Euclidean, and if space-time geometry is Euclidean this tensor must vanish everywhere.

Since this statement deals with a tensor, it must be true in all coordinate systems.

11.19 The Einstein-Ricci Tensor

In other words, the necessary and sufficient condition for a geometry to be Euclidean is that the Riemann-Christoffel tensor vanish. If this tensor does not vanish, then the geometry departs, to a greater or lesser extent, from Euclidean geometry at various points, depending upon the amount by which the tensor differs from zero at these points. Einstein was quick to realize that the law of gravity he was seeking would somehow have to be formulated in terms of such a tensor. He found, however, that his law could not be expressed directly in terms of the fourth-order Riemann-Christoffel tensor (which contains 256 components) but rather in terms of a second-order tensor, the **Einstein-Ricci tensor,** which he was able to obtain by a simple mathematical procedure from the Riemann-Christoffel tensor.

11.20 The Space-Time Metric Tensor

Although we shall not pursue this point, since to do so would involve us in tensor calculus, we shall indicate briefly how Einstein set up his law of gravitation in terms of the second-order Einstein-Ricci tensor. To begin with, we note that the theorem of Pythagoras is not the same in non-Euclidean as it is in Euclidean geometry. As we saw in Section 10.17, the distance between two nearby points (theorem of Pythagoras) in flat or Euclidean space-time (the space-time of special relativity) is given by

$$\Delta s^2 = \Delta x^2 + \Delta y^2 + \Delta z^2 - c^2\,\Delta t^2,$$

or $$\tag{11.1}$$

$$\Delta s^2 = \Delta x_1^2 + \Delta x_2^2 + \Delta x_2^2 + \Delta x_4^2,$$

where we have renamed the coordinates x, y, z, t, calling them x_1, x_2, x_3, x_4 (x_4 is placed equal to $\sqrt{-1}\,ct$), so that space and time are treated mathematically on the same footing.

This expression for the distance between two neighboring points is not valid in non-Euclidean geometry since space is not flat but curved. To take account of the curvature of space, we replace the above expression by the general expression

$$\begin{aligned}\Delta s^2 = {} & g_{11}\,\Delta x_1^2 + g_{12}\,\Delta x_1\,\Delta x_2 + g_{13}\,\Delta x_1\,\Delta x_3 + g_{14}\,\Delta x_1\,\Delta x_4 \\ & + g_{21}\,\Delta x_2\,\Delta x_1 + g_{22}\,\Delta x_2\,\Delta x_2 + g_{23}\,\Delta x_2\,\Delta x_3 + g_{24}\,\Delta x_2\,\Delta x_4 \\ & + g_{31}\,\Delta x_3\,\Delta x_1 + g_{32}\,\Delta x_3\,\Delta x_2 + g_{33}\,\Delta x_3\,\Delta x_3 + g_{34}\,\Delta x_3\,\Delta x_4 \\ & + g_{41}\,\Delta x_4\,\Delta x_1 + g_{42}\,\Delta x_4\,\Delta x_2 + g_{43}\,\Delta x_4\,\Delta x_3 + g_{44}\,\Delta x_4\,\Delta x_4\end{aligned} \tag{11.2}$$

This differs from the previous expression by the presence of cross-product terms, such as $\Delta x_2\,\Delta x_3$, and the coefficients g_{ik}. We see that there are just 16 of these coefficients (i and k separately can be any one of the four integers, 1, 2, 3, 4) and we can show that they are the components of a second-order tensor. Since the distances between points in our curved space-time depend upon the tensor g_{ik}, this tensor is called the **metric tensor**. This metric tensor is important because it determines completely the geometry of space-time. If we know the value of this metric tensor, we also know the nature of space-time geometry and hence the properties of the gravitational field.

11.21 The Energy-Momentum Tensor T_{ik}: Einstein's Gravitational-Field Equations

Einstein's great contribution was the discovery of a set of equations (his famous gravitational-field equations) the solutions of which give the values of the g_{ik}'s, and therefore the value of the gravitational field at any point in space-time. These equations are Einstein's generalization of Newton's law of gravitation.

To obtain these equations, as we have noted above, Einstein first introduced the Ricci tensor. This is a second-order tensor R_{ik} which depends entirely on the g_{ik}'s and which is a measure of the way in which space-time at a given point departs from Euclidean space as a result of the presence of a gravitational field.

Now, we know that the presence of a gravitational field is associated with the presence of matter in our space-time. Hence the tensor R_{ik} must also be related in some way to this matter. However, since R_{ik} is a tensor we can establish such a relationship only if we represent matter by such a tensor. Einstein succeeded in finding such a tensor T_{ik}, which turns out to represent not only matter but also any energy that may be present, so that we may call this the *energy-momentum or energy-matter tensor*.

This tensor is an extension of the energy-momentum four-vector discussed in Section 10.24. To set up an equation to represent the geometry in the presence of matter, one might be inclined to place $R_{ik} = T_{ik}$—that is, to say that the curvature tensor is determined by the matter-energy tensor—but this would be wrong since T_{ik}, the energy-momentum tensor, obeys the principle of conservation of energy and momentum, whereas the tensor R_{ik} does not obey a conservation principle. However, there is a tensor related to R_{ik} that does obey a conservation principle, and this obviously is the tensor which must be equated to T_{ik}.

To obtain this tensor we first introduce the quantity R, which is the measure of the actual curvature at a point and is an invariant (the same in all coordinate systems). This curvature invariant, which plays the same role in four-dimensional space as the Gaussian curvature does at a point on a two-dimensional surface, can be derived by simple mathematical manipulations from the tensor R_{ik}. Einstein subtracted the tensor $\frac{1}{2}g_{ik}R$ from R_{ik} and thus obtained the Einstein-Ricci tensor that obeys the conservation principle and hence may be equated to T_{ik}. By doing this he arrived at his gravitation-field equations:

$$\boxed{R_{ik} - \frac{1}{2} g_{ik}R = -\frac{8\pi G}{c^4} T_{ik}}, \qquad (11.3)$$

where G is the Newtonian gravitational constant and c is the speed of light. It is interesting to see that the speed of light now enters into the law of gravity. The coefficient G/c^4 has to be introduced to make the physical dimensions the same on both sides of the equation, and the numerical factor 8π is introduced to make Einstein's equation and Newton's agree numerically for very weak gravitational fields for which Newtonian gravitational theory is a good approximation.

This single equality really represents ten different equations, since i and k can take on all values from 1 to 4 independently (there are only ten, and not sixteen, equations since R_{ik}, g_{ik}, T_{ik} are symmetric, that is, $R_{ik} = R_{ki}$, $g_{ik} = g_{ki}$, $T_{ik} = T_{ki}$). These ten equations constitute Einstein's law of gravitation, and their solutions give the general relativity counterpart of Newton's classical law of gravitation. Since from these solutions we also obtain the components, g_{ik}, of the metric tensor, we are able to determine the geometry of space-time at each point in a gravitational field, and hence to see how such a field is related to the departure of the geometry at a point in space-time from Euclidean geometry.

11.22 The Invariant Path of a Particle in Curved Space-Time: The Geodesic

To study the motion of a particle, such as a planet, in a gravitational field, we must have more than Einstein's field equations. Just as in the classical theory we must have a law of motion $F = ma$ in addition to Newton's gravitational law of force, we must also have a law of motion in this case. We cannot use Newton's law of motion $F = ma$ in general relativity since it is not invariant to a transformation of coordinates. We can obtain the counterpart of this law of motion in general relativity theory by considering a particle moving from an initial point A_i to a final point A_f through a series of neighboring points A_1, A_2, etc. along some permissible path in the gravitational field (as shown in Figure 11.8). Obviously this path is determined by the gravitational field and hence by the geometry of space-time. The law that governs this path, and is therefore the counterpart of the law $F = ma$, can be obtained as follows from Einstein's generalization of Newton's first law of motion given in Section 11.8. Let Δs_1 be the distance from A_i to A_1, Δs_2 the distance from A_1 to A_2, etc. where the points A_1 and A_2 are to be taken very close together. Let us now sum all these distances to obtain the total path length from A_i to A_f:

$$\text{total path length} = \Delta s_1 + \Delta s_2 + \cdots + \Delta s_f. \quad (11.4)$$

Since the Δs's all depend on the g_{ik}'s as shown by Equation (11.2), we can find this sum as soon as Einstein's field equations are solved, since these solutions give us the g_{ik}'s at each point of space-time. The law of motion is now the following:

A particle in a gravitational field moves only along that path for which the total path length is a minimum—that is, along a geodesic.

Since this law depends on the tensor g_{ik}, it is the same in all coordinate systems and thus has the property that is required of a law of nature. Although Einstein originally introduced this law of motion as distinct from his field equations, it was shown later that it can be derived from the field equations themselves.

If one solves Einstein's field equations for the gravitational field surrounding a point mass (such as the sun, for example), one obtains Newton's law of force to a first approximation. However, Einstein's expression for the gravitational force contains terms in addition to the Newtonian expression which turn out to be negligible under *most conditions*. But precisely these terms give effects which, though numerically small, are, from a conceptual point of view, drastically different from what one can obtain from Newton's law of gravitation. Although one has to solve Einstein's equations in their entirety to see just how Einstein's law of gravitation differs from Newton's law (this was first done for a planet moving in the gravitational field of the sun by K. Schwarzschild, in 1917), it is possible to derive some of the physical consequences of Einstein's theory by calling upon the principle of equivalence. This is, indeed, how Einstein first predicted some of the surprising results of his theory.

11.23 Applying the Principle of Equivalence to a Body in a Gravitational Field

We know that, according to the principle of equivalence, all the effects one can observe in a small region of a gravitational field can be reproduced by an appropriately accelerated system. Thus, if the sun's mass is M, we can reproduce the effects of its gravitational field on a body at a distance r from the sun's center by supposing that the sun is not present and that the body is in an elevator that has an acceleration GM/r^2 (where G is the gravitational constant) in a radial direction away from the center of the sun, as shown in Figure 11.9. By the same token, we can eliminate the gravitational field of the sun at any point for a given observer by imagining this observer to be in a small elevator that is in free fall towards the center of the sun.

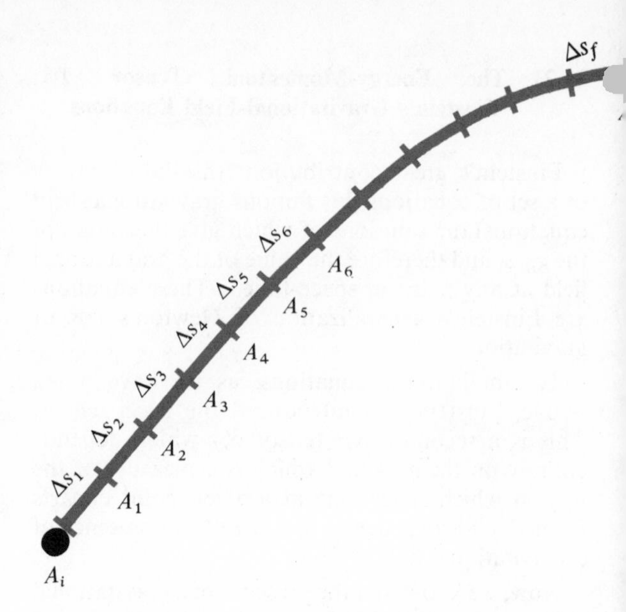

FIG. 11–8 *The total path length from an initial point A_i to a final point A_f is the sum of the infinitesimal line elements, Δs_i, from A_i to A_f.*

We now consider the second case in more detail and suppose that this elevator has at each point the speed that it would have if it had fallen to this point from an infinite distance (that is, the speed of escape towards the sun at each point). As far as the person in the freely falling elevator is concerned, there is no measurable gravitational field present, even though he is in the gravitational field of the sun. All his instruments behave as though he were in a region of space where there is no gravitational field.

Therefore, to determine the effect of a gravitational field on an instrument, all we need do is compare the instruments of the observer in the freely falling elevator with instruments that are at

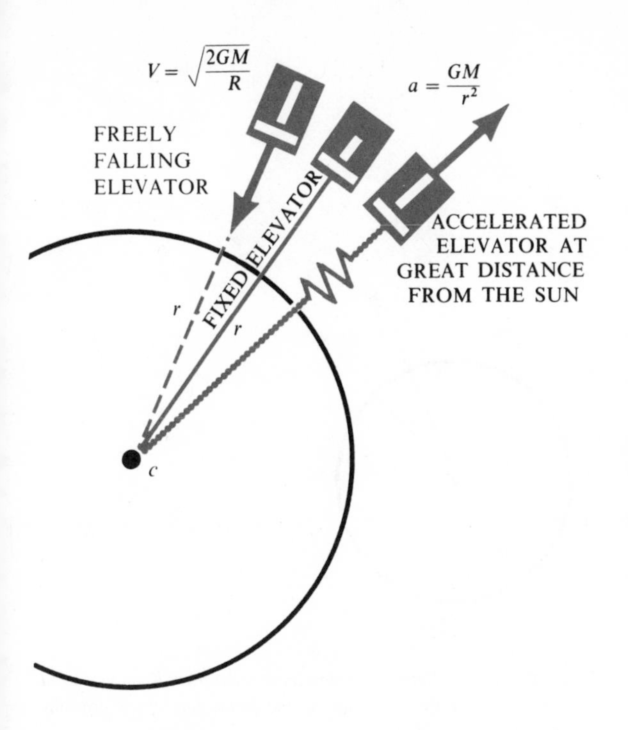

FIG. 11-9 *Lengths compared by an observer in a freely falling elevator with those of an observer in a fixed elevator in the gravitational field of the sun. To the observer in the freely falling elevator, radial lengths in the fixed elevator appear shorter, whereas transverse lengths are unchanged. The elevator to the right is at a very great distance from the sun, being accelerated away from the sun with an acceleration equal to* $(GM)/r^2$.

rest at the same point in the gravitational field. If our observer in the elevator is at the distance r from the center of the sun we can find out how lengths and time are affected by the gravitational field at this distance from the center of the sun by having the observer in the freely falling elevator compare his clocks, rods, and masses with those of an observer fixed at the distance r from the center of the sun.

11.24 Contraction of Time in a Gravitational Field

From the results of the special theory of relativity we know the falling observer finds that the clock of the fixed observer is slower than his clock, and that rods of the fixed observer pointing towards the sun's center are shorter than his rods (see Figure 11.9). This follows at once from the Einstein contraction of space and time for moving bodies (see Section 10.14). To see how big this contraction effect is at the distance r from the center of the sun, we note that the freely falling observer, having fallen from an infinite distance, is moving towards the center of the sun with a speed equal to $\sqrt{2(GM/r)}$ (see discussion on the speed of escape, p. 100).

As far as this falling observer is concerned, therefore, it appears to him that the clock fixed in the gravitational field of the sun is moving away from the sun with a speed equal to $\sqrt{2(GM/r)}$. *In other words, to the falling observer, for whom there is no gravitational field, it appears that the clock at rest in the gravitational field is retarded with respect to his clock, by an amount equal to* $\sqrt{1 - (v^2/c^2)}$, *where* v^2 *is equal to* $2(GM/r)$, *so that the retardation of a clock at a distance r from the center of the sun is equal to*

$$\sqrt{1 - 2\frac{GM}{rc^2}}.$$

From this we see that the closer the clock gets to the center of the sun (that is, the larger the gravitational field in which it is placed) the slower it goes. This applies not only to mechanical clocks, but to any device that beats and can therefore be used as a clock—for example, the human heart. From this we conclude that a person living in an intense gravitational field ages more slowly than one in a weak gravitational field. *The retardation of time in a gravitational field has important astronomical consequences, because an atom placed in an intense gravitational field vibrates more slowly, and hence emits a spectrum that is different from that emitted by the atom that is not subject to a gravitational field.*

11.25 The Einstein Red Shift

We find that the spectral lines emitted by atoms in intense gravitational fields are displaced towards the red end of the spectrum, just as they would be (as a result of the Doppler shift) if the atoms were moving away from us. This effect, which is known as the **Einstein red shift**, was verified experimentally by Pound and Rebka, using what is known as the **Mössbauer effect**. The frequency of gamma rays emitted by a radioactive nucleus near the surface of the earth was compared with the frequency of the radiation emitted by the same kind of radioactive nucleus at a greater distance from the surface of the earth, and it was found that this difference agrees with the results obtained from Einstein's theory.

Another way of obtaining the formula for the Einstein red shift is to compare the frequency of radiation emitted by an atom at a point in a gravitational field with the frequency of radiation emitted by a similar atom in field-free space. We know that if the frequency of emitted radiation is ν_0, the energy of a single photon emitted from the atom is $h\nu_0$ (where h is a universal constant which we shall discuss later). (See Figure 11.10.) We may now treat this photon as though it had a mass $h\nu_0/c^2$, according to the Einstein energy-mass relationship. If we picture the atom as being on the surface of a body such as the sun, whose mass is M and whose radius is R, then when the photon has reached a point at a great distance from the surface of the sun, it has an amount of potential energy, relative to the surface of the sun, equal to

$$\frac{GM}{R}\frac{h\nu_0}{c^2}$$

(see the discussion of potential and kinetic energy, pp. 98, 100). This gain in the potential energy of the photon must occur at the expense of its original energy $h\nu_0$, so that its new energy is now

$$h\nu_0 - \frac{GM}{R}\frac{h\nu_0}{c^2}.$$

If the frequency of the photon as measured at this great distance from the sun is ν, we must therefore have (since the energy of the photon at this great distance is $h\nu$)

$$h\nu = h\nu_0 - \frac{h\nu_0}{c^2}\frac{GM}{R}$$

or (11.5)

$$\nu = \nu_0\left[1 - \frac{GM}{c^2R}\right]$$

which gives the effect of the gravitational field on the frequency of radiation emitted by atoms in this field. This same expression can be obtained as a first approximation from the formula for the slowing down of a clock given in our previous discussion.

We shall refer to this formula again when we discuss the application of the Einstein red shift to the determination of the radii of small stars. As we can see from the expression, the effect becomes more and more pronounced as the radius of the surface on which the atom is vibrating gets smaller and smaller, so that the effect becomes most pronounced for stars that are very small (white dwarfs).

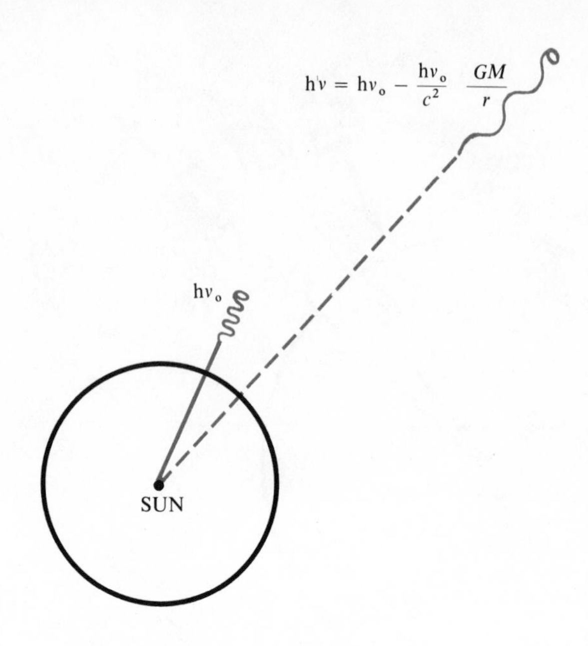

FIG. 11-10 *A photon becomes reddened when it moves from the strong part (near the center) to the weak part of the gravitational field of a body.*

11.26 The Contraction of Lengths in a Gravitational Field

Not only clocks but lengths are affected by gravitational fields. Again, let us consider the man in the freely falling elevator, who is equivalent to a person who experiences no gravitational field. *If he holds up a unit length in the direction in which he is falling (that is, radially to the sun) and compares it with a unit length parallel to his, but fixed in the gravitational field at the same distance from the center of the sun (that is, the distance r), he finds that this fixed length is shorter than his length.* Again, since it appears to him that the fixed length is moving past him with a speed $\sqrt{2(GM/r)}$, he finds that this fixed length is contracted by an amount $\sqrt{1 - 2(GM/rc^2)}$, according to the results from the special theory of relativity (see Section 10.13).

11.27 The Nature of the Geometry in a Gravitational Field

This at once leads the observer to conclude that the geometry of space-time in the gravitational field

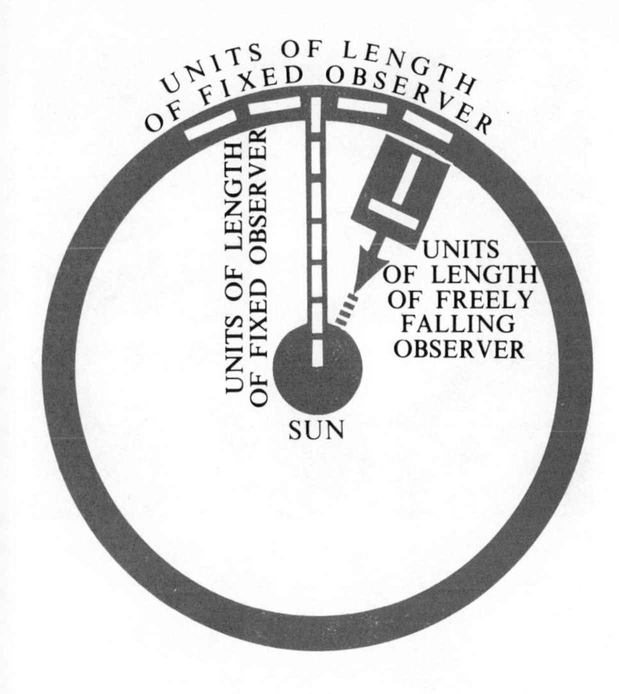

FIG. 11–11 *The fixed observer in a gravitational field measures the radius of a circle with a unit of length that is contracted as compared with a unit not in a gravitational field (the unit of the freely falling observer), whereas he measures the circumference with a unit of length that is unchanged.*

is non-Euclidean. We can see this as follows: if our falling observer compares a unit length that is parallel to the floor of the elevator with a parallel fixed length in the gravitational field (Figure 11.11) *he finds that there is no contraction because his motion is at right angles to these two lengths.* In other words, if a circle were drawn around the sun in the gravitational field, and if an observer fixed in the gravitational field measured first the radius of this circle and then its circumference, and compared these two measurements, he would find the circumference not equal to 2π times the radius (Figure 11.11).

This follows from what we have already said. *The unit lengths placed along the radius of this circle are all contracted, whereas the unit lengths placed along the circumference show no effect since they are at right angles to the motion of the falling elevator.* From this it follows that more unit lengths can be placed along the radius of a circle than would be the case if the circle were out at a great distance from any gravitational fields (that is, if the freely falling observer were measuring the radius of the circle). We see from this analysis that the circumference of a circle in a gravitational field is less than 2π times the radius of the circle. This means that the geometry in a gravitational field is of the elliptical type (see discussion on non-Euclidean geometry, Section 11.9).

Although we have analyzed the geometry in a radial gravitational field by considering a circle with the source of the field (the sun) at the center of the circle, we can apply a similar analysis to a small circle around any point in the field. However, the analysis in this case is more complicated, and we shall not give it here. Note that for a small circle around a point in the gravitational field the departure from Euclidean geometry is small and difficult to detect because the geometry in a small enough region of space even in an intense gravitational field can be approximated by Euclidean geometry. This is equivalent to saying that a small piece of curved surface (for example, the surface of the earth in the neighborhood of a point) is approximately flat.

11.28 The Increase of Mass in a Gravitational Field

One more effect of a gravitational field may be mentioned at this point. *The mass of a body increases as the gravitational field surrounding it increases, so that mass itself can no longer be defined independently of gravitational fields.* Again we see that this follows from the analysis made by the observer in the freely falling elevator. If he compares the unit mass in his elevator with one that is fixed in the gravitational field, he finds that the latter is greater by the factor

$$\frac{1}{\sqrt{1-\frac{v^2}{c^2}}}=\frac{1}{\sqrt{1-2\frac{GM}{c^2r}}},$$

for exactly the same reasons that lengths and time are contracted.

11.29 The Bending of Light in a Gravitational Field

We now consider how the gravitational field of a body such as the sun influences a ray of light. To do this, we replace the gravitational field of the sun by an elevator that has an acceleration radially away from the center of the sun equal to that which a freely falling body would have at that point. As we have noted, all the effects that an observer experiences in such an elevator without the sun's

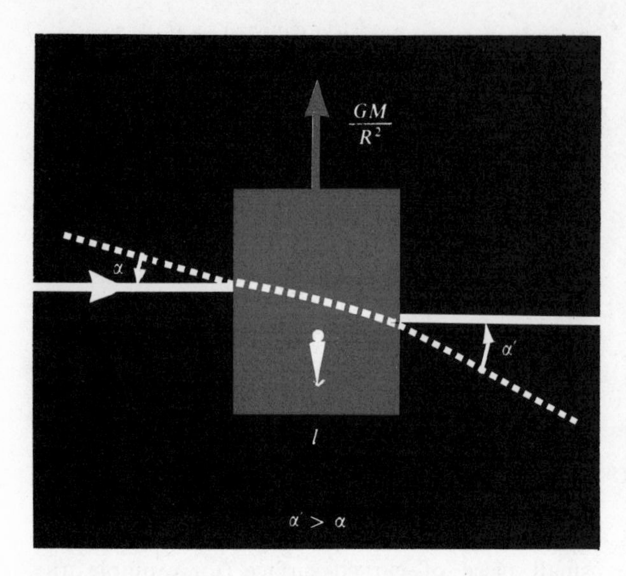

FIG. 11–12 *The bending of a ray of light in a gravitational field is illustrated by the principle of equivalence applied to a ray of light as seen by an observer in an accelerated elevator. The aberration of light is used to derive the effect.*

being present, are the same as those which are experienced by an observer fixed in the gravitational field (as shown in Figure 11.12) at a distance r from the center of the sun.

We consider a beam of light entering the elevator in a direction parallel to its floor. As seen by the observer in the elevator, this beam deviates from its direction as seen by a fixed observer by an angle α equal to v/c, because of the aberration of light (see p. 68), where v is the speed of the elevator at the moment the beam enters it from the left, and c is the speed of light. By the time the beam has moved across the elevator, the speed of the elevator has changed because of its acceleration. Since the acceleration of the elevator is GM/r^2, the change in the speed of the elevator is $(GM/r^2)t$, where t is the time it takes the light to cross the elevator.

If the width of the elevator is l, this time is obviously l/c, so that the speed of the elevator at the moment the beam leaves the elevator is $v + (GM/r^2c)l$, since v was the speed of the elevator when the light entered the left-hand side. From this it follows that the light, on leaving, suffers an additional aberration effect and thus makes an angle α' equal to

$$\left(v + \frac{GM}{r^2}\frac{l}{c}\right)\frac{1}{c}$$

with the direction of the beam when it entered (as seen by the man in the elevator). Thus the total

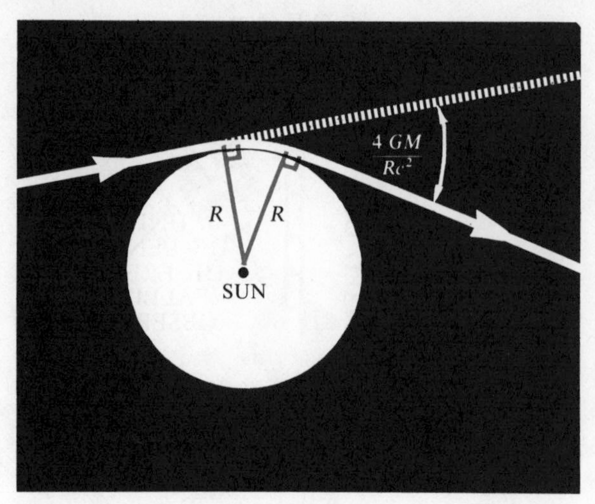

FIG. 11–13 *The actual bending of a ray of light as it grazes the photosphere of the sun. The total angular change in the direction of the beam is* $(4GM)/Rc^2$.

amount of bending suffered by the beam equals the difference between this aberration and its aberration on entering. In other words, the total bending of the beam while it is passing through the elevator is equal to

$$\left(v + \frac{GMl}{r^2c}\right)\frac{1}{c} - \frac{v}{c}$$

or

$$\frac{GMl}{r^2c^2}.$$

We see that the beam undergoes an amount of bending equal to GM/c^2r^2 per unit length of its journey.

If we consider a beam of light coming from a very great distance and just grazing the surface of the sun, we can obtain the total amount of its bending by finding the sum of all such effects over its entire path. When this is done, we find that the total angle through which the light is bent is equal to $4GM/Rc^2$, where R is the radius of the sun, and M is its mass, as shown in Figure 11.13. If numbers for M, G, and R are put into this formula, we find that the total angle of bending is 1″.75, which agrees quite well with the observational results obtained from eclipse expeditions.

11.30 The Precession of the Perihelion of Mercury

Now that we have considered some of the consequences of the general theory of relativity, we are in a position to see how it accounts for the precession of the perihelion of Mercury. We saw in our discussion of planetary motion (Chapter 9) that even when the perturbations on Mercury of all the bodies in the solar system are taken into account, using Newtonian gravitational theory, there is still a discrepancy in its motion that cannot be accounted for. The major axis of the planet's orbit does not remain fixed, but precesses slowly in the same direction as the planet revolves around the sun by an amount that is 43″ per century in excess of that predicted by Newtonian theory. In other words, the perihelion of Mercury advances continuously by this amount.

We may understand how this happens from the point of view of relativity theory, if we take into account the non-Euclidean nature of the geometry of space-time in the neighborhood of Mercury. We have seen that this geometry is elliptical. *This means that the circumference of a circle drawn around the sun having a radius equal to the radius of Mercury's orbit* (*which we may take as a circle for the present discussion*) *is smaller than 2π times the radius of this circle.*

We now consider Kepler's third law applied to the planet Mercury, keeping in mind the non-Euclidean character of space-time. We know from this law that the period of this planet is equal to some constant times the $\frac{3}{2}$ power of its mean distance from the sun. If the geometry governing the motion of the planet were Euclidean, the period of the planet would equal some other constant multiplied by the $\frac{3}{2}$ power of the circumference of the orbit, since in Euclidean geometry the circumference is equal to 2π times the radius. From this it follows that if a planet started from a given point in its orbit, it would come right back to this point again after one sidereal period, as determined from Kepler's law applied to this orbit.

However, we must now modify these considerations because the circumference of the orbit of the planet is less than 2π times its radius, *so that the distance traveled by the planet in one sidereal period is now greater than the circumference as determined by Euclidean geometry.* In other words, if the planet starts from a given point in its orbit, say its perihelion, it will overshoot this point after one period, as determined by Kepler's third law. Hence the motion of the planet is just as though this perihelion point itself had moved forward, as shown in Figure 11.14. This, then, accounts for the motion of the perihelion of Mercury.

Of course we cannot give the magnitude of the effect by this argument, but it shows why an effect of this sort should be present. An exact mathematical analysis of the orbit using Einstein's theory gives excellent agreement with observation.

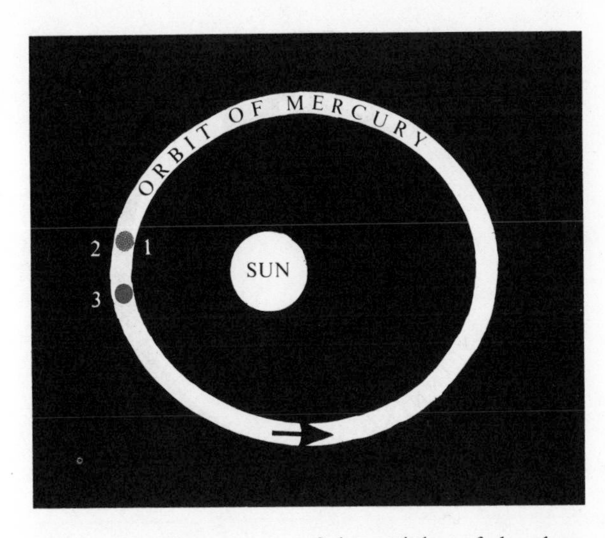

FIG. 11–14 *The precession of the perihelion of the planet Mercury arises because the relationship between this period and the circumference of its orbit is not governed by Kepler's third law but by elliptical geometry. Thus the planet overshoots the perihelion. After one period, the planet does not end up at point 2 but at point 3, which is the new perihelion position.*

11.31 The Space-Time Geometry of Rotating Systems

We finally consider the nature of the geometry for an observer situated on a rotating system. We know from what we have already said that such an observer experiences inertial forces (centrifugal and Coriolis) that cannot be distinguished from gravitational forces. Hence, it follows that such an observer finds himself governed by a geometry that is different from Euclidean geometry. That this is, indeed, the case is at once clear if we remember that rods and clocks placed in various positions in the rotating system are affected because of their motion. If our spinning system is a disk, as shown in Figure 11.15, and *our observer on the disk places a unit rod along the radius of the disk, its unit length is unaffected, because its motion resulting from the rotation of the disk is at right angles to its own length. However, if the same rod is placed along the circumference of this disk, it will suffer an Einstein contraction, because it will be moving parallel to its length.* Owing to this effect, the man on the spinning disk finds that the circumference of the disk is greater (more unit lengths can be placed along it because of their contraction) than he would if the disk were not spinning.

FIG. 11–15 *On a rotating disk, unit lengths laid off radially remain unchanged, but unit lengths laid off transversely along the circumference of the circle are shortened. This results in hyperbolic geometry.*

This means that for the man on the disk the circumference of a circle is greater than 2π times its radius, since unit lengths are not affected when placed along the radius. Thus, the geometry for the man spinning with the disk is hyperbolic. We can derive all the effects that are attributed to the centrifugal and Coriolis forces from an analysis of this non-Euclidean geometry which governs his space-time.

Although in the last two chapters we have considered the theory of relativity only as it applies to a region in the neighborhood of the sun, it has had enormous influence on our ideas of cosmology. In a later chapter we shall consider this point in detail when we discuss the cosmological problem.

11.32 Mach's Principle of Relativity

In discussing the general theory of relativity we saw that there is no experiment that an observer can perform to differentiate between gravitational forces and inertial forces, such as arise in accelerated systems (e.g., a rotating disk). However, the general theory of relativity recognizes the difference between these two types of forces insofar as their origins are concerned. *The gravitational forces originate from masses and cannot be completely eliminated by transforming to different coordinate systems, whereas the inertial forces are the result of accelerations and can be transformed away.*

However, there is an extension of the general theory of relativity which goes back to Ernst Mach (1838–1916) who ascribed inertial forces as well as gravitational forces to the distribution of masses of bodies in the universe: this is known as the **Mach principle.** We may illustrate it by considering the forces (centrifugal, Coriolis, centripetal, etc.) experienced by a body on a rotating disk. According to the Mach principle these forces are present only because the disk is rotating with respect to the distant stars, and the inertial forces we ascribe to this rotation are forces that originate from the interaction of the rotating disk and these distant stars. In other words, according to Mach's principle, if no other matter were present in the universe, a person on a rotating disk would experience no inertial forces, and, indeed, it would be impossible for him to detect any kind of acceleration.

We may therefore restate the Mach principle as follows: all the effects experienced by an observer on a rotating disk (or in any kind of accelerated system) are the same as the effects he would experience if the disk were not rotating and the universe were spinning in the opposite sense about the observer. In other words, according to Mach's principle there is no observational difference between the rotation of a body with respect to the fixed stars and the rotation of the stars with respect to the fixed body, so that the inertia of a body is determined by the distribution of matter throughout the universe. This principle has recently been incorporated in certain cosmological theories; see Chapter 31.

EXERCISES

1. If a foot ruler were laid along a radius pointing to the center of the sun at a distance of one million miles from the center, how would its length compare with its length if it were placed at right angles to the radius?

2. (a) By how much would a clock slow down if it were on the surface of Jupiter as compared to its rate in interstellar space far away from any massive body? (b) How large is the Einstein red shift on the surface of the sun?

3. Compare the world line of a beam of light where there is no gravitational field with its world line near the sun as it moves radially away from the sun.

4. One can show that in a gravitational field emanating from a mass point m, the space-time interval (or the line element) ΔS between two neighboring events at a distance r from the mass point is given by:

$$\Delta S^2 = \frac{\Delta r^2}{1 - \dfrac{2Gm}{c^2 r}} + r^2[\Delta\theta^2 + \sin^2\theta(\Delta\phi)^2] - \left(1 - \frac{2Gm}{c^2 r}\right)c^2\,\Delta t^2$$

where Δr is the distance between the two events along the line of sight, $\Delta\theta$ is the difference in declination, $\Delta\phi$ is the difference in right ascension, Δt is the time interval between the two events, G is the gravitational constant, and c is the speed of light. Use this expression to show (a) how much a length placed radially in this field shrinks, (b) how the speed of light changes in a gravitational field, and (c) what happens to a clock in a gravitational field.

12

PROPERTIES OF THE PLANETS*

12.1 Symmetry of the Solar System Relative to the Sun

When we consider the solar system as a whole, we are at once struck by certain symmetrical and regular features which dominate the entire structure. Since it is best to consider the physical characteristics of the individual planets in the light of these regularities, we shall first briefly describe them, and then collect all the data into a single table, from which we can proceed to a more detailed discussion of each planet separately.

The feature that is most apparent when we look at the solar system as a single structure is its symmetry with respect to the sun, as shown in Figure 12.1a. If we examine this structure in detail, we find first that all its principal bodies (the planets) circulate around the sun in the same direction and *that their orbits all lie in planes that are very nearly coincident with the sun's equatorial plane* (see Figure 12.1c). Moreover, all the planets, except Uranus, rotate about their axes in the same direction as they revolve in their orbits (which is also in the same sense as the sun itself rotates) and these axes of rotation stand almost perpendicular to the plane of the ecliptic. We also find that the satellites associated with these planets, with a few minor exceptions, revolve about their respective planets in the same direction as the planets themselves revolve around the sun. Furthermore, the periods of rotation of the planets, except for those of Mercury and Venus, range from 10 to about 25 hours, so that again we see that the variation is not great.

12.2 Distances of Planets: Bode's Law

There is also a regularity in the distances of the planets as we move outwardly from the sun. This was first written down in its numerical form by Titius of Wittenburg, and discovered independently by Bode in the year 1772. This Bode-Titius law may be expressed by stating that if the earth's distance from the sun is taken as unity (that is, if we express the distances in astronomical units) the mean distance from the sun of any one of the other planets can be found from the formula

$$\boxed{d_n = 0.4 + 0.3(2^n)}, \qquad (12.1)$$

where n is to be placed equal to $-\infty$ for Mercury, 0 for Venus, 1 for earth, 2 for Mars, 3 for the asteroids, etc. This formula gives the distances for the planets shown in Table 12.1. We have given the true distances in the last column, and we see that except for Mercury, Neptune, and Pluto the agreement is quite good, so that here, too, we discern a remarkable regularity in the geometry of the solar system.

TABLE 12.1

PLANETARY DISTANCES ACCORDING TO BODE'S LAW

Planet	Distance according to Bode's law	True distances in astronomical units
Mercury	0.4	0.39
Venus	0.7	0.72
Earth	1.0	1.00
Mars	1.6	1.52
Gap (the asteroids)	2.8	2.9
Jupiter	5.2	5.20
Saturn	10.0	9.55
Uranus	19.6	19.2
Neptune	38.8	30.1
Pluto	77.2	39.5

12.3 Bode's Law and the Asteroid Gap

If we compare the distances given by Bode's law with the distances as measured, we find that there appears to be a gap between Mars and Jupiter,

FIG. 12–1 *(a) The symmetry of the solar system as related to the sun. The orbits of terrestrial planets are in red and the major planets in white. The orbits are not to scale. (b) Relative sizes of the planets compared to the sun, whose rim is shown.*

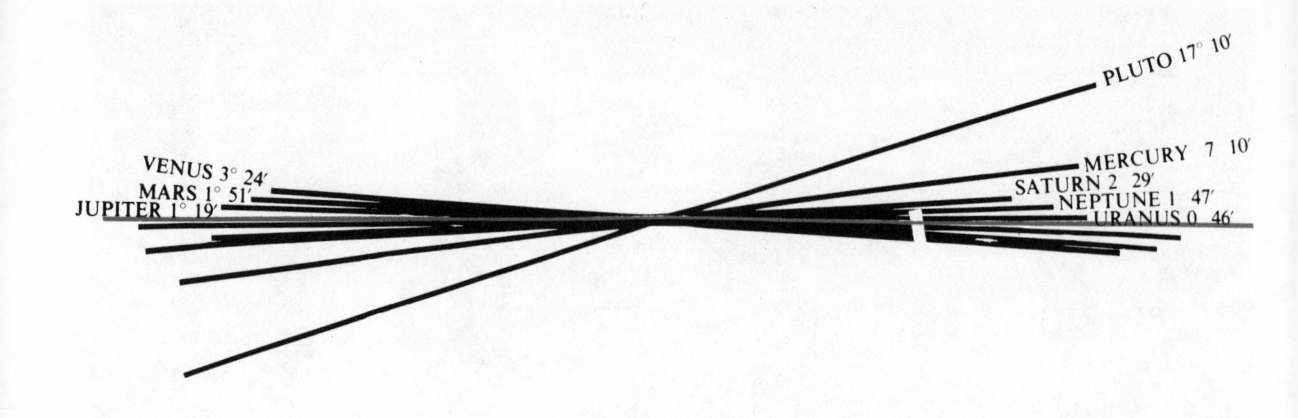

FIG. 12–1 *(c) The tilts of the planes of the planetary orbits are all very small.*

which is not predicted by Bode's law. One might suppose this to be a flaw that discredits the law itself. It is not, however, since we now know that this spatial gap is occupied by a swarm of tiny planets, the asteroids. If one calculates the mean distance from the sun of these asteroids (we shall see that this is given essentially by the mean distance of the most massive of these bodies, the asteroid Ceres), one finds that it corresponds very closely to the value obtained from Bode's law.

In other words, Bode's law tells us that we can expect to find something in the nature of a planet at this mean distance, and we do. We shall see that this gap represents more than a mere geometrical division in the solar system, for it separates the planets into two groups having different physical characteristics and dynamical properties.

It should be noted that one can obtain a numerical relationship similar to the Bode-Titius law if, instead of the earth's distance, the mean distance of some other planet is taken as unity. In that case, of course, different numbers have to be used in the formula, but the same regularity prevails.

12.4 The Masses of the Planets

We find regularity not only in the distances but also in the masses of the planets. Here, however, there is a remarkable division between the four innermost planets, Mercury, Venus, Earth, Mars, and the four outer ones, Jupiter, Saturn, Uranus, Neptune (we disregard Pluto, which requires special consideration). As we move away from the sun, we find that the masses increase gradually until we reach Mars, at which point there is a dip (the mass of Mars being about one-tenth that of the earth). If we take the four innermost planets together, and include the mass of the moon as well, we find that the total mass is about twice that of the earth's mass, with masses ranging from about $5\frac{1}{2}$ per cent of the earth's mass for Mercury to 82 per cent for Venus. We see that these four innermost planets contribute very little to the total planetary mass of the solar system.

When we pass to planets beyond the asteroid gap, there is a sudden and sharp increase in the masses starting with Jupiter, and then a gradual falling off. The masses in this group range from about 318 times the mass of the earth for Jupiter to about 15 times for Uranus. We shall exclude the planet Pluto, since this object seems to be in a category by itself.

12.5 Sizes, Densities, and Oblatenesses of Planets

Having divided the planets into two groups according to their masses, we find that the same grouping holds for their diameters, densities, and oblatenesses. For the four inner planets, which we shall refer to as the **terrestrial planets**, the equatorial diameters range from about 38 per cent of the earth's diameter for Mercury to about 97 per cent for Venus. The equatorial diameter of Mars is slightly larger than one-half that of the earth. In the case of the outermost planets, the **Jovian planets** as we shall call them (again except for Pluto), the equatorial diameters range from $3\frac{1}{2}$ times the earth's diameter for Neptune to 11.2 times for Jupiter.

It is clear from the masses and diameters of the terrestrial and Jovian planets that the former are considerably denser than the latter. In fact we find that whereas in the case of the terrestrial planets the densities range from 4 gm/cm^3 for Mars to $5\frac{1}{2}$ gm/cm^3 for the earth, in the case of the Jovian planets the densities range from 0.7 gm/cm^3 for Saturn to 2.3 gm/cm^3 for Neptune (Pluto excluded). The most massive planet, Jupiter, has a density of 1.33 gm/cm^3. In view of these facts we must conclude that the chemistry of the terrestrial planets is quite different from that of the Jovian planets, and that this is related to the way in which these two groups of planets were formed during the early stages of our solar system.

We also note that whereas the terrestrial planets are hardly oblate at all, the maximum being 0.005 for Mars, the Jovian planets are quite oblate, the maximum being 0.096 for Saturn. This difference in oblateness (see Section 3.11) arises because of the lesser density of the Jovian planets and because of their more rapid spinning.

TABLE 12.2

NATURAL SATELLITES OF PLANETS

Planet	Satellites	Distance from planet	Sidereal period	Synodic period			Orbit incl.	Orbit eccentricity	Radius	Mass	Apparent magnitude, m_v, at mean opposition
		10^3 km	days	d	h	m	° (R = retrograde)		km	10^{24} g	
Earth	Moon	384	27.32	29	12	44		0.054 9	1 738	73.5	−12.7
Mars	1 Phobos	9	0.3		7	39	1.1	0.021	6		+11.5
	2 Deimos	23	1.26	1	06	21	1.6	0.002 8	3		+12.5
Jupiter	1 Io	422	1.77	1	18	29	0	small	1 670	73	+ 5.5
	2 Europa	671	3.55	3	13	18	0	and	1 460	47.5	+ 5.8
	3 Ganymede	1 070	7.15	7	04	00	0	vari-	2 550	154	+ 5.1
	4 Callisto	1 883	16.69	16	18	05	0	able	2 360	95	+ 6.3
	5	181	0.49		11	57	0.4	0.003	70		+13
	6	11 470	250.59	264			28	0.158	50		+14
	7	11 740	259.7	276	10		26	0.206	10		+18
	8	23 500	737	630			33 R	0.40	10		+18.5
	9	23 700	758	645			25 R	0.27	8		+19
	10	11 850	255	272			28.5	0.135	7		+19
	11	22 560	692	596			16.5R	0.207	8		+19
	12	21 200	631	551			33 R	0.16	6		+19
Saturn	1 Mimas	186	0.94		22	37	1.5	0.020 1	300	0.04	+12.1
	2 Enceladus	238	1.37	1	08	53	0.0	0.004 4	300	0.07	+11.7
	3 Tethys	295	1.89	1	21	19	1.1	0.0	500	0.65	+10.6
	4 Dione	377	2.74	2	17	42	0.0	0.002 2	500	1.0	+10.7
	5 Rhea	527	4.52	4	12	28	0.3	0.001 0	700	2.3	+10.0
	6 Titan	1 222	15.95	15	23	15	0.3	0.029 0	2 440	137	+ 8.3
	7 Hyperion	1 481	21.28	27	07	39	0.5	0.104	200	0.31	+14.5
	8 Iapetus	3 560	79.33	79	22	05	14.7	0.028 3	500	1	+11
	9 Phoebe	12 950	550.41	523	15	36	30 R	0.163 3	100		+14
Uranus	1 Ariel	192	2.52	2	12	30	0	0.003	300	1.2	+14
	2 Umbriel	267	4.14	4	03	28	0	0.004	200	0.5	+15
	3 Titania	438	8.71	8	17	00	0	0.002 4	500	4	+13.8
	4 Oberon	586	13.46	13	11	16	0	0.000 7	400	2.6	+14.0
	5 Miranda	128	1.41					<0.01	100	0.1	+16.9
Neptune	1 Triton	353	5.88	5	21	03	20.1R	0.0	2 000	140	+13.6
	2 Nereid	5 600	360				27.5	0.76	100	0.03	+19.5

12.6 Natural Satellites of Planets

There is a marked difference in the numbers of natural satellites associated with the two groups of planets. Whereas there are just three satellites attached to the terrestrial planets (the earth's moon and the two tiny satellites of Mars), there are altogether 28 satellites associated with the Jovian group (12 for Jupiter, 9 for Saturn, 5 for Uranus, 2 for Neptune), several of which are more massive and larger than our own moon. This difference in the number of satellites may very well be associated with the difference in speeds of rotation of the planets in the two groups and may have its origin in the genesis of the solar system itself.

12.7 The Elements of a Planet's Orbit

It is convenient, in discussing the characteristics of an individual planet, to divide the data into two parts: (1) the characteristics of the planet's motion, and (2) the physical characteristics of the planet itself. In Chapters 8 and 9 we devoted a good deal of space to the analysis of the nature of orbits in general. However, to describe a particular planet we must specify more than just the nature of the orbit, which is given essentially by the eccentricity and the size of the semimajor axis. We must also specify just how the orbit is orientated in space with respect to some frame of reference, and for this we must know the following three quantities:

1. orientation of the orbit in its own plane, which is given by the direction that the semimajor axis makes with a fixed direction in this plane;
2. the tilt that this plane of the orbit makes with some fixed plane (let us say the ecliptic); and
3. the angle that the line of intersection of these two planes makes with some fixed direction in space.

In other words, the orbit of a planet is completely specified by the following five numbers:

1. semimajor axis, a;
2. eccentricity, e;
3. inclination of the plane of the orbit to the plane of the ecliptic, i;
4. the angle that the major axis makes with the line of intersection of the plane of the ecliptic and the plane of the orbit, ω;
5. the angle that the line of intersection of the two planes makes with the line from the sun to the vernal equinox—called the longitude of the ascending node, Ω.

These quantities are all shown in the diagram in Figure 12.2. If we also want to know just where a planet is in its orbit, we must know two more things about it, namely

6. the planet's sidereal period, P;
7. the epoch, E, that is, at what moment it was at its perihelion.

The quantity P, the planet's sidereal period, is in a somewhat different category from the others because it can be calculated by Kepler's third law if the mean distance of the planet from the sun, the gravitational constant, and the mass of the sun are known.

In our description of the individual planets we first give some data associated with the motion of the planet and then go on to the physical characteristics.

THE TERRESTRIAL PLANETS

12.8 Mercury: Its Orbital Characteristics and Its Motion

Although the planet Mercury is difficult to observe, because its angle of separation from the sun (elongation) is never very great, recorded observations of it go back to 264 B.C. The best time for observing this planet in the early evening is in March or April, if it is then at maximum eastward elongation, and the best time for viewing it as a morning star (just before the sun rises) is in September or October if it is then at its maximum westward elongation.

Since Mercury is the planet closest to the sun, its mean distance being 36 000 000 miles (58×10^6 km), its average speed in its orbit is greater than that of any other planet, varying from 35 mi/sec (56 km/sec) at perihelion (28 000 000 mi, or 46×10^6 km from the sun) to 23 mi/sec (37 km/sec) at aphelion (43 400 000 mi, or 69.5×10^6 km from the sun). It is easy to compute from these figures that Mercury's sidereal period is 88 days, and that its synodic period is 116 days.

From its perihelion and aphelion distances we find that the eccentricity of the orbit of Mercury is 0.206, which is larger than that of any other planet. The inclination of the plane of the orbit of Mercury with respect to the ecliptic is 7°, and the longitude of its ascending node is a little more than 47°. *The period of rotation of this planet, according to recent*

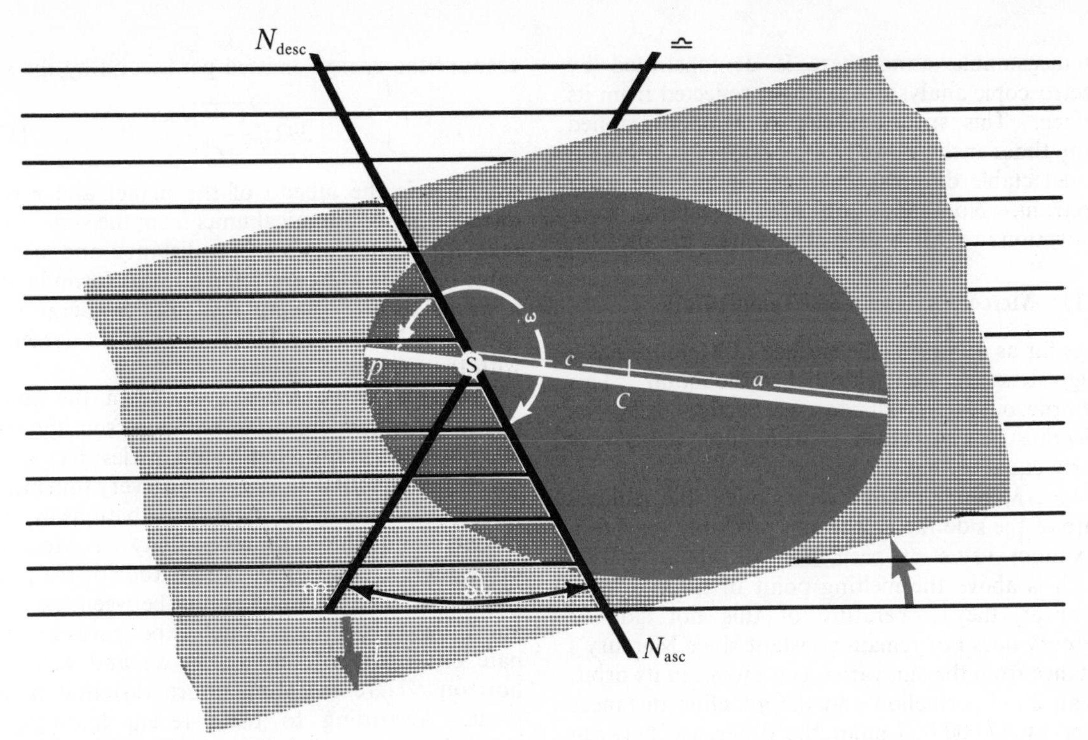

FIG. 12–2 *The elements of a planet's orbit are shown: the semimajor axis of the orbit, a; the eccentricity* $e = c/a$; *the longitude (measured from the Vernal Equinox, ♈) of the ascending node, ☊; inclination of the orbit to the ecliptic, i; the angle from the ascending node to the perihelion point, ω;* N_{asc} *is the ascending node, and S is the position of the sun.*

radar observations, is equal to about 60 earth days.

12.9 Mercury's Diameter and Mass

Since we can measure the apparent diameter of Mercury (that is, the angle that the diameter of its disk subtends at the eye, which ranges from 5″ to 13″ as the distance between Mercury and the earth varies) and also the distance of the planet from the earth at any moment, we can find the true diameter, which is about 3 100 miles (4 840 km).

Finding the mass of a planet is more difficult than finding its size because we must use its gravitational effect on some other body. If a planet has a satellite revolving around it, we can get the planet's mass simply by applying Newton's law of gravitation to the motion of the satellite (actually Kepler's third law). We have, in fact, from Kepler's third law as applied to the orbit of the satellite,

$$P^2 = \frac{4\pi^2 a^3}{GM},$$

where P is the period of the satellite, a is its mean distance from the planet, and M is the planet's mass. Here we have neglected the mass of the satellite compared to the mass of the planet. From this equation we have for the mass of the planet

$$M = \frac{4\pi^2 a^3}{GP^2}.$$

But if it does not have a satellite, we must analyze the way in which the planet perturbs the orbits of other bodies. In the case of Mercury this is difficult because it is so close to the sun, but it can be done. It has been found that the mass of this planet is somewhat more than 5 per cent of the mass of the earth. From knowledge of the mass of a planet and its diameter we can compute its density, which for Mercury is about 5.46 gm/cm³.

12.10 The Atmosphere of Mercury

Because of its small mass, Mercury has a fairly small speed of escape, 2.4 mi/sec. For this reason, and because of its high surface temperature, it cannot retain an atmosphere. That Mercury has no measurable atmosphere is demonstrated by spectroscopic analysis of sunlight reflected from its surface. This spectrum has been photographed many times at the Lowell Observatory, and there is no detectable difference between it and the solar spectrum. No bands or lines resulting from absorption in a Mercurian atmosphere are present.

12.11 Mercury's Albedo and Temperature

As far as we know, the surface of Mercury has a rough structure, similar to that of the moon. This is borne out by its albedo (see Section 6.29)—its reflecting power—which is 0.06, and hence very nearly equal to that of the moon.

Since Mercury rotates very slowly, the temperature of the side facing the sun probably reaches a maximum value at perihelion of about 400°C, which is above the melting point of tin and lead. However, the temperature of this hot side of Mercury does not remain constant since Mercury's distance from the sun varies as it moves in its orbit. Because the perihelion and the aphelion distances differ by 17 000 000 miles the difference between the amounts of radiation received by Mercury at these two positions may be as great as 50 per cent. At its closest approach to the sun, Mercury receives about 12 times as much radiation as the earth does, but at its greatest distance from the sun it receives only about 6 times as much. Hence the temperature of the hot side of Mercury falls to about 285°C at aphelion, and the temperature there when Mercury is at its mean distance from the sun is about 340°C. The values cited above were obtained by E. Pettit who used thermocouples (see Appendix A.36) to measure directly the thermal radiation emitted by the hot side of Mercury. These thermocouples, which had the shape of capital N, were formed by two pure bismuth wires connected diagonally by a wire consisting of bismuth and 5 per cent tin.

We can also determine the temperature mathematically by computing the amount of radiation the planet receives per second from the sun, and subtracting from this the amount reflected by the planet per second. This difference, which is the amount of radiant energy absorbed by the planet per second, determines how hot the planet gets.

In Chapter 16 we show that the absolute (Kelvin = centigrade + 273°) temperature to which a surface is raised when it absorbs radiation is proportional to the fourth root of the absorbed radiation. This leads to the following formula for the absolute surface temperature of a planet taking into account the energy emitted per second by the sun:

$$392\sqrt[4]{\frac{1-A}{r^2}}, \qquad (12.2)$$

where A is the albedo of the planet and r is its distance in astronomical units from the sun. If we take for r Mercury's mean distance and use the value 0.06 for Mercury's albedo, this formula gives a mean surface (facing the sun) temperature of 616°K for Mercury. This is in excellent agreement with the measured value.

The side of Mercury away from the sun at any moment, according to readings from the radio telescope at Parkes, New South Wales, has a temperature of about 0°C, suggesting a very thin atmosphere. Because of the slight perturbations by other planets, and the high eccentricity of Mercury's orbit, there are librations in longitude of the planet which give rise to a small region between the sunlit and the darkened faces which experiences an alternate oscillation of the sun above and below its horizon. Here the temperature variation is very great. According to these recent findings, we must discard our previous notion that Mercury keeps one face to the sun, while the other is in perpetual darkness.

12.12 Transits of Mercury

An interesting observational feature of Mercury is an occasional solar transit. It can then be observed as a small black spot moving slowly across the apparent disk of the sun. Transits of Mercury occur about 13 times per century. The latest one in this century occurred on November 7, 1960; the next will occur on May 9, 1970. From these transits we have greatly increased our knowledge of Mercury's position.

12.13 Venus: Its Orbital Characteristics and Motion

Venus, the planet next in order from the sun, dominates the evening sky after the sun has set (when no moon is present) when it is visible as an evening star. Since it is by far the brightest of all planets, it has played a great role in astronomy. When it is near its greatest eastern elongation, it is visible to the naked eye in the late afternoon, even before the sun has set.

The mean distance of this planet from the sun is 67 260 000 miles (108.21×10^6 km) and its orbital velocity is about 22 mi/sec (35 km/sec), and this is very nearly the same at all points in its orbit, since the orbit is almost circular. Since its orbital eccentricity is only 0.007 (it is smaller than that of any other planet), *Venus' distance from the sun varies only by about* 1 *million miles.* From Venus' velocity and the radius of its orbit we can calculate its sidereal period around the sun. The synodic period of Venus is 584 days, or 1 year and 7 months, and its sidereal period is just 224.7 days. The inclination of the plane of its orbit to the plane of the ecliptic is about 3°24′ and the longitude of the ascending node of its orbit is 75°47′.

12.14 Venus: Speed and Direction of Rotation and the Orientation of Its Axis

Because there are no fixed markings on the visible disk of Venus (a cloud layer surrounds the solid surface), it is difficult to measure its period and direction of rotation. Independent radio-telescopic investigations (the 1 000 ft fixed instrument in Puerto Rico, staffed by Cornell University, and another at the Jet Propulsion Laboratory of the California Institute of Technology) indicate that the period of rotation is once every 247 ± 5 days. As seen from above the north pole of Venus, the direction is clockwise (an exception to the counterclockwise direction of rotation of all other planets). The axis is tilted about 6° with respect to the perpendicular to the the plane of orbit.

12.15 Physical Characteristics

Measurements of the physical characteristics of Venus show it to be more nearly like the earth in its general structure than any of the other planets. Its apparent diameter is 64″ when it is closest to us at the time of inferior conjunction (26 000 000 miles or 42×10^6 km distant from the earth) and only 10″ at superior conjunction (160 000 000 miles or 260×10^6 km distant from the earth). Except for the moon, a few asteroids, and some comets now and then, no other celestial objects get as close to us as this planet. From the measured values of its apparent diameter and distance at a definite time we find that its diameter is 7 560 miles (12 200 km) so that its surface area is 95 per cent of the earth's and its volume is 88 per cent of the earth's volume.

We can determine the mass of Venus by observing its gravitational effects on the orbits of the earth and of Mars. We find from observations that its mass is about 82 per cent of the earth's mass and its density is 5.1 gm/cc or about 92 per cent of the earth's.

12.16 Phases of Venus and Its Albedo

As the planet changes its position relative to the sun and the earth, its appearance changes just the way the moon's does, so that it passes through successive phases ranging from a thin crescent phase to a full phase. These variations in the appearance of Venus were observed first by Galileo, with the aid of his homemade telescope, and their similarity to the phases of the moon convinced him of the correctness of the Copernican theory. If the Ptolemaic theory were correct, Venus would show only the crescent phase. See Figures 12.3 and 12.4.

Since Venus is completely surrounded by a cloud, its very high albedo is 0.75.

12.17 Atmosphere of Venus

Because of the dense opaque cloud surrounding Venus, we know very little about its surface, but we do have some information about its atmosphere. We have obtained this information by spectral analysis of the solar radiation it reflects. Most of this sort of work was started at the Lowell Observatory and then carried on more extensively at the Mount Wilson Observatory. The spectroscopic evidence clearly points to the presence of large quantities of carbon dioxide in the atmosphere of Venus, as compared to the quantities of this gas found in the earth's atmosphere. According to the calculations of Dunham, the amount of carbon dioxide present in that part of the atmosphere which can be observed from the earth is equivalent to a layer 400 meters thick completely surrounding the planet at standard temperature and pressure conditions (that is, at a temperature of 0°C and a pressure of 1 atmosphere). This is about 250 times the quantity present in the earth's atmosphere.

As for water vapor and oxygen, spectral evidence indicates that, in the part of Venus' atmosphere that lies above the clouds, there appears to be less than one one-thousandth the quantity of these gases than there is in an equivalent layer at the same height above the earth's surface. The spectral evidence for the presence of other gases is much too weak to enable us to reach any conclusions.

The conclusions in the previous paragraph are based on spectral data in the ordinary optical part of the spectrum, but recent evidence, obtained February 21, 1964, and based on infra-red spectroscopy, indicates the presence of a considerable amount of water vapor in Venus' atmosphere.

FIG. 12–3 *Venus in blue light. This shows the crescent phase. (200-inch photograph from the Mount Wilson and Palomar Observatories)*

Until recently, it was impossible to analyze the infra-red spectrum of Venus because of the absorption of infra-red radiation by our own atmosphere (a result of terrestrial water vapor). Balloon flights up to heights of 87 000 feet above the earth's surface have enabled astronomers to obtain the infra-red spectrum of Venus in the 11 300 Angstrom region, which corresponds to the infra-red absorption of the water molecule. In a balloon expedition from Holloman Air Force Base, New Mexico, J. S. Strong and his colleagues found (by means of special infra-red spectroscopic equipment) that there is a fairly intense absorption band at 11 300 Angstroms. They identified this band by comparing it with the water vapor band at 11 287 Angstroms. From the intensity of this absorption band in the spectrum of Venus (the height of the balloon eliminated the obscuring effect of terrestrial water vapor), Strong and his co-workers concluded that above the clouds the amount of water vapor is equivalent to a liquid layer 0.003 9 inches thick. They further concluded that the water vapor in the entire atmosphere of Venus may be as large as that in our own atmosphere and that the clouds surrounding Venus may consist of water droplets. That this infra-red absorption arises from Venus is indicated by the fact that its Doppler shift corresponds to the calculated Doppler shift of Venus.

12.18 The Nature of the Cloud Surrounding Venus

We know from the high albedo of Venus that the surface is completely surrounded by an opaque cloud, but we have little evidence about the nature of this cloud. Speculations concerning its chemical nature have ranged from those of Wildt, who argued that it was made up of droplets of formaldehyde, to those of Menzel and Whipple, who have presented a strong case for water vapor clouds. The argument against fine water droplets as the principal constituent of the clouds is based on spectroscopic evidence. As Dunham's analysis has shown, the upper limit of water vapor in all of Venus' atmosphere cannot exceed 2 to 5 per cent of that in the earth's atmosphere above the Mount Wilson Observatory. Moreover, the color of these clouds is yellowish, whereas we know that water clouds are almost perfectly white reflectors. (See Section 12.17.)

Recent estimates of temperature conditions on the surface of Venus based on radio emissions from the planet are also difficult to reconcile with water clouds since the very high temperatures found should result in a good deal of water vapor. Nevertheless, Menzel and Whipple argue that if Venus were completely covered with water (which they suggest), the atmosphere there would be exceptionally free of all kinds of dust or nuclei and the special cloud formations possible under those conditions would correspond to those found on Venus. The first corroborative spectroscopic evidence for water vapor in Venus' atmosphere was obtained with spectroscopes sent to heights of more than 78 000 feet by a balloon on November 28, 1959.

Wildt's formaldehyde hypothesis is based on a photochemical process in which solar energy is used by water vapor and carbon dioxide to synthesize formaldehyde. In this way Wildt sought to account for the absence of water vapor in the atmosphere.

It has also been suggested that the cloud layers may be composed of various oxides (essentially dust of the sort that we have here on earth) and possibly also salts, such as sodium chloride. Kuiper has used polarization experiments in the infra-red region of the reflected light to develop a theory that the

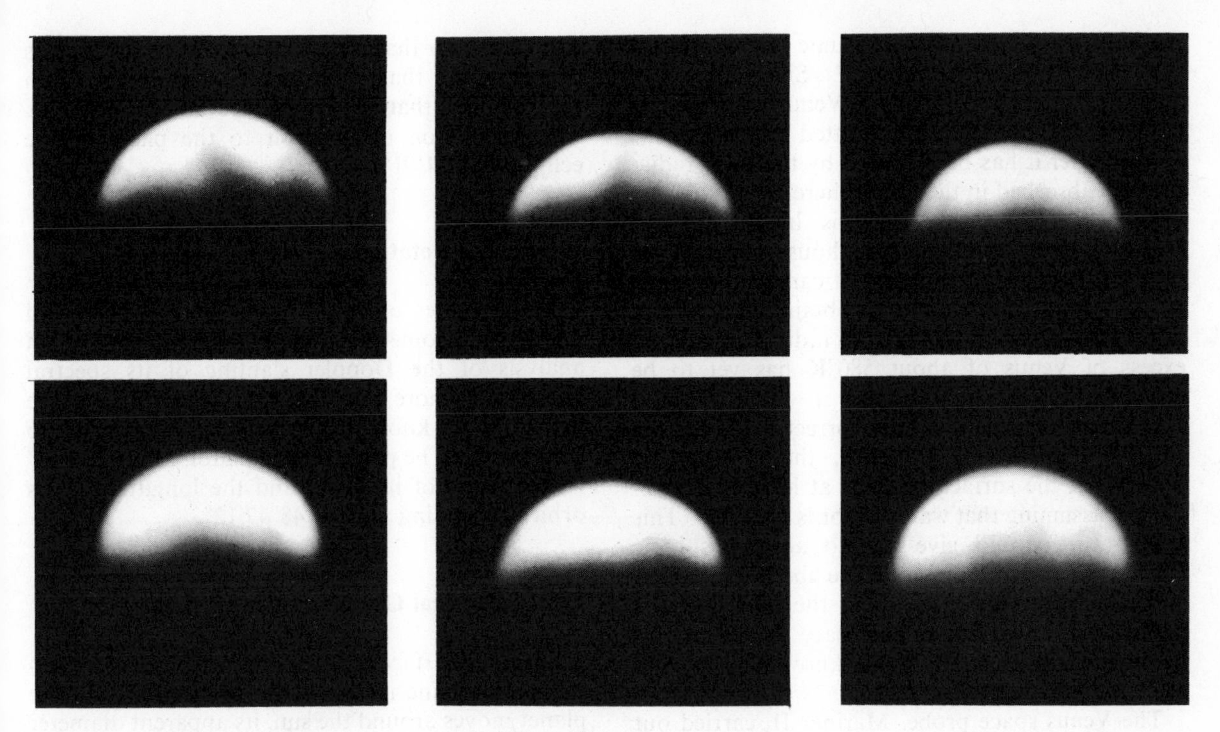

FIG. 12–4 *Venus, six photographs, all taken in ultraviolet light from June 6 to July 1, 1927. The remarkable homogeneity of the cloud is evident.*

clouds are composed of carbon and oxygen atoms strung into large molecules.

12.19 The Temperature of Venus

Although we don't know the precise chemical nature of the clouds, we do know that they are in fairly violent motion, probably as a consequence of the intense heating that takes place on the side facing the sun. Because of this motion, the surface temperature is probably uniform over the entire surface—provided that it is due entirely to radiation from the sun.

However, recent measurements of radio waves emitted by Venus in the centimeter wavelength range (at about 3 cm and 10 cm) show that the surface of Venus is much hotter than can be accounted for by solar radiation alone. If we use Equation (12.2) and place $A = 0.75$ (Venus' albedo) and $r = 0.72$ (Venus' mean distance from the sun in A.U.) we obtain about 300°K for the surface temperature of Venus. This value, which is derived on the assumption that the temperature of Venus is due entirely to absorbed solar radiation, is in pronounced disagreement with the temperature obtained from the intensity of radio waves emitted by the surface.

In using radio waves in this way one assumes that their emission is due to the random chaotic thermal motions of the molecules on the planet's surface and that therefore their intensity accurately indicates surface temperature. The most recent observational evidence shows that although the disk temperature, as determined from centimeter radio waves, varies slightly with the wavelength, the values obtained are all very close to each other. The average value of the disk temperature obtained from centimeter wavelengths is about 580°K.

In the millimeter radio wavelength range (8 mm and 8.6 mm) the temperature range is 300°K to 400°K, which is considerably smaller than for the 3-cm to 10-cm range mentioned above. This may be accounted for if we suppose that the centimeter radiation is emitted close to the surface where the temperature is very high, but that the millimeter radiation is emitted from the upper layers of Venus' atmosphere.

A surface temperature for Venus as high as 580°K cannot be accounted for by the solar radiation alone, even if we assume that there is a very high **"greenhouse effect."** Since there is a good deal of carbon dioxide in Venus' atmosphere, the infra-red rays that are emitted by the planet's surface after it has been heated by the solar radiation are absorbed in the atmosphere by this carbon dioxide so that the surface is heated further. However, even with this greenhouse effect, if we take into account the amount of carbon dioxide in Venus' atmosphere we cannot obtain temperatures in excess of 350°K, so that the radio temperature excess of Venus of about 580°K has yet to be accounted for.

If we accept 580°K as the correct value for the surface temperature of Venus, the atmospheric pressure at the surface must be at least 10 atmospheres, assuming that water vapor is present. This temperature would give rise to an atmospheric pressure of 30 atmospheres in the absence of water.

We can say very little about the nature of the surface of Venus, but recent space probes of this planet are adding greatly to our knowledge and will answer many questions.

The Venus space probe, Mariner II, carried out in 1962 by NASA in the U.S. has corroborated the conclusions drawn from the radio waves about the temperature on the surface of Venus. The data from this probe give a temperature of about 800° Fahrenheit, and the surface appears to be a vast waterless desert swept by winds traveling at hundreds of miles per hour. The atmospheric pressure at the surface is about 30 times that on the earth's surface. The data also indicate that there is hardly any water vapor and (very surprisingly) that carbon dioxide appears to be missing. The clouds themselves appear to consist of hydrocarbons. Finally, the Mariner II data show that there is no measurable magnetic field around Venus and that it may really have a retrograde rotation with a period of about 250 days.

12.20 Mars: Its Orbital Characteristics and Motion

It is easy to recognize the planet Mars at night with the naked eye, because it is characteristically reddish and fairly bright. Although its mean distance from the sun is 141 000 000 miles (227.9 × 10^6 km), its eccentricity is fairly large, 0.093, so that its distance from the sun varies by more than 26 000 000 miles (41 500 000 km). When it is in opposition, and also at perihelion, its distance from the earth may be less than 35 000 000 miles (56 × 10^6 km). Its sidereal period is 687 days, or 1.88 years, so that its average speed in its orbit is slightly more than 15 mi/sec (24 km/sec). From this we find that its synodic period is 780 days. The inclination of its orbit to the plane of the ecliptic is 1°51′01″.1.

12.21 The Rotation of Mars

Mars rotates on its own axis, just as the earth does, but somewhat more slowly. From an analysis of the Doppler slanting of its spectral lines, and, more accurately, a study of surface markings, we know that its sidereal day is equal to $24^h37^m23^s$. The plane of its equator is tilted 23°59′ to the plane of its orbit, and the longitude of its orbit's ascending node is 48°47′12″.

12.22 Physical Characteristics of Mars

Since the surface of Mars is clearly visible, we can easily determine its geometric properties. As the planet moves around the sun, its apparent diameter ranges from 3″.5 at conjunction to 25″ at the most favorable opposition. From the apparent diameter and the distance at any moment, we find that its true diameter is 4 191 miles (6 760 km), so that its surface area is 29 per cent of the earth's surface area, and its volume is 15 per cent of the earth's volume.

By applying Kepler's third law to the motion of the two satellites of Mars, we find that its mass is 0.108 of the earth's mass. If we divide the mass of Mars by its volume we find its mean density is 3.97 gm/cm^3. Since the surface gravity of a planet varies directly as its mass and inversely as the square of its radius, we can find this quantity without too much trouble. It is 0.376 times that on the earth, so that a 100-lb man will weigh only 37 lb on Mars. We can also find, using our knowledge of its mass and radius [see formula (6.11), p. 100)], that the speed of escape on the surface of Mars is 3.15 mi/sec (5 km/sec).

12.23 The Nature of the Martian Atmosphere

That Mars has an atmosphere is evident at once from spectroscopic data, and also from the scattering of light near its surface. In addition to its atmosphere we can observe continuously changing cloud formations and also vast dust storms. There is no hydrogen or helium in the atmosphere of Mars because of its low speed of escape, but this speed is sufficiently high to enable Mars to retain the heavier and therefore more slowly moving molecules of oxygen, nitrogen, carbon dioxide, and water vapor. Although carbon dioxide is the only gas that has been definitely identified thus far, most astronomers are inclined to agree with Kuiper that water vapor is present. He showed from a study of infra-red spectral lines in the radiation from Mars that the polar caps consist of frozen water. Polarimetric observations of the light reflected from the polar caps have now confirmed this. The amount of water vapor that appears to be present in the Martian atmosphere is about equal to the total amount that we find in the part of the earth's atmosphere that lies beyond 8 km above the earth's surface. Although this is 100 times smaller than the amount present in the entire earth's atmosphere, it is sufficient to account for the clouds and the hoarfrost that have been detected by polarimeters.

Although we have no specific evidence for the existence of free oxygen in the Martian atmosphere, we must not conclude from this that oxygen is absent. Because of the strong oxygen absorption lines originating in the earth's atmosphere, the weak oxygen lines from Mars (if they exist) are almost entirely obscured, so that we must suspend judgment until data can be obtained with spectroscopes mounted in observatories orbiting the earth above our atmosphere. However, from preliminary studies, we can say that the amount of oxygen in the atmosphere of Mars is probably less than 0.1 per cent of that in the earth's atmosphere covering an equal surface area.

The gases that are present in the atmosphere of Mars probably consist of atoms that are difficult to detect spectroscopically, that are too heavy to escape from the planet, and that are not too active chemically. This leaves us with nitrogen and the rare gases, argon, neon, and krypton, which are probably present—if at all—only in very small abundances, as on the earth.

If we consider the Martian atmosphere as a whole and picture it as a shell of homogeneous gas at standard temperature and pressure conditions (0°C and at a barometric pressure of 30 inches of mercury), it then has a thickness of about 2 km. If we assume that the molecular weight of the Martian atmosphere is about 32, this gives a mass of about one twenty-fifth of the mass of the earth's atmosphere. If we accept the latter figures, it follows that the atmospheric pressure at the surface of Mars is about one-third of that on the earth.

12.24 Cloud Formations on Mars

Three groups of colored clouds have been observed in the atmosphere of Mars:

1. the **white clouds**, which may consist of tiny ice crystals, and which rise up to heights of 15.5 miles;
2. the **blue clouds**, recorded in blue and violet light, but invisible in red or yellow light, which lie below the white clouds; and
3. the **yellow clouds**, visible in red and yellow light, but not as clearly in blue or violet light, which are much lower than the blue ones, and rise to an altitude of about 3 miles.

12.25 Surface Markings: The Variation of Polar Caps

The surface of Mars is characterized by two distinctive variable features: the **polar caps** and the **green** areas. Whether observed photographically or visually, the polar caps are the most conspicuous feature of Mars, and one has little trouble detecting their seasonal changes. As the winter season advances in the northern hemisphere, the northern polar cap grows irregularly until it covers almost half the hemisphere. With the approach of spring in this hemisphere (corresponding to March 21 on the earth) the northern polar cap begins to shrink quite uniformly until, at the Martian summer solstice, it is reduced to a few hundred miles in diameter. As the summer advances, the cap diminishes still further, becoming much smaller. This whole cycle of variations starts again at the beginning of the Martian fall season.

The situation in the southern hemisphere is quite similar, except that during the height of summer the white polar cap vanishes entirely. The difference in the northern and southern summer is due to the difference in the distance of Mars from the sun at perihelion and aphelion positions. This sequence of events occurs quite regularly from Martian year to Martian year, and other white areas show similar variations, disappearing and reappearing at the same times each year.

12.26 The Nature of Polar Caps

Since measurements have shown that the temperature during the summer season at the poles may rise above the melting point of ice, it is reasonable

FIG. 12–5 *The large Martian crater that fills most of this picture is about 75 miles across. The small, clearly defined crater seen in the upper right rim of the larger crater is about 3 miles in diameter. Craters of all sizes in between can be seen in the picture, which has been described as one of the most remarkable scientific photographs of this age. It was taken by the Mariner IV television camera at 5:30:33* P.M. *PDT on July 14, 1965. Distance between the surface area photographed and the camera was 7800 miles. Showing Atlantis, between Mare Sirenum and Mare Cimmerium, the picture covers an area 170 miles from east to west and 150 miles from north to south. (From Office of Public Education and Information, California Institute of Technology Jet Propulsion Laboratory, National Aeronautics and Space Administration, Pasadena, Calif.)*

to assume that the polar caps are some kind of snow and not carbon dioxide, as had been supposed at one time. From the rate at which the polar caps grow and diminish in size, we must conclude that their maximum thickness is not more than a few inches. Because of the low atmospheric pressure on Mars it is probable that the polar caps are essentially hoarfrost that consists of very small grains. Because of the low atmosphere pressure the hoarfrost probably sublimates (becomes gaseous without first melting) directly into the gaseous atmosphere under the sun's radiation. According to Dollfus what is then left behind is a kind of crystalline structure similar in appearance to opal glass and full of small pores and cavities. The evidence for this is based on comparison of the polarization of light reflected here on earth from sublimating hoarfrost with the polarization of light reflected from melting ice and snow. The variations in the polar caps that we have just described are accompanied by variable clouds which undoubtedly have their origin in the condensation of water vapor rising from the sublimating caps.

12.27 The Red and Green Surface Markings

The reddish areas that give Mars its conspicuous red color are probably covered with fine metallic oxides that are similar in nature to those which give the Painted Desert in Arizona its magnificent coloration. The greenish surface markings on Mars show systematic changes with the seasons such that many astronomers have concluded they are some sort of vegetation. That some types of plant life can exist on Mars is a reasonable assumption; carbon dioxide and water vapor give the planet at least the minimum conditions for photosynthesis. Although no conclusive spectroscopic evidence has been obtained for the presence of chlorophyll, Kuiper's studies have led him to the conclusion that the amount of water present in the atmosphere of Mars "is compatible with a lichen cover" and that the color types seen on Mars are similar to those of lichens. Polarization of the sunlight reflected from these dark green areas differs considerably from the polarization of sunlight reflected from seed plants that have been tested in this way. However, the polarization of sunlight from certain flowerless microplants (with no specific leaf or stem structure) is similar to the Martian polarization of these green regions.

12.28 Vegetation on Mars

The theory that the green markings are vegetative is based on the way in which they change with Martian seasons. We find that these dark green markings are most prominent in a given hemisphere during the spring and summer seasons, when the polar caps are shrinking. These regions gradually grow smaller in the fall or winter or fade out entirely, and in some cases a distinct change in coloration is visible, with the green giving way to a russet brown at the height of the winter season. Since most of these changes occur periodically from Martian year to Martian year, we may reasonably conclude that they are the result of the growth and dying out of vegetation. (See Figure 12.6 for the surface markings as photographed in red and blue light.)

Recently E. Öpik has given an excellent argument for supposing that these dark regions are vegetation. If they were barren areas, the yellow dust storms, continuously occurring on Mars, would have covered them under thick layers of yellow dust during the past few billion years. That they are not covered in this way, but instead show continual seasonal variation, is a clear indication of their regenerative powers.

12.29 Martian "Canali"

When Schiaparelli, the Italian astronomer, in 1877 announced his discovery of a great abundance of fine, dark, straight lines crossing the reddish surface of Mars, he started a controversy concerning intelligent life on that planet which, to this very day, is still raging. Although Schiaparelli used the Italian word *canali* (literally meaning anything from grooves to man-made channels) in describing his observations, various astronomers, including Lowell in this country, accepted this as evidence of the presence of highly intelligent beings on Mars. Today, very few astronomers agree with Lowell that these markings are a continuous network of canals. W. H. Pickering has pointed out that any series of small markings arranged along some direction gives the impression of a continuous straight line when viewed from great distances. Such irregular features may be visible evidence of craters, fault lines in the planet's crust, or rift valleys like those found here on earth.

12.30 Climatic Conditions on Mars: Radio Emission

Since Mars is about half again as far from the sun as the earth is, it receives from the sun less than one-half as much heat per second as the earth does. We may, therefore, expect the surface of Mars to be much colder than the surface of the earth. Moreover, because of the very thin atmosphere the variations in temperature are much more extreme and occur more rapidly on Mars than on the earth. The temperature of this planet's surface is known from radiometric observations made at Lowell and Mt. Wilson observatories. Although in the polar regions the temperature rarely rises above the freezing point of water, it reaches a value of 50° to 60°F in the equatorial regions at noon. In general, we find that the dark greenish areas are warmer than the reddish ones, but violent fluctuations take place over the entire surface. The temperature at the equator will drop to well below the freezing point at sunset and fall still lower during the night, so that the nights are probably extremely cold. From theoretical calculations and from extensive observations we know that the mean temperature of the surface of Mars taken over a period of time is of the order of −23°C. These values for the surface temperature of Mars are in agreement with the values obtained from the intensity of radio waves emitted. Two observations in the 3-cm wavelength region have been made, and the intensity of this radiation is in reasonably good agreement with what can be expected from a surface having the temperature of Mars and emitting thermal radio waves.

From what we have said, it is clear that Mars, at best, is an inhospitable planet for intelligent life as we know it. At present there is no evidence for the existence of intelligent beings there. However, knowing how great the variations are in living forms and under what a great variety of conditions

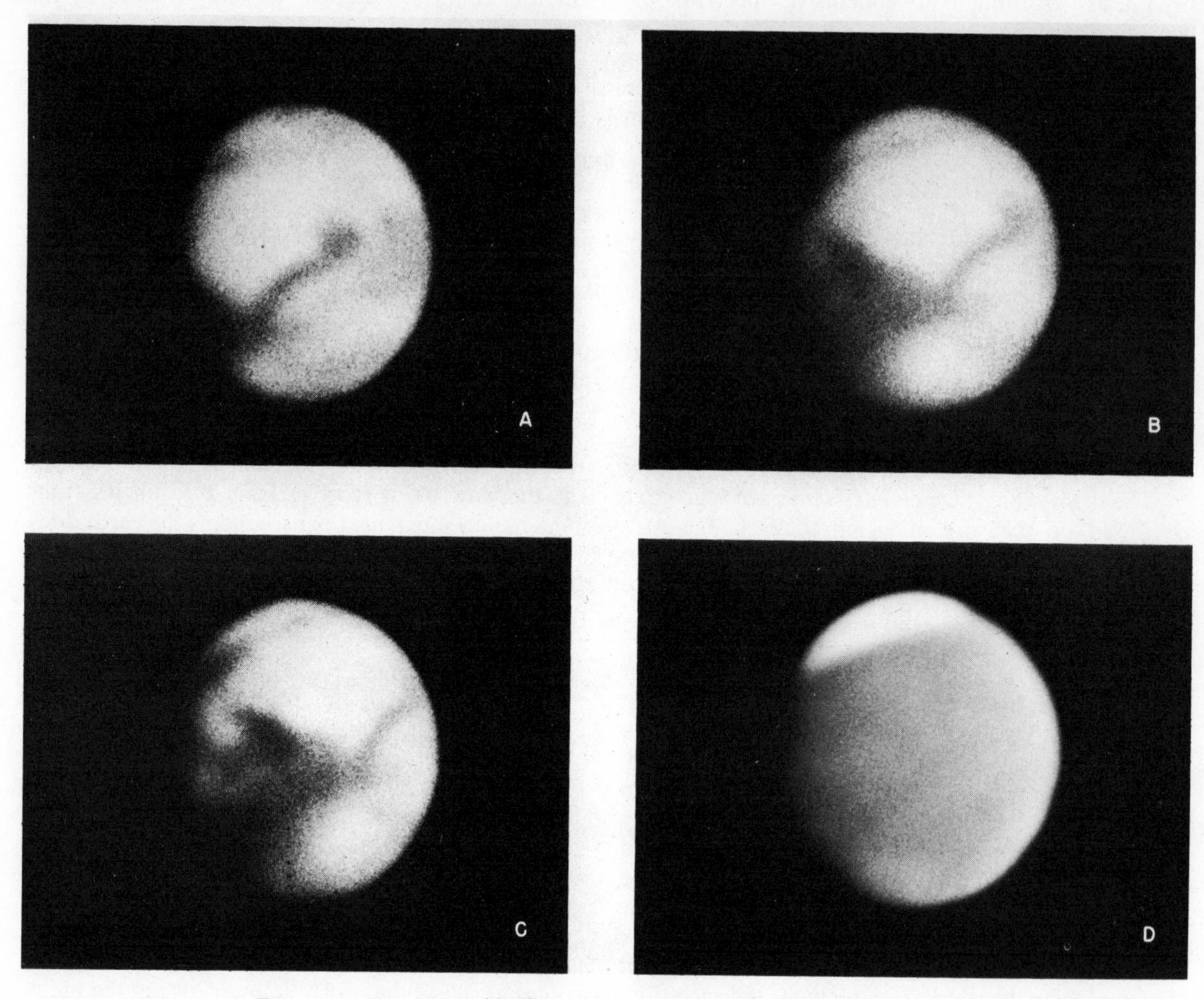

FIG. 12–6 *Mars: views (a), (b), and (c) taken in red light showing surface features and effect of rotation; (d) taken in blue light showing only the atmosphere. (100-inch photographs from the Mount Wilson and Palomar Observatories)*

intelligent life exists here on earth, we cannot entirely exclude the possibility of such life on Mars.

12.31 Martian Satellites: Phobos and Diemos

Mars is accompanied by two small objects which were first discovered by Hall in 1877. Because of their extremely small size, we can see them only through very large telescopes when Mars is in or near opposition. These satellites are remarkable because of their smallness and because they are so close to Mars itself.

Phobos, the inner satellite, is only 5 800 miles (9.3×10^3 km) from the center of Mars, and revolves in its orbit once every 7^h39^m, so that its period of revolution is less than one Martian day. This means that Phobos rises in the west and sets in the east as seen from Mars. From the apparent brightness of this object as measured from the earth, we find that its diameter is about 7 miles (12 km). As seen from a point on the surface of Mars, Phobos would appear about one-third the diameter of the full moon, as seen from the earth.

The more distant satellite, Diemos, is over 14 600 miles (23.5×10^3 km) from the center of Mars, and has a period of 30^h18^m. It is not nearly as bright as Phobos, and from brightness measurements we find that its diameter is only about 4 miles (6 km).

THE ASTEROIDS

12.32 The Discovery of the Asteroids

If we accept Bode's law as giving a reasonably correct picture of the arrangement of the planets relative to the sun, we must expect to find a planet at a distance of 2.8 astronomical units from the sun. But no planet is there, and for a long time this large gap between Mars and Jupiter was extremely puzzling to astronomers. However, we now know that this is not really a gap, but is a region in which there are virtually thousands of small or minor planets.

On January 1, 1801, Giuseppi Piazzi, while making routine observations, noticed what appeared to be a small star where none had been the day before. He then noted, on the following nights, that this object changed its position relative to the background stars in the manner of planets. Since Piazzi was not able to keep this object in view continuously, he sent records of his observations to other astronomers, but these communications were so slow in transmission that the object was lost sight of, apparently irretrievably. However, the great mathematician Gauss applied to Piazzi's data his newly developed method of determining the orbit of a planet from three distinct observations. Even though these data were rather meager, Gauss was so skillful that he obtained for this object an orbit lying between Mars and Jupiter, and the orbit was so accurately calculated that just one year after Piazzi's first observation the object was relocated within $\frac{1}{2}°$ of the position Gauss had predicted. This minor planet, to which Piazzi gave the name of **Ceres** (from the patron goddess of Sicily), is the largest and most massive of the asteroids, and its mean distance to the sun is just 2.77 astronomical units (in excellent agreement with Bode's law).

In 1802 Olbers discovered a second asteroid, **Pallas**, for which Gauss also obtained an orbit lying between Mars and Jupiter. Thereafter, in the years 1804 and 1807, two further asteroids, Juno and Vesta, were discovered, but a fifth and much fainter one, Astrae, was not discovered until 1845. Since then, however, with the introduction of very large telescopes, the rate of discovery has increased greatly, and at the present time reliable orbits of more than 1 500 have been determined. Over 3 000 asteroids have been discovered, and it is estimated that there may well be over 30 000 that can be photographed.

TABLE 12.3
ASTEROIDS

Asteroid	Number	Radius	Mass	m_{pg}	Rotational period	Period	Orbital data a	e	i
		km	g		h m	d	A.U.		°
Ceres	1	350	60×10^{22}	4.0	9 05	1 681	2.8	0.079	10.6
Pallas	2	230	18×10^{22}	5.1		1 684	2.8	0.235	34.8
Juno	3	110	2×10^{22}	6.3	7 13	1 594	2.7	0.256	13.0
Vesta	4	190	10×10^{22}	4.2	5 20	1 325	2.4	0.088	7.1
Hebe	6	110	20×10^{21}	6.6	7 17	1 380	2.4	0.203	14.8
Iris	7	100	15×10^{21}	6.7	7 07	1 344	2.4	0.230	5.5
Hygiea	10	160	60×10^{21}	6.4	18?	2 042	3.2	0.099	3.8
Eunomia	15	140	40×10^{21}	6.2	6 05	1 569	2.6	0.185	11.8
Psyche	16	140	40×10^{21}	6.8	4 18	1 826	2.9	0.135	3.1
Nemausa	51	40	9×10^{20}	8.6		1 330	2.4	0.065	9.9
Eros	433	7	5×10^{18}	12.3	5 16	642	1.5	0.223	10.8
Davida	511	130	3×10^{22}	7.0		2 072	3.2	0.177	15.7
Icarus	1566	0.7	5×10^{15}	17.7		408	1.1	0.827	23.0
Geographos	1620	1.5	5×10^{16}	15.9		507	1.2	0.335	13.3
Apollo		0.5	2×10^{15}	18		662	1.5	0.566	6.4
Adonis		0.15	5×10^{13}	21		1 008	2.0	0.779	1.5
Hermes		0.3	4×10^{14}	19		535	1.3	0.475	4.7

The brightness in column 5 is given in photographic magnitudes (see Section 15.12) for the asteroid at a unit distance from the sun and at a unit distance from the earth. The radius R of the asteroid and its photographic magnitude are related by the empirical formula $\log R = 3.746 - 0.2\, m_{pg}$.

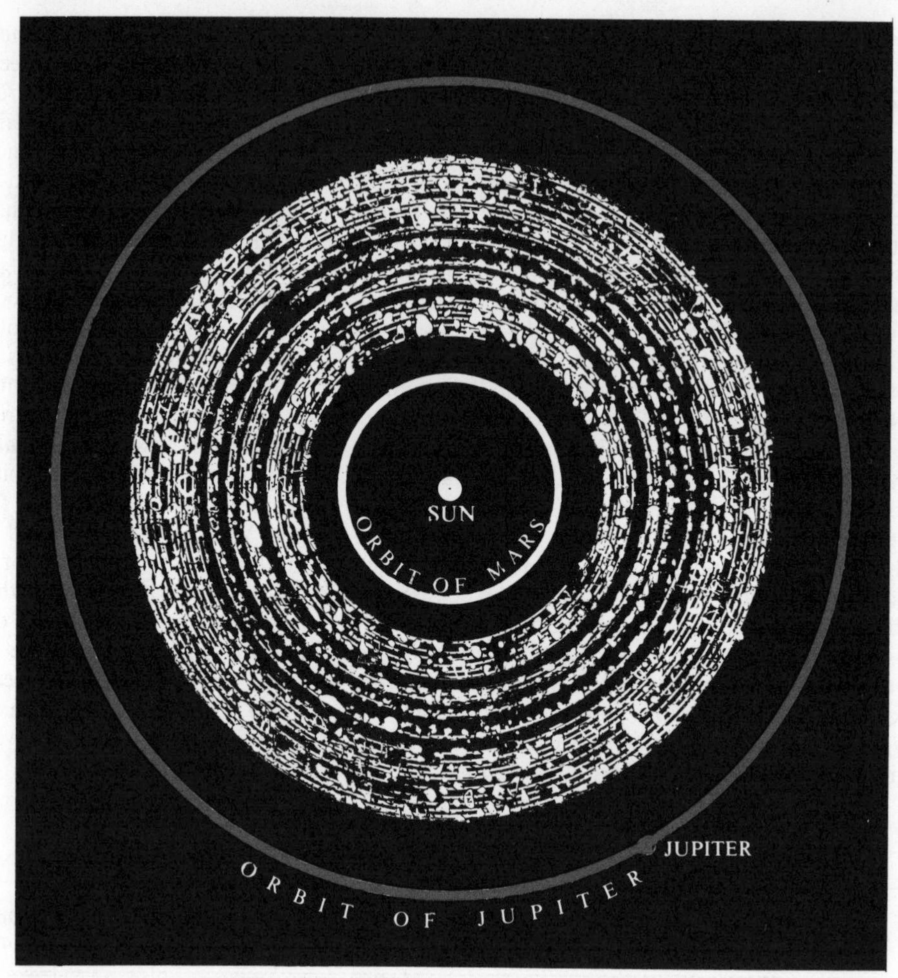

FIG. 12–7 *The Kirkwood gaps in the belt of the asteroids.*

12.33 The Mean Distance of Asteroids: The Kirkwood Gaps

Although the mean distance of the asteroids from the sun, taken altogether, is of the order of 2.8 astronomical units, there is a wide range in their individual distances, with **Eros** (the closest one) having a mean distance of 1.46 A.U., and **Hidalgo** (the farthest) a mean distance of 5.71 A.U. The asteroids form a wide band between Mars and Jupiter, with most of them clustered around the mean distance of 2.8 A.U.; there are, however, several marked gaps in this band where no asteroids are found.

Kirkwood explained these gaps as being due to the perturbing effects of Jupiter on the orbits of the asteroids. This is quite understandable, since these gaps occur just where the period of an asteroid moving in the gap would be exactly commensurable with that of Jupiter. In other words, no asteroid can be found having a mean distance corresponding to a sidereal period that is one-half, one-third, two-fifths, or a similar fraction of Jupiter's period, as shown in Figure 12.7.

12.34 Diameters and Masses of Asteroids

Even the largest of these asteroids are so small that their disks can be measured only with the most accurate micrometrical devices. Ceres, with a diameter of about 434 miles (700 km), is the largest, and it is followed by Pallas, 285 miles (460 km); Vesta, 236 miles (380 km); Juno, 120 miles (190 km). From an analysis of the apparent brightnesses and diameters, it appears that Ceres has the same albedo as the moon, and that Pallas reflects about the same as Mars does, whereas Vesta is as good a reflector as Venus.

All of these objects have very small masses. Ceres, the most massive, has a mass of about one eight-thousandth that of the earth. A person standing on Ceres would weigh one-thirtieth of what he does on the earth, and an object traveling with half the speed of a rifle bullet would escape from this asteroid. The total mass of all the asteroids is of the order of one three-thousandth the mass of the earth, with Ceres and Pallas accounting for about half of this mass.

Because of the small masses of these bodies, their internal gravitational forces are much too weak to give them a spherical shape, and the likelihood is that most of them have highly irregular structures. This is borne out by the periodically fluctuating brightness that is observed for many of them. This means that these asteroids, besides being highly irregular, are rotating in some way or other. The measurements indicate that in most cases the rotation is in the same sense as the revolution of the asteroid around the sun.

FIG. 12–8 *Trail of Eros. Photographed by Van Biesbroeck with the 40-inch refractor, February 16, 1931. Exposure time, 1 hour, 30 minutes (Yerkes Observatory photograph)*

12.35 The Orbits of the Trojan Asteroids

The asteroids have served an important function in astronomy, since it is possible to determine their orbits with great accuracy and to check this against Newtonian gravitational theory. The agreement is excellent, and some interesting results have been obtained, the most remarkable of which has to do with the motion of a group of these asteroids that are about the same distance from the sun as Jupiter is, and therefore have periods nearly equal to Jupiter's. These asteroids, twelve in number, form the famous **Trojan group**.

The remarkable quality of their motion was first established by Lagrange, who showed, by applying gravitational theory, that if a body is moving about the sun in the same orbit as a planet, but in such a way that the sun, the planet, and the body are at the three vertices of an equilateral triangle, this equilateral configuration remains unchanged as the body and the planet revolve around the sun. Moreover, this configuration is stable so that if the body is slightly disturbed, it returns to the equilibrium position. Seven of the Trojan asteroids are close to the vertex of the equilateral triangle that precedes Jupiter, and five are at the vertex of the equilateral triangle that follows Jupiter.

12.36 Using the Asteroids to Measure the Astronomical Unit

Some of the asteroids serve another important function in astronomy because they come so close to the earth that we are able to measure their distances with extreme accuracy. This is especially true of Eros (see Figure 12.8), whose mean distance from the sun is 1.458 A.U., and whose orbital eccentricity is 0.223. At its closest approach to the earth, it is just about 14 000 000 miles (22.4×10^6 km) away, so that we can determine its parallax and its distance in miles with great accuracy. Since the dimensions of the orbit of Eros are well known in astronomical units, from its period and Kepler's third law we can calculate its distance from the earth at the moment of closest approach in astronomical units and compare this with the actual distance from the earth in miles or kilometers as measured at this moment. From this comparison we can thus find the value of the astronomical unit in miles (or kilometers). This is one of the most accurate methods for determining the mean distance from the earth to the sun. However, more recent procedures using radar beams bounced off Venus give results that promise to be even more accurate.

12.37 Jupiter: Its Orbital Characteristics and Motion

The asteroid belt separates the terrestrial planets from the major planets, of which the first from the sun is Jupiter. The largest and most massive of all the planets, it is next to Venus in brilliance. Unlike Mercury and Venus, Jupiter can be above the horizon at all hours of the night (i.e., its elongation can range continuously from 0 to 180°).

Jupiter's mean distance from the sun is 5.2 astronomical units, or 484 000 000 miles (778×10^6 km). Its orbit, with an eccentricity of 0.05, is midway in its ellipticity between those of the earth and Mars. Because of the eccentricity of its orbit, the distance of Jupiter from the sun increases by about 47 000 000 miles (76×10^6 km) as it moves from perihelion to aphelion. This planet may come as close as 370 000 000 miles (600×10^6 km) to the earth when it is in opposition, and may get as far away as 600 000 000 miles (960×10^6 km) at conjunction. Its sidereal period is 11.86 years, its synodic period 399 days, and its mean orbital velocity 13.06 km/sec.

Jupiter rotates more rapidly about its axis than any other planet, with a period of rotation ranging from 9^h50^m at its equator to 9^h56^m near the poles. The six-minute variation in going from the equator to the poles arises because different points on the Jovian surface rotate with different speeds. This means that the visible surface of Jupiter is not rigid. Although the speed of rotation is greatest at the equator, it does not fall off gradually as one moves to the poles of the planet, but rather in a series of sharp changes. There are distinct bands running parallel to the equator which have their own periods of rotation. Because of the rapid rotation of this planet a point on its equator moves with a velocity of 25 000 mph. The plane of Jupiter's equator is tilted 3° with respect to the plane of its orbit, which is almost parallel to the plane of the ecliptic, the tilt being 1°18′3″. The longitude of the ascending node is 99°26′20″.

12.38 The Diameter and Mass of Jupiter

The true diameter of the planet can be found from the measured values of its apparent diameter and its distance from the earth at a given moment. The apparent diameter varies from 50″ at opposition to 32″ at conjunction, with a mean value of 37″.8. This corresponds to a true equatorial diameter of 88 770 miles (143×10^3 km). However, since the oblateness of this planet is very pronounced, because of its rapid rotation, the polar diameter of 83 010 miles (133×10^3 km) is considerably smaller. The oblateness is 1/15.4, and the mean diameter is 86 850 miles (140×10^3 km).

It is quite easy to determine the mass of Jupiter with very high accuracy because of its many satellites and its perturbations of the orbits of the asteroids. From an analysis of all the observations we find that the mass is 317.8 times that of the earth. This value for the mass leads to a surprisingly small value for the density. From Jupiter's diameter, its volume is known to be more than 1 300 times that of the earth, so that the mean density is 1.33 gm/cm^3—about the same as that of the sun.

12.39 Acceleration of Gravity on the Surface of Jupiter

From the mass and the radius of Jupiter, we find, using Newton's law of gravity, that the gravitational acceleration on its surface would be 2.64 times that on earth if Jupiter were not rotating. Because of Jupiter's rotation the actual acceleration of gravity at its equator is less than this value. However, because of the planet's rapid rotation and its oblateness, the acceleration of gravity increases at the poles. On taking these things into account we find that the weight of a body at the poles of Jupiter is 15 per cent greater than at the equator, whereas on the earth the difference is only about 0.5 per cent.

12.40 Surface Appearance of Jupiter

The most striking feature about Jupiter, when viewed through a telescope, is the series of bands across its surface parallel to its equator. These vary in dimensions and positions from time to time, and contain within them irregularities that come and go relatively quickly. They are probably atmospheric phenomena. The most pronounced of these features is the famous **Red Spot**, which, although its appearance has changed considerably since its discovery in 1879, is fairly permanent in its position and dimensions. It is about 30 000 miles (48.4×10^3 km) long and 7 000 miles (11.2×10^3 km) wide, and of a dark reddish hue, although this coloration is fading out.

FIG. 12–9 *Jupiter, in blue light, showing large red spot, satellite Ganymede, and shadow above. The light of the sun is coming from the right. This photograph also shows the pronounced flattening of the planet and the band structure of its atmosphere. (200-inch photograph from the Mount Wilson and Palomar Observatories)*

Since its discovery, this spot has tended to get rounder, and although it appears to be fixed in position, it is really not attached to any solid surface below. Since the period of rotation has diminished by about 5 seconds since its discovery, we know it must be revolving faster than the surrounding atmosphere (and possibly faster than the surface underneath). We may conclude, therefore, that the Red Spot has circled the planet twice in a twenty-year period. It has also shifted its position in latitude, and thus we must conclude that it belongs to the planet's atmosphere. The bands themselves are probably also gaseous or cloudlike, as suggested by the variations in their rotational periods. See Figure 12.9.

12.41 Jupiter's Atmosphere

The nature of a planet's atmosphere depends upon the speed of escape from its surface and on the temperature conditions. We know from Jupiter's mass and radius that the speed of escape on its surface is slightly more than 37 mi/sec (60 km/sec). We also know from radiometric measurement and the analysis of the amount of sunlight that Jupiter receives per second from the sun (about one twenty-fifth of that which the earth receives) that the surface temperature is about 135°K. These figures show that if any of the known gaseous atoms or molecules are in Jupiter's atmosphere, they cannot be moving fast enough to escape.

We may therefore conclude that there are great abundances of hydrogen and helium on this planet, since these are the most abundant elements in the universe and were undoubtedly in the original planetary material. Helium is present in its free form, since it is an inert gas, whereas hydrogen, being very active chemically, is present both as H_2 molecules and in compound forms. Indeed, hydrogen compounds such as methane and ammonia are observed in Jupiter's atmosphere. However, there is so much more hydrogen than any other element in our universe and so little of it has escaped from Jupiter, that it is reasonable to assume that a great deal of hydrogen is present on Jupiter, not in combination with other atoms, but in free form, since these other atoms could hardly have been present initially in sufficient abundance to take care of all the hydrogen.

If any free oxygen was present on Jupiter originally, it must now be completely bound up in the form of water and metallic oxides. This does not mean that there is any water vapor in the atmosphere; the temperature is too low for that. The water on Jupiter is probably present beneath the atmosphere in the form of a layer of ice, perhaps as much as 15 000 miles thick. It is also safe to assume that there is no free nitrogen in the atmosphere, because all of the nitrogen must have combined with free hydrogen to form ammonia. The presence of ammonia was verified spectroscopically by Wildt in 1932, who showed that the absorption bands in the orange and red in Jupiter's spectrum arise from methane and ammonia. The presence of methane in Jupiter's atmosphere can easily be understood in terms of the chemical affinity of carbon for hydrogen.

Since carbon, nitrogen, and oxygen are probably much less abundant in Jupiter's atmosphere than hydrogen, there is good reason to suppose that great quantities of hydrogen are still present in the form of the molecule H_2. The coloration observed in the various bands in the atmosphere has not been explained, but Wildt suggests that this may be due to various compounds of sodium, potassium, or iodine dissolved in liquid ammonia.

12.42 Radio Emission from Jupiter: Centimeter Radiation

Since 1955, when 3-cm wavelength radiation was first observed, radio telescopes have yielded some interesting data on atmospheric conditions on Jupiter. The intensity of the radiation observed in 1955 indicated a surface temperature in agreement with the radiometric value given above. However, since then steady radio emission from Jupiter has been observed over a wide range of wavelengths (from about 3 to 70 cm) with an unusually intense signal at 10.3 cm. Although the surface temperature calculated from the short-wavelength end of the radio spectrum agrees with the temperature values determined from thermocouple measurements and that calculated from the absorbed solar radiation, the radio temperature values increase as we go to the long-wavelength end of the spectrum. Thus, using the intense 10.3-cm radiation, temperatures ranging from 315° to 640°K have been measured.

If we accept as correct the surface temperature measured by thermocouples and calculated from the absorbed solar radiation, then only the radiation in the 3-cm wavelength region is thermal in nature, i.e., arises from the random thermal motions of the molecules on the surface of Jupiter. We must therefore find some other way to account for the intense long-wavelength radiation. If Jupiter has a sufficiently intense magnetic field, of the order of 7 gauss, then charged particles (such as electrons, protons, and ions) can be trapped in belts around Jupiter similar to the Van Allen Belt around the earth. Under these conditions these charged particles spiral around the magnetic lines of force with sufficiently high speeds to emit the type of radio spectrum that has been observed. It is probable that this synchrotron mechanism (see Section 28.28) can account for the intensity of the radio emission in the 10- to 70-cm range observed from Jupiter.

12.43 Radio Emissions from Jupiter: Meter Radio Waves

In addition to centimeter radio waves, intense and spasmodic radio emission has also been observed in the 15-meter range. This remarkable radiation consists of pulses of such intensity that Jupiter is by far the strongest radio source in the sky in this wavelength range. When Jupiter radiates in this wavelength region it emits more than 6 pulses per minute at a peak power greater than 4×10^{-13} erg/sec per frequency interval received on one square centimeter of the earth's surface. If we consider the wavelength region from 15 to 30 meters there is a minimum of 13 pulses per minute. This is truly an enormous amount of radiation power and is about equal to that of the "quiet sun." This means that the radio-power output of Jupiter per unit surface area is about 100 times greater than the radio emission per unit surface of the quiet sun. We may compare the energy emitted by a single radio burst from Jupiter with the energy released in the San Francisco earthquake of 1906 and the Krakatoa volcanic eruption of 1883. Various hypotheses have been presented to account for this radio emission, including earthquakes, volcanic eruptions, and vast electrical storms. Nonetheless, in spite of this vast amount of non-thermal radio energy Jupiter emits, its thermal radio-energy emission is 10^7 times greater.

12.44 Internal Structure

Although we know very little about the internal structure of Jupiter, we may conclude from its oblateness that it is much more highly concentrated towards the center than the earth is; if it were not, its oblateness would be much greater. This reasoning has given rise to various theories in which the interior is pictured as consisting of a highly concentrated core (probably of heavier elements such as iron and rock) surrounded by a very thick layer of lighter material (probably compressed water or ice) with an outer layer of gases in a liquid state. In a more recent theoretical investigation, Miles and Ramsey have calculated models for Jupiter containing 76 per cent hydrogen by mass. They picture hydrogen as being in a solid state in the interior under extremely high pressure, with the heavier elements not only concentrated in a central core, but also, in part, distributed uniformly throughout the hydrogen.

12.45 Satellites of Jupiter

In 1610, immediately after Galileo had built his telescope, he directed it at Jupiter, and was amazed to see four tiny objects attending this planet. These were the first new heavenly bodies discovered by man since antiquity. They are known as the Galilean satellites. Besides these four larger bodies revolving about Jupiter, there are eight others. Although the first four can be easily seen, even with no more than good binoculars, the others are so tiny that it was not until 1892 that Barnard, using a large telescope, discovered a fifth Jovian satellite (the one closest to the planet).

The diameters of the four Galilean satellites range from about the moon's diameter to that of Mercury, and their masses range from 0.64 to 2.1 times the moon's mass. Their orbits are very nearly circular and lie practically in the plane of Jupiter's equator (the maximum inclination is about $\frac{1}{2}°$); their periods range from about 2 to slightly more than 16 days. The innermost of these four satellites is at a distance of 262 000 miles whereas the outermost is at a distance of 1 171 000 miles. Because the cross section of Jupiter's shadow where it is traversed by these satellites is very large, eclipses occur at each revolution for the three inner Galilean satellites (**Io, Europa, Ganymede**) and with great frequency for the outermost one (**Callisto**).

12.46 Roemer's Determination of the Speed of Light

If someone on the earth were to observe a periodic phenomenon (such as the recurring eclipse of a satellite by its planet) the time between the number of events (and therefore the number of events in a given interval) would always be the same if the earth were either fixed with respect to the phenomenon or were moving with uniform speed away from it or towards it. However, the frequency of the events (or the interval between them) would be different in each of these three cases. If the observer were moving away from the periodic phenomenon, he would observe fewer events per interval of time than if he were fixed with respect to the phenomenon; and if he were moving towards it, he would observe more events per interval of time. This is nothing more than the Doppler effect (see Section 5.14) applied to periodic events. In the year 1675, Roemer, a Danish astronomer, applied this effect to the analysis of the eclipses of the inner satellites of Jupiter and thus was able to calculate the speed of light.

To see how this is done we consider the earth and Jupiter so situated in their orbits that the radius vector from the sun to Jupiter and the radius vector from the sun to the earth make an angle θ with one another. Let the time between two successive eclipses of the innermost satellite as observed from the earth be t, assuming the earth and Jupiter fixed in their orbits. What is the actually observed interval and how does it compare with t? To answer this question we note that the earth is sometimes moving away from Jupiter, sometimes towards it, and sometimes transverse to its direction. This time interval changes as follows: it equals t at opposition and conjunction of Jupiter (E_{opp} and

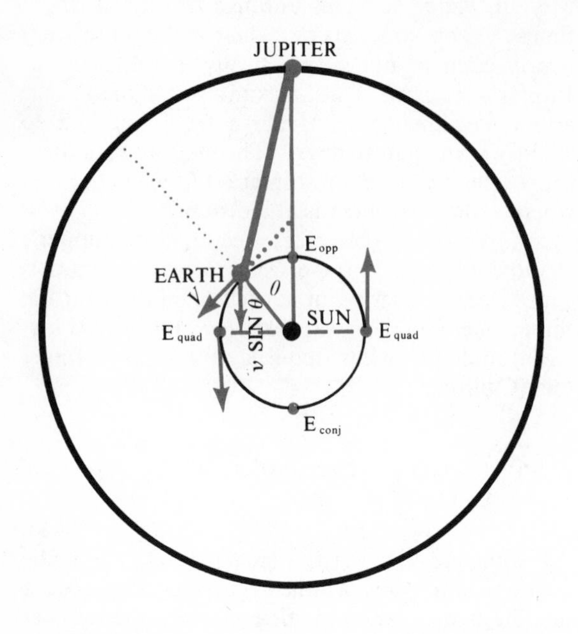

FIG. 12–10 *The delay in the occurrence of eclipses of Jupiter's moons as seen by an observer on the earth is due to the earth's motion away from Jupiter.*

E_{conj} for position of earth); it is a maximum when the earth is receding from Jupiter in quadrature ($E_{quad\ 1}$ for position of earth) and a minimum when the earth is approaching Jupiter in quadrature ($E_{quad\ 2}$ for position of earth). This interval therefore varies continuously and depends at any one moment not only on the motion of the earth but also on the motion of Jupiter. It thus also depends on the shape of the earth's orbit and on the shape of Jupiter's orbit. However, in our analysis we shall suppose that Jupiter is not moving and that the earth's orbit is a circle. (Jupiter's motion is so slow compared to the earth's that we may neglect it in this simple analysis.) Moreover, we shall assume that the direction from the earth to Jupiter is parallel to the direction from the sun to Jupiter. (This is a good enough approximation for our purpose.)

With the earth in position E (as shown in Figure 12.10), the speed with which it is receding from Jupiter is $v \sin \theta$, where v is the average speed of the earth. Thus, the time between two successive eclipses at this position is equal to t (the interval between successive eclipses when the earth and Jupiter are fixed) plus the time it takes the light to travel the additional distance that the earth has receded from Jupiter during the time interval. This additional time that must be added on to t is just $\left(\frac{v}{c} \sin \theta\right)t$, where c is the speed of light. Thus, the time interval between successive eclipses in position E is just

$$t + t\frac{v}{c}\sin\theta = t\left(1 + \frac{v}{c}\sin\theta\right).$$

Since $\sin\theta$ takes on all positive and negative values between $+1$ and -1 (including zero) as the earth moves around its orbit, this formula shows exactly how the observed interval between eclipses changes.

Let us now suppose that an eclipse occurred at a time T when Jupiter was in opposition (angle $\theta = 0$). How long after the time T will the next eclipse be seen by an observer on the earth? If the earth were fixed with respect to Jupiter, it would be just the time interval t later. But the motion of the earth changes this. Since v is the average speed of the earth, the angle θ changes from its value 0 to a value vt/A, where A is the astronomical unit, and the next eclipse will thus occur at a time that is the average of the times $t + T$ and

$$T + t\left(1 + \frac{v}{c}\sin\frac{vt}{A}\right),$$

since θ changes continuously. Suppose that an eclipse occurs when the earth is on the other side of the sun so that Jupiter is in conjunction. How long after T will this occur? (Of course, in this position the eclipse cannot be observed from the earth, but we can still talk about it.) This time will be

$$T + \left[nt\left(1 + \frac{v}{c}(\sin\theta)\right)\right]_{\substack{\text{average}\\ \text{from } \theta=0^\circ\\ \text{to } \theta=180^\circ}}$$

$$= T + nt + nt\frac{v}{c}(\sin\theta)_{\substack{\text{average}\\ \text{from } \theta=0^\circ\\ \text{to } \theta=180^\circ}},$$

where n is the number of eclipses from opposition to conjunction. The last term on the right-hand side of this expression is just the total delay between the first and last eclipse introduced by the earth's motion, and this total time delay is about 1 000 seconds. This time delay was first detected and measured by Roemer. Since

$$nt\,v(\sin\theta)_{\text{average}} = 2A,$$

where $2A$ is the total distance that the earth has receded from Jupiter during the n eclipses, we must have

$$\frac{2A}{c} = 1\,000.$$

From such an analysis Roemer computed the speed of light.

12.47 Outer Jovian Satellites

The Jovian outer satellites are all very tiny (the diameter of the largest is of the order of 80 miles or about 130 km) and extremely faint, with masses too small to be accurately measured. Eight of the satellites move about Jupiter in the same direction as Jupiter moves about the sun, but four of them, apparently the very outermost ones, move in retrograde orbits. The orbits of the outer satellites are highly inclined to the plane of Jupiter's equator and have fairly large eccentricities. It is quite probable that some of the outer satellites were captured by Jupiter long after the solar system had been formed.

12.48 Saturn: Its Orbital Characteristics and Motion

Saturn, when visible to the naked eye, is prominent among all the stars and can be recognized by its steady, yellowish brilliance. This planet was the most distant known to the ancients, and with its rings forms the most beautiful image of all the planets when viewed through a telescope. Because the orientation of its rings with respect to the earth changes from year to year, Saturn appears to vary in brightness by as much as 70 per cent every 15 years.

The mean distance of Saturn from the sun is about 887 100 000 miles ($1\,427 \times 10^6$ km) and because the eccentricity of the orbit is 0.056, its distance from the sun increases by 100 000 000 miles as it goes from perihelion to aphelion. At its closest approach to the earth Saturn's distance is 747 000 000 miles, and its greatest distance is 1 028 000 000 miles. Its sidereal period is almost $29\frac{1}{2}$ years and its synodic period is 378 days. Saturn's orbital plane is inclined to the ecliptic $2\frac{1}{2}°$, and the longitude of the ascending node is 112°47′10″.

Because the surface of Saturn is marked by bands (similar in structure to those found on Jupiter, but without any permanent features), it is difficult to determine the period of rotation with any accuracy. However, by analyzing the Doppler tilt of its spectral lines, Moore showed, in 1939, that the period of rotation increases as one moves from the equator to higher latitudes. It appears that there is a current at the equator that is moving at a speed of 880 mi/hr (1 420 km/hr). From all available data we may conclude that the period of rotation ranges from 10^h14^m at the equator to 10^h38^m at the poles. The plane of the equator is inclined 26°45′ to the plane of the orbit.

12.49 Physical Characteristics of Saturn

Because of the high speed of rotation this planet's oblateness amounts to about 10 per cent, so that its mean diameter differs considerably from its equatorial diameter. From its known distance from the earth at any moment, and its apparent diameter at that moment, we find its true equatorial diameter is 74 000 miles, whereas its true polar diameter is 66 400 miles, so that its true mean diameter is 71 600 miles (116×10^3 km). Its surface area is thus 81 times that of the earth, and its volume is 769 times the volume of the earth.

We can determine the mass of this planet by observing its satellites, which are nine in number, and also by observing the perturbations which it produces in Jupiter's motion. These methods give a value 95 times the earth's mass, and hence an over-all density of only 0.68—about three-fourths the density of water—so that Saturn would float if it were placed in a large enough ocean. From the mass and the radius of this planet we find that the acceleration of gravity on its surface is about 11 per cent greater than on the earth's surface, but as we go from Saturn's equator to its poles this value increases by 30 per cent. From the mass and radius of Saturn, we find that the speed of escape from its surface is 23 mi/sec (37 km/sec), so that this planet has probably retained all the elements it had when it was formed.

12.50 Appearance of Saturn

The general appearance of Saturn (see Figure 12.11) is very much like that of Jupiter, except that its belts are not as clearly defined and seem to change more slowly. The center region is marked by a brilliant yellowish hue which becomes greenish as one moves to the poles. There is no doubt that these bands and colorations are due to the presence of gases in their liquid state and probably ammonia crystals with traces of the alkali metals dissolved in them. Since radiometric measurements and an analysis of the heat received from the sun give surface temperatures as low as −150°C, it is safe to assume that most of the ammonia has been frozen

FIG. 12–11 *Saturn and ring system. The bands in its atmosphere can also be seen. (100-inch photograph from Mount Wilson and Palomar Observatories)*

out of the atmosphere, and probably only such gases as methane, argon, helium, and hydrogen are present. The clouds themselves, just as in the case of Jupiter, are probably composed of small ammonia crystals.

12.51 Radio Emission from Saturn

Very weak radio waves from Saturn in the 3.75-cm region were detected by Drake and Ewen in 1958, and more recently 3.4-cm radiation was picked up with the aid of the 85-ft reflector at the University of Michigan. An analysis of the intensity of this radiation (which is thermal in origin) gives a surface temperature for Saturn of about 160°K, which is in good agreement with the temperature obtained from measurements of the infrared radiation from this planet. It also agrees with Kuiper's calculations based on the absorption of solar radiation.

12.52 The Internal Structure

From the oblateness of Saturn we may conclude that its internal structure is fairly similar to that of Jupiter, with a high central condensation in both planets. Miles and Ramsey have extended their analysis to Saturn, and have concluded that it probably contains 62 per cent to 69 per cent, by mass, of hydrogen. According to this picture, the oblatenesses of both Jupiter and Saturn, resulting from their axial rotations, suggest central cores of heavy elements with masses about ten times that of the earth. The remainder of Saturn's mass, according to this model, can be assumed to be hydrogen in a solid state. On the other hand, Wildt has suggested a model consisting of a central rocky core 26 000 miles in diameter, an intermediate layer of ice under pressure 8 000 miles thick, and an outer mantle of gas in liquid state 15 000 miles thick.

12.53 The Rings of Saturn

Saturn is the only planet with a system of rings. This concentric system consists of three rings, A, B, and C (A refers to the outermost or exterior ring), which are flat, extremely thin, and lie precisely in Saturn's equatorial plane. Although Galileo was the first to detect these rings, he was not aware of their nature because they were oriented in such a way that he saw them only edgewise (see Figure 12.12). They remained a perplexing problem until Huygens explained their nature in 1655.

These rings were thought to be a single structure until D. Cassini, in 1675, discovered a division between the rings A and B. The rings' triple character was established in 1850 by Bond when he discovered the hazy or crepe ring that is closest to the planet. The radius of the entire ring system measured from the center of Saturn to an exterior

FIG. 12–12 *Saturn. Rings on edge showing condensations, December 12, 1907. Drawn by Barnard.*

point on the outer ring is slightly more than 85 000 miles (137×10^3 km). This outer ring is 10 000 miles wide, and Cassini's division between the A and B rings is 3 000 miles. This division appears to be uniform all the way around with no observable variation.

The middle ring B is about 16 000 miles (26×10^3 km) wide and is considerably brighter than the outer ring, being equal in brightness to the equatorial regions of the planet itself. The innermost ring C is separated by about 1 000 miles (1.6×10^3 km) from the B ring, and is referred to as the gauze or crepe ring because it is quite faint and semitransparent. This ring is 11 500 miles (19×10^3 km) wide, so that the entire ring system has a width of 41 500 miles (68×10^3 km). This leaves a space of about 7 000 miles between the innermost part of C and the surface of the planet. The entire Saturnian structure, including the rings, is about the same size in its equatorial plane as Jupiter. Although it is difficult to measure the thickness of the rings, the estimates do not exceed 10 miles, and recent measurements indicate no more than inches or fractions of an inch.

12.54 The Appearance or Phases of the Rings

Since the rings lie exactly in the equatorial plane of the planet, they are seen from the earth in different phases as the earth and the planet revolve around the sun. During the $29\frac{1}{2}$-year period of the planet there are times when the plane of the rings of Saturn passes through the earth, and since it takes about a year each time for Saturn to sweep through this position, the rings of Saturn are presented to the earth edgewise at least once during each of these two periods. Actually, from the earth we can see the rings edgewise as many as three times during the period that Saturn's plane of rings is crossing the plane of the earth's orbit, as shown in Figure 12.13.

When Saturn is halfway between these positions on either side of the sun, observers can see the rings in their entirety, the upper surfaces of the rings being visible at one time, and the lower surfaces at the other.

12.55 The Nature of the Rings

We know from spectroscopic data and photometric studies, as well as from theoretical analysis, that the rings are not continuous solid or liquid structures, but rather swarms of individual particles, each of which may be considered a satellite revolving around the planet. Direct observation shows that one can look right through the rings and see objects, such as stars, behind them. Moreover, measurements of the variations of the albedo of the rings for different orientations with respect to the earth are consistent only with a particle structure and not with a continuous structure.

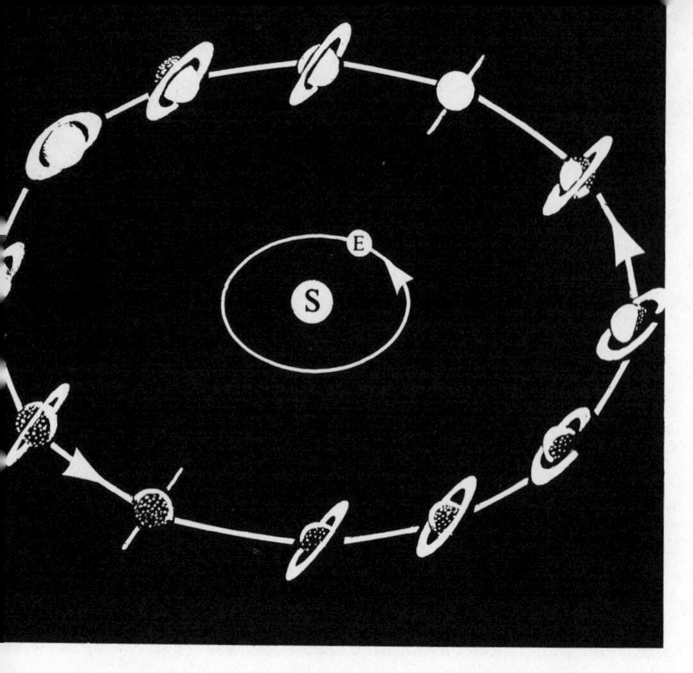

FIG. 12–13 *The orientation of the rings of Saturn in its various positions in its orbit as seen by an observer on the earth.*

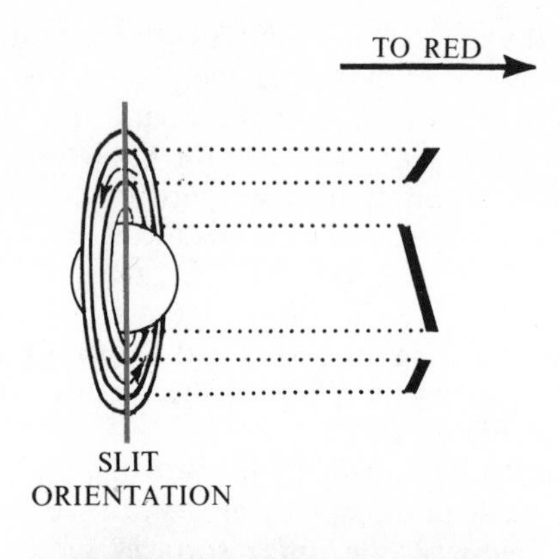

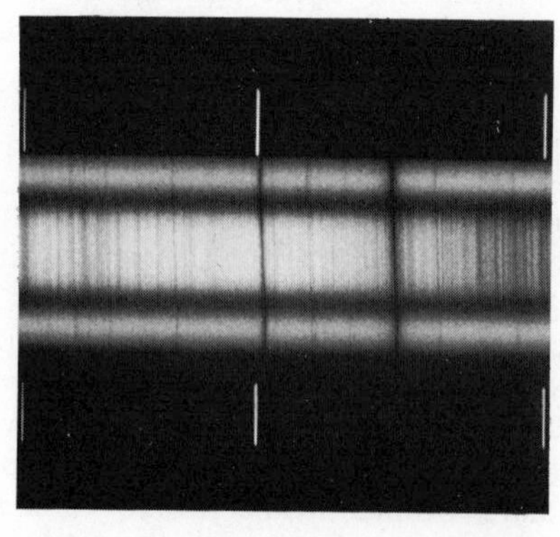

FIG. 12–14 *The lines in the spectra of the rings of Saturn are broken, indicating the particle structure of the rings, and that these particles move around Saturn in Keplerian orbits.*

From spectroscopic evidence it is clear that the inner parts of the rings are revolving around the planet faster than the outer region, as shown in Figure 12.14. If the rings are viewed through a spectroscope with the slit of the spectroscope lying in the plane of the rings, the spectral lines of the entire planet including those of the rings show a broken structure with the corresponding ring segments tilted with respect to those arising from the main body of the planet. The upper part of each segment is displaced towards one end of the spectrum (say the red) and the lower part of the line displaced towards the other (the violet). This shows clearly that the rings are not continuous solid or liquid structures, since they do not rotate as a solid disk does, but rather according to Kepler's third law.

This spectroscopic evidence of the discontinuous character of the rings is in agreement with the theoretical analysis of James Clerk Maxwell. He showed that neither solid nor liquid rings could continue to revolve around a planet since they would be highly unstable. Any kind of perturbing influence, such as Saturn's satellites exert, would cause such rings to undergo violent oscillations and ultimately break into pieces. On the other hand, one can show—and this also was done by Maxwell—that rings consisting of many particles will remain stable, because any slight disturbance is damped out quickly provided the mass of the rings is small compared with the mass of the planet. We know that this is, indeed, the case since all of the rings together have a mass about 25 per cent of that of the moon. The divisions in the rings are of interest, because one can show, as Kirkwood has done, that they are due to the perturbations of the satellites of the planet.

We may understand the origin of the rings if we recall that Roche (see Section 6.37) had shown that if a satellite approaches too close to a planet, the tidal action of the planet tears the satellite into pieces, provided the satellite is not too dense. For a satellite whose density is the same as Saturn's this limiting distance (the **Roche limit**) from the center of the planet, below which the tidal action becomes destructive, is 2.44 times the planet's radius. Since all the rings lie within this distance, we may reasonably assume that their origin was a satellite that was torn to pieces by Saturn's tidal action.

12.56 Satellites

Nine satellites have been discovered revolving around Saturn, ranging in diameter from 1 500

FIG. 12–15 *Mars, Jupiter, Saturn, and Pluto. First three photographed with the 100-inch telescope; Pluto with the 200-inch telescope. (From Mount Wilson and Palomar Observatories)*

miles (2440 km) for the largest to about 62 miles (100 km) for the smallest. Six of these satellites can be observed with moderately small telescopes but the others require fairly large instruments. The masses of some of these satellites can be determined from the way they perturb each other's orbits, and since **Titan's** mass has been calculated as being twice the moon's mass, the speed of escape from its surface is large enough for it to support an atmosphere. We know that this is, indeed, the case since Kuiper detected the bands of methane in the spectrum of Titan in 1944. Thus this satellite is unique, since no other is known to possess an atmosphere. All the satellites but one revolve around the planet in the same direction that the planet revolves around the sun. Phoebe, the satellite farthest from Saturn (over 8 000 000 miles), was found by E. C. Pickering, in 1898, to have a retrograde motion.

Figure 12.15 shows photographs of Mars, Jupiter, Saturn, and Pluto.

12.57 Uranus: Its Orbital Characteristics and Motion

Uranus, which was unknown to the ancients, is the first planet from the earth that is invisible to the naked eye. It was discovered in the year 1781 by the great British astronomer, Sir William Herschel, who detected its disk while searching the sky in the region of Gemini with a seven-inch telescope which he himself had constructed. He thought this object to be a comet, but several months of observations and calculations made clear that its orbit, being nearly circular, could not be that of a comet. The newly discovered planet was first called Georgius Sidius, after George III, of revered memory; it was later given the name of Uranus at the suggestion of Bode. Interestingly enough, it had been observed on many previous occasions, but it had then been thought to be a star.

Uranus moves in a fairly eccentric orbit (eccentricity = 0.047) so that its distance from the sun increases by 84 000 000 miles as it goes from perihelion to aphelion. Its mean distance, however, is 19.18 astronomical units, or 1 786 000 000 miles. The plane of the orbit of Uranus is practically coincident with the plane of the ecliptic, the inclination being 46′ 21″. It moves with a speed of about 4 mi/sec (6.8 km/sec) so that its sidereal period is slightly more than 84 years, and its synodic period is 369.66 days.

12.58 Diameter and Mass of Uranus

The planet's apparent mean diameter is 3″.75, and this, taken with its mean distance from the earth, gives for the mean true diameter a value of 30 000 miles (48×10^3 km). However, these measurements are subject to considerable error since Uranus is so faint and appears so small that its disk is difficult to observe with accuracy. Because the oblateness of this planet is equal to about 1/14, the equatorial diameter is considerably larger than the polar diameter.

We can determine the mass of Uranus by studying the motions of its satellites, and also by comparing the observed perturbations it induces in Saturn's orbit with the computed perturbations. Its mass is 14.5 times that of the earth, so that its density is 1.60 times that of water. From its mass and mean diameter we find that the speed of escape from this planet is somewhat more than 13 mi/sec. The mean acceleration of gravity on Uranus is 91 per cent of that on the earth.

12.59 Rotation of Uranus

It is obvious from its large oblateness that this planet must be rotating rapidly; spectroscopic data show its period of rotation to be 10^h49^m. The axis of rotation lies almost in the plane of the ecliptic. As shown in Figure 12.16 the equatorial plane of Uranus is inclined about 98° with respect to the plane of its orbit. Hence the planet as viewed from the earth has retrograde rotation.

12.60 The Atmosphere of Uranus

Since the albedo of Uranus is the same as that of Saturn and Jupiter, we may assume that except for

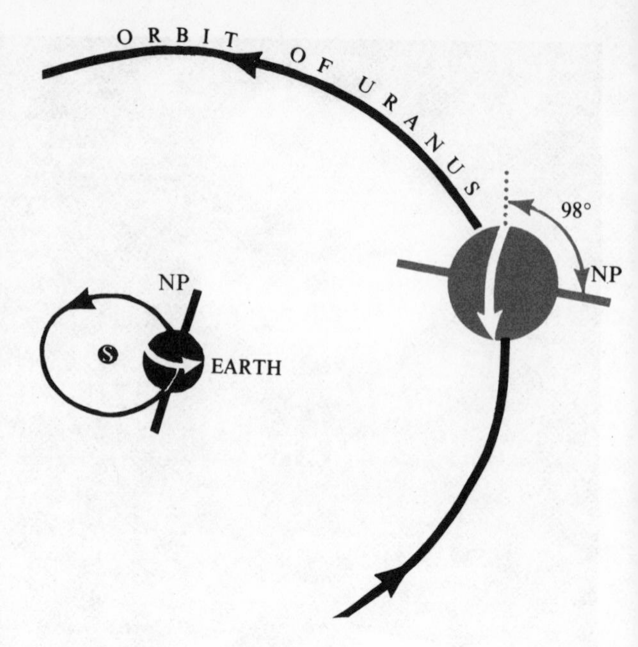

FIG. 12–16 *The tilt of the axis of Uranus with respect to the normal to the plane of its orbit is larger than that of any other planet.*

variations resulting from temperature differences the atmospheric conditions are the same for all three planets. The spectrum for Uranus shows bands in the red, orange, and green, such as we find in Jupiter's and Saturn's spectra, so that methane is undoubtedly present in the atmosphere. Just as in the cases of Saturn and Jupiter, there are heavy clouds probably composed primarily of ammonia crystals.

12.61 Satellites of Uranus

Uranus is accompanied by five known satellites, all of which revolve in the same direction about it as the planet rotates. These are extremely faint, and their diameters range from a few hundred miles to six or seven hundred miles for Titania, the largest. The orbits of these satellites are very nearly circular, and lie in one plane, which is coincident with the equatorial plane of the planet.

12.62 Neptune: Its Discovery

With the observations made on Uranus by Herschel and his followers, and the twenty or more additional observations that had been unknowingly made by Lemonnier previous to its discovery, it was possible to compare an accurately observed orbit with a calculated one in which the perturbations of all the known planets were taken into account. The comparison of these orbits showed clearly that some other disturbing influence was present since the planet departed from its predicted positions by an amount far in excess of what one might reasonably consider as being due to observational error. By the year 1845 the difference between the computed and the observed positions amounted to very nearly 2′ of arc, and although it was possible to rearrange the elements of the orbit of Uranus over the entire span of time from 1782 to 1845, so that at no point did the observed and computed orbit differ by more than 20″ of arc, this discrepancy was still too large to be acceptable. As a consequence, it was generally agreed that an unknown planet was present, and that its perturbing influence was responsible for the discrepancy.

Both the French astronomer U. J. Leverrier and the British astronomer J. C. Adams independently calculated the position of the unknown planet. Following Leverrier's directions, Galle, at the Berlin Observatory, found it within 1° of the position as given by Leverrier. The discovery of this object, the planet Neptune, created tremendous excitement since it was the greatest single achievement of mathematical astronomy since the time of Newton, and indicated clearly that in the hands of experts, a combination of science and mathematics can be a tool of great power.

12.63 Orbital Characteristics and Motion of Neptune

Neptune, although invisible to the naked eye, can be seen as a faint greenish object even with ordinary binoculars. Its mean distance from the sun is 30.07 astronomical units, or 2 797 000 000 miles ($4\,500 \times 10^6$ km). The orbit, which is inclined 1°47′ to the ecliptic, is very nearly circular (its eccentricity being about 0.009). Neptune moves with a speed of 3.3 mi/sec (5.4 km/sec) in its orbit, so that its sidereal period is 165 years and its synodic period 367 days. The planet rotates once every 15^h40^m (from spectroscopic data) in the same direction that it revolves about the sun, and the plane of its equator is tilted 28° with respect to the plane of its orbit.

12.64 The Diameter and Mass of Neptune

From all the observations of its apparent diameter, Neptune's true diameter is known to be about 28 000 miles (44×10^3 km). Since the orbits of its two satellites are known, and its perturbing action on Uranus has been studied with great care, we know with fairly high accuracy that its mass is about 17.20 times that of the earth. Using its mass and radius, we find that the acceleration of gravity on its surface is slightly larger than it is here on earth, but the speed of escape is almost twice the speed of escape from the earth.

Neptune superficially looks very much like Saturn and Uranus, so we may conclude that its atmosphere is much the same as theirs. Its internal structure is probably similar to those of the other Jovian planets since its mean density is 2.3 times the density of water—very close to their mean densities.

12.65 Satellites of Neptune

Neptune has two satellites, Triton and Nereid, the latter discovered by Kuiper in 1949. Although no oblateness can be observed on Neptune's disk, the plane of Triton's orbit is undergoing continual precession, and this can be accounted for only by a bulge at the planet's equator. Triton moves around Neptune in an almost circular orbit having a radius of 220 000 miles. Its motion is retrograde, and its orbit is inclined 20° to the orbit of the planet. It is probably somewhat larger than our moon, and its mass is estimated as being about one-fortieth of the earth's mass. It is probable, therefore, although there is no evidence for it, that Triton possesses an atmosphere.

12.66 Pluto

Just as Neptune was discovered by comparing the computed and observed orbits of Uranus, so Pluto was discovered in 1930 after calculations made by Gaillot and by Percival Lowell (the first two letters of the planet's name are the same as Lowell's initials) clearly indicated that both Neptune and Uranus were being perturbed by some more distant planet. Although Pluto was discovered by Tombaugh at the Lowell Observatory, after Lowell's death, the honor for the discovery must also go to Lowell who had begun searching for this planet almost a quarter of a century before. After its discovery it was found that Pluto had been photographed as far back as 1914 but had escaped detection as a planet.

The mean distance of this planet is $39\frac{1}{2}$ astronomical units, or 3 675 000 000 miles ($5\,900 \times 10^6$

FIG. 12–17 *Pluto: two photographs showing motion of planet relative to the sun in twenty-four hours. (200-inch photographs from Mount Wilson and Palomar Observatories)*

km), the sidereal period is slightly more than 248 years, and the synodic period is 367 days. Since the eccentricity of its orbit is 0.25, Pluto's position at perihelion is 35 000 000 miles within Neptune's orbit, whereas at aphelion it is 1 800 000 000 miles beyond Neptune's orbit. For this reason, and because of its small mass, Kuiper and other astronomers have concluded that Pluto was once a satellite of Neptune. The plane of its orbit is tilted 17°10′ with respect to the ecliptic, so that at no time does it get closer than 240 000 000 miles to Neptune.

From the perturbing effects of Pluto on Neptune and Uranus, its mass is thought to be 80 per cent of the earth's mass, so that it is probably similar to the terrestrial planets in constitution. This would be borne out if its diameter were the same as the earth's, but it is much too small and far away for any reliable measurement of its disk to be made at the present time, although Kuiper has found a value of 0″.23 of arc for Pluto. However, if we assume that its mean density lies between 3 and 10 times the density of water, its diameter must lie within 20 per cent of the earth's diameter. Figure 12.17 shows Pluto's apparent motion on the celestial sphere.

EXERCISES

1. How would Bode's law look if you expressed it in terms of Venus' mean distance from the sun? In terms of Jupiter's mean distance?

2. (a) Compute the mass of Mars from the motion of one of its satellites. Use the data given in the text. (b) Do the same thing for Jupiter.

3. Calculate the Roche limit for (a) Jupiter and (b) Saturn.

4. By how much does Jupiter's rotation reduce the weight of a body on its equator?

5. If our moon were revolving around Jupiter at the same distance from Jupiter's center as from the earth's center, would its tidal action on Jupiter be larger than on the earth? If so, how much larger would it be?

6. If the solar system were all alone in the universe, which one of its measurable characteristics would remain unaltered in the course of time?

TABLE 12.4

PLANETARY ORBITS

Planet		Semimajor axis of orbit		Sidereal period	Synodic period	Mean daily motion	Mean orbital velocity	Eccentricity e	Inclination to ecliptic i	Mean longitude of ascending node, ☊	Mean longitude of perihelion, $\bar{\omega}$ measured from ☊	Planet at 1950 Jan 0.5 L
		A.U.	10^6 km	Days	Days	″	km/s		° ′ ″	° ′ ″	° ′ ″	° ′ ″
Mercury	☿	0.387 1	57.91	87.969	115.88	14 732.42	47.90	0.206	7 00 10.6	47 08 43	75 53 54	33 10 06
Venus	♀	0.723 3	108.21	224.700	583.92	5 767.67	35.05	0.007	3 23 37.1	75 46 50	130 09 51	81 34 19
Earth	⊕	1.000 0	149.60	365.257		3 548.19	29.80	0.017			101 13 11	99 35 18
Mars	♂	1.523 7	227.94	686.980	779.94	1 886.52	24.14	0.093	1 51 01.1	48 47 12	334 13 6	144 20 07
Jupiter	♃	5.202 8	778.3	4 332.587	398.88	299.13	13.06	0.048	1 18 31.4	99 26 20	12 43	316 09 34
Saturn	♄	9.540	1 427	10 759.20	378.09	120.46	9.65	0.056	2 29 33.1	112 47 10	91 05	158 18 13
Uranus	♅	19.18	2 869	30 685	369.66	42.23	6.80	0.047	0 46 21	73 28 50	169 03	98 18 31
Neptune	♆	30.07	4 498	60 188	367.49	21.53	5.43	0.009	1 46 45	130 40 50	43 52	194 57 08
Pluto	♇	39.44	5 900	90 700	366.74	14.29	4.74	0.249	17 10	109 00	224	165 36 09

PHYSICAL ELEMENTS OF PLANETS

Planet		Semidiameter (equatorial) at 1 A.U.	Semidiameter (equatorial) at mean C or O †	Radius (equatorial) R_e		Ellipticity $\frac{R_e - R_p}{R_e}$	Volume	Reciprocal mass (including satellites)	Mass (excluding satellites) $\mathscr{M}$	Density ρ	Surface gravity	Escape velocity	Rotation period (equatorial)	Inclination of equator to orbit
		″	″	km	⊕ = 1		⊕ = 1	1/☉ = 1	⊕ = 1	g/cm³	cm/sec²	km/s		° ′
Mercury	☿	3.34	5.45	2 420	0.38	0.0	0.055	6 050 000	0.054	5.4	360	4.2	88^d	
Venus	♀	8.43	30.5	6 100	0.96	0.0	0.88	408 600	0.815	5.1	870	10.3		23 ?
Earth	⊕	8.80		6 378	1.00	0.003 4	1.000	328 700	1.000	5.52	982	11.2	$23^h\ 56^m\ 4^s.1$	23 27
Mars	♂	4.68	8.94	3 380	0.53	0.005 2	0.150	3 089 000	0.108	3.97	376	5.0	$24^h\ 37^m\ 22^s.6$	23 59
Jupiter	♃	98.47	23.43	71 350	11.19	0.062	1 318	1 047.38	317.8	1.33	2 600	61	$9^h\ 50^m.5$	3 05
Saturn	♄	83.33	9.76	60 400	9.47	0.096	769	3 497.6	95.2	0.68	1 120	37	$10^h\ 14^m$	26 44
Uranus	♅	32.8	1.80	23 800	3.73	0.06	50	22 930	14.5	1.60	940	22	$10^h\ 49^m$	97 55
Neptune	♆	30.7	1.06	22 200	3.49	0.02	42	19 100	17.2	2.25	1 500	25	15^h	28 48
Pluto	♇	4.1	0.11	3 000	0.47		0.1	400 000?	0.8?	uncertain			$6^d.39$	

† C = inferior conjunction (Mercury and Venus only), and O = opposition.

13

COMETS AND METEORS*

COMETS

In the year 1910 there appeared in the sky two celestial objects, rivaling the moon in brilliance and extending for great distances. One was Halley's Comet, which reappears every 76 years; the other was an unknown comet which appeared unexpectedly. These two were spectacular representatives of a large group of bodies that move in our solar system in orbits ranging from elliptical to hyperbolic.

13.1 Orbits of Comets

Those comets which move in elliptical orbits are permanent members of the solar system and reappear periodically, whereas those that move in hyperbolic orbits appear only once and then leave forever. However, it is extremely difficult to tell whether a body is moving in a highly elongated elliptical orbit or in a parabolic or hyperbolic orbit when it is close to perihelion, which is the only time we can study the orbit of a nonrecurrent comet. It is almost impossible to say whether or not any of these apparently nonrecurrent comets are really members of the solar system. It might be that they are, but that they move in closed orbits with very long semimajor axes.

Actually, an assignment of a calculated orbit to a comet is not very satisfactory because the computed orbit is one that the comet would have if its motion, at any particular moment, were determined entirely by the sun, so that the perturbations of the planets are not taken into account. *The action of the planets as the comet approaches the sun may be such as to change the orbit drastically*, and we know that many of the initial orbits associated with comets moving within Jupiter's orbit were undoubtedly greatly affected by Jupiter.

These considerations are particularly important because recent theories have suggested that all comets in our solar system are really permanent members of it, but with the bulk of them having enormously long periods so that they spend most of their time far away from the sun, forming a halo around the solar system at distances of the order of 150 000 astronomical units. This idea has been suggested by J. H. Oort, who argues that the comets are part of the solar system because they share the motion of the sun through the Milky Way. As a matter of fact, E. Strömgren has shown that no recorded comets have orbits that are definitely known to have been hyperbolic before they entered the region of the planets. This greatly strengthens Oort's theory. We shall return to this point when we discuss the origin of comets.

13.2 Appearance of Comets

When a comet gets close enough to the earth to be observed (it may still then be quite far away from the sun) it looks something like a star surrounded by a kind of luminous fog, but then as it approaches the sun, it acquires a tail, which glows with a hazy light. The tail has a fairly definite structure, and in almost all instances points away from the sun, as shown in Figures 13.1 and 13.2.

Although almost a dozen comets are discovered each year, most of them are much too faint to be seen with the naked eye, so that these visitors to our part of the solar system are much more numerous than most people think. In spite of the large number of comets that have come into the solar system (over 1 000 have been listed) the average person cannot be expected to see more than half a dozen during a lifetime, and no more than two or three of these will be brilliant ones.

Among all the known comets there are only two, the Schwassmann-Wachmann Comet and the Oterma Comet, that can be observed every year. When a comet is discovered, and if it is remarkable enough to warrant a proper name, it is named after its discoverer or in honor of the astronomer who analyzed its orbit. As examples of this we have Halley's, Encke's, Donati's, and Brooks' comets. If a comet is relatively insignificant, it is usually first designated by the year in which it is discovered —for example, 1963b (where the letter represents the order of discovery)—and then when the orbit is computed a roman numeral replaces the letter to represent the order of the perihelion passage of the comet.

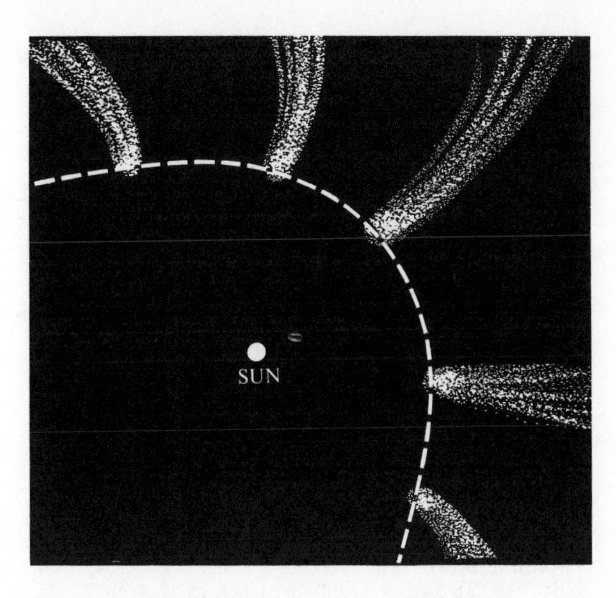

FIG. 13–1 *The tail of a comet points away from the sun at all points of the orbit of the comet.*

13.3 Periodic Comets

Of the periodic comets about 60 have periods ranging from 100 to 10 000 years, with about 30 of them in the 100- to 1 000-year range, and the other 30 with periods between 1 000 and 10 000 years. There are 69 periodic comets with periods less than 100 years, and about 36 of these have actually been observed to return up to as many as 41 times.

FIG. 13–2 *Brooks' Comet, 1911 V, October 19. Tail formed of sheet-like streams.* Astrophysical Journal, *XXXVI, Pl. II.* (*Yerkes Observatory photograph*)

TABLE 13.1

COMETS

Comet	Recent perihelion date	Period	Angle from ascending node to perihelion ω	Longitude of ascending node, Ω	Inclination i	Eccentricity e	Perihelion distance q	Semi-major axis a, of orbit
		y	°	°	°		A.U.	A.U.
Encke	1961.10	3.30	185	335	12.4	0.847	0.339	2.21
Grigg-Skjellerup	1957.09	4.90	356	215	17.6	0.704	0.855	2.89
Temple (2)	1957.10	5.28	191	119	12.5	0.545	1.38	3.0
Kopff	1958.05	6.3	160	120	5	0.556	1.51	3.4
Giacobini-Zinner	1959.82	6.5	172	196	30.8	0.72	0.94	3.5
Schwassmann-Wachmann (2)	1961.68	6.53	358	126	3.7	0.384	2.155	3.50
Wirtanen	1961.29	6.67	343	86	13.4	0.543	1.62	3.55
Reinmuth (2)	1960.90	6.7	45	296	7.0	0.46	1.93	3.6
Brooks (2)	1960.46	6.75	197	177	5.6	0.50	1.76	3.6
Finlay	1960.67	6.85	321	42	3.5	0.705	1.07	3.6
Borrelly	1960.45	7.01	351	76	31.1	0.604	1.450	3.67
Faye	1955.17	7.42	201	206	10.6	0.565	1.655	3.80
Whipple	1955.91	7.42	190	189	10.2	0.356	2.450	3.80
Reinmuth (1)	1958.23	7.67	13	124	8.4	0.478	2.03	3.90
Oterma	1958.44	7.89	355	155	4.0	0.144	3.39	3.96
Schaumasse	1960.29	8.18	52	86	12.0	0.705	1.195	4.05
Wolf (1)	1959.22	8.42	161	204	27.3	0.396	2.505	4.15
Comas Solá	1961.26	8.57	40	63	13.5	0.577	1.775	4.19
Väisälä (1)	1960.35	10.5	44	135	11.3	0.635	1.745	4.79
Schwassmann-Wachmann (1)	1957.36	16.1	356	322	9.5	0.132	5.53	6.4
Neujmin (1)	1948.96	17.9	347	347	15.0	0.774	1.54	6.8
Crommelin	1956.80	27.9	196	250	28.9	0.919	0.744	9.2
Olbers	1956.45	69.6	65	85	44.6	0.930	1.18	16.8
Pons-Brooks	1954.39	70.9	199	255	74.1	0.955	0.775	17.2
Halley	1910.30	76.2	112	57	162.3	0.967	0.587	17.8

Of the 69 comets with periods less than 100 years, about 45 are definitely associated with Jupiter since their aphelia lie close to Jupiter's orbit. We therefore surmise that these comets originally had very long periods, but that their orbits were drastically altered by Jupiter's perturbations. As a group, they constitute what is known as Jupiter's family of comets, and their periods lie between 5 and 7½ years. Among the 69 comets mentioned above, about 49 have periods lying between 3 and 9 years.

13.4 Some Famous Comets

Of all the comets that have been observed Halley's Comet is probably the most famous, and records of its appearance have been traced back to 240 B.C. It is named in honor of Halley because he was the first astronomer who proved that this comet could be identified with those that appeared in 1531, 1607, and 1682, and he predicted it would reappear in 1758. When it returned in 1910, its tail was so long that the earth passed right through it. This comet, as well as others with similar periods, has a highly inclined orbit of large eccentricity and has a retrograde motion. See Figure 13.3.

Although in general comets move around the sun in precise accordance with the law of gravitation, Encke's Comet, which has a period of 3.3 years, appears to be an exception. Since the date of its discovery in 1819, Encke's Comet has been observed each time it has returned, and it has been found that even after the perturbations of all the planets are taken into account, there is still a steady decrease in its period which cannot be due to the gravitational

Fig. 13–3 *Halley's Comet, May 12 and 15, 1910. Tails 30° and 40° long. Photographed from Honolulu with 10-inch focus Tessar lens. (From Mount Wilson and Palomar Observatories)*

attraction of any one body. From 1819 to 1914, its period decreased by $2\frac{1}{2}$ days, corresponding to a reduction of its mean distance from the sun by 275 000 miles. This phenomenon can only be explained by assuming that the comet periodically moves through some resisting medium, such as a swarm of meteors.

The periodic comet Schwassmann-Wachmann, discovered in 1925, is remarkable for the near-circularity of its orbit, its eccentricity being 0.14, which is quite small for a comet. The orbit itself lies entirely between Jupiter and Saturn. Among its interesting characteristics are its sudden changes in brightness which, O. Struve suggests, are due to bombardment by corpuscles, such as ionized particles, emitted in bursts by the sun.

The comets that appeared in 1668, 1843, 1880, 1882, and 1887 have a common characteristic which seems to indicate that they are parts of what was once a single comet which was torn into pieces by the tidal action of the sun. This common characteristic is that their orbits are almost identical. Of this group the comet 1882 II is of particular interest because it is the most brilliant and one of the brightest comets ever observed. At its last appearance it was plainly visible during the daytime, and passed

FIG. 13-4 *Donati's Comet, September 29, 1858. Drawing by G. P. Bond of Harvard College Observatory. Shows detail in head. (Yerkes Observatory photograph)*

right through the sun's corona at a distance of about 300 000 miles from the sun's surface. Even when it was within 4° of the sun it could still be seen in the sky by blocking out the sunlight with one's hand.

As an example of the way in which a planet can affect the orbit of a comet, we may note that the Brooks Comet, in 1886, suffered a change in its period from 29 to 7 years when it passed close to the planet Jupiter. In the same way Lexell's Comet of 1770 suffered a reduction of $2\frac{1}{2}$ days in its period when it came close to the earth. We shall come back to these two effects when we discuss the masses of comets.

13.5 Physical Characteristics of Comets

Although comets are among the most spectacular objects and are the largest bodies in our solar system, they are physically rather inconsequential since, as we shall see, they contain very little mass in spite of their large bulk. Most of the mass of the comet is contained in its head, which consists of the **coma**, a hazy nebulous halo of transparent matter that is luminous, and also, in most cases, of a **nucleus**, which is a fairly concentrated bright region near the center (see Figure 13.4). In addition to these two parts of a comet there is apt to be a **tail** as well, which is a stream of bright luminous matter that always points away from the sun regardless of where the comet is in its orbit. The tail is a rather complex structure, and is probably formed by material ejected from the head of the comet as it approaches the sun.

The head itself is not at all simple in structure, but often consists of concentric shells from which jets of material are constantly emitted. At times small spikes may develop pointing towards the sun rather than away from it. This was particularly true of the Arend-Rowland Comet (see Figure 13.5), 1956h, which in addition to a tail developed a spikelike structure 20° to 30° in length, which pointed to the sun.

13.6 Spectral Features and Chemical Nature of Comets

When a comet is first visible, its brightness is due primarily to reflected sunlight, but as it gets closer to the sun, the intense solar radiation excites the gases so that they become self-luminous. A spectral analysis of the light from a comet shows, in addition to the reflected solar spectrum, bright bands of various carbon and hydrogen molecules such as C_2, CH, CH_2, NH, NH_2, OH, CO, N_2^+, and OH^+. However, these bands are not permanent since the molecules tend quickly to combine with others to form various carbon gases, which are driven into the tail by the pressure of radiation from the sun as well as by collisions with streams of corpuscles from the sun. When the comet gets quite close to the sun, the spectrum shows bright lines of sodium, iron, and nickel.

The nucleus of the comet itself is a very porous structure consisting, according to Whipple, of large chunks of ices of methane, ammonia, and water, with meteoric materials—particles of iron, nickel, and calcium—embedded in them. As the comet approaches the sun, the volatile matter evaporates and is swept back into the tail. According to Whipple, as much as 1/200 of the mass of a comet evaporates during each perihelion passage. Whipple has attempted to explain the internal spiral motions of some comets by means of a jet propulsion mechanism involving the emission of evaporated material from the comet's head.

13.7 The Formation and Direction of the Comet's Tail

Since the sun is constantly emitting high-speed corpuscles from its surface, we surmise that these, together with the sun's radiation, sweep away from the sun any small bits of matter they encounter. Thus, if small particles and gases are released from

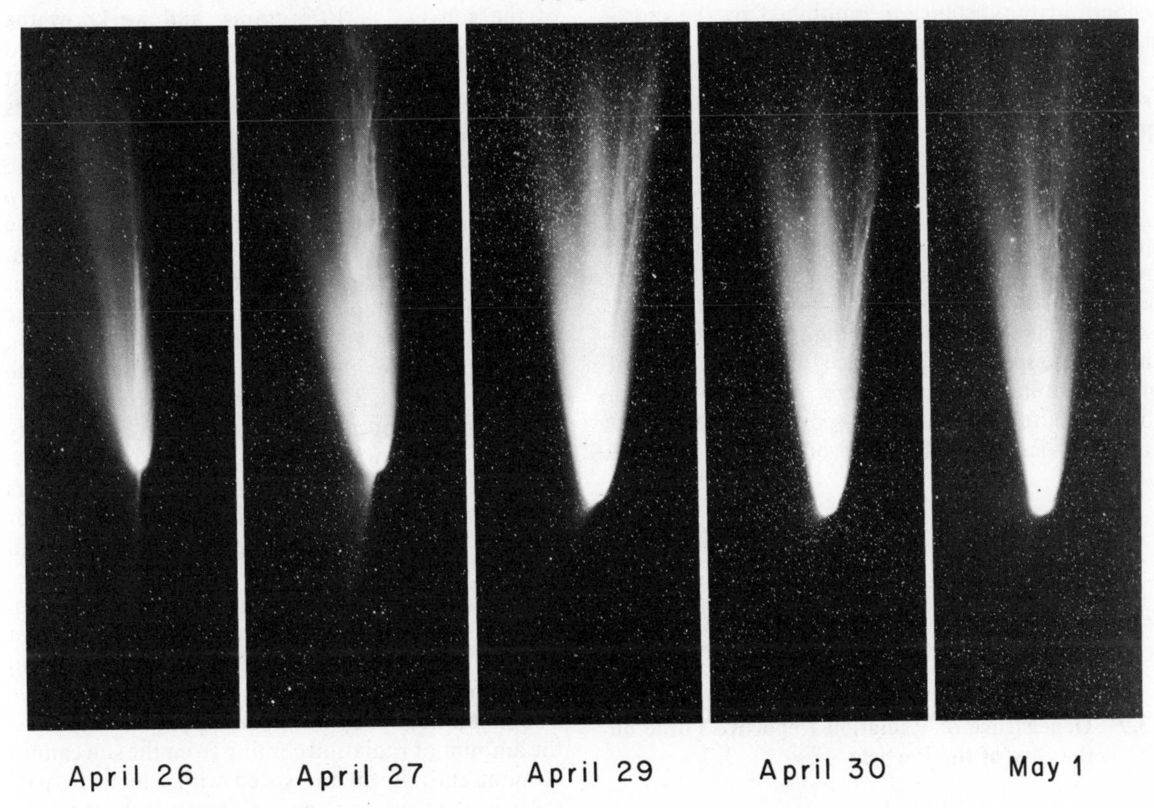

FIG. 13–5 *Arend-Rowland Comet, 1957. Five views taken with a 48-inch Schmidt telescope. The changes in the structure in the tail as this comet approached the sun are clearly visible. Of particular interest is the jet from the head of the comet on April 26. (From Mount Wilson and Palomar Observatories)*

the head of the comet, they are driven away from the sun by this action, and this repelling action accounts for the formation of the tail.

That tails of comets are composed of matter streaming away from the head is corroborated by photography. Successive photographs of tails of comets show bright condensations moving rapidly away from the head. Thus, Curtis was able to photograph such a condensation in the tail of Halley's Comet. It was moving away from the head at an average speed of about 45 mi/sec when first seen, and then increased its speed to about 60 mi/sec a day later. This kind of acceleration at a point in a comet's tail can be due only to repulsive forces emanating from the sun; these forces may be 100 times larger than the sun's gravitational attraction at that point.

13.8 Radiation Pressure

We can see from the following simple analysis that the force of repulsion arising from the radiation pressure on a small enough particle can be much greater than the gravitational attraction of the sun on this particle. Let the density of the particle be ρ, and let its radius be r, so that its mass is $\frac{4}{3}\pi r^3 \rho$. If this particle is at a distance d from the center of the sun, the gravitational attraction towards the sun is, according to Newton's law,

$$F_{\text{attraction}} = \frac{GM_{\odot}}{d^2}\left(\frac{4}{3}\pi r^3 \rho\right),$$

where $M_{\odot}$ is the mass of the sun.

On the other hand the repulsion arising from the radiation pressure depends upon how much radiation strikes this particle per second. This is clearly equal to the amount of radiation from the sun passing through one square centimeter per second, at the distance d, multiplied by the cross-sectional area of the particle. If we let $L_\odot$ be the total luminosity of the sun (the radiation emitted from the sun in all directions per second); then this quantity is just

$$L_\odot \frac{r^2}{d^2},$$

since $L_\odot/(4\pi d^2)$ is just the amount of radiant energy from the sun passing through one square centimeter per second at the distance d. We obtain the force exerted on the particle by dividing the above expression by the speed of light (since force is the amount of momentum transferred per second, and we obtain the momentum of radiation by dividing the radiant energy by its speed, the speed of light) so that this force is

$$\frac{1}{c}\frac{L_\odot r^2}{d^2}.$$

For a particle having a cross section of 1 square centimeter this force at the distance of the earth from the sun is 4.5×10^{-5} dynes.

13.9 Dependence of Radiation Repulsive Force on the Size of the Particle

If we now compare the expression for the radiant repulsive force with that of the gravitational attractive force, which we can do by dividing the former by the latter, we obtain

$$\frac{\text{force of radiation}}{\text{force of gravitation}} = \frac{\dfrac{L_\odot r^2}{cd^2}}{\dfrac{\frac{4}{3}\pi G M_\odot \rho r^3}{d^2}} = \frac{L_\odot}{\frac{4}{3}\pi G M_\odot c r \rho} = \frac{L_\odot}{\frac{4}{3}\pi G M_\odot c}\left(\frac{1}{\rho}\right)\left(\frac{1}{r}\right). \quad (13.1)$$

We see from this that for a particle of given density the repulsive radiative force is larger than the gravitational force if the radius r of the particle fulfills the following inequality:

$$\boxed{r < \frac{L_\odot}{\frac{4}{3}\pi G M_\odot c}\left(\frac{1}{\rho}\right)}. \quad (13.2)$$

We see, then, that the smaller r is, the greater the radiation force is compared with the gravitational force. Since $L_\odot$ is 4×10^{33} ergs/sec, and the mass of the sun is 2×10^{33} grams, and we know the value of the gravitational constant G and the speed of light c, we can calculate the constant coefficient on the right-hand side of this inequality. Its numerical value in cgs units is 2.5×10^{-4}. Hence, we know that at any distance d from the sun the radiative repulsive force on a particle of density ρ gm/cm^3 is larger than the gravitational attractive force if the radius is less than $(2.5 \times 10^{-4})/\rho$ cm.

Thus, if a particle has the density of rock (about 3 gm/cm^3) it is repelled by the radiation from the sun if its radius is smaller than 0.8×10^{-4} cm. Since from the evidence of the acceleration of the material in the tail streaming away from the head of a comet it appears that the repulsive force may be anywhere from 30 to 100 times the sun's gravitational attractive force, we know that the diameters of the particles in the tail cannot be much greater than 2.5×10^{-6} cm, if their density is about equal to that of rock.

Although the foregoing analysis of how radiation from the sun can repel small particles may account to some extent for the formation of a comet's tail and its pointing away from the sun, it does not give a complete answer to this puzzling problem. Thus, the amount of radiation coming from the sun cannot account entirely for the speed with which the particles in the tail are streaming away from the head of the comet nor for the apparent rigidity of some of the tails.

However, we now know that most of the features of the tail are due to a **solar wind** which consists of a steady stream of ionized plasma (charged particles) flowing away from the sun. Although such a corpuscular stream had been predicted by various astronomers, particularly Biermann, no direct evidence for it was available until the solar plasma experiment on the Mariner II space probe established definitely that such a solar wind exists. We shall discuss the details of the Mariner II solar plasma experiment in Section 14.10.

13.10 Sizes of Comets

It is impossible to give definite figures for the sizes of comets since they vary enormously in structure and size among themselves, and the structure and size of a single comet change considerably as it approaches and recedes from the sun. However, we can say that the head of a comet may range anywhere from a few thousand miles to a few hundred thousand miles, being largest when the comet is still about 100 000 000 miles from perihelion. The nuclei of even the very largest comets are quite small, seldom exceeding 1 000 miles. The largest part of the comet is the tail, which is hardly ever less than 5 000 000 miles long when the comet is near the sun, and may, as in the case of Halley's Comet, extend out to a distance of 100 000 000 miles. Since the tail is fan-shaped, it may be many millions of miles across at the greatest distance from the head.

13.11 The Effect of the Tidal Action of the Sun on Comets

As we have noted, the heads of comets change shape as they approach the sun, and this is undoubtedly due to the tidal action of the sun's gravitational field which tends to tear them apart. An excellent example of this process is Biela's Comet, which has a period of $6\frac{3}{4}$ years. In its four recorded appearances in 1772, 1806, 1826, and 1832, it moved along in its orbit as a single body without any change in its structure. But in 1846 it became a pear-shaped comet, and then separated into two parts which traveled together but which independently developed nuclei and tails. These two components were again seen in 1872 separated by a distance of 1 500 000 miles, but they have never been seen since then. Since we know that their orbits could not have been affected by any of the other planets, their invisibility must be due to some other mechanism. This division of a comet into two parts also occurred in Taylor's Comet (1916 I) which has a period of $6\frac{1}{2}$ years.

13.12 Masses and Densities of Comets

Whereas the dimensions of comets are very large, their masses are extremely small and not measurable. This is evident from Brooks' and Lexell's comets, which do not measurably affect the orbits of planets they closely approach. From such observations it seems most likely that no comet has a mass greater than 1/1 000 000 of the earth's mass. (Even so, this maximum is roughly 6×10^{18} kg—which by terrestrial standards is an enormous mass.)

Because of their small masses and large volumes the densities of comets are extremely small; this is borne out by their transparency. The density is probably largest in the nucleus, becoming negligible towards the end of the tail. In the nucleus the density may be about 1/1 000 000 of the density of air at sea level; and in the tail of a comet such as Halley's the density is so low that 2 000 cubic miles has less material in it than one cubic inch of ordinary air. Because comets are so very tenuous, we must conclude that they are able to move in their orbits in high agreement with Newton's law of gravity only because interplanetary space is practically empty. If there were any appreciable amount of matter in the form of dust in our solar system, comets would not move the way they do.

13.13 Origin of Comets

Less than four decades ago it was still believed that comets originated in interstellar space. Today we know that they do not; if they did, many of them would have to be traveling at speeds exceeding 26 mi/sec (42 km/sec) (the speed of escape from the sun) as they pass perihelion. Moreover, most of them would approach the sun from the direction of Hercules, towards which our entire solar system is moving. But neither of these situations holds. It has, therefore, been suggested by Oort, who made use of the work of E. Strömgren, that the majority of the comets are members of the solar system lying in a shell at a distance of about 150 000 A.U. from the sun. From a consideration of the number of comets that we detect each year, and from the assumption that all the comets move in orbits with their major axes distributed in all possible directions pointing away from the sun, and that the perihelion distances of their orbits have a wide range, with only a few being less than 2 A.U., it is clear that only a very few of the total number of comets that exist are ever seen. Oort and Van Woerden estimated that there may be more than 100 000 000 000 comets in the cloud that forms a shell around our solar system.

METEORS

13.14 The Nature of Meteors

"Falling stars" or "shooting stars" catch almost everyone's imagination, although they are called stars only by poetic license. Actually most of them are tiny objects (often no larger than a grain of sand); their brightness develops as they rush into our atmosphere and are vaporized by intense heat produced by atmospheric friction.

If we could see all of the earth's atmosphere from some point on the ground, we would find that billions of tiny particles stream very rapidly into the earth's atmosphere each day and vanish in a burst of light. All these objects taken together are called **meteors.** But occasionally these objects are so large and massive that they are not completely vaporized. They then strike the earth's surface and are known as **meteorites**. The very bright meteors that light up the entire landscape are known as **fireballs.**

13.15 The Number Entering the Earth's Atmosphere

That meteors are objects that enter our atmosphere is clear from the speed with which they move across the sky and also from the measurements of their heights made from different stations on the earth's surface. These objects, then, may be part of the swarm of particles in our solar system that are moving in their own orbit around the sun and that are swept into the earth's atmosphere. They may also be particles from interstellar space. If we consider that from a given point on the earth's surface only a small portion of the earth's atmosphere can be seen, it is clear that the number of meteors that strike the earth's atmosphere is millions of times greater than the average of four to eight per hour that the single observer can see. It is estimated that the total number of meteors that enter the earth's atmosphere in 24 hours and are bright enough to be seen is 200 million. All of these taken together have a mass of about 10 tons. If we keep in mind that there may be ten times as many that are too faint to be seen, the total mass of meteoric material that enters our atmosphere per day may be something of the order of 100 tons.

13.16 Detecting Meteors by Radar and Photography

Today, with improved radio techniques, it is possible to detect meteors by the effects they have on radar waves from the earth when these meteors strike the so-called *E* layer of the atmosphere about 60 miles above the earth's surface, the height at which most meteors are observed visually or on photographic plates. When the meteor strikes this layer, it induces a sudden increase in the ionization (knocking the electrons out of atoms) at the point of collision, changing the way this *E* layer reflects radar waves at this point. Thus, by continually monitoring this *E* layer with radar waves, we can pick up meteors that might otherwise be undetected. At the present time, however, the best way of studying meteors is by photographic methods, and extensive programs for carrying out photographic observations have been developed.

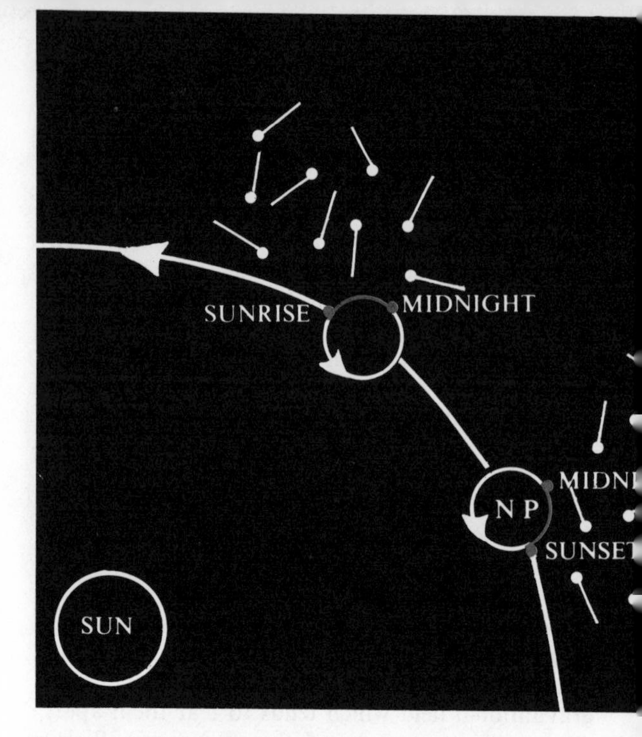

FIG. 13–6 *The viewing of meteors from sunset to midnight, and from midnight to sunrise.*

13.17 The Best Time for Observing Meteors

The best time for viewing meteors is between midnight and sunrise because the observer must then look in the general direction in which the earth is moving if he wants to see them. He therefore sees the meteors (moving with a speed less than the earth, $18\frac{1}{2}$ mi/sec) that the earth is overtaking, as well as those that are moving towards the earth. In order to see meteors from sunset to midnight the observer must look in a direction pointing away from the earth's motion, so that he sees only those meteors which are moving directly towards the earth with speeds greater than that of the earth, as shown in Figure 13.6.

13.18 Heights of Meteors

From observations of the heights of thousands of meteor trails made at stations over the entire earth's surface we know that, on the average, meteors first become visible at a distance of about 60 miles (96 km) above the surface of the earth, regardless of their apparent brightnesses. The heights at which these meteors disappear depend upon how big they are, the brightest disappearing at heights of about 40 miles and the fainter ones at heights of about 52 miles. An interesting aspect of this is that the average height at which meteors are observed over their entire trail is greater if the meteor is faint and moving with a high speed. An exception is found in the case of the fireballs, which come very close to the earth's surface, sometimes penetrating to within a height of about 15 miles.

13.19 The Velocities of Meteors

The velocities of meteors have been carefully studied by Öpik and Hoffmeister, separately, who have concluded that these objects must be traveling with speeds $2\frac{1}{2}$ times the speed of the earth in its orbit. In other words, according to these observers meteors at the same distance as the earth is from the sun have speeds that are greater than the parabolic speed with respect to the sun at that distance (the speed of escape at that distance) and must, therefore, be visitors from interstellar space. This is not true for all meteors but is true only on the average. Öpik has estimated that only one-third of all the meteors do not have speeds of escape, and are therefore true members of our solar system.

This is in disagreement with the estimates of Whipple who measured the speeds of 144 meteors and found only 15 among them with speeds greater than 26 mi/sec, which is the speed of escape from the solar system at the distance of the earth. Since he ascribes some of these hyperbolic speeds to possible errors of observation, he concludes that there is no reliable evidence of a single meteor moving with the speed of escape. His results are supported by McKinley's investigations which dealt with 1 100 meteors, among which he found only 32 with speeds above the speed of escape. This work would lead us to conclude that most meteors have their origin in the solar system itself and not in interstellar space.

13.20 Kinetic Energies, Masses, and Composition of Meteors

From the amount of light that is emitted by the trail of the meteor we can determine the amount of kinetic energy that the meteor had on entering the atmosphere if we assume that a known fraction of its kinetic energy is transformed into radiant energy. From this and our knowledge of the meteor's speed we can determine its mass, since kinetic energy equals half of its mass times the square of its speed. This leads to a mass of only a few milligrams for most meteors.

From the spectrum analysis of meteor trails as well as from the chemical analysis of meteorites we find that meteors are composed of iron, nickel, calcium, silicon, etc., but in varying quantities. Whereas calcium and silicon are found to dominate in one group of meteors, iron and nickel stand out in the other. It thus appears that there is a group of stony meteors and another group of iron-nickel meteors.

13.21 Meteor Showers

If a swarm of particles revolves in an orbit (along which they are spread out) which pierces the plane of the ecliptic at a point close to the earth's orbit, then periodically the earth will pass through or close to the swarm and an unusually large number of the particles in the swarm will enter the earth's atmosphere. An observer on the earth will then see a meteor shower. Since all meteors in this shower are moving parallel to each other in space (their orbits around the sun are all parallel to each other), it will appear to an observer watching these meteors that they are all diverging from a single point in the sky, an effect that is due to perspective. This point is called the radiant of the shower, and the shower itself is designated by the constellation in which the radiant appears to lie. Thus, the radiant of the Leonid shower, which occurs in November, lies in the constellation of Leo, etc.

Although during such a shower the number of observed meteors per hour increases considerably above what is usually observed, the effect in most cases is not very dramatic. However, occasionally there may be a tremendous increase in number, and meteors then appear in great profusion. Thus, in the Leonid shower of November 13, 1833, meteors were so abundant that at some stations as many as 200 000 per hour were seen. With radar techniques it is possible to observe such showers in the day-time, and some have been recorded at the Jodrell Bank Station in England.

From an analysis of orbits of some of the showers we know that the particles in such a swarm must have originated from some previous comet. Thus

Table 13.2

METEOR SHOWERS

Meteor shower	Maximum shower	Associated comet
Quadrantids	Jan. 3	
Lyrids	Apr. 21	1861 I
η Aquarids	May 4	Halley (?)
δ Aquarids	July 30	
Perseids	Aug. 12	1862 III
Draconids	Oct. 10	1933 Giacobini-Zinner
Orionids	Oct. 21	Halley (?)
Taurids	Nov. 4	Encke
Andromedids	Nov. 10	Biela
Leonids	Nov. 16	1866 I temporary
Geminids	Dec. 13	
Ursids	Dec. 22	Tuttle
Permanent daytime streams		
Arietids	June 8	
ξ Perseids	June 9	
β Taurids	June 30	Encke

the orbit of the Leonids appears to be almost identical with that of the faint comet 1886 I, and the Perseids are moving in an orbit practically the same as that of the comet 1861 I. In this way it is possible to associate most of the meteoric showers with comets that disintegrated and whose material was spread out over their entire orbits by the action of the sun. Table 13.2 gives data for the various showers and their relationship to the known comets.

13.22 Meteorites *

If a large mass of material, either of stony or metallic nature, enters the earth's atmosphere, and if the collision is not too violent, it sometimes survives the vaporizing action of atmospheric friction and strikes the earth with explosive force. If the impact is violent enough, a meteoric crater is formed and a good part of the meteorite disintegrates into tiny particles that are dispersed in all directions. If the entire meteorite is not destroyed in this collision, the residue can be recovered for study. Sometimes one finds meteorites occurring in groups as though they had entered together or as though a single one had been broken into pieces during the collision. Since 1900 an average of six meteorites a year have been observed to enter the atmosphere and have been recovered, but obviously many more have remained undiscovered. The masses of meteorites range from a few grams to many tons.

There are characteristic surface features that enable one to recognize these objects almost immediately. The surface of the meteorite, if found soon after the fall, is very glossy and dark in color, indicating that it was probably formed from the fused material that remained while the meteor was moving through the atmosphere. The surface itself is quite irregular and in some cases covered with deep pits as though quantities of material had been vaporized out of it. Many of the meteorites are of stone; others are masses of alloys of iron, nickel, and cobalt. When meteorites are heated they emit gases, and we have found about 75 of the known elements present in these objects.

13.23 Zodiacal Light

In addition to comets, meteors, and meteorites, which we can observe directly, there is a great deal of dust in our solar system which produces a glow in the sky that is referred to as **zodiacal light** (see Figure 13.7). Since the dust lies practically in the plane of the ecliptic, the Zodiacal light itself can be seen under proper conditions in the western horizon after sunset as a luminous band extending upward from the horizon along the ecliptic. For a northern-hemisphere observer the ecliptic is most nearly perpendicular to the horizon at sunset when the sun is near the vernal equinox, and in this hemisphere therefore the Zodiacal light can be seen best after sunset in the spring. The band of light is broad near the horizon, but becomes much narrower as one approaches the Milky Way. The effect can also be seen in the eastern horizon, just before sunrise, and it is most pronounced during the fall.

Zodiacal light is due to the scattering of sunlight by the particles of dust in this band of matter around the sun. It has recently been suggested by Russian astronomers that zodiacal light is due to a tail of dust (similar to a comet's tail) following the earth as it moves around the sun. This dust trail also explains the interesting effect known as "gegenschein" (see Figure 13.8) or "counterglow" —a faintly luminous diffused spot in the ecliptic that is about 180° away from the sun's position.

Fig. 13–7 *This photograph of the Milky Way from Cassiopeia to Canis Major shows the Zodiacal light. Taken by Osterbrock and Sharpless with the Greenstein-Henyey 140° camera, October 1950. Exposure 15 minutes. (Yerkes Observatory photograph)*

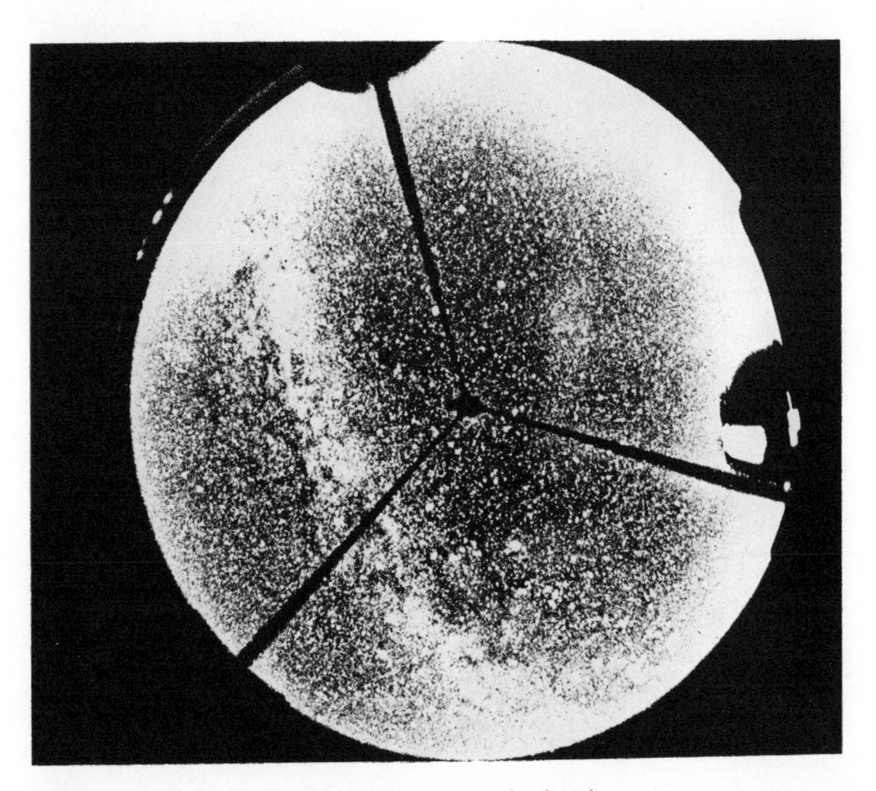

Fig. 13–8 *The Gegenschein, photographed by Osterbrock and Sharpless with the 140° camera, October 10, 1950. Exposure 15 minutes. (Yerkes Observatory photograph)*

EXERCISES

1. What is the approximate total (kinetic plus potential) energy per unit mass of one of the comets in the halo of comets surrounding the solar system?

2. (a) How large is the acceleration of gravity (arising from the sun) at the outer regions of the halo of comets? (b) What is the maximum speed a comet can have in the halo and still remain in the solar system? (Use the approximate distance of the halo given in the text.)

3. Consider a comet that is moving away from the sun towards Alpha Centauri. How much faster is it moving at its aphelion (in the halo) than it would be if no stars other than the sun were present?

4. What is the period of a comet that belongs to the halo of comets?

5. Assume, on the average, that the perihelion distance of a comet in the halo is about 1 A.U. and that its aphelion distance, on the average, is that of the halo. How does its orbital angular momentum per unit mass compare with that of the earth per unit mass?

6. How close to the sun must a comet approach to suffer fission? (Assume the density of the comet to be about 10^{-6} gm per cc.)

7. How large is the average radiation force on the earth? On Mercury?

8. If a one gm meteor strikes the earth's atmosphere with a speed of 20 miles per sec, how much heat is generated when it is completely slowed down by the atmosphere?

9. Assume that meteors are moving with the same speed in equal numbers in all directions relative to a stationary earth. How many more on the average would an observer count per unit time after midnight than before midnight if one assumes that the mean speed of the earth is $18\frac{1}{2}$ miles per sec?

10. Analyze the effect of the continual bombardment of meteors on the orbital motion of the earth.

14

THE SUN

The dominant body in our solar system is the sun, controlling, because of its great mass, the motions of all the planets and the other bodies in this region of space, and influencing, by means of the radiation that it pours out to them, the physical and chemical conditions on the planets. The sun has an added importance to astronomers on the earth since it is the only star that is close enough to be studied in detail, and may thus be compared observationally with the stellar models that are derived from theoretical analysis. We shall discuss the purely stellar properties of the sun in the later sections of the book in which we deal with over-all properties of stars. At this point we consider only those phenomenological aspects of the sun that are pertinent to questions about the solar system.

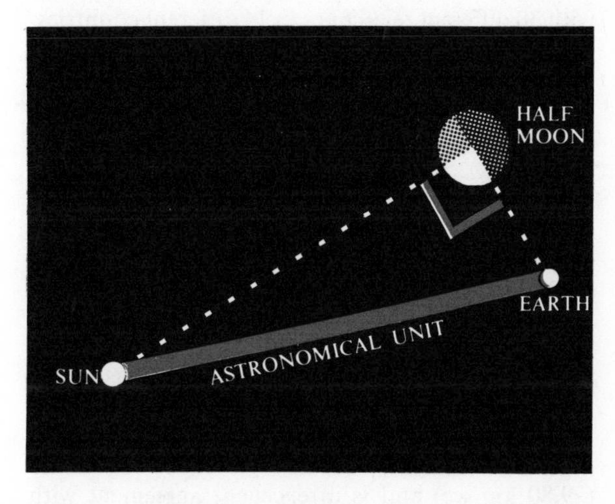

FIG. 14–1 *Aristarchus' procedure to determine the distance from the sun to the earth by observing the precise occurrence of the half moon.*

14.1 The Solar Parallax or the Mean Distance of the Sun from the Earth

Of all the data about the sun that we must have in order to understand the structure and the dynamics of the solar system, *the mean distance of the sun from the earth* (*the* **astronomical unit**) *is probably the most important.* The reason is that all other distances in our solar system are expressed in terms of the astronomical unit. Moreover, only when we have this datum can we determine such things as the dimensions, the mass, the density, and the luminosity of the sun.

Measuring this distance or, what is the same thing, finding the sun's geocentric parallax is extremely difficult, and involves measurements of the most delicate and accurate kind. The first attempt was made by Aristarchus of Samos, who argued correctly that at exactly half moon the line from the earth to the moon makes a right angle with the line from the sun to the moon, so that the distance from the sun to the earth is the hypotenuse of a right triangle. This procedure, however, is not very accurate since it depends on knowing precisely when the half moon occurs, and this, in general, is difficult to determine. (See Figure 14-1.)

The direct procedure for determining the parallax of the sun by observing the center of its disk from two widely separated points on the earth's surface fails because the apparent shift of this point on the sky, as seen from the two stations on the earth, cannot be determined since the stars are not visible when the sun is out. However, this method may be used during a total eclipse of the sun.

The most feasible method for determining the solar parallax is to measure the geocentric parallax of some planet or asteroid (see Section 6.9). From the parallax of such an object we can find its mean distance from the sun in miles, and since this mean distance in astronomical units is known from Kepler's third law, we can find the value of the astronomical unit. Very extensive measurements of this sort have been made on the asteroid Eros, and after ten years of data gathering and the analysis

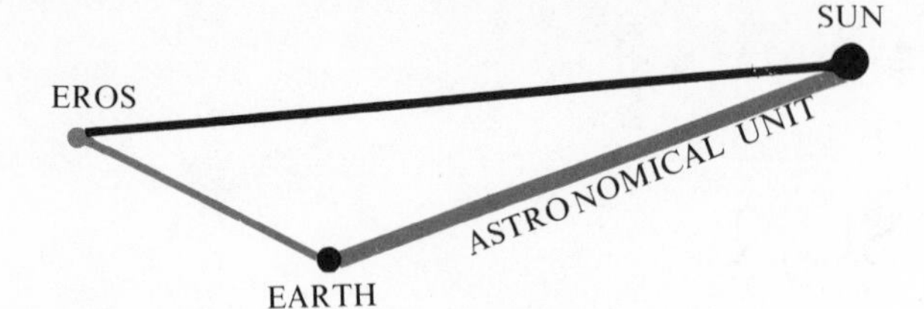

FIG. 14–2 *Determining the parallax of the sun by accurately measuring the distance of Eros when it is at its closest to the earth.*

of these data (2 847 photographic plates, taken with 30 different telescopes in 14 different countries, were processed) we find that the best value for solar geocentric parallax is

$$8''.794 \pm 0''.002 = 4.263 \times 10^{-5} \text{ radians}$$

(E. Rabe 1950)

as shown in Figure 14.2.

Other methods of finding the value of the astronomical unit are based upon determining the speed of the earth in its orbit (using the properties of light such as aberration and the Doppler shift). In addition there are certain other gravitational methods that depend upon perturbations. The best value for the mean distance is 92 956 020 ± 11 000 miles ($1.495\,985 \times 10^8$ km). This leads to a value of the earth's orbital velocity of $18\frac{1}{2}$ mi/sec (29.80 km/sec) and is in excellent agreement with the value determined from the aberration of light.

14.2 Dimensions of the Sun

Since the sun is not a solid body but gaseous all the way down to the center, the question of its diameter is somewhat complicated by the fact that one cannot say precisely where the atmosphere of the sun ends and the main body of the sun begins. However, there is a bright surface which stands out sharply on a photographic plate and which is the disk that one sees when the sun is just faintly visible through thin clouds. This is the surface (referred to as the **photosphere**) that we have in mind when we speak of the diameter of the sun. As seen from the earth, its average apparent size is about 32′ of arc. From this and our knowledge of the value of the astronomical unit, we can find the true radius of the sun from the equation

$$A\alpha = R, \tag{14.1}$$

where α is the angle (expressed in radians) subtended by the sun's radius at the earth, A is the astronomical unit, and R is the radius of the sun, as shown in Figure 14.3.

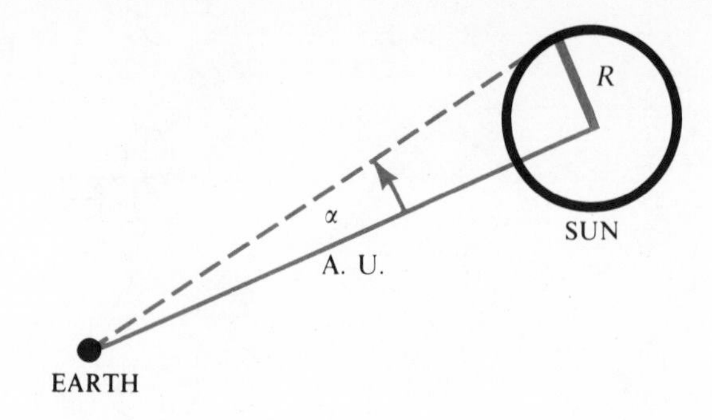

FIG. 14–3 *The relationship between the radius of the sun, the angle that the sun subtends at the eye, and the distance of the sun from the earth.*

If, in this equation, we substitute for α (about 16′) its value in radians, we then find that the diameter of the sun is about 865 400 miles (13.920×10^5 km), or somewhat more than 109 times the earth's diameter. From this value we note that the area of the sun's surface is more than 12 000 times the earth's area, and its volume exceeds the earth's volume by 1 300 000 times.

14.3 The Mass of the Sun

We can easily determine the mass of the sun, since we know the earth's orbit with very great accuracy. If we apply Kepler's third law to the earth, we have

$$P_{\oplus}^2 = \frac{4\pi^2 A^3}{GM_{\odot}},$$

or

$$M_{\odot} = \frac{4\pi^2 A^3}{GP^2},$$

where $P_{\oplus}$ is the earth's period ($365\frac{1}{4}$ days). By introducing the numerical values for G, A, and P into this equation, we find that $M_{\odot} = 2 \times 10^{33}$ gm, or approximately 335 000 times that of the earth.

In using this equation to find the mass of the sun, we must be very careful to insert a very accurate value for A because this quantity appears to the third power. Thus, a small percentage of error in A leads to three times this percentage of error in the sun's mass. However, the ratio of the mass of the sun to the mass of the earth can be found in another way: namely, by considering the perturbations of nearby planets, and then the above equation leads to a very accurate determination of the astronomical unit.

14.4 The Mean Density of the Sun

Knowing the mass and the radius of the sun we can find its mean density which is 1.4 gm/cm³. Of course this value does not apply to any particular point inside the sun, and, as we shall see, the density increases considerably as one goes down to the center of the sun. However, we shall find that the temperature also increases so that even at the center, where the density is quite high, the sun is still in a gaseous state.

14.5 The Surface Gravity of the Sun

From the mass and radius of the sun we find that the acceleration of gravity at its surface, given by $GM_{\odot}/R_{\odot}^2$, is 28 times that on the surface of the earth, so that an object weighing 100 lb on the surface of the earth would weigh 2 800 lb on the surface of the sun, and a pendulum placed on the surface of the sun would oscillate about five times as rapidly as the same pendulum does on the earth's surface. From the mass and the radius we can also compute the speed of escape from the surface of the sun by using the expression

$$\sqrt{\frac{2\,GM_{\odot}}{R_{\odot}}}.$$

On substituting the appropriate values into this expression, we find that the speed of escape from the sun is about 386 mi/sec (6.17×10^2 km/sec), or slightly more than 55 times the speed of escape from the earth.

14.6 The Luminosity of the Sun: The Solar Constant

The sun differs from all other bodies in the solar system in one very important respect: *it is a self-luminous body*. Although it emits vast quantities of radiation per second in all directions, the amount that we receive here on the earth per second is less than 1/2 000 000 000 of this. We can best express the amount of radiation that reaches the earth from the sun by considering how much of it, per unit time, strikes one square centimeter of a surface placed just outside the earth's atmosphere and at right angles to the sun's rays. We find that such a surface receives 1.95 cal/min. This quantity, 1.36×10^6 ergs/cm² sec, is called the **solar constant**.

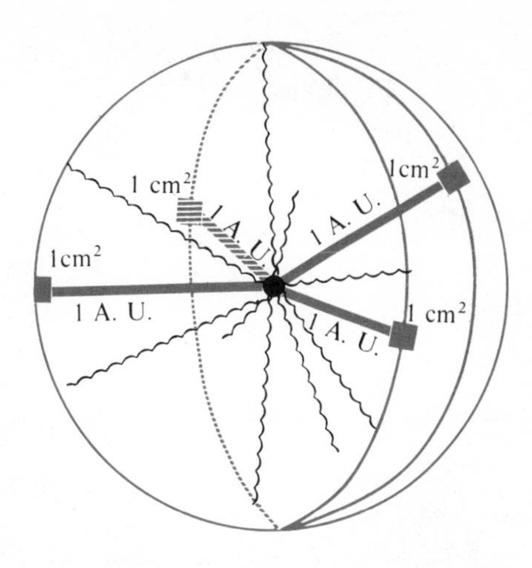

FIG. 14-4 *Each square centimeter of the surface of a sphere, with a radius equal to 1 A.U. drawn around the sun, receives 1.36×10^6 ergs of radiant energy every second.*

This constant, which was originally obtained by making measurements on the surface ot the earth and then estimating the necessary corrections that would have to be introduced because of absorption of the earth's atmosphere, can now be determined with great accuracy from data gathered by artificial ,atellites.

From our knowledge of the solar constant and the value of the astronomical unit, we can determine the luminosity of the sun, i.e., the total energy emitted per second by the sun in all directions. To do this we note that each unit area of the surface of a sphere surrounding the sun and having a radius equal to the distance of the earth from the sun receives exactly (1.36×10^6 ergs) of radiant energy per second, as shown in Figure 14.4. We can, therefore, find the total amount of energy emitted by the sun per second in all directions by finding the total area of such a sphere and multiplying this area by the solar constant. In other words, the total luminosity of the sun is $4\pi A^2 1.36 \times 10^6$ ergs/sec, where A, the astronomical unit, is to be expressed in centimeters. This product is 3.8×10^{33} ergs/sec, which therefore is the luminosity of the sun.

This means that each square centimeter of the sun's surface emits energy at the rate of 6.25×10^{10} ergs/sec. Expressed another way, the power equivalent of one square yard of the sun's surface is 70 000 horsepower. Of course, the radiation that we detect with the eye is only part of the total

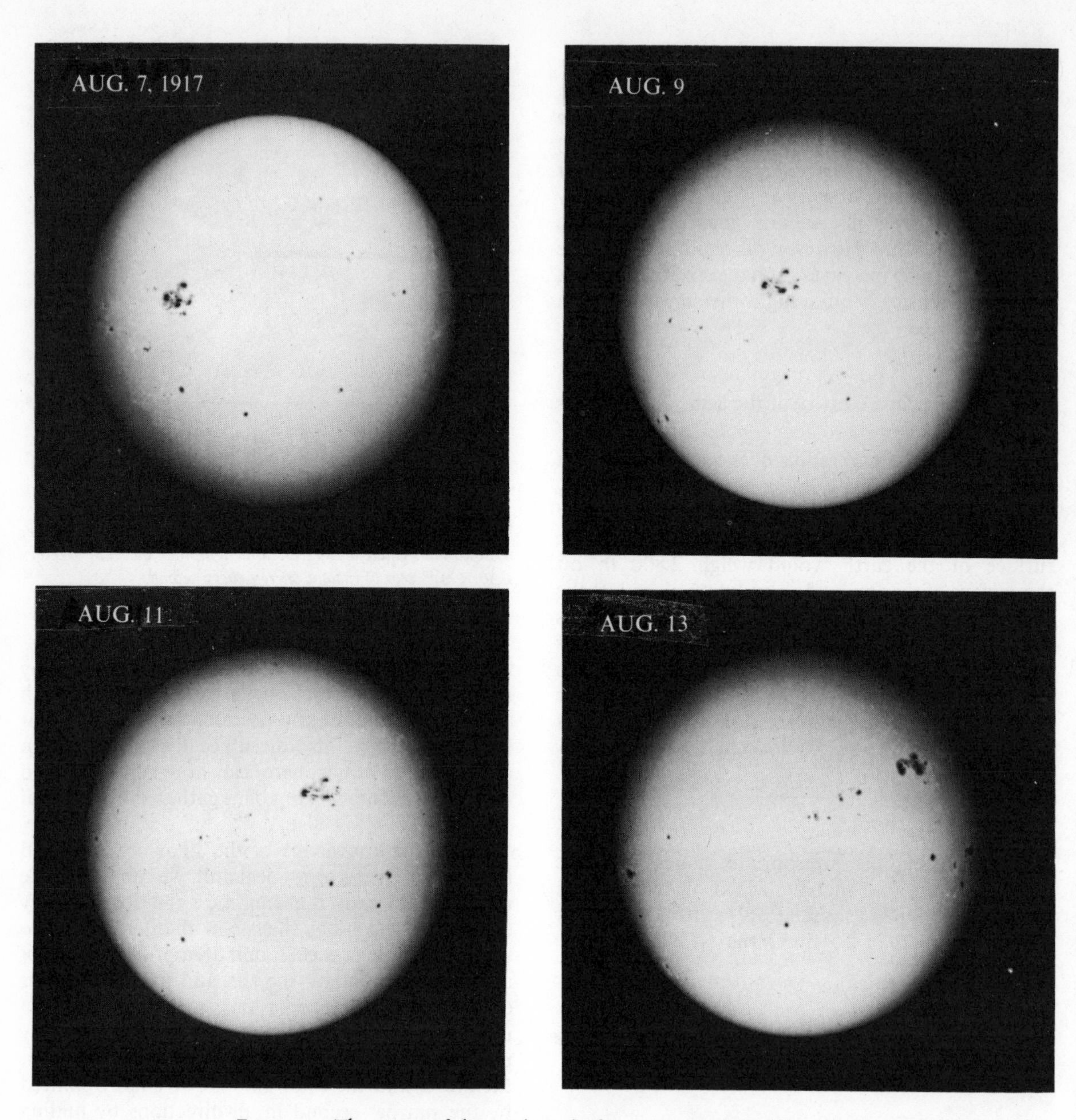

FIG. 14-5 *The rotation of the sun shown by four exposures of the disk. Photographed by Miss Calvert with the 12-inch telescope on August 7, 9, 11, and 13, 1917.* (*Yerkes Observatory photograph*)

radiation emitted by the sun since there is a good deal of energy in the infrared, in the ultraviolet, and in the radio parts of the spectrum. However, we shall discuss these points when we consider the general properties of stellar radiation.

14.7 Rotation of the Sun

Since there are markings such as sunspots (see Section 20.7) which remain visible on the surface of the sun for fairly long periods of time, we can determine the period of rotation by studying these spots. The rotation of the sun can also be detected and measured spectroscopically. Since one edge of the sun's disk is approaching us and the other edge is receding, the lines in the solar spectrum (see Chapter 20) show a Doppler effect which varies as we move across the sun's surface. An analysis of this gives the period of rotation of the sun. As seen from the earth, these spots travel from east to west (the same direction in which the planets revolve around the sun), and take about four weeks to line up with the earth again, as shown in Figure 14.5. Since this is the synodic period of the sun's rotation, we can find the sidereal period from it by using formula (6.1). This turns out to be different for different solar latitudes, a fact that has been known for a long time. In 1860, Carrington discovered that there is a systematic variation in the sun's period of rotation as one moves from the equator to the poles. The fastest rotation occurs at the equator, where the mean sidereal period is about 25 days. A point on the sun's equator has a speed of rotation of about 2 km/sec.

At a solar latitude of 30° the period of rotation increase to 25.85 days, and to 27.84 days at 40°. At the poles the period of rotation is estimated to be about 34 days. We shall discuss the means of accounting for this when we consider the general properties of stellar rotation in a later chapter.

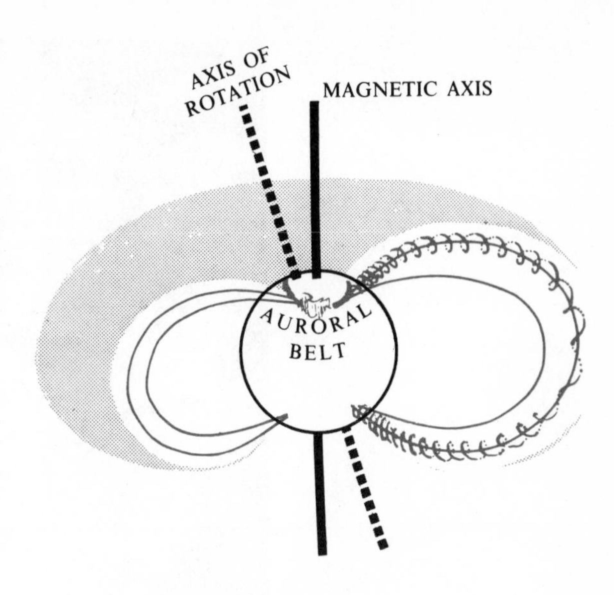

FIG. 14–6 *A charged particle circling around the earth's magnetic lines of force.*

14.8 Solar Influence on the Earth

Aside from the dynamical effects that the sun, in virtue of its great mass, has on the earth, and the climatic effects resulting from the solar radiation striking the earth's surface, other important atmospheric phenomena are definitely associated with the sun. A study of the photosphere of the sun shows a continuous variable activity ranging from sunspots to minor and irregular disturbances which develop and die out in a matter of hours or days. These variable phenomena affect the earth's atmosphere in a way that is most discernible in disrupted radio communications.

Since the ionization in the ionized layers of the earth's atmosphere is due primarily to the ultraviolet radiation coming from the sun, disturbances in these layers occur whenever variations in the solar radiation are strong enough. Changes in these ionized layers have sudden and dramatic effects on radio transmission and reception on the earth's surface, so that it is easy to detect the important outbursts on the sun's surface by recording these radio disturbances.

It is clear from the nature of these disturbances that solar outbursts such as flares (see Section 20.17) are accompanied by intense ultraviolet emissions and storms of charged corpuscles, such as electrons, protons, and other ions. When such charged corpuscles have the right amount of kinetic energy, they are trapped in the earth's magnetic field between the magnetic poles, and thus form the belts of ionized particles that were first discovered by van Allen. Not only are the ionized layers affected by solar phenomena, but the magnetic field of the earth itself is influenced by the charged particles streaming from the sun.

We generally find that a day after a flare has been observed on the surface of the sun the earth's magnetic field undergoes erratic variations. These variations are called magnetic storms, and at times they are so violent that they induce electric currents in the earth itself which interfere with long-distance telephone and telegraph communications. Flares are generally associated with sunspots, and the most violent ones occur in the neighborhood of large sunspot groups.

The corpuscles emitted by the sun are also responsible for the aurora borealis or northern lights (in the southern hemisphere called aurora australis), which can be seen either as isolated rays and streamers or as luminous arches across the northern sky. These aurorae are seldom seen more than 40° away from the magnetic poles and tend to occur with maximum frequency at about half that distance. The rays and streamers constituting the northern lights can be observed to follow the magnetic lines of force, so that there is no doubt that these phenomena are associated with the charged solar corpuscles that follow the magnetic lines of force in helical paths. Charged corpuscles of extremely high energy can pass through the magnetic lines of force that surround

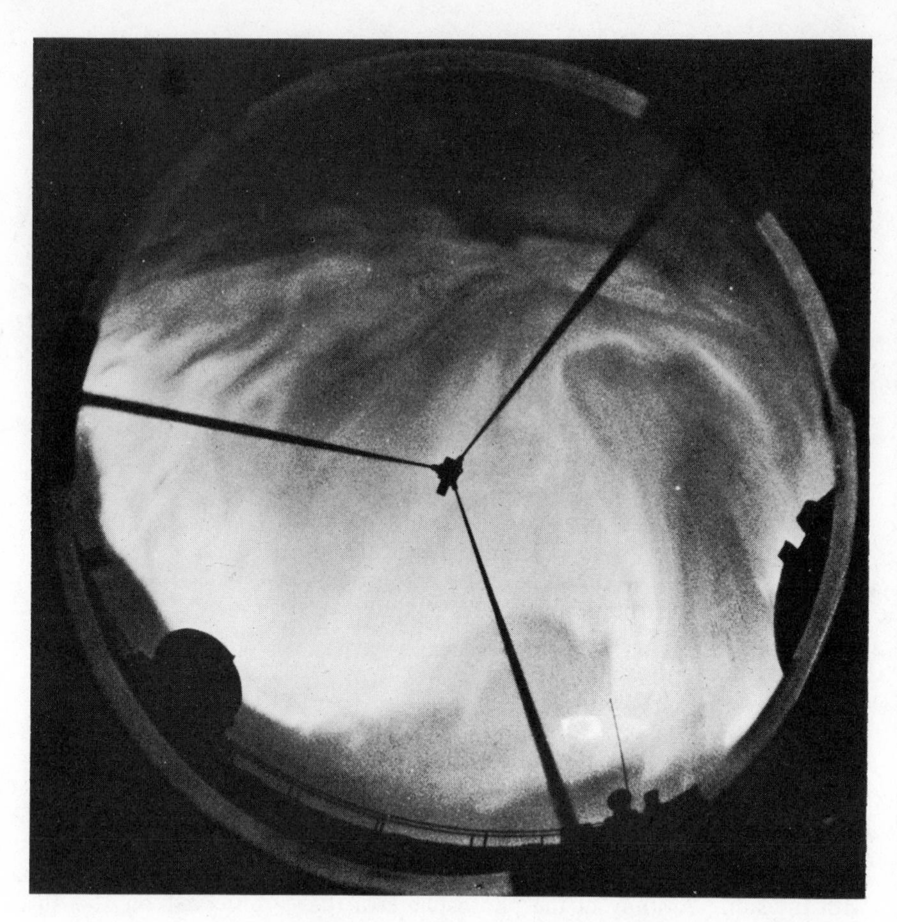

FIG. 14-7 *The great Aurora of August 19–20, 1950, photographed by Sharpless with the Greenstein-Henyey 140° camera. Exposure 15 minutes. (Yerkes Observatory photograph)*

the earth, but a charged particle that does not have enough energy to do this spirals around a line of force and strikes the earth's atmosphere near either the north or the south magnetic pole (see Figure 14.6). As a result the molecules and atoms in the earth's atmosphere are set vibrating, emitting radiation which we observe as an aurora. See Figure 14.7.

14.9 Solar Cosmic Rays

In addition to solar corpuscles of rather small energy [they travel at speeds of 2 000 mi/sec (3.2×10^3 km/sec)], which are responsible for the magnetic storms and aurorae, there are solar corpuscles that come into the earth's atmosphere with energies anywhere from 100 000 to millions of times as great. These high-energy particles are called **cosmic rays**, and although the bulk of the cosmic rays come from interstellar space, we now know that the sun also emits them in quantity. During the last 20 years particularly brilliant solar flares have been observed to be accompanied an hour or two later by great increases in cosmic-ray intensity. It is clear from this that the sun sends cosmic-ray particles to the earth in addition to charged particles of small energy.

14.10 Solar Wind

In Section 13.9 we discussed the effect of the solar wind on particles in a comet, and indicated that definite evidence for this solar corpuscular stream was obtained from Mariner II. Data from the Mariner II solar plasma experiment were received continuously from October 29 to 31, 1962, and an analysis of these data has shown that there is a continuous and measurable radial flow of plasma

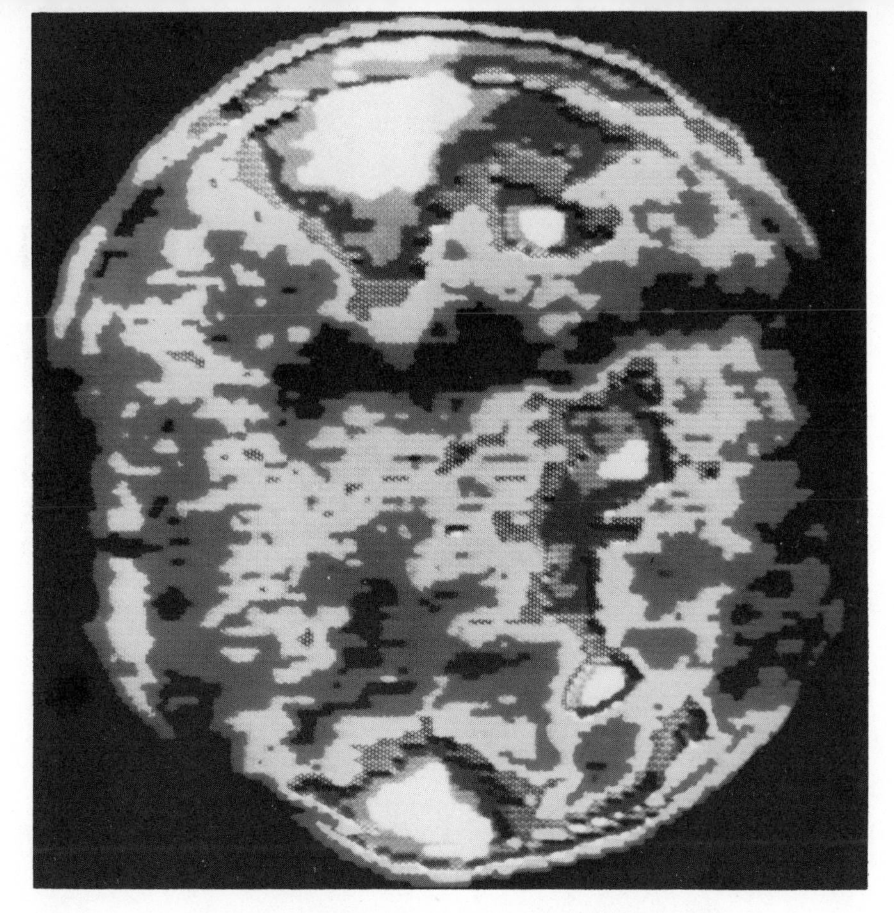

FIG. 14–8 *Relative intensity of solar regions is shown in tones of gray in this TV-monitor display of data from the X-ray/ultraviolet spectroheliograph on Orbiting Solar Observatory 7. White is greatest intensity. (From the National Aeronautics and Space Administration)*

from the sun. This plasma consists of streams of charged particles (probably protons; see Section 18.10) with velocities ranging from 314 km/sec to 1 250 km/sec. The mean velocity of these charged particles is in the neighborhood of 625 km/sec. This is a greatly simplified description of a very complex phenomenon. The actual flow of the charged particles in the solar wind is quite irregular and consists of bursts of varying intensity and thickness. These bursts are clearly related to storms on the surface of the sun. We shall see later in our discussion of the formation of stars that this outward flow of matter from a star's surface plays an important role in the star's evolution. Ultraviolet and X rays are also emitted from the sun (Fig. 14.8).

EXERCISES

1. If all the planets were lined up at their mean distance from the sun on the same side of the sun, what acceleration would the sun experience?

2. If all the planets in the solar system were to fall into the sun, how and to what extent would the rotation of the sun be altered?

3. If you were on Jupiter, how large would your solar constant be?

4. If two sunspots are on the same meridian on the sun but one is at the equator and the other at latitude 40°, how long will it be before the two spots are on the same meridian again?

5. If a clock were on the surface of the sun, by how much would it be slowed down as compared with a similar clock far away from any ponderable bodies?

6. By how much would a ruler shrink on the surface of the sun if it were held in a vertical position?

7. How much does the gravitational field on the surface of the sun alter the speed of light?

PART TWO

STELLAR PROPERTIES AND THE STRUCTURE OF STARS

15

STELLAR DISTANCES, LUMINOSITIES, AND COLOR

15.1 Stellar Astronomy and Astrophysics

When astronomers, starting with Sir William Herschel, undertook detailed and serious studies of the stars, investigations grew into two seemingly independent branches of astronomy. *On the one hand we have* **stellar astronomy**, *the older of the two branches, which is concerned primarily with the way stars are distributed in space, and how they arrange themselves to form such groups as galaxies, globular clusters, and so on. On the other hand, the more recent branch,* **astrophysics**, *is concerned with the properties of stars as individuals and with how stars maintain themselves in equilibrium under the vast forces that are at work in their interiors.*

For a long time these two branches of investigation proceeded quite independently of each other since it appeared that the position of a star in the Milky Way, and how it happened to be moving, were in general unrelated to its internal structure. However, during the last few decades, it has become more and more evident that the two branches are quite intimately related, and that one cannot hope to understand stellar dynamics or galactic structure without a deep understanding of the astrophysics of individual stars. Although there was already some indication of this in the analysis of stellar motions in the early 1930's, it was only after Baade had discovered the existence of various stellar populations that the intimate relation between astrophysics and stellar astronomy was clearly evident. The correlation between these two branches of astronomy is still more dramatically revealed in the recent theoretical work on stellar evolution and on the formation of heavy elements in stellar interiors.

Superimposed on these considerations, we have the more general cosmological questions, and here, too, we shall see that the answers to very important problems can be obtained only through a thoroug understanding of the astrophysics of individu stars. It is clear that if we are to have a prop understanding of galactic structure and cosmolog we must begin by studying the star as an individua For it is only when we know individual characte istics of stars in detail that we are able to unde stand the communal properties of stars an ultimately the structure of the universe.

15.2 Stellar Properties

Just as we find a wide variation among individua in any category (for example, human being animals of any species, etc.) so we find suc variations among stars. In order to see how w may correlate various stellar features that diff from star to star, we must consider these feature in some detail. We shall try to study them in logical order so that the knowledge and measur ment of one property may assist us in understandin those that follow. Ultimately, we shall present a analysis of each feature in sufficient detail, so tha altogether, they give us as complete a picture of star as we can get.

The order in which we propose to consider thes characteristics is the following:

1. distances;
2. brightnesses (magnitudes):
   (a) apparent,
   (b) absolute—luminosities;
3. colors—color index;
4. surface temperatures;
5. spectral classes;
6. atmospheres;
7. radii;

8. masses and densities;
9. interior structure and chemistry;
10. evolution of stars;
11. properties of variable stars.

'e shall find it necessary as we go along to use the ws of physics just as we did in the section on the lar system; and just as we proceeded there, we all introduce these laws and explain them on the sumption that the student has had no previous ontact with physics. In this chapter we begin by onsidering distances, brightnesses, and colors, nce we can introduce these stellar characteristics using the physical concepts we have already scussed.

.3 The Definition of Stellar Parallax

Although the distance of a star from us is a rely accidental characteristic, and has no relationip to the star's intrinsic properties, it is, neverthess, one of the most important things we can learn out a star, since it is only after we have discovered ow far away a star is that we can measure many of s other properties.

In discussing the motion of the earth (Chapter 5) e mentioned that one method of convincing urselves that the earth is moving is by observing e apparent shifting of nearby stars as the earth volves around the sun. We shall now see how ne can use these observations to find the distances f stars that are not too far away. In Figure 15.1 t S be the position of the sun, E_1 and E_2 two early opposite points on the earth's orbit, and P e position of a star. If the observer views the ar first from position E_1 and then from position $_2$, it appears to shift against the distant background stars by the angle α. Half of this angle α, xpressed in seconds of arc, is called the parallax, ", of the star. *We therefore define the trigonometric arallax of a star as the angle that is subtended at e star by the mean distance between the earth and e sun, i.e., by the astronomical unit.*

5.4 Using Photographic Plates to Measure Parallaxes

This angle is not measured visually, but by eans of telescopes and photographic plates. et us suppose that at position E_1 a telescope is irected at the star in question, so that an image of is formed exactly at the center of the photo-

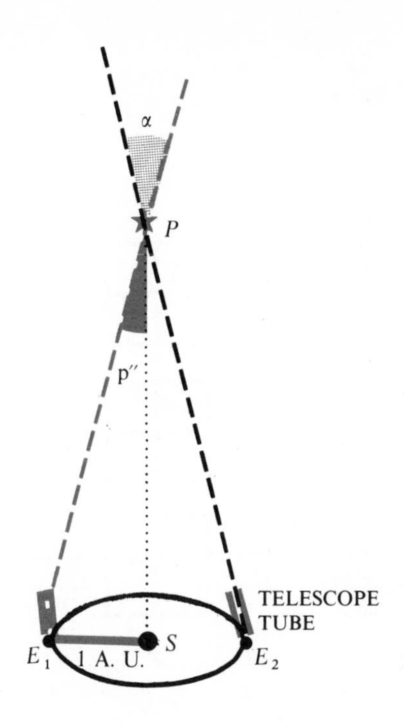

FIG. 15–1 *The parallax of a star is determined by observing it as photographed from two diametrically opposite positions in the earth's orbit.*

graphic plate (i.e., the light from the star enters the telescope parallel to the optic axis of the telescope). We suppose now that without changing the orientation of the telescope, an image of the star is again obtained on a second photographic plate when the earth is in position E_2. Since the direction of the optic axis of the telescope in position E_2 is parallel to its direction in position E_1, a ray of light from the star entering the telescope in position E_2 makes an angle α with the optic axis, and *forms an image of the star on the photographic plate at a point displaced by a small amount from the center of the photographic plate as shown in Figure 15.2. Since this angle α is twice the parallax of the star we can find the parallax by measuring this angle.*

We do this by first measuring the amount OS by which the image on the second photographic plate is displaced with respect to the center of this plate (i.e., with respect to the image of the star on the first photographic plate). If we then divide this displacement OS by the focal length F of the telescope (which is just the distance of the photographic plate behind the objective lens), we obtain the angle α in radians. If we then take one-half this value and change to seconds of arc, we have the parallax of the star. It should be noted that the

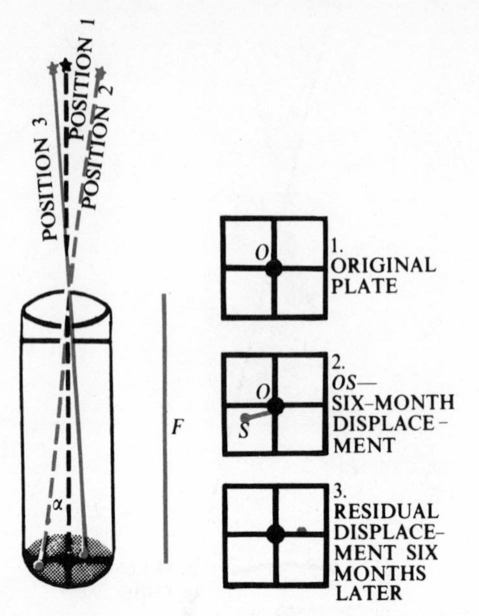

FIG. 15–2 *The positions of a star image on three different photographic plates (taken at 6-month intervals) are used to measure the parallax of the star.*

longer the focal length of the telescope, the bigger the displacement OS on the photographic plate. It is, therefore, advantageous to work with long-focal-length telescopes if one wants to make precise parallactic measurements. The modern procedure in this field is based on a thorough analysis of the entire problem of parallactic measurements made in 1903 and thereafter by Frank Schlesinger, who worked with the 40-inch Yerkes refractor.

15.5 Three or More Sets of Photographs Required for Parallax Measurement

Two photographs taken six months apart will not give us the correct parallax of a star if the star moves with respect to the solar system while the earth goes from E_1 to E_2. Therefore, at least one more photograph must be taken when the earth is back in position E_1, and this third photograph must be compared with the first one. If the star image is not back at the center of this third plate, we know that part of the displacement OS is due to the motion of the star, the motion of the entire solar system, or both. Half of this residual displacement of the star image on the third photographic plate (the displacement occurring in six months) must be subtracted from OS to obtain the correct parallax. *Actually, for each star whose parallax is to be determined, accurate work requires five or more sets of photographs.* Each of these sets consists of several plates taken six months apart.

15.6 The Relative Stellar Parallax

Another point must be made in connection with this procedure. Since different photographic plates are used, perhaps having different dimensions or inserted differently, and since the photographs are taken at different times of the year when temperatures are different (so that thermal effects may change properties of the telescope and the photographic plates), small spurious displacements may arise. Moreover it is impossible in practice to point the telescope exactly parallel to its initial direction after six months. Because of these errors we do not compare one photographic plate directly with another, but rather measure how much the given star is displaced with respect to certain very distant background stars on the *same* photographic plate.

These background stars, although they may appear to be close, are really at great distances from us and hence have very small parallaxes. If there are any spurious displacements in the image of the parallax star resulting from the errors described above, the distant background star images are displaced by the same amount. Hence, the errors can be detected. *By measuring the position of the parallax star with respect to these distant background stars, we eliminate these errors. The parallax we obtain in this way is called the* ***relative parallax****.* In order to obtain the absolute parallax we must have some way of determining the parallax of these distant background stars. We shall see later how this can be done. *By adding the background parallaxes to the relative parallax of the star, the star's* ***absolute parallax*** *is then obtained.*

One must also be careful in parallax work to take account of atmospheric refraction that occurs when the light from the star enters the atmosphere, since this affects the position of the star on the photographic plate. We can do this by always taking the photographs when the given star is at the same hour angle (that is, in the same part of the sky).

15.7 Errors in Parallax Measurements

What we have outlined above is a simplified version of the actual procedure used by the observational astronomer. The actual photographic sets are not taken precisely at six-month intervals but on four or five successive nights at both positions, E_1 and E_2. In parallax measurements the observer must also take note that the sky changes its appearance in six-month intervals and that certain stars that are visible from position E_1 will not be visible from E_2 because they will then be in the daylight zones. Therefore, for parallax measurements one generally chooses the time of the year when the stars being considered are near quadrature.

Since all physical measurements are burdened with errors arising from the fallibility of the scientist himself and from the imperfections in the apparatus used, *each measured parallax will have a definite error associated with it. From the measured parallaxes of many stars we find the average probable parallax error is of the order of 0″.005.* In the case of the nearest star, α Centauri, which has an absolute parallax of 0″.76, this error is less than 1 per cent, but the percentage error goes up rapidly as we go up to the more distant stars. Thus if a star has a measured parallax of 0″.01, the average error is 50 per cent, so that we must suppose that the true value of the parallax has a 50 per cent chance of being somewhere between 0″.015 and 0″.005. *It is clear from this that when we go to stars whose parallaxes are smaller than 0″.01 the percentage error is much too large for the measurement to be of much value.* We shall see, however, that there are indirect ways of obtaining the parallaxes of these distant stars which depend on first determining their intrinsic luminosities. Because of the enormous difficulties and the great precision involved, direct measurements of parallaxes have been made for no more than about 6 000 stars, although at the present time the parallax program is being pursued intensively at many observatories. Since images of many stars are obtained on a single photographic plate, we can, under favorable circumstances, use the same sets of photographic plates to determine the parallaxes of several nearby stars at the same time. The most extensive list of directly measured stellar parallaxes is given in the *General Catalogue of Trigonometric Parallaxes* of the Yale Observatory. In Table 15.1 we give a list of the parallaxes of nearest stars with some additional data included.

15.8 Stellar Distances

Once we know the parallax of a star, we can determine its distance by simple geometry. Since the parallax of even the closest star is very small, we may treat the lines from the star to the earth and the star to the sun (i.e., the lines EP and SP, as shown in Figure 15.3), either one of which may be taken as the distance d of the star, as though they were the radii of a circle having the star P as its center. We may then treat the distance of the earth from the sun (i.e., the astronomical unit A) as a small arc of this circle, and hence write

$$\frac{\alpha}{2} d = A,$$

or

$$d = \frac{A}{\alpha/2},$$

where $\alpha/2$ is the parallax expressed in radians. Since we wish to express this distance in terms of the parallax as measured in seconds, we must replace $\alpha/2$ by its value in terms of p''. But this is just $p''/206\,265$ *since there are 206 265 seconds in one radian.* Hence, the distance of a star is

$$\boxed{d = \frac{206\,265 A}{p''}}. \tag{15.1}$$

Note that in this formula the parallax is expressed in seconds of arc, and the distance in whatever units we choose for the astronomical unit, A. This formula means that if the star has a parallax of 1 second of arc, its distance from the sun is 206 265 times the astronomical unit.

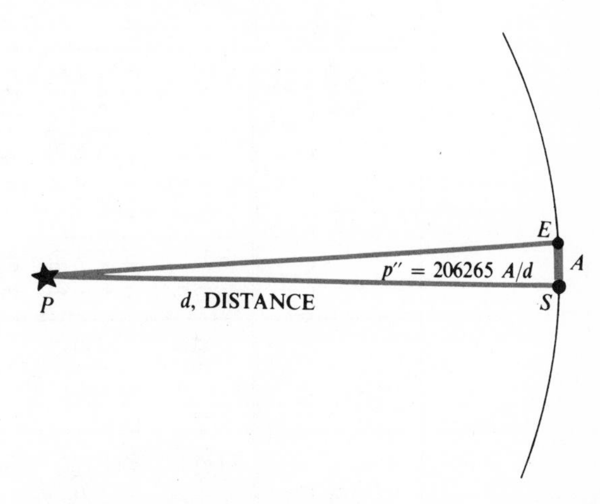

FIG. 15–3 *The relationship between the parallax, the distance, and the astronomical unit is shown.*

Table 15.1

FIFTY NEAREST STARS

Star		Epoch 1900 α	Epoch 1900 δ	App. mag. m_v	Color index	Abs. mag. M_v	Spectral class*	Proper motion μ	Direction of proper motion, θ, clockwise from N	Parallax p''	Radial velocity v_r	Mass $\mathscr{M}$	Radius $\mathscr{R}$	Notes†
		h m	° ′					″/y	°	1 000 p''	km/s	$\mathscr{M}_\odot$	$\mathscr{R}_\odot$	
Grm 34 = +43° 44	A	0 13	+43 27	8.1	1.55	10.3	M2	2.89	82	278	+ 14			A = sp db
(= CC19)	B	0 13	+43 27	11.1	1.78	13.3	M5	2.89	82	278	+ 21			AB = 38″
v. Maanen = Wolf 28		0 44	+ 4 55	12.3	0.55	14.2	wF5	2.949	155.4	235	+260?			
L726–8 [13]	A	1 34	−18 28	12.4	1.9	15.3	M6e	3.36	80	375	+ 29	0.044		a = 5″, P = 200y
(= UV Cet)	B	1 34	−18 28	12.9	1.9	15.9		3.36	80	375		0.035		B = Flare st.
τ Cet		1 39	−16 28	3.5	0.72	5.7	sdG8	1.934	296.5	275	− 16			
ε Eri		3 28	− 9 48	3.7	0.87	6.1	K2	0.990	271.7	303	+ 15			
o^2 Eri = 40 Eri	A	4 11	− 7 49	4.4	0.81	5.9	K1	4.08	213	201	− 42	0.11		A db? 0″01, 3y
	B	4 11	− 7 49	9.6		11.2	wA	4.08	213	201	− 42	0.44	0.018	BC a = 6″9
	C	4 11	− 7 49	11.1		12.6	M5e	4.08	213	201	− 45	0.21	0.43	P = 248 y
Kapteyn = −45° 1 841		5 08	−44 59	8.9		10.9	sdM0	8.73	131	251	+242			Nearest subdwarf
Ross 614	A	6 24	− 2 44	11.3		13.2	M5	1.00	131	249	+ 24	0.14		a = 1″0
(= CC390)	B	6 24	− 2 44	14.8		16.8		1.00	131	249		0.08		P = 16.5 y
Sirius = αCMa	A	6 41	−16 35	− 1.5	0.01	1.4	A1	1.32	204	375	− 8	2.31	1.8	a = 7″6
	B	6 41	−16 35	8.7		11.5	wA5	1.32	204	375		0.98	0.022	P = 49.9 y
Luyten = +5° 1 668		7 22	+ 5 32	9.9		12.0	M4	3.75	171	266	+ 26			db?
Procyon = αCMi	A	7 34	+ 5 29	0.4	0.40	2.6	F5	1.25	214	287	− 3	1.75	1.7	a = 4″5
	B	7 34	+ 5 29	10.8		13.1	wF	1.25	214	287		0.64	0.01	P = 40.6 y
Grm 1 618 = +50° 1 725		10 05	+49 57	6.6	1.37	8.3	M0	1.464	249.8	222	− 27			
AD Leo = +20° 2 465		10 14	+20 22	9.4	1.55	11.0	M4	0.429	258	212	+ 10			Flare st. invis. comp.
Wolf 359		10 52	+ 7 36	13.7		16.8	M6e	4.71	235	425	+ 13			
Lal 21 185 = +36° 2 147		10 58	+36 38	7.5	1.51	10.4	M2	4.78	187	398	− 87	0.35		Invis. comp. $\mathscr{M}$ = 0.03
Ross 128		11 43	+ 1 23	11.1		13.5	M5	1.38	153	298	− 13			

* w = white dwarf, sd = subdwarf.

† P = period, a = semimajor axis of secondary relative to primary, db and trip = double and triple star with invisible components, sp = spectroscopic.

Star	Epoch 1900 α	Epoch 1900 δ	App. mag. m_v	Color index	Abs. mag. M_v	Spectral class*	Proper motion μ	Direction of proper motion, θ, clockwise from N	Parallax p''	Radial velocity v_r	Mass $\mathscr{M}$	Radius $\mathscr{R}$	Notes†
	h m	° ′					″/y	°	1 000 p''	km/s	$\mathscr{M}_\odot$	$\mathscr{R}_\odot$	
Wolf 424 A	12 28	+ 9 34	12.6		14.4	M6	1.87	276	230	− 5			}0″.5
B	12 28	+ 9 34	12.7		14.5	M7	1.87	276	230				
+15° 2 620	13 41	+15 26	8.5	1.44	10.0	M3	2.30	129	201	+ 15			
Proxima Cen	14 23	−62 15	10.7		15.1	M5e	3.85	262	763		0.1		Flare st. nearest st.
α Cen A	14 33	−60 25	0.0	0.69	4.4	G2	3.69	281	752	− 22	1.09	1.23	}a = 17″.7
B	14 33	−60 25	1.4		5.8	K4	3.69	281	752		0.88	0.87	}P = 80.1 y
−12° 4 523 = CC995	16 25	−12 25	10.1	1.60	12.0	M4	1.18	182	244	− 13			
−46° 11 540	17 21	−46 47	9.4	1.5	11.0	M4	1.04	147	213				
−44° 11 909	17 30	−44 14	11.1		12.7	M5	1.16	217	209				
Barnard = +4° 3 561	17 53	+ 4 25	9.5	1.75	13.2	M5	10.27	356	545	−108			db. 0″.06, 1.1 y
+59° 1 915 A	18 42	+59 29	8.9	1.54	11.2	M4	2.28	324	283	+ 1			}17″
(= Σ2 398) B	18 42	+59 29	9.7	1.58	12.0	M5	2.28	324	283	+ 14			
Ross 154	18 44	−23 54	10.6		13.3	M5	0.72	103	350	− 4			Flare star
61 Cyg A	21 02	+38 15	5.2	1.21	7.5	K5	5.20	52	292	− 64	0.59		}a = 24″.6, 720 y,
B	21 02	+38 15	6.0	1.4	8.4	K7	5.20	52	292	− 63	0.50		}? invis. trip.
−39° 14 192	21 11	−39 15	6.7	1.42	8.7	M0	3.47	251	256	+ 20			
−49° 13 515	21 27	−49 26	8.9		10.6	M3	0.81	185	219	+ 18			
ϵ Ind	21 56	−57 12	4.7	1.05	7.0	K5	4.69	123	286	− 40			
Krüger 60 = DO Cep A	22 24	+57 12	9.8	1.63	11.8	M3	0.86	246	250	− 24	0.27	0.51	}a = 2″.4, 45 y
(= +56° 2 783) B	22 24	+57 12	11.4		13.4	M4e	0.86	246	250	− 28	0.16		}B = Flare st.
L789–6	22 33	−15 52	12.6		14.9	M6e	3.260	45.2	296	− 60			
−21° 6 267 A	22 33	−21 08	9.3		11.0	M2	0.46	99	219	− 8			}23″
B	22 33	−21 08	11.0		12.7	Me	0.46	99	219				
−15° 6 290 = Ross 780	22 48	−14 47	10.2	1.60	11.7	M5	1.11	123	206	+ 9			
−36° 15 693	22 59	−36 26	7.4	1.50	9.6	M2	6.90	79	273	+ 10			
Ross 248	23 37	+43 39	12.2	1.8	14.7	M6e	1.82	176	316	− 81			
−37° 15 492	23 59	−37 51	8.6	1.48	10.3	M3	6.08	113	219	+ 24			

If A is expressed in miles, the distance of the star is given in miles as

$$d\,(\text{in miles}) = \frac{206\,265'' \times 93\,000\,000}{p''}. \quad (15.2)$$

Thus a star with a parallax of 1 second of arc would be at a distance of 19 000 000 000 000 miles.

Since stellar distances expressed in miles are too cumbersome to use, we introduce a larger unit. If we take the astronomical unit itself as our distance unit, i.e., if we place $A = 1$ in our formula, we obtain the distance of a star in astronomical units as

$$d\,(\text{in astronomical units}) = \frac{206\,265}{p''}. \quad (15.3)$$

We see from this that the astronomical unit is also too small to be used conveniently for stellar distances.

15.9 The Light Year and the Parsec

There is a natural unit which is considerably larger: the ***light year****—the distance that light travels in one year.* We can obtain the distance of a star in light years by expressing A itself in light years. Since light takes about 8 minutes to reach us from the sun, A is a very small fraction of the light year (it is the same fraction of a light year that 8 minutes is of one year) and our formula for the distance of a star in light years becomes

$$d\,(\text{in light years}) = \frac{3.26}{p''}. \quad (15.4)$$

Another unit of distance that is in common use in astronomy is the **parsec**, *which is defined as 206 265 times the astronomical unit.* Hence the distance of a star in parsecs is

$$d\,(\text{in parsecs}) = \frac{1}{p''}. \quad (15.5)$$

This is the simplest formula to use for stellar distance since we obtain *the distance in parsecs by inverting the parallax expressed in seconds.*

15.10 The Distances of Naked-Eye Stars

An examination of Tables 15.1 and 15.2 shows us first that most of the naked-eye stars are more than 10 parsecs away from us, and that within a distance of $3\frac{1}{2}$ parsecs there are about 20 known stars altogether, 10 of which are in the form of pairs (5 double stars). Of these 20 stars, 8 are visible to the naked eye, but most of the visible stars in the northern and southern hemispheres are 100 or more light years away from us. The closest star to us (excluding the sun) is α Centauri, which is really a double star, about which a third member, Proxima Centauri, revolves.

15.11 The Brightness of a Star

Now that we know how to determine the distance of a star, we can measure its luminosity, i.e., how much energy in the form of radiation it pours out per second in all directions. However, in order to do this we must first introduce a method of measuring a star's brightness.

Here we must differentiate between two ideas: the **apparent brightness**, which refers to the way a star appears to us and which does not take its distance from us into account, and the **intrinsic** or **absolute brightness**, which refers to the way a star would appear to us if it were at some standard distance.

15.12 Apparent Brightness and the Magnitude Concept

The apparent brightness is a measure, on some acceptable scale, of the amount of radiant energy per second from the star that strikes a square centimeter of a sensitive surface (the retina of an eye, a photoelectric cell, a photoplate, etc.) placed at right angles to the direction of the star. The scale that we use in measuring the apparent brightness of stars is not based on an energy unit such as the erg, but rather on a system in which certain stars themselves are taken as standards. In other words we introduce a scale in which all the stars are compared with certain standard stars.

This procedure goes back to Hipparchus, who catalogued over 1 000 naked-eye stars about 120 B.C. In his catalogue Hipparchus arranged the stars according to how important they appeared to him and he classified those which appeared brightest as the most important. Thus, he spoke of stars of the first magnitude (meaning the first importance), second magnitude (second importance), and so on,

TABLE 15.2

ONE HUNDRED VISUALLY BRIGHTEST STARS

Star		Epoch 1900 α		Epoch 1900 δ		App. mag. m_v	Color index	Abs. mag. M_v	Spectral class	Proper motion 1 000 μ'' yr	Direction of proper motion, θ, clockwise from N	Distance d	Rad. veloc.* v_r	Notes†
		h	m	°	′						°	pc	km/s	
Alpheratz	α And	0	03	+28	32	2.1	−0.07	−0.5	B9p III	211	139	31	−12 v	sp. var, 96.7 d, Mn star
Caph	β Cas	0	04	+58	36	2.3	+0.34	+1.5	F2 IV	557	109	14	+11.7	
Ankaa	α Phe	0	21	−42	51	2.4	+1.07	+0.2	K0 III	444	154	27	+75 v	sp. and vis. db, 3 849 d
Schedar	α Cas	0	35	+55	59	2.2	+1.16	−1.3	K0 II–III	59	121	50	− 3.7	irreg. var.
Diphda	β Cet	0	39	−18	32	2.0	+1.01	+0.8	K0 III	234	80	18	+13.1	
Cih	γ Cas	0	51	+60	11	2.2	−0.2	−0.9	B0e IV	28	94	40	− 7 v	vis. db 2″, irreg. var.
Mirach	β And	1	04	+35	05	2.1	+1.62	+0.2	M0 III	213	122	24	+ 0.3	
Polaris	α UMi	1	23	+88	46	2.0	+0.6	−4.5	F8 Ib	45	95	200	−17 v	sp. db, 30 y, var. 4.0 d
Achernar	α Eri	1	34	−57	45	0.5	−0.17	−2.2	B5 IV	96	108	35	+19	
Almach	γ And	1	58	+41	51	2.2	+1.3	−2.3	K3 II–III	70	138	80	−11.7	tr. vis. and sp. db, 10″, 55 y
Hamal	α Ari	2	02	+22	59	2.0	+1.17	+0.3	K2 III	241	127	22	−14.2	
Mira	o Cet	2	14	− 3	26	2.0	+1.5	−1.0	M6e III	234	181	40	+63	var. 332 d, vis. db (close)
Menkar	α Cet	2	57	+ 3	42	2.5	+1.16	−1.0	M2 III	75	187	50	−25.9	
Algol	β Per	3	02	+40	34	2.1	−0.05	−0.5	B8 V	8	125	31	+ 5 v	ecl., sp. tr., 2.87 d, 1.87 y
Mirfak	α Per	3	17	+49	30	1.8	+0.48	−4.1	F5 Ib	36	132	150	− 2.4	
Aldebaran	α Tau	4	30	+16	19	0.8	+1.55	−0.8	K5 III	203	160	21	+54.2	vis. db, 31″
Capella	α Aur	5	09	+45	54	0.1	+0.81	−0.6	G8, G0	436	168	14	+30 v	sp. db, 104 d
Rigel	β Ori	5	10	− 8	19	0.1	−0.05	−7.0	B8 Ia	2	135	270	+22 v	qu., vis. db 9″, sp. db 22 d
Bellatrix	γ Ori	5	20	+ 6	16	1.6	−0.22	−4.1	B2 III	17	200	140	+18.2	+
El Nath	τ Tau	5	20	+28	32	1.7	−0.13	−2.9	B7 III	179	169	80	+ 8.0	
Mintaka	δ Ori	5	27	− 0	22	2.2	−0.21	−6.0	O9.5 II	3	146	450	+17 v	ecl., tr., vis. db 53″, sp. db 5.7 d
Arneb	α Lep	5	28	−17	54	2.6	+0.22	−4.8	F0 Ib	6	30	300	+24.7	
Alnilam	ε Ori	5	31	− 1	16	1.7	−0.18	−6.8	B0 Ia	1	180	500	+26.1	
Alnitak	ζ Ori	5	36	− 2	00	1.8	−0.21	−6.2	O9.5 Ib	7	131	400	+18.1	tr., vis. db 2″.5, sp. db 19 d
Saiph	κ Ori	5	43	− 9	42	2.1	−0.16	−7.1	B0.5 Ia	6	149	700	+20.8	
Betelgeuse	α Ori	5	50	+ 7	23	0.4	+1.85	−5.9	M2 I	30	74	180	+21 v	semi-reg. var. 5.8 y, db 40″
Menkalinan	β Aur	5	52	+44	56	1.9	+0.04	−0.2	A2 V	49	264	26	−18 v	ecl. sp. db 3.96 d
Mirzam	β CMa	6	18	−17	54	2.0	−0.23	−4.5	B1 II–III	4	270	200	+34 v	sp. var. 0.25 and 42 d
Canopus	α Car	6	22	−52	38	−0.7	+0.16		F0 I–II	24	56		+20.4	
Alhena	γ Gem	6	32	+16	29	1.9	0.00	−0.5	A0 IV	66	136	30	−12 v	sp. db 2 175 d

* v = variable, + = distance increasing, i.e., red shift.

† db = double, tr = triple, qu = quadruple, sp = spectroscopic, vis = visual, var = variable.

Star		Epoch 1900				App. mag. m_v	Color index	Abs. mag. M_v	Spectral class	Proper motion 1 000 μ'' yr	Direction of proper motion, θ, clockwise from N	Dis-tance d	Rad. veloc.* v_r	Notes†
		α		δ										
		h	m	°	′						°	pc	km/s	
Sirius	α CMa	6	14	−16	35	−1.4	−0.01	+1.41	A1 V	1 324	204	2.7	− 8 v	vis. db 49.9 y, 9″
Adhara	ε CMa	6	55	−28	50	1.5	−0.17	−5.0	B2 II	4	135	200	+27.4	vis. db 8″
Wezen	δ CMa	7	04	−26	14	1.8	+0.63	−7.0	F8 Ia	5	292	600	+34.3	
Aludra	η CMa	7	20	−29	06	2.4	−0.07	−7.1	B5 Ia	8	294	800	+40.2	
Castor	α Gem	7	28	+32	06	1.6	+0.05	+0.8	A1, A5	202	237	14	+ 3 v	vis. tr., each sp. db
Procyon	α CMi	7	34	+ 5	29	0.4	+0.41	+2.7	F5 IV–V	1 247	214	3.5	− 3.2	vis. db 4″ 41 y
Pollux	β Gem	7	39	+28	16	1.2	+1.01	+1.0	K0 III	624	264	10.7	+ 3.4	nearest giant
Naos	ζ Pup	8	00	−39	43	2.2	−0.27	−7.3	O5	33	282	800	−24	
	γ Vel	8	06	−47	03	1.9	−0.25	−4.2	WC7	8	290	160	+35	vis. db 41″
Avior	ε Car	8	20	−59	11	1.9	+1.2	−3.1	K0, B	30	296	100	+11.5	db, composite sp.
	δ Vel	8	42	−54	21	1.9	+0.04	+0.1	A0 V	89	170	23	+ 2.2	vis. tr. 3″, 69″
Suhail	λ Vel	9	04	−43	02	2.2	+1.7	−4.3	K5 Ib	25	281	200	+18.4	
Miaplacidus	β Car	9	12	−69	18	1.7	−0.01	−0.4	A0 III	186	301	26	− 5	
Scutulum	ι Car	9	14	−58	51	2.2	+0.18	−4.2	F0 Ib	21	266	180	+13.3	
	κ Vel	9	19	−54	35	2.5	−0.16	−3.0	B2 IV	12	275	120	+22 v	sp. db 117 d
Alphard	α Hya	9	23	− 8	14	2.1	+1.43	−0.7	K4 III	35	333	35	− 4.3	
Regulus	α Leo	10	03	+12	27	1.3	−0.11	−0.8	B7 V	248	269	26	+ 3.2	vis. tr. 177″, 3″
Algeiba	γ Leo	10	14	+20	21	2.0	+1.2	−0.5	K0p III	349	119	32	−36.7	vis. db 400 y, high vel.
Merak	β UMa	10	56	+56	55	2.4	−0.02	+0.6	A1 V	88	71	23	−12.0	
Dubhe	α UMa	10	58	+62	17	1.8	+1.06	−0.6	G9 III	138	240	30	− 8.9	close vis. db
Zosma	δ Leo	11	9	+21	04	2.6	+0.12	+0.8	A4 V	204	135	23	−21	
Denebola	β Leo	11	44	+15	08	2.1	+0.08	+1.6	A3 V	509	256	13	− 1	
Phecda	γ UMa	11	49	+54	15	2.4	0.00	−0.1	A0 V	94	88	32	−12	
Gienah	γ Crv	12	11	−16	59	2.6	−0.09	−2.4	B8 III	161	275	100	− 4.2	
Acrux	α Cru	12	21	−62	33	0.8	−0.26	−3.7	B1, B3	44	228	80	− 6 v	vis. db 5″, each sp. db
Gacrux	γ Cru	12	26	−56	33	1.7	+1.58	−2.5	M3 III	272	176	70	+31.3	optical pair
Muhlifain	γ Cen	12	36	−48	25	2.2	−0.01	−1.7	A0 IV	198	266	60	− 7.5	vis. db 85 y
Mimosa	β Cru	12	42	−59	09	1.3	−0.25	−4.3	B0 III	51	240	130	+20.0	
Alioth	ε UMa	12	50	+56	30	1.8	−0.02	−0.2	A0p	115	95	25	−10	sp. db 4.15 y, Cr-Eu star
Mizar	ζ UMa	13	20	+55	27	2.1	+0.03	0.0	A2 V	129	103	26	− 9 v	tr., vis db 14″, sp. db 20 d
Spica	α Vir	13	20	−10	38	1.0	−0.23	−3.1	B1 V	53	229	65	+ 1 v	ecl. sp. db 4.0 d
	ε Cen	13	34	−52	57	2.3	−0.23	−3.6	B1 IV	35	229	150	+ 5.6	
Alcaid	η UMa	13	44	+49	49	1.9	−0.19	−2.3	B3 V	120	260	70	−10.6	
Hadar	β Cen	13	57	−59	53	0.6	−0.24	−5.0	B1 II	36	220	130	−12 v	vs. db 1″.2
Menkent	θ Cen	14	01	−35	53	2.1	+1.02	+0.9	K0 III–IV	738	225	17	+ 1.3	

Star		Epoch 1900 α		Epoch 1900 δ		App. mag. m_v	Color index	Abs. mag. M_v	Spectral class	Proper motion 1 000 μ'' yr	Direction of proper motion, θ, clockwise from N	Dis-tance d	Rad. veloc.* v_r	Notes†
		h	m	°	′						°	pc	km/s	
Arcturus	α Boo	14	11	+19	42	−0.1	+1.24	−0.2	K1 III	2 285	209	11	− 5.2	
	η Cen	14	29	−41	43	2.4	−0.21	−3.0	B2 V	48	224	120	− 0.2	db, composite sp.
Rigil Kent	α Cen	14	33	−60	25	−0.3	+0.71	+4.2	G2, K1	3 678	281	1.3	−22 v	vis. tr. 80.1 y, 2s.2
	α Lup	14	35	−46	58	2.5	−0.22	−2.5	B1 V	33	220	100	+ 7.3	
Izar	ε Boo	14	41	+27	30	2.4	+0.93	−0.6	K1, A	49	280	40	−16.5	vis. db 3″, composite sp.
Kochab	β UMi	14	51	+74	34	2.0	+1.49	−0.6	K4 III	31	280	33	+16.9	
Alphecca	α CrB	15	30	+27	03	2.2	−0.02	+0.5	A0 III	156	130	22	+ 2 v	ecl. sp. db 17.36 d
Dzuba	δ Sco	15	54	−22	20	2.3	−0.14	−4.0	B0 V	35	200	180	−15	
Acrab	β Sco	16	00	−19	32	2.5	−0.09	−4.0	B0.5 V	27	200	200	− 7 v	vis. tr., 14″, 1″
Antares	α Sco	16	23	−26	13	0.9	+1.83	−4.7	M1, B	30	192	130	− 3.2	vis. db 3″
	ζ Oph	16	32	−10	22	2.6	0.00	−3.4	O9.5 V	22	35	160	−19	
Atria	α TrA	16	38	−68	51	1.9	+1.43	−0.4	K2 III	43	148	29	− 3.6	
	ε Sco	16	44	−34	07	2.3	+1.15	+0.6	K2 III–IV	664	248	22	− 2.5	
Sabik	η Oph	17	05	−15	36	2.4	+0.05	+0.8	A2.5 V	96	23	21	− 1.0	close db
Shaula	λ Sco	17	27	−37	02	1.6	−0.23	−3.2	B1 V	33	180	90	0	sp. db 5.6 d
	θ Sco	17	30	−42	56	1.9	+0.38	−4.0	F0 Ib	12	110	150	+ 1.4	
Ras-Alhague	α Oph	17	30	+12	38	2.1	+0.15	+0.9	A5 III	262	154	17	+13	
	κ Sco	17	36	−38	59	2.4	−0.21	−3.3	B2 IV	30	203	140	−10	
Eltanin	γ Dra	17	54	+51	30	2.2	+1.54	+0.8	K5 III	23	197	40	−27.7	
Kaus Australis	ε Sgr	18	18	−34	26	1.8	−0.02	−1.7	B9 IV	137	198	50	−11	
Vega	α Lyr	18	34	+38	41	0.0	0.00	+0.5	A0 V	346	36	8.1	−13.7	
Nunki	σ Sgr	18	49	−26	25	2.1	−0.20	−2.4	B2 V	62	173	80	−11	
	ζ Sgr	18	56	−30	01	2.6	+0.09	−0.4	A2 IV	20	270	40	+22	vis. db 21 y
Altair	α Aql	19	46	+ 8	36	0.8	+0.22	+2.3	A7 IV–V	658	54	4.9	−26.2	
Peacock	α Pav	20	17	−57	03	1.9	−0.20	−2.9	B3 IV	87	177	90	+ 2 v	sp. db 11.8 d
Sadir	γ Cyg	20	19	+39	56	2.2	+0.66	−4.8	F8 Ib	1		250	− 7.5	
Deneb	α Cyg	20	38	+44	55	1.3	+0.08	−7.2	A2 Ia	3		500	− 4.6	
Gienah	ε Cyg	20	42	+33	36	2.5	+1.03	+0.6	K0 III	483	47	24	−10 v	
Alderamin	α Cep	21	16	+62	10	2.4	+0.23	+1.5	A7 IV–V	159	72	15	−9	
Enif	ε Peg	21	39	+ 9	25	2.4	+1.56	−4.6	K2 Ib	26	92	250	+ 4.9	vis. db 82″
Al Na'ir	α Gru	22	02	−47	27	1.8	−0.14	−0.2	B5 V	197	141	25?	+11.8	
	β Gru	22	37	−47	24	2.2	+1.62	−2.6	M3 III	133	99	90	+ 1.6	
Fomalhaut	α PsA	22	52	−30	09	1.2	+0.09	+1.9	A3 V	367	117	7.0	+ 6.5	
Scheat	β Peg	22	59	+27	32	2.5	+1.7	−1.4	M2 II–III	234	55	60	+ 8.7	
Markab	α Peg	23	00	+14	40	2.5	−0.04	0.0	B9.5 III	73	125	32	− 3.5	

using the word magnitude to designate how bright a star appears. In about the year 180 A.D. Ptolemy extended Hipparchus' work and *from that time on the idea of magnitude as related to brightness became more and more firmly entrenched in astronomical literature.* Today we still use the same idea, although it is now a precisely defined term instead of being a concept based upon the crude kind of visual estimates made by Hipparchus and Ptolemy.

The modern magnitude scale dates from 1854 when the British astronomer Pogson, following the work of Sir John Herschel, showed that Ptolemy's magnitude scale according to which the numbers 1, 2, 3, 4, 5, 6 are assigned to the six groups into which Ptolemy had divided the naked-eye stars *is not really a linear scale* (i.e., a scale in which a sixth-magnitude is one-sixth as bright as a first-magnitude star) *but is actually a logarithmic scale to the base 10. What this means is that the numbers that represent the magnitudes of stars are not proportional to the brightness of the stars themselves, but rather to the logarithm (to the base 10) of their brightnesses.*

15.13 The Precise Magnitude Scale

To introduce the modern magnitude scale it was therefore necessary to have a precise way of relating the magnitude of a star to its brightness. This was done by noting that an average first-magnitude star (of Ptolemy's) is about 100 times brighter than an average sixth-magnitude star. Pogson therefore proposed that the magnitude scale be made precise and universal by *defining a first-magnitude star as being exactly 100 times brighter than a sixth-magnitude star.* The zero point of the new scale was then set by choosing as typical sixth-magnitude stars (on this new scale) those which appeared as sixth-magnitude stars in the famous German catalogue, the *Bonn Durchmusterung*, that was then in wide use. With these stars established as sixth-magnitude stars on this new Pogson scale, all stars 100 times brighter are automatically first-magnitude stars. Thus, for example, Spica, Aldebaran, and Altair are all typical first-magnitude stars, whereas Polaris has exactly the apparent magnitude of +2. In fact, the apparent magnitude +2 of Polaris has sometimes been used as a standard for assigning magnitudes.

15.14 The Brightness Ratio for Stars of Different Magnitudes

We shall now see how to assign magnitudes to stars that are not exactly 100 times fainter than (that is 1/100 as bright as) the standard first-magnitude star. We can do this by determining how much two stars differ in brightness if they differ by one magnitude. To ascertain this from our scale let b_1 be the apparent brightness of a star of the first-magnitude, b_2 the apparent brightness of a star of the second-magnitude, etc., so that b_m is the apparent brightness of a star of the mth magnitude. The factor b_m/b_{m+1} is then the number of times an mth-magnitude star is brighter than a star of magnitude $m + 1$. *Let us call this factor, a, so that a star of the first magnitude is a times as bright as one of the second magnitude; a star of the second magnitude is a times as bright as one of the third magnitude, etc.* In other words we have

$$\frac{b_1}{b_2} \times \frac{b_2}{b_3} \times \frac{b_3}{b_4} \times \frac{b_4}{b_5} \times \frac{b_5}{b_6} = \frac{b_1}{b_6} = 100,$$

$$a \times a \times a \times a \times a = a^5 = 100.$$

We see from this that a is equal to the fifth root of 100 ($\sqrt[5]{100}$) or, what is the same thing,

$$\boxed{a = 10^{2/5}}, \qquad (15.6)$$

which is very nearly equal to 2.512. Thus a star of any magnitude is somewhat more than $2\frac{1}{2}$ times brighter than the star of the next higher magnitude. *Note that the larger magnitudes are associated with the fainter stars.*

Let us now consider two stars, one of which is of the mth apparent magnitude, and the other of the nth apparent magnitude, and let b_m and b_n be their apparent brightnesses respectively. What is the ratio of these brightnesses? We know that for each magnitude step the brightness decreases by a factor of $10^{2/5}$, and since in going from m to n there are exactly $n - m$ steps, the apparent brightness of the nth-magnitude star is smaller than that of the mth-magnitude star by the factor

$$\overbrace{10^{2/5} \times 10^{2/5} \times \cdots \times 10^{2/5}}^{(n-m) \text{ factors}} = 10^{(2/5)(n-m)}.$$

We thus have

$$\boxed{\frac{b_m}{b_n} = 10^{(2/5)(n-m)}}. \qquad (15.7)$$

15.15 Relationship between Brightness and Magnitude

Since the power to which 10 must be raised to give any particular number is called the logarithm of that number to the base 10, we have

$$\log_{10} \frac{b_m}{b_n} = \frac{2}{5}(n - m). \qquad (15.8)$$

If we now consider the star of the nth magnitude as being a kind of standard to which we may refer all other stars, we have

$$\boxed{\log_{10} b_m = -0.4m + B}, \qquad (15.9)$$

where b_m refers to the apparent brightness of any star of apparent magnitude m, and B is a constant. This gives us the relationship between apparent brightness and the apparent magnitude. In Table 15.2 a list of the visually brightest stars and their apparent visual magnitudes is given.

15.16 Faintest Measurable Magnitudes

With our modern large telescopes, such as the 200-inch Mt. Palomar reflector, it is possible to photograph objects down to the 24th magnitude. In fact, the faintest star that has been recorded by the 200-inch telescope has an apparent magnitude of +23.9. By means of photoelectronic devices, such as photomultiplier tubes, used in conjunction with telescopes mounted in high-altitude balloons and orbiting observatories, it will be possible ultimately to go several magnitudes beyond this. This has already been demonstrated by the work of M. Schwarzschild and his Princeton group, who used a 12-inch refracting telescope mounted in a balloon flying at a height of 82 000 ft in September 1957 to photograph the solar surface. Much larger telescopes are being mounted in space balloons and on orbiting observatories.

Although visual methods for determining the stellar magnitudes were used almost exclusively in the past, today these have been replaced by methods depending upon photography and photoelectric devices. The eye used as a photometric instrument has the disadvantage that it cannot keep a permanent record, that it is subject to fatigue, and that it does not respond with greater intensity to cumulative light. Thus, no matter how long we may look at a very faint source, our eye does not store up the effect of the light to make the source appear brighter after a given time. But since a photographic plate can do this, it is possible to record very faint sources photographically by exposing the plate a sufficiently long time. At the same time the photographic plate gives us a record that can be kept as long as desired. The ultimate limitations on a photographic plate are determined by the graininess of the plate and the background brightness of the night sky; very faint stars cannot be photographed because the entire plate becomes fogged by the background light after a sufficiently long exposure.

15.17 Photographic and Visual Magnitudes

Since ordinary photographic plates are relatively insensitive to the reds and the yellows, the apparent brightness of a given star as recorded by the eye and as recorded by a photographic plate are, in general, not the same. *The eye sees red and yellow stars as brighter than blue-white stars even if they all send the same amount of light to the eye per second. Just the opposite is true of the photographic plate.* Hence, we must be careful to distinguish between magnitudes determined photographically and those measured by the eye. For this reason we shall speak of *the photographic magnitude* (m_{ph}) *of a star when its apparent magnitude is determined photographically, and its visual magnitude* (m_{vis}) *when its apparent magnitude is determined visually.* Generally these two numbers will differ; we shall see later how this difference can be used to measure the color of a star. Figure 15.4 shows the same star field taken with photographic and photovisual filters.

Although we shall be using the concept of visual magnitude right along, we must not think of this quantity as being determined by an actual visual measurement. Indeed, visual magnitudes are determined photographically by using special photographic plates that have the same color response as the eye does, or by using appropriate filters in conjunction with photoelectric cells. Instead of using a photographic plate, we can use a photoelectric cell with the right kinds of filters to give both the photographic and the visual magnitudes. Since we can construct very sensitive photoelectric cells and photomultiplier tubes we can measure magnitudes of even faint stars (fourteenth to fifteenth magnitude) with medium-sized telescopes.

15.18 The Accuracy of Magnitude Measurements

To obtain the magnitude of a star from photographic plates we measure either the size or the blackness of the image of the star formed on the

FIG. 15-4 *Color index shown by two photographs of the same star field taken with photographic and photovisual filters.* (*Yerkes Observatory photographs*)

plate, since both of these quantities depend upon the amount of starlight striking the plate. Since these two quantities also depend upon the type of photographic plate used and the exposure time, astronomers use standardized plates and carefully record the length of time each plate has been exposed. Today special instruments called photometers are used to measure the stellar images very precisely; with such devices, among which the Eichner iris-diaphragm photometer is outstanding, we can determine the apparent magnitudes of stars with errors no larger than 0.03 magnitude.

For higher sensitivity and precision, photoelectric cells and photomultiplier tubes are now employed in place of photographic plates. In the photomultiplier tubes there are rows of sensitive surfaces arranged in such a way that an electron emitted from the first surface, when light impinges on it, cascades down to the other surfaces so that more and more electrons are knocked out of each successive surface as the cascade proceeds (see Figure 15.5). In this way a stream of electrons starting from one surface can be magnified millions of times after encountering a dozen or more such surfaces. This leads to an enormous amplification of the original light beam.

15.19 Bolometric Magnitudes

Since neither the eye, nor a photoelectric device, nor a photographic plate is sensitive to the entire range of colors in the radiation coming from stars, none of these instruments gives us a complete picture of the energy emitted by the star. To obtain such a picture we must have some way of measuring the radiation in regions outside the sensitivity of these three devices. Thus, since stars such as the sun emit radio waves, we can take this type of radiation into account only by means of

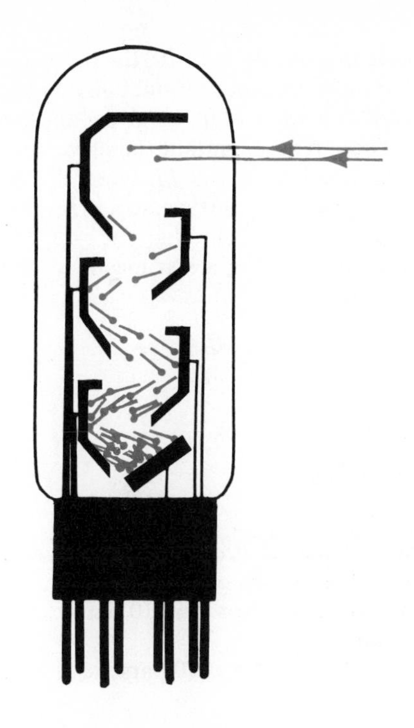

FIG. 15–5 *In a photomultiplier tube an electron ejected from one surface by a photon cascades down and triggers off a stream of electrons from the other surfaces.*

radio-telescopes. To emphasize the fact that the ordinary photographic and visual magnitudes do not give us complete information about a star's radiation, *we introduce the* ***bolometric magnitude***, *which is physically more important than either of the other magnitude scales, since it gives us information about the entire range of radiation.*

This magnitude scale is arbitrarily set in such a way that *the visual magnitude of the sun is placed equal to its bolometric magnitude.* For a star that differs from the sun, the bolometric magnitude is then obtained by an appropriate correction, which depends on the surface temperature of the given star. *The more a star differs in temperature from the sun, the larger is the correction that must be made in the visual magnitude to obtain the bolometric magnitude.* This correction ranges from -0.73 magnitude for stars whose absolute surface temperature is 4 000°, to -2.7 magnitude for stars whose absolute surface temperature is 25 000°.

15.20 The Apparent Magnitudes for Some Well-Known Celestial Objects

Now that we have introduced a magnitude scale, it is instructive to give a few examples of the apparent magnitude of some well-known objects. Among the stars Ptolemy included in his first-magnitude group are some that are brighter than the first magnitude according to the modern scale, so we must consider this scale as extending into negative numbers. Since the sun has the greatest apparent brightness of all objects in the sky, it has the smallest apparent visual magnitude. On the Pogson scale this is -26.78, according to J. Stebbins. On the same scale the full moon has an apparent visual magnitude of -12.6, whereas Venus and Jupiter at their brightest have apparent visual magnitudes of about -4.

The brightest star in the sky is Sirius with an apparent magnitude of -1.5, which is about $3\frac{1}{2}$ magnitudes less than the apparent magnitude of Polaris, and is therefore about 25 times as bright as Polaris. Since α Centauri is about 6.3 times brighter than Polaris, it is 2 magnitudes brighter than Polaris and therefore has an apparent magnitude of 0. In this way by comparing various stars with Polaris and remembering that each magnitude difference corresponds to a brightness ratio of about 2.512, we can assign apparent magnitudes to all the visible stars.

15.21 Standard-Magnitude Stars

Although in discussing the apparent magnitudes of stars such as Sirius and α Centauri we used Polaris as our standard, we must use a large number of standard stars if we desire a sufficiently high accuracy. Such standard-magnitude sequences have been established in various limited regions of the sky and have been used as controls by observers everywhere. Thus the Mount Wilson North Polar Sequence (which consists of a group of circumpolar stars whose magnitudes have been very well determined, and in which the magnitude variations occur in very small steps) has been used as a standard for visual and photographic magnitudes for many years. In addition to this sequence others have been set up to be used with various filters and in photoelectric photometry.

15.22 The Luminosity and the Absolute Brightness of a Star; Absolute Magnitude

The first step in determining the luminosity of a star is to assign it an apparent magnitude m in the manner described above, but since this apparent magnitude is a measure only of apparent brightness, we cannot obtain the luminosity from it without some additional information. It is clear that the apparent magnitudes of stars, which range from −26.78 for the sun to +24 for the faintest stars visible to the 200-inch telescope, are determined by two factors: (1) *the distances of the stars, and* (2) *their luminosities, or what is the same thing, their intrinsic or absolute brightnesses.* Therefore, to find the luminosities of stars we must know not only their apparent magnitudes but also their distances or their parallaxes.

If all the stars in the sky were at the same distance from the earth, then their apparent brightnesses or their apparent magnitudes would measure their true brightnesses. If under these conditions a given star appeared to be twice as bright as another star, it would, indeed, be intrinsically twice as bright, and hence its luminosity would be twice as large. However, since stars are not all at the same distance, we may draw no such conclusions from the apparent magnitude data.

15.23 The Absolute Magnitude of a Star

We may imagine the distance factor eliminated by picturing all the stars as being placed at some standard distance from the earth. If this could be done and if all the stellar magnitudes could then be remeasured, a comparison of the reassigned magnitudes of any two stars would give us a comparison of their true brightnesses. *We may, therefore, refer to these reassigned magnitudes as the absolute magnitudes.*

To make this explicit, we must agree upon a definite distance at which we are to picture all the stars as being placed. *By international agreement this has been taken as* ***10 parsecs****, and we shall, therefore, define the absolute magnitude, M, of a star as the apparent magnitude it would have if it were placed at a distance of 10 parsecs from us.*

15.24 Assigning Absolute Magnitudes to Stars

Let us consider all the stars as they are now and imagine how they would look if they were all placed 10 parsecs away from us. Associated with each star as it appears to us now, we have three numbers b_m, the apparent brightness; m, the apparent magnitude; and d, the distance. At 10 parsecs each star would have associated with it three other numbers, B_M, the absolute brightness; M, the absolute magnitude; and 10, the distance in parsecs. Finding the absolute brightness and the absolute magnitude of a star of given apparent brightness and apparent magnitude therefore means replacing b_m and m by B_M and M, when the true distance d is replaced by 10 parsecs. In other words, the problem of finding the apparent brightness and magnitude of a star means carrying out the transformation

$$\left.\begin{matrix} b_m \\ m \end{matrix}\right\} \rightarrow \left\{\begin{matrix} B_M \\ M \end{matrix}\right.$$

for

$$d \rightarrow 10.$$

This is shown in Figure 15.6.

It is clear that

for stars at distances **greater** than 10 parsecs, m is **larger** than M;
for stars at distances **less** than 10 parsecs, m is **smaller** than M; and
for a star that is **exactly at** 10 parsecs, $m = M$.

15.25 The Relationship between m and M

To find the exact relationship between m and M we must know how the apparent brightness of a

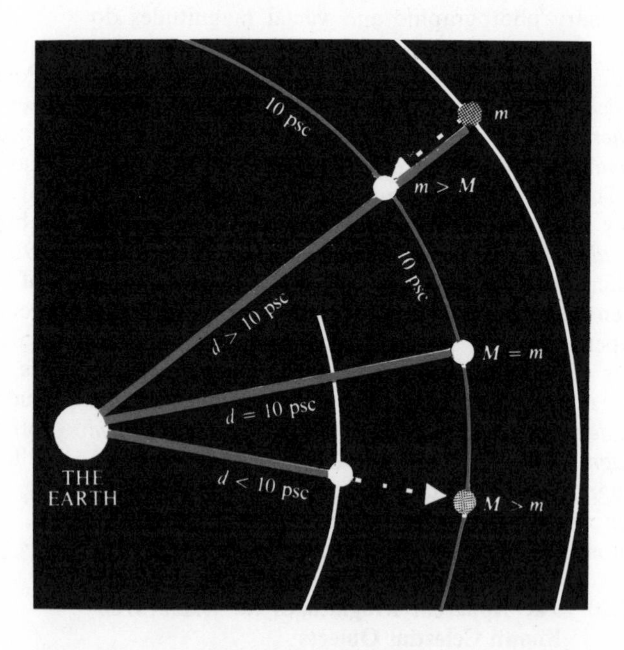

FIG. 15–6 *A star of apparent magnitude, m, at its true distance, d, has an apparent magnitude, M (called its absolute magnitude), when seen at a distance of 10 parsecs.*

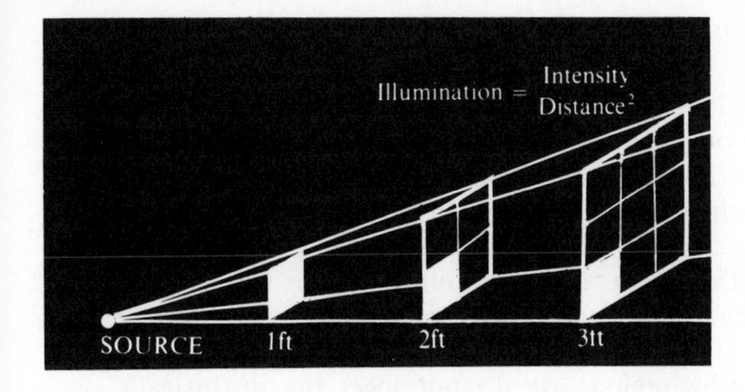

FIG. 15–7 *The intensity of light from a point source falls off inversely as the square of the distance as shown by the larger area over which light is distributed as the distance increases.*

star changes when the distance changes. The apparent brightness of a celestial body varies inversely as the square of its distance from us, because its light moves uniformly in all directions, spreading out in concentric spheres. Thus, the intensity falls off inversely as the square of the distance from the star because the energy in the light is spread uniformly over larger and larger spherical surfaces. Figure 15.7 shows the inverse-square law.

If we keep this in mind, it is easy to find M, if m and d are known. Let us consider a star (as shown in Figure 15.6) at distance d and suppose that it is brought to a distance of 10 parsecs. If we compare its apparent brightness, b_m, at the distance d with its apparent brightness, B_M (its absolute brightness), at 10 parsecs, we have from the inverse-square law described above

$$\boxed{\frac{b_m}{B_M} = \frac{10^2}{d^2}}, \tag{15.10}$$

where d is to be measured in parsecs.

On the other hand, we may look upon the star when it is at a distance d and the same star when it is at 10 parsecs as being two different stars having magnitudes m and M respectively, and we may then ask how two such stars compare in apparent brightness. We gave the answer to this question in the previous section [see Equation (15.7)], where we saw that since each magnitude step represents a brightness ratio of $10^{2/5}$, two stars having magnitudes m and M will have a brightness ratio of $10^{(2/5)(M-m)}$, there being just $M - m$ steps in going from m to M. In other words, we must have

$$\frac{b_m}{B_M} = 10^{(2/5)(M-m)}.$$

If we compare this with our previous equation for b_m/B_M, we see that

$$10^{(2/5)(M-m)} = \left(\frac{10}{d}\right)^2,$$

or

$$10^{(M-m)/5} = \frac{10}{d},$$

so that

$$d = 10^{(5-M+m)/5}.$$

From this it follows that

$$\log_{10} d = \frac{5 - M + m}{5}$$

or

$$M = m + 5 - 5\log d.$$

Since $d = 1/p''$, (note that $-\log d = +\log p''$), we see, finally, that

$$\boxed{M = m + 5 + 5\log p''}. \tag{15.11}$$

From this equation we can determine the absolute magnitude of any star if its apparent magnitude and parallax are known.

15.26 Stellar Luminosities

Since we shall express the luminosities of all stars in terms of the sun's luminosity, $L_\odot$, *we must begin by finding the absolute magnitude* $M_\odot$ *of the sun, for once we know this, we can find the luminosity of any star from a knowledge of its absolute magnitude by means of the relationship* [see Equation (15.8)]

$$\log\left(\frac{L}{L_\odot}\right) = 0.4(M_\odot - M),$$

where L and M represent the luminosity and absolute magnitude of the given star. Since we know that $L_\odot = 3.90 \times 10^{33}$ ergs/sec, we can at once find L from this equation if we know $M_\odot$.

To find the absolute magnitude of the sun, we use the equation $M = m + 5 - 5\log d$, with $m = -26.78$ and $d = 1/206\,265$, since this is the distance of the sun in parsecs; we thus have

$$M_\odot = -26.78 + 5 + 5\log 206\,265,$$

or

$$M_\odot = +4.71.$$

Hence if the luminosity L of a star is expressed in solar units we have

$$\log_{10} L = -0.4M + 1.884.$$

15.27 Luminosities of Some Well-Known Stars

From this value of $M_\odot$ we know that a star whose absolute magnitude is 3.8 has a luminosity of about 2.512 times that of the sun, or has a luminosity of 9.90×10^{33} ergs/sec, etc., and a star whose absolute magnitude is 5.8 has a luminosity of $(3.90/2.512) \times 10^{33}$ ergs/sec, etc. As a specific example, we consider the luminosity of the star Sirius. Since its apparent magnitude (it is a binary) is -1.44 and its parallax is 0.375″, our equation for the absolute magnitude gives us the value $+1.4$. In other words, this star is just a little more than 3.3 magnitudes intrinsically brighter than the sun. Hence its luminosity is 20 times greater than that of the sun.

As another example we may take the star Rigel, which is the blue-white star in the lower right-hand corner of Orion. Since its apparent magnitude is $+0.11$, and its parallax (which cannot be determined trigonometrically but is found from its spectral class; see Section 18.1) is about 0.004″, its absolute magnitude is about -7, so that this star is about 12 magnitudes more luminous than the sun. Since 10 magnitudes correspond to a factor of 10 000 in luminosity, 12 magnitudes correspond to a factor of over 60 000, so that Rigel emits radiant energy at a rate of more than 2×10^{38} ergs/sec.

Using the same procedure, we can determine the absolute magnitudes and hence the luminosities of other well-known stars. These data are contained in Tables 15.1 and 15.2, one for the nearest stars and the other for the brightest stars. What is interesting about the data in these tables is the great abundance of very faint stars and the scarcity of the intrinsically luminous stars. Thus, we find that among the 40-odd stars closer than 5 parsecs only three are intrinsically brighter than, or as bright as, the sun, and twenty are 1 000 or more times intrinsically fainter (less luminous) than the sun.

Indeed, a very luminous star like Rigel is so rare that we have to go out almost 700 light years from the earth to pick up one such star, as compared to a distance of $4\frac{1}{2}$ light years to pick up another star (α Centauri, $M = 4.38$) like the sun. Considering the amount of space we have to survey to pick up a Rigel compared to the amount of space we must survey to pick up an α Centauri, we see that stars like the sun are at least three million times more numerous in our Milky Way than are stars like Rigel.

15.28 Naked-Eye Stars Give an Incomplete Picture

These tables indicate that the way the stars appear to our naked eye is not a true picture of their distribution since we see only distant stars that are intrinsically very bright and faint stars that are very close to us. It is clear that as we go farther and farther from the earth, our eyes tend to favor the very bright stars so that we miss more and more of the fainter ones. If we look all around us in the nearby regions we do not meet the intrinsically bright stars (because they are so rare) and pick up only the very faint ones. We must keep these facts in mind if we wish to get a true picture of the distribution of stars in the Milky Way. We shall come back to a discussion of this point in a later chapter. An interesting point which the tables emphasize and to which we shall come back later is the large number of stars that belong to physically connected groups (such as double or triple systems).

We see from the tables that stellar luminosities vary over an enormously wide range, going all the way from a star like Wolf 359 ($M = 16.8$), which is about 50 000 times intrinsically fainter than the sun, to a very luminous star like Rigel, which we have already discussed. The tables also give the spectral classes of the various stars; we shall discuss this in the section on stellar spectra.

15.29 Color and Color Index

The total luminosity of a star, although in itself a very important property, does not contain nearly all of the information that we can gain from the radiation coming from the star. A great many data can be deduced by analyzing the light and determining its component colors in detail. However, even without doing this we can obtain other important and useful data by noting that there are stars like Betelgeuse and Antares which are reddish in color, stars like Rigel which are blue-white, and still others like Sirius which are white. In fact, we find that the colors of stars range all the way from red to bluish.

Just as it is possible to measure the luminosity or the magnitude of a star, it is also possible to introduce a precise numerical definition of color. To do this we use the fact that given photometric instruments, such as the eye or a photographic plate, react differently to different colors. Since the eye is more sensitive to the reds and yellows than it is to blue, *a bluish star and a red star of the same intrinsic luminosity and at the same distance from us do not have the same apparent visual magnitudes.* The reddish star has a smaller visual magnitude than the bluish star. On the other hand, just the opposite is true for a photographic plate. The image of the bluish star formed on the photographic plate is larger and blacker than the image of the reddish star for the same exposure time. In other words, for a given star, the photographic and the visual magnitudes differ, in general, as we have already noted in our discussion of apparent magnitudes (Section 15.17).

We make use of this difference to define the color of a star. We do this by introducing the **color index** (see Figure 15.4) which we define as the difference between the magnitudes determined in two different colors or by means of two photosensitive devices that have different color responses. The color index is generally given as the blue minus the visual magnitudes ($B - V$); however, we shall define it as the difference between the photographic and the visual magnitudes:

$$\boxed{I = m_{ph} - m_{\text{vis}}}. \qquad (15.12)$$

The redder a star, the larger this number is, and the bluer a star the smaller this number. Since a white star may be defined as one to which the eye and the photographic plate respond equally, astronomers have chosen the photographic and visual magnitude scales to make I equal to zero for such a star. I is therefore greater than 0 *for stars that are redder than this, and less than* 0 *for stars that are bluer than this.*

The color indices of some well-known stars are given in Tables 15.1 and 15.2. Vega is considered to be a white star since its color index on this scale is 0.00 whereas Rigel, with a color index of -0.05, is a blue-white star, and the sun and Capella with color indices equal to $+0.81$ are yellow stars. Arcturus, an orange star, has a color index of $+1.24$, and Betelgeuse and Antares with color indices of $+1.85$ and $+1.83$ respectively are red stars.

EXERCISES

1. If an astronomical observatory were operating on Jupiter, how would the parallaxes of the stars differ from those obtained on the earth?

2. How much brighter does Sirius appear than (a) Polaris, (b) Canopus, (c) α Centauri, and (d) Betelgeuse?

3. Suppose a magnitude scale were introduced on which a fifth-magnitude star were taken to be 81 times brighter than a first-magnitude star. How many times brighter would an mth magnitude star be than an nth one?

4. If we used the magnitude scale in problem 3, what would the relationship be between absolute magnitude M and apparent magnitude m?

5. Calculate the total absolute magnitude of a binary system if the absolute magnitudes of the individual components (which cannot be seen separately) are 0 and 2.

6. (a) How much would the luminosity of the sun have to be increased if its absolute magnitude were to be changed to -1? (b) Assume that our Milky Way contains 10^{11} stars, each, on the average, as luminous as the sun. What is its absolute magnitude?

7. What would the apparent magnitude of Rigel be if it were one astronomical unit away from us?

8. What is the significance of the constant B in Equation 15.9 of the text?

9. What would the relationship between the apparent and absolute magnitudes be if the standard distance were 30 parsecs instead of 10?

16

THERMODYNAMICS AND PROPERTIES OF RADIATION; STELLAR SURFACE TEMPERATURES

16.1 The Temperature of a Body

The only information we can get about stars at the present time must come to us from the radiation they emit. We have already seen that this enables us to assign a distance, a luminosity, and a color to a star. These are all quantities that can be measured directly, but there are other stellar characteristics that we must infer from the radiation by means of some indirect method based upon the known laws of radiation. One such important property is the *surface temperature of the star*, which, as we shall see, we can determine by properly analyzing the star's color. However, before we consider how to do this, we must have a clear understanding of the temperature concept itself.

To most people "temperature" simply means a number on a scale called a thermometer, but to the astronomer and the physicist temperature is associated with a definite physical state of a body, and the number on this thermometer in contact with the body is but one of various ways of measuring that state. That there are several ways of doing this is important in astronomy for it is clearly impossible to obtain the temperature of a star by bringing it in contact with a thermometer. Since we must work with the radiation from the star, we proceed by first determining how the temperature of a body is to be related to the radiation it emits.

16.2 Heat: The Breakdown of the Principle of Conservation of Mechanical Energy

Most people generally identify temperature with heat, and use the two terms almost synonymously. However, these are two distinctly different concepts, and in order to avoid confusion we must introduce them both in a precise way. We begin by considering the concept of heat itself. We start with the principle of conservation of energy, which we discussed in Section 6.22. We saw there that if a body is moving without being subjected to any frictional forces, its total mechanical energy remains constant. However, if frictional forces are present, this is not true, and there appears to be a breakdown of the principle of conservation of energy. Thus, if a pendulum is left to itself after it has been set swinging, it finally loses all the mechanical energy that was imparted to it and comes to rest.

Before the year 1842 this was interpreted as a failure of the principle of conservation of energy, but we now know that even in the case of a pendulum coming to rest this principle still holds if we keep in mind that the mechanical energy that manifests itself in the motion of the pendulum has been transformed into some other kind of energy. In other words, we may still maintain the principle of conservation of energy if we allow for other kinds of energy in addition to kinetic and potential.

16.3 Heat, Internal Energy, and the Conservation of Energy: The First Law of Thermodynamics

If we examine the pendulum in detail as it is coming to rest, we discover that at each moment energy in some form is flowing out of it into the surrounding air. *If we add to the mechanical energy that the pendulum has at any moment the amount that it has already transferred to its surroundings by this process, we again find that the total quantity of energy remains unchanged.* What this means is that we may still retain the principle of conservation of energy if, in addition to the mechanical energy present in the bob of the pendulum (as shown in Figure 16.1), we take into

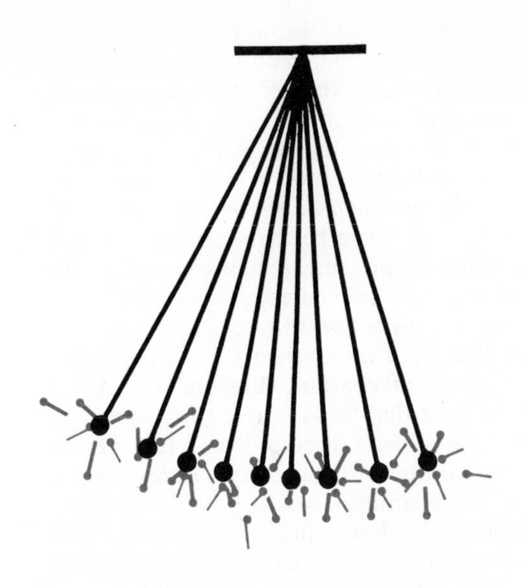

FIG. 16–1 *The bob of a pendulum loses mechanical energy because it is continually being bombarded by molecules of air moving at random.*

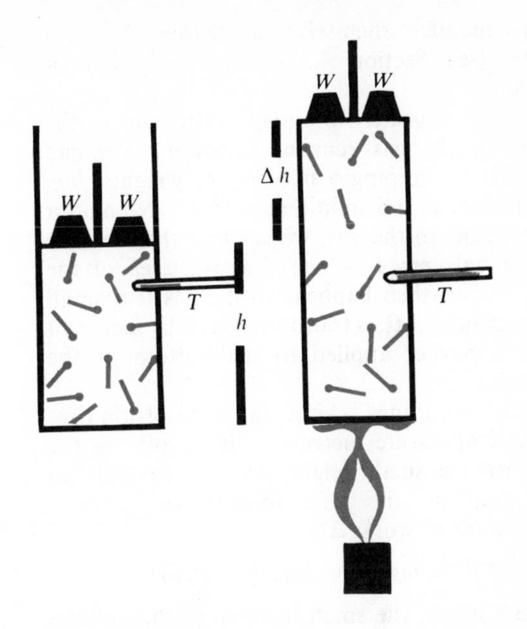

FIG. 16–2 *When heat is supplied to gas in a container kept at a constant pressure (shown by weights on the piston) the volume, V, increases, and the temperature, T, rises.*

account the amount of energy that has flowed out of the pendulum and into the surrounding atmosphere.

Let us consider in more detail the energy that may at any moment flow out of or into a system and also any hidden energy the system may have that we cannot classify as bulk kinetic or potential energy in the usual sense of these terms. We can get some idea of what is involved by noting that as the bob of the pendulum swings back and forth it is continuously being bombarded by the molecules in the atmosphere, and hence robbed of its energy, so that the mechanical energy that was previously concentrated in the small volume of the bob of the pendulum is now distributed over large numbers of molecules moving at random in various directions. Thus, the energy that was lost by the pendulum and is now hidden in the surrounding atmosphere is associated in some way with the random motion of the molecules in this atmosphere.

If we now call the energy that is flowing into or out of a system ***heat****, and the hidden energy that is present in the form of the random motion of its molecules* ***internal energy****,* we can extend the principle of conservation of energy by saying that *the heat that flows into a system must equal the increase in the system's internal energy plus the increase in its mechanical energy* (which will show up as work done by the system against external forces). *This principle is referred to as the* ***first law of thermodynamics****.* Together with another law (the second law) it forms the basis of thermodynamics.

We shall see later that *the* ***temperature*** *of a body is a measure of how much internal energy in the form of random molecular motion it has, or, more accurately, how much kinetic energy, on the average, each of its molecules has.*

16.4 First Law of Thermodynamics Applied to a Gas

We can formulate the first law of thermodynamics precisely by considering a gas in a cylinder with a piston, as shown in Figure 16.2. We also suppose that a thermometer is kept in contact with the gas so that its temperature can be read off at any time. In general, the condition of the gas itself is represented by three parameters: its pressure, P, its volume, V, and its absolute temperature, T. *The absolute temperature scale is defined as the centigrade scale plus 273°. The* ***absolute zero*** *is 273.15° below zero centigrade.* The centigrade scale is frequently referred to as the Celsius scale, after the scientist who first described it. We shall see that these three numbers may not be chosen independently since they are connected by the **equation of state of the gas.** We shall come back to this later and merely note here that any changes that take place in a gas manifest themselves as changes in these numbers (see Section 4.5 on the definition of pressure).

We now consider a situation in which the pressure in the gas remains constant. We can insure this by keeping a fixed set of weights, mg, on the piston, as shown in Figure 16.2. No matter what happens to the gas, we see that the pressure always equals mg/A, where A is the area of the piston. What then happens to the gas if a small amount of heat, ΔQ, is transferred to it by means of a Bunsen burner applied to the bottom of the cylinder?

We find two things taking place: the piston rises and the temperature increases. If we picture the piston rising a small height, Δh, we see that the flow of heat into the gas enables the gas to do a small amount of work, ΔW,

$$\Delta W = mg\,\Delta h = PA\,\Delta h = P\,\Delta V,$$

where ΔV is just the small increase in the volume of the gas.

On the other hand, since the temperature rises, not all of the heat goes into work. Some of it remains trapped in the gas as internal energy, ΔU. The first law of thermodynamics states that *there must be a balance between the small amount of heat, ΔQ, that flows into the gas, and the sum of the small amount of work done by the gas and the small increase in the internal energy of the gas.* This may be expressed in mathematical form as follows

$$\boxed{\begin{aligned}\Delta Q &= \Delta U + \Delta W \\ &= \Delta U + P\,\Delta V\end{aligned}}\,. \qquad (16.1)$$

As we can see, this statement of the first law of thermodynamics is nothing more than an extension of the principle of conservation of energy to thermodynamic systems.

16.5 The Specific Heat of a Gas: The Specific Heat at Constant Volume and the Specific Heat at Constant Pressure

The quantity ΔU, which represents the increase in the internal energy, must be related to the increase ΔT in the absolute temperature of the gas. In fact we may write

$$\Delta U = C\,\Delta T,$$

where C is called *the* ***specific heat*** *of the gas, and by definition is the amount of heat needed to increase the temperature of one gram of the gas by one degree.* However, here we must be a bit more careful, because there are two different specific heats of a gas, depending upon whether its volume or its pressure remains constant as heat is added to it.

If heat is added to a gas while the volume remains constant (the piston is kept immovable) we speak of *the specific heat at constant volume and write it as C_v.* On the other hand, if heat is added to a gas while its pressure remains constant (the volume of the gas then increases) we speak of its *specific heat at constant pressure and write it as C_p.* It is clear that C_p is larger than C_v, because while the gas expands at constant pressure, it is using some of the heat that flows into it to do work (as in the example already discussed) and hence more heat is needed to raise its temperature a given amount. The quantity $C_p/C_v \equiv \gamma$, which is called the **ratio of specific heats**, plays an important role in the equilibrium of gaseous configurations, and hence is essential to an understanding of the internal structure of stars.

16.6 The Equation of State of a Perfect Gas

Although in discussing the first law of thermodynamics we made use of the properties of a gas, we did not write down the equation of state of the gas. However, to apply these properties to an understanding of conditions inside a star we must have an equation that connects the quantities P, V, and T. Since in general this equation for an actual gas is quite complex, we shall suppose that the gases we deal with inside stars are **perfect** or **ideal gases**. The equation of state then becomes very simple. We shall be able to derive the equation of state of a perfect gas from the fact that its internal energy depends only upon its absolute temperature and not on its volume. This means, simply, that *the molecules in a perfect gas (except when they are colliding with each other, which lasts only a very short time) do not affect one another.* In other words, *except during collisions there are no forces between these molecules.* In the interiors of stars where the temperatures are very high, the state of the gases approximates that of an ideal gas to a very high degree of accuracy so that the simple equation of state for ideal gases applies.

We shall derive the expression for a perfect gas in two steps. We again consider the cylinder of gas whose state is defined by the quantities P, V, T. Keeping the temperature constant, we reduce the volume by very slowly lowering the piston. If the volume of the gas is now reduced to one-half its original value, it is clear that the pressure of the gas against the walls of the container is twice as great. The reason is that the molecules of the gas have only half as much space to move in, and since they still move with the same average speed as before, they collide with the walls twice as often. *Note that the average speed of the molecules remains the same since the temperature is unaltered.* This relationship between P and V when T is constant is known as **Boyle's law**, and may be written in the form

$$\boxed{PV = \textbf{constant}}\,. \tag{16.2}$$

If we allow the temperature of the gas to change while we change its volume, Boyle's law no longer holds since the pressure of the gas changes with the temperature even if its volume remains fixed. In the same way we find that even though the pressure is kept constant, the volume of the gas changes if its temperature changes. We must now incorporate this fact (i.e., that P, V, and T all change together) into a single expression.

We know experimentally that if our gas is at 0°C and we raise or lower its temperature by 1°C, keeping its volume constant, its pressure increases or decreases by a definite amount, and if we increase or decrease the temperature by 2°C, keeping the volume constant, the pressure increases or decreases by twice the amount, etc. *In the same way, if we keep the pressure fixed, allowing the volume to change as the temperature increases or decreases, we find experimentally that the volume changes by the same amount per degree change of temperature.* This is known as the law of **Charles and Gay-Lussac.**

These phenomena can be incorporated into a single expression by making use of the experimental fact that no matter what happens to the pressure, volume, or absolute temperature of a perfect gas, the quantity PV/T remains unaltered. Thus we may write for **the equation of state of a perfect gas**:

$$\boxed{\frac{PV}{T} = R}\,, \tag{16.3}$$

where R is called the **gas constant.**

16.7 The Dependence of the Equation of State of a Perfect Gas on the Quantity of Gas

We must be careful in using this equation to specify the amount of gas that we are dealing with since the constant R increases as this amount increases. It is customary to refer to R as the gas constant if the amount of gas is **1 mole**, i.e., *if its mass in grams is numerically equal to the molecular weight* (see Section 18.7) *of the gas.* If the molecular weight of the gas is μ (for a mixture of gases μ is the mean molecular weight) and its mass is m then we are obviously dealing with m/μ moles of gas, and the equation of state of the gas is

$$\frac{PV}{T} = \frac{m}{\mu} R$$

or

$$PV = \frac{m}{\mu} RT. \tag{16.4}$$

It can be shown that the gas constant, R, is related to C_p and C_v. In fact, it turns out that $R/\mu = C_p - C_v$.

Since we often apply the equation of state to points inside a star, it is convenient to write this equation in terms of quantities that can be evaluated at a point. The pressure and the temperature are such quantities, but the mass and volume are not. However, we can eliminate these by introducing the density, ρ, at a point inside a star. If we divide both sides of the equation of state by the volume, we have

$$P = \frac{m}{V}\frac{R}{\mu} T$$

or

$$P = \rho \frac{R}{\mu} T$$

since m/V is just the density.

16.8 The Equation of State and the Boltzmann Constant

We may alter this equation by introducing instead of R, which refers to 1 mole of gas, *the universal constant k, which is known as* ***Boltzmann's constant****, and which may be considered as the gas constant for a single molecule.* Since in one mole of gas there are N_0 molecules (where N_0 is equal to 6.02×10^{23}, and is known as **Avogadro's number**) we shall place $R = N_0 k$, and thus obtain for the pressure in a perfect gas at a point

$$P = \frac{N_0 k}{\mu} \rho T.$$

Since the number of molecules in 1 mole of gas is the same regardless of the gas we are dealing with, and 1 mole of hydrogen (the atom not the molecule H_2) has a mass of 1 gram, it follows that N_0 is

1 gram/H, where H is the mass in grams of a single hydrogen atom. We thus have, finally, for the pressure at a point inside a perfect gas of molecular weight μ, and of density ρ, the expression

$$P = \frac{k\rho T}{\mu H}. \quad (16.5)$$

This formula shows us immediately how important the mean molecular weight of a star's constituent gases is for its internal structure. We shall come back to this point in more detail when we discuss the internal structure of stars.

16.9 The Kinetic Definition of the Absolute Temperature

From this equation of state we can now get some insight into the meaning of absolute temperature by noting that the pressure can also be related to the molecular motions in a gas. We can do this by considering a rectangular container (as shown in Figure 16.3) of length L and cross section A. The pressure of the gas against side A can be expressed in terms of the force exerted against the side by the molecules that hit it every second. If all the molecules have the same mass m, and the average speed of the molecules is v, then the average momentum of a single molecule is mv. Hence, when a molecule moving parallel to the side L hits the wall A, it brings to the wall, on the average, an amount of momentum mv.

After hitting the wall, the molecule has exactly the same speed in the opposite direction, so that its momentum is $-mv$, which it must have obtained from the wall. The wall, therefore, receives an amount of momentum mv, and gives up an amount $-mv$, resulting in a net gain of momentum by the wall of $2mv$, in the direction of the initial momentum of the molecule, as shown in Figure 16.3. The net force against the wall exerted by the gas is therefore measured by the number of such momentum transfers that occur per second, since we know that force, according to Newton's second law, is the time rate of change of the momentum of a body.

Since a molecule after hitting the wall has to travel a distance $2L$ (to the opposite wall and back) before it can hit the wall again, it spends a time $2L/v$ between two successive collisions with the same wall. Hence it hits this wall $v/2L$ times per second. Therefore, every molecule moving parallel

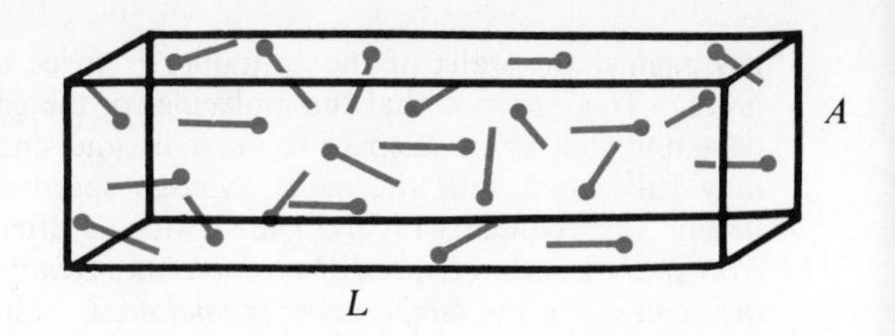

FIG. 16–3 *The gas pressure on the walls of a container is due to the continual transfer of momentum to the walls by the randomly moving molecules.*

to L transfers each second to the wall an amount of momentum equal to

$$\frac{v}{2L}\,2mv.$$

If we now suppose that there are n randomly moving molecules in the container, the effect is as though one-third of them or $n/3$, on the average, were at any moment moving parallel to L. Thus the total momentum transferred to the wall A per unit time is

$$\left(\frac{n}{3}\right)\left(\frac{v}{2L}\right)(2mv)$$

or

$$\left(\frac{n}{3}\right)\left(\frac{mv^2}{L}\right). \quad (16.6)$$

To find the pressure exerted on the wall we must divide this force by the area of the wall so that we get

$$P = \frac{1}{3}\frac{nmv^2}{AL} = \frac{1}{3}\frac{nmv^2}{V},$$

where $V = LA$, or the volume of the container. We may write this as

$$PV = \frac{nmv^2}{3},$$

but we also know that $PV = RT = nkT$ (see Section 16.7), so that if we compare these two equations we see that

$$kT = \frac{mv^2}{3},$$

or finally,

$$T = \frac{mv^2}{3k} \quad (16.7)$$

$$= \frac{2}{3k}\left(\frac{1}{2}mv^2\right).$$

Hence:
$$\frac{3}{2}kT = \frac{1}{2}mv^2. \qquad (16.8)$$

We therefore have obtained a more profound interpretation of the temperature which shows us how the absolute temperature is related to the average kinetic energy of a molecule in a gas. Thus, *the absolute temperature T is a measure of the average concentration of the internal energy per molecule in a given mass of the gas.* In a sense, it is the internal energy per molecule. What is ordinarily called heat in a body is associated with the total internal kinetic energy of the body.

16.10 The Principle of the Equipartition of Energy

There is another profound conclusion that we may draw from Equation (16.8). Since $v^2 = v_x^2 + v_y^2 + v_z^2$, where v_x, v_y, v_z are the components of the average velocity of the molecules in a rectangular coordinate system, we may write from 16.8

$$\frac{3}{2}kT = \frac{1}{2}m(v_x^2 + v_y^2 + v_z^2),$$

but $v_x^2 = v_y^2 = v_z^2$ since the molecular motions are random. Hence

$$\frac{3}{2}kT = \frac{m}{2}3v_x^2$$

or

$$\frac{1}{2}kT = \frac{1}{2}mv_x^2 = \frac{1}{2}mv_y^2 = \frac{1}{2}mv_z^2. \qquad (16.9)$$

This tells us that in a gas of absolute temperature, T, the kinetic energy associated with each degree of freedom of the motion of the molecules (there are just three degrees of freedom since there are just three mutually perpendicular directions in space) has exactly the same amount of energy, namely, $\frac{1}{2}kT$; this fact is known as the **principle of the equipartition of energy**. Although we have established this principle for the case of three degrees of freedom, it holds for any number of degrees of freedom, for example rotational or vibrational degrees of freedom of a body.

16.11 Isothermal and Adiabatic Process in a Gas

If a gas undergoes changes in which *the temperature remains constant at all times*, it is governed by Boyle's law, and we speak of an **isothermal process.** Another important process occurs in a gas when the gas is completely insulated and can neither give heat to nor gain heat from its surroundings. Such a process is called an **adiabatic process**, and when it occurs the pressure and volume change together according to the equation

$$PV^{\gamma} = \textbf{constant}, \qquad (16.10)$$

where γ is the ratio of specific heats, C_p/C_v. Adiabatic processes are important in stellar interiors because the gases there may be in states of turbulence or convection, and as each small quantity of gas rises and falls it undergoes adiabatic changes.

16.12 The Need for a Second Law of Thermodynamics

The first law of thermodynamics is important because it shows us that energy is always conserved, even though it may take on different forms. *However, there is nothing in the first law that places any restriction on the way in which mechanical energy can be transformed into heat, or heat back into mechanical energy.* Nevertheless, there is already an indication in this first law that there is some kind of restriction in nature as to the way one type of energy can be changed into another. We observe from the first law that some of the heat that flows into the gas in the cylinder becomes internal energy of the gas. Part of this internal energy becomes **unavailable** for doing work for us. The reason is that the internal energy is of a **disorganized** kind since it is the energy of the random motions of the molecules of a gas. *The exact statement of this fact, namely, that heat cannot be transformed into work under all conditions, is the core of the second law of thermodynamics.*

To express the second law in a precise form, we again analyze the behavior of a gas in a cylinder undergoing various processes. If we are dealing with a perfect gas, we may at any moment describe its state by specifying any two of the three quantities P, V, and T. We allow the state of the gas to change and observe the way these three quantities are altered.

16.13 The Change of a Gaseous System from an Initial to a Final State

Let the gas start out from a precise initial state defined by the set of values P_i, V_i, and T_i and

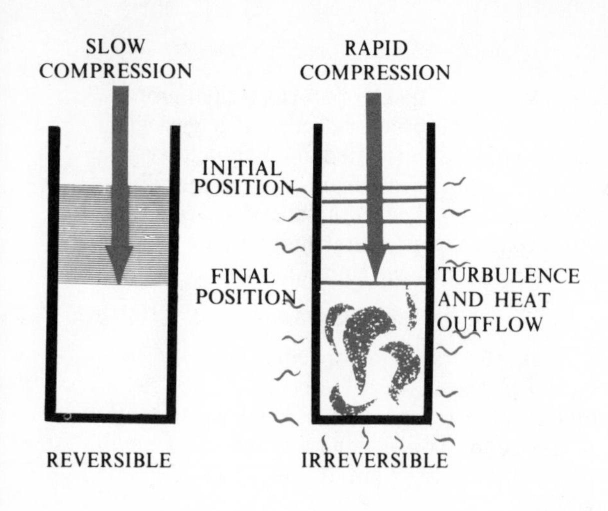

FIG. 16–4 *A gas undergoing a change from an initial to a final state in two different ways: one reversible, the other irreversible.*

reach some precise final state defined by the set of values P_f, V_f, and T_f. Now, *even though the initial state is precisely defined and the final state is precisely defined, the system can go from the initial state to the final state in a great variety of ways. In other words, there is no unique path for the system to take in going from the initial to the final state.*

As an example we may picture the gas in a cylinder as going from a given state in which P, V, and T have definite values to a state in which the volume is one-half as large and the temperature is unchanged (Figure 16.4). This may occur in two ways: on the one hand we may lower the piston very quickly, thus generating a large amount of heat as a result of the work done on the gas. We can then allow this heat to flow off until the temperature is equal to its initial value. On the other hand we may lower the piston very slowly, so that the temperature always remains the same (an isothermal process) and any small amount of heat generated flows out of the gas very slowly. (We do this by keeping the cylinder in contact with a reservoir of heat that is always at the same temperature as the gas in the cylinder.) The final result, as far as the gas is concerned (i.e., the final state of the system), is the same in both cases; yet one process is associated with a rapid outflow of heat, the other with a very slow transfer of heat at each state of the compression.

16.14 Reversible and Irreversible Processes

These two processes differ in one other remarkable and profound characteristic. In the slow process the system is always in **equilibrium**, so that if we stop the process at any moment, everything reverses itself; the piston rises again to its initial position. In the case of this slow compression, the work we do in pushing the piston down does not increase the internal energy of the gas since its temperature remains constant at all stages of the compression. This means, by the first law of thermodynamics, that there is a slow (and hence reversible) transfer of heat (this is just the work we do during the compression) from the gas to the reservoir. Hence, if we allow the piston to rise again very slowly to its initial position, the same amount of heat will flow slowly from the reservoir back to the gas again and we thus retrieve all the work that we put into the system during the compression. This is so since the temperature of the gas (and hence its internal energy) remains constant during this slow expansion so that the heat that flows back must just equal the work done. Thus, since we get back the work that we put in, and since the piston is back to its initial position and its internal energy (temperature) is the same, everything is as it was initially *and this slow compression is reversible.*

This is certainly not true in the rapid compression case, for even if we leave the system to itself after the piston has been violently pushed down, and then allow heat to flow slowly back from the reservoir into the gas until the piston is back to its original position (P, V, T are then equal to their initial values), the amount of heat that has returned to the cylinder, and hence the work done by it in raising the piston, is the same as in the second case (slow compression case), *but it is not equal to the amount of heat that flowed from the hot compressed gas into the reservoir*, and hence it is not equal to the work done during the rapid compression. In fact it is less than this, so that some of the heat that flowed into the reservoir (that is, some of the work done during the rapid compression) remains as unavailable internal energy of the reservoir, even though the gas in the cylinder is back to its initial state again.

The reason for this is that we do more work during the rapid compression than we do during the slow compression. During the slow compression we work against an average pressure determined entirely by Boyle's law, which is the average of the initial and final pressures as determined entirely by the initial and final volumes of the gas. During the rapid compression, however, the temperature of the gas immediately rises and the average pressure of the gas during the entire compression is higher than that in the slow compression. Hence we do more work during the rapid compression and all of this is not returned to us when the piston comes slowly back to its initial position. *Thus the rapid compression is irreversible.*

These examples show us that two types of processes can occur in isolated systems—a **reversible process**, and an **irreversible process**. The importance of the second law of thermodynamics lies in the fact that it recognizes that *events in nature can be separated into reversible and irreversible processes.*

16.15 Statement of the Second Law of Thermodynamics in Terms of Irreversible Processes

We may state the second law either in terms of the irreversible process chosen by the physicist Clausius, who expressed the second law as follows: *if two isolated bodies are in contact, heat always flows spontaneously from the hotter to the colder body,* or by the equivalent irreversible process of Lord Kelvin who expressed the second law as: *in an isolated system, heat can never be completely transformed into work; some of the heat always becomes unavailable internal energy.* These statements express the fact that there are processes in nature that can go only in one direction (they are irreversible).

We may express this law somewhat differently by saying that all irreversible changes that occur in an isolated system result in a decrease of the total available energy in the system (that is, energy that can do work).

To formulate the second law mathematically, we again consider a system going from a given state A to a given state B. As we have seen, there is an infinitude of paths that such a system can take, and yet regardless of which path it takes, the final and initial states of the system are the same. There must therefore be some intrinsic characteristic of the system whose change depends only upon the initial and final states, and not on the path that the system follows. What is the nature of this quantity? It is certainly not the total heat that the system gives up or receives when it goes from its initial to its final state since this can have almost any desired value. The work that is done on or by the system also depends on the path the system takes.

16.16 The Concept of Entropy

There are, however, two quantities that are independent of the path, and that depend only

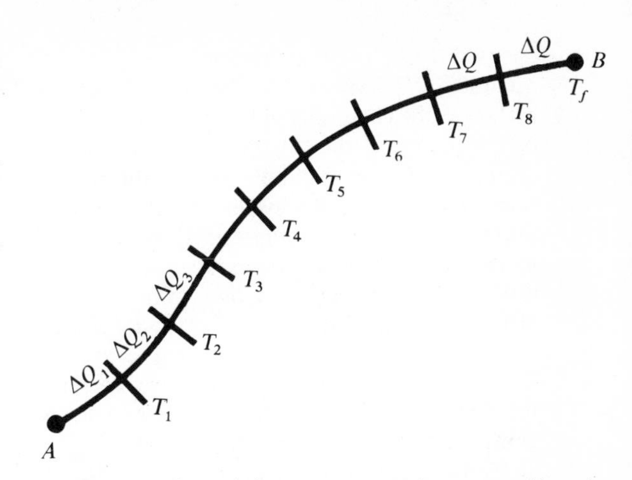

Fig. 16–5 *The total change in entropy along a reversible path from A to B is the sum of all the infinitesimal quantities $\Delta Q_i/T_i$ over this path.*

upon the initial and final states of the system. One of these is the internal energy of a system, which we have already considered, and the other is a quantity called the **entropy** of a system, which was first introduced by Clausius. To define this quantity, S, we consider a system changing along a reversible path from an initial state, A, to a final state, B. We divide this path into many (n in number) small steps and suppose that the system gains (or loses) a small amount of heat ΔQ_i, during the ith step. Let the temperature of the system during this ith step be T_i, as shown in Figure 16.5, and let us consider the quantity $\Delta Q_i/T_i$.

We now add all these quantities for each step; in other words, we consider the sum

$$\frac{\Delta Q_1}{T_1} + \frac{\Delta Q_2}{T_2} + \cdots + \frac{\Delta Q_n}{T_n},$$

where T_1 is the temperature of the system in the initial state A, and T_n is the temperature of the system in the final state, B. We define this sum as the change, $S_B - S_A$, in the entropy of the system when the system goes from its initial to its final state *along a reversible path.*

The concept of entropy is extremely important; therefore we shall further consider and clarify the above definition of the change of the entropy of a system. We may first note that *the above sum truly represents the change in the entropy of a system only if each of the above steps is* ***infinitesimal***, because it is only then that the system can lose or gain an infinitesimal amount of heat, ΔQ, which is a necessary condition for the change to be reversible. If a system loses or gains a finite amount of heat, Q, during any step in its path, its temperature changes and this step is not reversible, so that the quantity Q/T (where T is the temperature of the system at the beginning of the step) is not a measure of the change of the entropy as defined above.

For this reason the above sum evaluated along a reversible path of the system must consist of an infinite number of infinitesimal terms. Mathematicians call such a sum an **integral**, and it can be evaluated according to certain rules. We should also note that if a system loses an infinitesimal amount of heat ΔQ at the absolute temperature T (a reversible loss of heat) its entropy **decreases** by the amount $\Delta Q/T$. In the same way if the system gains an infinitesimal amount of heat ΔQ at the absolute temperature T, the entropy of the system **increases** by an amount $\Delta Q/T$.

Clausius chose this definition of the change of the entropy of the system because he was able to show that the sum $\Delta Q_i/T_1 + \cdots + \Delta Q_n/T_n$ for a system going from some initial to some final state depends on the path that is taken by the system if the path is irreversible, but that this sum is the same for all reversible paths. In other words, this sum, taken along any reversible path of the system, depends only on the initial and final states of the system (not on the reversible path that is chosen) and hence represents the change in some intrinsic quantity of the system that depends only upon the initial and final states. Clausius called this intrinsic quantity the **entropy** of the **system**. Although we have defined the change of entropy between two states of a system as the above sum taken along a reversible path between the two states, it defines the change in entropy regardless of how the system goes from one state to the other. In other words *the change in entropy of a system between two fixed states is independent of the path (reversible or irreversible) the system takes to go from one to the other.*

16.17 The Application of the Clausius Sum and Examples of Entropy Change

We can best illustrate the difference between the Clausius sum

$$\left(\frac{\Delta Q_1}{T_1} + \frac{\Delta Q_2}{T_2} + \cdots + \frac{\Delta Q_n}{T_n}\right)$$

for reversible and irreversible processes that take the system from the *same initial state to the same final state* by analyzing a simple but very instructive example. We again consider a perfect gas in a cylinder at pressure P_i, volume V_i, and temperature T. These define the initial state of the system. The initial pressure is given by the appropriate weight W_i on the piston in the cylinder as shown in Figure 16.6(a). We now consider this gas in some final state defined by P_f, V_f, and T, where $P_f < P_i$, and $V_f > V_i$ so that the gas has undergone an expansion. The final pressure is given by a smaller set of weights W_f on the piston as shown in Figure 16.6(b). Note that the initial and final temperatures are the same.

There is an infinite number of processes, both reversible and irreversible, by means of which the system can go from the initial to the final state. Of these we shall consider just two, one reversible and the other irreversible, at all stages of which the temperature of the system remains equal to T (isothermal processes). Of course there are non-isothermal processes which are also reversible and irreversible, but for these it is more difficult to calculate the Clausius sum.

The first isothermal process we shall consider is the reversible one in which we allow the gas in the cylinder to expand very slowly, keeping T constant. We do this by slowly reducing the weights on the piston of the cylinder (at just the proper rate to keep the pressure always inversely proportional to the volume according to Boyle's law) while the cylinder is kept in contact with a large heat reservoir at temperature T. During this process the gas does work against the piston (by pushing it up) and is kept from cooling off by a steady but very slow transfer of heat from the reservoir to the cylinder. Because of this slow transfer of heat, the process is reversible at each stage.

We can now calculate the Clausius sum

$$\frac{\Delta Q_i}{T_i} + \frac{\Delta Q_2}{T_2} + \cdots + \frac{\Delta Q_f}{T_f}$$

as follows: We first note that $T_i = T_2 = \cdots = T_f = T$, since the process is isothermal and is carried out at the constant temperature T. We may therefore write the sum as

$$\frac{1}{T}(\Delta Q_i + \Delta Q_2 + \cdots + \Delta Q_f).$$

We now note that because the temperature of the gas remains constant, its internal energy does not change during the process. Hence, according to the first law of thermodynamics, the total heat transferred from the reservoir to the gas (the quantity in the parenthesis above) is just the work done by the gas in lifting the piston. Thus the

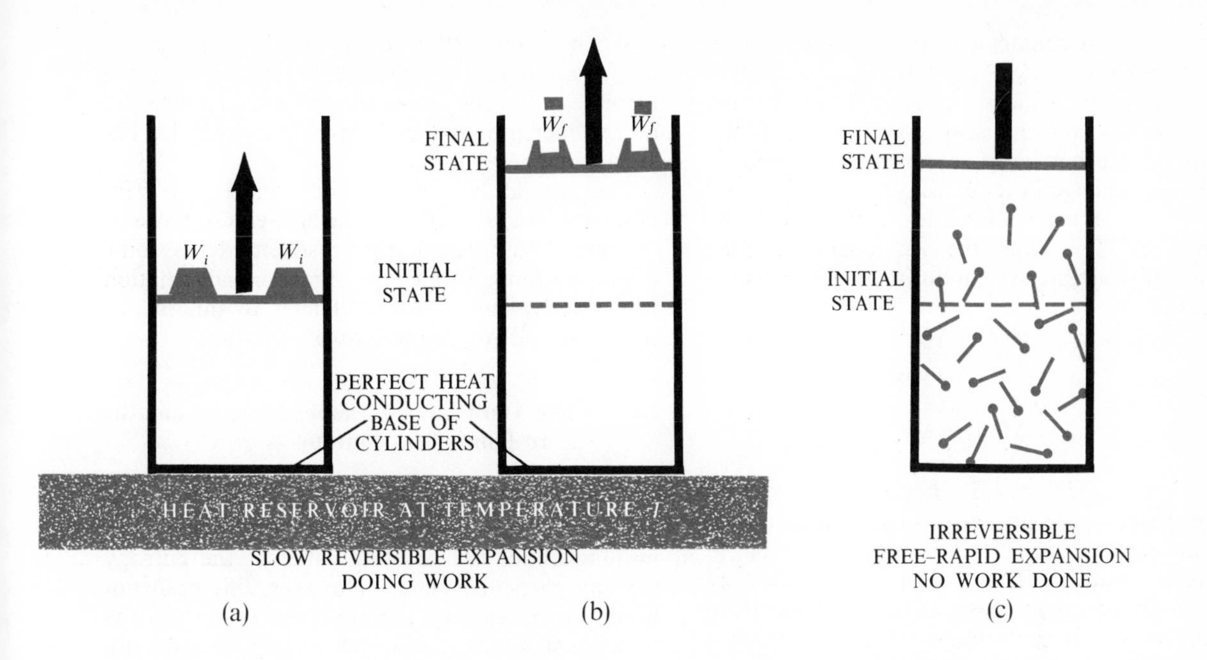

FIG. 16–6 *When a gas expands reversibly (very slowly as it does work against a piston) its change of entropy is given by the Clausius sum [(a) and (b)]. When the gas expands freely (irreversibly) as in (c), without doing any work, the Clausius sum does not give the change in entropy, for it is smaller than this change.*

Clausius sum, and hence the change in entropy of the gas as it expands from its initial to its final state (both at the same temperature), is

$$S_f - S_i = \frac{\text{work done on piston}}{T}. \quad (16.11)$$

Since this quantity is positive, the entropy of the gas increases. Note that in this case the Clausius sum and the entropy change are equal because the Clausius sum is evaluated along a reversible path.

It should be noted that although this quantity giving the change in entropy was computed by considering a special reversible process, it is the entropy change no matter how the system gets from its initial to its final state. *But only for reversible processes does one find that the Clausius sum gives the entropy change.* It should also be noted that in the reversible process considered above the heat reservoir loses exactly as much entropy as the gas gains so that the change of entropy of the entire system (gas and reservoir together) is zero. This is as it should be, because the total process in the combined system is reversible and *in an isolated system the entropy remains constant for reversible processes.*

Let us now consider an irreversible process that takes the gas from its initial to its final state and see that the Clausius sum calculated for such a process does not give the change in entropy of the gas. We allow the gas to expand freely from its initial volume V_i to its final volume V_f without its pushing against a piston (that is, without its doing work). See Figure 16.6(c). This process is irreversible because the gas, if left to itself, cannot compress itself back to its initial volume.

Since the gas does no work during this free expansion, its internal energy and hence its temperature remain constant. *This means that it takes in no heat during any stage of the free expansion.* Hence the Clausius sum for this process is zero and certainly not equal to the change in entropy. There are, of course, infinitely many irreversible processes, each one of which consists of a number of separate free expansions (irreversible) and a number of slow work-performing expansions (reversible). *In all of these the Clausius sum is smaller than its value for a completely reversible process which gives the change in entropy. Thus the change in entropy is the maximum value of the Clausius sum and this is attained along a reversible path (process).*

As another example of the application of Clausius' definition to the change in the entropy of a system we consider a system consisting of a hot body of absolute temperature T_1 in contact with a cooler body of absolute temperature T_2 ($T_1 > T_2$). If a small amount of heat ΔQ flows from the hot body to the cooler one, then from the above discussion we see that the entropy of the hot body decreases by an amount $\Delta Q/T_1$ (the change in its entropy is $-\Delta Q/T_1$) and the entropy of the cooler body increases by the amount $\Delta Q/T_2$. Thus the *net* change in the entropy of the entire system during this irreversible process is

$$\Delta S = \frac{\Delta Q}{T_2} - \frac{\Delta Q}{T_1}$$

$$= \Delta Q \left(\frac{1}{T_2} - \frac{1}{T_1}\right).$$

Since $T_1 > T_2$, $1/T_2 > 1/T_1$; hence the parenthesis in the above expression is positive, which means that the net change in the entropy of the entire system is positive. This means that the entropy of the entire system increases as the heat flows from the hot body to the cold body. *In general we may say that the entropy of a system will always increase if there is any net flow of heat from one part of the system to any other part.*

It should be noted in the example just cited that although the first body gives up its heat reversibly and the second body takes up the heat reversibly, the process as a whole is irreversible because there is a *net* flow of heat as a result of a finite temperature difference.

As another instructive example, we note that whenever friction is involved the total entropy increases. This is so because friction causes heat to flow into a body so that there is an immediate increase in entropy, but there is no compensating decrease because there is no heat loss by any object. The heat gained by the body comes from work done.

16.18 The Significance of the Second Law of Thermodynamics

An important property of the entropy is the following: *no matter what happens in an isolated system, its entropy can never decrease. If the system undergoes none but reversible processes, the entropy remains constant, but any irreversible process is always accompanied by an increase in the entropy. In other words, whenever energy becomes unavailable (which is the earmark of an irreversible process), entropy increases. This means that the entropy of the universe as a whole is constantly increasing.*

We may relate this to the increase in the disorganization in the universe by stating that the increase in the entropy of the universe is a measure of the increase in the disorganized (that is, unavailable) energy in the universe!

The second law of thermodynamics is a very powerful tool for analyzing the behavior of physical systems. With its aid, it is possible to derive some important fundamental laws of matter and radiation with relative ease. We shall refer to this in our further discussion of the laws of radiation.

16.19 The Third Law of Thermodynamics and the Zero Point of the Entropy

Our discussion above defines only the change of entropy between two states of a system; it tells us nothing about the absolute value of the entropy for any particular state. However, this need not bother us because we can always refer the entropy to some standard state which we can all agree on. In any case, only differences in entropy are important in considering the changes in the state of a system.

As long as we limit ourselves to the first and second laws of thermodynamics we have no way of introducing an absolute definition of the entropy. However, there is a third law of thermodynamics, originating in the quantum theory and first expressed by Nernst (also known as the **Nernst heat theorem**). This law, which enables us to assign an absolute value to the entropy of a system in any state, states that *the entropy of any system goes to zero when its absolute temperature goes to zero.* Since this theorem gives us the entropy for what we shall call the **zero state** of the system (entropy zero at absolute temperature 0°), we can find the entropy of any other state by computing the step-by-step change of the entropy along a reversible path from the zero-entropy state to a final given state according to Clausius' method given above.

Thus knowledge of the entropy of a system gives us an insight into the state of the system.

16.20 Basic Properties of Radiation

Now that we have some understanding of heat, temperature, and entropy, we shall see that we can get at the surface temperature of a star by studying the radiation that this star emits. It is clear from the most cursory examination of the way hot bodies emit radiation (for example, a very hot piece of metal) that both the color and the rate of radiation depend in some way upon how hot the surface of the body is. Hence, if we know the relationships (physical laws) that exist between the surface temperature of this hot body and the color and the amount of radiation it emits per second, we can find its surface temperature. Before we consider these relationships we shall review some basic ideas of radiation.

Although Newton propounded a corpuscular theory of radiation and lent the great weight of his genius to its promulgation, it was contradicted by experimental evidence and was ultimately replaced by the wave theory as presented by Huygens (a Dutch contemporary of Newton's). Thus, Newton's corpuscular theory predicted that light should travel faster in a dense medium than in a less dense medium. We know, however, from the experiments of Fizeau, Michelson, and others, that just the opposite is true. Moreover, the Newtonian corpuscular theory finds it difficult, if not impossible, to account for interference and diffraction of light (i.e., the bending of light around corners). Therefore, we shall utilize the wave theory of radiation in the first part of our discussion.

According to the wave theory light consists of transverse electromagnetic vibrations which are propagated in a vacuum at 3×10^{10} cm/sec. As shown in Figure 16.7, such vibrations are characterized by three numbers:

1. the distance between two successive crests, the **wavelength**, λ;
2. the number of crests passing a given point per second, the **frequency**, (ν); and
3. the speed of light, c.

Since the speed, c, is the distance traveled by a crest in a unit time, we must have

$$\boxed{\lambda\nu = c}\,. \tag{16.12}$$

The electromagnetic characteristic of light was first proposed theoretically by James Clerk Maxwell.

The wavelength of light determines what we recognize as **color**. But although our eyes respond only to a small range of wavelength, from $\lambda =$ 7 500 Å (red light) to about 3 900 Å (violet light) (where Å, the **angstrom**, is a unit of length equal to 10^{-8} cm), the spectrum of electromagnetic radiation ranges from the very largest radio waves having wavelengths of many miles, down to the very penetrating gamma radiation having wavelengths of less than one angstrom. If we could spread out all the radiation in the universe according to wavelengths, we would have a continuous variation from the very long to the very short, as shown in Figure 16.8.

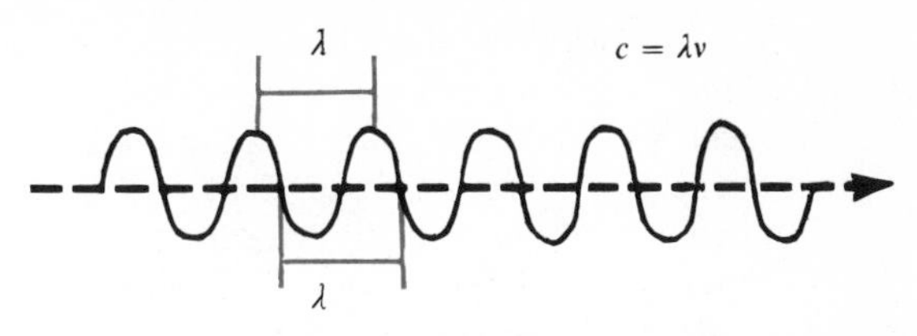

FIG. 16–7 *The relationship between wavelength, frequency, and speed of a wave of light.*

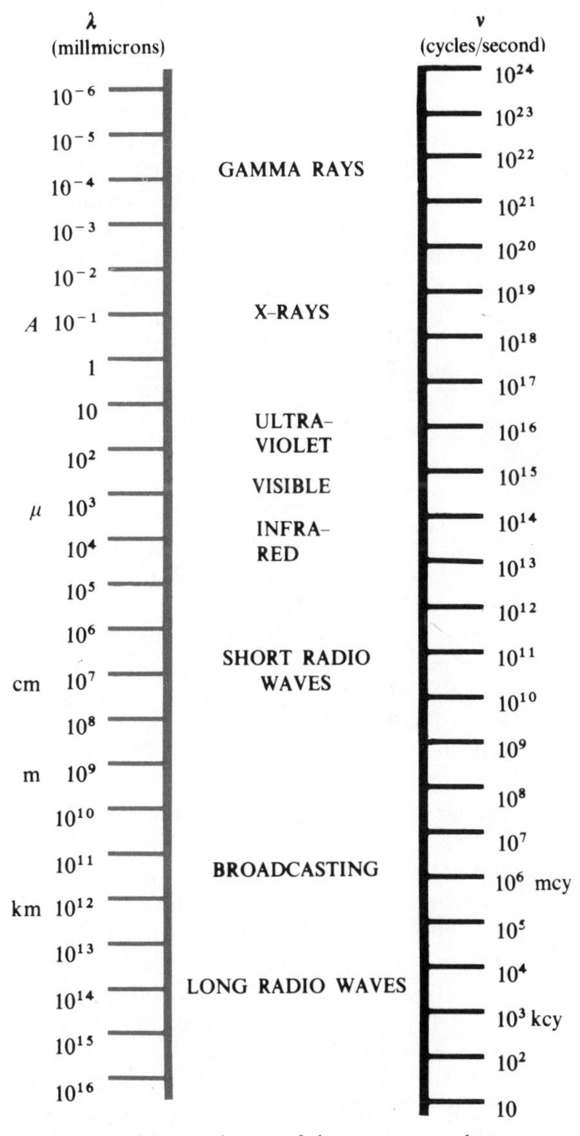

FIG. 16–8 *The spectral range of electromagnetic radiation.*

Since the wavelength and frequency must change together, it follows that the long waves have small frequencies and the short waves have large frequencies. It is easy to see that red light has a frequency of about 4×10^{14} vibrations per second, whereas violet light has a frequency of about 8×10^{14} vibrations per second.

16.21 The Nature of a Surface and the Radiation It Emits

Since we shall be concerned with the radiation emitted by stars, which are hot bodies, we must ask at this point what the general relationship is between the temperature of the surface of a hot body and the way that surface emits energy. As it turns out, this question has meaning only if we first specify the nature of the surface, because two qualitatively different surfaces raised to the same temperature do not emit energy at equal rates. We may understand this better if we consider the way a body absorbs radiation because, as we shall see, the emission and absorption of radiation by a surface are related.

16.22 Emission and Absorption of Radiation by Various Surfaces: Black Bodies

We consider a flat surface on which there is incident a monochromatic beam of radiant energy, as shown in Figure 16.9(a). Two things may happen to this beam. On the one hand, it may be **reflected**. This means that it bounces off the surface without any change in its character, i.e., the wavelength (the color) of the reflected beam is exactly the same as that of the incident beam. On the other hand, the beam, or part of it, may be **absorbed**. This means that the radiant energy at incidence is transformed into internal energy of the surface (i.e., random molecular motion). Now these two processes, reflection and absorption, are mutually exclusive. The more radiant energy the body absorbs, the less it reflects, and vice versa.

A surface that reflects all radiation striking it (regardless of the wavelength) is called a **perfect reflector**. Obviously such a surface is white (here we are dealing with diffuse reflection and not specular reflection, as in the case of a mirror). A surface that completely absorbs all wavelengths of radiation falling on it is called a **perfect absorber**. Obviously, such a surface is black; indeed, it is referred to as a **black body**.

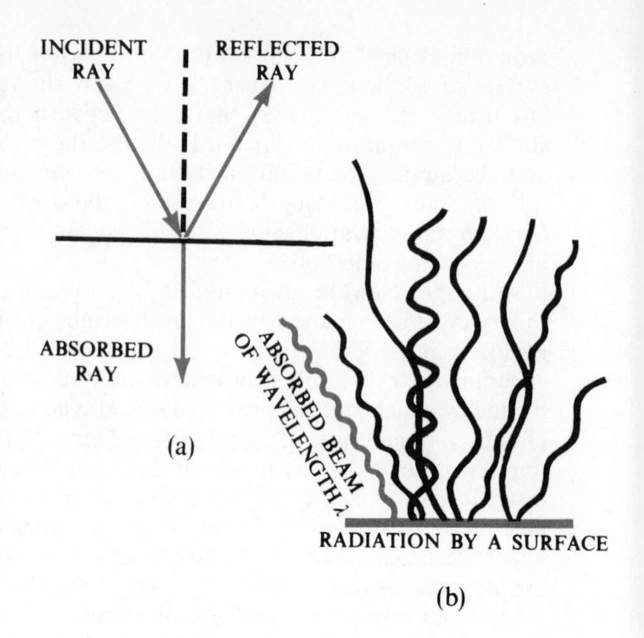

FIG. 16–9 *The behavior of a beam of light on striking a surface. (a) In addition to the incident beam there are a reflected and an absorbed beam. (b) The surface begins to radiate energy as soon as it absorbs energy. The radiated energy is spread over the entire spectrum.*

We note that a perfect absorber is necessarily a **perfect radiator** because the more energy a body absorbs per second, the hotter it gets and the more energy it re-radiates per second. It should be emphasized, though, that re-radiation of energy by a surface is not the same as reflection. If light of a given wavelength is reflected, it comes back to us exactly as light of that same wavelength; but if it is first absorbed and then re-radiated, it comes back to us as radiation consisting of a mixture of many wavelengths. In other words, an absorber takes the energy of a specific wavelength and then spreads it out over all the wavelengths in various concentrations as shown in Figure 16.9(b).

16.23 Black-Body Radiation

Since a black body is the most efficient emitter of radiation for a given temperature, and since we are interested in the universal laws of the emission of radiation, we shall consider the laws of black-body radiation, i.e., how the radiation emitted by a perfect black body (perfect absorber) depends upon the absolute temperature of the black body. This was one of the most important problems facing physicists of the late nineteenth century, and because classical physics failed to find an answer

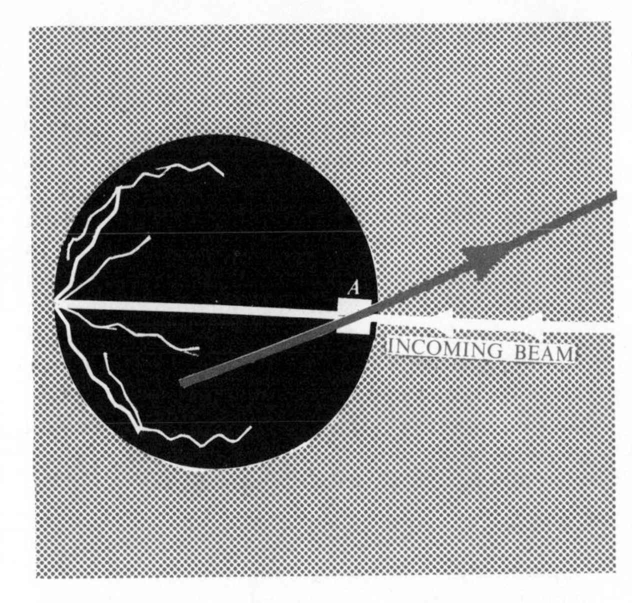

FIG. 16–10 *A small opening A in the wall surrounding a hollow region (hohlraum) behaves like a black body.*

to it the **quantum theory** was born. The problem is essentially the following:

> given a black body of unit area and absolute temperature T, how is the radiant energy this body emits per second distributed over various wavelengths?

In other words, if $\mathscr{I}_{\lambda T}$ is the intensity of the radiation emitted in the wavelength λ, how does this depend upon the wavelength of the emitted radiation and the absolute temperature of the body?

We may first note that we can very easily approximate a black body to any desired degree of accuracy by considering a hollow region surrounded by absorbing walls with a small opening in one wall. See Figure 16.10. *This small opening behaves like a perfect black body because a beam entering A is completely absorbed by the walls of the cavity. The emitted beam (in red) shows black-body radiation coming from the opening. Therefore if the walls of this container are heated, the radiation coming from the small opening is black-body radiation.* We may associate with this radiation a definite temperature T, which is the temperature throughout the hollow region, and hence of the radiation which is in thermodynamical equilibrium with the walls of the container. We may therefore speak of the temperature of the black-body radiation emitted from this small opening. Essentially what we have here is a furnace with a small window, so that we may say that radiation emitted through a small window of a furnace is black-body radiation.

16.24 The Problem Associated with Black-Body Radiation

The question that now arises is the following: If the temperature inside the hollow region is a definite value, T, what is the nature of the radiation that is emitted from one square centimeter of opening every second? In other words, if we consider black-body radiation of temperature T and analyze it spectroscopically, how much of each particular color is present, or how does the intensity of this radiation depend upon its wavelength? The answer was known experimentally since all one had to do to find it was to analyze the radiation spectroscopically. This can be represented graphically (see Figure 16.11) by means of curves that show the relationship between intensity and wavelength for different absolute temperatures. The problem which the physicists of the nineteenth century had to solve is the following: using the laws of classical physics (mechanics, thermodynamics, electricity, etc.) can one derive the algebraic expression for these curves theoretically? As it turned out, classical physics was unable to solve this problem, and it became clear by the end of the nineteenth century that only a radical departure from all previous existing theories could lead to a solution.

The two attempts made by outstanding physicists of that period led to algebraic expressions that agreed with the curves at one end of the spectrum but deviated from them at the other end. Before we see how the correct algebraic expression was finally obtained, we shall indicate how far classical physics is able to go in giving some insight into the nature of this radiation.

16.25 Wien's Displacement Law

If we examine each of the plotted curves representing actual black-body radiation shown in Figure 16.11 we observe that they rise fairly steeply to a maximum at the short-wavelength side, and then fall off to zero gradually. In other words, for any given absolute temperature, there is one color (one wavelength) for which the intensity of the black-body radiation is a maximum. We may also observe that as the temperature rises (in other

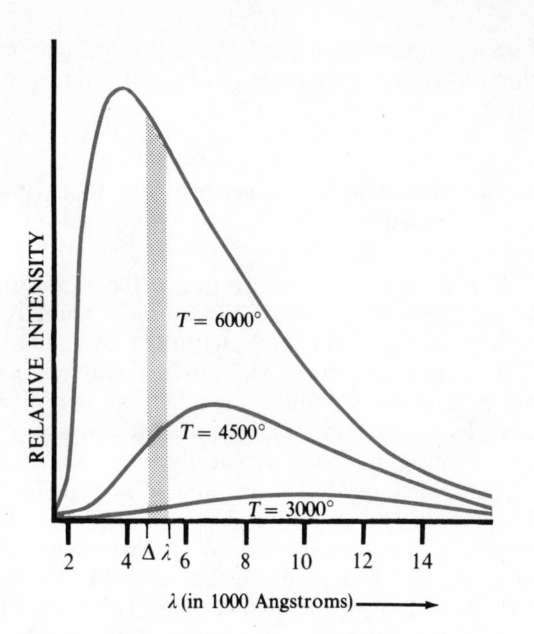

FIG. 16–11 *A graph of the intensity of the radiation emitted by a black body as a function of the wavelength. The shift of the wavelength of maximum intensity with increasing temperature is shown. The small shaded segment represents the energy emitted per second in the wavelength interval* $\Delta\lambda$.

words, as we go to the curves higher up on the intensity scale) the wavelength associated with this maximum gets smaller and smaller. What this means observationally is that the hotter a furnace is, the greater is the intensity of the blue-violet colors in the radiation as compared with the red-yellows. This, of course, agrees with our knowledge that as an object is heated it first takes on a deep red hue and then becomes yellowish and then finally whitish as its temperature rises.

The dependence of the wavelength of maximum intensity on the absolute temperature can be found by using the classical laws of physics. This was done by the physicist Max Wien at the end of the nineteenth century. *He was able to show from the laws of classical physics that the wavelength of maximum intensity varies inversely with the absolute temperature*; in other words,

$$\lambda_m = \frac{C}{T}, \tag{16.13}$$

where C is a universal constant whose value 0.289 8 cm deg can be measured in the laboratory. This formula is known as **Wien's displacement law.**

16.26 Stefan-Boltzmann Law

Still further information can be obtained about black-body radiation from the laws of classical physics. If we examine one of the radiation curves and consider a small area of width $\Delta\lambda$ under it, we see that this represents the radiation emitted per second in the wavelength λ over the range $\Delta\lambda$ as shown in Figure 16.11. Therefore, the area under the entire curve represents the radiation emitted per second in all colors (all wavelengths). How does this depend upon the absolute temperature of the black body? Again it was found that this question can be answered by using the laws of classical physics. The answer was first given experimentally by the Austrian physicist Stefan, and later derived theoretically by the great physicist Boltzmann who applied the second law of thermodynamics to black-body radiation enclosed in a container. He thus obtained what is called the **Stefan-Boltzmann** law in the following form:

$$\boxed{\epsilon = \sigma T^4}, \tag{16.14}$$

where ϵ is the total energy (in all wavelengths) emitted per second from 1 square centimeter of the black body and σ is a universal constant called the **Stefan constant** (its value can be measured in a laboratory and is $5.669\,2 \times 10^{-5}$ erg cm^{-2} deg^{-4} sec^{-1}). This formula means that if the absolute temperature of a black body is doubled, each square centimeter of its surface emits 2^4 or 16 times as much energy per second as previously, and if it is tripled 3^4 or 81 times as much as previously, etc.

These two formulae, Wien's displacement law and the Stefan-Boltzmann law, are as far as classical physics is able to go. The complete formula giving the intensities for all wavelengths cannot be derived from classical physics. Nevertheless, even without the complete formula it is possible to use the Stefan-Boltzmann law and the Wien displacement law to determine the effective surface temperatures of stars.

16.27 The Effective Surface Temperatures of Stars

We can use the Wien displacement law and the Stefan-Boltzmann law to determine the surface temperatures of stars if we suppose that stars emit radiation like black bodies. Although we shall see that this is not exactly true, it is a sufficiently good approximation to enable us to apply the laws

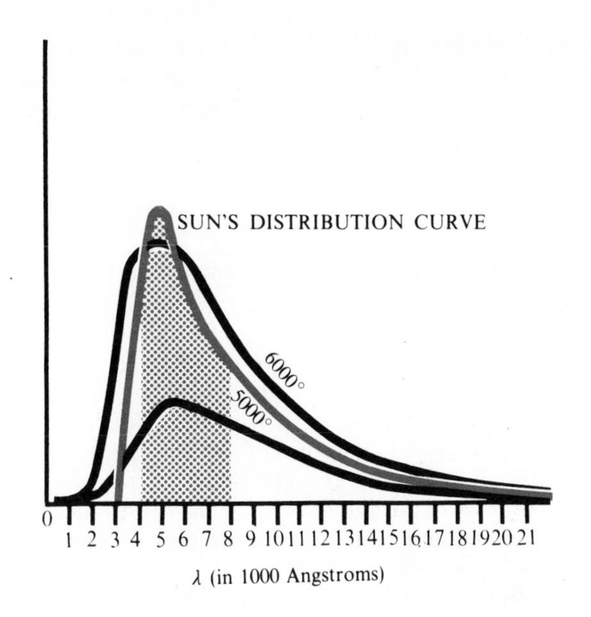

FIG. 16–12 *The radiation emitted by the sun compared with the radiation emitted by a black body of the same temperature.*

of black-body radiation to stars. The reason is that stellar radiation originates in the very deep interior of the star and has therefore been absorbed and re-emitted so often by the time it gets to the surface that it is very nearly black-body radiation.

To find the surface temperature (that is, the temperature of the photosphere of a star) by using Wien's displacement law, all we have to do is determine the wavelength of the most intense color in the radiation from the star. This gives us the quantity λ_m, and since we know the constant C in Wien's law we can solve for the absolute temperature T. We might refer to this as the **color temperature** of the star. This quantity, however, is only approximately related to the actual physical temperatures of the star's surface since a star does not radiate exactly like a black body. Although the departure from black-body radiation is small, it is significant. This is shown in Figure 16.12, where the radiation from the sun is compared with the curves for black-body radiation. We shall come back to the idea of the color temperature later.

A physically more meaningful number associated with the surface temperature can be obtained if we consider the radiant energy emitted per second in all colors (all wavelengths) from one square centimeter of the star's surface. If the surface of a star radiated exactly like a black body, the energy ϵ emitted per second from one square centimeter would obey the Stefan-Boltzmann law: $\epsilon = \sigma T^4$. If we know the size and distance of a star we can find ϵ from observations and solve for T. Although the quantity T obtained from this equation is not the real temperature of the surface of the star, we may nevertheless refer to it as the **effective temperature** which we define as follows:

The effective temperature, T_e, of a star is the temperature to which a black body (furnace) would have to be raised so that the radiant energy emitted per second from one square centimeter of its surface would equal that emitted per second from one square centimeter of the star's surface.

In general, although the effective temperature and the color temperature defined above are close to each other they are not equal.

If we had an algebraic formula that for a given temperature could give us the entire distribution of radiation for all colors, we could then find the surface temperature of a star (assuming it to be a black body) by determining which one of these formulae best fits the actual radiation emitted in all colors by the star. But, as we have seen, classical physics is unable to give us such a formula. To obtain one we must consider the changes introduced in classical physics that led to the **quantum theory of radiation.**

16.28 Planck's Radiation Law for Black Bodies

At the close of the nineteenth century it was obvious that classical physics was powerless to solve the problem of black-body radiation and that some drastic changes would have to be introduced. In the year 1900 Max Planck succeeded in obtaining a formula for $\mathscr{I}_{\lambda T}$ which agrees in every detail with the experimental data for black-body radiation. But he did so by introducing an entirely new concept of radiation.

Up until the time of Planck's work, it was assumed that radiation is entirely wavelike. In other words, it was assumed that radiation spreads out continuously and occupies every element of space however small. This means that if we subdivide a hollow region containing radiation into infinitesimally small spatial elements, each such spatial element contains a small bit of radiation as shown in Figure 16.13.

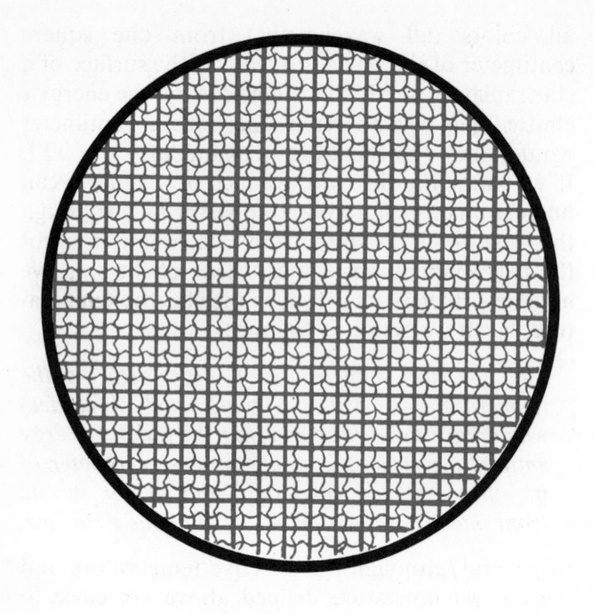

FIG. 16–13 *According to the classical wave theory of radiation, every small element of volume (however small it may be) contains a non-zero amount of radiation energy.*

This would, indeed, be the case if radiation were a wave phenomenon, spreading out uniformly in all directions. This assumption Planck had to deny to obtain the correct radiation formula. In fact, he had to suppose that black-body radiation is emitted and absorbed in little pellets, which he called **quanta of energy** (now **photons**). Later, Einstein was able to show that not only is radiation emitted and absorbed in the form of pellets, but that it has this characteristic at all times, even when it is in a hollow enclosure. This means that the radiation is not distributed continuously in a given region, but is concentrated in pellets, or photons, which have **corpuscular** or particle-like properties.

In order to derive the correct radiation formula Planck had to assign to each photon an amount of energy, E, that is different for different-colored photons. He found it necessary to assume (now an established fact) that the blue photons are more energetic than the red photons, and he expressed this by means of his famous formula

$$\boxed{E = h\nu}, \tag{16.15}$$

where E is the energy of the photon, ν is its frequency and h is a universal constant known as **Planck's constant of action.** Its numerical value, which can be measured in the laboratory, equals 6.625×10^{-27} gm cm^2/sec. *We see from this formula that the larger the frequency of a photon (i.e., the bluer it is) the more energy it contains.*

16.29 The Fundamental Constant of Action

The existence in nature of a fundamental constant of action, h, shows that if we could magnify processes sufficiently (i.e., if we could actually see subatomic processes) we would see that they do not occur continuously but in a kind of jerky, spasmodic fashion, and that no process can occur that involves an amount of action less than h. We shall define the concept of action precisely in Section 18.18, although here we may just note that it is a measure of the way a body or a system of particles is moving about.

Using the concept of the photon and Equation (16.15) Planck derived the following radiation formula for the intensity, $\mathscr{I}_{\lambda T}$, of the emitted radiation of wavelength λ:

$$\mathscr{I}_{\lambda T} = \frac{2hc^2}{\lambda^5}\frac{1}{e^{hc/k\lambda T} - 1}, \tag{16.16}$$

where c is the speed of light, h is Planck's constant of action, k is Boltzmann's constant, and e is the base of the natural logarithm (2.718). This formula gives us the amount of energy in each color emitted per second from one square centimeter of a black body at a temperature T.

16.30 The Significance of Planck's Radiation Law

Planck's formula can be expressed in words as follows: if the absolute temperature of the surface of the black body is T, then the intensity of the energy emitted per second in the wavelength λ from one square centimeter of this body is obtained by taking twice Planck's constant, multiplying this by the square of the speed of light and dividing the product by the fifth power of the wavelength of radiation being considered, and then dividing this entire quotient by the quantity $(e^{hc/k\lambda T} - 1)$.

Actually, of course, just as one cannot have a physical point or an infinitely thin line, one cannot speak of the radiation emitted in one particular wavelength since it is physically impossible to pick out, by means of instruments, radiation having a single definite wavelength. *The only thing that has physical meaning is the intensity of the energy emitted in the wavelength range from λ to $\lambda + \Delta\lambda$,* *where* $\Delta\lambda$ *is a small but measurable range of wavelengths.* When we use Planck's formula with this understanding, we must multiply it by $\Delta\lambda$ to give the total intensity of energy emitted. Thus, we may speak of the energy emitted in the wavelength range from 5 880 Å to 5 880.1 Å, in which case $\Delta\lambda$ equals 0.1 Å or 10^{-9} cm. We must therefore multiply Planck's formula by this value to obtain the intensity of the energy emitted per second by one square centimeter of a black body *in this wavelength range.*

What we measure physically is not the intensity itself (see Figure 16.14), which is the energy emitted per unit time by a unit area into a unit solid angle in a direction exactly at right angles to the unit area, but rather the ***flux***, *which is the radiation leaving the surface per unit time in all directions.* We can obtain this from Equation (16.16) by simply multiplying by π. In other words $\pi\mathscr{I}_{\lambda T}\,\Delta\lambda$ is the amount of radiation emitted per second in all directions (in the wavelength range from λ to $\lambda + \Delta\lambda$) by one square centimeter of a black body having an absolute temperature T.

Returning to the previous example, we may note that the amount of energy emitted per second in all directions from one square centimeter of the sun's surface (assuming that it radiates like a black body) in the wavelength range from 5 880 Å to 5 880.1 Å is given by (using Planck's formula)

$$\frac{(2\pi)(\overset{h}{6.62\times 10^{-27}})(\overset{c^2}{3\times 10^{10})^2}(\overset{\Delta\lambda}{0.1)\times 10^{-8}}}{\underset{\lambda^5}{(5\,880\times 10^{-8})^5}\left[\underset{e}{2.718}^{\frac{(6.62\times 10^{-27})(3\times 10^{10})}{(1.38\times 10^{-16})(5\,880\times 10^{-8})(5\,500)}}-1\right]}\ \frac{\text{ergs}}{\text{sec}}$$

where we have used 5 500° for the absolute temperature of the surface of the sun. If we carry out the indicated arithmetic, we find a value of 4×10^9 ergs/sec. In the same way we can calculate the amount of energy radiated per second by one square centimeter of the sun for any wavelength.

16.31 The Brightness of a Star in a Given Wavelength

With the aid of Planck's formula we can obtain the color temperature of stars. To see how, we note that from Planck's formula, assuming that stars radiate like black bodies, we can determine the amount of energy coming to us per second from a star in any particular wavelength range just as we did above for one square centimeter of

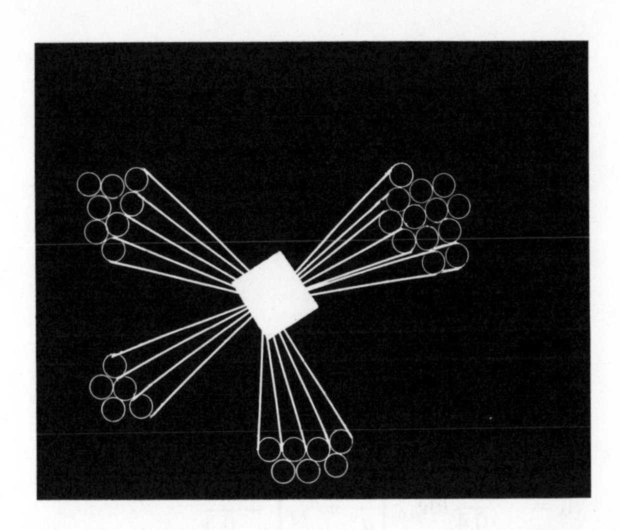

FIG. 16–14 *The radiation flux is the energy emitted from a surface in all directions.*

the sun's surface. The total amount of energy emitted by the star per second in the wavelength range λ to $\lambda + \Delta\lambda$ in all directions is obtained by multiplying the energy emitted from a unit area by the total area of the star. If R is the radius of the star, its surface area is $4\pi R^2$, and the total energy emitted per second in the wavelength range λ to $\lambda + \Delta\lambda$ is

$$4\pi R^2 \times \pi\mathscr{I}_{\lambda T}\,\Delta\lambda$$

area times flux per unit area

If the star is at a distance d, we then obtain the radiant energy coming to us from the star and striking one square centimeter of the earth's surface per second by dividing the previous expression by $4\pi d^2$ (the inverse-square law). In other words, we receive from the star per second, on a unit area at right angles to the star's direction, an amount of radiation

$$\frac{R^2\pi\mathscr{I}_{\lambda T}\,\Delta\lambda}{d^2}.$$

Since this is the amount of energy in the wavelength range from λ to $\lambda + \Delta\lambda$ that strikes one square centimeter of the earth's surface per second, we may refer to $R^2\pi\mathscr{I}_{\lambda T}/d^2$ (the energy received per unit wavelength range) as the apparent brightness of the radiation for the wavelength λ. In other words, we shall write

$$b_\lambda = \frac{R^2}{d^2}\,\pi\mathscr{I}_{\lambda T} \qquad (16.17)$$

where b_λ is the apparent brightness for the wavelength λ.

16.32 The Color Temperature of a Star

We can now use this formula to obtain the surface temperature of the star as follows: we first take the logarithm of this expression to the base 10 and thus obtain, according to Equation (15.9), an expression for the apparent magnitude for the wavelength λ

$$-0.4m_\lambda + B_\lambda = \log_{10} b_\lambda = \log_{10}\left[\frac{R^2}{d^2}\,\pi\mathscr{I}_{\lambda T}\right],$$

where B_λ is related to the apparent magnitude of some standard star for the wavelength λ, or

$$\begin{aligned} m_\lambda &= -\frac{1}{0.4}\log_{10}\left[\frac{R^2}{d^2}\,\pi\mathscr{I}_{\lambda T}\right] + \frac{B_\lambda}{0.4} \\ &= -5\log_{10} R + 5\log_{10} d \\ &\qquad - 2.5\log_{10}\mathscr{I}_{\lambda T} + C_\lambda, \qquad (16.18) \end{aligned}$$

where C_λ is a number that need not concern us here but which we can calculate for each value of λ. The importance of this formula is that it enables us to assign an apparent magnitude to each color of the radiation coming from the star.

We now consider two distinct colors: the one to which a photographic plate is sensitive (wavelength λ_{ph}) and the other to which the eye is sensitive (wavelength λ_{vis}). We then have a quantity $m_{\lambda_{ph}}$ and another one $m_{\lambda_{vis}}$ expressed by means of Equation (16.18) in the same way. Since by definition (see p. 267) the difference between these two quantities is just the color index, $m_{\lambda_{ph}} - m_{\lambda_{vis}} = I$, we obtain the color index expressed in terms of Planck's formula by using Equation (16.18) for each term on the left-hand side of this equation for I

$$\begin{aligned} m_{\lambda_{ph}} - m_{\lambda_{vis}} &= I = -2.5\log_{10}\mathscr{I}_{\lambda_{ph}T} \\ &+ 2.5\log_{10}\mathscr{I}_{\lambda_{vis}T} + (C_{\lambda_{ph}} - C_{\lambda_{vis}}) \\ &= 2.5\left[-\log_{10}\frac{2hc^2}{\lambda_{ph}^5}\frac{1}{e^{hc/k\lambda_{ph}T}-1}\right. \\ &\left. + \log_{10}\frac{2hc^2}{\lambda_{vis}^5}\frac{1}{e^{hc/k\lambda_{vis}T}-1}\right] \\ &+ (C_{\lambda_{ph}} - C_{\lambda_{vis}}). \end{aligned}$$

Note that the terms containing d and R drop out in the subtraction since they are the same for $m_{\lambda_{ph}}$ and $m_{\lambda_{vis}}$. If we now substitute the appropriate values for λ_{ph} and λ_{vis} and the values for the various constants in this formula, we obtain by simple arithmetic the color index expressed entirely in terms of the surface temperature of the star:

$$\textbf{color index} = I = \frac{7\,200 - 0.64T}{T},$$

where we have chosen λ_{ph} equal to 4 300 Å, and λ_{vis} equal to 5 400 Å, and where we have dropped small numerical terms. If we solve this equation for T, we obtain

$$\boxed{T = \frac{7\,200}{I + 0.64}}. \qquad (16.19)$$

Because of the numerical approximation we used to derive this formula, it is not applicable to very hot stars because then the exponential in Planck's formula is not large compared with 1.

By substituting the value of the color index of the star into this formula, we can calculate the quantity T, which is called the **color temperature** of the star. Tables 15.1 and 15.2 show the color indices from which the surface temperatures of some well-known stars can be calculated. Essentially what the color temperature represents is the best fit we can get of a Planckian black-body radiation curve to the observed star's radiation curve. In Table 16.1 the color and effective temperatures of five stars are given.

TABLE 16.1

COLOR AND EFFECTIVE TEMPERATURES OF FIVE STARS

Star	Spectral class	T_{color}	T_e	T_{color}/T_{eff}
τ SCO	B0 V	21 000°	25 000°	0.84
β Per	B8 V	8 500	12 800	0.66
β Aur	A2 IV	7 300	9 700	0.76
α Lyr	A0 V	7 100	11 000	0.65
α CMa	A1 V	7 100	10 300	0.69

The color temperatures compared to the effective temperatures of five stars of early spectral classes. The intensities of the Balmer lines Hβ, Hγ, and Hδ were used to obtain these color temperatures.

16.33 Radiation Pressure

We have already noted in discussing the effect of the sun's radiation on a comet's tail (see Chapter 13, pp. 233–234) that radiation exerts a pressure just the way a gas does. Let us now suppose that we have black-body radiation of a given temperature T in a hollow region and that this radiation is in equilibrium with its surrounding walls. We can get at the pressure exerted by the radiation on these walls by considering how much radiation is absorbed by each square centimeter of the wall per second. The amount of energy striking one square centimeter of the wall per second is the amount contained in a cylinder of unit cross section c centimeters long (where c is the speed of light). If u is the density of the radiation (that is, the amount of radiation in each unit volume), the total radiant energy striking one square centimeter of the wall per second is just uc (energy density times volume).

But the momentum of radiation is its energy divided by the speed of light (see Section 10.20). Therefore, the momentum transferred to the wall per second by the radiation striking it is uc/c, or just u. But since on the average only one-third of the radiation is moving in a given direction, the amount of momentum transferred per second to any wall is $u/3$. This therefore is the pressure of the radiation. In other words, the pressure of black-body radiation in a container is just one-third of the energy density of this radiation. Now it can be shown from the Stefan-Boltzmann law that the energy density u of black-body radiation is given by

$$u = \frac{4\sigma T^4}{c}$$

$$\boxed{= aT^4}, \qquad (16.20)$$

where $a = 4\sigma/c$ and σ is defined on p. 282. From this we see that the pressure of radiation is just given by

$$\boxed{p_{\text{radiation}} = \frac{1}{3} aT^4}. \qquad (16.21)$$

EXERCISES

1. (a) If 100 calories of heat are supplied to a gas in a heat-insulated cylinder and the piston in the cylinder then slowly rises, by how much does the internal energy of the gas increase if the piston does 2×10^8 ergs of work? (b) If the cylinder in (a) were rigid without a movable piston, what would the increase in internal energy of the gas be?

2. If a gas were enclosed in a completely insulated container so that no heat could flow into it or out of it (adiabatic container) by how much would its internal energy increase if its volume were decreased by 100 cc at an average pressure of 10^6 dyn/cm^2?

3. A perfect gas at 0°C and at pressure of 76 cm of mercury occupies a volume of 100 cc. What is its pressure if its temperature rises to 100°C and its volume is increased to 150 cc?

4. What are the values of C_p and C_v for gaseous hydrogen if $C_p/C_v = \frac{5}{3}$? (The value of the gas constant R for one mole of gas is 8.3×10^7 ergs per mole per degree.)

5. The absolute temperature at the center of the sun is approximately 15×10^6 degrees and the central density is about 100 gms/cc. If we take the mean molecular weight μ at the center to be about $\frac{4}{3}$ (completely ionized helium), what is the central pressure?

6. What would be the effect on the mean temperature throughout a star of replacing some of its hydrogen by heavier atoms?

7. (a) Use the data given in problem 5 to find the average speed of the helium nuclei near the center of the sun. (b) Do the same thing with free electrons there.

8. During the adiabatic compression of a perfect gas for which $\gamma = \frac{5}{3}$, the volume is reduced to half its initial value. What happens to the pressure? To the absolute temperature? (Hint: replace P in Equation (16.10) by its value in terms of V and T as given in the gas equation.)

9. A body at absolute temperature 300° is brought in contact with one at absolute temperature 325°. One-tenth of a calorie flows from the hotter to the colder body. What is the net change of entropy of the system? (Assume that the interchange of the heat has no measurable effect on the temperature of either body.)

10. (a) The surface temperature of the sun is about 5 500° absolute. At what rate is its entropy decreasing? (Hint: use its luminosity.) (b) If the earth absorbed one-half of the solar radiation striking it, at what rate would its entropy be increasing? (Assume the average absolute temperature of the earth to be 300°K.)

11. If we assume that stars radiate like black bodies, where in the spectrum (at what wavelength) is the maximum intensity of the radiation from (a) Betelgeuse, (b) Arcturus, (c) the sun, (d) Sirius, and (e) Rigel?

12. If the surface absolute temperature of the sun were to increase by 20 per cent, how much more luminous would it become?

13. What is the radiation pressure at the center of the sun if one uses the data given in problem 5?

14. How much energy does a photon have if its wavelength is 1 000 Angstroms?

15. Consider black-body radiation at an absolute temperature of 10^3 degrees inside a container that is perfectly insulated on the outside and that has perfectly reflecting walls on the inside. How much radiation energy per cubic centimeter is there in the container?

16. (a) Assume that the sun radiates like a black body at a temperature of 5 500°K. How much energy is emitted by one sq cm of the solar surface in the continuous spectrum between the wavelengths 7 500 Å and 7 503 Å? (b) How much between the wavelengths 3 500 Å and 3 503 Å? (c) How much between the wavelengths 5 500 Å and 5 503 Å?

17. (a) One mole of a perfect gas is allowed to expand isothermally until its volume is increased five-fold. What is the change in entropy in the gas when it goes from its initial to its final state? (b) If the absolute temperature of the gas is 300°K and the gas goes from its initial to its final state along a reversible path, how much work does the gas do during the expansion?

18. What is the temperature of a black body for which the maximum in its radiation distribution curve is: (a) In the yellow line, 5 890 Å, of sodium? (Use the Wien displacement law.) (b) In the red line 6 563 Å of hydrogen? (c) In the violet line 3 990 Å of hydrogen?

17

SPECTRAL CLASSIFICATION OF STARS*

In the two preceding chapters we saw that we can obtain a great deal of information about the surface conditions of stars by studying their radiation. However, we concerned ourselves there only with the over-all properties of the radiation and not with its detailed characteristics. We saw that we may treat stars as though they were black bodies and that by analyzing the intensity distribution of the radiation among the various wavelengths we can find the surface temperatures of stars. However, we can get a good deal more information from radiation if we take into account its detailed spectral features. Before we can do this we must briefly consider the various types of spectra that can be obtained from radiating bodies.

THE SPECTRAL ANALYSIS OF RADIATION

17.1 The Continuous Spectrum

We shall see in Appendix A on astronomical instruments that *when light passes from a vacuum into a dense medium such as glass, it slows down, and its speed in the dense medium depends upon its color* (*that is, its wavelength*). *The long wavelengths travel more rapidly than the short ones.* Because of this phenomenon it is possible to spread white light out into its constituent colors by passing it through a glass prism. The prism bends the short wavelengths (slower light) more than it does the longer wavelengths (faster light). When we do this, we obtain what we call a **spectrum** of the radiation. The nature of the spectrum depends upon the physical and chemical nature of the source. *If we are dealing with a solid, a liquid, or even a dense gas that is brought to incandescence, we obtain a* ***continuous spectrum****, in which the colors change continuously from deep red to deep violet.* Actually the spectrum extends beyond the red into the infrared and longer wavelengths, and beyond the violet into the ultraviolet region and shorter wavelengths, but the eye cannot detect these parts of the spectrum.

The continuous spectrum is almost exactly the same for all substances and therefore gives us almost no indication of the chemical nature of the substance. The intensities with which the various colors appear in this type of spectrum depend only upon the temperature of the body and how closely it approximates a black body, and not on the chemical nature of the body.

17.2 The Bright-Line Emission Spectrum

If instead of considering radiation emitted by a solid, a liquid, or a dense gas, we consider the radiation emitted by a highly rarefied (low-pressure) gas, the type of spectrum we obtain is quite different. The easiest way to achieve this is to place a small amount of gas in a previously evacuated glass container and then to pass an electric discharge through it as shown in Figure 17.1. Because some of the molecules and atoms in the glass tube are always electrically charged, they will start moving with increasing speed towards one end of the tube when the gas is subjected to a sufficiently high voltage. *Because of this motion they collide with other atoms, and as these collisions become more*

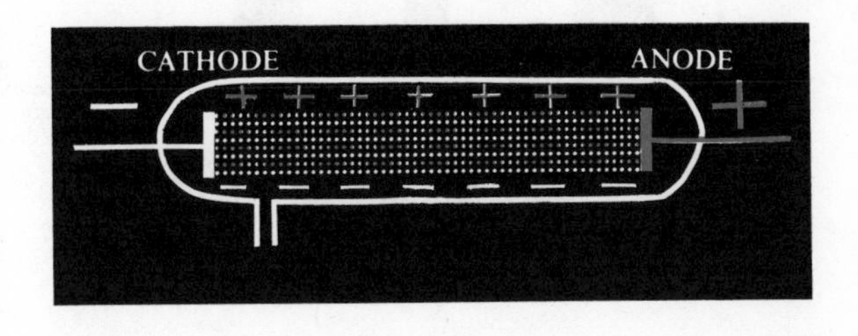

FIG. 17–1 *An electric discharge through a gas in a tube.*

and more violent, the colliding atoms are excited as a result of the energy transferred to them by the collisions. They then re-emit the energy so that the whole tube begins to glow (*as in a mercury arc, a neon tube, etc.*).

If we analyze this radiation spectroscopically, we obtain a spectrum that is quite different from that emitted by a hot solid. Instead of the continuous array of colors, *we obtain only a discrete set of lines of different colors separated by dark spaces. Such a spectrum is called a* ***bright-line emission spectrum****, and the lines present are entirely characteristic of the atoms in the tube.* Each line in such a spectrum can be specified by giving its wavelength. Thus, in the case of the sodium spectrum there are two very close yellow lines having the wavelengths 5 890 Å and 5 896 Å. In the case of hydrogen some of the lines are

H_α	6 563 Å	H_δ	4 102 Å
H_β	4 861 Å	H_ε	3 970 Å,
H_γ	4 340 Å		

and in the case of neon, 5 852 Å, 5 882 Å, 5 954 Å, 5 976 Å. *If the gas inside a tube is a mixture of different types of atoms, the spectrum of this mixture contains the lines that are characteristic of each type of atom that is present.* Thus, the bright-line spectrum is a very powerful tool for the chemical analysis of sources of light if these sources are in the form of attenuated gases.

If the gas in the tube is a compound rather than an element, we have a **molecular spectrum**, and we find that the colors are not concentrated in sharp lines, but spread out over bright bands (each band actually consists of numerous lines very close together). We therefore speak of a **bright-band spectrum** rather than a bright-line spectrum.

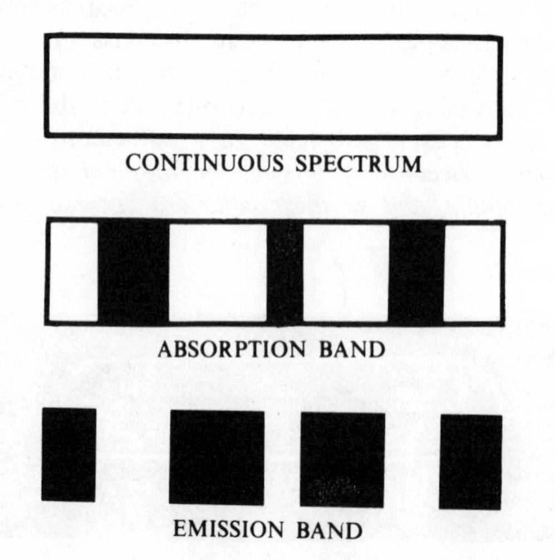

FIG. 17–2 *A schematic diagram of emission and absorption bands.*

17.3 The Absorption Spectra

We must consider still another type of spectrum which is of great importance to astronomers. Let us suppose that we have a hot incandescent solid surrounded by a cool gas. Without the cool gas we would obtain the usual continuous spectrum, but with the cool gas we find that the continuous spectrum is broken up by a series of fine, sharp dark lines. *These lines are present in the spectrum at precisely the positions that would be occupied by the bright lines in the spectrum of the gas if this cooler gas were radiating in a tube all by itself.* This is called an **absorption spectrum** and is as characteristic of the cool surrounding gas as the bright-line emission spectrum is. In fact, we assign to these dark lines, which are called **absorption lines**, the same wavelengths as we assign to the bright lines of the emission spectrum. If the cool gas is composed of molecules, the absorption spectrum is a band spectrum (a set of dark bands separated by bright regions). (See Figure 17.2.)

17.4 The Way Different Spectra Arise

In order to make use of spectroscopic analysis in astronomy we must first explain how the above-mentioned three types of spectra arise, and we must decide which one is most important in astronomy. For the moment let us picture an atom as consisting of electrically charged particles that are associated with specific restricted modes of vibration (like the strings in a violin) by their selective absorption of energy in the form of electromagnetic radiation. When we discuss atomic structure accurately and in detail in Chapter 18, we shall see that the charged particles in an atom (the electrons) are not really vibrating in the classical sense but have associated with them restricted stationary wave patterns, each of which corresponds to a definite frequency and hence, according to Planck's fundamental formula (16.15), to a definite energy. The atom really emits or absorbs radiation whenever one of its component charged particles (electrons) changes from one stationary wave pattern to another (as if we changed from the vibrating D string to the E string of a violin); we shall see later just how this process occurs. An atom becomes excited when one of its charged particles acquires more energy and changes from a lower to a higher mode of vibration. It then returns to a lower mode of vibration by emitting electromagnetic radiation—which we detect as emission lines in the spectrum of an atom.

Since electrical charges in different types of atoms are bound to the atom by forces that differ in strength from one type of atom to another, the frequencies of the restricted modes of vibrations also differ. Hence different types of atoms emit radiation of different frequencies or wavelengths. Thus, each type of atom is characterized by a fixed set of spectral lines arising from its various modes of vibration.

Although the foregoing is only a crude representation of what happens when the atom emits radiation, nonetheless it may help us understand why the various types of spectra arise.

Since each atom has its own characteristic modes of vibration and hence its own spectral lines, it is at first difficult to understand why a continuous spectrum arises at all. *However, we must remember that in the case of a solid, a liquid, or a very dense gas the atoms are crowded tightly together so that they are not free to vibrate independently of each other.* In other words, as soon as one atom in a solid begins to vibrate, its neighbors are set vibrating also until vibrations occur among whole groups of atoms and the individual characteristics are lost. We thus get all kinds of vibrations, one on top of the other, giving rise to a continuous spectrum.

This is not so for atoms in an attenuated gas in a tube. To a great extent, each atom in the tube vibrates independently of its neighbors, and we observe its individual radiation characteristics in the form of sharp lines. As more and more atoms are forced into the tube, more and more interference occurs between neighboring atoms, and the characteristic lines become fuzzy and broad. *Finally, when the atoms in the gas have been packed together closely enough, the lines become so broad that they merge into each other, and a continuous spectrum arises.*

When the gas in the tube consists of molecules, the individual atoms of the molecule are not free to vibrate independently of each other. What happens in that case is that the entire molecule is set vibrating and rotating in various ways, and innumerable close lines concentrated in bands are emitted.

To understand the origin of the absorption spectrum, we must keep in mind that when radiation of a particular kind passes an atom, the atom can alter its modes of vibrating by absorbing this radiation just as a sound wave of a particular frequency sets a tuning fork vibrating, if the tuning fork has a pitch that exactly corresponds to the wavelength of the passing sound. If we now consider a column of atoms in a cool gas through which continuous-spectrum radiation from a hot body is passing, the atoms in the column will change their modes of vibration (see Figure 17.3) in response to the radiation of the proper wavelength. *The atoms that have absorbed radiation then re-emit it in all directions, so that an observer analyzing the column of radiation finds it deficient in this particular wavelength by comparison with radiation that does not pass through cooler gas.* The column of atoms in the cool gas has scattered in all directions the radiation of the wavelength in question; thus an absorption line appears. An absorption line may be produced in another way when an atom actually removes radiation (a photon) of a specific frequency by absorbing it and then, instead of re-emitting it immediately in a different direction (**scattering**), collides with a neighboring atom and transfers to it this energy of excitation, sending the atom off with increased kinetic energy. This process is called **absorption**.

It is clear from this analysis that radiation coming from the stars should be characterized by an absorption spectrum. And, indeed, it is. *Stellar spectra generally consist of bright continuous backgrounds threaded by fine dark lines which are characteristic of the atoms in the cool atmospheres of the stars.* The continuous part of the spectrum

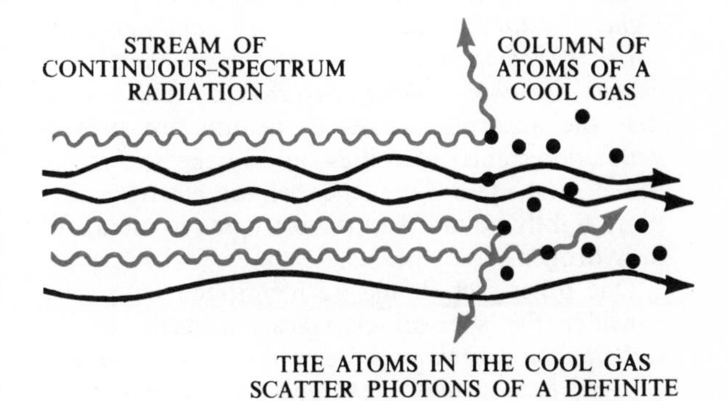

FIG. 17-3 *A stream of radiation sets an atom vibrating which, in turn, re-emits in all directions the radiant energy that it has absorbed from the stream. This process is called scattering.*

arises from the star's photosphere, and the absorption lines are due to the scattering and the absorption of radiation by the atoms in the cool atmosphere. If molecules are present in the star's atmosphere, characteristic absorption bands, in addition to the absorption lines, are present in the spectrum.

17.5 Differences in Stellar Spectra

Different intensities of the outwardly flowing radiation from a star's hot photosphere produce different absorption lines in the star's spectrum, because the atoms in the cool atmosphere react or respond differently to different intensities of radiation. We shall consider this in more detail later, but the explanation, briefly, is as follows.

The charged particles (i.e., electrons) are not equally tightly bound in different atoms, such as hydrogen and helium, and thus they are not equally free to respond to radiation of a given frequency and thus to go from one mode of vibration to another. The more tightly bound an electron is inside an atom (to the nucleus by electrostatic forces), the more energy must be transferred to the atom from the radiation to compel the electron to a higher mode of vibration. The more energetic the radiation from the star—that is, the higher the star's surface temperature—the more effective this radiation will be in changing the charged particles' modes of vibration, thereby producing different absorption lines. For this reason stars having different surface temperatures are characterized by the absorption lines of different atoms. *In general, the more tightly bound the charged particles of an atom are, the higher the star's surface temperature—and hence the hotter the radiation from the photosphere—must be to change the atom's modes of vibration.* This means that the absorption lines of helium are present with appreciable intensities only in the spectra of the hottest stars since the helium electrons are more tightly bound than the electrons in metals or in hydrogen.

One other fact should be mentioned before we consider the spectral classification itself: *if the radiation passing through the atmosphere is energetic enough, it may tear a charged particle out of an atom entirely. We then say that the atom is **ionized**. If such an ionized atom interacts with the radiation passing it, we obtain absorption lines of the ionized atom, which are quite different from those of the un-ionized atom.* Since hydrogen has only one charged particle in its atom, ionized hydrogen can have no absorption lines, for when it loses its single charged particle, the atom (the proton) has no modes of vibration.

17.6 Spectral Classification of Stars

With these points in mind, we can now describe the spectral classifications astronomers accept as characterizing the majority of stars. In spite of their vast range in luminosities and surface temperatures, stars can be grouped into a very few spectral classes. *Of course these classes are not sharply differentiated and the spectral characteristics vary continuously as we go from the very hottest to the very coolest stars.* But since there are clear distinctions from one class to the next, this type of classification has great practical use. We shall find that as we go from one class to the next, there is a continuous variation in the intensities of some of the outstanding spectral lines.

The work of classifying stars spectroscopically was begun in 1885 by E. C. Pickering at the Harvard College Observatory, and was carried on with great success later by Annie J. Cannon. The Henry Draper Catalogue, which consists of nine volumes and was completed in 1924, gives spectral classification of 255 000 stars. *There are seven main spectral classes to which the letters O, B, A, F, G, K, M are assigned, and each of these is divided into ten subgroups (e.g., B0, B1, B2,. . ., B9).* To these seven main groups three minor groups, R, N, S, which are related to the K and M classes, have been added.

In order to differentiate among stars of different luminosities in the same spectral class, Morgan, Keenan, and Kellman have introduced a luminosity criterion, thereby spreading out the spectral groups into a two-dimensional classification. This classification, which assigns a luminosity class to each spectral type, is referred to as the MKK system. Roman numerals are used for luminosity class designations according to the following scheme:

Ia—bright supergiants
Ib—faint supergiants
II—bright giants
III—normal giants
IV—subgiants
V—normal dwarfs or main-sequence stars.

The application of the MKK system to the brightest stars is given in Table 15.2 of the 100 brightest stars.

17.7 Spectral Class Characteristics

Table 17.1 gives the essential features of each spectral class together with the associated color and temperature and some typical examples of stars of each class. In case of O stars and some B stars, weak emission lines brighter than the background continuous spectrum are found as well as absorption lines. In general, the presence of emission lines means that the atmospheres of these stars are fairly hot. However, weak emission lines are also found occasionally in stars of spectral classes M, N, and S which have cool atmospheres. This may indicate that the atmosphere has a hot upper level something like the solar chromosphere but more extensive (see Section 20.13). We should note that an absorption line is not absolutely black and devoid of radiant energy, but is dark merely by comparison with the background continuous spectrum. If the background spectrum were suddenly removed, then all of the absorption lines would appear as bright lines. What we find as we inspect Table 17.1 is that *the spectral sequence, as we go from the O type stars down to the M type stars, is at the same time a temperature and a color sequence.* As we move from the O to the M type stars, we go from very hot stars having temperatures of more than 50 000°K to cool stars having temperatures of about 3 000°K or less. At the same time we observe that the color varies from deep blue-white down to red.

We see, moreover, that the hot stars such as the O and B stars are characterized by the lines of highly ionized atoms. Thus, in the case of the O type stars, the lines of singly ionized helium and doubly ionized nitrogen and oxygen are present. As we move from the O to the B type stars [see Figure 17.4(a)], the lines of neutral helium become more pronounced, and the lines of the highly ionized atoms become less intense. In both

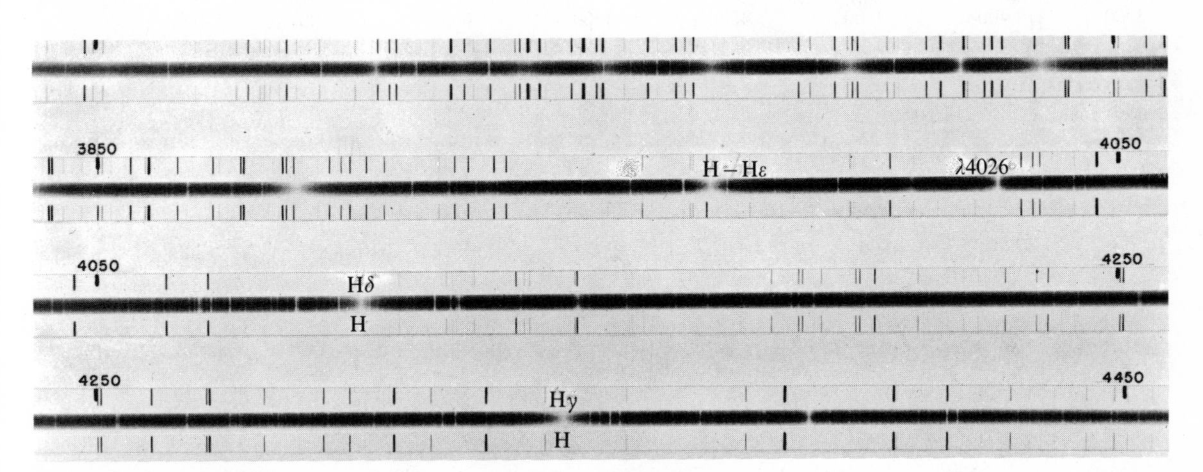

FIG. 17–4(*a*) *High-dispersion stellar spectra. Z Sco B0, λ3 250–λ4 650.*

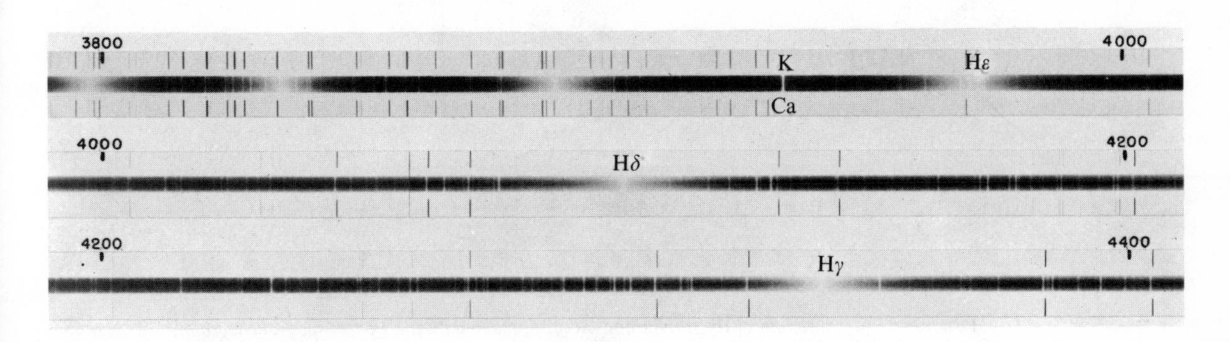

FIG. 17–4 (*b*) *High-dispersion stellar spectra. Alpha Canis Majoris. dA2. λ3 800–λ4 400. (From Mount Wilson and Palomar Observatories)*

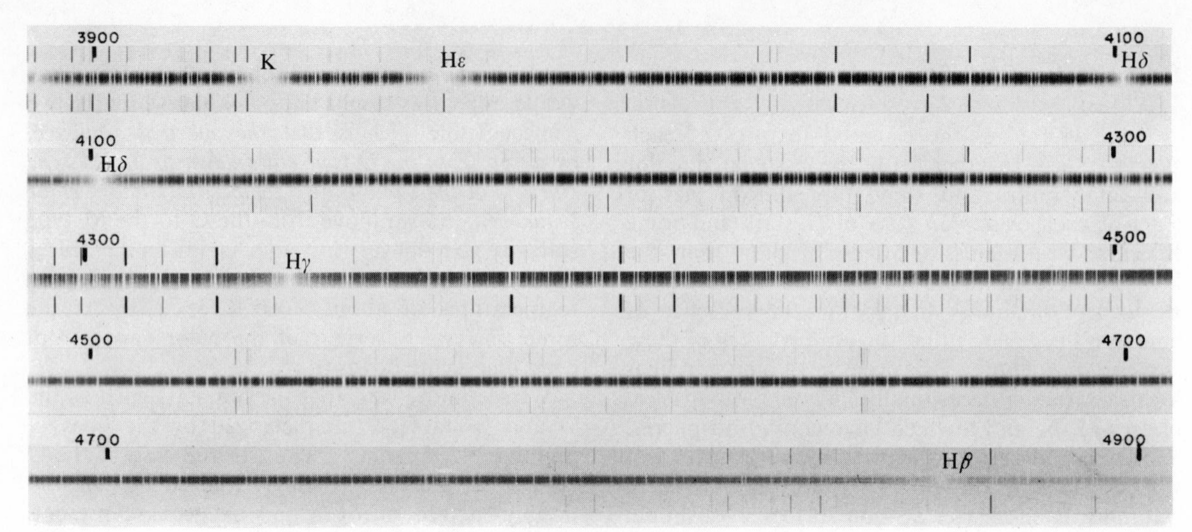

FIG. 17–4 (*c*) *High-dispersion stellar spectra. Alpha Canis Minoris. dF5. λ3 900–λ4 900. (From Mount Wilson and Palomar Observatories)*

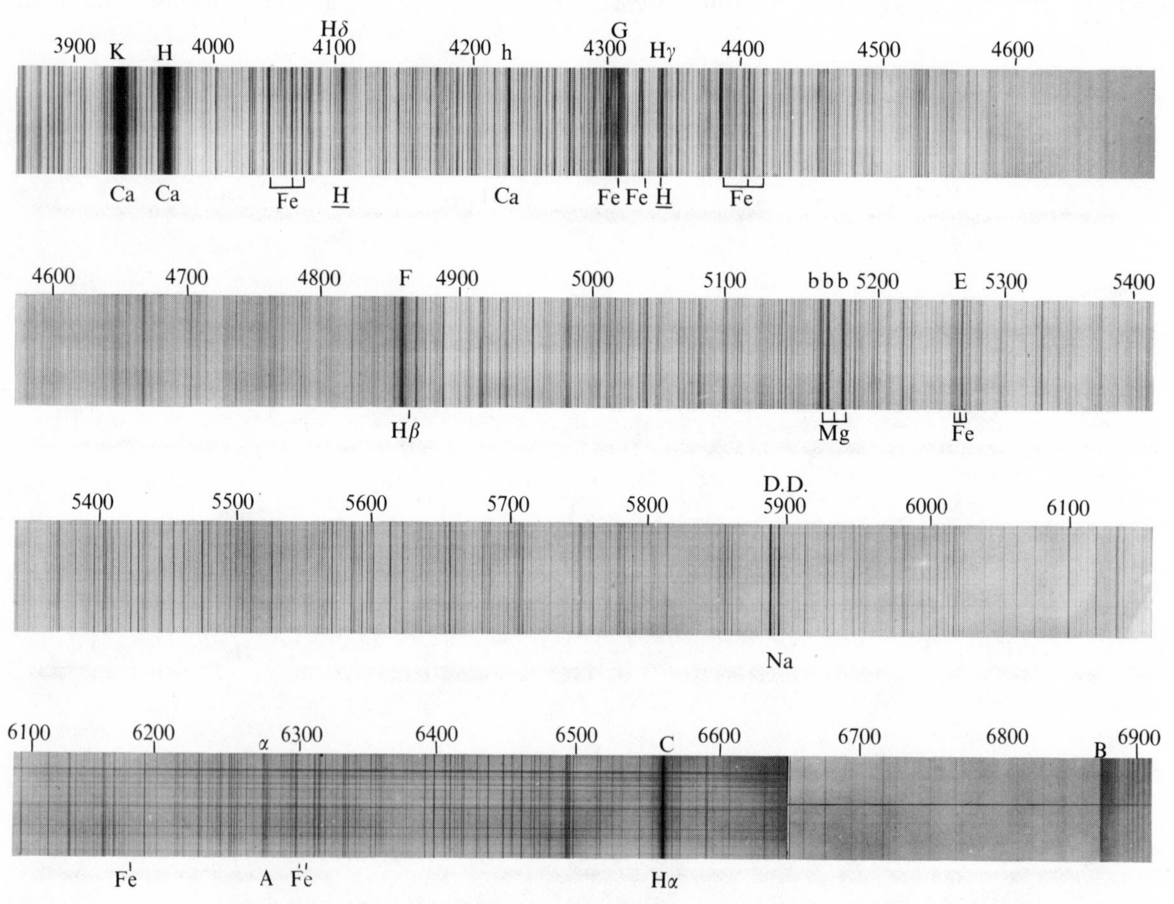

FIG. 17–4 (*d*) *Spectrum of the sun λ3 900–λ6 900, taken with the 13-foot spectroheliograph. This shows the greatest abundance of the spectral lines. The H and K lines of calcium are particularly prominent. The sodium, magnesium, and iron lines are also very strong, as are the Balmer lines of hydrogen. (From Mount Wilson and Palomar Observatories)*

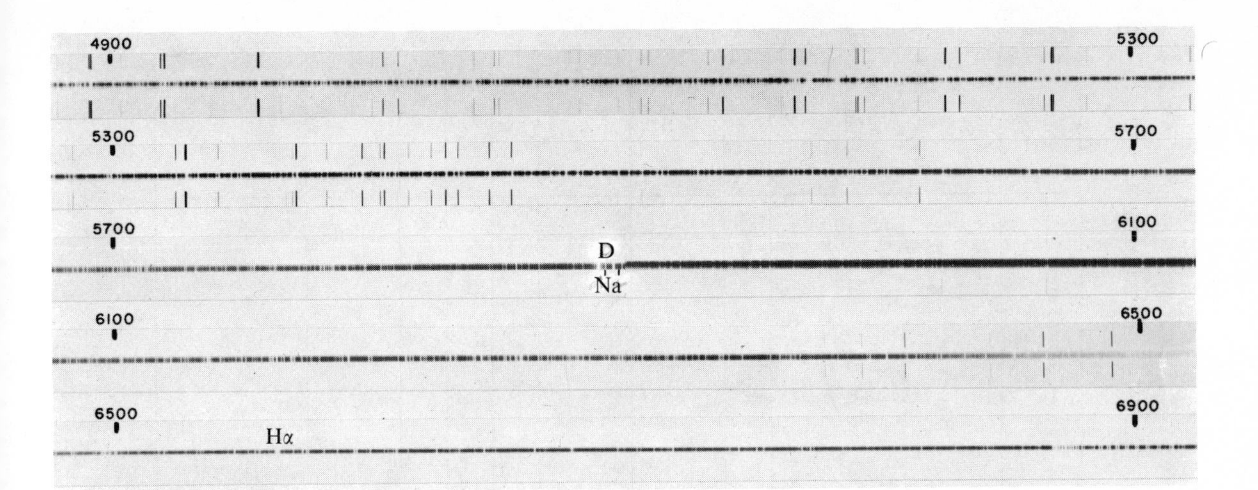

FIG. 17-4 (*e*) *High-dispersion stellar spectra.* α *Bootes.* *gKO.* *λ4 900–λ6 900.* (*From Mount Wilson and Palomar Observatories*)

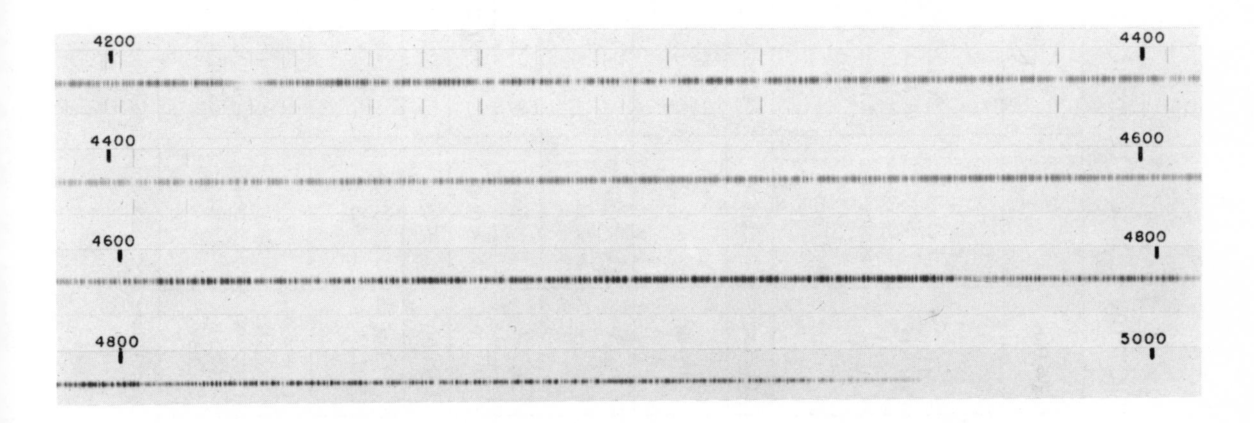

FIG. 17-4 (*f*) *High-dispersion stellar spectra.* β *Bootes.* *gM2.* *λ4 200–λ5 000.* (*From Mount Wilson and Palomar Observatories*)

classes the lines of hydrogen are present, but are comparatively weak. As we pass from the B stars to the A stars [see Figure 17.4(b)], the helium lines become weaker and weaker. The hydrogen lines become stronger and reach their fullest intensity for AO type stars. The hydrogen lines diminish in intensity as we move into the F class (see Figure 17.4(c)), and here the metallic lines, particularly those of ionized calcium, begin to dominate. The ionized calcium lines grow stronger and then begin to diminish in intensity as we move into the G class. Although still present, the hydrogen lines are weaker and the helium lines have disappeared. Now the lines of the neutral metals are dominant [see Figure 17.4(d)]. Molecular bands also begin to show up. These characteristics are greatly intensified for the K type stars [see Figure 17.4(e)]. Finally, in M stars [see Figure 17.4(f)] the lines of the neutral metals are quite strong, and bands of molecules, such as TiO (titanium oxide), begin to dominate. In Figure 17.5 the spectra of the principal spectral types are shown. These photographs show clearly the gradual variation as we go from the O to the M type stars.

The line spectra of the R and N stars resemble those of the K and M stars, but their molecular band spectra are dominated by carbon molecules such as CN, CH, etc. The S stars differ from the M stars in showing molecular bands of zirconium oxide instead of titanium oxide.

TABLE 17.1

SPECTRAL AND OTHER CHARACTERISTICS OF STARS

Spectral class	Wolf-Rayet	O	B	A	F	G	K	M	Carbon stars R	Carbon stars N	S
Most prominent lines	HeII emission Lines CII, NIII NIV OIII, OIV	HeII HeI HI OIII NIII CIII SiIV	HeI HI CII OII NII FeIII MgIII	HI (Balmer) Ionized metals	HI CaII TiII FeII	H and K Lines of CaI, FeI, TiI, MgI Some molecular bands Balmer Lines	H and K of CaII, HI (Balmer) Molecular bands	Molecular bands of TiO, CaI	Bands of C_2 CN CH Carbon stars	Similar to R in bands. Lines like K	Bands ZrO YO, LaO Lines of Technitium
Colors	Blue	Blue-white	Blue-white	White	Creamy	Yellow	Orange	Red	Red	Red	Red
Color index B-V		O5 −0.45	−0.20	0.00	+0.40	+0.60	+1.00	+1.5			
B.C. Bolometric correction		4.6	2.5	0.5	0.1	0.00	0.3	1.8			
M_v		−6	−3.7	+0.7	+2.8	+4.6	+5.2	8.9			
T_e		35 000	21 000	10 000	7 200	6 000	4 700	3 300			
T_{color}		70 000	38 000	15 400	9 000	6 700	5 400	3 800			
Percentage of stars in galaxy*			2.5% 0.03%	26.7% 0.6%	11.0% 0.2%	16.7% 9.3%	35.4% 39.1%	7.6% 50.0%			
Typical examples	Fainter component of the binary HD193576	ζ Puppis	ζ Persei Rigel γ Lyr	Sirius Vega Deneb δ Cas	Canopus (FO) Procyon (FS) β Cas	Sun Capella β Her	Arcturus Aldébaran ϵ Peg	Betelgeuse Antares η Gem			

* The first line in the percentage of stars in the Galaxy is the apparent percentage and the second line is the real percentage.

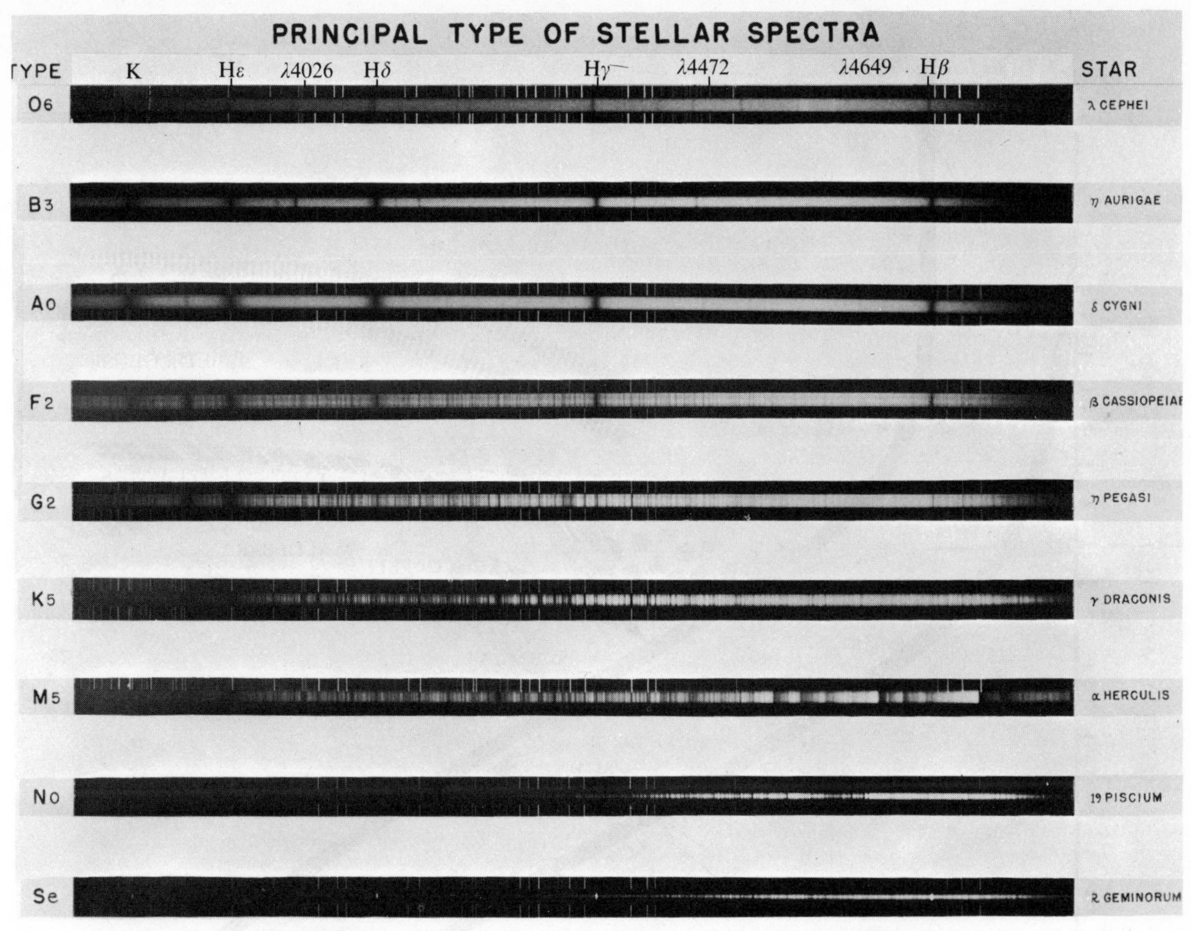

FIG. 17–5 *Principal types of stellar spectra: O, B, A, F, G, K, M, N, and S.* (*From Mount Wilson and Palomar Observatories*)

17.8 The Significance of the Spectral Classes

It is important to note that the intensities of the various lines in the spectral sequence above do not necessarily indicate the abundances (in the atmosphere of the star) of the atoms that give rise to the absorption lines. Of course, other things being equal it is true that the more abundant a particular atom is in the atmosphere, the greater is the intensity of its absorption lines, *but the variations that we observe in the intensities of the absorption lines as we pass from the hot to the cool stars are due primarily to the variation in temperature.* As we shall see later, hydrogen and helium are the most abundant substances in stars, but the absorption lines of helium are not present in the spectra of cool stars, because the electrons in helium are tightly bound and only very hot radiation can affect their modes of vibration. We shall consider these points in more detail in Chapter 19.

THE HERTZSPRUNG-RUSSELL DIAGRAM

17.9 Giants and Dwarfs

Whereas the spectrum of an atom characterizes that atom completely, this is not true of stars. Two stars may belong to the same spectral class and yet differ considerably in internal structure and other characteristics. The reason is that the spectral sequence is, as we have noted, a temperature sequence. But the surface temperature of a star is related to two characteristics, the star's luminosity and its radius. This is at once obvious if we consider that two stars having the same surface temperature (and hence belonging to the same spectral class) may have different luminosities because one is bigger than the other. The reason is that the temperature determines the rate at which

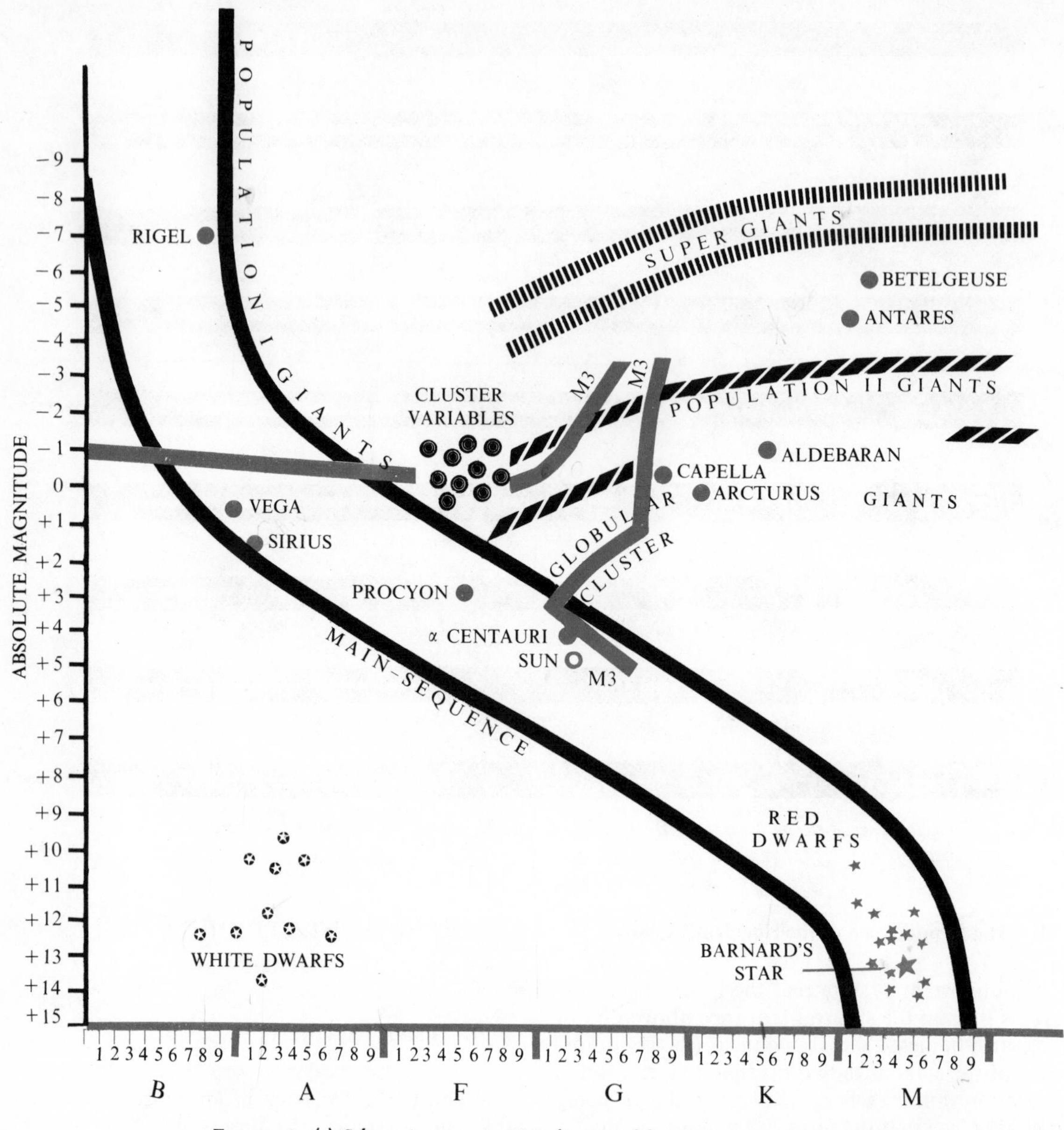

FIG. 17–6 *(a) Schematic composite H-R diagram of the various types of stars. The diagram of the globular cluster M 3 is also shown.*

each square centimeter of the star's surface emits energy, and the total luminosity is measured by this rate times the total number of square centimeters in the surface. Because of this relationship involving the luminosity, the radius, and the surface temperature, stars of the same spectral class can fall into two different groups. There are two such main groups.

This is best illustrated by means of a diagram which stemmed from work begun by Hertzsprung in 1911 and later carried on by Russell. These two investigators discovered that *the cooler stars (yellow or red stars) consist of two groups, one composed of intrinsically faint stars, and the other of stars of high luminosity.* Hertzsprung called the faint stars **dwarf stars**, and the stars of high luminosity **giant stars**. The results of Hertzsprung's, Russell's, and subsequent investigations

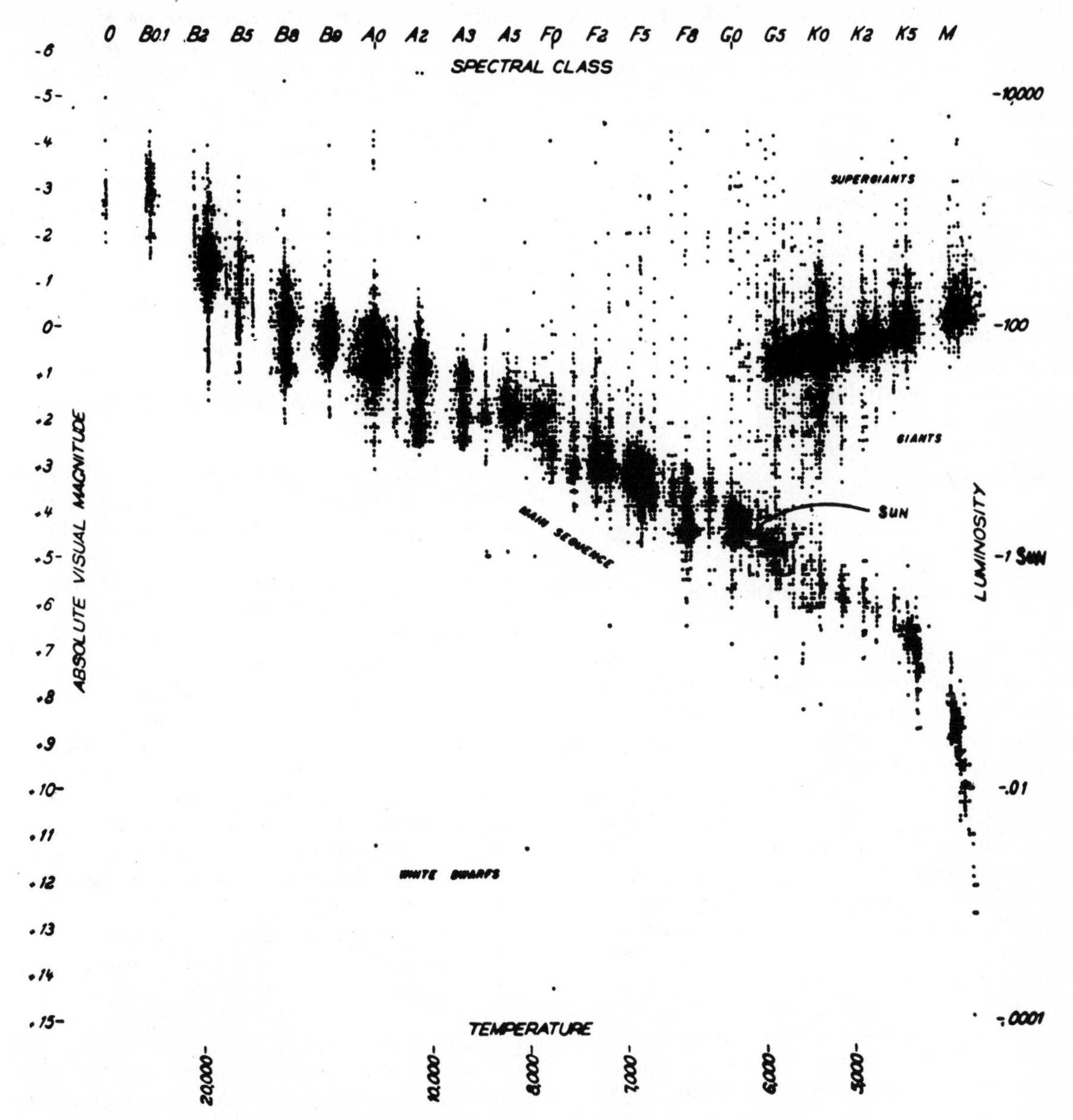

FIG. 17–6 (*b*) *H-R diagram of stellar absolute magnitudes plotted against spectral types.* (*O. Struve*, Stellar Evolution. *Princeton, N.J.: Princeton University Press, 1950.*) (*Yerkes Observatory photograph*)

are best represented on a graph known as the **Hertzsprung-Russell (H-R) diagram**. If we plot the absolute magnitudes of a large group of randomly chosen stars along the ordinate axis and their spectral classes along the abscissa, we obtain the diagram shown schematically in Figure 17.6(a). Some typical stars are labeled in this diagram. In Figure 17.6(b) the H-R diagram obtained for a few thousand actual stars is shown.

We see that most of the stars fall on a curve that runs diagonally from the hot stars of high luminosity down to the cool, faint red stars. This group of stars is called the **main sequence**. Another group of intrinsically bright stars (the **giants**) lies on a branch along which the luminosity increases slightly as we move to the right from the hot to the cool stars. The average absolute magnitude of these luminous stars is about zero, and we see that they are far fewer than those on the main sequence. This group of stars, which is called the **giant branch**, merges with the main sequence for FO type stars. In addition to these two main branches, in the lower left-hand corner there is a large group of stars called **white dwarfs**, and another very small group almost parallel to, but above, the giant branch called the **supergiants**.

The stars on the lower right-hand part of the main sequence are referred to as **red dwarfs**, and those in the upper left-hand corner are called **blue giants**. We see that the sun may be characterized as a yellow main-sequence star, Sirius as a white main-sequence star, Rigel as a blue giant, Capella as a yellow giant, Arcturus as an orange giant, and Antares and Betelgeuse as red supergiants. The nearest known star to the sun, Proxima Centauri, is a red dwarf.

17.10 Discussion of the H-R Diagram

The first thing that impresses us as we examine the Hertzsprung-Russell diagram is that the main sequence and the giant branches are not well-defined lines, but rather broad bands. Part of this can be accounted for by the errors we make in carrying out the measurements needed to place a star on the diagram. But the width of the branches cannot be entirely explained in this way. *Some of it arises because when we take a random sampling of stars over the entire sky, we are dealing with stars that differ greatly in their intrinsic characteristics.* It is reasonable to assume that this kind of intrinsic variation from star to star is partly responsible for the width of the branches. In particular, we shall see later that large ranges in the masses and in the initial (when first formed) chemical compositions of the stars included in the H-R diagram greatly affect its structure.

We also note that in taking a random sampling of the stars in the sky, *we tend to favor the luminous stars and to underplay the fainter ones.* Thus, in Figure 17.6(b) there are far too many giants and supergiants as compared to the other groups. *In particular, we should note that the number of white dwarfs actually present in our universe must be much greater than indicated by the small number given in the diagram.*

The concentration of the stars into distinct groups in the H-R diagram shows us that there are definite patterns according to which stars are constructed; otherwise we should find stars in equal numbers all over the diagram. *The concentration of the stars into distinct branches tells us that definite structural laws govern them and that as stars evolve, they occupy different positions along the branches in accordance with these laws.* We shall see in Chapter 23 that in taking a sampling of many stars we include stars at various stages of evolution. Hence, the Hertzsprung-Russell diagram is a powerful tool enabling us to study the evolution of stars.

17.11 The H-R Diagram for Nearby Stars

To abstract from the H-R diagrams all the information we can about the evolution of stars, we must work with a diagram that is free of large variations in both the masses and the initial compositions of the stars. *If we could obtain a diagram of stars all of which began their lives with the same chemical compositions, but with different masses, we could then see how the mass of a star affects its evolution.* To do this we must work with groups of stars which we have reason to believe were chemically similar at their birth. We shall see in Chapter 27 that there are several hundred such groups in the Milky Way called **open** or **galactic clusters**, and for these the H-R diagrams do not show the scattering into wide branches that we find in the case of the random sampling of stars given above.

Before we study these clusters, however, we shall consider the H-R diagram for all known stars that lie very close to us. This diagram is shown in Figure 17.7, where we have incorporated the data for stars lying within 10 parsecs of us. Although it contains stars varying in mass and in initial chemical composition, so that the evolutionary picture is obscured, it is still important because it gives us a much more accurate picture of the relative abundances of giant and dwarf stars than the standard H-R diagram does.

We see from this diagram that the red dwarfs are the most abundant stars, followed by the white dwarfs. Giants, supergiants, and bright main-sequence stars are extremely rare. That the main-sequence branch is not a thin line, but already shows stars departing from it, arises from the fact that even within 10 parsecs we must expect to find differences in the initial chemical compositions and in the masses of the stars and hence in their rates of evolution.

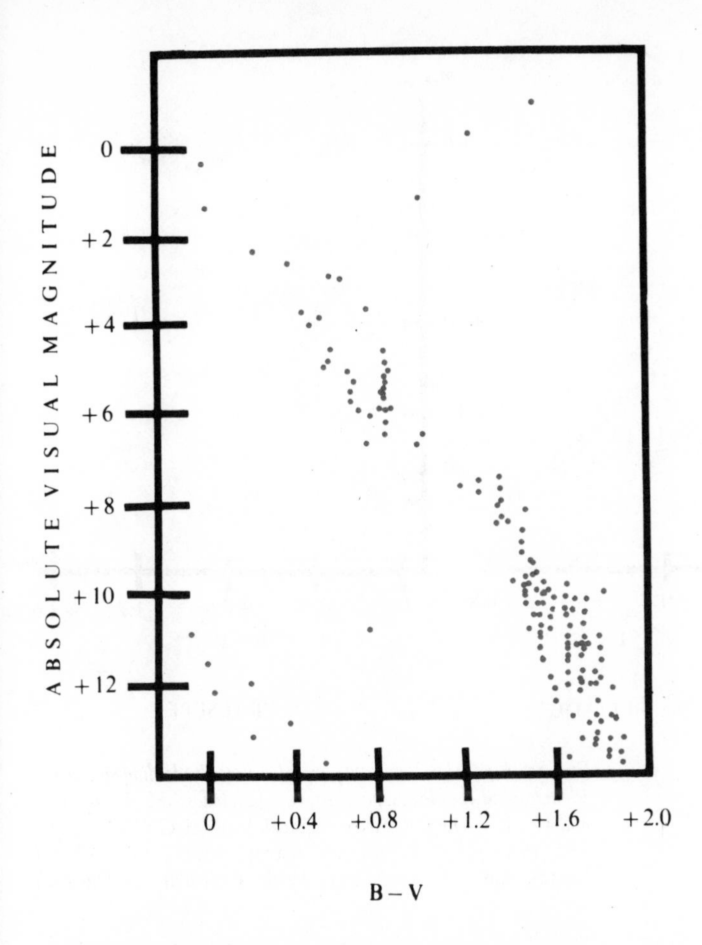

FIG. 17–7 *H-R diagram of stars lying within 10 parsecs. (Adapted from H. L. Johnson and W. W. Morgan,* Astrophysical Journal *CXVII (1953), 313. © 1953 by the University of Chicago)*

17.12 The H-R Diagram for Open or Galactic Clusters

To obtain observational evidence for the evolution of stars, we must consider the Hertzsprung-Russell diagram of chemically homogeneous groups of stars such as open or galactic clusters. A great deal of observational work has been done in this field by such investigators as Johnson, Morgan, Sandage, and Eggen, who applied photoelectric procedures to bright galactic clusters such as the Pleiades, the Hyades, and the Praesepe. *Since all the stars in a cluster are at very nearly the same distance from us, errors in the absolute magnitudes of individual stars do not enter into the picture: we may work with the apparent magnitudes of the stars in a cluster since these give an accurate account of the range in the intrinsic luminosities of the stars.* In fact, what one does in setting up the H-R diagram for such a cluster is to plot the apparent magnitudes against the color indices. Then when the distance of the cluster becomes known one can convert the apparent magnitude into absolute magnitude by means of Equation (15.11), Section 15.25. It is convenient to use the color index rather than the spectral class since the color index is a precisely determined number and the spectral sequence is itself a color sequence. In Figure 17.8 the H-R diagrams—or, as they are very often called, the **color-luminosity** or **color-magnitude** diagrams—of the Pleiades and the Praesepe clusters are shown.

What is remarkable about these diagrams is that

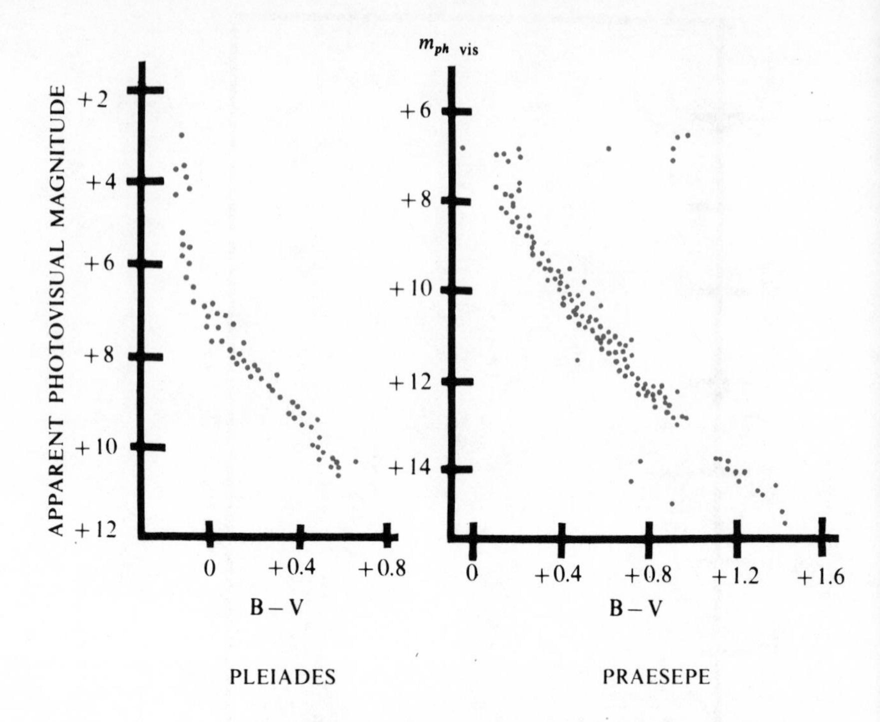

FIG. 17–8 *Color-luminosity or color-magnitude diagram of the Pleiades and the Praesepe clusters. (Pleiades: H. L. Johnson and W. W. Morgan,* Astrophysical Journal *CXVII (1953), 313; Praesepe: H. L. Johnson,* Astrophysical Journal *CXVI (1952), 640. © 1952, 1953 by the University of Chicago)*

all the stars fall (with a few exceptions) along a thin line. This line, as we observe, has the shape of the main-sequence branch in the general H-R diagram, but if we look at the two examples given, we see that *it does not terminate at the same left-hand point for both clusters.* In the case of the Pleiades, the main sequence extends all the way to the left of the color index -0.2, whereas for the Praesepe cluster it stops at the color index $+0.2$. In terms of the absolute magnitude, we find that the Pleiades cluster has stars as bright as $M_v = -3$, whereas the Praesepe cluster has no stars brighter than $M_v = 0.5$. If we disregard these differences, we find that the bulk of the stars in both clusters lie on almost parallel lines which correspond to the main sequence of the usual H-R diagram. In fact, if we superpose one diagram on the other, the two main-sequence lines almost coincide, and we find a departure occurring only in the upper left-hand part of the diagram. Although we have given detailed examples for only two open clusters, the same general situation applies to the others.

At this point we must make another observation. As we move up the main sequence of an open cluster and approach the point where it terminates, we observe that there is a sharp turn to the right, away from the main sequence. Since their termination points differ, this turning away from the main sequence occurs at different values of the color index for different clusters. It must, then, be related to an important property of the cluster itself.

The various features we have discussed about the H-R diagrams of open clusters can best be illustrated by a composite diagram of ten different galactic clusters, as shown in Figure 17.9. Since all the main sequences of these clusters coincide, they are represented by a single line that ranges from color index $+1.2$ to $+0.4$. As we go beyond this point, the line for each cluster deviates to the right at a point which can be used to classify the cluster. We shall see that this point bears an important relationship to the age of the cluster. Note that the stars in some of the clusters do not all lie on one continuous line. Thus, we find that the stars of some of the clusters such as $h + \chi$ Persei, M 41, M 11 lie on small branches far to the right of the main sequence. We have also included

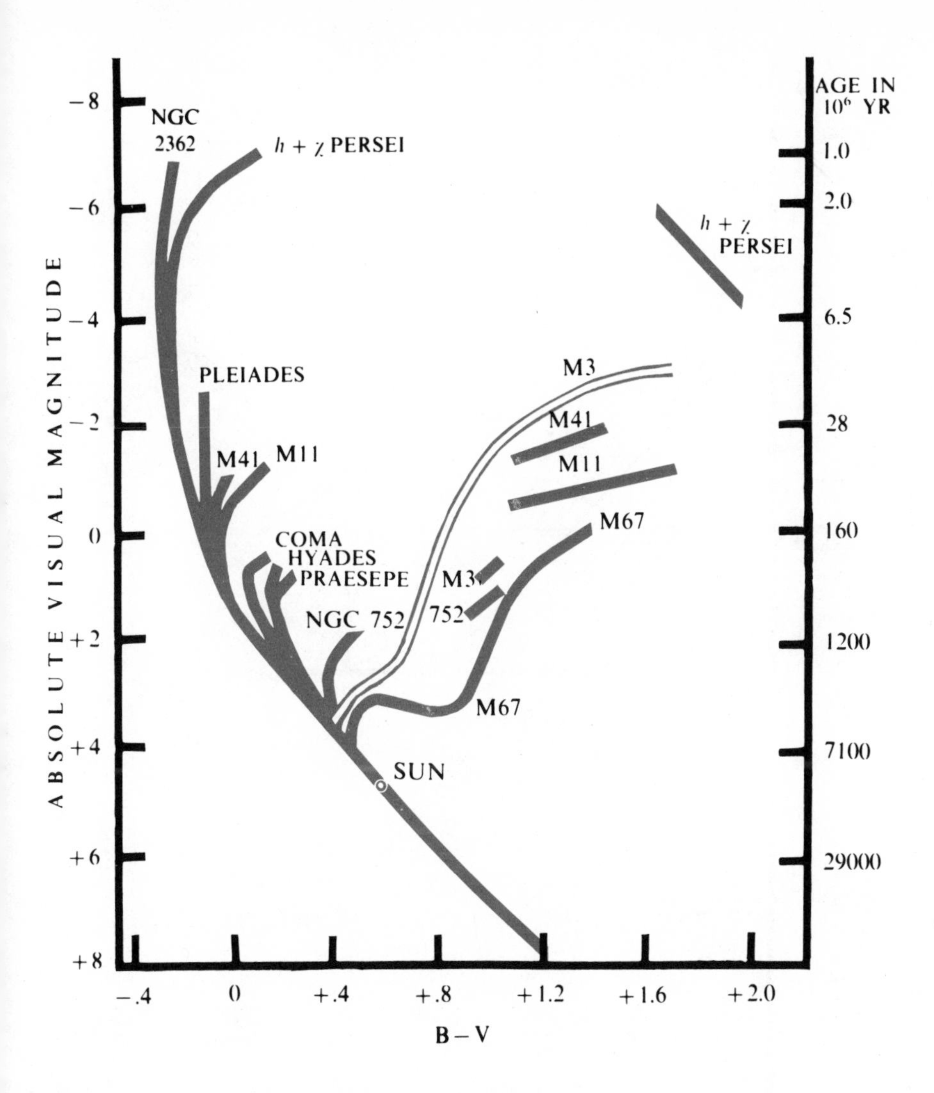

FIG. 17–9 *A composite H-R diagram of various open clusters. (After Sandage)*

among the ten examples the cluster M 67, whose detailed color-luminosity diagram is given in Figure 17.10.

We shall come back to the H-R diagrams of open clusters and their relationship to the evolution of stars in Chapter 27 after we have discussed the internal constitution of stars from the astrophysical point of view. However, before we leave the H-R diagrams for galactic clusters we must call attention to galactic cluster NGC 2264, which has been studied by M. F. Walker. In the H-R diagram (Figure 17.10) in which this cluster is analyzed, the heavy line gives the position of the main sequence. We see that most of the stars in this cluster lie above the main sequence and that there are a few stars off at a great distance to the right. We shall see later in Chapter 27, Section 27.28, that Clusters M 67 and NGC 2264 (which we analyze in detail there) are examples at the extreme ends of the evolutionary sequence, the former being a very old cluster and the latter a very young one.

17.13 The H-R Diagram of Globular Clusters

Another example of a stellar group consisting of stars that all had the same initial chemical

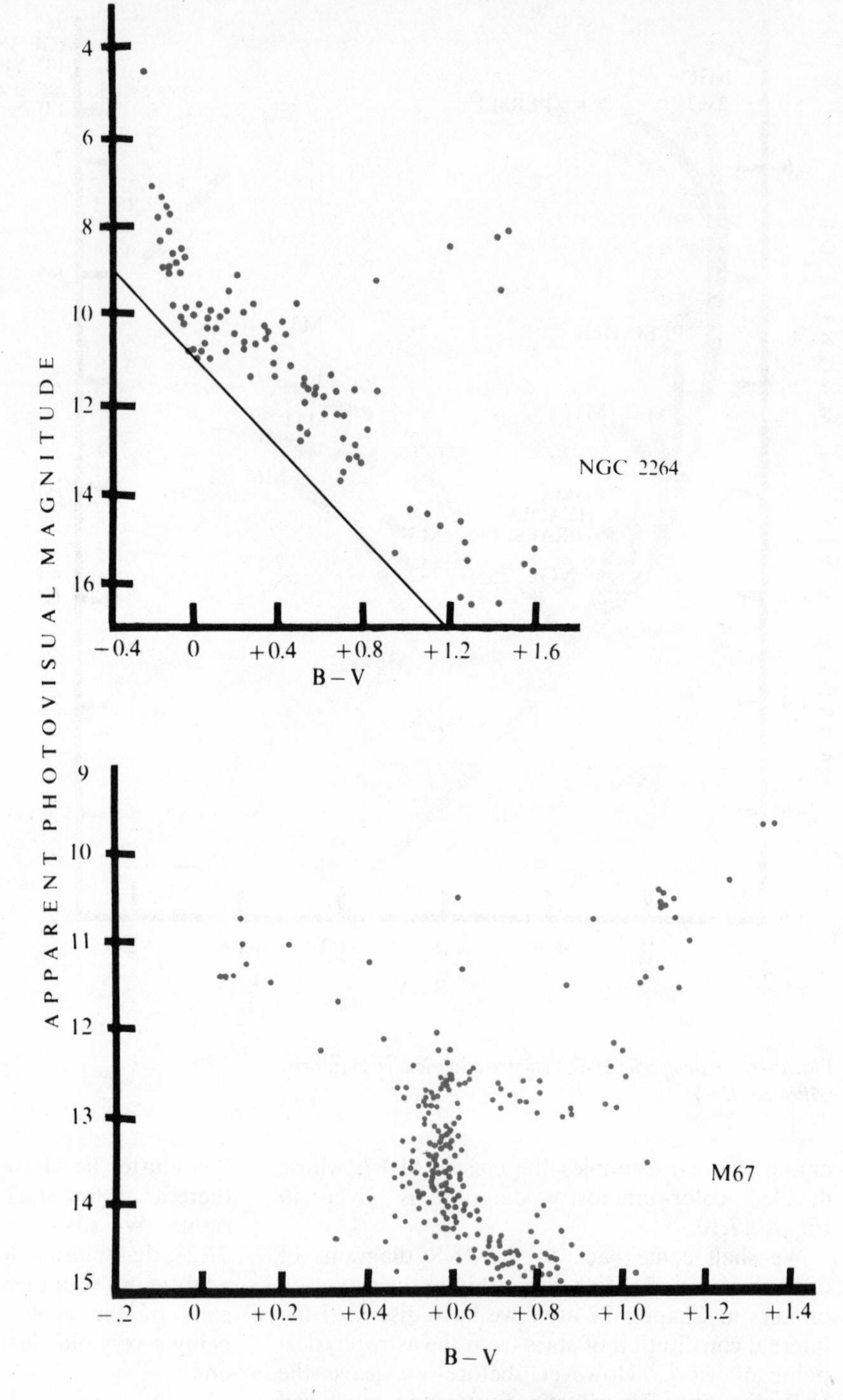

FIG. 17–10 *A detailed H-R diagram of two extreme galactic clusters, NGC 2264 (a very young cluster) and M 67 (a very old cluster). (NGC 2264: M. Walker,* Astrophysical Journal, *Suppl. No. 23, 1956; M 67: H. L. Johnson and A. R. Sandage,* Astrophysical Journal *CXXI (1955), 616.* © *1955, 1956 by the University of Chicago)*

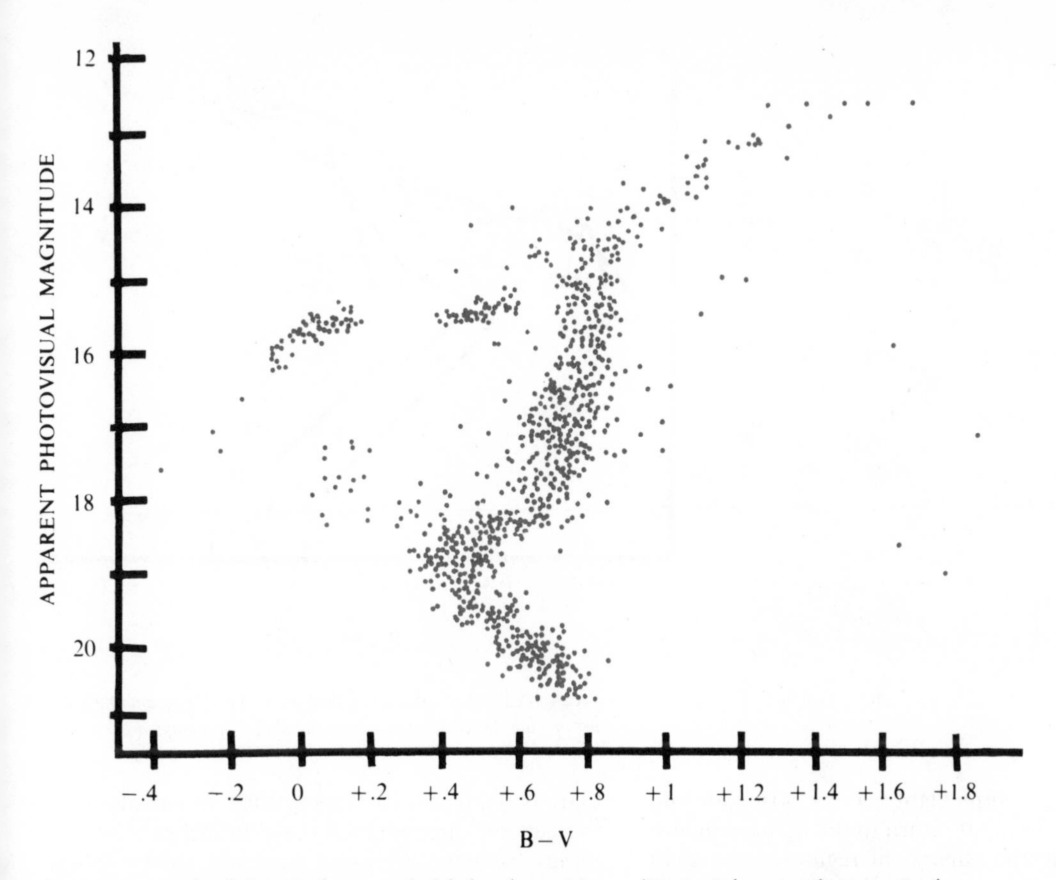

FIG. 17–11 *A detailed H-R diagram of globular cluster M 3.* (*H. L. Johnson and A. R. Sandage,* Astrophysical Journal *CXXIV (1956), 379. © 1956 by the University of Chicago*)

composition is a **globular cluster**. *These clusters are huge symmetrical conglomerations of stars, containing anywhere from a few hundred thousand to a few million stars, and lying at great distances from us in a kind of halo around the center of the Milky Way* (we shall discuss this point in more detail in Chapters 27 and 28). In the H-R diagram of globular cluster M 3 (Figure 17.11) we see that the stars in this cluster lie not only along the main sequence extending from color index +0.8 to +0.4, but also along branches that depart from the main sequence and run up to the giant and supergiant regions. Although most of the globular clusters have H-R diagrams similar to this one, there are slight differences which are best presented by again considering a composite diagram.

To carry out this comparison we must be able to superpose one diagram on top of another, but we can do this only if we have some common point or region which we know to be of the same absolute magnitude (note that in these diagrams, just as for open clusters, we plot apparent magnitude against the color index). If we again consider the H-R diagram of the globular cluster M 3, we see that there is a gap in the upper branch lying between color index −0.2 and +0.5. Such a gap occurs in the H-R diagrams of all globular clusters. Stars are really found to lie in this gap, but since these stars, which are called **cluster-type variables**, undergo continuous changes in luminosity, they are not included in the diagram. We do know, however (from evidence which we shall give later), that all these stars have very nearly the same average absolute magnitude, about +0.6. Hence, if we wish to make a composite H-R diagram of globular clusters, we must superpose these diagrams so that the gaps for all the clusters coincide.

When we do this we obtain the composite picture shown in the next H-R diagram, Figure 17.12. We see that all the globular clusters have main-sequence branches that do not quite coincide, and also branches that move off the main sequence

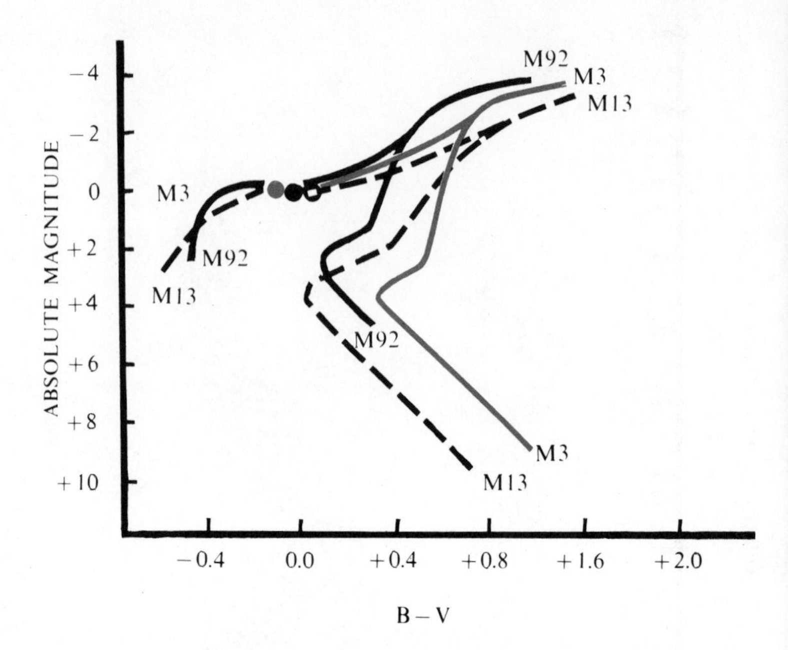

FIG. 17–12 *A schematic comprehensive H-R diagram of a number of globular clusters in which the RR Lyrae gaps are matched.*

into the giant and supergiant regions. In addition, there are branches that return to the main sequence from the giant and supergiant regions. We shall see later that there is good evidence for believing that as stars evolve (age), they move along in the general directions of these branches. All the things we have discussed here are shown in the composite schematic H-R diagram, Figure 17.6(a).

EXERCISES

1. If an error of 25 per cent is made in the parallax of a main-sequence star, how far off the main sequence will it be placed in the H-R diagram?

2. How does the size of a red giant star differ from that of a white giant?

3. (a) What can you say about the relative abundances of the following stars from an examination of the H-R diagram: red dwarfs, white dwarfs, blue-white giants, red supergiants, giants? (b) What would tend to make your analysis in (a) based solely on the H-R diagram wrong?

4. We shall see later that stars evolve from the main sequence into the giant branch. Why then are there so few stars found in the region between the main-sequence and the giant branches?

18

ATOMIC STRUCTURE AND THE ORIGIN OF STELLAR SPECTRA

We have seen that the variation in the spectral class, as we go from star to star, is due to a combination of things, the most important of which is the surface temperature of the star. This is most clearly depicted in the H-R diagrams which we discussed in detail in the last chapter. Although the over-all variations that occur in stellar spectra are due primarily to the variation in temperature, other factors have a significant influence on the fine details of the spectra. These fine details are important in astronomy since by studying them we can determine the absolute magnitudes of stars. We can best grasp this by referring again to the general H-R diagram (Figure 17.6(a)).

18.1 Absolute Magnitude and the H-R Diagram

In principle it should be possible with the aid of the H-R diagram to obtain absolute stellar magnitudes. We may illustrate this by an example. Let us suppose that we determine the spectral classification of a star and find it to be similar to that of the sun. We know then that it must lie on the H-R diagram on a line perpendicular to the horizontal axis at the point *G*2. If it were not for this diagram, we would have no way of knowing where along this line the star lies. However, since the H-R diagram teaches us that this star can lie either on the main-sequence, the giant, or the supergiant branch, we are already well along the way to knowing what its absolute magnitude must be: it must be approximately equal to that of the sun, to that of Capella, or to that of a yellow supergiant. If we had no further data to guide us, and still wanted to make a reasonable guess about the magnitude of this star, we could place it equal to +4.8 (the sun's absolute magnitude) and be sure that this would have a 90 per cent chance of being correct.

This type of procedure is useful for studying the statistical properties of large groups of stars. However, for any particular star we cannot be sure that assigning an absolute magnitude in the manner above is correct. Fortunately, there are fine differences in the spectra of main-sequence, giant, and supergiant stars of the same spectral class which enable us to assign a star of a given spectral class to one of these three categories. Investigations into these fine spectral differences were begun by Kohlschütter and Adams in 1913, at Mount Wilson Observatory, and have been carried on extensively since then, so that astronomers are now able to set up fairly precise spectral criteria for differentiating among main-sequence stars, giants, and supergiants.

These procedures allow us to assign absolute magnitudes to stars and from these absolute magnitudes to determine their parallaxes. This process for finding the parallax of a star is called the **method of spectroscopic parallaxes**. However, before we can describe this method in detail, we must analyze the way in which absorption lines are produced by the atoms in the atmosphere of a star. And to do this, we must first briefly discuss the structure of the atom.

18.2 The Introduction of Quantum Theory into Atomic Structure

In our discussion of the properties of radiation we saw that it was necessary to discard classical ideas. We can no longer look upon radiation as being just a wave phenomenon, but instead, following Planck, we must introduce the quantum concept in which radiation is pictured as consisting of photons, each one of which, for a given frequency ν, has a given amount of energy $E = h\nu$. While this profound revolution was taking place in our concept of radiation, important changes were also occurring in our picture of the structure of matter itself.

When Planck developed his quantum theory, he did so without introducing any assumption about the structure of the atom itself. He simply assumed that the energy was emitted and absorbed by some kind of microscopic oscillators in the black body which he pictured as vibrating more or less as a violin string or any elastic body does (so-called simple harmonic oscillations). However, although this picture was sufficient to enable Planck to derive his radiation law, it cannot tell us how any single atom emits or absorbs radiation. To understand how spectral lines are formed, we must have a model of an atom that is capable of emitting and absorbing radiation in agreement with spectral evidence. It is clear that the model of the atom we accept must incorporate within it the quantum theory and hence must follow Planck's departure from classical concepts.

18.3 Properties of Electrical Charges

Since we know from the work of Maxwell that radiation is an electromagnetic phenomenon and we also know that it is emitted and absorbed by atoms, it is clear that the atom itself must be composed of electrically charged particles. This was known for a long time: the Greeks had already discovered that it is possible to induce an electrical charge on matter such as a piece of ebony by rubbing it with another bit of matter, such as cat's fur. These elementary investigations demonstrated that there are two types of electrical charge; we designate one as negative and the other as positive. Obviously there is an equal amount of positive and negative charge in all neutral matter.

The important characteristics of an electrically charged particle differentiating it from neutral matter are that:

1. it responds to other electrical charges placed in its vicinity, and
2. it reacts to passing radiation.

The knowledge of these two properties of an electric charge, which have been studied extensively, is basic to an understanding of the structure of the atom and to the way in which the atom emits and absorbs radiation.

18.4 Coulomb's Law of Force between Electrical Charges

When two electric charges of the same sign are placed together, they repel each other, and when two charges of opposite sign are placed near each other, they attract each other. The law governing this repulsion and attraction of electric charges was first discovered by Cavendish and independently by Coulomb. It is now known as Coulomb's law of force since his results were published first. This law can be stated as follows:

If in a vacuum a point charge of magnitude Q_1 (Q_1 is the amount of electrical charge) is placed at a distance from a like point charge of magnitude Q_2, the two charges repel each other with a force (along the line connecting them) which is equal to the product of the two charges divided by the square of the distance between them.

We may express this algebraically as follows:

$$F = \frac{Q_1 Q_2}{r^2}. \tag{18.1}$$

This is known as Coulomb's inverse-square law of force, and is the basis for our understanding of electrostatics. Using this law of force we define the unit of electric charge as follows:

If two equal charges in a vacuum are placed 1 cm apart, and if each charge exerts a force of 1 dyne on the other, then the magnitude of each of these charges is 1 electrostatic unit.

18.5 Positive and Negative Charges in Atoms

That atoms are composed of both positive and negative electric charges in equal quantities was already known in the middle of the nineteenth century as the result of Faraday's experiments on electrolytic solutions. Faraday showed that when a salt is dissolved in water, each molecule breaks up into negative and positive ions. Thus, in a solution of sodium chloride, each molecule NaCl is dissociated into a positive ion Na^+ and a negative ion Cl^-. By passing an electric current through such a solution, Faraday demonstrated that the amount of electrical charge on any ion is always an integral multiple of the charge on the sodium ion.

We can understand why this is so if we picture electric charges in nature as occurring in integral multiples of a fundamental indivisible unit. Faraday's investigations prove that this fundamental indivisible unit exists, both as a positive charge and as a negative charge, and that regardless of what sort of ion we may have in a solution (whether it be positive or negative) it always has 1, 2, 3, etc. units of this fundamental charge, and never a fraction of it ($\frac{1}{2}$, $\frac{1}{3}$, $\frac{1}{4}$, etc.). Although Faraday's work demonstrates that atoms consist of equal quantities of positive and negative charge, it tells us nothing about the way these charges are arranged to give a neutral atom.

One more notable result that Faraday obtained is the ratio of the charge e on a given ion to the mass M of that ion. Faraday discovered that this ratio e/M varies from ion to ion, and has the largest value for the hydrogen ion. This agrees with the fact that the mass of the hydrogen ion is smaller than that of any other ion.

18.6 Radiation by Accelerated Charges

Although Coulomb's law of force tells us how one charge is influenced by the presence of another charge, it does not tell us how such a charge reacts to radiation. This gap in the classical picture of the electrical nature of matter was filled by Maxwell, who, on the basis of Faraday's experiments, developed the electromagnetic theory of light. Maxwell showed that light is an electromagnetic disturbance that is propagated in the form of a wave through empty space. This means that a beam of light consists of energy in the form of rapidly oscillating electric and magnetic fields. If a charged particle is placed in such a field, it oscillates in time with these varying fields.

What happens is that the electromagnetic field of the passing radiation exerts a force on the charge just the way a gravitational field imparts an acceleration to a mass. *Since the charged particle acquires kinetic energy when it oscillates in response to the radiation passing it, the charge must have acquired this energy from the radiation itself. In other words, the charged particle absorbs some of the passing radiation.*

Just as a charged particle can absorb radiation from an electromagnetic field, it can also emit radiation. If we set a charged particle oscillating by mechanical means, this charged particle emits radiation at a very definite rate, which can be computed from Maxwell's theory. If the acceleration of the charged particle is a, and if e is the electric charge on it, then Maxwell's classical theory of radiation shows us that the charged particle emits energy at a rate equal to

$$\frac{2}{3}\frac{e^2a^2}{c^3} \quad \text{ergs/sec,} \qquad (18.2)$$

where c is the speed of light.

If we take all of these properties of charged particles into account we have a complete mechanism for setting up a model of an atom that can both emit and absorb radiation. Neither Maxwell nor the other physicists of his day were in a position to do this since nothing was known about the physical properties of the individual charges constituting an atom; however, it was not long before the investigations of men such as Sir J. J. Thomson led to just this kind of information. We shall consider their findings after taking note of one other important discovery, which lies more in the domain of chemistry than of physics.

18.7 The Mendeleev Periodic Table: Atomic Number and Atomic Weight

The introduction of the atomic theory of matter early in the nineteenth century made it relatively easy to understand the laws of formation of chemical compounds, even though there was no precise theory of the way atoms combine to form these compounds. The greatest advance in nineteenth-century chemistry occurred with the discovery by Mendeleev of the **periodicity** in the chemical properties of the elements as one passed from one element to another. Mendeleev presented his discovery in the form of the Periodic Table of Elements. To understand this table, we must introduce two fundamental concepts: (1) the atomic number, and (2) the atomic weight.

If we arrange all the elements in order, starting with the lightest and going to the very heaviest, the atomic number of an element is its position in this array. Thus, we assign the atomic number 1 to hydrogen, the atomic number 2 to helium, 3 to lithium, and so on. We shall see later that the atomic number is related to the number of electrically charged particles inside a neutral atom.

The atomic weight is the measure of the mass of a single atom on some arbitrary but convenient scale. The scale that has been in use until the present time is one in which the oxygen atom is arbitrarily given the atomic weight 16. On this scale hydrogen has the atomic weight 1.008, and helium the atomic weight 4.003.

We can now understand Mendeleev's great discovery by arranging some of the elements starting with hydrogen in the sequence of increasing weight, as shown in Table 18.1. The subscript attached to the symbol of each element is its atomic number and its superscript is its atomic weight. Mendeleev's great contribution consisted in the discovery that all the elements in a given vertical column possess similar chemical properties. Thus, Li, Na, K, etc. have similar properties and constitute the group called the alkali metals. Fl, Cl, Br, etc. constitute another group called the halogens, and He, Ne, Ar, etc. belong to the family of noble gases. What is important about this arrangement of elements is that there is a repetition of chemical properties in a cycle of eight elements. This is not true for the entire chart of elements; as we go to heavier elements there is a change from a cycle of eight to a cycle of eighteen, etc. All the elements are shown in Table 18.2.

TABLE 18.1
CHEMICAL PERIODICITY OF SOME OF THE LIGHTER ELEMENTS

${}_{1}H^{1}$	${}_{2}He^{4}$	${}_{3}Li^{3}$	${}_{4}Be^{9}$	${}_{5}B^{11}$	${}_{6}C^{12}$	${}_{7}N^{14}$	${}_{8}O^{16}$	${}_{9}Fl^{18}$
	${}_{10}Ne^{20}$	${}_{11}Na^{23}$	${}_{12}Mg^{24}$	${}_{13}Al^{27}$	${}_{14}Si^{28}$	${}_{15}P^{31}$	${}_{16}S^{32}$	${}_{17}Cl^{35}$
	${}_{18}Ar^{40}$	${}_{19}K^{39}$	${}_{20}Ca^{40}$	${}_{21}Sc^{45}$	${}_{22}Ti^{48}$	${}_{23}V^{51}$	${}_{24}Cr^{52}$	

18.8 Electrical Discharges through Gases at Low Pressure

Although Mendeleev's discovery aided the chemist greatly in understanding the chemical behavior of the elements, it did little to further the physicist's knowledge of the structure of the atom itself. It soon became clear that only by penetrating physically into the atom and analyzing its constituent particles could one construct a model of the atom that would correctly account for its behavior. Experimental investigations of this nature were carried out by a group of outstanding physicists such as Sir J. J. Thomson of Great Britain, Jean Perrin in France, and Kaufmann and Goldstein in Germany. Their investigations stemmed from work that had been begun by Plückert and Geissler in Germany and Sir William Crookes in Great Britain on the conduction of electricity through gases at low pressure.

We have already seen in our discussion of the emission spectrum that when a high enough voltage is placed across a tube containing a gas at low pressure, the entire tube glows. This, as we have noted, is the result of a flow of electrical charge through the tube. If the pressure in the tube is reduced more and more until hardly any gas remains, the glow disappears and is replaced by a remarkable fluorescence on the walls of the glass tube opposite the cathode (the negative high-voltage terminal).

This fluorescence can be shifted around by applying a magnet to the tube. Moreover, it can be blocked off by placing an object between the cathode and anode. These phenomena led physicists to believe that some kind of radiation emanates from the cathode. This radiation was given the name **cathode rays** by Goldstein, and it already appeared then that cathode rays consist of negatively charged particles. Later, Goldstein made small holes in the cathode of the tube and discovered what he called **canal rays** consisting of a stream of some sort flowing away from the anode. [See Figure 18.1(b).]

18.9 The Discovery of the Electron

By applying electric and magnetic fields to the cathode rays, it is possible to demonstrate that these rays do, indeed, consist of negatively charged particles which we now call **electrons**. If the cathode-ray beam is placed in an electric field, it is distorted from a straight line into a parabola (it behaves like a stream of water leaving the nozzle of a hose held horizontally above the ground). If now a magnetic field at right angles to the electric field is imposed upon the beam, it can be brought back into a straight line again. Sir J. J. Thomson was the first to apply two such fields to a cathode ray and to show that one can thereby determine the

Table 18.2
THE CHEMICAL ELEMENTS

Element	Symbol	Atomic number	Atomic weight (chemical scale)
Hydrogen	H	1	1.008
Helium	He	2	4.003
Lithium	Li	3	6.939
Beryllium	Be (Gl)	4	9.013
Boron	B	5	10.812
Carbon	C	6	12.012
Nitrogen	N	7	14.007
Oxygen	O	8	16.000
Fluorine	F	9	18.999
Neon	Ne	10	20.184
Sodium	Na	11	22.991
Magnesium	Mg	12	24.313
Aluminum	Al	13	26.982
Silicon	Si	14	28.09
Phosphorus	P	15	30.975
Sulphur	S	16	32.066
Chlorine	Cl	17	35.454
Argon	Ar (A)	18	39.949
Potassium	K	19	39.103
Calcium	Ca	20	40.08
Scandium	Sc	21	44.958
Titanium	Ti	22	47.90
Vanadium	V	23	50.944
Chromium	Cr	24	52.00
Manganese	Mn	25	54.940
Iron	Fe	26	55.849
Cobalt	Co	27	58.936
Nickel	Ni	28	58.71
Copper	Cu	29	63.55
Zinc	Zn	30	65.37
Gallium	Ga	31	69.72
Germanium	Ge	32	72.60
Arsenic	As	33	74.924
Selenium	Se	34	78.96
Bromine	Br	35	79.912
Krypton	Kr	36	83.80
Rubidium	Rb	37	85.48
Strontium	Sr	38	87.63
Yttrium	Y	39	88.908
Zirconium	Zr	40	91.22
Niobium	Nb (Cb)	41	92.91
Molybdenum	Mo	42	95.95
Technetium	Tc (Ma)	43	99
Ruthenium	Ru	44	101.07
Rhodium	Rh	45	102.91
Palladium	Pd	46	106.4
Silver	Ag	47	107.874
Cadmium	Cd	48	112.41
Indium	In	49	114.82
Tin	Sn	50	118.70
Antimony	Sb	51	121.76
Tellurium	Te	52	127.61
Iodine	I (J)	53	126.909
Xenon	Xe (X)	54	131.30
Caesium	Cs	55	132.910
Barium	Ba	56	137.35
Lanthanum	La	57	138.92
Cerium	Ce	58	140.13
Praseodymium	Pr	59	140.913
Neodymium	Nd	60	144.25
Promethium	Pm (Il)	61	147
Samarium	Sm (Sa)	62	150.36
Europium	Eu	63	151.96
Gadolinium	Gd	64	157.25
Terbium	Tb	65	158.930
Dysprosium	Dy (Ds)	66	162.50
Holmium	Ho	67	164.937
Erbium	Er	68	167.27
Thulium	Tm (Tu)	69	168.941
Ytterbium	Yb	70	173.04
Lutecium	Lu (Cp)	71	174.98
Hafnium	Hf	72	178.50
Tantalum	Ta	73	180.955
Tungsten	W	74	183.86
Rhenium	Re	75	186.3
Osmium	Os	76	190.2
Iridium	Ir	77	192.2
Platinum	Pt	78	195.10
Gold	Au	79	196.977
Mercury	Hg	80	200.60
Thallium	Tl	81	204.38
Lead	Pb	82	207.20
Bismuth	Bi	83	208.988
Polonium	Po	84	210
Astatine	At	85	211
Radon	Rn	86	222
Francium	Fr (Fa)	87	223
Radium	Ra	88	226.05
Actinium	Ac	89	227
Thorium	Th	90	232.047
Protactinium	Pa	91	231
Uranium	U (Ur)	92	238.04
Neptunium	Np	93	237
Plutonium	Pu	94	239
Americium	Am	95	241
Curium	Cm	96	242
Berkelium	Bk	97	243
Californium	Cf	98	244
Einsteinium	Es	99	—
Fermium	Fm	100	—
Mendelevium	Md	101	—
Nobelium	No	102	—
Lawrencium	Lw	103	—

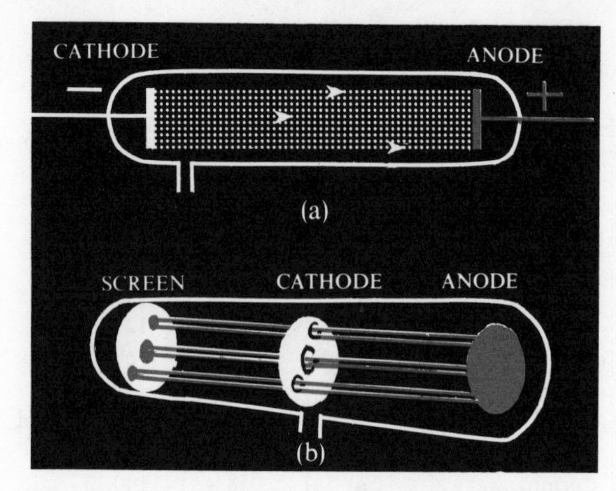

FIG. 18–1 *(a) Cathode-ray tube with electrons being emitted by the cathode. (b) Canal rays. Positively charged particles moving from the anode through holes in the cathode.*

quantity e/m (the ratio of the charge e to the mass m of the electron). He did this by setting down the mathematical condition that must be fulfilled when the effect of the electric field is just canceled by the effect of the magnetic field.

Thomson discovered that this ratio is always the same regardless of the chemical nature of the cathode. *This means that the electron is a fundamental constituent of all matter and must therefore be one of the building blocks of all atoms.* Thomson further discovered that the ratio e/m for the electron is about 1 836 times that of e/M found for the hydrogen ion by Faraday. Since the charge on an electron is equal in magnitude to the positive charge on the hydrogen ion, it follows that the masses of all electrons are equal and about 1 836 times smaller than that of the mass of the lightest known atom, hydrogen. It is clear from this that the electrons are not atoms in the way that, for example, hydrogen and helium are (there is no place for them on the periodic chart of elements) but are constituents of atoms.

18.10 The Discovery of the Proton

Since an atom is a neutral structure, it contains an equal positive charge for every electron (negative charge). Such positive charges constitute the particles in Goldstein's canal rays that we mentioned previously. The e/M for the particles in the canal rays are not the same in all cases since the canal ray consists of positive ions for which the charge and mass may vary. The largest value for e/M for a canal ray is obtained when the canal rays consist of hydrogen ions. We may therefore take the hydrogen ion as being the fundamental positive charge within an atom, just as the electron is the fundamental negative charge. *The hydrogen ion is called the* ***proton.***

18.11 The Charge and the Mass of the Electron

Following the discovery of the electron, Millikan in 1909 measured its charge and found it to be 4.802×10^{-10} esu (electrostatic units). This means that if 2.1×10^{9} electrons could be concentrated in a point they would exert a repulsive force of 1 dyne on the same number of electrons concentrated at a point 1 cm away.

From the knowledge of e/m for the electron and the measured value of the charge e, we find that the mass m of the electron is 9.108×10^{-28} gm. This means that if a force of 1 dyne were applied to an electron continuously, the electron would suffer an acceleration of about 10^{27} cm per sec per sec (assuming the laws of classical mechanics held for the electron).

18.12 The Rutherford Model of the Atom

With the discovery of the electron and the proton it appeared that little remained to the task of constructing a model of the atom. All one had to do, it seemed, to set up such a model was to find an equilibrium configuration in which one could properly arrange electrons and protons. The mathematical and theoretical machinery for doing this was available since the great physicist Lorentz had already applied Maxwell's electromagnetic theory to matter and had developed a theory of the electron. Only one thing remained to be determined experimentally before this electron theory could be applied to a model of the atom, and that was the way in which the electrons are distributed in relationship to the positive charge in the atom.

Sir J. J. Thomson believed that the positive charge was distributed like a fluid over a large core in which the electrons were embedded. He developed a theoretical model of the atom on this basis. This model, however, was in contradiction to the experiments of Lord Rutherford and his collaborators, who subjected a thin gold foil (see Figure 18.2) to a bombardment by the very

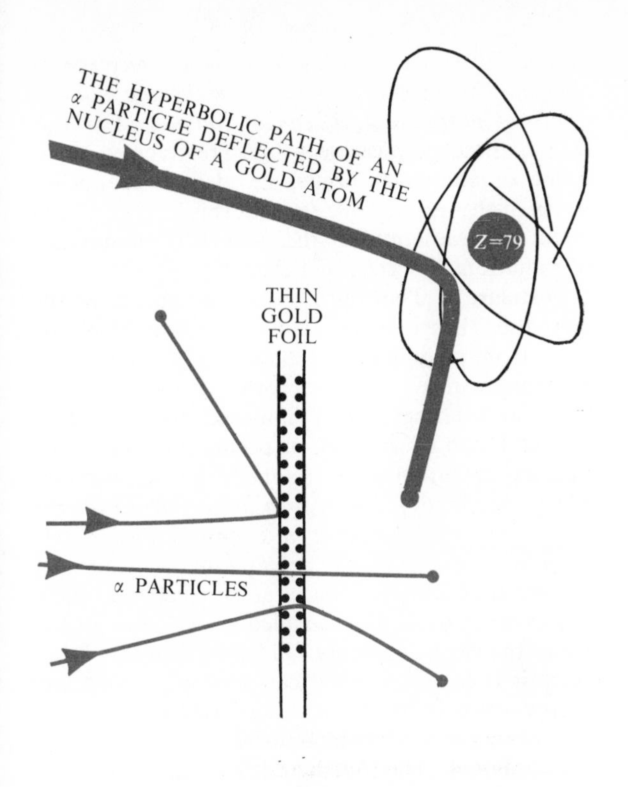

FIG. 18–2 *The bombardment of a thin gold foil by α particles, showing the sharp deflection of the α particles by the nuclei of the gold atoms. An enlargement of one gold atom, showing how the nucleus deflects the α particle in a hyperbolic orbit.*

energetic positively charged particles (*so-called* **alpha particles**, *which are doubly ionized helium atoms*) emitted by radioactive atoms such as uranium.

Rutherford's group showed by this means that *the positive charge in an atom is concentrated in a tiny massive nucleus and that the negatively charged electrons lie in a kind of cloud surrounding this nucleus.* When an alpha particle strikes the gold foil, it generally passes right through without being deviated very much since there are relatively vast spaces between positively charged nuclei of the gold atoms. However, every now and then an alpha particle passes very close to the nucleus of a gold atom and is then violently repelled by the nucleus' positive charge. This is exactly what Rutherford found. On this experimental basis, Rutherford suggested the famous planetary model of the atom, consisting of a central positively charged massive core surrounded by electrons revolving around this nucleus in orbits, just the way planets revolve around the sun.

18.13 The Contradiction between the Rutherford Model and Classical Physics

Although the Rutherford model of the atom accounts very nicely for Rutherford's α-particle experiments, it seems to have one fatal flaw, which at first appeared to make it completely unacceptable. In order to prevent an electron from falling into the nucleus of the Rutherford model, we must suppose that it is moving in a circular or elliptical orbit around the nucleus. This means, of course, that it is constantly being accelerated the way a planet is [see Equation (3.6)]. But as we have seen, when a charged particle is accelerated, it must, according to the laws of Maxwell and Lorentz, radiate energy in the form of electromagnetic waves at a rate proportional to the square of the acceleration. If this law were valid for an electron moving in an orbit around the nucleus of an atom, it can be shown that the electron would lose all of its energy in a fraction of a second and fall into the nucleus.

Most physicists, therefore, rejected the Rutherford model because it could not, according to the classical laws of electricity and magnetism, lead to stable atoms. Thus, it appeared that the only model of the atom that was in accord with the facts adduced by Rutherford's experiments was in violent contradiction to the basic laws of electricity and magnetism, as developed by Maxwell and others. This difficulty was overcome by Niels Bohr, who was able to keep the Rutherford model by imposing upon it Planck's quantum theory.

18.14 The Bohr Theory of the Atom

We can best understand what Bohr did by considering the hydrogen atom, to which he first applied his theory. According to the Rutherford model the dynamical properties of such an atom, consisting of a single electron and proton, are similar to those of a two-body gravitational system, like the sun and a planet (see Figure 18.3). The electron must then be pictured as moving in a planetary orbit around the proton (the nucleus of the hydrogen atom)—but with one important difference as compared to the motion of the planet around the sun. In the case of the latter the total energy of the system is conserved, so that once the planet is in a given orbit, it always remains in that orbit and the system maintains itself in the given configuration forever.

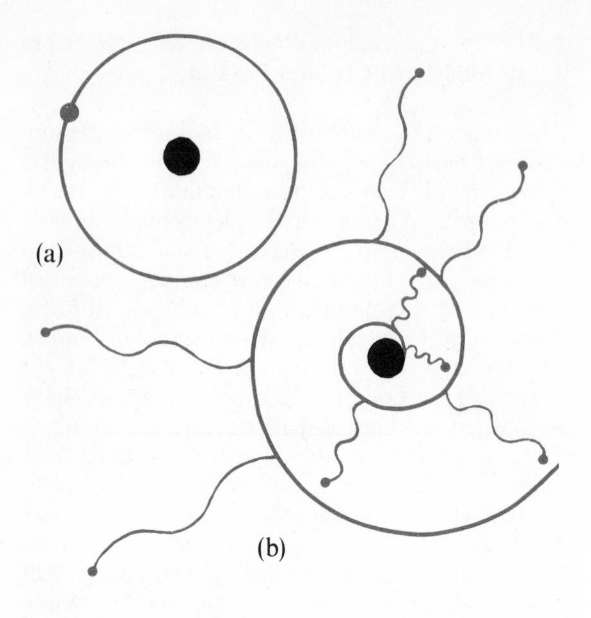

FIG. 18–3 *The Rutherford model of the atom, showing the electron (a) in a planetary orbit, and (b) spiraling into the nucleus as it radiates energy classically.*

This is not the case for the electron and the proton if the classical laws of physics are retained. The electron must lose energy continuously according to these laws, and so no stable orbits appear possible. It is here that Bohr made his great contribution. He imposed certain new and revolutionary conditions on the behavior of the electron inside an atom, which lead to stable orbits. These conditions are four in number, and were introduced by Bohr simply as a working hypothesis without any prior experimental or theoretical justification.

18.15 The Bohr Postulates

Postulate 1. In order to prevent the hydrogen atom from collapsing and the electron from falling into the proton, Bohr first postulated that the electron may move only in certain permissible orbits. In other words, *not all sizes of orbit are allowed to the electron but only a discrete set of sizes.* This, of course, is a drastic departure from classical mechanics since, as we know, there is nothing in Newtonian theory which in any way restricts the size (semimajor axis) of a planet's orbit around the sun.

Postulate 2. Bohr's second postulate was introduced to keep the electron from radiating energy continuously while moving in one of the permitted orbits. It states that *as long as an electron is revolving about the proton in one of the allowed orbits of postulate 1, it cannot radiate.* For this reason these permissible orbits are referred to as **stable** or **stationary orbits**, or also as **stationary states** of the atom. We see that this postulate is in direct contradiction to the Maxwell-Lorentz law of radiation of accelerated charges.

Postulate 3. The third postulate concerns the way the electron may emit or absorb energy. Since from postulate 2 it follows that as long as the electron is in a permissible orbit it does not emit or absorb energy, Bohr imposed the condition that emission or absorption occurs only when the electron is jumping from one orbit to another. *When the electron jumps to a larger orbit (farther away from the proton) it must* ***absorb*** *energy, and when it jumps to a smaller orbit (closer to the proton) it must* ***emit*** *energy.* Since, as we shall see, there is a smallest orbit, the so-called **ground state** of the atom, the electron cannot fall lower than this orbit. Once it is in this orbit it can no longer emit any energy since it has no smaller orbit to fall into. This allows stable atoms to exist.

Postulate 4. The fourth and final postulate gives the quantitative description of the emission or absorption of energy by an electron and introduces the Planck theory directly. *It states that whenever an electron jumps from one orbit to another, it emits or absorbs energy, as the case may be, in the form of a single photon, and the energy of the photon emitted or absorbed is exactly equal to the difference of the energy of the electron in the two orbits.*

To be precise, if the energy of an electron in its initial orbit is E_i, and the energy in its final orbit is E_f, then the energy of the photon emitted when an electron jumps from orbit i to orbit f is just $E_f - E_i$ (if this quantity is negative the photon is emitted and if it is positive the photon is absorbed). Since according to Planck's theory the energy of a photon is $h\nu$, it follows that

$$h\nu = E_f - E_i.$$

Hence, the numerical value of the frequency of the emitted (or absorbed) photon is

$$\boxed{\nu = \frac{|E_f - E_i|}{h}}, \qquad (18.3)$$

where the vertical lines mean the absolute value. This last postulate enabled Bohr to calculate the frequencies of photons emitted or absorbed by electrons jumping from one orbit to another, and thus, as we shall see, to account for the spectral emission and absorption lines of hydrogen.

18.16 Analysis of the Bohr Postulates

A few words about the fourth postulate will make it even clearer. *The first thing to observe is that only one photon is associated with each jump of the electron, regardless of whether the electron jumps to a neighboring orbit or to a distant orbit.* Note that with the Bohr theory an electron jumping from a given orbit is not restricted to landing in the next nearest orbit. Thus, an electron may cascade down from a given orbit to the ground state emitting a series of photons of different frequencies, or it may jump down directly to the ground state by emitting a single photon whose frequency is the sum of the frequencies of the photons emitted in the cascade process (see Figure 18.4). In the same way the electron may jump to a higher orbit by absorbing a single photon of the right frequency, or by absorbing a succession of photons whose frequencies add up to the frequency of the single photon.

The second thing to note is that when an electron jumps in a single step from a lower to a higher orbit, it does so only by absorbing a photon of one single frequency and no other. This means that the electron is very selective in its reaction to a stream of radiation passing it. If this stream contains photons of many different frequencies, the electron moving in one of the permissible orbits absorbs a photon only if that photon has just the right amount of energy (just the right frequency) to bring the electron to some higher permissible orbit.

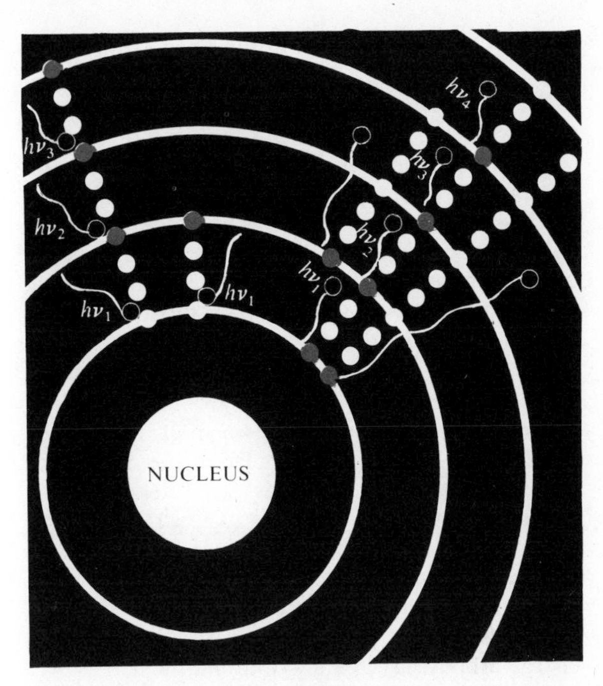

FIG. 18–4 *An electron cascading down to the ground level of an atom emits a photon during each transition from level to level, according to the Bohr model.*

18.17 The Hydrogen Spectrum: The Balmer Lines

By imposing these four postulates, Bohr derived the observed wavelengths of the lines in the hydrogen spectrum, and in particular he accounted for the famous **Balmer lines** (*the series of lines that lie in the visible range of the hydrogen spectrum*). In the year 1885, Balmer, investigating the four visible lines in the hydrogen spectrum that were known at that time, discovered a very interesting regularity in the frequency spacings of these lines which he expressed by means of a simple formula involving the squares of the integers. He went on from this to predict that his formula would give the frequencies of other undiscovered lines in this series. This proved to be the case. The Balmer formula for the frequencies for these lines is

$$\nu_{\text{Balmer}} = A\left(\frac{1}{4} - \frac{1}{n^2}\right), \tag{18.4}$$

where A is a universal constant, and n ranges through all the integers beginning with 3. Bohr's great achievement was to derive this formula for the frequencies of the Balmer lines, and at the same time to express the constant A in terms of Planck's constant h, the charge e of the electron, and the mass m of the electron.

In spectroscopic work it is convenient to use the reciprocal of the wavelength, $1/\lambda$, which is referred to as the **wave number**, $\tilde{\nu} \equiv 1/\lambda = \nu/c$, so that the Balmer formula is written as

$$\tilde{\nu} = \frac{A}{c}\left(\frac{1}{4} - \frac{1}{n^2}\right), \tag{18.5}$$

where c is the speed of light. The universal constant A/c is referred to as the **Rydberg constant**, and is generally designated as R in spectroscopic formulae.

18.18 The Energy Levels of the Bohr Orbits

We can easily derive the Balmer formula the way Bohr did by considering the case of circular orbits for the electron in the hydrogen atom, and by applying to the motion of this electron the laws of classical electricity and mechanics with Bohr's modification. We note that to obtain the frequency of the photon emitted by an electron jumping from a higher to a lower orbit, we must first know how much energy the electron has in each of these orbits, for it is the difference in these two energies that determines the frequency.

We consider the nth orbit (we shall number the orbits 1, 2, 3,..., starting from the lowest or ground-state orbit). If v_n is the velocity of the electron in this orbit, and m is its mass, the kinetic energy of the electron in this orbit is $\frac{1}{2}mv_n^2$. Besides kinetic energy, the electron also has potential energy because it is in the electrostatic field of the proton. This is quite similar to the potential energy a body has in a gravitational field.

Since e is the magnitude of the charge of both the electron and the proton, the potential energy of the electron in the nth orbit is $-e^2/r_n$, where r_n is the radius of the nth orbit. The potential energy is taken as negative since the state of zero potential energy is defined as one in which the electron and proton are infinitely distant from each other, and a positive amount of work must be supplied to move the electron infinitely far away from the proton. Combining these two expressions for the kinetic and potential energy of the electron in the nth orbit, we obtain for the total energy in the nth orbit

$$E_n = \frac{1}{2} mv_n^2 - \frac{e^2}{r_n}.$$

Since Bohr's postulates say nothing at all about the velocity of the electron in an orbit, we must eliminate v_n from the above expression. We can do this by noting that the centrifugal force acting on the electron moving in its circular orbit must be exactly balanced by the electrostatic force exerted by the proton on it. Since the former is just mv_n^2/r_n (see Section 3.19) and the latter is just e^2/r_n^2 (see Coulomb's law, Section 18.4) we have

$$\frac{mv_n^2}{r_n} = \frac{e^2}{r_n^2}$$

or

$$mv_n^2 = \frac{e^2}{r_n}. \tag{18.6}$$

If we substitute this into the formula for the total energy, we have

$$E_n = \frac{1}{2}\frac{e^2}{r_n} - \frac{e^2}{r_n}$$
$$= -\frac{1}{2}\frac{e^2}{r_n}.$$

This, then, is the expression for the total energy of the electron in the nth orbit. However, we cannot express this total energy in terms of measurable quantities until we specify r_n in terms of these quantities.

18.19 The Concept of Action and the Bohr Quantum Condition

Since Bohr postulated that only a discrete set of orbits is permissible for an electron, it follows that only a discrete set of values of r_n is permitted. To find this discrete set, some kind of restriction must be imposed upon the size of the orbits. Bohr derived the restrictions that had to be imposed by using Planck's discovery of the existence in nature of a fundamental unit of action, h. He concluded that an electron may move in an orbit around a proton only if the action associated with the electron in this orbit is an integral multiple of this fundamental unit of action. This, in a sense, is similar to what Planck did to derive his law of radiation when he restricted the "harmonic oscillators" which he pictured as existing in a black body to a discrete set of oscillations (see Section 16.28).

In classical physics the action associated with a particle as it moves from an initial point to a close

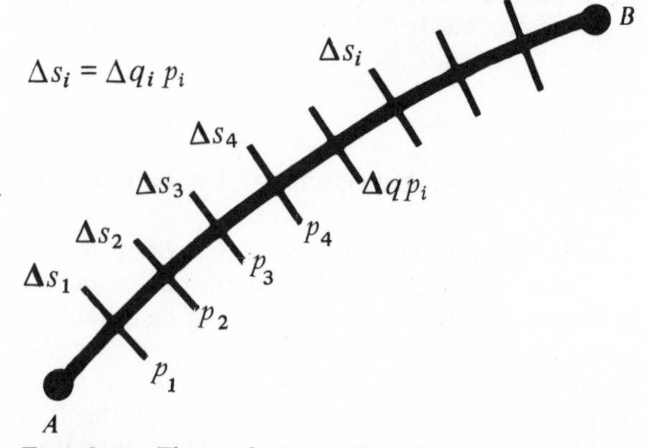

FIG. 18–5 *The total action along the path of a particle from A to B is the sum of the action elements $\Delta q_i p_i$, where Δq_i is the infinitesimal distance traveled by the particle and p_i is its momentum at i.*

neighboring point is defined as the momentum mv of the particle times the distance it moves in going from the initial to the neighboring point as measured along the path of the particle. In other words, the action is

$$mv(\Delta s),$$

as shown in Figure 18.5, where Δs is the small stretch of path between the two points. If the particle moves over a long stretch of path between two widely separated points, we may assign a total action to such a path by dividing it up into little strips Δs_1, Δs_2, Δs_3, etc. and adding the action for each of these strips.

If we consider the electron in the nth circular orbit, the action along any arc of this orbit is mv_n times the length of the arc since v_n is the same no matter where the electron is in its orbit. It follows, then, that the total action of the electron associated with the nth orbit is $mv_n 2\pi r_n$, i.e., the momentum times the circumference. Bohr imposed the condition that this action must then be n times the fundamental unit of action, h, since Planck showed that action in nature occurs only in multiples of h. In other words we must have

$$2\pi r_n m v_n = nh$$

or

$$r_n = \frac{nh}{2\pi m v_n}.$$

18.20 The Radii and the Energies of the Bohr Orbit

But from Equation (18.6) we see that $v_n = e/\sqrt{mr_n}$ so that

$$r_n = \frac{nh}{2\pi e}\sqrt{\frac{r_n}{m}}.$$

This finally leads to (squaring both sides)

$$r_n = \frac{n^2h^2}{4\pi^2 e^2 m}. \qquad (18.7)$$

We obtain the radius of the lowest orbit by placing $n = 1$; this is referred to as the Bohr radius, or the radius of the first Bohr orbit:

$$r_1 = \frac{h^2}{4\pi^2 e^2 m} = 0.53 \times 10^{-8} \quad \text{cm}.$$

If into the expression for the total energy of the electron in the nth orbit we substitute the value for r_n taken from Equation (18.7) we obtain

$$E_n = -2\pi^2 \frac{me^4}{n^2h^2}. \qquad (18.8)$$

18.21 The Principal Quantum Number n

We see that we have expressed the total energy of the electron in the nth orbit in terms of the measurable constants, m, e, and h, and the integer n. By assigning to n the numbers 1, 2, 3, etc. we obtain the energies of the electron in the various orbits. In particular, for $n = 1$, we obtain the energy of the electron in the ground state which is just $-2\pi^2(me^4/h^2)$. *The electron inside the hydrogen atom cannot have an energy less than this.*

The integer n is referred to as the **principal quantum number** of the electron and is used to specify its various permissible orbits. Spectroscopists have found it convenient to assign letters of the alphabet to the orbits associated with the different values of the principal quantum number. Thus the orbits having principal quantum numbers 1, 2, 3, etc. are called the *K*, *L*, *M*, etc. orbits or shells, respectively, as shown in Figure 18.6.

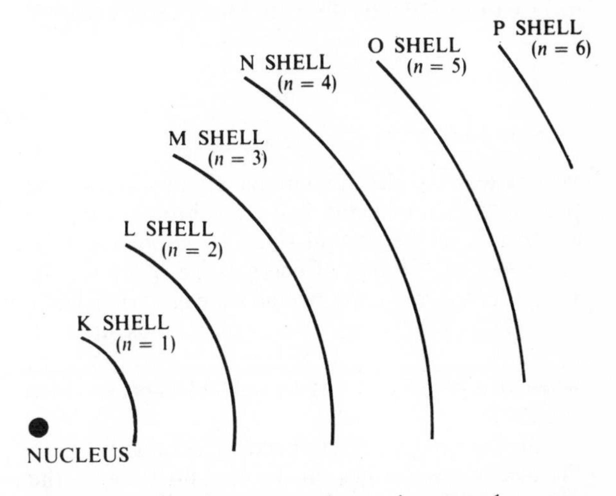

FIG. 18–6 *The orbits corresponding to the principal quantum numbers $n = 1$, $n = 2$, $n = 3$, etc. are called the K, L, M, etc. orbits.*

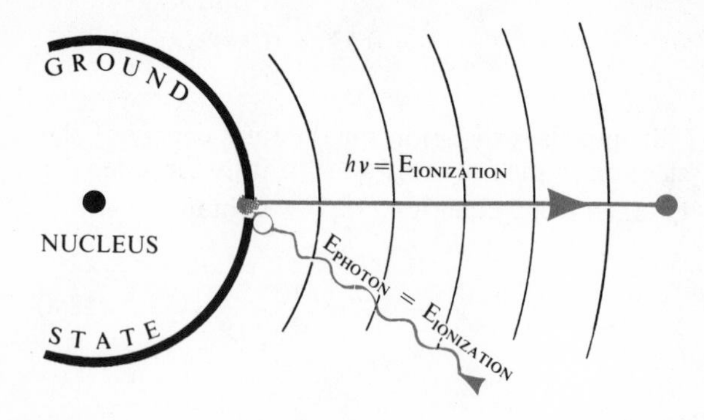

FIG. 18–7 *A photon whose energy is equal to, or greater than, the ionization energy of an electron in an atom removes the electron completely from the atom when it is absorbed by the electron.*

18.22 The Ionization Energy of the Hydrogen Atom

We note that the energy of the electron in a discrete orbit is always negative. This simply means that work must be done (positive energy has to be piped into the atom) before the electron can be torn out of the atom (i.e., before it can be brought to a state of zero energy and hence to an infinite distance from the proton). Since the energy of the electron in the ground state, using Equation (18.8) with $n = 1$, is just

$$E_1 = \frac{-2\overbrace{(3.142)^2}^{\pi} \times \overbrace{(9.108 \times 10^{-28})}^{m} \times \overbrace{(4.802 \times 10^{-10})^4}^{e}}{\underbrace{(6.62 \times 10^{-27})^2}_{h}}$$

$$= -2.178 \times 10^{-11} \quad \text{ergs},$$

we can tear the electron completely away from the proton (i.e., ionize the hydrogen atom, when its electron is in the ground state) by supplying to it $+2.178 \times 10^{-11}$ ergs of energy. In other words, since the electron in the ground state has a deficiency of 2.178×10^{-11} ergs of energy, it ends up with zero energy, and it is no longer bound to the atom when it is given just this amount of energy. (See Figure 18.7.)

This energy can be supplied in various ways to the electron when it is in the ground state of the atom. Thus, if the hydrogen atom suffers a collision with some other body, the electron may acquire enough of the kinetic energy of the colliding body to be torn out of the atom. On the other hand, it may acquire this amount of energy by absorbing a photon of just the right frequency, which we can find by equating $h\nu$ to 2.178×10^{-11} ergs, so that

$$\nu = \frac{2.178 \times 10^{-11}}{6.62 \times 10^{-27}} = 3.3 \times 10^{15} \quad \text{vibrations/sec.}$$

This corresponds to a photon having the wavelength 9×10^{-6} cm or approximately 900 Å. Since such a photon lies deep in the ultraviolet region of the spectrum, the hydrogen atom with its electron in the ground state cannot be ionized by ordinary light.

18.23 The Electron Volt as a Unit of Energy

There is another way of representing the energy required to ionize the hydrogen atom when its electron is in the ground state. If we place a free electron between two plates of a charged condenser, the electron is repelled by the negative plate and attracted by the positive plate. As it falls towards the positive plate, its kinetic energy increases; we may therefore use as a new unit of energy the kinetic energy an electron acquires when it falls from the negative to the positive plate of a condenser in which the plates are maintained at a potential difference of 1 volt. This unit of energy is referred to as **1 electron volt**, and is equal to 1.602×10^{-12} ergs. Using this unit of energy, we may say that the **ionization energy** of hydrogen is 13.595 electron volts (ev), or, as it is often referred to, the **ionization potential** of the hydrogen atom with its electron in the ground state is 13.595 volts.

If a photon or a colliding body has an energy greater than this ionization energy, the electron in the ground state of the hydrogen atom absorbing this photon, or acquiring all the kinetic energy of the colliding body, has more than zero energy, so that even after it has been torn out of the atom, the electron will continue moving with a certain amount of kinetic energy. This excess energy is just the difference between the energy of the photon absorbed by the electron and the ionization energy required to bring the electron up to zero energy.

18.24 The Excited State of an Atom

If the photon or collision supplies to the electron in the ground state less than the ionization energy, the electron cannot leave the atom. If the energy supplied to it is just right, the electron jumps into an orbit having a larger value of n. We then say that the atom is no longer in the ground state, but is in an excited state defined by the principal quantum number n. The energy of the electron in this excited state is just given by Equation (18.8). The electron does not remain in this excited state very long, but after an interval of about 10^{-8} seconds it jumps down to some lower state, ultimately coming back to the ground state in a series of such jumps. As it does so, it emits photons of various wavelengths.

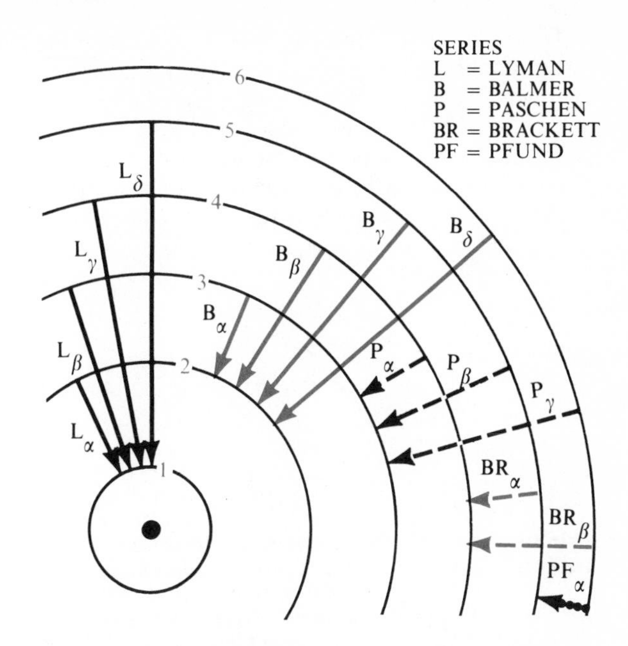

FIG. 18-8 *The various series in the hydrogen spectrum arise when excited electrons in hydrogen atoms jump to lower levels, emitting photons of frequencies corresponding to the energy differences of the levels.*

18.25 Bohr's Derivation of the Frequency of the Hydrogen Lines

We can now use the Bohr formula to determine the frequency of the emitted or absorbed photons. Let us suppose that the electron is in the nth excited state, and it jumps down to the Mth state. In doing so it must emit an amount of energy $E_n - E_M$ in the form of a photon, whose frequency, ν_{nM}, is just $(E_n - E_M)/h$. If we substitute into this formula the expressions for E_n and E_M [note we get E_M from Equation (18.8) by replacing n by M] we get

$$\nu_{nM} = \frac{\left[-\dfrac{2\pi^2 me^4}{n^2h^2} - \left(-\dfrac{2\pi^2 me^4}{M^2h^2}\right)\right]}{h}$$

$$= \frac{2\pi^2 me^4}{h^3}\left[\frac{1}{M^2} - \frac{1}{n^2}\right]. \qquad (18.9)$$

For the wave numbers $\tilde{\nu}_{nM} \equiv \nu/c$ we have

$$\tilde{\nu}_{nM} = \frac{2\pi^2 me^4}{ch^3}\left[\frac{1}{M^2} - \frac{1}{n^2}\right]. \qquad (18.10)$$

18.26 The Balmer Lines and Other Spectral Series

This formula (with slight modifications that were introduced later) is the basic formula for the study of the hydrogen spectrum. By assigning different values to the integers M and n, we can obtain the frequencies of various lines in the hydrogen spectrum. Since these lines can be grouped into different sets or series, as they are called, each of which lies in a different part of the spectrum, we can apply this formula to study these spectral series.

We first consider the series we obtain if we analyze the light coming from excited hydrogen atoms in which the electrons jump from higher orbits to the ground state. This is shown schematically in Figure 18.8. The wave numbers of the spectral lines associated with these photons are given by placing $M = 1$ in Equation (18.10):

$$\tilde{\nu}_{n1(\text{Lyman})} = \frac{2\pi^2 me^4}{ch^3}\left[1 - \frac{1}{n^2}\right] \qquad (n > 1).$$

If we allow n to take on the values of 2, 3, 4, etc. we obtain the wave numbers for a series of lines called the **Lyman series**.

The line associated with the transition of the electron from the second orbit to the first orbit is called the **Lyman α line**, and has a wavelength of 1 260 Å. Since all the other lines in this series have wavelengths smaller than this, we see that the Lyman lines lie in the ultraviolet region of the spectrum.

If we place $M = 2$ in Equation (18.10), and allow n to take on all integral values larger than 2, we obtain the formula for the wave numbers of the **Balmer series**: (see Figure 18.8)

$$\tilde{\nu}_{n2(\text{Balmer})} = \frac{2\pi^2 me^4}{ch^3}\left[\frac{1}{4} - \frac{1}{n^2}\right] \qquad (n > 2).$$

If we compare this formula with the empirical formula found by Balmer, we see that the Rydberg constant is equal to $2\pi^2 me^4/ch^3$.

If we place $n = 3$ in this formula, we obtain the wave number of the line associated with the photon emitted when the electron jumps from the third orbit (the M orbit) to the second orbit (the L orbit). This line in the red part of the hydrogen spectrum is referred to as the H_α line, and has a wavelength of 6 561 Å. Thus we see that the Balmer lines lie in the visible part of the spectrum.

Just as we obtained the formula for the Lyman series and the Balmer lines, we obtain the **Paschen series**, involving transitions of the electron from higher states to the third or M orbit, by placing $M = 3$ and allowing n to take on all integral values larger than 3. Other series that have been studied are the **Brackett series**, $M = 4$ and $n > 4$, and the **Pfund series**, $M = 5$ and $n > 5$.

EXERCISES

1. (a) Calculate the electrical force in dynes between an electron and a proton separated by a distance of 10^{-8} cm. (b) Compare this with the gravitational force between these two particles. (See Appendix C for the charge on an electron and for the masses of an electron and a proton.)

2. (a) If an electron were moving with a constant speed equal to one per cent of the speed of light in a circle of radius 10^{-8} cm, at what rate, according to Maxwell's electromagnetic theory of radiation, would it be radiating energy? (b) At this rate of radiation, how long would it take the electron to lose all its kinetic energy?

3. (a) If an electron were completely annihilated and its mass were transformed into energy, how many ergs would be released? (b) If the energy in (a) were all in the form of a single photon, what would the frequency and the wavelength of this photon be?

4. (a) Assume that the electron in a hydrogen atom moves around the proton in a circular Bohr orbit. What is its speed in the first Bohr orbit? (b) If the electron obeyed the classical law of electricity and magnetism, at what rate would it be radiating energy from this orbit?

5. (a) Use the Bohr formula to calculate the frequencies and wavelengths of the first five Lyman lines. (b) Do the same thing for the first five lines in the Paschen series.

6. Compare the wavelength difference between the first two lines in the Balmer series and the wavelength difference between the twentieth and the twenty-first lines.

7. Assume that in a singly ionized helium atom the remaining electron moves around the nucleus according to the Bohr theory. Write down the expression for the frequencies of the spectral lines that correspond to the Lyman, the Balmer, and the Paschen series in the neutral hydrogen atom.

8. Assume that the ionized helium atoms in a stellar gaseous atmosphere are excited by collisions. How much higher must the temperature in such an atmosphere be on the average than the temperature of an atmosphere in which the collisions are energetic enough to excite the Balmer lines of hydrogen? (Assume that in both atmospheres the theorem of equipartition of energy applies.)

19

ATOMIC STRUCTURE AND STELLAR ATMOSPHERES

19.1 The Emission of Radiation by a Glowing Gas: The Partition Function

From our derivation of the Bohr formula for the hydrogen spectrum, we can now see just why a gas under low pressure and subjected to a high voltage glows and emits the bright-line spectrum characteristic of the atoms in that gas. We can see that if a previously evacuated tube is filled with a small amount of this gas and a high voltage is placed across it, the atoms in the gas suffer collisions with the various charged particles that are always present in such a tube and that are moving from one end of the tube to the other. The high voltage placed across the tube sets the charged particles moving, and they acquire a good deal of kinetic energy. If they then collide with hydrogen atoms the collisions may be violent enough to throw the electrons from the ground states of the atoms to excited states. A moment later the electrons jump back again, emitting photons of definite frequencies, and these same atoms are ready to be excited again by other collisions (see Figure 19.1).

If at any moment large numbers of atoms suffer collisions that throw their electrons into the same excited state, and if, 10^{-8} seconds later, these electrons jump down to the same lower state, a batch of photons of the same frequency are emitted. If this process is going on for all the atoms in the tube and if all possible excitations are taking place simultaneously, with subsequent transitions of the electrons to all possible lower states, photons of all the various frequencies mentioned above in the discussion of the Lyman, Balmer, etc. series will constantly be emitted. What is happening, then, in a tube filled with hydrogen glowing in this fashion, is that electrical energy is being constantly transformed into radiant energy by means of collisions that excite the hydrogen atoms.

We may give a more explicit description of this process as follows. Let there be altogether N hydrogen atoms inside the tube across which there is a constant high voltage. (This number of atoms is always the same since we are neither adding nor subtracting gas.) At any moment an instantaneous photograph of all these atoms would show their electrons in all the various permissible excited states or orbits. A certain number, N_1, of the hydrogen atoms in the tube would have their electrons still in the ground state (these atoms would be either those which had suffered no collisions or those for which the electron had just returned to the ground state after having been excited). We would find a certain number, N_2, with their electrons in their first excited state, a certain number,

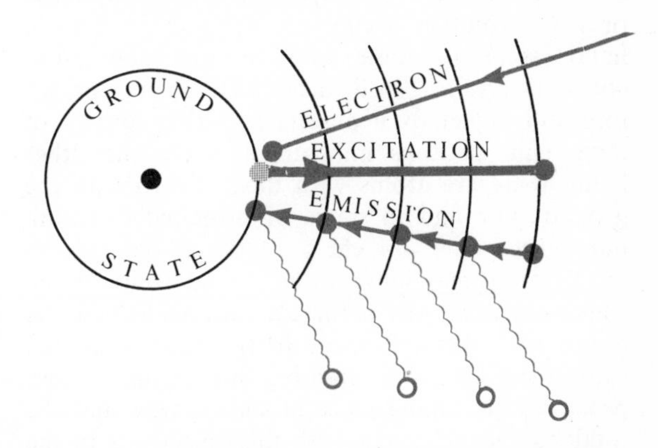

FIG. 19–1 *An electron is thrown into a higher level by a collision and immediately thereafter jumps to lower levels emitting corresponding photons.*

N_3, with their electrons in their second excited state, etc., and finally a certain number, N_i, of atoms in a state of ionization, i.e., with their electrons completely torn away.

Since all of these atoms taken together must give us the total number of atoms, we have

$$N = N_1 + N_2 + N_3 + \cdots + N_i. \quad (19.1)$$

This sum divided by N_1 is called the **partition function** of the atoms among the excited states.

We see, then, that all the atoms in the tube are distributed among the various excited states. This partition of the total number of atoms into excited groups is always the same regardless of when we look at the gas in the tube (i.e., N_1, N_2, etc. do not change) although any one particular atom is found in one group at one moment and in some other group at another moment.

19.2 The Formation and Intensities of Emission Lines

To clarify this point, we consider all the atoms that have their electrons in the second excited states, i.e., in orbits of principal quantum number $n = 3$. If left alone, and if these were the only atoms present, the electrons of some of them would jump down to the first excited states, and the electrons of the others would jump down to the ground states with the corresponding photons being emitted in each case. The electrons in the first excited states would then also jump down to the ground states with emission of photons; there would therefore be no atoms left in the excited states and the tube would cease to glow.

But since these atoms are not the only ones present, two other phenomena occur. On the one hand, some of these same atoms may suffer collisions resulting in their electrons' being thrown into still higher excited states, so that not all of them emit photons immediately. On the other hand, there are atoms with their electrons in the ground state and in states of principal quantum number $n = 2$, $n = 4$, etc. so that in all these atoms the electrons may jump to the second excited level. This means that just as there are atoms leaving the group with their electrons in the second excited state, there are atoms entering this group. These processes are taking place in such a way that the total number of atoms with their electrons in the second excited state remains constant. In other words, we have a state of statistical equilibrium in which the numbers N_1, N_2, etc. remain constant even though atoms are continuously moving from one group to another and thus emitting the observed spectral lines.

The only way in which the partition numbers, N_1, N_2, etc. can be altered is by changing the voltage across the tube. As the voltage increases, atoms with their electrons in higher states of excitation become more numerous. If, for any voltage, we know the partition numbers N_1, N_2, etc. we can calculate the intensities of the various spectral lines. It is clear that the more atoms there are in a particular state, the more intense are the various spectral lines associated with the photons emitted when the electrons jump down from this state because there are more such photons emitted per second.

19.3 The Formation of Absorption Lines in Stellar Atmospheres

In general, it is quite difficult to determine the partition numbers, N_1, N_2, etc. for the gas in a glowing tube, but in the case of the atoms in the atmosphere of a star this can be done without too much difficulty. The situation here is just the opposite of that for the hydrogen gas in the tube, for in stellar spectra we are concerned with absorption rather than emission lines. However, since absorption lines are formed by comparatively complicated processes, explaining the detailed structure of the dark lines in stellar spectra is a task requiring great ingenuity and a very skillful use of mathematics and physics. Nevertheless, we can obtain a good insight into the way absorption lines are formed by considering the simple process of **scattering** which contributes most to line formation.

To understand this process, we again consider hydrogen and imagine a star whose atmosphere consists almost entirely of this gas. We need not be concerned with what lies under the atmosphere, but merely suppose that the black-body radiation streaming through the atmosphere comes from a very hot underlying surface (the photosphere of a star). This radiation, as we have already noted, would give a continuous spectrum if there were no atmosphere present, but the hydrogen atoms in the atmosphere act as an obstacle to some of the photons in the radiation. Since the atoms in stellar atmospheres, such as the sun's, are *not* tightly packed together, most of the photons in the streaming radiation pass right through the atmosphere without encountering any atoms, so that the

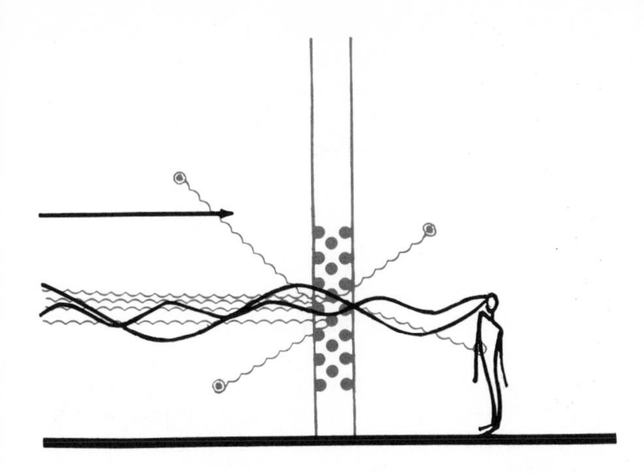

FIG. 19–2 *The atoms in a slab of material scatter photons of a particular frequency in all directions so that these scattered photons do not reach the eye of the observer.*

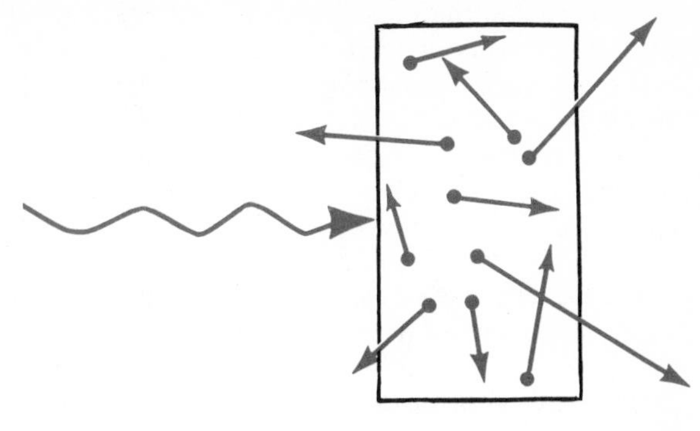

FIG. 19–3 *The atoms in a slab of atmosphere have random velocities so that the Doppler shift in the radiation as seen by each atom is different.*

continuous part of the spectrum is dominant, and the absorption lines are weak in comparison. *To see how a particular absorption line is formed, we must follow those photons which are intercepted by atoms.*

Let us consider, then, a stream of photons of a particular frequency, ν_{nm}, striking a small slab of atmosphere, as shown in Figure 19.2. If this slab were not present, the observer standing off to the right would detect all of the photons in the stream, and he would find no break in the continuous spectrum at this frequency. If, however, an atom in the slab has its electron in an excited state of quantum number m (the $m - 1$ excited state) at the moment that one of these photons collides with it, the electron in this atom is in a position to absorb this photon by jumping to the excited state n.

If a large number of atoms are in this same initial excited state m, the electrons in all these atoms can absorb the photons of frequency ν_{nm} and end up in the higher excited state n. Of course, since the electrons cannot stay in this higher excited state very long, they must jump down to lower states again, re-emitting photons of either the same or other frequencies. *However, these re-emitted photons do not reach the observer, but are scattered away from the incident beam in all directions.*

The reason is that when an atom emits a photon, there is an equal probability for the photon to be emitted in any direction as shown in Figure 19.2. We see, then, that in the process of absorbing and re-emitting radiation the atoms scatter photons of frequencies ν_{nm} out of the original beam, so that the observer detects fewer photons of this frequency than he would if the slab of hydrogen atoms were not present. The observer therefore *finds a dark space in the continuous spectrum at the point corresponding to the wavelength* $c/\nu_{nm} = \lambda$.

19.4 The Intensities of Absorption Lines

Although this scattering process is easy enough to understand, its actual application to the analysis of absorption lines is rather complicated. To begin with, it is quite clear that stellar atmospheres consist of more than one slab of atoms. Many layers, hundreds of kilometers thick and under various complex physical conditions, contribute to the absorption lines.

Moreover, the atoms in any layer are not standing still waiting for photons to strike them. They are in rapid random motion, so that the frequencies of the incident photons as seen by these atoms may not be exactly ν_{nm} because of the Doppler effect (see Section 5.14) as shown in Figure 19.3. Again, although we have assumed that the atoms in our slab are not packed closely together, they do, nevertheless, affect each other to some extent, and this has an effect on the way they absorb photons of a given frequency.

We must also take into account other conditions which make the actual problem of line formation a difficult one. However, assuming that scattering is the most important single contributing factor, we can give a brief description of what determines how strong a particular absorption line is. We shall also see how the surface temperature of a star determines which spectral lines are most prominent for that star.

To calculate the intensity of a given absorption line, say the one corresponding to the frequency ν_{nm}, *we must first determine how many atoms there are in the initial excited state m, for only such atoms can give rise to such a line. If the number of such atoms in the atmosphere is large, the corresponding absorption line is intense; on the other hand it is weak if there are few such atoms present.*

But the only way we can have atoms in this excited state in the atmosphere of a star is if the atoms collide among themselves with sufficient energy, or if there are energetic enough photons in the radiation streaming from the photosphere of the star to toss the electrons from the ground into the excited state m. These two processes can thus prepare the atoms for their role in the formation of a given absorption line. Since the violence of collisions between atoms in the atmosphere of the star and the energy of the most abundant photons coming from the photosphere are determined by the star's surface temperature, we can begin to see how important a role the temperature plays in the formation of the spectrum of a star. We shall illustrate this by considering the Balmer lines.

19.5 The Partition Numbers for the Balmer Lines in the Atmosphere of a Star

Since the Balmer absorption lines can be formed only if the electrons of the hydrogen atoms jump from the first excited state (principal quantum number $n = 2$) to higher excited states by absorbing the appropriate photons, we can determine the intensities of the Balmer lines in the spectrum of a star only if we know at any moment how many hydrogen atoms in the atmosphere have their electrons in this first excited state. *There is a simple way of doing this if we suppose that the atoms are all colliding with each other, and that these collisions have brought about a steady state, so that no matter when we look at the atmosphere, the number of atoms in any excited state is always the same* (in other words, if we have a situation similar to the one we have described in the case of the glowing hydrogen in the tube). Again, just as in that case, we must find the partition of the total number of hydrogen atoms among the various excited states. However, in this case, it is not an electric voltage that determines the numbers N_1, N_2, N_3, etc. but rather the absolute temperature T, for, as we have noted, the excitations of the atoms are due to collisions among themselves, and these are more or less violent depending upon how high the temperature is. There is a simple exponential formula which enables one to calculate the ratios N_2/N_1, N_3/N_1, etc., i.e., to compare the number of atoms in any excited state with those in the ground state. A knowledge of these numbers is basic for determining the intensities of the absorption lines.

19.6 The Boltzmann Excitation Formula and the Intensities of Absorption Lines

This formula, which was first derived by Boltzmann, shows clearly how important the temperature is in determining the distribution of atoms among the various excited states. We consider two different states, A and B, of the atom, where B is the state of higher excitation (both B and A are here considered as excited states, although A may be taken as the ground state if desired), and we let $\chi_{AB} = E_B - E_A$ be the energy difference between these two states (the electron must acquire this amount of energy to go from state A to state B).

Boltzmann discovered that for a given absolute temperature, T, under which the atoms in the gas are colliding with each other, the number of atoms in state B as compared with the number in state A, that is, the ratio N_B/N_A, is given by the formula

$$\frac{N_B}{N_A} = g_{AB}e^{-\chi_{AB}/kT}, \qquad (19.2)$$

where g_{AB} is a physical quantity which depends upon the nature of the states A and B, and hence is determined by the structure of the atom, e is the base of the natural logarithms, and k is the universal Boltzmann constant. *The negative sign in the exponent tells us that the greater the energy difference between a given excited state and the ground state the fewer the atoms in that excited state for a given temperature T. The form of the exponent shows us that as the temperature increases, the negative exponent becomes smaller, so that more and more atoms are then found in the excited state.* This is so because collisions become more violent and frequent as the absolute temperature increases.

We can apply Boltzmann's formula (19.2) to determine the ratio N_2/N_1 of the number of hydrogen atoms in the first excited state to the number in the ground state. We must know this ratio if we are to determine the intensities of the Balmer lines. We have

$$\frac{N_2}{N_1} = g_{12}e^{-\chi_{12}/kT},$$

where χ_{12} is the additional energy the electron must acquire to jump from the ground state, 1, to the first excited state, 2.

Since in the case of the hydrogen atom g_{12} is equal to 4, and χ_{12} is equal to 10.19 ev or 16.3×10^{-12} ergs, the formula becomes

$$\left(\frac{N_2}{N_1}\right)_{\text{hydrogen}} = 4e^{-(16.3 \times 10^{-12})/(1.38 \times 10^{-16})T},$$

where we have put in the value of the Boltzmann constant k. If we put different values of the temperature into this formula, we can see how the ratio of N_2/N_1 varies, and therefore how the intensities of the Balmer lines depend upon the temperature.

19.7 The Intensities of the Balmer Lines for Various Spectral Classes

In the case of the solar atmosphere, where we may assume the temperature to be about 5 700°K, this formula shows that, on the average, about four atoms in every billion are excited to the second level, where they are able to absorb the photons that correspond to the Balmer lines. Thus we see that the Balmer absorption lines are certainly not the most intense lines for stars like the sun. Nevertheless, we do find these lines present even in the solar spectrum; this shows us that hydrogen is extremely abundant in the atmospheres of stars. Indeed, we find the Balmer lines even in the spectra of stars such as Betelgeuse, a red star with a surface temperature of about 3 000°K. In other words, even though fewer than four out of every billion atoms take part in the formation of Balmer lines for stars as cool as the sun or cooler, we see that because hydrogen is the most abundant element in stellar atmospheres, the Balmer lines are present in the spectra of such cool stars.

As we go up the main sequence and consider stars with higher surface temperatures, the ratio N_2/N_1 increases and the Balmer lines become more intense. For a star such as Sirius, with a surface temperature of the order of 10 000°K, three atoms in every 10 000 in its atmosphere are excited to the second level, so that there are 100 000 times as many hydrogen atoms available for the formation of the Balmer lines in such stars as in the sun. Thus it becomes clear why the A type stars are characterized by intense Balmer lines.

Above the A type stars on the main sequence the formula shows that although fewer atoms are left with their electrons in the ground state, the number of atoms with electrons in the first excited state decreases relative to the number of ionized atoms. Thus the Balmer lines are most intense for A type stars (see Figure 19.4) and begin to diminish in intensity as we approach the B type stars. As the temperature increases, more atoms are thrown into the higher excited state by collisions and radiation, and the electrons are completely torn out of many of the atoms, and this means that no hydrogen absorption lines can be formed by such atoms. This later effect becomes increasingly important as we approach the O type stars. Again, by means of another exponential formula we can calculate the number of ionized atoms as compared with those that are un-ionized. Such calculations show quite clearly why the Balmer lines fall off in intensity as we go to the very hottest stars.

19.8 Saha's Ionization Formula and the Spectra of Ionized Atoms

In 1922, the Indian physicist M. N. Saha derived by means of thermodynamical and statistical considerations the ionization formula that is used in these calculations. The problem that Saha dealt with is similar to that which the chemist considers when he analyzes the dynamical equilibrium existing between dissociation and recombination in a given chemical reaction under a given set of conditions (e.g., pressure, temperature, abundances, etc.).

Thus, if C is a chemical compound, and A and B are the dissociation products, the chemist considers the process

$$C \rightleftarrows A + B,$$

where the arrow pointing to the right indicates dissociation, and the arrow to the left indicates recombination. *If, at any moment, under the given conditions, equilibrium exists, the rate of dissociation must exactly equal the rate of recombination, so that the numbers of molecules (or atoms, as the case may be) of types C, A, and B are constant (although by no means equal) as dissociations and recombinations go on simultaneously.* If the conditions are altered, either dissociation or recombination is favored, and either the amount of C or the amounts of A and B present at any moment increase.

Saha considered the ionization of atoms in a gas at a given temperature to be the same kind of phenomenon as dissociation in chemistry. If *a*

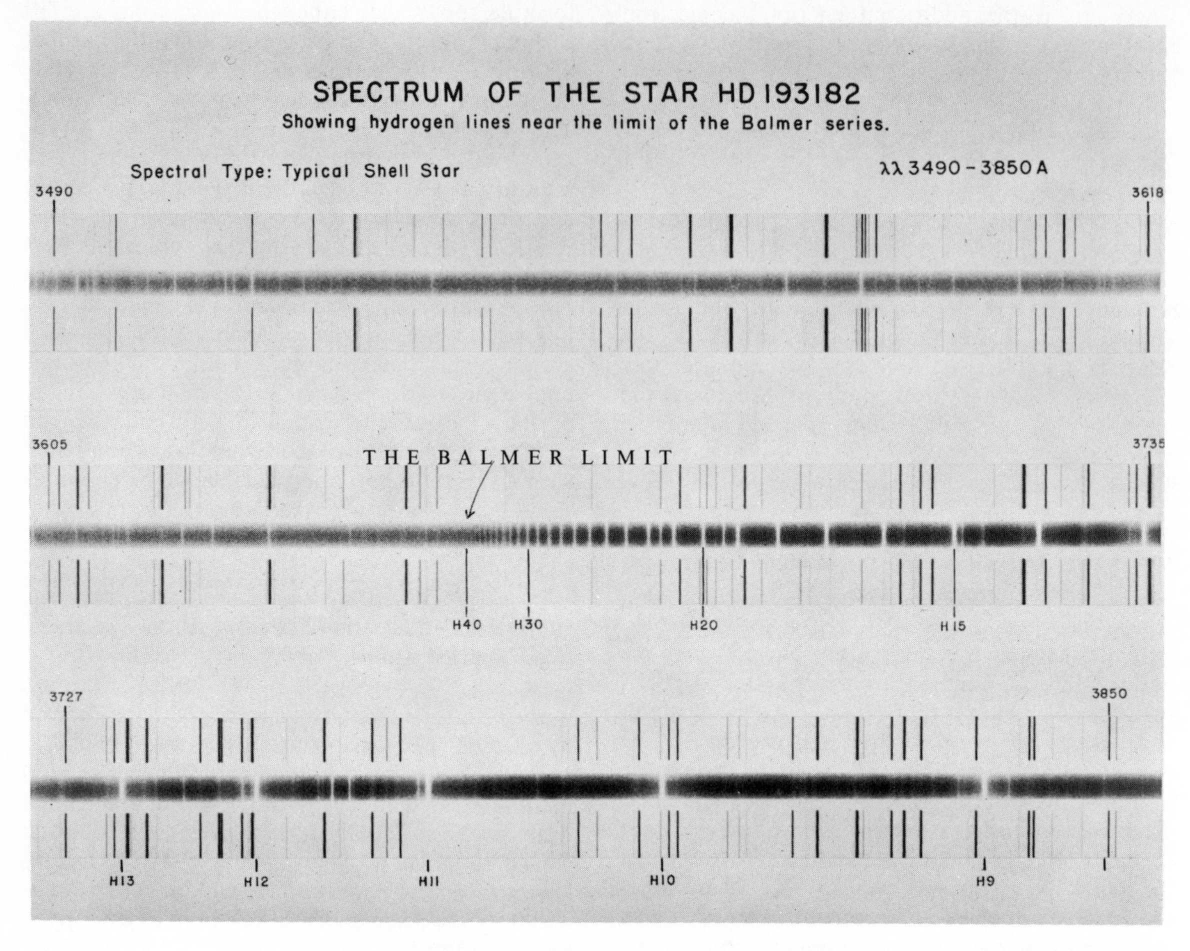

FIG. 19–4 *Spectrum of the star HD193182, showing the Balmer series down to and including the continuum. This photograph shows the way the lines converge to the limit. The lines appear bright because the plate is a negative.* (*From Mount Wilson and Palomar Observatories*)

represents the neutral atom (corresponding to C above) and e^- and a^+ represent the electron and the ionized atom, respectively (corresponding to A and B above), we have for the ionization process

$$a \rightleftarrows a^+ + e^-.$$

Since this process occurs as the result of collisions between atoms (or between an atom and a photon) the temperature plays an important part in the formula that determines the quantities of a, a^+, and e^- that are present at any moment.

Saha's formula enables us to determine the number of ionized atoms, N^+, that are present for a given absolute temperature, T. Although we shall not derive this formula, we can see from very general considerations how it depends on the temperature and other factors. To begin with, since ionization occurs only if the atom with its electron in the ground state is supplied with the necessary ionization energy χ_0 (either by a collision or by absorbing an appropriate photon) the number of ionized atoms depends on the factor $e^{-\chi_0/kT}$, just as in the case of the Boltzmann excitation formula.

Now if the atom acquired only the amount of energy χ_0, and no more, the electron on being torn out would be at rest relative to the ionized atom since there would be no excess energy to set it moving. In general, however, when atoms are ionized, the electrons that are torn out can have a range of kinetic energy depending upon how much energy above χ_0 was supplied to the electron. Obviously this increases with the temperature T, and hence there must be an additional temperature factor to account for all such cases. It turns out that this factor is of the form $\sqrt{T^3}$, or $T^{3/2}$.

It is also clear that the greater the number N_a of un-ionized atoms, the greater the number of ionized atoms, since each un-ionized atom has a chance of becoming ionized. Therefore, to find the total number of ionized atoms present, we must multiply these three factors together to obtain

$$N_a T^{3/2} e^{-\chi_0/kT}.$$

We must take account of one more factor before we can obtain the Saha formula. This factor, which tends to decrease the number of ionized atoms, is the number of free electrons (electrons not attached to atoms), N_e, present in the atmosphere. The more free electrons there are, the greater is the rate at which ionized atoms recapture electrons and thus become un-ionized. We must, therefore, divide the product above by N_e. Finally a numerical constant must be introduced, which depends upon the mass of the electron, Boltzmann's constant, and Planck's constant. If we put all these things together, we get Saha's formula for the number of ionized atoms per unit volume in a gas having an absolute temperature T:

$$N^+ = \textbf{constant} \times \frac{N_a}{N_e} T^{3/2} e^{-\chi_0/kT}, \qquad (19.3)$$

where N_a is the number per unit volume of un-ionized atoms in the ground state, and N_e is the number of free electrons per unit volume.

19.9 Saha's Ionization Formula Applied to Solar and Stellar Atmospheres

If we apply Formula 19.3 to the solar atmosphere, assuming that the temperature in the solar atmosphere is 5 700°K, we find that there is more than 30 times as much ionized as there is un-ionized magnesium. In other words, out of every 100 magnesium atoms three are neutral and 97 are ionized. We should therefore expect to find the lines of ionized magnesium more intense than those of un-ionized magnesium, and in the case of aluminum the situation is quite similar. If we go to a star like Sirius, we find that not only are practically all the calcium atoms singly ionized (one electron torn out) but most of them are doubly ionized as well (two of the electrons torn out of each atom).

Figure 19.5 shows the way calcium ionization changes as we go to higher and higher temperatures. The curve on the left shows that neutral calcium

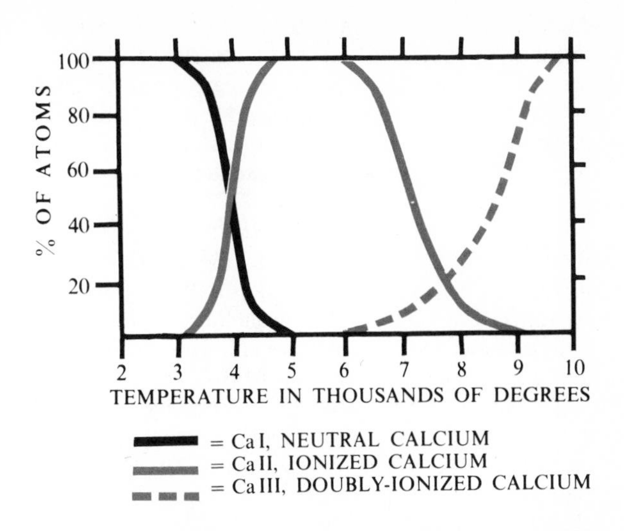

FIG. 19–5 *Schematic representation of the ionization of calcium in the atmosphere of a star as the temperature increases. (Adapted from Goldberg and Aller,* Atoms, Stars, and Nebulae. *Philadelphia: The Blakiston Co., 1943)*

(Ca I) falls off and practically disappears at 5 000°K, whereas the curve on the right shows that singly ionized calcium (Ca II) increases up to a temperature of 6 000°K and then falls off, almost disappearing entirely at 10 000°K as the amount of doubly ionized calcium (Ca III) begins to grow. In this way we can trace the ionization of various atoms, in particular hydrogen, and understand why lines of the neutral atoms give way to lines of the ionized atoms as we go up the main sequence.

19.10 The Variations of the Helium Spectrum along the Main Sequence

Just as we have analyzed the way the hydrogen spectrum changes as we go from spectral class to spectral class, we can do the same thing for the other elements. *But, of course, as we have noted, different elements do not behave the same way under the same conditions because their ionization potentials are different.* We may take as another example the case of helium, which next to hydrogen is the most abundant substance found in stars and interstellar space. The ionization potential of helium is 24.5 electron volts, as compared to the 13.5 for hydrogen, and the energy required to excite helium is also larger by about the same factor. Hence, the helium absorption lines are weaker in most of the stellar spectra than the hydrogen lines.

To excite helium atoms requires much more violent collisions and much more energetic photons than those required to excite hydrogen. Thus, we can expect helium lines to be absent from the absorption spectra of stars like the sun, and to be much weaker than the hydrogen lines in the spectra of stars like Sirius. However, as we consider stars with surface temperatures approaching 20 000°K, more and more excited helium atoms are found in these atmospheres and the helium absorption lines begin to approach their greatest intensities. At still higher temperatures, many helium atoms begin to lose one of their electrons by ionization. However, the helium atoms, even with one electron missing, are still able to produce absorption lines because the neutral helium atom has two electrons circling around its nucleus. Thus we find in the spectra of O type stars with surface temperatures of the order of 50 000°K that the lines of ionized helium are more intense than those of neutral helium.

19.11 Absorption Lines of Heavier Elements in Main-Sequence Stars

What we have done for hydrogen and helium we can do for the other elements whose lines appear in stellar spectra and determine for which spectral classes their absorption lines are most intense. As we go to the heavier elements such as iron and nickel, we find that the ionization potentials are of the order of 6 or 7 ev. These atoms can thus be easily excited, and even ionized, so that we can expect to find their most intense absorption lines in the spectra of the coolest stars. As we have already noted, this is, indeed, the case. Because these heavier elements can lose some of their electrons very easily (as indicated above for calcium and magnesium), the lines of these ionized atoms play a prominent role in stellar spectra.

We have already mentioned that molecular band spectra are prominent in atmospheres of the cool M type stars. This is so because in the atmospheres of such stars collisions between molecules are so weak that tightly bound molecules, such as titanium oxide, suffer hardly any dissociation at all. In addition to the band spectra we find in these stars the absorption lines of the most easily excited atoms, for which the ionization potentials are small.

19.12 The Profile of the Spectral Lines and the Abundances of Different Elements in Stellar Atmospheres

Although the techniques we have outlined above allow us to understand the way stellar spectra vary as we move along the main sequence, many refinements must be introduced into the theory before we can draw any conclusions from the spectral lines about the abundances of the elements in the atmosphere of a star. Of course, without going any further, we know that hydrogen and helium must be very abundant because their lines are found even in cool stars. But to determine, for example, the relative abundances of elements such as sodium and calcium, we must know the detailed shapes as well as the intensities of their absorption lines.

Actually, the absorption line is not just a thin dark streak across the spectrum, but it shows enormous complexity in its shape as one moves across it with a microphotometer. In some cases the line may be sharp and thin, and in other cases, broad and irregular, and this structure must be taken into account if we are to determine the number of atoms that give rise to the line. Such mathematical theories have been worked out in detail, and astrophysicists have been able to use them to determine the relative abundances of the various elements in stellar atmospheres. We shall give here, without further explanations, the results of such an analysis for the sun. We find that more than 81 per cent, by number, of atoms in the solar atmosphere are hydrogen, and more than 18 per cent are helium, with all the other elements contributing small traces to the remaining fraction of 1 per cent.

By means of the following simple analysis we can understand why the shape of an absorption line must be taken into account before we can determine the numbers of atoms in the atmosphere (say in a 1 cm^2 column above the photosphere) along a beam of radiation. Let us suppose that we are dealing with atoms that are capable of absorbing only one single frequency ν_0. If we have one such atom in our column, only one photon of frequency ν_0 is missing and the spectral line is not perceptible at all at the wavelength $\lambda_0 \equiv c/\nu_0$. If there are more and more of these absorbing atoms in the column, more and more of the ν_0 photons are absorbed and the line becomes darker and darker at the exact wavelength λ_0 until finally, if enough such atoms are present, all the given photons of frequency ν_0 are absorbed at this precise frequency and the line is then pitch black.

This, however, is an idealized situation because atoms absorb photons over a given range of frequency above and below the central frequency—just as a radio set can receive the signal slightly to the right or to the left of the optimum position of

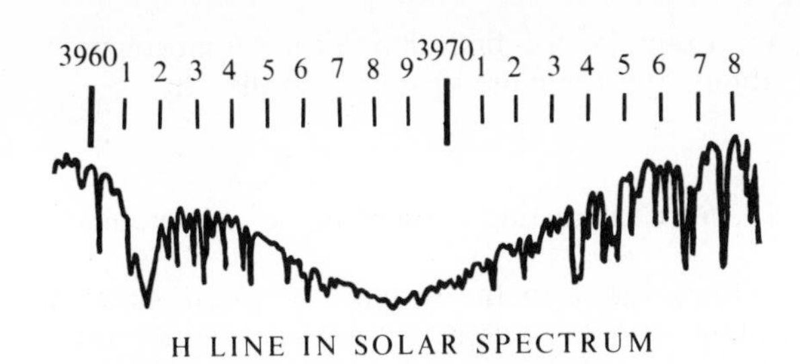

FIG. 19–6 *Actual tracings of absorption lines in the spectra of stars with a microphotometer show their profiles, which may be extremely complex. Line profile in the spectrum of the sun. (From Goldberg and Aller,* Atoms, Stars, and Nebulae. *Philadelphia: The Blakiston Co., 1943)*

TABLE 19.1
ABUNDANCE OF SOME OF THE CHEMICAL ELEMENTS IN THE SUN'S ATMOSPHERE
(N is the number of atoms above one square cm of the photosphere)

Element	log N	Element	log N
H	13.27	A	9.00
H^2	5.85	K	6.39
He	12.41	Ca	7.72
Li	2.57	Mn	6.55
Be	2.8	Fe	8.84
B	6.0	Ni	7.44
C	9.23	Cu	5.77
N	9.49	Zn	6.26
O	10.00	Sr	4.37
F	7.7	Zr	3.54
Ne	10.04	Pd	2.1
Na	7.69	Ag	2.0
Mg	9.05	Sn	2.2
Al	7.67	Ba	4.02
Si	8.80	Pt	2.60
P	7.08	Hg	4.35
S	8.52	Pb	3.8
Cl	8.30		

the tuning dial. For a variety of reasons—the Doppler effect, collisions among atoms, so-called natural broadening, etc.—as the radiation streams past more and more atoms of the given element in the column above the photosphere, photons with frequencies slightly greater and slightly less than ν_0 are absorbed, and the absorption line develops wings, as shown in Figure 19.6. Thus the line itself acquires a definite profile. Since the shape of this profile depends on the Doppler effect, on collision broadening, or on natural broadening, whichever of these processes are most important, we can determine how many atoms of the given element there are in the absorbing column of the atmosphere only if we know the shape of the absorption line and the relative importance of the various factors that are responsible for the shape. Since these theories have been worked out in detail, it is now possible to determine the abundances of various elements by a spectral-line analysis.

19.13 Experimental Determination of the Abundances of Elements in Stellar Atmospheres

Although theoretical techniques are extremely powerful for analyzing the atmospheres of the sun and stars, it is possible to do the same thing experimentally. One compares directly the intensity of a stellar absorption line of a particular atom with the same line formed in a so-called absorption chamber having a volume of 1 cm^3. If enough atoms under appropriate conditions are placed in such a chamber, we can obtain absorption lines of the same intensities as are found in stellar spectra. We then assume that the number of absorbing atoms in the absorption chamber is equal to the number of atoms of the same element in a column of the solar or stellar atmosphere extending from the photosphere to the top of the atmosphere, and having an area of 1 cm^2.

One must use this method with caution because conditions in the solar or stellar atmosphere may be quite different from those in the absorption chamber. Nevertheless, the method has been used successfully in finding the relative abundances of elements in solar and stellar atmospheres. Results obtained in this way agree well with results obtained theoretically. In Table 19.1 we have given the data for the sun, and there is good reason to believe that the same over-all abundances apply to most of the stars in our part of the Milky Way.

19.14 The Opacity of Stellar Atmospheres

Although we can gain a great deal of knowledge of the abundances of the chemical elements in stellar atmospheres by analyzing the spectral lines, before we can obtain a complete picture of the structure of a stellar atmosphere we must have some way of determining the physical conditions at each point in the atmosphere. Solving this problem is called **constructing a model stellar atmosphere**, and it involves the solution of certain complicated equations which we shall not discuss here. However, we shall consider here the empirical data that must be inserted in these equations before they can be solved. Among other things, the **coefficient of absorption** or the so-called **opacity** of the atmosphere, *which is a measure of how effectively the atoms in the atmosphere block the flow of radiation from the photosphere*, must be known.

The opacity of a stellar atmosphere is due to electrons (whether they are found in atoms or are free) reacting to the radiation flowing past them. *If these electrons are bound in atoms, absorption occurs by ionization processes which tear the electrons out of the atoms and set them in random motion as free particles. If the electrons are free particles moving in the neighborhood of the surrounding ions, they can also absorb radiation by experiencing an increase in kinetic energy.* By analyzing such processes, it is possible to determine the coefficient of opacity as a function of the depth in the atmosphere—that is, a function of temperature, pressure, and density. In general, the most effective absorbers are the metallic and the other heavy atoms because so many electrons surround their nuclei.

19.15 The Contribution of Hydrogen to the Opacity: The Negative Hydrogen Ion

Since hydrogen is very abundant, it contributes considerably to the opacity by an ionization process if the atmosphere is hot enough to throw many atoms into excited states. But, in cool atmospheres, where practically all of the hydrogen is in the ground state, only photons of the far ultraviolet can ionize the hydrogen, so that the contribution of hydrogen to the opacity by this process is small. However, hydrogen is very effective in screening the radiation by another process.

Since there are many free electrons in a stellar atmosphere, a neutral hydrogen atom can capture another electron and become a negative ion. Such negative ions are extremely effective absorbers of energy in the visual part of the spectrum because the two electrons attached to the single proton are rather loosely bound and therefore react to visible radiation very readily. The fact that stellar atmospheres in cool stars such as the sun are very opaque to ordinary radiation is due to this negative-ion absorption.

As we go to hotter atmospheres, the hydrogen atoms, in general, have their electrons in excited states from which ionization can occur much more readily, so that such atmospheres are considerably more opaque than the sun's atmosphere. Thus, for example, we find that Sirius' atmosphere is about 20 times more opaque than the sun's.

19.16 Constructing a Model Stellar Atmosphere

In addition to the opacity, we must have the following data to construct a model stellar atmosphere: the acceleration of gravity (derived from the mass and the radius of the star), the luminosity of the star, the number of free electrons per unit volume, and the chemical composition of the atmosphere. From this information, on the assumption that the atmosphere is in hydrostatic equilibrium (that is, at each point of the atmosphere the gas pressure supports the weight of the overlying layers) it is possible to determine the pressure and the density throughout and thus to obtain a complete picture. It is then possible to find the amount of energy flowing through the atmosphere in each part of the continuous spectrum and thus to determine how the absorption lines are formed.

19.17 The Method of Spectroscopic Parallaxes

In the introduction to this chapter, we stated that once we clearly understand the process by which absorption lines are formed in the spectra of stars, we can then introduce certain criteria to differentiate between main-sequence, dwarf, giant, and supergiant stars of the same spectral class. Even before the theory of spectral lines was understood, it had been established empirically that there are differences in certain lines in the spectra of stars belonging to the same spectral class as shown in Figure 19.7, but differing in absolute magnitude. In 1914, W. S. Adams and A. Kohlschütter discovered that one can use these differences as criteria for obtaining the absolute magnitudes of stars in the same spectral class. *The relationship that exists between the absolute magnitude of a star and its spectrum* is referred to as the **absolute-magnitude effect**, and the method based upon it for determining the absolute magnitude and, hence, the parallax of a star, is called the **method of spectroscopic parallaxes.**

19.18 Differences in the Absorption in the Spectra of Giants and Dwarfs in the Same Class

The absolute-magnitude effect manifests itself in changes in certain characteristics of the spectral

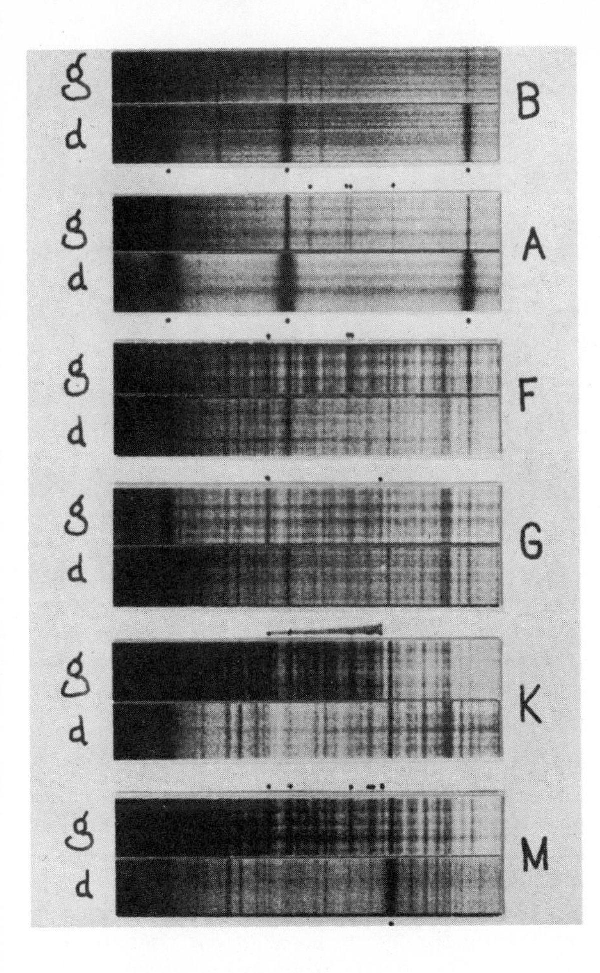

FIG. 19–7 *Spectral differences between giant and dwarf stars of classes B through M, by Morgan, 1950. In each case the lines in the dwarf spectrum are broader and fuzzier than those in the giant spectrum.* (*Yerkes Observatory photograph*)

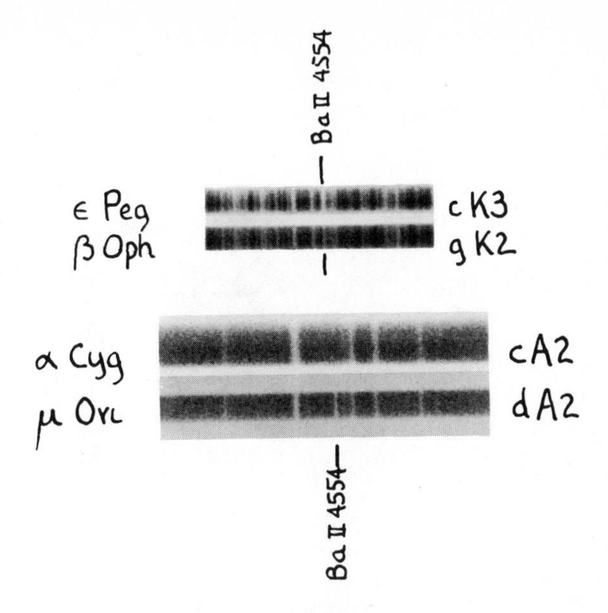

FIG. 19–8 *The behavior of Ba II 4554 in classes K and A. The line is stronger in the K supergiant ε Pegasi than in the giant β Ophiuchi. At class A, on the other hand, the line is not seen in the spectrum of the supergiant α Cygni illustrated, while it is a well-marked line in the spectrum of the dwarf μ Orionis.* (*Yerkes Observatory photograph*)

lines, two of which we shall discuss here. *First, we observe that when we go from dwarf to giant stars of the same spectral class, whereas the intensities of certain lines remain unchanged, the intensities of other lines, and particularly of ionized lines, change considerably.* This can be seen in Figure 19.8. *Second, we find that there is a considerable difference in the width and fuzziness of the majority of the lines.*

Before we discuss them in detail, we shall show that *both these effects are due to pressure differences in the atmospheres of dwarfs and giant stars belonging to the same spectral class.* Let us consider a dwarf star such as the sun, and one of its giant counterparts, a star like Capella. As we shall see later, *although the radii of stars vary by many factors when we go from the dwarf to the giant branch (keeping the spectral class the same) the masses change by a relatively small factor.* Thus, although the mass of Capella is only about four times that of the sun its radius is sixteen times larger. This means that the pressure in the atmosphere of Capella is much smaller than in the sun's atmosphere.

The ratio of the atmospheric pressures for stars in the same spectral class is given essentially by the ratio of the accelerations of gravity on their surfaces (their photospheres), since their temperatures are approximately the same. The reason is that the acceleration of gravity determines the weight of material above the surface (photosphere) of the star and, hence, the weight of the overlying atmosphere. This, in turn, means that if the acceleration of gravity is large, the atoms in the atmosphere are packed more closely together than if the acceleration is small.

Since the acceleration of gravity is equal to GM/R^2 on the surface of a star, where M is its mass, and R is its radius, the atmospheric pressure on the sun is greater than on Capella by the factor

$$\frac{GM_{\text{sun}}/R_{\text{sun}}^2}{GM_{\text{capella}}/R_{\text{capella}}^2} = \frac{M_{\text{sun}}/M_{\text{capella}}}{(R_{\text{sun}}/R_{\text{capella}})^2} = \frac{1/4}{1/256} = 64.$$

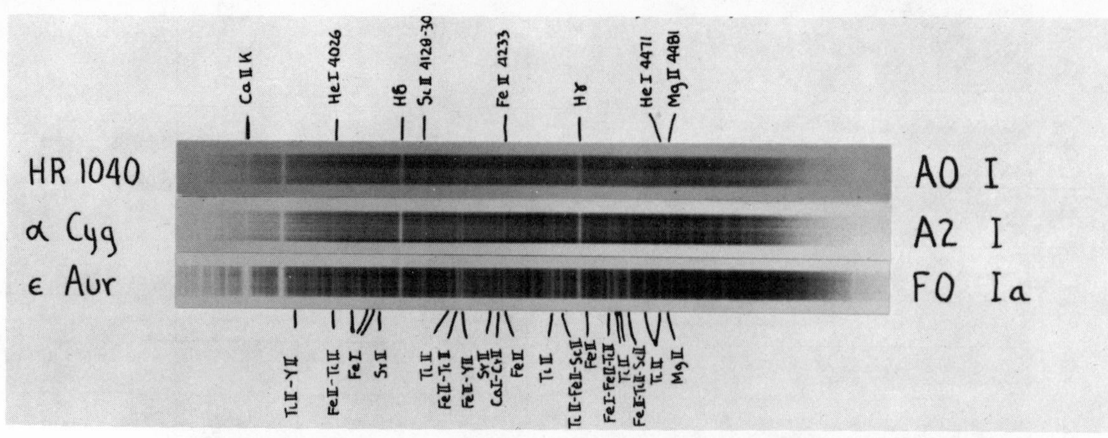

FIG. 19–9 *Supergiants AO to FO. The strongest lines of HeI are faintly visible in HR 1040. They are not visible on low dispersion spectra of α Cygni. The lines of FeII are very strong in the spectrum of α Cygni and are of about the same intensity in ε Aurigae. The lines of TiII and SrII are considerably stronger in ε Aurigae. The classification of the supergiants of type AO and later is a difficult problem. Their spectra differ so much from stars of lower luminosity that line ratios suitable for the latter cannot be used. Supergiants of classes A to M can best be classified by referring them to a normal sequence of high-luminosity stars whose spectra as a whole can be considered to define the class assigned.* (*Yerkes Observatory photograph*)

In other words, we may expect the pressure in the atmosphere of Capella to be about 100 times smaller than in the atmosphere of the sun. This means that if we compare a column of atmosphere extending from the photosphere of the sun up to the top of its atmosphere, with a similar column on Capella, the atoms and any free electrons that may be present are packed much more closely together (are more dense) in the sun than in Capella. From this fact we can draw immediate conclusions about the intensities of the absorption lines in the spectra of these two stars. Since radiation coming from the photosphere of the sun and passing through a column of atmosphere meets many more neutral atoms per path length of a given kind than similar radiation does on Capella, we see that, in general, the absorption lines of this neutral atom in the sun will be more intense than they are in the spectrum of Capella. This is so because the denser column of atmosphere is more effective in scattering and absorbing radiation than an attenuated column.

On the other hand the situation is different for the lines of an ionized atom. These are more intense in the spectrum of Capella than in the spectrum of the sun, and again because the atoms and free electrons in the solar atmosphere are packed more closely together than they are in Capella. If we look back at our discussion of what determines the intensities of the absorption lines of ionized atoms, we see that *one of the important factors is the number of free electrons per unit volume.* The greater this number, the fewer ionized atoms we have, and hence the weaker are the lines of ionized atoms. It follows that since the free electrons are much more closely packed together in the atmospheres of dwarfs than they are in the atmospheres of giants the lines of ionized atoms are in general stronger in giant stars. In Section 19.21 we shall qualify this statement somewhat, for although it is true that in general ionized lines of atoms are more intense in giant stars than in dwarf or main-sequence stars because of the free-electron factor, the situation for a dwarf and giant star in the same spectral class is more complex than indicated in the simple analysis given above. The formation of an ionized line and its intensity depend on the temperature and the ionization energy as well as the free-electron density, so that all three factors must be taken into account. Figure 19.9 shows the spectra of some supergiants from AO to FO.

19.19 Using Absorption Lines of Ionized Atoms as Criteria for Absolute Magnitude

As an example we may compare the ionized strontium lines in the spectrum of the sun with those of Zeta Capricorni (which is 6 000 times more

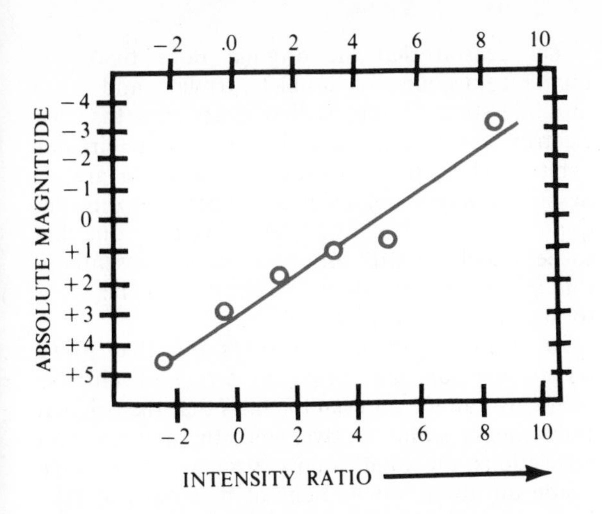

FIG. 19-10 *The absolute magnitude of a star as a function of the ratio of the ionized strontium 4 215 Å line to the neutral iron line 4 260 Å. (After D. Hoffleit, Harvard Observatory. From Goldberg and Aller,* Atoms, Stars, and Nebulae. *Philadelphia: The Blakiston Co., 1943)*

luminous than the sun). An analysis shows that they are much more intense in the latter than the former. It is possible, by studying many such spectra, to establish a spectral criterion for the absolute magnitude of a star in terms of the intensity ratio of the lines of certain ionized atoms to the lines of certain neutral atoms. Indeed, we can construct a graph showing such correlations, and then use this graph to find the absolute magnitudes of stars for which this intensity ratio can be measured. In Figure 19.10 the intensity ratio of the ionized strontium line 4 215 Å to the neutral iron line 4 260 Å is plotted against the absolute magnitude of known stars belonging to the same spectral class. We see that the curve is practically a straight line so that one can use this to determine the absolute magnitude, and hence the parallax, of a star of known apparent magnitude.

For let us suppose that we have a star of a given spectral class, for which the above intensity ratio of the two spectral lines is 5 on the abscissa scale. Moving across to the absolute-magnitude scale plotted along the vertical axis we see that this corresponds to a star whose absolute magnitude is zero. If the apparent magnitude of this star as measured directly is +5, we can find the parallax of the star from the formula

$$M = m + 5 + 5 \log p,$$

by solving for log p. We thus have

$$\log p = \frac{M - m - 5}{5}.$$

If we substitute $M = 0$, and $m = 5$, we find for log p the value -2, which means that the parallax of this star is 0″.01, or its distance is 100 parsecs.

19.20 The Accuracy of Spectroscopic Parallaxes

Just as we can use the intensity ratio of an ionized strontium line to a neutral iron line for spectroscopic parallaxes, we can use other such lines in the same way. This method has led to determination of the parallaxes of many thousands of stars. Its great advantage over the trigonometric method is that it requires the measurement of just a single photographic plate, and it can be carried out mostly by inspections so that the work proceeds very rapidly. Since the error in spectroscopic parallax work is of the order of 15 per cent of the value of the parallax itself, whereas trigonometric parallax measurements have an absolute error of 0″.005, the spectroscopic method becomes more and more reliable compared to the trigonometric method, as we go out to greater and greater distances. At a parallax of about 0″.03 both methods are equally reliable. For stars closer than this the trigonometric method is more reliable, and for stars more distant than this, the spectroscopic method gives better results.

As an example of the way in which the lines of the neutral atom vary as we go from giants to dwarfs, we may consider the neutral calcium line 4 227 Å in the dwarf star Lalande 21 185, and in the star Betelgeuse which is 300 000 times more luminous. Both of these stars belong to spectral class M2, but the neutral calcium line for Lalande 21 185, whose radius is half that of the sun, is much more intense than it is in the spectrum of Betelgeuse, which has a diameter 400 times that of the sun.

19.21 The Difference in Sharpness of Absorption Lines from Giant to Dwarf Stars

Just as the pressure in the star's atmosphere affects the relative intensities of the ionized and neutral lines, so, too, it has an effect on the sharpness and width of the spectral line. Since the atoms are much closer together in atmospheres in which the pressure is high (dwarf stars) than in atmospheres of giant stars where the pressure is low, the lines in the spectra of giants are, in general, sharper than they are in dwarfs. This is so because when atoms are closer together, there are frequent collisions among them, so that the spectral lines are wide and fuzzy instead of being sharp. An example of this effect is shown in Figure 19.7, in which the spectra of main-sequence stars are compared with the spectra of giant stars of the same spectral class. We see that whereas the lines are broad and fuzzy for the main-sequence stars, they are sharp and thin in the case of the giants.

Because of the difference in the ionization characteristics in the atmospheres of giants and dwarfs of the same spectral class, the surface temperatures of a dwarf and a giant are not quite the same. The reason is that we assign two stars to the same spectral class if their spectra are similar. This applies to the ionized lines as well as to the neutral lines of the spectra. But the only way the spectrum of a dwarf can have ionized lines similar to those in a giant star is if the dwarf star is somewhat hotter than the giant star to compensate for the greater concentration of free electrons in its atmosphere.

The basis for this argument is given in Equation (19.3). We see there that the electron density and the temperature work against each other in the ionization formula. This means that if we go from a dwarf to a giant star, with a resulting reduction in the electron pressure (that is, the electron density N_e), this can be offset by a smaller temperature T which appears both as a factor and in the exponent. We see further that the ionization energy χ_0 has an effect opposite to that of the temperature. This means that if in a giant star N_e and T are just right to favor the ionization of one type of atom, they may not be as favorable for another type of atom because of the difference in the ionization potentials.

19.22 Improvements in the Bohr Theory: Noncircular Orbits for the Electron

Even though we have analyzed the spectral lines in terms of the simple Bohr theory of the atom, we find that this theory is hardly adequate to account for the many details in the spectrum. In particular, we discover that certain lines that first appeared to be single are really double, and in some cases are composed of three or more individual lines. Such fine and hyperfine structures of the spectral lines are of great importance in our analysis of the conditions in stellar atmospheres, ultimately enabling us to distinguish between giants, dwarfs, and main-sequence stars, as we have just seen in our discussion of the method of spectroscopic parallaxes.

The reason that the original Bohr theory is unable to account for doublets, triplets, and other multiple spectral lines is that it assumes that the electron revolves around the nucleus in circular orbits. Thus, the properties of the atom are all described in terms of a single number, the principal quantum number, n, and this allows only one line to be associated with any two orbits. Essentially this means that we allow the electron in the atom to have only one degree of freedom.

As a matter of fact, it turns out that if we try to develop the theory in terms of circular orbits we cannot get even the lines that the original Bohr theory seems to give, since there are certain restrictions on what jumps electrons can make inside an atom, which Bohr at first did not take into account. In the case of circular orbits these restrictions would limit the electron to going from a given orbit to the next nearest orbit and to no others, so that only one Balmer line could then be emitted. For all of these reasons, it is clear that noncircular orbits must be admitted; this was, indeed, the first improvement made in the Bohr theory.

19.23 The Angular Momentum of an Electron in an Atom: The Azimuthal Quantum Number, l

Since the general problem of the motion of the electron in the atom treated by the semiclassical method of Bohr is the same as that of the motion of a planet, we are naturally led to elliptical Bohr orbits for the electron. This immediately introduces an additional degree of freedom, since we know that if we have an elliptical orbit of given size (given semimajor axis), the shape (i.e., the eccentricity) of the orbit may vary over a wide range. Now, just as the size of the orbit (in the case of the electron the principal quantum number, n) determines the total energy of the electron moving in it, so the shape of the orbit, or the eccentricity, e (see Section 8.14), determines the angular momentum of the electron. We may naturally ask whether the eccentricity, too, has a quantum number associated with it just as the energy has the quantum number n associated with it.

It is fairly easy to see that this is so, and for the following reason. When the electron moves in an elliptical orbit, its behavior differs in one very important respect from its behavior in a circular orbit. Whereas in the latter its speed is always the same so that its mass does not alter, in the elliptical orbit its speed and hence—according to the special theory of relativity—its mass vary continuously. *Because of this continuous change in mass, the entire elliptical orbit of the electron precesses around the nucleus with a definite frequency.* This means that we must assign to the electron an additional small amount of energy, namely, the energy of the precession of its orbit, which depends upon the shape of the orbit.

Since this precession is periodic, it, too, is characterized by a quantum number, the azimuthal quantum number l. We shall see that just as the quantum number n determines the size (the semi-major axis) of the electron's orbit and the energy of the electron in its orbit (not counting the energy of precession) the quantum number l specifies the shape of the orbit or the angular momentum of the electron. *It can be shown that for a given value of n not all orbital shapes (i.e., values of angular momentum of the electron) are permissible, but only those for which l has the integral values 0, 1, 2, etc. up to n − 1.* In other words, if the principal quantum number of an electron inside an atom is n, there are exactly n orbits of the same size but of different shapes in which the electron can move. For a given value of l the orbital angular momentum of the electron is $lh/2\pi$. It actually is $\sqrt{l(l+1)}(h/2\pi)$ if we take all the refinements of the theory into account, but for our purpose it is sufficient to speak of the angular momentum as $lh/2\pi$.

19.24 Sublevels Arising from Quantum Number l

The introduction of the azimuthal quantum number means that the various energy levels introduced by Bohr are not really single levels, but are composed of sublevels which depend upon the angular momentum of the electron. Thus instead of there being one line in the hydrogen spectrum when the electron jumps from an orbit of principal quantum number n to an orbit of principal quantum number m, there is a multiple of lines very close together since the electron can jump from one of the sublevels associated with n (and having a definite value of l) to some sublevel associated with m, as shown in Figure 19.11. Because there are definite rules which do not allow all such jumps from sublevel to sublevel, not all lines that one might expect from all the sublevels present are found in the spectrum.

It is easy to list all the sublevels for a given n since we know that l can have all integral values

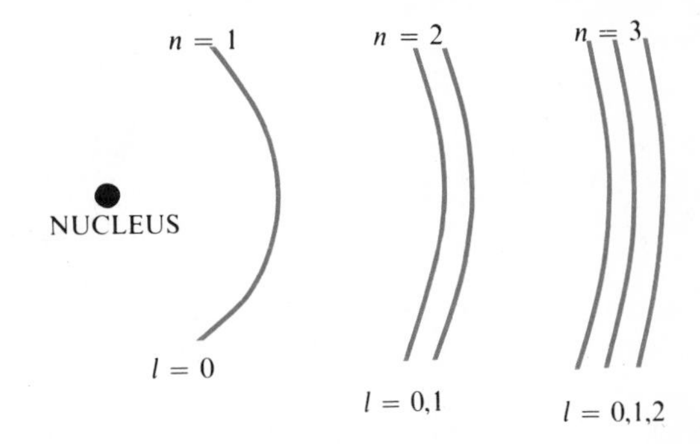

FIG. 19–11 *Schematic diagram showing sublevels associated with a given principal quantum number.*

from 0 up to and including $n - 1$, and no others. Thus for $n = 1$, l can have only the value 0, so that there are no sublevels for the ground state. For $n = 2$, there are the two sublevels with azimuthal quantum numbers $l = 0, 1$; for $n = 3$, three sublevels with quantum numbers $l = 0, 1, 2$; etc.

19.25 The Orientation of the Orbit of an Electron in a Magnetic Field: A Preferred Direction in Space

Since a particle like an electron ordinarily has three degrees of freedom, we should expect to find still a third quantum number associated with its motion inside an atom. That this must be the case is immediately clear if we note that the orbit of the electron in the atom is not completely specified if we give only its size and shape. If we picture the orbit as lying in a plane, we see that this plane can be orientated in all directions in space, as shown in Figure 19.12.

Although it is obvious that various orientations of this plane can be introduced, it is not clear why the orientation should have a quantum number associated with it since the introduction of a quantum number means that not all orientations are permissible. But all directions of this plane in space seem to be equally permissible, so long as there is nothing in space to single out some preferred direction. However, when a magnetic field is present, not all directions are on the same footing.

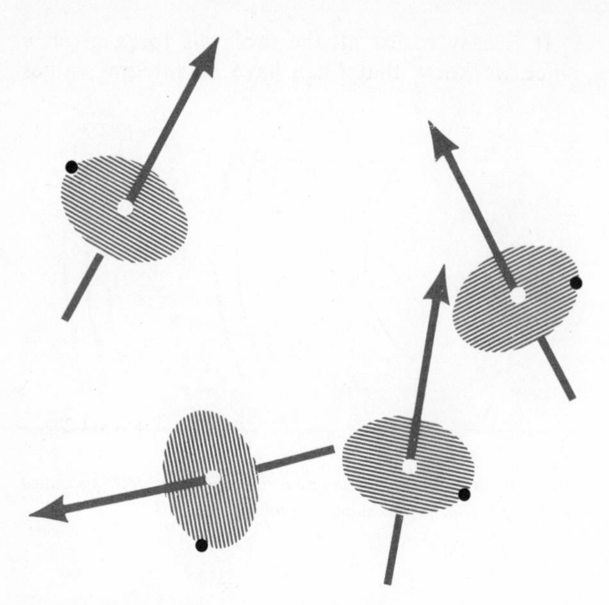

Fig. 19–12 *The orbital angular momentum vector of the electron (the normal to the plane of its orbit) can be oriented in various directions.*

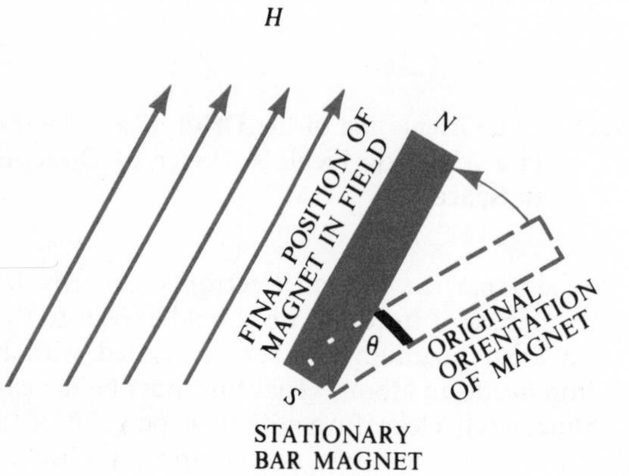

Fig. 19–13 *A non-spinning magnet aligns itself parallel to the magnetic field.*

Let us suppose that a magnetic field of magnitude H points in a given direction, and that a bar magnet is placed in it with its length making some angle θ with the direction of the field, as shown in Figure 19.13. *If the magnet were left to itself, it would align itself with the field unless it were spinning around its own axis.* If it were spinning, however, it would precess about the field with a definite frequency maintaining the angle θ, just the way a top spinning on the earth's surface

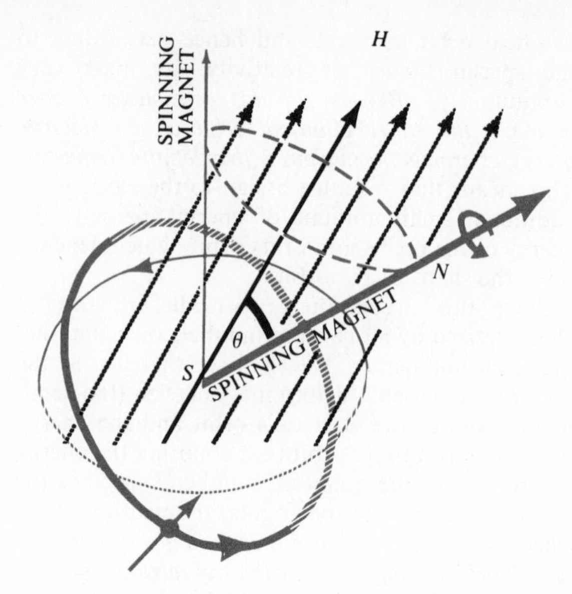

Fig. 19–14 *A spinning magnet (an electron moving around in its orbit) precesses around the magnetic field.*

precesses around the earth's gravitational field. *Thus, the presence of a magnetic field establishes a preferred direction in space.*

19.26 The Magnetic Quantum Number, m: The Ordinary Zeeman Effect

When an atom is placed in a magnetic field, it behaves the way a spinning bar magnet does. The electron is a charged particle and hence its motion about the nucleus turns the atom into the equivalent of a small bar magnet with its length perpendicular to the plane of the orbit. Not only is it a bar magnet, but it is a spinning magnet because of its orbital motion (angular momentum). Thus, when the atom is placed in a magnetic field, the plane of the electron's orbit precesses, as shown in Figure 19.14.

Since this precessional motion is periodic, we must assign to it a new quantum number, m. This is called the **magnetic quantum number** of the electron. It signifies that when an atom is placed in a magnetic field only those orientations θ given by the quantum numbers m are permitted for the electron's orbital plane with respect to this field. *The rule for the magnetic quantum number m is the following: for a given value of l it can only take on those integral values ranging from $-l$ to $+l$.* The reason for this is that m is the projection of l along the magnetic field so that $m = l \cos \theta$. But $\cos \theta$ can range only from -1 to $+1$ so that m must range (always as an integer) from $-l$ to $+l$. Thus, the existence of the magnetic quantum number m means that in a magnetic field each of the sublevels l associated with a given principal quantum number n is itself split up into $2l + 1$ subsublevels.

We may illustrate this by taking again the principal quantum numbers $n = 1, 2$. Since for $n = 1, l = 0$, the only possible value for m is also 0, and we see that even in a magnetic field the ground state has no sublevels. For $n = 2$, we have the two l values 0 and 1. For $l = 0$, m is 0, but for $l = 1$, there are the three values -1, 0, and $+1$ for m (note that $2l + 1 = 3$ for $l = 1$). As we go to higher values of the principal quantum number, the subsublevels become increasingly numerous when a magnetic field is present.

We see from these considerations that *the spectral lines of an atom break up into complicated multiplets in a magnetic field, a phenomenon which is known as the* **Zeeman effect**. This is of great importance in astronomy since the Zeeman effect enables us to determine the strengths of magnetic fields in stellar atmospheres. If one uses the Bohr theory as described up to this point to explain the Zeeman effect, one finds that each line in the spectrum of an atom should break up into three lines when the atom is placed in a magnetic field. This is called the normal Zeeman effect. But the actual Zeeman pattern (the anomalous Zeeman effect) in most cases is more complicated than this and cannot be explained by the Bohr theory.

19.27 The Spinning Electron and the Spin Quantum Number, s: The Anomalous Zeeman Effect

Although it may seem that three quantum numbers should be sufficient to describe the motion of an electron completely, since generally we associate just three degrees of freedom with this motion, we know that a fourth quantum number must be introduced. We can surmise this from the theory of relativity, since it tells us that four numbers must be introduced to describe events in nature because time must be treated together with space. We must therefore seek another periodic motion of the electron if we are to give the electron a fourth quantum number. This turns out to be the spin motion of the electron, which was first introduced into the Bohr theory of the atom by Goudsmit and Uhlenbeck in 1925.

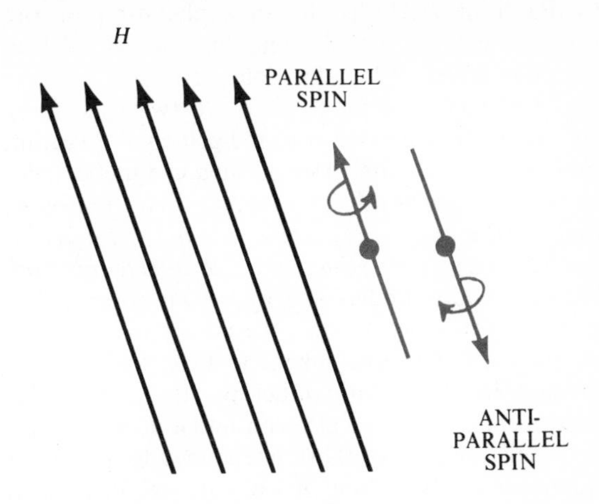

FIG. 19–15 *The spin angular momentum vector of an electron aligns itself either parallel or antiparallel to a magnetic field.*

It was only by assuming that the electron is spinning about its own axis, the way the earth does, that Goudsmit and Uhlenbeck were able to explain certain refinements in the spectra of atoms that had not previously been satisfactorily accounted for. To explain these observations they found it necessary to assign to the electron a spin angular momentum equal to $\frac{1}{2}\frac{h}{2\pi}$, and hence a **spin quantum number**, s. *We can account for the observed features of the spectral lines for various atoms only if the spin quantum number, s, takes on just two values, namely* $+1$ *or* -1. This means that if we picture the spinning electron as being the equivalent of a small magnet, and place it in a magnetic field, it will line up either with the magnetic field or against it. No other orientations of the spin axis of the electron in a magnetic field are possible. This is shown in Figure 19.15.

Because the electron is spinning, the Zeeman pattern of lines in the spectrum of an atom when it is in a magnetic field is much more complex than it would be if the electron had no spin. In fact, it is because this so-called anomalous Zeeman pattern is observed and cannot be explained in terms of the three quantum numbers, n, l, m, that it is necessary to assign a spin quantum number to the electron.

19.28 The Additional Subsublevels of an Electron in an Atom Resulting from the Spin

The introduction of the fourth quantum number, together with a very important principle discovered by Pauli in 1927, led to an explanation of the arrangement of the elements in the Mendeleev periodic table. We may note first that because the electron is spinning, there are twice as many subsublevels associated with the principal quantum number n when the atom is in a magnetic field, as we had previously mentioned. The reason is that *when a magnet is parallel to a magnetic field it has less energy than when it is antiparallel to this field.* We can readily understand this if we observe that we have to do work when we take a magnet that is lined up with the magnetic field and turn it around into the antiparallel position. For this reason the spin of the electron introduces two new sublevels for each of those introduced by the other quantum numbers, l and m: one sublevel for parallel alignment of electron spin and magnetic field and the other for antiparallel alignment.

It is easy to count the number of subsublevels associated with any quantum number n. Observe first that for each l value associated with n, there are $2l + 1$ subsublevels arising from the magnetic quantum number m, because that is exactly how many integral m values we can assign for this l value (as we have noted, m must range from $-l$ to $+l$). But l can take on all values from 0 to $n - 1$. Therefore, we must assign to the quantity $2l + 1$ the values 0, 1, etc. up to $n - 1$, and add the total number. In other words, to obtain the total number of subsublevels, not counting the contribution of the spin quantum number, we must evaluate the sum (for a given n)

$$\underset{l=0}{(2 \times 0 + 1)} + \underset{l=1}{(2 \times 1 + 1)} + \underset{l=2}{(2 \times 2 + 1)} + \cdots + \underset{l=n-1}{(2 \times [n - 1] + 1)}.$$

The sum is exactly equal to n^2. Not counting the effect of the spin, then, there are n^2 subsublevels associated with the quantum number n, where $n = 1$ for the ground state, etc. If we now consider the effect of the spin, we must multiply this by 2. *This means that for a principal quantum number n there are altogether $2n^2$ subsublevels.* Thus the ground state—i.e., the K shell—has two levels, the L shell, with principal quantum number 2, has 8, the M shell has 18, etc.

19.29 The Pauli Exclusion Principle

Now suppose that we have the nucleus of a given atom of atomic number N, and that we want to arrange the electrons around this nucleus so as to obtain the neutral atom in its ground state (that means, of course, that we must have N electrons appropriately arranged in orbits to compensate for the positive charge N on the nucleus). Since all physical systems in nature tend to fall to states of lowest energy, we might expect all the electrons to fall into the K shell.

However, this does not occur. Hence, we have the great variety of elements found in nature and the periodicities in their chemical properties. *It is because of Pauli's exclusion principle, mentioned above, that all electrons do not fall into the ground state. This principle states that only as many electrons ($2n^2$) can be accommodated in a shell of given principal quantum number n as there are subsublevels associated with that shell. In other words, there can only be 1 electron in each subsublevel.* This means that if the atom we are speaking about were to be placed in a magnetic field so that all the subsublevels were separated from each other, each electron would then be moving in its own subsublevel and, thus, would have its own distinct energy, different from that of any other electron.

There is another way of expressing this principle which is the one most commonly found in books. To get at this formulation of the Pauli exclusion principle we note that since each subsublevel has associated with it four distinct quantum numbers, n, l, m, s, so also does each electron. From this it follows that even when a neutral atom is in its ground state, so that all the electrons are in their lowest possible orbits, the electrons must be arranged in such a way that *no two electrons can have the same set of quantum numbers, n, l, m, s, assigned to them.*

19.30 The Explanation of the Periodicity of the Elements Using the Pauli Exclusion Principle

From this principle we can see why the chemical properties of atoms are periodic as we move along the Mendeleev table. First note that chemical properties are determined by the number of electrons in the outermost shell of an atom. As we go from the lighter to the heavier atoms, the additional electrons must go into higher shells because the Pauli exclusion principle closes the lower shells to them. These outer shells in the heavier atoms will have the same electronic configurations as the outer shells have in lighter atoms so that chemical properties are repeated.

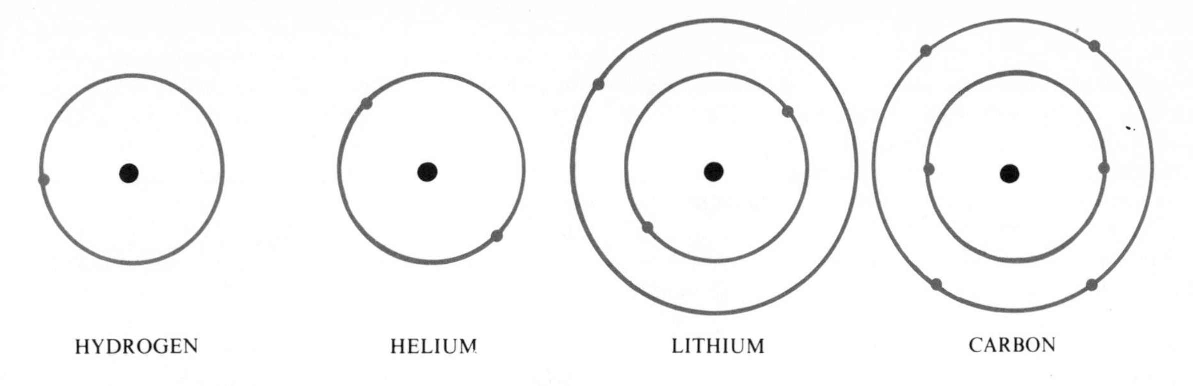

FIG. 19–16 *The arrangements of electrons in their orbits of some of the lighter elements: H, He, Li, C.*

We may illustrate this with a few of the lighter elements (as shown in Figure 19.16). Since hydrogen has only one electron, the electron goes into the lowest or K shell. In helium there are two electrons because the nucleus has two protons, and since there are just two subsublevels associated with the K shell both of these electrons are in this shell. But they move in such a way that their axes of spin are antiparallel to each other. Thus, one of these electrons has the quantum numbers $\overset{n}{1}, \overset{l}{0}, \overset{m}{0}, \overset{s}{1}$, and the other the quantum numbers $\overset{n}{1}, \overset{l}{0}, \overset{m}{0}, \overset{s}{-1}$. *Since these two electrons close the K shell according to the Pauli exclusion principle, there is no room for any other electron to jump into this arrangement and the two that are already there are very tightly bound. Because of this, helium is an* ***inert*** *or* ***noble*** *gas.*

If we go to lithium, we must arrange three electrons outside the nucleus, and since the third electron cannot find room in the K shell, it must circulate around the nucleus in the more distant L shell (principal quantum number $n = 2$). This electron is rather loosely bound so that lithium is chemically active. Now let us jump to neon, with atomic number 10. This atom must have 10 electrons properly arranged around the nucleus; two of these must be in the K shell to close that shell, and the other 8 must be in the L shell, which is therefore also closed.

It is easy to see what quantum numbers we must assign to these 8 electrons. Each of them has a set of quantum numbers of the form 2, l, m, $+1$, or 2, l, m, -1, where l has the value of 0 or 1, and m the appropriate value associated with it. Thus, there are electrons inside the neon atom with such quantum sets as 2, 0, 0, -1; 2, 0, 0, $+1$; 2, 1, 0, -1; 2, 1, 0, $+1$; etc. We can see that only 8 such distinct sets are possible. Since the outer electrons in neon form a closed shell just as the 2 electrons in helium do, we know that neon must have the same chemical properties as helium—which is, of course, the case. Neon, Helium, Argon, Krypton, Xenon, and Radon are noble gases. In 1962, the first noble-element compounds were produced, and the term inert has since become a relative expression.

If we go to the next element in the atomic chart, sodium, we have 11 electrons to arrange, and since 10 of them form the two closed shells K and L, the remaining one must lie in the M shell (principal quantum number $n = 3$), and we see, of course, that sodium with its external electron must behave chemically the way lithium does. Thus, sodium and lithium are found in the same chemical family. In this way it is possible to go through the whole table of chemical elements and to see why the spectral characteristics of various chemical elements in stellar atmospheres depend upon the temperatures in the way they do.

19.31 The de Broglie Wave Theory of Matter: The de Broglie Wavelength

Although the Bohr theory of the atom served magnificently to explain the most obvious features of atomic spectra and to give the physicist a deep insight into the atom's structure, it was clear by the early 1920's that it was burdened by many logical inconsistencies and could not give correct explanations of the intensities of spectral lines, or indicate why certain lines were missing. In addition to this the Bohr theory made use of certain physical concepts, such as orbits inside an atom, that were beyond the scientist's observational domain. For these and other reasons leading physicists of the time, quite independently of each other, began to develop atomic models built upon new and revolutionary concepts.

The first of these great developments occurred as the result of the work of Prince Louis de Broglie, who was greatly stimulated by the particle-wave dualism that exists in radiation. As we know, depending upon the nature of the phenomenon, we may describe radiation either in terms of its wave properties or in terms of its photon-corpuscular properties. De Broglie was led by these considerations to suggest that a particle such as an electron must also possess a particle-wave dualism, but of such a nature that only its particle aspect is immediately obvious to us; its wave character is obscured in most phenomena.

He was greatly strengthened in this belief by the knowledge that even when a particle is at rest, it possesses an amount of energy mc^2. Arguing that this energy must have a frequency associated with it through the relationship $mc^2 = h\nu$, he proceeded by purely formalistic methods to derive a formula for the wavelength of a particle that depends on the momentum of the particle. *If the momentum of any particle is p, de Broglie's formula for the wavelength is exactly*

$$\lambda = \frac{h}{p}. \tag{19.4}$$

What is amazing about this expression is that Planck's constant appears in it, and that it decreases with the momentum of the particle.

This idea of de Broglie's received little attention, and it would probably not have influenced the development of physics at that time but for the experiments of Davisson and Germer. These two American physicists, carrying out experiments with electrons deflected from metal surfaces, discovered inadvertently that these reflected electrons arranged themselves in patterns, just as though they had wave properties and definite wavelengths. By studying these patterns, Davisson and Germer were able to measure the wavelengths that one would have to assign to the electrons to account for the pattern, and found that these were exactly given by de Broglie's formula.

19.32 Schroedinger's Wave Equation for an Electron

Once de Broglie's theory was accepted, physicists began looking for some way of representing the behavior of electrons without treating the motion of an electron inside an atom according to the usual laws of classical mechanics, for these had been set up for particles and not for waves. What was needed was a so-called wave equation, i.e., an equation that would describe the motion of the electron in terms of a wave amplitude the way Maxwell's wave equation describes the properties of light in terms of electromagnetic wave amplitudes. *Such a wave equation for the electron was finally discovered by Schroedinger and is now known as the Schroedinger equation.* However, the wave amplitude of the electron which one obtains by solving the Schroedinger equation is not a real number, but a complex number involving $\sqrt{-1}$.

19.33 The Probability Amplitude, ψ, of the Electron

There is, therefore, a fundamental difference between the wave amplitude that we must assign to an electron according to the theories of de Broglie and Schroedinger and the wave amplitude that we assign to such wave phenomena as sound, light, and the mechanical vibrations propagated along taut metal wires when they are plucked. For whereas the latter are real physical phenomena, the wave amplitude of the electron has both a real and an imaginary part.

The interpretation that must be given to this wave amplitude was first stated by the physicist Max Born, in 1927. *He showed that the wave amplitude (which is designated by the Greek letter ψ, and is referred to as the* ***Schroedinger wave function****) associated with an electron is related to the probability of finding the electron at a certain point in space if its energy is known.* The actual probability itself is the product of ψ and the quantity ψ^* which we get when we replace $\sqrt{-1}$ by $-\sqrt{-1}$ wherever it appears in ψ. This product is called the square of the absolute value of ψ and is written as $\psi\psi^*$.

19.34 The Application of Schroedinger's Equation to the Motion of an Electron in an Atom

When Schroedinger obtained his wave equation, he applied it immediately to the motion of an electron around a proton (the hydrogen atom) and was able to obtain, by this single procedure, all the results that had been so painstakingly extracted from the corrected Bohr theory during the previous thirteen years. Indeed, what Schroedinger showed was that the three quantum numbers, n, l, m, appear automatically in the solution ψ_{nlm} (the wave function) of his wave equation, purely as a mathematical consequence of his theory. Because of this he was immediately able to obtain the energy subsublevels of the atom and the correct spectral lines, with the additional result (which went beyond the Bohr theory) that the correct intensities of the spectral lines also appeared. Moreover, the Schroedinger wave equation shows that the orbital angular momentum is not given by $l(h/2\pi)$ but by $\sqrt{l(l+1)}(h/2\pi)$. This remarkable achievement brought the wave formalism irrevocably into atomic theory and banished forever the idea of particles as distinct from waves, replacing it by the concept of the wave-particle dualism.

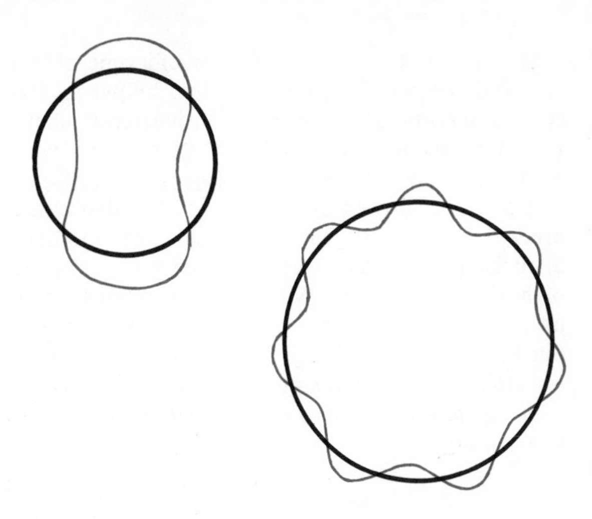

FIG. 19–17 *The circumference of a Bohr orbit is equal to an integral multiple of the de Broglie wavelength of the electron in that orbit.*

19.35 The de Broglie Wavelength and the Bohr Orbits

Since the de Broglie–Schroedinger wave concept of an electron is so at variance with our ordinary idea of a particle, it is natural to ask how the Bohr theory, based upon a particle electron, even as a first approximation, fits into this wave picture. We can get some idea of this if we recall (limiting ourselves to circular orbits) that the Bohr orbits are those for which (see p. 317)

$$2\pi r_n m v_n = nh,$$

where r_n is the radius of the nth orbit, v_n is the speed of the electron in this orbit, and n is the principal quantum number. Since $2\pi r_n$ is the circumference C_n of this orbit, and mv_n is the momentum p_n of the electron in this orbit, we have then

$$C_n p_n = nh,$$

or

$$C_n = n\frac{h}{p_n} = n\lambda_n,$$

[see the de Broglie equation (19.4)], where λ_n is the de Broglie wavelength of the electron in the nth orbit. We can now see that the Bohr orbits have the following significance: *in the Bohr theory, only those orbits are permitted to an electron inside an atom whose circumferences can exactly accommodate an integral number of de Broglie wavelengths of the electron*, as shown in Figure 19.17.

19.36 The Heisenberg Uncertainty Principle

While Schroedinger was developing his wave equation in Vienna, an entirely different approach to the solution of the problem was taken by a group of German physicists led by Heisenberg. He pointed out that the Bohr theory was built upon certain erroneous ideas because it depended upon concepts such as the orbit of an electron that could not be observed experimentally and, therefore, could not have precise physical meanings. He showed that *if one tries to ascertain the motion of an electron in an orbit, this very process destroys the orbital motion.*

Let us suppose that we wish to observe an electron moving in a neutral hydrogen atom. We can do this only by showering the atom with photons, with the hope that one of these will hit the electron moving around the proton, and bounce off it and into our detecting device. In this way we can, indeed, observe the electron at some particular moment, but because of the way electrons interact with photons, we know that the electron will be disturbed in this process, and hence we cannot know at the moment that we observe it how the electron is moving in its orbit, or even if it is in this orbit at all.

In fact, if we wish to observe the electron with greater and greater accuracy, we must use light of shorter and shorter wavelength (i.e., more energetic photons) so that the disturbing effect upon the motion of the electron becomes greater and greater. Indeed, to determine the position of the electron with extreme accuracy, we would have to use a photon so energetic that it would probably tear the electron completely out of the atom, leaving us with no orbit at all.

Heisenberg stated this in the form of a general and very famous principle, which is now known as the **uncertainty principle** or the **principle indeterminancy.** By analyzing the interaction of an electron with a photon under the most general conditions, Heisenberg showed that trying to determine precisely where an electron is at any moment disturbs it to such an extent that we lose knowledge of how it is moving. Moreover, if we determine its state of motion (i.e., momentum) we destroy our knowledge of its position. *This means that we cannot know simultaneously, with unlimited precision, both the position of an electron and its motion or momentum.*

19.37 Heisenberg's Formula for the Uncertainty Principle

To arrive at Heisenberg's principle in the form in which he stated it, let us suppose that we determined the position, q, of an electron with an uncertainty Δq, and that we also know its momentum, p, with an uncertainty Δp. In classical physics, before the quantum theory was introduced, it was assumed that given sufficiently accurate instruments, Δq and Δp could both be simultaneously reduced to as small values as desired. But according to Heisenberg, this is not the case. If one reduces Δq, then Δp must increase by the same factor and vice versa. This means that the product $\Delta q\, \Delta p$ cannot be reduced below a certain value. Heisenberg's great achievement was in finding this minimum value and expressing it by means of the inequality

$$\Delta q\, \Delta p \geq \frac{h}{2\pi}.$$

What is remarkable about this relationship is that the accuracy of our measurement cannot increase indefinitely, but is limited by Planck's constant. In a sense we can grasp this if we recall that Planck's constant represents the smallest amount of action that we can have in nature, so that whenever we try to follow a particle in its motion, we can never subdivide its action (the product of p and q) below this unit of action, h.

19.38 The Quantum Mechanics

By insisting that a physical theory of an atom deal only with observable quantities, Heisenberg developed a new kind of mechanics, the **matrix mechanics**, which, although mathematically different from Schroedinger's wave mechanics, is able to give the energy levels of an atom as accurately as the Schroedinger wave theory does. For some time these two theories were developed independently of each other, but then it was shown by the British physicist Dirac that both the Schroedinger wave mechanics and the Heisenberg matrix mechanics are different aspects of a more general mechanics which has since been called the **quantum mechanics**.

Although the Bohr orbits cannot have precise meaning in quantum mechanics, since, as we have seen, the electron cannot be assigned to a precise orbit, they do have an important significance. It can be shown, by using the solutions (ψ functions) of the Schroedinger equation for electrons in an atom, that the Bohr orbits are those regions in the atom where the electrons have the greatest probability of being found. Thus, although it is no longer permissible to speak of the Bohr orbits as precise paths for electrons inside an atom, it is still permissible to use these orbits to describe what is taking place since they present *the most probable positions* for the electrons.

19.39 The Incompleteness of the Schroedinger Wave Equation

As we have noted, the Schroedinger wave function, ψ, contains the three quantum numbers n, l, m as a mathematical consequence of the theory, so that these no longer have to be introduced separately as they were in the Bohr model of the atom. However, the spin quantum number does not appear, and since we know that this quantum number must be present to complete the description of the behavior of the electron, we know that there is still something missing from the Schroedinger equation.

It was at once obvious to most physicists that the reason for the incompleteness is that the Schroedinger equation does not conform to the special theory of relativity. As we noted in Section 10.6, the laws of nature must be so formulated that they have the same mathematical structure if we pass over from one observer (one coordinate system) to another moving with uniform speed with respect to the first. This is not true of the Schroedinger equation. In other words, the Schroedinger equation is not invariant to a Lorentz transformation.

19.40 Dirac's Relativistic Wave Equation

Although physicists knew what was lacking, they were unable immediately to see how to correct the Schroedinger equation and to bring it into conformity with the special theory of relativity. But this was finally achieved by Paul Adrien Maurice Dirac, who discovered how to set up a proper relativistic wave equation for the electron. This led to one of the most remarkable developments in physics—one with important applications in astronomy as well. When Dirac correctly merged the wave mechanics with Einstein's special theory, he obtained a much more complicated wave equation for the electron than Schroedinger's. In fact, he discovered that in order to describe the electron properly, four different wave functions obeying four different interrelated wave equations are needed.

At first this result appeared to give more than was required of the theory, for it seemed that in order to describe a spinning particle such as an electron having only two possible states of spin (parallel and antiparallel to a magnetic field) only two different wave functions obeying two interrelated wave equations would be needed. This appeared all the more logical since the four wave functions of Dirac break up quite naturally into two groups of two. The two wave functions in one of these groups describe the two possible states of spin of an ordinary electron and lead to the same spin magnetic properties of the electron that Goudsmit and Uhlenbeck had proposed. The two other wave functions, however, describe the two states of spin of an electron with negative energy. Thus, although Dirac had succeeded in deriving the spin quantum number of the electron from the special theory of relativity, it appeared that his wave equations gave more than was required—or desired.

The reason that Dirac obtained four wave equations can be understood if we realize that in the special theory of relativity the total energy of a freely moving particle is obtained from the relationship

$$E^2 = c^2p^2 + m_0^2c^4,$$

where p is the momentum of the particle, m_0 is its rest mass, and c is the speed of light (see Section 10.20). If we solve this equation for the energy, we must take the square root of the right-hand side, and we know that we then obtain both a positive and negative root. In other words, treated merely as a mathematical expression, the energy in the special theory of relativity may be either positive or negative.

We know that negative energies for a freely moving particle (a particle not moving in a gravitational or an electromagnetic field) have no physical meaning in classical physics. Hence these negative roots of the energy equation are simply discarded in classical physics where the wave properties of electrons are neglected and where electrons are treated as particles. The double set of wave equations that Dirac obtained is merely a recognition of the existence of these positive and negative sets of energy values which appear in the special theory of relativity.

19.41 The Positron and Antimatter

However, it soon became clear that whereas no harm is done in ordinary relativity theory (i.e., when wave properties are neglected) by discarding the negative energy values, one is led into serious error in trying to do this with the Dirac wave equations. In other words the complete set of four equations must be retained in the Dirac theory even though only two of them describe electrons of positive energy and the other two describe electrons with negative energy. This means that the Dirac theory requires the existence of electrons in states of negative energy. The reason electrons must be placed in these negative-energy states is the following: If the negative-energy states were empty, all the ordinary electrons in our world (all the positive energy electrons) would jump down into these empty negative-energy states in one catastrophic explosion and a vast release of energy. However, if each state of negative energy has one electron in it, no other electron can fall into this state because the Pauli exclusion principle limits each state to only one electron.

At first this appeared to be a severe drawback to the theory for it seemed to introduce into physics concepts that had no physical meaning or reality. However, as Dirac pointed out, since such negative-energy electrons can never be detected anyway, and hence can in no way influence observed physical phenomena, retaining them in the theory would lead to no real conflict with experimental results, and therefore cannot validly be objected to.

But it soon became clear that these negative-energy electrons are more than a convenient mathematical fiction, for it was shown that although an electron of this sort cannot be observed as long as it remains in a state of negative energy, it can (by absorbing a sufficient amount of positive energy in the form of a photon of energy equal at least to $2mc^2$) be thrown into a state of positive energy and hence be made observable. If this happens, two visible particles each of mass m are automatically created. One of these is the electron itself, which is now visible because of its positive energy, and the other is the absence of the electron (a hole) from the invisible sea of negative-energy electrons all around us. (See Figure 19.18.)

This hole, being the absence of negative energy and of negative charge, behaves just like an ordinary particle of positive charge. The so-called ***Dirac-holes*** *created in this way are called* **positrons** and were first detected in 1934 in cosmic rays. Today they are created daily in laboratories throughout the world. When a positron meets an ordinary electron the two particles completely annihilate each other with a burst of energy (two gamma rays) equal at least to $2mc^2$ being emitted. For this reason the positron is referred to as the **anti-electron**.

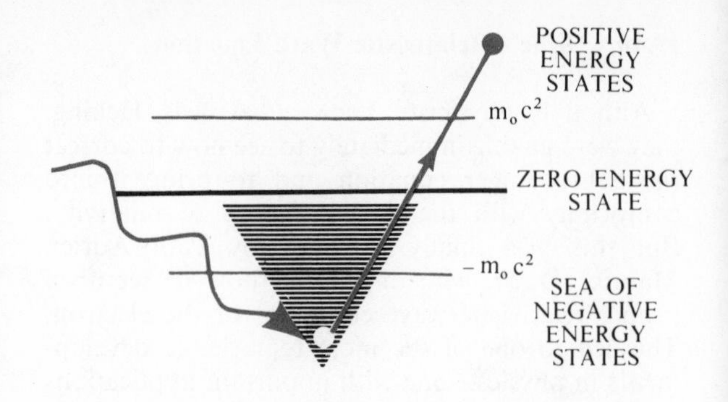

FIG. 19–18 *If a photon having enough energy is absorbed by the negative-energy electron, a hole in a sea of negative-energy states and a positive-energy electron (both becoming visible) are created.*

More recently **anti-protons** and **anti-neutrons** have also been discovered; thus there is every reason to suppose that there are regions in our universe containing **anti-matter**, which has the same physical and chemical properties as ordinary matter, but which would annihilate ordinary matter on coming in contact with it.

EXERCISES

1. (a) Let N_1 be the number of excited hydrogen atoms with their electrons in the L orbit (the second Bohr orbit) and N_0 the number of unexcited hydrogen atoms. Use the Boltzmann excitation formula to compare the ratio N_1/N_0 for the solar atmosphere and for Sirius' atmosphere. Do the same thing for (b) Rigel and (c) the sun. Use the following atmospheric temperatures for the sun, Sirius, and Rigel, respectively: 5 700°K, 10 000°K, 15 000°K.

2. Use the Saha ionization formula to compare the number of ionized hydrogen atoms relative to neutral hydrogen atoms in (a) the sun's atmosphere, (b) Sirius' atmosphere, and (c) Rigel's atmosphere. Use the temperatures of problem 1 for these stars and take for N_e (the number of electrons per cc.) the following values respectively: 4×10^{13}, 2.8×10^{24}, 3×10^{15}.

3. If an electron were contained in a box 10^{-8} cm on a side, what would the uncertainty in its momentum be? How fast would it be moving about in the box and what kinetic energy would it have because of the uncertainty in its momentum?

4. If an electron and a proton are moving with the same speed, how does the wavelength of the electron compare with that of the proton?

5. List all the subsublevels (that is, give their quantum numbers) for the case of the principal quantum number $n = 3$ if the hydrogen atom is in a magnetic field.

6. In the text some causes are given for the broadening of spectral lines. Explain these in more detail.

20

THE ATMOSPHERE OF THE SUN*

20.1 The Absorption or Fraunhofer Lines in the Solar Spectrum

We can illustrate many of the atmospheric characteristics and spectral features of stars which we have discussed in a general way (see Chapter 19) by considering the sun and its atmosphere in some detail. An analysis of the light from the sun shows an absorption spectrum—a continuous background crossed by thousands of dark absorption lines (they are called **Fraunhofer lines**, in honor of the German physicist who first plotted more than 700 of them). The great variations in the intensities and in the widths of the sun's absorption lines are a source of considerable information about the structure and physical conditions in the solar atmosphere; they indicate that constructing a model of this atmosphere is a very complex problem.

Although at the present time more than 25 000 solar absorption lines have been identified with the lines of known atoms here on earth, there are still thousands of them in the ultraviolet and infrared regions that are of unknown origin. *One of the difficulties in identifying lines in the ultraviolet region (lines of wavelength less than 2 500 Å) arises because the ozone in the upper layers of the earth's atmosphere is a strong absorber of ultraviolet radiation.*

In addition to the visible lines, there are also present molecular bands arising from hydrocarbons, such as CH, and from cyanogen, CN. Even though the temperature in the solar atmosphere is too high for most compounds to exist, these and a few others are not completely dissociated.

Recent work with high-altitude rockets equipped with instruments to operate above the ozone layer shows that the continuous spectrum becomes much weaker below 1 700 Å, and that the absorption lines disappear almost completely. For wavelengths down to about 500 Å there is a rather weak continuous spectrum upon which are superimposed many strong, bright lines which arise in the solar atmosphere. In addition to ultraviolet and infrared radiation, the sun's spectrum contains radio waves, to which radio astronomers are now paying a great deal of attention.

20.2 The Solar Photosphere: Limb Darkening Effect

If we consider the continuous radiation in the solar spectrum and compare it with that in an absorption line, we see that the former comes to us from some layer below the atmosphere without interference from any atoms. *We may therefore say that when we look at the sun by means of radiation in the continuous part of the spectrum we are really peering down into this lower region.* This is not the case when we look at the sun by means of the radiation in an absorption line. In other words, the atmosphere is not equally transparent to all frequencies of radiation. Hence the depth to which we can peer into the atmosphere depends upon the wavelength of the light coming to us. Since the appearance of the sun and hence of its surface, as we see it through a dark filter, is determined primarily by the radiation of the continuous spectrum, we say that the continuous spectrum originates from this surface. We call this surface the **photosphere** of the sun. The gaseous material lying above this is the solar atmosphere, which consists of two tenuous layers, the **chromosphere** and the **corona**.

Although we have assumed in our discussion that the continuous radiation from stellar surfaces is black-body radiation, a careful study of the photosphere of the sun shows us that this is not so. In Figure 20.1 we can see that as we move across the photosphere from its center to its edge, the brightness falls off perceptibly. This is called the **limb darkening effect**. *This means that the sun*

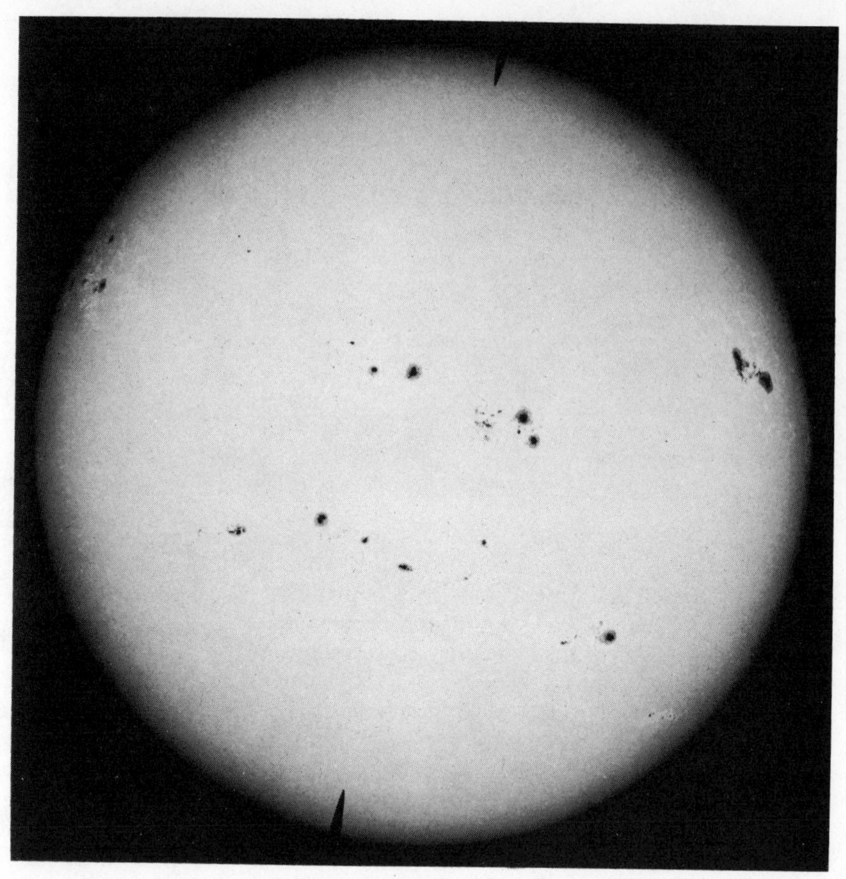

Fig. 20–1 *Photograph of sun near sunspot maximum showing 23 sunspot groups. Taken July 13, 1937, this photograph shows the limb darkening of the sun as we go from the center of the sun's disk to its edge. (From Mount Wilson and Palomar Observatories)*

does not radiate like a black body—for if it did, it would be uniformly bright over its entire surface.

The reason for the limb darkening is that the radiation coming to us from the center of the sun's disk originates in a deeper and therefore hotter layer below the atmosphere than the radiation coming to us from the edge of the sun. This is made clear in Figure 20.2, where we have three different rays, the ray C coming from the center of the disk and the rays E_1 and E_2 from the upper and lower edges.

Now let us suppose that because of the absorption of the sun's atmosphere, AB is the maximum depth to which we can peer into the sun. It is clear from this that the center ray comes to us from a layer at which the temperature is T_A and whose distance from the center is OA. If we consider the edge ray E_1, it, too, must originate from a point A' in the interior of the sun which is at the same depth AB (as measured along the ray from the top of the atmosphere) as the point A is. Therefore, as we can see from the figure, the point A' is farther from the center than the point A, so that the temperature $T_{A'}$ associated with this layer of atmosphere is less than T_A. Thus, the radiation along the rays E_1 and E_2 appears less intense than the radiation along A. Note that a ray starting from point P is absorbed by the solar atmosphere and hence does not reach the earth. Although this accounts for most of the limb darkening, there is a residual effect which cannot be accounted for in this way and involves complicated spectral theory.

20.3 The Application of the Limb Darkening Effect

By analyzing the solar limb darkening, first using the simple explanation given above, it is possible

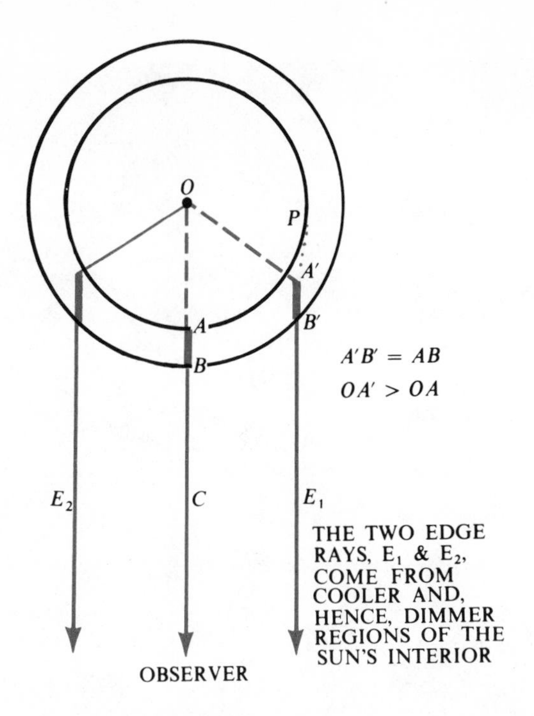

FIG. 20–2 *The limb darkening effect occurs because when we look at the rim of the sun, the absorption in the sun's atmosphere prevents us from seeing as deeply into the sun as we can when we look at its center.*

to get an idea of how the temperature in the atmosphere varies as we go from the upper layers down to the photosphere. For it is clear that as we move from the center of the disk out to the limb, we are looking at atmospheric layers that are closer to the top of the atmosphere, and hence cooler, as shown in Figure 20.2. An analysis of this sort shows that as we move from a height of about 400 km above the photosphere down to a depth of about 400 km below the photosphere, the temperature rises from some 4 200° to about 8 000°K.

An interesting feature about the limb darkening is that it is not equally pronounced over the entire continuous spectrum. The effect is smaller in the red part of the spectrum than it is in the blue part. The reason is that the atmosphere is much less opaque to red light than it is to blue light so that the red light from any part of the disk comes to us from points that are all at nearly the same distance from the center of the sun whereas the blue does not. From our knowledge of atomic structure we can see why the solar atmosphere is more opaque to blue light than to red. We know that most of the solar atmosphere consists of hydrogen and that only very energetic blue photons can excite hydrogen atoms. In other words, *since the blue photons can interact with hydrogen atoms whereas the red cannot, the former are absorbed more readily.*

The limb darkening, as we have noted, arises because the radiation coming from the photosphere suffers absorption in its contacts with the various atoms in the atmosphere. In other words, the atoms are, to some extent, opaque to this radiation, and this opaqueness, which is measured by the **coefficient of opacity**, is related to the limb darkening. Since we can calculate this coefficient from our knowledge of atomic theory and express it in terms of the density of the atmosphere at any point, we can, by measuring the limb darkening, determine the density at various depths of the atmosphere. Again, we find that as we go from the top of the atmosphere to regions a few hundred kilometers below the photosphere the density increases from about 10^{-7} gm/cm^3 to about five times this value. Since one may assume that the solar atmosphere behaves like a perfect gas, it is possible from our knowledge of the temperature and the density to determine the pressure at various points and hence to construct a model of the sun's atmosphere.

20.4 Solar Granules

Although a photograph of the sun taken in ordinary light seems at first sight to show the photosphere as a smooth surface, we find under good seeing conditions that the surface is really resolved into innumerable small bright granules (see Figure 20.3). These irregular white hexagonal structures, which look like rice grains and are in a constant state of turbulence, cover about 60 per cent of the sun's surface. The most intensive studies of these granules have been carried out by Dr. Martin Schwarzschild and his associates, who succeeded in obtaining photographs of the solar photosphere by means of unmanned balloons 80 000 feet above the earth's surface.

From a study of these photographs Schwarzschild obtained considerable information about the structure and behavior of the granules. The average diameter of a granule is of the order of 1 500 km, although some as small as 500 km are found. The granules are very transient structures which last on the average about 8.6 minutes, and then dissolve into the background of the photosphere. The average temperature in the granules is about 100°K higher than that of the surrounding photosphere, but there are fluctuations of the order of ±92°K. The most reasonable explanation for

FIG. 20–3 *Solar granulations: a highly magnified section of the sun's surface. (Photograph from Mount Wilson and Palomar Observatories)*

these structures is that they are turbulent cells of hot gases that bubble up from below the surface of the photosphere and penetrate into the solar atmosphere.

20.5 The Origin of the Granules: The Solar Convective Zone

Although no completely acceptable theory has been advanced to account for these granules, it is reasonable to suppose that they are the result of some turbulent process going on below the photosphere. We shall see later that there are good theoretical reasons for believing that whereas the atmospheric and photospheric layers on the sun are held in balance by a process that is called **radiative equilibrium** (i.e., a process in which most of the energy is transported out of the sun by means of radiation) the layers of material at a certain depth below the photosphere in the sun are kept in equilibrium by a convective process in which most of the energy is transported by an up-and-down motion of the solar material. This region, which extends for thousands of miles below the photosphere, is called the sun's **hydrogen convective zone**, and probably this underlying turbulent motion causes a continual formation and destruction of the granules on the photosphere somewhat like the bubbling in a pot of boiling water.

20.6 Spectroheliograms of the Photosphere

The general appearance of the sun changes as we examine its radiation in different parts of the spectrum, and takes on an especially interesting appearance if it is photographed with the light emitted in one particular absorption line. Such photographs are called **spectroheliograms**, and today we can obtain them in almost any part of the spectrum by means of special filters with very narrow transmission bands. Another method is to use a **slit spectrograph** in which only the light from one part of the solar disk is allowed to enter. If a second slit is placed behind the first one (where the spectrum is formed to admit only the light in one absorption line) and if the sun's disk is allowed to trail across the first slit while a photographic plate behind the second slit follows the sun at the same rate, a complete photograph of the sun's disk in this particular wavelength is obtained. Such photographs in the Balmer H_α and the ionized calcium line are shown in Figure 20.4(a), (b), and (c).

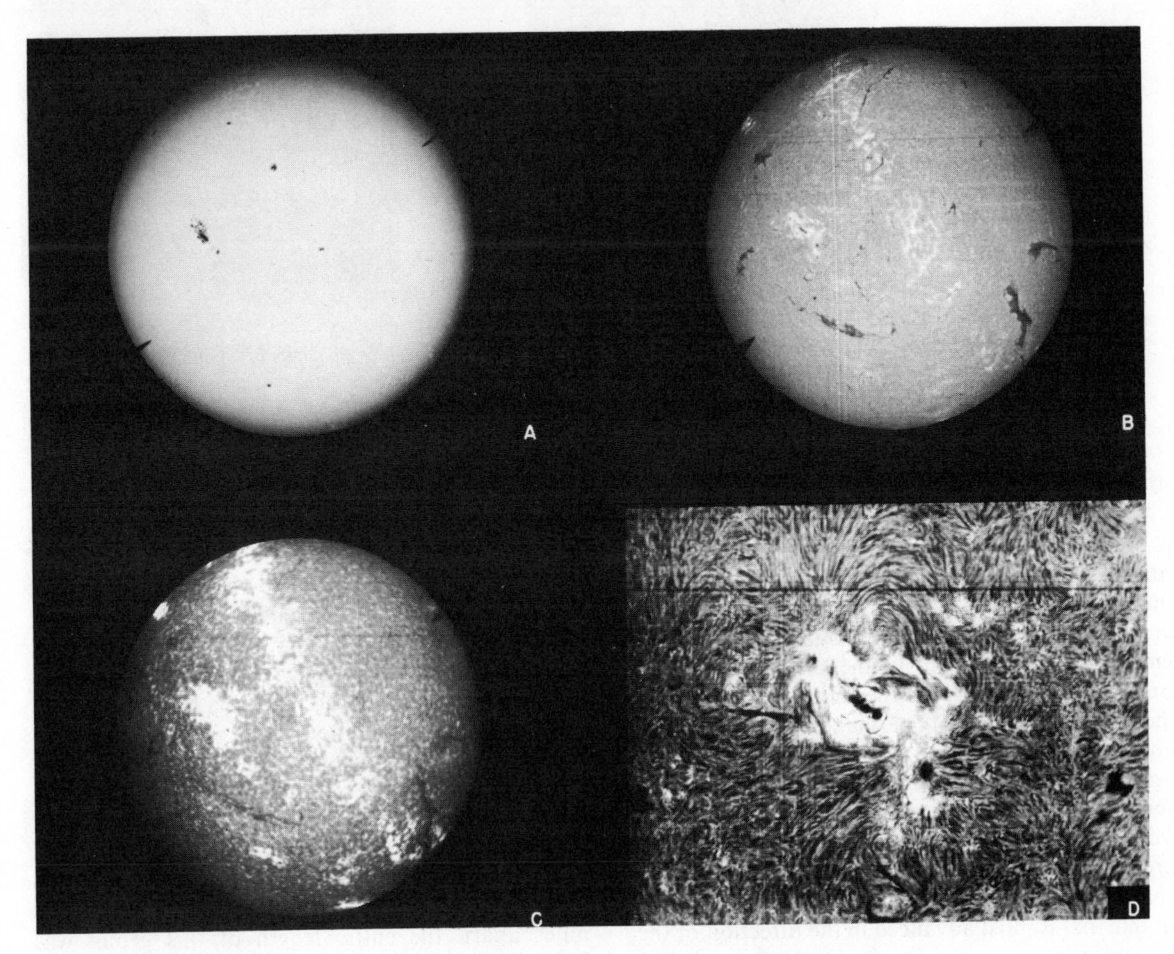

FIG. 20–4 *(a) Two photographs of the sun, one in ordinary light, the other in red hydrogen light (Hα), taken August 12, 1917. (b) Sun: (A) ordinary photograph, (B) hydrogen (Hα), (C) calcium spectroheliogram, (D) enlarged hydrogen spectroheliogram. September 15, 1949. (From Mount Wilson and Palomar Observatories)*

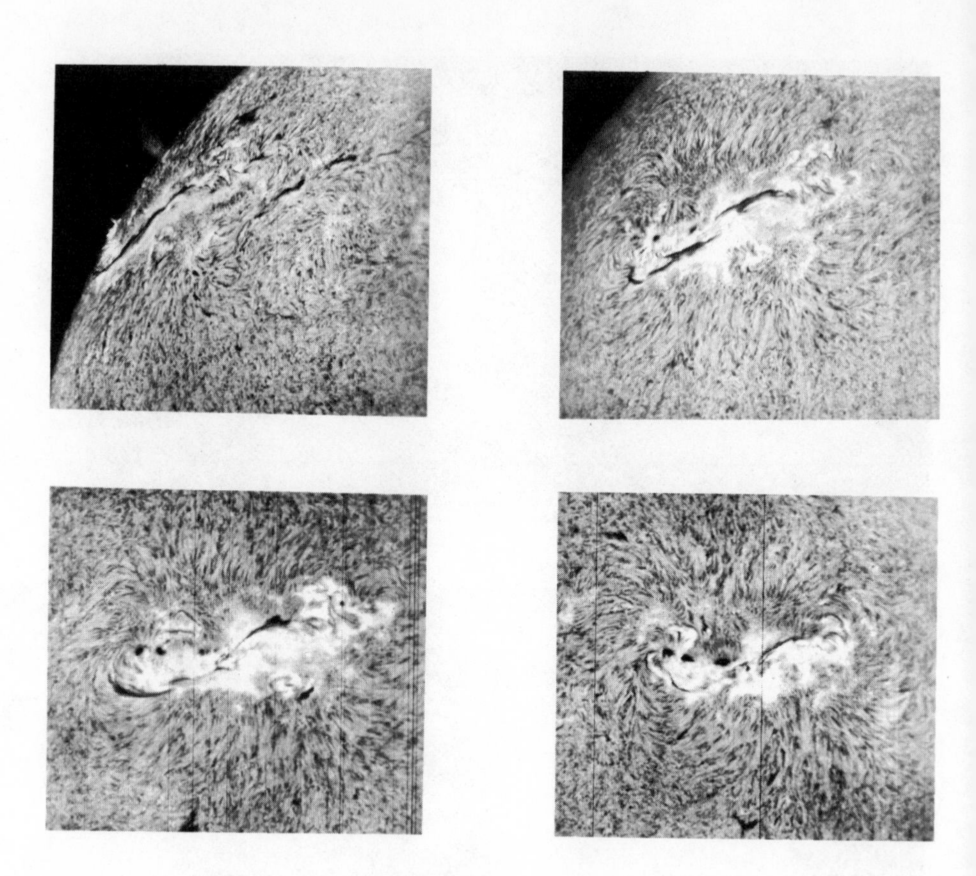

FIG. 20–4 (*c*) *Series of photographs of a section of the sun taken in red hydrogen light* ($H\alpha$). *Taken August 3, 5, 7, and 9, 1915.* (*From Mount Wilson and Palomar Observatories*)

20.7 Sunspots

An important variable feature of the photosphere, covering a relatively small area (compared with the total surface area of the sun), is the **sunspot**, which consists of a dark central area or **umbra**, surrounded by a somewhat brighter region, the **penumbra**, from which filaments originate and converge radially towards the center of the umbra. The sunspots are not permanent features, but grow and disappear (in a definite pattern). When a group of sunspots first appears, the individual members are small dark regions or pores (scattered over 3° or 4° in solar longitude) about 1 000 miles across which then grow quite rapidly and begin to separate. In such a group, the leading spot (i.e., the one that is most advanced in the direction of the sun's rotation) is generally the larger one. When the spots have reached their maximum size, the leading spot, which takes 8 or 9 days to grow, is generally about 10° in advance of the following spot which reaches its largest size in about 3 or 4 days.

After reaching their maximum size, the spots begin to decay rather slowly, generally disappearing in a few weeks, but sometimes lasting a number of months. In general, the following spot disappears first by breaking up into smaller spots which diminish in size and then vanish. The advancing spot remains by itself, but grows smaller and smaller, finally disappearing entirely after a few weeks.

Although average spots have umbras 20 000 to 30 000 miles in diameter, some unusually large spots occasionally develop such as those observed from January to May 1946. The leading spot was 62 000 miles long and 35 000 miles wide, and the following spot was 90 000 miles long and 60 000 miles wide. Since the two spots were about 40 000 miles apart, the entire length of this group was 200 000 miles, and it had a maximum area of 5 700 million square miles. This group of spots lasted more than 99 days and survived four rotations of the sun. An exceptionally large spot is shown in Figure 20.5.

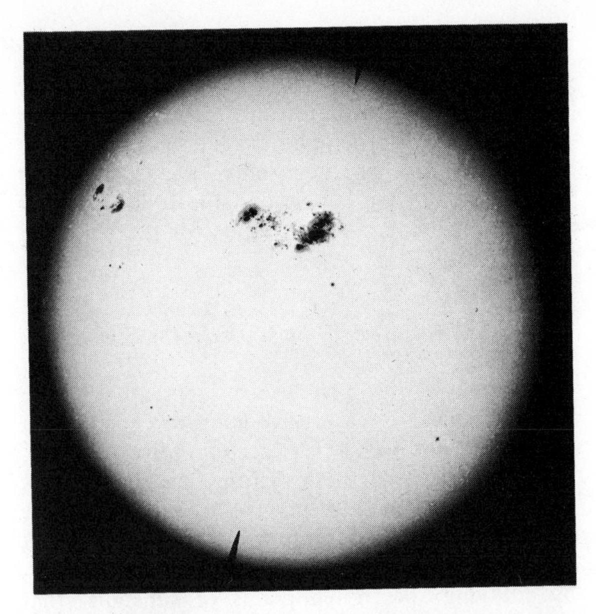

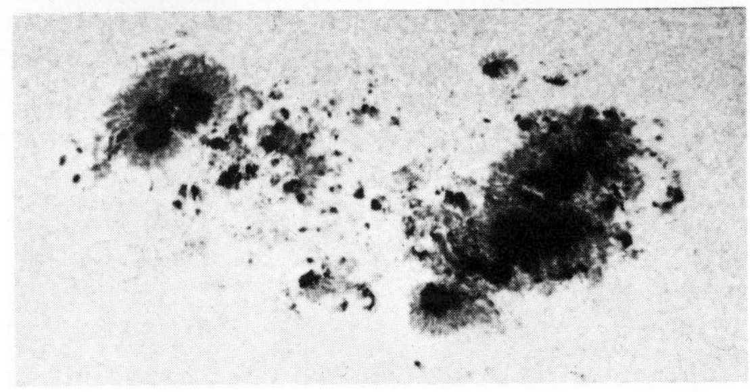

FIG. 20–5 *Whole solar disk and an enlargement of the great spot group of April 7, 1947. (From Mount Wilson and Palomar Observatories)*

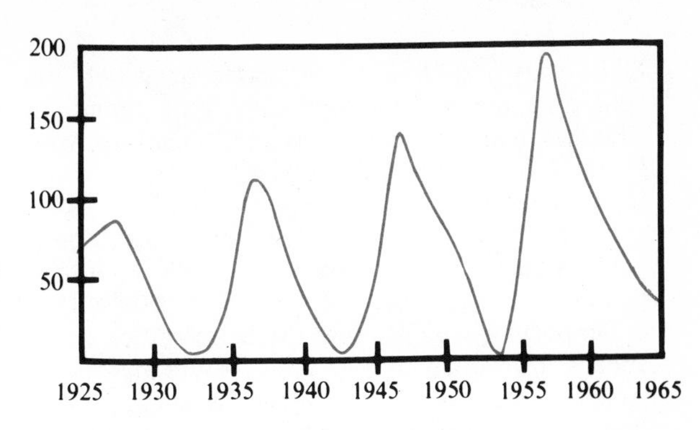

FIG. 20–6 *The frequency of sunspots varies from year to year as shown by the graph, where the number of spots is plotted against the year.*

20.8 The Distribution of Sunspots on the Photosphere and Their Periodicity

Sunspots, which occur both in groups and as scattered individuals, are not distributed over the entire surface of the sun, but are generally found between heliographic latitudes of $\pm 5°$ and $\pm 30°$. They are rarely found at the solar equator or at latitudes greater than $\pm 45°$, and they occur with equal frequency above and below the equator. The groups generally start in the upper latitudes, but the later groups are generated at lower and lower latitudes with the number of spots reaching a maximum when the group is formed at a latitude of about 15°. When the spot group is formed near the equator, sunspot activity is at a minimum, but before this group fades away, a new group, marking a new cycle, begins to form in the upper latitudes.

One of the most remarkable features of the sunspot groups is their cyclical growth and decay with their alternating periods of minimum and maximum activity. In Figure 20.6 the number of sunspot groups observed annually is plotted against the years. It can be seen that the value of the maximum in each cycle has increased almost steadily from 1925 to the present time. From the figure we see that the average interval between times of maximum activity in a group is about 11.1 years, although this may vary from time to time by as much as a year or more. In general, although we may expect spots to occur every year, occasionally there will be years (such as 1913) in which very few or no spots at all occur.

There appears to be a fairly constant ratio between the numbers of individual spots not belonging to groups and the total number of spots during any one cycle. This ratio varies from about 0.7 at sunspot minimum to 0.87 at sunspot maximum. The ratio of the total number of groups to the total spot number varies from 0.097 at minimum activity to about 0.083 at maximum.

20.9 The Temperature of Sunspots

Even though the spots appear black to us, we know that they emit light and only appear dark by comparison with the photospheric background. The reason is that the temperature of the spot itself is lower than the temperature of the surrounding photosphere by more than a thousand degrees. The temperature of the spot can be compared with that of the photosphere by applying the laws of black-body radiation to the rates at which the spot and the surrounding area emit radiation. This has been done by Pettit and Nicholson, who found a flux ratio of 0.47, which means that the temperature of the spot is about 4 700°K. Actually the temperature in the spot varies from about 4 300°K in the umbra to about 5 500°K in the penumbra.

Supporting evidence that the temperature of the spot is lower than that of the surrounding photosphere comes from their respective spectra. We find that the lines of ionized atoms are less intense in the sunspot spectrum than they are in the spectrum of the surrounding photosphere, whereas just the opposite is true for bands of molecules such as titanium oxide (TiO). See Figure 20.7.

20.10 Magnetic Fields of Sunspots

By studying the Zeeman pattern (see Section 19.26) of the lines in the spectrum of a spot we can learn a good deal about the magnetic fields associated with sunspots. This work was undertaken in 1908 by G. E. Hale, who showed that the magnetic intensity in a spot varies with the spot's size, ranging from 100 gauss (a gauss is a unit of magnetic-field intensity) in a small spot to about 4 000 gauss in a large spot. Thus, these sunspots behave like magnetic poles. Since we know that magnetic poles always have opposite polarity (a north pole is always found together with a south pole) we know that sunspots that occur in pairs should also have opposite magnetic polarity. This is found to be the case.

What is remarkable, however, about the polarity of the sunspot pairs, is that if the leading spot of a pair in one of the hemispheres (say the northern hemisphere) has a given polarity (let us say it behaves like a north pole), the leading spots of all pairs in the given hemisphere have this same polarity during that sunspot cycle and just the opposite polarity holds for all the pairs in the other solar hemisphere during that cycle. However, when a new cycle begins, the polarities are all reversed, so that in the northern hemisphere the leading spot behaves like a south pole and the following spot like a north pole and vice versa for the southern hemisphere.

20.11 The Motions of the Gases in the Sunspots

Various theories have been presented to explain these remarkable features of sunspot activity, but as yet none has been able to account for all of them. However, most of these theories are based upon the fact that there seems to be a considerable amount of vortical motion within the sunspot itself. This was first observed by Hale, who found, by studying spectroheliograms of sunspots, that gaseous vortical filaments apparently flow into the spots. Later, through similar studies, Evershed showed that material is also flowing out of the spots at about 2 km/sec tangent to the solar surface.

If we suppose that the vortical structure of the spot indicates a vortex of ionized gases swirling around just as the air in a tornado does here on earth, we can account for the magnetic features of the spot in terms of such motion. According to one such theory developed by Bjerknes and Alfvén, most of the vortical motion occurs below the photosphere and connects the advancing spot in the pair with the following spot, forming a horseshoe structure. The magnetic field is thus accounted for by the electric currents within this horseshoe structure generated by the rapidly spiraling streams of ionized gases.

A detailed and complex theory has recently been given by H. W. Babcock. It postulates that the sunspots arise from a combination of the over-all

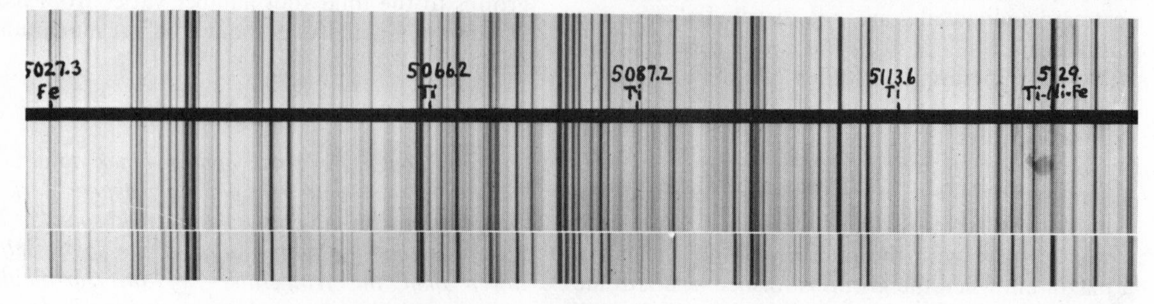

FIG. 20–7 *The spectrum of a sunspot with the spectrum of the photosphere for comparison. The difference in the indicated lines can be seen. In addition, the spot spectrum (center) shows a greater multiplicity of lines. Taken with horizontal spectrograph in Snow Building by Fox. (Yerkes Observatory photograph)*

weak solar magnetic field and the differential rotation of the sun as one moves from the equator to the poles. Babcock pictures the magnetic lines of force of the over-all magnetic field of the sun as emanating from the sun's north magnetic pole, where they have a mean intensity of 1 or 2 gauss, and re-entering at the south magnetic pole. The outer portions of these lines of force extend out to great distances but the submerged parts of each are just beneath the surface of the sun, so that these submerged lines form a kind of thin magnetic shell just below the photosphere.

According to Babcock the differential rotation of the sun twists these lines and draws them out in longitude until each line is wound around the sun about five times. This greatly amplifies (by a factor of about 45) the magnetic field right below the surface at a latitude of about $\pm 55°$. Ultimately, this concentrated magnetic energy bursts out of the photosphere to form a pair of spots. With this theory Babcock has succeeded in explaining many of the observed features of sunspot phenomena.

20.12 Faculae

One of the other variable features of interest on the photosphere of the sun is the appearance of **faculae**, which are much larger in surface area than the sunspots, and which generally occur in the regions associated with the sunspots before the latter make their appearance. They remain visible after the spots disappear, and have an average life of about 15 days. Although these regions are brighter than the surrounding photosphere, the difference is too slight to enable one to see faculae except when they are near the limb of the sun. The limb darkening effect provides a favorable contrasting background. The faculae undergo rapid changes in appearance in comparison with variations of the spots, but in over-all development they appear to follow the changing sunspot cycle of 11 years.

20.13 The Chromosphere

The part of the sun's atmosphere called the **chromosphere**, which is primarily responsible for the formation of the absorption lines, consists of gases that appear to be distinct from the photosphere, although there can be no doubt that the gases from the photosphere and the chromosphere merge. The most interesting feature of the chromosphere is the intricate pattern of numerous small prominences extending out from the photosphere and ranging in lengths from about 1 000 to 3 000 miles. The chromosphere was first seen during a total eclipse of the sun, and for many years such eclipses provided the only opportunities for observing it. Today, however, we can photograph the chromosphere with a telescope in which the bright disk of the sun is obscured by a diaphragm. Such a telescope, called a **coronagraph**, was first developed by the French astronomer Lyot, in 1930, and is now in common use in observatories all over the world. It is now possible to use the coronagraph with a camera that takes pictures continuously so that one can actually see moving pictures of the solar atmosphere.

If during an eclipse we watch the sun carefully, and observe it just at the moment of totality, the chromosphere is visible to us as an intensely red, narrow band around the sun. This is so because most of the energy comes to us in the form of the H_α line resulting from the de-excitation of great quantities of hydrogen atoms that are present in the solar atmosphere. Although the usual solar spectrum consists partly of dark lines, and these dark lines arise from the absorption by atoms in the chromosphere, the chromosphere itself yields a bright-line emission spectrum. We can obtain such a spectrum by photographing the chromosphere with a slitless spectrograph (a prism in front of the objective of the telescope) at the moment of totality of a solar eclipse.

20.14 The Flash Spectrum of the Chromosphere

We can obtain a spectrum of the chromosphere in this way because the chromosphere is such a thin band that it behaves essentially like a slit. When we obtain such a spectrum, each line itself is an image, in that particular wavelength, of the chromosphere. We then have what is called a **flash spectrum**. We obtain a bright-line spectrum because the very intense continuous radiation from the photosphere is not present, so that the energy emitted within each line can be detected. We must recall that the dark lines in the absorption spectrum are not really devoid of radiation but are dark by comparison with the bright background.

The bright lines in the flash spectrum are, for the most part, the same lines that we find in the absorption spectrum, but there are some conspicuous differences which are worth noting. Thus,

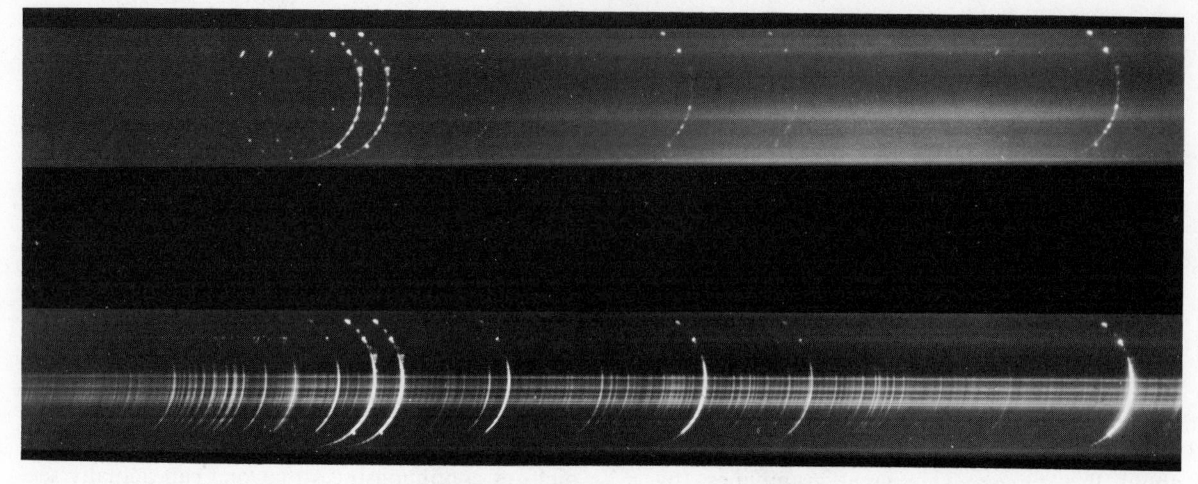

FIG. 20–8 *Sun, spectrum of the "flash" taken at the total eclipse of January 24, 1925, Middletown, Connecticut. The bright emission lines which are in the shape of arcs are images of the part of the sun's atmosphere above the moon's disk.* (*From Mount Wilson and Palomar Observatories*)

although the helium lines are missing in the absorption spectrum of the sun, in the flash spectrum they are very much enhanced. This is also true for the lines of ionized metals as well as for the lines of the neutral atoms which originate from high states of excitation. The lines in the flash spectrum are also different in detail from their absorption counterparts. Thus, whereas only four of the Balmer lines are visible in the latter, the entire Balmer series is present in the flash spectrum. Among the ionized atoms in the flash spectrum there are also present the two strong lines of singly ionized calcium.

20.15 The Distribution of Elements at Various Heights in the Solar Atmosphere

An examination of the flash spectrum (Figure 20.8) shows that it is possible, by analyzing the length of the lines, to see how high above the photosphere various elements are found. As we have noted, each line is an image of the chromosphere extending beyond the moon's disk. The closer the layer of atoms is to the disk of the moon (i.e., the closer these atoms are to the photosphere) the smaller is the arc of the chromosphere in which they lie, and hence the smaller is the image of that part of the photosphere formed by their spectral lines. This is shown in Figure 20.9, where it can be seen that atoms lying in layers closer to the photosphere cut out smaller arcs of the chromosphere.

Not only can we learn something about the distribution of the elements in the sun's atmosphere, but we can also discover something about the difference between the chromosphere and the photosphere. As we go higher and higher in the atmosphere, in other words, as we look at the longer spectral images in the flash spectrum, we find that these are the lines of the ionized metals and those elements requiring greater excitation energies. Thus, one of the lines of ionized helium is found at a height of 10 000 km above the photosphere. This

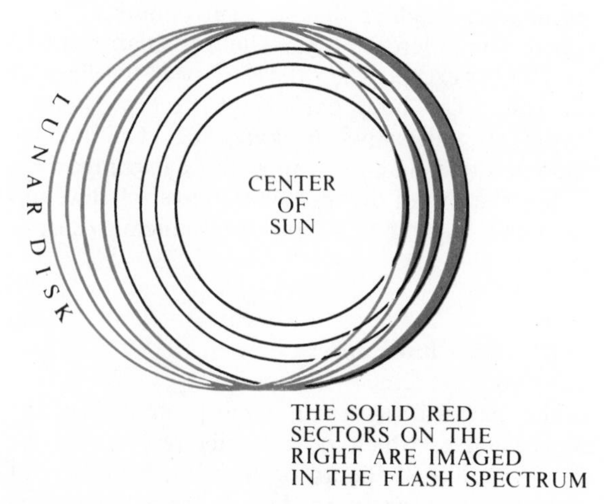

FIG. 20–9 *The higher a layer of atmosphere is above the sun's disk, the longer the strip cut out of it by the moon's disk.*

means that the temperature in the upper atmosphere of the sun must increase.

The heights at which different elements lie above the photosphere vary. We have found that ionized calcium extends up to a maximum height of 9 000 km and helium up to a maximum height of 7 500 km. Such metals as iron in both ionized and neutral states are found out to a distance of 2 500 km. The actual height to which the chromosphere rises is about 5 000 miles (8 000 km) but, as we have seen, it is not of uniform physical and chemical composition.

20.16 Variation of Temperature in the Solar Atmosphere

As we move away from the photosphere where the temperature is about 6 000°K (as given by the laws of black-body radiation) to the upper photosphere (just below the chromosphere) the temperature as determined by the absorption lines falls to about 4 600°K. From that point on, the temperature increases to values of about 8 000°K at a height of some 600 km above the photosphere. The actual kinetic temperature corresponding to the molecular motions may even be higher since we know, from recent spectra obtained with rockets at heights of a few hundred kilometers, that the ultraviolet hydrogen Lyman alpha line and other lines with wavelengths less than 1 900 Å are emitted from the chromosphere.

20.17 Flares

In our discussion in Chapter 15 of the relation of solar activity to terrestrial phenomena, we mentioned as one of the most important influences that of flares on the solar surfaces. As we noted there, these are associated with sunspot activity. When a flare occurs, it stands out in the chromosphere as a region of intense brightness (see Figure 20.10) which rises rapidly to a maximum and remains visible for an hour or so, although the actual maximum brightness lasts only for a few seconds. This phenomenon is only vaguely understood, and we know only that a flare represents a great concentration of solar energy resulting from some kind of imbalance within a sunspot group.

The flare that occurred on May 8, 1951, was ejected from the photosphere at 700 km/sec, and reached a maximum height of some 60 000 km. From such measurements, we know that in an intense flare the energy emitted in the optical part of the spectrum is of the order of 5×10^7 ergs/sec (somewhat more than a calorie per second) for each square centimeter of the flare. Thus, if a flare had an area of 1 cm^2 and lasted for a thousand seconds, it would emit 5×10^{10} ergs, or roughly 1 200 calories of visible radiation. If we consider the fact that certain flares may occupy an average of 1 per cent of the solar disk, the total energy emitted in a single flare may be as high as 10^{31} ergs.

Because of this remarkable emission of energy in a flare, the explanation of this phenomenon has presented considerable difficulty. In a recent theory Gold and Hoyle seek to account for solar flares by showing that the energy necessary to cause these flares can be stored in the neighborhood of a sunspot only by a particular class of magnetic fields whose lines of force have the general shape of twisted loops protruding out of the photosphere. If two such loops of opposite twist meet they destroy each other and the stored-up energy is released in a very short time. Gold and Hoyle therefore consider these twisted magnetic loops as a mechanism for the formation of flares.

20.18 Prominences

If the edge of the sun is viewed during the time of total eclipse, or if it is photographed with a coronagraph, an amazing complexity of red structures called **prominences** is observed shooting out of and returning to the solar disk. These are not distributed uniformly over the solar disk, but are concentrated in two zones, one associated with the neighborhood of sunspots, and the other in latitudes of 45°. Although most prominences change their shapes rapidly, there are some, so-called **quiescent prominences**, in which changes are relatively minor so that these prominences may last several days.

In general, prominences are flat, wispy, and cloudlike. They range in length from a few hundred to hundreds of thousands of miles. A maximum recorded height of over 1 050 000 miles for a prominence was observed June 4, 1946. Although, in general, the material in prominences rushes out at less than the speed of escape, speeds as high as 450 km/sec have been observed. In most cases the velocity range is from a few to a few hundred kilometers per second. The intricate structures of prominences, and the way the material rushes away from and returns to the solar disk, indicate that local magnetic fields play an important

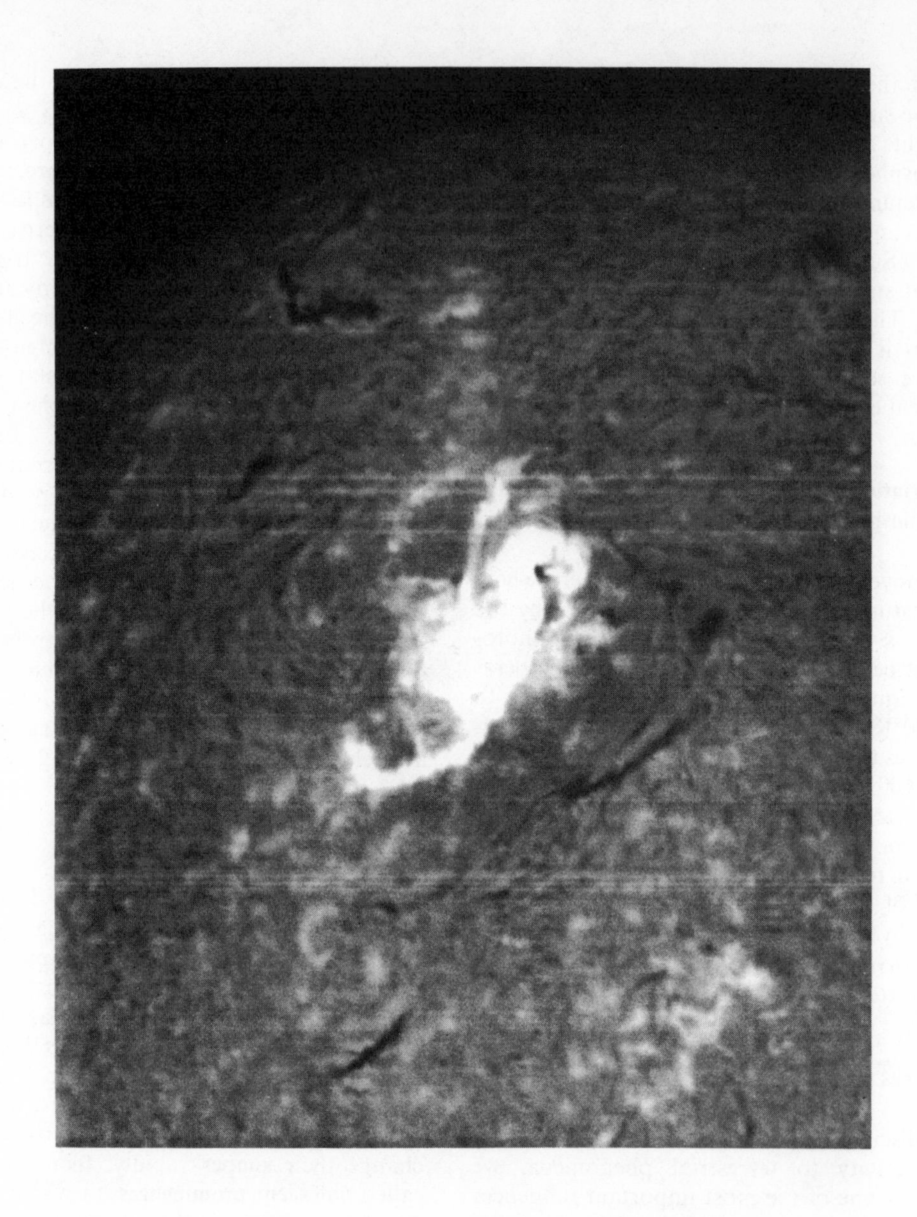

FIG. 20–10 *Solar flare photographed in red light of the hydrogen α line, July 16, 1959. (From Mount Wilson and Palomar Observatories)*

role in their behavior. See Figure 20.11(a), (b), (c), and (d).

20.19 The Over-All Magnetic Field of the Sun

We have already seen that the sun possesses local magnetic fields which are associated with sunspot activity, but for a long time it was questionable whether a polar magnetic field, such as we have on the earth, is present in the sun. Early in the century Hale had detected a magnetic field by means of careful analysis of the Zeeman pattern in the solar spectrum at sunspot minimum. He found a very weak field whose magnetic poles were within 6° of the poles of the sun's rotation. Later, however, much weaker fields were detected, and for a time it was thought that there is no general field at all. As a result of the very accurate magnetic scanning of the sun's surface by H. W. and H. D. Babcock, we now know that there is a

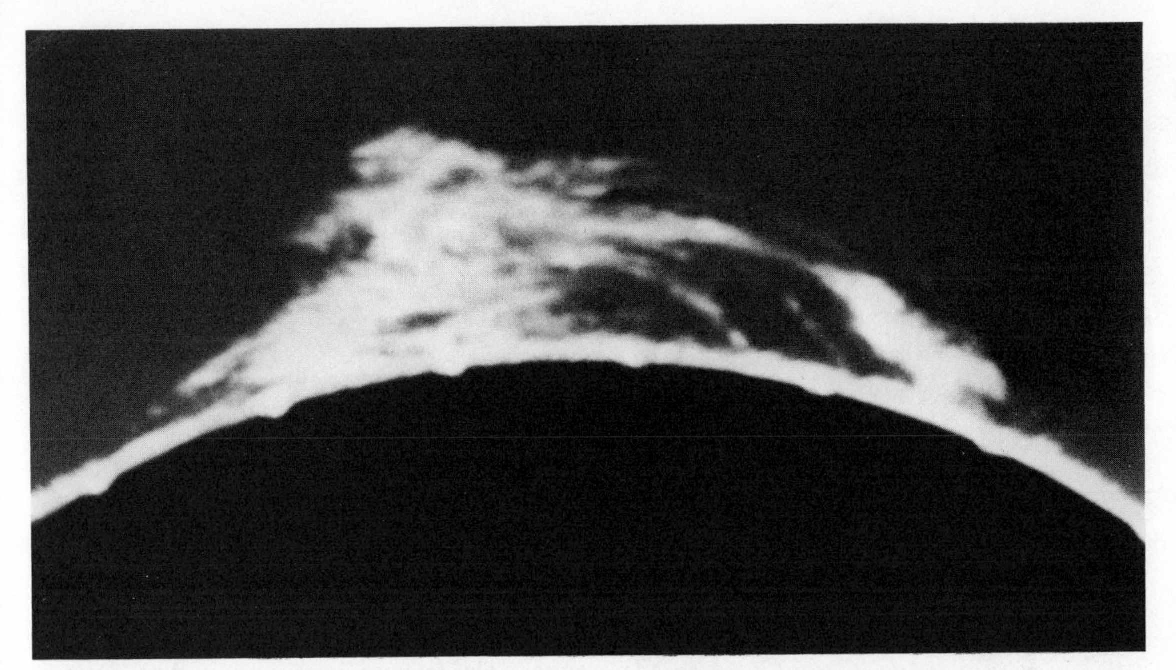

FIG. 20–11 *(a) Large prominence 132 000 miles high. Photographed in light of calcium, August 18, 1947. (From Mount Wilson and Palomar Observatories)*

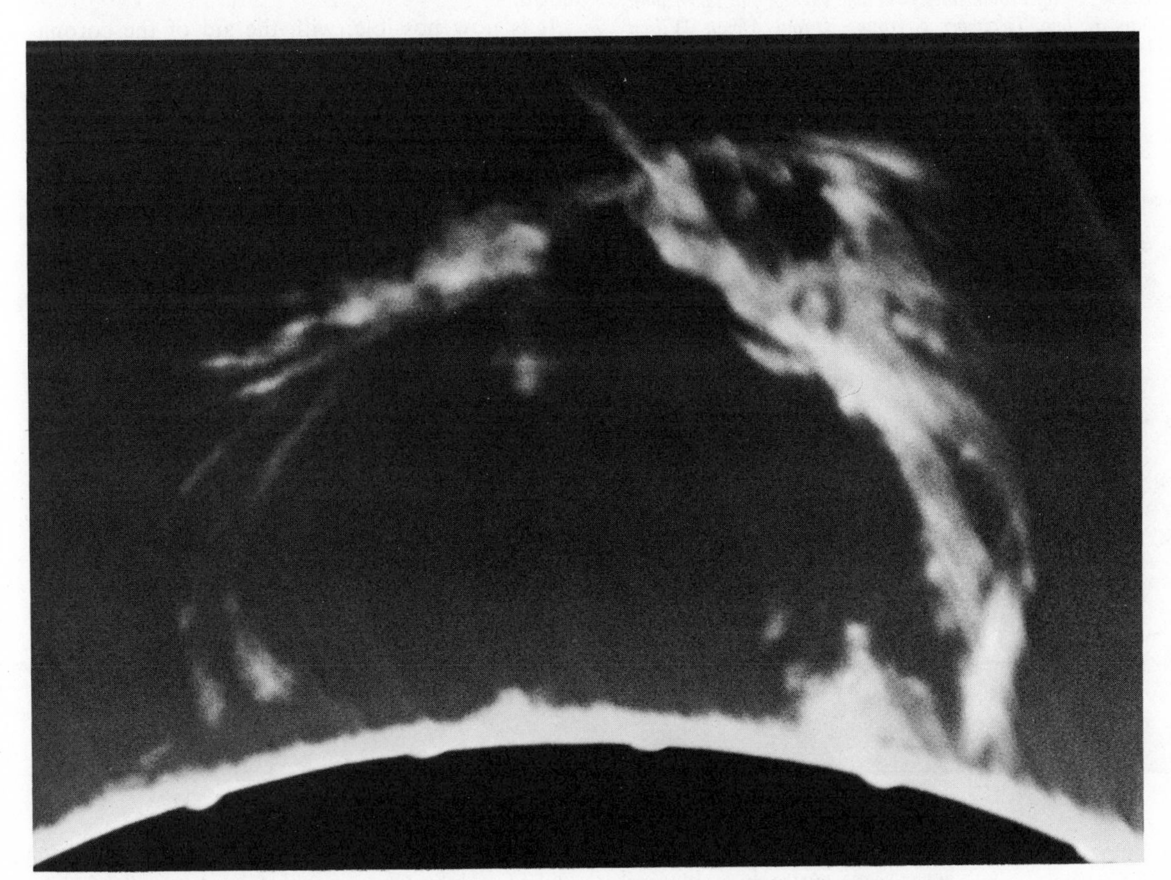

FIG. 20–11 *(b) Prominence 205 000 miles high (almost the distance from the earth to the moon), photographed in violet light of the calcium K line, July 2, 1957. (From Mount Wilson and Palomar Observatories)*

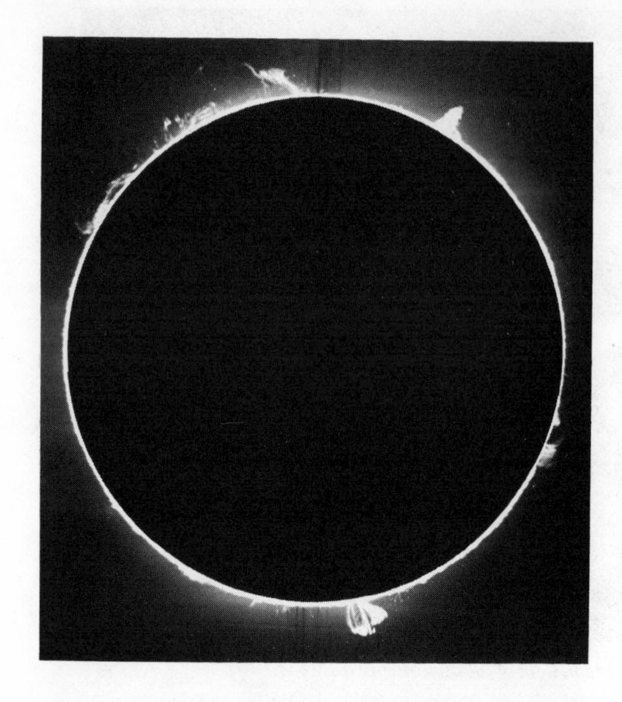

FIG. 20–11 (*c*) *Prominences. Whole edge of sun taken with calcium K line, December 9, 1929.* (*From Mount Wilson and Palomar Observatories*)

general magnetic field of the order of a few gauss having opposite polarity at the two poles. One of the amazing features of the sun's general magnetic field is that it appears capable of reversing its polarity, as indicated by the magnetographic observations made in 1959 by H. D. Babcock. A magnetogram of the sun is shown in Figure 20.12.

20.20 The Corona

At the moment of totality of a solar eclipse, we can see, surrounding the sun, a large luminous halo called the **corona** (see Figure 20.13), which varies in shape from a uniform spherical structure to an elongated one, depending upon sunspot activity. At sunspot maximum the corona has its most symmetrical structure, and at sunspot minimum it is quite asymmetrical, having a much greater extension in the equatorial direction with short spikes appearing near the north and south poles. This tenuous outer envelope emits a pearly white light and is about half as bright as the full moon.

It is now possible, with the aid of the coronagraph, to observe the inner or brighter parts of the

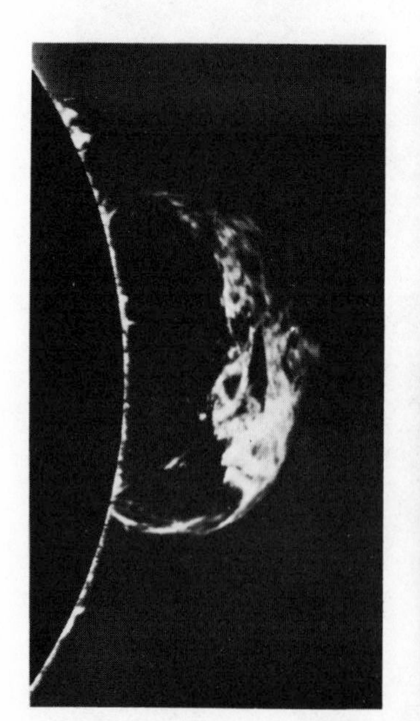

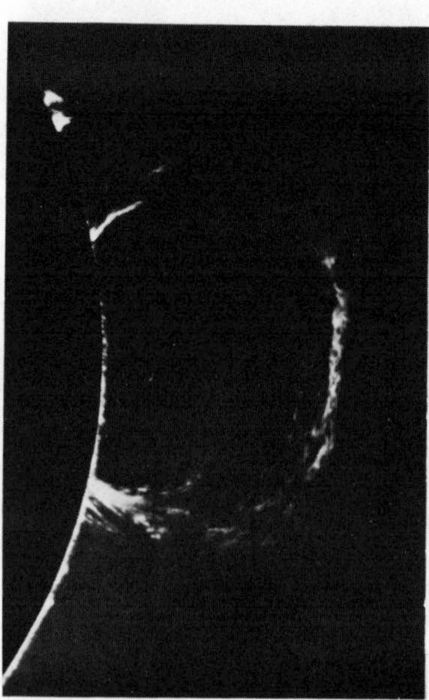

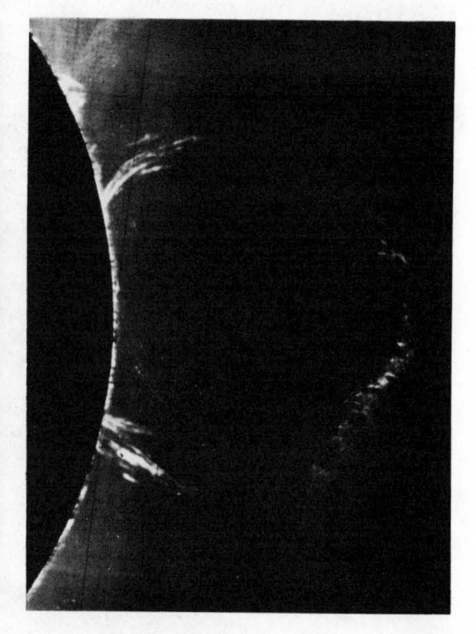

FIG. 20–11 (*d*) *Prominence of July 15, 1919. The three views show the increase in height of 200 000 km in one hour.* (Astrophysical Journal, *50, Pl VI.*) (*Yerkes Observatory photograph*)

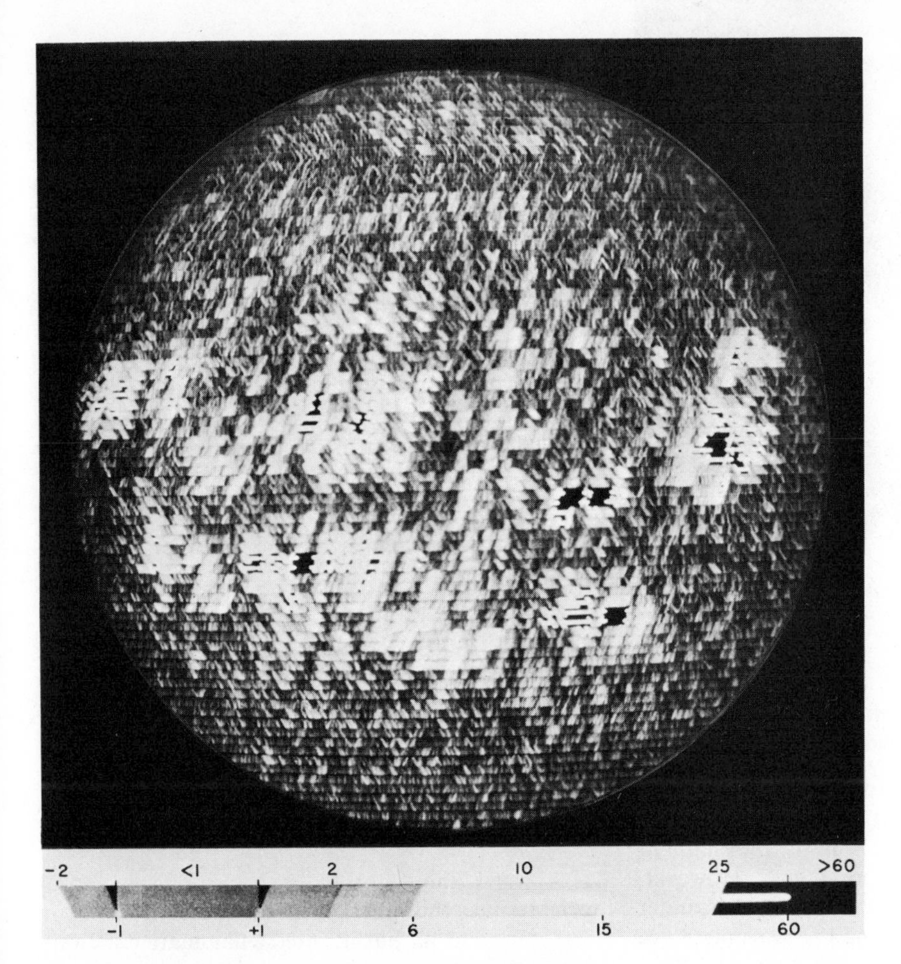

FIG. 20–12 *Magnetogram showing the intensity and polarity of weak magnetic fields in the atmosphere of the sun, July 21, 1961. This magnetic map of the sun's disk shows the location, field intensity, and polarity of weak magnetic fields. The records are made automatically by a scanning system that employs a polarizing analyzer, a powerful spectrograph, and a sensitive photo-electric detector for measuring the longitudinal component of the magnetic field by means of the Zeeman Effect. The calibration strip at the bottom of the picture shows how the recording line slants to right or left to indicate magnetic polarity, and how it changes in brightness and form to indicate seven different levels of magnetic field intensity in gauss. The extended magnetic areas on the solar disk are characteristically bipolar, and usually produce sunspots as well as other solar activity.* (*From Mount Wilson and Palomar Observatories*)

corona on any clear day. It is easy to see that the structure of the corona is greatly affected by the conditions of the underlying photosphere and chromosphere. The radiation in the corona comes to us in the form of a continuous spectrum with the same general characteristics as that of the sun. But there are changes as we move away from the solar disk out to a distance of about 300 000 miles. The spectrum of the inner corona is purely continuous, but emission lines appear in the middle part of the corona and remain visible throughout this portion. Finally, we find the absorption lines reappearing in the light from the corona's outer regions.

FIG. 20–13 *Solar corona photographed at the total eclipse of June 8, 1918. Green River, Wyoming. (From Mount Wilson and Palomar Observatories)*

20.21 The Velocities of Electrons in the Corona and the Electron Temperature of the Corona

Since the continuous part of the spectrum is the same as that of the sun, what we see when we look at the inner part of the corona is the sunlight scattered to us by the electrons there. However, since there are no absorption lines in this inner coronal spectrum, the scattering process must have obliterated them because of the Doppler shift. When electrons are in motion and scatter light, the scattered light (as viewed by an observer relative to whom the electrons are moving) suffers a change of wavelength. Since the electrons in the corona have random velocities, the Doppler shift is towards the red end of the spectrum in some cases and towards the blue end in others, so that each spectral line is broadened. The complete obliteration of the absorption lines means that the electrons in this part of the corona have very high random velocities. These high electron velocities tell us that the inner corona has a very high electron temperature, probably in excess of 300 000°K. This does not mean that a thermometer placed there would actually register a temperature of 300 000°K since there are not enough electrons per unit volume in the corona colliding with the thermometer every second to keep it at that temperature; the correct interpretation is that the electrons are moving as fast as they would if they were in a furnace having a temperature of 300 000°K.

20.22 The Chemical Elements in the Corona

The emission lines we find in the mid-corona are due primarily to very highly ionized calcium, iron, and nickel atoms which have lost anywhere from 9 to 15 electrons. The observed lines themselves are those referred to by spectroscopists as **forbidden lines**. This means that under ordinary laboratory conditions we do not observe them. In the corona, however, where the number of atoms per unit volume is extremely small, the forbidden electron transitions in an atom that give rise to these lines do occur.

Since we know how much energy is required to remove electrons from atoms such as iron and nickel, it is easy to compute how energetic collisions must be between the atoms, or between electrons and the atoms, to bring about this high state of ionization. We can thus determine what the electron temperature is in this part of the corona. We find from such calculations that electrons are moving with kinetic energies of about 1 000 ev, which is equivalent to a temperature of about 1 million degrees.

20.23 Metastable States of Atoms and Forbidden Lines

A word about forbidden transitions is appropriate since this phenomenon occurs not only in the solar corona but in interstellar matter as well. If an atom is excited by a collision with a photon, the electron, after remaining in the excited state for a millionth of a second, in general, cascades down to the ground state. But it may happen that during this cascade process the electron lands in some lower state from which it cannot immediately jump down into the ground state because such a transition is forbidden by the rules of the quantum mechanics. Such excited states are called **metastable states**, and the lifetime of an electron in such a state may be very long (seconds or even hours) compared to the normal lifetime of 10^{-8} seconds. Metastable states are known in which the lifetime may be as much as several hours.

When an electron after its long life in this metastable state jumps down to the ground state, the photon that is emitted gives rise to a forbidden line. Under ordinary conditions in the laboratory forbidden lines do not occur because collisions of the atoms among themselves or with the walls of the container are so frequent that the electrons are thrown out of these metastable states before they have a chance to jump down. In the highly tenuous corona, however, an atom is left to itself long enough for the electron to jump down to the ground state and emit the forbidden line.

20.24 Transparency of the Corona

We may note that the high temperature we find in the corona is not in disagreement with the low surface temperature obtained by applying the laws of black-body radiation to the sun. The reason is that whereas we may consider the surface of the sun as radiating approximately like a black body, this is not so for the corona because the corona is extremely tenuous, and hence allows radiation to pass right through it without affecting its continuous character in any way. The only time we may treat a hot gas as a black body is when the gas is so opaque to radiation that it has the effect of smearing any radiation of a given color out into the entire continuous spectrum by continually absorbing and re-emitting it. The corona does not do this. In other words, the corona is so transparent to the radiation coming from the photosphere that it does not impose its own physical characteristics on this radiation.

The story is somewhat different, however, for radiation having wavelengths above 50 cm, and we can examine the corona in this wavelength region by means of radio telescopes. In terms of this long-wavelength radiation the corona does behave somewhat like a black body, and the temperature found by analyzing such radiation agrees with that derived from the photosphere.

20.25 Total Mass of the Solar Atmosphere

By bringing together all the available data on the structure of the solar atmosphere we find the total amount of matter it contains to be of the order of 10^{17} tons. Although this is a vast amount of material, it is only the twenty-billionth part of the total solar mass. We also can see from this how very opaque the solar atmosphere is, for this total amount of mass represents only a few grams of matter for each column of solar atmosphere having a cross section of 1 cm^2, yet it is sufficient to restrict sharply the depth to which we can peer into the sun.

21

STELLAR RADII AND MASSES

If we look back at the list of stellar properties in Section 15.2, we see that up to this point we have discussed all but radii and masses. Once we have the radius and the mass of a star, it is an easy matter to determine its average density. We have not discussed radii and masses until now because it is possible to understand everything that has gone before, even the structure of stellar atmospheres, without having to know these quantities. Of course, in our considerations of the differences between giants and dwarfs we did make use of the fact that the masses of these stars are not very different, whereas their radii vary considerably, but more specific information was not necessary.

However, before we can properly study the internal structure of the star (i.e., before we can get a total picture of the star) we must have a more precise knowledge of radii and masses, for, as we shall see, the radius, mass, luminosity, and chemical nature of a star are all intimately related so that varying any one of them results in a change in the others. Therefore at this point we shall survey the observational methods that are available for determining radii and masses, and also see what the data are.

21.1 The Spurious Disk and the Direct Measurement of Stellar Radii

At first glance, it appears to be a relatively easy matter to find the radius of a star since we know that from the star's distance and its apparent size (i.e., its angular diameter) we should be able to determine its true diameter. We can, indeed, do this for the sun, as shown in Figure 14.3. If α'' is the angular diameter of the sun in seconds as seen on the earth, d its distance, and D its true diameter, then it follows that

$$D = d\frac{\alpha''}{206\,265}. \qquad (21.1)$$

Since the average angular diameter of the sun at the mean distance of the earth is $2(959''.63)$ its true radius is 6.96×10^{10} cm.

However, we cannot apply this method to a star because even with the most powerful telescopes the stars show no visible diameter. In other words, stars appear as mere points. It is true that when we look through a telescope, we see a small disk as the star image, *but this disk (surrounded by fainter rings) is not an image of the true disk of the star, but rather the result of the wave nature of the light itself and the diffraction suffered by light when it passes through an aperture (e.g., the objective of the telescope).*

Appendix A of this book gives a discussion of astronomical instruments, and we see there that when light passes a sharp edge, it is slightly bent. This phenomenon, called **diffraction**, is responsible for the formation of the spurious disk on a photographic plate (it may be mistaken for the true disk of the star). If the spurious disk, as formed by the objective of the telescope, is larger than the image of the true disk, *it is impossible to tell what the apparent diameter of the star is, since all we have available is the diameter of the spurious disk.*

From our discussion of the diffraction of light we know that the angular diameter of the spurious disk in seconds of arc is equal to $2.52 \times 10^5(\lambda/a)$, where λ is the wavelength of the light that enters the telescope, and a is the aperture of the telescope, both expressed in the same units. This is equal to $4''.5/a$ for visible light, if a is expressed in inches. Even for the telescopes of greatest aperture the spurious disk is large enough to obscure the image of the true disk for all but one or two very large stars close to our solar system.

Thus, if we consider the 200-inch telescope, the diameter of the spurious disk is $0''.03$, which is larger than the apparent diameter of all but a few supergiant stars such as Betelgeuse and Antares. If we assume that the parallax of Betelgeuse (α

Orionis) is 0″.012 (although because of errors the actual parallax may be as small as 0″.005) and that its diameter is about 300 times the sun's diameter, we find that its apparent angular diameter is about 0″.03. This is just at the limit of measurability by direct methods. However, because of the errors involved in the determination of the parallax, it is much too risky to use this method in measuring the diameter of stars in general.

21.2 Radii from Luminosities and Effective Temperatures

The direct method described above cannot be used at all to measure the diameters of smaller stars such as Vega since their apparent diameters are smaller than their spurious disks by a factor of about 10 or more. It does not do to try to look at the spurious disk under magnification to see if the image of the true disk can be observed, for all we accomplish by using a high-powered eyepiece is to magnify the spurious disk itself. How then are we to determine the diameters of stars in general?

There are various indirect methods for finding a star's diameter, one of which we shall discuss now. These methods use the relationship between the luminosity of a star, its surface temperature, and its radius. If we assume that the surface of a star radiates energy like a black body, we then know from the Stefan-Boltzmann law that each square centimeter emits energy at a rate σT_e^4, where σ is the Stefan-Boltzmann constant, and T_e is the effective temperature of the star (see Section 16.27). From this and a knowledge of the total luminosity we can find the radius since we know that the total luminosity must equal this expression multiplied by the surface area. In other words, we must have

$$\text{luminosity of star} = L = 4\pi R^2 \sigma T_e^4. \quad (21.2)$$

We can now solve this expression for R to obtain the star's radius.

Instead of doing this directly, which involves inserting the numerical value for $4\pi\sigma$, we can simplify things by expressing the radius in terms of the solar radius. To do this we note that when we apply the above formula to the sun, we get

$$L_\odot = 4\pi R_\odot^2 \sigma T_{\odot e}^4. \quad (21.3)$$

All we need do now is divide this equation into the previous equation to obtain

$$\frac{L}{L_\odot} = \left(\frac{R}{R_\odot}\right)^2 \left(\frac{T_e}{T_{\odot e}}\right)^4,$$

since $4\pi\sigma$ cancels out. We now solve this for $R/R_\odot$ to obtain

$$\frac{R}{R_\odot} = \frac{\sqrt{(L/L_\odot)}}{(T_e/T_{\odot e})^2}. \quad (21.4)$$

Thus, if we know the luminosity of the star expressed in solar luminosities, and take its square root and divide this by the square of the effective temperature of the star's surface expressed in terms of the sun's effective temperature, we obtain the radius of the star expressed in solar radii.

21.3 Typical Stellar Radii of Main Sequence and Giants from Luminosities

As examples we shall consider stars belonging to various branches of the H-R diagram. To apply our formula to the red dwarf, Barnard's star (spectral class M5 and absolute visual magnitude 13.2), we must know its total luminosity which we can obtain from its bolometric magnitude. This takes account of the infrared and ultraviolet radiation in addition to the visible radiation emitted by the star. By adding the bolometric correction −0.7 (see Section 15.19) to its absolute magnitude we obtain 12.5 for the bolometric magnitude of Barnard's star. In other words, this star is 7.7 magnitudes or 1 285 times less luminous than the sun. *Since its effective surface temperature is approximately one-half that of the sun, we find from Equation (21.4) that its radius is one-eighth that of the sun. This is approximately the case for most of the red dwarfs.* As we go up the main sequence, however, the stellar radii increase as the following examples will show.

If we consider Sirius, whose effective surface temperature is about twice that of the sun, we find that its bolometric magnitude is +0.8. It is, therefore, 4 magnitudes brighter, or about 40 times more luminous than the sun. Our formula thus gives us for the radius of this star in solar units

$$\frac{R_{\text{Sirius}}}{R_\odot} = \frac{\sqrt{40}}{(2)^2} = \frac{6.4}{4} = 1.6.$$

The blue giant, Rigel, has an absolute visual magnitude of −7 and a bolometric magnitude of −8, so that it is 12.8 magnitudes brighter, or more than 64 000 times more luminous than the sun. Since its surface temperature is about three times that of the sun, the formula gives

$$\frac{R_{\text{Rigel}}}{R_\odot} > \frac{253}{9} \cong 28$$

for this star. Thus, as we go from the faint red dwarfs to the very luminous blue giants on the main sequence, the radii increase about 120 times.

We can use the same method to find radii of giants, supergiants, and white dwarfs. Giant stars, such as Capella, have radii of the order of 16 to 20 times the sun's radius, whereas red supergiants, such as Betelgeuse (α Orionis) and Antares, have radii of the order of 300 to 400 times the sun's radius. As an example of one of the largest known stars we have the large companion of the binary Epsilon Aurigae with a surface temperature of about 1 200°K and a radius about 2 000 times that of the sun.

21.4 Radii of White Dwarfs from Luminosities

The smallest stars known are the white dwarfs, which occupy the lower left-hand corner of the H-R diagram. Since the surface temperatures of these stars are approximately that of Sirius, they must, indeed, be very small since they are many magnitudes fainter than Sirius. As an example we may consider Sirius B, which is a white dwarf with an absolute visual magnitude of +11.4, or a bolometric magnitude of about 10 (since the bolometric correction for a star of this sort is about −1.4). From this we see that the sun is about 5.2 magnitudes or approximately 100 times more luminous than this star. If we apply Equation (21.4) and note that the temperature of this star is about twice the effective temperature of the sun, we obtain a radius of about 1/40 that of the sun. Two other typical white dwarfs are 40 Eradini B, with a radius 0.019 of the sun's, and Van Manaan's star, with a radius 0.007 of the sun's. From these considerations we see that the radii vary drastically as we go from white dwarfs to supergiants.

21.5 Direct Measure of Radii: Using the Interferometer

Although we have determined stellar radii by using the laws of black-body radiation, it is possible to measure the radii of supergiants directly by using the **interferometer**. This is an instrument invented by Michelson, which makes use of the interference of light (the very thing that gives rise to the spurious disk when the image of the star is obtained in a telescope) to measure the diameter of a star. Although the theory of the instrument is explained in detail in Appendix A, we shall here describe its application to the measurement of a star's diameter.

As we have seen, when light from a distant point enters the objective of a telescope, the image obtained is a disk whose diameter depends upon the aperture of the objective. If we cover the objective with a screen having a small opening at some distance from the center of the objective, the image we then obtain is still a disk, except that the size of the disk is greatly increased. If we now punch another hole of equal size in this screen at the same distance from the center of the objective, but diametrically opposite the first hole, we find that the diffraction disk is no longer uniformly bright, but rather crossed by a number of straight, equidistant, dark lines (**fringes**). These dark fringes, which are perpendicular to the line connecting the two openings, are produced by the interference of the two separate beams of light entering the objective through openings in the screen, as shown in Figure A.33. We see from this figure that the spacing, y, between two adjacent bright and dark lines is given by the formula

$$\frac{y}{f} = \frac{1}{2}\frac{\lambda}{s},$$

where f is the focal length of the objective, λ is the average wavelength of the effective light coming from the star, and s is the distance between the two small holes in the screen.

Fringes occur for the following reason: as we have seen, a diffraction pattern (i.e., a spurious disk) is formed when the light passes through a single small opening. Hence, with two small openings, we obtain two superimposed disks. Since the light beams forming the two bright disks come from different openings they are, in general, out of phase so that at various points over these two superimposed disks there occurs either complete cancellation of light, giving rise to the dark fringes, or reinforcement, giving rise to bright fringes.

Let us now suppose that we have two point sources of light at a great distance, whose angular separation in radians as seen by an observer on earth is α. Let us also suppose that we allow light from these points to fall on our screen with the two openings. In each case we get a spurious disk with light and dark fringes. The light beams from the two point sources together then give two superimposed disks (each with fringes) whose positions with respect to each other depend on the

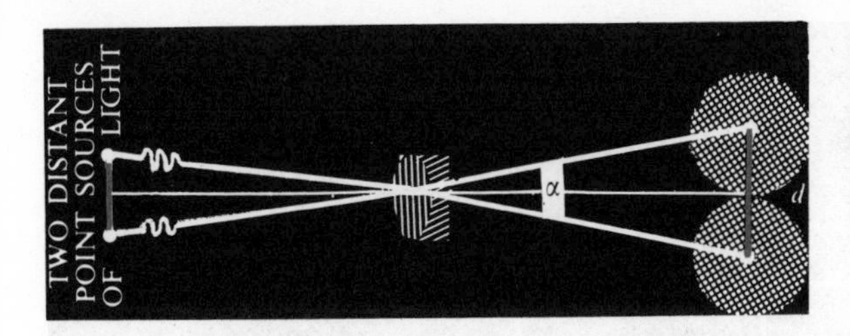

FIG. 21–1 *Two very distant point sources of light are imaged as two spurious disks in a telescope.*

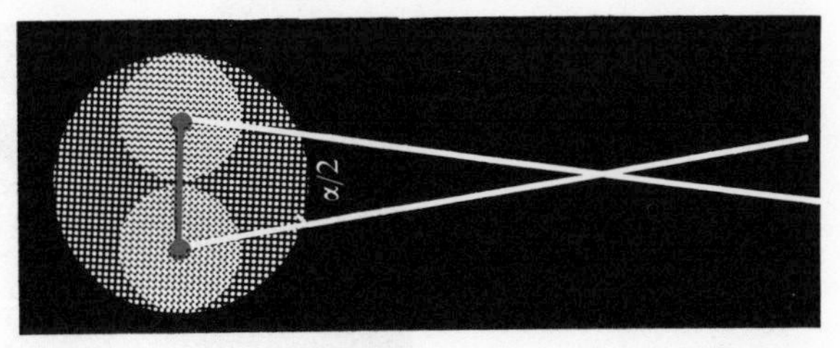

FIG. 21–2 *The light from a star may be pictured as coming from two point sources, each point representing half of the star.*

separation s between the openings in the screen. The distance between the centers of these spurious disks in the focal plane of the telescope depends upon the angular separation α of the two point sources and on the focal length f of the objective. If d is the distance between the centers of the two spurious disks, then it is clear from the diagram (Figure 21.1) that $d = f\alpha$. *We thus have two sets of fringes superimposed upon each other, and it is possible by increasing the distances, s, to arrange things so that the bright fringes in one spurious disk fall exactly on the dark fringes of the other spurious disk. We then have a uniformly lit image.* When s is chosen so that this happens, we find that

$$\boxed{\alpha = \frac{1}{2}\frac{\lambda}{s}}. \tag{21.5}$$

21.6 The Application of the Interferometer to Measurement of Giant Stars *

We can now apply this analysis to the determination of the diameter of a single star. To do this we may picture the total light of the star as coming from two half-disks whose centers in angular measure are separated by $\alpha/2$, as shown in Figure 21.2. We then have essentially the same situation as in the case of the two point sources described above. We find that as we separate the apertures in our screen, the bright fringes in the superimposed spurious disks get broader and broader, and the dark fringes narrower and narrower, until finally, when the openings of the screen are separated far enough, the diffraction disk of the star is uniformly illuminated.

If we can achieve this, we can then apply Equation (21.5)—with a slight correction. If the actual angular diameter of the star is α, we do not use $\alpha/2$ for the separation of the two half-disks, but rather 0.41α, since it is the separation between the effective centers (i.e., the centers from which we may picture the entire light as coming) that we must use. We then find, on applying the formula, that the angular diameter of the star is

$$\alpha = \frac{1.22\lambda}{s_0}, \tag{21.6}$$

where s_0 is the value of s for which the disk is uniformly illuminated. This holds if the actual surface of the star is uniformly luminous, but if it exhibits the same sort of limb darkening as the sun, the angular diameter is given by $\alpha = 1.43\lambda/s_0$.

Even in the case of the very large stars, the distance s_0 by which the two apertures in the screen have to be separated before the fringes disappear is greater than the diameters of the objectives of the largest telescopes now in use. However, it is possible to overcome this difficulty by replacing the apertures in the screen by two mirrors mounted on long steel tracks. Such tracks were first used with the 100-inch telescope, arranged so that the mirrors could be separated to a maximum distance of 20 feet, as shown in Figure 21.3. With this instrument, the diameters of such stars as Betelgeuse, Arcturus, Antares, and Aldebaran were measured as indicated in Table 21.2. These measurements completely confirm the radii obtained from the laws of black-body radiation.

21.7 White Dwarfs

Since from the laws of black-body radiation we have found that white dwarfs are extremely small stars, it is desirable to have some independent way of checking this result. One method is to use one of the consequences of the general theory of relativity, namely, the Einstein red shift (see Section 11.25). We know that the frequency of the radiation emitted by an atom in an intense gravitational field (i.e., on the surface of a very

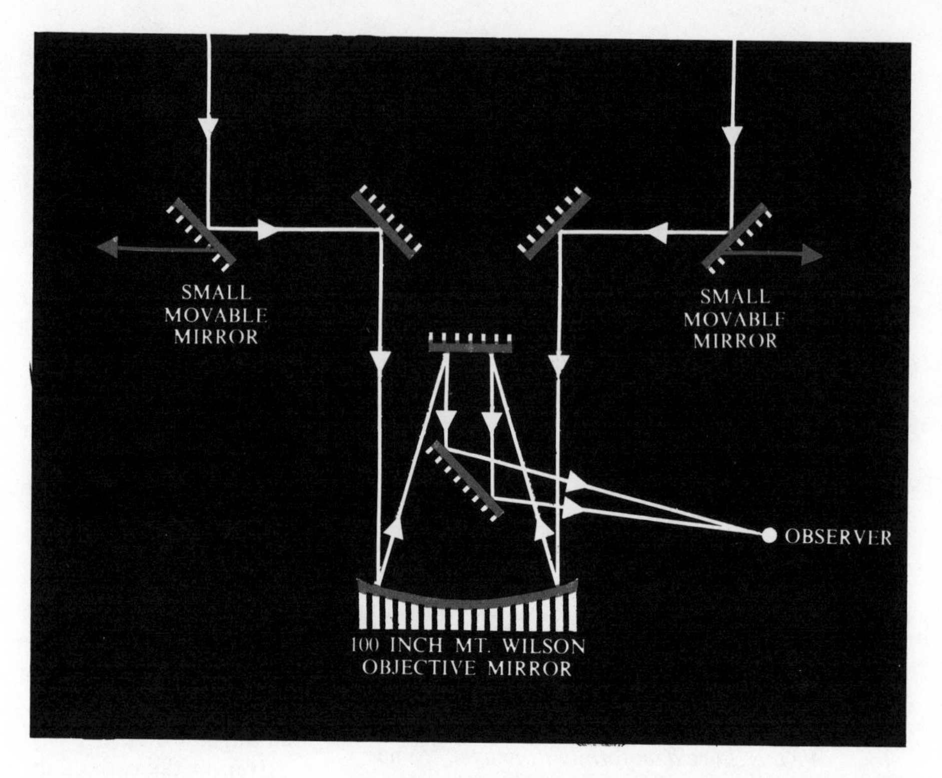

FIG. 21–3 *An interferometer consisting of two movable mirrors on tracks mounted on 100-inch reflector.*

small, massive star) is smaller than the frequency of similar radiation emitted by the same kind of atom in empty space. The amount $\Delta\lambda$ by which the wavelength is shifted in the gravitational field is given by the formula

$$\Delta\lambda = \lambda \frac{G\mathcal{M}}{c^2 R},$$

where λ is the wavelength of the light, G is the gravitational constant, c the speed of light, $\mathcal{M}$ the mass of the star, and R its radius.

If there is some independent way of determining the mass of a star, we can find its radius from this formula by measuring the Einstein red shift in its spectral lines. It is most pronounced for white dwarfs, since their masses are about the same as that of the sun, while their radii are about $0.01 R_{\odot}$. For such stars, this relativistic red shift is equivalent to a Doppler red shift of about 30 km/sec.

Since the Doppler red shift (arising from the radial velocity of a star) and the relativistic red shift occur together, we cannot use the latter to determine the radius of a star unless we know how big the ordinary Doppler shift is. This can be determined if the white dwarf is a component of a **binary system** (two stars revolving about a common center) as is the case for Sirius B and 40 Eradini B. We can then find the ordinary Doppler shift from the spectrum of the main component of the binary. Thus if we measure the ordinary Doppler effect in the spectrum of Sirius A (the bright component of the binary), and subtract this from the total red shift in the spectral lines of Sirius B, we find a residual red shift equal to 20 km/sec. This gives the radius of the star if its mass is known. We shall see that the masses of binary stars such as Sirius B and 40 Eradini B can be found from Kepler's third law of motion. When we do this, we obtain radii for white dwarfs which are in agreement with the values found from the laws of black-body radiation.

21.8 Stellar Masses

In addition to the radius of a star we must also know its mass before we can analyze its internal structure. But in most cases the masses of stars cannot be obtained by direct measurement. As we know, the mass of a body can be found only if we can measure the acceleration induced in it by a known external force or if we have some way of measuring its gravitational interaction with some other body. Since it is impossible to subject a star to a force of our own choosing, we can obtain its mass only by observing the orbit of some object in the star's gravitational field. We can do this if the star is a component of a binary, and only for such systems can the masses of the two components be obtained directly.

21.9 Kepler's Third Law Applied to Binaries

Although we shall discuss binary systems in detail in a later chapter, here we shall consider only their orbital characteristics. From these, by applying Kepler's third law of planetary motion in its correct form (as derived from Newton's law of gravitation; see Section 3.31), we shall be able to find the masses of the two components. We saw in Chapter 9 that if the sidereal period of a planet around the sun is P, and the mean distance of the sun is a (the semimajor axis), then

$$P^2 = \frac{4\pi^2 a^3}{G(\mathscr{M}_\odot + \mathscr{M}_p)},$$

where $\mathscr{M}_p$ is the mass of the planet. We may now apply this formula to a binary system since the two stars move around a common center of mass like the sun and any planet. (The sum of the masses of the two stars replaces the sum of the masses of the sun and the planet in the denominator.)

In Figure 21.4 the orbits of the two components of the binary system are shown relative to their center of mass. We note that the line connecting the two stars must always pass through this center of mass—a point dividing this line into two parts, the ratio of whose lengths to each other is always the same. From this it is clear that *when star 1 has returned to its initial point after one complete revolution, star 2 must also be back at its initial position, so that the period P assigned to any binary system applies to both stars.*

If we now apply Kepler's third law to a binary system, we have

$$P^2_{\text{binary}} = \frac{4\pi^2 a^3}{G(\mathscr{M}_1 + \mathscr{M}_2)}$$

or

$$\mathscr{M}_1 + \mathscr{M}_2 = \frac{4\pi^2 a^3}{GP^2_{\text{binary}}}, \tag{21.7}$$

where a is now the mean distance between the two stars (the semimajor axis of the orbit of one star relative to the other). It is clear from this formula that *if we can measure the position of the center of mass relative to the two stars, and if we can determine the true orbit of one star with respect to the other (so that the mean distance a can be found) we can then, by measuring the period of the binary system, determine the mass of each star separately.* For if the distance of the first star from the center of mass is d_1 and the distance of the second star from this point is d_2, then $\mathscr{M}_1/\mathscr{M}_2 = d_2/d_1$, so that this equation together with Kepler's third law enables us to find the masses $\mathscr{M}_1$ and $\mathscr{M}_2$ separately.

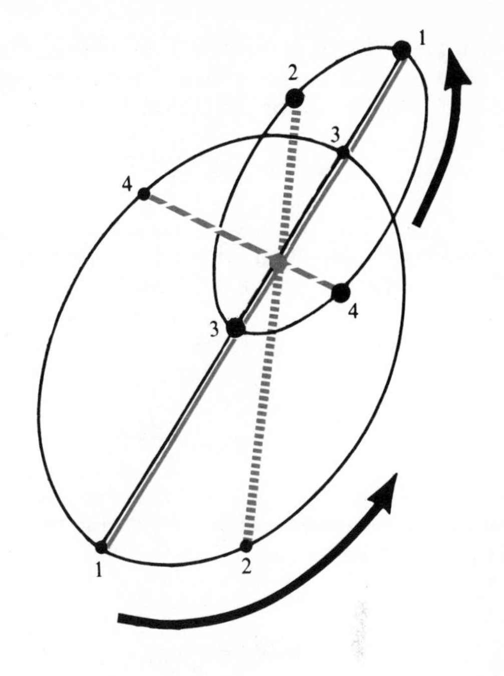

FIG. 21–4 *The line from one component of a binary to the other must always pass through the center of mass of the system as the two stars revolve about each other.*

Since we shall always give the mass of a star in terms of the sun's mass, it is convenient to express Kepler's third law for the binary system in terms of the earth-sun system. We do this by dividing Equation (21.7) by Kepler's third law for the earth, with the period expressed in years:

$$\mathscr{M}_\odot = \frac{4\pi^2 A^3}{G},$$

where A is the astronomical unit, and where we have neglected the mass of the earth compared with the sun. We then obtain the formula

$$\frac{\mathscr{M}_1 + \mathscr{M}_2}{\mathscr{M}_\odot} = \frac{(a/A)^3}{P^2_{\text{binary}}},$$

where the period P of the binary is to be measured in years. If d is the distance of the binary system we know from Equation (15.1) that

$$d = 206\,265 \frac{A}{p''},$$

where p'' is the parallax of the binary system. On solving this equation for A and on substituting into the previous equation, we have for the sum of the masses of the two components (expressed in solar units, i.e., placing $\mathcal{M}_\odot = 1$),

$$\mathcal{M}_1 + \mathcal{M}_2 = \frac{\left(206\,265 \frac{a}{d} \frac{1}{p}\right)^3}{P^2_{\text{binary}}}.$$

Since a/d is just the mean angular separation of the two components expressed in radians, and there are 206 265 seconds in one radian, we finally obtain the formula

$$\boxed{\mathcal{M}_1 + \mathcal{M}_2 = \frac{\alpha''^3}{p''^3 P^2}}, \qquad (21.8)$$

where α'' is the mean angular separation of the two components expressed in seconds of arc.

21.10 The Masses of Visual Binaries

In the case of a visual binary (where the two components can be seen) the position of the center of mass can be found and the true orbit can be obtained from the observed apparent orbit. It is then possible to find the individual masses of the two components. This has been done for more than 100 binary systems with periods ranging anywhere from 5.7 years to 100 years and up. If the two components of a binary system cannot be seen separately, the problem is a more difficult one, but the elements of the orbit can still be found if the spectra of the two stars can be observed. We shall discuss this in more detail in Chapter 26.

To illustrate how Kepler's third law is applied to a binary system, we shall consider the visual binary α Centauri, whose parallax is 0".760. The mean angular separation between the two components is 17".60 and their period is 80.1 years. We thus obtain for the sum of the masses

$$\mathcal{M}_1 + \mathcal{M}_2 = \left(\frac{17.60}{0.760}\right)^3 \frac{1}{(80.1)^2}$$
$$= 1.95\mathcal{M}_\odot.$$

Since observation shows that the center of mass divides the line connecting the two stars in the ratio of 1 : 1.25, the individual masses are 1.08 and 0.87 solar masses. We see from this that the mass of α Centauri is about equal to the sun's mass.

Table 21.1 gives the masses for some well-known visual binary systems. What is remarkable about these values is their small range compared to the broad ranges of luminosities and radii. *Thus, although the absolute bolometric magnitudes of these stars go from about 1.3 to about 12.3 (a range of 11 magnitudes or a luminosity range of about 25 000), the masses range from about 0.15 to about 3.5 times the sun's mass.* From this we see that the mass is a relatively stable characteristic of a star and that even a slight change in it results in a drastic change in the star's structure.

Although all the data for the observable binary stars show that the masses of stars do not change appreciably as we go from one branch of the H-R diagram to another, there are exceptions. Occasionally we find stars with unusually large masses. An example of such a star is HD47129, a spectroscopic binary whose orbital elements have been analyzed in detail by J. S. Plaskett. He found that the total mass of this binary is 100 times the sun's mass, and that each component has a mass about 50 times that of the sun. Such massive stars, however, are very rare, and probably unstable, as is true in this case.

21.11 Mass-Luminosity Relationship *

A careful analysis of the data in the table of stellar masses (Table 21.1) shows that there is a correlation between the absolute bolometric magnitude and the mass of a star. We see from the table that *the more massive a star is, the more luminous it is. This is known as the* ***mass-luminosity relationship*** *and can be used to determine the masses of stars which are not components of a binary system.* The mass-luminosity relationship can best be illustrated by plotting the masses of individual stars in our table against their absolute bolometric magnitudes. This is shown in Figure 21.5, where the absolute bolometric magnitudes are plotted along the vertical axis, and the masses along the horizontal axis. We see that except for a few stars in the lower left-hand corner (the white dwarfs) most of the points fall along a well-defined curve which is a graphical representation of the mass-luminosity relationship.

As we shall see later, a theoretical mass-luminosity relationship for stars having similar internal

TABLE 21.1
ELEMENTS OF SOME WELL-KNOWN VISUAL BINARIES

Star	Semi-major axis a	Period P	Parallax p	Apparent visual magnitude, m_v 1	2	Spectral class 1	2	Absolute bolometric magnitude, M_{bol} 1	2	Mass $\mathscr{M}$ 1	2
	″	years	″							$\mathscr{M}_\odot$	
η Cas	11.99	480	0.170	3.44	7.18	G0 v	K5	4.54	7.51	0.94	0.58
o^2 Eri B, C	6.89	247.9	0.201	9.62	11.10	B9	M5e	10.26	9.5	0.45	0.21
ξ Boo	4.88	150.0	0.148	4.66	6.70	G8 v	K5	5.41	6.70	0.85	0.75
70 Oph	4.55	87.8	0.199	5.09	8.49	K0 v	K4	5.56	6.85	0.90	0.65
α Cen A, B	17.66	80.1	0.760	0.09	1.38	G4	K1	4.40	5.65	1.08	0.88
Sirius	7.62	49.9	0.379	−1.47	8.64	A1 v	wA5	0.80	11.22	2.28	0.98
Krü 60	2.41	44.6	0.253	9.80	11.46	dM4	dM6	9.11	9.97	0.27	0.16
Procyon	4.55	40.6	0.287	0.34	10.64	F5 IV–V	wF8	2.59	12.62	1.76	0.65
ζ Her	1.35	34.4	0.104	2.91	5.54	G0 IV	dK0	2.94	5.52	1.07	0.78
85 Peg	0.83	26.3	0.080	5.81	8.85	G2 v		5.26	7.18	0.82	0.80
Ross 614 A, B	0.98	16.5	0.251	11.3	14.8	dM6+		9.9	11.9	0.14	0.08
Fu 46	0.71	13.1	0.155	10.01	10.39	M4	M4	8.26	8.64	0.31	0.25

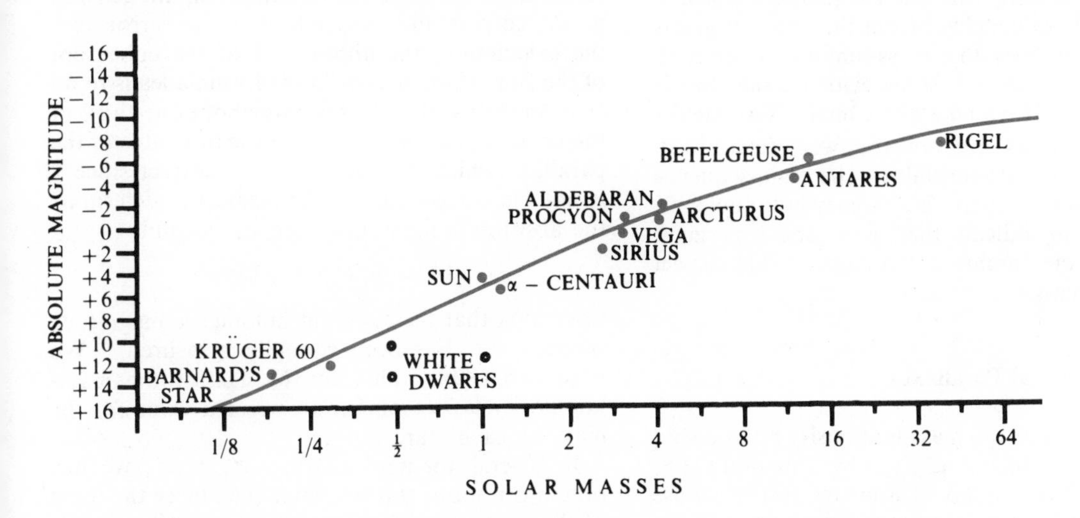

FIG. 21–5 *The mass-luminosity relationship. The absolute bolometric magnitude is plotted against the mass of the star. The white dwarfs are shown in the lower left-hand corner.*

structures (stars built according to the same model) can be derived from the basic physical conditions governing their equilibrium. Since most of the stars in the Milky Way belong to the main sequence and thus probably have the same internal structure as the sun does, we may assume that they obey the same mass-luminosity relationship and hence fall somewhere along the curve shown on the graph. Because the white dwarfs have a structure different from that of the sun, they do not obey the same mass-luminosity relationship. In fact, we may take the departure of a star from the mass-luminosity relationship as an indication that it has a different internal structure from that of the sun.

The mathematical formula that best fits the mass-luminosity curve given in the figure is

$$\log\left(\frac{L}{L_\odot}\right) = 3.5 \log\left(\frac{\mathscr{M}}{\mathscr{M}_\odot}\right).$$

This means that the luminosity of a star increases with the 3.5th power of the mass. By using the curve obtained from the empirical mass-luminosity relationship as shown in Figure 21.5, we can determine the masses of main-sequence stars from their luminosities.

21.12 The Masses of Giant Stars

We have seen that the masses of main-sequence stars (see Figure 21.5) fall on a fairly well-defined mass-luminosity curve. But we cannot be sure that this is true of all the giant stars. Unfortunately, there are not enough data on the masses of giant stars to enable us to draw any definite conclusion. At present the mass of only one giant star, the binary Capella, and the masses of two subgiants, Zeta Hercules A and Zeta Hercules B, have been accurately determined. The Capella system has a mass of about $3.02\mathcal{M}_\odot$, whereas the two subgiants have masses of about $1.1\mathcal{M}_\odot$ and $1.08\mathcal{M}_\odot$ respectively. Although the masses of these three stars fall on or lie close to the mass-luminosity curve for the main-sequence stars, it is unreasonable to conclude from this that all giants and subgiants obey this mass-luminosity relationship. There is, indeed, some evidence that this is not the case. If we take giant binary systems for which the masses are not too well determined, we find that they do not fall on the main-sequence mass luminosity curve but somewhat above it. This seems to indicate that giant and supergiant stars are more luminous than one would expect from their masses.

21.13 Dynamical Parallaxes

With the aid of the mass-luminosity relationship and Kepler's third law as given by Equation (21.8) we can find the parallax of a binary system whose orbit is well known. From this equation we see that the parallax of a binary system must be very accurately known if the error in the mass is to be small. The reason is that the sum of the masses of the two stars varies inversely as the cube of the parallax. This means that *any small error in the parallax gives rise to an error in the mass that is about three times as large, so that a 10 per cent parallax error means a 30 per cent mass error.* This naturally leads us to the idea that it might be more favorable to reverse the procedure and to determine the parallax of the binary system from a knowledge of both its orbit and the sum of the masses of its components.

If we solve for the parallax from Equation (21.8) we have

$$p'' = \frac{\alpha''}{[P^2(\mathcal{M}_1 + \mathcal{M}_2)]^{1/3}}.$$

Since α'' and P can be measured quite accurately for a binary system of known orbit, the parallax of this system can be found if $(\mathcal{M}_1 + \mathcal{M}_2)$ is known. It might seem at first that this leads us nowhere since in general $(\mathcal{M}_1 + \mathcal{M}_2)$ is not known. Fortunately, three things enable us to overcome this difficulty:

1. we know the mass-luminosity relationship;
2. the masses of most stars that obey the mass-luminosity relationship do not differ much from the sun's mass;
3. in the above formula for the parallax the sum of the masses appears to the one-third power.

We first assume in accordance with point 2 that $(\mathcal{M}_1 + \mathcal{M}_2)$ is equal to 2 (i.e., twice the mass of the sun) and substitute this value into the above formula. Although this assumption, in general, is not correct and thus introduces an error into the calculation, the appearance of the cube root of the sum of the masses in the formula leads to an error in the parallax that is only about one-third of the error in $(\mathcal{M}_1 + \mathcal{M}_2)$. We use this value of the parallax (which, of course, is not correct since 2 is not the correct sum of the masses) to determine the absolute magnitudes from the formula

$$\mathrm{M} = \mathrm{m} + 5 + 5 \log p''$$

(assuming that the apparent bolometric magnitude of each star has been carefully measured). We now check this value for M against the mass-luminosity law to obtain a corrected value for the mass of each star.

In general, the new value for $(\mathcal{M}_1 + \mathcal{M}_2)$ we thus obtain is not our starting value (i.e., twice the mass of the sun) but differs from it by a small amount. However, we can now use this new value for $(\mathcal{M}_1 + \mathcal{M}_2)$ in place of the value 2 and go through the entire procedure again, thus finding a corrected parallax. This again leads to a slight correction in $(\mathcal{M}_1 + \mathcal{M}_2)$ and hence to a still more accurate parallax. We can go on repeating this procedure until no new corrections are obtained. We then say that the parallax we have obtained is the **dynamical parallax** of the binary system. We find that in most cases this iterative process need only be carried out three or four times to lead to a self-consistent value for the masses.

Because masses of most stars do not vary over a wide range, the parallaxes of binaries found in this

TABLE 21.2

MASSES, RADII, LUMINOSITIES, AND MEAN DENSITIES OF SUPERGIANTS, GIANTS, MAIN-SEQUENCE STARS, AND DWARFS

Spectral class	log $(\mathcal{M}/\mathcal{M}_\odot)$			log $(R/R_\odot)$			log $(L/L_\odot)$			log $\bar{\rho}$ (in gm cm^{-3})		
	super-giant	main sequence		super-giant	giant	dwarf	super-giant	main sequence		super-giant	main sequence	
O5	+2.2	+1.60				+1.25		+5.5			−2.0	
B0	1.7	1.23		+1.3	+1.2	0.88	+5.5	4.1		−2.1	−1.3	
B5	1.4	0.85		1.5	1.0	0.60	4.9	2.8		−2.9	−0.8	
A0	1.2	0.55		1.6	0.8	0.42	4.4	1.9		−3.5	−0.6	
A5	1.1	0.34		1.7		0.25	4.1	1.3		−3.8	−0.3	
F0	1.1	0.25		1.8		0.13	3.9	0.8		−4.2	0.0	
F5	1.0	0.15		1.9	0.6	0.08	3.8	0.4		−4.5	+0.1	
		giant	dwarf					giant	dwarf		giant	dwarf
G0	1.0	+0.4	+0.03	2.0	0.8	+0.02	3.8	+1.5	+0.1	−4.9	−1.8	+0.1
G5	1.1	0.5	−0.03	2.1	1.0	−0.03	3.8	1.7	−0.1	−5.2	−2.4	+0.2
K0	1.1	0.6	−0.09	2.3	1.2	−0.07	4.0	2.0	−0.4	−5.7	−2.9	+0.3
K5	1.2	0.7	−0.16	2.6	1.4	−0.13	4.3	2.3	−0.8	−6.4	−3.4	+0.4
M0	1.2	0.8	−0.32	2.7		−0.20	4.5	2.6	−1.2	−6.7	−4	+0.5
M5			−0.65			−0.5		3.0	−1.1			+1.1
M8			−1.1			−0.9			−3.3			+1.7

way are surprisingly accurate, with errors on the average of about 5 per cent and seldom exceeding 15 per cent. H. N. Russell and Charlotte Moore used this method to obtain the dynamical parallaxes of more than 2 000 visual binaries.

21.14 Stellar Densities

From the mass and radius of a star it is now possible to determine its average density:

$$\rho_{\text{average}} = \frac{\mathcal{M}}{\frac{4}{3}\pi R^3}.$$

Since, as we have noted, the radii of stars vary enormously, whereas the masses are relatively the same, we see that the mean densities vary widely as we go from white dwarfs to red supergiants. As a basis for comparison we note that the mean density of the sun is 1.4 gm/cm^3. From this we can get an idea of the densities of the stars in various other branches in the H-R diagram.

Since the diameters of white dwarfs are about 1/100 that of the sun, whereas their masses are only slightly less than that of the sun, their average densities are of the order of several hundred thousand times that of the sun. As examples we have Van Manaan's star, which has a mean density of about 10^6 gm/cm^3, and the companion of Sirius with a mean density of about 65 000 gm/cm^3.

As we go up along the main sequence, the densities do not differ too much from that of the sun, but decrease slowly. For red dwarfs on the main sequence, the average density is about 25 gm/cm^3, whereas for main-sequence B and O type stars the average densities are of the order of 0.01 to 0.02 gm/cm^3.

When we leave the main sequence, however, and go to giant and supergiant stars, the densities change drastically, ranging from about 3×10^{-3} gm/cm^3 for giants like Capella, to 3×10^{-7} gm/cm^3 for red supergiants like Betelgeuse. A list of masses, radii, and densities is given in Table 21.2.

Although among the observable stars those of greatest density are the white dwarfs, recent theoretical investigations by Ambartzumian, Salpeter, and others have shown that stars of enormously greater densities can and probably do exist. These stars, which are too faint to be observed optically, have densities going from hundreds of tons per cubic centimeter (**neutron stars**) to billions of tons per cubic centimeter (**hyperon stars**), and have diameters of only a few tens of kilometers. We shall discuss them in detail in Chapter 24.

EXERCISES

1. How far away from the sun would we have to go before its apparent disk were as small as the spurious disk in the 100-inch telescope?

2. If a star were 100 times as luminous as the sun but its temperature were about the same as that of the sun, how would its radius compare with that of the sun? (Assume that they both emit energy like a black body.)

3. By what distance must the mirrors of the Michelson interferometer be separated to eliminate the fringes in the spurious disk of Betelgeuse in the yellow light of sodium?

4. Compare the Einstein red shift for atoms in the gravitational field on the surface of the white-dwarf companion of Sirius with that for atoms on the photosphere of the sun.

5. The mean angular separation between the two components of a binary system is 2″ of arc and the period of the system is 3 years. If the parallax of this system is 0″.5, what is the sum of the masses in solar units of the two components of the system?

6. Use the empirical mass-luminosity relation to find the masses of a main-sequence star whose absolute magnitude is -1.

7. The following data were obtained observationally for a visual binary: $\alpha'' = 1''.5$, $P = 0.85$ years. Use the empirical mass-luminosity relation to obtain the best dynamical parallax for this system.

8. Use the empirical mass-luminosity relationship to show that if the absolute magnitude of a binary system is M, then

$$M = M_0 - 26.25 (\log \alpha'' - \log p'' - (2/3) \log P)$$

where M_0 is the absolute magnitude of the sun, and α'', p'', and P have their usual meanings.

9. Show that if there is a small error Δp in the observationally determined parallax of a binary system, the percentage error in the calculated mass is $300\, \Delta p/p''$, where p'' is the observed parallax of the system.

10. If another star with a mass equal to twice the sun's mass were revolving around the sun at a mean distance of one astronomical unit, what would its period be?

22

STELLAR INTERIORS

We have seen that it is possible from a spectral analysis and the laws of black-body radiation to determine the photospheric conditions of stars and also the chemical constitution and physical nature of their atmospheres. However, ultimately the solution of the problem of a star's atmosphere must involve a complete analysis of the structure of the star, particularly its deep interior. This is so because the temperature of the photosphere and the physical conditions in the atmosphere are determined by the flow of radiation from the deep interior and by the gravitational fields originating from the matter in the star.

22.1 Surface Conditions of a Star as Related to Its Interior

Although we say that a star like the sun radiates a certain amount of energy per second because its surface temperature is about 6 000°K, it is more proper to say that the sun's photospheric temperature is about 6 000°K because 4×10^{33} ergs of energy pass through its surface every second. *If, for some reason or other, the conditions in the interior of any star change, the star must readjust itself to accommodate the new outward flow of energy, which means that the effective temperature of the star, its color, its spectral characteristics, its radius, and the structure of its outer atmosphere must also change.*

In addition to the outward flow of energy, the mass is important in determining the surface characteristics and the atmospheric structure, since we know that the mass for a given radius alters the gravitational field and hence the pressure and density of the atmosphere. In the previous chapter we saw how sensitive the over-all stellar conditions are to any change in the mass of a star. The problem, therefore, of the structure of stellar atmospheres is part of a much more general problem—that of **internal structure and constitution of stars.**

Offhand, it seems that determining the internal structure of a star is the most difficult thing in astronomy since there is no direct way of getting information about the interior. At best, because of the extreme opacity of stellar atmospheres, the depth to which we can peer into a star is extremely small (a few hundred kilometers into the photosphere), amounting to only the tiniest fraction of the radius of the star. Nonetheless, if we make certain assumptions about the conditions inside a star, we can solve this problem to a surprisingly high degree of accuracy.

22.2 Statement of the Problem of Stellar Structure

The problem that confronts us is the following: *given a distribution of matter whose mass equals the known mass of a star, and given the chemical composition of this matter, what must be the conditions (pressure, temperature, density, acceleration of gravity, outward flow of energy, etc.) at each point within this distribution if it is to be in equilibrium, if it is to occupy the given volume, and if it is to radiate energy at the given rate?* In other words, is it possible to specify physical conditions at each point inside a star if we know its mass, its chemical composition, its radius, and its luminosity? We shall see that this can be done, and as a matter of fact, that we can do so knowing only the mass and the chemical composition of a star.

22.3 Assumptions Introduced To Solve the Problem of Stellar Interiors

We shall see that this problem can be solved if we make the following assumptions:

1. *The laws of physics that govern the interior of a star are the same as those discovered here on the earth.*

2. *The star is stable, i.e., neither blowing up nor collapsing.* This means that at each point in the star the material is in dynamical equilibrium (all the forces acting on any element, that is, a small volume of its matter, are balanced so that the element suffers no net acceleration). This does *not* mean that there is no motion of the matter at a point inside the star; it simply means that there is no net inward or outward flow of matter.

3. *The star is in a steady thermal state.* This means that the rate at which energy is produced in the deep interior is exactly equal to the rate at which it is radiated from the surface, so that there is a slow but steady outward flow of energy from the interior of the star. It also means that the temperature at each point in the star remains practically constant for very long periods. If this were not so, energy would collect at some layer inside a star, the temperature of this layer would rise, and an explosion would ultimately result. We shall exclude all such possibilities in these considerations.

4. *The star is a sphere.* This means that we consider only stars that are not rotating, since rotation introduces a flattening at the poles and hence a departure from spherical symmetry. By assuming that a star is a sphere, which is true to a high degree of accuracy for stars like the sun, we simplify the mathematics enormously, because of spherical symmetry. This means that conditions at any distance r from the center of the star are the same as those at any other point at this same distance r from the center. In other words, all the quantities with which we shall deal depend only upon one variable, the distance r from the center.

5. *The material inside a star, even at its very center, is in a perfect-gas state.* We shall see that this is more than an assumption since the conditions in a star's interior create this state of affairs.

By using these five assumptions we can now set down the equations that determine the conditions at any point inside a star. As we shall see, these equations really describe how the physical conditions change from point to point inside a stable star.

Use of the Assumptions. In order to write down these equations, we shall picture a star as being divided into a series of concentric shells (Figure 22.1), and concentrate our attention on the shell that is at the distance r from the center. Let the thickness of this shell be Δr (where the Δ represents a slight increment in r). Since we are interested in knowing the conditions in this shell, its thickness, Δr, is taken small enough so that there is no appreciable change in the physical conditions when we move from the bottom (point B) to the top (point A) of the shell. (In actually analyzing a star in this way, the thicknesses of these shells are not all taken equal.)

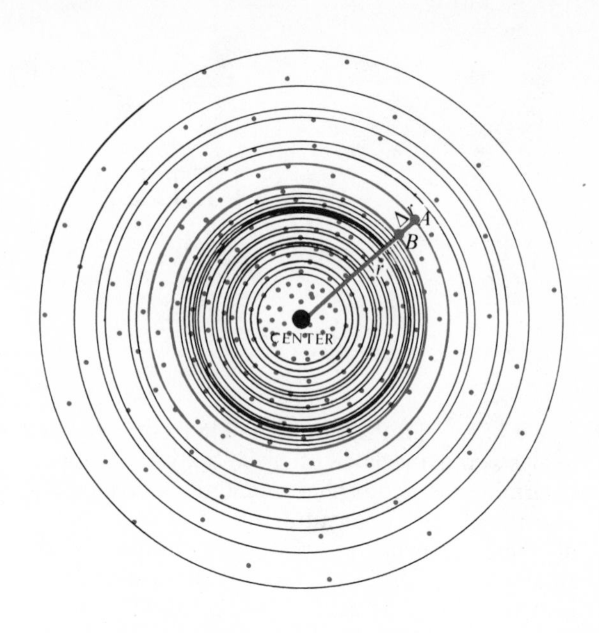

FIG. 22–1 *We picture a star as consisting of concentric shells with the conditions being the same at all points in a given shell.*

22.4 The First Equation of Stellar Interiors

If we now apply the second assumption to this shell, we can obtain the first of our equations. To do this, we consider an element of volume in this shell in the form of a small box of unit area and of height Δr. Its volume then is just Δr (height times base). If the density of the matter in the shell is ρ, then the mass of the matter in the box is just $\rho\,\Delta r$ (density times volume). If there were no force supporting this element of material, it would sink under its own weight, which is just its mass times the acceleration of gravity at this point [i.e., $(\rho\,\Delta r) \times g(r)$].

Now we know from Newton's law of gravitation that the acceleration of gravity $g(r)$ at this point arises from the mass, $M(r)$, of the material contained within the sphere of radius r. We may therefore express $g(r)$ as follows:

$$g(r) = \frac{GM(r)}{r^2},$$

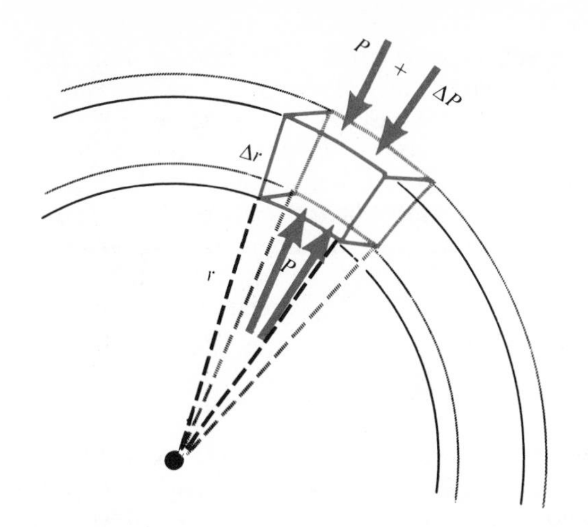

FIG. 22–2 *The forces acting on a small element of material lying in a shell at a distance r from the center of the sun.*

where G is the universal constant of gravitation, since in calculating $g(r)$ we may picture the mass $M(r)$ as being concentrated at the center. Hence this little box of stellar material is pulled downward by a force (its own weight) equal to

$$\underset{\text{weight of box of material}}{w} = \underset{\text{acceleration of gravity}}{\frac{GM(r)}{r^2}} \underset{\text{mass}}{(\rho\Delta r)}.$$

If there were no outward force to balance this weight at each point inside a star, the entire star would collapse under its own weight.

Since the material inside a star is in the form of a gas, *the gas pressure must support the weight.* Gas pressure is exerted both on the top and bottom of the box as shown in Figure 22.2, but the pressure on the bottom of the box is somewhat greater than on the top. The difference between these two pressures (called the **buoyant force**) supports the weight. If, then, P is the gas pressure on the bottom of the box, and $P + \Delta P$ is the pressure on the top of the box, ΔP represents the amount by which the pressure changes from the bottom to the top. As we shall see, *this change ΔP is negative since the pressure falls off as we move outwardly.*

Since the difference between the pressure at the bottom and at the top of the box must support the weight of the box (note that in this case since the pressure acts on a unit area, it is the same thing as the force exerted on the box by the surrounding gas) we must have

$$\underset{\text{buoyant force on the box}}{P - (P + \Delta P)} = \underset{\text{weight of the box}}{\frac{GM(r)}{r^2} \rho\, \Delta r}$$

or

$$\boxed{\Delta P = -\frac{GM(r)}{r^2} \rho\Delta r}. \qquad (22.1)$$

This is the first of our stellar-interior equations, and it says the following: If at a distance r from the center of a star we move outwardly a small distance Δr, the pressure diminishes by an amount given by the right-hand side of Equation (22.1). Thus, by using this equation we can find how the pressure changes from shell to shell inside a star, provided that we know the quantity $M(r)$ at each point (i.e., the way in which the mass is concentrated in the sphere of radius r) and also the density in each shell. Since these quantities are unknown, we must seek additional equations before our problem can be solved.

22.5 The Second Stellar-Interior Equation

We can at once find a second equation by noting that as we move outwardly from shell to shell, the quantity $M(r)$ must increase uniformly. In other words, there can be no break in the distribution of mass. To see how $M(r)$ increases when we pass through a shell of thickness Δr, all we have to do is compute the amount of matter in such a shell. Since the area of the shell at distance r from the center is $4\pi r^2$, and its thickness is Δr, its volume is just

$$\Delta V = 4\pi r^2\, \Delta r.$$

Since the density of the matter in the shell is ρ, the total mass of this shell is just

$$\underset{\text{mass}}{\Delta M(r)} = \underset{\text{volume}}{(4\pi r^2\, \Delta r)}\ \underset{\text{density}}{(\rho)}.$$

This is the amount by which $M(r)$ increases when a shell of thickness Δr is added to the star at a distance r from its center. This gives us our second equation.

$$\boxed{\Delta M(r) = 4\pi r^2 \rho\, \Delta r}. \qquad (22.2)$$

We now have two equations that govern the dynamical equilibrium of the star at each point.

However, we have three unknowns, P, $M(r)$, and ρ, so that at least one more equation must be found if we are to pursue our analysis.

22.6 Transport of Energy through Stellar Interiors

Since the two equations we have already obtained exhaust whatever information we can get from the second assumption concerning the dynamical equilibrium at a point inside a star, we must develop our third equation from assumption 3 concerning the thermal steady state at each point. It is clear that if conditions in a star must adjust themselves to allow a steady outflow of energy, there must be a continuous change of temperature from point to point to accommodate this outward flow. In other words, the third equation must tell us something about the **temperature gradient** (change in T per unit distance) in the interior of a star, and its relationship to the transport of energy.

There are three ways in which energy can be transported through a gaseous medium: **conduction**, **convection**, and **radiation**. Depending upon which of these processes is most important, we shall have one set of temperature conditions (**temperature gradient**) or another. We may dismiss conduction in the case of most stars since the coefficient of heat conductivity for gaseous matter is quite small even under the unusual conditions found in the interior of stars like the sun. However, since convection and radiation are both very important energy-transport mechanisms in stellar interiors, we shall consider each in turn.

22.7 Convective Transport of Energy: Convective Equilibrium *

When astronomers at the beginning of this century studied the problem of stellar interiors, they considered radiation as being of little importance in the transport of energy because energy is radiated in all directions with only a very small net outward flow. They pointed out that because a lower shell has a temperature that is only slightly higher than that of a shell a bit further out, the lower shell radiates only a bit more energy to the outer shell than the outer shell radiates to it. They felt that this effect is too small to account for the vast outpouring of energy from the surface of a star like the sun, and so they assumed that the most important mode of transfer is convection. *For such a process to occur, there must be a system of convective currents inside the star carrying the hot gases from the deep interior to the surface and back again.* This circulation of the gases is similar to the way the warm air from a furnace heats a house.

A star in which the transport of energy is due to convective currents is said to be in **convective** or **adiabatic equilibrium**. If this is true, then inside a star there is a very simple relationship between the pressure, P, and density, ρ, at each point:

$$P = K\rho^{\gamma}, \tag{22.3}$$

where γ equals C_p/C_v, the ratio of specific heats of the stellar gaseous material (see Chapter 16, p. 273), and K is a constant. For material at very high temperatures (no molecules, but only single atoms, are present) the quantity $\gamma = \frac{5}{3}$. However, as we shall see, there are certain regions in a star where the atoms (mostly hydrogen) are undergoing ionization. Under these conditions γ is less than $\frac{5}{3}$ since the atoms may then be treated as though they were complex molecules. If, then, the material in a star were governed entirely by convective equilibrium, Equation (22.3) would apply and we would have the third equation we are seeking. We could then solve our problem entirely, for we would have just three equations governing our three unknowns ρ, $M(r)$, and P. In fact, a good deal of work was done in the first part of this century with stellar models constructed according to these equations.

However, we now know that convection cannot be the main mechanism for the transfer of energy from a stellar interior to its surface because convective equilibrium can occur only under rather restricted conditions. *Convective currents can be maintained inside a star only if the internal temperature in any small box of stellar material (such as we considered in Figure 22.2) taking part in this convection is always out of step with the temperature of the surrounding region.*

In other words, if this box were to be given a slight outward push, thus causing it to rise to a cooler region, it could continue rising (convection would maintain itself) only if the interior of the box remained warmer than the surrounding region in spite of the fact that when the box rises and expands its interior cools off. So convection can be maintained inside a star only if the rising and therefore expanding material in a convective current cools off (because of its expansion) so slowly that its interior temperature is always higher than the surrounding temperature of the neighboring gases.

Since this condition is realized only where the temperature gradient (the change in temperature over a distance Δr) is fairly steep, most of the stellar interior is not in convective equilibrium, and the energy transport is primarily due to radiation. However, we shall see that as one gets closer to the center of the early type stars on the main sequence (i.e., *F* and earlier) the temperature gradient becomes so steep that convective equilibrium does apply. Thus, in general, we may expect the deep interior of a star to have two zones:

1. *a central core in convective equilibrium* governed by the three equations we have already obtained, and
2. *a radiative envelope* (consisting of most of the star) in which radiative equilibrium exists.

22.8 Radiative Transport of Energy and Radiative Equilibrium

To study the conditions in the major portion of the star we must, therefore, seek another equation to replace Equation (22.3). Since this new equation must describe the radiative transfer of energy from point to point inside a star, it will enable us to determine the temperature gradient from point to point. To find this equation we again consider a shell of matter at a distance r from the center, again focus our attention on the small box shown in Figure 22.2 and consider the radiation flowing through it.

We first introduce the quantity $L(r)$ which represents the total flow of energy per unit time through the lower interface of the shell at the distance r from the center of the star. The amount, therefore, that flows through the bottom surface of the box per unit time is

$$\frac{L(r)}{4\pi r^2},$$

since the box has a base of unit area. It is clear, then, that the temperature ΔT, as we move from the lower to the upper face of the box, must be sufficiently large to maintain this outward flow of energy.

We can get at ΔT by considering this radiant energy, which we may picture as being propelled outwardly through the box by the radiation pressure (just the way water pressure causes water to flow through a pipe). Let us suppose that the pressure of the radiation at the bottom of the box is $p_{\text{radiation}}$, at the top of the box is $(p_{\text{radiation}} + \Delta p_{\text{radiation}})$, so that the difference between the radiation pressure at the bottom and the top is just $-\Delta p_{\text{radiation}}$. This differential in pressure is responsible for the outward flow of the radiation.

The radiation pressure differential must be exactly balanced by (equal to) the resistance that the radiation meets as it moves out. We can, therefore, obtain our equation if we find all the factors that contribute to this resistance, and equate their product to $-\Delta p_{\text{radiation}}$. (Note that although we are speaking about radiation pressure here, we should really be talking about radiation forces. However, since the base of our box is of unit area, radiation pressure and force are numerically equal.)

22.9 The Third Stellar-Interior Equation

The first thing that contributes to this resistance to the flow of radiation is the inertia of the radiation itself. Since radiation, as we have noted (see Section 16.33), possesses momentum, a force must be exerted on it to push it through any material medium, and since the momentum associated with a given amount of radiative energy, E, is E/c, the pressure required to maintain a flow of radiation of amount E is proportional to E/c. It therefore follows that the decrease in the radiation pressure when we go from the bottom to the top of the box must be proportional to the quantity $L(r)/4\pi r^2 c$ since this is just the radiation momentum flowing through the bottom of the box per unit time. This, then, is the first of our resisting factors. *The second resisting factor is just the amount of material in the box itself.* It is clear that the more material there is in the box, the harder the radiation has to push to get through. Hence, the second resisting factor is $\rho\Delta r$ since, as we have seen, this is just the mass of the material in the box.

There is still a third resisting factor, and this arises because not all material reacts to radiation in the same way. As we have seen in discussing stellar atmospheres, different kinds of material respond differently to radiation. Thus, a given mass of metallic atoms is more resistant to the flow of radiation (i.e., more opaque) than the same mass of hydrogen. *Hence, our third resisting factor, which we shall call κ (the Greek letter kappa), is the **opacity** of the material.* Naturally this quantity depends upon the nature of the matter and the nature of the radiation, so that the value of κ must be obtained by properly averaging over all the various wavelengths, and properly taking into account the effects of the various types of atoms that are present.

We now multiply these three resisting effects together and equate them to $-\Delta p_{\text{radiation}}$. We thus get our third equation for those regions of the star where energy is transported by radiation:

$$\boxed{\Delta p_{\text{radiation}} = -\frac{\kappa L(r)}{4\pi r^2 c}\rho\,\Delta r}\,. \qquad (22.4)$$

Since we are interested in the temperature gradient instead of the radiation pressure we shall replace the left-hand side of this equation by its temperature equivalent.

We have seen that the radiation pressure is equal to $\frac{1}{3}aT^4$ (see Section 16.33). It can be shown from this that $\Delta p_{\text{radiation}}$ is equal to $\frac{4}{3}aT^3(\Delta T)$. We introduce this into the left-hand side of Equation (22.4) and, after dividing both sides by $\frac{4}{3}aT^3$, obtain

$$\boxed{\Delta T = -\frac{3\kappa}{16\pi ac}\frac{L(r)}{r^2}\frac{\rho}{T^3}\Delta r}\,. \qquad (22.5)$$

This equation determines how the temperature changes as we move outwardly through each shell of thickness Δr. The minus sign tells us that the temperature decreases as we move outwardly from the center to the surface of the star.

22.10 The Fourth Stellar-Interior Equation

The three stellar-interior equations (22.1), (22.2), (22.5) which we now have involve the five unknown quantities P, $M(r)$, $L(r)$, T, and ρ. In addition, there is present in the last of these equations the opacity κ, which also changes from point to point inside a star. It is clear, then, that we must seek additional relationships before our problem can be solved. We can obtain one such relationship immediately by considering how $L(r)$ changes as we go from one shell to the next.

Let us again consider the shell at distance r from the center so that $L(r)$, as previously defined, represents the flow of energy per unit time through the bottom of this shell. It is clear that, in general, the amount of energy flowing per unit time through the top of this shell is greater than $L(r)$, because some additional energy is generated in the shell itself. Thus, the energy flowing per unit time through the top of the shell is $L(r) + \Delta L(r)$, where $\Delta L(r)$ is just the amount of energy generated in the shell itself per unit time.

We can find an expression for $\Delta L(r)$ by noting that it depends on the product of two factors:

1. *the total material $4\pi r^2\rho\,\Delta r$ contained in the shell* since the more material there is, the greater is the energy generated; and
2. *the rate ϵ at which each gram of matter in the shell is generating energy.*

We therefore have

$$\boxed{\Delta L(r) = 4\pi r^2 \epsilon\rho\,\Delta r}\,. \qquad (22.6)$$

22.11 Eliminating the Density ρ from the Equations: The Stellar-Interior Equations in Final Form *

We thus have four equations for the five unknowns listed above, but we can eliminate one of these, the density ρ, by making use of the equation of state of the material in a star. We have assumed that the matter throughout a star is in the form of a perfect gas, so that we may write down at once the perfect-gas relationship between pressure, density, temperature, and mean molecular weight, μ, of the gas which we derived in Chapter 16, Section 16.8,

$$P = k\frac{\rho T}{\mu H},$$

where k is the Boltzmann constant and H is the mass in grams of the proton.

The presence of the mean molecular weight μ allows us to apply this formula to a gas consisting of a mixture of different elements such as we find in stellar interiors. If we solve this equation for the density ρ,

$$\rho = \frac{H}{k}\mu\frac{P}{T},$$

and substitute it into the four equations we have derived, we obtain the four equations in the final form in which they are used by astrophysicists:

$$\Delta P = -\frac{GH}{k}\mu\frac{M(r)}{r^2}\frac{P}{T}\Delta r, \tag{22.7}$$

$$\Delta M(r) = \frac{4\pi H}{k}\mu r^2\frac{P}{T}\Delta r, \tag{22.8}$$

$$\Delta T = -\frac{3H}{16\pi ack}\mu\kappa\frac{L(r)}{r^2}\frac{P}{T^4}\Delta r, \tag{22.9}$$

$$\Delta L(r) = \frac{4\pi H}{k}\mu\epsilon\frac{P}{T}r^2\Delta r. \tag{22.10}$$

In Section 21.11 we saw that stars, in general, and particularly those on the main sequence, obey an empirical mass-luminosity relationship. It is now possible to give a theoretical but crude basis for this from our interior equations. To see how, suppose that we alter some particular star (let us say the sun) by first enlarging it (or contracting it) by a given amount, and then multiplying the density at each point by a given factor. This would give us a larger (or smaller) star with a different mass and hence with a different luminosity. If we can see how this kind of change (called a **homology transformation**) changes the luminosity of a star, we can see the basis for the mass-luminosity relationship.

Our first equation tells us that when we carry out any change in the dimensions and density of a star, the pressure at a point varies as $M\rho\,\Delta r/r^2$. If we introduce an average pressure, $\bar{P}$, throughout the star, which we may picture as being defined as approximately half the central pressure, then P varies as $\bar{\rho}MR/R^2$ or $\bar{\rho}M/R$, where $\bar{\rho}$ is the mean density. But the mean density, $\bar{\rho}$, varies as M/R^3. Hence if we combine these last two expressions we see that the mean pressure varies according to the following relationship:

$$\bar{P} \propto \frac{M^2}{R^4},$$

where the symbol $\propto$ means *varies as*.

From the equation of state of a perfect gas we know that the mean temperature varies as $\bar{P}/\bar{\rho}$ or $(\bar{P}R^3)/M$. If then in this expression we use the previously obtained dependence of the mean pressure on mass and radius, we see that the mean temperature changes according to the relationship

$$\bar{T} \propto \frac{M}{R}.$$

We must now use the third of our interior equations to find out how the luminosity changes as a result of our homology transformation. Just by looking at it we see (from Equation 22.5) that the luminosity, L, varies as $(\bar{T}^4R)/\bar{\rho}$, or making use of the relationship for $\bar{\rho}$, as $(T^4R^4)/M$. If we now introduce into this the previously obtained dependence of $\bar{T}$ on M and R, we see that

$$L \propto M^3.$$

Although this is a crude derivation of the mass-luminosity relationship, we see that it agrees fairly well with the empirical result given in Section 21.11. In more advanced treatises on stellar interiors the mass-luminosity relationship is derived rigorously from the stellar-interior equations.

22.12 Initial or Boundary Conditions

With these equations the astrophysicist can determine the conditions at each point inside a star if he knows all the conditions at some particular point in the star to begin with. The reason we must have an initial set of data is that *these equations can tell us only how the quantities P, T, $M(r)$, $L(r)$ change from point to point.* We must therefore begin with a known set of values (values inside some particular shell) for a specific value of r. These known starting values for P, T, $L(r)$, $M(r)$ are called the **initial** or **boundary values** of the problem.

To illustrate, suppose we wish to find the conditions at each point inside the sun. We can begin by assigning to $M(r)$ and $L(r)$ the value $M(R_0) = 2 \times 10^{33}$ gm, and $L(R_0) = 4 \times 10^{33}$ ergs/sec at the surface, i.e., the point where $r = R_0$. If now we assign some reasonable values (determined spectroscopically or in some other way) to T and P at the surface, we can use our equations to tell us how the four quantities P, T, $L(r)$, $M(r)$ change as we move into the sun.

22.13 Mean Molecular Weight, μ

However, before we can carry out such a program, even with an initial set of values, we have to have some information about the three quantities μ, κ, and ϵ, which appear in our equations. For unless we can specify these quantities inside the star, we cannot make use of our equations. We shall first consider the mean molecular weight μ. Offhand, the attempt to determine this quantity might appear hopeless, but we shall see that because of the conditions in stellar interiors it is relatively easy to specify μ.

To obtain the mean molecular weight of a collection of different kinds of atoms or molecules in a cylinder here on earth, we take the total mass in the cylinder and divide by the total number of atoms and molecules it contains. Fortunately we do not have to do this in stellar interiors, because the temperature is so high that each atom is just about completely ionized. This makes the task of calculating the mean molecular weight fairly easy because when an atom is completely ionized, it contributes its external electrons to the total number of particles in the stellar gas. *We must therefore count each electron (just as we count each nucleus) in computing the average molecular weight.* However, although the electron must be counted as a distinct particle, it contributes nothing to the total weight of the gas so that *ionization reduces the mean molecular weight.*

22.14 Mean Molecular Weight of Completely Ionized Atoms

To illustrate this point, suppose we have an atom such as oxygen. The molecular weight of this atom is just 16, and if we have a gas consisting only of un-ionized oxygen (all the electrons bound to the oxygen nuclei) the mean molecular weight of this gas is 16. However, if the oxygen atoms are all ionized, this is not true. A completely ionized oxygen atom must be treated not as one single atom, but rather as nine different particles, each of which must be counted in taking the average molecular weight. In other words, the average molecular weight is now the total weight of these particles, 16 (since the electrons do not contribute to weight), divided by the total number of these particles which is 9. *This means the average weight of completely ionized oxygen (all electrons stripped away from the nucleus and moving as free particles) is approximately* 2 *instead of* 16.

It is easy to show that this is very nearly true for all the heavy atoms in the atomic table, starting with carbon. In fact, a heavy atom of atomic weight A contains, on the average, $A/2$ electrons in addition to its nucleus. Hence, when it is completely ionized, its mean molecular weight is very nearly $A/(A/2) \simeq 2$. We therefore lump together all the heavy atoms in a star as far as their contribution to the mean molecular weight goes. Atoms lighter than carbon must be treated separately since even when such atoms are completely ionized, their mean molecular weights are considerably smaller than 2. However, as we shall see later, hydrogen and helium are the only atoms lighter than carbon that can exist at the high stellar-interior temperatures. We need, therefore, only consider their mean molecular weights, in addition to those of the heavy atoms.

A completely ionized helium atom gives rise to three particles: the nucleus plus the two external electrons. The average weight of these three particles is therefore $\frac{4}{3}$ since the total mass is 4. This then is the contribution of helium to the mean molecular weight. From each completely ionized hydrogen atom we obtain two particles (a proton and an electron) so that the mean molecular weight of hydrogen is $\frac{1}{2}$.

22.15 Formula for Mean Molecular Weight

From this analysis we can now write down a simple expression for the mean molecular weight (of the completely ionized material in a star) which involves only the abundances of hydrogen, helium, and all the heavy elements taken together. Let us consider one gram of such a mixture and let X be the fraction of this gram that is hydrogen, Y the fraction that is helium, and Z the fraction that is everything else. It is clear then that we must have $X + Y + Z = 1$.

To get at the mean molecular weight of such a mixture, assuming complete ionization, we proceed as follows: we take the total mass (in atomic-weight units) and divide by the total number of particles. Since in the X grams of hydrogen there are X/H hydrogen atoms (H is the mass in grams of one hydrogen atom), and each of these gives rise to two particles on complete ionization, there are $2X/H$ particles arising from hydrogen. Since in the Y grams of helium there are $Y/4H$ helium atoms, and each of these gives rise to three particles on complete ionization, the number of particles arising from complete ionization of the helium is $3Y/4H$. Since in the Z grams of heavy material (assuming it to have atomic weight A) there are Z/AH heavy atoms, and each of these on complete ionization contributes $A/2$ individual particles, the Z grams of heavy substances give rise to $(A/2)(Z/AH)$ or $Z/2H$ individual particles. Thus, in this one gram of completely ionized substance there are

$$\frac{2X + (\frac{3}{4})Y + (\frac{1}{2})Z}{H} \quad \text{particles.}$$

To get at the mean molecular weight, we must now divide this into the total mass expressed in atomic-mass units. Since the total mass is just one gram, and one atomic-mass unit has a mass of H grams (the mass of the proton is one atomic-mass unit), the total number of atomic-mass units in one gram of matter is $1/H$.

If we divide this by the previous expression, we finally obtain for the mean molecular weight the formula

$$\mu = \frac{1}{2X + \frac{3}{4}Y + \frac{1}{2}Z}.$$

We can simplify this by using the equation $Z = 1 - X - Y$, so that we finally obtain

$$\mu = \frac{1}{2X + (\frac{3}{4})Y + (\frac{1}{2})(1 - X - Y)}$$

or

$$\boxed{\mu = \frac{1}{1.5X + 0.25Y + 0.5}}. \tag{22.11}$$

From this formula we can determine the mean molecular weight inside a star if we know its hydrogen-helium content.

We have thus eliminated μ from our equations and replaced it by a quantity involving X and Y. At first it might seem that this is no simplification since we have replaced one quantity by two. However, from our studies of the atmospheres of stars we know something about the abundances X and Y, so that we can always start our investigations by assigning specific values to these quantities based upon spectroscopic data. If this leads us into error, we can revise our values of X and Y and recalculate. This can be repeated as often as may be necessary to obtain a solution that corresponds to a particular star.

22.16 Variation of μ throughout a Star

Equation (22.11) for the mean molecular weight is based on the assumption that all the atoms are completely ionized. This is certainly true for the deep interior, but does not hold as we get close to the photosphere. However, a more accurate ionization formula for a gas with atoms in various states of ionization gives results that agree with the simple formula to a good approximation throughout most of the stellar interior.

It should be noted that in general the mean molecular weight is not the same at all points inside a star. *The increase of the temperature as we approach the center of a star affects the mean molecular weight in two ways. On the one hand, ionization of atoms near the center is different from what it is near the surface, and on the other hand, hydrogen is transformed into helium more rapidly near the center than elsewhere.* This, of course, results in a difference in the molecular weights between the inner and outer regions of a star. However, it may be that slow mixing takes place in stars, thus keeping the material fairly homogeneous throughout. If this does not occur, then the difference in molecular weight from point to point is an important factor in the structure of a star and must be taken into account in applying the equations. We shall indeed see that the way a star evolves as a result of the transformation of hydrogen into helium (and of the subsequent thermonuclear cooking) depends on whether or not mixing takes place.

22.17 The Opacity, κ, of Stellar Material

Now that we have eliminated the mean molecular weight as an unknown quantity, we must consider the opacity, κ. We shall see that we can express the opacity in terms of the temperature, the pressure, the hydrogen content, X, and the helium content, Y, so that it is not really an unknown quantity in our equations. To obtain such a formula for the opacity, we have to consider exactly how the atoms that are present hinder the radiation from flowing out to the surface of a star.

It is clear that the structure of a star will depend in a very sensitive way on the opacity, for if κ changes, the star must readjust all its parameters to allow whatever energy is generated in the interior to get to the surface without being blocked at any point. As an example of what can happen, suppose that the opacity becomes larger throughout a certain shell of the star. Some of the energy then is blocked in the shell and thrown back towards the interior. This results in a higher temperature in the interior and hence an increase in the rate at which energy is generated by the thermonuclear processes. This in turn causes a still greater outflow of energy, thus forcing the star to expand as a result of the increased pressure.

Let us see just what the opacity depends on. To begin with, any electrons bound to the nuclei of atoms contribute to the opacity because such electrons can absorb photons and leave their nuclei. This photo-ionization process contributes most to the opacity coefficient, κ. However, the electrons can contribute in two other ways to the opacity. If an electron is not bound to a nucleus, but moving in an open orbit while still under the influence of a nucleus, it can absorb radiation by increasing its kinetic energy. Essentially what happens is that the electron, moving in a hyperbolic orbit near a nucleus, absorbs a photon and shifts to a different hyperbolic orbit of greater total energy.

Finally, an electron can contribute to the absorption even if it is completely free and far away from any nucleus. If a stream of radiation passes such an electron, the electron oscillates in phase with this radiation the way a cork bobs up and down when a water wave passes it. The electron thus oscillating under the action of the photons passing it emits radiation in all directions, and thus scatters the radiation striking it. This scattering process, too, contributes to the absorption coefficient, κ.

22.18 Kramers' Opacity Formula

If only the first two processes are taken into account, we can write down the formula for the absorption which was first obtained by Kramers. This formula consists of a product of different factors. We first note there is a factor that merely depends upon certain physical constants, such as the charge on the electron, Boltzmann's constant, the mass of the electron, Planck's constant, the speed of light, etc. This turns out to be numerically equal to 4.34×10^{25}. Another factor that enters is the density of the material. Obviously the denser the material is, the greater is its opacity.

The temperature also enters, but its effect is opposite to that of the density. *If the temperature is high the radiation pressure is large, so that the radiation flows through the matter more easily than when the temperature is low.* Therefore, the temperature to some power must appear in the denominator of the opacity formula. The actual calculations show that this power is 3.5, so that the temperature factor is $T^{-3.5}$.

Still another factor is the abundance of atoms with electrons bound to their nuclei, because the more such atoms there are in the path of the radiation, the more effectively the radiation is blocked by photo-ionization. Since only the heavy atoms have a chance of retaining their electrons at high temperatures, this factor is just equal to the heavy-element abundance, $Z = 1 - X - Y$.

A fifth factor is the abundance of free electrons since we know that these, too, contribute to the opacity by the scattering process described above. Assuming that ionization is complete we can show that this factor is proportional to $1 + X$. The contribution of electron scattering to the absorption can be taken into account by a separate simple formula that must be used in conjunction with the Kramers' formula.

22.19 Eddington's Guillotine Factor, t

There is one other factor, the importance of which was first recognized by Eddington. If we take all the factors listed above and multiply them together, we obtain the formula first introduced by Kramers, but that formula really gives us the opacity on the assumption that every atom continues to contribute to the opacity no matter what happens to this atom. However, it is clear that because of ionization we have to be somewhat careful about counting atoms that can contribute to the opacity at any moment. If some particular electron in an atom has absorbed a photon and thus been torn out of the atom, that particular atom is no longer capable of absorbing radiation at the same rate as before until it has captured another electron to replace the one it lost. In other words, *there are moments right after the absorbing process when atoms cannot contribute fully to the opacity, because, as Eddington pointed out, they do not have their full quota of bound electrons.* To correct for this, he introduced the **guillotine factor**, t, which must divide Kramers' formula and which depends on the state of ionization of the atoms. Today the guillotine factor includes all corrections that are necessary to make Kramers' formula agree with the actual opacity.

There is still another correction that has to be introduced in Kramers' formula (of which he was not aware when he derived the formula). It arises because electrons have wave properties (see Section 19.31). This correction, which does not change Kramers' formula very much (it is approximately equal to 1), was first introduced into the opacity by Gaunt, and is known as the **Gaunt factor**, g.

If we take all of these factors into account, we finally obtain the opacity formula as it is used in most modern calculations of stellar interiors:

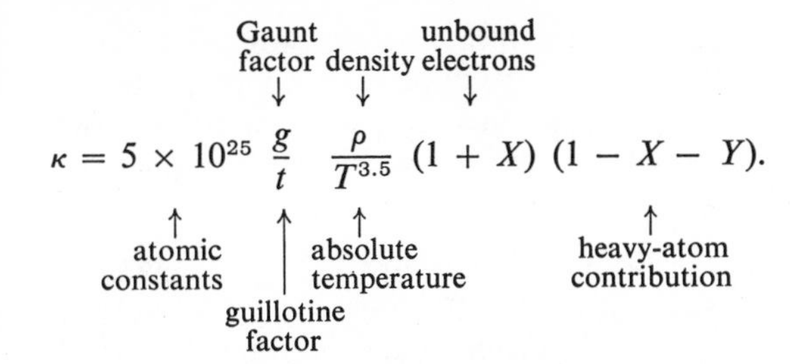

$$\kappa = 5 \times 10^{25} \frac{g}{t} \frac{\rho}{T^{3.5}} (1 + X)(1 - X - Y).$$

The opacity formula, and therefore the internal structure of a star, are quite sensitive to the guillotine factor, t, which depends not only upon how much heavy material is present, but also upon the relative abundances of the various kinds of heavy atoms. Thus, to take an example, if only half of the heavy material (all of which may account for only a few per cent of the star's mass) consists of oxygen (or atoms of about the same atomic weight) with the other half consisting of iron-like atoms, we obtain a value of the guillotine factor (and hence of the opacity) which is quite different from what it would be if 80 per cent of the heavy matter were oxygen, and 20 per cent were iron-like atoms.

22.20 The Guillotine Factor and the Opacity Tables *

Because of this sensitivity of the guillotine factor to the relative abundances of heavy elements, extensive calculations have been carried out by G. Keller and R. Meyerott, who derived opacity tables for various assumed mixtures of hydrogen, helium, and heavy atoms. These mixtures range all the way from groups of atoms containing 99 per cent hydrogen and 1 per cent all other elements to groups containing no hydrogen, 50 per cent helium, and 50 per cent heavy material. These tables give the opacity for 12 such groups of elements for temperatures ranging from 100 000°K to about 24 000 000°K.

To use the opacity formula in our stellar equations, we must first find an interpolation formula for t/g which gives an opacity that best fits the opacity values of Keller and Meyerott (or equivalent tables) for a wide range of temperatures. When this is done we find, in general, that different opacity formulae have to be used in different parts of the star, because of the complicated way the absorption changes as conditions change from point to point.

In the past, rather simple formulae such as

$$\frac{t}{g} = A[(1 + X)\rho]^n$$

have been used for the guillotine factor, where the coefficient A and the exponent n are chosen to make the formula fit the data in the tables as well as possible. Although such simple formulae are known to be inadequate for the detailed analysis of the interiors of individual stars, they have been useful in applying the interior equations to groups of stars in order to study their evolution over long periods of time.

If in the opacity formula we replace ρ by its value $(\mu H/k)(P/T)$ (from the gas equation) and introduce the interpolation formula for t/g, we then obtain an expression entirely in terms of unknowns P, T, and the hydrogen and helium abundances X and Y, so that the opacity itself is not really an unknown. However, we see again that we must know X and Y before we can solve our problem.

22.21 Energy Generation, ϵ, and the Ages of Stars

Although we have eliminated μ and κ as separate unknowns from our equations, the energy generation ϵ still remains to be considered. We shall now show that ϵ, just like these other quantities, can be expressed in terms of P, T, X, and Y, so that the entire problem can be reduced to finding the values of P, T, $L(r)$, $M(r)$ at each point inside a star of given mass for a given choice of X and Y. First, however, we shall consider briefly how energy can be produced inside a star where conditions are so vastly different from what they are here on the earth (except in the immediate vicinity of an exploding hydrogen bomb).

We may note (as an example) that the sun emits 4×10^{33} ergs per sec, or on the average 2 ergs per sec for each gram of its mass. Now we know from geological considerations and the study of uranium deposits in the earth that our solar system has been in existence for well over 3 billion years, so that the sun has been generating energy for at least 10^{17} seconds. This means that, on the average, each gram of matter in the sun must be capable of supplying (by some process or other) at least 2×10^{17} ergs. In ordinary chemical reactions (such as the burning of coal or oil) the total energy liberated per gram of matter is about 10^{12} ergs, which means that even the most efficient chemical fuel could not keep a star like the sun going for more than thirty or forty thousand years. It follows, therefore, that energy must be generated by some process other than chemical.

A possible mechanical process for generating the large amounts of energy radiated by a star is a slow contraction of the star, which releases the potential energy of the gravitational field. We know that a sphere of radius R and mass M must have radiated away a total gravitational energy of the order of GM^2/R per gram of matter in the sphere in contracting to its present size. For the sun this amounts to about 2×10^{15} ergs, which is 100 times less than what is required to account for the present age of the sun. In other words, the gravitational contraction gives an amount of energy 100 times too small and, hence, could have kept the sun going only for about 30 million years, instead of the billions of years required.

However, although gravitational contraction cannot be the principal source of energy for a star, it may be important at certain stages in a star's development. We shall see there are periods in a star's history when its core, depleted of thermonuclear fuel, must contract under its own weight and the weight of the overlying material. Gravitational energy is thus transformed into radiant energy. This contraction must be taken into account in considering the star's internal structure at that stage of its evolution.

22.22 The Need for Thermonuclear Reactions in Stellar Interiors

Since neither chemical nor gravitational processes can account for the total energy emitted by a star, we must look for the main source in nuclear transformations. Now we know that nuclear energy can be released either by **fission** (as in the case of early atomic bombs and atomic piles) or by **fusion** (as in the case of the hydrogen bomb), but from our knowledge of the very small abundances inside stars of the heavy elements that can undergo fission (uranium, etc.), we can dismiss the former process and concentrate on thermonuclear fusion. The only question we face, then, concerns the type of thermonuclear fuel that is involved in fusion processes inside stars. Here we must be guided by two considerations:

1. the fusion process must take place so slowly that the total nuclear energy is released gradually during billions of years;
2. the nuclear fuel must be so abundant that it can last for the required period of time.

Also involved is the question of whether the star is in its early stages of development or at an intermediate stage when its central temperature becomes very high. In the early stages of a star's development, the thermonuclear fuel must be sought among the light elements such as hydrogen. But since hydrogen is the most abundant substance in the universe it will still be one of the components of the nuclear reaction in the intermediate stages. After a certain fraction of hydrogen has been consumed in the early and intermediate evolutionary stages, we shall see that He^4 will then become the thermonuclear fuel.

A simple calculation shows us that thermonuclear fusion can adequately account for the energy released by stars. When four protons are fused to form a single He^4 nucleus, we see that 0.029 atomic units of mass are transformed into energy since the mass of He^4 (in atomic units) is 4.003, whereas the mass of four protons is 4.032 atomic units. In other words, for each gram of helium that is formed, about 0.007 grams of matter are transformed into energy. From Einstein's relationship $E = mc^2$, we find that this amounts to $0.008 \times (9 \times 10^{20})$ ergs, so each gram of matter that takes part in the thermonuclear fusion of hydrogen to helium releases about 7×10^{18} ergs. This is about thirty times more than is required to account for the energy per gram (2×10^{17} ergs) released by the sun since its formation. Thus this thermonuclear source of energy of the sun is certainly sufficient.

EXERCISES

1. Show that if one moves into a star through a thin shell that contains an amount of mass equal to $\Delta M(r)$, and which is at a distance r from the center, the pressure increases by an amount

$$\frac{GM(r)}{4\pi r^4}\Delta M(r).$$

2. Calculate the mass contained in a shell 10 km thick at a distance of 10^8 cm from the center of a star if the density in this shell is 20 gm/cc.

3. How much larger is the pressure at the bottom than at the top of the shell in problem 2 if the amount of mass of the star lying in the sphere surrounded by the shell is 10^{33} gm?

4. Find the amount by which the absolute temperature at the bottom of the shell in problem 3 is larger than at the top of the shell if the opacity κ in this shell is 20 cm^2/gm, the absolute temperature T in the shell is 2×10^6 °K, and the energy flux $L(r)$ is 3×10^{33} ergs/sec.

5. Find the radiation pressure and the gas pressure in the shell of problem 3 if the mean molecular weight of the material in the shell is 0.7.

6. What is the mean molecular weight of the material inside a star if ionization is complete and if the hydrogen content is 80 per cent and the helium content is 15 per cent?

7. Assume that the "guillotine factor" t/g is equal to 1 at a point inside a star where the density ρ is 100 gm/cm^3, the temperature is 1.5×10^7 °K, the hydrogen content X is 85 per cent and the helium content Y is 12 per cent. What is the value of the opacity κ?

8. Show that if the energy is transported by convection through a thin zone of a star, the gas pressure at the bottom of this zone is larger than that at the top of the zone by the amount $(\gamma P/\rho)\Delta\rho$, where γ is the ratio of the specific heats of the material in the zone, P is the gas pressure in the zone, ρ is the density of the material in the zone, and $\Delta\rho$ is the small difference between the density at the top and the bottom of the zone. [Hint: use Equation (16.10).]

9. What happens to the acceleration of gravity inside a star as one approaches the center of the star? Prove your answer.

10. Show that in the convective core of a star the change in pressure, ΔP, is related to the change in the temperature, ΔT, by the equation

$$\frac{\Delta P}{P} = \frac{\gamma}{\gamma - 1}\cdot\frac{\Delta T}{T}$$

where γ is the ratio of specific heats. [Hint: use Equation (16.10) and the gas equation.]

11. (a) Show that if a star is in gravitational equilibrium the quantity

$$P + \frac{GM^2(r)}{8\pi r^4}$$

decreases as one moves out from the center. (b) Use this result to obtain a lower limit for the central pressure in such a star.

23

NUCLEAR STRUCTURE AND STELLAR EVOLUTION

23.1 The Atomic Nucleus

Now that we have discussed the general conditions a thermonuclear reaction must fulfill to be a source of stellar energy, we must set down the expression for ϵ which is used in the stellar-interior equations. First, however, we must briefly discuss nuclear structure.

The study of the nucleus is much more difficult than that of the outer parts of the atom, because in the case of the latter, as we have seen, we deal with planetary electrons circulating around a central core. Our main concern there is with the interaction of individual electrons with the nucleus, even though these electrons interact with each other also. The effect of the nucleus on each electron is so large that the inter-electronic reactions may be disregarded to a first approximation.

This does not hold inside the nucleus, for *there is no central body in a nucleus, but rather a group of massive particles called* ***nucleons*** (neutrons and protons). Since each of these is equally important, we are at once involved in the *n*-body problem. Furthermore, whereas we know the nature of the force between the nucleus and the electrons (Coulomb's inverse-square law of force; see Section 18.4), we do not know the exact nature of the force between two nucleons, although we do know a great deal about its magnitude.

Until about 1933 physicists did not know what particles other than protons composed a nucleus, although they were inclined to believe that these were electrons. However, more and more evidence was accumulated during the 1920's against this idea. We may understand why electrons were first thought to inhabit the nucleus if we just consider the He^4 nucleus (i.e., the α particle). Since its atomic weight is 4, it was thought originally that there must be four protons in this nucleus to account for its mass. But since the He^4 nucleus has a total positive charge of 2, two of these protons must have their positive charges balanced by two electrons. Thus, the helium nucleus was pictured as consisting of 4 protons and 2 electrons moving in some complicated way. Another reason physicists thought it necessary to have electrons in the nucleus is that certain nuclei (such as strontium 90) are known to emit electrons spontaneously (β radioactive decay). These nuclei then change into nuclei of different atomic number.

However, with the development of quantum mechanics, it became clear that electrons cannot reside in the nucleus because their de Broglie wavelengths under ordinary conditions (see Section 19.31) are so much larger than the nuclei themselves. Electrons could stay inside nuclei only if their de Broglie wavelengths were very small. But this could be so only if the electrons had very high speeds—and then the nuclear binding forces would be too small to hold them inside the nucleus. We may also put it differently: if an electron were inside a nucleus (which means that we would know its position very accurately), its momentum would have to be so large (because of the uncertainty principle) that it could not be held in by the available forces. Another objection to electrons inside nuclei is that the total spin (the angular momentum) of certain nuclei, as measured experimentally, would not agree with the value obtained by adding up all the spins of the electrons and protons in the nucleus.

This situation remained a puzzling one until 1933, when the neutron was discovered as a particle emitted by certain nuclei bombarded by α particles. This suggested that only neutrons and protons exist inside nuclei. Using neutrons and protons, we can now construct a consistent model of any nucleus in which the constituent nucleons (the word nucleon is used to represent either a proton or neutron) are held together by nuclear forces vastly greater than the Coulomb electrostatic forces between electrons and nuclei.

23.2 The Neutron and the Nucleus

The neutron is slightly more massive than the proton, but it has no charge. Hence we can account for the structure of a nucleus like helium by introducing two protons and two neutrons, so that electrons are no longer needed. There is no difficulty in having neutrons inside a nucleus; the reason is that these particles (like protons) have very small de Broglie wavelengths since their masses are large compared to the mass of the electron.

In this neutron-proton model of the nucleus *we define its mass number, A (the integer closest to its atomic weight), as the sum of the total number of protons and neutrons in the nucleus. The atomic number, Z, is then equal to the number of protons.* Nuclei with the same Z but different A are called **isotopes** of the same element; and nuclei with the same A but different Z are called **isobars**.

23.3 The Half-Life of the Neutron and β Radioactivity *

It is possible to account for β radioactivity (the emission of electrons from certain nuclei as described above) by the fact that a neutron when left to itself (outside a nucleus) is unstable. *It spontaneously changes into a proton and an electron in about 12 minutes on the average, which is the half-life of the neutron.* In other words, if we start out with a certain number of neutrons, we have only one-half that number after 12 minutes, one-fourth the initial number after 24 minutes, and so on. Because a neutron behaves this way, it is clear that in certain nuclei (if these nuclei have too many neutrons as compared with protons) some of the neutrons will change spontaneously into protons and the resulting electrons will be emitted as β rays.

23.4 The Neutrino and β Radioactivity

When the neutron changes into a proton another particle—one that is almost impossible to detect directly—is emitted together with the electron. This must be so because without the emission of this elusive particle there could be neither **energy balance** nor **spin balance** when neutrons change to protons.

As far as the energy balance goes, the situation is as follows: since the masses of the neutron and of the proton are definite quantities and constants of nature, the amount of energy released when a neutron changes into a proton must be a constant (in mass units) and exactly equal (in mass units) to the difference between the mass of the neutron and the mass of the proton. In other words, in this decay of the neutron, if only an electron were emitted, the electron would always come out with the same energy (equal to the mass differential between the neutron and the proton). But this does not happen, for we find that the β rays (electrons coming from the decayed neutrons) are emitted with a whole range of energies. The only way, then, that an energy balance can exist is if some other particles having a complementary range of energy are emitted to make up this difference. Although the existence of these particles had been suggested by Pauli, the properties assigned to them are such as to make them practically undetectable, for they have neither charge nor rest mass. For this reason Fermi named these particles **neutrinos.** However, there are now experimental methods available for detecting these particles and studying their properties.

We come now to the spin balance as a reason for the existence of neutrinos. We know experimentally that the neutron has the same spin angular momentum, $\frac{1}{2}h/2\pi$, as the electron and proton. The quantity $h/2\pi$ is here taken as the unit of spin angular momentum (see discussion on spin, Section 19.27). If, then, a neutron changed only into an electron and a proton, without emitting any other particle, we would have a situation in which we start out with half a unit of spin (the spin of the neutron) and end up either with a total spin of zero (the resulting electron and proton come out with their spins opposite to each other, so that the total spin is 0) or with a total spin of 1 (the electron and proton come out with their spins pointing in the same direction as shown in Figure 23.1). In either case the final total spin would differ from the initial spin so that the spins could not balance.

This difficulty vanishes with the introduction of a neutrino if one assigns to it $\frac{1}{2}$ unit of spin. We then have three particles when the neutron decays, each with $\frac{1}{2}$ unit of spin, and it is always possible to arrange them to give a total final spin of $\frac{1}{2}$ unit, which is required because the initial spin (the spin of the neutron) is also $\frac{1}{2}$ unit.

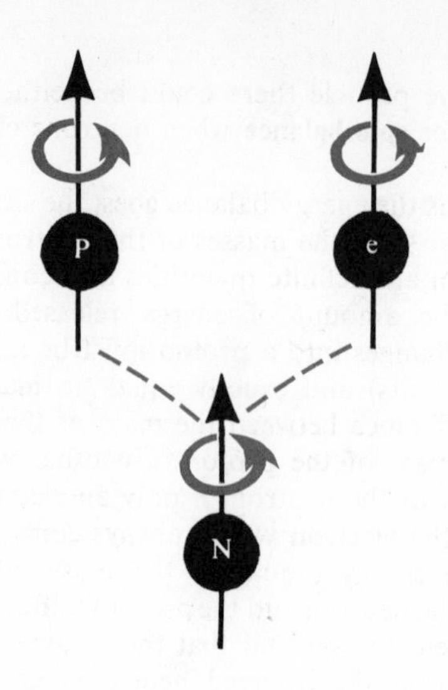

FIG. 23–1 *The decay of a neutron into a proton and an electron, in which the spins of all three particles are parallel. This means that the initial spin of the neutron cannot be balanced against the final spin unless another spinning particle is emitted.*

23.5 Nuclear Forces and the Nucleus

Now that we know the nucleus consists of neutrons and protons, the only question that remains is what kind of forces keep these particles (the nucleons) together. We can say two things about these forces at once. To begin with, they are much greater than the repulsive electrostatic forces, otherwise these repulsive forces between the protons would cause the nucleus to fly apart. Second, these nuclear forces must have a very short range as compared with the electrostatic Coulomb forces, so that they are brought into play only when two nucleons are within a few ten trillionths of a centimeter of each other. If this were not the case, the nucleus would not be a small, tightly bound structure, but rather a diffuse one with the nucleons separated by relatively great distances.

23.6 Nuclear Forces and Mesons *

By studying the way nucleons interact with each other, we have learned a great deal about these nuclear forces, and we now know that all nucleons exert equal forces on each other. Thus, the nuclear force between two neutrons is equal to that between a neutron and a proton and to that between two protons. We also know that two nucleons attract each other through the intermediary of a field of particles called **π mesons** of which there are three types, π^+ (positive charged), π^- (negative charged), π^0 (neutral). Each of these mesons has a mass of about 270 times the mass of an electron. If left to themselves, these π mesons decay (in about a 100-millionth of a second) into neutrinos and particles called **μ mesons.** A μ meson has a mass of about 207 times the mass of the electron. These μ mesons in turn decay into electrons and neutrinos in about a millionth of a second. When a meson decays, the mass it loses is converted into the kinetic energy of the resulting particles.

To understand how a neutron and a proton attract each other via the meson field, we may picture a neutron as momentarily giving rise to a negative π^- meson which is then instantaneously absorbed by the proton. In this rapid transformation, the neutron becomes a proton and the proton a neutron, so that the two particles behave as though they had exchanged places. Immediately following this the original proton (which is now a neutron) gives birth to a π^- meson which is then absorbed by the original neutron (now a proton) so that the original configuration is re-established. Immediately after this there is another exchange of a π^- meson, and this process continues with a π^- meson being thrown back and forth between two nucleons.

In this way the two nucleons are held bound to each other at a very close distance since it is only when the neutron and proton get very close together that a particle as massive as a π meson emitted (so-called **virtual emission**) by one of them can be absorbed by the other. It is because the π meson is a massive particle that the range of nuclear force is so very small and its magnitude is so very large.

We have discussed nuclear forces in the preceding paragraphs in terms of the π^- mesons but the neutral and positive π mesons also play a role in nuclear interactions. The existence of the π^+ meson is required to maintain proper symmetry in nature, and the π^0 meson is necessary to account for the nuclear forces between two like nucleons such as between neutron and neutron or between proton and proton. Since, as we have noted, all these nuclear forces are equal, the π^+, π^0, π^- mesons all have very nearly the same mass.

23.7 Nucleon-Meson Interactions

There are simple and elegant ways of representing the interactions between nucleons diagrammatically.

These are illustrated below (the solid lines represent nucleons, the dotted lines, mesons).

1. a proton emitting a π^+ meson or a neutron absorbing a π^+ meson (the proton becomes a neutron):

$$p \longleftrightarrow n, \pi^+, \qquad p \longrightarrow n + \pi^+$$

2. a nucleon (proton or neutron) emitting or absorbing a π^0 meson (here the nucleon remains unchanged):

$$\text{nucleon} \longleftrightarrow \text{nucleon}, \qquad \text{nucleon} \longrightarrow \text{nucleon} + \pi^0$$

3. a neutron emitting a π^- meson or a proton absorbing a π^- meson:

$$n \longleftrightarrow p, \pi^-, \qquad n \longrightarrow p + \pi^-$$

4. a neutron and a proton approach and interact by exchanging a π^+ meson (the proton becomes a neutron and the neutron becomes a proton):

p → n, exchanging π^+; n → p

5. a neutron and a proton interact by exchanging a π^- meson:

p → n, exchanging π^-; n → p

6. any two nucleons interact by exchanging a π^0 meson:

nucleon_1 → nucleon_1, exchanging π^0; nucleon_2 → nucleon_2

Since in each of these processes only one meson is involved (a single particle of the same mass is tossed from one nucleon to another in each case), the interaction forces must be the same. Although the range and magnitude of the force between two nucleons (and other properties of this force) can be understood in terms of these diagrams, the detailed characteristics of nuclear forces are so complicated as to be almost beyond analysis.

23.8 The Nature of Nuclear Forces

First of all the nuclear force cannot be represented mathematically by an inverse-square law. The force becomes repulsive when the two nucleons are closer than 10^{-14} cm, and then strongly attractive at distances greater than this, but less than 10^{-13} cm. Beyond this distance it falls off very rapidly but the exact mathematical form of this law is not known. Fortunately, the behavior of nucleons does not depend sensitively on the mathematical form of the law, but rather on how big the force is.

In addition to depending on the distance between the two interacting nucleons, the force also depends (but weakly) on the relative directions of spins of these two nucleons, and also on the directions that these spins take with respect to the line joining the two nucleons. Finally, we find that this force depends on how the total spin of the two particles is oriented with respect to their velocities.

Since the range of nuclear forces is very small, one nucleon attracts another only if the two are very close together. From this it follows that inside a nucleus each nucleon interacts only with its nearest neighbors so that the structural properties of the nucleus resemble those of a liquid drop. This "liquid drop" model has greatly assisted physicists in understanding the statistical behavior of nuclei.

23.9 Nuclear Binding Energies

With these preliminary ideas we can now understand how heavy nuclei are built up from light ones, but before we can do this we must introduce the concept of the binding energy of a nucleus. **Binding energy** *is the amount of energy required to disrupt a nucleus completely, i.e., required to separate all the nucleons so that they no longer attract each other, or, put differently, the amount of energy released when the nucleons come together to form a nucleus.*

The lightest nucleus H^1 is the proton itself. Following it, in order of increasing atomic weight, is the nucleus consisting of two nucleons, and since a combination of two protons is unstable, we have the only other possibility, a proton (represented by the symbol +) and a neutron (n). This structure, H^2, is called a **deuteron**, which we represent by ($+n$). The binding energy of the deuteron is about 2.23 Mev (million electron volts) (see Section 18.23). If we compare this with the ionization energy of hydrogen or helium (of the order of tens of electron volts) we see that the nuclear binding energies are about a million times larger. Because of this, great quantities of energy can be released in thermonuclear fusion.

As we go to heavier nuclei, the total binding energy of course increases. We shall consider the more significant quantity, the binding energy per nucleon (which also increases). For example, whereas He^4 has a binding energy per nucleon of about 7.1 Mev, nuclei like O^{16} and Ne^{20} have average binding energies per nucleon of about 8 Mev, and nuclei like Si^{28} and Ca^{40} have binding energies per nucleon of 8.6 Mev. The maximum binding energy per nucleon of 8.8 Mev is reached for Fe^{56}. For nuclei heavier than this the average binding energy per nucleon decreases; we shall see that this has an important bearing on the evolution of stars, and the formation of heavy elements beyond iron.

23.10 The Structures of Some Light Nuclei

To return to the structure of nuclei, we note that after the deuteron we have a three-nucleon system which may consist either of one proton and two neutrons, ($+nn$) (H^3), an isotope of hydrogen called **tritium**, or two protons and one neutron, ($++n$) (He^3), which is an isotope of helium. Tritium is radioactive with a half-life of 12 years and decays into He^3 by emitting a beta particle (electron) and a neutrino.

There is only one stable combination of four nucleons (the alpha particle or He^4 nucleus). It contains two protons and two neutrons ($++nn$). There is no known nucleus consisting of five nucleons so that at this point there is a gap in the table of nuclei. However, nuclei containing six nucleons do exist. Radioactive He^6 ($++{}^{nn}_{nn}$) (two protons and four neutrons) which decays in less than a second by emitting a β^- particle is an example of such a nucleus.

23.11 Heavy Nuclei

By adding more and more neutrons and protons, we obtain heavier and heavier nuclei and we find that the number of isobars associated with a given number of nucleons becomes larger and larger, because the number of possible stable combinations of nucleons increases. In general, the most stable nuclei are those in which the number of protons is about equal to the number of neutrons. However, because of the Coulomb repulsion between protons we can obtain stable nuclei as we go higher and higher in the atomic table only by introducing more neutrons than protons to offset the increasing Coulomb repulsion. If there are either too many neutrons or too many protons in a nucleus, the nucleus is not stable and radioactive decay sets in. *If there are too many protons, the nucleus decays by emitting a* ***positron*** *(β^+) and if there are too many neutrons, it decays by emitting an* ***electron*** *(β^-).*

23.12 Graphical Representation of the Nucleus

These simple ideas about nuclei now enable us to indicate how one nucleus can combine with another to form a more complicated compound nucleus. We can best see what happens with the aid of a graphical representation. We shall first consider a nucleus without taking into account its positive electrostatic charge. Clearly, if a nucleon (either a proton or a neutron) came within the nuclear force range of such a nucleus, the nucleon would immediately be absorbed because of the large nuclear force exerted on it. In other words, it would fall right into the nucleus.

We may illustrate this graphically by picturing such a nucleus (disregarding the electrostatic charge of the protons) as a deep well into which any proton or neutron can fall on close approach. This is shown in Figure 23.2. The depth of the well is related to the total binding energy of the nucleus; the horizontal lines within the well represent the energy levels in which the individual nucleons in the nucleus may be pictured as oscillating back and forth.

This picture of the nucleus is incomplete since it disregards the Coulomb repulsion between the protons in the nucleus and also the electrostatic field surrounding the nucleus. Since such an electrostatic field repels a positive charge approaching the nucleus, we can represent it graphically by a hill (Coulomb barrier) which gets steeper and steeper as we approach the nucleus and reaches a maximum value at the point where the well begins.

Thus, if we are to take both the well and the hill into account we may represent a nucleus graphically as shown in Figure 23.3. The height of the

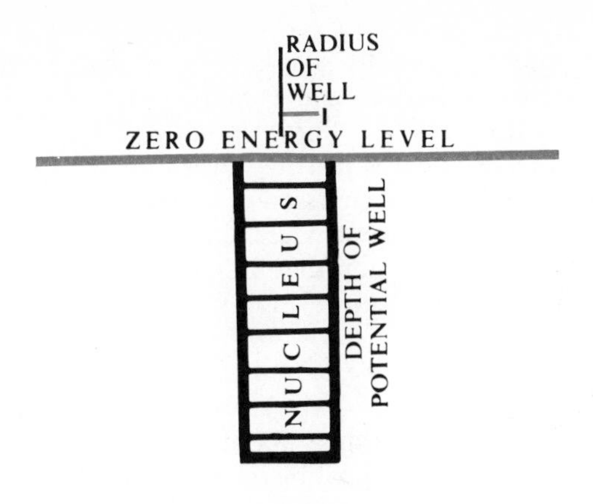

FIG. 23–2 *If a nucleus had no electric charge on it, it could be represented by a deep well below the zero energy level.*

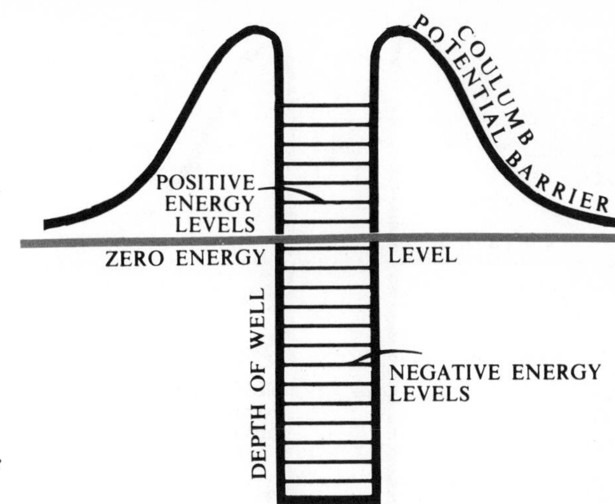

FIG. 23–3 *A schematic representation of the Coulomb potential barrier surrounding the nucleus and merging into the nuclear potential well. The lines within the well represent the energy levels of the particles (neutrons, protons) bound in the nucleus.*

Coulomb repulsion barrier depends upon the total charge on the nucleus (i.e., the total number of protons) so that it must be made higher and higher as we go to heavier and heavier nuclei. Now if we look at this picture, we realize that a neutron can interact with a nucleus much more readily than a proton does because as far as a neutron is concerned, it has no Coulomb barrier to contend with before it can combine with the nucleus. *In other words, as far as a neutron is concerned, it sees a nucleus according to our first picture (i.e., merely as a well). But as far as a proton is concerned, the Coulomb barrier is a reality and the proton must first penetrate this barrier or rise above it before it can fall into the well of the nucleus.*

From these considerations, it is clear that whereas slow neutrons can be used to bring about nuclear transformations, very energetic protons are required because the protons must be moving fast enough to climb completely up the hill or else to tunnel through it at some point. We shall return to the question of tunneling through the Coulomb barrier a bit later. Since a heavy nucleus has a high Coulomb barrier surrounding it, two heavy nuclei have much more difficulty interpenetrating and forming a new compound nucleus than two light nuclei do.

23.13 Penetrating the Coulomb Barrier

From Figure 23.3 showing the Coulomb barrier and the well we can draw a few important conclusions. We notice that the figure is divided into two parts by the zero energy level. A particle inside the nucleus but below this level has a total amount of energy (kinetic plus potential energy, where the potential energy arises because of nuclear forces) that is negative. Such a particle has no chance of leaving the nucleus and simply oscillates back and forth in one of the possible discrete energy levels. In other words, a nucleon with less than zero energy is a captive inside a nucleus and moves in a closed orbit just the way a planet whose total energy is less than zero (see Section 8.33) is a captive of the sun and moves in a closed orbit in the solar system. However, if a particle in a nucleus has a total amount of energy that is positive, it oscillates back and forth in one of the levels above the zero energy line. Because such a particle has wave properties (as we have already seen in Chapter 19) it must be represented by a wave, one part of which (with a very large amplitude) lies inside the nucleus, and the other part of which (with a very small amplitude) lies outside as shown in Figure 23.4.

Since the amplitude of the wave of this nucleon is much larger inside the nucleus that it is outside, the nucleon spends most of its life inside the nucleus. But because part of the nucleon wave lies outside the nucleus (with a very small amplitude) there is a very small but finite chance of finding the nucleon outside the nucleus. In other words, if the total energy of a particle inside a nucleus is positive, there is always a finite probability that the particle will penetrate the Coulomb barrier as it

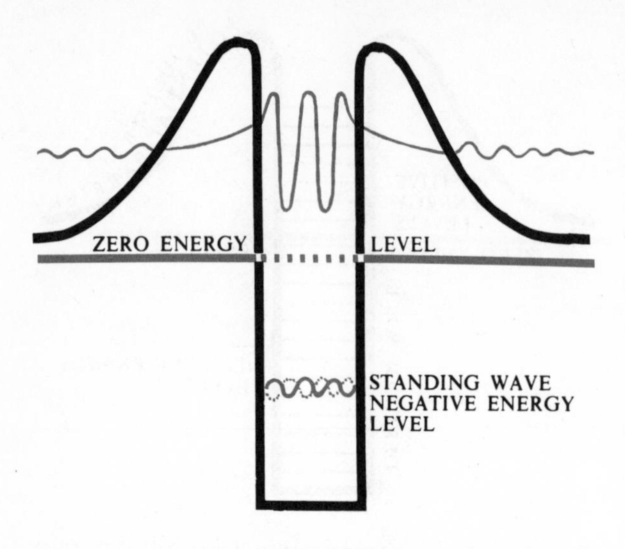

FIG. 23–4 *The Schroedinger wave pattern of two particles of a nucleus, one in an energy level below the zero energy level, the other in an energy level above the zero energy level.*

oscillates back and forth and leave the nucleus. The narrower the barrier is, the better is the chance for penetration, so that a particle in a nucleus with a lot of energy has a good chance of escaping.

23.14 α-Particle Emission

By using the fact that particles can thus tunnel through a Coulomb barrier, G. Gamow was able to account for the radioactivity of heavy nuclei like uranium. In such a nucleus, in which many particles are crowded together, there are some that are forced to oscillate in very high energy levels. *If, among these, two protons and two neutrons get close enough together they may coalesce to form an α particle which then escapes by tunneling through the barrier.* This accounts for the emission of α particles by heavy radioactive nuclei.

23.15 The Problem of Nuclear Fusion

Since in stars we are interested not in the decay of a nucleus, but rather in the fusion of two or more nuclei, we must consider the reverse process of the emission of α particles by nuclei. In other words, we must consider the following problem: given a nucleus, represented by a Coulomb barrier superimposed on a deep well, what is the chance that another nucleus similarly represented will penetrate the barrier and combine with the first nucleus to form a third compound nucleus? Since this problem is just the reverse of the radioactive problem (the emission of α particles from a complex nucleus) we can use the same procedure to answer this question that Gamow used to solve the α-particle problem. This will enable us to find the quantity ϵ.

However, before we calculate the quantity ϵ on the basis of what we have already said, we shall state a few obvious things about the nuclear transformations that are possible in stars like the sun. We see at once that although it is very easy to bring about nuclear transformations with neutrons, these do not play much of a role inside stars (except under certain unusual conditions which we shall discuss in Section 23.49) because neutrons decay so quickly that ordinarily none is present in the interior of stars. We also, in general, dismiss interactions between very heavy nuclei because their Coulomb barriers are so high that these nuclei would have to be moving at very high speeds for interactions to occur, and even at temperatures of tens of millions of degrees such nuclei have speeds that are much too small to bring about any appreciable nuclear transformations.

We are, therefore, left with nuclear reactions involving the lightest elements, and of these hydrogen is the most important because it is the most abundant. What then are the possible nuclear reactions inside stars where hydrogen is one of the interacting nuclei? We shall see that there are two such sets of reactions which can account for the great quantities of energy released by stars like the sun, both of which lead to the fusion of four protons to form a helium nucleus.

23.16 The Proton-Proton Chain *

*The first set, which starts as a direct interaction between two protons themselves, is called the **proton-proton chain**. Although this set of reactions proceeds very slowly at the relatively low temperature found in stars like the sun, it is the primary mechanism for energy release in these stars and in the stars lying below the sun on the main sequence.*

The proton-proton chain consists of three steps. In the first step, two protons coalesce to form a deuteron (D) with the emission of a positron and a neutrino. In this step, 1.44 Mev of energy are emitted although not all of this energy leaves the star as electromagnetic radiation, since a small

TABLE 23.1

ELEMENTARY PARTICLES

Classification			Name	Symbol	Charge	Mass M_{ph}		Spin	Life in free space	Decay
						electron	atomic weight (physical scale)		seconds	
Bosons			Photon	γ	0	0	0.0	1	infinity	stable
Bosons	Mesons		π meson (pion)	π^+, π^-	+1, −1	273.2	0.149 9	0	2.55×10^{-8}	$\mu + \nu$
				π^0	0	264.3	0.145 0	0	2.2×10^{-16}	$\gamma + \gamma$
			K meson (kayon)	K^+, K^-	+1, −1	966.8	0.530 5	0	1.22×10^{-8}	several modes
				$K^0, \bar{K}^0$	0	974	0.534 6	0	1.0×10^{-10} K_1^0 6×10^{-8} K_2^0	$\pi^+ + \pi^-$, $\pi^0 + \pi^0$
Fermions	Leptons		Neutrino	$\nu, \bar{\nu}$	0	0	0.0	$\frac{1}{2}$	infinity	stable
			Electron, positron	e^-, e^+	−1, +1	1	0.000 5	$\frac{1}{2}$	infinity	stable
			μ meson (muon)	μ^-, μ^+	−1, +1	206.79	0.113 5	$\frac{1}{2}$	2.212×10^{-6}	$e^{-,+} + \nu + \bar{\nu}$
	Baryons	Nucleons	Proton	$p, \bar{p}$	+1, −1	1 836.12	1.007 6	$\frac{1}{2}$	infinity	stable
			Neutron	$n, \bar{n}$	0	1 838.65	1.008 9	$\frac{1}{2}$	1.13×10^3	$p + e^- + \bar{\nu}$, $\bar{p} + e^+ + \nu$
			Λ^0 hyperon	$\Lambda^0, \bar{\Lambda}^0$	0	2 182.0	1.197 6	$\frac{1}{2}$	2.51×10^{-10}	$p + \pi^-, n + \pi^0$
			Σ^+ hyperon	$\Sigma^+, \bar{\Sigma}^+$	+1, −1	2 327.8	1.227 4	$\frac{1}{2}$	0.81×10^{-10}	$p + \pi^0, n + \pi^+$
			Σ^0 hyperon	$\Sigma^0, \bar{\Sigma}^0$	0	2 332	1.280	$\frac{1}{2}$	10^{-13}	$\Lambda^0 + \gamma$
			Σ^- hyperon	$\Sigma^-, \bar{\Sigma}^-$	−1, +1	2 341.7	1.284 7	$\frac{1}{2}$	1.6×10^{-10}	$n + \pi^-$
			Ξ^0 hyperon	$\Xi^0, \bar{\Xi}^0$	0	2 570	1.409	$\frac{1}{2}$?	1.5×10^{-10}	$\Lambda^0 + \pi^0$
			Ξ^- hyperon	$\Xi^-, \bar{\Xi}^-$	−1, +1	2 581	1.416 2	$\frac{1}{2}$?	1.3×10^{-10}	$\Lambda^0 + \pi^-$
Atomic particles			Hydrogen	H^1	0	1 837.13	1.008 1		infinity	
			Deuterium	H^2	0	3 671.43	2.014 7		infinity	
			Deuteron	D	1	3 670.43	2.014 2		infinity	
			α particle	α	2	7 294.0	4.002 7		infinity	

amount is carried off by the neutrino which passes unhindered through the star. This reaction proceeds very slowly, for although the potential barrier between two protons is not very high, one of the protons must turn into a neutron while the reaction takes place, and the probability for this is very small. In fact, we find (under conditions that exist in the deep interior of the sun) that a proton, on the average, will live for about 14 billion years before it is captured by another proton to form a deuteron. However, because of the great abundance of protons, even this slow reaction rate gives enough energy per second to account for the luminosity of the sun.

In the second step, the deuteron combines with another proton to form He^3, emitting a γ ray. In this reaction, which occurs very rapidly, about 5.5 Mev are released. The deuteron, on the average, cannot last more than 6 seconds in the deep interior of the sun. The He^3 formed in this second step can undergo a number of reactions, but the most probable one, which is the third step in our chain, involves another He^3 nucleus. We must realize that many reactions leading to He^3 are going on simultaneously so that there will be available many He^3 nuclei to interact with each other.

In the third step, two He^3's combine to form He^4, with the emission of two protons and a γ ray. In this reaction, about 12.9 Mev are released. On the average, two He^3 nuclei will react in this way once in about 10^6 years under the conditions existing in the sun.

The complete set of reactions in the proton-proton chain is given below:

$$_1H^1 + {_1H^1} \rightarrow {_1D^2} + \beta^+ + \nu \quad 1.44 \text{ Mev} \quad (14 \times 10^9 \text{ years})$$

$$_1D^2 + {_1H^1} \rightarrow {_2He^3} + \gamma \quad 5.49 \text{ Mev} \quad (6 \text{ seconds})$$

$$_2He^3 + {_2He^3} \rightarrow {_2He^4} + {_1H^1} + {_1H^1} + \gamma \quad 12.85 \text{ Mev} \quad (10^6 \text{ years})$$

(where the subscript is the atomic number, the superscript the atomic weight to the nearest integer, and where ν stands for the neutrino).

At least two sets of the first two steps in this chain must go on simultaneously if He^4 is to be formed since in the last step we need 2 He^3's. From this we see that the total amount of energy released for every helium nucleus formed is 26.2 Mev or 4.2×10^{-5} ergs (this is obtained by doubling the amount of energy released in the first two steps, adding the amount released in the last step and subtracting the small amount, 0.51 Mev, carried off by neutrinos). The proton-proton chain as a source of energy for main-sequence stars was first investigated by Hans Bethe and independently by C. F. Von Weizsäcker.

23.17 The Carbon Cycle

Bethe also investigated another series of nuclear reactions which lead to the transformation of four protons into He^4 with the release of energy. In this investigation Bethe showed that all light nuclei starting with lithium, but not including carbon, are quickly burnt up inside stars by coalescing with protons. He^4 is the end product in each case. These light nuclei, therefore, cannot be a source of energy except possibly during very early stages of a star's development. The carbon nucleus does not burn up but instead behaves like a catalyst. It takes part in a complete cycle of nuclear reactions but reappears after four protons have been transformed into He^4.

The carbon cycle consists of the following six nuclear reactions:

$$_6C^{12} + {_1H^1} \rightarrow {_7N^{13}} + \gamma \quad 1.95 \text{ Mev} \quad (1.3 \times 10^7 \text{ years})$$

$$_7N^{13} \rightarrow {_6C^{13}} + \beta^+ + \nu \quad 2.22 \text{ Mev} \quad (7 \text{ minutes})$$

$$_6C^{13} + {_1H^1} \rightarrow {_7N^{14}} + \gamma \quad 7.54 \text{ Mev} \quad (2.7 \times 10^6 \text{ years})$$

$$_7N^{14} + {_1H^1} \rightarrow {_8O^{15}} + \gamma \quad 7.35 \text{ Mev} \quad (3.2 \times 10^8 \text{ years})$$

$$_8O^{15} \rightarrow {_7N^{15}} + \beta^+ + \nu \quad 2.71 \text{ Mev} \quad (82 \text{ seconds})$$

$$_7N^{15} + {_1H^1} \rightarrow {_6C^{12}} + {_2He^4} + \gamma \quad 4.96 \text{ Mev} \quad (1.1 \times 10^5 \text{ years})$$

In the first step carbon combines to form N^{13} with the release of about 2 Mev in the form of a gamma ray. In the deep interior of a star like the sun a carbon nucleus will, on the average, move about for some 13 million years before it is converted into N^{13}. Since N^{13} is unstable, it decays to C^{13} in about seven minutes with the emission of a positron and a neutrino. Somewhat more than 2 Mev are released. The C^{13} then combines with a proton to give N^{14} and a gamma ray. The energy output is about $7\frac{1}{2}$ Mev, and the time for this reaction is about 3 million years. In the fourth step, N^{14} combines with a proton to give O^{15} and another gamma ray, releasing almost $7\frac{1}{2}$ Mev. This fourth step is the slowest of all the reactions, taking about 300 million years, and hence it determines the rate of the entire cycle. Since O^{15} is not stable, it decays into N^{15} in about 82 seconds, with the emission of a positron and a neutrino, releasing about 2.7 Mev. In the final step N^{15} absorbs a proton, but this does not lead to the formation of O^{16} as we might expect. Bethe has shown that, on the average, only one in 20 000 such reactions will give O^{16}. Usually what happens is that C^{12} reappears with an alpha particle and a gamma ray. The time for this reaction is about 100 000 years and the energy released is about 5 Mev.

The total energy released in a single carbon cycle (one alpha particle formed) is 25 Mev or 4×10^{-5} ergs. The reason this is somewhat less than the energy released by the proton-proton chain is that the two neutrinos in the carbon cycle carry off somewhat more energy than do those in the proton-proton chains.

23.18 The Rate of Energy Generation, ϵ: The Collision Factor

Now that we have described the types of nuclear reaction that occur in main-sequence stars, we can write down an expression for the rate of energy generation, ϵ, that appears in our equations. We shall not derive this formula rigorously, but rather show from general considerations how we can get at the various factors it contains. We start by noting that if two nuclei (such as protons, or a proton and a carbon nucleus) are to interact, they must first collide with each other. Hence, one of the factors that must appear in ϵ is the number of times per second, on the average, a nucleus of one kind (a proton) collides with a nucleus of the other kind that takes part in the reaction. (Note that not every such collision necessarily gives a nuclear reaction.)

To get at this number, we let the sum of the diameters of the two colliding nuclei be d, and let one of the nuclei, type 1, be moving with a speed v relative to the other, type 2. If we watch the type 1 nucleus, it will have moved relative to the other a distance v in a unit time. We may therefore picture a cylindrical volume equal to v times the area, πd^2, as being tunneled out by the two nuclei together as they approach and suffer a grazing collision, as shown in Figure 23.5. It is clear from the figure that the total number of collisions per unit time that can occur between this single type 1 nucleus and all type 2 nuclei is equal to the total number of type 2 nuclei in this tunnel. If N_2 is the number of type 2 nuclei per unit volume, the number of collisions that will occur in a unit time between one nucleus of type 1 and all type 2 nuclei (having relative speed v) is therefore

$$\pi N_2 v d^2.$$

Now if we suppose that there are N_1 nuclei per unit volume of the first kind, each one of these suffers this number of collisions in a unit time. Hence, the total number of collisions per unit time per unit volume of the two kinds of interacting nuclei will be

$$\pi N_1 N_2 v d^2.$$

Note, this holds only for those nuclei of the two kinds that are moving with a speed v with respect to each other.

If X_1 represents the abundance by weight of nucleus 1, X_2 that of nucleus 2, and ρ the density at the point where the nuclear reaction is going on,

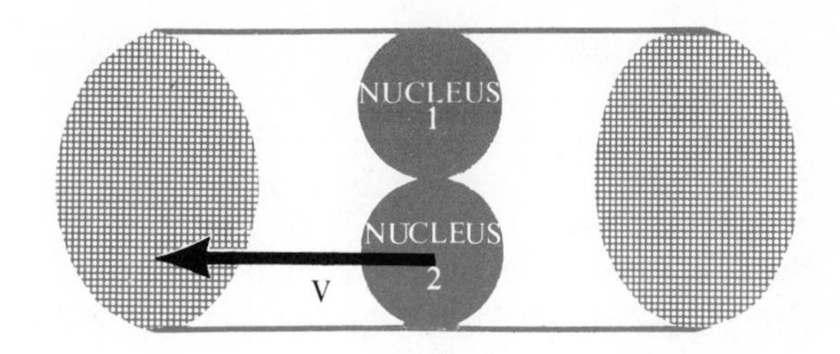

FIG. 23-5 *Grazing collision of two interacting nuclei moving with a velocity v with respect to each other. The volume swept out by the collision cross section in a unit time is that of the cylinder shown in the figure.*

then N_1 is proportional to ρX_1 and N_2 to ρX_2, so that the number of collisions occurring per second is, from the previous formula,

$$\pi \rho^2 X_1 X_2 v d^2.$$

If we take into account the fact that nuclei have wave properties according to the de Broglie picture, we must replace d by its appropriate quantum-mechanical wavelength. Now, as we have seen, the wavelength of a particle is equal to its momentum divided into Planck's constant so that we must replace d by h/mv, where m, the so-called **reduced mass**, is a quantity which depends upon the masses m_1, m_2 of the two colliding nuclei. The reduced mass m, which is equal to $\dfrac{m_1 m_2}{m_1 + m_2}$, must be used in this analysis because the collision must be referred to the center of mass of the two nuclei. Only the kinetic energy of each nucleus relative to the center of mass of the two nuclei is effective in a nuclear interaction; the motion of the center of mass does not count. This introduces the reduced mass of the two nuclei. We, therefore, have for the number of collisions occurring per second per unit volume:

$$\pi \rho^2 X_1 X_2 \frac{h^2}{m^2 v}.$$

23.19 Maxwell-Boltzmann Factor in ϵ

The above is the number of collisions per unit time that occur for nuclei moving with a relative velocity v with respect to each other, but we know that the relative velocities of nuclei inside stars depend upon the temperature. Maxwell and Boltzmann independently showed many years ago that at a given temperature the velocities of molecules in a gas are not all the same but are distributed about a certain mean value, with some moving slowly and some quite rapidly. From the Maxwell-Boltzmann formula, we can show that the number of nuclei that have relative velocities in the neighborhood of v at the absolute temperature, T, is just proportional to

$$\frac{v^2}{T^{3/2}} e^{-(1/2)(mv^2/kT)},$$

where m is the reduced mass and k is Boltzmann's constant. This, then, is the second factor that must be introduced in the formula for ϵ.

23.20 The Gamow Factor, $\mathcal{G}$, in ϵ

If the two interacting nuclei did not repel each other because of their electrostatic charge, we could obtain the total number of effective collisions per second giving rise to a nuclear reaction by multiplying these two factors together and summing over all relative velocities. *But, as we know, before two nuclei can combine they must not only collide but must also penetrate the Coulomb barrier. We must, therefore, introduce into our formula a third factor to take this into account. This factor, $\mathcal{G}$, was first discovered by Gamow and reduces the number of effective collisions.* It is proportional to

$$e^{-(4\pi^2 Z_1 Z_2/h)(1/v)},$$

where v is the relative velocity of the two nuclei, Z_1 is the electric charge on nucleus 1, Z_2 the electric charge on nucleus 2, and h is Planck's constant. We see that this factor reduces the number of effective collisions drastically as Z_1, Z_2, or both, are increased. In other words, for a given temperature, T, nuclear reactions become less and less probable as we go to heavier and heavier nuclei.

23.21 The Nuclear Factor, Γ, in ϵ

There is still another factor, the last one, the necessity for which we may understand from the following argument. Even after two nuclei collide and interpenetrate their respective Coulomb barriers, a nuclear reaction may still not take place. For it may well be that no compound nucleus can be formed. In other words, *a nuclear reaction will take place only if the final product nucleus is a possible one.* We must therefore introduce a fourth factor to take account of this. Although this factor depends strongly on the nuclear forces, it also depends on the structure of the two colliding nuclei and (but only weakly) on their relative velocity. Although this factor Γ can be computed to a fair approximation for very light nuclei, it is almost impossible to compute it for heavy nuclei, so that most of our information about it must come from laboratory experiments.

23.22 The Complete Formula for ϵ

If we take all of these factors into account and multiply them together, we find that the total number of reactions per unit time per unit volume (for the given relative velocity) is proportional to the following product:

$$\underbrace{\left(\frac{\pi h^2}{m^2}\frac{\rho^2}{v} X_1 X_2\right)}_{\text{collision factor}} \underbrace{\left(\frac{v^2}{T^{3/2}} e^{-(1/2)(mv^2/kT)}\right)}_{\text{Maxwell-Boltzmann factor}} \underbrace{\left(e^{-(4\pi^2 Z_1 Z_2/h)(1/v)}\right)}_{\text{Gamow factor}} \underbrace{(\Gamma)}_{\text{nuclear factor}}.$$

This gives the number of reactions per unit time per unit volume for pairs of nuclei having one particular relative velocity. If we want to get a formula for the number of reactions per unit time per unit mass for nuclei having all possible relative velocities, we must first divide this expression by the density, ρ (since this is just the mass per unit volume), and then add all such expressions for all possible relative velocities. This is done mathematically by using the integral calculus. It is clear that the final formula depends on the types of nuclei that are interacting, their concentrations, the density of the stellar matter, the absolute temperature, and on Γ.

In the case of the proton-proton chain the first step proceeds so slowly that Γ cannot be measured experimentally and must therefore be calculated from theory. However, for all the other reactions in both the proton-proton chain and the carbon cycle, experimental values for Γ are available, so that the final formula for ϵ can be written down.

If ϵ_{pp} is the energy released per unit time per gram of material at absolute temperature T for the proton-proton chain, and ϵ_{cc} is this quantity for the carbon cycle, we find, after carrying out all the above indicated operations, that to a very good approximation we can express each of these quantities in terms of the absolute temperature to some power. Thus, we have

$$\epsilon_{pp} = A\rho X^2 T^n, \qquad (23.1)$$

$$\epsilon_{cc} = B\rho X Z_{cn} T^m, \qquad (23.2)$$

where A and B are almost constants (they change slightly with T), X, as usual, is the hydrogen content, Z_{cn} is the carbon-nitrogen fraction of the heavier material Z (this is about $\frac{1}{3}Z$ for stars like the sun), and n and m are exponents that change slightly with the temperature of stellar material.

It turns out that n changes from about 6 to 3.5 as T varies from 4×10^6 °K to 24×10^6 °K, and m goes from 20 to 13 as T changes from 12×10^6 °K to 50×10^6 °K. For the deep interiors of stars like the sun, therefore, where we can expect the temperature to be anywhere from 12×10^6 to 15×10^6, $n = 4$, and $m = 20$. For this same temperature A is equal to 1.05×10^{-29}, and B equals 1.6×10^{-142}. Taking these things into account, we can now express the rate of energy generation per unit mass for both the proton-proton chain and the carbon cycle in the interiors of stars like the sun as follows:

$$\epsilon_{pp} = 1.05 \times 10^{-29} \rho X^2 T^4, \qquad (23.3)$$

$$\epsilon_{cc} = 1.6 \times 10^{-142} \rho X \frac{Z}{3} T^{20}. \qquad (23.4)$$

23.23 Carbon Cycle vs. Proton-Proton Chain

From these expressions we see at once that because B is very small compared to A, the carbon cycle does not contribute measurably to the energy generation for stars like the sun. However, as we go up on the main sequence the carbon cycle becomes more and more important, finally dominating. *The reason is that T in the deep interior of a star increases as we go up the main sequence, and this increase has the twentieth-power effect on the carbon cycle but only the fourth-power effect on the proton-proton chain.* Thus, if T is increased by 1 per cent, the rate of the carbon cycle increases by over 20 per cent whereas the rate of the proton-proton cycle increases by about 4 per cent. We shall see that the sun lies slightly below the point on the main sequence where the carbon cycle and proton-proton chain contribute equally to the energy generation. Above this point (F stars and earlier ones) the carbon cycle dominates and below it the proton-proton chain contributes most.

23.24 How the Interior Equations Are Used

Now that we have expressed the energy generation ϵ in terms of the density ρ, the absolute temperature, and the abundances X and Y, we can solve the problem of stellar interiors. We first replace the density ρ by $\left(\frac{\mu H}{k}\frac{P}{T}\right)$ from the gas law, in the expression for ϵ, and then introduce it into the last of our four interior equations (22.10). We also introduce into these four equations the expressions we found for μ and κ. Our four equations then contain just the four unknowns P, T, $M(r)$, $L(r)$, as well as the hydrogen and helium abundances X and Y. If, then, we have some independent way of determining X and Y, we are left with just four unknowns at each point inside the star, and these can be found by solving our four equations.

Since our four equations tell us only how our four unknowns *vary* from point to point inside a star, we cannot use these equations to find the conditions at each interior point unless, to begin with, we know the values of the four quantities at some particular point. Then our equations enable us to obtain the values of the unknowns at neighboring points.

To see this we note that each of the four equations tells us how one of our unknowns P, T, $L(r)$, $M(r)$ changes as we move a distance Δr from the point at which we assume all four quantities to be known. Thus (to illustrate), the third equation (22.9) tells us that if, from a point at a distance $r = 10^{10}$ cm from the center of a star, at which the pressure, P, is known to be 10^{12} dynes/cm², the temperature, T, 10^6 °K, the outward energy flow, $L(r)$, 10^{33} ergs/sec, the mean molecular weight, μ, 0.5, and the opacity, κ, 4 units, we move out towards the surface by an amount $\Delta r = 10^7$ cm (100 km), the temperature will decrease by an amount obtained by multiplying $\frac{3H}{16\pi ack}$ (which is numerically equal to 3.24×10^{-6}) by

$$\frac{0.5 \times 4 \times 10^{33} \times 10^{12} \times 10^7}{10^{20} \times 10^{24}}.$$

In other words, the temperature decreases by 648°K if we move out 100 km from the given position. Of course, if we move in towards the center the same distance from the same initial point the equation tells us that the temperature will increase by 648°.

By using the three other equations and the same

initial data [plus the known values of $M(r)$ and ϵ at the starting point] we can determine the decrease in P and the increase in $M(r)$ and $L(r)$ as we move out 100 km. The numbers we have used were arbitrarily chosen to illustrate the procedure, but it is clear from this numerical example why our equation can give us a complete picture of the interior of a star only if we know the values of P, T, $M(r)$, $L(r)$ at some point to begin with. Thus, if we do know these quantities at some point, we can go from this point to another nearby point and then to a third point and so on, always moving in steps of size Δr. In this way, starting from the given data at some interior point we can push out towards the surface of a star and in towards its center.

It is clear that the bigger the step Δr is, the less accurate will be the calculated values of P, T, $M(r)$, $L(r)$ at the various points (because our formulae were derived on the basis of infinitesimal steps), but then we have less numerical work to do to find these values. In order to obtain very accurate results, we have to make Δr very small—and this, of course, means extensive calculations.

23.25 Solving the Interior Equations for a Star

With these ideas in mind, let us see how we would determine the conditions in the interior of a star like the sun. The results we obtain will, of course, depend on our choice of X and Y, and what assumption we make about the mean molecular weight throughout the star. As we have indicated previously, if we assume that the mean molecular weight is constant throughout the star, we are dealing either with the star in its early history (before enough time has elapsed for much of its initial hydrogen to have changed into helium, so that its chemical composition is still very nearly the same throughout) or else with a star which may be well advanced in age, but in which a mixing process keeps it chemically homogeneous. For stars well along in age and in which mixing does not occur, the mean molecular weight is not the same at all points. Although this is true for all the stars like the sun, it is too complicated to consider in an elementary text such as this; we shall, however, consider the general evolution of such stars later on. We discuss at this point only chemically homogeneous stellar models since for our purposes they provide an accurate enough picture of the way we use our equations to determine a star's structure.

Let us suppose, then, that we start out by adopting for the entire star a definite set of values for the hydrogen and helium abundances X and Y. From these we calculate the mean molecular weight, μ, for the entire star.

We divide the star into a series of concentric shells each having a thickness Δr, as shown in Figure 22.1. (Note that this thickness need not be the same throughout the star, and, indeed, in actual calculations it is generally log Δr that is the same and not Δr. However, for our present discussion, it simplifies things to keep Δr the same for all shells.) From the assumed values of X and Y and the knowledge of T and P in any shell we can calculate the opacity and energy generation in that shell.

We shall suppose then that we start our calculation at some point inside the star, let us say at the nth shell which is a distance r_n from the center, and at a certain depth below the surface. Since we know nothing at all about the conditions in this shell, we must assume a set of values for P, T, $L(r_n)$, $M(r_n)$ and, using the adopted values for X and Y, we calculate κ and ϵ in this shell. If we substitute these values, as well as the value for μ, into the four equations (remembering that G, H, k, a, c are known universal constants), we can use our equations (as previously described) to obtain ΔP, ΔT, $\Delta M(r)$, and $\Delta L(r)$, as we move a distance Δr either towards the surface or towards the center of the star. *In other words, our equations tell us the amounts by which P, T, $L(r)$, $M(r)$ change as we go from one shell to the next.* Thus, from the numerical values assumed for these quantities in the nth shell, we find a set of numerical values in shell $n + 1$ (going out towards the surface) and in shell $n - 1$ (going in towards the center). *In this way, from our initial set of values for the nth shell, we can find values for the four physical parameters everywhere in the star, always evaluating κ and ϵ at each point first (which we can do from the new values of T and P that we obtain).* In evaluating ϵ we take both the proton-proton chain and the carbon cycle into account to be sure that we do not miss anything, even though one of these may contribute very little, relatively speaking.

We see from our equation that since ΔP is negative for increasing r (i.e., for Δr positive or going out towards the surface) the pressure decreases outwardly and increases inwardly. The same thing is true for the temperature since ΔT is negative for positive Δr. On the other hand $M(r)$ increases as we go outwardly and decreases as we go in, which is at once obvious since we add shells of material as we go towards the surface and take shells away as we go in towards the center. The same is true for $L(r)$.

23.26 Surface and Center Conditions *

To see what happens as we move towards the surface, *we first define the surface of a star as that shell at which both the pressure and the temperature simultaneously go to zero.* Actually we identify this with the photosphere of the star, for even though the temperature there is a few thousand degrees, we may still take it as being zero compared to the millions of degrees at the center. If the pressure and temperature both go to zero at some distance, $r = R$, from the center, our calculations then give us the definite values $L(R)$ and $M(R)$ at this point. These, then, are the calculated values of the total luminosity L and total mass M of the star, whose calculated radius is then R.

However, since we started with an assumed set of values for X and Y, we cannot know that we have obtained a correct model of a star until we go into the center with our equations to see whether or not conditions there can be properly fulfilled. *As we go in towards the center, P and T increase, while $M(r)$ and $L(r)$ decrease, so that the center must be defined as the point where the latter two are simultaneously equal to zero.* If this happens for $r = 0$ for the assumed values of X and Y, our model correctly depicts the structure of a star. At this point (the center of the star) both T and P have definite values which are referred to as the **central values of temperature and pressure**.

We see that if everything works as indicated, we obtain a solution to the problem of stellar interiors. With our equations, we have constructed a model of a star of radius R which has a computed luminosity, L, a computed mass, M, a given chemical composition, μ, and definite central values of pressure, temperature, and density. This model, of course, need not correspond to a known star. It merely shows us that a star having the assumed chemical composition and the computed properties can exist in principle; that is, in accordance with the laws of physics.

23.27 Stars with Convective Cores

There is one other point we should keep in mind in applying Equations (22.7) through (22.10). These equations tell us how the four variables change as long as the energy is transported by radiation, but if convection sets in, we must replace our third equation by the much simpler one (see Section 22.7 and Equation 22.3) which describes how the temperature gradient is determined by the convective transport of energy. Now this will occur (if at all in the deep interior) when we get close enough to the center of the star. Therefore, as we move in, we must examine the conditions to see at which point convection takes over. We find, in general, that for stars higher up than the sun on the main sequence, there is a convective core which may extend out to 10 per cent of the radius. For stars below the sun on the main sequence, there appears to be no convective core, or a very small one, and even the sun itself is in a doubtful category.

23.28 Constructing the Model for a Known Star

The procedure outlined above works in principle, but for a random choice of X and Y the chance that everything will come out as indicated is rather small. If then, we find that the surface and central conditions are not fulfilled for our first choice of X and Y, we must take a different combination of these abundances and start all over again. We repeat this (each time getting closer to the necessary surface and central conditions) until a solution is obtained, so that the computations, in general, are extremely involved and protracted. This process of trial and error finally leads to the X and Y of a model or, put differently, it tells us what the X and Y of a model must be for it to obey the stellar-interior equations.

However, if we apply these equations to some particular star—for example the sun—we are in a more favorable position because we know its total mass, its luminosity, and its radius. In other words, we know the surface values $L(R_\odot)$, $M(R_\odot)$, and $R_\odot$, and instead of starting at some point in the interior, we can start at the surface (i.e., the outermost shell) and move in towards the interior with our equations. Of course, we still have to start with an assumed set of values for X and Y.

23.29 A Chemically Homogeneous Model of the Sun *

Since we know L, M, and R for the sun, we can put these surface values directly into our equations for a given choice of X and Y, and then see whether $L(r)$ and $M(r)$ go to zero simultaneously as r goes to zero (i.e., as we go to the center). Since we know something about the chemistry of the solar atmosphere, we can be guided by this knowledge in choosing X and Y to begin with. Therefore, the most recent investigations have led to models for the sun in which the hydrogen and helium contents are found to be quite similar to those obtained from the spectroscopic analysis of the solar atmosphere.

The interior of the sun has been investigated more extensively than that of any other star. In one series of calculations (I. Epstein and L. Motz) in which the mean molecular weight was assumed to be constant throughout, a solution was found with a hydrogen content of 93 per cent, a helium content of 6.5 per cent, and with the heavy elements contributing about 0.5 per cent. The central temperature in this model is 13 million degrees and the central density is 100 gm/cm^3. According to this model the sun possesses a convective core extending out to about 8 per cent of the radius and containing about 4 per cent of the mass. The proton-proton chain contributes all of the energy, 29 per cent of which is generated within the core. See Figure 23.6.

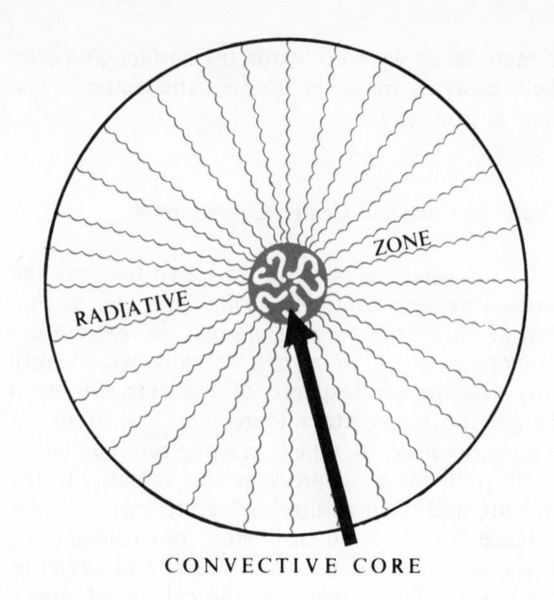

FIG. 23–6 *A model of the sun showing the convective core and the radiative envelope.*

Since the sun is approximately 5 billion years old, and since thorough mixing of the solar gases does not occur, this model is probably incorrect. We know that thorough mixing does not take place because we find elements like lithium in the solar atmosphere in quantities that could not possibly exist in the interior, such elements being burnt up very quickly in thermonuclear reactions at high temperatures.

23.30 An Inhomogeneous Model of the Sun

For the foregoing reasons a better model of the sun has been computed by M. Schwarzschild and his colleagues in which the mean molecular weight is not assumed to be constant throughout.

The basis for such a chemically inhomogeneous model is the following. We know that originally (about 5 billion years ago) the sun was chemically homogeneous and consisted primarily of hydrogen. In time, however, because of the different temperatures in different shells, hydrogen was transformed into helium at different rates inside the sun, so that there now exists a gradation in the hydrogen content ranging from zero in the central core (a helium core) to its original value near the surface.

To construct such an inhomogeneous model of the sun, Schwarzschild and his collaborators began with a homogeneous model consisting almost entirely of hydrogen and thus obtained an initial set of temperature, pressure, and density values at each interior point. Using these values they then computed the rate of energy generation and hence the transformation rate of hydrogen to helium at each point. By assuming that this transformation has been going on at the same rate for the past 5 billion years, they then computed the present helium content at each point and in this way obtained different mean molecular weights for different zones inside of the sun.

To simplify the calculations they computed appropriate molecular weights for five large zones inside the sun rather than for many narrow shells. Using the techniques described previously, they then constructed a model for the sun in its present state, which takes into account the evolution of the sun's chemical composition. It also takes into account the convective zone (which arises from the ionization of hydrogen) below the photosphere of the sun (see Section 20.5). This model, of course, depends on the assumed homogeneous chemical composition of the original sun, but since this is the same as the present chemical composition of the solar atmosphere, the initial values of hydrogen and helium are accurately known from spectroscopic data (80 per cent hydrogen, about 18 per cent helium, and 2 per cent heavy elements).

23.31 The Properties of the Chemically Inhomogeneous Solar Model

With these initial data, Schwarzschild obtained an inhomogeneous solar model with the following properties. There is a convective zone below the photosphere which extends down to a point where the temperature is about 100 000°K. The bottom of this zone is at a distance from the center equal to about 82 per cent of the radius. Below this zone, of course, the chemical composition is different from what it is on the outside, with the hydrogen content gradually diminishing until we come to a core that has a small hydrogen content. It may be noted that according to this model there is no convective core at the center of the sun. The mass of the sun above the outer convective zone equals only 3 per cent of the total mass.

As we go down to the interior starting from the bottom of the convective zone, the hydrogen content diminishes gradually until it reaches a value of about 30 per cent at the boundary of the central core. This indicates that in the almost 5 billion years of the sun's existence more than one-half its hydrogen in the central regions has been transformed into helium. In this model the central density is about 132 gm/cm^3, and the central temperature about 15×10^6 °K. This central temperature is somewhat higher than that found for the homogeneous model, and indicates that the carbon cycle may have to be taken into account and may contribute a measurable amount to the total luminosity of the sun. See Figure 23.7.

If this model is a fairly good approximation to the present sun, and represents, fairly accurately, its previous history, we can say something about the evolution of its luminosity, its effective temperature, and its radius during the 5 billion years of its life. We find that the present luminosity is about 1.6 times the initial luminosity, the radius is about 1.04 times the initial radius, and the effective temperature is now 1.1 times the initial effective temperature. In other words, in the early Pre-Cambrian era the sun was a smaller, cooler, less luminous, and redder star. Schwarzschild has concluded from this that the average temperature on the earth must then have been close to the freezing point of water, and that this must have played an important part in the evolution of life.

This model also indicates that the sun has moved slightly off the main sequence and is advancing towards the giant branch. We shall see that this parallels the evolutionary tracks of all chemically

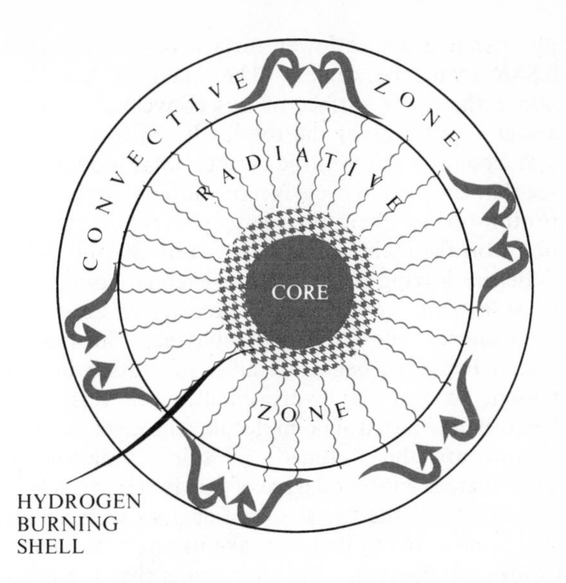

FIG. 23–7 *A chemically inhomogeneous model of the sun showing a hydrogen-poor helium-rich core, a hydrogen burning shell, a radiative zone, and a convective envelope.*

inhomogeneous stars (i.e., stars in which mixing does not occur).

23.32 Another Chemically Inhomogeneous Model of a Star

The star ζ Hercules A, which has evolved off the main sequence more than the sun has, has been investigated fairly thoroughly by Dr. Eva Novotny. This star has the following properties: spectral class G0, $M = 1.07M_\odot$, $R = 1.97R_\odot$, and $L = 5.4L_\odot$. According to the computed model, the hydrogen content varies from 73 per cent at the surface down to 0 in the core, which contains nearly 7 per cent of the total mass and occupies 2 per cent of the total radius. The heavy element content of this star is just 0.5 per cent so that its helium content decreases from 99.5 per cent in the core to 26.5 per cent near the surface. Its central temperature is 25.5×10^6 °K, and its central density is 600 gm/cm^3. The core of this star, according to this model, is contracting slowly and releasing gravitational energy which contributes about 0.5 per cent to the total luminosity. The carbon-nitrogen cycle contributes 83 per cent and the proton-proton chain 17 per cent.

The core itself has practically the same temperature throughout so that no convection occurs. There is, however, a convective zone below the photosphere whose lower surface is at a distance $0.84R$ from the center. The mass of the star above the lower surface of this convective zone is about 1 per cent of the total. From this model, and a consideration of the various solar models, we see that the carbon cycle is more important than the proton-proton chain for energy generation at some point on the main sequence between G0 and G2.

Before leaving these inhomogeneous models, we must explain briefly why a convective zone lies immediately below the photosphere. The reason is that hydrogen begins to suffer ionization in this region, so that the hydrogen atom in this zone behaves like a diatomic molecule (the electron and proton are the "atoms" of this "molecule"). Thus there is another degree of freedom (ionization) in addition to the translational degrees of freedom (see Section 16.10) that can take its share of energy (ionization energy). But this means that the ratio of specific heats γ is smaller in this hydrogen ionization zone than it is above or below it. But Equation (22.3), taken together with the gas equation (16.5), then tells us that the temperature changes taking place in a small mass of gas as it moves up or down are so small that it cannot adjust itself (in terms of temperature) to the surrounding regions. It therefore continues to move so that convection is set up.

23.33 Models of Families of Stars

Thus far we have been considering the detailed internal structure of individual stars such as the sun and ζ Hercules A, from which we can get some insight into the evolution of these stars. But it is still possible to understand the broad evolutionary picture without a detailed analysis of individual stars. We work instead with a series of models to describe some particular family or group of stars that have evolved together but at different rates (e.g., stars in globular and open clusters, etc.). We can apply our interior equations to such a series of models just as we do to an individual star except that we no longer try to make any one of our models fit all the data of one particular star. Since we know from what we have already done that the mass of the star and its mean molecular weight, μ, determine its internal structure, we can calculate models of families of stars by allowing the mass, the mean molecular weight, or both, to vary over a range of values that best correspond to the family of stars being considered.

Thus if we wish to compare our series of calculated models with a particular family of stars such as an open cluster (e.g., the Hyades cluster) to see whether our theoretical models correctly predict the evolution of stars in the cluster, we must compute a family of models having different masses but the same initial chemical composition as the stars in the cluster had when they were first formed from the original gaseous mixture. The stars in such a family all have the same chemical composition in their outer envelopes, but differ in their mean molecular weights in homologous zones because of different rates of evolution.

Although there are many features missing in the picture of the evolution of stars, enough has been learned from observational data (such as H-R diagrams) and theoretical calculations of models to enable us to give a broad discussion of stellar development. Some of the following material is based on an extrapolation of our presently known models into regions where little work has been done and hence is of a rather speculative nature. Some of the details will undoubtedly change but the general picture will probably remain the same.

23.34 Thorough Mixing and Stellar Evolution*

Since the paths of evolution in the H-R diagram for the stars in any family of given initial chemical composition are different, depending upon whether or not thorough mixing takes place, we shall consider the two cases separately. Such mixing may occur through rotation; this introduces other complications which we shall not mention here.

If a star begins initially as an amorphous mass of gas consisting almost entirely of hydrogen, contraction occurs during its early years. During this period of contraction the star lies to the right of the main sequence at a point depending on its original mass. Because of its contraction its interior temperature increases and it approaches the main sequence, reaching it (as shown in Figure 23.8) when its central temperature is about 10^7 °K, and thus sufficiently high to set off the proton-proton chain which then generates enough energy to prevent the star from contracting further.

But as this thermonuclear process goes on, the mean molecular weight of the star increases since more and more hydrogen is transformed into helium. Because of this the interior temperature at each point increases (as can be deduced from the interior equations), and this increases the rate of the proton-proton reaction. Helium is then

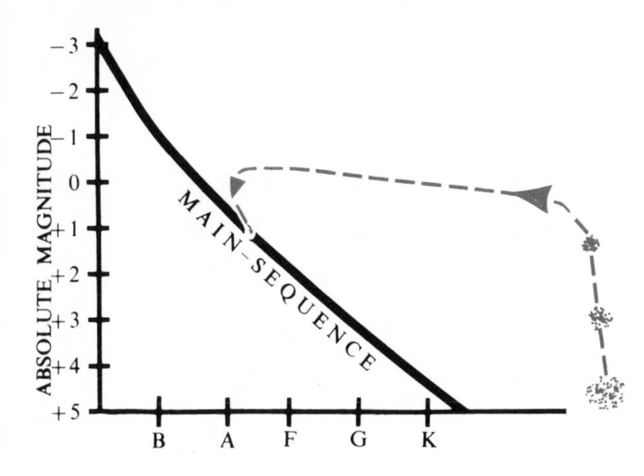

FIG. 23–8 *The contraction of a star that has no outer convective zone from initial pre–main-sequence amorphous state onto the main sequence.*

formed at an even faster rate, leading to a still greater increase in the mean molecular weight, and hence a still greater internal temperature increase, etc., so that evolution proceeds even faster. It follows from this that as long as the chemical composition remains uniform throughout the star (because of mixing), the star moves up along the main sequence of the H-R diagram. This occurs very slowly at first, but more and more rapidly as the hydrogen is used up. In fact one can show from the equations that the radius of the star increases with about the cube root of the mean molecular weight μ, the temperature with about the square of μ, and the luminosity with about the eighth power of μ.

If we assume that a thoroughly mixed star consisted initially of hydrogen only, it cannot have evolved far beyond the sun's position on the main sequence since under such conditions only the proton-proton chain can operate and this cannot release energy fast enough to account for the luminous B and O stars. The reason for this is that there is no carbon present and therefore no way in which the carbon-nitrogen cycle can proceed. However, if a star starts out with some heavy matter (e.g. carbon, nitrogen, and some metals) it will evolve past the sun on the main sequence, assuming that it remains thoroughly mixed.

A star of this sort will move up the main sequence with its central temperature steadily increasing until the carbon cycle becomes operative. It will then radiate energy even more rapidly, and evolve along the main sequence at a still faster pace.

23.35 The Evolution of Chemically Inhomogeneous Stars

Although there are probably numerous stars that evolve in this way with their chemical composition always remaining uniform, it appears, from the many investigations that have been carried out recently, that the great majority of stars are chemically inhomogeneous and hence do not evolve along the main sequence during their entire lifetimes, but leave it at a certain stage of their development.

In discussing the evolution of such stars we shall differentiate between those that began their lives as pure hydrogen stars (with none, or very little, carbon or other heavy material present) and those that were formed from a gaseous mixture containing heavy atoms as well as hydrogen. In general, of course, since hydrogen is the most abundant material in the universe, all stars consist mostly of hydrogen in their initial phases. *Nevertheless, we must differentiate between the evolution of the initially pure hydrogen stars, which we refer to as Population II stars, and the others, the Population I stars.*

23.36 Population I and II Stars

Since our galaxy was initially composed almost entirely of hydrogen (the primordial matter in the universe) the pure hydrogen stars (Population II stars) began their lives at that time and are now the very oldest. These stars have numerous characteristics that differentiate them from the younger Population I stars. Main-sequence Population II stars extend on the H-R diagram from late spectral class A (color index +0.3) into class M (color index +1.74). They are found somewhat above the main sequence and also among the yellow and red giants and supergiants but there are no blue-white giants among these stars. Some Population II stars are found below the main sequence and these are referred to as subdwarfs. *We find that the atmospheres of these stars are deficient in metals as compared to stars like the sun.* Other characteristics that differentiate Population II stars are their large space velocities relative to stars near the sun and their distributions in the Milky Way. Population I stars lie all along the main sequence (as well as along the giant branch) and are found in the spiral arms of the Milky Way but not in its nucleus, which consists entirely of Population II stars.

The early stages of evolution are about the same for both sorts of stars. They both start as cool masses of gas and after their initial contraction appear as red stars to the right of the main sequence. In this early stage both Population I and II stars are luminous primarily because of gravitational contraction, and only after they have reached the main sequence and thermonuclear reactions are well under way do their evolutionary paths diverge. In both cases, however, the mass of the star determines how rapidly it contracts onto the main sequence, at what spectral class it begins its main-sequence life (the longest part of its life), how rapidly it moves up along the main sequence, and at what point it leaves the main sequence.

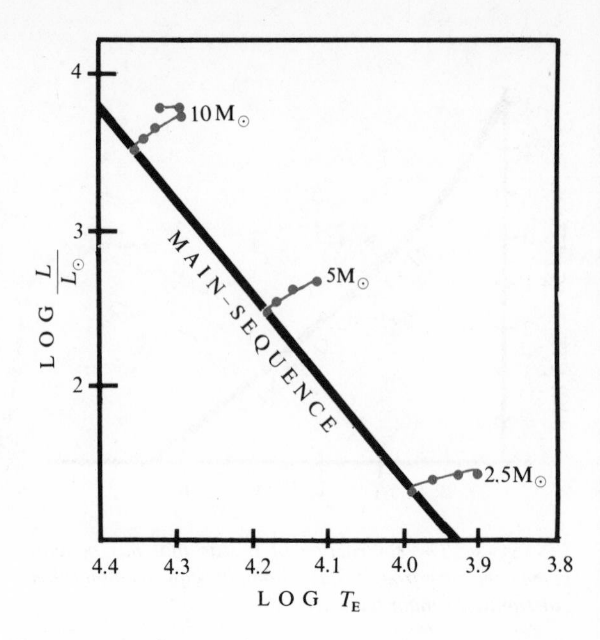

FIG. 23-9 *Evolution tracks in the H-R diagram for three upper main-sequence stars. (After M. Schwarzschild, Princeton University)*

23.37 The Evolution of Population II Stars without Internal Mixing

The only thermonuclear reactions that can occur in Population II stars during their main-sequence state are those in the proton-proton chain since, as we have said, these stars started out with very little, if any, carbon. Because this chain goes on rather slowly the star spends a long time (which varies with the initial mass of the star) on the main sequence but ultimately the star leaves it and moves on to the giant branch unless thorough mixing keeps it homogeneous. If the star remains chemically inhomogeneous it must leave the main sequence after a certain part of its hydrogen has been transformed into helium.

If we consider such a star whose mass is $1.2M_\odot$, we can follow its evolution with the aid of models calculated by M. Schwarzschild and his collaborators. They demonstrated that these stars evolve away from the main sequence along tracks that are perpendicular to the main sequence and which bring the stars to various points on the present globular-cluster tracks in the H-R diagram as shown in Figure 23.9, where the calculated tracks for stars with 2.5, 5, and 10 solar masses are also plotted. We see that there is a distinct difference between evolutionary tracks of two stars with different initial masses. The H-R diagram of a typical globular cluster shows how closely the calculated track of a star having a mass of $1.2M_\odot$ follows the track of an actual globular-cluster star (see in particular Sections 27.25, 27.26, 27.27).

During the early part of the main-sequence phase there is a gradual depletion of hydrogen at the center of the star. The resulting increase in mean molecular weight causes a rise in the central temperature, which, together with the increased density, accelerates the thermonuclear reactions. Thus, the star becomes more and more luminous, while at the same time the hydrogen in the core is transformed into helium at an ever increasing rate. If there is enough carbon present, the carbon-nitrogen cycle takes over when the temperature in the core becomes high enough (of the order of 20–30 million degrees) and then the transformation proceeds even more rapidly. During this phase, the gravitational contraction stops completely or proceeds very slowly.

23.38 The Chandrasekhar Limit and the Evolution of Population II Stars

When the core consists entirely of helium, which is thermonuclearly inert at these temperatures, the star begins a different phase of its evolution in which its initial mass is the determining factor in what takes place. Since at this stage the helium core generates **no** nuclear energy, it must be at a constant temperature throughout (i.e., it must be **isothermal**). The energy is now released primarily through the burning of hydrogen by the proton-proton chain, the carbon cycle, or both in a thin shell surrounding the core, as shown in the schematic diagram in Figure 23.7. *The density of the core is now so high that it can support the weight of the material lying* *above it as well as its own weight, provided that this total weight is not too large. Chandrasekhar has shown that this is so if the total mass of the star, when it consists entirely of helium, is not more than $1.44M_\odot$.*

If the total mass of a Population II star exceeds this limiting value by too much (say, it is $2M_\odot$ or larger), the helium core is unable to support the overlying weight of the star, and contraction proceeds fairly rapidly in the core as well as throughout the rest of the star. For this reason the evolution of a Population II star of large initial mass is different from that of a star with initial mass smaller than $2M_\odot$, as shown in Figure 23.10. Thus, Population II stars with masses equal to $M_\odot$ or slightly larger remain on the main sequence for about 5×10^9 years, and then begin to leave it, but only very slowly as compared to more massive stars.

In the more massive stars the helium cores contract so quickly under the weight of the overlying material that the central temperatures increase rapidly to about 10^8 °K, at which point new thermonuclear reactions take place. Helium begins to burn by a process in which three helium nuclei are fused to form a carbon nucleus. For stars of mass $1.2M_\odot$ this stage is also finally reached but only after a longer time.

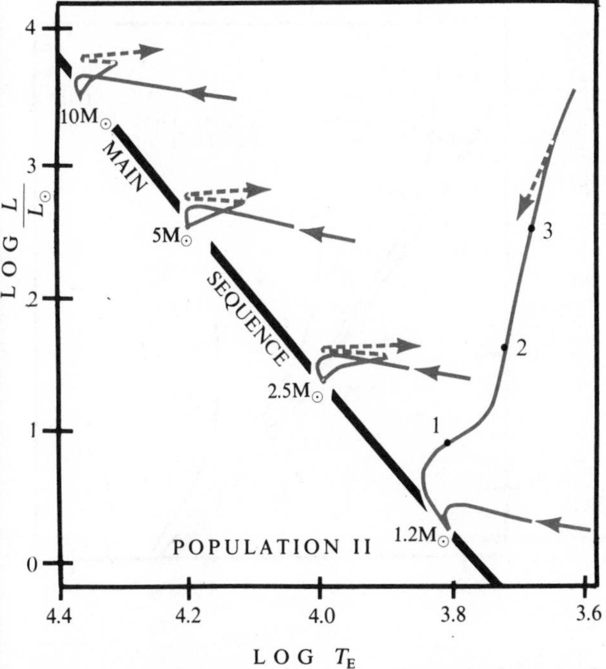

FIG. 23–10 *Evolutionary tracks in the H-R diagram of Population II stars of masses greater than 2.5 $M_\odot$ and with masses less than 2.5 $M_\odot$.*

23.39 Evolution of Population II Stars off the Main Sequence

The isothermal helium core of a star of mass $1.2M_\odot$ grows steadily as the hydrogen in the shell surrounding it slowly burns. Just before such a star leaves the main sequence it consists of three main zones: a noncontracting but growing isothermal helium core, a thin hydrogen-burning shell surrounding this core, moving steadily out towards the surface, and a hydrogen-rich envelope beyond this shell having the original composition of the star.

When a star of this sort leaves the main sequence, it becomes larger and larger so that the photosphere becomes cooler. However, at this point a deep convective zone sets in below the photosphere and tends to slow down the expansion of the envelope. The evolution now proceeds rapidly as more and more hydrogen is burned and the helium core grows. As the mean molecular weight increases, the star becomes progressively more luminous, quickly passing through the subgiant stage by moving upwards and to the right of the main sequence, designated by positions 1, 2, 3 on the H-R diagram. According to Schwarzschild's calculations, the star at this stage is 1 000 times more luminous than it was when it reached the main sequence, and the helium core contains about 40 per cent of the entire mass of the star.

To support such a large luminosity the hydrogen-burning rate increases considerably and the helium core becomes so massive that it begins to contract under its own weight in spite of the fact that it is already very dense (density of the order of thousands of grams per cubic centimeter). This contraction releases gravitational energy and thus contributes to the luminosity of the star. At this point the core contracts so rapidly that the central temperature rises to several hundred million degrees and the helium burning mentioned in connection with the more massive stars starts. This releases such large quantities of energy that the star begins to move into the supergiant branch.

23.40 Postgiant Evolution of Population II Stars*

Since the calculations of evolutionary models have not been carried out beyond this point, we can discover what happens as a star grows still older only by studying H-R diagrams of globular clusters. If we assume that a star continues to evolve along the H-R path of globular clusters, it appears from the plotted observational data that after becoming a supergiant the star must move downward and to the left towards the main sequence following the left-hand branch of the H-R globular-cluster diagram. In this branch, as we have already noted in Section 17.13, there is a break representing the RR Lyrae cluster-type variables, which therefore probably represents a short stage in the final evolution of a Population II star.

Once a Population II star returns to the main sequence, it does not stay there, but probably ends its life as a white dwarf, dropping down to the lower left-hand corner of the H-R diagram. But it can do this only if by some kind of ejection process it first gets rid of a sufficient amount of mass, so that the residual white dwarf can exist in equilibrium. During these final evolutionary stages, great changes take place in the interior of the star resulting in a gradual build-up of the heavy elements. Thus, the material ejected from the star before it becomes a white dwarf is rich in heavy elements, and in time becomes the dust and gaseous matter from which Population I stars, such as the sun, are born.

23.41 The Evolution of Population I Stars

Since Population I stars are second- or third-generation stars, formed from the dust and gaseous material ejected in the final stages of Population II stars (first-generation stars), they consist initially of a homogeneous mixture of such atoms as C^{12}, O^{16}, Mg^{24}, up to Fe^{56} in varying amounts, with hydrogen still the most abundant element. Lighter elements such as deuterium, He^3, He^4, Li^7, Be^9, B^{11} are also present. In the initial stages of a Population I star with a mass about equal to that of the sun, the evolution proceeds just as in the case of a Population II star, with the luminosity arising primarily from gravitational contraction which continues until the central temperature reaches a few million degrees. The very lightest elements, such as deuterium, He^3, Li^7, etc., are then transformed into He^4 by thermonuclear fusion with hydrogen. This is a short phase in the star's evolution just before it reaches the main sequence. During this phase, gravitational contraction stops, but it then starts again and continues until the central temperature is high enough to trigger the proton-proton chain. How soon this happens depends upon the initial mass of the star. A star as massive as the sun reaches the main sequence in about 5×10^7 years, but more massive stars evolve from the nebulous stage more rapidly.

When a Population I star reaches the main sequence, its future evolution (as in the case of a Population II star) depends upon whether it remains chemically homogeneous as a result of mixing, or becomes progressively more and more inhomogeneous with an isothermal helium core and a hydrogen-rich envelope. If the star remains chemically homogeneous it evolves by moving up the main sequence, with the carbon-nitrogen cycle taking over the main burden of generating energy after it passes the G spectral class. Such stars generally develop convective cores when enough energy is generated in the core by the carbon-nitrogen cycle.

On the other hand, if the star is not thoroughly mixed, it leaves the main sequence by moving to the right and somewhat upwardly as shown in Figure 23.10. *The time it takes a Population I star to leave the main sequence depends on its initial mass. The more massive it is, the more quickly it passes through its main-sequence stage and begins to move towards the giant branch.* A star having the mass of the sun stays on the main sequence for about 6×10^9 years and remains close to it, moving

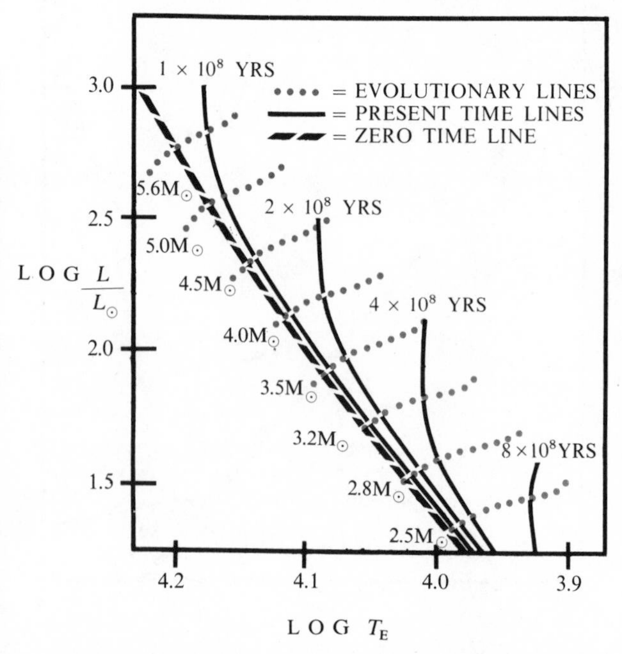

FIG. 23-11 *Time lines in the H-R diagram in the evolution of stars from one time line to another. The initial main sequence is shown as the zero time line. (After M. Schwarzschild, Princeton University)*

slowly away from it, for an equal amount of time. *But the point at which a star leaves the main sequence is determined by the amount of mass in its isothermal helium core as compared with its total mass.*

23.42 Evolutionary Lines and Time Lines on the H-R Diagram

The Population I star of given mass cannot have an isothermal core and remain in equilibrium, unless the mass of the core is smaller than a certain limiting value first discovered by Chandrasekhar and Schönberg. If the mass of the core exceeds this value the star must change its structure. *They showed that a star with an initial hydrogen content of about 80 per cent changes its structure and leaves the main sequence when about 12 per cent of its hydrogen has been transformed into helium, i.e., when the helium core consists of 12 per cent of the total mass.* The way in which stars with isothermal cores but with different total masses leave the main sequence is shown in the dotted evolutionary lines which cut across the solid **time lines**. (Note that a star does not evolve along a time line.) Thus Figure 23.11 is to be interpreted as follows: *The distance from the main sequence (measured along an evolutionary line) of a point of intersection of any time line with this evolutionary line represents the time it takes a star of given mass to evolve from its zero time line (the initial main sequence) to this later time line.*

To illustrate we consider two Population I stars with 2.8 and 3.5 solar masses, respectively, starting out from the main sequence. The evolutionary lines for these stars show that the more massive star suffers a greater change in its spectral class in a given time. Thus, whereas in 3×10^8 years the effective temperature of the $3.5M_{\odot}$ star decreases by about 2 000°K (going from about 12 000°K to about 10 000°K), the effective temperature of the $2.5M_{\odot}$ star decreases from about 9 770°K to 9 330°K.

We note that the time lines in our diagram correspond to the observed H-R diagrams of open clusters. We saw in Section 17.12 that stars in open clusters fall along lines which coincide on the main sequence, but which turn off to the right at different points. Since all the stars in a given cluster originated at the same time (from the same gaseous material) they are all the same age. However, the way each star evolved away from the main sequence depended upon its initial mass. Thus, the stars of small mass (lying in the lower right-hand part of the H-R diagram) have not moved off the main sequence at all, during the life of the cluster, whereas the more massive stars (in the upper part of the H-R diagram) will have turned off to the right.

If we superimpose the observed H-R diagram for clusters on the calculated time tracks given above it is possible to assign ages to the various open clusters. We see, for example, that the older clusters are those for which the turn-off point occurs lower down on the main sequence.

Thus, just as the observational data of globular clusters confirm the theory of evolution of Population II stars, the observed H-R data for the open clusters confirm the theory of evolution for Population I stars.

23.43 The Evolution of Population I Stars beyond the Main Sequence

As far as we know, a Population I star, after it has left the main sequence, evolves the way a Population II star does. In the case of the former, however, the carbon cycle plays the prominent role in releasing energy in the shell surrounding the helium core. There is also a difference between the heavy elements formed in two stellar populations during their giant and supergiant evolutionary stages. Since Population II stars consist only of hydrogen, they cannot synthesize heavy nuclei of all kinds, but only those which can be constructed with α particles. Hence they are deficient in certain metals (which will be discussed in more detail later). This is not true for Population I stars because they are formed from enriched material containing heavy atoms so that almost all types of heavy nuclei can be formed.

23.44 The Evolution of Population I Stars High Up on the Main Sequence*

Since the Population I stars are formed from dust and gaseous matter containing carbon and nitrogen, the carbon cycle controls the massive Population I (early spectral class) stars while they are on the main sequence. Stars such as these are very luminous with high effective temperatures and hence belong to the O and B spectral classes. These blue-white giants are among the youngest stars in our galaxy, some of them being a mere 30 or 40×10^6 years old. Since they are extremely luminous, they are consuming their hydrogen fuel at a very fast pace, and hence are evolving rapidly. Thus, the Population I stars, as distinct from the Population II stars, are found to lie everywhere along the main sequence with the very youngest ones in the upper left-hand corner.

THE FORMATION OF HEAVY ELEMENTS IN POPULATION II STARS

One of the most perplexing problems that faced astrophysicists—the problem of the formation of heavy elements—has now been solved as the result of the study of stellar evolution Since there is a gap in the atomic isotopes at mass 5, it is impossible for heavy elements such as Fe, Ni, etc. to be formed from hydrogen at the temperatures existing in the interiors of stars like the sun. At these temperatures there is no way to go from pure H beyond He^4 with thermonuclear reactions. Since until recently astrophysicists were unable to construct stellar models with high interior temperatures, they dismissed the possibility that heavy elements could be built up inside stars.

23.45 The Triple Helium Reaction and the Formation of C^{12}*

We know, however, that the temperatures in the helium cores of stars reach very high values when these stars pass from the main sequence to the giant stages. If the mass of the initial star is less than $1.5M_\odot$, the temperature in the helium core remains below 10^8 °K for a long time, but as this core grows with the increasing transformation of hydrogen to helium, it begins to contract and ultimately the core temperature exceeds 10^8 °K. If the initial mass of the star is greater than $2M_\odot$, the core contracts very rapidly (almost catastrophically) and very high temperatures are reached very quickly. Until temperatures of the order of 10^8 °K are reached, the helium core releases energy only by gravitational contraction. But at this point the helium thermonuclear reaction (first discovered by Salpeter) takes place. It may be considered in two steps:

$$\begin{aligned} &He^4 + He^4 \rightleftarrows Be^8 \\ &Be^8 + He^4 \rightarrow C^{12}_{\text{excited}} \rightarrow C^{12} + \gamma \end{aligned}$$

In the first step two helium nuclei combine to form the unstable nucleus Be^8 which, when left to itself, breaks down into two helium nuclei again. The fact that this process is reversible is shown by the lower arrow towards the left. However, for temperatures in excess of 10^8 °K equilibrium exists in the helium core between Be^8 and α particles. At any moment there is present among the α particles in the core a certain number of Be^8 nuclei; in fact there is one Be^8 nucleus for every billion α particles.

Because of this abundance of Be^8 (even though very small) the second reaction proceeds with the formation of an excited C^{12} nucleus. This happens because the Be^8 nucleus captures an α particle and forms an excited C^{12} nucleus if the captured particle has just the right amount of energy. The excited C^{12} nucleus then settles down to a stable state with emission of a γ ray. During this process about 7.4 Mev of energy is released. Thus, this triple α-particle reaction begins to contribute to the luminosity of the star in the giant stage of its existence.

When helium burning sets in at the tip of the red-giant branch, a runaway thermal process (the helium flash) occurs in the contracting core. The triple α process causes rapid heating in the case without any expansion, owing to the high degeneracy there (the great abundance of free electrons that conduct heat very quickly). The resulting rapid increase in temperature in the core accelerates the helium burning, and the helium flash occurs with a great release of energy in a very short time (at 10^{14} ergs/gm/sec). This released energy does not make the star extremely luminous (more luminous than a whole galaxy, as we might at first expect) because the outer layers of the helium core blanket the energy so completely that only a tiny fraction reaches the hydrogen-rich envelope. The flash, which lasts for a short time, cools the helium core. The triple α process then proceeds at a slow, steady pace until a carbon core is formed.

When the temperature in the core, as a result of the contraction, is high enough, the rate at which energy is released in the nuclear transformation is high enough to bring the core contraction to a halt or to slow it down considerably. If we suppose that the central density at this stage is of the order of 100 000 gm/cm³ (which is a very likely density according to calculated models), then at 100 million degrees this triple α-particle fusion releases about 600 ergs/gm/sec.

23.46 Formation of Elements beyond C^{12} by α-Particle Capture

When a sufficient amount of C^{12} has been formed, other helium-burning reactions proceed simultaneously with the triple α process. Thus C^{12} captures an α particle to form O^{16}; O^{16} captures an α particle to form Ne^{20}; Ne^{20} is ultimately transformed by He^4 to Mg^{24}, and so on. Of course, these reactions do not go on equally fast because the Coulomb barrier increases with atomic weight. However, in the temperature range of $1\text{–}2 \times 10^8$ °K, the reaction rates (except for the Ne^{20} reaction that leads to Mg^{24}) are of the same order of magnitude.

This means that if all the C^{12}, O^{16}, Ne^{20}, Mg^{24}, that we now find in our universe were formed in this type of cooking process in giant stars, their atomic abundances ought to be similar except for Mg^{24}. That this is so has been pointed out by Seuss and Urey who have found the following atomic abundance ratios for these nuclei in our solar system:

$$C^{12}:O^{16}:Ne^{20}:Mg^{24} = 1:6:2:0.2.$$

This is also in agreement with Aller's estimate based upon spectroscopic abundances of these elements in the atmospheres of young stars such as the sun. He finds the ratios 1:5:8:1. We cannot expect complete agreement between these two sets of ratios because the nuclei formed in our solar system have probably experienced a second state of hydrogen burning.

23.47 Formation of Elements beyond Mg^{24}

At this stage there is still some helium in the core, but in time all of it is transformed into C^{12}, O^{16}, Ne^{20}, and some Mg^{24}. The core then becomes isothermal since the temperature is too low for nuclear transformations to occur among these heavier nuclei, and it again contracts until temperatures of the order of 10^9 °K are reached. The intense γ rays that are present at these temperatures are energetic enough to break Ne^{20} down into O^{16} plus an α particle, which is then available for capture by a Ne^{20} nucleus to form Mg^{24}. In other words, Ne^{20} is slowly transformed into O^{16} and Mg^{24}. In each of the above α-particle capture processes about 5 Mev is released.

As the amount of Mg^{24} increases, still heavier elements are formed since at 10^9 °K Mg^{24} can capture the α particles released in the neon dissociation. This reaction transforms Mg^{24} to Si^{28}. The heavy elements such as S^{32}, Ar^{36}, Ca^{40}, (in decreasing abundances) are also synthesized by the α particles released in the dissociation of nuclei by γ rays (similar to the Ne^{20} dissociation).

23.48 Heavy-Nuclei Interactions and Fe^{56}

These α-particle reactions, resulting in the formation of more and more Ca^{40}, continue as long as the temperature in the core is between 10^9 °K and 3×10^9 °K. When the temperature exceeds 3×10^9 °K, because of further core contraction, the α process no longer contributes much to heavy-nuclei formation; at these high temperatures, with densities ranging from 10^5 to 10^9 gm/cm^3, the heavy nuclei interact thermonuclearly among themselves. These reactions go on at a steady rate so that statistical equilibrium exists among the various nuclei.

This α-particle build-up of nuclei as well as the build-up by heavy-nuclei interaction stops with the formation of Fe^{56} and a new stage in the development of the star begins. The reason that further contraction in the core (resulting in higher temperatures) does not bring about the formation of nuclei heavier than Fe^{56} is that this nucleus has the maximum binding energy per nucleon. This means that the bombardment of the Fe^{56} nuclei by protons, α particles, or other nuclei does not build up nuclei heavier than Fe^{56} because the Fe^{56} nuclei are disrupted by such collisions at these high temperatures. The only way we can build up nuclei beyond Fe^{56} is by having a source of neutrons since neutrons can enter the nucleus of iron at low speeds and bring about a nuclear transformation without disruption.

At this stage the star consists of a slowly contracting gaseous iron core (in which no nuclear reactions occur) surrounded by a layer of nuclei such as Mg^{24}, Si^{28}, S^{32}, Ar^{36}, Ca^{40}, in which the α process is still operating. This zone in turn is surrounded by a helium-burning shell containing primarily He^4, but also some C^{12}, and O^{16}. This in turn is surrounded by a thin shell in which hydrogen burning is producing He^4, and out beyond this is the radiative envelope of the star consisting of the original hydrogen-rich material.

Although the star remains in equilibrium in this stage, probably as a supergiant, for a fairly long time (just how long is still speculative, but probably of the order of 100 million years), this may be preliminary to an explosive stage. The reason is that up to a certain point the build-up of iron proceeds slowly, but when the temperature is sufficiently high and enough elements lighter than iron have been synthesized, large quantities of iron may build up in a matter of seconds. If the core now contracts further, resulting in an increase in temperature, the violent collisions between Fe^{56} and other nuclei break the iron up into α particles. The temperature may now be so high that this will occur very violently and matter will be ejected explosively from the outer regions of the star preliminary to the star's becoming a nova.

23.49 R and S Neutron Capture

In the evolution of the initial hydrogen star from the main sequence to the giant branch the heavy elements up to Fe^{56} with atomic weights that are multiples of 4 (e.g., O^{16}, Si^{28}, Ca^{40}) are formed. To explain the formation of nuclei beyond Fe^{56} and of isotopes such as N^{14} whose atomic weights are not multiples of 4, we have to consider nuclear reactions involving neutrons and also those involving protons in second- or third-generation stars (Population I) which started their lives with such nuclei as C^{12}, O^{16}, Ne^{20} already mixed in with hydrogen.

Nuclei beyond Fe^{56} can be built up only if there are sources of free neutrons in the interiors of stars. If these sources supply enough neutrons, not only are nuclei beyond Fe^{56} built up, but the various isotopes between neon and iron not in the $4n$ series are also formed.

Here we must differentiate between two types of neutron capture: the **S (slow) process**, and the **R (rapid) process**. In the S process, the capture occurs so slowly that there is time enough between successive captures of neutrons for the nuclei to settle down to stable isotopes by emitting electrons in the form of β rays. In this slow process, the nuclei formed are those which do not have an excess of neutrons and hence are stable.

Since different nuclei capture neutrons at different rates, the nuclei that are built up in greatest abundances by the slow process are those which do not have a great affinity for neutrons, i.e., those that have a small neutron-capture cross section. Although formed from nuclei that have a large neutron-capture cross section, they themselves are not affected by neutrons and hence become more and more abundant. Since all the various nuclei that are present at each stage of a star's evolution compete for whatever neutrons are around, the types of nuclei that are built up by neutron capture depend on the abundances of the nuclei that are already present and on their respective neutron-capture cross sections.

If the abundances of such light nuclei as Ne^{20}, etc., which compete with iron for neutrons, are not too great this S process is sufficient to build up nuclei beyond Fe^{56} and, in fact, right up to Bi^{209} but not beyond this nucleus. When Bi^{209} captures a neutron, it becomes radioactive with the emission of an α particle. This α-particle emission occurs so rapidly that there is no time for further neutron capture, so that the S process stops its heavy-elements build-up at Bi^{209}.

23.50 Neutron Sources in Stellar Interiors

Before we discuss the neutron R process, which accounts for the formation of elements beyond Bi^{209}, we may mention briefly the various neutron sources in stellar interiors. In stars containing such isotopes as C^{13}, O^{17}, Ne^{21}, Mg^{25}, Mg^{26}, etc. neutrons are released by nuclear transformations involving these nuclei and α particles, as illustrated by the interaction of Ne^{21} and an α particle:

$$Ne^{21} + He^{4} \rightarrow Mg^{24} + \textbf{neutron}.$$

Of course, as we have noted in our discussion of Population II stars, such nuclei cannot be formed in the helium-burning process if no mixing takes place. However, it is possible that some of the C^{12}, O^{16}, Ne^{20}, etc. is forced out of the core into the hydrogen-burning zone of the star, in which case the isotopes of odd atomic weight are formed from C^{12}, etc. by proton capture.

If this is so in Population II stars, or if we are dealing with second- or third-generation (Population I) stars in which proton reactions are available for the formation of the odd isotopes, then there are enough neutrons released by α-particle capture for the synthesis of heavy elements up to Bi^{209} by the S process. It has been shown in particular that the interaction of Ne^{21} with α particles can lead to about 14 neutrons per Fe^{56} nucleus so that a sufficiently strong neutron source is available for formation of these very heavy nuclei by the S process.

23.51 The Formation of Elements beyond Bi^{209}

To account for the formation of heavy elements beyond Bi^{209} we require a process other than the S process since, as we have noted, the slow capture of neutrons by Bi^{209} leads to radioactive nuclei. A rapid neutron process has, therefore, been suggested for the synthesis of heavy elements right up to Cf^{254}. *Since this R process requires intense beams of neutrons, it can only occur where such beams are created in short periods of time, and it appears that this can take place only in certain supernovae (S-Novae) explosions.* The observational evidence for the build-up of elements such as Cf^{254} by the R process (Type I supernovae, discussed in Section 24.39) comes from these supernovae light curves, which correspond to a half-life of 55 days (i.e., the intensity of the radiation emitted by these S-Novae diminishes by a factor of 2 every 55 days). This is equal to the half-life of the radioactive nucleus Cf^{254}.

We can account for this half-life of the S-Novae light curve by assuming that the major part of the energy released by Type I supernovae in this later stage comes from the spontaneous decay of Cf^{254}. But this means that great quantities of Cf^{254} somehow or other are built up very rapidly from nuclei such as Fe^{56} during the time of the explosion of these supernovae. This can only occur by the R neutron capture process. It can, indeed, be shown that if sufficiently intense neutron fluxes are available, great quantities of heavy elements such as Cf^{254} can be built from Fe^{56} in a matter of 10 to 100 seconds.

There is direct experimental evidence for this in the data first obtained in the thermonuclear test at Bikini in November 1952, for it was then observed that Cf^{254} occurred in great quantities in the debris of the hydrogen bomb. Since this element could have been formed only by the irradiation of U^{238} and lighter elements (such as iron in the casing of the H-bomb) by the intense neutron beams that were released, we may conclude that if the neutron beam is intense enough, heavy elements up to and above U^{238} can be built up very quickly from nuclei as light as Fe^{56}. From remnants of supernovae such as the Crab Nebula, we know that vast amounts of energy are released during the early stages of the explosion (sufficient to raise the temperature even in the outer envelope close to 10^9 °K), so that α-particle reactions involving such nuclei as Ne^{21} proceed at a very rapid pace. This creates intense neutron beams throughout the explosive regions of the supernovae, making the R process possible. It has been shown by G. R. and E. M. Burbidge that the mean time for the production of neutron beams during this early stage of the novae is of the order of 10 seconds.

23.52 Formation of Heavy Elements in Population I Stars

Thus far we have discussed the formation of heavy elements in Population II stars. The same analysis applies to Population I stars, with the following difference. *Since in the initial stages of Population I stars elements up to Fe^{56} are already present (but only those in the 4n series of isotopes) the types of nuclei built up are much more varied than those formed in Population II stars.* In particular, as a result of proton-capture after the Population I star has left the main sequence and passed into the giant branch, nuclei such as N^{14}, F^{19}, Na^{23}, etc. are synthesized. Of course, after these stars have become novae or have ejected their matter into space by some other means, the interstellar gas and dust are further enriched by the addition of a greater variety of isotopes. It is from this sort of material that third-generation stars are formed and it may very well be that our sun is such a star.

23.53 Chemical Differences between Population I and II Stars

The difference between the way heavy elements are built up in Population I and II stars enables us to differentiate between these populations now by spectroscopic analysis. Because Population II stars started out initially as pure hydrogen stars, whereas Population I stars began with an admixture of the metals, we can expect the Population II stars to show a metal deficiency relative to hydrogen as compared to Population I stars. Although the analysis of chemical abundances is a very difficult and complicated task, enough evidence has been adduced to show that this is indeed the case, and it is now possible to use the metal-hydrogen abundance ratio as an indication of the age of a star. *The young stars have higher metal-to-hydrogen ratios in their atmospheres than the old stars.*

23.54 Nuclear Transformation on Stellar Surfaces

Under certain conditions neutrons can combine with nuclei on the surface of a star to synthesize isotopes above hydrogen. We know that elements such as lithium, beryllium, etc. are found in the atmosphere of the sun in much greater abundances than can possibly occur in stellar interiors. These isotopes cannot be stored up in the interior of stars since they quickly combine with protons at the high interior temperatures. It therefore seems that these nuclei must be formed quite near the surface of a star by nuclear processes involving high-energy protons and α particles originating in hot spots such as flares on the photosphere.

If these protons and α particles acquire enough energy, as the result either of a flare or of the sudden build-up of a magnetic field, they collide with heavy nuclei to create beams of neutrons. As these neutrons diffuse through the stellar atmosphere, they are captured by protons and α particles to form deuterium and lithium, and the lithium in turn is built up into beryllium and boron by further neutron capture.

One more important bit of evidence for the formation of heavy nuclei by neutron capture is available in the presence of the element technetium, Tc, in stellar atmospheres. The lines of this element were first observed by P. Merrill in absorption-line spectra of S stars. Since the longest-lived isotope of this element has a half-life of only 300 000 years, it is clear that unless these stars are very young and were formed no more than a half-million years ago from material already containing Tc (which is possible but highly improbable) the Tc must have been and still is being formed on the surface of these stars by a neutron-capture process.

Taken altogether, we can see that the continuous thermonuclear processes and neutron capture in stellar interiors can account for all the known elements in our universe. At the same time it is possible to explain the existence of Population I and Population II stars on a chemical basis. We may draw one further interesting conclusion about the state of the matter in our universe at the present time. As we know, most of it (either inside stars or in the interstellar gas and dust) is hydrogen with perhaps an admixture of about 10 per cent heavier material. Hence, stellar evolution and element formation have not progressed very far and the universe still has a long way to go before even half of all the original hydrogen is transformed into heavy elements.

EXERCISES

1. Compare the rates of energy generation by the carbon cycle and the proton-proton chain at a point inside a star like the sun, where the temperature is 1.5×10^7 °K, the hydrogen content X is 0.85, and the heavy element content Z is 0.03.

2. Repeat problem 1 for a star high up on the main sequence at a point in its interior where the temperature is 2.5×10^7 °K and the hydrogen and heavy element contents are the same as in problem 1.

3. How many protons must be coalesced into helium every second to release the energy that is radiated per second by the sun?

4. Since under ordinary conditions the sum of the masses of two He^4 nuclei is smaller than the mass of Be^8, the latter decays spontaneously into the former. However, if the temperature of the helium gas is high enough, the mass of Be^8 can be lower than the sum of the masses of the two helium nuclei because of the dependence of mass on velocity. At what temperature will the sum of the He masses be equal to the mass of Be^8? (Use the equipartition theorem and take the mass of He to be 4.002 8 and that of Be^8 to be 8.007 85.)

5. Use the expression for the rate of energy generation by the carbon cycle and the proton-proton reaction to show that the carbon cycle takes place in a core close to the center of a star, whereas the proton-proton chain operates at much greater distances from the center.

6. Why is one more likely to find a convective core in a star when the carbon cycle is dominant than in one where the proton-proton chain dominates?

7. Assume that the sun consists of 85 per cent hydrogen. How long can it continue radiating at the present rate by changing its hydrogen to helium?

24

VARIABLE STARS AND STARS WITH UNUSUAL CHARACTERISTICS

24.1 The Definition of Variable Stars

Since, as we have seen, stars evolve and undergo changes that eventually become discernible, every star is variable in principle. However, since the evolutionary time scale is very long, the day-to-day variations in the luminosities of stars arising from this evolution are not detectable. In this chapter we shall limit ourselves to stars in which the variations are large enough and occur in short enough periods of time to be detectable. Even so, we shall be a bit more selective and exclude from our consideration small local changes such as occur in the sun's atmosphere and photosphere. We shall also disregard the small continual variations in local stellar magnetic fields, and the minor variations arising from rotations, etc., which probably occur in most stars.

Astronomers have designated as ***variables*** *those stars which exhibit a measurable variation in their brightness or luminosity not caused by the fluctuations in our own atmosphere.* Although the term "variable" originally referred only to luminosity, today we include among such stars those with a measurable variation in their spectra, their magnetic properties, etc. even though their luminosities may not change very much. Variable stars deserve a special chapter because they do not fit into the evolutionary scheme outlined in the last chapter, and because there is an important correlation between many of their properties and the structure of the galaxy so that they serve as a useful tool in analyzing stellar systems. Moreover the analysis and interpretation of their structure and the cause of their variations are among the most challenging and interesting problems in contemporary astronomy.

Strictly speaking, variable stars include a group of binary stars (eclipsing binaries) in which the variations in the apparent brightness are due to the periodic eclipse of one star by its companion. Such variables are called **extrinsic** or **geometric variables**, and we shall discuss them in Chapter 26 on ordinary binary systems. *Here we shall limit ourselves to* ***intrinsic variables****, in other words, to variables in which there is a real change in luminosity arising from causes within the star itself.* It is convenient here to give a schematic diagram (Table 24.1) representing the classes of intrinsic variables which we discuss in this chapter.

24.2 Classes of Intrinsic Variables *

We divide these stars into two groups: **irregular variables**, and **regular** or **periodic variables**. *There are three types of irregular variables*:

1. **Completely irregular variables**: these are stars that undergo slow, nonperiodic, and completely unpredictable fluctuations; they are generally red giants undergoing rather small luminosity variations over a brightness range rarely exceeding 2 magnitudes, and on the average considerably smaller than this.
2. **Semiregular variables**: these stars have mean periods but the periods themselves change considerably.
3. **Explosive variables**: the various types of **novae**.

The regular or periodic variables are divided into two large subgroups, which we may refer to as:

1. **Long-period variables** (periods of about 100 days or more), and
2. **Short-period variables** (periods of a few hours to 100 days).

TABLE 24.1
SCHEMATIC DIAGRAM OF INTRINSIC VARIABLES

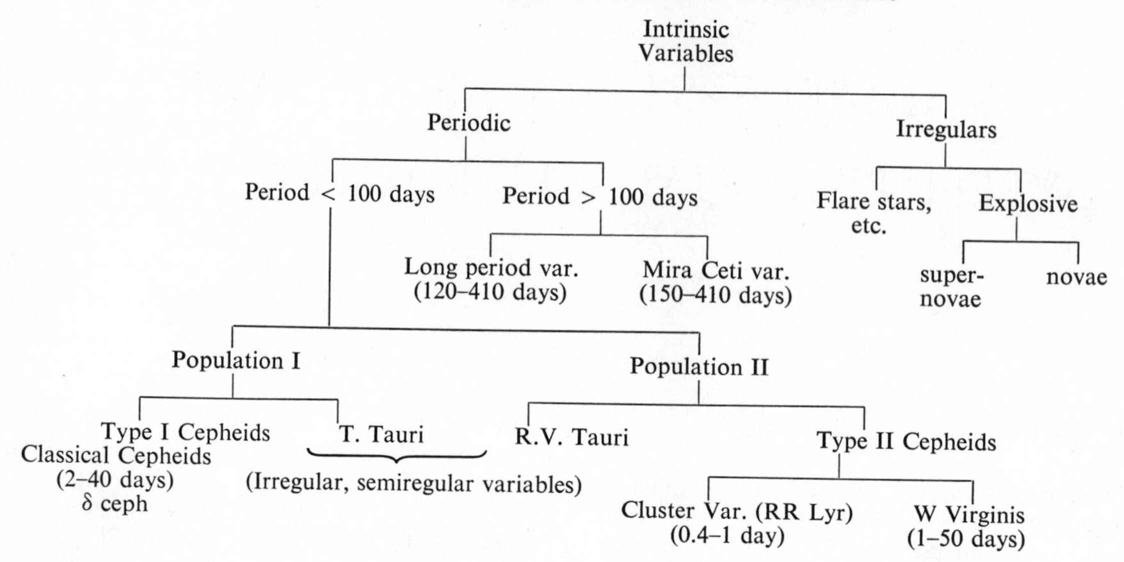

Tables 24.2A, 24.2B, and 24.3 list the variable subgroups and some of their characteristics.

24.3 The Historical Background of Variables

Variable stars were studied as long ago as 134 B.C. when Hipparchus first observed a nova, which led him to prepare the first catalogue of stellar magnitudes. But the modern study of variables did not begin until the year 1596, when David Fabricius began a systematic study of the star Mira (omicron Ceti). Little work was done in this field until 1784, when J. Goodericke made the important discovery that δ Cephei (which gives its name to the Cepheid variables as a group) is a regular variable. The irregular variable α Hercules was first discovered by Herschel in 1795, and the semiregular variable R Scuti by Piggott, who in the same year also discovered the peculiar variations in the star R Coronae Borealis.

Although it was difficult to make reliable observations of variables during this early period because photography had not yet been introduced into astronomical work, Chambers was able to list 113 variable stars by 1865. Photography completely revolutionized the study of variable stars; in the hands of E. C. Pickering of Harvard the rate of variable-star discovery increased enormously. By 1915 about 1 700 stars were definitely known to be variables. The latest variable-star catalogue lists over 14 700 variable stars. By means of photography S. I. Bailey was able to discover the very short-period RR Lyrae type stars, and its use also led Miss Henrietta Leavitt to her famous **period-luminosity law** for Cepheid variables.

24.4 Light Curves of Variables

To understand the behavior of a variable star and to gain some insight into its structure, we must have an accurate light curve showing how its luminosity changes from moment to moment. We do this visually, photographically (see Figure 24.1(a) and (b)), photoelectrically, or by a combination of all three. In actual measurements comparisons are made continuously between the brightnesses of the variable stars and those of a group of standard neighboring stars. When an accurate set of data is obtained for some variable, we plot the apparent magnitude of the star (along the ordinate axis) against the time, generally in days (along the abscissa). Although the observations themselves, which are the difference between the apparent magnitude of the variable and that of the comparison star, are often plotted against the time, in most cases the apparent magnitudes of the variable (obtained from the observations) are plotted directly. The final curve gives the variation of either the visual or the photographic magnitudes, and in some cases the variation in some particular part of the spectrum is also

TABLE 24.2
(A) CEPHEID-TYPE VARIABLES

I.A.U. symbol	Name	Population	Period	$m_v(10)$
			days	
Cδ	Classical Cepheids (δ Cep)	Extreme I	2 to 40	5.2
RR	Cluster variables (RR Lyr)	Extreme II	0.4 to 1	10
	Dwarf Cepheids	I	0.06 to 0.3	10
δSc	δ Scuti stars		0.08 to 0.19	8
CW	W Vir type	II	1 to 50	
βC	β CMa, β Cep type	I	0.15 to 0.25	5.3

(B) SELECTED CEPHEID VARIABLES

Star	Range of magnitude	Mean absolute magnitude M_v	Period	Range of spectral class
			days	
SU Cas	6.05–6.43	−2.5	1.95	F2–79
TU Cas	7.90–9.00		2.14	F5–G2
Polaris	2.08–2.17		3.97	F7 (double star)
δ Cep	3.71–4.43	−3.5	5.37	F4–G6
β Dor	4.24–5.69		9.84	F2–F9
ζ Gem	3.73–4.10	−4.5	10.15	F5–G2
X Cyg	3.53–8.09	−5.1	16.39	F8–K0
1 Car	3.6 –4.8	−6.5	35.52	F8–K0
U Car	6.30–7.55		38.75	F8–K5
SV Vul	8.43–9.40		45.13	G2–K5

studied. In Figure 24.2 typical light curves of different variable stars are shown.

24.5 Classical Cepheids—Characteristics of Their Light Curves

Since Cepheid variables, particularly the classical Cepheids, play an important role in the analysis of the Milky Way, we shall discuss these stars first. Although their periods range from about 1 to 50 days, most of them have periods in the neighborhood of 5 days. The light curve of δ Cephei (the prototype of all classical Cepheids) is quite smooth and shows that, in general, the rise in brightness of Cepheid variables is much steeper than its decrease. As we go from the ultraviolet to the red end of the spectrum for any particular Cepheid, the light curves flatten out, indicating that the range of variation in luminosity is much wider in the ultraviolet than it is in the infrared.

The changes in the spectral class of δ Cephei as it runs through a complete cycle are shown on the curve of the integrated brightness (all colors included). From the light curves of δ Cephei and η Aquila, another classical Cepheid (with a period of 7.1 days), we see that there is a variation of about 1 magnitude from minimum to maximum (although this may occasionally be as high as 1.5 for some classical Cepheids). The mean absolute magnitudes of classical Cepheid variables range from −2 to −5, with δ Cephei having a mean absolute magnitude of −3.5. The light curve of η Aquila has a slight hump on the descending branch which seems to be typical of classical Cepheids with periods of 7 to 8 days.

24.6 Spectral Classes of Cepheid Variables

As a Cepheid variable passes through its cycle of variations, its spectral class changes from about G2 at minimum to F2 at maximum as shown by the points on the curve in Figure 24.4. Taken as a group, these classical Cepheids are yellow giants or supergiants, with mean spectral classes (see Figure 24.5) lying between F0 through G0, with the G0 Cepheids being the more luminous. Even though no more than about 500 classical Cepheids have been observed in our galaxy, they are quite prominent in the sky because of their great luminosity. Some that are visible to the naked eye are listed in Table 24.2B.

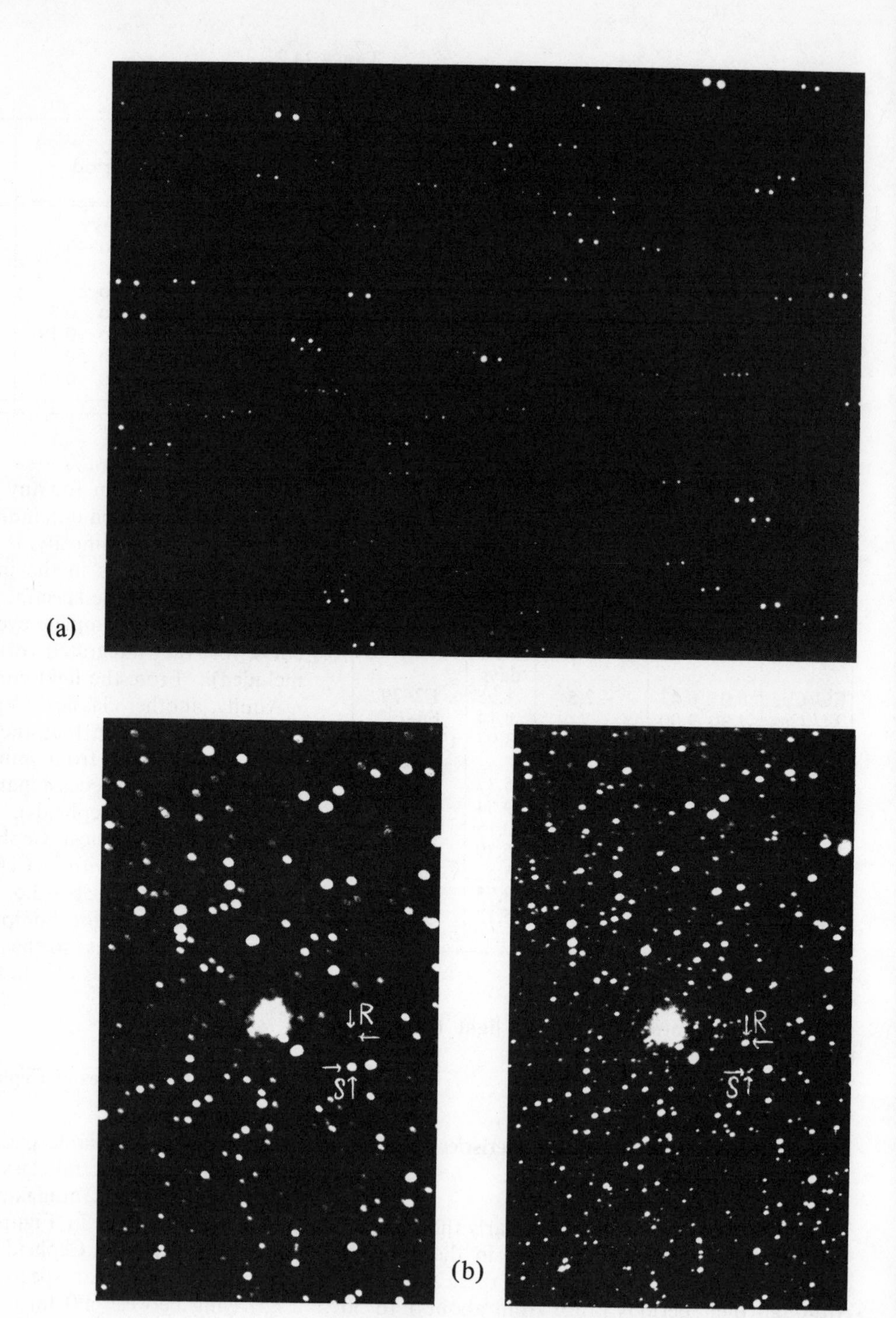

FIG. 24–1 (*a*) *Variable star WW Cygni at maximum and minimum brightness.* (*From Mount Wilson and Palomar Observatories*) (*b*) *Variable stars R and S Scorpii. A double print showing notable changes in brightness. Photographed by Barnard with a 10-inch Bruce lens.* (*Yerkes Observatory photograph*)

Table 24.3

CEPHEID CHARACTERISTICS AS A FUNCTION OF PERIOD, P

$\log P$	$\overline{M}_v$	$\overline{M}_B$	Spectral class		Change in magnitude $\Delta M_v = \Delta m_v$	Color index	Change in color index	$\log \frac{\mathcal{M}}{\mathcal{M}_\odot}$	$\log \frac{R}{R_\odot}$	$\log \frac{L}{L_\odot}$
			at maximum	at minimum						
in days										
Classical Cepheids										
0.4	−2.6	−2.2	F5	F8	0.4	+0.42	0.13	0.8	1.4	3.0
0.8	−3.5	−2.9	F6	G3	0.8	+0.60	0.32	0.9	1.8	3.5
1.2	−4.4	−3.6	F7	G8	1.0	+0.76	0.55	1.1	2.1	3.8
1.6	−5.3	−4.4	F8	K1	1.4	+0.88	0.67	1.3	2.5	4.2
Cluster variables (RR Lyrae)										
−0.4	+0.6	+0.7	A5	F1	1.3	+0.15	0.35	0.3	0.7	1.9
0.0	+0.5	+0.7	A7	F3	0.6	+0.25	0.1	0.4	1.0	1.8
W Virginis stars										
0.6	−1.8	−1.3	F3	F8	0.6	+0.5	0.2	0.7	1.6	2.6
1.0	−2.7	−2.0	F5	G1	0.7	+0.7	0.4	0.8	1.9	3.0
1.4	−3.5	−2.7	F7	G4	0.9	+0.8	0.6	1.0	2.2	3.4

24.7 Velocity Curves of Variables

In addition to the light curve, we can also plot the velocity curve by analyzing the light spectroscopically and measuring the periodic shifting of the spectral lines towards the red and violet ends of the spectrum. *From this Doppler shift we can easily determine the radial velocity of the star's surface, and we find that it oscillates back and forth (relative to the observer) in a way that is directly related to the change in luminosity.* In fact the velocity curve is the mirror image of the light curve as shown in Figure 24.6. *We see that the surface of this particular star alternately approaches and recedes from us (at maximum speeds, both of approach and recession, of 20 km/sec) as though the entire star were undergoing pulsations.* The velocity curve also shows the remarkable fact that the surface is expanding most rapidly at the moment that the luminosity is greatest. The luminosity has its mean value when the star is most highly contracted and again when it is at its greatest expansion. The luminosity is at a minimum when the star is contracting most rapidly.

24.8 Population Properties of Cepheids

Until quite recently, all Cepheids were thought to be similar and hence were grouped together, but the recent intensive studies of Population I and Population II stars show that Cepheids (like all other stars) belong to one or the other of these groups. This is of great importance in establishing the period-luminosity law (which we shall discuss presently) and therefore has a great effect on our knowledge of distances obtained by using this law. The distribution of the classical Cepheid variables in our Milky Way shows clearly their Population I characteristics. They lie in the plane of the galaxy out near the spiral arms. But there is another group of Cepheids, Type II or W Virginis stars, which are Population II stars, and which, like these, are found in globular clusters and towards the center of our galaxy. As a group these stars also are very luminous, lying close to the giant branch but somewhat below the classical Cepheids. The mean spectral classes of W Virginis stars lie between F2 and G2 and their absolute magnitudes range from −0.5 to −2.5.

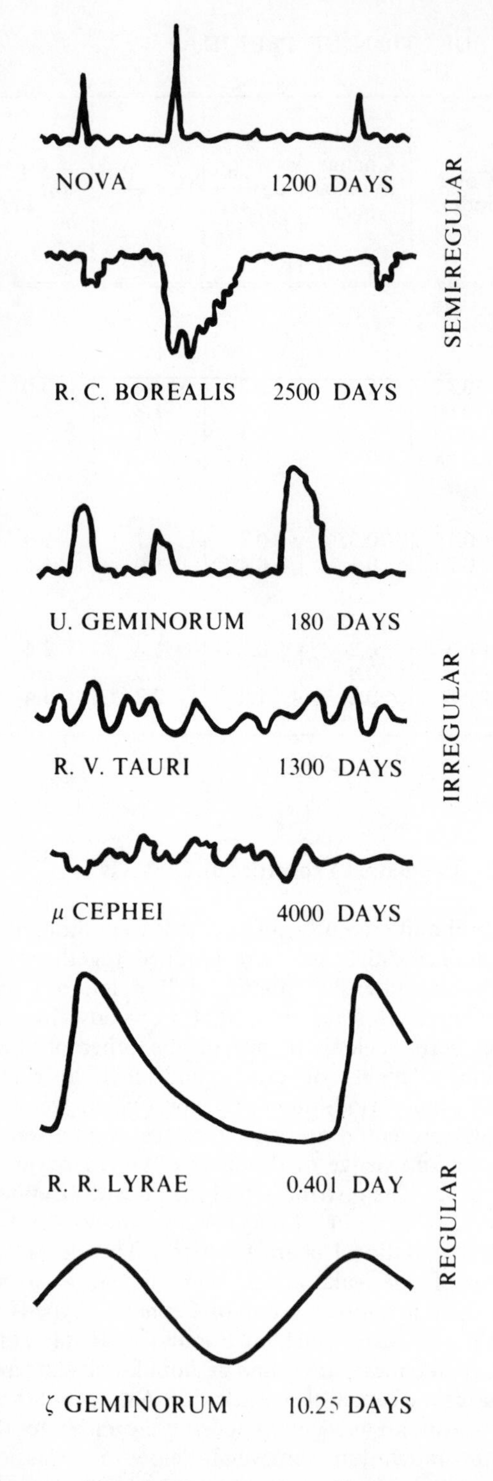

FIG. 24-2 *Typical light curves of variable stars. (From W. Becker,* Sterne und Sternsysteme. *Dresden: T. Steinkopf, 1950, pp. 105, 125)*

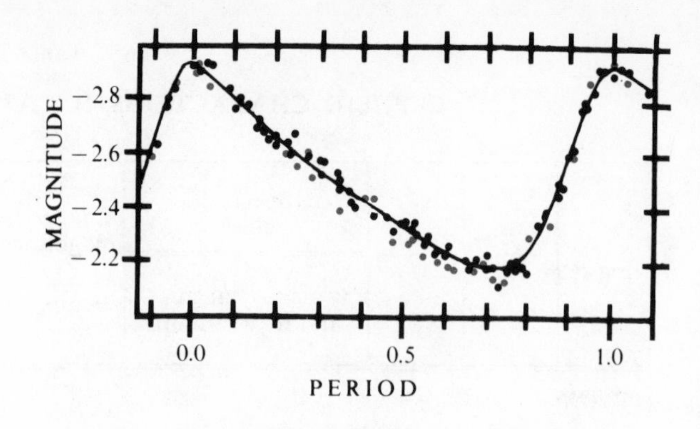

FIG. 24-3 *The light curve of δ Cephei. (Stebbins, black, and Danjon, red.) (From Goldberg and Aller,* Atoms, Stars, and Nebulae. *Philadelphia: The Blakiston Co., 1943)*

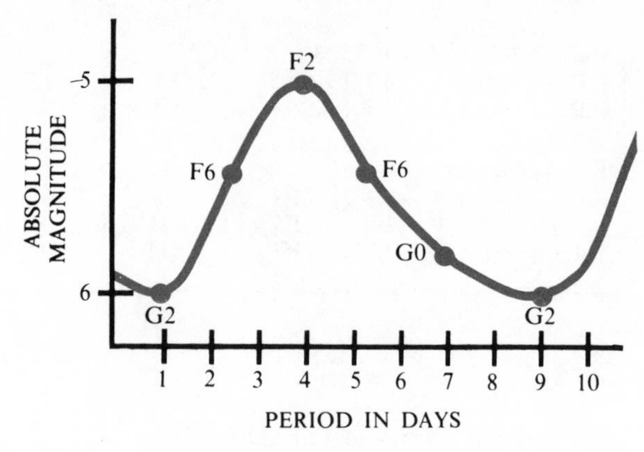

FIG. 24-4 *The variation of the spectral class of a Cepheid as its luminosity changes.*

The periods of W Virginis stars range from 2 to 75 days but most of them cluster around 17 days. Figure 24.7 shows a typical W Virginis star light curve. They are characterized by rather bumpy downward branches and rather flat maxima as compared to those of classical Cepheids.

24.9 Spectral and Other Features of W Virginis Stars

An interesting characteristic of some W Virginis stars was pointed out by A. H. Joy, who noted that of ten such stars he studied in globular clusters, seven show bright emission lines. He found that the hydrogen emission lines are at their greatest intensity when these variables are at maximum luminosity. Another interesting feature of the spectrum, which is quite marked in W Virginis stars for a very short time, just as the luminosity

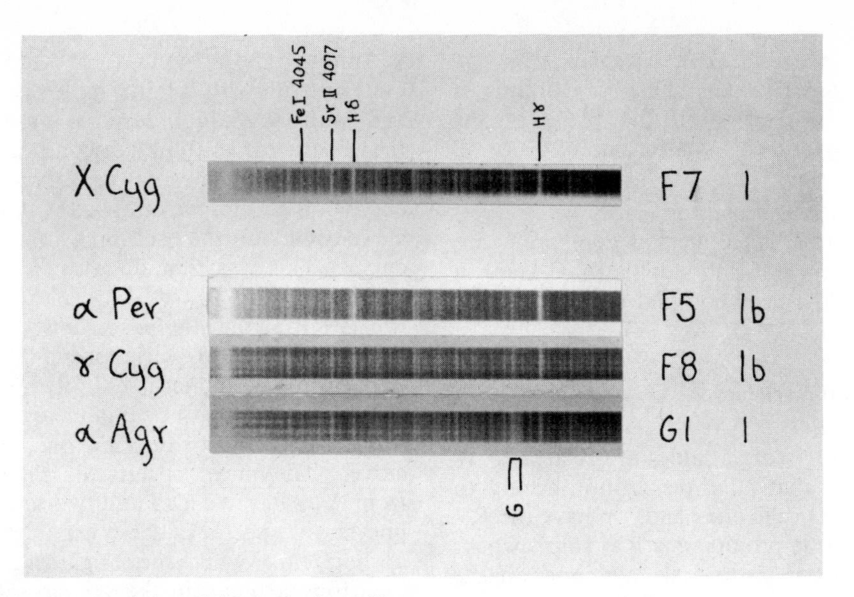

FIG. 24–5 *Changes in spectral class of the Cepheid variable χ Cygni. The plate of χ Cygni was taken near maximum light. The spectral type can be determined by interpolating between the standard supergiants illustrated. Two useful criteria of type are the intensity of the H lines and the appearance of the region of the G-band. It is of the greatest importance in classifying and studying the spectral variations of groups of stars like the Cepheids to use stars of similar luminosity for comparison. (Yerkes Observatory photograph)*

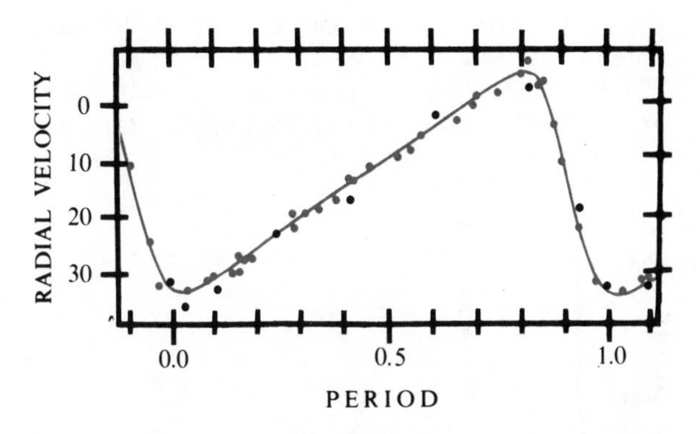

FIG. 24–6 *The velocity curve of a Cepheid variable. (Jacobsen, red, Henroteau, black.) (From Goldberg and Aller,* Atoms, Stars, and Nebulae. *Philadelphia: The Blakiston Co., 1943, p. 132)*

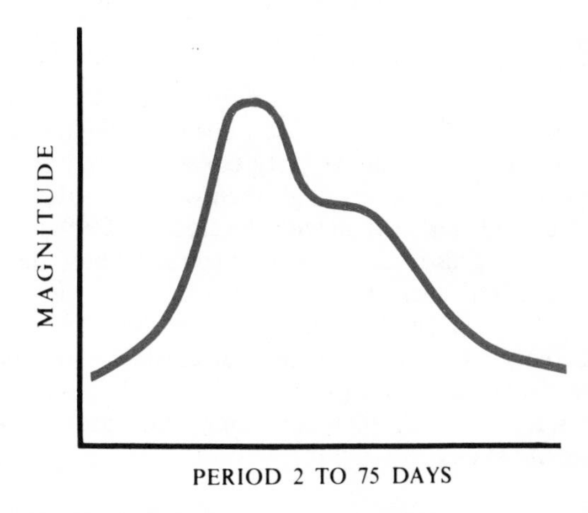

FIG. 24–7 *A typical W. Virginis light curve.*

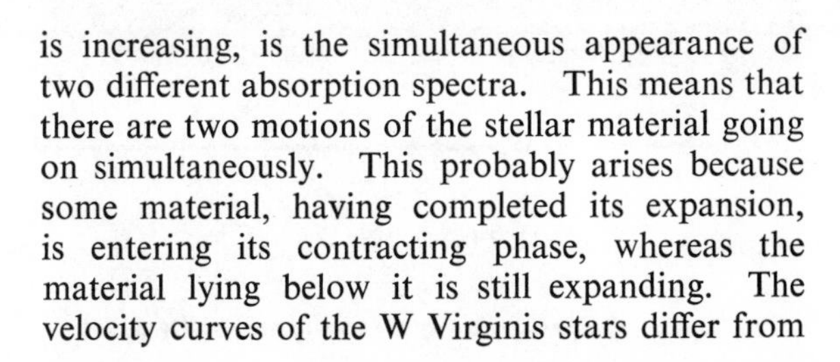

is increasing, is the simultaneous appearance of two different absorption spectra. This means that there are two motions of the stellar material going on simultaneously. This probably arises because some material, having completed its expansion, is entering its contracting phase, whereas the material lying below it is still expanding. The velocity curves of the W Virginis stars differ from those of classical Cepheids in that the former are *not* the mirror images of the light curve, although the two curves are closely correlated.

Although the W Virginis stars, taken as a group, are found mostly in globular clusters, some (like W Virginis itself) lie in our galaxy. These have two of the characteristics that are common to all Population II stars found in our galaxy:

1. *They lie at considerable distances from the plane of the galaxy.* Thus, W Virginis is found 1 750 parsecs off the plane of the Milky Way and this is also true of 30 or 40 other such stars.
2. *They all have high space velocities with respect to the sun* (discussed later) as compared with classical Cepheids and Population I stars in general in the neighborhood of the sun.

24.10 RV Tauri Variables *

Two other classes of variables are related to W Virginis stars in that all three groups belong to Population II. On the one hand, we have the RV Tauri variables (the prototype is RV Tauri) whose periods range from 60 to 100 days, and on the other hand, the very short-period RR Lyrae stars. Although the RV Tauri stars are not quite in the class of regular variables, forming a sort of bridge between the periodic Cepheids and semiregular variables, we shall discuss them here because they appear to be closely related to the W Virginis stars. While we can assign mean periods to the RV Tauri stars, their light curves are unstable and show irregularities in shape and period.

They range in spectral class from late G type to early M type, and their mean absolute magnitude is about -3. Their velocity curves have characteristics that definitely align them with Population II Cepheids rather than with the classical Cepheids, and this is also true of their spectra. Thus, they show bright lines and a doubling of their absorption lines in similar parts of their light curves. Moreover, they occur in globular clusters and they are distributed in our galaxy the way the Population II stars are and, like the latter, they are high-velocity stars.

24.11 RR Lyrae Variables *

The RR Lyrae stars are Population II variables with periods between 0.3 and 0.9 days. Although there is some uncertainty about their absolute magnitudes, which according to various estimates lie between 0.0 and $+0.8$, they all appear to have about the same absolute magnitudes; the most recent data give $+0.6$ as the best value. From their positions in the H-R diagram, we see that they fit very nicely into H-R diagrams for globular clusters, and this is where most of them are found.

Although the light curves of RR Lyrae stars are quite smooth with a fairly rapid rise in luminosity and with a gradual decrease on the downside, the velocity curves show a discontinuity. However, the changes in the radial velocity follow the light curve in a general way. The break in the velocity curve occurs on the ascending branch of the light curve. This was first discovered by Struve and Sanford.

The RR Lyrae stars also show the same peculiarities in spectral variations that the type II Cepheids do. Balmer lines of hydrogen appear for a short time during the rapid increase in luminosity of the star and just before the break in the velocity curve. The mean spectra of these stars range from A to F. We also find two sets of absorption lines in the spectra of these variables occurring for a short time while the brightness is increasing, indicating a break in the velocity curve. One of these sets of lines arises from surface material approaching us and the other set from material that is rushing away. Figure 24.9 shows the spectrum of RR Lyrae. The light curve of an RR Lyrae star shows that its luminosity varies by about 1 magnitude during any one cycle.

Although some RR Lyrae stars lie in our neighborhood of the galaxy, we find them in increasing numbers as we go towards the center and away from the plane of the galaxy (i.e., they lie in the halo part of our galaxy, which we shall discuss later). Finally, since their space-velocity characteristics are similar to those of the W Virginis stars, we

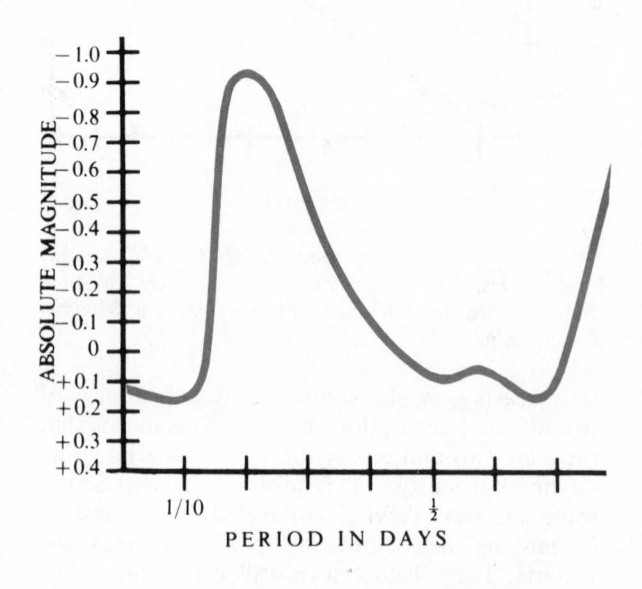

FIG. 24–8 *A characteristic light curve of an RR Lyrae star.*

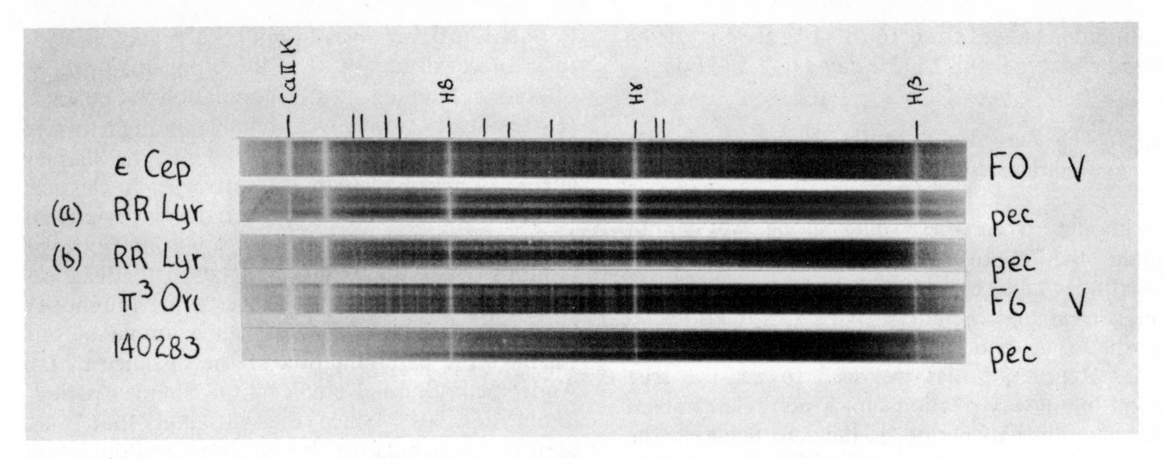

FIG. 24–9 *The cluster-type variable RR Lyra. The spectrum of RR Lyra is peculiar and cannot be located accurately in a spectral classification system. The intensity of the H lines is similar to the normal dwarf ε Cep (FO) at the phase when spectrogram (a) was taken. The numerous metallic lines present in the spectrum of ε Cep are faint or invisible on plate (a). At the same time, the K line has an intensity similar to an early A star. At the phase when plate (b) was taken the H lines had become weaker and the K line stronger. The H lines are similar to π^3 Ori (FG), while the metallic lines are about as strong as in class FO. Spectrum (a) bears a resemblance to the high-velocity dwarf HD 140283. (Yerkes Observatory photograph)*

must classify RR Lyrae variables as Population II stars. Because they are found in great numbers in globular clusters they are also called **cluster-type variables.**

24.12 The Period-Luminosity Law of Cepheid Variables

Because a relationship exists between their luminosities and periods, Cepheid variables are used to measure distances of groups of stars lying outside our galaxy. This remarkable relationship was discovered at Harvard by Miss Henrietta Leavitt, who was investigating the Small Magellanic Star Cloud (an irregular galaxy lying about 150 000 light years away from us) which contains numerous Cepheid variables. *Since these are all at about the same distance from us, any correlation between their periods and their apparent magnitudes must represent a relationship between period and absolute magnitude.* Since Miss Leavitt did not know the distance of the Small Magellanic Cloud, she correlated the periods of the Cepheid variables to their apparent magnitudes and thus discovered that *the longer the period of a variable is, the smaller its apparent magnitude* (*i.e., the brighter it appears to us*).

Her observations are given in Figure 24.10, where the apparent magnitudes are plotted along

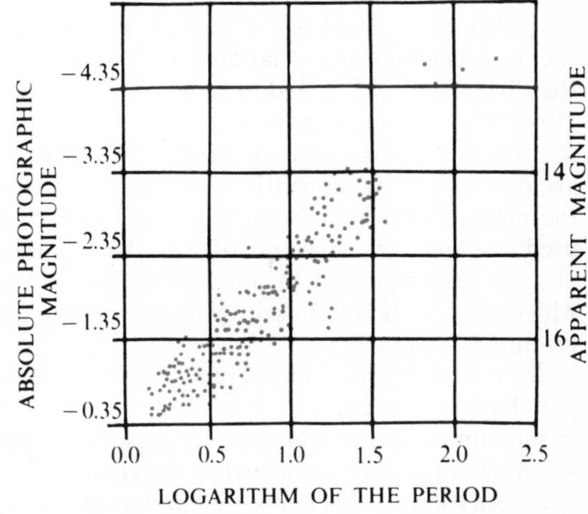

FIG. 24–10 *Period-luminosity curve plotted from the observations of Miss Leavitt. (After Shapley, in Goldberg and Aller,* Atoms, Stars, and Nebulae. *Philadelphia: The Blakiston Co., 1943, p. 134)*

the ordinate axis, and the logarithms of the periods in days along the abscissa. Since these variables are all at very great distances, their apparent magnitudes are very large, ranging from +13 to +17, even though they are intrinsically very bright stars. From this material, Miss Leavitt concluded that a given Cepheid variable is about 2 magnitudes brighter than another if the period of the brighter star is 10 times longer than that of the fainter. From the plotted data we see that the apparent magnitude changes from 16 to 14 as the log of the period changes from 0.5 (3.2 days) to 1.5 (32 days).

24.13 The Zero Point of the Period-Luminosity Relationship

Although from these data alone we cannot obtain the relationship between the absolute magnitudes and periods of these stars, it is nevertheless clear that the curve giving this relationship has the same shape as the curve for the observed data. Hence, all that we need to get the true period-luminosity relationship, which relates absolute magnitude to period, is the **zero point** on the curve. In other words, since we already know the shape of the true curve, all that remains to be learned is the position of one of its points on the absolute magnitude scale. Put differently, we have to know what absolute magnitude one of these observed apparent magnitudes corresponds to, and we can discover this by measuring the distance of a single variable.

Harlow Shapley set out to find the zero point of the period-luminosity relationship by measuring the distance of a Cepheid variable in our galaxy. However, even the nearest Cepheid variable is so far away that its parallax cannot be measured accurately by trigonometric methods. Moreover the method of spectroscopic parallaxes cannot be used because the spectra of variables change. Shapley therefore had to use a statistical method which lacks accuracy since it involves working with groups of stars.

Since Shapley was unaware of the two stellar populations, the curve he obtained includes both Population I and Population II Cepheids, which we now know obey somewhat different period-luminosity laws. There are two period-luminosity laws since Population I Cepheids are, on the average, 1.5 magnitudes brighter than Population II Cepheids, as shown schematically in Figure 24.11, where we have also included the RR Lyrae type stars. The two types of Cepheids lie on period-luminosity curves that are parallel to each other, but separated by 1.5 magnitudes. The RR Lyrae stars do not appear to obey a period-luminosity law, or, if they do, the variations of their luminosities with the period are too small to be discernible.

24.14 The Cepheid-Variable Distance Scale

Since the period-luminosity curve derived by Shapley is based primarily on data obtained from Population II variables in our own galaxy, it leads to incorrect distances when it is applied to groups of stars containing mostly Population I Cepheids. His results, however, were reliable enough to give the first period-luminosity law. Thus, the Shapley curve assigns an absolute magnitude 1.5 to classical variables which, like δ Cephei, have periods of $5\frac{1}{3}$ days. This leads to too small an intrinsic luminosity for these stars and hence to distances that are also too small when the period-luminosity law is applied to them. As an example of this we have the determination of the distance of the Andromeda Spiral Nebula by the Shapley period-luminosity law. Since the variables that were used in this nebula for this purpose are Population I, the distance of 750 000 light years that was found is too small by a factor of 2.

It was, indeed, by investigating variable stars in the Andromeda Nebula that Baade discovered the two Populations of Cepheids. He pointed out that if the Andromeda Nebula were really 750 000 light years away, we should be able to photograph some of its RR Lyrae variables, because they are luminous enough (he took 0 for their absolute magnitude) to be photographed with the 200-inch telescope at that distance. However, these variables did not show up on Baade's plates, and he argued, correctly, that the Andromeda Nebula is at a much greater distance than 750 000 light years. He concluded that there are two types of Cepheids in the Andromeda Nebula which obey different period-luminosity laws, and that the RR Lyrae stars belong to one type and the classical Cepheids to another. Since the observed Cepheid variables in the Andromeda Nebula are similar to the classical Cepheids in our own galaxy (Population I stars), we must apply the revised period-luminosity curve to these variables to obtain the correct distance of this galaxy—which we now know is about 2 200 000 light years.

With the aid of the period-luminosity law for Cepheid variables we can measure the distances of groups of stars beyond our Milky Way, such as galaxies, provided we use either the Population I or the Population II Cepheids they contain and do not mix up these two groups. Since the period of a Cepheid gives us its mean absolute magnitude and since we can measure its mean apparent magnitude directly, we can obtain its distance in parsecs from the formula $M = m + 5 - 5 \log d$. If a group of stars, such as a globular cluster, contains RR Lyrae variables, we obtain its parallax by placing $M = +0.6$ in the formula and using the observed value of m. This is one of the most

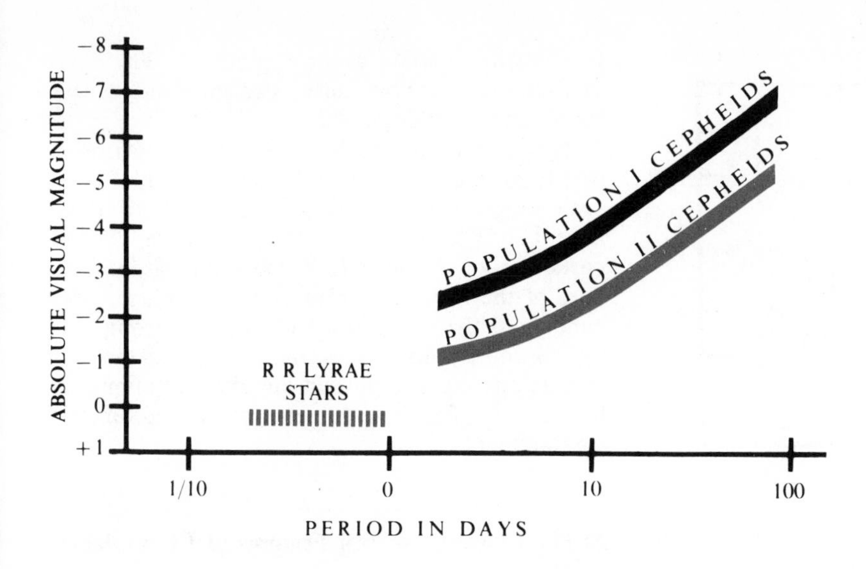

FIG. 24–11 *A schematic diagram showing the period-luminosity laws for Population I and Population II Cepheids. In the lower left-hand corner of the diagram the RR Lyrae variables are shown. The Population I Cepheids are 1.5 magnitudes more luminous than the Population II Cepheids.*

useful methods of finding the distances of globular clusters.

From recent work done with Cepheid variables in the Large Magellanic Star Cloud, Richard van der Riet Wooley, the Astronomer Royal, has found that even within the same class of variables there may be different period-luminosity laws which depend on the group of stars in which the Cepheids lie and hence presumably on their origin and evolution. He discovered that the period-luminosity law for Cepheid variables in the Large Magellanic Star Cloud is somewhat different from the period-luminosity relationship for Cepheids in the Small Magellanic Star Cloud. If we write the period-luminosity law in the form

$$\text{mean brightness} = A + B \log \text{period},$$

we find that although A is the same for Cepheid variables in both the Large and Small Magellanic Star Clouds, B changes from about 2.9 in the Large Cloud to 2.4 in the Small Cloud.

24.15 Long-Period Variables

Variables with periods of 100 days or more lie in a borderline region in which it is difficult to introduce a precise classification. We shall, therefore, consider in detail only the broad categories and mention briefly some of the subgroups. The *M*, or **Mira type**, stars are regular variables (although slight variations in their periods occur from time to time), with periods ranging from about 150 to 450 days. Their luminosities vary by more than 2.5 magnitudes during a single cycle. They straddle the giant branch (spectral classes M, R, N, S) of the H-R diagram, the more luminous ones (as given by M_{vis}) having the shorter periods. Their mean absolute visual magnitudes range from about -2.2 for 150-day periods to about 0.3 for 400-day periods. It should be noted that since all long-period variables are very far away, no reliable trigonometric parallaxes are available, and these absolute-magnitude estimates are based on statistical parallaxes and are therefore not too reliable. In Figure 24.12 the light curve of Mira Ceti (o Ceti) itself is plotted as an example of how the luminosity of a variable of this type changes.

A second group of long-period variables has properties similar to the Mira type stars. These are called LP (long-period) variables, and they differ from the Mira type stars in that their luminosities vary by less than 2.5 magnitudes during a single cycle. These variables also lie near the red end of the giant branch in an H-R diagram and their mean visual absolute magnitudes range from $+1$ to -2, with the long-period stars being fainter than the short-period ones. There are so many

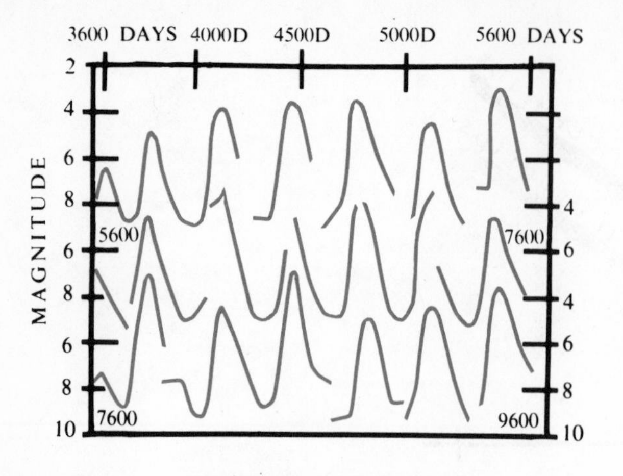

FIG. 24–12 *The light curve of Mira Ceti. (After L. Campbell, "Popular Astronomy": Goldberg and Aller,* Atoms, Stars, and Nebulae. *Philadelphia: The Blakiston Co., 1943)*

small differences as we go from one variable to another in these two groups that we cannot detail here all their characteristics. Instead we shall describe the broad features of both groups by giving a fairly detailed account of Mira Ceti itself.

TABLE 24.4
LONG-PERIOD VARIABLES

Spectral class at max		T_e max	T_e min	$\log \frac{\mathcal{M}}{\mathcal{M}_\odot}$	$\log \frac{R}{R_\odot}$	$\log \frac{L}{L_\odot}$	$\overline{M}_v$	$\overline{M}_{bol}$	Change in magnitude ΔM_v
	in deg	°K	°K						
K5e	2.0	3 700		1.02	1.9	3.3	−2.5	−3.4	
M0e	2.1	3 400	3 200	1.09	2.1	3.5	−2.5	−3.8	2
M5e	2.4	2 800	2 200	1.11	2.3	3.6	−0.6	−4.0	4
M8e	2.7	2 300	1 700	1.23	2.7	3.9	+0.3	−5.0	6
R8e	2.4	2 800	2 200	1.01	2.1	3.2	−1.7	−3.2	5
N0e	2.5	2 400	1 900	1.05	2.3	3.4	−1.2	−3.6	4
N5e	2.7	2 100	1 800	1.15	2.7	3.7	0.0	−4.4	3
S, Se	2.5	2 500	1 900				−1		6

24.16 Mira Ceti

Mira (the Wonderful) Ceti was the first star recorded by Fabricius as a variable. Its apparent magnitude at maximum brightness generally lies between 3 and 4, although it may reach the 2nd magnitude or become even brighter, as was the case in 1779 when its apparent magnitude was 1.2. At minimum brightness it falls to the 9th magnitude. Although its average period is 330 days, the interval between successive maxima ranges from 320 to 370 days.

Its spectral class varies from M6 at maximum to M9 at minimum; this appears to hold for all such long-period variables. Nearly all of them have bands in their spectra, with those of TiO particularly strong. The low-temperature arc lines of the metals are also found. Both the TiO bands and the metal arc lines become stronger as the luminosity diminishes, indicating that the surface temperature falls considerably at minimum brightness. This is in agreement with radiometric measurements.

24.17 Unusual Spectral Features of LP Variables

One of the most interesting features in the spectra of these long-period variables is the appearance of bright emission lines of hydrogen at maximum. This phenomenon is so characteristic of these stars that *the presence of bright hydrogen lines together with band spectra is an identifying mark of long-period variables*, and has been used in the past to pick up such stars. The hydrogen emission lines, which are weak at minimum luminosity but become very intense at maximum, probably originate deep down below the photosphere, and vary in intensity because of the star's pulsation. At minimum luminosity, when the star is most compressed, the relatively dense atmosphere obscures the emission lines. At maximum luminosity,

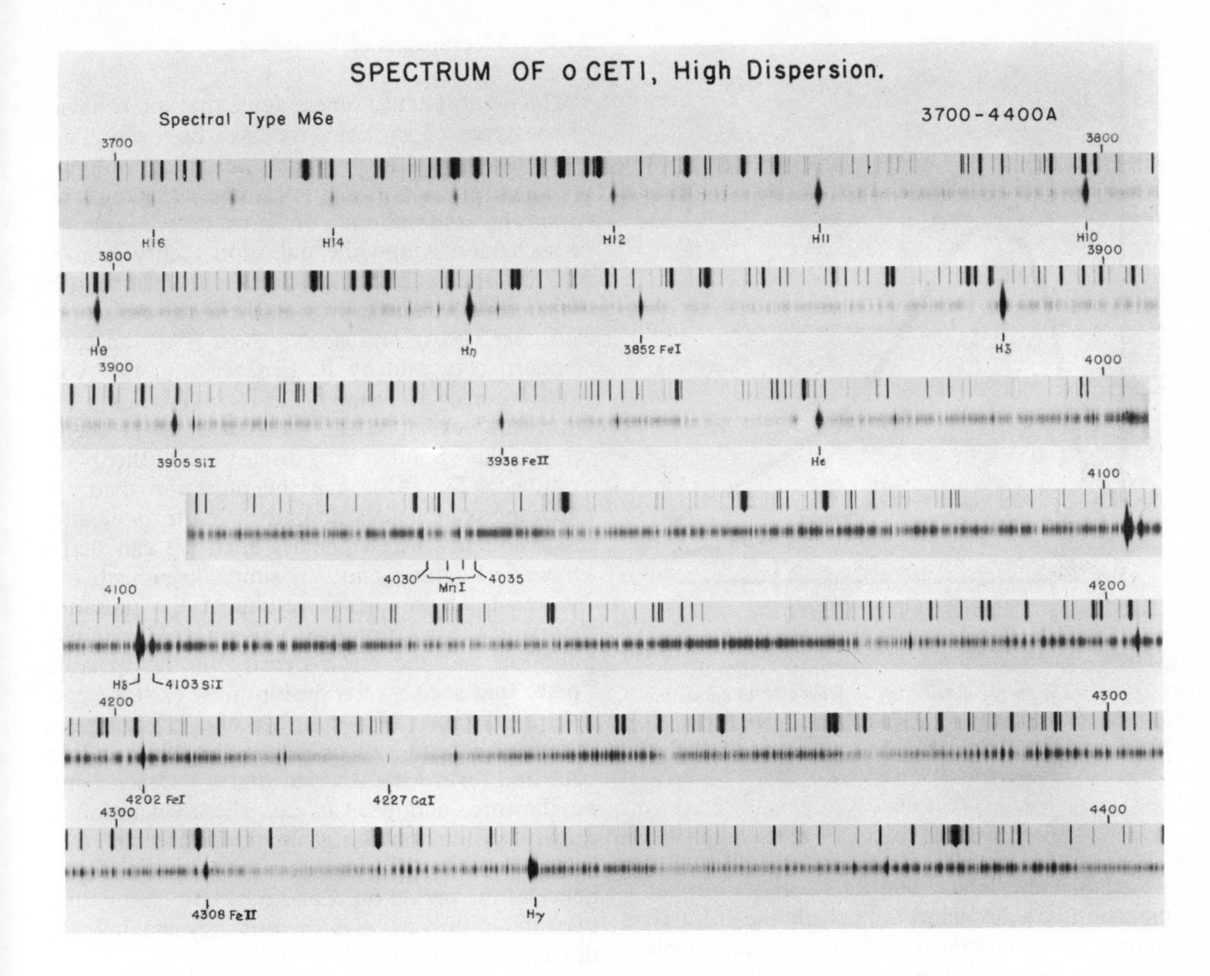

FIG. 24–13 *The spectrum of Mira Ceti, high dispersion. (Photograph from Mount Wilson and Palomar Observatories)*

however, the atmosphere is sufficiently attenuated to allow the emission lines to stand out strongly. See Figure 24.13 for the spectrum of *o* Ceti (Mira).

24.18 Energy Distribution Curve of Mira

The sharp drop at minimum luminosity in the surface temperatures of these stars as measured both radiometrically and spectroscopically (in the case of Mira Ceti, from 2 600° to 1 900°, a change by a factor of 1.37) leads to an apparent contradiction between the observed luminosity and the theoretical energy distribution, as we see in Figure 24.14. The Stefan-Boltzmann law tells us that each square centimeter of Mira at maximum luminosity, when the temperature is 2 600°K, should radiate $3\frac{1}{2}$ times more rapidly than at minimum luminosity, when its temperature is 1 900°K. If, then, only its surface temperature changed during a cycle, Mira would be only about $3\frac{1}{2}$ times brighter at maximum than at minimum.

However, this is not the case. It is 5 magnitudes or 100 times brighter (both visually and photographically) at maximum. There is thus a 25-fold discrepancy between the observed luminosity and that calculated from the Stefan-Boltzmann law. Part of this is due to Mira's larger radius at maximum luminosity, about a 20 per cent increase, which gives an area that is larger by a factor of 3/2. But this accounts for only a small fraction of the discrepancy and we must look elsewhere to explain it all.

A large part of the remaining discrepancy is due to the change in shape of the energy distribution curve. At a temperature of 2 600°K there is much

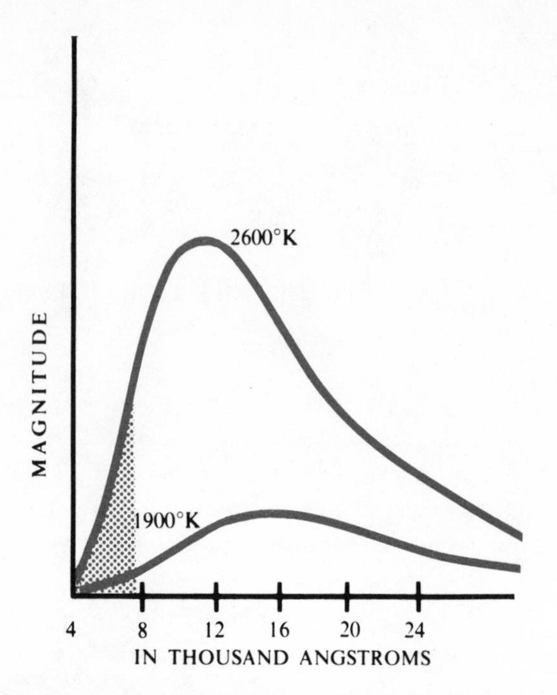

FIG. 24–14 *The energy distribution of Mira Ceti at maximum and minimum brightness. The shaded area is in the visible part of the spectrum.*

more energy in the visual and photographic part of the spectrum (as shown in the shaded part of Figure 24.14) than there is at 1 900°K, so that a proper comparison can be made only if all the infrared radiation at minimum is added to the visible radiation. However, this still does not account for the discrepancy completely. Some of it still remains and may be due to the intensification of the absorption bands and lines when the star is faint. If the total radiation (most of it is in the infrared) is taken into account at both maximum and minimum, the luminosity during a single cycle of Mira changes by about 1 magnitude and the discrepancy between the observed and calculated luminosity is reduced considerably.

All red variables like Mira Ceti are giants with fairly large radii. Interferometer measurements give a mean radius about 300 times that of the sun for Mira. Mira is a double star with a faint B companion. This was first suggested by A. H. Joy, who discovered that there is a tenth-magnitude B-type spectrum associated with Mira's characteristic M supergiant spectrum. This B-type companion was finally detected visually by A. G. Aiken, who showed that it is itself a variable star with a period of about 14 years and a luminosity range of about 2 magnitudes.

24.19 Mechanisms of Variables

Today most astronomers agree that the behavior of the types of variables we have been discussing is best accounted for by radial pulsation, and most models for such variable stars have been constructed on this assumption. The most detailed account of such models and the pulsation theory is given by P. le Doux and T. H. Walraven in Volume 51 of *Handbuch der Physik*, by S. Rosseland, whose book *Pulsation Theory of Variable Stars* is the accepted standard text, and by R. F. Christy in the April 1964 issue of *Reviews of Modern Physics*. The latter has developed the most satisfactory theory up to this point. We discuss this theory in Section 24.22. Although the pulsation theory is much too complicated and advanced for a detailed discussion in this elementary text, we can derive some of its consequences by simple arguments.

One of the important results of the pulsation theory is the relationship between the period of pulsation and the mean density of the variable. To see that such a relationship must exist we consider a unit mass of the surface of a pulsating star of radius R. This element of matter oscillates back and forth with a period that can depend only on the force acting on it (i.e., the acceleration of gravity, g) and on the geometry of the star (i.e., its radius, R). We therefore obtain the dependence of the period on R and g by combining these two quantities to give a quantity having the dimensions of time.

24.20 Relationship between Period and Mean Density of Variables

From this it follows that the period is proportional to $\sqrt{R/g}$. This holds if the pulsations are the result of the interplay between the gas pressure inside the star and the gravitational force. Physical considerations show that such a relationship should hold because the larger a variable is, the more slowly it should pulsate, and the greater the acceleration of gravity is, the more rapidly it should pulsate (in analogy with a pendulum).

Since $g = G\mathscr{M}/R^2$, we see that on substituting this in the above formula, the period is proportional to (where $\mathscr{M}$ is the mass)

$$\sqrt{\frac{R^3}{G\mathscr{M}}}.$$

If we now introduce into this equation the mean density $\bar{\rho} = \mathscr{M}/\frac{4}{3}\pi R^3$, the period is proportional to $1/\sqrt{\bar{\rho}}$. This fundamental relationship between the period and the mean density of periodic variables is confirmed by observations and is one of the strongest arguments in favor of the pulsation theory. It is usually written in the form

$$P\sqrt{\bar{\rho}} = C,$$

where C is a constant.

If we determine this constant for one Cepheid variable, we may apply the equation with the same constant to all other variables. Since the application of the formula shows that a variable star with a mean density equal to that of the sun (1.4 gm/cm^3) has a period of 2 hours, the constant C is equal to $\sqrt{5.6}$. We may therefore write the period-density relationship in the form

$$P\sqrt{\bar{\rho}} = \sqrt{5.6}, \qquad (24.1)$$

provided that P is measured in hours and the density in gm/cm^3. Thus, the period of δ Cephei which has a mean density of 5.6×10^{-4} gm/cm^3, should be about 100 hours, which is in good agreement with the observed value of 5.37 days.

The period-density relationship tells us at once that a period-luminosity law must hold as well. This follows because the relationship tells us that, in general, long-period stars are less dense and hence larger than short-period stars (assuming the masses to be about equal). But the larger a star is, in general, the brighter it is, so that the period-luminosity law follows.

24.21 The Phase-Lag Problem

Although the pulsation theory accounts for most of the features of the light curve, the **phase-lag problem** has only recently been clarified. To understand the nature of this problem we consider the relationship of the luminosity of a star to its internal structure. We know that when a star is compressed, its internal temperature rises, and it becomes more luminous because it generates energy at a greater rate. It follows from this that a pulsating star should be most luminous when it is most compressed (when the variable's radius is at a minimum). *But the observed velocity curves of variables show that maximum luminosity occurs one-quarter of a period later than the state of highest compression, and actually coincides with the moment of maximum velocity of expansion.* This contradicts the pulsation theory if we suppose that all points in the interior of the Cepheid variable oscillate together (i.e., in phase), for if this is so, the theory predicts that maximum luminosity should occur at maximum compression.

To overcome this difficulty, M. Schwarzschild has suggested that the pulsations in the interior and at the surface of a Cepheid variable do not occur in unison, but are out of phase. He has worked out the details of this theory and has obtained luminosity and velocity curves that are in satisfactory agreement with the observations.

24.22 The Origin of Pulsations *

One problem that has only recently been solved in connection with the pulsation mechanism is how such pulsations arise and how they are sustained with such steady periods over long time intervals. We can best understand the difficulties involved in this problem by considering the processes that go on inside a stable star like the sun. As long as the energy-generation mechanism of a star operates smoothly, no instabilities resulting in pulsations can occur since the energy balance is continuously maintained to keep the star in dynamical equilibrium.

If, however, a sudden change were to occur in the energy-generating mechanism (for example, a sharp transition from the proton-proton chain to some other type of nuclear transformation) dynamical instabilities could occur. Thus, since the outflow of energy would be interrupted during this transition, the star would contract quite rapidly and its internal temperature would rise considerably. This higher temperature could then trigger a different set of nuclear transformations resulting in a greatly increased energy outflow. This, in turn, would again cause the star to expand under the increased pressure engendered, and its internal temperature would drop, thus cutting off the new set of nuclear reactions. The star would then contract again and repeat the same cycle.

Although this mechanism appears to account for the pulsations of stars, there are certain difficulties associated with it. The first of these is the dissipation of the pulsation energy into unproductive heat because of frictional forces. If a sphere of gas, such as the sun, were compressed by some vast force, it would begin to pulsate, but these pulsations would be quickly damped out by the friction. However, if this contraction triggered off a new source of energy, this additional energy might just be sufficient to keep the pulsations going provided certain conditions were fulfilled. To begin with, this additional energy would have to be released in a time that is close to the period of pulsations of the variable. If we look at the time cycles of various types of thermonuclear transformations that are available (see Section 23.17) we see that there are no thermonuclear chains that can occur quickly enough to keep the pulsations going.

One possibility might be the formation of radioactive nuclei during the contracting phase of a variable. If enough such nuclei were formed, and if their decay period (the half-life) were about equal to that of the pulsation, then the energy released by these atoms during their radioactive decay could, in principle, keep the star pulsating. However, to account for the known periods of Cepheid variables, large abundances of radioactive nuclei with these periods would have to be produced in stellar interiors—and radioactive nuclei with the required periods are not known to exist. Another problem concerns the stability against pulsations of stars like the sun and, in general, all the main-sequence stars. This problem has been investigated by Eddington, le Doux, Cowling, and others. They have shown that even if the energy generation depends very sensitively on the temperature, as in the case of the carbon cycle, pulsations cannot maintain themselves after they are started in some way or other.

This objection, however, does not apply to the variable stars which lie in the red-giant and supergiant part of the H-R diagram. It may well be that these pulsations can occur at an evolutionary stage in a star's development when a change in its energy-generating mechanism takes place. In fact, a giant star may begin to pulsate when a sudden contraction of its helium core occurs.

In the previous paragraphs we discussed Cepheid pulsations in terms of the triggering of new energy mechanisms by stellar contraction to keep the pulsations going, which would involve the pulsation of the entire star; but there is still another process that can account for pulsating variable stars which involves only the outer regions of a star—in particular, the hydrogen and helium II ionization zones. Starting from a preliminary analysis of instabilities in the outer ionization zones given by the Russian astrophysicist S. A. Zhevakin, and a more detailed analysis by Cox, Christy has obtained models for pulsating variables that agree very well with the observations (light curves and velocity curves), by showing that the hydrogen and the He II ionization zones can, under proper conditions, begin to oscillate and continue to do so stably.

The analysis is based on the following idea: In a zone of a star where ionization of some atom sets in, the temperature must fall because not all the energy flowing into this zone from the interior goes into kinetic energy of molecular motion (internal energy, see Section 16.3); some of it ionizes the atoms. The material in this zone of the star will therefore begin to collapse and will continue to do so until ionization is approximately complete. At this point, recombination of the ions and electrons will set in, and, with the release of the ionization energy, the temperature in the zone will begin to rise and the zone will expand. Christy used the interior equations (22.7 through 22.10) to analyze such ionization zones, but in these equations he allowed the variables $M(r)$, $L(r)$, P, T, and r to change periodically with the time. He then solved these time-dependent equations to obtain his pulsating models.

24.23 Semiregular and Irregular Variables

There are two groups of irregular variables, one consisting of stars in which the variations occur rather slowly (such as Betelgeuse) and the other composed of stars in which the variations occur catastrophically in short periods of time: these are the **novae** and **supernovae**.

The stars in the first group are all red giants or supergiants (belonging to the spectral classes M, R, N, S) with mean absolute magnitudes ranging from about -1 to -3. Their luminosity variations are quite small (seldom exceeding 2 magnitudes, and in most cases less than 1). According to A. H. Joy, the spectral variation during a single cycle of any one of these variables is usually no more than 1 or 2 subdivisions of a spectral class. Although the irregular variables have no periodicity (the variations occurring in a rather unpredictable way from moment to moment) the semiregular variables have a periodicity of sorts, generally longer than 100 days.

From their distribution in the galaxy and from their velocities it appears that the M-type regular and semiregular variables having high luminosities are members of Population I, whereas those of

TABLE 24.5

TYPES OF IRREGULAR AND SEMIREGULAR VARIABLES

Desig.	Type and features	Population	Mean period	Spectral class	Absolute magnitude M_v	Change in abs. magnitude ΔM_v	Apparent magnitude m_v	Mean galactic latitude
			days					
RV	RV Tau, UU Her, Irreg. min. alternating depth	II	75	G to K	−2	1.3	7.4	23
SR a, b, c, d	Long-period semireg. including μ Cep, δ Ori	I to II	100	G to M N	−1	1.6	5.4	22
I	Irregular			K to M N	−0.5	1.3	4.4	22
RW	T Tau, RW Aur. Ass. with neb. em. lines	I Extreme		dFe dGe	+5	3	11	14
RCB	RCrB. Sudden decr. of brightness	I		G, K R	−5?	4	10.5	14
UG	SS Cyg, U Gem (Sudden periodic incr. of brightness)		60	B, A	8 ± 3	3.6	13	25
Z	Z Cam, CN Ori (Sudden periodic incr. of brightness)		20	F	10 ± 3	3.2	13.5	22
	SX Cen. Superimp. long and short periods		{30, 800}	F to M		{1.2, 2.0}	13.5	15
UV	Flare stars, UV Cet			dM3 to 6e	15	2	13	

comparatively low luminosity (absolute magnitudes of about −1) are rather heterogeneous, with some of them belonging to Population II. The semiregular and irregular variables belonging to spectral class N and R give every evidence of being Population I stars.

24.24 Explosive Variables

The explosive variables consist of all stars that undergo very rapid luminosity changes in relatively short periods of time. Although the changes differ considerably in amplitude from one member of this group to another, ranging from a few magnitudes, in the case of stars such as U Geminorum, to about 16 magnitudes in the case of supernovae, there are sufficient similarities to warrant our discussing them together. However, we shall consider the novae and supernovae separately, and first treat the less violently explosive variables consisting of the Wolf-Rayet stars, the shell stars, the U Geminorum, the Z Camelopardalis, the flare, and the R Coronae Borealis stars.

24.25 The Wolf-Rayet Stars

Although Wolf-Rayet stars are not ordinarily classed as variables, their spectral lines indicate that these stars contain a central core surrounded by a rapidly expanding shell of gas, as shown in the schematic drawing in Figure 24.15. No more than a few hundred of these stars have been observed, all at very great distances, with absolute magnitudes (as estimated from an analysis of these stars in nearby galaxies, such as the Magellanic Star Clouds) ranging from −4 to −8. From proper motion studies of these stars in our galaxy, however, they appear to have a mean absolute magnitude of about −3.4, so that they are similar to O-type stars of the same effective temperature.

An analysis of their spectra, which contain atomic emission lines, shows (see Figure 24.16) that these stars have surface temperatures ranging from 60 000 to 100 000°K, from which, together with their luminosity, we must conclude that their radii are about twice that of the sun. The emission lines (they are not bands), which are extremely wide (about 60 to 100 Å in some cases) and many

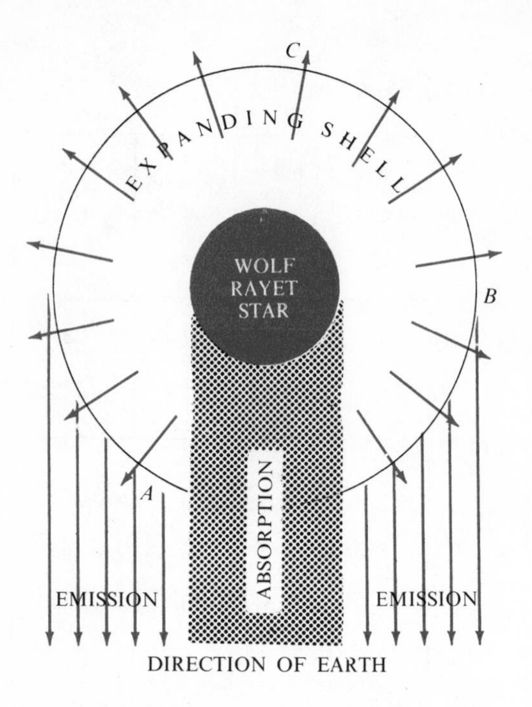

FIG. 24–15 *Schematic drawing of the expanding shell of a Wolf-Rayet star.*

times brighter than the continuous background spectra, are due to doubly ionized helium and highly ionized atoms of carbon, nitrogen, oxygen, and silicon. These lines are at times accompanied at their violet end by absorption lines. The presence of wide lines in the spectra of these stars leads to the conclusion that they are surrounded by a shell of radially expanding gas. For if we consider light coming towards us from such a shell (see Figure 24.15), we observe that it consists of contributions by atoms in the parts of the shell A, coming towards us, the parts B, that are neither approaching nor receding, and the parts C, receding from us.

The different Doppler shifts arising from all the different radial motions of these various parts of the shell are superimposed, giving us the observed broad spectral lines. The absorption component at the violet end of the emission line also indicates an expanding shell since it arises from radiation coming from the core of the star, which is then absorbed in the part of the shell that is approaching us. The width of the spectral lines indicates that the material in the shells of Wolf-Rayet stars is streaming away from the main body at speeds in excess of 1 000 km/sec, although certain difficulties in this interpretation have led some astronomers to reject these high velocities of expansion. (See Figure 24.16.)

The most extensively studied Wolf-Rayet star is the fainter component of the eclipsing binary HD 193576. Its mass is 12 times that of the sun, and it is accompanied by an O6 component having a diameter 10 times that of the sun, and a mass equal to $28M_\odot$. The Wolf-Rayet member of this binary has a small dense core having a diameter twice that of the sun, and an effective temperature of about 80 000°K. Surrounding this is an envelope of 7 solar diameters, and beyond this envelope is the expanding shell, which appears to be detached and whose diameter is 16 times that of the sun. Since other Wolf-Rayet stars have been discovered to be binaries, it is the opinion of some astronomers that all stars in this group are binaries, and that the only reason we do not observe them as such is that the planes of their orbits do not pass through the earth, so that eclipses cannot be observed.

24.26 Shell Stars

Besides the Wolf-Rayet stars, there are others with envelopes that are expanding much more slowly. Among these are the P Cygni stars, certain shell stars of which Pleione is a good example, and certain types of class B emission stars. The absorption lines in the spectrum of a P Cygni star are displaced towards the violet, indicating that the part of the atmosphere in which these lines originate is approaching us. Since the emission lines are broadened, the analysis of the previous section applies here too, and we may conclude that the atmospheres of these stars are expanding. The widths of these emission lines show that the shells of these stars are expanding at speeds of up to 200 km/sec. That stars like P Cygni are related to novae is suggested by the novae-like outburst in the seventeenth century suffered by P Cygni itself, which thereafter became a 5th-magnitude star.

The class B emission stars are those B-type stars with equatorial rotational velocities of 250 km/sec. This high rotational speed induces instabilities because it results in an equatorial centrifugal acceleration of the same order of magnitude as, or larger than, the gravitational acceleration. Thus a B-type star of $10R_\odot$ having an equatorial speed of 300 km/sec experiences at its equator a centrifugal acceleration of 1.43×10^3 cm/sec^2, whereas the gravitational acceleration is of the order of 10^3 cm/sec^2. Because of this, material streams out of the star in the equatorial plane and an expanding

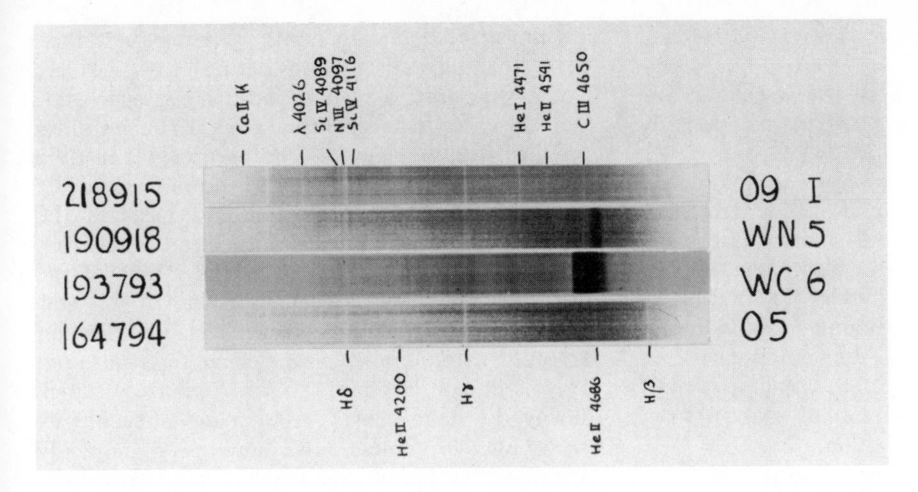

FIG. 24-16 *Two Wolf-Rayet stars. The spectral types of the two Wolf-Rayet stars were determined by Sanford and Wilson (Ap, J 90, 237, 1939). HD 218915, a supergiant of class O9, has a spectrum similar to HD 190918, except for the emission lines. HD 164974 has a pure absorption spectrum which is very early in type. (Yerkes Observatory photograph)*

atmosphere is formed. The rapid rotation itself causes the lines of the spectrum to be broadened since one edge of the star is approaching and the other receding.

The Pleione stars, among which are such stars as 48 Libra, and γ Cassiopeiae are shell stars with bright hydrogen lines. They are apparently stable for long periods of time, but then appear to undergo changes resulting in the disappearance of the bright hydrogen lines. Later, the bright lines may reappear together with narrow dark lines which may first grow quite strong and then finally weaken again. The analysis seems to indicate that a shell develops in these stars in which the material is streaming out at speeds of about 50 km/sec.

24.27 U Geminorum Stars

The U Geminorum stars are characterized by fairly large and rapid increases in brightness followed by slow declines. In this respect they are similar on a reduced scale to novae. These stars can best be understood by studying the two well-known members of this group, U Geminorum and SS Cygni. They both alternate between states of quiescence and rapid changes in luminosity during which they vary by about 4 magnitudes. Their maximum luminosities generally last anywhere from 3 to 20 days (average 8 days) and the periods of quiescence may be as short as 20 or as high as 200 days. These stars are of spectral class B and A and are generally bluish-white in color. At minimum luminosity the spectra of these stars are characterized by wide emission lines, which, as the stars become more luminous, merge gradually into the continuous background. At maximum brightness these lines reappear as absorption lines.

Using the most recent spectroscopic data for U Geminorum stars, R. Kraft has shown that most of the bright stars in this group are exceedingly close binaries with very short periods of a few hours or less. He surmises that all U Gem stars are binaries.

24.28 Z Camelopardalis and R Coronae Borealis Stars

The Z Camelopardalis stars are similar to the U Geminorum stars in general appearance and behavior and are probably related to them. Their luminosities change by 2 to 3 magnitudes in periods of 10 to 20 days. They are characterized by relatively long periods of quite regular changes which are suddenly replaced by rapid fluctuations of small amplitude.

The R Coronae Borealis stars are almost opposite in behavior to U Geminorum stars in that they are at maximum luminosity for very long periods of time compared to their periods of minimum luminosity. Their absolute magnitudes at maximum luminosity are always the same, but they vary considerably at minimum luminosity. In fact, the amplitudes at minimum luminosity are completely random over a range of 6.5 magnitudes. The luminosities of these stars may drop suddenly by as many as 6 or 7 magnitudes.

24.29 Flare Stars

A group of stars, the **flare stars**, which show extremely rapid luminosity changes, has recently been found among the class M dwarfs. They are characterized by sudden increases in luminosity which may last only a few minutes, but sometimes as long as half an hour. All known flare stars, among which are Proxima Centauri and U Ceti (the faint component of the double star 60 Kruger), have absolute magnitudes ranging from 11 to 16. Moreover, many of them are components of binaries, and it may well be that the flare phenomenon is related to the gravitational tidal action of the two components on each other.

An interesting spectroscopic feature of these stars, which seems to be related to the flare phenomenon, is the presence of fairly strong emission lines of H and C II during the quiescent phase of the star. Although at maximum luminosity it is difficult to obtain a reliable spectrum because of the short duration of the flare, it has been observed that the background continuous spectrum in the ultraviolet region is greatly enhanced and that the hydrogen lines are intensified. There is also evidence that both neutral and ionized helium lines are present at maximum luminosity, which indicates a surface temperature of at least 10 000°K. Because of the speed with which flares occur they bear a striking resemblance to solar flares, but on a considerably larger scale involving a rise in temperature from 3 000°K to 10 000°K. It is also estimated that a flare involves about 1 per cent of the stellar surface.

NOVAE

Novae and supernovae differ from other explosive variable stars in the magnitude of the explosion involved. Whereas stars such as U Geminorum, shell stars, etc. have shells that are expanding at several hundred km/sec, the energy released suddenly by a nova is so great that the material is ejected from the star at speeds of thousands of km/sec. Since the explosive energy released in supernovae is thousands of times greater than that released by ordinary novae, we shall discuss these two groups separately.

24.30 Ordinary Novae

An ordinary nova increases by about 12 or 13 magnitudes in anywhere from a few hours to a few days (see Figure 24.17(a)). We may best illustrate what happens by considering Nova Aquilae which was observed as an ordinary A-type star, and then in 1918 became a nova 13 magnitudes brighter. A sequence of photographs taken over a period of ten years (1922–31) shows the scale of the explosion and enables one to measure the speed of the expanding shell of ejected material. We find an expansion rate for the radius of the shell of 1″ per year. The luminosity of this star increased from an absolute magnitude of +5 to an absolute magnitude of −8 during this outburst. This is direct evidence of the luminosity range during the nova stage. After a nova reaches its maximum brightness, its luminosity suddenly diminishes along an exponential curve that is very nearly smooth or along one with small irregular fluctuations as shown in the two diagrams (Figure 24.17(b)).

The data obtained from Nova Aquilae show that the luminosity of a nova decreases fairly rapidly within the first few days after the maximum is attained. After that the decrease is gradual, and a few years elapse before the nova returns to its original luminosity. Nova Aquilae became three magnitudes fainter in the first 8 days after maximum.

That the nova outburst is accompanied by a rapidly expanding shell of gas is evident not only directly from photographs but also from the variations in the spectra of these stars. In the case of Nova Aquilae (see Figure 24.18(a) and (b)) a faint greenish envelope surrounding the star was seen telescopically a few months after the initial outburst; by 1940 this envelope, expanding at 2″ per year, could be clearly detected on photographic plates. Although in the case of Nova Aquilae this expanding shell is spherical and concentric with the star itself, this is not true for all novae, as can be seen from the photograph of Nova Persei (see Figure 24.19). Here we find an asymmetrical expanding shell as though the explosion were one-sided. From an analysis of the rate of expansion of novae shells, it appears that they leave the novae at the moment of maximum luminosity. In Table 24.6 we list the apparent magnitudes at maximum and minimum luminosities of certain novae observed since 1892.

24.31 Fast and Slow Novae

Although the luminosity curves of novae are quite similar in their general characteristics, we distinguish between **fast** and **slow novae**. *The fast*

FIG. 24–17 *(a) Nova Herculis of 1934 before and after outburst.* *(Yerkes Observatory photographs)*

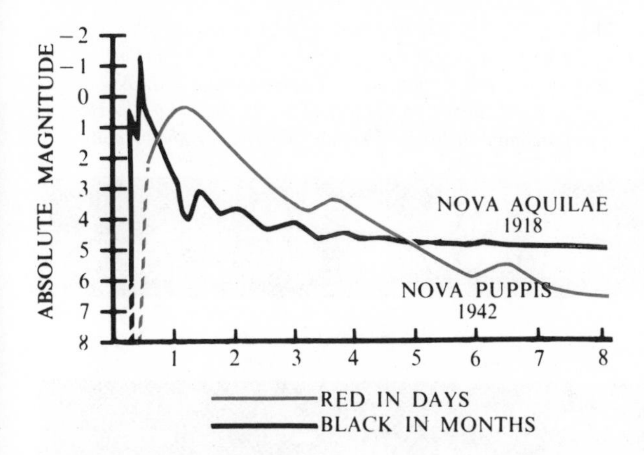

FIG. 24–17 *(b) Two types of light curves of novae. (Nova Aquilae (black), 1918 (after L. Campbell, Harvard); Nova Puppis (red), 1942). (C. Payne Gaposchkin,* Introduction to Astronomy. *New York: 1954)*

nova undergoes a complete change in luminosity, returning to its original prenova stage in a matter of months, or at most a few years. The initial rise to maximum luminosity, involving a change of 12 to 13 magnitudes, is extremely rapid, occurring within a few hours or a day or two. The nova then loses about 3 magnitudes rather smoothly in a few days, and continues to diminish in brightness quite slowly while undergoing a series of fluctuations. These are followed by a final smooth decline to the initial luminosity.

In the slow novae, which may take as long as a month to rise to maximum, the same variations occur, but over a number of years or even centuries. In general, slow novae at maximum are 1 to 2 magnitudes fainter than fast ones, and their decline in brightness is much more gradual. Slow-novae fluctuations are more pronounced than those of fast novae (of the order of 1 to 2 magnitudes) and their light curve after maximum is quite different.

24.32 Recurrent Novae

Certain types of novae (three of which have been studied fairly extensively) are recurrent, and some astronomers believe that this is true of all ordinary novae, but with very long periods, possibly thousands of years or more. T Pyxides is an example of a recurrent nova with a short period. In 1890, 1902, 1920, and 1944 it changed from a fourteenth- to less than an eighth-magnitude star, undergoing similar variations on each occasion, with light curves that leave no doubt that these were nova outbursts.

24.33 Spectra of Novae *

The spectacular and pronounced changes that occur in the spectrum (see Figures 24.18(a) and (b)) of a nova are consistent with the observed expansions of a surrounding shell of gas. Hardly any spectral data are available for the prenova stage of a star, but it appears likely that before becoming a nova the spectrum is the same as it is after the star has settled down in its postnova blue-white state. The observational data in most cases begin when the nova is about 2 magnitudes fainter than maximum, at which stage the spectrum consists of a strong continuous background dominated by absorption lines, with some faint bright lines also present.

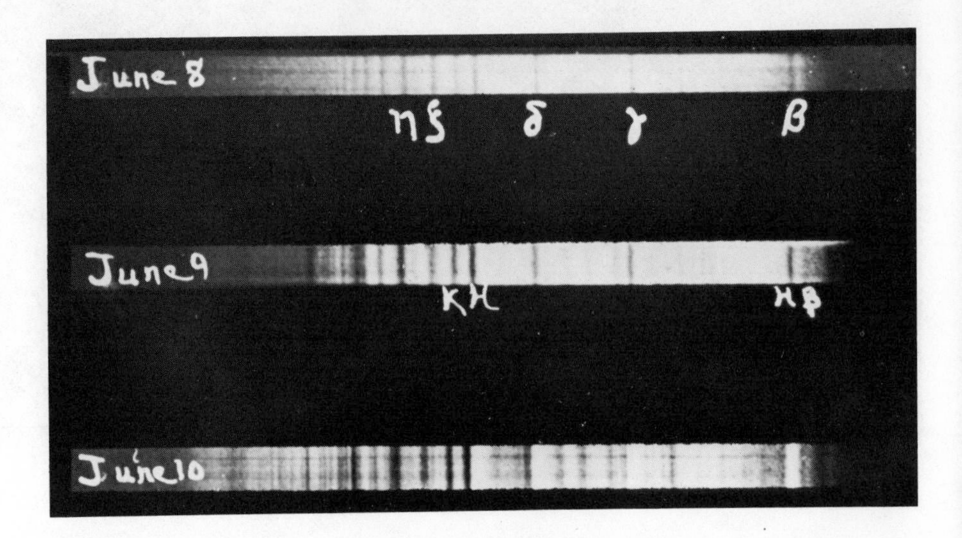

FIG. 24–18 *(a) The spectrum of Nova Aquilae No. 3 taken by Parkhurst with 30° objective prism, June 8, 9, and 10, 1918, without comparison spectrum. The broadening of the absorption lines and the sudden appearance of bright lines (e.g. the Balmer lines) are quite striking. (Yerkes Observatory photograph)*

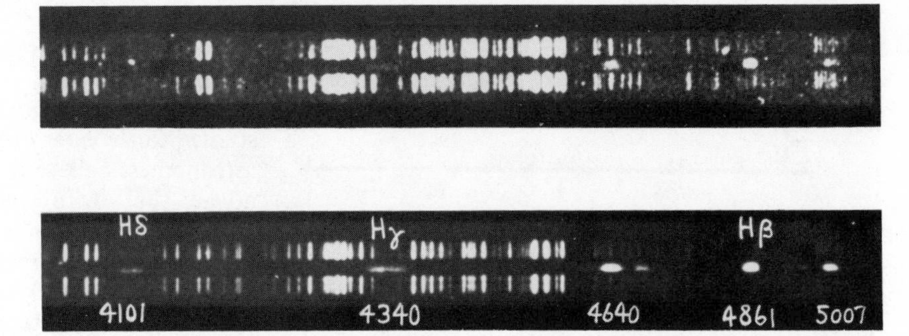

FIG. 24–18 *(b) This photograph of Nova Aquilae taken with the Bruce spectrograph, September 1 and 5, 1927, shows the very pronounced broadening of the bright hydrogen Balmer lines Hδ, Hγ, Hβ. (Yerkes Observatory photograph)*

During the initial rise to maximum luminosity the spectrum remains about the same, but with the absorption lines shifted towards the violet. Since, except for the Doppler shift, the spectrum during this stage does not change very much, we may conclude that although the atmosphere is expanding rapidly it is still intact and dense enough to give rise to the characteristic absorption lines. At this point the ionization and excitation characteristics of the spectral lines indicate temperatures ranging from 10 000°K to 20 000°K. However, as the nova approaches maximum luminosity, the spectrum changes to that of an F supergiant. The magnitude of the violet shift up to this point indicates a radial velocity of about 1 000 km/sec.

At maximum luminosity there is an abrupt change and the continuous F supergiant spectrum is replaced by a forbidden bright-line spectrum, characteristic of planetary nebulae (see Section 28.20). Right after maximum the continuous spectrum fades, and at first broad, bright lines appear with absorption components on their violet edges. This occurs because the expanding shell of gaseous material (which is now much larger than the star itself) becomes highly rarefied and emits bright lines which are broadened in the manner described in the section on Wolf-Rayet stars (see Section 24.25). The displacement of the faint absorption components towards the violet is due to the motion towards us of that part of the shell lying between us and the main body of the star.

The absorption lines are those of hydrogen (weak), ionized metals, oxygen, nitrogen, and

TABLE 24.6
SELECTED GALACTIC NOVAE

Nova	Year	Galactic coordinates		Apparent magnitude			$m - M$	Postnova spectral class	t_3	Type
		l^{II}	b^{II}	pre-nova	max	post-nova				
		°	°						day	
Tau (Crab)	1054	184	− 5		−6			e neb†		SN I
Cas (Tycho)	1572	120	+ 2		−4.1		8	e neb		SN I
Oph (Kepler)	1604	4	+ 6		−2.2					SN I
η Car	1843	287	− 1		−0.8	7.9	7	pec	3 000	N
V841 Oph 2	1848	7	+17	> 10*	3	12.7	10	O con	300	N
T Aur	1891	177	− 1	> 13	4.0	14.8	10.2	O e	120	N
GK Per 2	1901	151	−10	13.5	0.2	13.3	8.5	O e	12	N
DN Gem 2	1912	183	+15	15	3.5	14.6	11.5	O e	34	N
V603 Aql 3	1918	33	0	10.6	−1.1	10.6	7.3	O e	7	N
V476 Cyg 3	1920	87	+13	> 15	2.0	16.1	10.6	O e	14	N
RR Pic	1925	271	−25	12.7	1.2	9	7.6		150	N
DQ Her	1934	72	+26	14.3	1.4	13.8	7.6		105	N
CP Lac	1936	102	− 1	15.3	2.1	14.9	10.2		9	N
CP Pup	1942	253	− 1	17	0.2		10.6	O e	8	SN?
Recurrent nova					max	min				
T CrB (P = 79 y)	1866 1946	42	+18		1.9	10.6	9.8		6	
WZ Sge (P = 32 y)	1913 1946	58	− 8		7.3	15.8	14.4		33	
RS Oph (P = 35 y)	1898 1933 1958	20	+10		4.3	11.6	12.7		10	

* > = fainter than.
† e = emission line.

A simple empirical rule relates the maximum brightness of a nova (that is, its maximum M_v) to the speed of its decline in days, given as t_3 (the time the nova takes to decline 3 magnitudes from the maximum) in column 7:

$$M_v(\max) = -11.3 + 2.3 \log t_3.$$

helium (essentially resembling the spectrum of an F supergiant). At this stage, too, since the shell is quite attenuated, the radiation from a second shell moving rapidly outward from the star is observed. This consists of a second set of absorption lines displaced towards the violet and superimposed on the previous spectrum.

As the shells expand further and further and become more and more attenuated, the bright metal lines become weaker and weaker, but those of gases like hydrogen and helium remain fairly strong. To account for this we note that the high-temperature radiation coming from the main body of the star easily tears away the outer electrons of the metals but does not ionize hydrogen and helium. Because the gases in the shells are highly attenuated, the metallic atoms find it difficult to recapture electrons, and therefore they radiate and absorb in the deep ultraviolet to form lines that are not visible.

When the gaseous shells expand sufficiently, the continuous spectrum almost disappears and only emission lines remain, among which are the prominent nebular green lines of OIII, NeIII, and NiII. These are forbidden lines (see Section 20.23 on metastable states) and they are emitted only because the gases in the expanding shells are so rarefied that collisions between atoms are infrequent. Although the continuous spectrum is no longer evident, it is still present because the main body of the star continues to emit radiation; however, the emission lines from the gaseous shells

FIG. 24–19 *Expanding nebulosity around Nova Persei, 1901. This photograph clearly shows the asymmetry of the outburst. The irregular structure of the expanding shell is apparent.*

are so strong that the continuous background spectrum is obscured. In the final stage, the nebular emission lines of the shells disappear and only a continuous Wolf-Rayet type spectrum of the star itself remains. This continuous spectrum is practically free of absorption but contains broad, intense emission lines.

R. Kraft, following on the discovery by M. Walker that Nova DQ Her is a very close binary with a period of 4 hrs. 39 min., has recently analyzed the spectra of ten old novae and has demonstrated that seven of them are certainly very close spectroscopic binaries (see Section 26.14) with periods ranging from 1 hr. 22 min. for WZ Sge to 127.6 days for TcrB. He has advanced the hypothesis that being a member of a close binary such as those he studied is a necessary condition for a star to become a nova.

24.34 Energy Released and Mass Ejected during Nova Stages

The energy released (about 10^{45} ergs) during the entire nova stage, as measured by the area under the light curve, represents only a small fraction of the total amount available from the nuclear reaction inside the star itself. The amount of mass ejected during this explosion is but a small fraction (10^{-5}) of the entire mass of the star. This supports the suggestion that novae can go on exploding this way over and over again. If we suppose that the nova ejects about the same amount of material each time it explodes, its mass will be reduced to about that of a white dwarf after about 10 000 explosions. If we assume that the period between these outbursts is of the order of 10^5 years, a nova becomes a white dwarf in about a billion years.

Since the total amount of energy emitted in 10 000 explosions (assuming that each explosion releases about 10^{45} ergs) is larger than the amount available from the total conversion of hydrogen to helium, whether by the proton-proton chain or the carbon cycle, it is clear that higher nuclear processes such as the transformation of helium to carbon or the build-up of heavier nuclei (as described in Sections 23.45 and 23.46) is involved in this transition from nova to white dwarf. This agrees with our assumption that the nova phase of a star occurs near the end stage of the star's evolution.

24.35 The Number of Novae

About 250 novae have been recorded in our own galaxy by S. I. Bailey, and it is estimated that 25 novae brighter than the 9th apparent magnitude occur in our galaxy each year. This estimate is strengthened by the 67 ordinary novae that were found in the Andromeda Nebula between 1909 and 1926, mainly towards the nuclear unresolved region. Hubble, working at the Mount Wilson Observatory, showed that about 25 or 30 occur in this nebula every year, although during the last 50 years we have observed only about 8 of them as naked-eye objects.

If 25 or 30 stars in a given galaxy become novae every year, there must have been about 100 billion such occurrences during the last $4\frac{1}{2}$ billion years, and this again indicates that novae must be recurrent. Otherwise it would follow that every star in our galaxy must already have been a nova to account for the total number. However, we know from the evolution of stars that most of the stars on the main sequence have not yet passed the nova stage. Of course it may be that most of the novae are now invisible, but this too leads to far too many white dwarfs to agree with the observed data if novae were not recurrent. If we suppose, however, that on the average each nova accounted for 1 000 explosions during its lifetime, the total number of stars that have passed through the nova stage and become white dwarfs during the past 4 billion years would be 10^8, which is in better agreement with the observed number of white dwarfs.

24.36 Distribution of Novae

Novae are strongly concentrated towards the galactic plane, with most of them lying within 10° of the galactic equator. They are found in greatest abundance towards the center of our galaxy in the longitudinal quadrant centered in the neighborhood of Sagittarius (galactic longitude 327° or 0° in the new galactic coordinate system). From this we conclude that they are Population II objects. This conclusion is strengthened by the following arguments given by Baade: they are found in globular clusters such as M 80, which are members of the halo around our galaxy; and they are also found in certain dwarf-type galaxies which consist of Population II stars only.

From the few observations we have of stars before they become novae it appears that these objects lie below the main sequence and have absolute magnitudes of about +5. Since in both the prenova and postnova stages they are O-type stars, with surface temperatures as high as 50 000°K, most of them probably have much smaller diameters than the sun. From the luminosity and temperature in pre- and postnova stages we find that these diameters range from 5 to 30 per cent of the sun's. Their mean parallax, taken from five observed novae, is of the order of 0".01, which shows that most of these novae are at great distances from us.

24.37 Supernovae

Supernovae are stars that explode much more violently than ordinary novae. Whereas the latter increase in luminosity by a factor of about 5×10^5, a supernova at maximum luminosity is 10^8 to 10^9 times more luminous than it was before exploding. These objects have absolute magnitudes at maximum luminosity that may be as small as −19, according to the corrected distance scale for Cepheid variables, and often at maximum brightness are as luminous as the entire galaxy in which they occur.

Although over 50 supernovae have been observed, all but three of them were discovered in galaxies outside our own (see Figures 24.20(a) and (b)), and it is estimated that they occur in our galaxy once every three or four centuries. The three that have been recorded are SN (supernova) Tauri (now the Crab Nebula) which attained an apparent magnitude of −5 and was visible in full daylight; SN Casseopeiae 1572, which was observed by Tycho Brahe (and was also visible in full daylight, becoming brighter than Venus); and SN Ophiuchi 1604, Kepler's supernova, the remnants of which were recently discovered as a nebulous structure emitting radio waves similar to the Crab Nebula.

An example of a supernova in another galaxy is that of August 1885 in the Andromeda Nebula. At maximum brightness its apparent magnitude was +6 and it was then about as luminous as the entire nebula. This supernova became 10^{-4} times fainter after 6 months and then disappeared entirely.

FIG. 24–20 (*a*) *The spiral galaxy NGC 7331 before and during the maximum of the supernova of 1959. Two photographs taken with the Crossley reflector. The supernova, as can be seen, lies in the dark dust lanes of the galaxy. If the dark dust lanes of this galaxy are closer to us than the other (the bright) side, the spiral arms are trailing and not unwinding.* (*Lick Observatory photograph*)

24.38 Energy Released and Matter Ejected by Supernovae

The amount of energy released (10^{49} ergs) by a supernova indicates that these explosions are of an entirely different order of magnitude from those of an ordinary nova. This energy is a large fraction of the total thermonuclear energy available in such stars. At maximum luminosity supernovae radiate as much energy in one day as the sun does in one million years. Moreover, the very broad emission lines observed in the spectra of these objects show that the material is ejected at speeds up to 12 000 km/sec. The amount of material ejected during a supernova outburst is anywhere from 1 to 10 per cent of the total mass. This also indicates that the mechanism responsible for the supernova outburst is quite different from that of ordinary novae. After the outburst the vast amount of ejected material becomes gaseous nebulosity surrounding a small hot residual core (which may become a neutron star, see Section 24.50) such as we find in the Crab Nebula, the structure of which we shall discuss in Chapter 28. Figure 24.20(c) is a photograph of the spectrum of the supernova in NGC 6946.

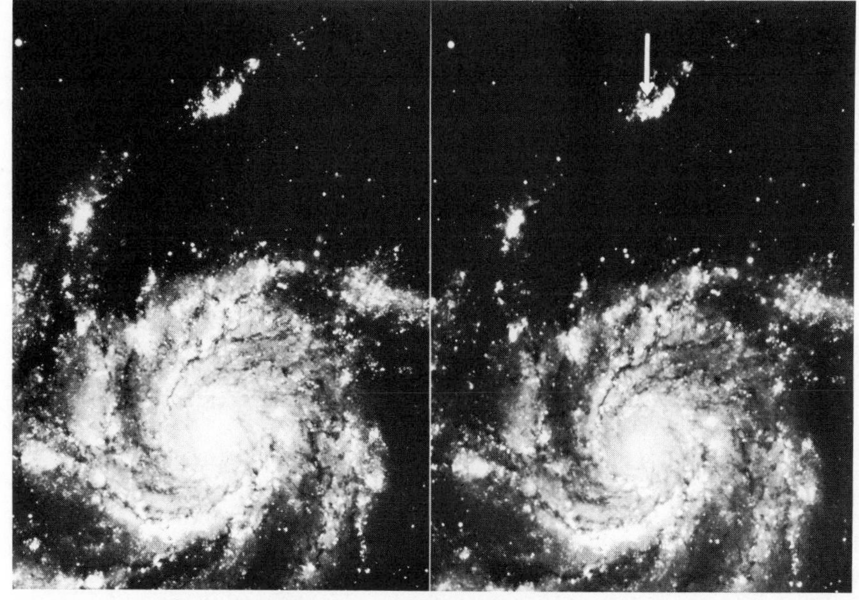

FIG. 24–20 (*b*) *NGC 5457 Type Sc galaxy in Ursa Major. Photographed with supernova June 9, 1950, and without supernova February 7, 1951. The many separate arms in this galaxy can be traced from branchings of two principal dust arms that begin in the nucleus. The outer arms are broken up into clumps or associations of stars, in one of which the supernova occurred. The distance modulus of this galaxy is 27.0, so that the average width of an arm is 380 parsecs. The largest H II regions are 120 parsecs in size, and the largest star associations are 620 parsecs by 190 parsecs. (From Mount Wilson and Palomar Observatories)*

24.39 Type I and Type II Supernovae *

According to R. Minkowski, who studied their spectra, and Baade, who studied their light curves and maximum luminosities, there are two groups of supernovae, Type I and Type II.

Type I supernovae attain luminosities 10^8 times that of the sun, and in an outer galaxy one of them was observed with a maximum luminosity 6×10^8 times that of the sun. At maximum, the absolute photographic magnitudes of Type I supernovae cluster around -16. The light curves of Type I supernovae are similar to those of ordinary novae and show an exponential decay with a half-life of 55 days (as we mentioned in Chapter 23 on the formation of the heavy elements). From this it appears that in the end stages of supernovae of this type the energy release comes primarily from the radioactive decay of Cf^{254}. Baade, who

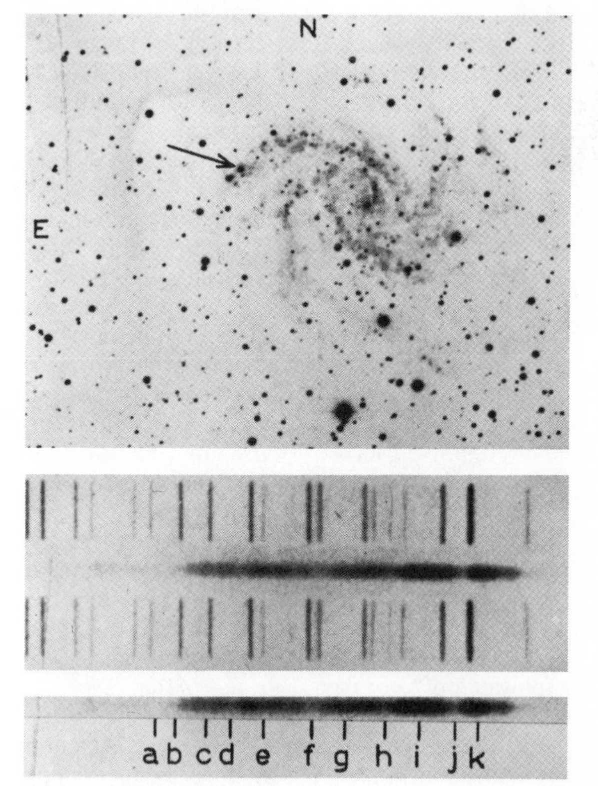

FIG. 24–20 (*c*) Top. *Arrow points to a supernova in the spiral nebula NGC 6946, photographed on July 10, 1948, with the Crossley reflector, exposure one hour.* Center. *Supernova spectrum photographed July 7–8, 1948, exposure four hours. The farthest left comparison line is 3610 of Pd, the farthest right, 5086 of Cd.* Bottom. *Lines and letters designate dark-line features. (Yerkes Observatory photographs)*

classified Type I supernovae as Population II stars, has shown that these stars remain at maximum brightness for about one week, and then decline about 3 magnitudes in 20 or 30 days. Following this their brightnesses diminish at the rate of 0.02 magnitudes per day.

At maximum brightness Type II supernovae are about 10 to 20 million times brighter than the sun and thus are not quite as luminous as Type I. The one that was observed in 1940 in the outer galaxy, NGC 4725, is a good example of this group. The light curves of Type II supernovae (which Baade classified with Population I stars) are different from those of Type I supernovae. Thus, the 1940 supernova, which attained a maximum luminosity 30 million times that of the sun, faded slowly at first, declining 3 magnitudes in about 100 days, then more rapidly. Unfortunately, since supernovae were classified before the division of stars into two populations, supernova designations I and II do not correspond with the population designations.

24.40 The Origin of Novae and Supernovae

No adequate explanation has yet been found to account for either novae or supernovae, although various theories have been suggested. Since in the case of ordinary novae only a comparatively small amount of material is ejected, most theories assume that the explosive phenomenon originates not very far below the photosphere in a region where convection ordinarily occurs in stars like the sun. Biermann has suggested that in stars that have depleted all their hydrogen, convection does not set in properly in this region and the photosphere collapses. According to his theory, the energy coming from the deep interior, instead of inducing convection, is absorbed by the rapid ionization of the heavy elements under these conditions (absence of hydrogen) and hence is not available to support the convection zone. The outer layers of the star must then collapse, releasing in a single outburst all the energy that was trapped in ionization. This ionization energy, together with the great quantities of gravitational energy released, triggers the explosion of ordinary novae. After such an explosion takes place, an intermediate equilibrium stage follows and the star remains quiescent until the ionization process has trapped enough energy to trigger another explosion. The requirement that no hydrogen be present for the Biermann process to occur fits in with the idea that novae are at the end stage of their evolution, at which point all the hydrogen has been transformed into heavy nuclei.

It is much more difficult to account for supernovae because of the vast amounts of energy they suddenly release. This action appears to be the result of a gravitational collapse of most of the star on an enormous scale. One possible explanation is that the protons in the nuclei of the core of the star are rapidly transformed into neutrons. According to Zwicky this releases enormous quantities of energy. Another theory, the work of Ambartsumian, deals with hyperon stars, which we shall discuss in the Section 24.51.

Recently, S. Colgate and R. White have shown that for all stars of mass greater than 1.5 $M_{\odot}$ (the Chandrasekhar limit, see Section 23.38) catastrophic collapse can occur when the star develops a dense neutron core. When the outer material of the star collapses onto the core, high temperature shock waves occur. The shock wave has a temperature of 10^{12} degrees, a density of 10^{15} gm/cc, and a thickness of about 10 km. The neutrinos emitted in this collapse heat the outer regions of the star, causing an explosive ejection of matter similar to that found in supernovae.

24.41 White Dwarfs

As we have seen, a star at the end stages of its evolutionary development gets rid of great quantities of matter by passing through a nova stage and ejecting this matter violently by a series of explosions. The residual star, called a **white dwarf**, is then in its final, very slowly changing, stage of life. More than 250 white dwarfs have been listed.

White dwarfs are characterized by two features that differentiate them from main-sequence stars. First, lying as they do in the lower left-hand part of the H-R diagram, they are highly underluminous for their color. Second, they do not obey the ordinary mass-luminosity relationship, but fall below this curve. The faint component of Sirius, a typical white dwarf, has a mass about equal to that of the sun, and yet a luminosity which is only $0.003L_{\odot}$. *We may summarize these points by stating that the white dwarfs have the same masses as ordinary stars, but very small radii.*

24.42 Color, Spectra, and Absolute Magnitude of White Dwarfs

Although as a group these stars are called white dwarfs, they have color indices ranging from −0.6

to +1. There are white dwarfs that are very blue, some that are yellow and red, and probably others that are black and therefore undetectable. Most of those that have been observed belong to spectral class A, although some F and G type "white" dwarfs have also been discovered. Even though we assign white dwarfs to the same spectral classes as main-sequence stars, there are many small spectral differences between main-sequence stars and white dwarfs of the same class. It must be noted that the spectral class is not as clearly defined for a white dwarf as it is for a main-sequence star.

Whereas the spectral lines of main-sequence stars are conspicuous and fairly sharp, those of a white dwarf are very broad and sometimes entirely missing. Certain white dwarfs, such as Kuiper's, have no detectable spectral lines, but exhibit pure, continuous spectra. However, the careful work of J. Greenstein has shown that some of the white dwarfs that were thought to have only continuous spectra do have absorption lines which are very difficult to observe.

The absolute magnitudes of the white dwarfs range from +9 to +16, although most of them cluster around +12. Their masses are difficult to measure; very little is known about them except for those of Sirius B and 40 Eridini B, which are components of binary systems. However, a relationship derived by Chandrasekhar between the mass and radius of a white dwarf is used to calculate the masses of these stars.

24.43 Numbers of White Dwarfs

Although not many white dwarfs have been observed, the total number is probably very large as deduced from star counts by Ambartsumian and Shajyn, and from stellar evolutionary considerations by Salpeter. All things considered, we conclude the number of white dwarfs may, indeed, be equal to the number of all other stars in our galaxy. However, Luyten's estimates are at variance with these conclusions. Using other star counts, he estimates that there are 55 white dwarfs in a sphere around the sun having a radius of 96 parsecs, or 1 white dwarf per 1 400 psc^3. Since there are about 8 000 ordinary stars in such a sphere, Luyten concludes that 1 per cent of all stars are white dwarfs. However, among the 100 nearest stars, 7 are known to be white dwarfs, so that Luyten's estimate appears to be too small, at least in our part of the galaxy. If we keep this in mind and remember how difficult it is to detect white dwarfs, it is easier to accept the large estimates than the small.

24.44 The Theory of White Dwarfs

Since white dwarfs are enormously dense, we can understand their structure only if we know the properties of matter at very high densities and pressures. Chandrasekhar has shown that a star can reach the white-dwarf stage and remain stable as a white dwarf only if its mass (after it has become a white dwarf) is less then $5.75M_{\odot}/(\rho/N_eH)^2$, where ρ is the density of the stellar matter, N_e is the number of free electrons per cm^3, and H is the mass in grams of the hydrogen atom. The quantity (ρ/N_eH) is essentially the mean molecular weight that must be assigned to the material in the white dwarf. Note that the nuclei are not included in computing the mean molecular weight because nuclei move very slowly compared with the free electrons and hence contribute very little to the pressure exerted by the gaseous material in white dwarfs. Stars more massive than this can reach the white-dwarf stage only by ejecting matter, and from this it follows that white dwarfs are the result of an explosive ejection of material (i.e., nova stage of a star).

Having ejected enough material, a star becomes a stable white dwarf only after it has completed its gravitational contraction and reached a very high density. Properties of matter at these very high densities have been studied by many physicists. Fermi and Dirac separately have derived the basic theory for the behavior of such matter consisting of particles like electrons (i.e., particles that have a spin of $\frac{1}{2}$ and hence obey the Pauli exclusion principle).

24.45 Degenerate Matter

Matter under very high pressure is said to be in a **degenerate state** provided its temperature is below a certain value determined by a criterion first given by Fermi. This criterion is fulfilled in white dwarfs, so that the material in these stars is degenerate. Since white dwarfs have used up all their nuclear fuel, there is nothing (that is, no sufficient outflow of energy) to prevent them from contracting, which they continue to do until all the atoms have been squeezed so close to each other that their electron shells begin to interpenetrate.

At this evolutionary stage, most of the star consists of heavy atoms with only a small external fringe of hydrogen and helium. This outer fringe of a white dwarf cannot be considered as an ordinary atmosphere, because its density is enormous as a consequence of the very large surface gravity (the surface gravity of Sirius B is 30 000 times that of the earth). There can be no hydrogen in the interior of white dwarfs because the nuclear burning of this hydrogen at the interior temperatures would make these stars much more luminous than they actually are.

24.46 Pressure Ionization

As the white dwarf contracts gravitationally, the interpenetration of the electron shells continues until the nuclei themselves are as close to each other as the radius of the lowest electron shell. Hence the electrons can no longer be identified with any particular nucleus and must be counted as free. This process of separating electrons from their nuclei is called ***pressure ionization****.*

When the pressure ionization is complete, we have a gas of electrons moving about in a matrix of a much heavier and more sluggish gas of positive nuclei, so that the material of a white dwarf has the physical properties of a metal. At this stage the stellar material consists of negatively and positively charged particles; it is neutral as a whole because the number of negative charges just equals the number of positive ones. Under these conditions, just as in the case of a metal, the energy released from the interior is transported out to the surface by conduction rather than by radiation or convection.

24.47 Degenerate Electron Gas

The electron gas itself is now in a completely degenerate state, the properties of which we can explain as follows:

We first picture a gas of particles without spin, and imagine this gas as cooling off and contracting at very low temperatures and under very high pressures. All of the particles in this gas then sink to their lowest energy levels so that there is no motion at all. This is so because these particles are not governed by the Pauli exclusion principle and there is nothing to prevent them from falling into the zero energy state. However, this is not true for a gas of electrons such as we have in white dwarfs. Even though the pressure may be very high and the temperature low, all the electrons in such a gas cannot be brought to rest because the Pauli exclusion principle prevents them from all having zero energy at the same time. As more and more electrons are squeezed together in each cubic centimeter, more and more of them must move at higher and higher speeds because the Pauli exclusion principle requires that no more than two electrons (with opposite spins) in each element of volume have the same energy. Thus, in an ordinary white dwarf such as Sirius B, with about 10^{28} electrons per cm^3, most of the electrons in any small element of volume are moving about quite rapidly no matter how low the temperature may be, because only two electrons can have zero energy and the others must lie, two by two, in higher and higher energy states. This means that there are some electrons with speeds corresponding to millions of degrees even though the temperature of the entire group of electrons is low. *An ordinary white dwarf, then, is a gaseous configuration consisting of a matrix of closely packed heavy nuclei through which a degenerate electron gas moves. Towards the surface of a white dwarf the degree of degeneracy drops off; and at the surface itself, the atoms are not ionized and the material is in the ordinary gaseous state.* In the case of a star like Sirius B, only 75 per cent of its material is in a degenerate state.

24.48 The Equation of State of Degenerate Matter

The behavior of a degenerate gas differs from that of an ordinary gas, primarily in its equation of state. Whereas in an ordinary perfect gas the pressure is proportional to the absolute temperature, in the case of the completely degenerate electron gas the pressure, to a first approximation, is independent of the temperature. However, at higher approximations it depends upon higher powers of the temperature.

Since there is no limit to how dense matter can become under sufficient gravitational contraction (that is, if the total mass is large enough), the white dwarfs continue to contract and their densities continue to grow. As long as the density remains under 10^8 gm/cm^3 (100 tons per cm^3), we may still speak of a white dwarf in the usual sense since even at this high density the nuclei retain their identities, and the degenerate electron gas maintains the star in its state of equilibrium. However, if the pressure of the degenerate electron gas does not compensate for the gravitational force, the white dwarf becomes unstable and contracts until there are more than 10^{32} electrons per cm^3. Many electrons in each cubic centimeter will then be moving at enormous velocities (because of the Pauli exclusion principle)—indeed, at very nearly the speed of light. A new process then sets in because the energy conditions are favorable for the transformation of protons into neutrons by inverse β decay.

24.49 Inverse Beta Decay Process

Under these conditions the electrons having very high energies begin to combine with protons inside nuclei. These protons are changed into neutrons, and neutrinos are emitted, according to the scheme

$$p + e^- = n + \nu.$$

In this way more and more protons inside nuclei are transformed into neutrons. But this process cannot continue indefinitely because, as we have noted (Section 23.11), nuclei become unstable if they contain too many neutrons compared with protons. When a critical number of protons in any nucleus has been transformed into neutrons, this nucleus decays into a lighter nucleus emitting separate protons and neutrons. Since this happens to many nuclei at the same time, this process leads to stars with densities much greater than 10^8 gm/cm^3. These stars then consist of a mixture of three degenerate gases: neutron, proton, and electron. This is true only in the core of the star where the densities are very high. The outer layers still consist of ordinary nuclei and degenerate electrons, and the material near the surface itself contains un-ionized atoms.

24.50 Neutron Stars *

As the star continues to contract, electron densities ranging from 10^{34} to 10^{38} electrons/cm^3 are finally reached; by this time most of the nuclei have been disrupted by the inverse β decay process. *There are now thousands of times more neutrons than protons present* (*note that the number of protons always equals the number of electrons since the star is always neutral*) *and we have what is called a* ***neutron star***. Oppenheimer and Volkoff have shown that a neutron star can exist in equilibrium if its mass lies between 0.3 and $0.7M_\odot$. The radii of such a star is of the order of 20 km and its surface gravity is 2×10^{11} times the gravity on the earth's surface.

The neutron gas in such stars is degenerate and exerts an enormous pressure which maintains the star in equilibrium under its gravitational attraction. During this process of the transformation of a white dwarf into a neutron star, enormous amounts of energy are released. If all of it were released suddenly, the star could not be stable but would explode violently; hence, this process must go on gradually. The energy cannot be released as radiation since it would have to be spread out over a time much longer than the assumed age of our galaxy (about 14×10^9 years), and no neutron stars would now exist. But a white dwarf can evolve into a neutron star nonexplosively well within 14×10^9 years without disturbing the star in any way because the *neutrinos emitted in the inverse β decay process can carry off most of the energy.*

Because of the enormous gravitational fields surrounding neutron stars, the passing light from other stars is drastically curved (a gravitational lens). We should therefore be able to detect neutron stars by looking for ghost stars in their neighborhoods. The images of more distant stars formed by the gravitational lens are shown in Figure 24.21.

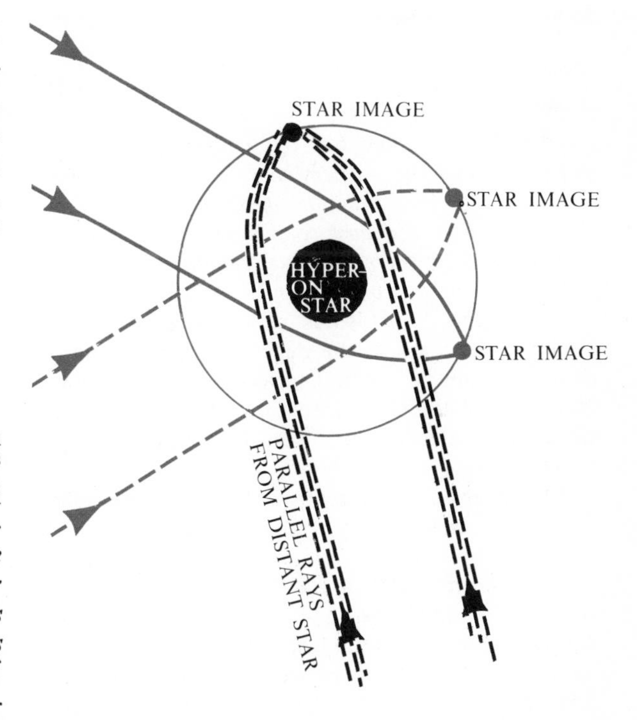

FIG. 24–21 *A small dense hyperon star behaves like a gravitational lens, forming images of distant stars on a spherical surface.*

Neutron stars can also be detected by the kind of radiation they emit. Since the surface temperatures of neutron stars are very high, the laws of radiation (see Chapter 16) show that these stars emit X-rays. Recently, H. Friedman and his group have detected X-rays from the Crab Nebula (they worked with rockets above the earth's atmosphere) so that the star at the core of this nebula may be a neutron star. Chiu and Salpeter have given a theoretical analysis that agrees with the observations.

The core temperatures of neutron stars are of the order of 2×10^9 °K and their surface temperatures are of the order of 10^7 °K. The wavelength of the emitted radiation is about 3 angstroms. The core, which consists of a degenerate neutron gas, and which has a diameter of about 20 km, is surrounded by a nondegenerate layer one meter thick. In this one-meter layer the temperature drops from 2×10^9 to 10^7 degrees; this layer is surrounded by a photosphere that is only one centimeter thick. These stars emit about 10^7 ergs per second, mostly in the X-ray region.

24.51 Hyperon Stars *

Neutron stars of large enough mass ultimately evolve into **hyperon stars**, the theory of which has recently been worked out by Ambartsumian and others. *A **hyperon** is a very short-lived particle with a mass greater than the mass of a nucleon. It does not exist under ordinary conditions, but if enough energy is available hyperons are created momentarily in the neighborhood of a nucleus.* When a neutron star contracts to densities of the order of 10^{15} gm/cm^3, there are in the degenerate gas so many neutrons moving at speeds close to that of light, that most of them are transformed into hyperons which are stable at these very high densities and energies.

The hyperons are stable at these high densities because the nucleons they ordinarily would decay into obey the Pauli exclusion principle, and hence have nowhere to go since all the possible lower energy states are already filled. As the star continues to contract, more and more neutrons are changed into hyperons until there are equal numbers of hyperons and neutrons. At this stage some μ and π mesons (see Section 23.6) and a completely degenerate electron gas are also present, but the number of electrons is thousands of times smaller than that of neutrons and hyperons. The free mesons are stable at this stage because their decay into electrons is forbidden by the Pauli exclusion principle.

A hyperon star has a mass about equal to that of the sun, but its radius is only a few kilometers. It consists of three regions: the hyperon core, which is the densest region and contains almost the entire mass (there also are present here degenerate neutrons, some π mesons, and a few electrons); a surrounding layer composed primarily of degenerate neutrons (no hyperons), a few electrons (1 part in 1 000), and protons; and the outer regions containing the same kind of matter as in white dwarfs, namely, nuclei and electrons. At the very surface there is an outer shell, probably no more than a few meters thick, containing ordinary atoms.

It should be noted that a hyperon star is stable only if it is left to itself. If it suffers a collision with some object, the energy released by the collision allows all the hyperons to decay into nucleons, and a vast explosion occurs with a tremendous release of energy. Ambartsumian has suggested that this may be the origin of the supernovae.

The analysis we have just outlined shows why black dwarfs are the ultimate stages of a star's evolution. If we picture a star as contracting to its ultimate stage, its entire matter will be completely degenerate and therefore unable to emit any energy. For if any particle (e.g., electron, hyperon, etc.) were to emit any energy it would have to fall into a lower state. But this is impossible because according to the Pauli exclusion principle there is no room in any lower energy state for the particle to fall into after it emits the energy. Thus, these black dwarfs can be quite hot and yet not visible.

24.52 Rapidly Rotating Stars *

Although, strictly speaking, stars that are rotating rapidly are not variables, we include them in this chapter because of their unusual spectral characteristics, and also because a star in very rapid rotation may become unstable and undergo intrinsic changes, even to the extent of suffering fission. Although F. Schlesinger first detected the rotation of a star by analyzing the motion of an eclipsing binary, today rotating stars are recognized by the unusual widths of their spectral lines.

There is no direct evidence that there are any stars that are not rotating. We do, however, know that not all stars are rotating *rapidly*. Indeed, in some cases (such as the sun) the rotation is extremely slow (2 km/sec). Among the main-sequence stars there is a correlation between spectral class and the speed of rotation. In Figure 24.22 this correlation is shown, and we see that the highest rotational speeds are found in O stars. Rotational speeds diminish from a maximum of 500 km/sec for Oe and Be stars (where the e stands for emission). There is a very sharp drop in rotational speeds for stars later than F5.

Since the axes of rotation of stars are distri-

TABLE 24.7
SOME TYPICAL WHITE DWARFS AND THEIR CHARACTERISTICS

Star	Proper motion μ''	Parallax p''	Spectral class	Apparent magnitude m_v	Absolute magnitude M_v	B-V	$\log \frac{\mathscr{M}}{\mathscr{M}_\odot}$	$\log \frac{R}{R_\odot}$	Density $\log \frac{\bar{\rho}}{\bar{\rho}_\odot}$
v. Maanen	3.00	0.″238	w F8	12.35	14.24	+0.61	0.1	−1.95	6.0
L 870.2	0.67	0.″065	w A5	12.82	11.89	+0.32		−1.9	
40 Eri B	4.07	0.″200	w A2	9.50	11.02	+0.03	−0.37	−1.82	
Sirius B	1.32	0.″376	w A5	8.40	11.30	+0.4	−0.01	−1.70	5.1
Procyon B	1.25	0.″291	w F8	10.80	13.1	+0.5	−0.29	−1.90	5.4
R 627	1.01	0.″081	w A5	14.24	13.79	+0.25	−0.2	−2.3	6.7

buted at random in space, and we do not know for any particular star what this direction is, our observations based upon the Doppler effect do not give us v_e, the equatorial velocity, but rather $v_e \sin i$, where i is the angle of inclination of the axis of rotation to the line from the earth to the star. However, we can obtain the mean equatorial velocity for various groups of stars by multiplying by an appropriate statistical factor.

The observed data themselves are based on the broadening of the spectral lines. The light contributing to the spectrum of a star comes to us not only from the center of the disk but also from the two edges of the star, one of which is approaching, and the other receding because of the rotation. The absorption lines are therefore broadened by the Doppler shift, just as in the case of a shell star. However, the spectra associated with rotating stars do not show any of the other characteristics of shell stars, so that it is easy to distinguish between these two types.

The maximum rotational velocities occur between B5 and B7 stars and they decrease as we move in either direction along the main sequence. Among these B-type stars, velocities of rotation as large as 500 km/sec have been found, but it is estimated that over 50 per cent of the B-type stars have rotational velocities of about 100 km/sec or greater, with 2 per cent having velocities greater than 300 km/sec.

The most interesting characteristic of rotating stars is the sharp drop in rotational velocity that occurs below F5, and Struve has indicated that this may have important cosmological consequences and may even be related to the formation of planetary systems. B-type rotational stars in many instances exhibit lines of hydrogen superimposed on the broadened absorption lines. Aside from a correlation to spectral class, there is no other important group characteristic (such as population type) to which rotations seem to be related.

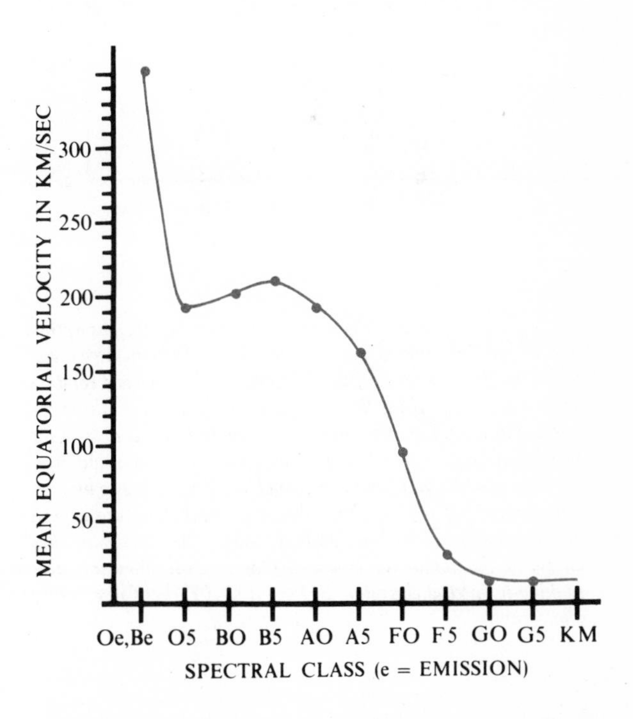

FIG. 24–22 *Correlation between spectral class and speed of rotation of stars. (Data from C. W. Allen,* Astrophysical Quantities, *2d ed. London: Athlone Press, 1963, p. 204)*

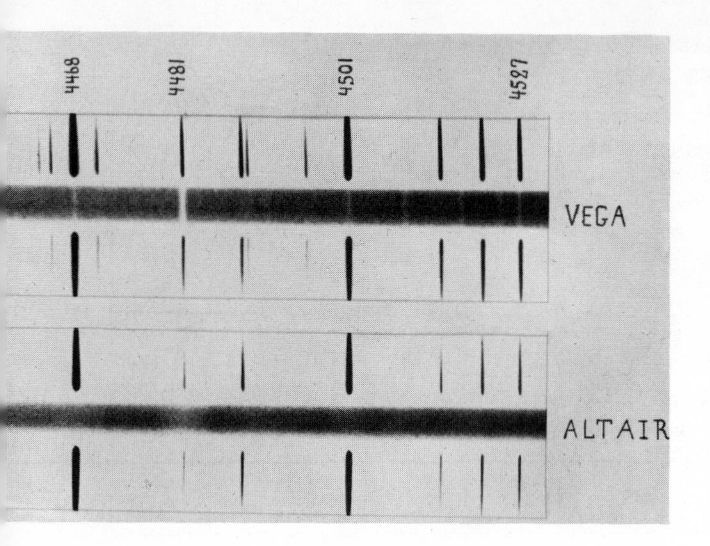

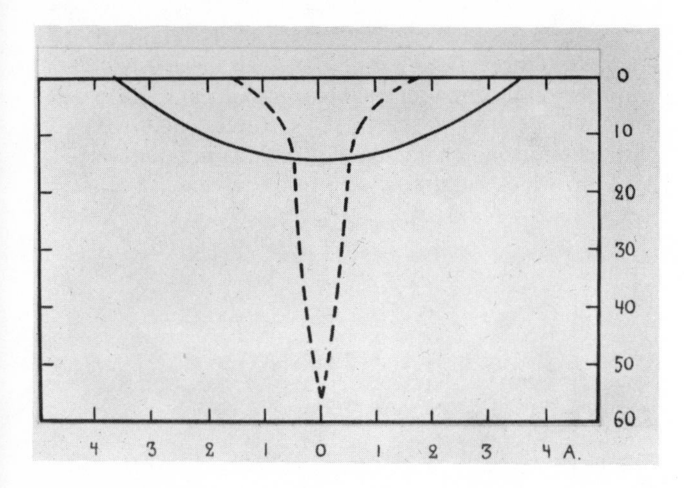

FIG. 24–23 *Spectra of Vega and Altair, showing the absorption line λ 4481 of ionized magnesium. It is sharp in Vega and diffuse in Altair because of the difference in the observed rotation of the two stars. If i is the angle between the line of sight to the star and its axis of rotation, then V sin i can be calculated from the line broadening, where V is the equatorial speed of rotation of the star. For Vega, V sin i is zero and for Altair it is 240 km/sec. The contours of the two lines shown at the bottom are typical of nonrotating (solid-line profile) stars. A. Slettebak has analyzed the rotational broadening of spectral lines for stars ranking in spectral class from BO to GO. (Yerkes Observatory photograph)*

The rotational velocities (these can be found from the broadening of the spectral lines; see Figure 24.23) have also been measured for giant stars and we find that in the range from F to G these giants are rotating more rapidly on the average than the main-sequence stars of the same spectral class. This is in agreement with the evolution of stars outlined in Chapter 23. We saw there that stars evolve by moving off the main sequence towards the right into the giant branch. In other words, giant F stars have greater rotational velocities than main-sequence F stars because the former evolved from rapidly rotating A and B stars on the main sequence. Fairly extensive calculations have shown that the expected rotational change of a star evolving from the main sequence agrees fairly well with observed data.

One of the surprising features about stars in general is not that we find large rotational velocities but rather that most stars appear to have rotational velocities much too small to be consistent with the conservation of angular momentum. If a star evolves from a mass of gaseous material and dust, it must have started out initially with a great deal of angular momentum because this dust cloud is part of the rotating galaxy. Since the velocity of rotation of our galaxy in the neighborhood of the sun is about 220 km/sec, we can show that if a star contracts, without any loss of angular momentum, from a cloud having a radius of 2 parsecs down to a sphere having a radius about 5 times that of the sun, then the rotational velocity at the equator is very close to the speed of light. Since no known star is rotating at such extremely high speed there must be some mechanism that dissipates the angular momentum of the star during the time it contracts down to its final size. Some such mechanism that makes use of the interaction of the contracting gas cloud with other gas clouds or with surrounding magnetic fields has been proposed.

24.53 Magnetic Stars

In addition to localized magnetic fields of a few thousand gauss related to sunspots, there is associated with the sun a small general magnetic field of the order of a few gauss. A small number of stars, however, have been discovered with general magnetic fields in their reversing layers of the order of several thousand gauss. Most of the observations of the general stellar magnetic fields have been made by H. W. Babcock, who has investigated and catalogued 338 such stars. He analyzed the Zeeman (see Section 19.26) pattern of 150 of these because their spectra have lines that are sharp enough to permit detection of fields larger than 500 gauss. Of these stars, 89 have measurable general magnetic fields exceeding that of the

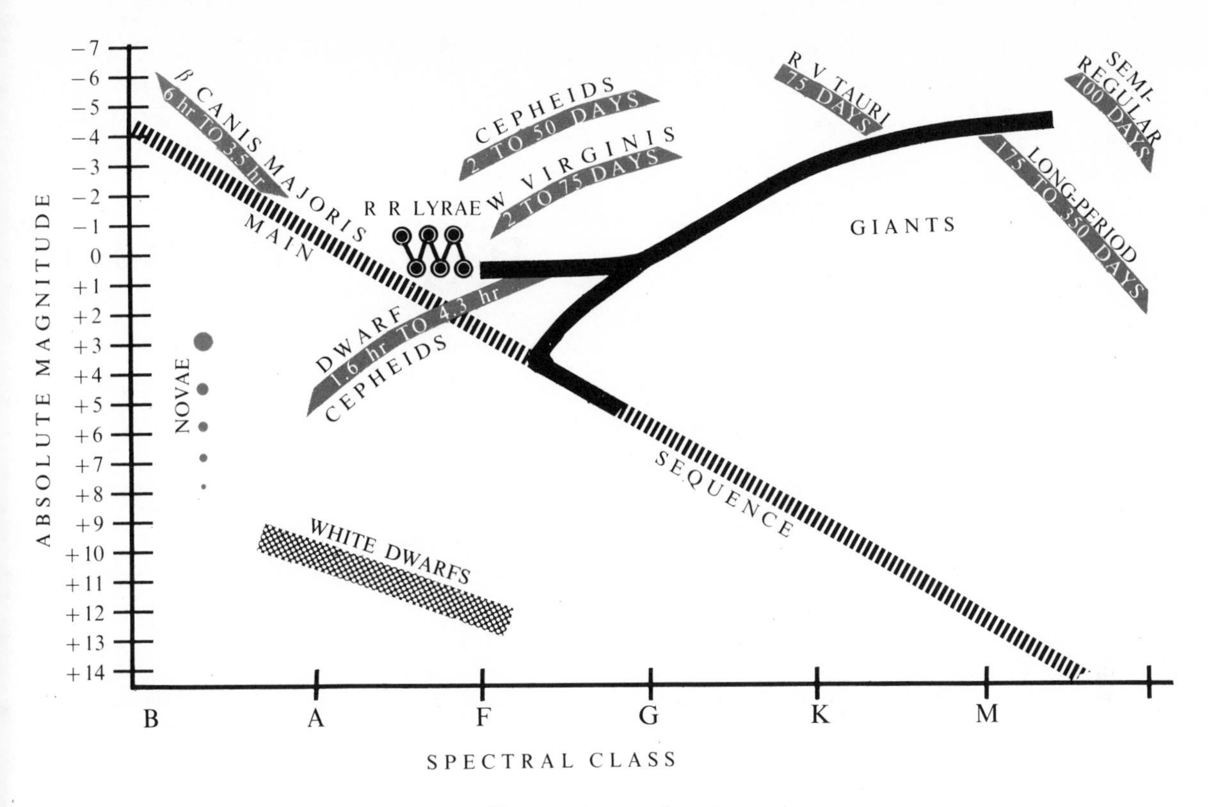

FIG. 24–24 *H-R diagram showing types of variable stars relative to the main sequence.*

sun. Although the remaining 188 stars have lines that are too broad for definite conclusions, Babcock lists 66 as showing some Zeeman effect. Thus, at the present time 155 stars show evidence of large magnetic fields.

Most of the stars with large magnetic fields have peculiar spectroscopic features. Although the great majority of them are A stars, a few are M giants and S stars, and there is even one cluster-type variable in this group. One of the interesting features of these peculiar stars is that the absorption lines of different elements indicate different values for the effective magnetic fields. This can only mean that these lines are produced in different layers above the photosphere. The rate at which H (the strength of the magnetic field) changes from line to line for any one element is in agreement with this interpretation, for it indicates there are large differential velocities in the gases of the stellar atmospheres at different levels.

Babcock's analysis of the stellar Zeeman patterns shows that the general magnetic fields in stars vary with time to the extent of a few thousand gauss. In some of these magnetic variable stars the polarity itself reverses periodically, with the north and south magnetic poles interchanging. In fact, Babcock has developed a theory of solar sunspots (see Section 20.11) in which the alternating polarities of these stars are related to change in polarity of the sun's general magnetic field. Babcock lists 22 of these irregular variables with alternating polarities and 15 whose magnetic fields vary irregularly but without change of polarity. He concludes from an analysis of the changing magnetic fields that the variations are due to large-scale intrinsic hydromagnetic fluctuations on the surface of these stars.

24.54 Spectral Variables

Among these magnetic variables, there are some with variable spectra, which as a group are referred to as **spectral variables**. The intensities of the absorption lines in the spectra of these stars change regularly with periods that are very nearly equal to the periods of rotation of these stars. Since spectral variations are in step with the magnetic cycles of the stars, they may be explained by assuming that the magnetic field is attached rigidly to the star and rotates with it. In most of their other characteristics, these peculiar magnetic A stars are similar to the normal Population I stars and are frequently found in open cluster. They lie slightly above the main sequence (see Figure 24.24).

PART THREE

STELLAR SYSTEMS AND THE STRUCTURE OF THE MILKY WAY

25

STELLAR MOTIONS

25.1 Motions of Stars and Stellar Aggregates

Thus far we have been discussing stars as individuals without paying too much attention to their group characteristics. Since, however, stars occur in aggregates ranging from binaries to vast structures that we call galaxies, we cannot have a complete understanding of their characteristics, behavior, and evolution until we analyze these stellar aggregates. To understand how stars collect into stable systems such as the Milky Way, and continue as a group, we shall have to study the dynamical properties of such groups. But before we can do this, we must have a correct kinematical picture of such stellar systems—i.e., we must have the observational data that properly describe how the stars are moving about and how they are distributed.

We shall first consider stellar motions in our galaxy, because not only will an understanding of these lead us to the solution of the dynamical problem, but it will also give us direct information about the way stars are distributed in the Milky Way. As an example of how knowledge of stellar motions enables us to analyze the distributions of stars, we may consider the neighborhood of the sun. If we want to survey all the stars within 10 parsecs of the sun, we must have some way of selecting them from among the vast number of stars that appear on our photographic plates. However, it is physically impossible to do this by measuring the parallaxes of all these stars and eliminating those that are too small. Instead *we pick out the nearby stars by analyzing one component of their motion with respect to the sun.*

When we look at the stars our first impression is that they are fixed relative to each other since, unlike the planets, they do not seem to change their positions over long periods of time. This is why the idea of fixed stars is still accepted by uninformed people. However, accurate observations show that stars, in general, are moving relative to each other, and that the shapes of some constellations will be altered in a few hundred years.

Stellar motions were first observed by Halley, in 1718, when he detected that Arcturus and Sirius had changed their places on the celestial sphere by a full degree and by a half degree, respectively, since the time of Ptolemy. Hundreds of observations since Halley's time indicate that the stars are moving about on the celestial sphere with speeds relative to the sun which are about the same as those of the planets in their orbits. Because stars are at vast distances, it is difficult to detect their motions visually, but we can measure them with modern instruments.

25.2 The Frame of Reference for Stellar Motions

Before we consider these measurements, we must agree on the meaning of the motion of a star. Since we have seen in the discussion on relativity theory that motion in the absolute sense has no meaning, we can discuss stellar motion only if we first introduce a frame of reference. If we compare the positions of a star (R.A., δ) on the celestial sphere at the beginning and end of a given time interval we determine the motion of the star with respect to the earth. But since the earth itself is moving around the sun, and since its axis is precessing and nutating, the observed displacement of the star on the celestial sphere is due mostly to aberration, precession, nutation, and its annual parallactic shift. *Since we are not interested in these effects, which are due to the earth's motion within the solar system, we shall eliminate them by referring the motions of the stars to the sun.*

Let O in Figure 25.1 be the position of an observer on the sun (which is now our frame of reference), and let A be the position of a star as seen from the sun. If the star moves from A to B in one year, we shall call the distance AB the

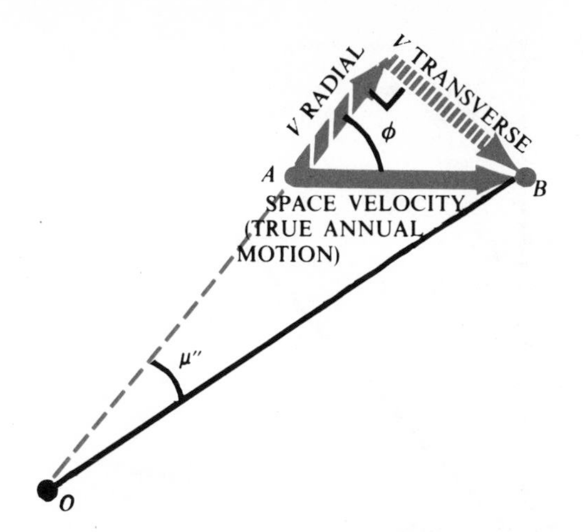

FIG. 25–1 *The annual motion of a star, relative to the sun, decomposed into its transverse and radial components.*

annual motion of the star, and define the space velocity V of the star relative to the sun as this distance per year. To determine this velocity we break it up into two mutually perpendicular components:

1. the radial velocity, V_R, along the line of sight, and
2. the transverse velocity (or tangential velocity), V_T, perpendicular to the line of sight.

Since V_R and V_T are perpendicular to each other, we have

$$V^2 = V_R^2 + V_T^2. \tag{25.1}$$

Moreover, if the space velocity V makes an angle ϕ with the radial velocity (with the line of sight), we also have

$$V_R = V \cos \phi,$$

and

$$V_T = V \sin \phi.$$

Although we cannot measure V directly, we can measure V_R and V_T separately, and then compute V from Equation (25.1).

25.3 Proper Motion and Transverse Velocity

We do not measure V_T directly, but rather the apparent annual angular displacement μ'' of the star relative to the sun. If we measure μ'', which is called the **proper motion** of the star and is expressed in seconds of arc per year, we can find V_T if we know the distance d of the star from the sun. We have

$$V_T = d\,\frac{\mu''}{206\,265},$$

where the divisor 206 265 is introduced to convert μ'' from seconds into radians. Since this equation gives us the transverse velocity in distance per year, we must alter it somewhat if we want to obtain V_T in km/sec. If we replace d in the formula by its equivalent 206 265 (A/p'') [see Equation (15.1)], where p'' is the parallax of the star, and A is the astronomical unit, we obtain

$$V_T = \frac{\mu'' A}{p''}. \tag{25.2}$$

If we now express A in kilometers, and divide this by the number of seconds in a year, we obtain the transverse velocity of the star relative to the sun in kilometers per second:

$$\boxed{V_T = 4.74\,\frac{\mu''}{p''}\ \text{km/sec}}\,. \tag{25.3}$$

25.4 Measuring the Proper Motion

In order to obtain the transverse velocity, we must first measure the proper motion, μ''. This is relatively easy since all we have to do is compare the position of a star on the sky some time in the past with its position today (see Figure 25.2 which shows the change in position of Barnard's star). Even though the stars are so far away that proper motions are, in general, very small, all we need do is wait long enough (even for very distant stars) for their displacements to become measurable. This must ultimately happen because stars continue moving in the same direction without interruption.

In practice, the catalogue positions of stars (R.A., δ) which have been carefully corrected for aberration, nutation, and the constant of precession, are analyzed from year to year, and the proper motions are determined from the differences that are found in these positions. A geometric way of determining the proper motions is to take all the observations of a given star over a sufficiently long period of time, and to plot both the R.A. and declination separately against the time of observation. We thus obtain two straight lines whose slopes give the proper motions, as shown for Barnard's star in Figure 25.3.

The total proper motion, μ'' (which is expressed in seconds of arc), is then obtained by properly combining these two values. To do this we must also know the position angle in the sky towards which the proper motion is taking place—that is,

FIG. 25–2 *Barnard's star in Ophiuchus, with greatest known proper motion. Double photograph shows marked change in position in 22 years. (Yerkes Observatory photograph)*

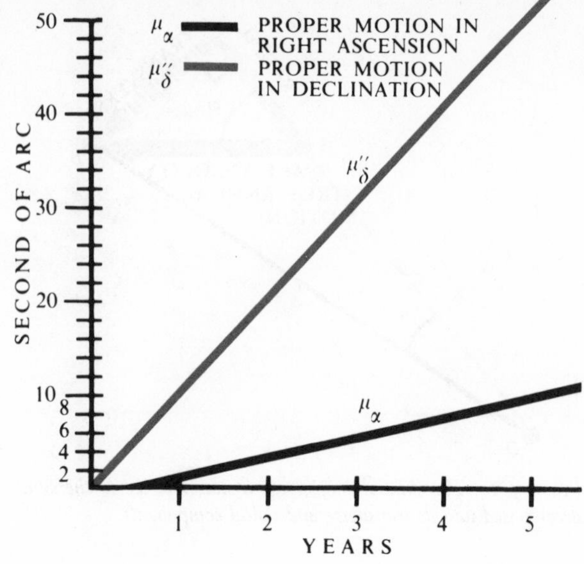

FIG. 25–3 *The right ascension and declination of Barnard's star plotted separately against the time of observation.*

the angle ψ between the direction on the celestial sphere of the proper motion itself, and the northward direction as shown in Figure 25.4. If we consider the eastward direction as well on this sphere (at right angles to the northward direction), we can break up the proper motion into its northward component $\mu \cos \psi$, which is just its proper motion in declination, μ_δ'', and into its eastward component, $\mu \sin \psi$, which, however, is not equal to the proper motion, μ_α, in R.A.

The reason for this is that the proper motion in R.A. is measured along the equator, which is a great circle, whereas the eastward displacement, $\mu \sin \psi$, is measured along a small circle. In Figure 25.4, which shows the annual displacement of a star on the celestial sphere, we see that $\mu'' \sin \psi$ is just equal to $\mu_\alpha \cos \delta$.

Since P_1P_2 represents the total annual proper motion, we have

$$\mu = \sqrt{\mu_\alpha^2 \cos^2 \delta + \mu_\delta''^2}.$$

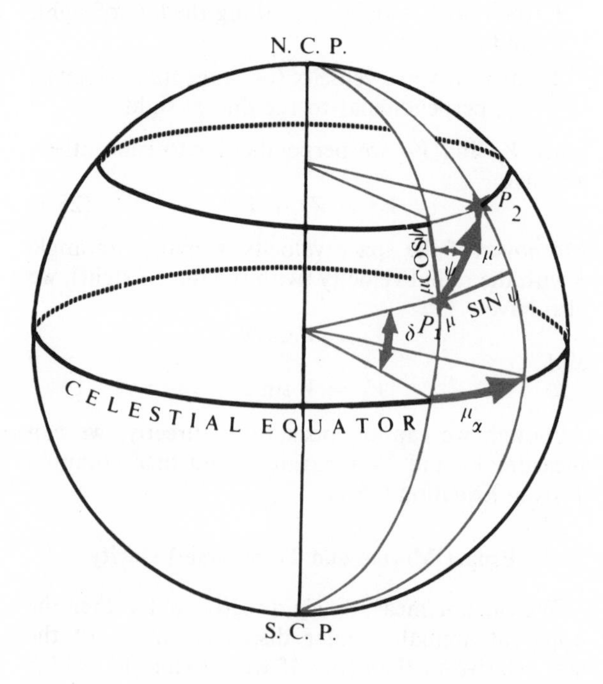

FIG. 25–4 *The proper motion of a star is decomposed into proper motion in declination and proper motion in right ascension.*

Here μ_α is expressed in seconds of time per year, whereas μ_δ'' is expressed in seconds of arc per year, *so that in using this formula we must be careful of units.* The accuracy with which the proper motion of a star can be measured depends on the interval of time separating the two observations of the star. The longer this interval is, the greater the accuracy will be. In Boss' Preliminary General Catalog, the proper motions are given in each coordinate with errors that are no greater than 0.″005. Note that if both μ_α and μ''_δ are accurately known, the position angle ψ as well as the total proper motion can be found.

25.5 Relative and Absolute Proper Motion

The proper motions of bright stars are derived from R.A. and δ measurements made with the meridian circle (see Appendix A, Optical Astronomical Instruments) at two widely separated times, but in the case of faint stars, we must use photographic plates. We compare two photographs of the same region of the sky, taken with the same telescope (to reduce error) from 5 to 10 years apart. The relative proper motions (relative to the stars in the field) can be found with great accuracy by comparing the positions of displaced stars with those of stars that have remained the same or have shifted by very small amounts. From these relative proper motions, we can then obtain the absolute proper motions by measuring the absolute proper motion of a few bright stars in the same field very accurately with the meridian circle. These reference stars may then serve as standards for future work.

The use of photography in proper-motions work is of great importance because it permits us to pick out the numerous stars with large proper motions. This is done most effectively with the "**blink microscope**" which enables us to view alternately two photographic plates of the same star field taken at different times. *If we properly align these plates in the microscope, the two images of the star field appear superimposed on the retina of the eye, so that any star that has altered its position in the field appears to jump back and forth during this blink procedure.* This effect is most pronounced if the second photograph is taken long after the first one. With this procedure we can easily detect proper motion as small as 0.″1.

25.6 Large Proper Motions as a Guide to Nearby Stars

We are interested in stars with large proper motions because such stars are probably nearby; hence we should include them in any program dealing with the distribution of stars in our part of the Milky Way. The proper motion of a star depends not only on how large its actual space velocity is, but also on how close it is to us. Since, as we shall see, most stars have about the same space velocity, large proper motions, in general, signify nearby stars.

A tenth-magnitude star in Ophiuchus, now known as Barnard's star (see Figure 25.2), has the largest known proper motion (10.″27 per year). Its apparent motion among the stars alters its position by the moon's apparent diameter every 180 years. However, this star, as well as that discovered by Kapteyn ($\mu'' = 8.''7$ per year), and a third star, $-36°15693$ ($\mu'' = 6.''90$), have unusually large proper motions. In all, only 50 known stars have proper motions larger than 2″ per year, whereas over 300, according to Luyten, have proper motions larger than 1″ per year. The average of the proper motions of all the naked-eye stars is of the order of 0.″1 per year. Luyten, by using the blink method at the South African and South American Harvard Stations, has estimated that there are 80 000 stars with proper motions greater than 0.″05 per year.

25.7 Mean Proper Motions and Spectral Class

Some interesting correlations have been found between proper motions and spectral classes of stars. B-type stars, in general, have small proper motions, whereas F and G type stars have the largest mean proper motions. The mean proper motions for B and O type stars range from about 0.″02 to 0.″03 per year, while F and G type stars have mean proper motions lying between 0.″17 to 0.″18. The mean proper motions of A and M type stars range from 0.″06 to about 0.″07, and K stars have mean proper motions of about 0.″12 (Boss' Preliminary General Catalog).

25.8 Transverse Velocity

From the proper motion and the parallax of a star we can determine its transverse velocity with the aid of Equation (25.3). If in this formula we substitute the measured proper motions and the parallaxes for various stars, we find that the transverse velocities are of the order of 20 to 30 km/sec for stars in the neighborhood of the sun. However, there are some stars (Barnard's star, for example) with unusually high transverse velocities.

From the annual proper motion of Barnard's star, 10″.25, and its parallax, 0″.54, we obtain its transverse velocity:

$$V_T = \frac{(4.74)(10.25)}{0.544} = 89.31 \text{ km/sec}.$$

Since the transverse velocity is one component of the space velocity, we see from this that the space velocity of this star relative to the sun is also very large.

25.9 The Doppler Shift and Radial Velocity

A star's radial velocity is found by measuring the Doppler shift of its spectral lines (see Chapter 5 on the motion of the earth). We saw in Chapter 5 that the wavelength of light from a moving source as measured by an observer fixed on the earth is altered by an amount that depends upon the relative speed of recession or approach of source and observer. (Note that the effect is the same whether we picture the source as moving and the observer as fixed, or vice versa—the only thing that matters here is the **relative radial motion**.)

We can easily derive the relationship between the radial speed and the change in wavelength by a simple argument. We consider a wave coming from a source that is at rest relative to an observer (no motion of approach or recession). If the frequency of this wave is ν_0 (the subscript 0 indicates that the observer has zero velocity relative to the source) the observer detects exactly ν_0 waves, each of length λ, entering his eye every second.

Now consider what happens when the observer moves towards the source at constant speed v. His eye now receives more than ν_0 waves per second because even if the wave were standing still, his motion itself would carry him past a certain number of wavelengths in a unit time. This then must be added to the number ν_0 he would receive if he were not moving towards the source. Since he moves a distance v in a unit time, and since the length of a wave is λ, the number of waves he passes in virtue of his own motion is exactly v/λ, or $(v/c)\nu_0$, because $\lambda_0 = c/\nu_0$. If we add this to the frequency ν_0, we obtain the frequency of the light as measured by the moving observer.

If the observer is moving away from the source, this quantity must be subtracted from ν_0 to give the observed frequency. We may therefore write for the measured frequency of light from a source relative to which the observer is moving radially with a speed v,

$$\nu = \nu_0\left(1 \pm \frac{v}{c}\right), \qquad (25.4)$$

where the plus sign is to be taken when the source and the observer are approaching (higher frequency), and the minus sign when they are receding from each other. This formula for the Doppler shift is correct as long as $v \ll c$ (source moving slowly with respect to observer). But if v is large, this formula must be corrected by a relativistic factor (see Section 31.7).

Since we measure the wavelength and not the frequency, we shall convert this formula to one involving wavelengths by using $\nu = c/\lambda$, which gives us

$$\frac{1}{\lambda} = \frac{1}{\lambda_0}\left(1 \pm \frac{v}{c}\right),$$

or, by inverting,

$$\lambda = \frac{\lambda_0}{\left(1 \pm \dfrac{v}{c}\right)} \cong \lambda_0\left(1 \mp \frac{v}{c}\right).$$

Since we measure here the change in wavelength $\Delta\lambda = \lambda - \lambda_0$, rather than the wavelength itself, we introduce this expression into the formula to obtain the following equation for the radial velocity of the star (where we have placed λ_0 equal to λ, since the two are nearly equal):

$$v = \mp \frac{\Delta\lambda}{\lambda} c. \qquad (25.5)$$

The following convention is adopted when this formula is used:

1. if the star is receding (spectral lines shifted towards the red), λ is larger than λ_0, so that $\Delta\lambda$ and hence v are positive;
2. if the star is approaching (spectral lines shifted towards the violet), $\Delta\lambda$ and v are negative.

This formula enables us to determine the radial velocities of stars directly regardless of their distances, as long as their spectral lines can be observed. We may illustrate this procedure with the following example. In the spectrum of a given star the line of wavelength 5 000 Å is displaced by 0.5 Å towards the red. What is the radial velocity of the star? On substituting 0.5 Å for $\Delta\lambda$, 5×10^3 Å for λ, and 186 300 mi/sec for c, we obtain

$$v = \frac{0.5}{5 \times 10^3} \times 186\,300 \text{ mi/sec}$$
$$= 18.6 \text{ mi/sec}.$$

This tells us that the star is receding from us along the line of sight at 18.6 mi/sec.

25.10 Accuracy in Measuring Radial Velocities

The difficulty in this procedure lies in achieving a sufficiently high dispersion (separation of close lines) so that the displacements of the stellar spectral lines relative to those from a source on the earth can be accurately measured. If the star is bright enough, we can do this by letting its light pass through a number of prisms successively. But with faint stars only one prism can be used and the dispersion is necessarily small. If the star has a number of intense sharp lines, the accuracy is quite high and the error is of the order of 0.5 km/sec. This implies that the error in measuring the wavelength is not greater than 1 part in 600 000. However, many stars have fuzzy lines so that the error in their calculated radial velocities may be as high as 5 to 10 km/sec.

25.11 Radial Velocities of Stars After Correcting for the Motion of the Earth

Since the Doppler shift gives us the radial velocity of the star relative to the earth, this must be corrected for the earth's rotation and its revolution about the sun. When this is done, we obtain the radial velocity relative to the sun. In Chapter 5 we discussed the earth's motion around the sun and saw that the Doppler shift of stars resulting from this motion can be measured. In fact, this is one method used to determine the solar parallax, i.e., to find the value of the astronomical unit.

After correcting for the earth's motion, we find, on the average, that the radial velocities of stars in the neighborhood of the sun are about ± 20 km/sec. Of the 15 000 radial velocities that have been measured, about 32 per cent are smaller than 10 km/sec, 27 per cent lie between 10 and 20 km/sec, and the remainder are larger than 20 km/sec. About 6 per cent of the stars that have been studied have radial velocities of 60 km/sec or larger. The measurements of radial velocities lag far behind those of proper motions since radial velocities must be measured for each star individually, whereas proper motion can be measured en masse on photographic plates containing many stars. Because of this about 330 000 proper motions have been obtained compared to only about 15 000 radial velocities.

If we combine the radial and transverse velocities of a star, we can obtain its space velocity, v, relative to the sun. Again we find, on the average, that most space velocities are of the order of 20 to 30 km/sec. However, there are a few very high-velocity stars, which we shall discuss in detail later on. As an example of one, we have the bright star Arcturus, which has a space velocity of 135 km/sec.

25.12 Solar and Stellar Motions in a Nonsolar Frame of Reference

The space velocities we have been discussing are measured in a coordinate system, with its origin at the sun. The stellar motions that we obtain this way, however, are misleading as far as the kinematics of the sun are concerned, for they are derived on the assumption that the sun itself is not moving among the stars. To analyze the motions of the stars relative to the Milky Way itself (which we must do if we are to get a correct picture of the structure of our galaxy) we must re-examine the stellar motions, taking into account the sun's motion relative to the stars as a group. To do this we must shift the frame of reference (i.e., our coordinate system) from the sun to some other point. The ideal thing, of course, would be to use the geometric center or possibly the center of mass of the entire galaxy as our coordinate origin, since we would then have the velocity data that are pertinent to the dynamics and the structure of the galaxy. But this is impossible since we have no means of determining this ideal point.

We therefore consider the stellar motions in a frame of reference determined by the positions of the nearby stars themselves. Because the stars are moving, this frame of reference is not a permanent one and changes over long periods of time; but since we are interested in the motions of the stars at the present time, and we know that only slight changes occur in stellar positions from year to year, our frame of reference based upon the stars themselves has as much permanence as we require. However, this frame of reference has one feature that may lead to difficulties; it depends to some extent on the stars we use to establish it. If we use just the nearby stars, the motions we obtain are different from those we obtain when we use the very distant stars. Moreover, a frame of reference based on the red stars gives us results different from those based on the blue stars, and so on. In other words, *since a frame of reference based on a particular selection of stars is not absolute in our galaxy, the results we obtain vary to some extent as we change from one frame to another.* However, if we stick to one frame of reference, the results we obtain are consistent. We shall, therefore, use a frame of reference determined by the observable stars within 20 parsecs of the sun.

25.13 The Solar Apex and Antapex: Peculiar Motions of Stars

If the sun is moving with respect to these nearby stars (i.e., within our stellar frame of reference), we should be able to detect this motion by carefully analyzing either the proper motions of these stars or their radial velocities. This has been done for many stars within the 20-parsec radius and has led to a clear picture of the sun's motion. The point of the celestial sphere towards which the sun (in this frame of reference) appears to be moving is called the **solar apex**, and the point opposite to it on the celestial sphere is called the **antapex**. To determine the solar motion completely, we must find the position (R.A., δ) of the apex, and the speed of the sun towards this point. If we know the solar motion, we can subtract it from the space velocity of any star and thus obtain what is called its **peculiar motion**. This is nothing more than the velocity of the star in a frame of reference determined by the nearby stars themselves.

25.14 Determining the Local Solar Motion from Proper Motions

To see how we determine the local solar motion from proper motions, we consider the average proper motion of a group of stars in any small region of the sky. If the sun were not moving, and if the stars in this group were moving at random with different speeds, the average of the proper motions of all these stars would vanish or be very nearly equal to zero. However, because of the sun's motion, this is not the case. The average proper motion of the group depends upon the angle between the direction of the solar motion and a line from the sun to the group.

If the group of stars is at the apex or at the antapex, the average proper motion vanishes since the motion of the sun is towards these stars or away from them, as the case may be (i.e., the solar motion is radial with respect to these stars), and it has no effect on their transverse velocities. Actually, if we look at the stars near the apex they appear to diverge in all directions from it, and if we look at stars near the antapex they appear, in general, to be converging towards it, as shown in Figure 25.5.

However, if the direction to this group of stars is at some nonzero angle to the solar motion, this motion causes each star to appear to drift in the opposite direction (i.e., opposite to the sun's

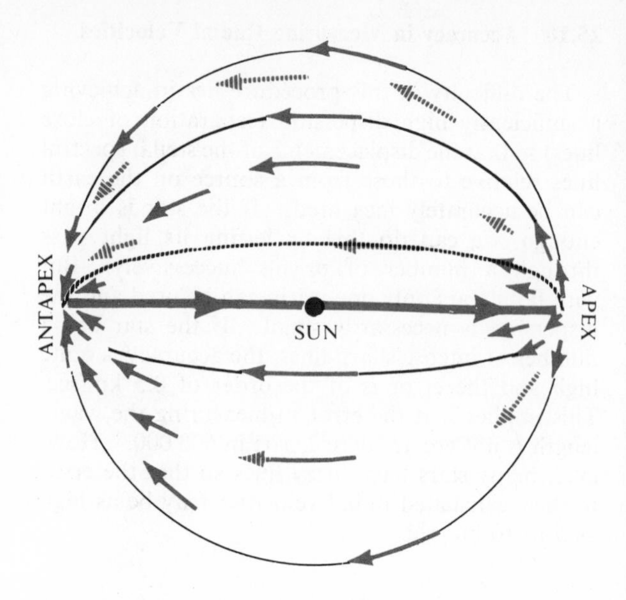

FIG. 25–5 *The effect of the solar motion on the proper motions of the stars. The mean proper motion of stars cancels out at the apex and antapex on the celestial sphere, but there is a mean drift at all other points which is a maximum at points lying at right angles to the direction of the solar motion, as seen from the sun.*

motion) so that the proper motions of the stars in this group no longer average out to zero. There is thus a residual effect opposite to the solar motion which is most pronounced for stars at right angles to the solar motion, as shown in Figure 25.5, and which gets smaller and smaller as we approach the apex and antapex. The actual procedure for determining the local solar motion is thus based upon the analysis of the mean proper motion of groups of stars lying along any great circle on the celestial sphere that passes through the apex and the antapex, as shown in Figure 25.5.

25.15 Procedure for Calculating the Right Ascension of the Apex and Antapex

Since the local solar motion (a vector quantity which can be resolved into three mutually perpendicular components: along the line of sight to the star; parallel to increasing R.A.; and parallel to increasing declination) affects the apparent proper motions of a star in right ascension and declination separately, we can find the position of the apex by analyzing proper motions in right ascension (R.A.) and declination (δ) separately. This is obvious, as can be seen in Figure 25.6, since the solar motion along the line from the apex to the antapex has components in R.A. and δ.

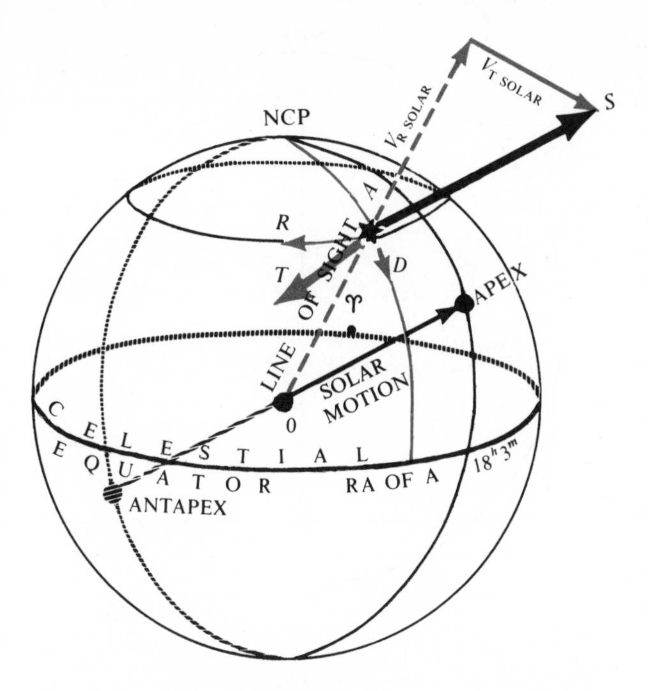

FIG. 25–6 *Schematic diagram showing the contribution of the solar motion to the proper motion of a star at point A on the celestial sphere. A S is the solar motion, A R is the contribution to the proper motion in right ascension (corrected by the appropriate cosine), A T is the contribution of solar motion to the total proper motion, and A D is the contribution of the solar motion to proper motion in declination.*

The component of the solar motion in δ gives the stars on the celestial sphere an apparent southward or northward drift which can only affect their observed proper motions in declination, and the solar motion in R.A. gives the stars an apparent eastward or westward drift which affects their apparent proper motions in R.A., but not in declination.

To determine the R.A. of the apex, we divide the celestial sphere into eight north-south running sectors, each 3 hours wide. From any catalogue (such as Boss' Preliminary Catalog) we select for each segment several hundred stars without regard to their declination (they may be north or south of the celestial equator, as shown in Figure 25.7), and consider their apparent proper motions in R.A. Note that the same number of stars is taken for each segment. Essentially, we consider the projection of each star onto the equator by means of a great circle passing through the star and the N.C.P. We then note the annual apparent proper motion of this projection along the equator as also shown in Figure 25.7.

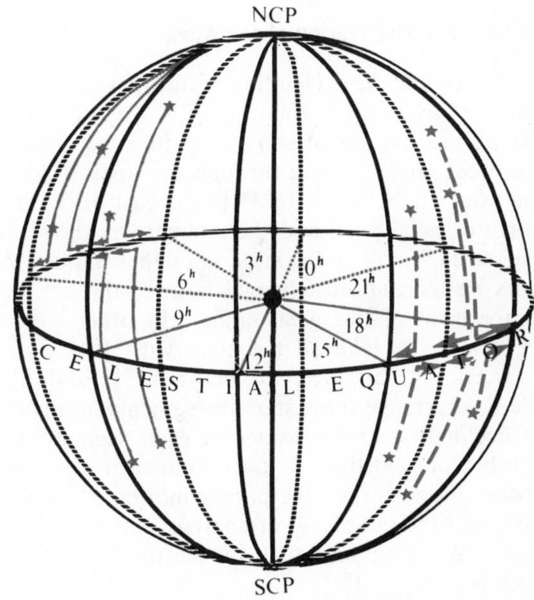

FIG. 25–7 *The analysis of proper motion of stars in right ascension to determine the apex of the solar motion.*

The projection of each star in any segment appears to be moving either to the east or west along the equator (positive or negative R.A.). We now take all stars with negative apparent motions in a given segment, and represent their number by a single arrow on the equator having a length proportional to this number. In the same way we represent the number of stars with positive apparent motions by an arrow of appropriate length in the opposite direction. The sum of these two arrows, the **resultant apparent proper motion** (taken vectorially, as shown by the red arrow), gives the preponderance of proper motion, either to the east or west, resulting from the solar motion.

If we proceed in this fashion, and assign a resultant arrow to each segment, we find that these arrows vary in length from segment to segment. The two segments at 0^h and 12^h have arrows of maximum length, and the two segments at 6^h and 18^h have arrows of about zero length. The number of stars with apparent negative and positive motions in R.A. for each segment is indicated in Figure 25.8. We find from this type of analysis that the R.A. of the apex (one of the points where the resultant arrow is zero length) is 18^h3^m, or $271° \pm 2°$.

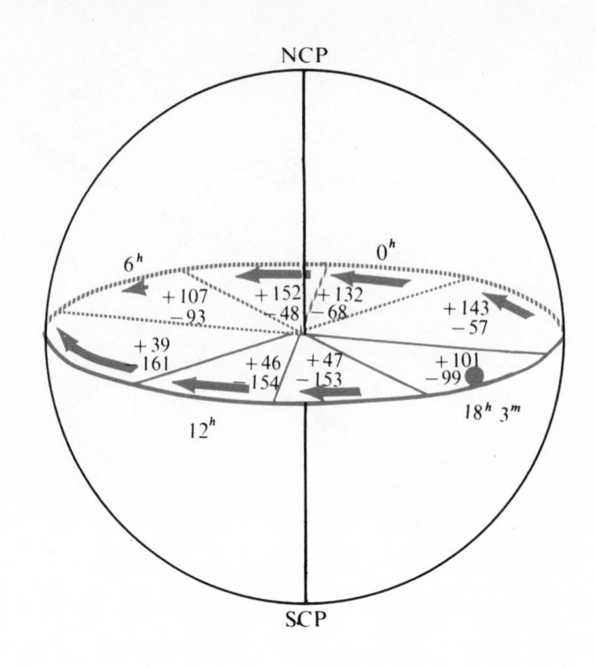

FIG. 25–8 *We obtain the right ascension of the apex and the antapex by finding the segments where the mean proper motion vanishes. (After Russell, Dugan, Stewart,* Astronomy. *Boston: Ginn & Co., 1938, vol. 2, p. 658)*

25.16 The Declination of the Apex

We can find the declination of the apex (assuming that we have already found its R.A.) by analyzing the proper motion of stars lying in groups along the great circle passing through the apex (that is, the hour circle that is 18^h3^m to the east of ♈), the north celestial pole, and the antapex, as shown in Figure 25.9. To get a large enough sampling of stars we include those lying $\frac{1}{2}$ hour to the left and to the right of this great circle. In other words, we essentially take all the stars with R.A. lying between $17\frac{1}{2}^h$ and $18\frac{1}{2}^h$ or between $5\frac{1}{2}^h$ and $6\frac{1}{2}^h$. We now arrange these stars in segments according to declination and consider in each segment the numbers of northward and southward apparent proper motions (i.e., the proper motion in δ), and carry out the same kind of analysis that we did for R.A. We then find that the declination of the apex is $+30° \pm 1°$.

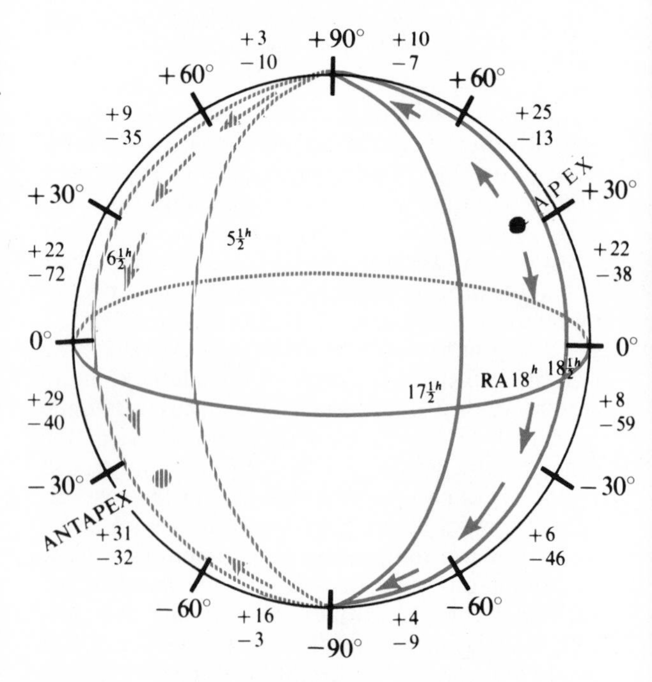

FIG. 25–9 *The declination of the apex is found by analyzing proper motions in declination. (After Russell, Dugan, Stewart,* Astronomy. *Boston: Ginn & Co., 1938, vol. 2, p. 660)*

25.17 The Dependence of the Apex on Selected Stellar Samplings

Although the calculated position of the apex depends to some extent on the numbers of stars we use and to an even greater extent on the groups of stars we use, the results are all in substantial agreement, and the figures given are those generally accepted now. If instead of taking random samples of neighboring stars, we choose special groups of stars, there are appreciable differences in the calculated results for the local solar motion. As an example of this, we find that if we take only B-type stars, the calculated position of the apex is R.A. 274°, δ + 30. For stars between late B and early F we find R.A. 265°, δ + 24; for F to G, R.A. is 270°, δ + 27; K and M spectral classes give R.A. 280°, δ + 34. The solar motions relative to globular clusters, high-velocity stars, or RR Lyrae stars differ considerably from those found from a random sampling. We shall see later that this is due to the structure of the galaxy itself, and to the motion of the solar system relative to the galactic center.

25.18 The Use of Radial Velocities for Determining the Solar Motions

From the proper motions themselves we can find the direction of the local solar motion, but not its magnitude. To find the magnitude, we must use the radial velocities of the neighboring stars (from which, of course, we can also find the direction). The problem here is much easier than in the case of proper motions since all we need do is consider stars near the apex and near the antapex. In Tables 25.1 and 25.2 the positions of the apex in galactic coordinates (see Chapter 1) (λ is galactic longitude, and β is galactic latitude) are given for various groups of stars. The values of the solar motion in km/sec determined from radial velocities are also given as well as the numbers of stars studied in each group.

Stars near the apex should, on the average, be approaching us with a speed equal to the solar motion and those at the antapex should be receding with the same radial velocity, so that the Doppler shift for these groups of stars should give us the data we need. Indeed, this would be the case if the radial velocities gave us the same direction for the solar motion as did the data from proper motions. All we would then have to do would be to find the average radial velocity of recession of stars near the antapex or the average radial velocity of approach of the stars near the apex, which would be close to but not necessarily equal to the former. The mean of these two averages would then give us the magnitude of the solar motion. But this procedure cannot be followed because the radial velocities, in general, give us a direction for the solar motion different from that found from proper motions.

We then proceed as follows: since we may not assume that we have the direction of the solar motion, we set up a three-dimensional coordinate system, with one of its axes pointing towards the north celestial pole, another towards the point R.A. 18^h, δ 0°, and the third axis directed towards R.A. 12^h, δ 0 (autumnal equinox), as shown in Figure 25.10. We now resolve the solar motion into

TABLE 25.1
SOLAR APEX AND MOTION DETERMINED FOR VARIOUS SPECTRAL CLASSES

Spectral class	α	δ	λ	β	V km/sec
B0–B5	274°	+30°	25°	+19°	21
B8–A2	265	24	16	24	16
A5–F2	265	24	16	24	16
G5	270	27	21	21	19
K0–K2	274	30	25	19	20
K5–M	279	36	30	16	25

TABLE 25.2
SOLAR MOTION AND APEX DETERMINED FROM DIFFERENT GROUPS OF STARS

Group	No. of stars used	λ	β	V km/sec
Giants	—	25.6°	16.8°	13.5
Dwarfs	—	26.9	5.9	30.3
O Stars	49	49	25	35.7
Double stars	3 850	49	36	—
Supergiants	205	26	15	19.7
Cepheids	156	—	—	20
Open clusters	29	50	—	30
Interstellar Ca II	261	22	20	21.1
Mira variables	305	47	10	54
RR Lyrae	64	53	12	130
High velocity	598	44	6	66.9
Globular clusters	50	55	0	200
Subdwarfs	79	76	1	153

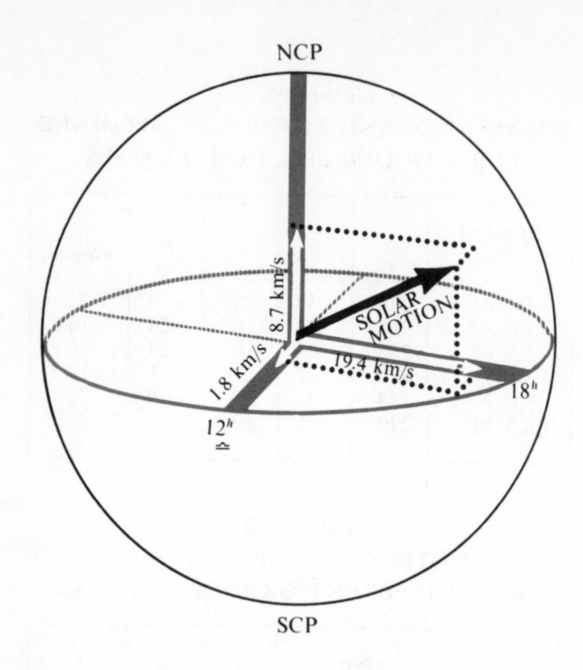

FIG. 25–10 *To obtain the velocity of the solar motion we analyze the radial velocity of stars in a three-dimensional coordinate system.*

three components along these three axes and determine them separately by analyzing the radial velocities of stars along these three axes, as was first done by Campbell and Moore in 1925 for 2 034 stars. As an example of this procedure, using some of the stars in Campbell's Catalogue, we find that the component of the solar motion along the axis towards the north celestial pole is 8.7 km/sec. The component along the axis R.A. 18^h, $\delta\ 0°$, is 19.4 km/sec, and the component along the axis R.A. 12^h, $\delta\ 0$, is 1.8 km/sec. We thus obtain for the local solar velocity

$$V_{\text{solar}} = \sqrt{(8.7)^2 + (19.4)^2 + (1.8)^2} = 21.3 \text{ km/sec},$$

as taken from Russell, Dugan, and Stewart.

That there is a small component of the solar velocity along the axis R.A. 12^h, $\delta\ 0$, indicates that the direction of the solar motion obtained from radial velocities differs slightly from that obtained from the proper motions. Campbell and Moore, using all the data of 2 034 stars, found a solar motion of 19 km/sec with the apex at R.A. 17^h54^m, $\delta + 27°.2$. This means that relative to nearby stars the sun is moving at a speed of about 20 km/sec towards the constellation of Hercules.

Just as the calculated position of the apex depends upon the stars chosen, so too does the velocity of the solar motion, as determined from radial velocities of different groups of stars as given in Table 25.2. We should note the high values for the solar motion obtained from RR Lyrae stars, globular clusters, high-velocity stars, and subdwarfs. As we see, these are so much larger than the others that they must be considered separately, which we shall do later.

25.19 The Peculiar Motion of a Star—Its Radial Component

Since the solar motion is the velocity of the sun in a frame of reference determined by the neighboring stars, we can obtain the motion of any one of these stars itself (its so-called **peculiar motion**) with respect to this stellar frame by adding the calculated solar motion to the observed space velocity of the star. If we consider the space-velocity vector, V_s, of a star as defined previously and shown in Figure 25.11, and represent the solar motion as a vector Ω pointing towards the apex, we see at once how the observed motions of the stars are to be corrected for the solar motion to obtain the peculiar velocities.

If θ is the angle between the direction of the solar motion and the line of sight to the star, then the component of the solar motion along the line of sight is just $\Omega \cos \theta$. Hence if the star's observed radial velocity, V_R, is increased by this amount, we obtain the radial component, $V_{R_{\text{peculiar}}}$, of its peculiar velocity,

$$V_{R_{\text{peculiar}}} = V_R + \Omega \cos \theta. \qquad (25.6)$$

25.20 The Transverse Component of Peculiar Velocity

The correction for the transverse component of the star's velocity is a bit more complicated because the observed space velocity of the star does not necessarily lie in the plane determined by the solar motion and the line of sight to the star, as shown in Figure 25.12. Since, in general, the space velocity of the star does not lie in the plane defined by the solar motion and the line of sight to the star, this is also true of the transverse velocity. We therefore decompose the observed transverse velocity of the star into a component V_T'' at right angles to this plane and a component V_T' lying in this plane. It is clear now that the contribution of the solar motion to the component V_T' is simply that part, $\Omega \sin \theta$, lying in the direction of V_T', and as we see,

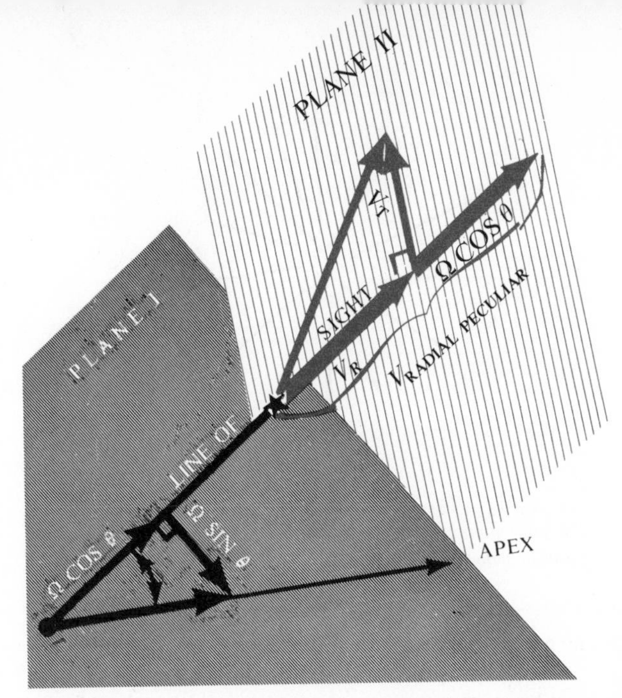

FIG. 25–11 *If we correct the observed motion of a star for the solar motion, we obtain the peculiar velocity of a star. In this figure the correction is given for the radial component of the stellar motion.*

this must be added to V_T' if V_T' is measured positive towards the apex. Hence, $V_T' + \Omega \sin \theta$ is the component of the peculiar transverse velocity lying in the plane determined by the solar motion and the line of sight to the star. There is still, of course, the component V_T'' of the transverse velocity at right angles to this plane, but that is unaffected by the solar motion. However, these two components, $V_T' + \Omega \sin \theta$ and V_T'', must be added vectorially to give the total peculiar transverse velocity:

$$V_{T_{\text{peculiar}}} = \sqrt{(V_T' + \Omega \sin \theta)^2 + (V_T'')^2}. \tag{25.7}$$

Since we measure the proper motion, and not the transverse velocity directly, it is important to see what the peculiar proper motion of a star is after its observed proper motion is corrected for local solar proper motion. We can get at this by noting that the addition of $\Omega \sin \theta$ to the observed transverse velocity of a star whose parallax is p'' would lead to the addition of $p''(\Omega \sin \theta)/4.74$ to its observed proper motion, and this would be its **peculiar proper motion**. Hence the solar motion has the effect of making the component of the observed proper motion parallel to the solar motion (if this component is measured positive towards the apex) smaller by this amount than this

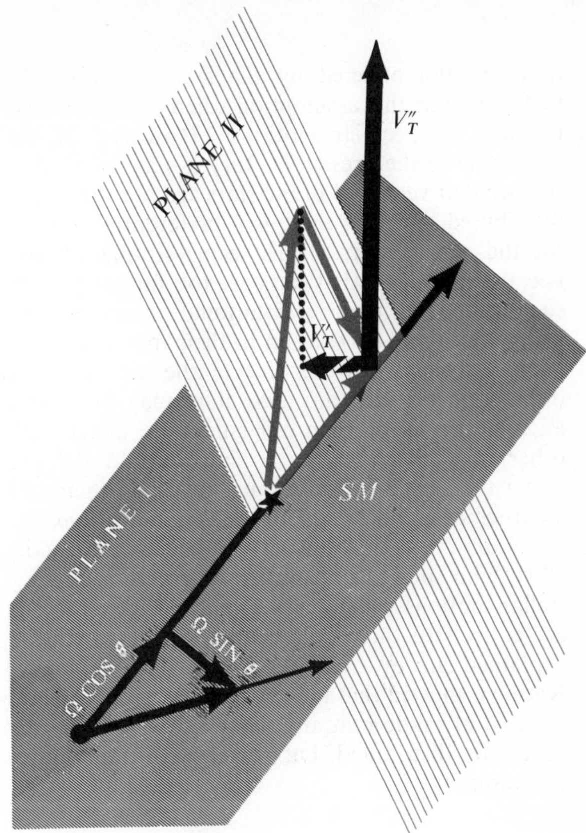

FIG. 25–12 *The correction of the transverse component of the star's motion for the solar motion.*

component of the peculiar proper motion. Thus if μ' is the component of the observed proper motion lying in the plane *SM*, then this component of the peculiar proper motion is just

$$\mu' + \frac{p''\Omega \sin \theta}{4.74}.$$

The peculiar velocities of the stars give us a much deeper insight into the way they are moving with respect to each other than we get from their motions with respect to the sun. *The apparent displacement of a star arising from the solar motion is called its* ***parallactic motion***. We shall see that by measuring the parallactic motion of a group of stars we can find its mean or average parallax.

25.21 Preferential Motions of Stars as Indicated by Peculiar Velocities

It might seem from what we have already said that peculiar motions of the stars are random since they are not affected by the sun's motion. In fact, we made this assumption when we introduced the idea of peculiar motions. However, at the beginning of the present century it was found that the peculiar velocities of the stars are not randomly distributed, but that there is a preferential motion for the nearby stars. This seems strange at first because it appears that stars are too far away from each other for their gravitational interaction to bring about any correlation in their motions.

Nevertheless, however feeble the gravitational interaction between any two stars may be, taken all together the stars in our galaxy do affect each other in such a way as to introduce a definite pattern in their motions and to keep them all together as a galaxy. *This preferential motion of the stars* was first discovered by Kapteyn in 1904, who showed that there is correlation among stellar motions which, as we shall see, is related to the entire structure of the Milky Way, and can be explained only by a dynamical picture of our galaxy. Kapteyn showed that the peculiar velocities of stars selected at random at any point in the sky are not randomly distributed, but rather have preferential directions.

25.22 Kapteyn's Two Star-Stream Concept

Kapteyn announced this in the form of the two star-stream hypothesis. In order to see what this means, let us first suppose that we could observe in a given region of the sky a compact group of stars having completely random peculiar velocities. How would these stars appear after a sufficiently long period of time? If the sun were not moving at all, each star would appear to move out from the center of the group quite independently of the other stars (because of their random proper motions) and in time we would have not a concentrated globule of stars, but rather a spherical cluster spread out over a given area of the sky, with the slowly moving stars concentrated near the center and the rapidly moving ones out towards the circumference.

Let us now analyze the peculiar motions of the stars in this group by a special velocity diagram. From the center of the group we draw arrows of various sizes in all directions. Each arrow is to represent by its length the number of stars in the group with peculiar velocities (for example, peculiar proper motions) in the direction of the arrow. Thus, the more stars in the group there are with proper motions in a given direction, the

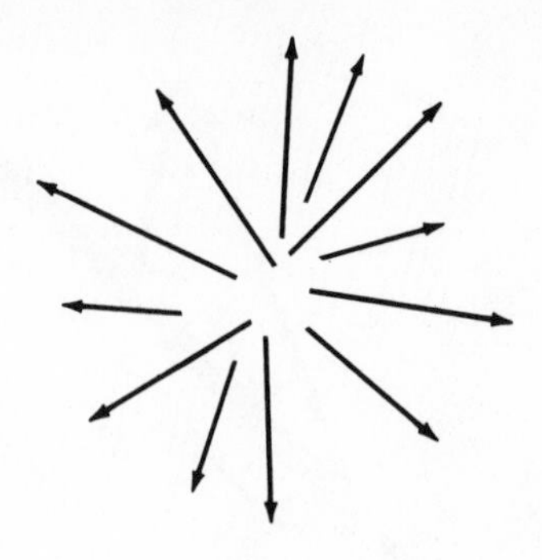

FIG. 25-13 *The velocity diagram of stars, taking into account the solar motion. In other words, the velocity diagram of the peculiar velocities of stars.*

larger is the arrow in that direction. If there were no solar motion, and if the stars were moving at random, the curve connecting the ends of the arrows radiating out in all directions from the center of the group would be a circle, as shown in Figure 25.13. However, because of the parallactic motion of the stars in this group (arising from the solar motion) we would find, instead of a circle (again assuming random peculiar velocities), a kind of oval distribution with the greatest number of stars moving towards the antapex, because the whole cluster appears to drift in that direction.

As was first pointed out by Kapteyn, the actual observations do not agree with this model. He discovered that the curve giving the distribution of peculiar proper motions is not oval, but rather irregular in shape with two lobes, as though the stars were streaming in two directions. This is found to be the case no matter in what part of the sky the proper motions of groups of stars are analyzed.

If the two preferred directions obtained in each part of the sky are plotted on the celestial sphere, they all appear to converge towards two points (one for each stream) which Kapteyn called the **apparent vertices**. A typical star-stream diagram is shown in Figure 25.14. To explain these preferential directions of motion, Kapteyn introduced the idea that there are two streams of stars in our galaxy intermingling freely with each other, but moving in two distinct directions, as given by the two convergent points. By analyzing the peculiar velocities in various parts of the sky, Kapteyn showed that the directions of these two drifts (as observed from the moving sun) form an angle of 100°, and that their velocities are very nearly equal.

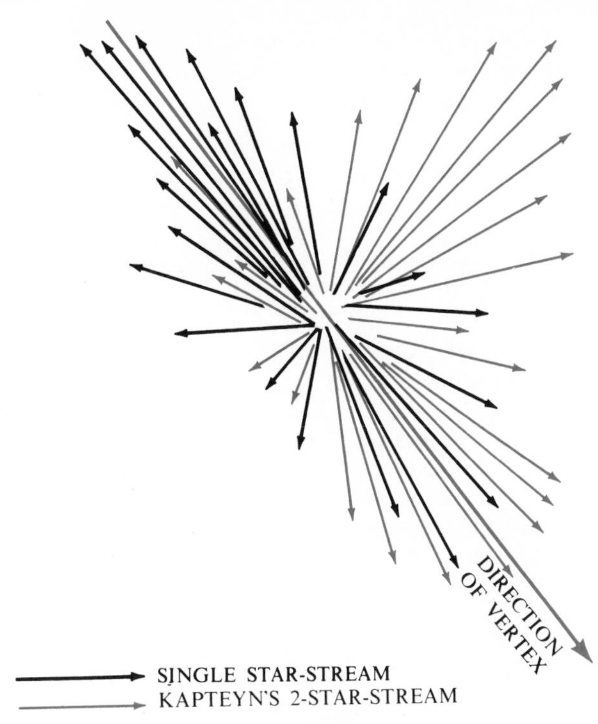

FIG. 25–14 *A typical star-stream diagram.*

25.23 The Ellipsoidal or Single-Stream Hypothesis

Although at first astronomers were inclined to assign a physical significance to these two streams as though the stars in our galaxy were, indeed, separated into two different physical groups, we now know that this is but a mathematical scheme for representing the observations and that one can just as well represent them by another scheme. The simplest and physically most meaningful of these is the one introduced by Karl Schwarzschild, the **ellipsoidal** or **single-stream** hypothesis. This scheme is now universally accepted since it is in accord with the actual velocities of the stars in our galaxy.

To understand how the single-stream or ellipsoidal theory arises, we may imagine observing the group of stars introduced above, not from the sun, but rather from the centroid of the group itself (in other words, after the parallactic motion has been eliminated). We then find that Kapteyn's two star streams are not tilted 100° to each other, but are moving along the same line but opposite to each other on the celestial sphere. This line, then, forms an axis of symmetry of the observed peculiar velocities of the stars. Starting from this idea Schwarzschild introduced the alternative hypothesis that instead of two groups of stars there is a single group of stars streaming preferentially along this axis. Although stars, in general, prefer to move along this line of the true vertices, there is a certain amount of drifting perpendicular to this line of symmetry so that the actual distribution of velocities defined above is an ellipsoidal one.

25.24 The Vertices of the Preferential Motion

From an analysis of the proper motions of stars in various parts of the sky, we find that the preferential direction of this star streaming lies in the plane of the Milky Way, and pierces the celestial sphere at the two points R.A. 6^h15^m, $\delta + 13$ [a point in Orion]; R.A. 18^h15^m, $\delta - 13$ [a point in Scutum], which are called the **vertices** of the **preferential motion**. We see that the line connecting these two vertices passes through the center of the Milky Way.

The average velocity of drift along the line of the true vertices is 41 km/sec, whereas the drift at right angles to it in the plane of the Milky Way is of the order of 25 km/sec. A third component of the drift at right angles to the plane of the Milky Way is 18 km/sec. We see from this analysis that the stars, in general, seem to be drifting in the plane of the Milky Way almost towards or away from its center with a smaller component of drift at right angles to this plane. Although preferential star streaming was first found by using proper motions, we shall see that the same results are obtained from radial velocities. The greatest drift effect is found in stars of class A and is almost entirely absent in B stars.

We shall see that the Schwarzschild single-stream picture can be derived from the dynamics of the entire galaxy, and that this motion arises from the gravitational interaction of the stars in the neighborhood of the sun with the great mass of stars in the central core of the Milky Way. We shall, in fact, see that the Milky Way itself is rotating and that the stars in the region of the sun are moving in very nearly Keplerian orbits about the center of the galaxy. Lindblad proved that the observed star streaming is due to the departure of these orbits from circularity and to their slight tilt with respect to the plane of the Milky Way.

25.25 The K Effect

At this point it is important to mention a very curious effect which was first detected in 1910 by

Kapteyn and Frost, but which was first clearly defined with the aid of careful observations by Campbell. It is now referred to as the **K effect.** Kapteyn and Frost discovered from the radial velocities of B-type stars at the apex and antapex that whereas the sun appears to be approaching the former at a speed of 18.4 ± 1.4 km/sec, it appears to be receding from the latter at 28.4 ± 1.4 km/sec. Since these two velocities are not equal, it follows that the mean peculiar radial velocity of B-type stars does not vanish so that these stars appear to be receding from us at a speed of 5 km/sec. This is the K term for B0 stars and is just the deviation of the two radial velocities given above from their average value.

Campbell, using a larger sampling of stars and taking only those at the apex and antapex, found that this K effect is primarily associated with the bright late O and early B-type stars. It is a recession of these stars in all directions of the order of +4 km/sec. The K effect is negligible in other spectral classes, as shown in Table 25.3.

TABLE 25.3
KAPTEYN EFFECT DETERMINED BY SPECTRAL-CLASS SELECTION
(Sampling for Stars with Apparent Magnitude Less Than 6)

Spectral class	K effect
	(km/sec)
B0	+5.3
A0	+1.4
F0	0.0
G0	−0.5
K0	0.0
M0	+0.4

Although the explanation of the K effect has not been given and is still one of the unsolved problems of stellar motion, Plaskett has pointed out that this term is associated with stars that are not at very great distances. He showed that this term vanishes for B stars that are fainter than magnitude 7.5 and that it is considerably smaller for faint O-type stars than for bright ones. A more puzzling effect, but one similar to the K term, is that associated with very faint stars from the ninth to the thirteenth apparent magnitude. They seem to have a velocity of approach of about 4 km/sec, and Trumpler has also shown that an effect of this sort exists for certain open clusters.

25.26 Dependence of Peculiar Velocities on Galactic Longitude

The investigation of the K effect has led to the discovery of what is dynamically a more important effect, namely, the dependence of the peculiar radial velocities on galactic longitude. This phenomenon was first discovered in 1915 by Gyllenberg, who was investigating the dependence of the K effect on the distribution of stars with respect to the center of the galaxy. He found that the average of the peculiar velocities of B-type stars lying in a given galactic longitude depends on this galactic longitude. Later it was discovered that this holds not only for B-type stars but for other stars as well, and Oort was able to show in 1927 that this effect can be represented by the formula

$$V_{R_{\text{peculiar}}} = A \sin^2 (\lambda - \lambda_0),$$

where A and λ_0 are certain constants, and λ is the galactic longitude of the group of stars whose mean peculiar radial velocities are being analyzed. We shall see later that this effect can be accounted for by the rotation of the galaxy.

25.27 Asymmetry of Stellar Motion

Although in our discussion of the velocities of stars we originally assumed that the stars were moving at random, we now see that this is certainly not the case and that there is a preferential motion in the plane of the galaxy. However, disregarding this drift, the motions on the whole are isotropic and almost random for most stars in the neighborhood of the sun. Nevertheless, as was first pointed out by Boss, there is a remarkable asymmetry in the motions of certain stars with peculiar velocities in excess of 62 km/sec.

We can best represent this by a velocity diagram for these stars (given in Figure 25.15) which were first studied and thoroughly analyzed by Oort. The diagram shows that almost without exception, stars with peculiar velocities in excess of 53 km/sec are moving in the same general direction in the plane of the galaxy away from Cygnus and towards galactic longitude 273°. It is as though these high-velocity stars as a group were moving almost at right angles to the line of the true vertices and in a

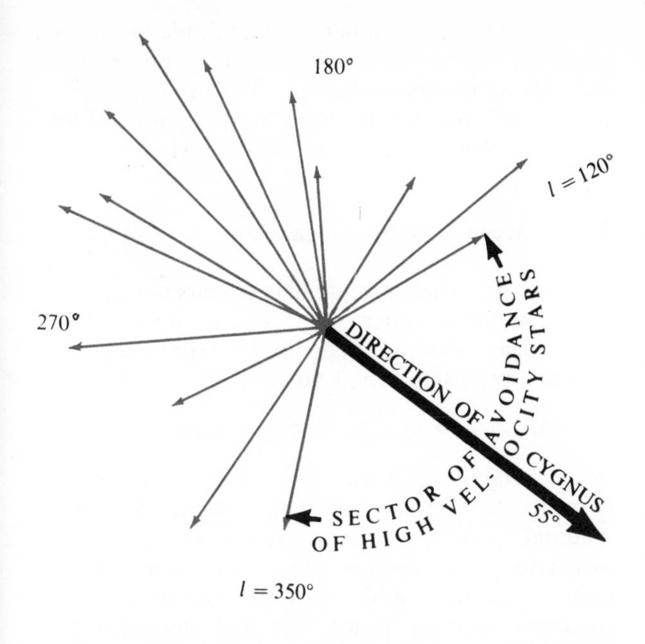

FIG. 25–15 *A schematic diagram showing the asymmetry of stellar velocities. All the high-velocity stars appear to be moving in one sector.*

general direction opposite to the solar apex. Although this effect was first discovered for high-velocity stars, it applies to other groups of objects as well, among which are globular clusters, RR Lyrae type stars, and Population II stars in general. We shall see that this asymmetry can be explained in terms of the galactic structure and its rotation.

25.28 Secular Parallaxes of Stars: Mean Parallax

We have seen that if we add the solar motion to the observed proper motion of a star, we obtain its peculiar proper motion, which consists of a component at right angles to the solar motion and a component ω'' parallel to the solar motion, where

$$\omega'' = \mu' + \frac{p''\Omega \sin \theta}{4.74},$$

and where μ' is the component of the actual observed proper motion parallel to the solar motion and Ω is the solar motion itself. Since this expression contains the parallax of the star, it appears at first that we should be able to determine this quantity by measuring ω''. In fact, we find on solving for the parallax

$$p'' = \frac{4.74(\omega'' - \mu')}{\Omega \sin \theta}. \qquad (25.8)$$

This apparent displacement of a star resulting from the solar motion is referred to as the **secular parallax.**

However, we cannot measure ω'' for a single star. The reason is that we can only measure the star's observed displacement which is a combination of the solar motion and the star's actual motion. In other words, it is impossible to determine from the observed proper motion of an individual star the part that is due to the solar motion. Nevertheless, we can use this formula to determine the mean parallax for a large number of stars with very nearly random peculiar proper motions. This procedure works particularly well for groups that are homogeneous in character, such as Cepheid variables, high-velocity stars, open clusters (such as the Hyades), long-period variables, and the like. In fact, Shapley used this procedure to determine the parallax of δ Cephei in order to find the zero point of the period-luminosity law for Cepheid variables. (See Section 24.13.)

In a large homogeneous group there are, in general, as many stars with proper motions in the direction of the solar motion as opposite to it, so that ω'' is as likely to be positive as negative. ω'' therefore drops out of the equation when the average value of p'' is taken:

$$p''_{\text{average}} = \frac{-\mu'_{\text{average}} \times 4.74}{\Omega \sin \theta}. \qquad (25.9)$$

(Note that $\sin \theta$ does not have to be averaged since θ is the same for all stars in the group.) This would be the correct formula for the mean parallax if all the stars were close together and at the same distance from the apex, but if we consider the mean parallax of stars for varying angular distances θ from the apex, the formula has to be corrected slightly by replacing $\left(\frac{\mu'_{\text{average}}}{\sin \theta}\right)$ by $\left(\frac{\mu' \sin \theta_{\text{average}}}{\sin^2 \theta_{\text{average}}}\right)$,

so that we finally obtain for the **mean parallax** of a group of stars at different positions in the sky

$$p''_{\text{average}} = \frac{-4.74(\mu' \sin \theta)_{\text{average}}}{(\sin^2 \theta)_{\text{average}}}. \qquad (25.10)$$

If a sufficiently large number of stars is taken to compute this average, the accuracy is fairly high even if all the peculiar proper motions ω'' do not cancel out exactly. However, if the peculiar proper motions ω'' are small enough, and their average does cancel out, the results are quite good. In our discussion of trigonometric stellar parallaxes we saw that we measure the relative parallaxes of nearby stars with respect to distant background stars and that to obtain the absolute parallax of a nearby star we must first measure the parallax of these background stars. Since we can now use the procedure just outlined to find the mean parallaxes of the background stars, we can obtain the absolute parallaxes of the nearby stars by adding this mean parallax to the measured relative parallaxes as shown in Figure 25.16.

25.29 The Application of the Method of Secular Parallaxes

As an example of the application of the method of secular parallaxes we may consider the mean parallaxes of stars of different apparent magnitudes. As we go to fainter and fainter stars, the mean parallax decreases. However, the mean parallax also depends on the position of the stars relative to the plane of the Milky Way. We find a systematic larger mean parallax for stars of a given apparent magnitude near the galactic pole than for stars of the same apparent magnitude near the plane of the galaxy. The mean parallax also depends on the spectral class. We find it increases from 0".007 for B-type stars to 0".012 for F-type stars, but then decreases to 0".008 for M-type stars.

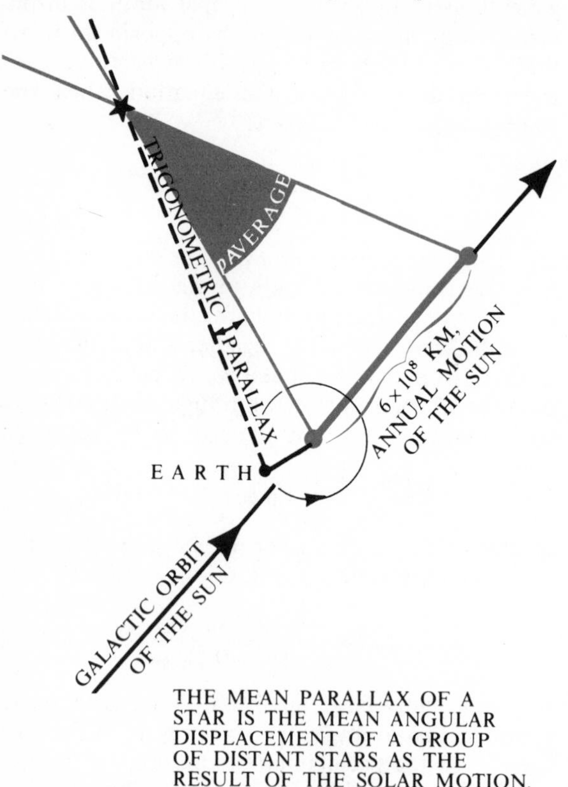

FIG. 25–16 *To obtain the absolute parallax of a star, we must correct it by the mean parallax of distant background stars obtained from the solar motion.*

25.30 Mean Absolute Magnitudes

If we know the mean parallax of stars in a group, and their mean apparent magnitude, we can find their mean absolute magnitude. However, the formula one has to use is **not**

$$M_{\text{average}} = m_{\text{average}} + 5 + 5 \log p''_{\text{average}}$$

but a slightly different one which is properly corrected for the absorption of light by interstellar material. Since mean parallaxes are generally associated with distant stars, this correction is quite important and must be included. The corrected formula which was first suggested by Oort on a semi-empirical basis, is the following:

$$M_{\text{average}} = m_{\text{average}} + 4.5 + 5 \log p''_{\text{average}} - A, \tag{25.11}$$

where A is the correction introduced to take account of interstellar absorption.

The mean parallaxes of stars of different apparent magnitudes and different spectral classes (as given in Tables 25.4 and 25.5) show us that most of the naked-eye stars are certainly considerably more luminous than the sun, as can be seen from the following argument. Since their mean parallaxes are generally smaller than 0".1, they are farther than 10 parsecs away, so that their absolute magnitudes are smaller than their apparent magnitudes. But their apparent magnitudes for the most part are smaller than 5 (i.e., less than the absolute magnitude of the sun) so that their absolute magnitudes are also less than that of the sun.

Finally we may note that it is possible to obtain the mean parallax of a group of stars having the same proper motion and a given apparent magnitude by means of the following formula:

$$\log p''_{\text{average}} = -0.69 - 0.0713m + 0.645 \log \mu''. \tag{25.12}$$

This formula, developed empirically by Kapteyn and given a theoretical basis by K. Schwarzschild, gives results that are considerably in error when applied to individual stars but are fairly good for groups of stars.

TABLE 25.4
MEAN PARALLAXES OF DISTANT STARS ACCORDING TO SPECTRAL CLASS

Spectral class	Mean μ''	Mean parallax	No. of stars
B	0".025	0".006 6	490
A	0.046	0.009 0	1 647
F	0.076	0.012 5	656
G	0.053	0.008 0	444
K	0.057	0.009 0	1 227
M	0.052	0.007 9	222

TABLE 25.5
MEAN PARALLAXES OF STARS OF DIFFERENT APPARENT MAGNITUDES

Apparent magnitude	Mean parallax	Apparent magnitude	Mean parallax
1	0".083	8	0".007 2
2	0.056	9	0.005 3
3	0.030 4	10.0	0.004 0
4	0.002 7	11.0	0.002 8
5	0.017 1	12.0	0.002 2
6	0.012 8	13.0	0.001 6
7	0.009 6		

EXERCISES

1. Calculate the transverse velocities of the ten nearest stars listed in Table 15.1, using their proper motions.

2. (a) Use the listed radial velocities of the ten stars in problem 1 to calculate their space velocities, using the results of problem 1 for the transverse velocities. (b) Calculate the average space velocity of these ten stars.

3. Use the formula for the mean parallax to find the mean parallax of the ten B-type stars for which the data are available.

4. Use the formula for the mean parallax to find the mean parallax of four O-type stars for which data are available.

26

MULTIPLE STELLAR SYSTEMS

26.1 The Statistical Importance of Binaries

Now that we have discussed the motions of stars, we are in a position to investigate the way stars are arranged in aggregates, from the simplest to the most complex groups. Since, in general, the distances between stars are very large, the gravitational effect of one star on another, taken at random, is too small to be observable, and only the sum of all these interactions is important in considering the structure of large assemblies such as our own galaxy. However, within any galaxy there are groups in which the stars are so close together that they are all dynamically related and form a single structure moving as a unit through that galaxy.

The simplest such structure is the **binary system**, in which two stars revolve about a common center of mass according to Kepler's laws. These stars are certainly important in stellar statistics since an analysis of stars in the neighborhood of the sun shows that probably as many as 50 per cent of the stars in our galaxy are components of binary systems. Of the known 254 stars closer than $10\frac{1}{2}$ parsecs, 127 are components of 61 different multiple systems. In other words, it appears that about one-half of all observed stellar objects are multiple systems consisting of two or more stars. Astronomers have now catalogued 30 000 such multiple systems.

26.2 Double Star Systems

We divide double stars into three groups:

1. **Visual doubles**. These consist of two components that can be seen as individual stars with available optical telescopes. We shall include in this group not only **visual binaries** (i.e., pairs in which the two stars are held together gravitationally) but also **optical pairs** in which the two stars are nearly in the same line of sight on the celestial sphere but are really at a very great distance from each other. As we shall see, there are very few of these, but we must be aware of them in considering the statistics of double stars.
2. **Spectroscopic binaries**. These are double stars in which the angular separation between the components is so small (either because the two stars are physically very close to each other or because the entire system is at a very great distance from us) that they are beyond the resolving power of our telescopes and cannot be seen as individuals. However, the binary nature of these stars can be detected spectroscopically because of the periodic displacement of their lines resulting from the Doppler shift.
3. **Eclipsing binaries.** These are systems in which the plane of the orbit of the two stars passes very nearly through the earth, or lies very nearly edgewise to us. These systems may, of course, be spectroscopic binaries and even visual binaries. However, in most cases the components of these binaries are too close together to be resolved by telescopic methods and in some cases too faint to show up as spectroscopic binaries.

26.3 Criterion for Visual Binaries

Before we can say whether two stars that appear close together really form a binary system, we must have a workable criterion. This criterion is based on Aitken's suggestion that the smaller the linear separation between two stars, the greater the probability of their being members of the same binary system. We may put this differently by saying that *the smaller the angular separation between two stars is and the smaller their apparent magnitudes are, the greater is the probability that they are components of a binary system.* Aitken introduced the following empirical threshold formula to differentiate between optical pairs and binary systems:

$$\log a'' = 2.8 - 0.2m, \tag{26.1}$$

where a'' is the angular separation in seconds of arc between the two stars and m is the apparent magnitude of the two stars taken together. *If a'' is less than this, we may assume the stars to be a visual binary, but if a'' is greater than this, we may classify them as an optical pair.*

This gives a limiting value of 6″ for a 10th-magnitude star. Thus, this formula tells us to count all stars brighter than the 10th magnitude as binaries if their angular separation is less than 6″. For stars of the 6th magnitude the angular separation must be less than 40″. Using this criterion Aitken found that over 17 000 stars lying between declination −30 and +90 are visual binaries. He concluded that down to the 9th magnitude one star out of every 18 is a visual binary. Of these pairs about 64 per cent have components separated by less than 1″.0 and 23 per cent by less than 0″.5. He went on to point out that among the stars brighter than the 6th magnitude 1 out of 9 is a visual binary.

Although we know more than 17 000 physical double stars (about 40 000 visual doubles are known but all are not physical doubles), the orbits of fewer than 300 have been determined by observation. The reason is that almost all the visual doubles are systems with very long periods (because the two components are far from each other; see Kepler's third law, Section 8.27) and not enough observations have been made in the available time to compute the orbits accurately.

26.4 Methods of Measuring

The observations of visual doubles are made with a special eyepiece in which the east-west direction and the north-south direction are taken as coordinates in a polar coordinate system, as shown in Figure 26.1. One fixes the origin of the coordinate in the eyepiece on the primary star (the brighter star and in general the one with the smaller motion) and locates the position of the secondary star with respect to it by specifying its position angle θ and its distance ρ'' expressed in seconds of arc. Note that the actual displacement ρ'' is given in linear measure and converted to angular measure by dividing by the focal length of the telescope. The measurement is made by a micrometer eyepiece having two parallel wires, one of which is movable,

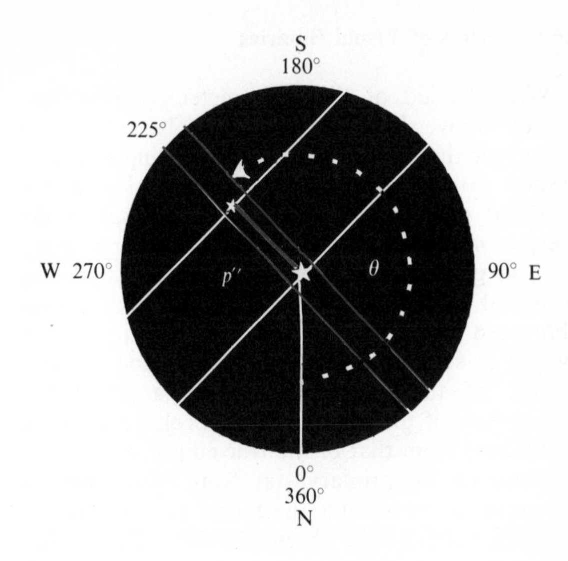

FIG. 26–1 *Coordinates in an eyepiece used to obtain the relative orbit of a binary. The origin of the coordinate is taken as the primary star.*

as shown in Figure 26.1. Although photographs can be used if ρ'' is large enough, in general visual methods must be employed for very close binaries.

26.5 Statistics of Visual Binaries

Although one may definitely conclude that a visual double is a binary system only if the orbital motion of one star relative to the other is observed, a very strong statistical argument can be presented to show that most visual double stars are physically connected. If we just consider the 6 500 naked-eye stars in both hemispheres of the celestial sphere, we can show statistically that the chance against finding a single pair that is not physically connected is about 80 to 1.

Sometimes it is possible to conclude that two stars form a visual binary even though an orbit is not available. This is true for a system with a large proper motion in which the angular separation between the two stars remains about the same. If the proper motions of the two components are quite different in magnitude and direction, we may conclude that the stars are not physically connected. Although in dealing with two stars that appear close together and that have small proper motions the evidence that they form a binary is difficult to adduce, it is reasonable to assume on the basis of statistics that such stars are probably binaries if they show no relative motion over a long period.

26.6 Orbits of Visual Binaries

With the aid of the micrometer eyepiece discussed above it is possible by observation to determine the orbit of the rapidly moving star with respect to the primary. *One then obtains the apparent orbit (which is always an ellipse) of the component relative to the primary.* But the primary star, in general, does not lie at the focus of this observed ellipse because this ellipse is the foreshortened projection of the true elliptical orbit (which is seen obliquely) onto the plane perpendicular to the line of sight. As a result the eccentricity of the observed (i.e., projected) ellipse is different from that of the true ellipse, so that the distance of the primary star from the center of the projected orbit is not, in general, proportional to the eccentricity of the observed orbit. In the diagram of Figure 26.2 the observed apparent orbit of a visual binary is shown. In Figures 26.3(a) & (b) the photographs show the orbital motions of some visual binaries.

An extreme example of the deviation of the primary star from the focus of the observed apparent elliptical orbit is given by a binary system with a true circular orbit which is seen almost edgewise. In this case the secondary star appears to oscillate back and forth with respect to the primary, and the observed orbit is highly eccentric, although the primary star is at the center of this apparent orbit.

Since the apparent orbit is the projection of the true orbit, Kepler's second law of areas applies to the apparent just as it does to the true orbit. But the third law does not, since, in general, the semi-major axis of the true orbit is foreshortened when it is projected on to the plane of the sky. The projection of the semi-major axis of the true orbit is that diameter of the apparent orbit which passes through the primary star. *We may note that, in general, the focus of the projected ellipse does not lie on the diameter through the star.*

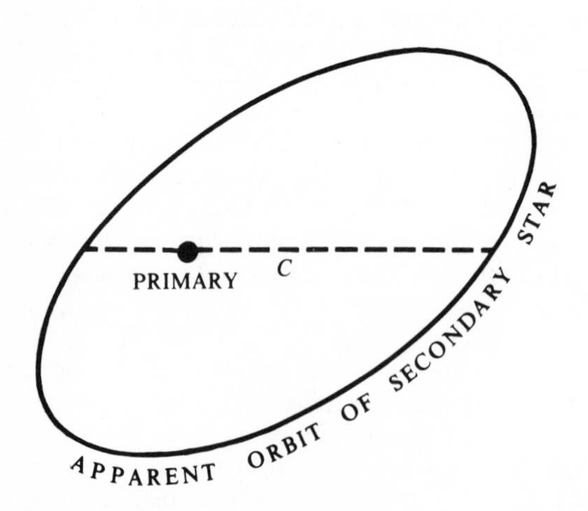

FIG. 26–2 *The observed apparent orbit of a visual binary.*

26.7 Periods of Visual Binaries

Since only the data of observed orbits are available to us, methods have been devised for determining the true orbit from these data. Observed orbits are available for only a small number of binary systems. Even though for many binaries the positions of the components have been determined from observations made during the past 100 years, only for a few of these have true orbits been calculated. For only about 235 binaries are the observations sufficiently reliable to allow the determination of true orbits. These have a large range of periods, the longest of which is 10 850 years for the system σ_2 Ursa Majoris. The components of this system have a mean separation of 500 A.U. On the other hand, the system δ Equulei has a period of 5.7 years, and 13 Ceti a period of 6.8 years. One of the shortest known periods, 1.8 years, is that of ξ Ursae Majoris.

The greatest number of observable orbits are those with periods between 25 and 100 years. Although there are undoubtedly systems with periods in excess of 11 000 years, they are extremely difficult to observe and, in general, orbits with periods of 200 years or more are not at all reliable. Unless they are very close to us, stars having periods less than 2 years are too close together to be studied as visual binaries and must be investigated spectroscopically.

26.8 The True Orbit

If a good apparent orbit is available for a visual system, a true orbit can be computed geometrically; it is described in terms of the same elements that are used in the description of planetary orbits (see Section 12.7). Thus, we have the period in years, the eccentricity of the true orbit, the semi-major axis of the true orbit, and so on. However, instead of introducing the inclination of the plane of the orbit to the plane of the ecliptic, as we did in the case of planets, *we introduce the angle of inclination of the plane of the orbit to the plane of the sky.*

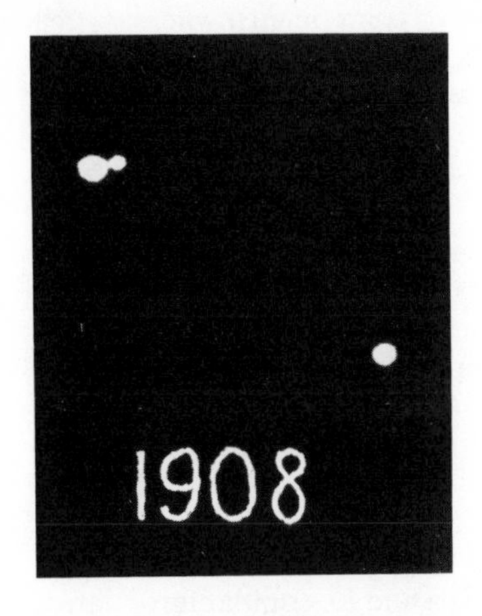

FIG. 26–3 (*a*) *Combination of three plates of the double star Krueger 60 showing conspicuous orbital motion (1908, 1915, 1920). Photograph by Barnard. (Yerkes Observatory photograph)*

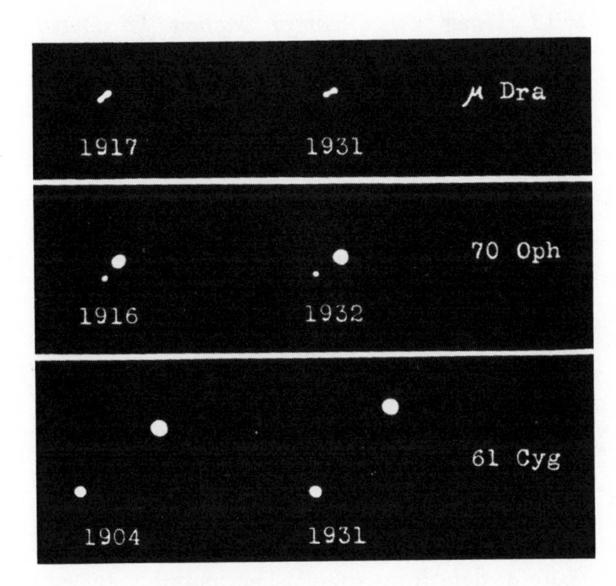

FIG. 26–3 (*b*) *Orbital motion of three visual binaries during 14, 16, and 27 years, respectively. (Yerkes Observatory photograph)*

The nodes are then defined as the points where the true orbit intersects the plane of the sky.

The line in the plane of the sky connecting the two nodes (called the line of nodes) and any line in the plane of the true orbit parallel to the line of nodes remain unshortened when projected onto the plane of the sky, but a line in the plane of the true orbit perpendicular to the line of nodes is foreshortened by the factor cos *i* (*i is the angle of inclination*) *on being projected onto the sky.*

26.9 Orbital Characteristics

Fairly good orbits have been computed for over 100 visual binaries by various methods of calculation, and they show a wide variety of characteristics. In general, the eccentricities are fairly large, with most of them clustering around 0.5; 10 per cent of the computed orbits have eccentricities less than 0.2 and 10 per cent have eccentricities larger than 0.85. Three known visual binaries, ξ Bootes, $O\Sigma$ 341, and γ Virginis have eccentricities greater than 0.89. The apparent separations of observed visual binaries have a wide range going from 0".16 to about 17", with 83 per cent of them less than 2".

There is a definite correlation between the number of visual binaries and spectral class, as shown in Table 26.1. 33 per cent of all known components of binary systems are G-type stars, whereas fewer than 2 per cent are B-type stars. Of course the small number of B-type stars may be a statistical result because of selection, since most B stars are at such great distances that we are bound to miss many visual binaries containing these stars.

TABLE 26.1

PROPORTION OF OBSERVED VISUAL DOUBLE PAIRS IN EACH SPECTRAL CLASS (APPARENT MAGNITUDE < 9)

Spectral class	O	B	A	F	G	K	M
Per cent visual doubles		2	8	6	9	4	2

26.10 Spectra of Visual Binaries

In discussing the spectra of visual binaries we must differentiate between systems in terms of whether the angular separation between the two components is large enough to enable us to see the spectra of both stars. In the case of a close pair the spectrum refers to the combined light. Since generally one star is brighter than the other, the spectrum is essentially that of the brighter component.

In Tables 26.2 and 26.3 the elements of the orbits and other characteristics of binary systems are given.

26.11 The Sum of the Masses of the Components of a Binary

In Chapter 21 we saw that if we can determine the orbit of a binary system, the sum of the masses of the two components can be found from Kepler's third law. In order to do this, all we need know is the true orbit of one component relative to the other. In Table 21.1 we have given the masses of some binaries, and we see that (as we have already mentioned in Section 21.11) the masses and luminosities of stars are correlated. Another interesting conclusion that can be drawn from this table is that binary systems do not differ much in mass. This is borne out by the fact that the periods and the true separations between the components increase together; the longer the period, the greater is the separation between the two components. In other words the sum of the masses of the two components remains very nearly the same from binary to binary.

26.12 The Ratio of the Masses of the Components of a Binary: The Motion Relative to the Center of Mass

Although the sum of the masses of the components of a binary system can be found from the relative orbit alone, we require additional information to find the ratio of the masses. We must know how each of the two components moves with respect to the center of mass of the system. We can do this if we determine the position (R.A., δ) of each component of the binary separately or if we determine the orbit of each component with respect to nearby stars. The center of mass of this system moves along the same straight line regardless of the position of the two components, whereas each component oscillates from side to side of this straight line. The two components are always on opposite sides of the center of mass at distances from it whose ratio is equal to the inverse ratio of their masses.

It was, in fact, because of Sirius' oscillation that Bessel, in 1834, suspected that it has a companion. After observing it for ten years, he concluded that it is a binary system with a period of 50 years. However, not until 1862 did the American telescope maker, Alvan Clark, observe the faint companion which we now know is a white dwarf. Observations of this component show that Bessel's determination of the period is correct. In Figure 26.4, the orbits of Sirius and its companion relative to the center of mass are shown.

26.13 Using the Observed Motion To Detect Binaries *

This method of picking up visual binaries applies to systems in which one companion is too faint to be observed visually. In such a system the proper motion of the bright companion is oscillatory instead of being linear. In general, it is difficult to pick up binaries of this sort since the apparent motion of the bright companion relative to the center of mass is quite small. But as we reduce the errors of our instruments and improve our measuring techniques, more and more such systems will be discovered.

As an example we may note that Van de Kamp has succeeded in reducing the mean error in the photographic position of a star to $\pm 0\overset{''}{.}01$. As a result he has found that certain stars that previously had been considered to move along a straight line really deviate from straight-line motion. Indeed, it is now possible by using these accurate methods to measure the oscillation imparted to a star by a retinue of planets. Thus, in the case of the double stars 61 Cygni, 70 Ophuichi, and CI 1244, small deviations from elliptical motions were observed which indicate the presence of other members in these systems with masses about equal to that of Jupiter. In 1963, Van de Kamp discovered a small invisible planetary companion to Barnard's star. The estimated mass of this planet is about $0.0015\mathscr{M}$, or 50 per cent larger than the mass of Jupiter.

In considering such data as dynamical parallaxes, luminosities, etc., of binary systems, we must keep in mind that selection plays an important part. Thus, for example, the mean parallaxes of binary

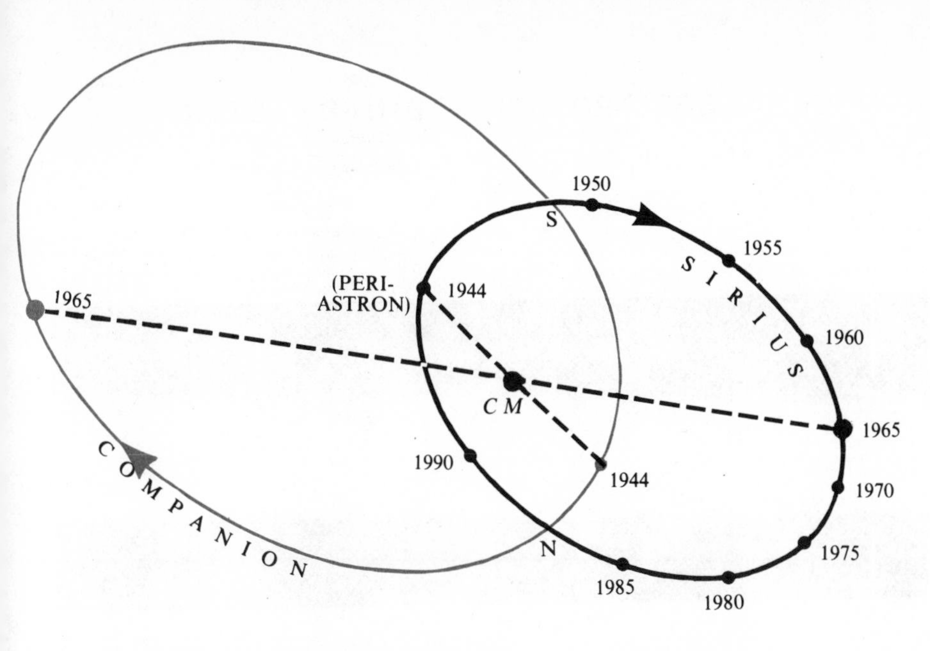

FIG. 26–4 *The orbits of Sirius and its companion relative to the center of mass.*

systems are, in general, greater than those of all other stars of the same mean apparent magnitude. This clearly arises because the resolving powers of our telescopes limit us to the nearby binaries.

26.14 Spectroscopic Binaries

The discovery of spectroscopic binaries is relatively simple since it is based on detecting periodic variations in the Doppler shift of the spectral lines (i.e., in the radial velocities) of these stars. In this way E. C. Pickering first discovered, in 1889, that Mizar (ζ Ursae Majoris) is a spectroscopic binary. More than 1 300 stars with variable radial velocities have been found and we must conclude that the majority of these are probably binaries. *In a binary system of this sort the two companions are so close together that they appear as a single star to the telescope, and unless their orbits are at right angles to the line of sight, their periodic Doppler shifts can be observed.* Although variations in radial velocities have been detected in many stars, sufficiently detailed observations of the spectral lines to allow us to eliminate all doubt about their binary character and to enable us to calculate orbits are available for only about 600 such systems.

If both stars of a spectroscopic binary are bright enough, we obtain two sets of lines that oscillate opposite to each other as shown in Figure 26.5. *Periodically, when the line connecting both stars is at right angles to the line of sight, one star is approaching and the other receding, so that the spectral lines of one star are displaced relative to those of the other by an amount proportional to the relative velocity of the two stars. A quarter of a period later, when the line connecting the two stars is parallel to the line of sight, only one set of lines is observed, and another quarter of a period later the two sets are again displaced but in opposite directions.*

As long as the two stars are equally bright, this periodic doubling of the lines is observed, but if one of the components is more than one magnitude brighter than the other, only the lines of the bright component can be detected, and these oscillate back and forth about mean positions.

26.15 Statistics of Spectroscopic Binaries

It is estimated that about 20 to 25 per cent of all the bright stars, and perhaps 35 per cent of the B-type stars, are spectroscopic binaries, and Otto Struve has gone as far as to suggest that all Wolf-Rayet stars are such pairs. On this basis we may conclude that there are probably 33 000 such systems down to the ninth apparent magnitude. The periods of most of these systems are less than 5 years, with the star VV Pup having the shortest known period, 1 hr. 40 min., although if we include the binaries belonging to the explosive variables, recently studied by R. Kraft (see Sections 24.37 and 24.34), there are even shorter known periods than

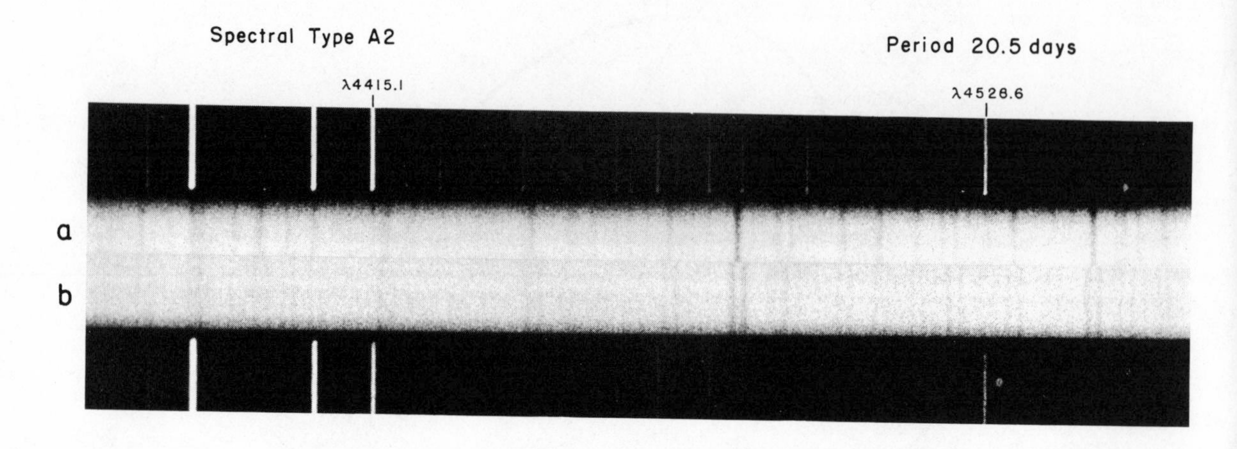

FIG. 26–5 *Spectrum of a spectroscopic binary star, showing shifts of lines towards violet and red on two exposures. The doubling of the lines is also shown. (From Mount Wilson and Palomar Observatories)*

this. The spectroscopic binary with the longest known period, 15.3 years, is ξ Hydra. The periods of 56 per cent of the known spectroscopic binaries are less than 10 days and the periods of 26 per cent lie between 10 and 100 days.

Both giants and dwarfs are found among spectroscopic binaries, with many mixed systems as well, consisting of a giant and a dwarf star. Spectroscopic binaries are also found among all spectral classes but there is an important relationship between period and spectral class (see Table 26.2). If we divide these binaries into two groups, one group with periods shorter than 20 days and the other with longer periods, we see that the short-period systems (i.e., those systems in which the components are close together) are found mostly in spectral classes O through F, whereas the long-period systems are found mostly in spectral classes F, G, and K.

26.16 The Velocity Curves of Spectroscopic Binaries

Before we can determine the orbital characteristics of a spectroscopic binary, we must derive its velocity curve from the spectral data. We do this by measuring the displacement of a given spectral line with respect to its normal position and calculating the radial velocity of the star at that moment. During a single cycle the displacement of the spectral line varies so that a complete set of radial-velocity values can be obtained during this cycle. If these radial velocities are then plotted against the time, the velocity curve is obtained. If the spectra of both components can be observed (the lines are periodically doubled) two velocity curves of exactly the same shape but of opposite phase are obtained. *The amplitudes of these two curves are different since the masses of the stars and*

TABLE 26.2

TWENTY SPECTROSCOPIC BINARIES

Name	Apparent magnitude	Spectral class	Period in days	Eccentricity e	Orbital velocity in km/sec	$a \sin i$ Millions of km	$\frac{m_1 \sin^3 i^*}{m_2 \sin^3 i}$	$\frac{m_2^3 \sin^3 i}{(m_1+m_2)^3}$	Parallax	Absolute magnitude
+6° 1 309	6.36	O8	14.41	0.04	206 247	40.9 48.9	76. 63.	13.2		
η Ori	3.44 eclipsing	B1	7.99	0.02	145 152	15.9 16.8	11.2 10.6	2.51	0".0057	−2.0 −2.0
β Sco	2.90 eclipsing	B1	6.83	0.27	126 197	11.4 17.8	13.0 8.3	1.26	0.0085 ±.0014	−1.7 −1.7
ψ Ori	4.66	B2	2.53	0.07	144 190	5.0 6.6	5.5 4.2	0.78	0.0057	−0.8 −0.8
α Vir	1.21 eclipsing	B2	4.01	0.10	126 208	6.9 4.4	9.6 5.8	0.83	0.014 ±.003†	−2.6 −2.2
ζ Cen	3.06	B2	8.02	0.50	312‡	29.8‡	16.4		0.0217 ±.0014	+0.5 +0.5
μ Sco	3.09 eclipsing	B3	1.45	0.05	480‡	9.5‡	16.5		0.0074 ±.0017	−2.2 −1.8
β Aur	2.07 eclipsing	A0	3.96	0.00	109 111	5.93 6.04	2.21 2.17	0.55	0.034 ±.006	+0.6 +0.6
ζ UMa A	2.40	A2	20.54	0.54	69.2 6.88	16.4 16.4	1.67 1.66	0.41	0.045 ±.002	1.4 1.4
o Leo	3.76	F5	14.50	0.0	54.0 63.1	10.8 12.6	1.30 1.12	0.24	0.026 ±.007	1.6 1.6
α Aur	0.21	G0	104.02	0.01	25.8 32.5	36.8 46.4	1.19 0.94	0.18		
29 CMa	4.90	Oe	4.39	0.16	218	13.0		4.58	0.0071	−0.8
ι Ori	2.87	Oe5	29.14	0.75	110	28.9		1.14	0.0051	−3.6
β Per	2.1 eclipsing	B8	2.87 688	0.04 0.13	44.1 10.0	1.73 93.0		0.025 0.070	0.027 ±.010	−0.7
β Ari	2.72	A5	107.0	0.88	32.6	22.9		0.042	0.064 ±.006	1.7
β Cap	3.25	G0	1375.3	0.44	22.2	377.		1.13		
κ Peg A	4.8	F5	5.97	0.03	40.5	3.36		0.040		
ξ UMa A	4.41	G0	665	0.41	7.0	58.3		0.018		
α Gem A	1.99	A0	9.22	0.50	13.6	1.49		0.0015		
α Gem B	2.85	A0	2.93	0.01	31.8	1.28		0.0097		

* Interferometer measures give $i = 50°$, $m_1 = m_2 = 3.7$.
† Parallax, if the star is a member of Kapteyn's Scorpius-Centaurus group.
‡ Relative orbital velocity and mean distance of the two stars, from measures on objective-prism plates.

Table data adapted from H. N. Russell et al., Astronomy. Boston: Ginn & Co., 1955, vol. 2, p. 706.

hence their velocities relative to their center of mass are different. If the two stars are of equal mass, the velocity curves are exact mirror images of each other. If the spectral lines of only one star are detectable, there is only one velocity curve.

In Figures 26.6 and 26.7, we see that the velocity curves of various binaries differ considerably in shape. This arises from the different eccentricities of the orbits and differences in orientations of the major axes with respect to the line of sight. We may illustrate this by considering three examples, one with $e = 0$, and two with $e = 0.5$. In these examples the straight horizontal line represents the motion of the center of mass of the system, the ordinate represents the velocity in km/sec, and the time is plotted along the abscissa.

In the first case, because the orbit is circular and hence completely symmetrical, the velocity curve is a sine curve. To determine the shape of the velocity curve for $e \neq 0$, we make use of the fact that in the true ellipse which the star describes about the center of mass, the star must sweep out equal areas in equal times in accordance with Kepler's second law of motion. *We also note that the points of intersection of the velocity curve and the straight line representing the motion of the center of mass correspond to the points on the orbit when the star is moving transverse (at right angles) to the line of sight, and hence has no radial velocity component. The points on the velocity curve that are farthest above and below this straight line are the points on the orbit where the star has its maximum velocity of recession and maximum velocity of approach, respectively.* These points on the true orbit, therefore, determine a line, the so-called **line of nodes**, which is at right angles to the line of sight. Since the points themselves are the nodes, the star has its maximum radial velocities at the nodes. The reason for this is given in the next paragraph.

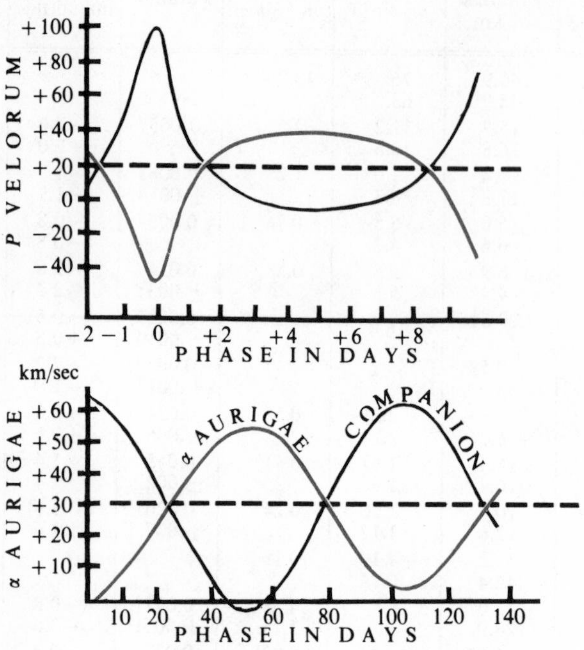

FIG. 26–6 *Typical velocity curves of some spectroscopic binaries. (After H. N. Russell et al.,* Astronomy. *Boston: Ginn & Co., 1955, vol. 2, p. 697)*

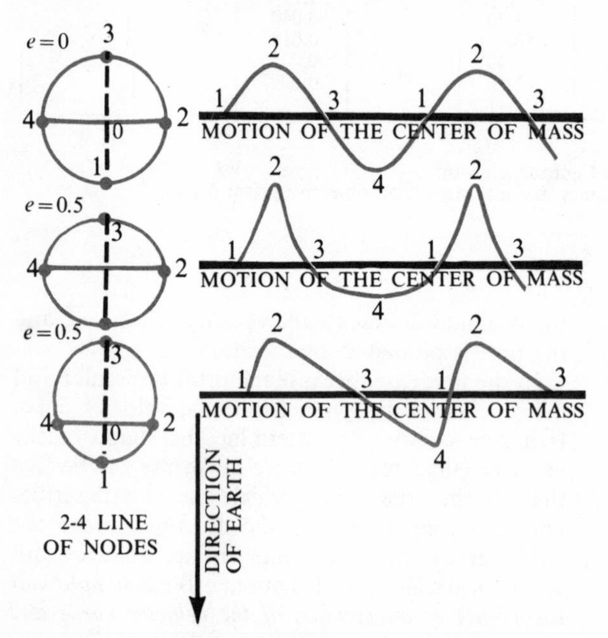

FIG. 26–7 *The shape of the velocity curve depends on the eccentricity of the orbit and the orientation of the major axis with respect to the line of sight. (After H. N. Russell et al.,* Astronomy. *Boston: Ginn & Co., 1955, vol. 2, p. 698)*

When the star is at either node, it must be exactly in the plane of the sky, and this must also be true of its companion because the center of mass of the binary lies on the line of nodes and hence also in the plane of the sky (the line connecting one component of a binary with the center of mass must pass through the other component as well). Thus, when the star is at either node, the line connecting it to the other component is perpendicular to the line of sight so that the gravitational force at this moment lies along this line of nodes and hence has no component along the line of sight. Therefore, the component of the acceleration of the star along the line of sight vanishes when the star is at either node. In other words, the acceleration changes its direction as the star passes across the plane of the sky as shown in Figure 26.8. If, then, we consider the component of the star's velocity perpendicular to the plane of the sky (i.e., the velocity along the line of sight or its radial velocity), we see that it increases up to the time the star reaches the node (acceleration positive) and decreases (acceleration negative) immediately after it passes the node, with the maximum velocity at the node itself.

26.17 Circular Orbits of Binaries

We can now use this analysis together with Kepler's second law to determine the shape of the velocity curve. If the orbit is circular, the velocity curve, as we have already noted, is sinusoidal regardless of the orientation of the orbit, as can be seen from Figure 26.9(a), where the line *cd* represents the line of nodes, and the points *a*, *b* the points of zero radial velocity. The true orbits and their orientation are shown to the left. The points on the velocity curve corresponding to the true orbital points *a* and *b* lie on the velocity line of the center of mass as shown on the figure, whereas points *c* and *d* are at maximum distances from this zero-velocity line. Since in the case of a circular orbit the star spends the same time in going from *a* to *c*, *c* to *b*, and *b* to *d*, the curve above and below the horizontal line must be identical and hence sinusoidal.

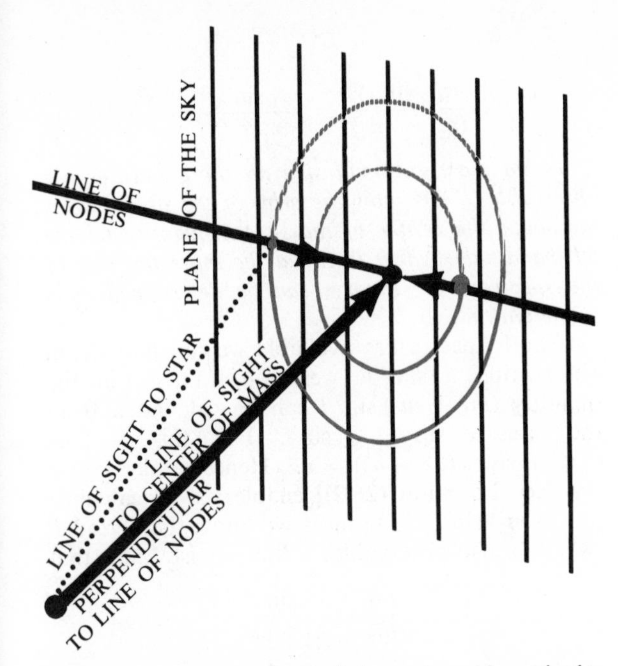

FIG. 26–8 *When a component of a binary is at either node, this star has no acceleration along the line of sight.*

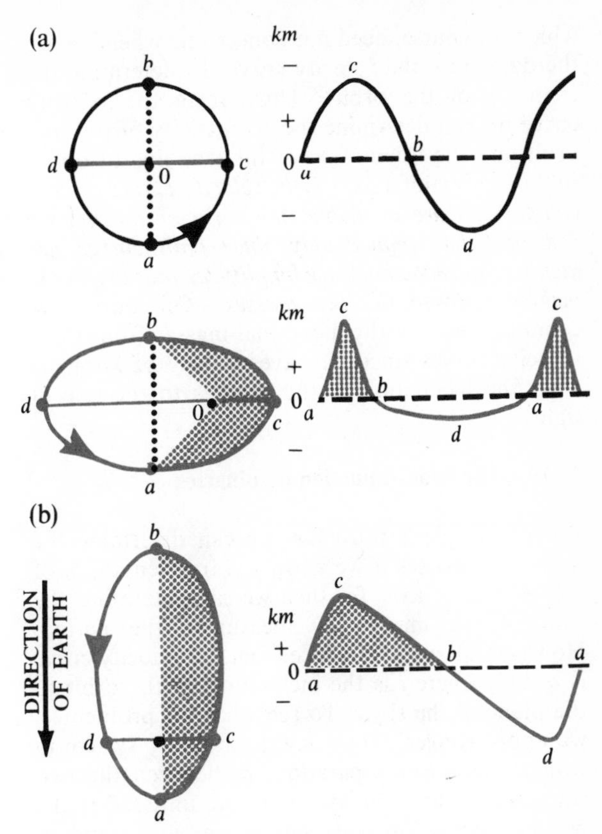

FIG. 26–9 *Applying Kepler's Third Law of Motion to the orbit of a binary to determine the shape of the velocity curve.*

26.18 Elliptical Orbits

In Figure 26.9(b) an elliptical orbit is shown in which the line of nodes coincides with the major axis, so that the semimajor axis points towards the observer. Since in this case the area of the segment *oacb* of the true ellipse is smaller than the area *obda* (where *a* and *b* are at the ends of the minor axis and *c* and *d* are at the ends of the major axis), the time interval from *a* to *b* along the time axis is smaller than the interval from *b* to *a* in the ratio that the area *oacb* is smaller than *obda.* This is shown on the diagram and we see that the part of the velocity curve above the horizontal time line (or zero-velocity line) is quite different from that below this line. It should be noted, however, that the area under the stretch *acb* of the velocity curve is exactly equal to the area above the stretch *bda.* The reason is that the area under *acb* represents the distance *a* to *b* along the orbit (that is, the arc *acb*) traveled by the star and the area above *bda* represents the distance *b* to *a* along the orbit. These two distances are equal by construction. Note also that since the area *aoc* is exactly equal to the area *ocb*, the time interval from *a* to *c* is exactly equal to the time interval from *c* to *b* and the interval from *b* to *d* is exactly equal to the time interval from *d* to *a.*

In the final example the orbit is oriented so that the line of nodes *cd* (now the minor axis) is perpendicular to the major axis (the line *ab*), and we observe now that the time interval from *a* to *b* is equal to the time interval from *b* to *a* since the area *oacb* is one-half of the area of the ellipse. We see that the upper branch of the velocity curve is the mirror image of the lower branch. However, the time from *a* to *c* is smaller than from *c* to *b* in the ratio in which the area *oac* is smaller than *ocb.* The same holds, of course, for the time intervals *bd* and *da.* If the orientation of the orbit is such that the semimajor axis is at some angle with respect to the line of nodes different from 0° or 90°, the shape of the velocity curve lies between the two shown. It is also clear that the amount by which the velocity curve differs from a sinusoidal curve depends on the eccentricity.

The height of the "zero"-velocity line above the abscissa line represents the velocity of the entire system (that is, the velocity of its center of mass) which, of course, need not concern us when we use the data from the velocity curves to determine the elements of the orbit. Thus, from the velocity curve we can determine the eccentricity of the true orbit and the orientation of the true line of nodes to the semimajor axis. *If the velocity curves of both components are available, the ratio of the masses can be found immediately since both curves are identical in shape but have amplitudes that are in the inverse ratio to the two masses.* Of course, we cannot determine the individual masses from these velocity curves since we have no way of knowing what the inclination of the orbit is to the line of sight.

26.19 The Mass Function of Binaries

From Kepler's third law we can determine the individual masses if we know the true length, a, of the semimajor axis, for then we can determine the sum of the masses by measuring the period. However, all we can deduce from the velocity curve is $a \sin i$, where i is the inclination of the orbit to the plane of the sky. To see what the problem is, we apply Kepler's third law to a binary system in which the mean separation, a, between the two stars is measured in astronomical units, P is the period expressed in years, and m_1 and m_2 the individual masses expressed in solar masses. We then have

$$m_1 + m_2 = \frac{a^3}{P^2}.$$

Since we cannot find a from the velocity curve, but only $a \sin i$, we introduce $\sin i$ into this equation by multiplying both sides by $\sin^3 i$ so that we get

$$(m_1 + m_2) \sin^3 i = \frac{(a \sin i)^3}{P^2}. \qquad (26.2)$$

Since a refers to the mean separation between the two components, and since, in general, only one spectrum is observable, it is convenient to introduce instead of a the mean distance a_1 from the center of mass of the system of the star whose spectrum is visible. Since

$$\frac{a_1}{a} = \frac{m_2}{m_1 + m_2}$$

we have

$$a = \frac{a_1(m_1 + m_2)}{m_2}.$$

If we substitute this in Equation (26.2), we have

$$(m_1 + m_2) \sin^3 i = \frac{(m_1 + m_2)^3 a_1^3 \sin^3 i}{m_2^3 P^2},$$

or

$$\frac{(m_2 \sin i)^3}{(m_1 + m_2)^2} = \frac{(a_1 \sin i)^3}{P^2} = \frac{s^3}{P^2}.$$

Since on the right-hand side all the quantities are observable (if we know the orbit of the star of mass m_1 about the center of mass), the quantity on the left-hand side, which is called the mass function of the spectroscopic binary, can be determined as a single entity.

If both spectra are available we can find (from observation) $a_1 \sin i$ as well as $a_2 \sin i$, so that the quantity $(m_1 + m_2) \sin^3 i$ can be calculated from the sum $a_1 \sin i + a_2 \sin i$. For this is just $(a \sin i)$ since $(a_1 + a_2) = a$. Hence Kepler's third law [see Equation (26.2)] enables us to calculate $(m_1 + m_2) \sin^3 i$ as soon as we know the period P (which is also observable). But we also know that

$$\frac{m_1}{m_2} = \frac{a_2 \sin i}{a_1 \sin i}$$

so that the mass ratio is known. Hence we can calculate $m_1 \sin^3 i$ and $m_2 \sin^3 i$ separately. If i is known, as for an eclipsing binary, m_1 and m_2 can be found separately. (See Table 26.2).

26.20 Determining the Inclination Statistically

We cannot determine i itself unless the system is a visual binary in which both components can be observed or is an eclipsing binary. In the case of an eclipsing binary we know that the angle i is very nearly 90°, and in many such cases the actual tilt can be determined accurately from the light curve. We can then obtain the masses of the two components with very little error; systems of this sort give us the most reliable stellar mass data.

It is possible, however, to obtain average values for $\sin^3 i$ even when we are dealing with a spectroscopic binary for which the spectrum of only one of the components is available. An analysis of velocity curves of many spectroscopic binaries shows that tilts of the orbits with respect to the plane of the sky are distributed at random. However, this is not equivalent to saying that all values of i are equally probable. For it can be shown that values of i near 90° are much more likely to be found than values near 0°.

We can see why this is so if we note that i is just the angle between the line of sight to the star and the perpendicular to the plane of the orbit as shown in the Figure 26.10. Let us consider, then, a certain region, S, of the sky whose direction is given

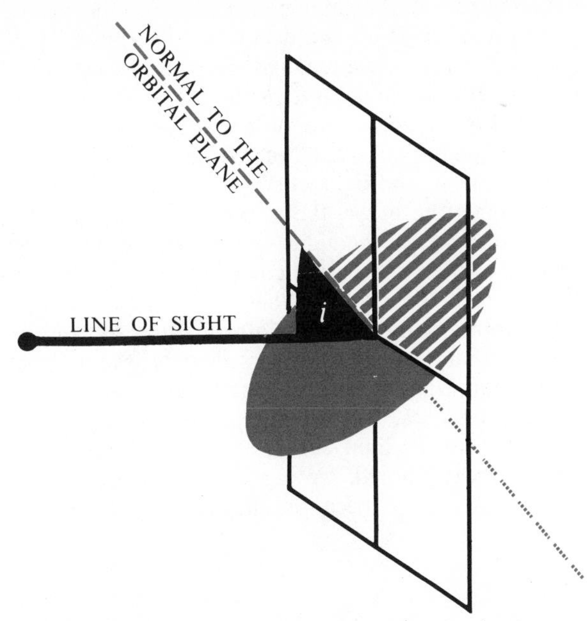

FIG. 26–10 *The tilt, i, of the orbit of a binary with respect to the line of sight.*

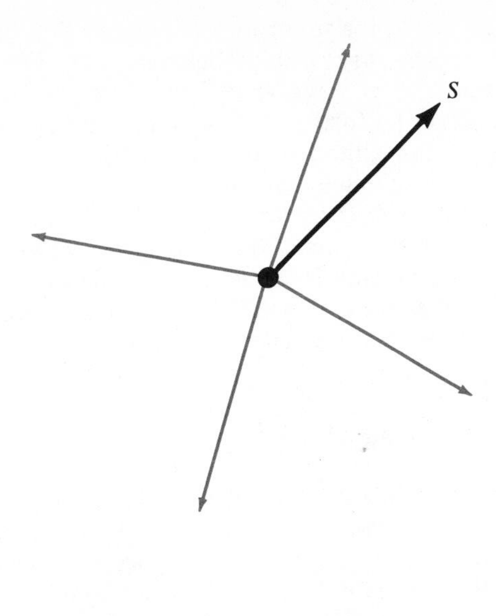

FIG. 26–11 *Vectors are used to represent the orientations of orbits of binaries.*

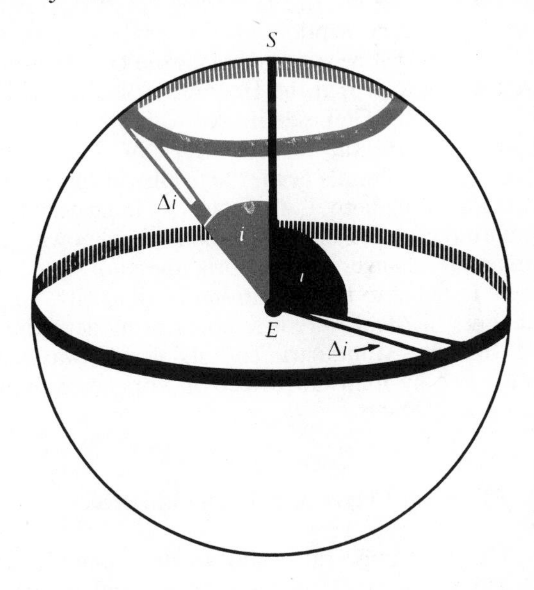

FIG. 26–12 *Even though the inclinations of the orbits of binaries are distributed at random, not all values of the tilt, i, of the orbit with respect to the plane of the sky are equally probable, as shown by the analysis of this figure.*

by a line from the earth drawn to that region. We may represent the plane of the orbit of any binary in this region by a vector starting from the earth as the center of the celestial sphere and terminating at the surface of the sphere as shown in Figure 26.11. Since the tilts of the orbits are distributed at random, the various orbital planes are represented by vectors radiating out from the earth in all directions. However, even though all possible directions are represented and no direction is to be preferred above any other, still it is clear that if we examine binary systems in the neighborhood of point S we find many more with a large angle i than with a small angle i.

To see this we consider all binary systems for which the tilt lies between some particular value of i and $i + \Delta i$, as shown in Figure 26.12. All such systems are represented by vectors whose end points lie in the small circular strip (on the surface of the celestial sphere) whose area can be shown to be proportional to $2\pi \sin i \, \Delta i$. Obviously this area represents the probability of finding binaries with the tilts of their orbital planes close to i (lying within a small neighborhood Δi of i). We see that because of the presence of $\sin i$ in this formula this number gets smaller as i gets smaller, so that the chance of finding orbits with i close to zero, i.e., orbits perpendicular to the line of sight, is much smaller than that of finding orbits with their planes parallel to the line of sight.

By this type of analysis we can show that as many values of i lie between 60° and 90° as lie between 0° and 60°, and that the mean value of $\sin^3 i$ is just $3\pi/16$. By using this mean value of $\sin^3 i$ we can obtain mean values for m_1 and m_2 (from the previously calculated values of $m_1 \sin^3 i$

and $m_2 \sin^3 i$) for spectroscopic binaries (for which both spectra are visible) belonging to different spectral classes. We find that these quantities range from 1.2 for F and G type binaries to 70 for O type, and although, in general, the two components of a binary are not of equal mass, the ratio of masses is very nearly 1, with the mean value of the ratio being about 0.8. However, we do find spectroscopic binaries in which the ratio of masses is as low as 0.28. Some of the data for spectroscopic binaries are given in Table 26.2.

26.21 Eclipsing Binaries

When the orbital plane of a binary system passes through the earth, or very nearly so, it may happen that the two stars eclipse each other periodically, and the apparent brightness of the system then varies in a predictable manner. Although as a group such stars are classed as variables, they are not intrinsic variables.

The best example of an eclipsing binary is β Persei or Algol (a corruption of the Arabic name Al Ghoul, hence it became known as the Demon Star). According to Kopal the Hebrews called it Satan's Head, and the Chinese referred to it as "Piled Up Corpses." In the constellation of Perseus it represents Medusa's head, the trophy of this Greek mythological hero. All of this indicates that before the nature of its variations was known, the apparent changes in the brightness of the star were imputed to the operation of evil spirits. The changes in Algol were first noted in modern times in 1670 by Montanari, and its period was first measured by John Goodricke of York, who found it to be $2^d20^h49^m$.

26.22 Light Curves of Eclipsing Binaries

The most important thing to determine for an eclipsing binary is its light curve, and we may illustrate the nature of such a curve by considering Algol. This binary remains at constant brightness for 2 days and 11 hours and then steadily declines in brightness for 5 hours until it is one-third its normal brightness. It then reaches its maximum brightness again after about 5 hours, and this cycle repeats itself. The variations that give the principal features of the light curve are the result of the periodic eclipsing of the bright star by the faint component. In addition to these changes there is a slight dip in the light curve midway between any two successive principal eclipses. This is due to the eclipse of the faint component by the bright component. If all the details of the variation of the light are measured and plotted against time, we obtain the light curve of the binary itself.

In Figure 26.13 five different light curves are given including that of Algol. The deeper minimum in the light curve is called the **primary minimum**; the other, the **secondary minimum**. An examination of the light curves shows that they vary in shape and period considerably from binary to binary. By studying these light curves we can determine many of the orbital features as well as the geometries of the components themselves. Since eclipsing binaries, in general, are also spectroscopic binaries, we can obtain velocity curves in addition to the light curves and from these, as we have already noted, we can determine the masses very accurately because $\sin i$ is approximately equal to 1.

26.23 The Elements That Determine the Light Curves

It is clear from a consideration of the orbit of an eclipsing binary system that the shape of its light curve is determined by the following factors:

1. the shape of the relative orbit;
2. the sizes of the components of the binary;
3. the orientation of the major axis of the orbit with respect to the line of sight;
4. the ratios of luminosities of the two components;
5. the departure from sphericity of the two stars;
6. the sizes of the relative orbits;
7. the reflection effect; and
8. the limb darkening.

Most of these influences on the shape of the light curves are present in the examples we have given. However, to understand these things more fully, we shall consider some of them in more detail.

The simplest case is one in which the orbit is exactly circular, the plane of the orbit is exactly parallel to the line of sight, both stars are equally luminous and both are of the same size. The primary and secondary minima are then identical and are equally spaced in time. The period is equal to twice the time between two successive minima and, since both stars are equally bright, the minima are exactly equal. Since the total eclipse occurs when one star exactly covers the other, the

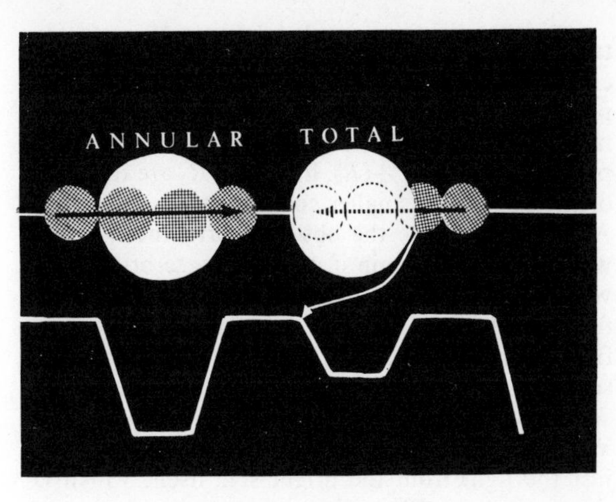

FIG. 26–14 *The shape of the light curve depends upon the relative luminosities of the two components.*

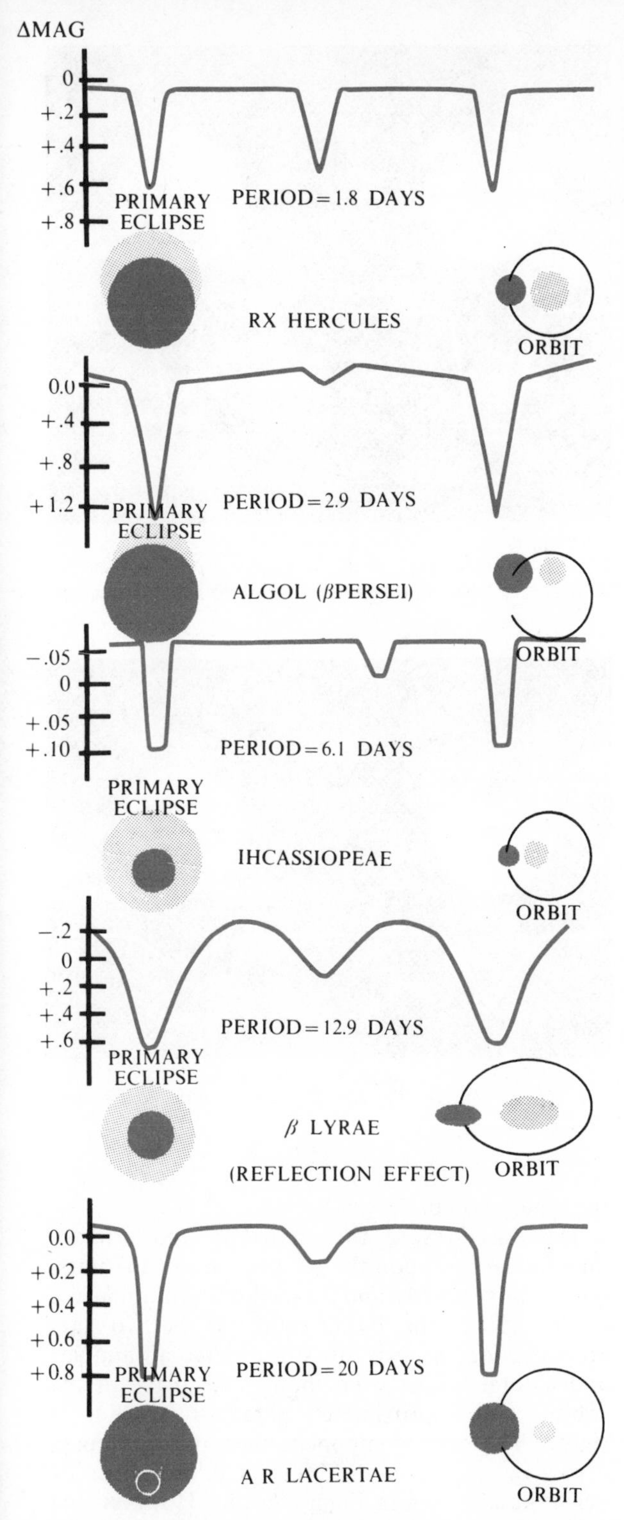

FIG. 26–13 *Light curves of some typical eclipsing binaries, including Algol. The ordinate represents the change in apparent magnitude from a standard value indicated as zero. (Adapted from O. Struve et al.,* Elementary Astronomy. *New York: Oxford University Press, 1959)*

minimum is a sharp point because totality lasts only for a moment. Although in a circular orbit the secondary minimum always occurs midway between the two primary minima, this can also occur for elliptical orbits provided the semimajor axis is properly oriented to the line of sight.

If we now consider a circular orbit again with its plane exactly parallel to the line of sight, but with one star hotter and larger than the other, as shown in Figure 26.14, the shape of the light curve, in general, is altered in two ways. Although the secondary minimum still lies midway between the two primary minima, it is not as deep as the primary minimum. Moreover, the secondary and primary minima are flat instead of being pointed. This is so because the primary minimum occurs when the small, cool star passes across the face of the large, hot star, and—as can be seen from the figure—the eclipse at this time is annular (and remains unchanged during the time the small star moves across the large one). The total light that reaches us then is both that coming from the surface of the faint star and that from the annular part of the bright star. On the other hand, during the secondary minimum the eclipse of the smaller, faint star is total. But it again lasts for the time it takes the small star to pass behind the large, brighter companion, so that the secondary minimum is also flat.

26.24 The Light Curve and Radii of Eclipsing Binaries

If the two stars have the same surface temperature, we can draw some conclusions about their radii from the depths of the minima in the light curve. The luminosity of each star is proportional to the square of the radius, and the maximum luminosity when both stars are visible is proportional to $R_1^2 + R_2^2$, if R_1 and R_2 are the radii of the larger and smaller components respectively. At primary eclipse the observed brightness is just what we would obtain if only the large bright star were visible, hence it is proportional to R_1^2. This follows because, although the smaller star is in front of the large one and blocks out some light, the smaller star still sends its own total light to us, supplying the amount of light that it blocks out. When the faint star is totally eclipsed we again see just the light from the bright star itself, which is proportional to R_1^2. Thus if the temperatures of these two stars are equal, the primary and secondary minima are also equal and this enables us to determine the ratio of the radii of the two stars. If we take the ratio of the brightness at maximum (between primary and secondary minima) to the brightness at either primary or secondary minimum we have, from what we said above,

$$\frac{b_{\text{max}}}{b_{\text{min}}} = \frac{R_1^2 + R_2^2}{R_1^2} = 1 + \frac{R_2^2}{R_1^2}.$$

We thus have the ratio of the radii of the two components in terms of the ratio of the brightnesses of the maximum and minimum of the light curve.

Although we have given the above analysis for two components of an eclipsing binary having the same surface temperature, we can apply the same kind of analysis to stars with different surface temperatures. If R_2 is smaller than R_1, we shall refer to the eclipse of the smaller star by the larger one as an **occultation**, and the eclipse of the larger by the smaller as a **transit**. If the apparent brightnesses of the two stars are b_1 and b_2 respectively, the total apparent brightness when there is no eclipse is $b = b_1 + b_2$. During the occultation, the brightness (b_{occ}) is just b_1. During the transit the brightness (b_{tran}) is just $b_1 - (R_2/R_1)^2 b_1 + b_2$, which is just equal to $b - (R_2/R_1)^2 b_1$. The reason for this is that during the transit the small star blocks out an area from the large star that is proportional to R_2^2, and replaces the light from this area by its own light. We therefore have, combining these results,

$$b_{\text{tran}} = b - (R_2/R_1)^2 b_1 = b - (R_2/R_1)^2 b_{\text{occ}}$$

or

$$(R_2/R_1)^2 = (b - b_{\text{tran}})/b_{\text{occ}}.$$

Since b, b_{tran}, and b_{occ} can be obtained from the light curve, we can find the ratio of the two radii of the binary components.

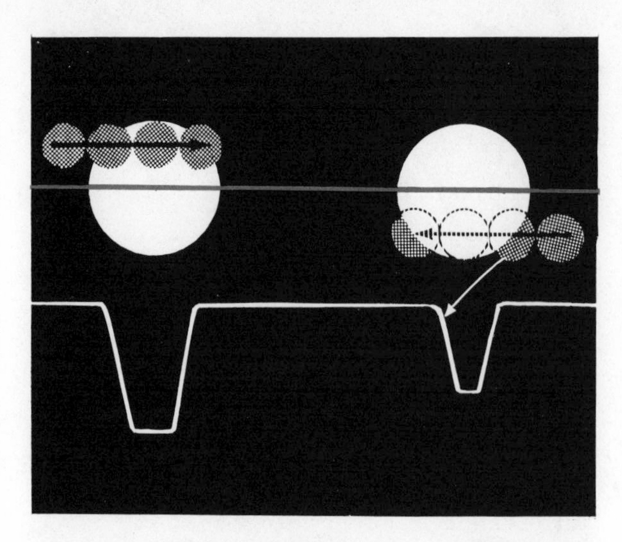

FIG. 26–15 *The shape of the light curve depends on how far off center the faint companion passes across the bright companion.*

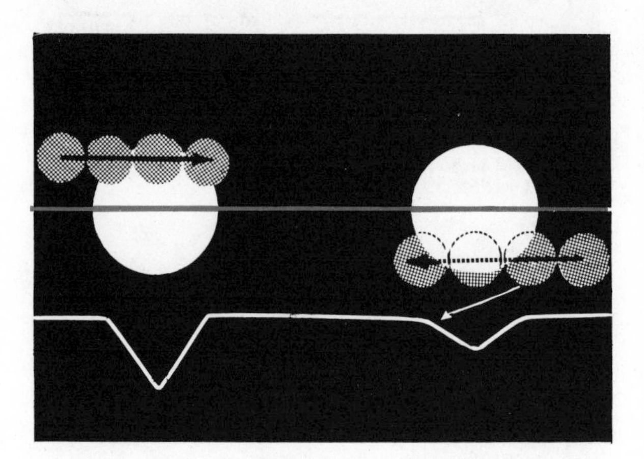

FIG. 26–16 *Partial eclipses occur if the tilt of the orbit is large enough.*

The durations of the secondary and primary minima depend upon the relative sizes of the stars, their relative speeds, and the inclination of the plane of the orbit to the line of sight. If the two stars are close together we can still observe an annular and a total eclipse even though the plane of the orbit is only approximately parallel to the line of sight. The faint component then does not pass across the center of the bright star but above or below it, as shown in Figure 26.15. However, the primary and secondary minima are still flat if the disk of the faint component passes completely behind the large star, i.e., if the two successive eclipses are **annular** and **total**. If the tilt of the orbit is sufficiently large, the eclipse is **partial**, as

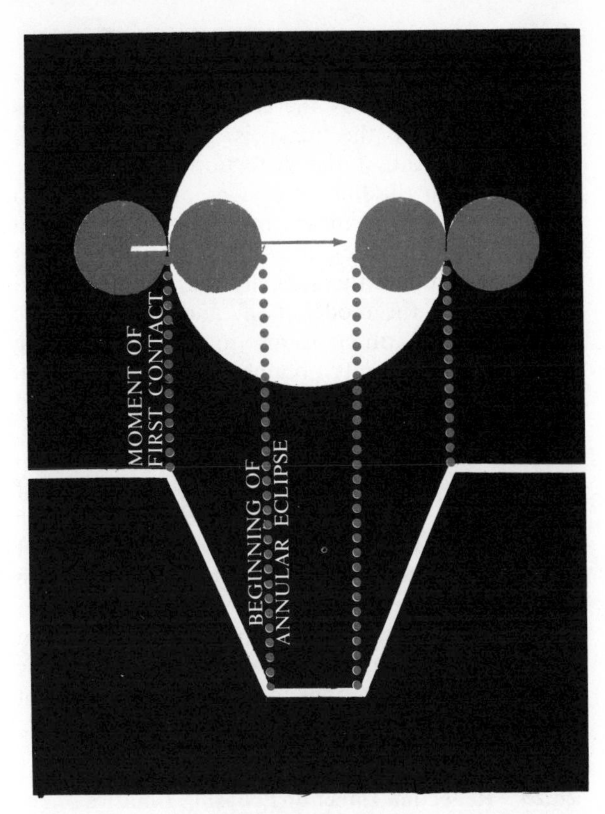

FIG. 26–17 *The radii of the components of a binary can be obtained from the light curve if the relative velocity of the two stars is known.*

shown in Figure 26.16. Therefore, the two minima will last but a moment (the moment at which the overlapping area of the two components is a maximum).

26.25 Determining the Radii of the Components of Eclipsing Binaries

If the velocity curve is known from spectroscopic data so that the relative velocity of one star with respect to the other can be obtained, we can determine the radius of each star separately. If we are dealing with an orbit whose plane passes through the line of sight, we know that the primary eclipse begins at the moment t_1 that the disk of the small star just touches the disk of the large star, and the annular eclipse begins at time t_2, when the disk of the small star has traveled a distance $2r_2$, if r_2 is the radius of the small star (see Figure 26.17). We thus have $(t_2 - t_1)v = 2r_2$, where v is the relative velocity of the two stars. Since the primary eclipse lasts until the moment t_3, when the advancing edge of the smaller disk just coincides with the farther edge of the large disk, it follows that

$$(t_3 - t_2)v = 2r_1 - 2r_2,$$

where r_1 is the radius of the larger component. Since the eclipse ends at a later time t_4 when the faint star is just past the large star, we have

$$(t_4 - t_3)v = 2r_2.$$

Adding all these together we obtain the total duration of the eclipse, $t_4 - t_1$, so that

$$(t_4 - t_1)v = 2(r_1 + r_2). \tag{26.3}$$

Thus if we know the relative velocity we can determine r_1 and r_2 separately from the shape of the light curve, provided the inclination of the line of sight to the orbit is equal to 90°. But if the orbit is not exactly parallel to the line of sight, a correction factor must be introduced. If the relative velocity v is not known, only $r_2/(r_1 + r_2)$ or the ratio of the radii can be found.

26.26 The Effect of Tidal Distortion of Eclipsing Binaries on Their Light Curves

In our analysis we assumed that the stars of the binary are exactly spherical, but this, in general, is not the case since the gravitational interaction between them causes tidal distortion in each one. This is quite pronounced if the stars are not too far from each other, and has two effects which we shall discuss separately. The first of these is related to the maximum brightness segment of the light curve, which in the case of spherical stars is represented by a straight line since then the combined brightness of the two stars when both are visible remains the same no matter where they are in their orbits. But if the stars are not spherical, their combined brightness varies as they move along their orbits even when both are visible; since the stars are ellipsoidal in shape, the area of the surface each presents to us changes. This gives rise to a rounded maximum and also to a rounded minimum as shown in the light curve of TX Cassiopeiae (see Figure 26.18).

26.27 The Apsidal Motion

The elliptical distortion has another important effect when the orbits are not circular. If we are dealing with elliptical orbits for eclipsing binaries, the position of the secondary minimum between the two primary minima depends on the orientation of the major axis of the relative orbit with respect

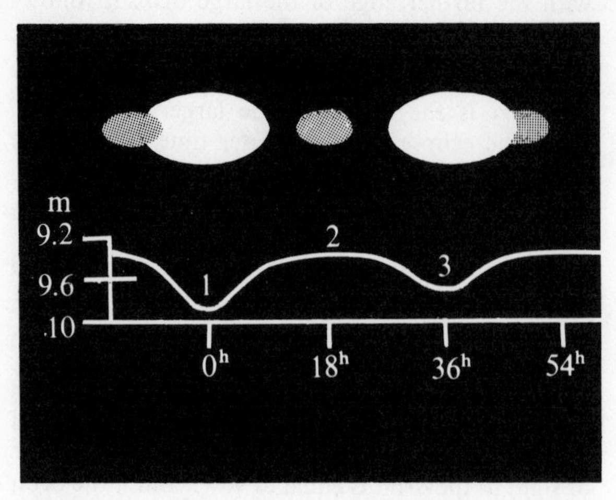

FIG. 26–18 *If the maxima and minima of a light curve are rounded, we deduce that the components of the binary are ellipsoidal. (Light curve of TX Cassiopeiae, after observations by R. J. McDiarmid, Princeton, N.J.)*

to the line of sight. If the stars were perfectly rigid spheres so that one star could not distort the other, the major axis of the relative orbits would always point in the same direction. But if the two stars are distorted by gravitational tides, the so-called **line of apsides** (the major axis) does not remain fixed in space, but rotates around at a definite period.

To analyze this we note that if the major axis points exactly towards the observer, the secondary minimum falls midway between the two primary minima on either side of it. That this is so can be seen at once from Figure 26.19(a), which shows a hypothetical orbit. It is clear from this figure that the interval of time from 1 to 3 is exactly equal to the time from 3 to 1, so that the secondary minimum is exactly halfway between the two successive primary minima occurring at 1. On the other hand, when the line of sight is perpendicular to the major axis, this is not the case (see Figure 26.19(b)). We see that the secondary minimum is now closer to one of the primary minima than to the other.

If the line of apsides rotates, then it is clear that this secondary minimum oscillates back and forth with a period equal to the period of rotation of the line of apsides. This is useful information; it can tell us a great deal about the internal structure of stars, and indeed it has been used to test various stellar models. When we construct a model of a star, as we have shown, we obtain information about the way the mass is distributed throughout the star. This distribution determines the degree of tidal distortion this star suffers in the presence of a companion. If the star has a high concentration of mass towards the center, its distortion is smaller than if the mass is spread out more uniformly. And if the distortion is smaller, the rate at which the line of apsides precesses is also smaller. Thus, by measuring the rate of apsidal motion we can check our stellar models. What we do is compute theoretically the rate of apsidal motion from the model itself, and then compare this with the observations made on eclipsing binaries. The results obtained in this way are in good enough agreement with the observations to convince us that our models are close to the truth.

As an example of eclipsing binaries with apsidal motion we have γ Cygni with an apsidal period of 46 years, GL Carinae with an apsidal period of 25 years, and CO Lacertae with an apsidal period of 40 years. Unfortunately most of the orbits with apsidal motions are of stars of spectral classes A5 and earlier, and with apsidal periods ranging from 40 years to 600 years, so that we cannot apply these data without reservations to a theoretical model of the sun, which is a G-type star.

26.28 Reflection Effect of Eclipsing Binaries

One more bit of information can be obtained from certain asymmetries in the light curves of eclipsing binaries. This is related to the brightness of the binary when both stars are visible. We consider two stars, one of which is large, faint, and cool, the other small, bright, and hot. As the small star approaches the large one, just before the large star is eclipsed, as shown in Figure 26.20, the light from its invisible hemisphere strikes the surface of the large cool star and is reflected to us so that the light curve at this point is at a maximum, and the same thing happens just after the primary minimum. Thus, the light curve is not constant between primary and secondary minima, but shows a maximum just before and after the primary minimum.

26.29 Binaries with Unusual Characteristics

It is easy to pick out eclipsing binaries by the variations in their apparent brightnesses; about 3 000 such systems are known. However, the data of only a comparative few, about 150, have been studied in sufficient detail to give reliable orbits. In Table 26.3, orbital data are given for some of the well-known eclipsing binaries, among which are some very unusual ones, because of the extreme

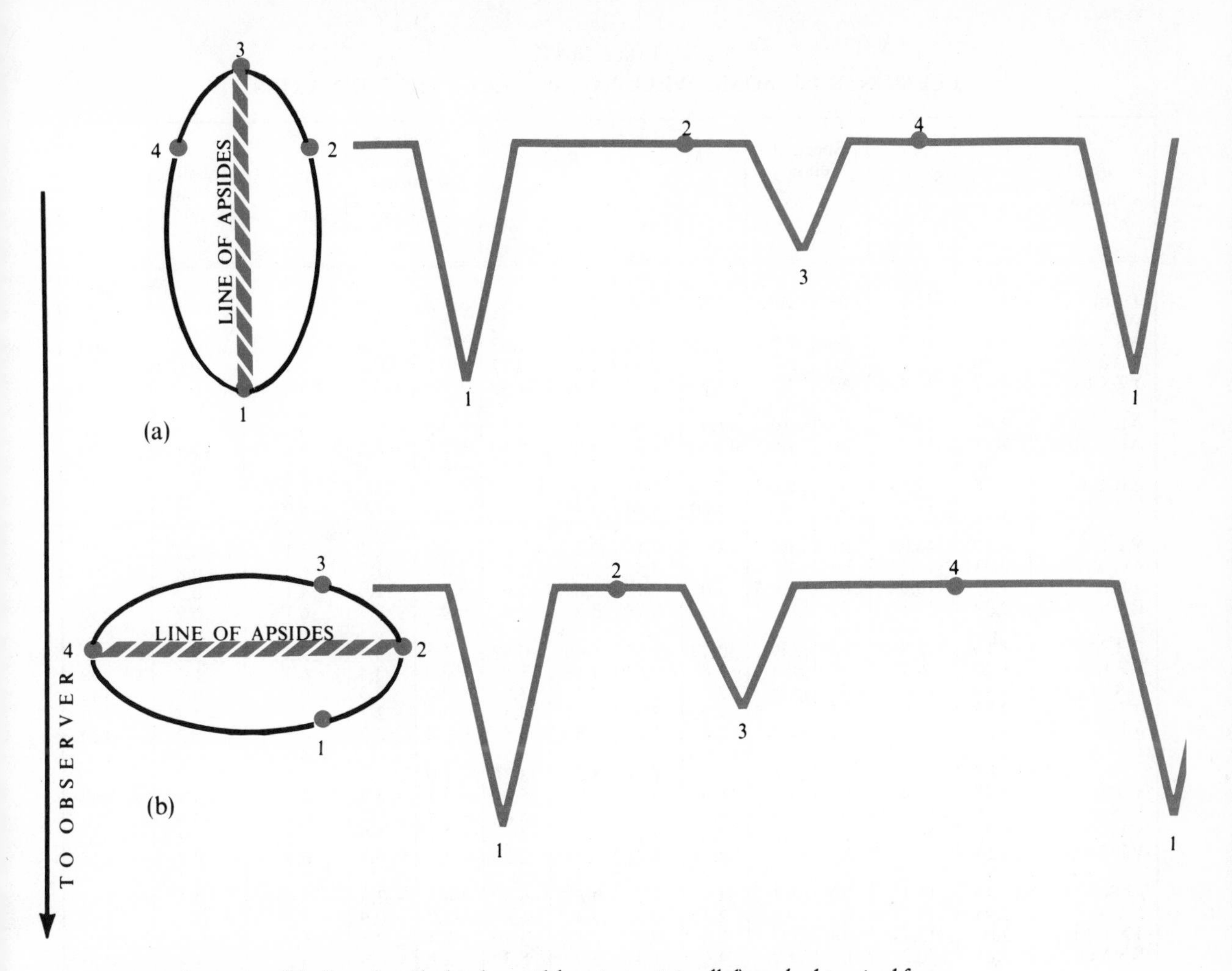

FIG. 26–19 *The rotation of the line of apsides (or the apsidal motion, as it is called) can be determined from the way the separation between the two minima in the light curve changes. (a) In the case shown here (the major axis points to the observer) the secondary minimum falls midway between the two primary minima on either side. (b) When the line of sight is perpendicular to the semimajor axis, the secondary minimum no longer lies between the the two primary minima.*

gravitational interaction between the two components. These are primarily supergiants, an example of which is the system ϵ Aurigae. In general, a binary of this sort consists of a cool supergiant star of spectral class M to F and a normal B-type star which is considerably smaller than the supergiant. Thus in the case of 31 Cygni, the supergiant is a K5 star and the small star is a B3 type, whereas ζ Aurigae is a K5 supergiant with a B8 component, and ϵ Aurigae is composed of an F supergiant and B component. However, β Lyrae is an exception with a giant B9 star and a small F-type companion. These giant binaries, which have periods ranging from 2 to 27 years, are characterized by peculiarities in their spectra.

From the photographs of the spectra of these stars one can see just what the nature of the eclipsing process is. When both stars are visible, one obtains an early type supergiant K to M spectrum superimposed upon a B-type spectrum which, in general, is very faint in the visible region. However, in the continuous part of the violet and ultraviolet spectrum, the B-type star completely dominates. Since the giant star has a very extended cool atmosphere, it plays a most important role in the light curve of the eclipsing binary, and in the changes observed in the spectra of these systems. As the hot B-type star passes behind the atmosphere of the cool giant star, the lines of ionized metals, particularly those of calcium, are generally

TABLE 26.3

ELEMENTS OF SOME WELL-KNOWN ECLIPSING BINARIES

Star variable	Period	Separation	Spectral class		Mass		Radius		Absolute bolometric magnitude		Apparent visual magnitude	Absolute visual magnitude	Distance
			1	2	1	2	1	2	1	2			
	day	$\mathscr{R}_\odot$			$\mathscr{M}_\odot$		$\mathscr{R}_\odot$						psc
σ Aql	1.95	15.1	B8	B9	6.8	5.4	4.2	3.3	−1.9	−0.9	5.1	−0.6	137
WW Aur	2.52	11.8	A7	F0	1.92	1.90	1.92	1.90	1.7	2.0	5.7	1.2	77
AR Aur	4.13	18.5	B9	A0	2.55	2.30	1.82	1.82	0.3	0.6	5.5	0.5	100
β Aur	3.96	17.5	A0	A0	2.33	2.25	2.48	2.27	−0.1	0.2	2.1	0.0	27
YZ Cas	4.47	19.4	A3	F5	3.3	1.6	2.75	1.49	0.4	3.1	5.6	0.7	90
AR Cas	6.07	34.8	B3	A0.5	11.9	3.0	7.1	2.3	−4.8	0.2	4.7	−3.0	350
AH Cep	1.77	18.7	B0	B0.5	16.5	14.2	6.06	5.50	−5.6	−5.2	6.6	−1.9	500
α Cr B	17.36	41.9	A0	G6	2.5	0.89	2.9	0.87	−0.1	5.4	2.3	0.7	22
AR Lac	1.98	9.13	G5	gK0	1.32	1.31	1.54	2.86	3.8	3.4	6.5	3.1	48
U Oph	1.68	12.8	B5	B6	5.30	4.65	3.4	3.1	−2.4	−1.9	5.9	−1.6	310
VV Ori	1.49	16.0	B1	B5	18	6.1	6.2	3.0	−5.3	−2.1	5.1	−3.4	500
AG Per	2.03	14.1	B5	B7	5.1	4.5	3.0	2.7	−2.3	−1.7	6.6	−0.8	300
ξ Phe	1.67	10.0	B7	A0.5	3.0	2.1	2.8	1.6	−1.2	1.0	4.1	−0.2	70
RS Sgr	2.42	10.1	B5	A5	1.4	0.94	3.2	2.6	−2.2	0.7	6.1	−0.9	250
R CMa	1.14	3.84	F0	gG9	0.49	0.11	1.06	0.97	3.3	5.5	5.9	3.3	33
RZ Cas	1.20	6.35	A0	gG1	1.80	0.63	1.53	1.80	0.9	3.4	6.3	1.5	90
U Cep	2.49	12.6	B8	gG8	2.9	1.4	2.4	3.9	−0.6	2.3	6.8	0.5	180
u Her	2.05	15.0	B3	B7.5	7.9	2.8	4.5	4.3	−3.8	−2.1	4.7	−2.4	260
δ Lib	2.33	11.6	A0	gG2	2.6	1.1	3.5	3.5	−0.8	2.2	4.8	−0.2	100
β Per(Algol)	2.87	15.7	B8	gK0	5.2	1.01	3.57	3.76	−1.0	2.7	2.2	0.0	27
V Pup	1.45	16.2	B1	B3.5	16.6	9.8	6.0	5.3	−5.1	−3.9	4.5	−3.5	400
U Sge	3.38	19.5	B9	gG2	6.7	2.0	4.1	5.4	−1.4	1.2	6.4	−0.6	250
V356 Sgr	8.90	46	B3	A2	12	4.7	5.0	13	−3.9	−3.2	6.8	−3.2	1 000
V505 Sgr	1.18	7.2	A1	gF8	2.33	1.21	2.27	2.26	2.7	0.3	6.5	0.7	145
μ^1 Sco	1.45	15.3	B9	gK1	14.0	9.2	4.8	5.3	−4.2	−3.3	3.1	−3.3	190
λ Tau	3.95	16.1	B3	A3	2.3	0.92	3.4	4.8	−3.2	−0.9	3.8	−1.8	132
TX UMa	3.06	13.7	B8	gG3	2.8	0.85	2.16	3.79	−0.4	2.1	6.9	0.6	180
RS Vul	4.48	20.9	B5	gF9	4.6	1.4	3.9	5.3	−2.6	0.9	6.9	−1.4	450

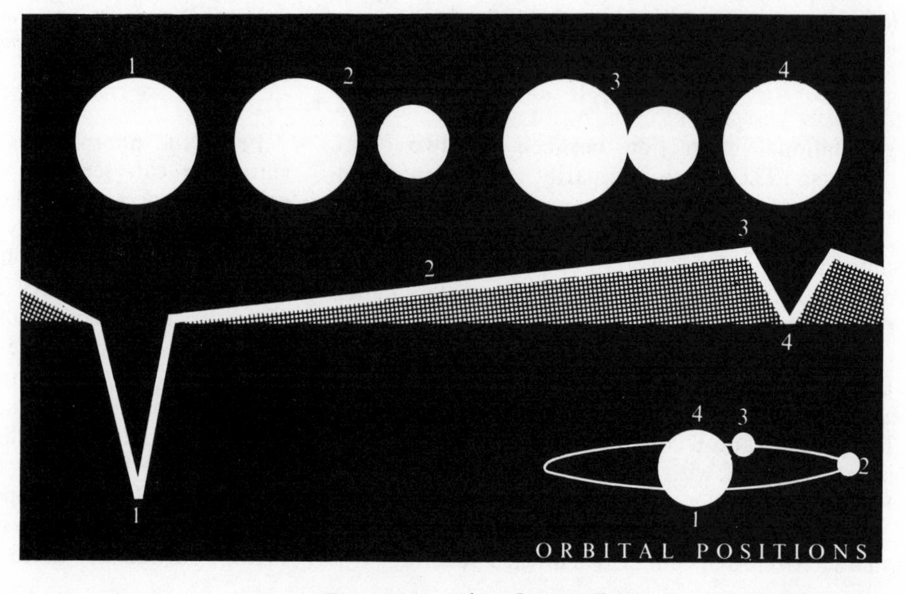

FIG. 26–20 *The reflection effect is shown in the shape of the light curve.*

enhanced as a result of the absorption in the K atmosphere of the ultraviolet radiation from the B-type star. As the B star moves further behind the giant these effects are greatly intensified with the ionized metallic lines becoming still more pronounced. But these lines finally disappear as the B star passes behind the main bulk of the K star, and the K spectrum is all that appears. When the B star reappears again, the variations in the spectra are similar, but in the reverse order to that we have just described.

26.30 Binaries in Which the Components Are Very Close *

Some very interesting phenomena occur in binary systems in which the components are very close together. These effects, which are evident in certain unusual spectral features, cannot be accounted for in terms of the normal types of binary orbits. Velocity curves and the spectral lines that one observes from such binary systems can be explained, as Otto Struve has pointed out, by supposing that the two components are so close together that huge streams of matter pass between them.

An example of such a system is β Lyrae, which is losing material continuously because of gaseous currents between the two components. The existence of these gaseous currents is indicated by the presence in the spectrum of this eclipsing binary of emission lines of varying intensity in addition to the normal spectra associated with the two stars. The B9 component of β Lyrae contributes 90 per cent of the light of the system but the F star is too faint to leave much impression on the spectrum. These stars are so close together that a jet of hot material passes continuously from the B to the F star, whereas the cool stream of gas goes from the F to the B star.

Struve has pointed out that this system is losing mass continuously at the rate of 10^{22} gm/sec, which causes the period of the entire system to increase slowly as the two stars revolve around each other. First one stream and then the other is cut off from sight, giving rise to a variation in the bright-line emission spectrum.

26.31 Multiple Star Systems

In some cases binary systems are accompanied by a third star to form a **triplet system**, and although these are quite rare compared to binaries, several well-defined triplets are known. α Centauri is one such system with two close companions and a distant component. The close pair are separated by 23.5 A.U. (period of 80 years) and the distant companion, Proxima Centauri, is separated from the other two by about 0.2 light years (period about 2 000 000 years).

Another complex system, α Geminorum (Castor), consists of three spectroscopic binaries all moving together as a unit. One of the binaries is of the second magnitude with a period of 9 days, and not far from it is another third-magnitude binary with a period of ·2 days. The fainter of these two spectroscopic binaries revolves around the other once every few hundred years. The third spectroscopic binary, a tenth-magnitude system, with a period of 0.8 days, is so far away from the other two that it revolves around them once every several thousand years.

In general, in such multiple systems we find two or more stars forming a compact system and one or more distant stars revolving around this central system.

EXERCISES

1. Plot the light curve of an eclipsing binary with the following characteristics: (a) the semimajor axis of the relative orbit, whose eccentricity is 0.3, is at right angles to the line of sight; (b) the diameter of the massive giant component is 10 times that of the faint component; (c) the surface temperature of the giant star is 8 000°K and that of the faint star is 4 000°K; (d) the average relative velocity of the two stars is 200 km/sec. Neglect the ellipticity, the reflection, and the limb-darkening effects.

2. Repeat problem 1 but take the surface temperature of both stars to be equal and take the major axis along the line of sight.

3. (a) Consider the relative orbit of a binary system in which the major axis is tilted 30° to the line of sight and the minor axis is perpendicular to the line of sight. If the eccentricity of the orbit is 0.5, what does this orbit look like to the observer? (That is, what is its shape projected on the plane of the sky?) (b) Show where the primary is on the major axis of the projected orbit. What would the angle of tilt have to be for the observer to see the projected orbit as a circle?

27

STAR CLUSTERS, STAR CLOUDS, AND THE MILKY WAY

27.1 The Milky Way Defined

Up to this point we have been discussing the properties of individual stars and, at most, small associations such as binaries. Our aim from now on, however, is to get a picture of the structure of the Milky Way itself or, as we shall refer to it, the Galaxy. The term Milky Way is generally applied to the faint luminous band circling the sky that can easily be seen with the naked eye on a clear moonless night. Its position in the sky is defined by certain familiar constellations such as Cygnus, Perseus, Aquila, Monoceros, Cassiopeia, etc. However, when we refer to the Milky Way as a galaxy we have in mind not only this faint band, which on examination with powerful binoculars or a telescope is revealed as consisting of thousands upon thousands of faint stars, but also all the other stars (visible to the naked eye and through telescopes) that appear all around us in the sky.

At first sight it appears that this luminous band of stars is something very distant and apart from the naked-eye stars that we see, but this appearance is deceptive and arises, as we shall see, because the stars in our Galaxy are not distributed uniformly in all directions around us, but lie in an extended disk which is quite thin in the neighborhood of the sun but thickens considerably towards the center of the Galaxy. The naked-eye stars seem to be distributed uniformly around us because we are in their very midst, but actually they are themselves part of the Milky Way (see Figure 27.1). Our purpose, then, is to discuss the nature of the distribution of stars in our Milky Way, and to determine just where our solar system lies in it. The only way we can discover this is by extensive star counts in all parts of the sky. However, as we shall see, we are greatly hindered in this in our part of the Galaxy by the presence of obscuration in the form of interstellar dust. This limits our counts of individual stars to a rather small neighborhood of the Milky Way, and if it were not for the existence of groups of stars which enable us in a sense to look beyond the obscuration, our picture of the Milky Way obtained from individual star counts would be a rather distorted one.

It is possible by combining the results of individual star counts, out as far as we can push them, with the results obtained from groups of stars such as open clusters (those lying in the Galaxy), star clouds, and globular clusters to obtain a fairly good picture of the structure of our Milky Way. In addition to this we shall be greatly aided in formulating this picture by investigating the distribution of interstellar matter itself.

27.2 Early Star Counts and the Structure of the Milky Way

Although, as we have noted, we cannot get a complete picture of the Galaxy from individual star counts alone, they are very useful in delineating certain of its features. We shall therefore begin by considering how complete a picture we can get by this approach. This procedure was one of the first adopted in investigating the Milky Way; it was used to construct a model of the Galaxy by Sir William Herschel, who surveyed the northern heavens with his telescope. Similar work was carried on in the southern hemisphere by his son, Sir John Herschel.

Using their star counts the Herschels pictured our Galaxy as having a thin, lens-like structure bulging at the center. However, they were unable

FIG. 27–1 *Constellation of Orion and surrounding region, showing some of the naked-eye stars. The bright nebulosity along and below the belt of Orion is clearly visible. The sizes of the images of Betelgeuse, a first (apparent) magnitude red star in the upper left-hand corner of the constellation, and of the blue-white star, Rigel, also of about the first (apparent) magnitude, show the difference between photographic and visual magnitudes. (From Mount Wilson and Palomar Observatories)*

to obtain a correct picture of it such as we have today and were under the impression that the solar system is at the center. Sir William Herschel recognized the importance of the plane of the Milky Way (the **galactic plane**) and pointed out that "the galactic plane is to our Milky Way what the plane of the ecliptic is to the solar system" and that all investigations about the structure of the Galaxy should start from that premise.

27.3 Kapteyn's Model of the Galaxy

The technique of star counts, enhanced by ingenious statistical methods, reached its greatest development in the hands of Kapteyn. In 1922, he pictured the Milky Way as an enormous cluster of stars shaped like a flattened ellipsoid of revolution (the so-called **Kapteyn universe**) with its diameter about five times larger than its central thickness and with the solar system at its center. Kapteyn estimated the total number of stars in the Milky Way to be 47 billion with a central density of about 45 stars per cubic parsec. Although considerable changes have been introduced in Kapteyn's picture, it is still essentially correct. However, we now know that the sun is not at the center of the Galaxy.

That the plane of the Milky Way must play a dominant role in any picture of the Galaxy is indicated by the concentration towards it of all kinds of objects. As we go to fainter and fainter stars, there is a greater and greater real concentration towards the plane of the Milky Way whether we consider spectral class O stars, M stars, Cepheid variables, binaries, clusters, or other groups. However, we shall see that there are exceptions to this, in line with the division of stars into the two populations we have already discussed. (The student must be careful to differentiate between a real and apparent concentration towards the plane of the Milky Way because of distance. Distant objects of both populations show an apparent concentration towards the plane.)

27.4 The Limitations of Star Counts

We begin our investigation by considering star-count data. We see at once that we are necessarily limited to an incomplete picture, for what we seek from these star counts (i.e., from the numbers of stars with known parallaxes, magnitudes, spectral classes, proper motions, etc.) is a picture of the distribution of stars throughout the Milky Way, as well as their motions and their luminosities. In other words, we seek the answers to the following questions: (1) Given a small volume of space anywhere in our Galaxy, how many stars lie in this volume? (2) How many of them lie in each absolute-magnitude range? (3) How many of them lie in a given velocity range? We seek, therefore, three distribution functions (which we shall define more precisely later) known as the **density function**, the **luminosity function**, and the **velocity function**.

The reason that we cannot answer these questions completely using only star counts is that we cannot get a complete count of stars unless we stay fairly close to our solar system. If we do this, we get a detailed picture of the number of stars in our neighborhood, but we get a distorted picture of the kinds of stars that are represented in our Galaxy. For we shall certainly fail to pick up such stars as Cepheid variables, blue-white giants, etc., if we stay close to our solar system. But if we do go to greater distances with our star counts, we are bound to miss larger and larger numbers of faint stars and therefore favor the intrinsically bright ones in picturing the kinds of stars constituting our Milky Way. Furthermore, as we shall see, the dust and gaseous matter we have already mentioned essentially cuts off stars beyond 6 000 light years in the plane of the Milky Way so that this, too, distorts our conclusions.

Nevertheless, when Kapteyn first undertook investigations of this sort and introduced his "selected-areas" program (which proceeded by counting the stars down to the seventeenth apparent magnitude in each of 206 selected, uniformly distributed areas in the sky), it was thought that if we could get the density and luminosity function for our region of space, we could then deduce from these the over-all picture. This was based on the hope that there is a regular pattern in the distribution of stars which can be deduced from the data in our neighborhood on the assumption that the sun is at the center of the Galaxy. We now know that this is not so and that the distribution functions obtained from star counts in a cube 10 light years on each side in the neighborhood of the sun are considerably different from those in the same size cube at a distance of 15 000 light years from our solar system in the direction of Sagittarius.

TABLE 27.1

(A) APPARENT DISTRIBUTION OF STARS WITH RESPECT TO GALACTIC LATITUDE

N_m = Number of stars per square degree brighter than apparent magnitude m

$\log N_m$

m	Galactic latitude				Mean 0° to 90°	Mean 0° to 90°
	0°	±30°	±60°	±90°		
	photographic magnitudes					visual
0.0						$\bar{5}.7$
1.0						$\bar{4}.4$
2.0		$\bar{4}.8$			$\bar{4}.96$	$\bar{4}.99$
3.0		$\bar{3}.27$			$\bar{3}.40$	$\bar{3}.55$
4.0	$\bar{2}.25$	$\bar{3}.85$	$\bar{3}.66$	$\bar{3}.60$	$\bar{3}.89$	$\bar{2}.11$
5.0	$\bar{2}.72$	$\bar{2}.32$	$\bar{2}.17$	$\bar{2}.11$	$\bar{2}.37$	$\bar{2}.60$
6.0	$\bar{1}.18$	$\bar{2}.78$	$\bar{2}.63$	$\bar{2}.58$	$\bar{2}.86$	$\bar{1}.07$
7.0	$\bar{1}.61$	$\bar{1}.23$	$\bar{1}.08$	$\bar{1}.03$	$\bar{1}.31$	$\bar{1}.54$
8.0	0.05	$\bar{1}.68$	$\bar{1}.52$	$\bar{1}.46$	$\bar{1}.75$	0.00
9.0	0.52	0.12	$\bar{1}.94$	$\bar{1}.88$	0.19	0.45
10.0	0.97	0.54	0.35	0.27	0.62	0.92
11.0	1.43	0.96	0.75	0.66	1.05	1.34
12.0	1.88	1.37	1.12	1.03	1.46	1.77
13.0	2.30	1.76	1.47	1.39	1.87	2.17
14.0	2.72	2.12	1.79	1.71	2.26	2.56
15.0	3.12	2.46	2.10	1.97	2.62	2.95
16.0	3.40	2.77	2.38	2.24	2.98	3.30
17.0	3.83	3.07	2.64	2.48	3.33	3.64
18.0	4.20	3.35	2.97	2.72	3.64	3.96
19.0	4.5	3.6	3.1	2.9	3.90	4.20
20.0	4.7	3.8	3.3	3.1	4.17	4.45
21.0	5.0	4.0	3.4	3.2	4.4	

In this table the logarithm of the number of stars, per square degree, N_m, brighter than apparent magnitude m is given for galactic latitudes 0°, ±30°, ±60°, ±90°.

(B) PERCENTAGE OF STARS IN EACH SPECTRAL CLASS

Spectral class	B	A	F	G	K	M
Per cent of stars $m_v \leq 8.5$	10	22	19	14	32	3

27.5 The Apparent Distribution of Stars Brighter Than a Given Apparent Magnitude

Even without using detailed star counts we can see, just from a general statistical analysis, that the density of stars appears to diminish as we go out to greater and greater distances from the solar system, but not equally in all directions. This conclusion is based on an analysis of counts of stars brighter than a given apparent magnitude m obtained by van Rhijn and Seares.

We see from Table 27.1, if we first consider the apparent bright stars in the sky (i.e., stars brighter than the ninth apparent magnitude), that the distribution appears to be fairly regular. Thus there are 530 stars brighter than $m = 4$, 1 620 brighter than $m = 5$, 4 850 brighter than $m = 6$, etc., so that the number increases by a factor of three for each apparent magnitude up to about $m = 9$. However, this ratio falls off as we go to fainter and fainter stars, so that there appear to be only about twice as many stars brighter than the eighteenth magnitude as there are brighter than seventeenth magnitude. This indicates that there is a fairly regular distribution of bright stars in all directions around the sun, but that this is not the case as we go to greater and greater distances.

27.6 Comparison of Observed and Theoretical Distribution

Considering these star counts alone we must conclude that the distant stars are more thinly scattered in all directions than the nearby stars. To see that this follows from the data in the table, we consider what the situation would be if the stars were uniformly distributed out to all distances from the sun. To simplify things we replace all stars by an equal number of stars of the same average luminosity.

Let r_m be the distance at which such a star has an apparent magnitude m, and let N_m be the total number of stars brighter than this apparent magnitude. In other words, N_m is the number of stars within a sphere of radius r_m with its center at the sun, so that

$$N_m = \rho(\tfrac{4}{3})\pi r_m^3,$$

where ρ is the density of the stars (number of stars per unit volume) which we assume to be constant at all distances from the sun. We now consider the number of stars $N_{(m+1)}$ in a sphere of radius $r_{(m+1)}$, where $r_{(m+1)}$ is the distance at which one of our average stars has the apparent magnitude $m + 1$. We thus have, since $N_{(m+1)}$ is the number of stars brighter than apparent magnitude $m + 1$,

$$N_{(m+1)} = \rho\tfrac{4}{3}\pi r_{(m+1)}^3 .$$

The ratio of $N_{(m+1)}$ to N_m is just

$$\frac{N_{(m+1)}}{N_m} = \left(\frac{r_{(m+1)}}{r_m}\right)^3.$$

Since stars are point sources of light, we know that the apparent brightnesses of two stars of the same luminosity vary inversely as the squares of their distances, so that one of our average stars at a distance r_m is brighter than one at a distance of $r_{(m+1)}$ by the factor $\left(\frac{r_{(m+1)}}{r_m}\right)^2$. But this ratio must equal 2.512 because a star at the distance r_m appears one magnitude brighter than one at $r_{(m+1)}$. In other words, $r_{(m+1)}/r_m = \sqrt{2.512}$. If we now substitute this into the expression above, we obtain

$$\frac{N_{(m+1)}}{N_m} = (\sqrt{2.512})^3 \cong 4. \qquad (27.1)$$

In other words, if the stars were distributed uniformly out to all distances from the sun, the ratios of star counts for successive apparent magnitudes (as obtained from van Rhijn's table) should be close to 4, and this holds regardless of the kinds of stars we are dealing with. The actual data, however, indicate that this ratio falls off from 3 to 2 as we go to the faint stars.

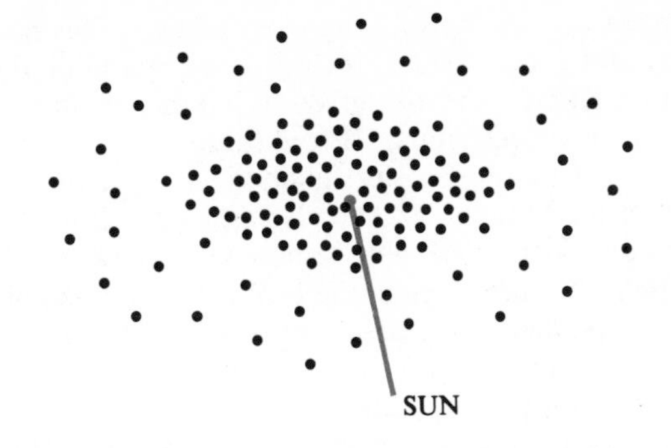

FIG. 27–2 *Kapteyn's Model of the Galaxy.*

Using data such as van Rhijn's, Kapteyn in the early years of his investigations concluded that there is a falling off of stars in all directions from the solar system and hence proposed the model of the Galaxy which we have already mentioned (see Figure 27.2). This has since been discarded because at the time van Rhijn and Kapteyn did their work, the existence of absorption from dust and gaseous material in interstellar space was not known. We now know that there is a much greater concentration of obscuring material in the plane of the Milky Way than in the directions at right angles to it, and that this leads to the impression that star counts fall off along the plane of the Milky Way just as they do at right angles to it, although not quite as rapidly.

As we shall see, the apparent falling off of stars at right angles to the plane of the Galaxy is due to a real thinning out of stars, because in our part of the Galaxy the stars form a thin layer and we lie close to the middle of it. However, the apparent thinning out of the stars in the plane of the Milky Way is a rather complicated matter and arises both because of the variation in the number of stars in various directions in this plane and because of the obscuration we have already mentioned.

27.7 The Local Neighborhood of the Milky Way: The Five-Parsec Region

Before we consider the over-all structure of the Milky Way, we shall try to get a picture of the structure in the neighborhood of the sun. We can best do this by measuring the parallaxes of stars with large proper motions since these have the greatest probability of lying close to us. This has been done by various investigators, particularly van de Kamp, who has given us a fairly complete list of the nearest stars. From these, we can get a picture of the luminosity function, at least in our part of the Milky Way, and also an idea of how the stars are distributed with respect to spectral class. These results are given in the H-R diagram (Section 17.11) plotted for stars within a radius of 10 parsecs around the sun.

Even though this material is taken from a very small sampling of space, it already gives us some very useful information about the distribution of stars in our part of the Milky Way, or at least about the luminosity function. We see that giants, supergiants, and stars belonging to the upper branch of the main sequence are completely absent, with no star present of luminosity greater than that of Sirius ($M = +1.5$). What is most remarkable is the great concentration of very faint red stars and the large number of white dwarfs. About 66 per cent of all these stars belong to class M and have absolute magnitudes ranging from $+10$ to about $+17$. Since these stars are distributed uniformly in all directions within the 5-parsec-radius sphere, the picture of this part of space is that of a uniformly dense sphere of faint red stars.

If we count the number of stars of a given absolute magnitude within this region, we see that this count increases up to about the fifteenth absolute magnitude so that the luminosity function appears to increase steadily with absolute magnitude. However, at the very faint end the luminosity function appears to fall off, although we cannot be sure that this is a real effect since there may be many stars that are so faint that we have not yet been able to detect them. We also note the relatively large numbers of white dwarfs ranging in absolute magnitudes from $+10$ to $+15$. Finally we see that the number of binary and multiple systems is quite high.

We may thus picture our neighborhood of space (up to 7 parsecs) as consisting primarily of very faint red stars, with an almost equal number of white dwarfs, but free of giants, supergiants, and upper main-sequence stars.

The number of stars per cubic parsec in our neighborhood is about 0.12, and since the masses of most of these stars are smaller than that of the sun, the mass-luminosity relationship indicates that the number of solar masses per cubic parsec in our neighborhood is about 0.05. We may also note that spectroscopic data show that most of these stars belong to Population I, although about 20 per cent are high-velocity stars and hence members of Population II.

27.8 The Eleven-Parsec Region

How far out from our solar system does the picture we have just drawn hold and at what point does it begin to deviate from the actual model of the Galaxy? The only way we can answer this question is by pursuing our star counts and applying our statistical methods to still greater distances. Of course, as we go further out, our trigonometric parallaxes becomes less and less accurate, and we also begin to overlook more and more of the faint stars. However, we can go out to about twice the distance, i.e., about 11 parsecs, without running into any serious discrepancies, although even in this close region we already find ourselves with incomplete data. The reason is that data such as parallaxes, spectral classes, proper motions, etc., are extremely unreliable for stars fainter than the fourteenth apparent magnitude. It follows from this that stars with absolute magnitude $+18$ or greater (such stars are known to exist since a star of absolute magnitude $+19.2$ has already been discovered) and lying between 5 and 11 parsecs from the sun have been overlooked in our survey.

Within this 11-parsec sphere 254 stars are known, and an analysis of these stars shows that they have a distribution similar to that for stars within the smaller sphere but with a considerable thinning out. Thus, the number of known stars per cubic parsec in this sphere is about 0.04 as compared with three times that number in the 5-parsec sphere. This already indicates that our star count is probably quite incomplete in this larger sphere. However, the luminosity function is about the same as in the smaller region of space with a very large concentration of stars in spectral class M. About 50 per cent of these stars are multiple systems and again we see that the white dwarfs are quite numerous, as shown in the H-R diagram in Figure 17.7. We observe from the diagram that the giant branch is represented by just one single orange star, Arcturus.

This diagram shows us just how incomplete our analysis is if we consider the white dwarfs. We have recorded 8 white dwarfs in this region, representing 3 per cent of the stars in the H-R diagram, whereas within 5 parsecs we recorded 5 white dwarfs representing about 10 per cent of the known stars. Thus approximately 70 per cent of the white dwarfs may be assumed to be missing in the 11-parsec picture, although it is probable that we are really picking up a mere 10 per cent of these stars. This situation becomes worse as we go out to greater and greater distances. There is, however, one important thing missing in the nearby picture that we do become aware of when we go to very great distances, and that is the presence of interstellar matter. For if we consider stars that are close to the sun, the amount of absorption resulting from interstellar matter is too small to be detected. However, the accumulation of interstellar matter between us and the distant stars results in enough absorption for us to detect it.

27.9 The Milky Way at Intermediate Distances from the Sun *

Now that we have seen what the picture looks like close to the sun, based on data obtained from accurately determined trigonometric parallaxes and proper motions, we shall go out to the more distant regions where trigonometric methods are unreliable and where we are forced to use spectroscopic parallaxes and general statistical methods. The trigonometric parallaxes do not allow us to go much beyond 100 parsecs, but spectroscopic parallaxes and mean parallaxes based upon the solar motion have been used to collect data up to 600 parsecs and beyond. Spectroscopic analysis enables us to determine the absolute magnitudes (and hence parallaxes) out to these distances with an uncertainty not in excess of half a magnitude.

(a) *Distribution at right angles to the plane of the Galaxy.* We shall consider the density function first and divide the material into two distributions—one at right angles to the plane of the Galaxy and the other in the plane of the Galaxy. This will reveal to us the distribution of absorbing material and its relationship to the plane of the Milky Way.

We discover that in the distribution at right angles to the plane of the Milky Way there is a marked difference between the picture we now obtain and that obtained from the nearby stars. Whereas the latter are distributed uniformly in all directions out to a distance of 10 parsecs, we now find that there is a definite drop in the star count as we move farther and farther away from the galactic plane. This is true for all types of stars and at an ever increasing rate as we move out to greater and greater distances. Thus, at a distance of about 1 500 parsecs, the number of stars at right angles to the plane of the Milky Way is less than 3 per cent of that in the plane of the Milky Way. Taking the data of F. H. Seares and combining all spectral classes, we obtain Table 27.2

TABLE 27.2
DECREASE IN STAR COUNTS $N(z)$ PERPENDICULAR TO THE PLANE OF THE MILKY WAY
According to Seares (Distance in parsecs)

Distance Z from plane	No. of stars $N(z)$ per cubic parsec	Z	$N(z)$
pc		pc	
1	1.00	631	0.13
79	0.90	794	0.09
100	0.84	1 000	0.07
126	0.73	1 259	0.05
200	0.50	1 585	0.03
316	0.31	1 995	0.02
398	0.24	2 512	0.01
500	0.17		

showing the mean concentration of stars (in number per cubic parsec) as a function of Z (the distance from the sun at right angles to the plane of the Milky Way) where the mean concentration is obtained by averaging stars on both sides of the plane of the Milky Way. *From this analysis of the star counts we see that the stellar distribution is symmetrical with respect to the plane of the Milky Way, but as seen from the sun there appears to be a greater concentration south of this plane than north of it. This might mean, according to W. Becker, that the solar system is not precisely in this plane but about* 14 *parsecs north of it. It might also mean, according to J. Schilt, that more of the obscuring dust is north of the sun so that more stars are seen to the south.*

The spectral analysis of these stars indicates also that the concentration of stars at right angles to the plane of the Milky Way has some dependence on spectral class and absolute magnitude. We find that the diminution of the concentration of stars as we move away from the galactic plane occurs sooner and proceeds more rapidly for the O and B stars and for the highly luminous main-sequence stars than it does for the later spectral classes and for those that are fainter. *In other words, we find that the blue and blue-white giants hug the plane of the Milky Way much more than the faint orange and red dwarfs.* Although data for this latter group can be obtained only out to about 200 parsecs, because they are so faint, we find that there is only a small drop in their concentration at these distances at right angles to the Galaxy.

As an example of the difference between the way in which stars of high luminosity and those of low luminosity fall off as we move away from the plane of the Milky Way, we note that out to a distance of 250 parsecs from this plane stars of absolute magnitude -4 diminish to about 3 per cent of their concentration near the sun, stars of absolute magnitude 0 fall off to 16 per cent, whereas those of absolute magnitude $+8$ fall off to about 50 per cent of their number in the plane. If we go out to about 1 500 parsecs, no stars brighter than $M = -2$ appear, and only 0.02 as many M-O stars appear as are found near the sun. However, the number of $M = +8$ stars at this distance is nearly 9 per cent of the number of similar stars near the sun (see Figure 27.3).

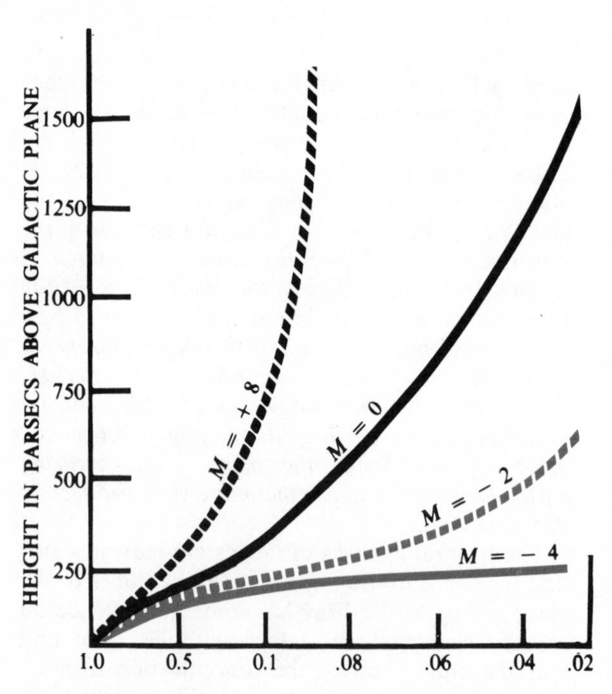

FIG. 27-3 *The distribution of stars of a given magnitude relative to the plane of the Milky Way.*

(b) *Distribution along the Plane of the Milky Way.* If we now consider the distribution of stars along the plane of the Milky Way, we run into a good deal of uncertainty, primarily because of the dust and gaseous material which begins to obscure stars very seriously beyond 1 000 parsecs. Again, because of the enormous amount of data that must be investigated (since there are such great numbers of stars in the plane of the Milky Way), we can only obtain an approximate picture of the situation. Nevertheless, at the present time we can draw some conclusions and obtain fairly reliable density and luminosity functions. Again, we shall consider only the density function at present. We shall treat the luminosity function for the entire distribution at a later time, combining the data for the stars in the plane of the Milky Way and those at right angles to it.

The observational evidence based on star counts, without taking into account absorption effects, shows a drop in the number of stars per cubic parsec in all directions in the plane of the Milky Way as we move away from the solar system, as though the sun were in the very center of a star-rich region. Thus, whether we look towards the center of the Milky Way (i.e., in the direction of Sagittarius), at right angles to it (i.e., in the direction of the solar motion), or in the direction opposite to the center, we find that whereas, at first, out to about 300 parsecs, the star counts remain fairly constant (estimating properly the number of very faint stars which we necessarily fail to pick up at these distances), the density of stars then falls off fairly sharply.

At a distance of about 700 parsecs on either side of the sun, along a line 60° from the galactic center, there is a fairly sharp minimum (a 40 per cent drop in the star count) followed by a rapid increase to a steady value at a distance of about 1 500 parsecs. Along the line towards the center of the Galaxy and along the same line away from the center there is a steady drop in the star count, but the drop away from the center is much steeper than that towards the center. This drop continues towards the center out to about 1 500 parsecs, after which the star counts remain fairly steady at about half their value near the sun. However, away from the center the drop extends out to 2 000 parsecs, reaching a value of 20 per cent of that in the sun's neighborhood, and then remains fairly steady.

If there were no obscuration, this would indicate that the sun is in a region of maximum stellar concentration, but if we take the absorption into account, this apparent maximum disappears, and we find that towards the center of the Milky Way the density function increases sharply out to 2 000 parsecs and then more slowly beyond that. It appears to reach a maximum at about 6 000 parsecs, after which it falls off. In the direction opposite to the center, there is at first a small dip in the star count, but then it remains fairly constant out to about 6 000 parsecs. This is indicated in the graph taken from the work of F. H. Seares, as shown in Figure 27.4.

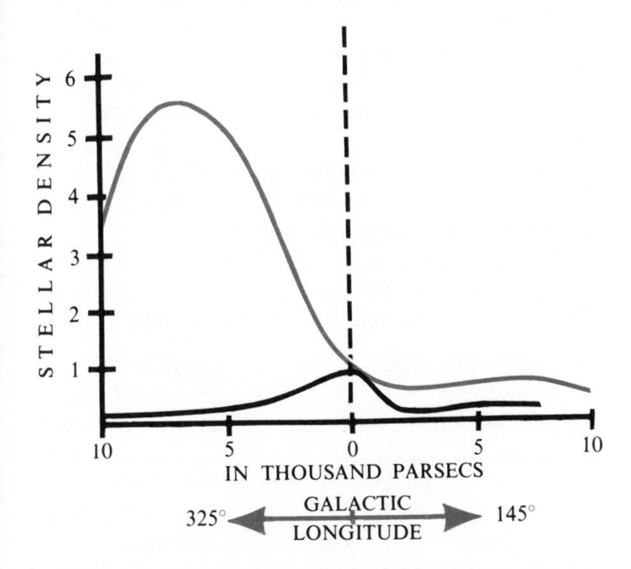

FIG. 27–4 *Graph showing the distribution of stars according to F. H. Seares. After M. Waldmeier,* Einführung in die Astrophysik *(Basel: Verlag Birkhäuser, 1948), p. 253.*

27.10 The Oort Model of the Galaxy: Surfaces of Constant Density

We cannot go much beyond this point with star counts since, as we have already indicated, the dust in the Milky Way limits us to regions within 6 000 light years. We can summarize what has been found in this small region of space by considering Oort's analysis which makes use of the concept of the **surfaces of constant density**. If we take a cross section of the Milky Way at right angles to its plane, we find, according to Oort, as shown in Figure 27.5, that the density function along any line parallel to the galactic plane is not constant but increases towards the center, and decreases as we move away from the center.

This means that out to about 2 000 parsecs the surfaces of constant density are not parallel to the plane of the Milky Way, but tilted with respect to it. We can best illustrate this by the diagram shown (after Oort), in which we plot star counts along two perpendicular directions: one in the plane of the Milky Way pointing towards the center (the abscissa) and the other at right angles to this plane.

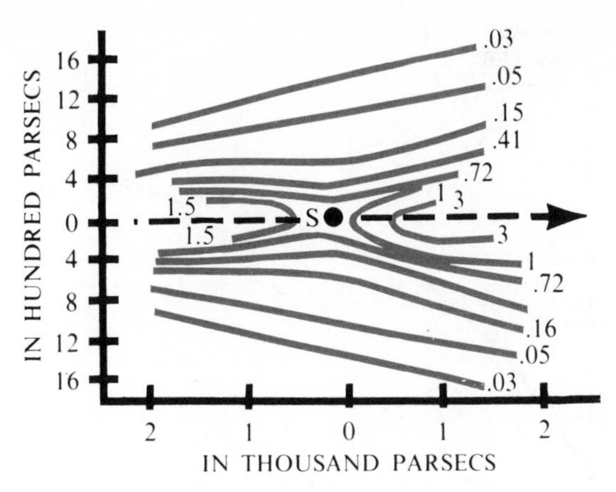

FIG. 27–5 *The Oort Model of the Galaxy showing the surfaces of constant star density.*

If we place the sun at the origin of this diagram, we find that at distances greater than 400 parsecs above and below the sun's position in the plane of the Galaxy the lines of constant density are fairly straight and diverge at a constant slope from the plane. At distances closer than 400 parsecs the curves of constant density are almost elliptical in shape and clearly indicate the great concentration of all types of stars towards the plane of the Milky Way.

27.11 The Luminosity Function

If we now use all the above data to obtain the luminosity function for stars out to 1 000 parsecs, we find that there is a steady increase in the number of stars per unit volume as we go to intrinsically fainter ones until we get to the fourteenth absolute magnitude (photographic) with what appears to be a slight dip at $M_{ph} = 9$. This is indicated in Figure 27.6 obtained from the data up to $M_{ph} = 14$. Thus we find within a cube around the sun, 300 parsecs on a side, that whereas there are about 5×10^5 stars of $M_{ph} = 14$ (all spectral classes taken into account) there are about 10^4 of $M_{ph} = +2$, and only a few hundred of $M_{ph} = -2$.

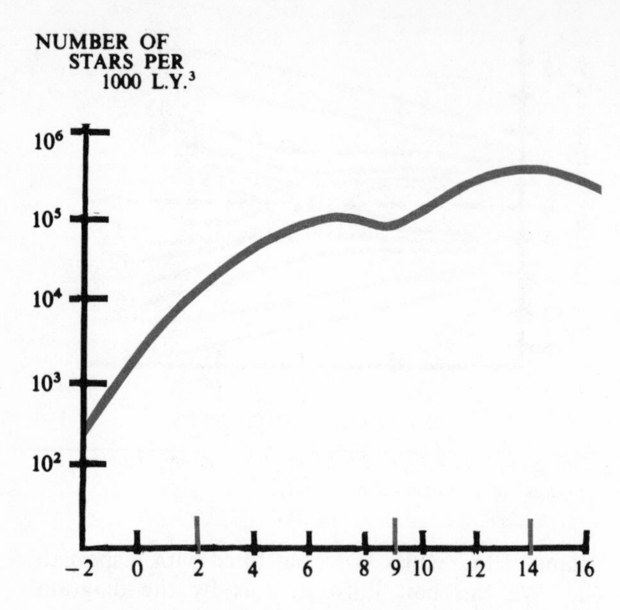

FIG. 27–6 *A graph of the luminosity function as obtained from star counts out to 1 000 parsecs. (After Bart Bok and P. F. Bok,* The Milky Way, *3rd ed. Cambridge, Mass.: Harvard University Press, 1957)*

We cannot be sure that this is an altogether correct picture and we do not know whether the drop in luminosity function after $M_{ph} = 14$ is a real one or is due to the lack of complete data. One thing, however, is clear: stars such as Rigel on the main sequence and red supergiants such as Antares and Betelgeuse are very rare indeed. We may note that whereas we can pick up another star like the sun (α Centauri, for example) within $4\frac{1}{2}$ light years (a little more than 1 parsec) we must go out almost 650 light years to pick up a star like Rigel. We must survey about $4\frac{1}{2}$ million times more space to pick up one Rigel than we must survey to pick up a star like the sun. Thus, the blue giants like Rigel (and also the red supergiants) are millions of times less numerous in our Galaxy than stars like the sun.

27.12 Distribution of Population I and II Stars

We also find that the two stellar populations are distributed differently in the Galaxy. Most of the stars in our galactic neighborhood are Population I stars and they all hug the plane of the Milky Way. We begin to meet more and more Population II stars towards the center of the Galaxy and we find them less concentrated towards the plane of the Galaxy than the Population I stars. Whereas the Population I objects in our Galaxy form a flattened system, Population II objects form a much more spherical distribution. We shall see that this is generally true of all galaxies. (See Section 30.4, Figures 30.2(a) and (b).)

27.13 Color Excess

One more feature about star counts within 2 000 parsecs is notable. As we push further and further out we find that the color indices of all the stars, regardless of spectral class, increase systematically with distance. This effect, which is called the **color excess**, is defined as the difference between the color index of a star of given spectral class near the sun and that of a star of the same spectral class far from the sun. This color excess is itself an indication of obscuration in the plane of the Milky Way, for we know that dust screens out the blue light by scattering it in all directions, thus making self-luminous objects at great distances appear redder than they actually are.

27.14 The Structure of the Galaxy at Great Distances: Star Clusters

If we consider the Galaxy on the basis of the data we have already presented, we may picture it as a thin layer of stars that extends uniformly out to several hundred parsecs, and then begins to thicken beyond 2 000 parsecs as we approach the center, as shown in Figure 27.7. To construct a picture of the Galaxy beyond 2 000 parsecs we must have recourse to objects that are still recognizable at great distances even though they are obscured by the dust just as the individual stars are. Star clusters are examples of such objects.

We must distinguish between two types of clusters which are different in their structures and also in their relationship to the Milky Way:

1. The **open clusters**, which generally consist of hundreds or at most a few thousand stars, and which are so spread out that the individual stars can easily be observed. These clusters are also referred to as **galactic clusters** because they lie close to the plane of the Milky Way and thin out very rapidly as we move away from the plane.
2. The **globular clusters**, which are stellar aggregates consisting of hundreds of thousands, and in some cases millions, of very closely

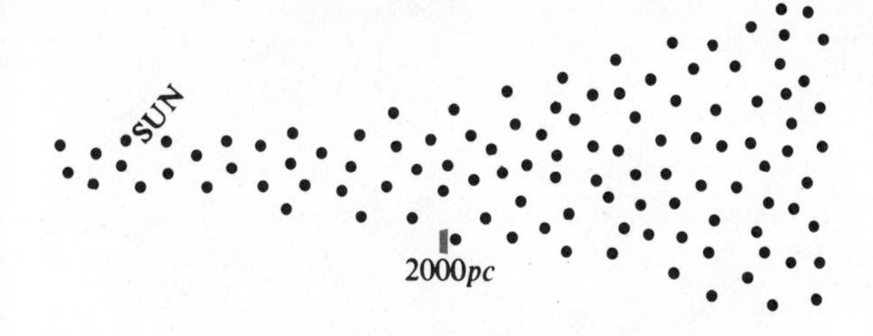

FIG. 27–7 *A preliminary contour of the Galaxy based on star counts out to 2 000 parsecs.*

packed stars. These clusters do not hug the plane of the Milky Way but are found distributed at right angles to it as well as towards the plane itself. Their shapes also differ from those of open clusters. Whereas in the latter the stars are rather irregularly distributed, they are arranged almost spherically in globular clusters.

27.15 Galactic or Open Clusters

The galactic or open cluster is a rather loose, irregularly arranged, aggregate of stars in which, nevertheless, the stars are close enough together to be recognizable to the naked eye as a cluster. Examples of such naked-eye clusters are the Pleiades, Hyades, Beehive (Praesepe), and the double cluster in Perseus (see the accompanying Figure 27.8). Although relatively few open clusters are visible to the naked eye (only about 30 are brighter than the fourth apparent magnitude, where the magnitude is obtained by lumping all the stars in the cluster together), many have been discovered with telescopes. There are 758 open clusters each designated with an NGC or an IC number in the 1958 Catalogue of Star Clusters and Associations of the Czech Academy of Science.

The apparent magnitudes of these clusters range from 0.6 for the cluster in the belt of Orion to 14.9 for NGC 6820. Their distances from the sun also vary considerably, ranging from about 40 parsecs for the Hyades to 3 300 parsecs for the faint cluster NGC 6846. Among the nearby clusters are the Pleiades and the Coma Berenices clusters at about 100 parsecs, the Praesepe cluster at about 200 parsecs, and the Orion Trapezium cluster at about 500 parsecs. Trumpler, who devoted a great deal of time to open clusters, estimated that most of the known clusters lie between 500 and 2 000 parsecs from the sun. The more distant ones blend into the general stellar background.

27.16 Dimensions of Open Clusters and Their Stellar Content

Another interesting feature of an open cluster is its apparent diameter. This dimension ranges from about 400′ in angular measure for a nearby cluster such as the Hyades (the Pleiades have an angular diameter of 20′, the Perseus cluster has an angular diameter of 40′, etc.) to about 0.5′ for the most distant ones such as NGC 6846. The Antares cluster has the unusually large angular diameter of 600′. We find, in general, that the more distant a cluster is, the smaller is its angular diameter, as we would expect if most clusters had about the same linear dimensions. The linear diameters are about 15 parsecs for the very largest to 1.5 parsecs for the very smallest, but Trumpler has found that most of them have linear diameters between 2 and 6 parsecs with the maximum number of them at about 2.5 parsecs.

From the distances of the clusters and their observed, integrated apparent magnitudes we can obtain their integrated absolute magnitudes. This has been done by I. M. Kopelow for about 300 clusters. He finds that although the absolute magnitude ranges from about 0 to -10, most open clusters have absolute magnitudes in the neighborhood of -3.5, which, as we shall see, is about 4 magnitudes fainter than the absolute magnitude of most globular clusters. The Hyades, the Pleiades, h and χ Persei, and the Praesepe cluster are among the most luminous with absolute magnitudes of about -8 or brighter. Kopelow also discovered an important relationship between the absolute magnitude of a cluster and the spectral class of its brightest stars. We shall discuss this point later in considering the H-R diagram of open clusters. The number of stars in these open clusters varies considerably, ranging all the way from 300 or more for the very populous ones, such as Persei, to 15 or 20 for the very small clusters.

27.17 Open Clusters and the Plane of the Milky Way

Open clusters have been extremely useful in investigating the Galaxy beyond 2 000 parsecs, because they are bright enough to shine through the obscuration even at very great distances.

FIG. 27–8 (*a*) *The Praesepe open cluster (NGC 2632) in Cancer. Taken by Barnard with the 10-inch Bruce lens. Its distance is 159 parsecs, and its linear diameter is about 4 parsecs. It contains about 100 stars and is about 2.5×10^8 years old. (Yerkes Observatory photograph)*

FIG. 27–8 (*b*) *NGC 5897 Star cluster, open type in Libra. (200-inch photograph from Mount Wilson and Palomar Observatories)*

FIG. 27–8 (*c*) *NGC 2682 Star cluster, open type, in Cancer. This is the cluster M67 at a distance of 830 parsecs. Its linear diameter is about 4 parsecs and it contains about 80 stars. It is one of the oldest known open clusters with an age of* 4×10^9 *years.* (*200-inch photograph from Mount Wilson and Palomar Observatories*)

FIG. 27–8 (*d*) *NGC 1432 Pleiades, an open cluster in Taurus. The diffuse nebulae around the individual hot stars can be seen. This cluster, which contains about 120 stars, is 126 parsecs away from us. It has a linear diameter of about 4 parsecs. The age of this cluster is* 6.3×10^7 *years.* (*100-inch photograph from Mount Wilson and Palomar Observatories*)

That they lie close to the plane of the Milky Way and that they increase in number as we approach this plane indicate that the picture we have drawn of our part of the Galaxy, as delineated by the star counts, also applies to regions beyond 2 000 parsecs. The great bulk of all the clusters that have been studied within 10 000 to 15 000 light years from our sun lie less than 500 light years away from the central galactic plane.

That open clusters hug the plane of the Milky Way out to about 15 000 light years indicates to us that out to this distance in the direction of Sagittarius the stars in our Galaxy lie in a thin layer parallel to the plane of the Galaxy. To go beyond this, of course, requires information which we must obtain from other sources. In this respect open clusters differ from globular clusters, which as we have already noted are distributed more or less in a spherical shell surrounding this plane. Since the way stars are distributed with respect to the plane of the Galaxy is an index to their population class, we may conclude that galactic or open clusters belong primarily to Population I whereas globular clusters consist primarily of Population II stars.

27.18 The Determination of Obscuration by Open Clusters

Trumpler's investigations of open clusters led to the first direct evidence of obscuration and interstellar absorption in the plane of the Galaxy. He observed that the linear diameter of galactic clusters seems to increase with distance as though the more distant clusters were actually larger than the nearby ones. Since it is unreasonable to assume that larger clusters are at greater distances, he explained this observation by pointing out that the distances assigned to the very distant clusters are too large because the obscuration increases their apparent magnitudes. If the distances of clusters are overestimated because the intervening dust makes them appear fainter than they would if there were no dust, and if we suppose that, on the average, clusters having the same general structure and composition have the same size, we then obtain larger and larger linear dimensions as we look at fainter and fainter clusters. This is so because we obtain the true diameter of a cluster by multiplying its apparent angular diameter by its distances as determined from the known absolute magnitudes of the main-sequence stars in it.

If d is the true distance of an open cluster and L is its linear diameter, then

$$L = \frac{d\,a'}{3\,438}, \qquad (27.2)$$

where a' is the angular diameter in minutes, and 3 438 is the conversion from minutes to radians. If there were no obscuration between us and the open cluster, we could determine d directly from the formula

$$M_v = m_v + 5 - 5 \log d$$

or

$$\log d = \frac{(m_v - M_v)}{5} + 1, \qquad (27.3)$$

where m_v is the apparent visual magnitude of the open cluster and M_v is its absolute visual magnitude. The quantity $(m_v - M_v)$ is called the **distance modulus of the cluster.**

The presence of dust increases this quantity because m_v is then observed to be larger than it would be if there were no absorption present. Thus the distance is overestimated by formula (27.3), so that the linear diameter of the cluster given by formula (27.2) is also too large. To correct for this we must subtract from the right-hand side of (27.3) a quantity A_d which is a measure of the interstellar absorption. This reduces the calculated distance. If the linear dimensions of typical clusters are known (and these can be found for nearby clusters from the proper motions of individual stars in the cluster, as we shall presently see), we can then compare their calculated linear dimensions with their known linear dimensions and thus obtain the amount of absorption.

27.19 Moving Clusters

Although at first sight it might appear that all the stars in a given constellation form a cluster, this is not the case; most constellations are not clusters at all. In certain cases where some of the stars of a constellation form a cluster, other members of the constellation may not belong to this cluster. The question therefore arises: How are we to determine whether or not some particular star belongs to a given cluster? For very distant clusters this presents little difficulty since the stars of the cluster appear so close together that all members are easily recognized, but in the case of nearby clusters the stars may be spread over a large region of the sky and a real problem arises. We can solve this problem by considering the motions of stars in the cluster.

Since the stars in a cluster are physically related to each other and remain together because of their mutual gravitational attractions, they move through space as a unit sharing a common motion. This is evident from a study of their proper motions, revealing their parallel paths and identical speeds, although, in general, most clusters are so far away that the individual stellar proper motions cannot be detected. If the proper motions of stars in a cluster can be observed, the cluster is called a **moving cluster.**

Since all members of such a cluster are moving together through space, the proper motions represented by directed arrows on the sky appear to converge to (or diverge from) a point on the celestial sphere. By using this convergent point it is possible to find all the stars belonging to the cluster even though they may be distributed all around us in the sky, since they must all have proper motions converging to this point. Thus it can be shown, for example, that all but two of the stars (α and η) in the Big Dipper have identical proper motions, and hence belong to the same moving cluster.

27.20 Parallaxes of Moving Clusters

Just as the proper motions for all stars in an open cluster are the same, so too are their radial and hence also their space and peculiar velocities. If the convergent point and the radial velocity of at least one member of the cluster are known, the parallax of the cluster can be found very accurately. To see this we consider the line OC from the earth to the convergent point, and the line OS to the star S in the cluster. The space velocity V_s of this star relative to the sun is then given by the vector $\mathbf{V}$ parallel to the line OC as shown in the diagram (Figure 27.9). If θ is the angle between OC and the line of sight to the star, then we know from the chapter on stellar motions that

$$V_R = V\cos\theta,$$

and

$$V_T = V\sin\theta = \frac{4.74\mu''}{p''},$$

where V_R is the radial velocity of the star, V_T is its transverse velocity, p'' is its parallax, and μ'' is its proper motion. If we divide the second of these equations by the first, we obtain

$$\frac{V\sin\theta}{V\cos\theta} = \frac{4.74\mu''}{V_R p''}$$

or

$$p'' = \frac{4.74\mu''}{V_R\tan\theta}. \qquad (27.4)$$

Since θ and μ'' can be measured directly, and V_R can be found from the Doppler shift, we can determine the parallax of any star in the cluster very

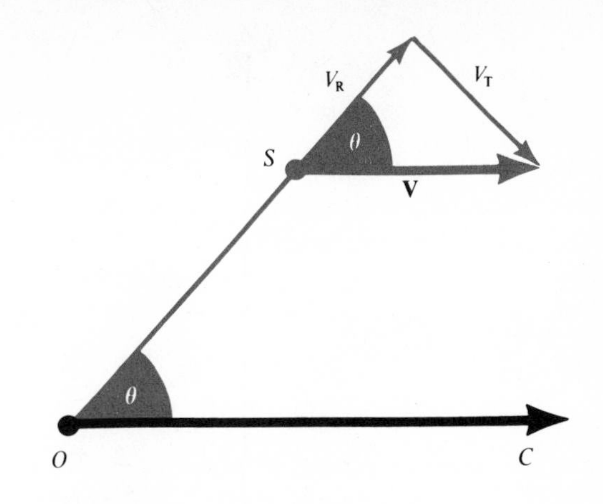

Fig. 27–9 *Determining the parallax of a moving cluster when its convergent is known.*

accurately. If we apply this procedure to the star δ Tauri in the Hyades cluster we find that $p'' = 0''.025$ since its proper motion is 0.115, its radial velocity is 38.6 km/sec, and θ is 29°.1.

27.21 The Hyades Cluster

An excellent example of a moving cluster that is relatively close to us (about 40 parsecs) is the Hyades cluster in Taurus. This cluster was first studied by Boss, who determined its convergent point from the proper motions of its members as shown in Figure 27.10, but the most detailed analysis of this cluster has recently been given by van Beuren. It is an approximately oblate spheroid (the smaller diameter is two-thirds that of the larger diameter) in which most of the stars lie at about 5 parsecs from the center (R.A. 4^h24^m, δ 15°45′) with some as far out as 10 parsecs. The equatorial plane of the cluster is approximately parallel to the galactic plane. It contains about 350 stars with a total mass equal to about $320M_\odot$, half of which is within 6 parsecs of the center. The coordinates of the convergent point of this cluster have been measured by various investigators since the time of Boss. The average value of all the determinations is R.A. 92°.9, δ 7°.6. According to van Beuren most of the stars in this cluster are G and K type main-sequence stars, with no stars bluer than A2 and with hardly any giants. The density of stars in the center of the cluster is more than three times that in the neighborhood of the sun.

Another example of a moving cluster is the Ursa Major group, which lies in a region of space to which our own solar system belongs and which

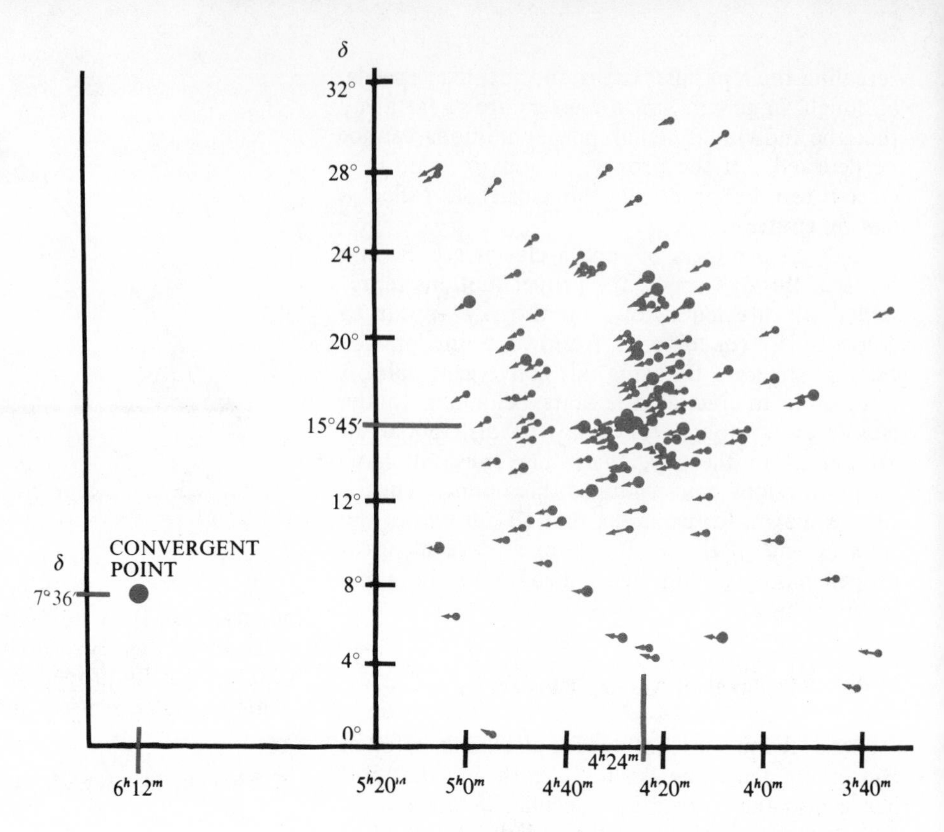

FIG. 27–10 *The convergent of the Hyades cluster as determined from the proper motions of its members. Adapted from a figure by Van Beuren, Leiden University.*

contains Sirius as one of its members. It is moving towards the point 30° away from the apex of the galactic star streaming. Two other known moving clusters are the Perseus Cluster and the Scorpio-Centaurus Cluster.

27.22 The Luminosity Function of a Cluster

If we examine the stars in a cluster and plot their H-R diagram, we can obtain some important data about their evolution and about the evolution of the cluster as a whole. We can also obtain important information about changes that occur in a cluster over a given period of time by analyzing the luminosity function of the cluster and comparing it with that of the general stellar field.

Considering the luminosity function first, we find that it changes markedly over a period of time. Thus the luminosity function of the cluster M 67 differs greatly from the over-all van Rhijn luminosity function for the stars in our Galaxy. We find that the maximum of this function for M 67 occurs at absolute magnitude +4, and that there is then a rapid decline to zero at absolute magnitude +9. The difference between these two luminosity functions indicates that the cluster must have changed considerably during its lifetime for it is clear that the stars of large absolute magnitude (i.e., the intrinsically faint stars) must have escaped from this cluster. Since this is true in general for clusters, it follows that open clusters become disrupted in time as a result of a continuous escape of the rapidly moving stars.

27.23 The Equilibrium and Disruption of Open Clusters: The Virial Theorem

To see what happens in a cluster we note that the members of the cluster can remain together only if the total kinetic energy of their random motions is equal to half of the total potential energy of the cluster. This important relationship, known as the **theorem of the virial**, is of extreme usefulness in studying the equilibrium of aggregates of objects, whether they be molecules, stars, or galaxies. We can see the justification for this theorem if we note that *the total kinetic energy of the random motions of the particles in an aggregate tends to cause the aggregate to fly apart, whereas the gravitational potential energy tends to keep the aggregate together. The aggregate can be stable only if the total potential energy exceeds the total kinetic energy.* The student will find it interesting to compare this statement with the energy conditions that we must impose on a planet for it to move in a closed orbit. (See Section 8.33.)

Since the potential energy of a cluster is of the order of $\frac{1}{2}G(M^2/R)$, where M is the total mass of the cluster, R is its radius, and G is the universal gravitational constant, and its total kinetic energy is $\frac{1}{2}M\overline{V}^2$, where $\overline{V}$ is the average random velocity of a star in the cluster, it follows from the virial theorem that we must have

$$\overline{V} = \sqrt{\tfrac{1}{2}(GM/R)}$$

if the cluster is stable.

If we now express the total mass of the cluster in solar units and the radius in parsecs, we see that the average velocity of a star in the cluster must not exceed

$$4.36 \times 10^{-2} \sqrt{M/R} \text{ km/sec},$$

if the cluster is to remain intact and not fly apart. Since the speed of escape of a single star at the surface of any cluster is equal to $\sqrt{2GM/R}$, we see that if a star near the extremities of a stable cluster has a speed equal to twice the average speed of stars in that cluster, it will escape from the cluster.

Because of the gravitational interactions among the stars in a cluster, there will in time be a distribution of velocities about the average value, with some stars moving more slowly and others with speeds sufficiently high to escape. *The reason is that as a star moves back and forth in a cluster, it can suffer considerable changes in its velocity and acquire enough energy under proper conditions to have a speed equal to or in excess of the speed of escape.* By this process clusters slowly lose stars.

We may note that the relationship between the average random velocity of the stars in a cluster and the size of the cluster can be used to determine its total mass. Thus, if the average random velocity is of the order of 2 km/sec and the radius is about 5 parsecs, the cluster has a total mass of about 10 000 suns.

27.24 The Relaxation Time of Clusters *

If a cluster is in equilibrium, the stars belonging to it move with various velocities distributed about an average value in such a way that each star, on the average, has the same total energy (potential energy + kinetic energy). This means that all stars at the same distance from the center of the cluster have, on the average, the same kinetic energy. Thus the more massive stars (hence the more luminous ones) at any given distance from the center are moving more slowly than the faint stars.

Such a cluster will remain in equilibrium unless some disturbing influence upsets it. This may be an encounter with another cluster, the gravitational tidal action of the core of the galaxy, or finally the escape from the cluster of one or more stars because of their high speeds. If any such event occurs, a certain time must elapse before an equilibrium distribution of velocities is re-established. *This interval, which is called the* ***relaxation time of the cluster****, plays an important role in the disintegration of the cluster.*

Of particular importance is the relaxation time of a cluster following the escape of a rapidly moving star. The escaping star carries away kinetic energy so that the relationship between the total kinetic energy and the total potential energy is upset and the cluster contracts, releasing enough gravitational energy to bring the kinetic energy and potential energy into balance again. The random motions of the stars in the cluster may be compared with the motions of molecules in a gas, and just as the latter supply the pressure that causes the gas to expand, the random motions of the stars in a cluster prevent the cluster from collapsing under its own gravitational attraction.

Spitzer has shown that the relaxation time for a cluster containing N stars, each having a mass M in solar units, is

$$8.0 \times 10^5 \frac{N^{1/2}R^{3/2}}{M^{1/2}(\log_{10} N - 0.5)} \text{ years},$$

where R is the radius of the cluster in parsecs. If we multiply this formula by the constant 133, we obtain the time for half the stars in the cluster to escape; this is known as the **half-life of the cluster**.

As an example we have the Pleiades with a relaxation time of 5×10^7 years and a half-life of 665×10^7 years. The data for a few clusters are given in Table 27.3. *Because of the escape of rapidly moving stars, which are also the faint stars, the luminosity function of the cluster departs more and more from the luminosity function of the general field of stars. In time, then, the cluster has a much greater concentration of luminous stars than the galaxy in general has.*

TABLE 27.3
CHARACTERISTICS OF SOME OPEN CLUSTERS

Name or designation	NGC or IC	Galactic coordinates		Distance	Diameter of concentrated areas		Number of brighter stars	Total apparent visual magnitude	Absorption A_v	Age
		l^{II}	b^{II}		Angular	Linear				
		°	°	psc	′	psc				$\times 10^6$ years
[19]	188	123	+22	1 000	14	4		9.3		10 000
M 103	581	128	− 2	2 100	7	4	30	6.9	1.5	25
	752	137	−23	400	45	5	60	6.2		1 600
h Persei	869	135	− 4	2 200	30	19	300	4.1	1.7	10
χ Persei	884	135	− 4	2 300	30	20	240	4.3	1.7	13
M 34	1039	144	−16	480	30	5	60	5.6	0.2	160
Perseus		147	− 7	155	240	11	80	2.2		10
Pleiades		167	−24	126	120	4	120	1.3	0.2	63
Hyades		179	−24	40.8	400	5	100	0.6	0.0	400
M 38	1912	173	+ 1	980	18	5	100	7.0	0.6	50
M 36	1960	174	+ 1	1 270	17	6	50	6.3	0.7	32
M 37	2099	178	+ 3	900	25	7	200	6.1		200
S Mon	2264	203	+ 2	800	30	7	60	4.3		2.0
τ CMa	2362	238	− 6	1 400	7	3	30	3.9	0.4	1.3
Praesepe	2632	206	+32	159	90	4	100	3.7	0.0	252
o Vel	I2391	270	− 7	170	45	2	15	2.6	0.2	20
M 67	2682	216	+32	830	17	4	80	6.5	0.2	4 000
θ Car	I2602	290	− 5	190	70	4	25	1.7		6.3
	3532	290	+ 2	430	50	6	130	3.3	0.0	100
Sco-Cen		320	+10	190	2 000	100	110	−0.8		3.9
Coma		228	+84	79	300	7	40	2.8	0.0	317
κ Cru	4755	303	+ 2	1 100	12	5	30	5.0	0.0	16
Ursa Maj		110	+50	22	1 000	7	100	−0.2	0.0	160
M 21	6531	8	0	900	12	3	40	6.8		6.3
M 16	6611	17	+ 1	2 000	8	5	40	6.6		1.3
M 11	6705	27	− 3	1 700	12	6	80	6.3	1.2	160
M 39	7092	92	− 2	255	30	2	20	5.1	0.2	252

TABLE 27.4
O ASSOCIATIONS

Association		Galactic coordinates		Number of stars	Distance	Associated features (NGC numbers)
		l^{II}	b^{II}			
		°	°		psc	
III + VII Cas	I	125	− 1	28	2 700	381, 366
I Per	I	135	− 5	180	1 900	h and χ Per
I Aur	I	173	0	15	1 100	χ Aur
I Ori	I	206	−18	1 000	500	1976, ϵ Ori
I Car	I	287	0	90	900	3293, I 2602
I Sco	I	343	+ 1	70	1 300	6231
IV Sgr	II	14	0	120	1 700	6561
II Cyg	I	76	+ 2	200	1 800	6871, I 4996, P Cyg
I Cep	II	101	+ 5	680	680	ν Cep

Table 27.5

T ASSOCIATIONS

Association	Galactic coordinates l^{II}	Galactic coordinates b^{II}	Diameter	Number of stars	Distance	Reference objects
	°	°	°		psc	
Tau T2	179	−20	4	11	170	T Tau
Aur T1	172	− 7	7	13	170	RW Aur
Ori T2	209	−19	4	399	400	T Ori
Mon T1	203	+ 2	3	141	800	S Mon, NGC 2 264
Lyr T1	60	+20	17	13	400	LT Lyr
CrA T1	0	−18	0.5	6	115	R CrA

Table 27.6

CONVERGENT POINTS OF SOME MOVING CLUSTERS

Cluster association or group	Convergent point relative to sun				Velocity	
	α	δ	l^{II}	b^{II}	rel. sun	corrected
	°	°	°	°	km/s	km/s
Hyades	94	+ 7	202	− 3	45	32
Orion	84	−17	220	−23	22	5
Praesepe	95	+ 4	206	− 3	41	27
Scorp-Cen	107	−46	256	−16	24	10
Coma Ber	121	−47	262	− 8	8	14

Table 27.7

CLUSTER AGE RELATIONS

Age in million years	1	10	100	1 000	3 200
Most luminous M_v on main sequence	−6.5	−3.7	−0.9	+1.9	+3.3
Earliest type on main sequence	O7	B1	B7	A5	A8
Smallest $(B - V)_{cor}$ on main sequence	−0.31	−0.23	−0.05	+0.30	+0.52

27.25 The Evolution of Open Clusters Using H-R or Color-Magnitude Diagrams

In our discussion of the evolution of stars as determined theoretically by stellar models in Chapter 23, we saw that stellar clusters can be used as a check on the theory since the H-R diagrams of clusters represent the time lines for the stars at their present evolutionary stage. If we picture all the stars in a cluster as having started out with the same chemical composition at the same time, but (because their masses and hence their luminosities were different) at different points along a zero main-sequence line, these stars will have evolved to their present position on the observed H-R (or C-M, color-magnitude) diagram of the cluster in the same time, but obviously along different tracks. This sort of analysis was started by A. Sandage and M. Schwarzschild and has been carried out in considerable detail by Sandage himself, who has given a complete discussion of the way the presently observed C-M diagram for M 67 has evolved from the original C-M diagram for this cluster when its stars were all on the main sequence.

27.26 The Analysis of Evolution of Open Clusters: M 67

The analysis proceeds in two steps: (1) the correlation of the present luminosity function of the cluster with the luminosity function as it would be if no stars had escaped from the cluster and if the stars in the cluster had not evolved at all; (2) the correlation of the observed cluster data with the data obtained from theoretical stellar models constructed by solving the stellar-interior equations (discussed in Chapter 23) under the assumption that the mean molecular weight of the cluster star changes as it evolves in the H-R diagram.

Sandage's method. To see how these procedures are employed we shall first outline Sandage's analysis of the evolution of stars in M 67 from the initial main-sequence distribution to the presently observed C-M diagram.

The observed C-M diagram for this cluster is given in Figure 27.11, where the absolute magnitudes are plotted against the effective temperatures of the stars rather than against their spectral classes or color indices. This distribution is shown by the heavy line which starts out at $\log T_e = 3.6$, $M_{bol} = 7.2$ and moves up along the main sequence to the point $\log T_e = 3.78$, $M_{bol} = 3.46$. This part of the C-M diagram is called the main-sequence part of the cluster although its upper end is somewhat to the right of the standard main sequence as compared to the lower part. The number of stars falls off very rapidly to zero as we go down the main sequence.

At about $M_{bol} = +3.46$ the observed C-M diagram, as given by the solid curve, turns to the right and extends horizontally to the point $\log Te = +3.66$. It then moves up sharply almost parallel to the absolute-magnitude axis, extending to point $M_{bol} = 0$. We may call this the subgiant branch of the cluster. In addition to this we have a branch starting at $M_{bol} = 1.2$, $\log T_e = 3.64$ and running to the left, almost parallel to the horizontal axis, to the point $M_{bol} = 1.78$, $\log T_e = 3.9$.

To see how the stars evolved we must have, in addition to the observed distribution, the line representing the original main sequence on which all the stars began their lives. We can obtain this **zero time line** from the theory of stellar structure (as developed in Chapter 23) if we know the original chemical composition of the cluster (i.e., at the time when the stars were first formed).

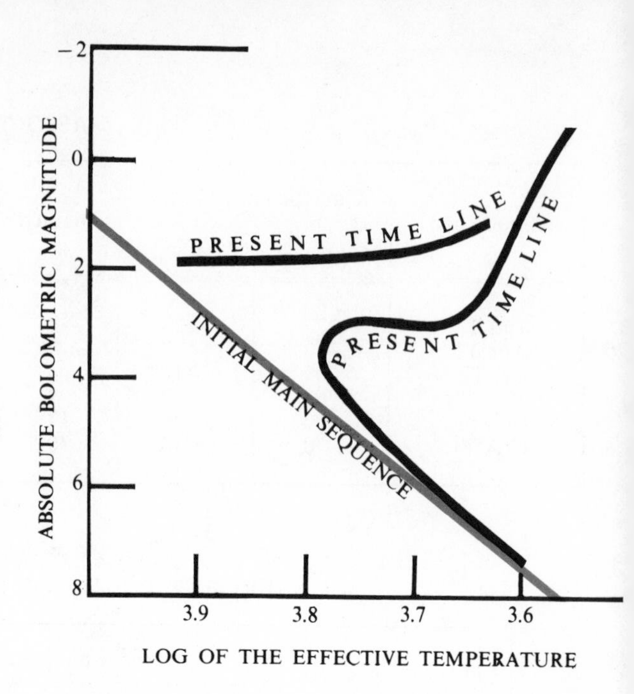

FIG. 27–11 *Schematic drawing of the H-R diagram of M 67 showing the zero time line (original main sequence) and the present time line (present distribution of M 67 stars). (After A. R. Sandage,* Astrophysical Journal, *CXXVI (1957), 329. © 1957 by the University of Chicago).*

Since the chemical composition of the present atmospheres is the same as the original chemical composition of the entire cluster, this information is available.

From this, and by using our interior equations, we can determine the zero time position in the C-M diagram for a star of given mass. The zero time positions for all the stars in the cluster lie on the straight line in the diagram that runs from the point $\log T_e = 3.6$, $M_{bol} = 7.2$ to the point $\log T_e = 3.9$, $M_{bol} = 2.5$. This, then, is the zero time line and is to be compared with the present time line described above. All stars in our cluster began their lives at the same moment on the zero time line, at points depending on their masses (the more massive stars higher up), and have spent the same time (present age of the cluster) in reaching their present positions on the observed C-M curve. However, their paths from the zero time line to the present time line are quite different. To carry through his analysis Sandage calculated these tracks.

He also had to know the luminosity function for the original main-sequence distribution so that he could determine how many stars lying in a given small stretch in the present C-M distribution originated from a definite segment on the zero time line. To obtain the original luminosity function

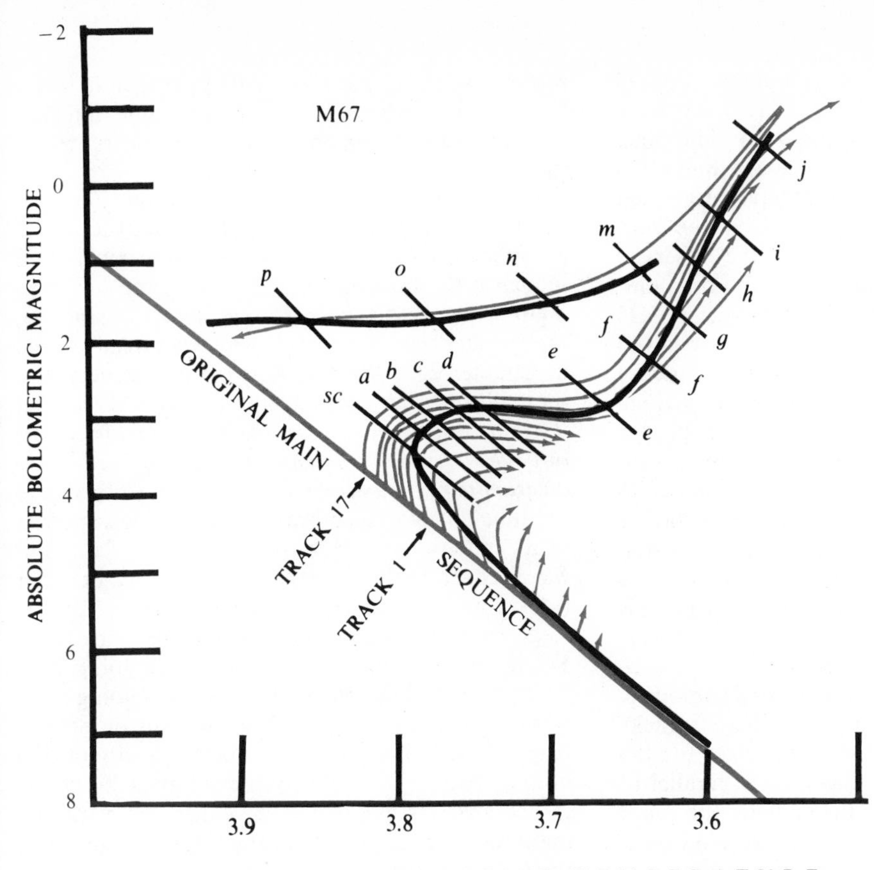

FIG. 27–12 *The evolutionary tracks for stars on the original main sequence or zero line to the present time line. The line marked SC represents the Schönberg-Chandrasekhar limit. (After A. R. Sandage,* Astrophysical Journal, *CXXVI (1957), 329. © 1957 by the University of Chicago)*

of the cluster he altered the present observed luminosity function by taking into account the rapidly moving stars that have escaped from the cluster during its lifetime.

27.27 Sandage's Analysis of the Evolution and Determination of the Age of M 67

We shall now see how the calculated evolutionary tracks and original luminosity function enabled Sandage to complete his analysis. We first note that the main-sequence part of the observed C-M diagram for the cluster coincides with the original main sequence up to about $M_{bol} = 6.8$ and then departs from it gradually up to about $M_{bol} = 3.46$, at which point it moves sharply into the subgiant branch. Up to this point, which we shall label *SC*, we can compute the evolutionary tracks by using statistical methods and the theory of stellar interiors. These tracks are shown in Figure 27.12, and we see that they are almost parallel to each other and to the absolute-magnitude axis. The stars with small absolute magnitudes moved from the zero time line to the present observed distribution along much longer paths and hence have evolved more rapidly and drastically than the intrinsically faint stars, which have departed only slightly from the main sequence during the age of the cluster. We see in particular that stars at the turn-off point and entering the subgiant branch on the observed C-M diagram (i.e., stars at the point *SC* with $M_{bol} = 3.46$, $\log T_e = 3.82$) have evolved from a point $M_{bol} = 4.4$, $\log T_e = 3.78$ on the zero time line. This is marked track 1.

To find which points on the subgiant branch correspond to the points on the original main sequence we must compare the present observable luminosity function with the computed original function since theoretical tracks have not been calculated for all points on the zero time line. Using the luminosity function Sandage finds, for example, that as many stars must have originated between $M_{bol} = 4.4$ and $M_{bol} = 4.3$ on the zero time line as there are stars now on the observed C-M diagrams between the point *SC* and the point *a*. It follows that a star at the point *a* ($M_{bol} = 3.26$ on the observed C-M curve) must have evolved from a star of $M_{bol} = 4.3$ on the main sequence and we can, therefore, draw its evolutionary track by connecting these two points by the proper evolutionary curve. Since all the calculated evolutionary tracks leave the zero time line parallel to each other, we can start this track parallel to track 1 as shown in Figure 27.12. Thus, by moving step by step in definite magnitude intervals on the observed C-M curve, we can relate points on this curve to the original zero-time main-sequence line by means of evolutionary tracks without having to compute these tracks in detail from stellar models.

Sandage in this way calculated 17 different tracks for M 67 and gave detailed mapping data for the seventeenth track. We see that it runs parallel to the observed C-M diagram throughout the giant and subgiant region and then turns back on itself and practically coincides with the upper left-hand horizontal branch of the observed C-M diagram.

27.28 The Time of Evolution and the Ages of Open Clusters *

Since all the stars in the cluster started their lives together, they must have spent the same time (the age of the cluster) along their evolutionary tracks from their original main-sequence positions to their present positions on the observed C-M track. Thus, the time it took a star on track 1 to go from the point 1 on the main sequence to the point *SC* on the observed C-M track is equal to the time it took a star on track 2 to go from 2_0 to 2_a on the observed distribution and also to the time it took a star on track 17 to go from 17_0 to point 17_p.

This time (the age of the cluster) can be computed by considering how long it takes a star of $M_{bol} = 4.4$ to use up 12 per cent of its hydrogen since we know from stellar models that this is the amount of fuel a star requires to go from the point 1 to the point *SC* on its evolutionary track. Since at *SC* $M_{bol} = 3.46$, whereas it started with $M_{bol} = 4.4$, we must take the average luminosity over the entire track and divide this into the total energy released when 12 per cent of its hydrogen is transformed into helium to obtain the time interval. This turns out to be about 5.1×10^9 years, which therefore is the age of this cluster.

Using this same procedure we can find the ages of other open clusters and we find that these ages are related to how high up along the original main sequence the turn-off point *SC* from the observed main sequence occurs. *The higher up this occurs, the younger the cluster is*, as we have already noted in Chapter 23. We also note, in connection with these clusters, as given in the data for M 67, that *the more massive a star is, the more quickly it runs through a given segment of the observed C-M diagram during its evolution.* We can, in fact, give the time any star spends along successive segments of its evolutionary track from Sandage's data, and we discover that whereas a star midway between track 1 and track 17 spends about 3.4 billion years to go from the original main sequence to the point *SC*, it spends about 330 million years in going from *SC* to *a*, and about 840 million years in going from *a* to *d*. It passes through the subgiant phase and reaches the giant branch at point *j* in about 530 million years. This is why the subgiant branch is rather sparsely populated as compared with the main-sequence segment and the turn-off segment in the C-M diagram.

The actual diagrams for four individual open clusters are given in Figures 17.8 and 17.10 (M 67, NGC 2264, the Pleiades, and Praesepe). A composite schematic diagram showing the turn-off points and ages for clusters is given in Figure 17.9. The observed C-M diagrams given above for open clusters, together with Sandage's analysis of the theoretical tracks for stars having masses less than $2M_\odot$, give us a fairly complete picture of the evolutionary tracks of Population I stars.

27.29 O and T Stellar Associations

In addition to the relatively tightly packed open clusters consisting of comparatively few stars, there are large, loosely knit groups known as **stellar associations** or **aggregates** which, according to Ambartsumian, are of great importance for the formation of new stars. These associations are of two types: the **O associations**, which contain as their brightest members O and early B type stars, and the **T associations**, which contain T Tauri variables. As Blauuw has pointed out, these stars have a strong clustering tendency and very few of them are found outside the associations. They are far too distant for proper motion measurements to show whether the individual stars are moving together, and it may well be that the mutual gravitational attractions among the members of the association are not strong enough to keep them together for very long, but these associations are not yet old enough to have dispersed to any extent. It is clear that the stars in these associations must be among the very youngest stars in our Galaxy and, indeed, both Ambartsumian and Blauuw have pointed out that O and B stars are actually being born in these associations at present. As examples of such associations we have the stars in the Orion region and also those in the Scorpio-Centaurus region.

Blauuw's observation of the clustering of O and B type stars is interesting in the light of the work of Morton S. Roberts, who has presented powerful arguments that all O and B type stars in our Galaxy originated in clusters and, indeed, in the same clusters. He first estimated the total number of O stars in our Galaxy from known star counts. He then estimated the total number of galactic clusters and multiplied this by the known number of stars per cluster. He found the two values for O stars in the Galaxy to be the same. He did the same thing for B-type stars with the same results. Going further, Roberts has suggested the possibility (and given quite strong evidence for it) that all Population I stars originated in open (galactic) clusters or associations.

The random motions of stars in O and T associations are quite different from those in open clusters because they are moving in such a way as to bring about the ultimate dissolution of the association. As Ambartsumian has stressed, these are unstable groups of stars which cannot have been in existence for more than tens of millions of years (i.e., for no more than about 1 per cent of the age of the Galaxy) so that the stars in them are only a few million years old.

Blauuw, in agreement with Roberts, has emphasized the very strong evidence that the majority of stars now being formed and recently formed in the plane of the Milky Way (i.e., the new Population I stars) were and are being formed almost exclusively in associations and hardly ever or never outside such associations. From an analysis of the luminosity function of the general stellar field and the luminosity function for stellar associations, Blauuw arrives at the conclusion that the rate of formation of stars is uniform at the present time and is not much different today from what it was billions of years ago. As we shall see, these aggregations of stars are associated with the gaseous matter and dust from which the new stars are being born.

27.30 The Ages of Associations

In general, Blauuw finds that stars born in these associations have random velocities below 30 km/sec, but certain of them are formed with unusually high speeds and move out to great distances from the plane of the Galaxy during their lifetime to form a kind of halo around the Galaxy. But, in general, most of the new stars formed have speeds that keep them well within 200 parsecs of the plane of the Galaxy.

The average initial random velocity of stars in an association enables us to determine the age of the association and the stars in it. Thus the stars in the association near ζ Persei, which are all within 30 parsecs of each other, have random velocities that indicate an average expansion rate of 12 km/sec for the association. This entire group must therefore have reached its present size in about 1.3 million years, which then represents the age of the stars in this group. This same sort of analysis leads to an age of about 4.2 million years for the association near the star 10 Lacertae and an age of about 70 million years for the Scorpio-Centaurus association.

As examples of stars that were born with high velocities there are the three stars, AE Auriga, μ Columba, and 53 Arietis, which are moving away from the Orion association at speeds in excess of 100 km/sec. Tracing them back we see that they must have been rifled into space about $2\frac{1}{2}$ million years ago as a result of an explosion in the Orion aggregation.

27.31 Globular Clusters

Although the distribution of galactic clusters enables us to trace the plane of the Milky Way out to distances of about 5 000 to 6 000 parsecs, we find that as we push farther and farther in towards the center of the Galaxy these open clusters merge into the distant star clouds, thus preventing us from completing our picture of the galactic structure. There is, however, another type of cluster, the **globular cluster**, which gives us important information about our Galaxy. As we have seen, *globular clusters differ in appearance, size, and star counts from open clusters (see Figure 27.13), but there are two*

FIG. 27–13 (*a*) NGC 5272 Globular star cluster in Canes Venatici. Messier 3. (200-inch photograph from Mount Wilson and Palomar Observatories)

other important differences as well. They do not hug the plane of the Milky Way as open clusters do, but form something of a halo around the Galaxy, and they are free of dust and gaseous matter and contain primarily Population II stars.

27.32 The Distribution of Globular Clusters

The globular clusters are very distant objects which lie mostly in the direction of the center of our Galaxy (in the direction of Sagittarius) at considerable distances from the galactic plane. Almost one-third of the total number can be photographed on a single plate. Instead of forming a flattened system (the way the Galaxy and the distribution of open clusters do) the globular clusters have a roughly spherical distribution as though they occupied a shell of space concentric with the Galaxy itself. As we shall see, Shapley used the globular-cluster distribution to determine the position of the galactic center.

Over one hundred globular clusters associated with our Galaxy have been identified and it is estimated that there may be another hundred hidden from us by the main bulk of the Galaxy. However, if we consider the number of globular clusters associated with other galaxies, this estimate may have to be increased to 1 000. These objects are faint and show neither measurable parallax nor appreciable proper motion, which indicates that they are at very great distances from us. We shall

TABLE 27.8
SELECTED GLOBULAR CLUSTERS

Cluster	NGC	Galactic coordinates l^{II}	b^{II}	Representative diameter angular	linear	Distance	Visual magnitude	Visual absorption	Observed no. of variables	Radial velocity in km/sec	Mass
				′	psc	kpsc					$10^4 \mathscr{M}_\odot$
47 Tuc	104	306	−45	7.6	10	4.6	4.01	0.2	11		
	2419	180	+25	1.9	32	58	10.7	0.3	36	+ 14	
ω Cen	5139	309	+15	14.2	20	4.8	3.57	1.1	164		
M 3	5272	42	+79	3.4	13	13	6.38	0.2	187	−150	21
M 5	5904	4	+47	4.5	12	9.2	5.93	0.0	97	+ 45	6
M 4	6121	351	+16	9.8	9	3.0	5.91	1.3	43		6
M 13	6205	59	+41	4.8	11	8.2	5.87	0.2	10	−228	30
M 19	6273	357	+10	3.5	7	7.3	6.88	1.3	4	+102	
M 22	6656	10	− 8	10	9	3.1	5.09	1.3	24	−148	700
Δ 295	6752	336	−26	15	24	5.6	6.2	0.6	1	− 3	
	7006	64	−19	1.2	17	48	10.68	0.3	40	−348	
M 15	7078	65	−27	2.8	11	13	6.36	0.2	93	−114	600

FIG. 27–13 *(b) NGC 6205 Globular star cluster in Hercules. Messier 13. (200-inch photograph from Mount Wilson and Palomar Observatories)*

consider the method of determining their distances later on. The brightest of these globular clusters, ω Centauri, is visible to the naked eye as a fourth-magnitude object in the southern hemisphere. There are about four other such naked-eye globular clusters, including M 13 (Hercules), 47 Tucanae, M 22 and M 5. The very faintest of these objects are detectable only in the most powerful telescopes.

27.33 Star Counts in Globular Clusters

The total number of stars in a globular cluster can be estimated only from star counts on photographic plates, and the results obtained this way are somewhat unreliable. If the photographic plate is exposed for only a short time, we miss most of the faint stars in the cluster, but if we expose the plate long enough to record many of the intrinsically faint stars, the dense concentration of stellar images at the center of the plate is so high that a fused image results. However, even with incomplete star counts it is clear that globular clusters must contain hundreds of thousands or even millions of stars. Using the 200-inch telescope, Sandage counted 44 500 stars in M 3.

27.34 Integrated Colors of Globular Clusters

The integrated colors (all stars contributing) of globular clusters are fairly uniform for clusters that are at high latitudes, but there is a reddening as we approach the galactic plane that is obviously due to obscuration in the plane of the galaxy. The color indices are all positive, indicating the *absence* from these clusters of blue and blue-white main-sequence stars. *The brightest stars in these clusters are all red giants with an average color index of* +1.4, *although occasionally stars of color index as low as* -0.3 *are found.* The integrated spectral classes of globular clusters range from A5 down to K, but most globular clusters have F8-type spectra.

27.35 Distances of Globular Clusters

Before a cluster can be analyzed in any detail its distance must be known. This can be obtained only indirectly by using some luminosity criterion for the individual stars in the cluster. The pioneer work in this field was done by Shapley, who obtained the distances of globular clusters by using both their RR Lyrae type variables and their brightest stars. Since we know that RR Lyrae type stars all have about the same absolute magnitude—about $+0.6$—we can obtain the parallaxes of such stars by measuring their mean apparent magnitudes and applying the standard formula. Since there is some uncertainty about the value $+0.6$ for the mean absolute magnitude of the RR Lyrae variables, the distances of the globular clusters based on this value are also uncertain. There is some evidence that the mean absolute magnitude of these variables may be as large as $+0.8$ so that the globular clusters' distances may have to be further corrected to this extent.

Another procedure used by Shapley to determine distances is based on his discovery that in all globular clusters containing variable stars, the brightest stars are about 1.5 photographic magnitudes brighter than the RR Lyrae stars. One can use this method to get the distances of globular clusters that are so far away that the RR Lyrae variables are too faint for reliable measurements, and of clusters containing no RR Lyrae variables.

Where Shapley was unable to obtain precise data for the brightest stars or for RR Lyrae variables, he correlated the apparent diameters of globular clusters of known distance with their parallaxes, and thus established a distances scale based upon the apparent size of the cluster. This would be a highly accurate method if all clusters were the same size, but since this is not the case errors are present which must be taken into account.

The distances that Shapley obtained by using these various methods are somewhat too large, because he did not take into account interstellar absorption. If we correct for this we find that ω Centauri (one of the close globular clusters) is at a distance of 4 800 parsecs; M 13 at 8 200 parsecs; M 3 at 13 000 parsecs; and 47 Tucanae at 4 600 parsecs. Most globular clusters are at considerable distances, the faintest ones as far out as 25 000 parsecs (in Leo and Sextans).

27.36 Dimensions and Stellar Concentrations

The absolute magnitudes of observed globular clusters range from -6 to about -10; for ω Centauri, for example, $M_{ph} = -9.9$. Their full diameters range from 7 to 120 parsecs. ω Centauri is estimated to have a linear diameter of 20 parsecs, whereas M 13 has a linear diameter of 11 parsecs. The great concentration of stars in the center is indicated by the size of the core, which in most clusters does not exceed 5 parsecs. It is estimated that the density of stars near the center of a globular cluster is thousands of times greater than that in the neighborhood of the sun. Thus in the 5-parsec core of a large globular cluster, there may be as many as 30 000 to 40 000 stars.

27.37 The Luminosity Function of Globular Clusters

An important feature of a globular cluster is its luminosity function, which indicates how many stars it contains in each absolute magnitude range. Because of the distances of these clusters it is difficult to pick up the very faint stars on photographic plates, and even the most accurate counts have not been able to go down to absolute magnitudes fainter than our sun. The most complete data on luminosity function have been obtained for M 3 by Sandage, who was able to get down to absolute magnitudes somewhat fainter than the sun. An interesting feature which he found—and which may be characteristic of other globular clusters—is a sharp peak at about $M = 0.0$. The luminosity functions of these clusters also indicate that most of the light does not come from the great mass of stars. Although 90 per cent of the light from M 3 comes from stars brighter than $M = 4$, 90 per cent of the mass of the cluster is due to stars fainter than this absolute magnitude. This indicates the presence in this cluster of a large number of white dwarfs, which Sandage estimates at about 50 000.

27.38 The Shape of Globular Clusters and Star Distribution

Although globular clusters appear quite spherical when first viewed, a careful analysis of their structure shows that they are ellipsoidal with discernible flattening. The most pronounced case is that of the cluster M 19, for which the ratio of the minor to the major axis is 0.4. This indicates that these clusters are spinning, and Shapley has found from an examination of 30 such clusters that the axes of rotation are distributed at random. There is also a considerable amount of variation in the concentration of stars towards the cluster centers, which has led to a classification of globular clusters according to these central concentration values. The range is from class I (the highest degree of concentration) to class XII, in which the stars appear to be rather loosely distributed. As examples, M 13 belongs to class V, and M 3 to class VI.

27.39 Proper Motions of Globular Clusters

Although these clusters are much too far away for reliable proper-motions measurements, some work has been done in this field, and proper motions have been measured for about eight or nine clusters. *All indicate that globular clusters have large space velocities.* This is borne out by their radial velocities, which can be measured much more easily by the Doppler shift. The radial-velocity work was begun by E. M. Slipher, who studied 17 globular clusters, and more recently continued by Mayall. If all the data are taken together, we find that the radial velocities range from about 300 km/sec of approach to 400 km/sec of recession. We conclude that most of the globular clusters have high peculiar velocities and must therefore be classed among the high-velocity objects. As we shall see, all this can be accounted for by the motion of the solar system around the center of the Galaxy together with the fact that the globular clusters themselves have elliptical orbits around this center.

27.40 The Equilibrium of Globular Clusters

We analyzed the equilibrium of an open cluster by considering the balance between its total kinetic and its total potential energy, and we can do the same thing for a globular cluster. We then find that $\overline{V}^2$ (the mean of the squares of speeds of the individual stars in the cluster) is $GM/\overline{R}$, where G is the gravitational constant, M is the total mass of the globular cluster, and $\overline{R}$ is a kind of average radius of the globular cluster which takes into account the fact that most of the mass is concentrated near the center. M. Schwarzschild and S. Bernstein have applied this formula to the globular cluster M 92 and found that its effective radius R is about 10 parsecs and that $\overline{V}^2$ is about 60 km/sec (note that this arises from the random motions of the individual stars within the cluster and has nothing to do with the total velocity of the cluster). From this they obtained for the total mass of the globular cluster the value $1.5 \times 10^5 M_\odot$.

27.41 C-M Diagrams for Globular Clusters and the Evolution of Population II Stars

We are able to test the theory of the evolution of Population I stars by using the Sandage technique of superimposing on the observed C-M diagram of open clusters the calculated evolutionary tracks obtained from the theory of internal structure of such stars, and we can do the same thing for Population II stars by superimposing on the C-M diagram of globular clusters the theoretical evolutionary tracks of Population II stars. We have already seen that these C-M diagrams of globular clusters are similar in their general appearance to those of open clusters, with certain significant differences in the region where the diagrams depart from the main sequence.

As an example of the procedure for studying the evolution of Population II stars we use Sandage's analysis of the globular cluster M 3. Again we picture the stars as having been originally distributed along the main-sequence line (see Figure 27.14) according to the original known luminosity function for these stars. The observed C-M diagram (note that the absolute bolometric magnitude is plotted against the log of the effective temperature) for the cluster is given by the heavy line which coincides with the original main-sequence line at absolute bolometric magnitude $M = 7$, and departs gently from it up to $M_{bol} = 3.46$, which again marks the sharp turn-off from the main sequence. We may consider the stars lying in this part of the observed C-M diagram as constituting the main-sequence part of this globular cluster.

We then have a stretch from the point SC to the point $M_{bol} = 2$, after which there is a very rapid rise almost parallel to the magnitude axis up to the giant branch at $M_{bol} = 0.0$. This region from SC to $M_{bol} = 0$ may be considered as the subgiant branch. The C-M line continues up into the

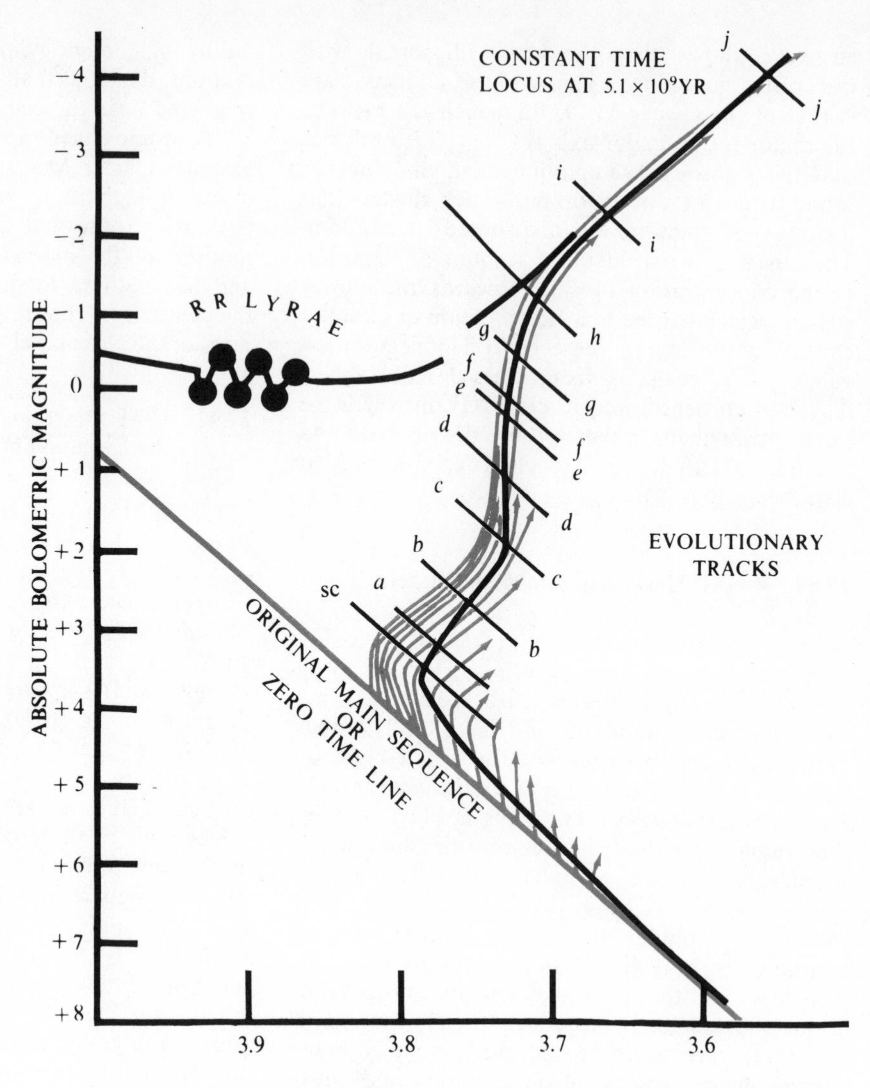

FIG. 27–14 *The zero-time line (original main sequence) and the present time line for the globular cluster M 3. Evolutionary tracks from the original main sequence to the present time line according to Sandage are also shown. (After A. R. Sandage,* Astrophysical Journal, *CXXVI (1957), 335. © 1957 by the Univerity of Chicago)*

supergiant branch at $M_{bol} = -4$, then falls back on itself, moving to the left into a horizontal branch at $M_{bol} = 0.0$. The typical RR Lyrae gap extends from $C.I. = 0.2$ to $C.I. = 0.0$.

This line may be taken as the constant-time line representing the present age of the globular cluster which, as we shall see, is about 6.5×10^9 years. Each point on it represents the present age of a star that has evolved from a point on the original main sequence (i.e., the zero-time line) in this period of time. The evolutionary tracks of the very faint stars ($M_{bol} > 4.5$) are very short and parallel to the magnitude axis, indicating a very slow evolution from the original main sequence to the presently observed main sequence with very small increases in luminosity.

As examples of evolutionary tracks we note that the star, $M_{bol} = 4.4$, on the original main sequence has evolved up to the point SC, $M_{bol} = 3.46$, and the star with $M_{bol} = 4.3$ on the original main sequence has evolved in the same time to the point a, $M_{bol} = 3.36$, etc. The various tracks are indicated in Figure 27.15.

Finally the star $M_{bol} = 3.8$ on the main sequence has evolved to the point j, with $M_{bol} = -4$, so that it is now a supergiant.

27.42 Time Tracks of Population II Stars and Ages of Globular Clusters

Just as with open clusters, we can analyze times taken by globular star clusters to evolve to their present observed positions on the C-M diagram, and how long they spent in each stage of their evolution. This can be done by calculating the amount of original nuclear fuel burnt by the star in going from stage to stage as it evolved. By using this sort of analysis, N. T. Woolf has shown that the age of the globular cluster M 3 cannot be more than 6.8×10^9 years. For if we consider a star that has used up all its fuel and is at the very end stage of its evolution (i.e., at the extreme left end of the horizontal branch of the C-M diagram) we know that it could not have emitted more energy during its lifetime (which is equal to the age of the cluster) than it had available in the form of unconsumed nuclear fuel at its birth.

If we now suppose that it originated at about $M_{bol} = 3.98$ and consisted then entirely of hydrogen, it is easy to show that it must have spent 70 per cent of its life in going from the point SC to the point j, and then down again to the end of the horizontal track. One can show from this sort of analysis, considering the star's original amount of nuclear fuel, that the star must have spent 4.6×10^9 years to go from the main sequence to SC, about 1.6×10^9 years to reach the point j, and about 2.3×10^9 years to pass down to and across the horizontal branch. This agrees with calculations of stellar models of Population II stars made by Haselgrove and Hoyle, and shows that the cluster cannot be older than the sum of these intervals, namely 8.5×10^9 years.

27.43 RR Lyrae Stars *

The RR Lyrae stars in a globular cluster represent a rather short phase in the evolution of Population II stars. Sandage has indicated that a star passes through the RR Lyrae domain in about 8×10^7 years and that during this time it undergoes changes in both period and luminosity. Whereas it has been customary to assign an absolute magnitude of zero to these stars, it is believed that a better figure is closer to $+0.6$. In any case, from the analysis of the M 3 globular cluster it appears that a star enters the RR Lyrae domain with a period of pulsation of about 1 day, leaves with a period of about 0.3 days, and is somewhat more luminous at the end of its variable phase than at the beginning.

27.44 Galactic Structure from Globular Clusters

The analysis of the structure of the Galaxy based on the distribution of open and globular clusters leaves us with a picture of a thin but well-populated layer of Population I stars, extending in all directions but increasing in stellar density towards the direction of Sagittarius. Both the open and globular clusters indicate that at a distance of 4 000 parsecs this thin layer begins to bulge out just where the open clusters merge into the dense star clouds. However, the globular clusters indicate that this structure, diagrammed schematically in Figure 27.15, continues far beyond 4 000 parsecs. This part of the Galaxy (shown in the figure) is also characterized by the presence of stellar associations in which new stars are being born from dust and gaseous matter.

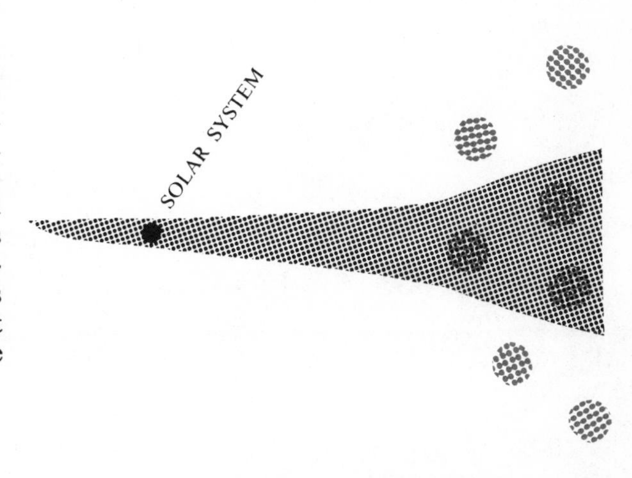

FIG. 27–15 *The schematic profile of the Milky Way beyond 4 000 parsecs as indicated by the distribution of globular clusters.*

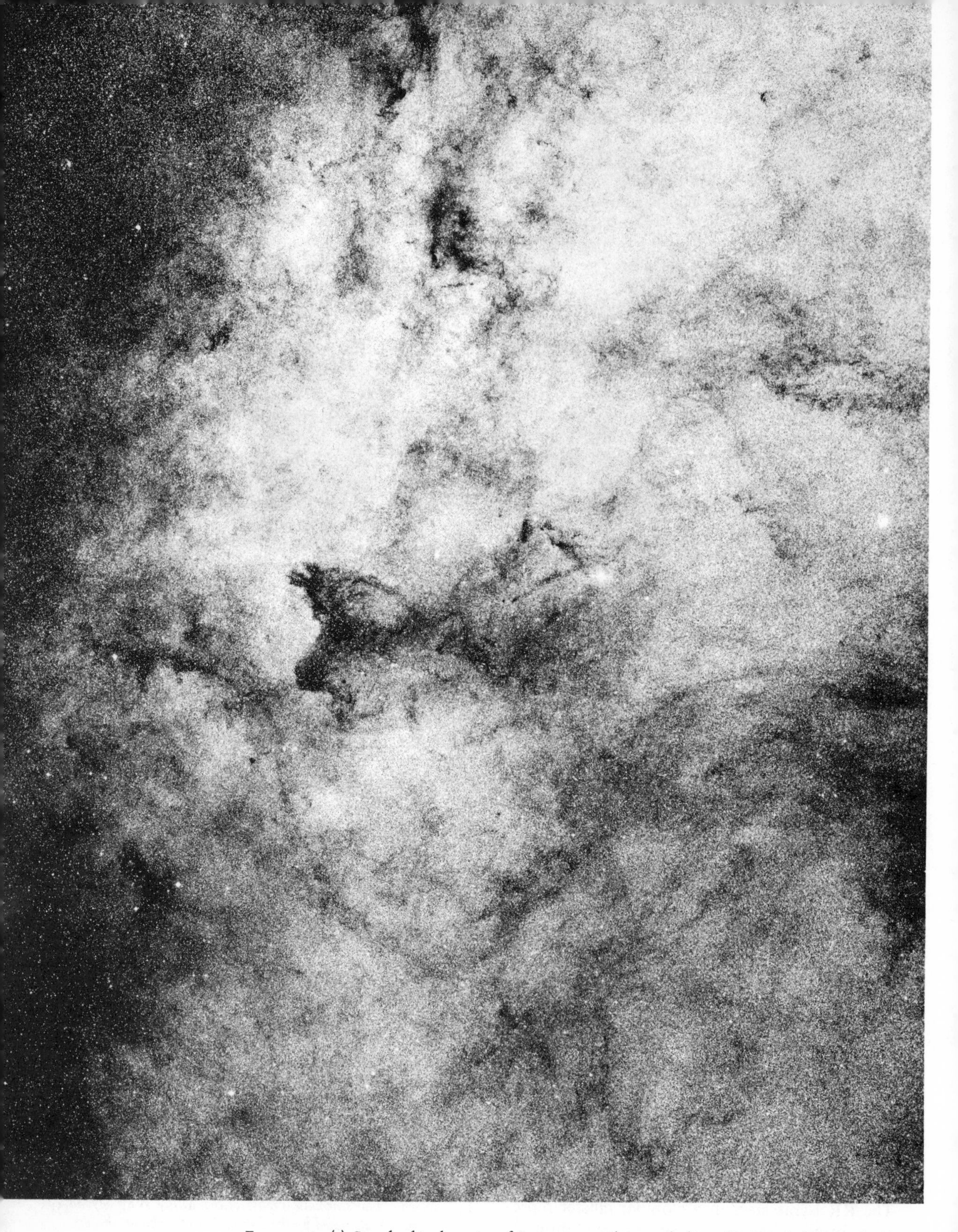

FIG. 27–16 (*a*) *Star cloud in the region of Sagittarius. Photographed in red light with the 48-inch Schmidt telescope. These are the richest clouds in the Northern Milky Way.* (*From Mount Wilson and Palomar Observatories*)

27.45 Star Clouds

We have seen, from star counts out to about 2 500 parsecs, from the analysis of open clusters out to 5 000 parsecs, and from the distribution of globular clusters out to still greater distances, that we can get a picture of the Galaxy up to the point where the nuclear bulge begins. However, we cannot penetrate into this nuclear region with star counts because of the obscuration, nor with open clusters because these clusters merge into the background of the distant stars. Globular clusters do not help us either because they surround the nucleus and are not found in the plane of the Galaxy itself.

However, at the point where we begin to lose sight of the open clusters we do find many luminous regions which, taken together, are part of the Milky Way itself but which nevertheless can be distinguished as separate, in the sense that they stand out as bright concentrated stretches. These **star clouds** consist of millions upon millions of stars; with the aid of a telescope these stars can be seen as very faint objects, but on photographic plates they are almost too numerous to be counted. (See Figure 27.16.) Star clouds extend almost unbrokenly from Cassiopeia to Cygnus and then from Cygnus to Sagittarius, in which constellation they are most dense; here they break up into two bands of clouds separated by a dark region called the **Great Rift.**

These star clouds consist of all groups of stars. The O, B, and A stars are so prominent that they can be used as indicators to determine the stellar density of the clouds. Besides these bright, early main-sequence stars, the star clouds contain other bright stars such as Cepheids, RR Lyrae variables, long-period variables, and eclipsing binaries which serve as beacons. Among the brightest of the star clouds are those in Cygnus and Scutum and the collection in Sagittarius. Although the clouds start at distances in excess of 3 000 parsecs, they undoubtedly extend into the very core of the Milky Way. The dimensions of clouds parallel to the Milky Way are generally of the order of 200 to 300 parsecs, and it is estimated that the star density in these clouds is about the same as in our own neighborhood.

FIG. 27–16 *(b) Cygnus region. R.A.* $20^h\ 10^m$*, declination* $+35°00'$*. The dark gaseous nebulosity can be seen threading this star cloud. (From Mount Wilson and Palomar Observatories)*

In the other direction from Cassiopeia, towards Perseus, there is a thinning out of the star clouds, but then, as we follow the Milky Way through Monoceros into the southern hemisphere, the clouds become very numerous and brilliant again through Carina, Musca, and Crux with great piling up in Scorpio. This region is broken up by dark masses of obscuration such as the **Coal Sack**. The general appearance of these star clouds in the Milky Way and particularly the rift in Sagittarius led very early in the analysis of the Milky Way to the concept of a spiral structure. Today, although few astronomers doubt that our Milky Way is a spiral galaxy, the actual proof, by tracing the arms, has not come from star counts but rather from radio signals emanating from the dust and gaseous matter between stars. In the next chapter we shall consider this interstellar matter and nebulosity.

EXERCISES

1. (a) What is the average velocity of the stars in a stable cluster consisting of 1 000 stars like the sun if the diameter of the cluster is 3 parsecs? (b) What is the speed of escape at the periphery of this cluster?

2. Calculate the average star velocities for five open clusters for which the necessary data are available in the table of open clusters. (Assume that these clusters are stable.)

3. Calculate the relaxation times for the clusters in problem 2.

28

INTERSTELLAR MEDIUM

28.1 Composition of the Interstellar Medium *

We have seen that the presence of interstellar dust prevents us from analyzing the nucleus of our Milky Way, and, indeed, cuts off most of the light from all but the brightest individual stars at distances beyond 2 000 to 2 500 parsecs. In addition to this dust, the existence of which was not definitely established until about 30 years ago as the result of Trumpler's work with open clusters (see Chapter 27), there are also present great quantities of interstellar gas, the evidence for which has been available for a long time. It can be observed directly in the form of fairly dense bright nebulous clouds associated with the hot O and B type stars, and can also be detected spectroscopically as a uniform distribution between the stars. The gas reveals itself in the bright nebulosities through certain characteristic emission lines, but the uniformly spread-out gaseous matter between stars is detected primarily by the absorption lines it introduces into the spectra of very distant stars.

Such absorption lines were first detected in 1904 by Hartmann, who observed that the sharp absorption lines of ionized calcium in the spectrum of the binary δ Orionis do not behave like the other stellar lines. Instead of oscillating back and forth in accordance with the Doppler effect for spectroscopic binaries, these lines always remain in the same positions (see Figure 28.1). E. M. Slipher concluded from this that these lines arise not from the gases in the stars themselves but from the interstellar gaseous medium. As we shall see, this type of spectral evidence allows us to determine the composition of the interstellar gas.

We thus have the following picture of interstellar matter, the various components of which we shall discuss separately:

1. *large masses of bright and dark nebulosity* (*this material consists of gas with particles of dust*), the evidence for which has been available for about a hundred years directly from telescopic observations and on photographic plates;
2. *the general interstellar gas*, the evidence for which has been available since 1904 in the spectra of distant stars in the form of stationary absorption lines of various metals, and more recently in the 21-centimeter radio signals from interstellar hydrogen;
3. *general interstellar dust*, the evidence for which has been available since 1932 from the study of star counts, open clusters, and the space-reddening (color excess) of distant stars. Although the amount of dust represents only about 1 per cent of the interstellar matter, it is responsible for almost all of the obscuration.

28.2 Bright and Dark Nebulosity

The most obvious evidence for the presence of interstellar dust and gaseous matter is found in the extended clouds of nebulosity. These are not only bright objects reflecting the light from stars close to them and also emitting their own radiation stimulated by the ultraviolet light from nearby hot close stars, but also dense dark clouds which reveal themselves by obscuring the star fields behind them. In addition to these extended gaseous masses of irregular shape there are also smaller localized bright nebulosities associated with individual stars; these are, first, the so-called planetary nebulae in which the gaseous matter forms large extended envelopes of approximately spherical shape around the stars; and second, the gaseous matter such as the Crab Nebula associated with supernovae. We shall discuss these various gaseous nebulosities separately, although bright and dark nebulae are essentially the same, differing only in that the bright ones are strongly illuminated by hot stars, whereas the dark ones are not.

INTERSTELLAR LINES

Clouds of atoms in space make their presence known by their effect upon transmitted light. They absorb small amounts of energy from the starlight passing through them, thereby producing absorption lines in the spectra of the most distant stars. The strength of such interstellar lines depends upon the number of absorbing atoms lying along the line of sight, and their velocities within the atomic cloud.

K LINE OF CALCIUM II

H LINE OF CALCIUM II

D LINES OF SODIUM I

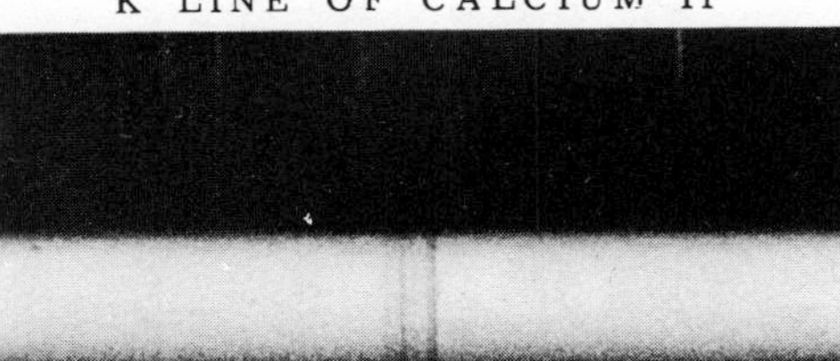

Five components are visible in the interstellar H and K lines in the spectrum of Epsilon Orionis. The displacements of these lines indicate heliocentric velocities for the five absorbing clouds of + 3.9, + 11.3, + 17.6, + 24.8, and + 27.6 kilometers per second, respectively.

Five components are also visible in each of the interstellar D lines in the spectrum of Epsilon Orionis. They yield the same radial velocities as the calcium lines.

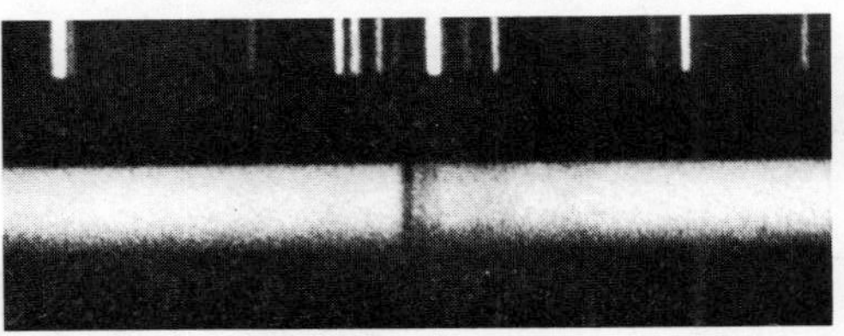

The star HD 172,987 is about 20,000 light years distant, and shows in its spectrum unusually strong, complex, interstellar calcium lines as shown above. The broad faint lines adjacent to the H line originate in the atmosphere of the star.

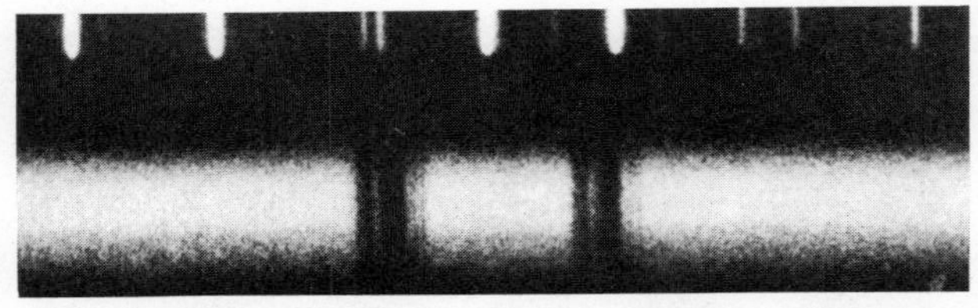

Each D line in the spectrum of 6 Cassiopeia is made up of two groups of lines, each group arising in separate clouds of sodium atoms in two different arms of our Galaxy whose radial velocities relative to the sun differ by about 30 kilometers per second.

The star HD 47240 shows in its spectrum three weak, highly displaced interstellar K lines. This star lies in the direction of, and beyond the gaseous nebula NGC 2237, in the constellation of Monoceros.

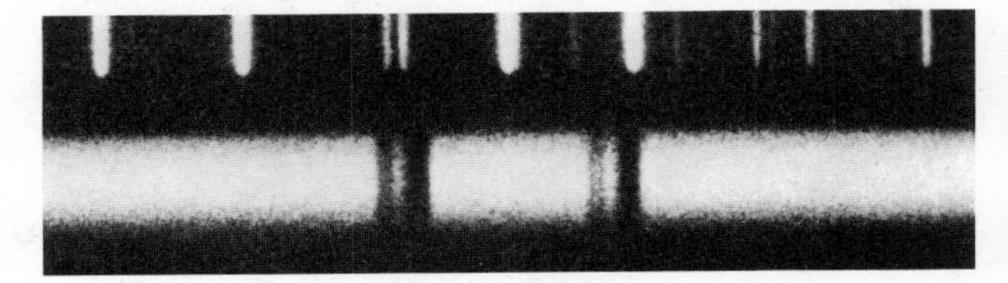

In the spectrum of HD 14134 is visible the same complex structure of the interstellar D lines that was shown in the spectrum of 6 Cassiopeia. This indicates that this star also lies in or beyond a second spiral arm of our Galaxy.

FIG. 28–1 *(a) Interstellar lines of calcium and sodium.* *(From Mount Wilson and Palomar Observatories)*

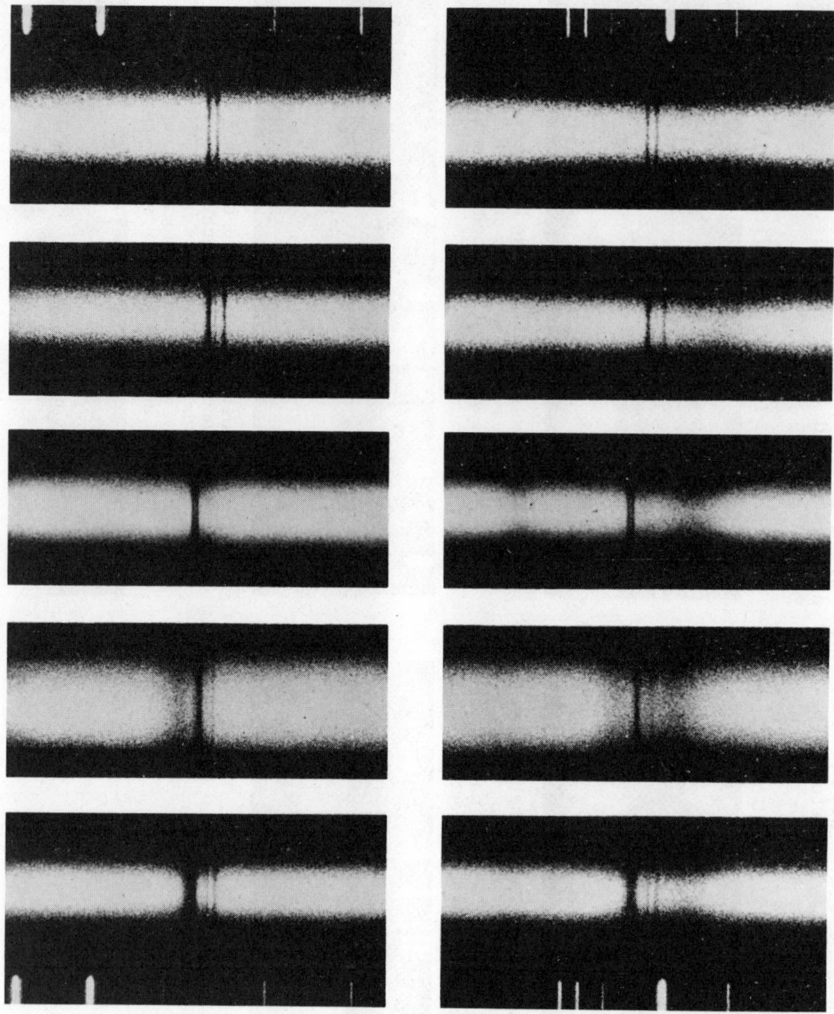

FIG. 28–1 (*b*) *Multiple stellar H and K lines of Ca II in the spectra of four stars give evidence of discrete clouds of interstellar gas. Interstellar lines with high dispersion. Calcium K (left) and H (right). From top to bottom: κ Aquilae, HD 167264, χ Aurigae, μ Sagittarii, HD 199478. (Yerkes Observatory photograph)*

28.3 Bright Nebulae—Their Spectral Features

The bright nebulae are generally diffuse masses of gaseous material (see Figure 28.2) exhibiting bright-line emission spectra or continuous spectra, which on closer inspection show the presence of faint dark lines. It has been found that the bright nebulae with the sharp emission lines are, in general, associated with stars of spectral class B1 or earlier, whereas those with continuous spectra are illuminated by stars later than B2. We now understand the mechanism that is responsible for these two types of spectra. Nebulae with sharp emission lines are illuminated by hot stars and thus receive sufficiently energetic ultraviolet radiation so that their atoms are excited and emit the observed spectral lines. Nebulae with continuous spectra, however, are close to the cooler stars and hence are illuminated by radiation that is not energetic enough to excite their atoms. The dust in these nebulae simply reflects the light which thus exhibits the spectrum of the cool star itself.

FIG. 28–2 *NGC 1432 Pleiades, diffuse nebulosity around Merope. Messier 45. (60-inch photograph from Mount Wilson and Palomar Observatories)*

FIG. 28–3 *NGC 1976 Great Nebula in Orion. The brightness of this nebula is due to four hot blue-white stars imbedded in its gaseous mass. Messier 42. (100-inch photograph from Mount Wilson and Palomar Observatories)*

We have an example of the type of radiation emitted by bright nebulae in the light coming from the great nebula in Orion (see Figure 28.3). The greenish-tinted light coming from this gaseous mass, in the midst of which the bright stars of the Trapezium shine out, is generally characteristic of this type of bright nebulosity. This typical coloration is due to the green lines λ 5 006.84 and λ 4 958.91, which are the most intense lines in the spectra of these objects.

Since these lines have never been produced in a laboratory, and hence at first could not be identified with any known spectral lines, it was thought they were due to a new kind of material called nebulium. But we now know that they are the spectral lines of doubly ionized oxygen and are present because the nebular material is extremely attenuated so that the oxygen atoms are widely separated and hence collide only rarely. We shall discuss this in a moment.

In addition to these lines of ionized oxygen the spectra of bright nebulae contain the lines of ionized helium, neutral helium, hydrogen, ionized carbon, and doubly ionized nitrogen.

28.4 The Green Lines of Nebulae and Metastable States of Atoms

To understand how the green lines of doubly ionized oxygen occur and why we do not observe them here on earth, we note first that they arise when electrons jump down to lower states from excited metastable states (see Section 20.23). As we have seen, a metastable state is one in which an electron has only a small probability of leaving as compared with the probability of its leaving an ordinary excited state. In other words, whereas an electron ordinarily jumps down from a nonmetastable excited state in 10^{-8} seconds, it may spend seconds, minutes, or even hours in a metastable state. Hence, if the electron in an atom is excited into a metastable state in the laboratory here on earth, it has little chance of jumping down to the ground state and thus emitting a photon

FIG. 28–4 *(a) NGC 6960 Filamentary nebula in Cygnus. (100-inch photograph from Mount Wilson and Palomar Observatories)*

because the excited atom suffers collisions with other atoms so frequently that the electron unloads its energy by such collisions rather than by the emission of a photon. Thus, we do not observe the line associated with the metastable state in laboratories here on earth.

However, when the atom with its electron in the excited metastable state is in a very attenuated gas, as is true in the case of nebulae, collisions are so rare that the electron has time to jump down to a lower state and unload its energy in the form of a photon. In such gaseous nebulae ionized oxygen atoms collect energy as they move about by absorbing ultraviolet photons coming from the neighboring hot stars, with the result that their electrons are torn off and sent speeding in all directions. These electrons are then captured by other ionized atoms. Some land in excited metastable states. In time these captured electrons fall to the ground states of the atoms and the emitted photons give rise to the observed spectral lines. Since these processes require very energetic radiation, it is easy to see why bright gaseous nebulae are visible only in the neighborhood of hot O and B type stars—for it is only stars of this sort that emit great quantities of radiation that is energetic enough to ionize oxygen doubly.

28.5 Examples of Bright Nebulae: The Orion Nebula

Bright emission nebulae are irregular in shape and vary considerably in size. In some cases they are no more than small wisps of gaseous matter (see the accompanying photographs, Fig. 28.4) having apparent diameters of a few minutes of arc and exhibiting themselves on photographic plates only after a long exposure. But in other cases they may be large gaseous distributions such as the Orion Nebula, covering several degrees of the sky.

The Orion Nebula has a central bright region with an apparent diameter of about 1°, and its total apparent extension is about 3°. Since its distance from us is estimated to be about 500 parsecs, its main body has a probable diameter of about 10 parsecs, but altogether it may extend 30 parsecs in all directions. The very densest portion of this nebula contains about 300 atoms/cm^3, i.e., about 10^{-22} gm/cm^3, and its average density is probably one-tenth of this or less. Its absolute magnitude is about -6.

The North American Nebula, which is about one-third the size of the Orion Nebula, is an example of a medium-sized bright nebula, and the Trifid Nebula with a diameter one-fifth that of the Orion Nebula is an example of a small nebula. (See Figure 28.5.)

It is possible to detect internal motions within these objects as well as their translational velocities as a whole. Thus in the case of the Orion Nebula radial velocities (after the solar motion has been subtracted) vary from point to point by as much as ± 8 to 10 km/sec. The nebula as a whole appears to be receding at about 17.5 km/sec. What we have said about the Orion Nebula applies to other bright nebulae as well.

Although the total number of these objects is difficult to determine, 325 of them have been catalogued, and H. D. Curtis has estimated that there may be as many as 1 000. Taking all things into account *we define nebulae of this sort as above-average concentrations of interstellar gaseous matter associated with hot O and B type stars.* Since the reflection nebulae differ from the emission nebulae only in their appearance, because they are associated with cool stars and reflect their radiation,

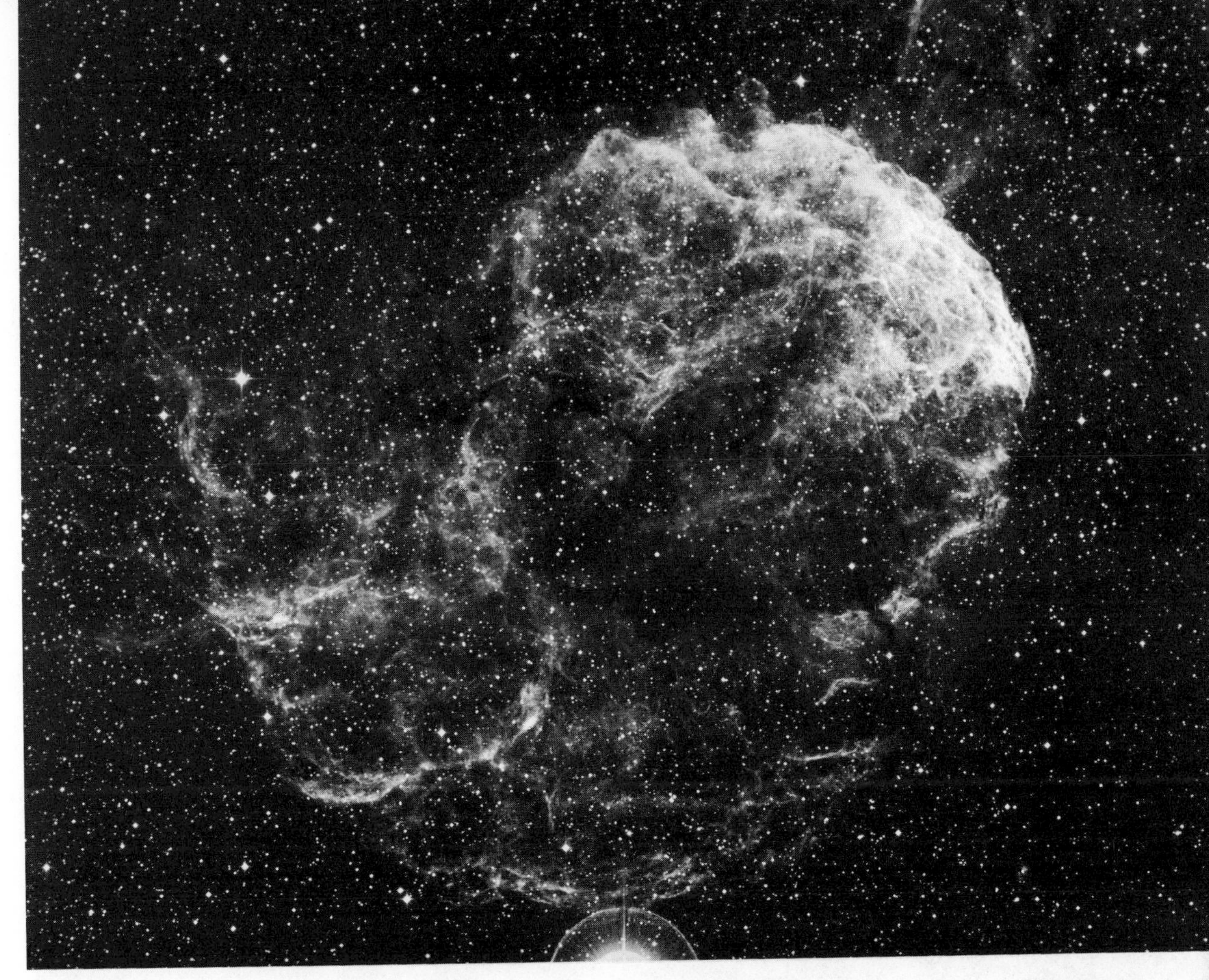

Fig. 28–4 *(b) IC 443 Gaseous nebula in Gemini, photographed in red light. The complex filamentary structure indicates that most of the visible radiation arises from the Balmer lines of hydrogen. (48-inch photograph from Mount Wilson and Palomar Observatories)*

we shall not discuss them further. What we have said about the emission nebulae applies to these as well.

28.6 Dark Nebulae

If there is an accumulation of interstellar matter similar to a bright nebula but not close enough to a star or a group of stars to reflect much of the stellar light or to be stimulated to emit its own radiation, this mass of gas is not visible to the naked eye. However, such structures can still be detected since they obscure stars lying behind them, often giving the appearance of dark empty patches in an otherwise unbroken field of stars. Sometimes, as in the case of the Horsehead Nebula (see Figure 28.8), the cloudlike structure of the dark nebula is immediately evident, and it is clear that there are stars lying behind it that cannot be seen. In many cases, however, this kind of structure is not at once apparent, and one detects the presence of such obscuring clouds only by noting the sudden change in the star count as one passes their boundaries. The obscuration of stars by these clouds is due to the presence of very fine dust, which occurs throughout interstellar space in our region of the Galaxy. (See Figure 28.6a).

Since such drops in star counts are generally found in the neighborhood of bright nebulae, it is clear that the dark nebulae are in many cases continuations of the bright nebulae into regions where the light from the neighboring hot stars has no effect. Often the dark nebulosities are dark lanes looking like rifts in the star fields as in Figure 28.6(b).

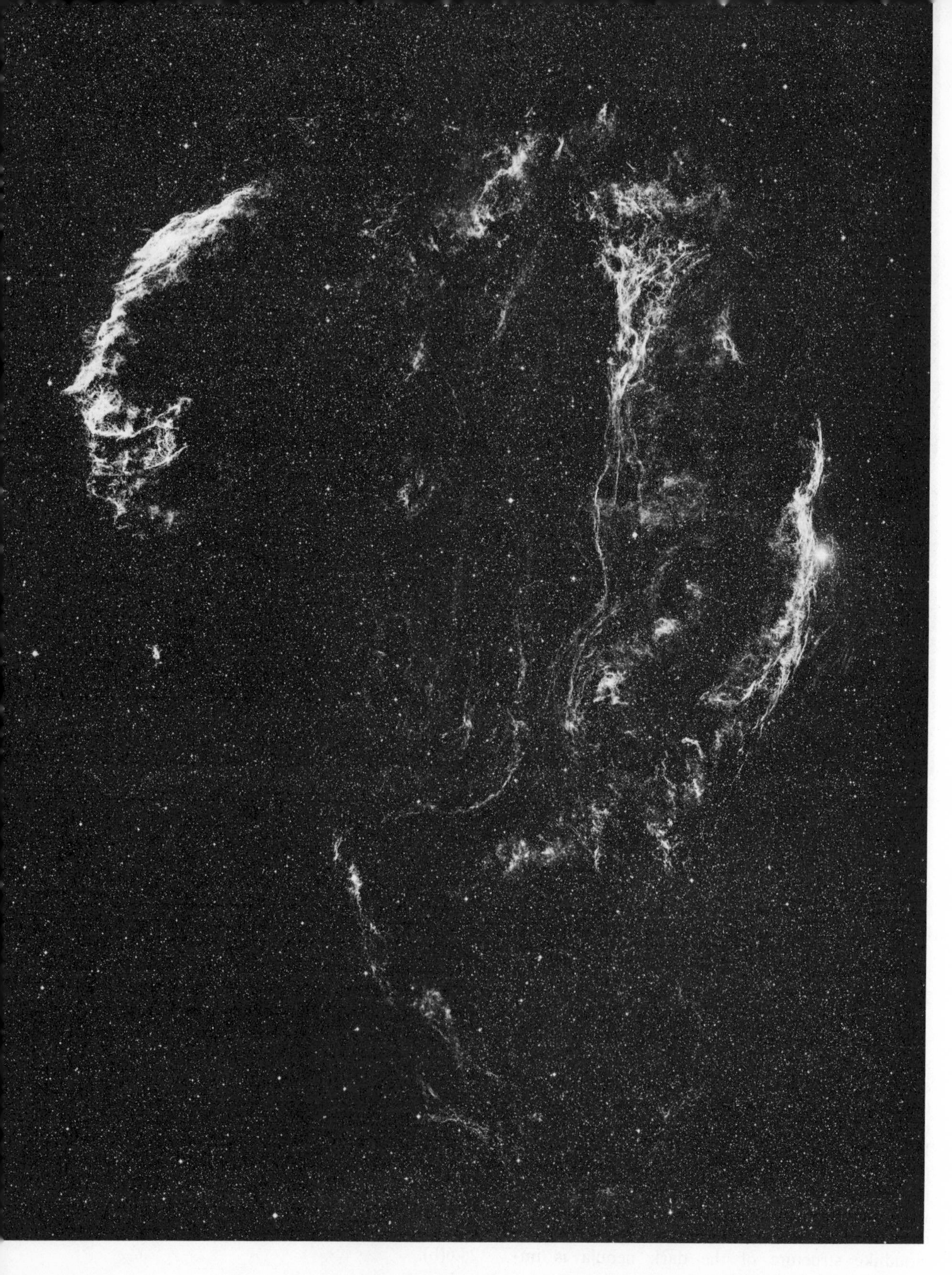

FIG. 28–4 *(c) Filamentary nebula in Cygnus, showing entire area of nebulosity. Photographed in red light. (48-inch Schmidt photograph from Mount Wilson and Palomar Observatories)*

FIG. 28–5 *NGC 6514 Nebula in Sagittarius. Messier 20, "Trifid" Nebula. Photographed in red light. Note the dark gaseous regions. (200-inch photograph from Mount Wilson and Palomar Observatories)*

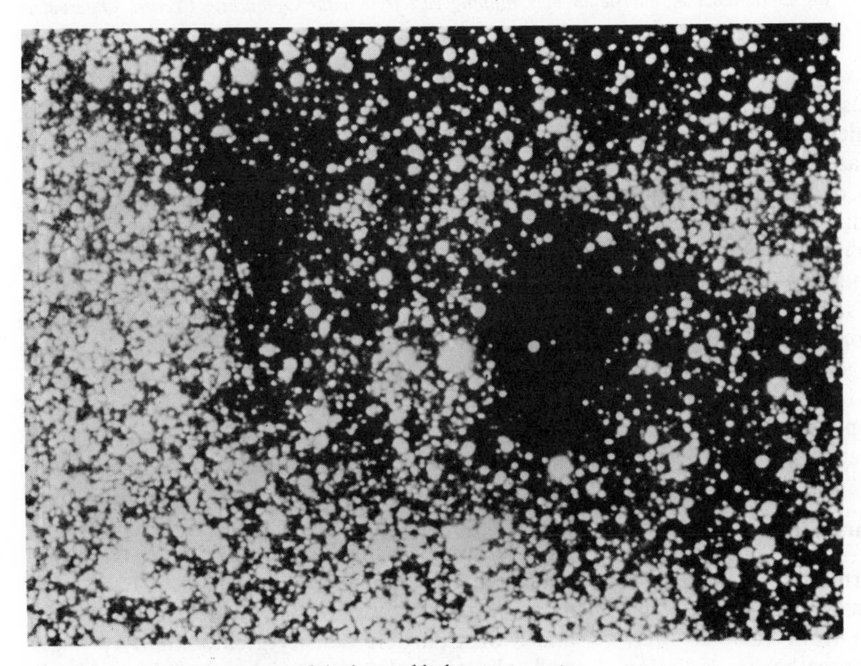

FIG. 28–6 *(a) Enlarged view of the largest black spot appearing in the star cloud in Sagittarius. See Figure 27.16(a). This shows in a striking manner the abrupt and sharply defined edge of the spot. (Yerkes Observatory photograph)*

FIG. 28–6 *(b) Dark lanes of interstellar dust running east from Rho Ophiuchi. (Yerkes Observatory photograph)*

At times the dark nebulous material appears in the form of small dark globules spattered throughout a bright nebulous region as though the material there were of unusually high density (as in Figure 28.7(a) and (b)). These nodules are most prominent in the Sagittarius, Ophiuchus, and Scutum regions of the Milky Way, and especially near M 3 (the Lagoon Nebula) where some 30 such objects can be seen. Many astronomers have suggested that these nodules, and in general dark nebulae such as the Coal Sack and Horsehead Nebula (Figure 28.8), are the initial phases of star formation, and hence are the first step in stellar evolution.

28.7 Dark Nodules as Primary Stages of Stars

We can account for the formation of these dark globules by picturing the dust particles in a gas cloud as absorbing radiation from various directions, and hence being driven together by radiation pressure into a globule which grows with time. As this globule moves about in the dark cloud, it sweeps up gaseous material and dust, and, depending on the density of matter in the cloud, doubles its mass in anywhere from 3×10^7 to 3×10^8 years. This goes on until the gravitational attraction within the globule itself is sufficiently large to bring about a very rapid evolution to the initial stages of a star. See the Herbig-Haro objects in Figure 28.9.

That these objects are approximately of the proper mass to initiate stellar formation has been demonstrated by Bok, who correlated their absorption of light with their masses. Many of them appear to have diameters between 10 and 25 thousand A.U. and absorb anywhere from 1 to 5 magnitudes of background light depending upon how large and dense they are. Bok finds that a globule having a total absorption of 5 magnitudes and having a diameter of 0.06 parsecs has a total mass of absorbing material greater than $0.002M_\odot$ and a dust-particle density greater than 10^{-21}gm/cm^3, whereas a globule that absorbs 1.5 magnitudes and has a diameter of 0.5 parsecs has a mass greater than $0.05M_\odot$ and a particle density of about 5×10^{-23} gm/cm.

G. Herbig has given a remarkable example of the actual formation of a star from nebulous material in the Orion Nebula. He has pointed out that the object FU Orionis, very close to the Orion Nebula, has changed from a sixteenth-magnitude object (when it was first photographed in 1936) to a tenth-magnitude object now. He estimates that this increase occurred in about 120 days and that this object has retained its present luminosity for a period of at least 28 years. Hayashi has analyzed events of this sort and demonstrated theoretically that nebular material of a sufficiently large mass and density will contract (gravitationally) slowly at first and then, at a certain stage, will collapse catastrophically into a star. FU Orionis probably had a radius about 300 times that of the sun before the collapse, but now has a radius only 20 to 25 times that of the sun.

FIG. 28–7 *(a) NGC 2237 Nebula in Monoceros, enlarged section. This gaseous nebula is particularly rich in dark pre-star nodules. Photographed in red light. (48-inch photograph from Mount Wilson and Palomar Observatories)*

28.8 The Interstellar Gaseous Medium

Thus far we have been discussing that portion of the interstellar medium which can be detected immediately in the form of bright or dark nebulosity. However, between us and star fields that appear to suffer no appreciable obscuration, we know that both gaseous material and dust are present but with densities considerably smaller than those of the gaseous nebulosities themselves. Most of this gaseous interstellar material can be detected through optical spectral lines; a more recent method uses the 21-centimeter radio emission line of hydrogen.

In addition to the absorption lines of ionized calcium, which were found in 1904, the sharp interstellar stationary lines of sodium have also been detected, particularly in the spectra of O-type stars. Other neutral and ionized atoms whose interstellar absorption lines have been found are Ca I, Ti II, and Fe. In addition to these observed lines there are also present certain unidentified absorption lines. Interstellar absorption bands have also been observed, including those of the molecules CH, CN, OH, and NH.

Certain emission lines which, in general, are very difficult to detect have also been discovered. Thus, the Balmer lines of HI and emission lines of OII, OIII, and NII are occasionally observed. It may at first appear puzzling that the absorption lines of hydrogen are not evident since, as we shall see, hydrogen is the most abundant substance in interstellar space. However, it is easy to understand this if we recall that neutral hydrogen in its ground state gives rise only to the Lyman absorption lines, which are in the ultraviolet part of the spectrum. Since the ozone in our atmosphere absorbs most of the ultraviolet radiation, the only way we have of detecting these interstellar hydrogen lines is by means of observatories orbiting above our atmosphere.

That the lines we have listed arise from the interstellar gas is indicated not only by their stationary character as compared with the Doppler displacements of the stellar lines, but also by the increase in their intensities with the increasing distances of the stars in whose spectra these lines are found. Thus we find that the Ca II lines are more intense in the spectra of stars at great distances from us than they are in nearby stars because clearly there is more gaseous material between us and the distant stars than between us and nearby stars.

FIG. 28–7 (*b*) *NGC 6611 Nebula in Scutum Sobieski. Messier 16. A few dark globules (pre-star stage) and small elephant trunks can be seen. Photographed in red light. (200-inch photograph from Mount Wilson and Palomar Observatories)*

Furthermore, we find that the intensities of the above lines are correlated to the reddening of stars (color excess), which we know is due to selective absorption of light from the stars by the dust that is intermixed with the intervening gaseous matter. We thus find that the more distant a star is, the greater is its color excess and at the same time the greater is the intensity of the interstellar lines in its spectrum, so that both of these characteristics must be associated with material between us and the stars.

28.9 The Cloud Structure of Interstellar Material

We know from the presence of bright and dark nebulosities that the gaseous matter is not spread out uniformly between the stars but tends to accumulate into dense clouds in certain regions. But even between observable nebulosities the interstellar gas appears to collect into cloudlike structures. These clouds seem to be fairly discrete and to have average sizes of about 8 parsecs. There are about 5×10^{-5} such clouds per cubic parsec, separated from each other on the average by about 40 parsecs, and moving about with speeds averaging about 7 km/sec. Their mean mass is about $400M_{\odot}$ and they contain on the average about 10 atoms/cm^3. Between these clouds the interstellar gas appears to be spread out quite uniformly (but much more thinly than in the clouds) with an average density of about 0.8 atoms/cm^3. The evidence for such clouds can also be found in the doubling of the interstellar spectral lines resulting from Doppler shifts.

FIG. 28–8 "*Horsehead*" *Nebula. Nebula in Orion south of Zeta Orionis, IC 434, Barnard 33. Photographed in red light. (200-inch photograph from the Mount Wilson and Palomar Observatories)*

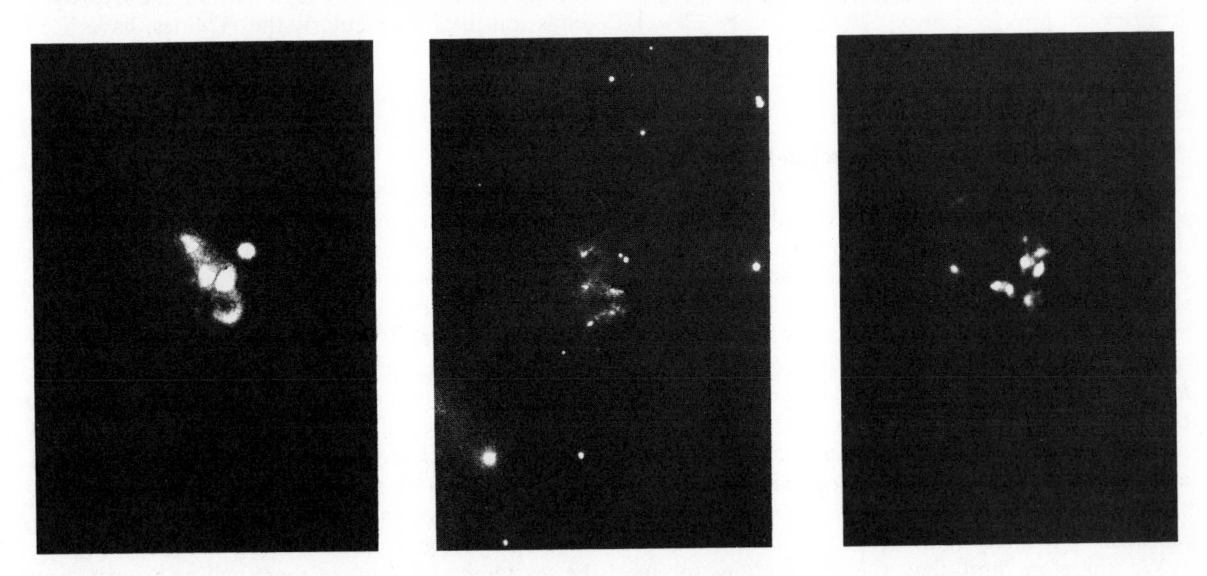

FIG. 28–9 *These Herbig-Haro Objects are examples of what astronomers believe to be stars evolving from interstellar dust. These objects change their shapes continuously, becoming more and more spherical. Such changes have actually been photographed. In these photographs the sphericity of each object can be clearly seen, showing the effect of gravitational contraction. (Lick Observatory photograph)*

All of this gaseous material hugs the plane of the Galaxy, and it appears that in the outer portions of the Galaxy as much matter is present in the form of interstellar gases as there is concentrated into stars. The clouds themselves, however, are rather filamentary and occupy about 7 per cent of the space in the neighborhood of the sun. If the material of the clouds were spread out uniformly along the galactic plane, its density would be about 0.6 hydrogen atoms/cm³ or about 10^{-24} gm/cm³.

An analysis of the interstellar absorption lines and the Balmer emission lines indicates that the relative abundance of elements in this interstellar gas is the same as in the stars, with hydrogen predominating. Table 28.1 shows the relative abundances of various chemical elements in the Galaxy. It clearly shows that hydrogen and helium dominate the chemistry of the universe.

TABLE 28.1
COSMIC ABUNDANCE OF ELEMENTS

At No. Z	Element	No. per 10^8	gms per ton	At No. Z	Element	No. per 10^8	gms per ton
1	H	839 000	456 000	12	Mg	42	680
2	He	159 000	421 000	14	Si	35	660
8	O	680	7 210	16	S	12	250
10	Ne	640	8 550	28	Ni	5	200
7	N	200	1 860	13	Al	3	54
6	C	130	1 030	11	Na	3	46
26	Fe	64	2 370	20	Ca	2	53

28.10 Strömgren Spheres: Hydrogen II Regions

The physical state of this interstellar hydrogen depends on whether or not it is in the neighborhood of hot stars. Bengt Strömgren has shown that there are two fairly distinct regions, HI and HII, associated with hydrogen in interstellar gaseous matter. *The hydrogen in the HI regions is un-ionized and relatively cool, whereas the hydrogen in the HII regions is ionized and hot. The HII regions, known as* ***Strömgren spheres****, are associated with O and B type stars and extend out from these stars to a point where the radiation suddenly becomes too weak to do much ionizing.*

We therefore find that the HI regions set in suddenly at a certain critical distance, S, from the hot star. Beyond this distance the ionization falls off very drastically. Thus, we find that whereas in a shell of radius $0.5S$ (as shown in Figure 28.10) all the atoms are ionized, and within a shell of radius $0.97S$, 94 per cent of the atoms are ionized, the percentage of ionized hydrogen atoms in a shell of radius $1.03S$ is only 33 per cent. The value of S depends upon how hot the exciting star is, and we find that it increases very rapidly with the surface temperature of the star. It ranges from about 0.5 parsecs for A0-type stars to 20 parsecs for B0-type stars having surface temperatures of 25 000°K, and it has a value of about 58 parsecs for O6 stars having surface temperatures of 46 000°K. Ionization extends beyond the values for S given here, but it then becomes so small that it may be neglected. It is in the HII regions that the hydrogen emission lines, particularly the H_α lines, have been detected.

FIG. 28–10 *The Strömgren sphere surrounding a hot blue-white star separates the ionized hydrogen from the un-ionized regions.*

28.11 Balmer Emission Lines from HII Regions

The extent of these HII regions (or Strömgren spheres) can be measured directly by observing the Balmer emission lines which are emitted when the protons and electrons in these regions recombine. When a proton in an HII region captures an electron, either in the M or in a higher level (see Section 18.26), the electron cascades down to the ground state and emits one of the Balmer photons, usually the H_α line. This photon then passes through the HI region (where un-ionized hydrogen is in its ground state) without being absorbed, because it is not energetic enough to excite this un-ionized hydrogen. These photons will, of course, be scattered and some of them will be absorbed by grains of dust, but they cannot be augmented by any emission within the HI region itself since the hydrogen atoms in this region are all in the ground state. Thus the H_α photons reaching us give a clear indication of the extent of the HII regions surrounding O and B type stars. From the analysis of the Balmer radiation it appears that in these HII regions there are about 2 to 5 protons per cubic centimeter.

28.12 Kinetic Temperatures of HI and HII Regions *

From a consideration of the speeds with which the atoms of hydrogen are moving around in the HI and HII regions we can assign to each of them a kinetic temperature. *By the **kinetic temperature** of a highly attenuated gas we mean the temperature that such a gas would have to have at ordinary densities and pressures for its atoms (or molecules) to move at the same average speed as do the atoms of the attenuated gas.* The atoms in the interstellar gas have high kinetic temperatures because the gas gains energy from the radiation reaching it from hot stars. Thus, when a hydrogen atom is ionized by a very energetic photon, the electron is kicked off in one direction and the proton rebounds with a certain amount of energy also. The proton thus acquires energy. In this way protons in the HII regions acquire greater and greater speeds as compared with the hydrogen atoms in the HI regions.

Calculations show that the kinetic temperature in the HII regions ranges from 5 000 to 10 000°K, whereas the kinetic temperature in the HI region is of the order of 100°K. We thus find that if we pass from an HII into an HI region there is a very drastic drop in the kinetic temperature. At the boundary between these two regions there is, therefore, a sharp pressure difference and hence the density difference. As a result there may be sudden incursions of cool dense material into the hot rarefied regions, giving rise to intense local compressions which may be the beginning of stellar formations. These are known as "**elephant trunks.**" [See Figures 28.11 and 28.7(b).]

28.13 The 21-Centimeter Line

As we have seen, the hydrogen in the HII regions can be observed directly by detecting the H_α emission line, but the neutral hydrogen in the HI regions cannot be detected by the usual spectroscopic methods. However, we do receive radiation from these HI regions which can be observed and measured by radio telescopes (see Appendix B, Radio Astronomy). The presence of neutral hydrogen is characterized by the emission of a sharp 21-cm line which was first predicted theoretically in 1944 by van de Hulst.

We can see how this line arises by introducing a small revision in our picture of the energy levels of electrons inside an atom. We saw in Sections 19.27 and 19.28 that each atomic energy level above the ground state generally consists of sublevels associated with different azimuthal and spin quantum numbers, and that these sublevels have slightly different energies. These sublevels give rise to the fine structure of spectral lines. However, we saw according to our atomic models that there is no fine structure (i.e., no sublevels) of the ground state. This is correct, however, only if we do not take into account the interaction between the spin of the electrons and the spin of the nucleus.

28.14 Electron-Proton Spin Interaction

If we do take this spin-spin interaction into account, we see that the usual ground level of the hydrogen atom really consists of two levels lying close together. To understand this we must picture the spinning electron as a small magnet, as we have previously done, possessing a magnetic moment and hence giving rise to a magnetic field. In the same way, since the proton is also spinning, it too behaves like a little magnet with a surrounding magnetic field. If the electron is in the ground state of the hydrogen atom, its spin may line up either parallel to the spin of the proton or antiparallel to it, as shown in Figure 28.12.

Since the parallel-spin case is similar to two magnets arranged with like poles opposite each other, we see that less work is required to separate the electron from the proton in the parallel than in the antiparallel case, because like magnetic poles repel each other. Thus the ground level of the electron-proton system has a somewhat higher energy in the parallel spin alignment than in the antiparallel. This is called the **hyperfine structure of the ground state**, and because the parallel ground level is higher than the antiparallel, transitions occur spontaneously from the former to the latter.

28.15 The Emission of the 21-Centimeter Line *

When such a transition occurs (i.e., when the electron in the ground state, with its spin axis parallel to the proton spin axis, flops over so that the two spins

FIG. 28–11 *Nebulosity in Monoceros. Situated in south outer region of NGC 2264. A large elephant trunk dominates this picture. Photographed in red light. (200-inch photograph from Mount Wilson and Palomar Observatories)*

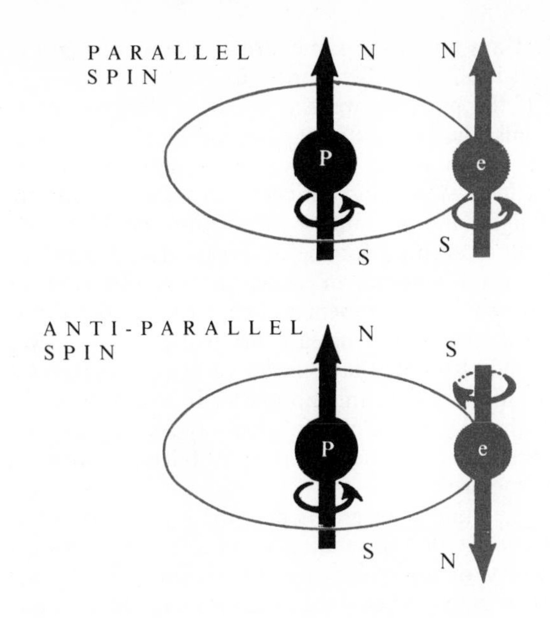

FIG. 28–12 *An electron in the ground state of an atom moves in such a way that either its spin is parallel to the spin of the proton or antiparallel to the spin of the proton. The first case corresponds to magnets lined with like poles opposite each other, and the second case corresponds to magnets lined with unlike poles opposite each other.*

are antiparallel), a photon having a wavelength of 21 centimeters is emitted. Since a proton has three times as many chances of capturing an electron with parallel spin as with antiparallel spin, there is in any given HI region three times as much neutral hydrogen with electrons in the higher sublevel of the ground state (parallel spin) as with electrons in the lower sublevel of the ground state (antiparallel spin). Therefore, as van de Hulst first pointed out, we should be able to detect this 21-cm line from HI regions because of the continuous transitions of electrons from the higher energy sublevel to the lower one. This radiation was first detected in 1951 by Ewen and Purcell.

The intensity of the 21-cm line depends on: (1) the number of hydrogen atoms per unit volume in the higher energy state, in which the spins of the electron and proton are parallel; (2) the average lifetime of the hydrogen atom in this state. The reciprocal of this lifetime is proportional to the probability that the electron spin will flop over and that a 21-cm photon will be emitted. This multiplied by the quantity (1) gives the amount of 21-cm radiation emitted from each cubic centimeter of space and hence the intensity of the 21-cm line.

The spontaneous transition of an atom from the parallel spin state with the emission of a 21-cm photon is forbidden by the laws of quantum mechanics and therefore occurs rarely, once every 11 million years. Also, a hydrogen atom with its electron in the anti-parallel state is only rarely excited to the higher state by absorbing a 21-cm photon. Thus radiative transitions between the two spin substates are rare. But atomic collisions do induce such transitions frequently. Two colliding atoms form an H_2 molecule momentarily, exchange electrons, and then separate. In this exchange, an event that occurs about every 10 years, the electrons may alter their spin directions in either way. Such collisions keep the ratio of the number of hydrogen atoms in the parallel spin state to the number in the ground state always the same (3:1). Atomic collisions thus keep the hydrogen atoms with their electron and proton spins parallel and the hydrogen atoms with these spins anti-parallel in a state of thermodynamic equilibrium, and just the right number of atoms are always in the upper energy state to give the observed intensity of the 21-cm radiation.

As we shall see, this line is not as sharp as we might expect because the hydrogen atoms are moving about at random, and, because of the Doppler shift, this movement gives rise to a slight spreading of the line. We shall see further that the Doppler shift of this line arising from the rotation of the Galaxy gives us important information about the Galaxy's spiral structure.

28.16 Neutral Hydrogen in the Spiral Arms of the Galaxy

With the aid of the 21-cm line we can now trace out the neutral hydrogen in the Milky Way. By analyzing the intensity of the line and its width we can determine the kinetic temperature of the hydrogen atoms in the HI region; we find it is about 100°K, in excellent agreement with other estimates. We also discover from the 21-cm line that the previous estimate of about 1 atom/cm^3 in the HI regions is correct.

By analyzing the profiles of the 21-cm line at various galactic longitudes, Oort and his coworkers have obtained important information about the rotation of the Galaxy and its spiral structure. If we look right towards the center of the Galaxy or 180° away from it in the galactic plane, the rotation does not affect the spectral lines, and we find that the profile of the 21-cm line is symmetrical in both cases. However, if we look away from the center in either direction, asymmetries appear in the profile which can be explained by the rotation of the Galaxy and by the distributions of neutral hydrogen in the spiral arms. We shall discuss this in detail in Section 29.7.

As we look towards the center of the Galaxy with our radio telescopes, the 21-cm radiation coming from hydrogen farther away than 5 000 light years is blocked out by absorption within the intervening hydrogen itself. This means that the 21-cm radiation that we receive from the direction of the center comes from HI regions having a thickness of 5 000 light years between us and the center of the Galaxy. However, if we look off to the side and away from the center, complete absorption of the 21-cm line does not occur because the line is broadened by rotational effects. As a result we can detect 21-cm radiation coming to us along off-centered directions from very great distances and hence can trace out the spiral arms by this means.

28.17 The Interstellar Dust *

We have seen that as we go out to stars at greater and greater distances, there is a very strong extinction of light resulting from the distribution of grains of dust in the interstellar gaseous medium. Although this dust probably accounts for no more than 1 per cent of the interstellar material, its absorption of light is so great that it is impossible to observe the parts of our Galaxy that lie beyond 6 000 light years. This dust is associated with the outer parts of the Galaxy (i.e., the spiral arms) and is characteristic of all galaxies having a spiral structure. As we have already remarked, the effect of this obscuring matter is most easily detected in the reddening of distant bodies, which indicates that the light coming from these bodies is extinguished selectively by the grains of dust. The long wavelengths pass through the dust more easily than the short wavelengths, which are scattered in all directions.

By analyzing how various wavelengths are scattered and absorbed by the grains of dust we can determine the grain size. We find from this type of analysis, and from the amount of reddening that occurs, that the extinction varies inversely with the wavelength. This means that light having a wavelength one-half that of red light is extinguished twice as effectively as red light. Applying optical theory we conclude that the grains of dust are about the same size as the wavelength of light itself (i.e., of the order of 10^{-5} cm).

If the particles were as small as electrons or as atoms, the light extinction would vary inversely as the fourth power of the wavelength, and if the particles were much bigger than the wavelength of light, the amount of absorption would be too small to account for the observations—that is, for the total amount of mass that is observed or believed to be present (since large particles just block the light instead of scattering it). We should also note that this law of extinction (the $1/\lambda$ variation) is the same no matter in which direction of the sky we look or which types of stars we investigate, so that the dust is fairly uniform in character.

The number of grains per cm^3 is of the order of 10^{-13} to 10^{-9} depending upon size, giving a dust density of approximately 10^{-26} gm/cm^3. In the HI regions where the surrounding gas kinetic temperature is of the order of 100°K, the temperature of the dust grains is of the order of 20°K to 50°K.

28.18 The Structure of Dust Grains *

We can get an idea about the shapes of the dust particles from the polarization of starlight. We say that a beam of light is **plane polarized** when the electric field of the beam always vibrates in one direction (i.e., the velocity of the beam and the electric-field vector always determine the same plane as shown in Figure 28.13). Now we have found that light coming from distant stars is

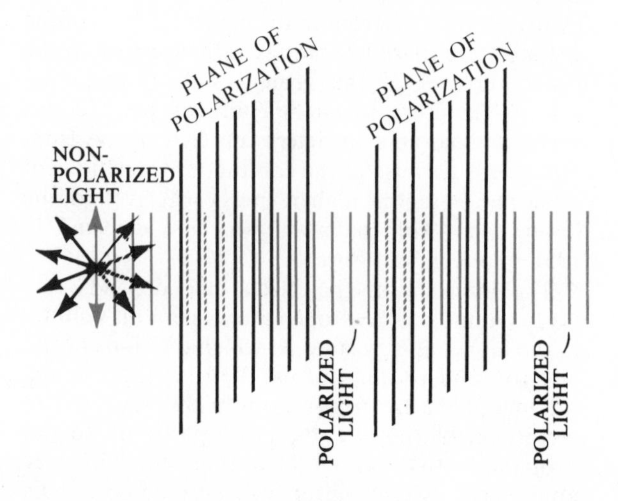

FIG. 28–13 *In a plane-polarized beam of light the electric vector always oscillates in the same plane.*

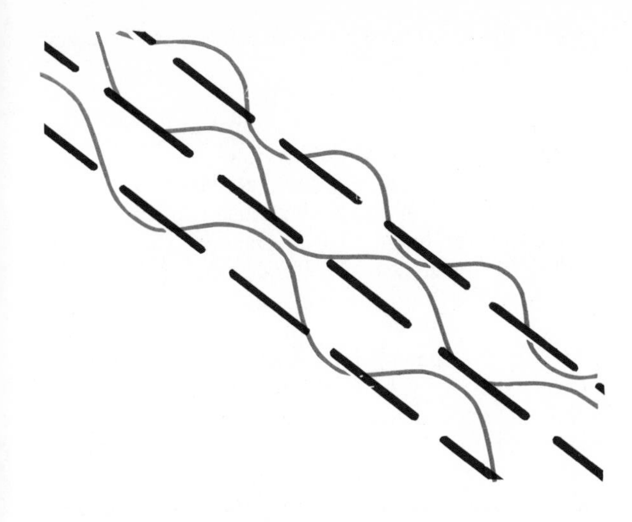

FIG. 28–14 *If the grains of dust in the Milky Way are needle-shaped and parallel to each other, the light passing through them becomes polarized.*

partially polarized when it reaches us. This indicates that the dust grains are elongated and are lined up in such a way as to allow light with its electric field perpendicular to these grains to pass through more easily than light vibrating parallel to them. In other words, we know that the grains are not spherical but are needle-shaped and are arranged end to end instead of being distributed at random.

If the grains were spherical, no polarization of starlight could occur, but with the arrangement of needle-like grains just described, shown in Figure 28.14, polarization does occur. An analysis indicates that the plane of polarization of the light from the stars is parallel to the plane of the Galaxy, so that the axes of the grains tend to be perpendicular to the galactic plane. The alignment is not complete but only partial, and in some parts of the Milky Way it is much more pronounced than in others.

The alignment of these grains indicates the presence of magnetic fields along the spiral arms of the Galaxy and leads us to the conclusion that at least some of these dust grains are composed of ferromagnetic substances such as the iron oxides. At one time astronomers thought that most of the dust consisted of metallic particles, but if this were so there would be a greater absorption of light than is observed. We now believe that the interstellar dust consists primarily of hydrogen compounds with ice crystals predominating. Van de Hulst has suggested that for every 100 grains of H_2O (ice) there are 30 grains of solid hydrogen, 20 grains of CH_4, 10 of ammonia, and 5 of magnesium hydride.

28.19 The Formation of Dust Grains

These dust grains are undoubtedly formed by the accretion of matter around molecules under appropriate conditions. As the grains move about, they either increase in size by collecting molecules and other smaller dust grains that may stick to them, or they are destroyed by collisions with other grains, or by the evaporation of atoms and molecules from their surfaces when they absorb radiation. All these processes go on simultaneously at rates that maintain a state of equilibrium so that the total number of grains remains fairly constant. It can be shown that, in the interstellar gas, grain sizes of the order of 10^{-5} cm are formed in about 100 million years, but the growth does not go beyond this point because a state of equilibrium is then reached.

28.20 Planetary Nebulae

In addition to the interstellar gaseous material and dust in the spiral arms of our Galaxy associated with Population I stars, there are also in the Galaxy gaseous nebulosities associated with Population II stars. These are the **planetary nebulae**, which look like planets on photographic plates; they have the greenish, disk-like appearance of the planet Uranus. See Figure 28.15.

A planetary nebula is an extended oval mass of gaseous material with a faint hot star at the center. Often the gaseous nebulous material exhibits a series of concentric shell structures with a fairly complicated turbulent formation (see Figure 28.16). The spectra of these objects contain not only the usual hydrogen and helium bright emission lines, but also the forbidden lines (transitions from metastable states) of OIII, etc., just as do the spectra of ordinary field nebulae.

The central stars are generally so hot that they emit most of their radiation in the deep ultraviolet region of the spectrum, but this ultraviolet radiation is so battered about from atom to atom in the nebulous envelope that it is degraded to visible radiation. This ultraviolet radiation gives the nebulous mass its luminosity. Since the central star emits far more ultraviolet than visible radiation, the nebula itself, which transforms the ultraviolet into visible radiation, appears much brighter than the central star.

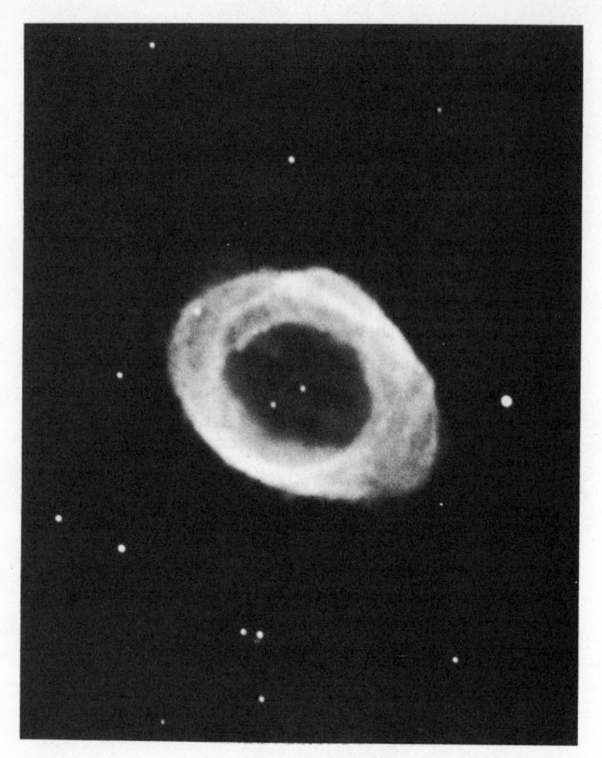

Fig. 28–15 *(a) NGC 6720 "Ring" Nebula. Planetary nebula in Lyra. Messier 57. (200-inch photograph from Mount Wilson and Palomar Observatories)*

28.21 The Distribution and Abundance of Planetary Nebulae

These objects have the same spatial distribution as the RR Lyrae variables and other extreme Population II stars. In fact, a planetary nebula has been found in a globular cluster and many lie in the central bulge of our Galaxy. In general, the bright (hence nearby) ones do not hug the plane of the Milky Way, but are found in high galactic latitudes as well as close to this plane; they are distributed rather symmetrically around the plane of the Galaxy. The faint and hence distant ones are concentrated towards the galactic center (i.e., the direction of Sagittarius). There are probably many more planetary nebulae in the nucleus of our Galaxy than are found in the spiral arms, but they cannot be seen because of the obscuring dust.

The total number of known planetaries exceeds 400, and they vary in apparent diameters from about 12′ for the nearby ones to a few seconds for the remote ones. The mean apparent diameter of the very distant ones (those that are concentrated towards the center of the Galaxy) is about 6″ of arc, and these are visible only in the light of the red hydrogen lines because of the obscuration which cuts out the blue and violet lines. The central star is generally a faint O-type star, with lines of

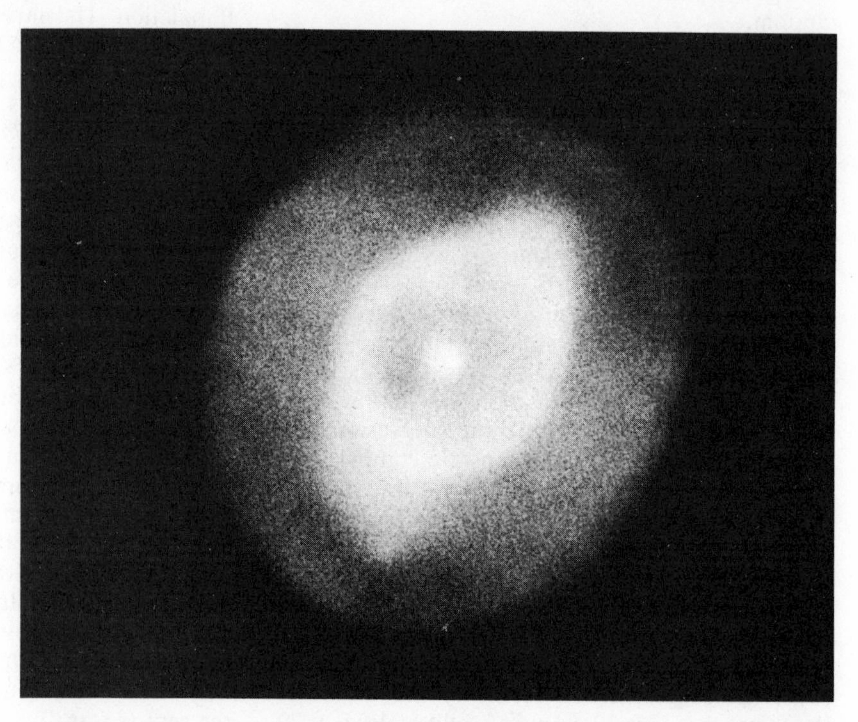

Fig. 28–15 *(b) NGC 3242 Planetary Nebula in Hydra. Photographed in red light. (200-inch photograph from Mount Wilson and Palomar Observatories)*

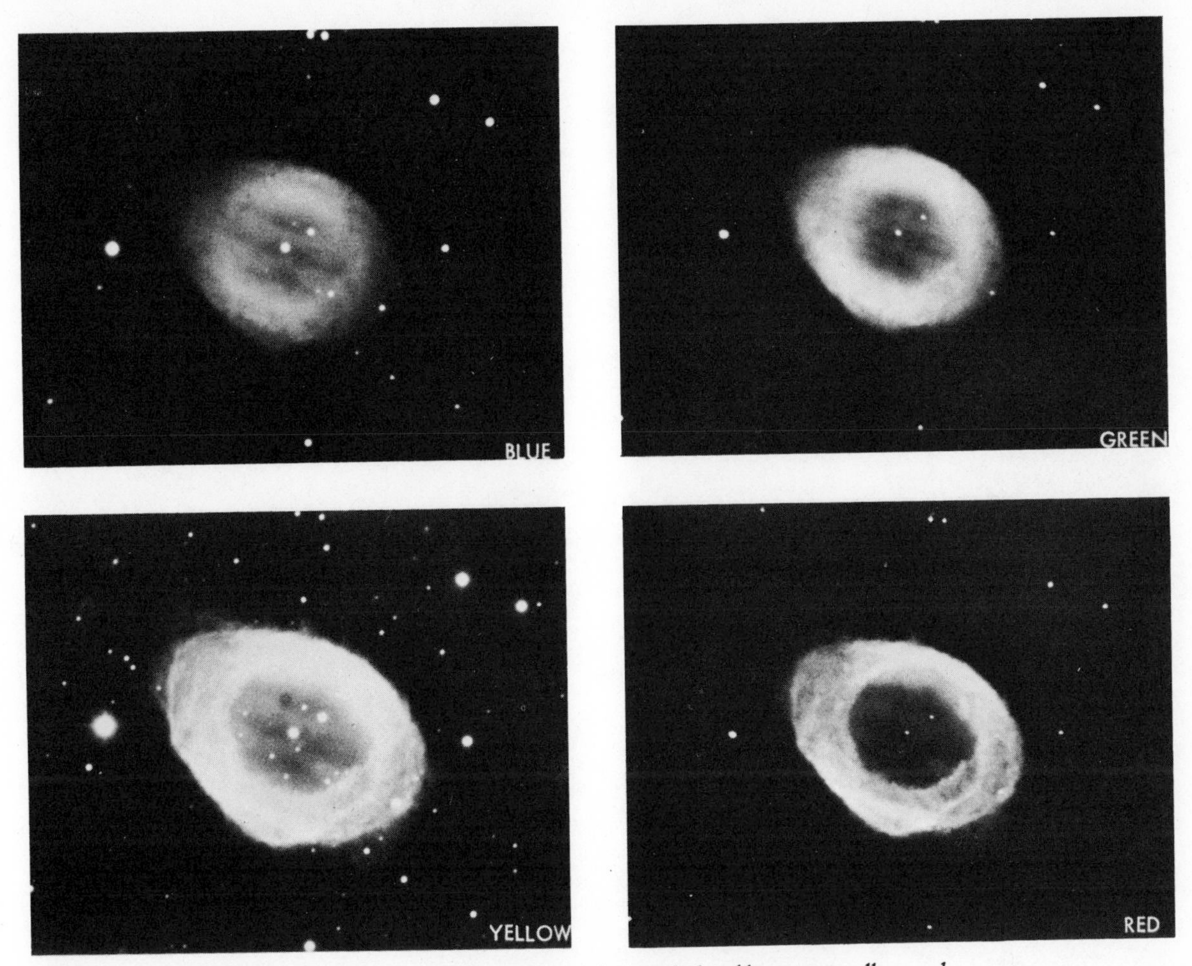

FIG. 28–16 *(a) NGC 6720 The "Ring" Nebula in Lyra. Photographed in blue, green, yellow, and red light. (200-inch photograph from Mount Wilson and Palomar Observatories)*

ionized helium prominent in its spectrum, or a Wolf-Rayet star with clearly visible bright lines. The general appearance of planetary nebulae as we follow them out to greater and greater distances indicates that they are probably of about the same size, varying in appearance only because of distance.

28.22 Luminosities and Diameters of Planetary Nebulae

The parallaxes of the nearby planetaries can be measured trigonometrically and are found to range from 0".04 to 0".01. In the case of the more distant planetaries we obtain a mean parallax of about 0".008. From the parallaxes and the apparent mean diameters we can find the actual diameters, which are of the order of 20 000 A.U. although there are variations above and below this value. We also find that the mean absolute photographic magnitude of the central star is about +3, with some as faint as +10 and some as bright as −2. Although most of these objects are too distant to give reliable proper motions, we conclude from the radial velocities, which on the average exceed 30 km/sec, that planetary nebulae must be included among the high-velocity stars.

28.23 The Structure of Planetary Nebulae and Internal Motions *

Although planetary nebulae appear to have a ring-like structure, the gaseous matter is really distributed in a thick shell transparent enough to allow the visible light from the central star to pass through it without much absorption, so that we can see the

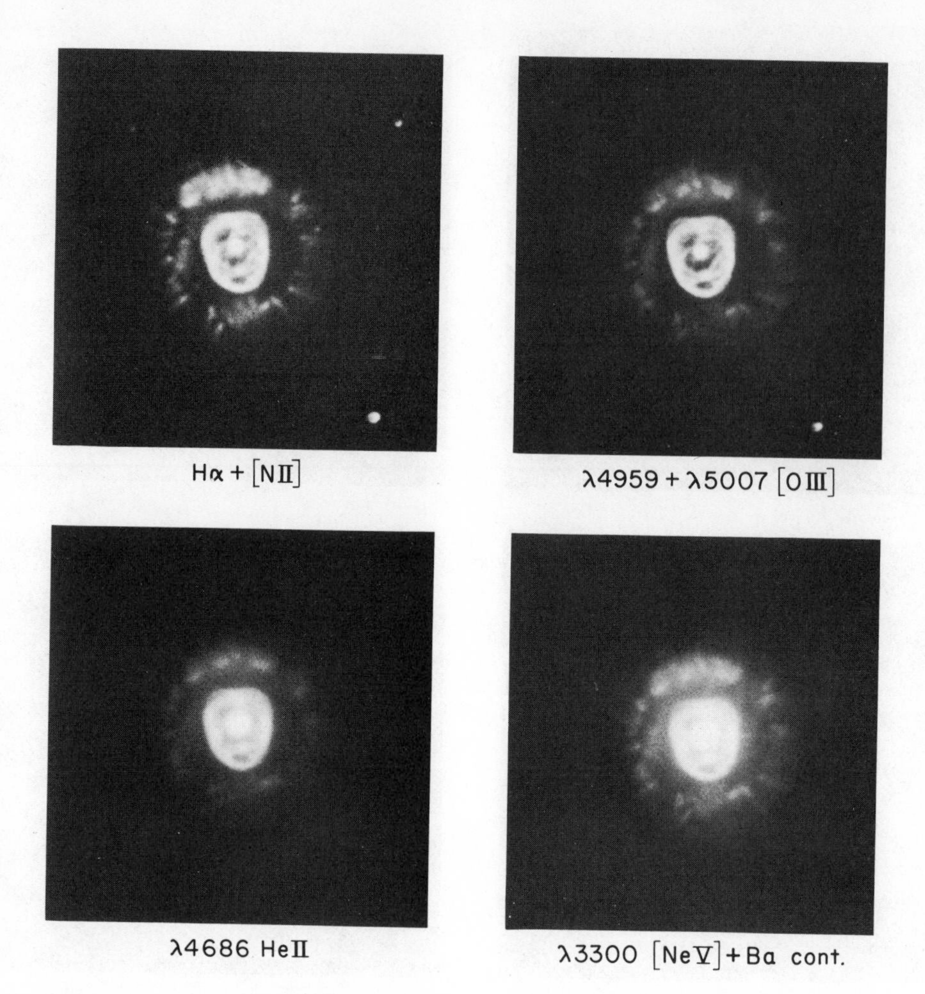

FIG. 28–16 (*b*) *NGC 2392 Planetary Nebula in Gemini. Photographed in red, yellow, violet, and ultra-violet light.* (*200-inch photographs from Mount Wilson and Palomar Observatories*)

central star almost unobscured. The ringlike appearance arises from reflected, scattered, and excited radiation coming from the parts of the nebulae that are at an angle to the line of sight, as shown in Figure 28.17.

The shell itself has a very complicated structure and in general is in a state of turbulence and internal motion resulting from the intense radiation of the central star. The internal motion and turbulence are indicated by the Doppler doubling of spectral lines and by other types of spectral irregularities (see Fig. 28.18). There also appear to be velocity differences within the nebula itself, ranging anywhere from 10 to 100 km/sec. There is reason to believe that this nebulous mass of material is expanding at speeds of the order of 20 km/sec. Such a speed of expansion is required

FIG. 28–16 (*c*) *NGC 3587 "Owl" Planetary Nebula in Ursa Major. Messier 97.* (*60-inch-photograph from Mount Wilson and Palomar Observatories*)

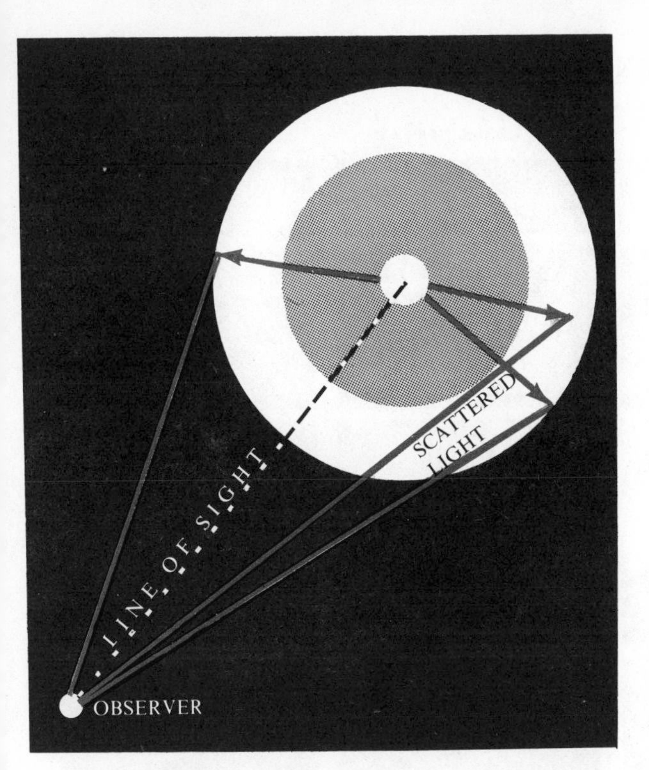

FIG. 28–17 *The ring-like appearance of planetary nebulae arises from the way the light of the central star is scattered by the gaseous envelope.*

by the great pressure of the ultraviolet radiation. There is also evidence that planetary nebulae are rotating, the rotation giving rise to additional internal motions. (See Figure 28.19.)

The densities in planetary nebulae are of the order of 10^3 atoms/cm^3, with a range from 100 atoms/cm^3 for the faint objects to 10 000 atoms/cm^3 for the very bright ones. These densities can be found from the number of electrons per cubic centimeter, which is determined by noting that the radiation in the continuous-emission spectrum comes from the recombination of free electrons and ions. The ultraviolet radiation from the central star first ionizes neutral atoms, thus ejecting electrons. These free electrons are then recaptured in the L level (second excited state) of the atom, emitting continuous visible radiation in the process. From the total continuous energy emitted in this way we know that there are anywhere from a few hundred to a few thousand electrons per cm^3, and that the electron kinetic temperature ranges from about 8 000 to 20 000°K.

Planetaries fall near the blue end of the horizontal branch of the H-R diagram of a globular cluster. This is borne out by the single planetary nebulae in M 15, which appears to be one magnitude brighter than the RR Lyrae stars in that cluster.

28.24 The Crab Nebula

In the last section we discussed the nebulous material found in expanding shells of gas around faint, hot, central stars, and an analysis of these shells indicates that this material was probably ejected violently from the central stars at some distant time in the past. From this it is reasonable to conclude that planetary nebulae may be the remnants of, or in some way related to, supernovae, although the evidence for this is not conclusive. There are, however, nebulosities associated with single stars which we know definitely were ejected from these stars during supernova explosions.

Three such explosions have been detected in our Galaxy in the past 1 000 years; the best example is the Crab Nebula in Taurus, which Chinese astronomers recorded as a supernova on the fourth of July in the year 1054. It gradually became as bright as the planet Venus, and it is still visible as the Crab Nebula, in which the gaseous material is expanding at a rate that confirms the date of the Chinese observations. This remarkable object (Figure 28.20) is now known to be a radio source, and has an apparent angular diameter of 5′ of arc. Since its distance is known to be 3 500 light years, it has a linear extension of about 2 parsecs. If it is viewed in the near infrared region of the spectrum (λ7 200 to λ8 400), it appears as an amorphous mass without any internal structure, but when photographed in λ6 300 to λ6 700 light it exhibits a remarkable filamentary structure.

28.25 The Structure of the Crab Nebula

According to Baade and Minkowski, the Crab Nebula consists of two interpenetrating structures: an external system of relatively slender filaments forming at the surface an envelope which is rapidly expanding at a rate of 1 300 km/sec, and an amorphous mass completely filling the interior region. The filamentary structure, which exhibits an emission spectrum consisting primarily of the lines of H_α and NII, accounts for only a few per cent of the total luminosity, whereas the amorphous mass, which contributes more than 80 per cent of the radiation, emits a continuous spectrum with no indications of bright or dark lines. The filamentary system is an almost elliptical structure; the amorphous material is an S-shaped mass.

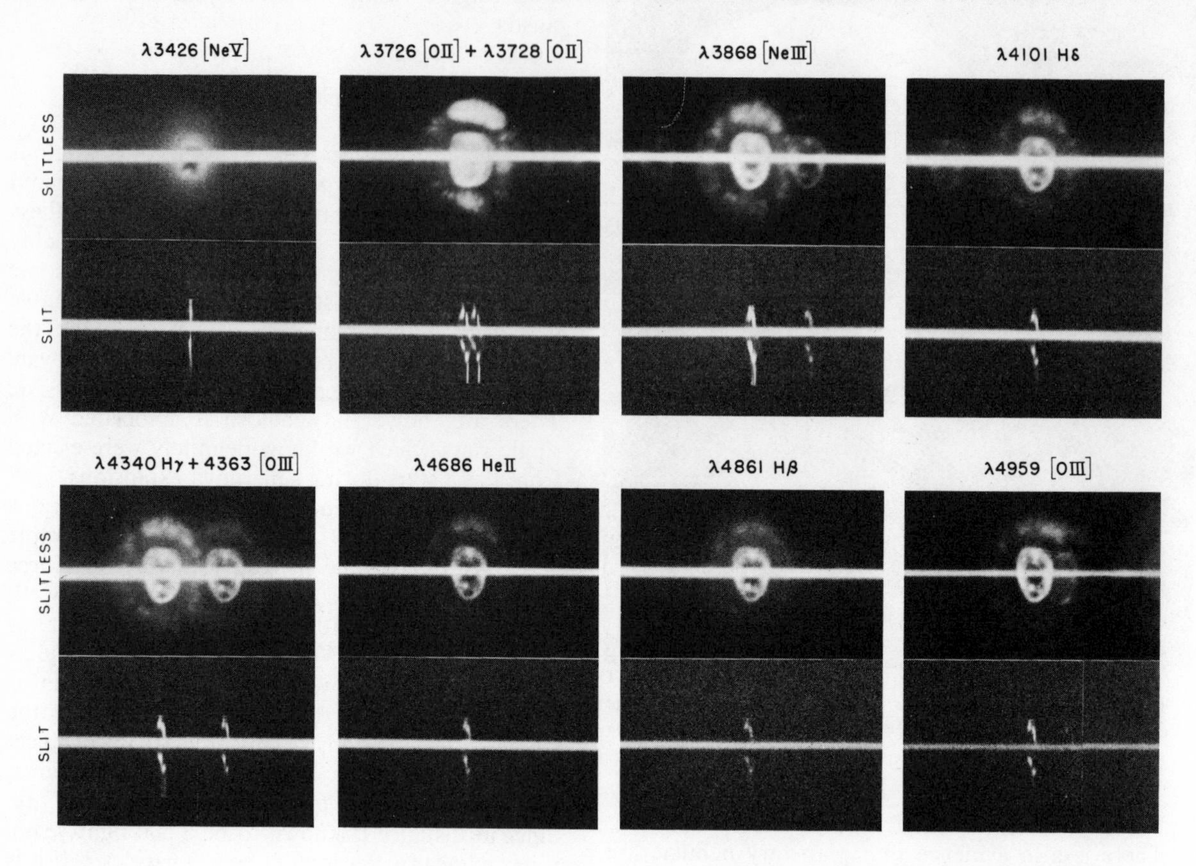

FIG. 28–18 *Spectra of the planetary nebula NGC 2392, slit and slitless. The slitless spectra show different image sizes depending upon the spectral lines in which the photograph is taken. The small size of the image in highly ionized NeV and in ionized HeII shows that the hot radiation is much more effective in ionizing atoms close to the star than at large distances. Note the pronounced irregularities in the spectral lines. (From Mount Wilson and Palomar Observatories)*

Since the usual OII, OIII, H, and He lines characteristic of planetary nebulae are observed in the spectrum of the Crab Nebula, it is natural to assume, as Baade and Minkowski did, that the luminosity of the gaseous mass is due to the ultraviolet radiation from the very hot and faint star at the center of this nebula, which is now thought by some astronomers to be a neutron star emitting X-rays. However, there are serious difficulties with this analysis since the total apparent photographic magnitude of the Crab Nebula is $+9$ while that of the central star is about $+16$. To assume then that the luminosity of the gaseous mass is due entirely to the ultraviolet radiation from the central star, we would have to account for a thousandfold increase when we go from the visible to the ultraviolet radiation from the central star.

28.26 The Problem of Optical Emission of the Crab Nebula

To understand the difficulty, we must first consider the mechanism responsible for the continuous spectrum emitted from the amorphous gaseous mass (see Figure 28.21). If we assume that the spectrum arises from the free electrons that emit continuous radiation as they slow down (as a result of interacting with electrical fields of nearby atomic nuclei) and also from the free electrons that emit continuous radiation as they recombine with nuclei, then the kinetic temperature of the electrons in this nebula would have to be of the order of 100 000°K, the electron density of the order of 1 000 electrons/cm^3, and the total mass of the order of $20M_\odot$ (which is very much larger than

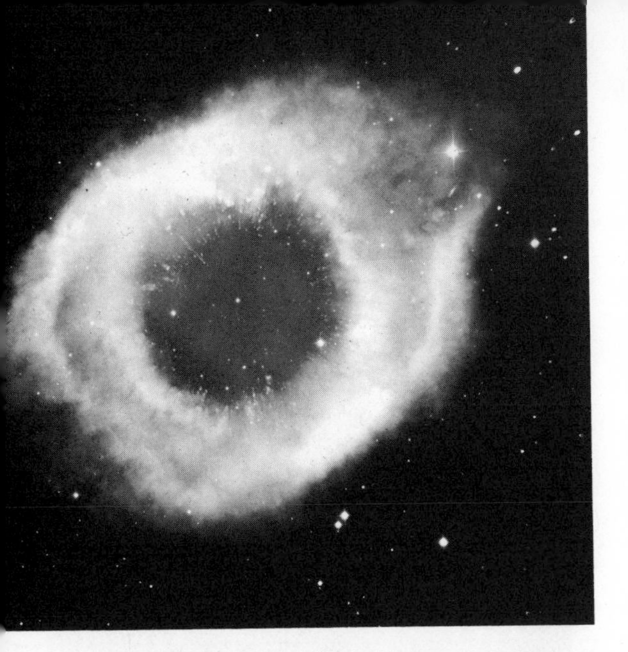

FIG. 28–19 *NGC 7293 Planetary Nebula in Aquarius. Photographed in red light. The expansion of the envelope and the internal motions are clearly indicated. (200-inch photograph from Mount Wilson and Palomar Observatories)*

the mass of typical bright planetary nebulae). This means that the mass of the star before it became a supernova would have had to be of the order of $20M_{\odot}$.

If these numbers are correct, the present total energy content (5×10^{47} ergs) of the Crab Nebula (i.e., the sum of the kinetic energies of all the electrons and ions and the ionization energy, the energy that would be released if all the electrons recombined with the ions) will be radiated away in a few thousand years at the present luminosity, which is 1.3×10^{36} ergs/sec (the absolute magnitude of the Crab Nebula is -2.2). It is clear, then, that the energy emitted by the nebulous material must be constantly replenished by some mechanism unless we suppose that the energy will all be dissipated in a few thousand years.

If we assume that this mechanism is the radiation from the central star itself, we run into difficulty because this central star, which emits a continuous spectrum with no visible absorption lines, has the color index of a late B-type star. It is therefore too cool (its surface temperature being of the order

FIG. 28–20 *NGC 952 Crab Nebula in Taurus. Remains of supernova of AD 1054. Messier 1, taken in red light. The filamentary structure is clearly visible and the amorphous structure which has only a continuous spectrum can be seen beneath it. There is some evidence that the central star is a neutron star because X-rays have been detected from this object. (200-inch photograph from Mount Wilson and Palomar Observatories)*

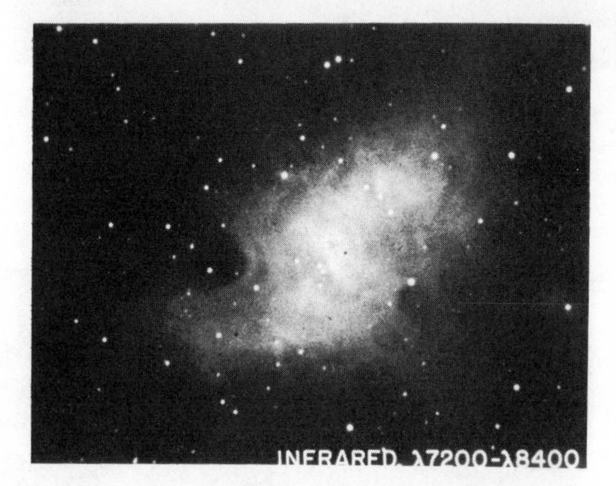

FIG. 28–21 *NGC 1952 Crab Nebula in Taurus. Messier 1. Four photographs in infrared, red, yellow, and blue. The detailed structure of the filaments in red light indicates that the radiation from the filaments is primarily due to the Balmer lines of hydrogen. (100-inch photographs from Mount Wilson and Palomar Observatories)*

of 20 000 to 30 000°K) to emit the required amount of ultraviolet radiation.

But if, in spite of this, we do assume that all the light coming from the nebulous mass originates in the central star ($M_{ph} = +4.8$), then we have to ascribe very unusual properties to this star if it is to emit ultraviolet radiation at a rate of 1.3×10^{36} ergs/sec, as required by the luminosity of the nebula. Taking into account the fact that the central star probably does not radiate like a black body in the visible spectral range, Minkowski found that the central star would have to have a luminosity about 30 000 times that of the sun, a radius of about $0.020R_{\odot}$, a mean density of 180 000 times that of the sun, and a surface temperature of 500 000°K to supply the nebula with energy enough to account for the emitted radiation. Shklovsky has pointed out that these data are at such variance with the appearance of the central star (which as we have seen has a color index indicative of a rather low temperature) that Minkowski's explanation is untenable. (See, however, the discussion of neutron stars, Section 24.50.)

28.27 The Problem of the Radio Emission

There is another difficulty with this explanation. As we have noted, the Crab Nebula, and indeed all the supernovae, are intense radio sources. The energy per unit frequency interval emitted in the radio wavelength regions is practically independent of the frequency of the emitted radiation and is about 400 times larger than the radiation per unit frequency interval emitted in the visible regions. From this Shklovsky has shown that the radio emission in the Crab Nebula is not due to the thermal motion of the ions (i.e., the random motion arising from the effective temperature of the nebula).

For if this radiation were thermal in origin, the kinetic temperature of the ions would have to be about 100-fold greater than that obtained from the observed optical radiation, and the number of electrons per cm^3 in the nebula would have to be greater than that observed optically by a factor of about 25. Moreover, if the radiation were thermal, its intensity per unit frequency interval would be about the same in the radio part of the spectrum as in the optical part, instead of being 400 times larger. Because of these difficulties associated with the usual Baade-Minkowski explanation, Shklovsky has proposed a new mechanism leading to what is called **synchrotron radiation** to explain the luminosity of the Crab Nebula.

28.28 Synchrotron Radiation from the Crab Nebula

According to this mechanism, very rapidly moving electrons (so-called ***relativistic electrons*** *because they are moving at speeds close to the speed of light, hence with kinetic energies considerably larger than m_0c^2, where m_0 is the rest mass of the electron) are decelerated in local magnetic fields in the nebula and emit radiation in this process.* In fact, Alfvén and Herlofson had already suggested this mechanism to account for the emission of radiation from discrete radio sources in space. Since this sort of radiation has been emitted by electrons in large synchrotrons here on earth, we can expect the same thing to occur in nebulous regions where high-velocity electrons and magnetic fields exist together. *Essentially, the electrons spiral around the magnetic lines of force and radiate energy of a frequency determined by the strength of the magnetic field.*

To account for the radiation emitted by the Crab Nebula, Shklovsky showed that magnetic fields of the required intensity do occur in localized regions of the nebula, and that there are present enough electrons with very high velocities to give the observed results. Since the mean internal motions in the Crab Nebula are of the order of 300 km/sec, there are eddy currents of ionized gases in various regions of it that give rise to chaotic localized magnetic fields. These magnetic fields under the Crab-Nebula conditions range anywhere from 3×10^{-4} gauss to 2×10^{-3} gauss, and although they are very small, radiation of the right frequency and intensity is emitted because there are enough electrons present moving at the proper velocities.

It can be shown that the Crab Nebula can emit the observed radio waves if its electrons have energies ranging from 2×10^7 to 2×10^{10} ev, whereas to emit the observed visible radiation its electrons must have energies of the order of 10^{11} or 10^{12} ev. Shklovsky finds that there are about 2×10^{50} relativistic electrons with energies between 2×10^7 and 2×10^9 ev, and that these are sufficient in number to account for the observed radio emission. He shows, in fact, that only a relatively small number of high-velocity electrons (of the order of $2 \times 10^{-6}/cm^3$) can give rise to very intense radio waves. The total number of relativistic electrons in the Crab Nebula with energies greater than 2×10^{11} ev is about 10^{47}, which is again sufficient to account for the observed optical radiation.

28.29 The Mass of the Crab Nebula

With the synchrotron-radiation mechanism it is no longer necessary to suppose that the mass of the gaseous material in the Crab Nebula is many times that of the sun. In fact, it appears now that the mass of the amorphous material that emits the continuous radiation is quite small and that if all its constituent electrons were moving with relativistic velocities, its mass would only have to be about $10^{-6}M_\odot$, to give the observed optical radiation. However, the mass must be considerably greater than this; otherwise the necessary magnetic fields could not be sustained. One can show that for such magnetic fields to exist, the mass of the amorphous material cannot be less than $0.05M_\odot$, and this is probably true for the filamentary material in the outer regions as well. It appears, then, that the total mass of the material ejected when the Crab Nebula originated in the supernova outburst is of the order of $0.1M_\odot$.

According to this picture, based upon the analysis of the Crab Nebula, it now appears that during a supernova outburst the original star is not completely disintegrated, but rather that a relatively small envelope is ejected just as in the case of ordinary novae, but at much greater energies. Thus a supernova and an ordinary nova appear to differ only in degree rather than in kind.

28.30 Polarization of Light from the Crab Nebula

That local chaotic magnetic fields are indeed present in the Crab Nebula is demonstrated by the

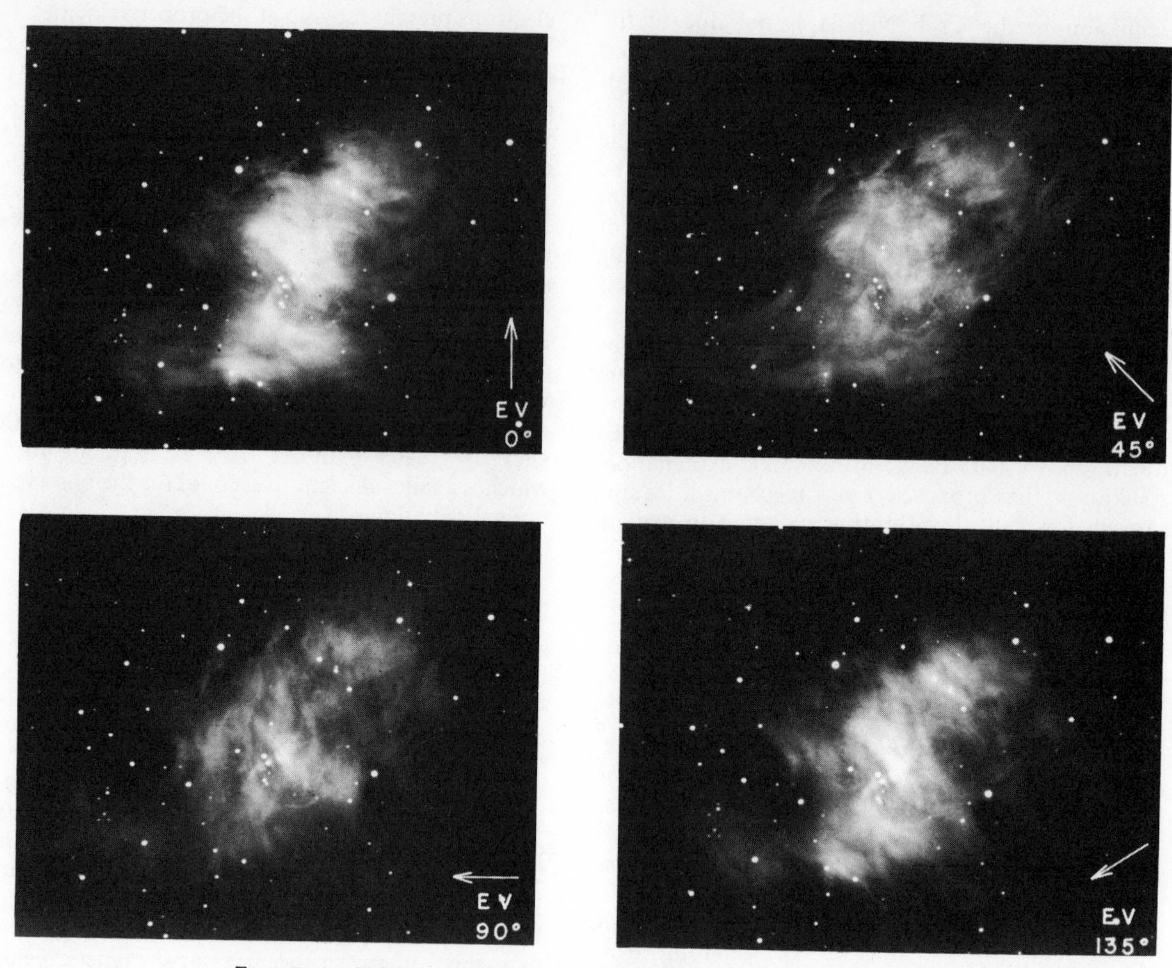

FIG. 28–22 *NGC 1952 The Crab Nebula in Taurus, photographed in polarized light. Direction of electric vector given. The arrow gives the direction (as seen by an observer on the earth) in which the electric vector of the light reaching the observer is vibrating. From the direction of polarization of the light we are able to determine the direction of the magnetic field in the gaseous matter.* (From Mount Wilson and Palomar Observatories)

polarization of light coming from this nebula. Synchrotron radiation is polarized in a definite way because the decelerated electrons in a synchrotron emit radiation principally in the direction of their motion. Thus, depending on what part of the Crab Nebula we observe, we find that the emitted synchrotron radiation is differently polarized depending on the orientation of the magnetic fields. If one studies the nebula with a polaroid filter, it is possible to trace out the magnetic fields, and one sees that as the filter is rotated, the brightness of various parts of the nebula changes so that the entire nebula alters its appearance continuously, as shown in the series of photographs taken with polaroid filter in different orientations (Figure 28.22).

28.31 The Origin of the Speed of Relativistic Electrons *

One of the problems presented by the synchrotron radiation mechanism is the origin of the large number of relativistic electrons that are needed to give the observed radio emission. The most likely explanation is that they rise from collisions between small local masses of gas moving at high relative speeds. If two such masses moving at a relative speed of 1 000 km/sec or more collide, temperatures of the order of 10^8 °K are generated at the interface of the collision and a certain number of particles acquire energies between 10^4 and 10^5 ev. The energies of these particles steadily increase as they collide over and over again with masses of gas carrying magnetic fields. In this way, by random collisions with turbulent magnetic masses of gas, charged particles can acquire very high energies.

Just as we can account for the luminosity of the Crab Nebula by the synchrotron mechanism, we can do the same thing for other discrete radio sources. In fact, Shklovsky has analyzed the discrete radio source corresponding to the supernova that occurred in A.D. 369 and which has properties similar to those of the Crab Nebula. The Tycho Brahe supernova has also been detected with radio telescopes, and it now appears that all supernovae are sources of intense radio waves.

29

THE STRUCTURE AND ROTATION OF THE GALAXY*

29.1 The Spiral Structure of the Galactic Disk*

We have seen that our solar system lies in a region of the Galaxy in which the stars are distributed in a thin layer about 1 000 light years thick, extending out to a distance about 15 000 light years towards the center (in the direction of Sagittarius) where the central bulge begins. This thin disk (which we may speak of as the outer portion of our Galaxy) consists of Population I stars lying in a matrix of gaseous matter and dust from which new stars are constantly being formed. Although most of the stars in this region are very faint red dwarfs, there are also hot blue-white stars, apparently among the youngest stars in our Milky Way.

Although we cannot push star counts much beyond 6 000 light years because of the obscuration, the distribution of open clusters enables us to trace the disk-like structure out to the 15 000 light-year distance already mentioned. Here we find vast accumulations of star clouds which indicate the direction of the center of the Galaxy, where great numbers of Population II stars such as the RR Lyrae variables and most of the globular clusters are distributed. We also find in this general direction a concentration of planetary nebulae, long-period and irregular variables, and novae, all of which mark the very center of the Galaxy itself. However, because of the obscuration we cannot obtain a detailed picture of the nucleus, although we can see that star clouds like those in Sagittarius define its edge and extend far into it. We shall see that we can surmise a good deal about the structure of the nucleus, first by considering objects such as galactic clusters that outline it, and then by analyzing extragalactic spiral nebulae, which we have reason to believe are similar in structure to our Galaxy. Before we do this, however, we must complete our picture of the spiral disk.

Although a detailed star count of the faint stars in the neighborhood of the sun does not give a definite picture of the spiral structure of our Galaxy, we obtain such a picture by tracing out to considerable distances the distributions of O and B type stars, stellar associations, the HII regions, type I Cepheids, and other objects associated with the spiral arms. In addition there is the analysis of the 21-cm line which gives us a clear picture of the distribution of hydrogen, again indicating the presence of spiral arms. Taken individually these objects and their related phenomena might not be decisive in this analysis, but when we find all of them occurring together in a given region of the sky, we must conclude that they trace out a spiral arm.

29.2 Three Galactic Spiral Arms

By correlating all the data from the distribution of these various types of galactic objects, we are able to delineate three distinct spiral arms.

The first evidence for spiral arms was obtained by Morgan, who analyzed the distributions of O and B type stars in the Northern Milky Way. These, together with the distribution of HII regions, give clear indications of two spiral arms, the **Orion arm**, in which the solar system itself is situated, and the **Perseus arm** at a considerable distance from the sun. The Orion arm appears to extend from about galactic longitude 40° to longitude 90°. It has been traced out over a distance of 5 000 parsecs from Cygnus through Cepheus, Perseus, Orion, and Monoceros. This arm, about 2 000 light years wide, runs at an angle of about 60° to the direction to the center of the

Galaxy and contains among other things the Orion Nebula, the Perseus nebulosity, and some HII regions. According to Morgan, the solar system lies about 200 light years within this arm near its inner edge—that is, on the concave side of the arm that faces the center of the Galaxy.

The second, or Perseus arm of the spiral is about 7 000 light years from the sun out towards the edge of the Galaxy, and is most easily recognized by the double cluster in Perseus. Morgan has traced this arm from galactic longitude 70° to a longitude of 140°. In general, it is similar to the first arm in structure and size. As we trace these arms out from the center moving from Cygnus towards Monoceros, they become wider in the plane of the Galaxy, which indicates that the spiral structure unwinds itself in that direction. We shall see that this is borne out by a study of the rotation of the Galaxy.

The concentration of O and B stars and HII regions that can be traced out through η Carinae is evidence for the existence of an inner arm in the region of Sagittarius and Scorpius. This arm includes the great star cloud in Sagittarius. One section of the arm can be observed between longitudes 300° and 350°, with another section lying between 250° and 263°. There appears to be a break of considerable extent in this arm between galactic longitudes 265° and 300°. Bok considers this to be a real gap in the spiral structure.

The outlines of these three arms are indicated in Figure 29.1, but it must not be supposed that they form continuous structures all around the center of the Galaxy. They are undoubtedly broken up by gaps, and there is also evidence that there are short spikes and spurs projecting from parts of the arms at various points, as shown in Figure 29.1.

It is clear that the evidence that we have presented here for the spiral structure is based on only a very small fraction of all the stars in our Galaxy, and many more data of the kind described here must be collected before we can construct anything like a complete picture. We have merely indicated here that there are three observable segments of spiral arms within 10 000 light years of the sun. To trace these arms all around the nucleus involves piecing together about 18 such segments. Since this is clearly impossible by means of star counts, because it would involve observing stars in the plane of the Milky Way out to distances of 60 000 light years or more, we must have recourse to some other observational data to complete the picture. Here we can make use of the emission of radio waves from HI regions (i.e., the 21-cm line).

29.3 Galactic Rotation

Since, as we have noted, neutral hydrogen is abundant in spiral arms, we can trace out these arms by studying the 21-cm lines. We can do this because the Galaxy is rotating so that the hydrogen in each of the arms is moving with different speeds around the center. For this reason the 21-cm line, in general, shows a Doppler shift that varies from arm to arm (see Section 29.7). Because of this relationship of the rotation of the Galaxy to its spiral structure, we shall consider the problem of galactic rotation in order to help us understand the structure.

In discussing the motions of stars in the neighborhood of the sun, and also the solar motion itself, we found it necessary to introduce a frame of reference determined by the neighboring stars themselves. We saw that in this frame of reference the sun is moving at 19 km/sec, at right angles to the line to the center of the Galaxy, towards the constellation of Hercules, and that the peculiar velocities of most of the stars in our neighborhood are of the order of 20 to 30 km/sec. We also noted that these peculiar motions are not random, but rather have a preferential direction parallel to a line towards the center of the Galaxy. Moreover, we saw that there are certain high-velocity stars moving in a general direction opposite to the solar motion, and that this high velocity is characteristic of stars such as the RR Lyrae stars (average speed relative to the sun is 130 km/sec) which are distributed in a subsystem of our Galaxy that is only slightly flattened towards the plane of the Milky Way. It is clear that to account for these apparently unrelated observations we must refer not to a frame of reference determined by nearby stars, but rather to a frame at the center of the Galaxy itself.

The difference between the observed motions of the more or less spherically distributed stars such as the RR Lyrae stars (and in general all high-velocity objects) and the observed motions of stars such as our sun and the O and B type stars that form a flattened system parallel to the plane of the Milky Way is due to the differential rotations of these systems about the center of the Galaxy. Indeed, as early as 1923 Lindblad first suggested the differential rotation of the Galaxy as a way to explain the asymmetry of the high-velocity stars. However, even before that time Charlier proposed a rotating galaxy on the basis of the observed proper motions. The proof for the rotation of the flattened systems of O and B type stars and, in

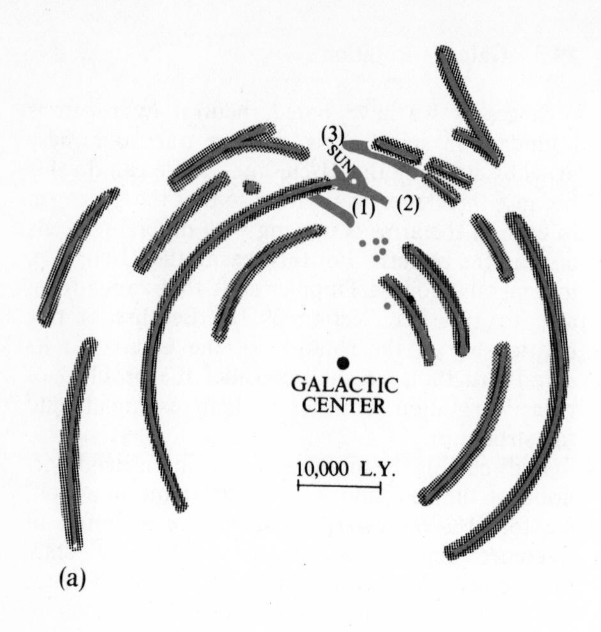

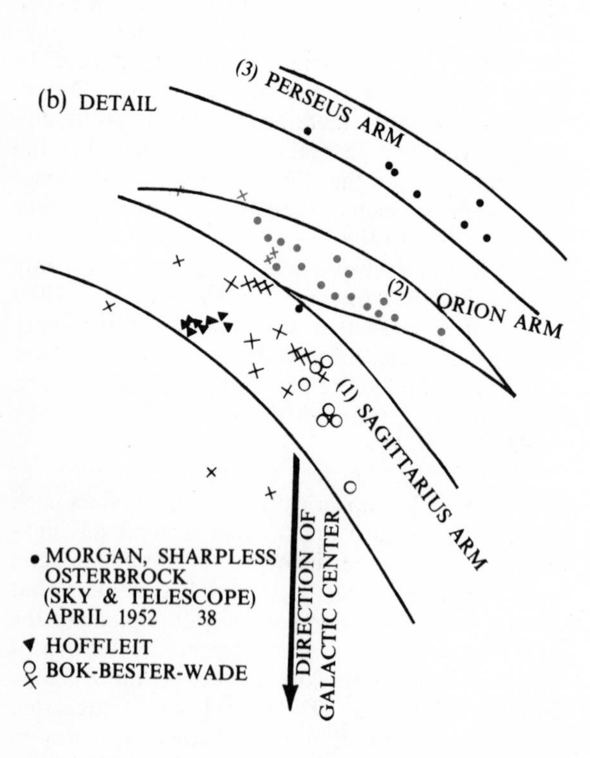

FIG. 29–1 *The schematic drawing of the three known arms of the Galaxy: the Perseus Arm, the Orion Arm, the Sagittarius Arm. These are not continuous structures, but are broken up by gaps and have short spikes and spurs. Adapted from the data of F. J. Kerr, Radio Physics Laboratory, Sydney, Australia.*

general, all Population I stars lying in the plane of the Milky Way was first given by Oort, who analyzed the radial velocities and proper motions of stars at distances between 1 000 light years and 10 000 light years from the sun.

29.4 Oort's Analysis of the Rotation of the Galactic Disk

Oort's analysis of the rotation of the galactic disk is based on the assumption (which we now know is correct) that the Galaxy is not rotating like a solid disk, but rather differentially according to Kepler's laws, with stars farther from the center moving more slowly than those close to it. We can understand why this is so if we picture the greater part of the mass of the Galaxy as concentrated in the nucleus, so that each star in the arms of the Galaxy revolves in an almost circular orbit around this nucleus in conformity with Newton's law of gravity.

To understand Oort's analysis, we consider the arrangement in Figure 29.2, where c is the center of the Galaxy, s is the position of the sun at distance $R_{\odot}$, and Q is the position of a star at a distance R from the center and at a distance r from the sun. The vector $\mathbf{V}(R_{\odot})$ is the velocity of the sun (assumed to be moving in a circular orbit around the center of the Galaxy) and $\mathbf{V}(R)$ the velocity of the star (also in a circular orbit). Oort proceeded by analyzing the radial velocity of the star relative to the sun. This is obviously equal to $V(R)\cos\alpha - V(R_{\odot})\sin\phi$, where α is the angle between the vector $\mathbf{V}(R)$ and the direction of the line of sight from the sun to the star and ϕ is the galactic longitude of the star. This follows since $V(R_{\odot})\sin\phi$ is just the component of the sun's velocity in the direction r and $V(R)\cos\alpha$ is the component of the star's velocity along this line, so that the difference between these two is just the radial velocity of the star relative to the sun.

Now if P is the time it takes the star to make one complete revolution around the Galaxy (i.e., its **galactic period**) and $P_{\odot}$ is the galactic period of the sun, we have

$$V(R) = \frac{2\pi R}{P} \quad \text{and} \quad V(R_{\odot}) = \frac{2\pi R_{\odot}}{P_{\odot}}.$$

It follows, therefore, that the radial velocity of the star relative to the sun is

$$V_r = 2\pi\left(\frac{R}{P}\cos\alpha - \frac{R_{\odot}}{P_{\odot}}\sin\phi\right).$$

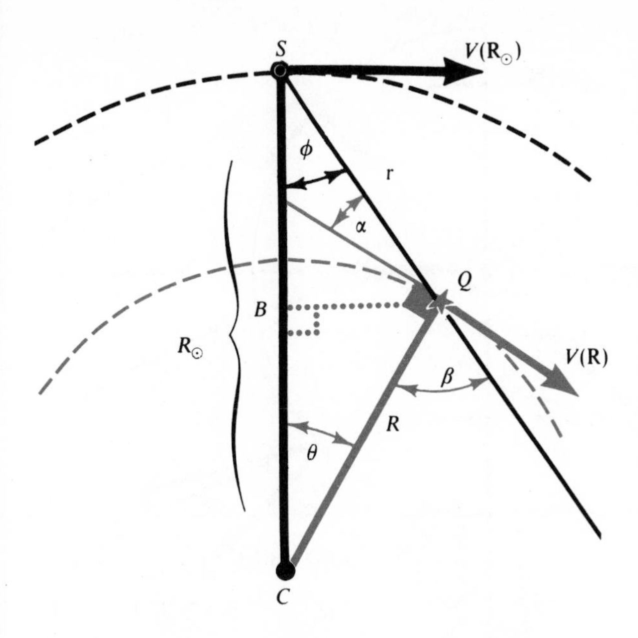

FIG. 29–2 *A schematic diagram used to derive the rotation of the Galaxy, based on Oort's analysis. C is the center of the Galaxy, S is the position of the sun, and Q is the position of a star.*

We see from the figure that $R \cos \alpha = R \sin \beta = R(\sin \theta \cos \phi + \cos \theta \sin \phi)$, since $\alpha + \beta = 90°$, where θ is the angular separation between the sun and the star as seen from the center of the Galaxy. We also have $R \sin \theta = QB = SB \tan \phi$, $R \cos \theta = BC$. Therefore we have

$$\begin{aligned} R \cos \alpha &= R(\sin \theta \cos \phi + \cos \theta \sin \phi) \\ &= (R \sin \theta) \cos \phi + (R \cos \theta) \sin \phi \\ &= (SB) \tan \phi \cos \phi + (BC) \sin \phi \\ &= (SB + BC) \sin \phi \\ &= R_\odot \sin \phi. \end{aligned}$$

If we introduce this into the expression for the radial velocity, we have

$$\begin{aligned} V_r &= 2\pi\left(\frac{R_\odot \sin \phi}{P} - \frac{R_\odot}{P_\odot} \sin \phi\right) \\ &= 2\pi R_\odot\left(\frac{1}{P} - \frac{1}{P_\odot}\right) \sin \phi. \qquad (29.1) \end{aligned}$$

If we are dealing with a star that is not too far away from the sun, $1/P$ and $1/P_\odot$ depends only on $R - R_\odot$ to a first approximation, so we may write for the radial velocity

$$V_r = 2\pi R_\odot(R - R_\odot)B \sin \phi,$$

where B is a constant of proportionality and essentially represents the way in which the angular velocity of rotation of the Galaxy changes with the distance from the center.

29.5 Oort's Constant of Rotation

Finally, we may note that if we use the cosine law in the triangle CSQ, and take account of the fact that $R_\odot$ is considerably larger than r (the distance of the star from the sun), we may place $(R - R_\odot)$ equal to $-r \cos \phi$. If we introduce this into the formula for the radial velocity and use the substitution $\sin \phi \cos \phi = \frac{1}{2} \sin 2\phi$, we finally obtain Oort's famous formula for the radial velocity of a star having a given galactic longitude ϕ and at a given distance r from the sun,

$$\boxed{V_r = Ar \sin 2\phi}, \qquad (29.2)$$

where A is called **Oort's constant of rotation.**

To see the physical significance of this formula, we consider the sun's orbit in the plane of the Galaxy with one line drawn from the sun to the center of the Galaxy as shown in Figure 29.3, and another drawn in the direction of the sun's motion around the center of the Galaxy. We now draw around the sun (in the plane of the Galaxy) a circle having a radius of a few thousand light years and note that the two lines divide the circle into four quadrants. *If we consider the stars in successive quadrants, we see from Oort's formula that their radial velocities relative to the sun change sign every time we pass from one quadrant to another because of the factor* $\sin 2\phi$. Thus, on the average, the radial velocities of the stars in the first quadrant are positive and those in the second quadrant are negative, and so on.

We can see that this is so without using the formula because stars in the first quadrant are (according to Kepler's laws) moving around the Galaxy faster than the sun, and hence on the average are receding from the sun, whereas the sun is overtaking stars in the second quadrant, and so on. This is shown by the vector diagram drawn for a typical star in each quadrant. It follows that there is a Doppler shift towards the red in the spectra of stars in the first quadrant, a shift towards the blue for stars in the second quadrant, and so on. Using this type of analysis, Oort showed that the flattened system of stars lying parallel to the plane of the Milky Way is rotating with a speed that decreases as we move away from the center.

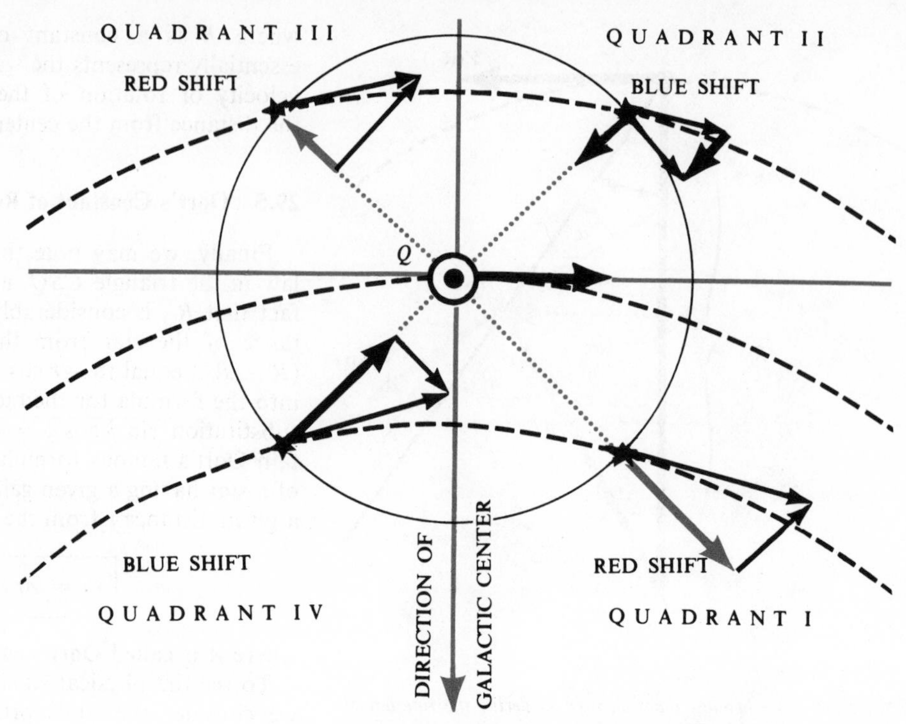

FIG. 29–3 *The schematic drawing of the orbit of the sun and the stars in its neighborhood around the center of the Galaxy. According to the Oort formula for the rotation of the Galaxy, the Doppler shift in the spectra of the stars alternates from blue to red as we go from one quadrant to the other. The velocity of any star relative to S is broken up into one component parallel to the line of sight from the sun, and one component perpendicular to the line of sight.*

29.6 The Numerical Value of Oort's Constant

From an analysis of the radial velocities of stars the value of Oort's constant is found to be about 18.2 ± 0.9 (km/sec) per kpsc, which taken together with other information implies that the sun is rotating around the center of the Galaxy with a speed of about 215 km/sec. The relationship of the rotational velocity to distance from the center is given in Table 29.1(B). Evidence gathered by the radio telescope at Parkes, Australia, substantiates previous indications that the central part of the Galaxy is rotating swiftly like a solid structure. We shall see that this model of the rotation of the Galaxy can account for most of the observed features of stellar motion in the neighborhood of the sun.

We can also derive Oort's constant from an analysis of proper motions and we obtain about the same value as above. It should be noted that the value of A depends on the groups of stars selected. Thus, for B-type stars $A \cong 18$ (km/sec) per kpsc, for Cepheid variables $A \cong 19$ (km/sec) per kpsc, and for K giants its value is about 17.5 (km/sec) per kpsc.

29.7 The Profile of the 21-cm Line and Galactic Rotation

Whereas the analysis of galactic rotation based on the velocity of stars limits us to a few thousand parsecs, the 21-cm line of neutral hydrogen enables us to study the rotation of the Galaxy and its spiral structure out to much greater distances. To derive the rotation of the Galaxy from the 21-cm line we must see how its profile varies as we analyze the 21-cm radiation from different parts of the Galaxy.

We have already noted that the shape of the 21-cm line depends on the radial velocity relative to the sun of the HI region emitting this line. The profiles of the 21-cm line show clearly the rotation of the Galaxy. To make this clear we give a simple analysis based on the assumption that the neutral

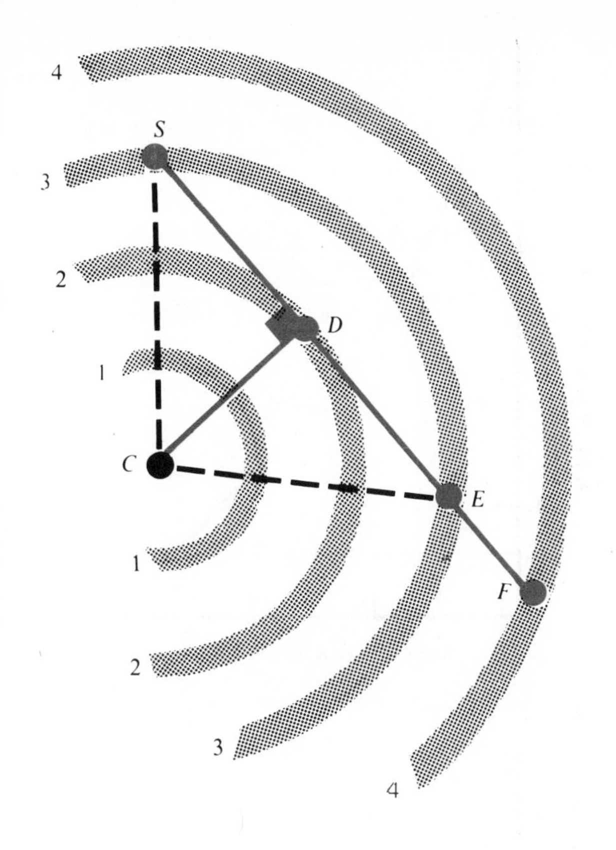

FIG. 29-4 *A schematic drawing of the spiral arms for an analysis of the Doppler shift in the 21-cm line.*

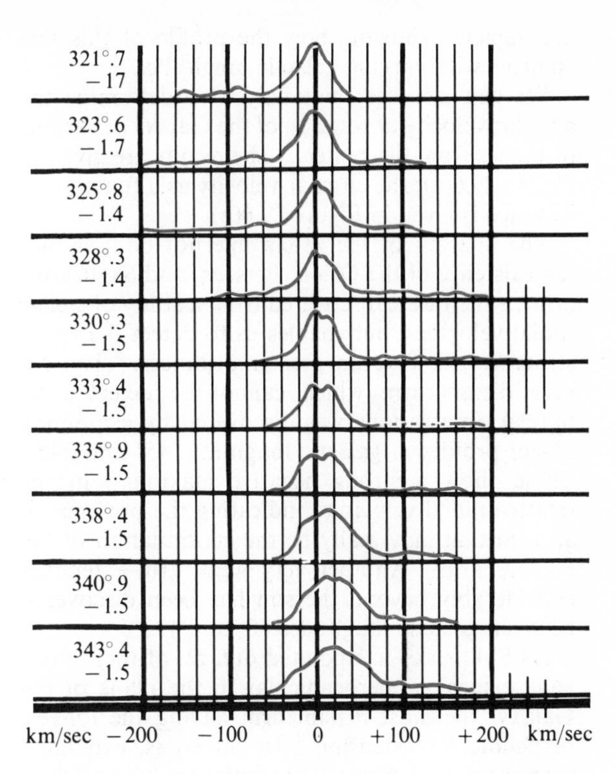

FIG. 29-5 *The variation in the profile of the 21-cm line with varying galactic longitudes. (After Kwee, Muller and Westerhout, from I. S. Shklovsky,* Cosmic Radio Waves, *Cambridge, Mass.: Harvard University Press, 1960)*

hydrogen is concentrated in distinct spiral arms and is revolving in circular orbits around the galactic center just the way the stars are, as shown in Figure 29.4. If we label these spiral arms 1, 2, 3, and 4, and consider the sun in, say, spiral arm 3, we may analyze the 21-cm radiation coming from the HI in the spiral arms 2, 3, and 4 along a line that is tangent to spiral arm 2 at point D. The radiation then comes from the hydrogen concentrated at the three points F, E, and D.

If we look at Equation (29.1) we see that the HI at D has the largest radial velocity since the value of the period of revolution for this HI is smaller than that for the HI at E or at F. This is so because the distance of D from the center is smaller than that for any other point along the line DF. Because of this the Doppler shift towards the red is a maximum for the 21-cm radiation coming from D. On the other hand, the Doppler shift for the 21-cm radiation coming from the HI concentrated at E is zero since E lies in the same spiral arm as the sun does, so that the period of this HI is the same as for the sun, and V_r as given by the formula vanishes. The period of the HI at F is greater than the period of the sun, since the distance F from the center is greater than $R_{\odot}$, and the radial velocity becomes negative so that the Doppler shift for this line is towards the blue.

29.8 Spiral Arms and the Peaks in the Profiles

Because of these different radial velocities the profile of the 21-cm line of the radiation coming along the direction FS is not sharp, but irregular, and contains three distinct peaks, one exactly at the 21-cm position corresponding to spiral arm 3 containing the sun, one on the red side of this position and corresponding to the inner arm, and one on the blue side corresponding to the outer arm. Because of the random thermal motion of the HI atoms the actual profiles are considerably more complicated than this simple picture indicates. Nevertheless, we can detect the three peaks and thus show that there are, indeed, three distinct spiral arms. In Figure 29.5 various 21-cm profiles are depicted showing how the profile of this line changes with varying galactic longitudes.

We can also use this analysis to determine the angular velocity of rotation of the Galaxy at various distances from the center. The results are given in Figure 29.6(a); the circular velocity as a function of distance is given in Figure 29.6(b).

The analysis of the 21-cm line not only verifies the existence of the Orion, Perseus, and Sagittarius arms, which were discovered from a study of stellar radial velocities, but enables us to detect the spiral structure out to even greater distances. Thus, a very distant arm which cannot be detected by optical observations was discovered by examining 21-cm profiles in galactic longitudes 45° and 345°. These show a strong negative maximum in the relative radial velocities, indicating the presence of an arm that may really be the continuation of the Perseus arm. Another very weak arm extending considerably beyond the sun has been discovered between galactic longitudes 60° and 120°.

As Shklovsky has pointed out, all of these arms have extensions perpendicular to the plane of the Galaxy, the most distant arm having the longest perpendicular extension. Its thickness, extending from galactic latitude 0° to latitude 30°, is about 800 parsecs. *Moreover, it appears from the 21-cm lines that the spiral arms are not all coplanar, but lie in planes that are slightly tilted with respect to each other.*

29.9 The Spiral Structure near the Galactic Nucleus

Although as we go in towards the center of the Galaxy, the 21-cm line is obscured to some extent because of the absorption by neutral hydrogen (not because of scattering by dust, since dust particles have hardly any scattering effect on wavelengths as long as 21 centimeters), we can still obtain information about the spiral structure close to the center by analyzing the Doppler shift in the 21-cm line. Because of this Doppler effect the radiation coming from arms close to the center has a wavelength slightly different from 21 centimeters. Hence it is not absorbed by the neutral hydrogen atoms lying between us and the distant arms near the nucleus so that we are able to detect the presence of HI in these arms. This is shown in the schematic diagram in Figure 29.7, and is based on the work of Shklovsky.

The analysis of 21-cm radiation from the region of the galactic center shows an expanding spiral arm at a distance of 3 kpsc from the center. This arm has a fairly complex structure near $l^{\mathrm{II}} = 348°$.

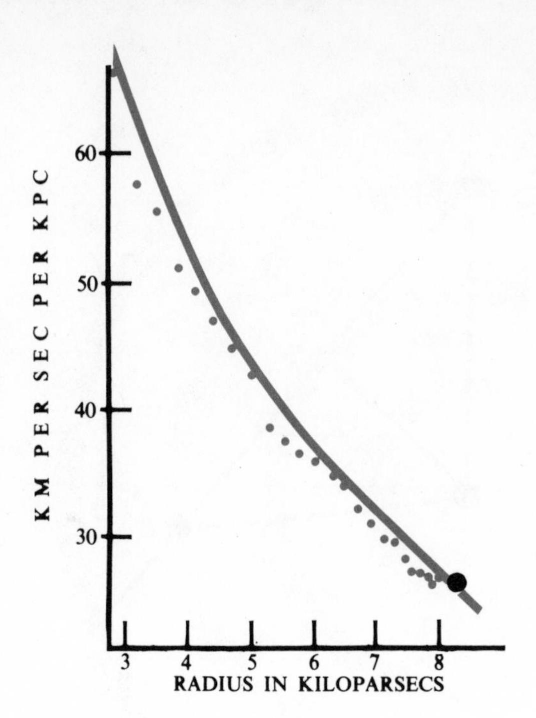

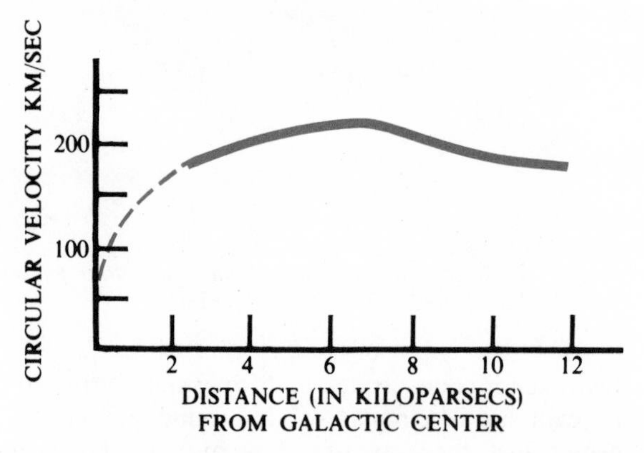

FIG. 29–6 *(a) The angular velocity of the Galaxy in km/sec per kpsc is plotted as a function of the distance from the center in kiloparsecs according to radio observations. (After Kwee, Muller, and Westerhout). (b) The circular velocity in km/sec is plotted as a function of distance from the center. (After Kwee, Muller, and Westerhout, from I. S. Shklovsky,* Cosmic Radio Waves. *Cambridge, Mass.: Harvard University Press, 1960)*

29.10 The Density Distribution of the HI and the Spiral Arms

Finally we are able to obtain the density distribution of HI in the Galaxy, and this, too, gives evidence of the spiral structure. This distribution of HI shown in Figures 29.8 and 29.9 is taken from the work of Oort, Kerr, and Westerhout. The first shows the density distribution of HI in

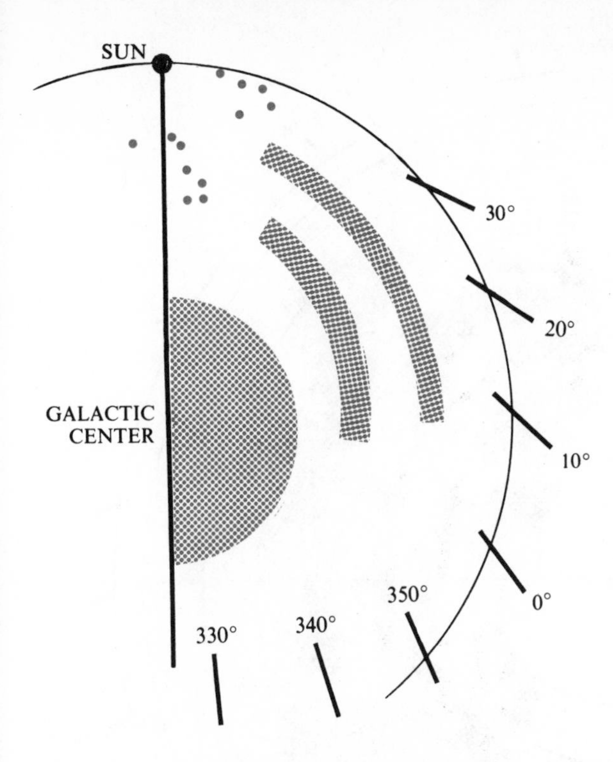

FIG. 29–7 *A sketch of the spiral structure in the inner parts of the Galaxy. (After Kwee, Muller, and Westerhout)*

the various spiral arms, with the Orion arm and the two outer arms clearly indicated. There is also a definite indication of the inner arm. The second figure is a contour map showing the projection of the maximum density of interstellar hydrogen in the galactic plane. In this second figure the inner structure is more clearly indicated. Shklovsky has pointed out that from this diagram one can distinguish four or five spiral arms within a distance of 9 kpsc from the center.

29.11 The 21-cm Absorption Line

Just as we analyze the 21-cm emission line, we can also analyze the 21-cm absorption line, and again trace out the spiral structure for the Galaxy. We can do this for fairly intense discrete sources of 21-cm radiation such as Cygnus A, Cassiopeia A, and Sagittarius A. From an analysis of both the Cygnus A and Cassiopeia A sources, as shown in Figure 29.10, we discover that there are two absorption maxima in the profiles of the 21-cm line associated with these two discrete sources, and these maxima clearly correspond to the Orion and the Perseus arms. It is also evident that Cygnus A

FIG. 29–8 *The spiral structure of the Galaxy as obtained from an analysis of the distribution of HI, using the 21-cm line. (After Oort, Kerr, and Westerhout, from I. S. Shklovsky,* Cosmic Radio Waves. *Cambridge, Mass.: Harvard University Press, 1960)*

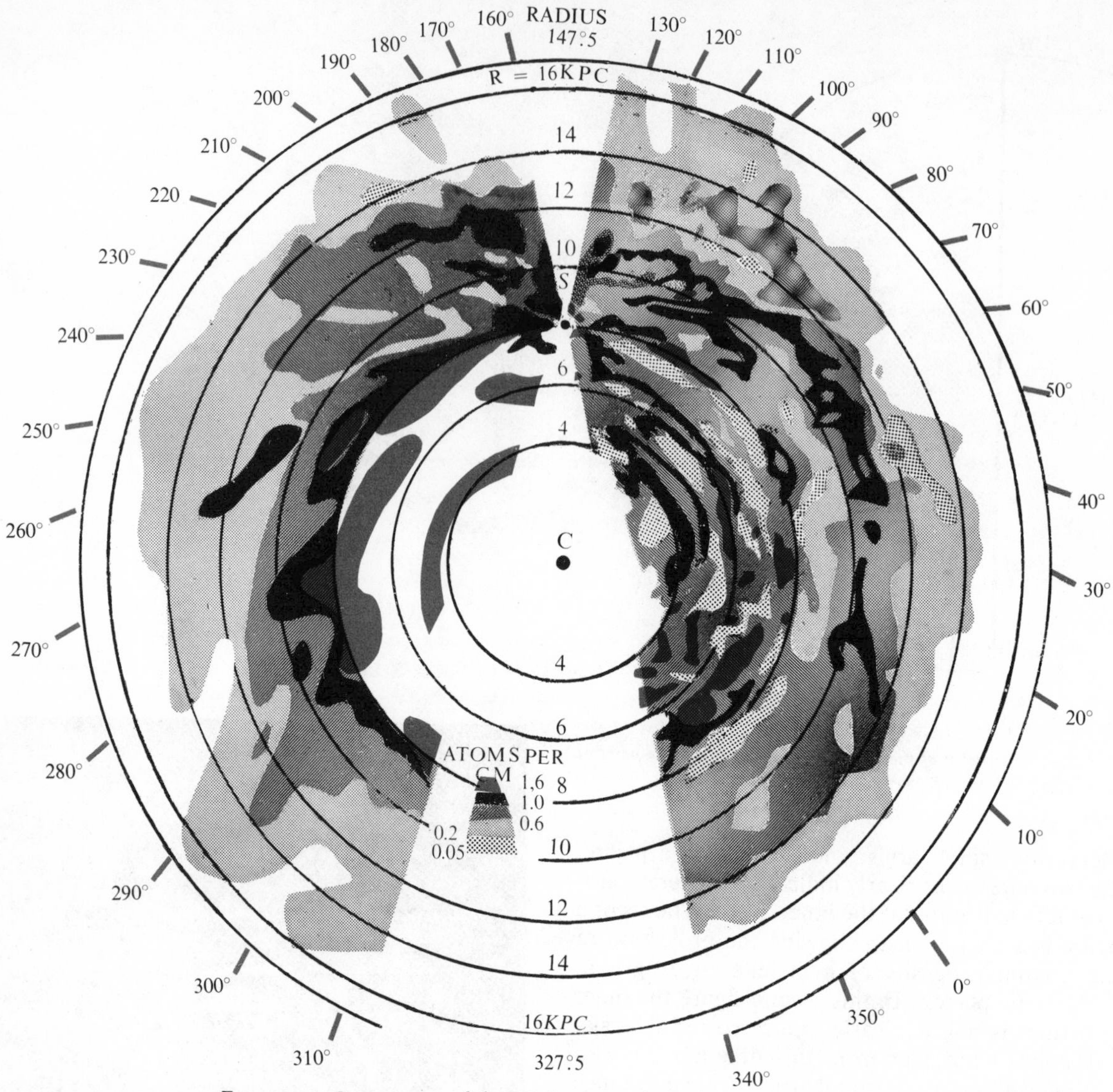

FIG. 29–9 *Contour map of the HI density in the Galaxy according to Oort, Kerr, and Westerhout.*

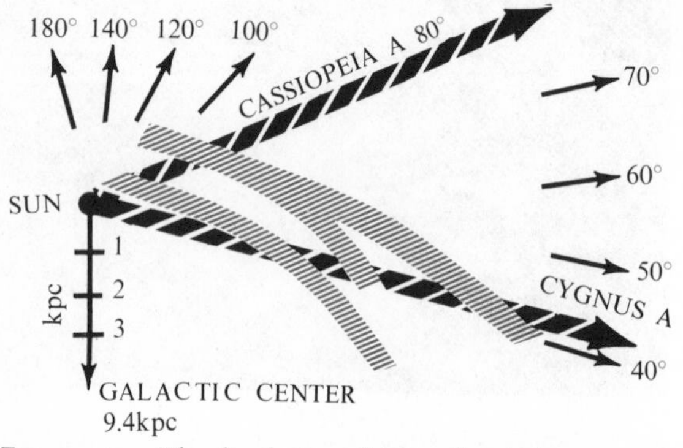

FIG. 29–10 *The distribution of interstellar hydrogen in the directions of Cygnus A and Cassiopeia A as determined from the absorption maxima of the 21-cm line.* (*From I. S. Shklovsky*, Cosmic Radio Waves. *Cambridge, Mass.: Harvard University Press, 1960*)

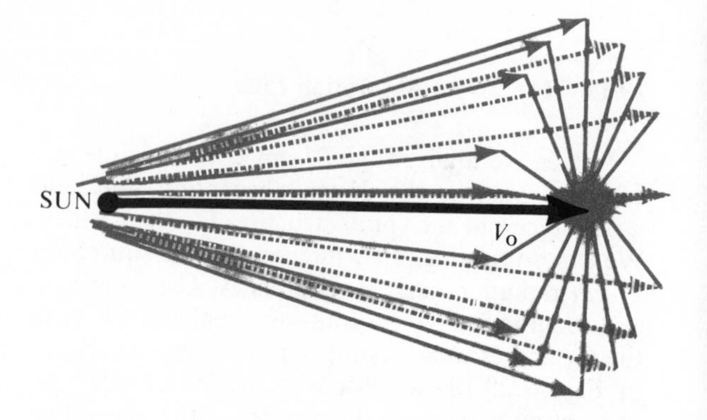

FIG. 29–11 *The slight departure of stellar motion from circular orbits in the neighborhood of the sun accounts for the preferential star streaming.*

is certainly at a distance from the sun greater than 9.5 kpsc since this is the distance of the Orion arm in the direction of Cygnus A. We may therefore conclude that this radio source is outside our Galaxy.

If we go in towards the center, we find from the absorption of 21-cm radiation from Sagittarius A that the innermost spiral arms are within 2 or 3 kpsc of the center and are moving towards the sun with a velocity of about 50 to 100 km/sec.

29.12 Consequences of the Rotation of the Galaxy —Star Streaming

The rotation of the Galaxy now enables us to understand the peculiarities of the stellar velocities in the neighborhood of the sun. We first note that the observed star streaming (see Section 25.23) and the asymmetry of stellar motions, particularly the motion of high-velocity stars, can be accounted for by the deviation from circular orbits of the majority of the stars in the neighborhood of the sun. This was first pointed out by Lindblad, who showed that *slightly elliptical orbits in the plane of the Galaxy or orbits somewhat inclined to this plane give rise to apparent drift towards or away from the galactic center as seen from the sun.*

To understand this, we consider the motion of the sun around the center of the Galaxy and assume that it is moving in a circular orbit. Its velocity is then given by the vector $\mathbf{V}_0$ as shown in Figure 29.11. Now if the stars in the neighborhood of the sun were all moving in circular orbits around the center of the Galaxy, their velocity vectors would all be parallel to that of the sun but of somewhat different lengths. However, since these stars are moving in elliptical orbits, their velocities deviate slightly in direction from that of the sun, and thus form a cone around the sun, as shown in Figure 29.11. Because of the preponderant distribution of stars in the plane of the Galaxy and because the axes of the elliptical orbits of the stars lie primarily in this plane, the number of velocity vectors that deviate in direction from the solar vector and lie in the plane of the Galaxy is much larger than the number of these vectors not in this plane.

If we connect the ends of the solar velocity vector to the ends of the other velocity vectors, we obtain the velocity distribution relative to the sun, and we see that most of these are distributed parallel to the plane of the Galaxy and not nearly as many at right angles to the plane. This gives rise to the apparent drift of stars in the plane of the Milky Way and to the ellipsoidal distribution of velocities which we have already discussed.

29.13 Galactic Rotation and High-Velocity Stars

We can now also see that the galactic rotation accounts for the observed motions of the high-velocity stars. It is clear that these stars are really moving slowly relative to the center of the Galaxy and not taking part in the general rotation the way the Population I stars are. This is borne out by the fact that there are no high-velocity stars (velocity relative to the sun exceeding 63 km/sec) moving in the direction of the rotation of the Galaxy relative to the sun. For if there were, their velocities around the center of the Galaxy would exceed the speed of escape from the Galaxy. (Their speed around the center of the Galaxy would be their speed relative to the sun plus the sun's speed around the Galaxy.) We can show, using a reasonable mass for the Galaxy, that if a star were moving parallel to the sun around the center of the Galaxy with a speed of at least 63 km/sec more than that of the sun, it would escape from the Galaxy. Since the speed of escape from the Galaxy in the neighborhood of the sun is about 300 km/sec, we see that there can be no high-velocity objects (stars, nebulae, clusters, etc.) moving in the same direction as the sun. *The observed high-velocity stars therefore are really not moving as fast as these appear to be but are lagging behind the sun and are really moving rather slowly around the Galaxy.* They appear to be moving rapidly only because the sun is moving away from them with its high orbital speed around the Galaxy.

As an example we take the RR Lyrae variables, which as a group are lagging behind and being overtaken by the sun (depending on their galactic longitude) with an observed average velocity of about 120 km/sec. Actually these stars are moving in the same direction as the sun around the Galaxy but in very elongated orbits, the inner parts of which lie in the nucleus of the Galaxy itself, and the outer parts near the sun. Thus, the RR Lyrae stars near the sun are at their apogalactic positions, and hence moving relatively slowly in their orbits (let us say at speeds of about 100 km/sec). Since the sun is moving at 220 km/sec, these stars and other Population II stars in the neighborhood of the sun appear to be moving opposite to the sun at a speed of 120 km/sec. In other words, the high-velocity stars are really Population II

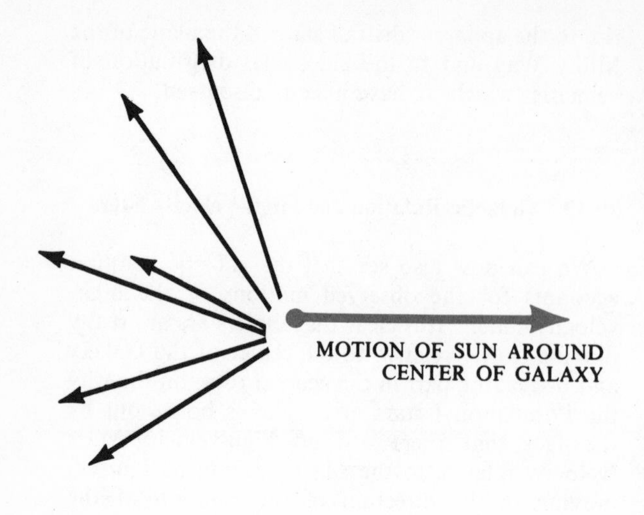

FIG. 29-12 *The distribution of high-velocity stars in the neighborhood of the sun. The asymmetry arises because the sun is running away from the stars.*

stars which move in highly elongated orbits and have their minimum speeds when they are near the sun, so that as seen from the sun they all appear to be moving rapidly, as shown in Figure 29.12.

29.14 Determining the Center of the Galaxy

One of the important problems in the structure of the Galaxy is the location of its center, which we have already indicated is at a distance of 9.5 kiloparsecs from the sun. This point can be located both optically and by means of radio astronomy. The first determination of the center of the Galaxy was made by Shapley, who pointed out that the position of the center can be found by noting that the globular clusters lie in a shell that is concentric with the nucleus of the Galaxy itself. By determining the center of the distribution of the globular clusters, Shapley demonstrated that the center of the Galaxy lies in the direction of Sagittarius about 9.5 kpsc from our solar system at galactic longitude 328°. Shapley's results were later confirmed by analyzing the counts of RR Lyrae variables outside globular clusters. These form a halo around the Galaxy the way globular clusters do and occur in great concentrations in Scorpius and Sagittarius.

More recent estimates of the position of the galactic center are based on studies of radio emission from the nucleus of the Galaxy. Radio emission from the nucleus in the 75-cm, the 50-cm, and the 22-cm spectral regions show high-intensity maxima of these wavelengths coming from a point at galactic longitude 327°.8 and galactic latitude −1°.4, which practically coincides with the galactic center as determined from the distribution of globular clusters and RR Lyrae stars.

29.15 The Radio Center of the Galaxy

These intense maxima have been identified with the discrete radio source Sagittarius A, which we have already mentioned. It now appears, according to Shklovsky, that Sagittarius A is the radio nucleus of our Galaxy. Studies of radio sources in the Galactic center have also been made in other wavelengths—for example, 1 390 megacycles per second (λ 21.6-cm). From such analyses it appears, according to Westerhout, that there is a nonthermal source at the center of the Galaxy (i.e., waves arising not from random motions of ions, but from some kind of macroscopic vibrations of the ionic plasma) having dimensions of 300 by 150 psc. Within this nonthermal source there is a thermal source 80 by 35 psc in area radiating at the same frequencies.

29.16 Ionized Plasma at the Center of the Galaxy

Sagittarius A itself appears to be a thermal source at the center of the Galaxy. We know that it must lie at the very center; otherwise it would shield us from the nonthermal radio emission. Since both thermal and nonthermal radiation arise from an ionized medium, it is clear that the very nucleus of the Galaxy itself is an HII region. From the intensity of the radio waves emitted by this region, we may conclude that the total mass of ionized hydrogen at the center of the Galaxy is of the order of $2.5 \times 10^5 M_\odot$, and that its density is 85 ions/cm^3.

The analysis of the 21-cm line by the Dutch astronomers indicates, according to Westerhout, that surrounding this HII central region and out to about 3 kpsc the gaseous matter consists primarily of neutral hydrogen, which is receding from the center at speeds ranging from 200 km/sec very close to the center, to 50 km/sec at 3 kpsc from the center. Surrounding this neutral hydrogen shell is another HII shell which extends out another kiloparsec.

29.17 The Source of Ionized Hydrogen in the Nucleus

Shklovsky explains this unusual distribution of hydrogen in the nucleus of the Galaxy by supposing that ionized hydrogen is first produced at the very center at a rate of one solar mass per year from the Population II stars and possibly from planetary nebulae, which are very abundant at the galactic center and which are near the end stages of their evolution. To produce such quantities of HII there would have to be 10^6 such planetary nebulae in a very small region near the center. The total central star count would then have to be of the order of several billion so that the number of stars per cubic parsec would have to be 10^5 times that in the neighborhood of the sun. This ionized hydrogen released in the center expands out to a distance of about 4 kpsc where it cools off and becomes neutral. Conditions in this cool gas are then favorable for the formation of young hot stars, and the radiation from these new stars then ionizes the surrounding neutral hydrogen to form the second HII region. The results we have just obtained indicate that there are unusual and peculiar features about the galactic center, the study of which Shklovsky believes will be a central problem in astrophysics and cosmology. The full nature of these peculiarities can best be seen in the detailed structure of the galactic nucleus derived from the 21-cm line by Oort and his coworkers.

29.18 The Structure of the Galactic Nucleus from the 21-cm Line

They find a dense core with a radius of about 100 parsecs surrounded by a rapidly spinning disk extending out to 300 parsecs. This disk in turn is surrounded by a well-defined rotating ring extending from 500 to 590 parsecs. Whereas the disk is rotating with a speed of 200 km/sec, the ring rotates with a speed of 265 km/sec. The hydrogen concentration in both the ring and the disk is about 1 atom per cm^3. Relatively high abundances of hydroxyl (paired hydrogen and oxygen) molecules have also been detected in this region. The data for this model of the galactic center are given in Table 29.1(A), where the distribution and the rotation are derived from radio emission and gravitational theory. We see from this table that about 60 billion solar masses lie within a radius of 8.2 kpsc of the center of the Galaxy. This agrees with the evidence obtained from dynamical considerations, for if we consider the motion of the sun around the Galaxy, we must have

$$\frac{GM}{R_\odot^2} = \frac{V_\odot^2}{R_\odot},$$

where M is the mass of the part of the Galaxy lying within a sphere of radius 9.5 kpsc from the center, $R_\odot$ is the distance of the sun from the center of the Galaxy, and $V_\odot$ is the velocity of the sun around the center. If we solve for M we obtain

$$M = \frac{V_\odot^2}{G} R_\odot.$$

If we introduce numerical values for $V_\odot$, $R_\odot$, and G, and express the mass of the Galaxy in solar masses, we obtain $9.5 \times 10^{10}\ M_\odot$, which is in agreement with what we found before.

29.19 The Galactic Disk

The study of radio emission in various frequencies ranging from a few megacycles per second to a few thousand megacycles (1 megacycle corresponds to a wavelength of 30 000 cm, 10 megacycles to 3 000 cm) and at various galactic latitudes shows that in addition to the distribution of stars, dust, and gas in the nucleus and the spiral arms, constituting the main bulk of the Galaxy, there is also a radio disk that envelops the Galaxy, forming a kind of shell about 400 light years thick, as shown in Figure 29.13. This shell contains some stars that extend out from the main body of the Galaxy, but consists primarily of an ionized plasma that gives rise to the radio emission. There are probably both discrete and continuous sources in this radio disk.

In addition to the ionized hydrogen plasma, the disk of the Galaxy also contains neutral hydrogen, which, according to Oort, lies in an extremely flat plane some 100 parsecs thick at the arm that is 3 kpsc from the center. The thickness of this layer of neutral hydrogen increases to 600 parsecs near the sun. Beyond 9 kpsc from the center this layer of hydrogen becomes irregular and is no longer confined to a plane. Kahn and Woltjer believe that these irregularities are due to the motion of the Galaxy through some kind of resisting intergalactic medium, but part of it may arise from the gravitational fields of the Magellanic star clouds. This distribution of the neutral hydrogen between the sun and the center of the Galaxy can be deduced from the Doppler profiles (see Section 29.7) of the 21-cm line on the assumption that the

FIG. 29–13 *The radio shell of the Galaxy.*

hydrogen is revolving around the nucleus of the Galaxy in circular orbits. However, circular orbits do not agree entirely with the observations, and the discrepancies can be explained only by assuming that the hydrogen is streaming outwardly (as well as rotating around the center) through the Galaxy with a speed that equals 7 km/sec near the sun but increases rapidly toward the center.

The neutral hydrogen is densest at a distance of 6 kpsc and 10 kpsc from the center of the Galaxy. It falls off very rapidly beyond 12 kpsc and is quite small at 4 kpsc. This is similar to the distribution that has been found in M 31. There appears to be a rapid expansion of this neutral hydrogen as we approach the center. Thus the 3 kpsc arm has an outward motion of 50 km/sec, and at 500 parsecs the speed of expansion is about 250 km/sec.

29.20 The Galactic Halo

As we move beyond this disk we come to the galactic halo or corona, which contains the globular clusters, RR Lyrae type stars, and other Population II stars. It is a vast ellipsoidal, almost spherical, region whose radius is considerably larger than the distance from the sun to the center of the Galaxy. Its volume is estimated to be 50 or more times larger than the main flattened body of the Galaxy. Since the radio emission from this halo seems to remain constant in both latitude and longitude, there appears to be little concentration of the material towards the core of the Galaxy, and it has been assumed by Oort that this halo is homogeneously filled with radio sources, with particularly intense emission in the 3.7-meter wavelength. J. E. Baldwin estimates that the radius of this halo is about 16 kpsc. The existence of such a radio corona is confirmed to some extent by a similar halo surrounding the Andromeda Nebula. Such a halo, of course, could account for a good deal of the over-all background radio spectrum detected by radio telescopes.

TABLE 29.1

(A) THE GALAXY

Diameter	= 25 kpsc
extended spherical system (including globular clusters)	= 30 kpsc
Thickness (at the center)	= 4 kpsc
Total mass	= $1.1 \times 10^{11}\ M_{\odot}$
Over-all density	= $0.10 M_{\odot}$ per psc^3
	= 7×10^{-24} gm per cm^3
Absolute magnitude (polar direction from outside the galaxy)	= −20.5
Distance of the sun from the galactic center	$\geq 8.2 \pm 0.8$ kpsc
Distance of the sun from the galactic plane	= 8 ± 12 psc north of the galactic plane
Velocity of escape:	
galactic center	= 450 km/sec
neighborhood of sun	= 290 km/sec
galactic rim	= 180 km/sec
Age of galaxy	= 1.5×10^{10} years
Number of stars in galaxy	~ 10^{11}

(B) ROTATIONAL VELOCITY IN GALAXY

Distance from center in kpsc	1	2	4	6	8.2
Rotational velocity in km/sec	150	180	210	225	215

The radiation coming from this halo indicates that the material responsible for the radiation is not in the form of discrete sources, but rather in the form of a continuous plasma. According to Shklovsky, the coronal radiation arises from a synchrotron process in which relativistic electrons are moving in weak magnetic fields. The magnetic fields can be accounted for by supposing that they are carried out into the halo at great distances from the galactic plane by ionized gas clouds moving at high velocities. An analysis of such gas clouds shows that the kinetic temperature within them may be of the order of 10^6 °K, but to account for the synchrotron radiation there would have to be electrons moving with speeds close to the speed of light intermixed with these clouds. Moreover, to obtain the observed intensity of radiation from the halo as analyzed by Baldwin, the density of such electrons per cm^3 has to be 4×10^{-12} if the intensity of the magnetic fields is of the order of 10^{-6} gauss, and 4×10^{-13} if the magnetic-field

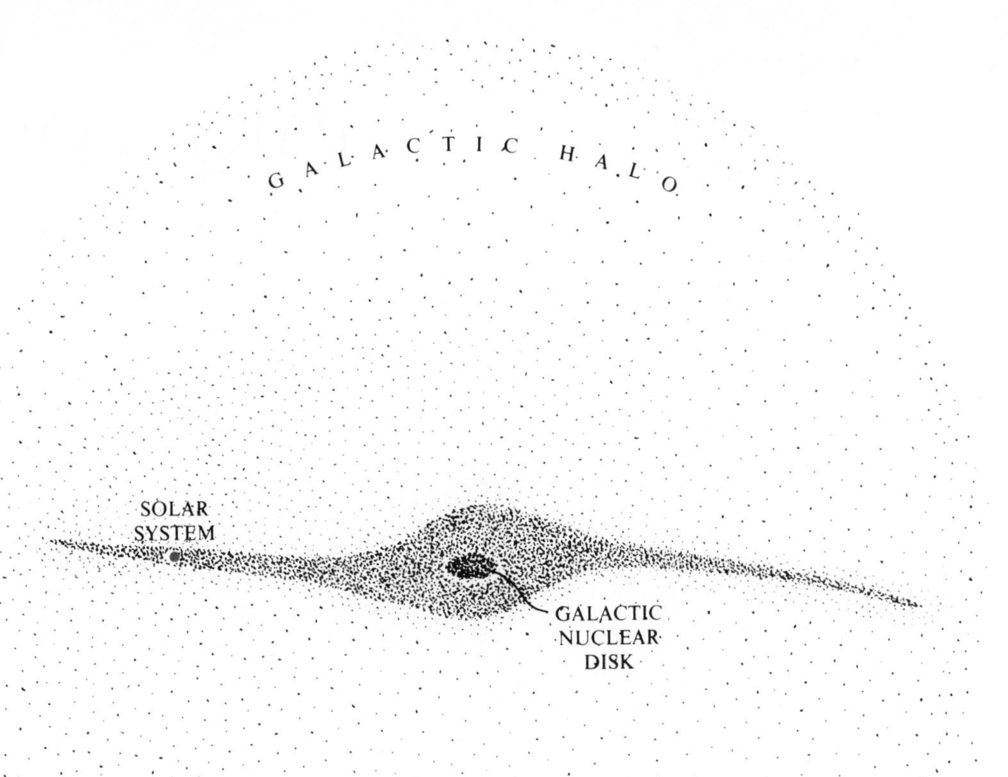

FIG. 29–14 *A schematic drawing of the Galaxy.*

intensity is 10^{-7} gauss.

These magnetic fields originate in the main disk of the Galaxy and have intensities that were first calculated by Chandrasekhar and Fermi. They showed by applying the theorem of the virial (see Section 27.23) that if there are magnetic fields present in the spiral arms of the Galaxy, these spiral arms can retain their structure only if the gravitational and magnetic forces are properly balanced. Under these equilibrium conditions, the magnetic fields in the spiral arms are of the order of 10^{-6} gauss. We shall see later that the interaction of charged particles with the moving magnetic fields in the arms of the Galaxy gives rise, at least to some extent, to the cosmic rays of the Galaxy.

Figure 29.14 shows a schematic drawing of the Galaxy as we picture its structure today. A composite photograph of the Galaxy taken by the Lund Observatory is to be found on the title pages. The Magellanic Clouds can be seen in the photograph.

PART FOUR

GALAXIES AND COSMOLOGY

30

GALAXIES*

If our knowledge of the structure of the Milky Way depended entirely on Milky Way data such as star counts, spectra, stellar motions, clusters, interstellar absorption, interstellar spectral lines, and so on, we could never be quite sure that our model correctly corresponds to the actual structure of the Galaxy. Since we are very close to the plane of the Galaxy itself, hemmed in by the obscuring dust, our results are necessarily incomplete and provide only a sketchy picture. If, however, we could stand away from the Milky Way and observe it at a distance, we could verify our results directly. We cannot do this with our own Galaxy, but we can do it with other galaxies which, we have reason to believe, are built like our own. There are, fortunately, galaxies so close to us that our optical and radio telescopes reveal a sufficiently detailed structure to allow us to confirm the model of the Galaxy which we described in the previous chapter.

30.1 The Andromeda Nebula

In the neighborhood of our Milky Way, there are some 17 galaxies close enough to show detailed structures, but only one of these, the Andromeda Nebula (M 31), has a spiral structure similar to our own, and is oriented in such a way that its detailed features can be analyzed. Although there are many varieties of galaxies ranging all the way from amorphous, almost spherical structures to those with highly concentrated nuclei and well-delineated spiral arms, we shall study Andromeda in some detail before we discuss galaxies in general. (See Figure 30.1.)

We shall see that M 31 is characteristic of spirals in general, and demonstrates the intimate relationship that exists between the structure of a galaxy and its composition. *The Andromeda Nebula is typical of spiral galaxies which contain both Population I and Population II stars, and in which the dust, gas, and the Population I stars are confined to the disk (the spiral arms). In such galaxies the nucleus is composed of Population II stars which pervade the flattened disk and which are found far out beyond the spiral structure. This differentiates the spiral from the nonspiral structures since the latter are completely free of dust and contain just Population II stars.*

30.2 The Distance and Geometrical Properties of the Andromeda Nebula

From the analysis of the type I Cepheid variables in the spiral arms of the Andromeda Nebula, we know that the distance of this galaxy is about 575 kiloparsecs. Its apparent optical diameter is about 2°.4 as revealed by a good photograph, but it is clear that this nebula is possibly twice as large as that since most photographs pick up only the blue-white giants in the spiral arms, undoubtedly missing many of the faint red stars. If we study the image of this nebula with the **microdensitometer** (a very sensitive electrical instrument that picks up images on photographic plates that are not visible to the naked eye), we find that the spiral disk of Andromeda is about twice the size of that on the usual photographic plates; and we now estimate that its apparent diameter is about 4°.8. Considering its distance, we see that its actual linear dimension is about 120 000 light years.

30.3 The Luminosity and the Number of Stars in the Andromeda Nebula

Since the plane of the Andromeda Nebula is not perpendicular to our line of sight, it appears to us foreshortened and elliptical in outline. However, its width is estimated rather speculatively to be only somewhat smaller than its length, about 100 000 light years. Its absolute magnitude is about

FIG. 30–1 *NGC 224 Great Nebula in Andromeda. Messier 31. Satellite nebulae NGC 205 and 221 also shown. (48-inch Schmidt photograph from Mount Wilson and Palomar Observatories)*

-19.9, so that it is about 25 magnitudes more luminous than our sun. This means that it contains at least 10^{10} stars, each as luminous as the sun. However, this is certainly an underestimate since the stars less luminous than the sun are difficult to detect, and probably contribute little to the visible radiation. It is therefore probable that the total number of stars in this galaxy is about twice that

FIG. 30–2 *(a) NGC 224 Great Spiral Nebula in Andromeda. Messier 31. Central region. The dust lanes can be traced right up to the nucleus, from diametrically opposite points of which two large spiral arms are seen to emerge. Red Population II stars form the nucleus. (100-inch photograph from Mount Wilson and Palomar Observatories)*

in our own Galaxy which contains about 10^{11} solar masses.

30.4 The Spiral Arms of the Andromeda Nebula

An examination of this nebula shows that it has a rather simple spiral structure with very little branching of the spiral arms, and is probably similar in structure to our own Galaxy. By studying Andromeda and resolving its central region into individual stars (see Figure 30.2(a)), Baade arrived at his concept of two different stellar populations, and reached the conclusion that the elliptical galaxies, which are free of dust and gas, contain only Population II stars, whereas the spirals, which do contain dust and gas, consist of both Population I and II stars as shown in Figure 30.2(b).

Baade, who studied the Andromeda Nebula in great detail, found that it is composed of seven spiral arms, the outermost of which is 21 kpsc from the center, and the innermost of which is 0.3 kpsc from the center. He has traced these arms both on the north and south sides of the nucleus and has shown that they contain the same kinds of stars, dust, gaseous material, Cepheid variables, open clusters, and so on, as our own Galaxy (see Figure 30.3). As we move in towards the center of the Andromeda Nebula, the outermost arm is defined by scattered groups of blue supergiants, with no obvious signs of gas or dust. Although there is no doubt that gas and dust are present in this outer arm, the radiation from this interstellar matter is too faint to be detected.

At distances of 12 to 9 kpsc from the center, where the fifth and fourth spiral arms are located, there is a maximum concentration of Population I stars with the blue supergiants dominating and with the dust becoming quite conspicuous. In the third and second spiral arms, at distances ranging from 5 to 2 kpsc from the center, HII regions become visible and the Population I giants are much less numerous. In the innermost spiral arm no Population I supergiants are visible at all, but the HII regions stand out clearly. As Baade remarks, "The association of dust and gas and Population I stars is most strikingly revealed by the HII regions of the Andromeda Nebula, which are strung out like pearls along the spiral arms. That these HII regions and with them their exciting O and B stars of high luminosity are deeply imbedded in the dust of the spiral arms is shown by their strong reddening." It is interesting to note that the HII

FIG. 30–2 *(b) Great Spiral Nebula in Andromeda. Messier 31. South preceding region showing resolution into stars. This region is similar to the spiral arms in our own Galaxy. It is rich in Population I blue giants, Cepheid variables, dust, gaseous nebulae, and HII regions. (100-inch photograph from Mount Wilson and Palomar Observatories)*

regions and emission nebulosities in the spiral arms of Andromeda, which Baade found by the hundreds on red-sensitive plates (these photographic plates are exceptionally sensitive in the H_α region), were completely undetected by Hubble, who carefully surveyed the fifth spiral arm using blue-sensitive plates.

30.5 Obscuration and Dust in the Plane of the Andromeda Nebula

That the dust in the Andromeda Nebula is concentrated in the spiral arms is clearly shown by the absence or slightness of reddening of the globular clusters as compared to the HII regions. These globular clusters, which are scattered throughout the nebula but are very definitely concentrated towards the nucleus, are not at all related to the spiral structure. There is some reddening of the globular clusters located on the far side of the nebula because the light from them must pass through the plane of the nebula and is therefore somewhat reddened. The few globular clusters that are heavily reddened are those that lie almost in the plane of the nebula itself, and hence are considerably obscured by the dust.

These observations of the globular clusters of Andromeda are consistent with the investigations of the Population II novae in this nebula. Of the 25 such objects that have been observed, only one shows appreciable reddening. This nova,

FIG. 30–3 *Stellar Populations I and II. To the left, the Andromeda Nebula photographed in blue light shows giant and supergiant stars of Population I in the spiral arms. The hazy patch at the upper left is composed of unresolved Population II stars. To the right, NGC 205, companion of the Andromeda Nebula, photographed in yellow light, shows stars of Population II. The brightest stars are red and 100 times fainter than the blue giants of Population I. The very bright, uniformly distributed stars in both pictures are foreground stars belonging to our own Milky Way system. (From Mount Wilson and Palomar Observatories)*

interestingly enough, lies in the fifth arm, so that its light must pass through the dust in the arm before it reaches us. The innermost arm of the Andromeda Nebula gives strong confirmation to the relationship we have outlined above between dust and the spiral structure of any galaxy. As we follow this arm into the nucleus itself, we find that whereas in its outer regions it is richly studded with Population I supergiants, these gradually diminish in number towards the nucleus and abruptly disappear. The spiral arm, however, continues into the nucleus itself as a lane of dust.

30.6 Stellar Formation from Dust and Gas in Spiral Arms

Baade characterized this situation in the following words: "We have today convincing reason to believe that dust and gas are the primary constituents of the spiral structure, and that the associated stellar Population I represents a secondary phenomenon, its stars being formed up to the present day in the dust and gas clouds of the spiral arms. This conclusion, originally solely based on observations in other galaxies, has been in recent years confirmed by investigations in our Galaxy which showed that dust and gas clouds like the Orion Nebula and I 5146 are the breeding places of young Population I stars."

From an analysis of all the material in the spiral arms of the Andromeda nebula we find that the spiral arms contribute less than 20 per cent of the total light from such galaxies. This means that the Population I stars in the spiral arms constitute only a small fraction of the total mass of the galaxy. Since this must also be true of the gas and dust, the amount of this material that is available for the future formation of stars is very small. Van de Hulst has shown by 21-cm line analysis of the neutral hydrogen both in Andromeda and in our own Galaxy that the amount of material available for the formation of new stars is not more than 2 per cent of the total mass of these two galaxies. We must conclude that the star-formation phase in spirals such as our Milky Way and Andromeda is just about over, and that most of the stars in these spiral systems were formed a long time ago. Of course, if Population II stars, which constitute the great bulk of these galaxies, eject enough material as they continue to evolve, this can replenish the supply of dust and gaseous material and the 2 per cent figure given above does not then indicate an absolute maximum.

That this picture of the disk is, indeed, correct is clear at once from an examination of red-sensitive plates that have been exposed for a long time. One finds on these plates that the spaces between the spiral arms are filled with dense sheets of Population II giants, whose numbers slowly decrease as we go outward. According to this analysis the spiral arms in these galaxies are thin layers of gas, dust, and O and B Population I supergiants embedded in the disk.

30.7 The Distribution of the Population II Stars in Andromeda

Baade was the first to point out that the nuclei of spirals such as Andromeda and our own Galaxy contain only Population II stars and are dust free. The stars in the nucleus of Andromeda are yellow and red giants, but one can trace these Population II stars out into the arms of the nebula also. These Population II stars actually are found everywhere in the flattened disk, and even in regions far beyond the outermost detectable spiral structure. By examining Population II stars along the minor axis of the nebula, Baade was able to follow these stars outward to a distance of 45′ from the center. The concentration of Population II stars as we go out from the center along the major axis diminishes at about the same rate as the brightness of the nebula, disregarding the arms. Along the major axis Population II stars have been traced out to 2° from the center, but probably will be found to go out much further when better red-sensitive plates are used. Baade pointed out that these Population II stars form quite an elongated system, and must be identified with the disk of the Andromeda Nebula itself and not only with the large halo surrounding it. The main disk, then, of the Andromeda Nebula (and presumably our own Galaxy) consists in part of Population II red and yellow giants.

Intermixed with these Population II giants there are undoubtedly ordinary giants similar to those found in open clusters such as M 67, and which have therefore evolved from Population I stars. These may, indeed, constitute the majority of stars in the disk and may account for most of the light from it (disregarding the spiral arms). But the most luminous disk stars are certainly the Population II red giants. These two types of giants can easily be differentiated because the Population II stars are poor in metals as compared to the ordinary giants.

30.8 The Radio Emission from the Andromeda Nebula

Just as we have gained a great deal of information about the structure of our own Galaxy by studying its radio spectrum (particularly the 21-cm line), we can do the same thing with the Andromeda nebula and we find that the situation is pretty much the same as in our Milky Way.

The 21-cm line can be detected along the major axis in a number of regions and these correspond to the spiral arms. This radiation can be traced out to almost 3° on either side of the nucleus; from it one can detect the law of rotation for these outer portions of the galaxy. What is of interest is the sharp maximum in the 21-cm line at about 8.7 kpsc from the center of M 31 corresponding to a similar maximum in our own Galaxy at 6.5 kpsc from our center. This is just where Baade has placed the fourth spiral arm in Andromeda in which the Population I stars, gas, and dust are at their maximum concentrations. We can also detect a radio halo around M 31 similar to the halo around our own Galaxy.

The radio data indicate that the sources of radio waves are distributed out to great distances from the plane of the galaxy and form an almost spherical system which extends more than 10 kpsc at right angles to the plane. The actual size of Andromeda as determined from radio waves is considerably larger than its optical size, and, according to Shklovsky, the part of this galaxy that contains radio sources has a diameter of about 100 kpsc.

30.9 Rotation of the Andromeda Nebula

That the Andromeda Nebula is rotating like our own Galaxy can be verified optically and from an analysis of its radio spectrum. The optical analysis is fairly simple since we can measure the Doppler shift of the spectral lines in the radiation

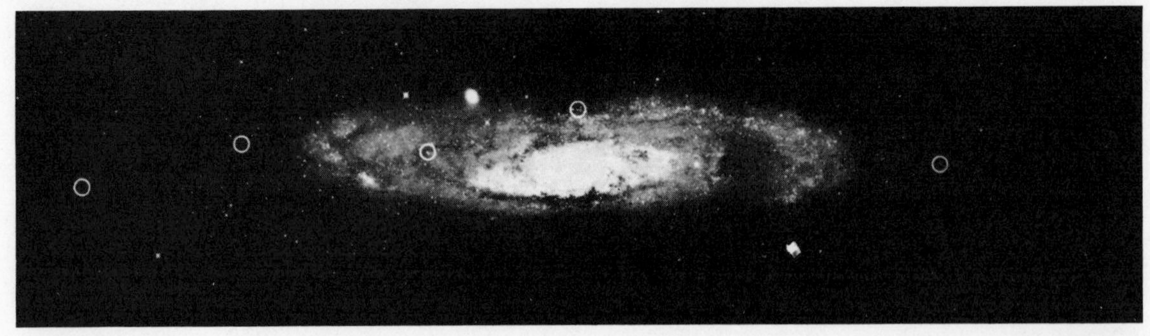

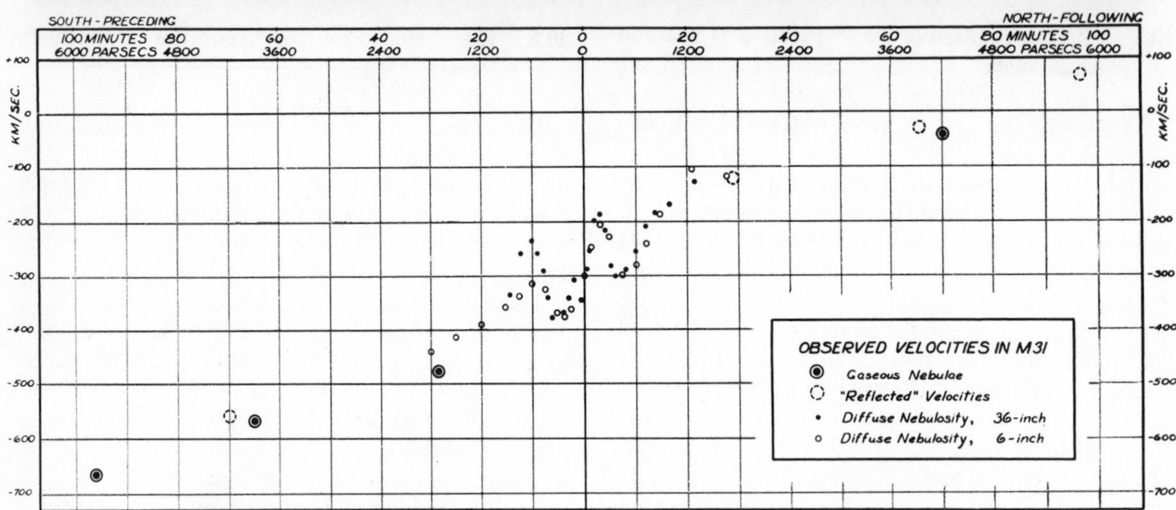

FIG. 30–4 *The Andromeda Nebula, photographed by Mayall, with observed velocities plotted below. Smoothed mean velocities of the unresolved central portion are represented by small dots for observations with the Crossley reflector, and by small open circles for observations with the 6-inch mirror. In the photograph, the bright-line emission nebulosities are encircled, and their velocities, with the exception of the one on the minor axis, are plotted as ringed dots. The velocities plotted are the actual observed velocities relative to the sun. To find the rotational velocities of various parts of Andromeda relative to its center, one must subtract from the observed velocities the observed velocity of the center of Andromeda relative to the sun.* (Lick Obs. Bull. *XIX* (*1940*), *41*.) (*Yerkes Observatory photograph*)

emitted by stars on either side of the nucleus of M 31.

The data are collected in the graph in Figure 30.4, where the radial velocities for various objects in the disk of the Andromeda Nebula are given. Out to an angular distance of 65′ to 70′ on either side of the nucleus the angular velocity appears to be uniform, as though the nucleus were rotating like a solid disk. From about 70′ out to 155′ on either side, the rotational velocity decreases with increasing distance, as is to be expected, since these outer portions are rotating around the nucleus according to Kepler's laws.

30.10 The Mass of the Andromeda Nebula

Using Newton's law of gravity and the observed rotational velocities, we find that the mass of Andromeda is about 2×10^{11} $M_\odot$. However, this is somewhat smaller than the mass obtained from an analysis of the radio emission. We find that the mean radial velocity of the interstellar hydrogen clouds in the Andromeda Nebula is about 8 km/sec (about the same as in our own Galaxy) and this leads to a mass of 3.4×10^{11} $M_\odot$, in other words, about three or four times the total mass of the Milky Way. The mass of interstellar hydrogen in M 31 as determined from an analysis of the 21-cm line is about 1.3 per cent of this, so that the ratio of gaseous matter to stars is about the same in Andromeda as it is in our own Galaxy.

30.11 General Properties of Galaxies

We have just given a detailed description of one external galaxy with spiral arms to show that these objects are similar to our own Milky Way. We shall see later that both Andromeda and our Milky Way are intermediate-type spirals. There are, however, other types of spirals which differ considerably from these two. In addition there are elliptical galaxies which have no spiral structure and which are now believed to be at least as numerous as the spirals.

It is convenient to classify galaxies according to a structural scheme introduced by Hubble (also independently by Lundmark, who developed it to its fullest extent). According to this scheme, the galaxies are divided into elliptical, spiral (normal and barred), and irregular structures. This standard classification is best represented by Hubble's forked diagram given in Figure 30.5(a). Although it has been enlarged by introducing more detailed subdivisions, the simple scheme still serves to differentiate the principal characteristics of various types of nebulae. This is to be compared with the photographs in Figure 30.5(b) and in Figure 30.11.

30.12 Elliptical Nebulae

The elliptical nebulae are divided into seven subgroups, starting with an almost spherical or globular structure, such as *NGC* 4486, which is an E0 type (see Figure 30.6(a)), and going down to the spindle-shaped lenticulated structures, such as E7, which resemble the large nuclei of ordinary spirals (Sa, Sb) (see Figure 30.6(c)). If these objects were seen head-on instead of edgewise, they would, of course, appear spherical, and hence one might be inclined to believe that all elliptical nebulae, even those that look spherical, are really elongated structures. According to this concept even the globular galaxies appear spherical only because we see them at right angles to their planes.

Although it is impossible to determine whether any particular globular galaxy is really spherical or simply a flattened system seen perpendicular to its plane, there are far too many globular galaxies to suppose that they are all flattened ellipsoidal systems. In fact, an analysis of the distributions of these objects given by Hubble shows that the observations agree best with the assumption that the elliptical galaxies contain equal numbers of the various ellipsoidal structures, from the spherical to the spindle, with their axes randomly distributed in space.

An elliptical galaxy exhibits no structural detail except that its nucleus appears considerably more condensed than its outer portions, which diminish

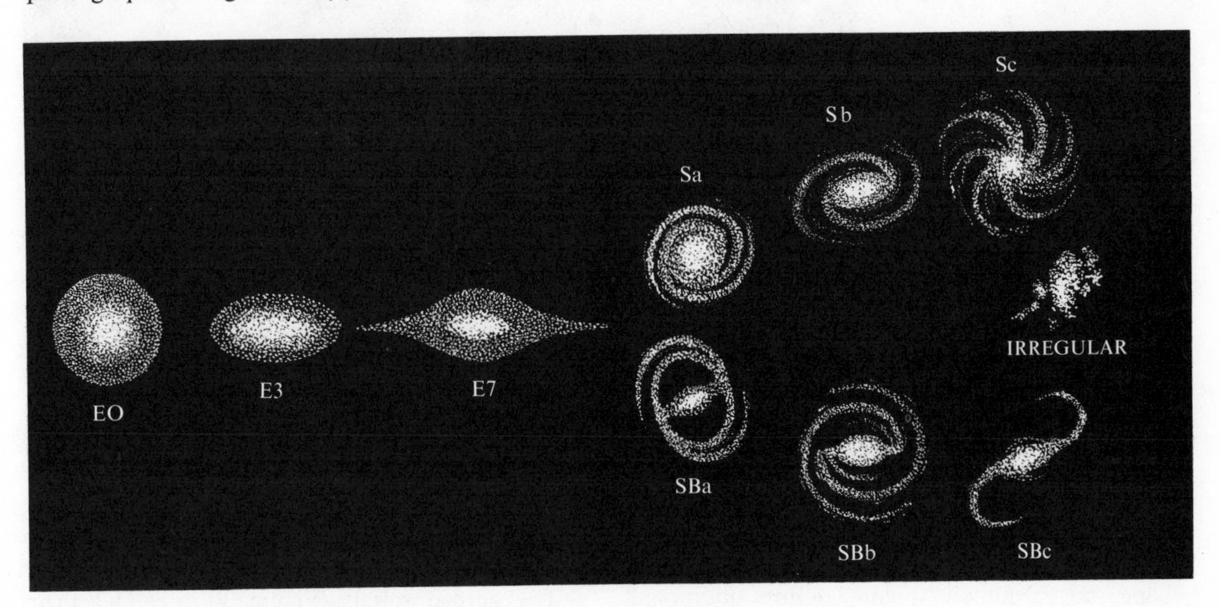

FIG. 30–5 *(a) Schematic diagram of the Hubble classification of galaxies.*

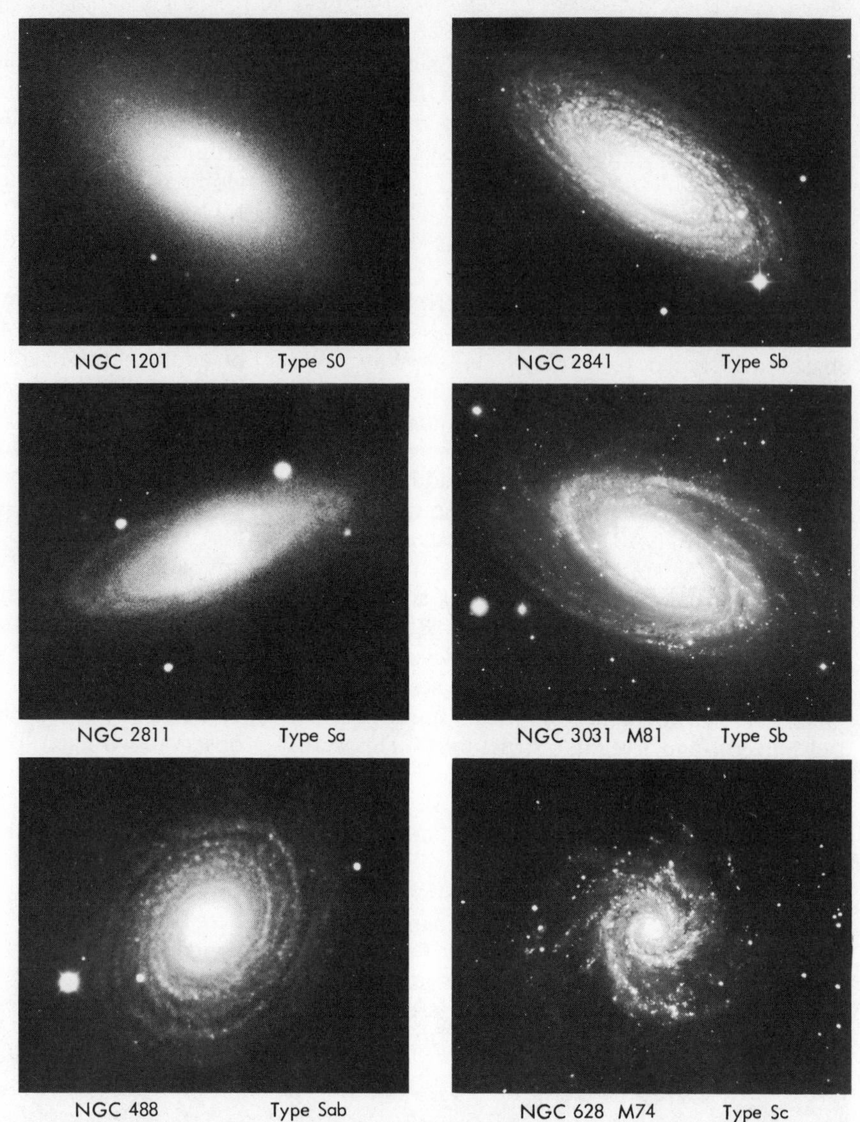

FIG. 30–5 *(b) Types of extragalactic nebulae, elliptical and spiral. Compare this with the Hubble scheme shown in Fig. 30–5 (a). (60-inch photograph from Mount Wilson and Palomar Observatories)*

in star concentration gradually as one moves out towards the edge. Each subtype of elliptical galaxy is characterized by its index, n, which is defined as 10 $[1 - (b/a)]$, where a and b are the apparent major and minor axes respectively. Among elliptical galaxies the smallest ratio of major to minor axis is 1:3 for the most highly flattened systems. We may note that the most highly flattened type of galaxy, the E7, is not elliptical in shape but definitely pointed. For additional examples of elliptical galaxies, see Figure 30.6(c) and (d).

Separating the elliptical galaxies from the spirals is a group of objects, the S0 galaxies, which have some of the characteristics of both elliptical and spiral systems. They look like highly flattened nuclei of spirals that have lost their spiral arms.

30.13 Normal Spirals

The **normal spirals** (Figure 30.7) exhibit their spiral-arm structure when their planes are perpendicular to the line of sight, i.e., in the plane of the

FIG. 30–6 *(a) NGC 4486 globular nebula (Eo peculiar) in Virgo. Messier 87. This is the brightest elliptical galaxy in the Virgo cluster and is one of the brightest galaxies known. If the distance modulus for the Virgo cluster is $m - M = 30.7$, then M_{ph} for NGC 4486 is -21.0. This is to be compared with $M_{ph} = -20.3$ for M 31 (Andromeda). Many globular clusters (over 500) are scattered over the entire image. The nuclear jet in this galaxy is shown in Figure 30–6(b). It can be traced out to 1 500 parsecs from the center and is the source of the radio noise as well as of blue synchrotron radiation. (From Mount Wilson and Palomar Observatories. Caption material reprinted from* The Hubble Atlas of Galaxies, *by Alan Sandage (1961), by permission of the Carnegie Institution of Washington and the California Institute of Technology.)*

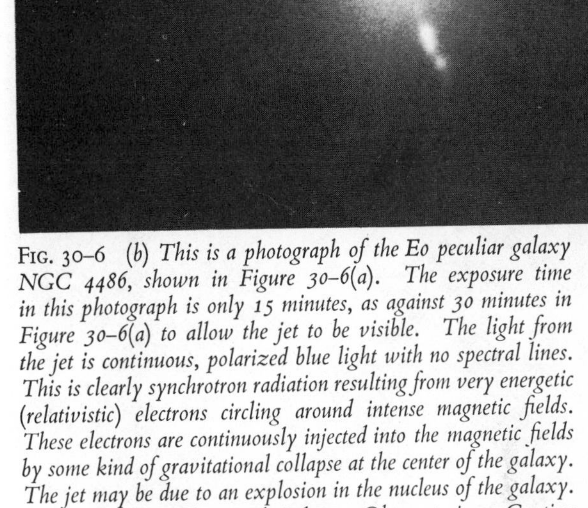

FIG. 30–6 *(b) This is a photograph of the Eo peculiar galaxy NGC 4486, shown in Figure 30–6(a). The exposure time in this photograph is only 15 minutes, as against 30 minutes in Figure 30–6(a) to allow the jet to be visible. The light from the jet is continuous, polarized blue light with no spectral lines. This is clearly synchrotron radiation resulting from very energetic (relativistic) electrons circling around intense magnetic fields. These electrons are continuously injected into the magnetic fields by some kind of gravitational collapse at the center of the galaxy. The jet may be due to an explosion in the nucleus of the galaxy. (From Mount Wilson and Palomar Observatories. Caption material reprinted from* The Hubble Atlas of Galaxies, *by Alan Sandage (1961), by permission of the Carnegie Institution of Washington and the California Institute of Technology.)*

FIG. 30–6 *(c) NGC 147 Nebula in Cassiopeia, shows resolution into stars. Members of Local Group. This is a normal E5 galaxy. Taken in red light. (200-inch photograph from Mount Wilson and Palomar Observatories)*

sky. However, when they are viewed edgewise they look just like the highly flattened E7-type galaxies except that they have a dark lane of absorbing matter running parallel to the major axis, as shown in the accompanying photographs, Figure 30.8. According to Hubble, the normal spirals are characterized by arms that emerge tangentially from the nucleus at two diametrically opposite points, and die out after one complete turn. In the regular (so-called classical) spirals the points at which the arms leave the nucleus are very well defined, and the arms themselves are sharply defined and quite symmetrical. Where the arms are not clearly delineated, but are broad and fuzzy, the outer edge of the nucleus seems to be divided into irregular and partly formed arms, and it is difficult to see from what points of the nucleus these arms originate. (See Figure 30.9.)

The relative importance of the nucleus varies (see Figure 30.10) and, according to its size as

FIG. 30–7 *(a) NGC 2903 Normal Spiral Nebula in Leo. The nucleus of this type Sc galaxy consists of eight intensely bright globules or knots that are probably HII regions. (200-inch photograph from Mount Wilson and Palomar Observatories. Caption material reprinted from* The Hubble Atlas of Galaxies, *by Alan Sandage (1961), by permission of the Carnegie Institution of Washington and the California Institute of Technology.)*

FIG. 30–6 *(d) NGC 205 Nebula in Andromeda, shown resolved. Satellite of Great Nebula. This is an elliptical galaxy with peculiar features. Since its outer envelope is similar to that of an So galaxy, this may be a transition case. The outer regions are highly resolved into stars, the brightest of which are all of the same apparent magnitude. Although generally free of dust, there are two faint dust patches near the center. 200-inch photograph taken in red light. (From Mount Wilson and Palomar Observatories. Caption material reprinted from* The Hubble Atlas of Galaxies, *by Alan Sandage (1961), by permission of the Carnegie Institution of Washington and the California Institute of Technology.)*

compared with that of the spiral arms, we have three subclasses, a, b, and c. Both the Milky Way and Andromeda are examples of Sb galaxies. These subclasses are also characterized by the degree to which the spiral arms are unwound. The width of an arm is related, within certain limits, to the total diameter of the nebula, ranging from one-tenth to one-fifteenth of the latter in various galaxies. The lengths of the arms also vary, ranging all the way from short structures to those which wind completely around.

The normal spirals contain the same kind of emission nebulae and absorbing material that are found in M 31 and in the Milky Way. This is clear from the photographs, which also show the presence of other Population I emission objects such as O and B blue-white giants.

30.14 Barred Spirals and Irregulars: The Magellanic Clouds

The barred spirals range from an SBa structure, which looks like the Greek letter θ, to the S-shaped SBc structure which is characterized by a very bright nucleus with a bar running through it, from the ends of which two spiral arms extend about one-quarter of the way around. Finally in the SBb spirals, which in structure lie between Sba and SBc, the nucleus and the bar running through it are very clearly defined, and the spiral arms—which extend almost completely around—originate at the ends of the bar almost at right angles to the bar itself (see Figure 30.11).

FIG. 30–7 (*b*) *NGC 3031 Spiral nebula in Ursa Major. Messier 81. This is an Sb-type galaxy that is similar to our own Galaxy and to Andromeda. Its distance modulus is 27.1, so that if the brightest stars in the central region have* $M_v = -3$, *their* m_v *is 24.1, and the central region cannot be resolved with the 200-inch telescope. Dust lanes can be tracked to within 35 seconds of arc from the center. The arms are resolved into numerous individual stars and HII regions. Twenty five ordinary novae have been observed in addition to Cepheids, and irregular blue and red variables, all brighter than* $M_{ph} = -4.5$. *(200-inch photograph from Mount Wilson and Palomar Observatories. Caption material reprinted from* The Hubble Atlas of Galaxies, *by Alan Sandage (1961), by permission of the Carnegie Institution of Washington and the California Institute of Technology.)*

FIG. 30–7 *(c) NGC 4736 Spiral nebula in Canes Venatici. Messier 94. This Sb spiral has an intense central region devoid of spiral structure. Its spiral arms are tightly wound. (200-inch photograph from Mount Wilson and Palomar Observatories)*

FIG. 30–8 *(a) NGC 4565 Spiral nebula in Coma Berenices, seen edge-on. Photographed on an unfiltered red-sensitive plate. (200-inch photograph from Mount Wilson and Palomar Observatories)*

FIG. 30–8 *(b) NGC 4594 Spiral nebula in Virgo, seen edge-on. Messier 104. The most pronounced feature is the heavy dust lane that bisects the galaxy. It is also remarkable because of the large nuclear bulge. This galaxy is either of type Sa or early Sb. The halo surrounding this galaxy contains globular clusters which are seen as bright condensations. Since the distance modulus of NGC 4594 is 30.6 the absolute magnitude of the brightest cluster in it is − 10.6. (200-inch photograph from Mount Wilson and Palomar Observatories. Caption material reprinted from* The Hubble Atlas of Galaxies, *by Alan Sandage (1961), by permission of the Carnegie Institution of Washington and the California Institute of Technology.)*

FIG. 30–9 (*a*) *Spiral galaxy in Triangulum (central region). Messier 33. This is a member of our Local Group and the nearest Sc galaxy. Its distance modulus is 24.5 according to Sandage. Stars down to* $M_{ph} = -1.5$ *can be detected. The spiral arms are completely resolved into bright stars, most of which are blue supergiants. In addition there are in the spiral arms, Cepheid variables, open clusters, novae, irregular variables, HII regions, and at least 3 000 red supergiants of* $M_v = -5$ *similar to those in the open cluster h and* χ *Persei. Globular clusters are also present. The integrated color index is 0.40. (200-inch photograph from Mount Wilson and Palomar Observatories)*

FIG. 30–9 (*b*) *NGC 598 Spiral nebula in Triangulum. Messier 33. Photographed in red light. (48-inch Schmidt photograph from Mount Wilson and Palomar Observatories. Caption material reprinted from* The Hubble Atlas of Galaxies, *by Alan Sandage (1961), by permission of the Carnegie Institution of Washington and the California Institute of Technology.)*

Although both Magellanic Star Clouds are generally classified as irregular galaxies, they really differ considerably in structure. Shapley has classified the Small Magellanic Cloud as a typical irregular dwarf galaxy, whereas de Vaucouleurs has shown that the Large Magellanic Cloud (see Figure 30.12) has a definite spiral structure in its outer portions. Moreover, while the Large Cloud is predominantly composed of Population I stars and contains an abundance of cosmic dust, the Small Cloud appears to be dust-free. To the unaided eye the Large Cloud has an angular diameter of 7° as compared to 3°.5 for the Small Cloud, but photography shows us that the Large Cloud has an extent of about 20° and the Small Cloud is also considerably larger than it appears. De Vaucouleurs has also demonstrated that a bridge of matter connects the Large Cloud to our own Galaxy.

The irregular galaxies show no definite spiral structure and no well-defined nucleus, and are best exemplified by the Magellanic Star Clouds. We may sum up these irregulars in the words of Hubble as "lacking both dominating nuclei and rotational symmetry."

30.15 The Relative Abundances of Different Types of Galaxies

When galaxies were first catalogued, it was thought that the spirals were far more numerous

Fig. 30–10 *(a) NGC 628 Spiral nebula in Pisces. Messier 74. This is a multiple-arm Sc-type spiral with well-defined dust lanes on the inside of the luminous arms. The nucleus is small and compact. The distance modulus is 30.0 and the approximate width of the arms is 1 000 parsecs. (200-inch photograph from Mount Wilson and Palomar Observatories)*

Fig. 30–10 *(b) NGC 5364 Sc Spiral nebula in Virgo. One of the most regular galaxies known. There is a thin bright completely closed ring around the core with some spiral structure inside it. The dust lanes are faint but visible. The redshift corresponds to 1 357 km/sec recession (corrected for solar motion) giving a distance modulus of 31.1. Hence, the average width of a spiral arm is 700 parsecs. (200-inch photograph from Mount Wilson and Palomar Observatories. Caption material reprinted from* The Hubble Atlas of Galaxies, *by Alan Sandage (1961), by permission of the Carnegie Institution of Washington and the California Institute of Technology.)*

than any of the others. However, more recent work indicates that the elliptical galaxies may be much more numerous than was originally thought, and some investigators believe that there are as many ellipticals as there are spirals. Using 600 bright galaxies, Hubble obtained the accompanying table (Table 30.1) of relative abundances of different types of galaxies. We must, however, realize that in this sort of frequency analysis selection and sampling play a very important role since we classify into groups only those galaxies whose structure can be clearly resolved, and these are the galaxies with large apparent diameters.

A survey of photographic plates of extragalactic objects shows that the greatest number of images is of very small objects whose structures cannot be discerned with sufficient accuracy to allow classification. In his analysis of the 600 objects, Hubble used only the very brightest nebulae listed in the Catalogue of Shapley and Ames, so that no question could arise as to the classification of these objects. We must, therefore, expect to find strong departure from these abundances in local samplings, and, indeed, we know that in many clusters of galaxies the number of ellipticals is greater than that of spirals. Within our own local group of galaxies consisting of 17 individual members, there are probably nine ellipticals.

Table 30.1
RELATIVE FREQUENCIES OF GALACTIC TYPES

Type	Frequency (per cent)
E0–E7	17
Sa, SBa	19
Sb, SBb	25
Sc, SBc	36
Irregular	2.5

30.16 The Apparent Diameters and Relative Sizes of Different Types of Galaxies

In Hubble's initial investigations of galaxies he suggested that the mean apparent diameter for

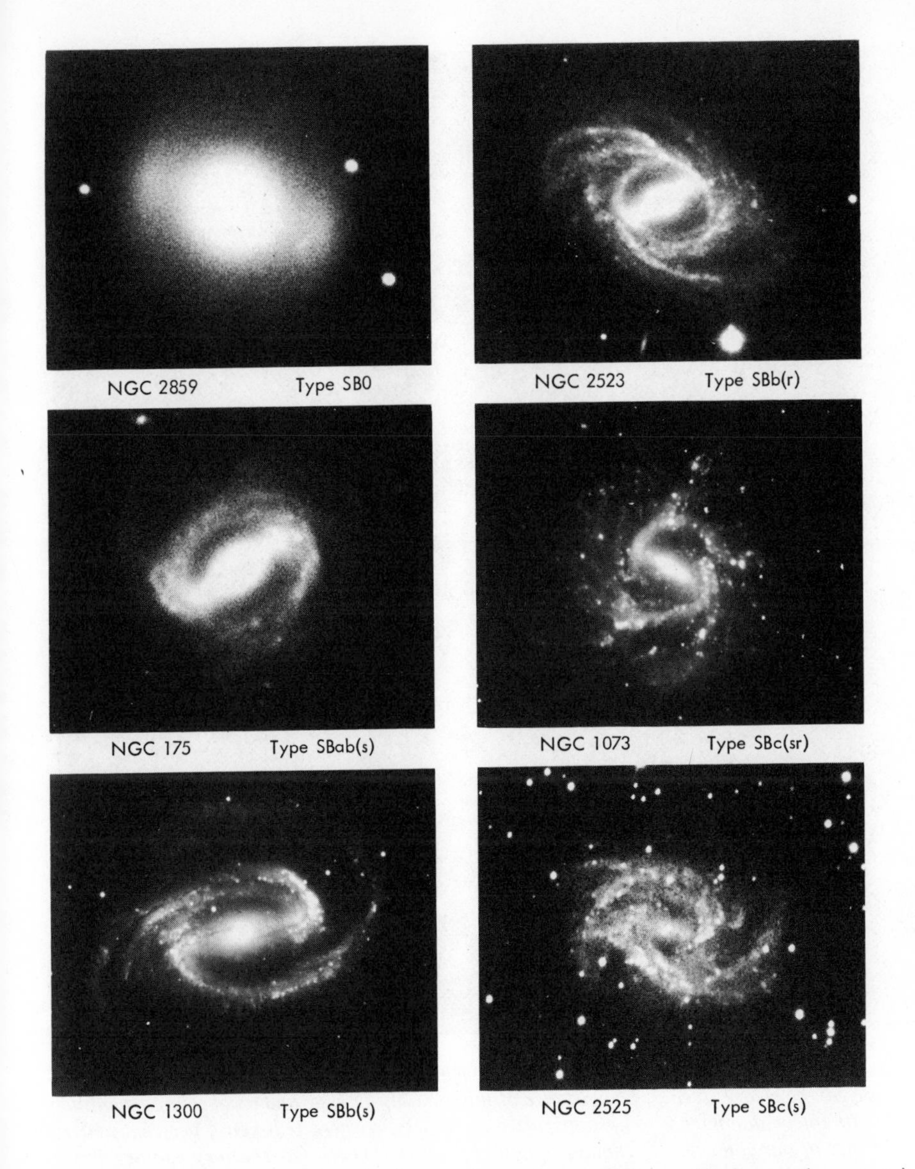

FIG. 30–11 *Classification of barred galaxies.* (*From Mount Wilson and Palomar Observatories*)

galaxies having the same apparent magnitudes increases as one goes from the elliptical to the spiral structures. According to his data the mean apparent diameter (for galaxies with $m_{vis} = 10$) increases from 2′ for the E0 class to 9′ for the Sc and SBc spirals. From this it appears that the spirals, on the average, are 2.4 times larger than the ellipticals.

This early estimate of Hubble's was revised later by Shapley, who found that the ratio of apparent diameters of spirals to those of ellipticals is 1.7, although the results he obtained depend on the apparent magnitudes of his samples, showing again that selection plays an important role in this type of analysis.

That there is a real difference between the sizes of spirals and ellipticals is indicated by an analysis of 85 galaxies belonging to the Virgo cluster, hence all at the same distance from us. De Vaucouleurs has shown that there is an increase in apparent diameter as one goes from the Sa to Sc spirals (4.′4 to 6′) and an increase from 2.′5 to 7.′3 in the barred spirals as one goes from the SBa to SBc, but in the ellipticals there is a decrease as one goes from E0 to E7. However, in the Virgo cluster of galaxies (see Figure 30.13) the spirals are, on the

FIG. 30–12 *The Large Magellanic Star Cloud—a member of the Local Group of galaxies. Its bar structure can be clearly seen. The distance of this cloud is 52 kpsc and its absolute magnitude is -18.7. Its linear diameter is 7 kpsc and its mass lies somewhere between 5 per cent and 17 per cent of that of our Galaxy. It contains about 10^{10} stars. (Lick Observatory photograph)*

average, larger than the ellipticals by a factor of 1.23. It appears from all available data that spirals are generally larger than elliptical galaxies.

30.17 Linear Dimensions of Galaxies

Where the distances of the galaxies can be determined, we can obtain the actual dimensions of the objects from microphotometric tracings although uncertainties both in distances and in the photometric tracings can give rise to serious errors. Since the faint outer portions of the distant galaxies do not leave enough of an impression on photographic plates to be detected even with sensitive microphotometry, we tend to underestimate the dimensions of these objects. The most reliable data, obtained from 27 bright galaxies, show that the ellipticals range from about 5 to 15 kpsc in diameter, whereas the spirals first increase in diameter from about 8 kpsc for the Sa spirals to more than 20 kpsc for the Sb, and then decrease down to about 10 to 7 kpsc for the Sc spirals. The irregulars are among the smallest, with a mean apparent diameter of about 2 kpsc. This sampling of 27 galaxies is greatly influenced by the presence in it of M 31 and M 81, which are both Sb types.

FIG. 30–13 *The Virgo Cluster of galaxies at a distance of about 11 megaparsecs. Its speed of recession is 1 150 km/sec and it contains about 2 500 galaxies, the 10th brightest of which has an apparent magnitude of 9.4.* (*Yerkes Observatory photograph*)

30.18 Absolute Magnitudes and Color Indices of Galaxies

Again using the 27 brightest galaxies we find that the absolute magnitudes range from about -14 for the ellipticals to -19.5 for the Sb types. The range in absolute magnitudes in the Sc spirals is -15.5 to -19, and the irregulars range from -12.8 to -16. If we consider all the galaxies taken together, the over-all range in absolute magnitude is 7 or more, from the faintest ellipticals and irregulars to the brightest Sb spirals. From an analysis of the absolute magnitudes of galaxies, or such a group as the Sb, it appears that there are few giant galaxies such as our Milky Way, M 31, and M 81.

Galaxies also vary in color; we find that the mean integrated color index varies from about 0.9 for ellipticals down to 0.37 for the irregulars. The Sb spirals have a color index of 0.7. It is clear that the difference between the colors of ellipticals and spirals is due to the presence of the O and B giants in the spiral arms of the latter (see Figure 30.2 (b)). This, of course, agrees with our knowledge that the ellipticals consist primarily of Population II stars, whereas the spirals contain Population I giants as well.

30.19 Mean Spectral Classes of Galaxies

As we move from the elliptical through the spiral structures, we find that there is only a slight variation in spectral types (although the spectrum of a galaxy is similar to that of a star, it does not correspond exactly to that of any single star), but the trend indicates a definite correlation with color. From the work of Humason, who investigated 546 galaxies, we know that the spectral class changes from about G3.7 through F6 for the Sc spirals. The average spectral class for all of them taken together is G1.4. An interesting property of the spectra of many nebulae is the presence of emission lines superposed on the continuum. This is particularly true of the late spirals and Magellanic-type irregulars, in which the H_α and H_β lines as well as the lines of bright gaseous nebulae are dominant. Mayall has also discovered that the line $\lambda 3727$, alone or with other emission lines, is found in the spectra of the nuclear regions of all types of galaxies.

30.20 Distances of Galaxies: Distance Moduli

We have been discussing such things as dimensions and luminosities of galaxies, but before we can measure these quantities we have to know the distances of the galaxies. It is clear that direct methods such as trigonometric parallaxes cannot be applied to extragalactic objects, and it is therefore necessary to apply indirect methods which are primarily photometric in nature. These photometric methods depend on our ability to resolve extragalactic galaxies into individual objects such as Cepheid variables (the objects most commonly used because of the period-luminosity law of Cepheid variables, see Section 24.14), globular clusters, open clusters, supergiants, novae, and planetary nebulae, whose absolute magnitudes we know from a study of such objects in our own Galaxy. This procedure is based on the assumption that all such objects have similar properties regardless of whether they are found in our Galaxy or any other galaxy. To apply this method we use the formula

$$m - M = 5 \log d - 5,$$

where m is the measured apparent magnitude of the object in the resolved galaxy, and M is the absolute magnitude as determined from an analysis of similar objects in our Galaxy. We have already discussed *the quantity* $(m - M)$, *called the distance modulus of the galaxy.*

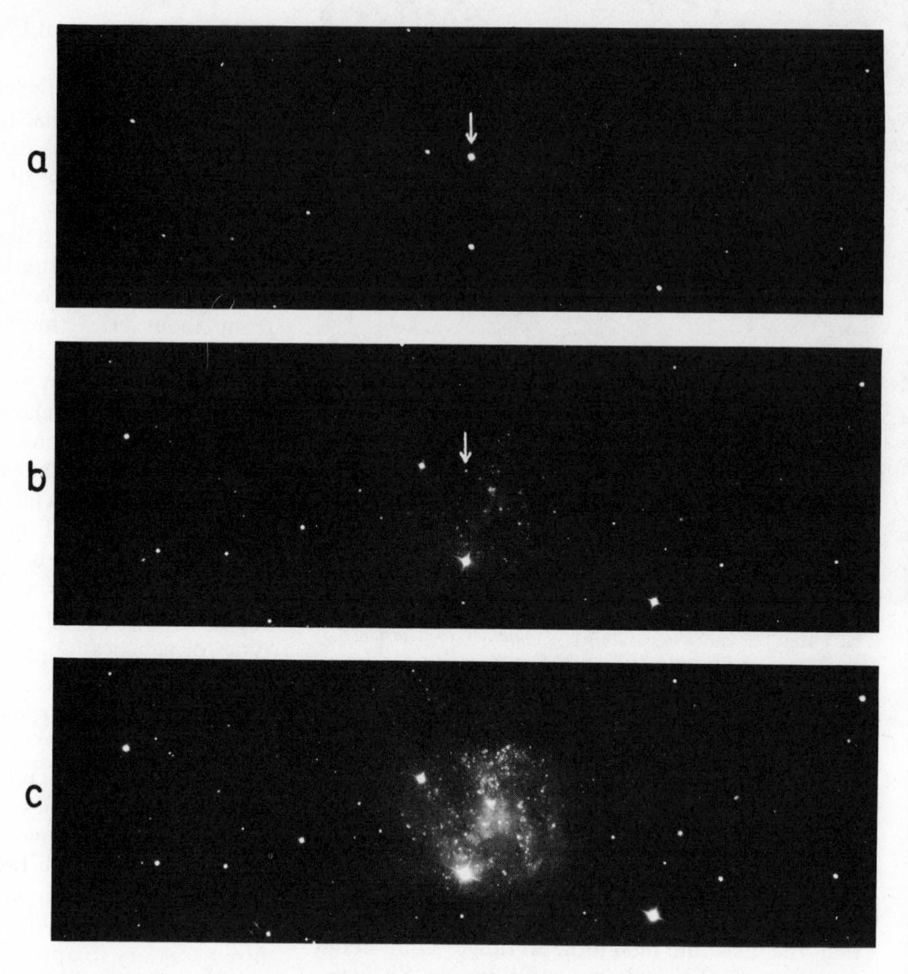

FIG. 30–14 *Supernova in IC 4168 (a galaxy in Virgo). Three views: (a) August 23, 1937, exposure 20 minutes. Maximum brightness; (b) November 24, 1938, exposure 45 minutes. Faint; (c) January 19, 1942, exposure 85 minutes. Too faint to observe. The appearance of the gaseous nebulosity in the lower plate is due to the long exposure time. (100-inch photographs from Mount Wilson and Palomar Observatories)*

30.21 Using Novae to Determine Distance Moduli

The method outlined above for finding the distance modulus for a galaxy is of primary importance since other less direct methods are based on it. These indirect methods make use of the over-all galactic dimensions and certain spectral laws. These secondary methods have to be applied to galaxies that cannot be resolved into individual stars. We may note that even before galaxies were resolved into individual stars, H. D. Curtis applied the photometric method to the novae in galaxies. He compared the average apparent magnitude at maximum luminosity of novae in the Andromeda Nebula with the known mean absolute magnitude at maximum luminosity of similar novae in our Milky Way. He thus obtained $(m - M)$ equal to 25 for M 31.

The use of ordinary novae for obtaining the distance moduli of galaxies in this way can be applied out to a distance of the order of 600 kpsc. But we can go to still greater distances by observing the known types of supernovae. This procedure has been successfully applied to about 35 galaxies. (See Figures 30.14 and 24.20.)

30.22 Other Luminosity Criteria for Determining Distance Moduli

If novae and supernovae cannot be used, we can use the period-luminosity law to obtain the distance moduli for galaxies whose Cepheid variables can be resolved. If the Cepheid variables in a galaxy are not bright enough to be detected, we still can have recourse to the brightest stars whose absolute magnitudes are known from studies in our own Galaxy. Of course, in galaxies where Cepheid variables can be detected and where in addition we have available very bright stars and novae, we can check one method against the other and, in general, we find that the values obtained for the distance moduli are in fair agreement—so that there is an inner consistency in all of these methods.

As we go to greater and greater distances, we find that the Cepheid method is the first that drops by the wayside, because of the comparatively low intrinsic luminosities of Cepheids. After that the order in which various objects cease being useful distance indicators is the following: irregular variables, brightest stars, planetary nebulae, the novae and supernovae. With supernovae it is possible to determine distances up to 5 million parsecs (see Figure 30.14). If no objects whose absolute magnitudes are known can be detected in a galaxy, the distance of the galaxy must be determined by a secondary method which makes use either of some known characteristic of the galaxy as a whole, such as its intrinsic luminosity or its actual size, or of a relationship between the distance and a dynamical property of the galaxy.

If we know that galaxies, on the average, have an absolute magnitude of -15, we can at once assign distance moduli to galaxies by measuring their apparent magnitudes. But there may be an error of as much as four magnitudes in this procedure since the galaxy may be this much brighter or fainter than average. This method can be made more accurate if we know the class to which a galaxy belongs, for we can then assign a more accurate value to its absolute magnitude on the basis of what we have already discussed. This procedure, which merely involves measuring the apparent magnitude of a galaxy, allows us to go out to distances of the order of about 100×10^6 psc.

In addition to the above procedures, we can also determine the distances by comparing the apparent sizes with the average linear diameters of galaxies as determined statistically. If none of these methods can be applied, it is still possible to find the distance by means of the spectral red shift, for we know observationally that the shift towards the red in spectra of distant galaxies increases with the distance in a regular way as expressed by Hubble's law, which we shall discuss later.

30.23 Radio Galaxies *

Finally, radio astronomy enables us to detect galaxies far beyond those that can be found with optical telescopes (see Appendix B). We have seen that there are radio sources in our own Galaxy and Andromeda, and we now know that this is true for practically all galaxies of spiral classes SBb, Sb, and later. The early-type galaxies, such as the ellipticals and the Sc, Sa, and SBa, show hardly any radio emission, and this is in agreement with what we know about the composition of these galaxies and the relationship of this composition to radio emission. These early types, having no spiral structure and practically no dust or gaseous material, do not emit radio waves because the radio-emission mechanism, which requires the presence of magnetic fields in gaseous nebulosities and dust clouds, is missing. However, the late-type spirals do have the necessary means for emitting radio waves. *It appears from this that a necessary condition for radio emission from a galaxy is the presence of Population I stars.*

The radio shape of a galaxy differs from its optical shape just the way the radio shape of our Galaxy and that of M 31 differ from their optical shapes. The galaxies in which radio emission in relationship to their optical emission is about the same as in our own Galaxy and M 31 are referred to as **normal galaxies**, but there is a group of radio galaxies in which the radio emission is as much as 1 000 times greater than that of a normal galaxy of the same visual apparent magnitude. In some cases no optical image has been obtained at all. These objects are referred to as **radio galaxies**.

An example of a radio galaxy is the source Cygnus A (one of the strongest discrete radio sources known (see Figure 30.15)). We now know from an optical investigation that this radio galaxy, more than 100 megaparsecs away from us, consists of two galaxies in contact. Apparently radio waves are emitted from colliding galaxies (or galaxies that are in fission) because these objects are not in equilibrium but in an extremely nonsteady state (see the photographs in Figure 30.16).

Fig. 30–15 *Cygnus "A" radio source. This intense radio source is seen, optically, to be two galaxies in collision. A faint halo surrounds these colliding objects. (200-inch photograph from Mount Wilson and Palomar Observatories)*

Fig. 30–16 *(a) NGC 2623 Galaxy, type Sc_{pec}, in Cancer. A source of radio noise seen as two colliding galaxies. (200-inch photograph from Mount Wilson and Palomar Observatories)*

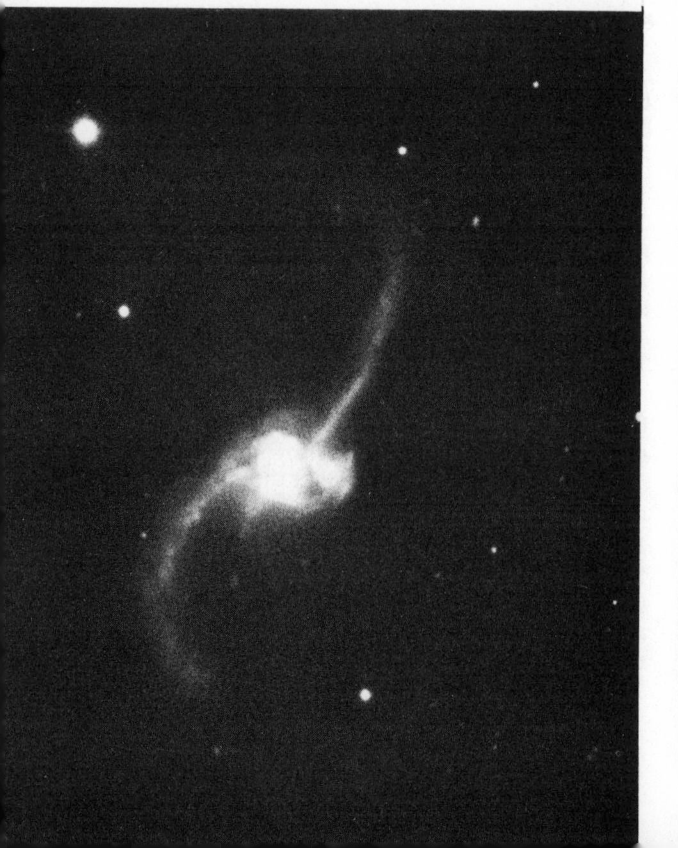

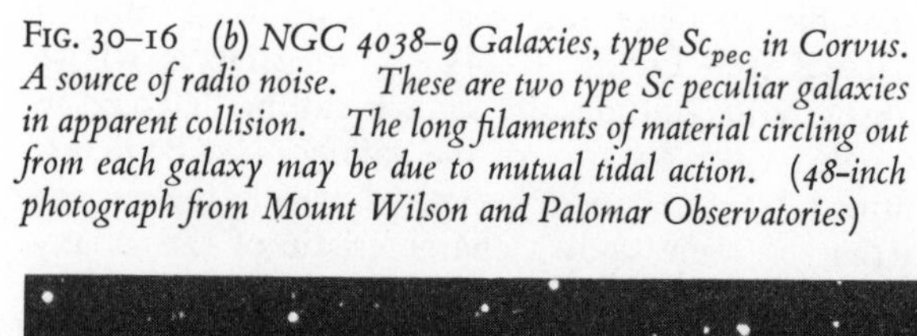

Fig. 30–16 *(b) NGC 4038–9 Galaxies, type Sc_{pec} in Corvus. A source of radio noise. These are two type Sc peculiar galaxies in apparent collision. The long filaments of material circling out from each galaxy may be due to mutual tidal action. (48-inch photograph from Mount Wilson and Palomar Observatories)*

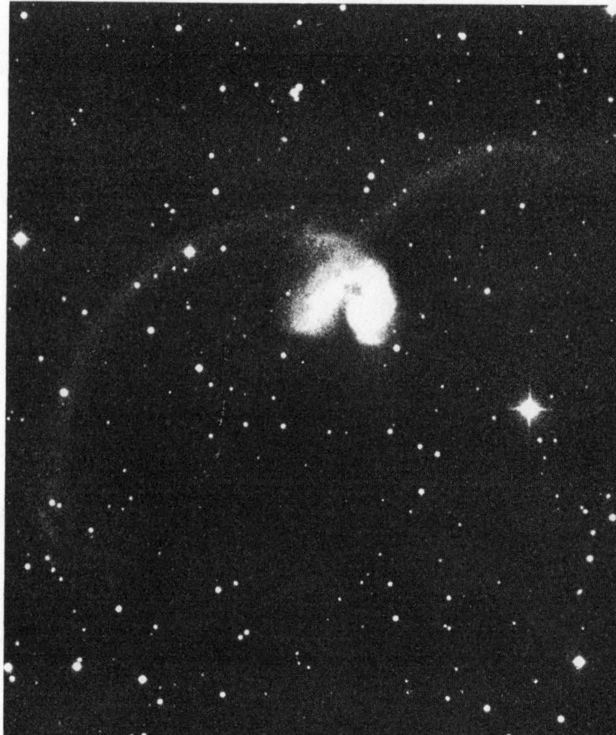

30.24 Radio Telescopes Extending the Observable Boundaries

We can use radio astronomy to push out to vast distances because of the great sensitivity of radio detecting devices and also because of the enormous power of extragalactic radio sources. Since radio telescopes can be made many times larger than optical telescopes, and since electronic amplification methods are much more readily available with radio telescopes than with optical telescopes, extremely weak radio sources can be detected.

With the large, modern, highly sensitive radio telescopes we can detect 10^{-17} (ergs/sec) per cm^2 for a frequency width of 1 megacycle per second. In the visual or photographic range such a source corresponds to a star of the thirtieth apparent magnitude. Since our most powerful telescope (the 200-inch) can only detect objects as faint as +23.5 apparent magnitude, we see that even with radio telescopes that are nowhere near their ultimate size and sensitivity we can detect objects that are about 600 times fainter and about 25 times more distant than those that we can detect optically. The great radio-emission power of certain extragalactic radio sources relative to their optical emission is also an important factor as can be seen from the following considerations.

Whereas the sun sends us 10^8 times more optical radiation than all other bodies in the universe combined (we exclude the moon because it is so close to us) the situation is quite the reverse as far as radio waves go. The total cosmic radiation flux from all sources at a wavelength of 10 meters exceeds that of the sun 100 000 times. As another example of this we may note that the radio flux from Cygnus A (300 000 000 light years) is as large as that from the sun (when the sun is quiet—no sudden flares). This means that the intrinsic radio luminosity of Cygnus A (which we can calculate from its distance) is 10^{26} times as great as that of the sun, whereas its optical luminosity is about 10^{11} times that of the sun. If we know the intrinsic radio luminosity of a radio galaxy or a normal galaxy, we can find its distance modulus from its radio flux.

Table 30.2 lists distances and other characteristics of some of the nearby galaxies.

FIG. 30–16 *(c) NGC 5128 Galaxy, unusual type in Centaurus. A source of radio noise, and one of the strangest objects ever observed. The inner region is that of an E0 galaxy, but the wide lane of dust is unique and wider and more chaotic than that of an Sb or Sc galaxy. The strongest radio emission comes from the dark band. This may be a collision of two galaxies or an explosion in the nucleus. (200-inch photograph from Mount Wilson and Palomar Observatories. Caption material reprinted from* The Hubble Atlas of Galaxies, *by Alan Sandage (1961), by permission of the Carnegie Institution of Washington and the California Institute of Technology.)*

30.25 The Distribution of Rotational Velocity in Galaxies

Since all galaxies belonging to the same structural class have similar dynamical properties, we can be fairly certain that all systems like our own Galaxy and M 31 are rotating in about the same way. This is borne out by an analysis of the spectra of these stellar systems from which radial velocities of their bright emission objects (supergiants, HII regions, planetary nebulae, etc.) can be obtained. However, only for the closest extragalactic galaxies can the velocities of the outer regions at great distances from the nuclei be obtained, and in such cases we find, just as for M 31, that whereas the nuclei revolve with constant angular velocity, there is a gradual decrease in the angular velocity as we move out to the furthest visible edges of these galaxies. For the very distant galaxies only the nuclei can be detected and we find that these nuclei rotate like solid disks with constant angular velocities.

As an example of the distribution of rotational velocity in a galaxy we have M 33, in which the region of constant rotation extends 15′ out from the nucleus and, indeed, right into the spiral arms themselves. At this limiting region the speed of rotation is about 120 km/sec. From this point on there is a gradual decrease of the rotational velocity out to the edge at 30′ where it is 65 km/sec.

The problem of determining the speed of rotation from the observed radial velocities is complicated

TABLE 30.2
GALAXIES

Galaxy	Type	Galactic coordinates		Diameter		Ellipticity	Distance	V	Color index	Absolute visual magnitude	Radial velocity (observed)	Mass $\log \frac{\mathcal{M}}{\mathcal{M}_\odot}$
		l^{II}	b^{II}	angular	linear							
Local group members		°	°	′	kpsc		kpsc				km/s	
Galactic system	Sb	—	—	—	20	—	8	—	0.8	−20.3	—	11.0
Large Magellanic Cloud	IrI	280	−33	470	7.8	0.89	52	0.1	0.45	−19.1	+280	10.1
Small Magellanic Cloud	IrI	303	−45	153	2.6	0.5	54	2.4	0.4	−16.8	+167	9.2
Andromeda Nebula M 31 NGC 224	Sb	121	−21	102	16.0	0.30	570	3.5	0.98	−20.9	−270	11.5
M 33 NGC 598	Sc	135	−32	34	5.7	0.64	600	5.8	0.55	−18.5	−190	10.1
M 32 NGC 221	E2	121	−22	5	0.8	0.74	600	8.2	0.9	−16.0	−210	9.6
NGC 205	E5p	121	−21	12	1.7	0.5	600	8.2	0.81	−16.0	−240	9.9
Sculptor system	E	284	−84	30			110	7	0.8	−13		8.5
Fornax system	E	237	−65	40			200	7	0.8	−15	+ 40	9
NGC 6 822	Ir	26	−19	15	1.7	0.7	400	8.7	0.5	−15.0	− 40	8.6
NGC 147	Ep	120	−14	9	1.0	0.63	400	9.6	0.90	−14.5		9
NGC 185	Ep	121	−14	6	0.7	0.83	400	9.5	0.93	−14.6	−340	9
IC 1 613	IrI	129	−60	12		1.0	600	9.6	0.5	−14.1	−240	7.9
Wolf-Lundmark system	E5	74	−73	10	1.5	0.5	500?	10.8	0.5	−13.5	− 80	8
Leo system I	E4	227	+49	12		0.6	400?					
Leo system II	E1	222	+68	11	1.3	0.9	400?	12	0.9	−12		
Possible Members												
IC 10	Sc	119	− 3	4							−340	
IC 342	Sc	139	+10	30							− 10	
NGC 6946	Sc	97	+11	7	2?	0.9	800?	9	0.8	−17	+ 40	
Leo system III	Ir	197	+54			1.0	1 000?	13	0.4			
Sextans system	Ir	246	+40			0.9		11	0.4		+370	
NGC 300	Sc	299	−80	20	6?	0.5	1 000?	8.5	0.5	−16	+250	
Selected outer galaxies							Mpsc				corrected for solar motion	
M 81 NGC 3 031	Sb	141	+41	20	16	0.51	3.0	6.9	1.02	−20.9	+ 80	11.1
M 82 NGC 3 034	IrII	141	+41	8	7	0.4	3	8.2	0.91	−19.6	+400	10.4
NGC 3 115	E7	248	+37	4.4	5	0.32	4	9.1	1.0	−19.4	+430	10.9
M 87 NGC 4 486	E1			4.0	13	0.85	11	8.9	0.97	−21.5	+1 220	12.6
M 104 NGC 4 594 Sombrero	Sa	299	+51	6.5	8	0.7	4.4	8.1	1.02	−20.4	+1 020	11.2
M 63 NGC 5 055	Sb			10	15	0.5	4	8.6	0.86	−20.1	+2 600	
NGC 5 128	Ep	310	+19	14	15	0.80	3.8	6		−23	+260	
M 51 NGC 5 194 Whirlpool	Sc	105	+68	9	9	0.58	2	8.3	0.63	−19.6	+550	11.0
M 101 NGC 5 457 Pinwheel	Sc	102	+60	20	23	1.00	3	8.1	0.6	−20.2	+400	11.2

by the fact that the plane of the galaxy is tilted with respect to the line of sight. Since the amount of this tilt is not known unless it is either perpendicular to or parallel to the line of sight, we do not know just how great the rotational velocity is without making some assumption about the shape of the galaxy. If we assume that the disk is perfectly round, we can determine the tilt of the galaxy by measuring the amount of foreshortening (i.e., by comparing the major and the minor axes on a photographic plate).

Thus, if b is the observed minor axis and a the major axis, $\sin i = b/a$, where i is the inclination of the plane of the galaxy to the line of sight. If then V_R is the radial velocity, measured spectroscopically for an object in the galaxy, the rotational velocity V_{rot} of this object around the center is just

$$V_{rot} = \frac{V_R}{\cos i}$$

or

$$V_{rot} = \frac{V_R}{\sqrt{1 - \frac{b^2}{a^2}}}.$$

If the plane of the galaxy is at right angles to the line of sight, we cannot determine the rotational velocity spectroscopically since there is then no differential Doppler shift of any object in the galaxy relative to the center.

Until about 1935 most of the data on the rotation of galaxies were based on an analysis of absorption lines and were thus limited to regions close to the nuclei. However, with the discovery of the fairly intense emission line λ3727 we can now investigate the internal motions out to greater distances from the nucleus, and we find, almost without exception, that the nuclei rotate like solid bodies, whereas the material in the arms (when it can be detected) rotates according to Kepler's laws. This has been verified for galaxies in all structural classes from E to I, and the results obtained with this emission line are in agreement with those obtained from the absorption lines.

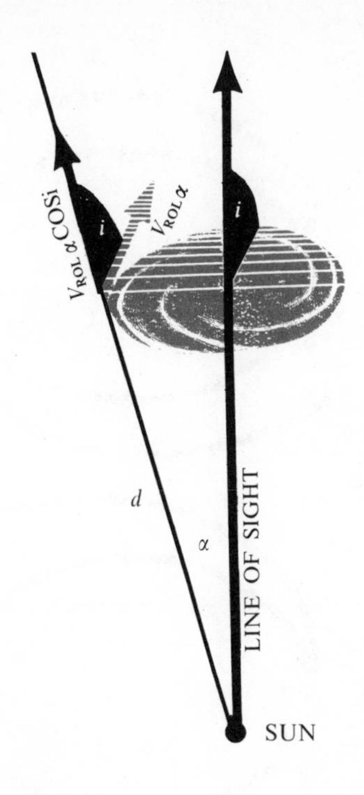

FIG. 30–17 *Determining the period of rotation of a galaxy by analyzing the rotational velocity at various distances from the center of the galaxy.*

30.26 Periods of Rotation of Galaxies

If we know the rotational velocity of a galaxy and its distance from us, we can obtain its period of rotation. To see this we note that if the measured rotational velocity at the angular distance α from the center of the nucleus is $V_{rot\ \alpha}$ (where the tilt is taken into account as described in the previous section) and d is the distance of the galaxy, the period is just $2\pi\alpha d/V_{rot\ \alpha}$, where α is measured in radians. If $V_{rot\ \alpha}$ is expressed in km/sec, α in sec of arc, and d in parsecs, the period is then just $29.7 \times d/V_{rot\ \alpha}$. Note that $d\alpha$ is the true distance from the nucleus to the object whose spectrum we are analyzing, as shown in Figure 30.17.

From an analysis of 30 objects Mayall has shown that the rotational periods increase as we go from the late elliptical galaxies to the Sc spirals. The rotational periods (as determined from the rotation of the nuclei, where the angular velocities are constant) vary from 5×10^6 years for E7 galaxies to 10^7 years for Sc types. From types Sa to Sc the rotational periods go from 10^7 to 8×10^7 years. Note that this applies only to the nucleus; periods of rotation for objects in the spiral arms are considerably longer. In the case of very late Sc types, both ordinary and barred, the periods are in excess of 100 million years.

30.27 The Motion of the Spiral Arms: Winding or Unwinding

An important question in connection with the rotation of galaxies is the direction of rotation with respect to their spiral arms: that is, whether the arms are trailing (i.e., winding up) or unwinding. Although offhand it seems that both types of rotation should and do occur, various theoretical models of galaxies predict either one or the other type of rotation but not both. Hence, knowing how a galaxy rotates relative to its arms is a useful way to check a particular model.

That the problem of determining observationally whether the arms are trailing or leading is not a

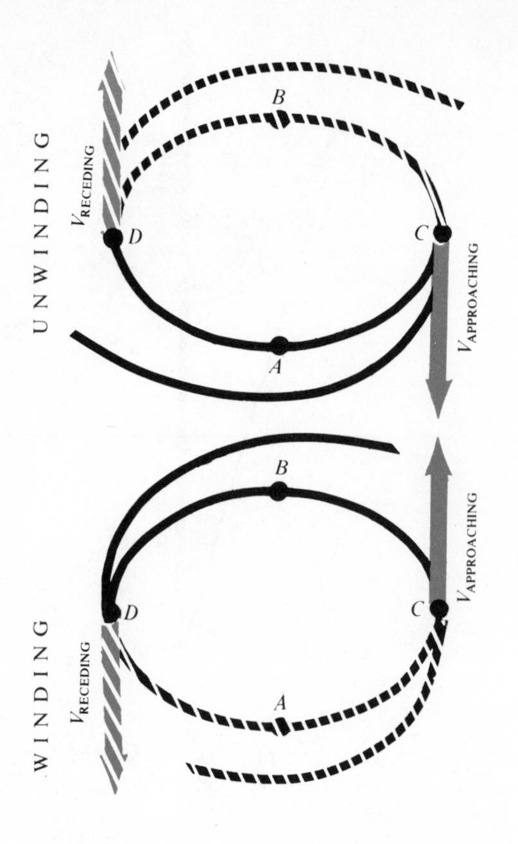

FIG. 30–18 *Determining the direction of rotation of a spiral galaxy by analyzing its tilt, as seen from the sun, in relationship to the observed velocities of its spiral arms.*

simple one is shown from a consideration of the diagram in Figure 30.18. We cannot know whether the arms are winding up or unwinding, even though we know that C is approaching and D is receding from us, because we do not know whether A or B is closer to us. If A is closer, then the arms are unwinding; if B is closer, the arms are winding up.

If the plane of the galaxy is at right angles to the line of sight, we have no way at all of determining the direction of rotation, but if the galaxy is tilted, so that we see it almost edgewise (i.e., $i < 10°$), we take as the side closer to us the one marked by the dark lane of obscuring matter. However, in such cases the arms cannot be clearly traced out, so we are left with some uncertainty as to whether they are trailing or not.

If the angle of tilt is sufficiently large to enable us to trace out the spiral arms unambiguously, the lane of obscuring material no longer stands out sufficiently to tell us which is the closer side. According to de Vaucouleurs, who has analyzed most of the available data, the balance of evidence favors the winding up of the arms (i.e., trailing) and agrees with the rotation of our own Galaxy as determined from radio data.

30.28 Masses of Galaxies

Since we cannot often observe the acceleration of one galaxy in the gravitational field of another, the only direct way we have of determining the mass of the galaxy is from an analysis of its internal motions such as rotational velocities. The simplest approach is the one we applied to our own Galaxy; we consider the rotational motion in the observed galaxy of an object far from the nucleus. If we picture this object as moving in a circular Keplerian orbit around the nucleus, the mass of the galaxy (see Section 3.19) is given as

$$M_{gxy} = \frac{RV_{rot}^2}{G},$$

where G is the gravitational constant, R is the distance of the object from the nucleus, and V is its rotational velocity.

If we express this in solar masses and introduce the period, P, of rotation of the galaxy, we have

$$M_{gxy} = 3.81 \times 10^{-5} PV_{rot}^3. \qquad (30.1)$$

This approximation gives us good results if the object whose velocity we are using to determine the mass is at a distance from the nucleus that is considerably larger than the radius of the nucleus itself. We may then treat the nucleus as though it were a concentrated mass acting almost like a point source of the galactic gravitational field. If, however, the object is close to the nucleus (i.e., close to the region of constant angular rotation), this formula holds only if the material in the nucleus has a spherical distribution. In general, the latter is not the case, and Kepler's formula has to be corrected because the nucleus itself is not spherically symmetric.

Such corrections have been introduced by Wyse and Mayall, and also by the Burbidges and Prendergast, who have analyzed a number of spirals and obtained their masses from their rotational velocities. Appropriate masses of some galaxies as obtained by the Burbidges and Prendergast follow: 2×10^{11} $M_\odot$ for NGC 3623 type Sa; 1.8×10^{10} $M_\odot$ for NGC 2146 SBc pec; 6×10^{10} $M_\odot$ for NGC 157 Sc; and about 2×10^{11} $M_\odot$ for NGC 3646 Sc, 41 kpsc. They have also found in the case of NGC 2146, for example, that the central density is 2.1×10^{-22} gm/cm^3, and that most of the mass of the nebula lies inside a diameter of 4.5 kpsc.

The accuracy of these mass determinations is influenced by the accuracy to which the galactic distances are known. These distances are determined from the red shift (i.e., velocity of recession, which we shall discuss later) on the assumption that the Hubble constant (see Section 31.4) has a value of 75 km/sec per megapsc. From all of the data obtained we see that, on the average, the masses of galaxies are of the order of 10^{11} $M_{\odot}$, as shown in Table 30.2.

30.29 The Mass-Luminosity Ratio for Galaxies

Since we have also listed the galactic luminosities in Table 30.2, we can see that there is a close correlation between the mass-luminosity ratio and the structural type. This ratio ranges from about 100 for ellipticals to about 1 for irregulars. We see that as we go from the irregulars through the spirals into the ellipticals this mass-luminosity ratio follows the same trend as the colors, the spectra, and other characteristics. The mass-luminosity ratio can, therefore, serve as an index of the galactic type.

Although, in general, the distribution of mass in a galaxy depends upon its structure, the work of the Burbidges and Prendergast indicates that in Sa and Sb types, most of the mass is concentrated in the region of constant angular velocity (i.e., in the nucleus). This is borne out in a detailed analysis of the Andromeda Nebula and is probably true of all spirals.

30.30 Galactic Evolution: Hubble's Theory

When Hubble first introduced his classification of galaxies and arranged them in the very nearly continuous sequence of structural types, astronomers were inclined to accept this sequence as an evolutionary scheme. Hubble himself believed that the elliptical galaxies represent the initial phases of galactic evolution and that the spirals represent the much older systems that evolved from ellipticals after the latter had reached a certain stage.

According to this theory galaxies were first formed from the original gaseous hydrogen as spherical structures in states of very slow rotation. In time, because of gravitational contraction, the speed of rotation increased and the systems became highly flattened, while at the same time the Population II stars reached the end stages of their evolution, ejecting into space great quantities of gaseous material containing heavy elements. From this gaseous material dust was formed and was swirled by the rapid rotation out into the outer regions, thus forming the spiral arm from which the Population I stars subsequently were born. As time went on, the arms grew, and the nucleus became smaller and smaller so that the spirals evolved into their final stage.

This theory meets with serious difficulty, however, on various grounds. To begin with, if this were, indeed, the path of evolution of a galaxy, the spirals would be the oldest galaxies and the ellipticals the youngest. We should therefore find a larger ratio of ellipticals to spirals for the very faint galaxies (i.e., the distant ones) than for the bright ones (i.e., the nearby ones)—for when we look at the faint galaxies, we are looking at the parts of our universe that are young; the light now reaching us from these galaxies started out a long time ago. We find, however, that instead of being larger, this ratio for very distant galaxies is about the same as for the nearby ones.

Another objection to the Hubble theory is that there is not enough angular momentum in the spherical and elliptical galaxies as we observe them now to give rise to the amount of flattening that is required to account for the spiral structures.

30.31 Oort's Theory of Galactic Evolution

For this reason certain astronomers have suggested an evolutionary scheme starting with the irregular galaxies and ending with the ellipticals. According to this picture, galaxies begin as amorphous distributions of dust and gas containing a fairly dense nucleus in which the oldest Population II stars are formed, so that the structure at this stage is rather irregular—like that of the Magellanic Star Clouds. As rotation speeds up, the outer masses of gas and dust are stretched into spiral arms and the old Population II stars collect into a spherical nucleus rotating like a solid structure. In time, all the gaseous material in the arms is collected into stars so that the spiral arms slowly disappear and a smooth elliptical structure is formed.

This theory also presents severe difficulties, since it is almost impossible to obtain a smooth globular form from the spiral structures by the process outlined above. Because of these difficulties, Oort has presented a kind of intermediate theory of galactic evolution in which ellipticals and spirals are pictured as evolving independently of each other instead of one from the other. In this theory the development of a galaxy is influenced by two things: (1) the presence or absence of dust, and (2) the presence or absence of rotation. According to Oort's evolutionary scheme, the galaxies were formed from the original thin expanding gas in our universe which varied in density from place to place. Because of the larger internal gravitational forces in the denser regions, these denser regions detached themselves from the expanding surrounding gaseous matter and began their galactic lives.

Within this irregular mass of gaseous matter there were large-scale internal random motions and turbulences resulting from the intermingling and collisions of eddies and currents of gas. Such internal collisions in this gaseous matter resulted in a loss of energy because of heating effects. This brought about a still greater contraction of large clouds, resulting in still more rapid internal motions and a greater variety of turbulences. The contraction of the cloud did not continue indefinitely because the denser regions in it were ultimately condensed into individual stars so that internal collisions became less and less frequent and internal energy was no longer radiated away. During this contraction phase, the system speeded up its rotation until a flat, disklike structure was ultimately formed.

Thus far this picture is similar to the Hubble evolutionary scheme, but now Oort introduces an important difference. He differentiates between an original gaseous structure (the **protogalaxy**) that started out with a great deal of rotation and one that began with a small amount of angular momentum. *The spherical galaxies are those which had very little rotation to begin with, whereas the highly flattened systems like the spirals began with a good deal of angular momentum. A protogalaxy having a large amount of rotational motion contracted perpendicularly to its place of rotation, but centrifugal forces prevented it from contracting in the plane.* Hence it ended up as a disk-like structure, and never became very dense, particularly in the outer regions where a great deal of gas still remained spread out and was not concentrated into stars. From this gas the spiral arms were then formed.

On the other hand, *in a protogalaxy in which there was only a small amount of rotation to begin with, the contraction was fairly uniform towards the center and no extended disk was formed.* Ultimately all the gaseous matter condensed into stars, and a globe or elliptical system resulted.

The Population II stars and such things as globular clusters, are, according to this picture, the objects that were first formed in the very early stages in the protogalaxy even before any great amount of contraction occurred. As contraction proceeded, these stars and other Population II objects retained their spherical distribution about the center, whereas the Population I stars originating from the gaseous material in the disk (after the disk had formed as a result of rotation) acquired a flattened distribution.

According to this picture, then, the difference between elliptical and spiral galaxies is in the amount of rotation that these objects had when they first began as protogalaxies. Those with large amounts of rotation became spirals, provided that enough interstellar matter to form spiral arms was present. If, however, the protogalaxy began with a small rotational motion and all its gaseous matter was condensed into stars, it necessarily evolved into an elliptical galaxy.

30.32 The Distribution and Number of Galaxies

Thus far we have been discussing the structures of individual galaxies without considering their distribution in space. We know that no matter in what direction we look away from the plane of the Galaxy (to avoid the obscuration, i.e., the so-called **zone of avoidance** 20° above and below the galactic plane) we find galaxies distributed apparently endlessly in all directions.

Down to the twentieth magnitude close to 500 such objects per square degree (this corresponds to 100 such objects over an apparent area equal to that of the full moon) in various parts of the sky have been counted on photographic plates. Since it is necessary to use large telescopes with great light-gathering power, only a small piece of the sky can be investigated at one time, and it is, therefore, impossible to survey the entire sky. Only random samplings can be taken in various directions. Thus, down to this magnitude we may estimate that there are 20 million galaxies distributed over the entire celestial sphere. With the 200-inch telescope, which brings us down to the twenty-third apparent magnitude, we estimate that there are over a billion such objects, and radio telescopes have revealed the existence of many billions more.

FIG. 30–19 *NGC 5194 Spiral nebula in Canes Venatici. Messier 51. Satellite nebula is NGC 5195. This magnificent spiral is dominated by dust lanes, the two most opaque of which lie on the inside of the two brightest arms. These arms and dust lanes describe almost perfect spirals as they emerge from the nucleus and unwind. The companion galaxy, NGC 5195, is an irregular type and its gravitational action probably accounts for the distortion of one side of NGC 5194. The distance modulus of this system is about 27.5. (From Mount Wilson and Palomar Observatories. Caption material reprinted from* The Hubble Atlas of Galaxies, *by Alan Sandage (1961), by permission of the Carnegie Institution of Washington and the California Institute of Technology.)*

30.33 The Clustering of Galaxies

Although no general pattern for this distribution has emerged, we know that there are local non-uniformities arising from a tendency of galaxies to cluster into groups. Hubble, who was one of the first observers to study such groups, expressed it as follows: "While the large-scale distribution of galaxies appears to be essentially uniform, the small-scale distribution is very appreciably influenced by the well-known tendency to cluster. The phenomena might be rightly represented by an originally uniform distribution from which nebulae have tended to gather about various points until now they are found in all stages from random scattering, through groups of various sizes, up to occasional great clusters."

Although in this vast distribution we find individual galaxies moving about by themselves and groups of multiple galaxies ranging from binary to more complex systems, we find in general that most of the galaxies, at least within 100 million light years or so, are not moving about as individuals but are members of loose groups which we shall call **clusters of galaxies**. These range in size from perhaps a dozen up to thousands of galaxies.

The existence of double or triple nebulae is well established. Some excellent examples are our own Galaxy with its satellite galaxies, the two Magellanic Star Clouds; M 31 and its two elliptical satellite galaxies; and the famous M 51 (the Whirlpool nebula) with a small component (see Figure 30.19).

30.34 The Coexistence of Galaxies

From an analysis of the Whirlpool nebula using a special type of composite photography, F. Zwicky showed that the main body of this galaxy consists of two coexisting galaxies superimposed on each other as shown in the schematic drawing, Figure 30.20. According to Zwicky, whereas the blue stars in this galaxy form a disrupted and very

FIG. 30–20 *Schematic drawing showing the coexistence of two galaxies in the Whirlpool Galaxy according to Zwicky.* (*From* Handbuch der Physik, *S. Flügge, ed., vol. 53. Berlin: Springer, 1959*)

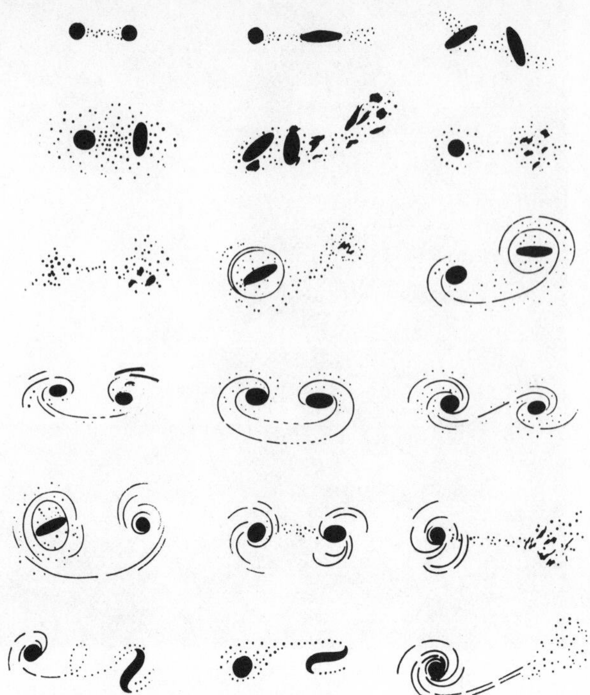

FIG. 30–21 *A schematic diagram taken from Zwicky showing the various types of bridged galaxies.* (*From* Handbuch der Physik, *S. Flügge, ed., vol. 53. Berlin: Springer, 1959*)

poorly organized spiral structure, the yellow-green stars have arranged themselves in a highly streamlined spiral with a smoothly distributed surface luminosity. Furthermore, this latter spiral is of the barred type while the other spiral is a normal structure. This coexistence of a barred and a normal spiral may throw an entirely new light on the evolution of such structures and on the whole classification scheme. It may very well be that many spirals must now be reclassified as coexisting galaxies of a type similar to the Whirlpool.

30.35 Galactic Binaries

If two galaxies appear to be close together, we may definitely conclude that they are really close together in space (i.e., gravitationally connected) only if their structures show distortions that arise from their mutual gravitational interaction, or if a bridge or filament of material connects the two. There are many such examples; Zwicky has analyzed the various types in detail. Figure 30.21 is a schematic diagram taken from Zwicky.

In recent years the wide-angle 48-inch Schmidt telescope has been used to survey extragalactic regions of the sky, and many thousands of gravitationally interacting double galaxies have been discovered, ranging all the way from those that are so close together that they may be considered as superimposed to those that are separated by two or three galactic diameters. One interesting feature about a galactic binary is that the two components do not necessarily belong to the same Hubble structural class; we find in fact that every type of galaxy from irregular to elliptical can form a binary system with every other type, and all such combinations have been observed.

The interrelationship in binary galaxies ranges from thin filaments of matter (which in some cases are extensions of spiral arms) extending from one galaxy to the other to a very intimate existence in which both galaxies interpenetrate each other. There are also cases where two galaxies are embedded in the same amorphous luminous cloud. The filamentary material between the two galaxies consists of very luminous stars, bright gaseous clouds, and also dust.

30.36 Masses of Galaxies in Binary Systems

The investigation of double galaxies has been useful in the study of galactic structures, for it is possible to determine the masses of the components of binary systems of galaxies by analyzing the radial velocities spectroscopically and thus determining the relative orbit of one with respect to the other. One treats the two galaxies as point masses moving in circular relative orbits and then simply applies Kepler's third law. This procedure for determining the masses of galaxies has been carried out extensively by Page and Holmberg, who first suggested this method. From an analysis of 35 double galactic systems Page has obtained a mean mass of $8 \times 10^{10}\ M_\odot$ for the component galaxies, and in a detailed investigation analyzing 26 double galaxies Holmberg found a mean mass of $6.5 \times 10^{10}\ M_\odot$.

An interesting point in connection with this has to do with the luminosities of galaxies. In principle we should be able to determine the mass of a galaxy from its luminosity if we suppose that large masses of matter in a galaxy are not concentrated in dark bodies. If we assume that the mass-luminosity ratio for a whole galaxy is about the same as the average M-L ratio obtained from an average of the stars in the neighborhood of the sun, we can then obtain the mass of a galaxy by multiplying the luminosity of the galaxy by this ratio. When we do this, we find that the masses of the most luminous galaxies are only a few billion times that of the sun. These objects are thus underluminous if we take as their correct masses the values determined dynamically in double galaxies or determined from the internal rotation for individual galaxies.

From these data of both Page and Holmberg we find that the M-L ratio for galaxies is much larger (of the order of 300) than it is for stars in our part of the Milky Way. Although these investigations probably contain fairly large errors, we must conclude that there are present in extragalactic nebulae great quantities of matter (with large M-L ratios) in the form either of gaseous clouds or of very faint massive stars such as white dwarfs, black dwarfs, or even hyperon stars (see Section 24.51).

30.37 Galaxies as Gravitational Lenses

There are two other methods that can be used to find masses of galaxies. One of these, which we shall discuss later, involves the analysis of the motions of the individual galaxies in a cluster. The other deals with galaxies acting as gravitational lenses. This question has been investigated in detail by Zwicky, who has concluded that compact galaxies should be able to bend light from a more distant galaxy gravitationally and thus bring it to a focus at a point between us and the focusing galaxy. This produces a combined picture of a tiny galaxy superimposed over the center of the light-bending galaxy. From the general theory of relativity (see Section 11.29) we know that the angle of deflection α (in radians) of a ray of light grazing a galaxy of mass M and radius R is $4GM/c^2R$, where G is the gravitational constant. This is just equal to $2.97 \times 10^{-28}\ M/R$.

From this it follows that a galaxy having a mass of $10^{10}\ M_\odot$ and a diameter of 1 000 light years bends light 4″ of arc. A lens nebula of this sort brings the light from another distant galaxy to a focus at a distance of about 5×10^7 light years. Because the photographic plate receives the light from the lens galaxy directly and also the light from the image formed by the lens galaxy, the final image on the plate is essentially a ring with a bright disk at the center, so that such objects should immediately stand out and should be easily differentiated from the images of ordinary galaxies. The discovery of such objects would, of course, not only serve as a means of measuring the masses of such galaxies, but would also be a remarkable verification of the general theory of relativity. Although no such images have been identified, there is every reason to believe they exist.

30.38 Aggregates of Galaxies

In addition to the binary systems we find that galaxies occur in higher-order multiplets such as NGC 6027 and Stefan's Quintet (see Figure 30.22), and we see from the over-all background of galaxies that they tend to collect into aggregates of various kinds. Zwicky has introduced the following classifications:

1. A **group**—a multiple system containing anywhere from a few dozen up to 100 members, in which the number of galaxies per unit volume is clearly greater than in the background region (see Figure 30.23). Although a group shows no definite concentration towards a central point it is readily distinguishable from the background.
2. A **cloud**—an irregular structure consisting of hundreds or even thousands of members. There is no concentration towards the center, but the number of galaxies per unit volume again is greater than in the background region, so that a cloud may be considered as a large group.

Fig. 30–22 (*a*) *NGC 3190 Group of four nebulae in Leo. NGC 3185, type SBa; NGC 3187, type SBc; NGC 3190, type Sb; NGC 3193, type E2.* (*200-inch photograph from Mount Wilson and Palomar Observatories*)

Fig. 30–22 (*b*) *A small cluster of nearby galaxies, Stephan's Quintet, NGC 7317–20. Three of the galaxies are barred spirals, two of which are intertwined and gravitationally connected. The largest galaxy in the group is an irregular spiral and the smallest member is an elliptical galaxy.* (*Lick Observatory photograph*)

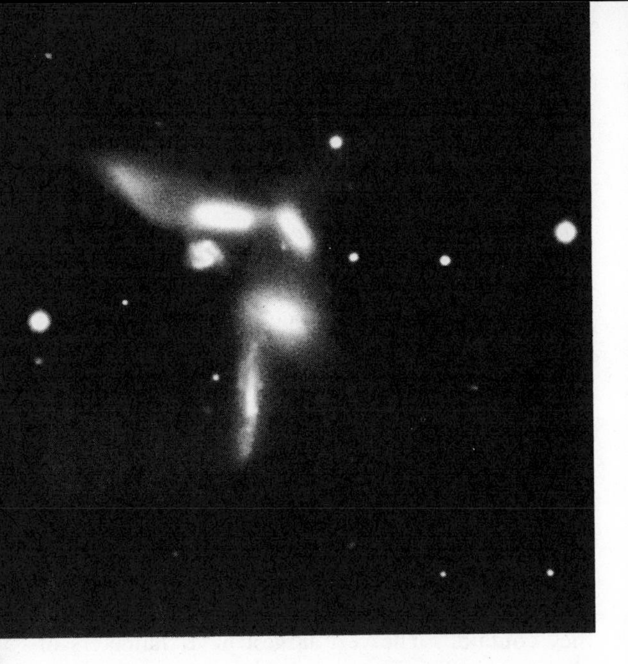

FIG. 30–22 *(c) NGC 6027 Group of five galaxies with unusual connecting clouds, in Serpens. These connecting clouds are evidence of the mutual gravitational interactions among these galaxies. (From Mount Wilson and Palomar Observatories)*

FIG. 30–22 *(d) NGC 3031 Galaxy in Ursa Major. Messier 81. Also shown are NGC 2976, 3034, and 3077. The galaxies NGC 3034, immediately above NGC 3031, and NGC 3077, in the lower right-hand corner, are irregular galaxies. (From Mount Wilson and Palomar Observatories)*

3. A **cluster of galaxies**—an aggregate of hundreds to thousands of galaxies with a definite concentration towards one or more points. In many cases there is a more or less spherical distribution with a fairly well-defined structure. Near the center of such a cluster the density of galaxies is anywhere from thousands to millions of times greater than that of the background region of the cluster. The richest clusters known contain about 10 000 member galaxies.
4. A **cloud of groups**—an association of many groups in which the density of galaxies between the component groups is greater than in the surrounding regions.
5. A **cloud of clusters**—an aggregate of many clusters ranging from double and triple clusters to hundreds of clusters, altogether containing hundreds of thousands of galaxies.

30.39 Clusters of Galaxies

In this text we shall discuss only clusters of galaxies since they have very interesting and unusual properties. These aggregates of galaxies were first detected by Sir William Herschel over a century ago, and have been under intensive investigation by astronomers working with large telescopes.

Many astronomers have concluded that clustering among galaxies is the rule rather than the exception, as is evident from Zwicky's analysis of the Coma cluster (see Figure 30.23(a)). In an area of 25 square degrees, in which most of the bright members of the Coma cluster are found, there are 165 field galaxies and 645 cluster galaxies brighter than apparent photographic magnitude +16.5. An analysis of the distribution of cluster galaxies and field galaxies shows that the ratio of the former to the latter is at least 23:1.

Field galaxies are probably those with such large kinetic energies that their velocities exceed the speed of escape from the average cluster, so that they are moving too fast to be held gravitationally to any one cluster.

FIG. 30–23 *(a) Part of cluster of nebulae in Coma Berenices. Distance about 40 million light years according to the old distance scale. Corrections give a distance of about 200 million light years. This cluster contains about 1 000 galaxies and is receding with a speed of 6 700 km/sec. (200-inch photograph from Mount Wilson and Palomar Observatories)*

30.40 Types of Galaxies in Clusters

The types of galaxies found in clusters depend upon whether the cluster is very compact and densely populated or mediumly compact. *The dense clusters appear to contain primarily elliptical or S0 systems with only a few spirals evident.* The open clusters such as the Virgo (see Figure 30.13) or the Ursa Majoris cluster contain all types of galaxies ranging from the globular to the irregular. We shall see that to some extent this difference can be accounted for by the more frequent collisions between galaxies in the very compact clusters than in the open clusters, which results in a sweeping out of the dust and gaseous matter from these galaxies into the intergalactic space in the cluster.

The dimensions of the clusters depend on their compactness and on the total number of galaxies they contain. The very largest have diameters of about 5 million parsecs, and the smallest have diameters of the order of a few hundred thousand

FIG. 30–23 *(b) Cluster of galaxies in Hercules. This cluster is extremely rich in spirals, ellipticals, and barred galaxies. (From Mount Wilson and Palomar Observatories)*

Table 30.3
SELECTED CLUSTERS OF GALAXIES

Cluster (or group)	Number of galaxies	Galactic coordinates		Diameter	Distance	Radial velocity	Galaxies per volume	m_v of 10th brightest galaxy
		l^{II}	b^{II}					
		°	°	°	Mpsc	km/s	$Mpsc^{-3}$	
Local Group	17				0.4	− 100	300	8
Virgo	2 500	284	+74	12	11	+ 1 150	500	9.4
Perseus	500	150	−14	4	58	5 400	300	13.6
Coma	1 000	80	+88	6	68	6 700	40	13.5
Hercules		31	+44		105	10 300		14.5
Cor. Borealis	400	41	+56	0.5	190	21 600	250	16.3
Boötes	150	50	+67	0.3	380	39 400	100	18.0
Ursa Maj II	200	149	+54	0.2	380	41 000	400	18.0
Hydra II		226	+30			60 600		18.6

parsecs. Of course, their apparent diameters depend on how far away they are, but clusters have been identified with apparent diameters ranging from 12° for the Virgo cluster down to 0°.2 for the very distant Ursa Majoris cluster. The shapes of the clusters vary considerably from the very rich compact clusters, which show almost perfect spherical symmetry, to the less compact irregular clusters.

We can best understand the nature of the cluster by considering one or more clusters in detail. To begin with, we have the Local Cluster, which consists of at least 17 members, the most important of which are our own Galaxy and the Andromeda Nebula. The details of the Local Cluster are given in Table 30.3. We see that this is a rather loose cluster with no definite concentration or well-defined nucleus. If we picture our own Galaxy as being in the center of this group, the diameter of the group is of the order of 1 000 kpsc (if one of the Leo systems and Sextans systems [NGC 300] are included in our local group of galaxies). As we see, it consists of all types of galaxies, as is to be expected in a non-compact cluster.

30.41 The Coma Berenices Cluster

The Coma Berenices cluster (Figure 30.23(a)) is an example of a compact spherical cluster. Its distance as first given by Hubble and Humason, using the old Hubble constant of recession, is about 14 megaparsecs, but on the new scale its distance is about 60 Mpsc. The apparent photographic magnitudes of the members of this cluster range from about +13.2 to +19.5, with +17.7 the value most frequently encountered.

With the 48-inch Schmidt telescope at Mt. Palomar, altogether 654 galaxies (excluding the field galaxies) brighter than the +16.5th photographic magnitude have been counted in this cluster out to 2°40′ from the center. Most of these galaxies are concentrated towards the center, which contains the two brightest members. However, the total number of galaxies in the cluster is considerably larger than this since many of the fainter galaxies in the center, which is extremely crowded, are completely obscured by the brighter, larger galaxies. An idea of the compactness of this cluster of galaxies can be obtained by considering the numbers lying within a narrow ring at a given distance from the center (again limiting ourselves to galaxies with apparent photographic magnitudes less than +16). Within a ring having a radius of 2.5 minutes of arc there are, according to Zwicky, about 2 777 galaxies per square degree, if we eliminate the background field galaxies of which there are about 7 per square degree. The actual number of galaxies counted in this ring with the 18-inch Schmidt telescope is 15. The galactic count falls off rapidly as we move away from the center and is 1 431 per square degree in a ring of radius 5′ of arc. In the outer regions at about 1°.5 from the center there are only about 13 galaxies per square degree.

The diameter of the Coma cluster is about 320′;

this corresponds to 4.4 million light years on the old distance scale and to 20 million light years with the corrected Hubble constant. If we take everything into account, the total number of galaxies in the Coma cluster may exceed 654 by a factor of 10 or more. In fact, using the 48-inch Schmidt telescope down to the +19th apparent magnitude and taking into account the field galaxies, of which there are 170 per square degree (on the average), Zwicky obtains a total count of 10 724 out to 6° from the center. The density of galaxies belonging to the Coma cluster (field galaxies are excluded) varies from about 87 per square degree in a thin ring around the center having a radius of 6° to about 2 578 per square degree in a thin ring having a radius of 5′ of arc. The cluster continues out beyond 6° with a count of about 51 galaxies per square degree. Using the old Hubble distance scale, we find that the diameter of this cluster out to 6° is about 9.5 million light years, and about 50 million light years with the revised Hubble constant. In Table 30.3 some data are given for the Coma cluster.

30.42 The Virial Theorem for Clusters

Spherically symmetric clusters of galaxies are stable structures in equilibrium under the mutual gravitational attractions of the members of the cluster. Since these clusters extend over very great distances, it is clear that Newton's law of gravitation is valid over distances of millions of light years. It is also evident from the structure of these clusters that the galaxies in such clusters are moving about in the same way individual stars move about within a galaxy. We may therefore assume that, on the average, galaxies in a cluster have velocities large enough to keep such clusters from collapsing, and yet not so large that the individual members can escape. In other words, for stable clusters to exist there must be a balance between the total kinetic energy and the total potential energy of the individual members. This can be stated in the form of the virial theorem, which we discussed in Section 27.23.

When applied to a cluster of galaxies this theorem states that twice the total kinetic energy of all the component galaxies of a stable cluster must equal the negative of the total potential energy. (Note that since the potential energy is itself negative, this quantity is therefore positive.) Since this relationship involves the average mass of the galaxies in the cluster, we can use it to obtain the masses of the galaxies if the average velocity of the galaxies within the cluster is known (i.e., the average random velocity of the galaxies relative to the center of the cluster itself, which is called the **mean dispersion of the velocities**). Zwicky was one of the first to apply this method to the Coma Berenices cluster (using about 600 of the brightest galaxies), and he obtained an average mass for the individual galaxies of $2 \times 10^{11}\ M_\odot$, which is many times larger than that obtained from the mass-luminosity ratio.

30.43 Zwicky's Method for Determining the Average Mass of Galaxies of a Cluster

To see what Zwicky did we now consider a sphere of radius R surrounding the center of a cluster of galaxies, and suppose that there are in this sphere N galaxies, each of average mass M. If the average velocity of these galaxies with respect to the center of the cluster is $\bar{v}$, then the total kinetic energy of this group of galaxies is

$$K.E. = \tfrac{1}{2}NM\bar{v}^2.$$

On the other hand, the total potential energy (if we assume a more or less uniform distribution of galaxies) is

$$P.E. \cong -\tfrac{1}{2}G\,\frac{N^2M^2}{R}.$$

To obtain the exact potential energy we should multiply this formula by a numerical factor of the order of 1, whose exact value depends on how the galaxies are distributed within the sphere; we shall assume this numerical factor to be exactly 1 in order to get some idea of the order of magnitude. If we apply the virial theorem we obtain

$$NM\bar{v}^2 = \tfrac{1}{2}G\,\frac{N^2M^2}{R}$$

or

$$M \cong 2R\,\frac{\bar{v}^2}{NG}.$$

To obtain the average mass of a galaxy from this equation we have to know the average velocity of the galaxies with respect to the center of the cluster (i.e., the mean dispersion of the velocities). But the only data we have are the radial velocities of the individual galaxies with respect to the earth as obtained from spectroscopic data (Doppler red shift). As we shall see, galaxies are receding from us at speeds that depend upon their distances from us, but this recession arises because of the motion of the entire cluster of galaxies, and the individual galaxies have a kind of random motion with respect to the entire cluster which we must know to find $\bar{v}^2$.

In the Coma cluster the average speed of recession of the 21 brightest galaxies is 6 600 km/sec but the cluster as a whole has a speed of recession of 5 550 km/sec. Hence the dispersion of the radial velocities (i.e., the mean radial velocity with respect to the cluster itself) is 1 050 km/sec. In other words, the typical velocity differs from the average by $\pm 1\,050$. If we use this quantity in our equation and consider a sphere around the center of the cluster of radius 1.7 million light years, or 1.61×10^{24} cm (using the distance scale based upon the old Hubble constant), which contains 670 brightest galaxies, we finally find that $M \cong 10^{11}\, M_\odot$.

This detailed discussion of a single compact cluster can serve as a model of similar clusters such as the Hydra and the Perseus clusters.

As an example of a much less compact cluster we have the Virgo cluster, whose characteristics are given in Table 30.3.

30.44 Galactic Collisions

In discussing radio galaxies we introduced the idea that colliding galaxies are strong radio sources and that Cygnus A is probably one such source. This appears to be borne out by optical telescopes, which indicate that we are dealing with two galaxies in close contact. Although, in general, a collision between two galaxies is an extremely rare event because of the great distances between neighboring galaxies as compared with their dimensions, such collisions are probably quite frequent in clusters of galaxies where the density is fairly high, as in the case of the Coma and Corona clusters (see Figure 30.24).

The problem of galactic collisions in clusters (see Figure 30.25) has been investigated by Spitzer and Baade, who have pointed out that these systems contain large numbers of S0 galaxies which have no spiral structures and no absorbing material and consist only of Population II stars. They have suggested that these clusters of galaxies may be the end products of collisions between their component galaxies since such collisions result in the sweeping out of any interstellar material in these galaxies and thus prevent the formation of Population I stars. We have already discussed how the kinetic energy of the colliding galaxies transferred to this interstellar material gives rise to radio emissions of such galaxies. From an analysis of the distribution of the galaxies in the Coma cluster Spitzer and Baade conclude that each galaxy collides with at least 20 other galaxies every 3×10^9 years provided that the galactic motions within the cluster itself relative to the center of the clusters are mostly radial. The collisions have relatively little effect on the stars in any galaxy but remove the interstellar matter. During such a collision the relative velocity between the two galaxies is of the order of thousands of km/sec.

30.45 The Dispersion of Velocities of Galaxies in Different Clusters

In discussing the masses of galaxies in a cluster we saw the importance of knowing the velocity dispersion, i.e., the mean random velocity of the individual galaxies. This varies from cluster to cluster, as Zwicky has pointed out, and increases with the size and population of a cluster. For small clusters it ranges from 250 to 500 km/sec, whereas for large clusters such as the Coma and Corona Borealis it is approximately equal to 2 000 km/sec.

30.46 The Systematic and Random Motions of Clusters of Galaxies

In the next chapter we shall consider in detail the general problem of the recession of galaxies. Here we merely note that the motion of a galaxy with respect to the earth depends upon whether it is one of the field galaxies or a member of a cluster. The field galaxies are receding from us with speeds that increase with their distances from us, but the speed of recession of a galaxy in a cluster is primarily due to the motion of the cluster itself, and we shall see that the cluster as a unit is receding from us with a speed that increases with distance. Aside from this systematic apparent recession of clusters of galaxies there are no detectable random motions of the clusters of galaxies relative to each other. Moreover, the clusters do not appear to be rotating nor do they seem to possess any other kinds of motion relative to the neighboring clusters.

30.47 The Number Density of Clusters of Galaxies

The number of clusters of galaxies is considerably greater than our original estimates had led us to

FIG. 30–24 *Cluster of nebulae in Corona Borealis. Distance about 120 million light years, according to the old Hubble Constant. On the new distance scale this cluster is about 5 times farther away. (200-inch photograph from Mount Wilson and Palomar Observatories)*

believe. It was originally estimated that with the 100-inch reflector we would pick up one cluster per 50 square degrees down to the 20.5th apparent magnitude, but the 48-inch Schmidt telescope has already resolved many more, and it is now possible to identify more than one cluster of galaxies per square degree on photographs taken with the telescope.

It is therefore reasonable to assume that we should find something of the order of 20 clusters per square degree on photographs taken with the 200-inch telescope. But the actual number of discoveries of clusters of galaxies on these plates is not very much larger than that obtained with the 48-inch. This can only mean that there is intergalactic dust and also intercluster obscuration in the form of dust which decreases the cluster count by actually cutting down the number of visible clusters and also by making it more difficult to identify a cluster.

Zwicky has pointed out an interesting feature in connection with clusters: the absence of superclustering. There appears to be no such thing as a globular supercluster of galaxies consisting of individual clusters of galaxies. Zwicky's analysis of the distribution of centers of clusters in unobscured regions of space shows that these centers are uniformly distributed in a random fashion.

30.48 The Problem of Stability of Clusters of Galaxies *

In this chapter we have considered clusters of galaxies which we assume to be in equilibrium under the mutual gravitational attraction of the individual galaxies. This means, as we have seen, that the total energy in a cluster is negative (i.e.,

FIG. 30–25 *Two nearby intertwined spirals, NGC 5432, NGC 5435, of type Sc. The bridge of matter between them connecting their spirals can be clearly seen.* (*Lick Observatory photograph*)

the total potential energy outweighs the total kinetic energy so that no single galaxy has a high enough velocity to escape from the system). We have also seen that if we then apply the virial theorem to these structures, we obtain for the average mass of the component galaxy in the cluster the unusually large value of the order of $2 \times 10^{11}\ M_{\odot}$. We thus arrive at extreme M-L ratios. Because of these large masses, Zwicky suggests that dwarf galaxies and also intergalactic material in clusters may account for a considerable portion of the total mass of a cluster.

As against Zwicky's suggestion, Ambartsumian has proposed the idea that many multiple systems of galaxies and large clusters are not in stable equilibrium, but possess a positive energy. This means that in these systems the kinetic energy outweighs the potential so that the individual members are moving with speeds in excess of the speed of escape and the systems themselves are disintegrating explosively. Hence the virial theorem does not apply and the true M-L ratios are smaller than the calculations show (in other words, the large M-L ratios are fictitious).

As an example Ambartsumian points out that when the virial theorem is applied to the Virgo cluster of galaxies, which contains about 1 500 members, the average mass of the component galaxies is about the same as the mass of our Galaxy. But since our Galaxy is a supergiant, and we know that the Virgo cluster has only a few such supergiants and consists mostly of dwarf galaxies (masses obtained from the luminosities of these galaxies lie between 1/10 and 1/100 of the mass of the galaxy), we obtain a serious discrepancy. According to Ambartsumian, we must therefore conclude that the virial theorem is not applicable to the Virgo cluster, and that its total energy is positive; it is therefore expanding and is not in equilibrium.

Ambartsumian applies the same reasoning to the Coma cluster and shows that the average mass of its members (about the same mass as that of our own Galaxy) obtained by the application of the virial theorem is again in contradiction with the observed luminosities of the members, which shows that they are not supergiant galaxies. He therefore concludes that many large clusters of galaxies are disintegrating systems because they have high velocities of dispersion.

30.49 Ambartsumian's Fission Theory

Using the same type of analysis Ambartsumian also suggests that certain multiple galaxies such as binaries, triplet systems, etc., are not stable but that the individual galaxies are receding from each other as though these galaxies had been formed from a single nucleus. In particular he argues that certain radio galaxies such as Cygnus A are not really colliding, but are rather examples of two galaxies in the process of formation by fission of a single nucleus, and that the two galaxies are presently separating from each other.

This point of view leads to a picture of the formation of galaxies similar to that of the formation of hot stars in the O and B associations. (See Section 27.29.) According to this theory galaxies are formed during an explosive expansion of a vast cloud of gaseous matter. However, Ambartsumian's analysis does not apply to all multiple systems and clusters, and it is safe to say that multiple galaxies and clusters can be divided into stable and unstable systems or disintegrating groups. Small groups having a trapezoidal configuration, such as Stephan's Quintet, are probably not stable and are expanding explosively. For the most part, however, close pairs of galaxies are probably stable. Loose irregular clusters, such as Virgo, are probably expanding but not violently and hence are unstable. However, most compact clusters, such as the Coma Berenices cluster, are probably stable.

30.50 Quasistellar Objects (Quasars): The Most Distant Objects That Have Been Observed Optically *

Although radio telescopes have enabled us to pick up radio galaxies out to distances of the order of 6 billion light years, recent evidence has been presented by J. Greenstein, T. A. Matthews, M. Schmidt, A. Sandage, and others that we may be able to use our optical telescopes here on the earth to detect objects as far out as 10 to 12 billion light years. This is based on their discovery that eleven very bright objects which we previously had thought were stars in our own Galaxy are really very distant objects (now called **quasars** from quasistellar objects) undergoing some vast explosion or implosion and hence releasing great quantities of energy in both the optical and radio regions of the spectrum. (See Figure 30.26.)

From the speeds of recession of these objects (as determined first by M. Schmidt from the red shift in their spectral lines) it is possible to determine their distances with the aid of Hubble's law (see Section 31.4). Greenstein and his associates found that one of these very bright objects is about 4 billion light years away from us. This means that for these objects to be visible here on earth they must be at least 100 times more luminous than our own Galaxy. But if this is the case we should be able to detect such objects out to 10 to 12 billion light years because of the vast amounts of energy they emit per second.

Although these objects were first thought to be colliding or exploding galaxies, there is very little direct evidence as to their exact nature, and it is now becoming more and more obvious that they are phenomena of a most unusual kind. Because of the tremendous rate at which they are pouring out energy (their luminosities are of the order of $10^{13}L_{\odot}$ or about equivalent to 10^7 supernovae) they are very short-lived and probably cannot exist (radiating at their present rate) for more than a few million years. If these are not colliding galaxies but single objects that are releasing energy explosively, they are clearly the most massive, and largest single, objects that have yet been observed, with masses ranging from 10^8 to 10^9 solar masses. Furthermore their linear dimensions must be of the order of a few light years. From this and from the rate at which these objects are radiating energy they must have surface temperatures that range from 10^4 to 10^5 degrees Kelvin.

If these are, indeed, individual objects releasing energy, we can account for their luminosity only by assuming that they are suffering a tremendous gravitational collapse exceeding anything hitherto known. The reason for this is that the total energy released by a quasar during its lifetime of 1 000 000 years is about 10^{60} ergs. This is far more energy than is available in thermonuclear reactions, since it corresponds to 10^6 solar masses and hence to the total mass of the radiating part of the quasar. Hence gravitational energy is the only adequate source. However, this gives rise to certain difficulties as far as the gravitational fields that are present on the surfaces of these bright objects are concerned. To release energy gravitationally at the rate indicated by their luminosities (of the order of 4×10^{46} ergs per sec), the speed of collapse at their surfaces would have to be of the order of $8 \times 10^{46}\, R^2/GM^2$ cm per sec, where R is the radius of the quasar, M is its mass, and G is the gravitational constant. This gives a speed of collapse of

FIG. 30–26 (*a*) *The Quasar 3C273, photographed by Sandage. Note the jet pointing away from the nucleus, which indicates the vast explosive release of energy. Since the length of the jet is about 100 000 light years, the explosion must have begun about 1 000 000 years ago. Although this object is about 2×10^9 light years away, it can be easily seen through a 6″ telescope. The identification of this object was made by Sandage from the original plate of this photograph.* (*From Mount Wilson and Palomar Observatories*)

the order of 10^9 cm per sec for objects having the mass and dimensions given. However, the acceleration of gravity at the surfaces of such objects is of the order of GM/R^2 or about 0.01 cm per sec^2. Hence it is difficult to see how such a small acceleration of gravity could lead to the kind of collapse that is required. In all of this one must keep in mind that we are seeing these objects as they were four billion years ago and it may be that the gravitational constant G was considerably larger then than the value we assign to it now. This, of course, would give a larger acceleration of gravity and hence permit the kind of collapse we now observe.

The linear dimensions of these objects have been determined from lunar occultation measurements. When a star is occulted by the moon, it does not disappear suddenly; instead, its light is visible for a small fraction of a second. Since we know how fast the moon is moving we can use this occultation time to measure the apparent diameter of the star if it is large enough. But this is rarely possible. In the case of our distant luminous objects, however, the occultation lasts long enough to be measured quite accurately with radio telescopes, and very reliable angular diameters can be obtained. This was done by C. Hazard, M. B. Mackey, and A. J. Skimmins of the Australian radio group.

Two quasars, 3C273 and 3C48 (3C stands for the Third Cambridge Catalogue of Radio Sources), have been studied extensively, and optical and radio spectra have been obtained for both. Not only are these quasars intense sources of optical radiation, but they also emit more intensely in the radio region than do other known radio sources. Whereas their optical luminosity is about 10^{46} ergs/sec, their radio luminosity is 10^{44} ergs/sec.

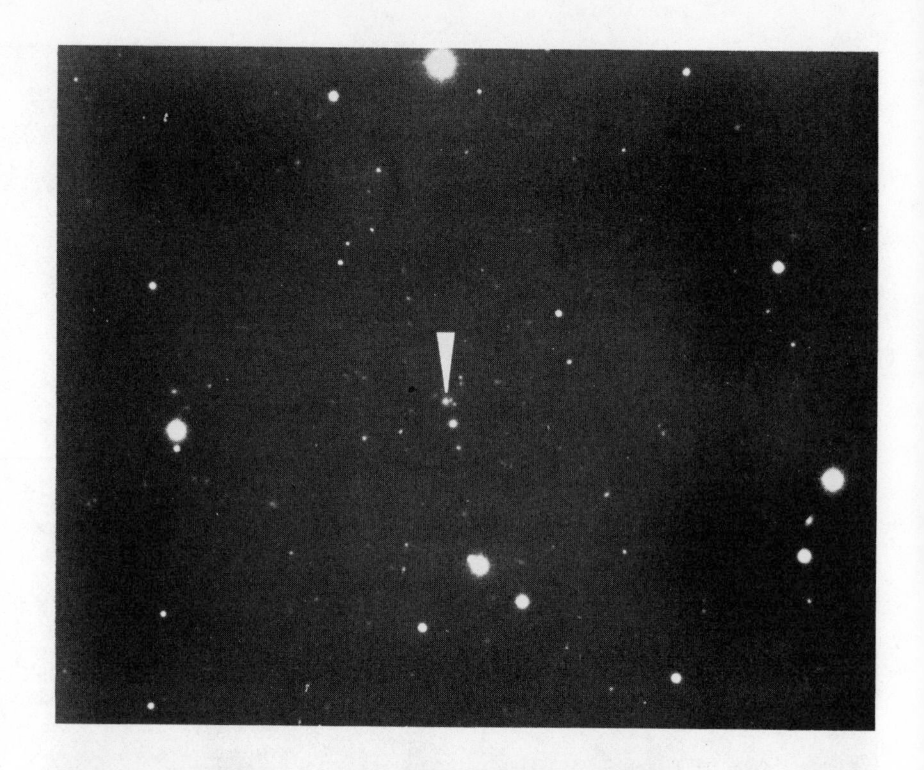

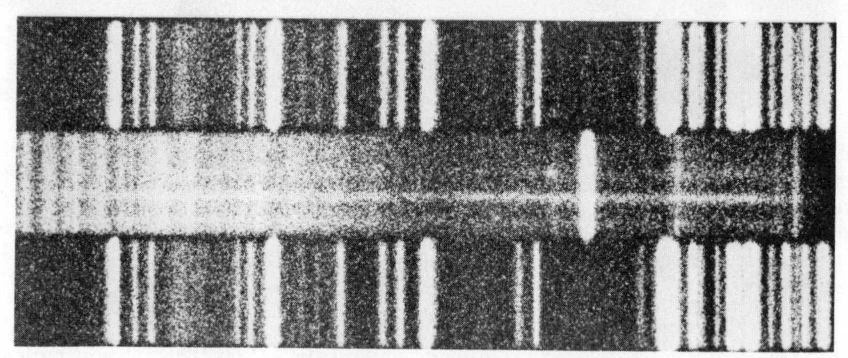

FIG. 30–26 *(b) The most distant object found to date of 1960, 3C295 in Bootes, measured by the 200-inch telescope. A source of radio noise. The spectrum of this object is also shown with the large shift of the lines to the red. This object also appears to be a quasar. (From Mount Wilson and Palomar Observatories)*

This is to be compared to the luminosity of a "radio galaxy," which is about 10^{43} ergs/sec.

The optical spectrum of quasars is quite different from that of any known star or galaxy. Although a few emission lines have been observed and measured (the Doppler shifts are of the order of 0.4 to 0.8), they do not correspond to the lines of any known atoms. They appear to correspond, however, to the strongly shifted Balmer lines of hydrogen. The optical spectrum is rich in synchrotron radiation and shows a considerable ultraviolet excess. The presence of synchrotron radiation indicates the presence of magnetic fields and relativistic (high-velocity) electrons.

To account for the radio-energy release by quasars, L. Gold and J. Moffit have proposed an ionic plasma vibration mechanism which leads to the right order of magnitude for the radio energy emitted by quasars.

One of the peculiarities in the luminosity of

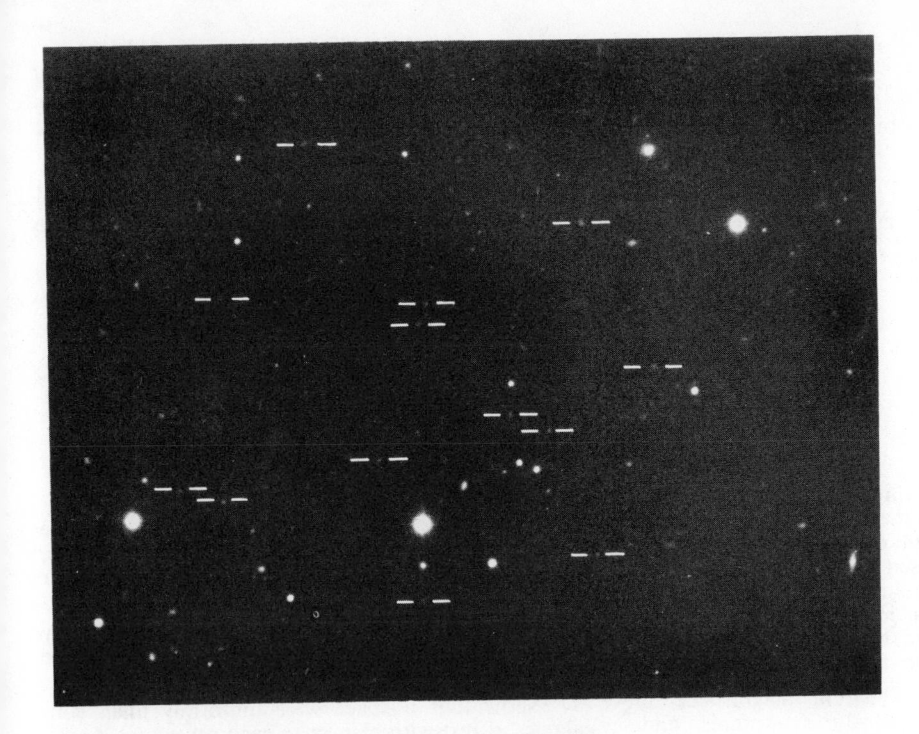

FIG. 30–27 (*a*) *Field of faint nebulae in Coma Berenices. Some of the faintest nebulae are indicated; distances are from 500 to 1 000 million light years (old Hubble constant). (200-inch photograph from Mount Wilson and Palomar Observatories)*

FIG. 30–27 (*b*) *At the center of this photograph is a faint cluster of galaxies in the constellation well over a billion light years away and near the limit of the observable universe. If the Great Andromeda Galaxy were moved out to the distance of this cluster in Pisces, it would look no larger and no brighter than the brightest members of this cluster. (200-inch photograph from Mount Wilson and Palomar Observatories)*

3C273 is its periodic variation. Using old sky patrol plates (on which 3C273 appears as a twelfth to thirteenth magnitude object), H. J. Smith and D. Hoffleit found a large light variation with an amplitude of about half a magnitude and a period of about 15 years. These plates were taken in the period from 1900 to 1960. It is extremely difficult to see how an object with a diameter of a few thousand light years can pulsate (its entire surface responding to mechanical or acoustical pulses) with a period of a few years.

The distances of these objects may be compared with the most distant clusters of galaxies. (See Figure 30.26.)

EXERCISES

1. Assume you are working with a telescope that can detect objects as faint as the twenty-first magnitude. Calculate the distances out to which the following types of objects can be used to obtain distances of galaxies (use the mean absolute magnitudes for each of the types listed): (a) Cepheid variables; (b) irregular variables; (c) blue-white giants; (d) red supergiants; (e) planetary nebulae; (f) novae; (g) supernovae.

2. What would be the apparent magnitudes of (a) the sun, (b) Sirius, (c) Rigel, (d) Antares, and (e) Vega if they were in the spiral nebula in Andromeda?

3. If our Galaxy were to collide with and pass right through the Andromeda Galaxy at a speed (relative to the Andromeda nebula) of 1 000 km/sec, how long would it be before our sun suffered its first collision with a star in Andromeda? (Assume that all the stars in Andromeda are about equal in size to the sun and that the average density of stars—the number of stars per cubic parsec—in Andromeda is about 10 times the density in the neighborhood of the sun.)

4. Consider a stable spherical cluster of 1 000 galaxies, each about the size of our own Galaxy. If the radius of this cluster is 10^6 light years, what is the average velocity of the galaxies in this cluster? (Assume the mass of each galaxy to be about equal to the mass of our Galaxy.)

5. Assume that the galaxies are distributed uniformly in the cluster in problem 4, and calculate (approximately) how many collisions, on the average, any galaxy in the cluster suffers with some other galaxy in 10^9 years.

6. Calculate the speed of escape from the periphery of the cluster in problem 4 and compare it with the average speed of the individual galaxies in the cluster.

7. Show that if space were uniformly filled with galaxies at rest with respect to each other and if each galaxy had the same intrinsic luminosity, then the number N of galaxies brighter than the apparent magnitude m that an observer would count all around him is given by the formula

$$\log_{10} N = 0.6m - 0.6M + 3 + \log_{10}\left(\frac{4\pi\rho}{3}\right),$$

where M is the absolute magnitude of each galaxy and ρ is the number of galaxies per unit volume.

8. Show that if all the galaxies in the uniform distribution of problem 7 were emitting radio waves at the same intrinsic rate, then the number N of such galaxies detected in all directions by a radio telescope down to the apparent radio brightness R is given by

$$N = \text{constant } R^{-3/2}.$$

(Hint: Assume the inverse square law for the apparent radio brightness of a radio source.)

31

COSMOLOGY

In the last two chapters we have been discussing individual galaxies and their clustering into groups of various sizes and structures. We have seen that galaxies and clusters of galaxies are found in all parts of the sky (except for the zone of avoidance because the dust in our Milky Way obscures galaxies in that region) and extend almost uniformly as individuals or in groups and clusters as far as our optical and radio telescopes can penetrate. The problem we must now consider is whether there is some over-all superstructure of this entire system.

We must also consider whether this superstructure (if it exists) is in a steady state (i.e., unchanging) or whether it is evolving in some fashion or other. *This phase of astronomy, which deals with the distribution and behavior of all the matter in the universe, is called* **cosmology**. Only recently, with the aid of the 200-inch telescope and radio telescopes, have we been able to collect enough data to begin to check some of the cosmological theories.

The problem of the cosmologist is to construct a model of the universe, based upon the known physical laws of nature, that is, in agreement with the observed data (i.e., with the observed distribution of galaxies in space and the way this distribution varies with time). Although several models of the universe have been proposed, available data do not permit us to choose the correct one among them. We do know, however, that certain models proposed in the past are incompatible with the observed distribution of matter and with the very faint amount of light coming to us from the night sky.

31.1 Olbers' Paradox *

As an example we consider the famous paradox proposed more than a century ago by Olbers, who was puzzled by the faintness of the starlight. Olbers assumed that the stars (this applies to galaxies and clusters of galaxies as well) are distributed uniformly throughout a Euclidean space (i.e., flat) that extends in all directions. He showed that if this were the case, *every bit of the night sky ought to be as bright as the sun* (the density of radiation at any point in such a universe ought to be infinite).

The reason for this is clear if we picture the earth as being surrounded by larger and larger equally thick concentric shells of stars, as shown in Figure 31.1. Since each star (or galaxy, or cluster of galaxies), on the average, has the same luminosity, every shell sends the same amount of radiation to the central point where the earth is. Although a star in a distant shell sends less light per second to the central point than a star in a nearby shell (in the inverse ratio of the squares of the distances of the shells), there are just enough more stars in the distant shell than in the nearby shell to compensate for the diminution in light intensity of a single star resulting from the distance factor. *Since in an infinite distribution there is an infinite number of shells, every point of the sky would appear as bright as the surface of any single star* (*let us say the sun*). *That this is not so* (*i.e., that the night sky is very faint*) *is the paradox.* Such an infinite distribution is also in disagreement with the known gravitational forces at a point. For if matter were distributed uniformly to infinity, the gravitational force at each point in such a distribution would be infinite, as was first pointed out in 1895 by Seeliger.

On the other hand, it is clear that the distribution of galaxies and clusters of galaxies cannot be uniform and finite in an infinite flat space. If the total number of galaxies were finite, the distribution could not be the uniform one that we now observe. There would either have to be a dense nucleus of galaxies and clusters of galaxies at some point in space, with the numbers of these objects thinning out as we move away from this nucleus, or the entire finite system would have become so dispersed

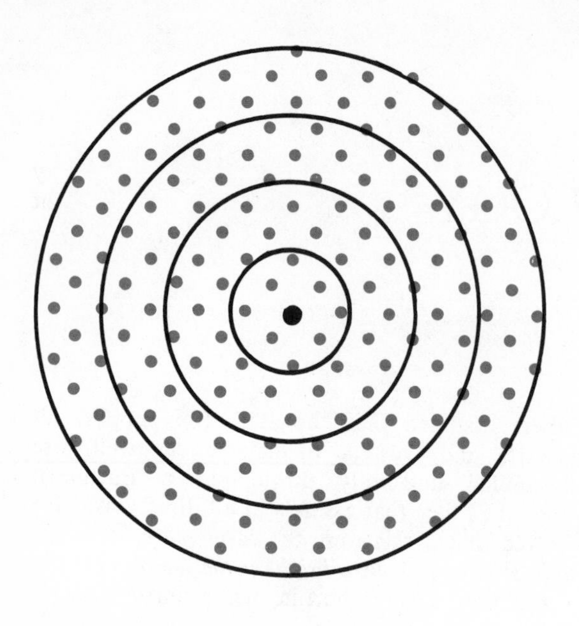

FIG. 31–1 *Schematic diagram showing concentric shells of stars with uniform distribution.*

in this infinite space during the billions of years it has been in existence that we would observe far fewer (in fact, hardly any) galaxies and clusters in our immediate neighborhood than we now do. It is clear, then, that both a static uniform infinite distribution of galaxies in an infinite Euclidean space and a finite distribution in an infinite space are incompatible with the actual observed distribution.

31.2 Charlier's Hierarchy of Galaxies

One attempt to overcome this paradox and still to retain an infinite distribution of galaxies in an infinite space was made early in this century by the Swedish astronomer Charlier, who proposed the idea of a superhierarchy of aggregates of stars and galaxies. Thus, he suggested that not only are stars concentrated in galaxies, and galaxies themselves in clusters, but that the clusters of galaxies form superclusters, and these superclusters in turn are concentrated into supersuperclusters, etc. However, as Zwicky pointed out, there is no evidence for aggregates beyond clusters of galaxies, although de Vaucouleurs has suggested that there are superclusters and that our Galaxy is a member of a vast supercluster with its center near the Virgo cluster.

If Charlier's hierarchy were a correct description of our universe, Olbers' paradox would disappear because we can always construct such a hierarchy to give the amount of radiation observed from the night sky and still have an infinite number of stars or galaxies in an infinite Euclidean space. We can do this by separating the superclusters by sufficiently large distances.

However, it is not necessary to introduce a Charlier hierarchical model of the universe to eliminate Olbers' paradox and the difficulties associated with the gravitational field. As long as cosmological theories were based on Euclidean geometry and the classical Newtonian mechanics, and as long as the universe was pictured as consisting of a static distribution, it was impossible to overcome this paradox. *But we now know that space is not Euclidean because of the presence of matter, and also that the distribution of galaxies is not static* (as is clear from the Doppler shift in the spectra of the distant galaxies). Both of these phenomena have greatly influenced our ideas of the structure of the universe and have led to uniform models in which Olbers' paradox disappears.

31.3 The Speeds of Recession of Distant Galaxies from Doppler Shifts

As we have already noted, the spectra of the galaxies are similar in their general features to those of individual stars. They contain certain well-defined absorption lines, the most prominent of which are the H and K lines of calcium in the violet regions of the spectrum. The most remarkable feature of these spectral lines is their large Doppler shift towards the red, which, except for some of the nearby galaxies, indicates large speeds of recession which increase with distance.

These large Doppler shifts were first detected in 1912 by Slipher, who analyzed the spectral lines in the Andromeda Nebula and showed that this galaxy seems to be approaching us at a speed of 128 km/sec. Slipher extended this analysis to the spectra of other galaxies and discovered that the fainter, and hence more distant, galaxies have still larger Doppler shifts and hence greater radial velocities. This indicates a systematic dependence of the Doppler shift on the distances of these nebulae. Thus, a few years after his work with the Andromeda Nebula Slipher analyzed the spectra of 15 spirals having very large Doppler shifts and showed that the radial velocities of these spirals are 20 to 30 times those of stars in our Galaxy, and in some cases as large as 1 800 km/sec. This was further confirmed in his work in 1924, when he

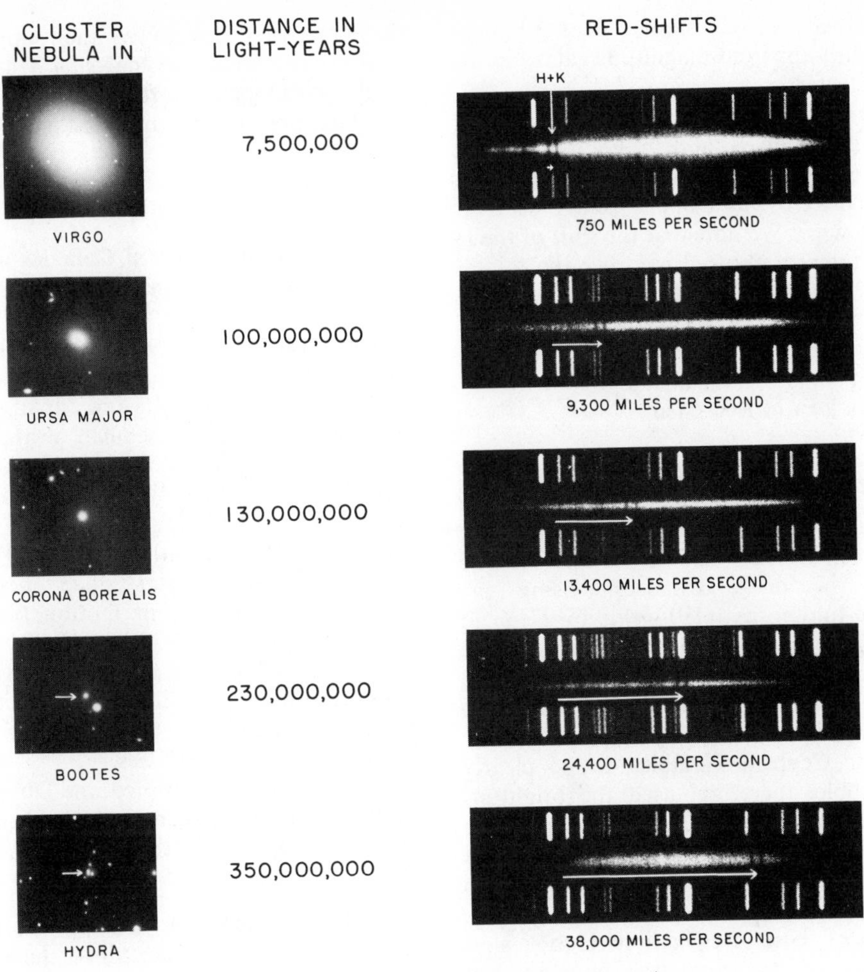

FIG. 31–2 *Relation between red shift (velocity) and distance for extragalactic nebulae. The distances given in this photograph are based on the old Hubble constant. They should be multiplied by approximately 5 to agree with the presently accepted value of the Hubble constant. (From Mount Wilson and Palomar Observatories)*

investigated the spectra of 43 galaxies and found additional radial velocities as high as 1 800 km/sec.

There are two remarkable things about Slipher's discovery:

1. in all but two of the galaxies the Doppler shifts indicate speeds of recession, and
2. the speed of recession increases with distance. (See Figure 31.2.)

The unexpected and unexplained size of the radial velocities of the distant galaxies, and the evident departure of these radial velocities from random motion, led Humason in 1929 to undertake a systematic investigation with the 100-inch telescope of the spectra of very distant nebulae (down to the eighteenth apparent magnitude). To photograph these spectra he used exposure times ranging from 50 to 100 hours, obtaining radial velocities for these distant galaxies which completely confirmed Slipher's results. In 1935 he obtained a velocity of recession of 42 000 km/sec for an eighteenth-apparent-magnitude galaxy in the Ursa Major cluster.

In Table 30.3 (p. 595) the radial velocities of some distant galaxies are given, and in Figure 31.2 the relationships of the apparent sizes and brightnesses of these galaxies to the Doppler shifts of their spectra are shown. We note that the shift of the H and K lines towards the red is larger the fainter the nebula is (the smaller its image is) on the photographic plate.

31.4 Hubble's Law of Recession

Since the relationship between distances of the nebulae and their speeds of recession can be established only if the distances of the nebulae are known, Hubble, in 1929, undertook a systematic determination of these distances. Using the Shapley period-luminosity relationship for Cepheid variables he determined the distances of some of the nearby nebulae for which Slipher had measured Doppler shifts. However, most of the nebulae whose spectra Slipher had analyzed were too far away for their Cepheid variables to be clearly detected. Hubble, therefore, used the brightest-star criterion (see Section 30.22) to determine the distances of these nebulae, and discovered that *their speeds of recession are strictly proportional to their distances.*

On the basis of these investigations Hubble suggested the idea of the expanding universe and formulated his law, now known as the **Hubble law of recession**. Humason was prompted by Hubble's hypothesis of the expanding universe to start his systematic study of the spectra of the very distant nebulae, and he did this work in conjunction with Hubble's determination of the distances. This work confirmed Hubble's law out to the most distant galaxies that could be measured in 1935. As we have already noted, these galaxies are receding at speeds in excess of 40 000 km/sec, a value that Hubble had already predicted on the basis of his law.

At these distances even the brightest stars in the galaxies cannot be observed separately and the distances must be determined from the appearance of the entire galaxy itself. One assumes that all the galaxies on the average have the same absolute magnitude and that therefore their distances vary inversely as the square roots of their apparent brightnesses; or, if we introduce the apparent magnitude of the galaxies, that the logs of their distances vary directly as their apparent magnitudes. Therefore, in analysing Hubble's law we relate the Doppler shifts (i.e., the speeds of recession) directly to the apparent magnitudes of the galaxies.

31.5 Radial Velocities of Galaxies within Clusters from Observational Data

The most extensive work in this field was done recently by Humason, Sandage, and Mayall, who investigated the red shifts and apparent magnitudes for 620 extragalactic nebulae at the Mt. Wilson and Palomar Observatories. This survey includes data for 26 clusters of nebulae, and we shall see that in this study Hubble's law applies to an entire cluster of galaxies rather than to the individual galaxies in the cluster since these individual galaxies have a random motion (i.e., a velocity dispersion) within the cluster itself. The analysis also includes the work done on 300 galaxies at the Lick Observatory. All the data on the red shifts were corrected for the solar motion with respect to the local cluster of galaxies.

As we have already noted, the Doppler shifts for very close galaxies such as Andromeda may show speeds of approach rather than recession. This is due to two things:

1. the recession does not apply to the members of the local cluster since the entire cluster moves through space as a unit, and each galaxy has a random motion within the cluster;
2. the sun itself is moving within our own Galaxy (because of the rotation of the Galaxy) at a speed of 220 km/sec so that we must subtract a component of this speed from the measured radial velocities of galaxies to obtain their true velocities of recession relative to our Galaxy.

The radial velocities obtained by these investigators (after correction for solar motion) range all the way from a few kilometers per second for nearby galaxies to a velocity of recession of over 61 000 km/sec for a member of the Hydra cluster. (See Figure 31.3.) To show the dependence of the velocities of recession of clusters on their distances and to illustrate the dispersion of velocities of the individual members of a cluster we give a few examples (all data corrected for solar motion). In the Virgo cluster the radial velocities range all the way from a velocity of approach of 452 km/sec to a velocity of recession of 2 492 km/

Fig. 31–3 Cluster of nebulae in Hydra. Distance about 360 million light years, according to the old Hubble constant. With the new Hubble constant this becomes about 1.5 billion light years. (200-inch photograph from Mount Wilson and Palomar Observatories)

sec; in the Perseus cluster the mean velocity of recession for the five galaxies studied is 5 600 km/sec; in the Coma cluster the radial velocities range from 4 655 km/sec to 8 422 km/sec; in the Hercules cluster velocities of recession range from 9 593 km/sec to 10 650 km/sec; and in the most distant cluster measured, the Hydra cluster, the velocities of recession range from 60 754 to 61 046 km/sec. The data for the various clusters are given in Table 30.3 (p. 595).

These data from Humason, Sandage, and Mayall, which indicate quite clearly how the speeds of recession increase with the distance, can be used to determine whether Hubble's law—that the velocity of recession increases linearly with the distance—holds all the way out to the most distant galaxies, or whether it breaks down if we go out far enough. If Hubble's law disagrees with the observational evidence, it is important to determine the nature of this discrepancy because this will tell us which of the various theoretical models of the universe is correct.

31.6 The Expressions of Hubble's Law in Terms of Observable Constants

Before we consider this point in more detail, we shall obtain a formula for Hubble's law which is best suited for comparison with observation. If the Doppler shift in a spectral line of wavelength λ is $\Delta\lambda$, then we know from our discussion in Chapter 25 that the velocity of recession is given by $v = c\Delta\lambda/\lambda$. Since Hubble's law states that the velocity of recession is proportional to the distance, we may write this law in the form

$$Hr = c\frac{\Delta\lambda}{\lambda} \tag{31.1}$$

where H is called Hubble's constant, and is expressed as a velocity per distance; in cosmological work it is generally given in km/sec/Mpsc (megaparsec = 1 million parsecs). If we let z represent $\Delta\lambda/\lambda$, we may express this law as

$$Hr = cz. \tag{31.2}$$

31.7 The Relativistic Doppler Shift and Hubble's Law

Before we can discuss this law and its application to the galaxies in more detail, we must say a few words about the applicability of the Doppler shift to this problem in the form we have introduced above. Since it follows from the theory of relativity that the speed of light is an absolute maximum for all bodies, it seems from the expression $v = c(\Delta\lambda/\lambda)$ that $\Delta\lambda$, the shift in the wavelength of a spectral line, can never exceed λ, for v would then be greater than c. However, this is not so, and only appears that way because the form we have used for the Doppler shift is not correct if relativity is properly taken into account. As long as the source of light we are dealing with is not moving very fast with respect to us, the simple form is correct. But when we are dealing with bodies moving with very high speeds we must use the so-called **relativistic Doppler effect**, which is given by the formula

$$\nu = \frac{\nu_0\left(1 - \frac{v^2}{c^2}\right)^{1/2}}{1 \pm \frac{v}{c}}$$

where ν_0 is the frequency of the radiation emitted by the source when the source is not moving with respect to the observer, and ν is the frequency of the radiation received by the observer. *If the source is moving with a speed v, the + sign applies when the source is moving away from the observer, and the − sign applies when the source is moving towards the observer.*

If we introduce the wavelength instead of the frequency we have

$$\lambda = \frac{\lambda_0\left(1 \pm \frac{v}{c}\right)}{\left(1 - \frac{v^2}{c^2}\right)^{1/2}}.$$

From this formula it follows that for a source that is receding from us (the + sign is used) with a speed v we must have

$$\frac{\Delta\lambda}{\lambda} = \frac{1 + \frac{v}{c}}{\left(1 - \frac{v^2}{c^2}\right)^{1/2}} - 1.$$

It is clear from this equation that $\Delta\lambda$ can take on all values up to ∞ as the speed of the receding object approaches the speed of light. Hence it is possible according to this formula for galaxies to be receding at such speeds that most of the radiation is shifted into the infrared and radio wavelength regions of the spectrum. Although this is the appropriate formula to use in discussing Hubble's law when the distances and hence the speeds of recession are very large, we shall retain the simple formula (31.1) in our present analysis for pedagogical reasons.

31.8 The *K* Term

If we plot cz against r for various galaxies, then according to Equation (31.2) we should get a straight line if Hubble's law is correct, and the slope of this line (the tangent of the angle which the line makes with the r axis) gives us Hubble's constant. Since r is not measured directly, it is more convenient to express this law in terms of the quantity that is actually measured (namely the apparent magnitude of the brightest star in the galaxy or the apparent magnitude of the galaxy itself).

From the relationship between the apparent magnitude, the absolute magnitude, and the distance d of an object we have $\log d = (\frac{1}{5})(m - M) + 1$. However, this formula cannot be applied directly to galaxies because of the obscuration and the red shift. Both of these effects cut down the amount of radiant energy we receive per unit time from the galaxies and therefore make these galaxies appear fainter than they would appear otherwise. *When the light leaves the galaxy it is not as red as when it reaches us, so that the red shift makes us overestimate the apparent magnitude.* For this reason, we must introduce a correction K for the apparent magnitude of galaxies, and we must therefore write

$$\log r = 0.2(m - K - M) + 1, \tag{31.3}$$

where r is the distance of the galaxy in parsecs. If we now introduce into this formula Hubble's law in the form

$$\log r = \log cz - \log H,$$

we have

$$\log cz = 0.2(m - K - M) + 1 + \log H,$$

or

$$\log cz = (0.2)m_{corr} + (\log H + 1 - 0.2M), \tag{31.4}$$

where m_{corr} is the corrected photographic magnitude.

31.9 Determining the Value of the Old Hubble Constant

Thus, if we plot the logs of the observed Doppler shifts (the quantities cz) against the observed corrected photographic apparent magnitudes, we should get a straight line if Hubble's law is correct. In applying this formula, however, we must be sure that we work only with objects in all the galaxies having the same absolute magnitude M. The value of the absolute magnitude that we must introduce in our formula depends on whether we work with Cepheid variables, the brightest star, supernovae, or the entire galaxy itself.

Since, as we have indicated in previous chapters, there is a good deal of uncertainty about these absolute magnitudes, and since there is uncertainty about the K correction term, the value of Hubble's constant is uncertain and depends on which criteria are used for K and M. Hubble originally used the value $M = -6.35$ for the brightest stars and obtained for H the value of 550 (km/sec) per Mpsc. This is referred to as the **old Hubble constant**.

Hubble also used the apparent magnitudes of field galaxies to obtain a velocity-distance relationship. In this case he worked directly with the mean velocities and mean distances of large numbers of galaxies. He used 29 field galaxies that could be resolved into stars and simply divided their mean velocity by their mean distance to obtain a value of 530 km/sec/Mpsc for H.

Finally, he worked with clusters of galaxies and obtained a velocity-distance relationship for clusters. His results are given in the graph in Figure 31.4. In working with the clusters of galaxies he chose in each cluster the fifth brightest galaxy, to which he assigned an absolute magnitude of -16.4. In the case of the field galaxies he had assigned the absolute magnitude -15.1. The results he obtained for cluster galaxies are not the same as those he found for field galaxies.

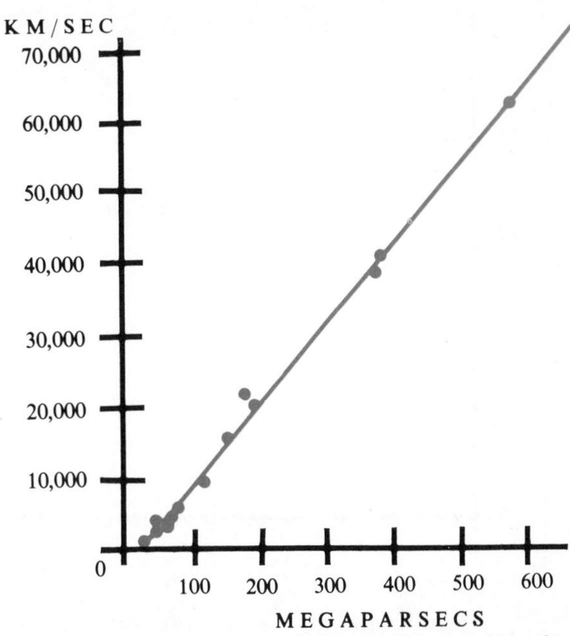

FIG. 31-4 *A graph showing the velocity-distance relationship for clusters of galaxies, based on the results of the work of Hubble.*

31.10 The Revised Hubble Constant *

When Hubble first did his work, the determination of the distances of galaxies was based on the distances of nearby galaxies, such as Andromeda, obtained from Shapley's period-luminosity law. Since, as we now know, this law is incorrect and has to be replaced by the revised period-luminosity law which takes into account the two Cepheid populations, the Hubble constant must also be changed. Since the correct period-luminosity law just about doubles the distance of the Andromeda Nebula, the general scale of distances of galaxies based on the analysis of the Andromeda Nebula is doubled, so that H itself is reduced, at least by a factor of 2, if no other changes are introduced.

However, improved absolute magnitude criteria for the brightest stars and also for field and cluster galaxies have led to an even smaller value of H. Thus, Humason, Sandage, and Mayall, from their analysis of the red shifts of about 800 galaxies, obtained a value of 180 km/sec/Mpsc. In certain galaxies, such as NGC 4321, they used the absolute magnitudes of resolved stars whose luminosities are known from similar stars in Andromeda. They also assumed that the brightest field and cluster galaxies are giants with luminosities about equal to that of Andromeda.

From their data they were able to show that

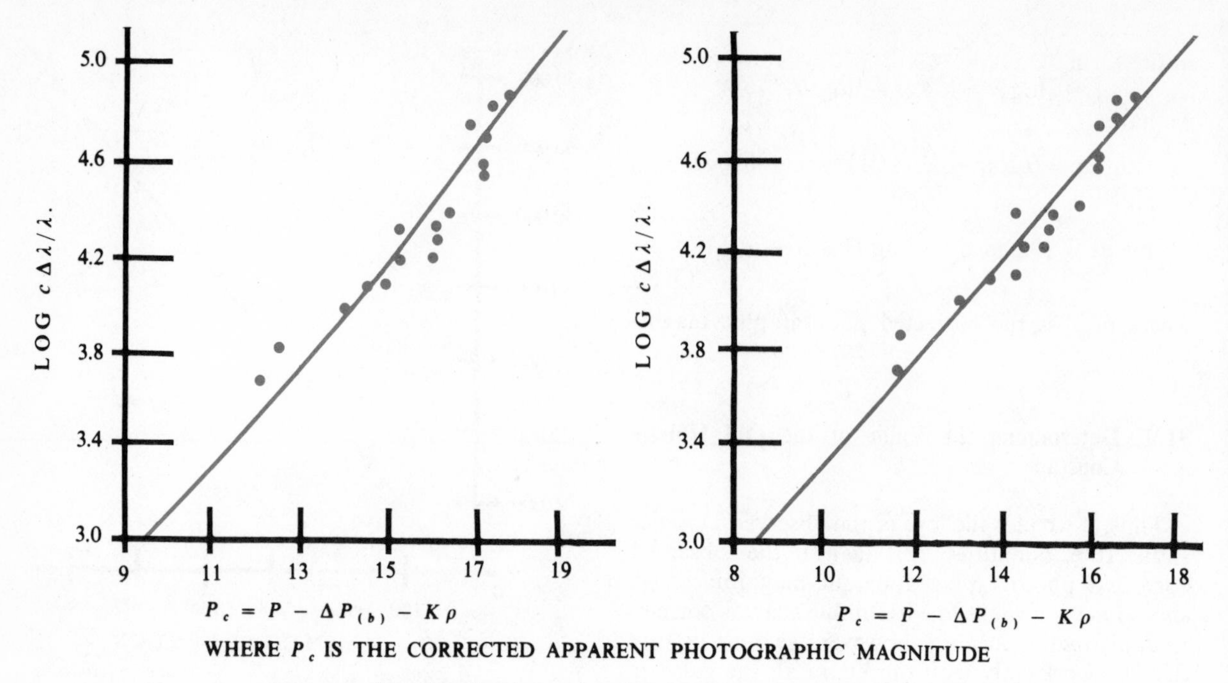

FIG. 31–5 *(a) The red-shift-ph magnitude relation for clusters of nebulae. The apparent photographic magnitudes have been corrected only for the latitude effect and for the selective effect of the red shift. The "energy" and "number" corrections are not included in the data but are introduced into the theoretical equations used for the interpretation. (b) Same as (a) for photovisual magnitudes. (From Humason, Mayall, and Sandage,* Astronomical Journal, *LXI (1956), 149)*

Hubble's law appears to break down for the very distant galaxies and clusters of nebulae. In other words, the velocity of recession appears to be proportional to distance only for small values of $\Delta\lambda/\lambda$. However, if $\Delta\lambda/\lambda$ exceeds 0.2—that is, if the velocity of recession is greater than 0.2 times the speed of light—the relationship between distance and velocity is not linear but is such as to indicate that there is a deceleration of the galaxies as we go out to greater and greater distances. Some of the data are given in the graphs in Figure 31.5.

The determination of H is extremely important in cosmology because it is one of the key numbers we must know to determine which particular theoretical model of the universe is correct. In addition to H we must know whether or not the galaxies at great distances are being accelerated, and if so, at what rate. In other words, we have to know whether the velocity-distance relationship departs from linearity, and if so, to what extent. For this reason a great deal of work has been done recently in determining H as accurately as possible and determining the distance-velocity relationship for very distant galaxies.

In 1958 Sandage proposed 75 km/sec/Mpsc for the Hubble constant on the basis of a revised analysis of the absolute magnitude of the brightest stars in resolved nebulae. Arguing that Hubble probably confused HII regions in resolved galaxies with the brightest stars, and taking into account the correct period-luminosity law, Sandage proposed a total correction of 4.1 magnitudes (2.3 for the P-L law correction, and 1.8 for the bright-star correction) to be added to the original Hubble distance modulus to offset these two errors. More recently astronomers have suggested that H be taken equal to 100 km/sec/Mpsc, so that at the present time there is a 25 per cent uncertainty in the value of Hubble's constant.

31.11 General Relativity Theory and Cosmology

Hoyle and Sandage have also investigated the question of deceleration. To understand how Hubble's constant and the deceleration enter into the cosmological problem, we must give a brief discussion of the theoretical basis of modern cosmological theories. All of these stem from the general theory of relativity, and in particular Einstein's field equations (see Section 11.21), which we take as our starting point. We saw there that if R_{ik} represents the Einstein-Ricci tensor (which is a measure of how the geometry of space-time at any point departs from Euclidean geometry) and g_{ik} is the metric tensor (i.e., the tensor that determines the distance between two neighboring points in space-time), then these field equations are

$$R_{ik} - \frac{1}{2} Rg_{ik} = -\left(\frac{8\pi}{c^4}\right) GT_{ik},$$

where c is the speed of light, G the universal constant of gravitation, R the Riemannian curvature at a given point, and T_{ik} the tensor that measures the density of energy-momentum and matter at that point. As we have seen, this is a set of ten equations, which to a first approximation (i.e., for sufficiently small masses and densities or at very great distances from any masses) leads to Newton's law of gravity. This set of equations also contains in it the law of conservation of energy-momentum and matter.

31.12 The Cosmological Principle or Postulate

Einstein was the first to see the importance of this set of equations for cosmology. He pointed out that one should be able to find the nature of the geometry of the whole universe if one could solve this set of equations for the correct distribution of energy and matter in the universe. In other words, if we knew the correct value for T_{ik} for the entire universe, we could then solve this set of equations for the g_{ik} everywhere and thus find the structure of our space-time.

Since, however, we do not know the exact distribution of energy and matter in the universe, we must proceed by making some reasonable assumption about it as Einstein did. *His assumption, which is now the basis for most cosmological theories, is called the* ***cosmological principle or postulate****; it states that, aside from random fluctuations that may occur locally, the universe must appear the same for all observers.* In other words, regardless of whether we look at the universe from our own Galaxy or from a galaxy in the Virgo cluster or a cluster a billion light years away from us, the over-all picture must be the same.

31.13 The Boundary Values of the g_{ik}'s for an Infinite Universe

Einstein applied this principle and his field equations to the universe by picturing all stellar and interstellar matter as being smeared out into a uniformly dense fog of material to fill the universe everywhere. Under these conditions, if one could know the density of this fog, one could then choose T_{ik} as constant and solve the field equations. However, since the field equations are differential equations (see Section 22.12) their solutions do not determine the g_{ik} completely, because, as we know, only the differences between values of the g_{ik}'s at two points can be found from differential equations. This means that the g_{ik}'s can be fully determined from the solutions of equations only if their values are known at the boundaries of the universe. Hence, if we assume that the universe is infinite, we must know the values of g_{ik}'s at infinity (i.e., the so-called **boundary values**) before we can determine them fully everywhere from the solutions of Einstein's field equations.

When Einstein attempted to carry out this program, he found he could not choose boundary values for the g_{ik}'s at infinity that were compatible with the solutions of his field equations as long as he assumed a uniform distribution of matter (i.e., a constant density for the smeared-out matter). This unsatisfactory situation prompted Einstein to discard the idea of an infinite universe and to replace it by a finite one. He reasoned that the presence of matter in the universe would introduce a constant curvature of space-time at each point and that this curvature, however small it might be, would cause the universe to curl around and ultimately close up like a sphere. This would give rise to a three-dimensional Riemannian (i.e., elliptical geometry) space of constant curvature.

31.14 The Cosmological Term Λ Introduced into the Field Equations

Now it is easy to show that the equation for such a universe cannot be obtained from Einstein's field equations as they stand, since the solutions of these equations give an infinite universe with inhomogeneities. Einstein overcame these difficulties by adding a small term to the left-hand side of his field equations. Of course one may do this only if this term does not destroy the character of the equations, and *since these equations involve tensors, the only term that can be added, if the tensor character is to be retained, must itself be a tensor. Moreover, since the field equations were set up to agree with the principle of conservation of mass and energy-momentum, any term that is added must not violate this principle.*

Einstein used one other condition to guide him in the choice of this term, namely that the universe must be static—that is, the universe must not vary with time. He imposed this last condition because when he did his work (1917) there was no observational evidence that the universe is expanding or that there are very large motions of stellar bodies. All observations then indicated that the stars are moving about rather slowly, so that Einstein felt justified in assuming that the universe is static and that the density of matter at each point and the distances between bodies remain constant.

Einstein showed that these three conditions are not violated if one adds the term Λg_{ik} to the left-hand side of the field equations. In order to keep the field equations about the same as they were without this term, Einstein chose Λ to be a very, very small constant. This small quantity, which has the dimensions of the reciprocal of a length squared (length^{-2}) is called the **cosmological** or **cosmical constant**. With this term present the field equations become

$$R_{ik} - \frac{1}{2} R g_{ik} + \Lambda g_{ik} = \frac{-8\pi}{c^4} G T_{ik}. \quad (31.5)$$

Because Λ is extremely small, the presence of this term has practically no effect on the gravitational forces between two bodies, so that Einstein's law of gravitation is essentially the same as it was before. However, this cosmological constant becomes more and more important at greater and greater distances.

31.15 The Einstein Model of the Universe

For positive values of Λ, and for a nonempty universe ($T_{ik} \neq 0$), Einstein obtained a solution of these equations that gives a uniform static model of the universe in which the density of matter is the same at all points and that has a constant positive Riemannian curvature. *The random velocities of matter in this universe cancel out to zero so that there is no net motion. The positive curvature means that this universe closes back upon itself and is therefore finite but boundless.* The Riemannian curvature of the Einstein universe at each point is just equal to Λ, and the radius of the universe is just equal to $\sqrt{1/\Lambda}$. One can show from the field equations that Λ is just equal to $4(\pi G/c^2)\rho$ (where ρ is the constant density of matter at each point of space). Einstein thought that this model represents the only possible solution of his field equations and that space itself (and all its properties) can exist only if matter is present. For this reason he felt that this static model of the universe incorporates the Mach principle of relativity which we discussed in Section 11.32, since it appeared that in this model of the universe space and matter are so inter-related that one cannot exist without the other.

31.16 The de Sitter Model of the Universe

However, another solution of Einstein's field equations was found by de Sitter for an empty universe, which shows that space can exist without matter. The de Sitter model of the universe has some very interesting properties which can be obtained directly from the field equations. If space is empty, so that $T_{ik} = 0$, the field equations reduce to

$$R_{ik} = \Lambda g_{ik},$$

(the term $-\frac{1}{2}Rg_{ik}$ combines with R_{ik} to give this result) or

$$R_{ik} - \Lambda g_{ik} = 0.$$

If we compare this with Einstein's original field equations for empty space,

$$R_{ik} = 0,$$

we see that the cosmological constant must be equivalent to a repulsion since it appears with a negative sign and hence tends to reduce the effect of R_{ik}, which we know leads to gravitational attraction. At small distances in this equation R_{ik} is the more important term, giving rise to an attraction, but at large distances Λg_{ik} outweighs R_{ik} and expansion occurs.

This means that if two test particles of negligible mass were placed in the de Sitter universe, they would recede from each other at greater and greater speeds as they separated to greater and greater distances because of the cosmological repulsion term Λ. We thus obtain an expanding universe from the de Sitter model. *We see then that Einstein's static field equations give either a universe completely filled with matter and without motion at one extreme of models, or a universe with motion but without matter at the other extreme.* By completely filled, we mean that the density of matter at each point (and hence the total amount of matter present) is just enough to give the calculated curvature—the size of this universe would have to change if matter were added or subtracted.

It is of historical interest to note that de Sitter's model was obtained before knowledge of the recession of the galaxies was available and therefore in a sense predicted this recession. That de Sitter's model requires empty space is not a serious objection to its practical application to our own universe because, as we know, the amount of matter in our universe is so small as compared with the amount of space that we may treat our universe as essentially empty.

31.17 Radius of Curvature in a Spherical Universe

Before we go on to a consideration of the time-dependent models that have replaced the Einstein-de Sitter models of the universe, we must understand that the radius of curvature of a spherical universe is not the distance from the earth to some distant point, but rather the radius of a three-dimensional hypersphere in a four-dimensional manifold. Just as the surface of our earth is a two-dimensional spherical surface in a three-dimensional space, so the Einstein universe is a three-dimensional spherical surface in four-dimensional space (one dimension of which is time). The radius then is the distance of the three-dimensional hypersurface from a point in four-dimensional space as shown in Figure 31.6, where the three-dimensional space is pictured as a thin shell.

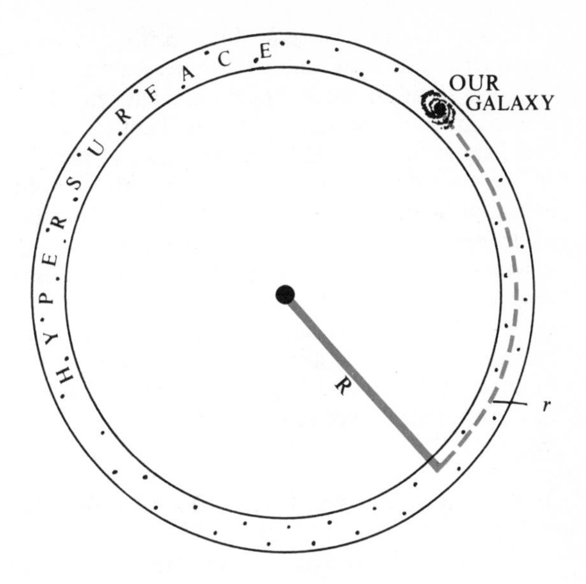

FIG. 31–6 *A schematic two-dimensional sketch of a three-dimensional hypersurface in a four-dimensional space. The distance of any point in space from our Galaxy is measured within the hypersurface itself and is given as r. The radius of the hypersurface in four-dimensional space is R.*

31.18 Time-Dependent Models of the Universe *

When Einstein did his work the validity of the concept of a static universe was taken for granted, but when de Sitter showed that the Einstein universe is essentially unstable and begins to expand when enough matter is taken out of it, mathematicians began to investigate nonstatic models of the universe. They did this by starting out with field equations in which the distances and densities are time-dependent quantities.

Work of this sort was first done by Friedmann and later carried on by Le Maitre, Robertson, and Eddington. The most general solution for all such model universes that are governed by the cosmological principle (i.e., conditions are everywhere the same) was given by H. P. Robertson and later, independently, by A. G. Walker. We shall introduce the observed data obtained from the receding galaxies into Robertson's general solution to see whether we can determine which of the nonstatic models of the universe best agrees with the observations. To understand how this is done, we again consider a Euclidean four-dimensional space with an observer (the earth) at the origin. We know from the discussion in Section 10.16 that the space-time interval (or the line element, as it is called) from the origin to a neighboring point whose coordinates in three-dimensional space are Δx, Δy, Δz is just

$$\begin{aligned}\Delta s^2 &= c^2\Delta t^2 - (\Delta x^2 + \Delta y^2 + \Delta z^2)\\ &= c^2\Delta t^2 - \Delta r^2,\end{aligned}$$

where Δr is the geometric distance from the origin to the point.

This line element holds only if no matter is present in space. But if matter is present, this line element does not apply, and we must replace it by a different one which Robertson has shown is the same for all four-dimensional spaces obeying the cosmological principle and has the form

$$\Delta s^2 = c^2\Delta t^2 - R^2(t)\Delta r^2.$$

Here $R(t)$ represents a scale factor which changes with time (as indicated by the t in parentheses); this gives a nonstatic and hence expanding or contracting universe. Because $R(t)$ determines the scale of a spherical universe at any moment t, it is generally referred to as the **radius of a closed spherical universe**.

To see the significance of this quantity, we again consider Figure 31.6, where we represent the spatial universe as a thin, three-dimensional shell in a four-dimensional continuum. In this picture $R(t)$ is the distance of the shell from the center of the shell in the four-dimensional continuum and Δr is the actual distance between two points within the shell itself, and therefore what we actually measure as distance. We observe that as $R(t)$ gets bigger, which represents an expansion of our closed shell, the spacing between any two points in the shell increases so that the distance from the earth to all objects gets larger and larger in agreement with the observed law of recession.

31.19 Time-Dependent Uniform Model Universes

We have illustrated various properties of universes by considering a closed, static, spherical universe with constant positive curvature, but whether this is the correct model of our universe can only be determined observationally. However, before we discuss the observational evidence we shall consider the other theoretical models that are consistent with the Robertson line element. Essentially, one has to find all the possible forms of $R(t)$ that are consistent with the cosmological principle by solving Einstein's field equations without the cosmological term (i.e., placing $\Lambda = 0$). We do not need the Λ term since we permit $R(t)$ to vary with time.

We begin by substituting the g_{ik}'s from the Robertson line element into the Einstein field equations (to see how the g_{ik}'s enter the field equation see Sections 11.20 and 11.21) and thus obtain a set of two differential equations for $R(t)$. If one solves these equations one obtains a time-dependent model of the universe which must then be checked against the observational data.

31.20 The Expansion Parameters of Time-Dependent Universes

To see how this is done we shall write down the two differential equations in a rather simple form by *introducing into them the velocity of expansion of the universe and the rate at which this expansion changes in time* (*i.e., the deceleration*).

Let us represent the rate at which R changes with time (i.e., $\Delta R/\Delta t$) by $\dot{R}$, which is essentially the speed of expansion, and let us represent the change of $\dot{R}$ with respect to time (i.e., $\Delta\dot{R}/\Delta t$) by $\ddot{R}$, which is essentially the acceleration or deceleration of the expansion at any point. We then find that R, $\dot{R}$, and $\ddot{R}$ are governed by the following two equations:

$$\frac{\dot{R}^2}{R^2} + 2\frac{\ddot{R}}{R} = -\frac{kc^2}{R^2}, \tag{31.6}$$

$$\frac{\dot{R}^2}{R^2} - \frac{8\pi G\rho}{3} = -\frac{kc^2}{R^2}, \tag{31.7}$$

where G is the gravitational constant, ρ is the density of matter at a point, c is the speed of light, and k is a quantity that according to the theory can take on only the values $+1$, 0, or -1.

We shall see that the density of matter in the universe and Hubble's constant together determine which of these three values k assumes. From the definition of Hubble's constant we have

$$\frac{\dot{R}}{R} = H. \tag{31.8}$$

If we now introduce the deceleration parameter q (which is a measure of how rapidly the expansion of the universe is slowing down) we have the following relationship (which is really the definition of q):

$$\frac{\ddot{R}}{R} = -qH^2. \tag{31.9}$$

The quantities q and H are the observational data that must be introduced into our equations before we can obtain a model of the universe.

31.21 The Deceleration Parameter and the Observational Data *

We can relate this deceleration parameter to the observational data obtained from the observed motions of galaxies and clusters of galaxies by going back to the magnitude-velocity relationship set down in Equation (31.4), which we shall write as

$$m_{corr} = 5 \log cz - 5(\log H + 1 - 0.2M). \tag{31.10}$$

This relationship holds only at a given instant of time if we picture the universe as being static at that moment. *However, if the universe is actually changing with time, because it is either expanding or contracting, then we must keep in mind that we see the universe at any moment not in its present state but the way it was at some past epoch. The farther away the object is that we are looking at, the farther in the past we see it.* We must also take into account that such things as absolute magnitude, distances, etc., are different from the way they now appear to us. If we incorporate these ideas into the above magnitude-velocity relationship we obtain according to Robertson

$$m_{corr_{bol}} = 5 \log cz + 1.086(1 - q - 2u)z + \text{constant}, \quad (31.11)$$

where z is the red shift $\Delta\lambda/\lambda$, and u is related to the time rate of change of the absolute magnitude of a galaxy. Since we can obtain the value of the deceleration parameter q from the observational data by using this equation, and we can obtain the value of H from the observational data by using Equation (31.10), we can introduce these values of H and q into the two differential equations (31.6) and (31.7) and decide which of the three values of k (i.e., which model) fits the actual universe.

We may point out that these two differential equations are related to certain properties of stellar systems which we have already discussed. Thus, if we consider the universe as a system of bodies in which we can neglect the total energy of the random motions of galaxies compared to the total kinetic energy associated with the velocities of recession, we can show that the first of these two equations is related to the virial theorem and the second to the theorem of conservation of energy.

31.22 The Dependence of the Curvature of the Universe on the Density and Hubble's Constant *

Depending upon the value of k we obtain three different uniform models of the universe:

Case 1: $k = +1$; this gives a closed, finite, elliptical universe of constant positive curvature and hence corresponds to the usual two-dimensional spherical surface.

Case 2: $k = 0$; this gives a flat, infinite, three-dimensional Euclidean space.

Case 3: $k = -1$; this gives an infinite, open, hyperbolic universe of negative curvature.

We may incorporate these three cases into a single statement by noting that the Riemannian curvature of any model universe is given by k/R^2, so that depending upon whether $k = +1$, 0, or -1 the curvature is positive, zero, or negative.

To see the significance of the different values of k we replace $(\dot{R}/R)^2$ by H^2 in Equation (31.7) and thus obtain

$$\frac{k}{R^2} = \frac{8\pi G\rho}{3c^2} - \frac{H^2}{c^2}. \quad (31.12)$$

We see from this equation that the curvature, k/R^2, of the universe depends on how dense the matter in the universe is as compared to the value of Hubble's constant. Since $8\pi G/3c^2 = 6 \times 10^{-28}$, we have

$$\frac{k}{R^2} = 6 \times 10^{-28}\rho - \frac{H^2}{c^2}. \quad (31.13)$$

Thus, whether the curvature of our universe is positive, negative, or zero depends on whether the square of Hubble's constant divided by the square of the speed of light is less than, equal to, or greater than 6×10^{-28} times the density of matter in the universe.

31.23 The Deceleration of the Expansion of the Universe

If we now subtract the first differential equation (31.6) from the second (31.7), noting that the first terms on the left-hand sides and the terms on the right-hand sides cancel out, we obtain

$$\frac{\ddot{R}}{R} + \frac{4\pi G\rho}{3} = 0, \quad (31.14)$$

or

$$-qH^2 + \frac{4\pi G\rho}{3} = 0, \quad (31.15)$$

from Equation (31.9).

We note immediately from Equation (31.14) by transposing the second term to the right-hand side that *$\ddot{R}$ is always negative (since R, of course, is a positive number) as long as there is matter in the universe. This means that the expansion is slowing down (decelerating) if we take the cosmological constant as* 0. *The reason is that the gravitational force of all the matter in the universe is constantly at work slowing down the receding nebulae.* We may state this differently by saying that the decelerating parameter q is always positive.

31.24 Observational Data and Model Universes

To determine from the observational evidence which of the three geometries applies to our universe, we must use the relationship between m and q. We can relate the observed parameter q to R (which determines the geometry of the universe) by replacing $(8\pi G\rho)/3$ in Equation (31.7) by $2qH^2$, which follows from Equation (31.15). We then obtain

$$H^2(2q - 1) = \frac{kc^2}{R^2}. \quad (31.16)$$

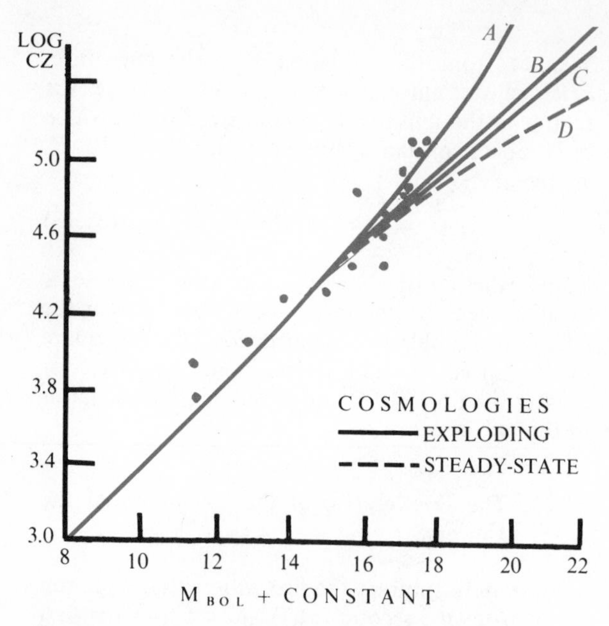

FIG. 31–7 *Curves A, B, and C on the graph correspond to an exploding universe (based on the work of Hoyle and Sandage). Curve D corresponds to the steady-state universe.* (Astronomical Society of the Pacific, *LXVIII (1956)*)

Thus, whether k is positive, zero, or negative depends on whether q is greater than, equal to, or less than $\frac{1}{2}$, and this is what we must determine observationally. Since q is related to the apparent magnitudes, m, and the Doppler shifts, z, of the galaxies through Equation (31.11) we can find q and thus determine the correct geometry of the universe by using this equation to plot the apparent magnitudes against the red shifts, z, for various assigned values of q, and then selecting that curve which best fits the observational red-shift data.

In Figure 31.7 three different curves, A, B, and C, taken from the work of Hoyle and Sandage are plotted for log cz and for the apparent bolometric magnitudes. Curve A represents the case $q = 2.5$; curve B the case $q = \frac{1}{2}$, and curve C the case $q = 0$. Curves A, B, and C correspond to an exploding universe (one that is evolving in time). Curve D, however, corresponds to the steady-state universe which we shall now discuss.

31.25 The Steady-State or Continuous-Creation Universe

The model for the steady-state universe arose originally from an extension of Einstein's field equations, suggested by Hoyle. It was derived independently by Bondi and Gold on the basis of what is now called the **perfect cosmological principle.** To see what Hoyle did, we recall that Einstein had to introduce the cosmological term into his field equations to obtain a finite universe.

In adding this term, Einstein was guided by certain criteria, an important one of which is the principle of conservation of mass and energy-momentum. Hoyle discarded this principle and introduced on the right-hand side of Einstein's field equations a tensor that is time-dependent, thus allowing the mass-energy-momentum tensor T_{ik} to change with time. *From this it follows that the total amount of matter in the universe is not a constant but increases.* For this reason, Hoyle's model of the universe is also referred to as the **continuous-creation universe**. It is also called the **steady-state** model because its observable features do not change with time, as we shall see in the next section.

31.26 The Perfect Cosmological Principle

Since Hoyle's treatment is highly mathematical, it is easier to get at the fundamental properties of this model of the universe by considering the Bondi-Gold approach based on the **perfect cosmological principle**. As we have seen, Einstein was guided in his work by what is called the cosmological principle, which requires that the universe be everywhere the same. *The perfect cosmological principle is an extension of this idea to time as well.* According to this principle as introduced by Bondi and Gold in 1948, the universe is not only the same everywhere, but it is also the same at all times for all observers. This concept gives rise to the steady-state picture of the universe since nothing in it ever changes. Here we must be careful not to confuse this model with a model in which there is no motion at all. The latter universe is static, whereas a steady-state universe permits a continuous flow of matter as long as the density of the matter at each point and its flow are always the same.

This model leads immediately to the concept of the continuous creation of matter. For if we consider the receding galaxies, we see that without continuous creation, the density of matter in the universe diminishes as the expansion continues. *To offset this, and to keep things the same at all times, it is necessary to postulate that matter is continuously created at each point of space.* Bondi, Gold, and Hoyle have shown that in the steady-state universe matter must be created at a rate of $3H\rho$ (where ρ is the mean density). If H equals 100 km/sec/Mpsc and if we take for the mean density of the universe the value 10^{-29} gm/cm^3, (assuming that all matter in the universe is uniformly smeared out) we find that the mean rate of continuous creation equals 10^{-46} gm/cm^3/sec. This is approximately 1 nucleon per 1 000 cm^3 per 500 billion years. This rate must be the same everywhere and at all times.

31.27 The Properties of the Steady-State Universe

To find the value of k that corresponds to the steady-state universe we note that the curvature of three-dimensional space for all uniform models of the universe is given by k/R^2. But by the perfect cosmological principle this datum, being an observed quantity, must be the same at all times and at all points. However, we know from what we have said that R varies with time, and for the steady-state universe $R(t)$ equals a universal constant times e^{Ht}, where t is the time and H is Hubble's constant. *It therefore follows that k/R^2 can be the same at all times (as required by the perfect cosmological principle) only if $k = 0$. This, then, is the value of k for the steady-state universe.* We note that this is the same as for the Euclidean universe, but the Euclidean universe is also an evolving or exploding one.

The difference between the steady-state model and the Euclidean expanding model lies in the value of q for these two cases. Whereas $q = \frac{1}{2}$ for the Euclidean exploding universe, as follows from (31.16), its value for the steady-state universe is -1. This result can be obtained from the exponential dependence of $R(t)$ on time and from the definition of q as given by Equation (31.9).

31.28 Applying the Observational Evidence: The Pulsating Universe *

To see which one of these four models, as given by theoretical curves A, B, C, and D in our figure, best corresponds to the observable universe, we must include in this figure the observational data. These are shown as little round dots as obtained by Sandage, Humason, and Mayall, and we see that the observations best fit curve A. We must therefore conclude that the best observational data at this time favor a closed, finite, spherical, and expanding universe (Riemannian, elliptical).

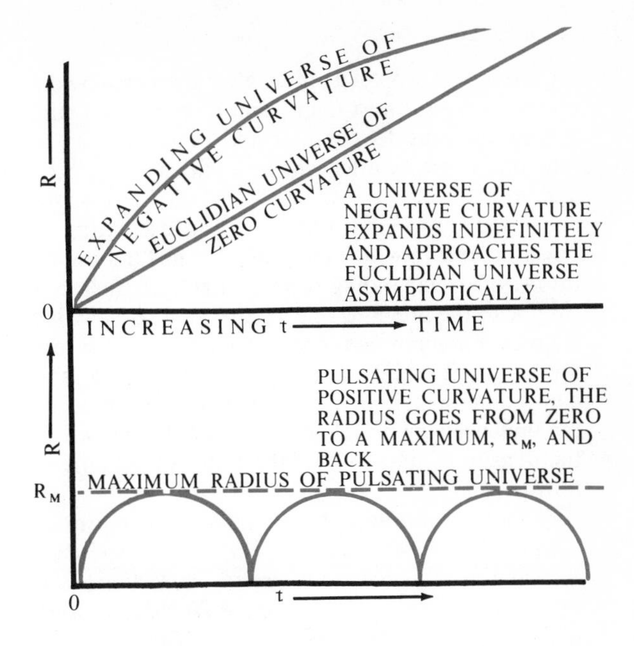

FIG. 31–8 *The radius of the universe plotted against time for different expanding models.*

This type of universe has an interesting property which we can determine from Equation (31.7). If in this equation we replace the mean density by the total mass of the universe divided by the volume of the universe (this quotient is of the order of M/R^3 multiplied by a universal constant) and transpose the second term on the left-hand side to the right-hand side, we obtain on multiplying through by R^2,

$$\dot{R}^2 = \frac{\text{constant}}{R} - kc^2,$$

where the constant contains the total mass of the universe and other factors.

Now the left-hand side, being the square of a number, is always positive. However, for $k = +1$ (which corresponds to our closed spherical universe), the right-hand side can be positive only if its first term always remains larger than its second term. This means that R cannot exceed a certain limiting value. In fact, if R equals the critical value 5×10^{10} light years, the right-hand side vanishes. If R were to exceed this value, then the right-hand side would become negative, which is impossible because the left-hand side is always positive. This means that a closed expanding universe must stop expanding after it has reached a critical size and must then begin to contract again. It will continue to contract and reach a state of maximum density after which it will expand again, and so on.

Thus, a universe of this sort is an oscillating or pulsating one in which the present is just a moment in one of its expanding stages. Each expansion is similar to the previous one, so that in a sense, there is no beginning or end in such a universe but rather an unending series of identical expanding and contracting stages. At the end of each contraction, the universe is in a highly condensed state in which the density of matter is of the order of billions of tons per cm^3.

If the age during any one cycle of a pulsating universe is defined as the time that has elapsed since the moment of highest condensation (zero time), this age is equal to the reciprocal of the Hubble constant. If H is of the order of 100 km/sec/Mpsc, then the present age of our universe (assuming it to be a pulsating one) is 10^{10} years.

The Euclidean model of the universe has properties similar to the Einstein-de Sitter model. In the case of the hyperbolic model the universe continues to expand indefinitely, thus approaching the Euclidean case as the density approaches zero. The curvature of the universe as a function of time for the three different models is plotted in Figure 31.8.

31.29 Radio Sources as Evidence for Cosmological Models of the Universe *

We can verify which of the various cosmological models best describes our universe by counting extragalactic radio sources. The most recent work in this field has been done by Ryle and Clark, whose observational evidence appears to contradict the steady-state model and to favor the exploding models. It is clear that if the universe is in a steady state, the number of galaxies containing radio sources must be the same everywhere, and must be the same now as it was billions of years ago. On the other hand, if the universe is evolving so that conditions are changing continuously, the number of galaxies and radio sources a few billion years ago was certainly different from what it is now. With radio telescopes it is possible to look far out into space, and to see the universe as it was billions of years ago.

If then we compare the number of radio sources in two equal volumes of space, one close to us (say, within 100 million light years) and the other at a great distance (5 or 10 billion light years from us), we should find a difference between these two counts, if the universe is evolving or exploding. The evidence that Ryle and Clark have obtained indicates that this is so and hence favors the closed spherical universe. The preponderance of evidence at the present time (taking into account the data obtained both with radio and optical telescopes) seems to eliminate the steady-state model.

There are still other observational criteria that can be used to decide between the steady-state and the evolving models of the universe. Thus the steady-state theory differs from the others in predicting that the angular diameters of very distant objects tend to zero with distance. There is also the question of differences between the physical characteristics of nearby galaxies and of very distant ones. If our universe is evolving, the very distant galaxies (as we see them now) appear younger, and we should therefore detect a difference in their apparent characteristics as compared to the nearby ones, which appear older. The characteristics of galaxies that should show such differences are color, shape, type, and stellar composition. Data of this sort cannot now give us a decisive answer as to which model of our universe is correct because the observations are not accurate enough. But in time, as observational techniques improve, this type of analysis will aid us greatly in determining the structure of our universe.

32

COSMOGONY, THE ORIGIN OF THE UNIVERSE: THE ORIGIN OF STELLAR SYSTEMS*

32.1 The Relationship of Model Universes to the Origin of Stellar Systems

The problem of the evolution of the universe and of systems such as galaxies, stars, and planetary systems is intimately related to the whole cosmological question, and its solution ultimately depends on whether we accept the steady-state model or the expanding (or evolving) model of the universe. In the case of the steady-state theory the formation and evolution of these systems is the result of a slow accretion of matter around nuclei of condensation, which is a continuous process. In time, as such clouds of matter grow, the chaotic motions within them give rise to a system of turbulences and then to a hierarchy of eddies which ultimately correspond to the various structures such as stars, etc., that we find in our universe.

However, in an exploding universe we must start with some highly condensed initial state in which the conditions were quite different from what they are now. In the very early stages of such an explosion the temperature was extremely high (of the order of trillions of degrees) but the temperature fell off very quickly as a result of the expansion, and after millions of years it was of the order of a few thousand degrees. This cooling off allowed the atoms to condense into molecules, and ultimately these molecules collected into clouds of fine grains of dust and gaseous matter.

This expanding mixture of fine dust and gaseous matter would presently be spread out as a uniform cloud filling the entire universe if no gravitational forces were present. However, the over-all gravitational force in this expanding mixture made such a uniform distribution impossible, for it can be shown that an expanding gaseous distribution subjected to internal gravitational forces is unstable and breaks up into local condensations of a certain limiting size. The size of these condensations must be such as to prevent the escape of particles from them. Gamow has analyzed this and has shown, taking into account the expansion of the universe, that the sizes and masses of these condensations (assuming a density of matter such as we now have in our universe) are about of the order of those of galaxies. We thus see that in the case of an exploding universe there is a natural mechanism for the fragmentation of the original material into galactic structures.

32.2 Fragmentation of the Expanding Cloud into Protogalaxies

What happened after the fragmentation into galaxies occurred was determined by the state of rotation (the amount and distribution of angular momentum) in these condensations. It can be shown that because of internal frictional forces a rotating gaseous system cannot continue to revolve unbrokenly (like a solid disk) but undergoes complicated turbulent motions consisting of eddies of various sizes moving about at various speeds. The larger an eddy is in such a turbulence, the larger its velocity is.

The conditions under which a smooth, unbroken rotational flow of matter breaks up into eddies were investigated many years ago by the physicist Reynolds, who set up a simple formula that defines a quantity, $\mathscr{R}$, called **Reynolds' number**. *This number, which is used as a criterion as to whether the smooth motion will break up into eddies or not,* is given by

$$\mathscr{R} = \frac{\rho v w}{\eta},$$

where ρ is the density of the fluid, v its velocity, w the width of the fluid filament, and η its viscosity. (Viscosity is a measure of the internal friction in the fluid. If η is large, the fluid is very viscous and flows with difficulty.) Reynolds showed that if the viscosity, the density, the speed, and the width of the stream are such that this number exceeds 1 000, the material must break up into eddies of various sizes. In the case of a uniform distribution of matter in a rotating structure as large as a galaxy, the conditions are always such that Reynolds' number is many times greater than 1 000, so that eddies must develop.

One of the consequences of having a system with eddies and turbulences is that local condensations occur in regions where eddies of various sizes are in contact or colliding with each other (see Figure 32.1). However, these local condensations generally disappear unless they are dense enough or large enough to remain stable and collect additional material as a result of the gravitational forces. These, then, are the condensations from which stars and also planetary systems ultimately evolve. This leads us to a consideration of the origin of the solar system and the various theories that have been developed to account for it.

FIG. 32–1 *Local condensations arising where turbulences are in contact with each other.*

32.3 The Origin of the Solar System

We have seen in our discussion of the planets that there are enough regular symmetrical characteristics in the solar system to indicate that the planets and our sun were formed together in a single process from the same primordial condensation arising from the turbulences discussed in the previous section.

32.4 The Kant-Laplace Hypothesis

The earliest theory introducing the concept that our solar system was formed as a condensation from a nebulous gaseous mass stems from the work of Immanuel Kant. Because this theory was later developed mathematically by Laplace, it is now referred to as the **Kant-Laplace hypothesis**. According to this theory the solar system began as a thin, lens-shaped, slowly spinning gaseous nebula with about the same mass as it now has.

As this gaseous mass contracted gravitationally, its rotational speed steadily increased because of the conservation of angular momentum. Laplace therefore argued that because of the centrifugal force arising from this rotation, a ring of the contracting nebula broke away from the central mass and condensed into a planet. The remaining nucleus continued to contract with additional rings breaking off from time to time and condensing into planets until the present solar system was formed [see Figure 32.2(a), (b), (c)].

There are serious arguments against this theory, most of which concern the distribution and total amount of angular momentum in this system. Thus, for example, it can be shown that if all the planets were now thrown into the sun, the speed of rotation on the surface of the sun would increase from its present value of 2 km/sec only to about 70 km/sec, and this speed is much too small to cause a ring to break away from the sun (the speed of rotation of the original extended nebular disk was even smaller than this).

Another argument against this nebular hypothesis deals with the distribution of angular momentum. In a slowly contracting system such as pictured by Laplace the bulk of the angular momentum always remains with the central nucleus, which contains the bulk of the mass. However, we know that the principal planets, which contain about 0.1 per cent of the entire mass of the solar system, contain about 98 per cent of the angular momentum, whereas the sun, which contains almost all of the mass, has about 2 per cent of the angular momentum. We may also note that Laplace's nebular hypothesis predicts planetary orbits that lie exactly in the equatorial plane of

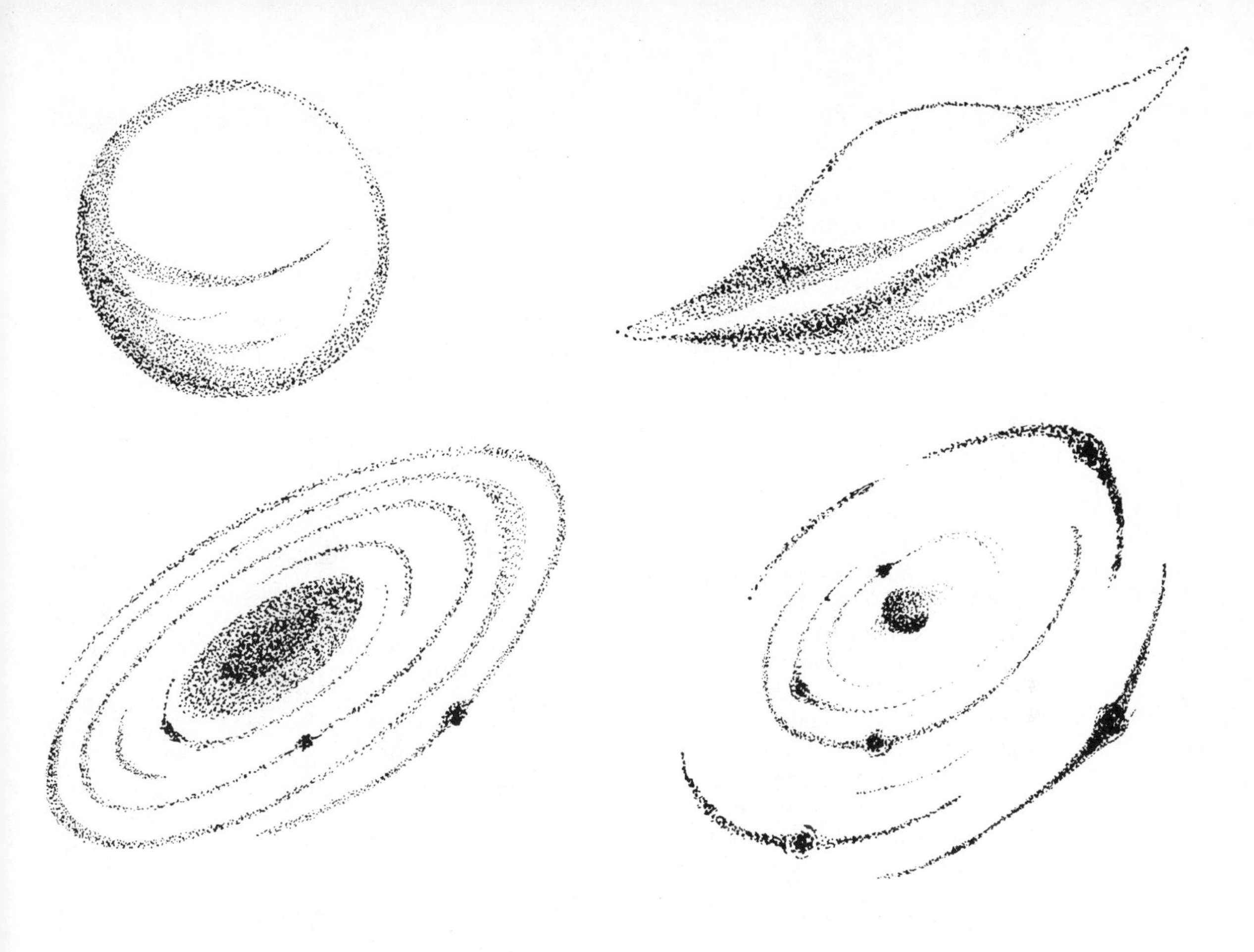

FIG. 32–2 *A schematic representation of the Kant-Laplace Theory of the Origin of the Solar System.*

the original solar nebula, and *there is no mechanism to account for the observed inclined planetary orbits.*

Finally, it was shown by Maxwell many years ago that Laplace's rings of matter could not condense into planets under their self-gravitational forces because these condensations would be destroyed by the inertial forces arising from the differential rotation between the inner and outer boundaries of each ring. Maxwell showed that a ring would have to be hundreds of times more massive than the planets are known to be before their gravitational forces could prevail against the disruptions arising from differential rotation.

32.5 The Tidal Theory

Because of these arguments astronomers discarded the Laplace hypothesis and developed other theories such as the **tidal theory**, which pictures the solar system as having been formed during a close encounter between the sun and another star. This theory, which had its origin in the work of Chamberlin and Moulton, was developed extensively by Jeans and Jeffreys, who proposed the idea that a star came close enough to the sun to raise huge tides on the sun's surface. These tides were then pictured as erupting and emitting large filaments of matter, which were then drawn out by the gravitational attraction of the passing star.

The passing star was pictured as imparting to the planets the great amount of angular momentum they now have. However, calculations show that it is impossible to bring a star close enough to the sun to tear enough material out of it to give rise to the planets as we know them, and at the same time to impart to these planets the required amount of angular momentum. Moreover, the material torn from the sun in this fashion could never condense into planets because it would consist of such hot gases that it would quickly diffuse into space. For these reasons the tidal hypothesis was discarded.

32.6 Modern Theories of the Origin of the Solar System

Because recent discoveries in astronomy have eliminated some of the serious objections to the nebular hypothesis, the current theories of the origin of the solar system start with the Kant-Laplace hypothesis, but with important variations. To begin with, we now picture the original nebulous gas from which our solar system was formed as consisting of a central nucleus (the **protosun**) which contained just about as much mass as the present sun does, and which ultimately evolved into our sun. This dense nucleus was surrounded by a tenuous shell of gaseous matter extending out to the extremities of our solar system and containing about $0.1M_\odot$. This diffuse gaseous shell is referred to as the **original solar nebula** (note that it does not include the protosun). All modern cosmological theories picture the planets as having been formed from this original solar nebula.

Since the original solar nebula contained primarily hydrogen and helium, it differed considerably from the present chemical composition of the planets. From this it follows that the planets now are but the mere remnants, chemically and physically, of the original solar nebula. The important difference between the original Kant-Laplace hypothesis and the present theory is that a much larger mass is assigned to the solar nebula in the present theory. It is now clear that with such a large mass in the solar nebula the difficulties in the distribution of the angular momentum can be greatly reduced.

To begin with, we note that a solar nebula having a mass of $0.1M_\odot$ could not be absorbed into the sun by gravitational attraction since the resulting rotational velocity would be so great—of the order of several thousand km/sec—that the centrifugal force would prevent it from coalescing with the sun. Hence, most of it would have to remain outside the main body of the sun and the great bulk of angular momentum would at all stages lie in the solar nebula rather than in the protosun. Thus, the difficulty of the distribution of angular momentum does not arise in this case.

However, there still remains the question of why the total mass of the planetary system now is so much smaller than the mass of the original solar nebula. We can explain this if we recall that the solar nebula originally consisted almost entirely of hydrogen and helium. We know that original hydrogen and helium atoms moved with speeds greater than the speed of escape from the planetary condensations so that most of these atoms slowly seeped out into space, carrying with them most of the mass and leaving the residual planetary condensations.

32.7 Formation of the Planets

We shall consider the theories of Weizsäcker and Kuiper to account for the formation of the planets in the original solar nebula. Weizsäcker started out by assuming a gaseous distribution consisting of a solar core and a surrounding nebula which he pictured as rotating. This rotation flattened the solar nebula into a disklike structure in the equatorial plane of the sun, and gave rise to turbulences and eddies of the sort described in Section 32.2. One can show that this must occur because the Reynolds number in such a solar nebula is many times larger than the critical value of 1 000. These vortices and eddies revolved around the solar core in Keplerian orbits. Weizsäcker showed that such a turbulent pattern of eddies cannot be a random pattern, but must arrange itself into a system of contacting vortices in shells, such as shown in Figure 32.3, with each shell containing the same number of vortices (Weizsäcker placed five vortices in each shell).

Weizsäcker pointed out that such a system of revolving shells of eddies is stable and that condensations of matter collect between successive shells where vortices are in contact. In these regions, which Kuiper calls the "roller bearings," the rotational motion of one vortex cancels that of the other so that accretion of matter occurs (see Figure 32.3). By arranging these bean-shaped vortices or eddies so that there are five in each shell Weizsäcker obtained a good approximation to Bode's law for the distances of the planets.

However, there are several difficulties associated with the Weizsäcker theory, among which are the following:

1. the lifetime of the solar nebula as derived from this theory is only about 1 per cent of the time it would take the sun to shrink down to its present size as determined from the mass-luminosity-radius relationship;
2. the rate of growth of the planets by accretion according to Weizsäcker's theory is much too slow to account for the formation of the planets;
3. the theory requires that the bean-shaped vortices maintain their beautifully symmetrical arrangement for unreasonably long periods of

FIG. 32–3 *Weizsäcker—1945. Vortices formed in the equatorial plane of a nebula of gas and dust rotating about the sun, according to Weizsäcker. Accretion would take place along the heavy concentric circles, to form planets and satellite systems with direct rotation and revolution. (Yerkes Observatory photograph)*

time in order for planets to grow in regions where the vortices are in contact.

Moreover, modern turbulence theory shows that such an arrangement cannot last for anywhere near the required length of time, and a much more complex system of vortices, such as shown in Figure 32.4, arises. In this arrangement the eddies are continuously forming and dissolving, with the larger eddies giving up their kinetic energies to the smaller ones.

32.8 Kuiper's Theory of the Origin of Planets

Heisenberg and Chandrasekhar have shown that, according to Kolmogoroff's theory of turbulences, this hierarchy of turbulences must have existed in the original solar nebula; thus one cannot obtain Bode's law directly since Weizsäcker's conditions do not apply. For this reason Kuiper has altered Weizsäcker's theory by taking into account the gravitational tidal action of the solar nucleus (i.e., the protosun) on this hierarchy of turbulences in the original nebula. In his theory Kuiper still introduces the system of primary vortices but does not arrange them according to the Weizsäcker scheme; instead he starts with a distribution of vortices that agrees with Kolmogoroff's theory of turbulences. Two questions now arise: (1) can such eddies give rise to nuclei of matter that will ultimately grow into planets by gravitational attraction? (2) can such nuclei occur at approximately the distances given by Bode's law? Kuiper shows that this is the case if we take into account the tidal action of the protosun and its Roche limit (see Section 6.37) within the solar nebula.

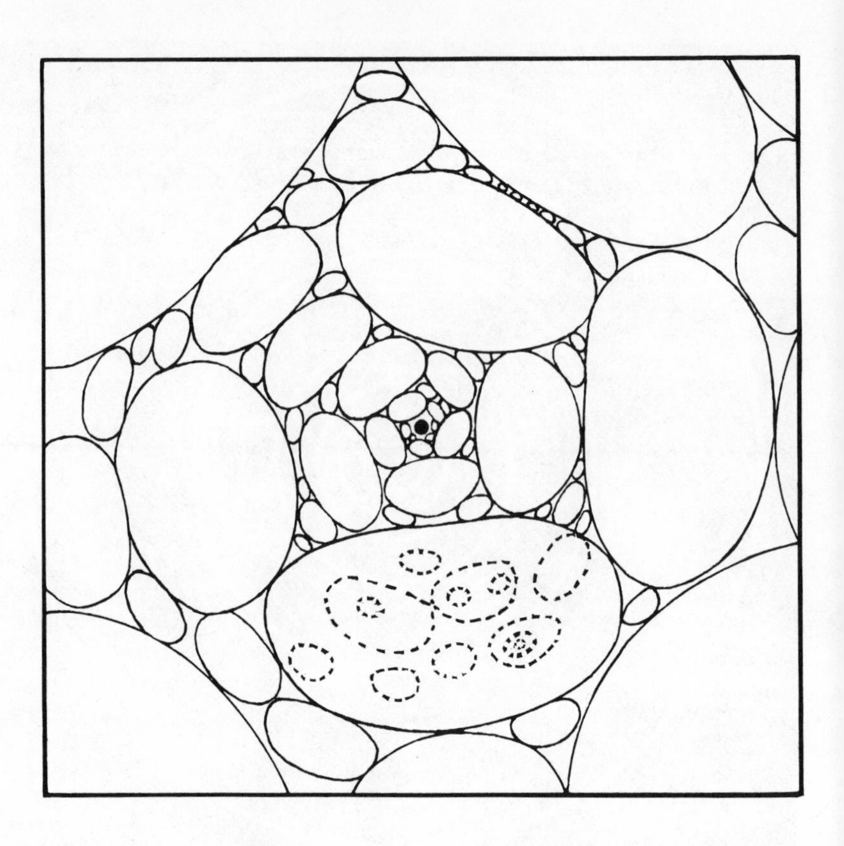

FIG. 32–4 *The turbulent pattern of vortices that arose in the primordial nebula about the sun according to the Kolmogoroff theory of turbulences. This pattern is the basis of Kuiper's theory of the origin of the solar system.* (*Yerkes Observatory photograph*)

Let us consider the solar nebula at a given distance from the protosun and inquire whether a core of matter suddenly forming at this distance can exist under its own gravitational attraction or whether it will be dissipated. We know that if the density of this core is so large that its distance from the protosun exceeds the Roche limit for that density, then the core (or the protoplanet, as we may now call it) is stable and continues to grow. If, however, the distance of this protoplanet is less than the Roche limit for the given density, the tidal action of the protosun will destroy it.

It follows, therefore, that planets were formed from the solar nebula only at those distances from the protosun at which the densities of matter exceeded the Roche-limit densities. Since the Roche-limit densities as derived by Kuiper are: at the distance of Mercury 10^{-5} gm/cm^3, at the distance of the earth 10^{-6} gm/cm^3, at the distance of Jupiter 10^{-8} gm/cm^3, at the distance of Neptune 10^{-10} gm/cm^3, protoplanets could have been formed in the original solar nebula if initial condensations having these densities occurred at these distances. These protoplanets could then grow by accretion. According to Kuiper this is how the planets began their lives.

32.9 The Importance of Turbulences *

We come now to the role of turbulences in Kuiper's theory. Let us suppose that in the original solar nebula (assuming that no turbulences were present) the densities at various distances from the protosun were in the neighborhood of those demanded by the Roche limit. Under these conditions the tidal action of the sun would tend to prevent protoplanets from forming. But the presence of turbulences acted as a trigger for the formation of the planets because such turbulences gave rise to density fluctuations in the solar nebula. These fluctuations resulted in densities at various distances larger than the critical densities of the Roche limit at these distances, so that stable protoplanets were formed and grew. Although this theory agrees fairly well with the structure of our solar system, it is by no means universally accepted by astronomers.

More recently H. C. Urey, Nobel Laureate, has proposed that the earth and the asteroidal meteoritic planets were formed at relatively low temperatures (less than the present central temperature of the earth) by the accretion of solids of which they are now composed. Urey's analysis was based on the present compositions of the planets and the remarkable difference in chemical composition between the terrestrial and the Jovian planets. His principal hypothesis, in agreement with Weizsäcker and Kuiper, is that the turbulent eddies described above were transformed into protoplanets (one for each planet) by accretion at a low temperature. As planetesimals fell into these protoplanets, water and ammonia first accumulated. In time the temperature on a protoplanet rose high enough to melt the iron ore chunks that fell onto the surface. Most of the metal in the form of iron-nickel alloys remained distributed throughout the silicates but ultimately fell into the core of the terrestrial planets while the volatile gases of low molecular weight escaped.

32.10 Cosmic Rays

In our analysis of the formation of the galaxies and the stars, we did not consider cosmic rays (see Section 14.9) which we now know contribute as much energy to interstellar space as electromagnetic radiation does. Each unit volume of interstellar space contains as much cosmic-ray energy as it does energy in the form of electromagnetic radiation. However, the average cosmic-ray energy per particle is far greater than the average electromagnetic energy per photon. At first sight it may seem strange that the energy density of cosmic rays (a nonthermal radiation) in interstellar space in the Galaxy is about equal to that of light (thermal radiation) since one would ordinarily expect the latter to be far larger. But light travels in straight lines (except for some slight bending by gravitational fields) and thus leaves the Galaxy after a few thousand years. The charged particles in cosmic rays, however, move in complete twisted orbits because of the galactic magnetic fields and are practically trapped in the Galaxy. Thus, these particles accumulate for millions of years.

Certain properties of cosmic rays are immediately evident from an analysis of the data. To begin with, if we consider cosmic rays with energies of the order of 10 Bev (billion electron volts, or, in the new system, Gev, gega electron volts) or larger, we find that they are remarkably time-independent if we disregard the occasional occurrences of unusually large solar flares. Below this energy there are fluctuations, as observed here on earth, but these can be accounted for by the interplanetary magnetic fields which modulate the low-energy cosmic rays but have little effect on the high-energy rays. The intensities of the primary beams of cosmic-ray particles (i.e., the beams just before they hit the earth's atmosphere) with energies ranging from about 20 Gev (these are energies measured for a single cosmic-ray particle) up to about 10^9 Gev per particle (the largest measured energies) have remained constant for the last thousand years. This is known from the carbon dating of ancient artifacts which have been uncovered.

Moreover, the cosmic-ray beams of very high energies are practically independent of direction (isotropic) and in particular are independent of the earth's rotation. For energies ranging from 10 to 30 Gev there is a variation of about 1 per cent in direction, but for energies greater than 10^4 Gev the variation is less than 0.1 per cent. Most of the variations for low-energy cosmic rays that we observe here on earth are due to solar flares.

We may therefore say that aside from the occasional intense cosmic-ray beams coming from solar flares, there are no visible sources of intense cosmic rays. We may not conclude from this, however, that the high-energy cosmic-ray sources are distributed symmetrically around the earth since this implies that the earth occupies a central position, at least relative to the cosmic rays. Since we may exclude this as highly improbable, we must conclude that the isotropic property of cosmic rays arises from the way they move about in our Galaxy.

32.11 The Cosmic-Ray Energy Spectrum

The number of primary cosmic-ray particles is not the same in all energy ranges but falls off as the energy per nucleon increases. This is best expressed in the form of an energy spectrum or distribution law which applies primarily to cosmic-ray particles with energies greater than 10^9 ev. This law gives the number of cosmic-ray particles of given energy striking one square centimeter of the earth's surface per second. We find that this number varies inversely as $E^{2.5}$ for these high-energy particles, where E is the energy per particle. We may express this as follows: if $N(E)$ is the number of primary cosmic-ray particles striking one square centimeter of the earth's atmosphere per second, and E is the energy per cosmic-ray particle, then

$$N(E) = KE^{-2.5},$$

where K is a constant that can be determined by observations. Note the negative exponent. Although in this energy law we have used 2.5 as our exponent, very accurate observations show that it is equal to 2.5 for cosmic-ray energies between 50 Gev and 10^2 Gev, but for larger cosmic-ray energies it ranges from 2.7 to 3.

32.12 Conditions Necessary for Cosmic Rays to Arise

The principal problem associated with cosmic rays is to account for their origin as well as for the energy spectrum given above. Today most cosmic-ray theories are based on the assumption that these particles acquire their energies from galactic sources.

According to Philip Morrison, to account for the stream of cosmic-ray particles observed in our part of the Galaxy the following conditions must be fulfilled:

1. *There must be a source or injecting mechanism that keeps a continual supply of charged particles, such as protons, flowing into the galactic mechanism that speeds these particles up to cosmic-ray energies.* Before these particles become cosmic rays they are probably ordinary protons that we find in interstellar space, in gaseous nebulosity, in hydrogen regions, or in stellar atmospheres.
2. *There must be a galactic mechanism that accelerates the particles to the various cosmic-ray energies.*
3. *There must be a stirring mechanism* (probably intimately related to the accelerating process) that distributes the particles in all possible directions so that the cosmic rays coming into our solar system are isotropic.
4. *A means must be available for keeping the particles stored in the accelerating mechanism while they are being accelerated to their cosmic-ray energies.* Since some particles take a much longer time than others to reach cosmic-ray energies, there must be different storage times for different particles.
5. Finally, *there must be a critical energy above which the beam of particles shows a definite and drastic change in its character.* This might possibly take the form of a change in the spectral character of the beam or show up as a nonisotropic distribution of the cosmic rays.

If any model is to explain the observed features of cosmic rays including the energy spectrum, it must include the above five processes.

32.13 The Fermi Theory of the Origin of Cosmic Rays

Various theories for the origin of cosmic rays have been suggested. According to Fermi's theory, the charged particles acquire their large energies by interacting with magnetic fields which are carried along by the interstellar clouds of matter in the spiral arms of our Galaxy. If a charged particle moves in a nonuniform magnetic field, it spirals around a line of force towards the region of increasing magnetic-field intensity. If the field becomes strong enough, i.e., the lines of force become sufficiently concentrated (see Figure 32.5), the motion of the particle in the direction of the lines of force becomes smaller and smaller until all the kinetic energy of the particle has been transferred to its circular motion around the lines of force, and it has lost all of its forward momentum; the particle is then reflected and moves back again until it reaches the opposite point of highly concentrated lines of force and is again reflected, so that it oscillates between the two magnetic mirrors without gaining or losing energy. This is the case if the system of magnetic lines, all taken together (i.e., the magnetic mirror itself), has no over-all uniform motion with respect to the charged particle. In other words, the charged particle and the magnetic mirror have no motion of approach or recession with respect to each other.

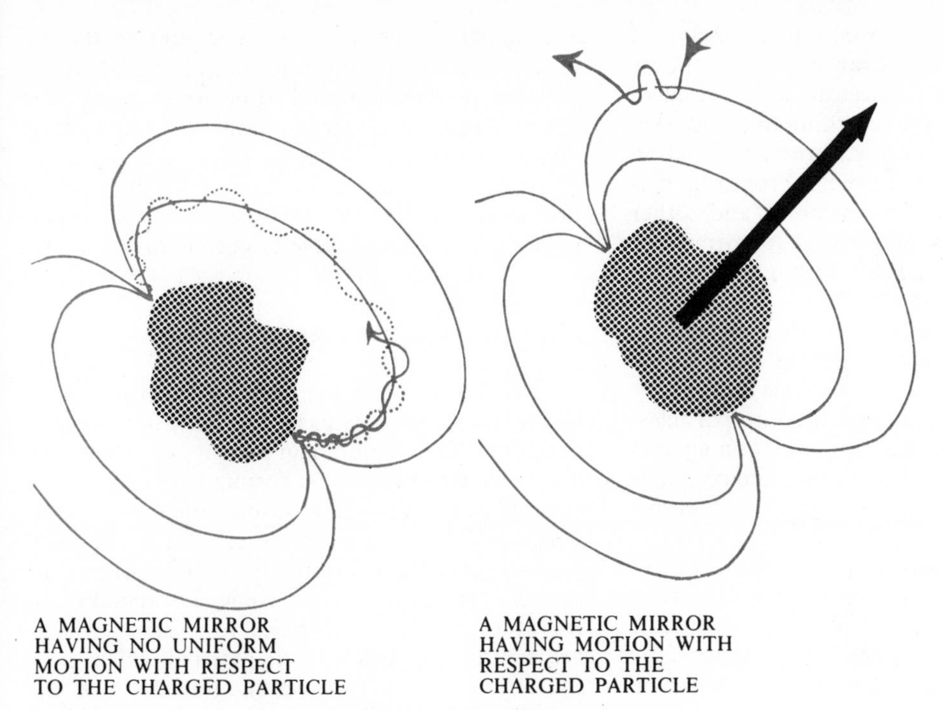

FIG. 32–5 *Magnetic mirrors in which charged particles are forced to oscillate back and forth as they circle around the lines of force.*

If, however, the magnetic mirror is moving with respect to the particle, the particle suffers a kind of collision with the mirror and acquires (or loses) some energy. In this way the mirror acts as an accelerating or decelerating mechanism which transfers some of the kinetic energy of moving cloud masses to charged particles. Fermi has shown that after colliding repeatedly in this way with clouds carrying magnetic fields, free protons and other charged particles acquire cosmic-ray energies. He has, in fact, derived from this model an energy spectrum for cosmic rays, at least for intermediate energy ranges, similar to the one that is observed.

32.14 Cosmic-Ray Sources

Although Fermi's theory has many interesting and admirable features, it still cannot account for the entire energy range of cosmic rays, and additional mechanisms for generating high-energy cosmic rays must be introduced. We now know that there are many different sources of cosmic rays and that we can account for their observed energy spectrum only if all contributing sources are taken into account. Thus, since the sun emits cosmic rays during solar flares, it is reasonable to suppose that most stars similar to the sun (as well as the usual flare stars) are sources of cosmic rays.

Not only are cosmic rays emitted from flares during the main-sequence phase of a star's life, but also during the mass-ejection processes that occur in the late evolutionary stages of stars, as discussed in Chapter 23. Thus it is certain that supernovae, and probably even ordinary novae, are intense sources of cosmic rays. There is reason to believe that such variables as the T Tauri stars, certain of the shell stars such as the Wolf-Rayet stars, and magnetic variables are all sources of cosmic rays. The data associated with such stars have been analyzed in detail by Philip Morrison, who has given very compelling reasons for accepting these stars as cosmic-ray sources.

32.15 The Crab Nebula and Other Galactic Cosmic-Ray Sources

The Crab Nebula is an important cosmic-ray source in our own Galaxy, and we may conclude that all such objects, which are remnants of supernovae, emit cosmic rays. From our discussion of the magnetic fields in the Crab Nebula we know that the electrons in this object are accelerated to very high speeds, and thus, according to Shklovsky, give rise to the continuous synchrotron radiation which we have already discussed. From this it is reasonable to assume that protons and other charged particles can be accelerated to cosmic-ray energies by the magnetic fields within such supernova remnants as the Crab Nebula.

Finally, since there is a great deal of kinetic energy per unit volume available in the Galaxy because of its rotation, the entire Galaxy itself may act as a mechanism to pump energy into particles. This is possible because there are general magnetic fields throughout the arms of the Galaxy, and energy can be transferred from the galactic reservoir to charged particles by a hydromagnetic process.

The most recent evidence, as presented in 1963 by de Shong, Hildebrand, and Meyer, seems to favor the remnants of supernovae such as the Crab Nebula as the primary sources of cosmic rays. The analysis is based on determining the ratio of electrons to positrons in the primary cosmic-ray beams (the beams just before they hit the earth's atmosphere). If the electrons in the primary cosmic rays are in the main part due to the collisions of original cosmic-ray protons with the protons in intergalactic space, there should be about as many electrons as positrons (but there will always be more electrons than positrons) in the beam since for every electron that is created during a proton-proton collision one positron is also created. Thus, if most of the cosmic rays arise from some kind of accelerating mechanism within the Galaxy, the ratio of electrons to positrons should range from about one to two. However, if the cosmic rays are primarily due to supernovae outbursts, the ratio of electrons to positrons in the primary cosmic-ray beams should be considerably higher than this because the electrons that are already present in the atmospheres of supernovae are accelerated to cosmic-ray energies during the explosion just the way protons are, whereas there are very few positrons around to be accelerated. The observational evidence, obtained from balloon flights, shows a much larger ratio of electrons to positrons than can be accounted for by proton-proton collisions. We thus conclude that supernovae are the main source of cosmic rays.

32.16 Extragalactic Cosmic-Ray Sources

Thus far we have been discussing cosmic rays that originate in our Galaxy, and these undoubtedly constitute the major component of those we observe. However, some cosmic rays originate in external galaxies by mechanisms similar to those in our own Galaxy. Furthermore, collisions between galaxies, and the explosive formation (fission) of multiple galaxies, release sufficient amounts of energy to accelerate particles to cosmic-ray velocities. In connection with external galaxies as sources of cosmic rays we should note that the contribution from the very distant galaxies becomes progressively less important because of their recession. Just as the recession of the galaxies enables us to eliminate the Olbers Paradox for the light coming from them, it reduces the importance of the distant galaxies as sources of cosmic rays.

If we consider all of these processes, we see that no single model accounts for the observed cosmic-ray spectrum and we must suppose that the same processes that are at work in the formation, evolution, and death of galaxies and stars are responsible for the cosmic rays. Although at present we have no direct evidence for the emission of cosmic rays from any single object except the sun, it may be that with artificial satellites and orbiting observatories we shall develop a cosmic-ray astronomy that will enable us to localize cosmic-ray sources fairly precisely. Until then we can only draw conclusions on the basis of reasonable theories.

PART FIVE

DEVELOPMENTS IN ASTRONOMY SINCE 1966

DEVELOPMENTS IN

The vast amount of data collected since 1966 through the manned space flights and the unmanned space probes evokes two responses in science: great satisfaction in the accomplishments, and urgency to learn more. After two decades of ample government funding, by 1973 the space and research projects were being severely cut back. The scientific community felt frustration and fear that the revolutionary progress would lose its momentum.

Here only the successes are visible—with the implicit assumption that there will be expansions, new developments, and great and small modifications in all domains of science in the future. The disappointments, bitter controversies, the lack of understanding or belief of colleagues, the jealousies and competition that exist are not revealed through the results.

This text describes the essential current knowledge in the field of astronomy. As students read and learn, they become an integral part of the continuum of adventure in science. They are able to know more about our universe than could the greatest astronomer who died before 1966. In the pages that follow we present the most recent developments in observational and theoretical astronomy to update the material in the main text. We have arranged this new material in accordance with the chapters and section headings of the main text so that the student can easily correlate the new with the old material.

CHAPTER 6: THE MOON

Although, as indicated in the main text, astronomers had discovered a great deal about the chemistry and physical features of the moon using earth-based telescopes, the knowledge obtained in the last decade from lunar probes, from spacecraft, and from landings on the moon is greater than that obtained in all preceding decades. We can present here only a brief resumé of this exciting work, which is still going on. The discoveries discussed stem from the following lunar probes and lunar landings.

The Ranger series 7, 8, and 9 in 1964–1965, which made the first "hard" landings (direct crash-landings) on the moon and, as they approached the moon, televised their photographs back to the earth. These photographs were accurate enough to show surface details as small as $\frac{1}{3}$ meter, which is 3,000 times as good as one can get with the best telescopes on the earth.

The Soviet Luna 9 in January 1966, which made the first soft landing on the moon and which, for three days, sent back to the earth the first close-up photographs of the moon.

The Surveyor series I, II, VI, and VII, which were launched between May 30, 1966 and November 1967. These all made successful soft landings on the moon and sent back many thousands of close-up photographs of the moon's surface, showing rocks and soil. These lunar probes also analyzed rocks and soil samples.

The five Lunar Orbiters launched between August 1966 and August 1967. These circled the moon in carefully calculated parking orbits from which they televised back to the earth numerous very detailed photographs of all faces of the moon. These pictures were used to choose landing sites for the Apollo astronauts.

The Apollo series of manned landings, which resulted in the return to the earth of hundreds of pounds of rocks, dust, and soil samples, and thousands of photographs.

The soft landing of a mobile Soviet laboratory, which explored the moon's surface for many days and sent back important information.

ASTRONOMY SINCE 1966

The findings were intriguing.

The **maria** (the so-called seas) are all chemically, physically, and geometrically similar to one another (the largest being about 1100 km across). The floors of all maria consist of a thick layer of grainy basalt that was probably formed from molten lava when it cooled. This basalt, which is called **mare basalt**, is similar to the basalt on the earth. The mare basalt contains small amounts of olivene and ilmenite, the oxides of iron and titanium, as well as the standard components, pyroxene and plagioclase feldspar. The basalt grains, which are about 0.02 mm in diameter, are clumped together into porous, compressible mats that have a density of about 1 gm/cm^3 and can easily support the weight of an astronaut with all his gear. All mare floors are heavily pitted, probably as a result of meteoritic bombardment. The dust covering the moon, which was formed by meteorite bombardment, consists of tiny glass beads. These were formed as molten droplets when the meteorites hit the lunar surface.

The highlands of the moon are also covered with basalt, but it contains only traces of ilmenite. This gives the highlands their light appearance as compared to the dark maria. Whereas the mare rocks have densities of about 3.3 gm/cm^3, the highland rocks have densities of about 2.9 gm/cm^3. This difference is due to the relative scarcity of iron and relative abundance of aluminum in the highlands.

The most abundant element in lunar rocks and soil is oxygen (about 58 per cent of all the atoms). The next most abundant element is silicon (about 20 per cent), which is present as silicon dioxide (sand). Then come aluminum, calcium, iron, magnesium, and titanium. No traces of water in any form or of organic compounds have been found.

The discovery that most of the moon's surface is basalt, which could only have come from molten lava, means that the entire surface of the moon was in a molten state at some time in the past. This was probably due either to intense meteoritic bombardment when the moon was first formed or to heating by radioactive material in the moon's crust, of which there is a fair abundance. The oldest rocks brought back from the moon are about 4.2 billion years old (these are from the highlands) and others are about 4 billion years old. From this information we can conclude that the rocks were formed after the first 600 million years in the moon's life, during which time it was in a molten state.

The moon was probably volcanic and subject to rather intense moonquakes for a period lasting some 700 million years. Since then it has been quite inactive, and the energy presently released per year in moonquakes, as monitored by seismometers left on the moon's surface, is less than one-billionth of that released by earthquakes.

There appear to be distinct layers of differing density in the moon. The densest material forms the moon's crust, which extends down to a depth of about 40 miles. The core of the moon, with a radius of about 400 miles, appears to be molten; it contains about 5 per cent of the moon's mass.

The chemical difference between the lunar rocks and earth rocks indicates that the earth and moon probably did not have a common origin. In particular, the moon could never have been part of the earth and been formed by being spun off the earth. The origin of the moon is still very much of a mystery, but the least objectionable theory is that the moon condensed gravitationally out of the same general material and at about the same time as the earth did. Thus the moon and the earth may be pictured as having originated as a double planet. The chemical differences between the lunar and terrestrial rocks and the difference between the moon's density and the earth's density are stumbling blocks in this theory but are not fatal to it.

6.13 The Mass of the Moon

Until space vehicles were placed in orbit around the moon or astronauts landed on its surface, the mass of the moon could be obtained only by measuring as accurately as possible the distance of the center of mass of the earth-moon system from the earth's center. Today the mass of the moon can be calculated very directly and precisely from the measured period of a space vehicle in a parking orbit around the moon and its mean distance from the moon's center, using Kepler's third law [Equation (8.29), with the mass of the moon replacing the sun's mass in the denominator]. The mass of the moon can also be calculated directly from the value of the acceleration of gravity measured by astronauts on the moon's surface. The formula (3.15) is used, with the quantities in that equation referring to the moon and not the earth.

6.33 The Explanation of the Tides

In this section we explained the tidal action of the moon on the earth by picturing the earth as falling in toward the moon. This may seem a bit difficult to understand—one pictures the moon as simply revolving around the earth. But we must remember that the moon is also pulling in on the earth, and the center of the earth is revolving around the center of mass of the earth-moon system in a small orbit, as described in Section 6.12. The earth's center is thus constantly being accelerated toward this center of mass and thus toward the moon. This is also true of a particle of water on the side of the earth facing the moon and of a particle of water on the side away from the moon. These particles of water experience centripetal accelerations toward the center of mass of the earth-moon system that are not quite equal to the acceleration of the entire earth toward the center of mass. One can think of the two high tides as being due to these differences in acceleration.

6.36 Tidal Friction and the Earth's Rotation

The relationship between the moon's distance from the center of the earth and the length of earth's day can be deduced from the fact that the total angular momentum of the earth-moon system must remain constant. We know that the orbital angular momentum of the earth and moon together is 4.82 times the rotational angular momentum of the earth. Taking the earth's rotational angular momentum as 1, we see that the total angular momentum (sum of orbital and rotational) of the earth-moon system is 5.82. The rotational angular momentum of the earth varies inversely as the length of the day d. And the orbital angular momentum depends only on the square root of the moon's distance p from the earth when the line from the earth to the moon makes an angle of 90° with the line from apogee to perigee. If we call this distance at the present time p_0, and call the present length of day d_0, then we must have

$$\frac{d_0}{d} + 4.82\sqrt{\frac{p}{p_0}} = 5.82,$$

where d is the length of the day when the moon's distance is p. This relationship permits one to calculate the length of day corresponding to a given lunar distance.

CHAPTER 7: ECLIPSES OF THE SUN AND THE MOON

7.9 The Importance of Eclipses

Until the advent of radio and radar astronomy, the prediction of the general theory of relativity that the path of a beam of light passing through a gravitational field is bent could be tested only during a total eclipse of the sun, as described in the main text. With radio and radar telescopes, astronomers can check this prediction at any time. There are two ways of doing this. One is to follow a celestial source of radio waves (such as a quasar) with a radio telescope as it is occulted by the sun. Just before such a source passes behind the sun, the radio waves will be deflected by the sun's gravitational field, and the apparent position of the source will be slightly altered.

The other way, which has been used extensively by I. I. Shapiro at MIT, is to bounce radar beams off Mercury and Venus when the sun lies between us and these planets. By analyzing the time it takes these beams to go from the earth to the planets and back, one can determine the effect of the sun's gravitational field on the propagation of the radiation. Both of these tests have given very strong support to Einstein's general theory of relativity.

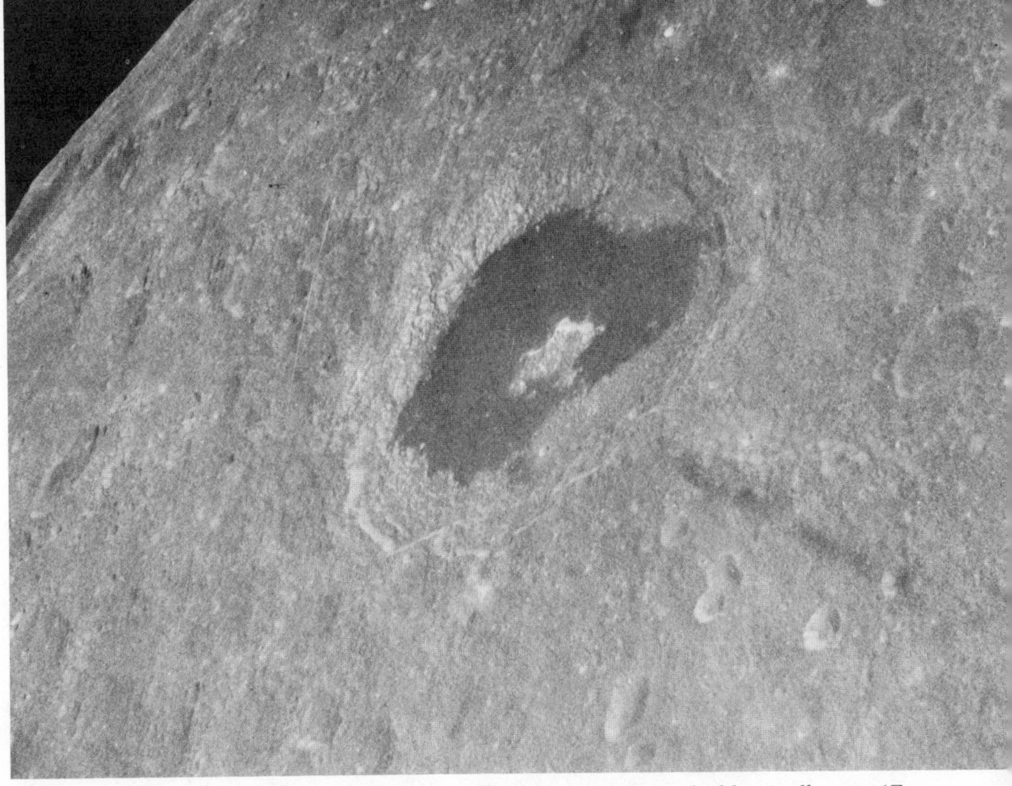

FIG. 6–44 *The crater Tsiolkovsky on the lunar far side, as photographed by Apollo 13.* (From the National Aeronautics and Space Administration)

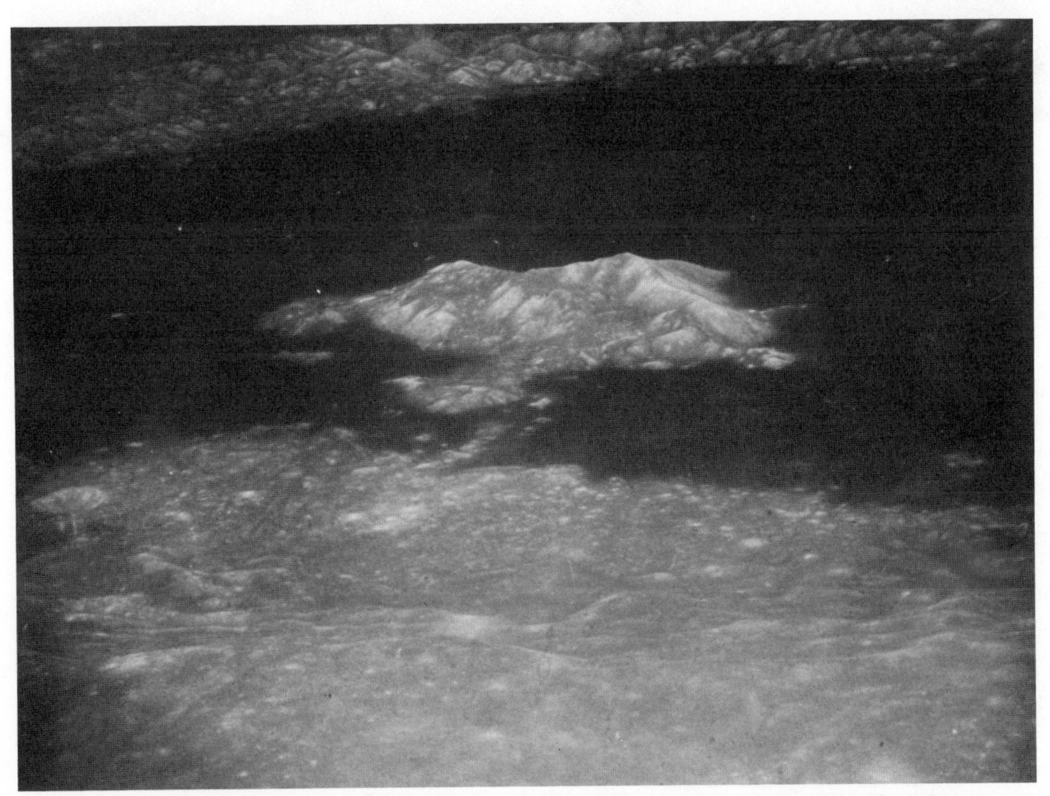

FIG. 6–45 *A closer view of crater Tsiolkovsky, photographed by Apollo 8. The flat floor of the crater is much darker than the surrounding lunar surface. The dark material is about 125 km across, measured from the near-to-far-side contacts in this view. The central peak, which stands as an "island" within the dark material, is about 40 km long.* (From the National Aeronautics and Space Administration)

FIG. 6–46 *Craters on the far side of the moon, photographed on the Apollo 12 mission.* (*From the National Aeronautics and Space Administration*)

CHAPTER 10: THE SPECIAL THEORY OF RELATIVITY

10.25 The Proper Time and the Aging of Fixed and Moving Observers

The full significance of the proper time can best be understood by first noting that Δs^2, the square of the space-time interval, between any two events is the same for all inertial observers. Consider then the inertial observer who, at the moment $t = 0$ as given by his clock, coincides with the first event just when it occurs. Let this observer be moving just fast enough to reach and coincide with the second event at the exact moment it occurs. The square of the proper time of this observer (the time given by his clock) multiplied by $-c^2$ is the square of the space-time interval between the two events.

When people in general discovered that one of the consequences of the special theory of relativity is that an observer who moves off into space and comes back to earth again ages more slowly than an observer fixed on the earth, they tended to consider it as the sheerest nonsense, and this feeling was shared by a number of top physicists. But there is now conclusive experimental evidence that a moving clock does slow down compared to a fixed clock and that the moving observer ages more slowly than the fixed observer. This evidence comes from two sources—the lifetime of certain short-lived massive particles that can be created momentarily in high-energy experiments, and the different rates of clocks attached to jet airplanes that circle the earth eastwardly and westwardly.

Various types of massive particles (called "strange particles" or baryons), which can be grouped into families, are created when protons collide violently with each other. Each such type of particle has a

FIG. 6–47 *Crater Lalande, a proposed future landing site on the moon. (From the National Aeronautics and Space Administration)*

definite lifetime which generally is very short and which is as characteristic of a given type of particle as are the spectral lines of a particular kind of atom. We find that the faster such a particle is moving when it is created, the longer it lives. Compared to such a particle at rest relative to us, its lifetime is longer by the factor

$$\frac{1}{\sqrt{1-\frac{v^2}{c^2}}}$$

(as deduced from the special theory of relativity), where v is its speed relative to us.

If a clock in a jet plane that is circling the earth eastwardly at a speed v relative to the ground is compared to an identical clock in a jet plane circling the earth westwardly at speed v relative to the ground, the eastwardly moving clock is found to run slower than the westwardly moving clock. The reason is that, owing to the earth's rotation, the eastwardly moving clock is traveling faster relative to a clock fixed on the earth's axis of rotation than is the westwardly moving clock. Actual experiments (carried out with clocks on two such jet planes) show that, in accordance with the formula for the slowing down of clocks deduced from the special theory of relativity, a clock moving eastwardly at a speed of 1 000 miles per hour relative to the ground loses 315 billionths of a second every time it goes around the earth compared to a clock moving westwardly at a 1 000 miles per hour relative to the ground. This is conclusive evidence of the slowing down of moving clocks.

FIG. 12–18 *This photomosaic of Mercury was constructed of 18 photos taken at 42-second intervals by Mariner 10 six hours after the spacecraft flew past the planet on March 29, 1974. The north pole is at the top and the equator extends from left to right about two-thirds of the way down from the top. Taken from a distance of about 210 000 km (130 000 miles), the pictures were computer-enhanced at the Jet Propulsion Laboratory. (From the National Aeronautics and Space Administration)*

CHAPTER 12: PROPERTIES OF THE PLANETS

So many new facts about the planets have been revealed by various space probes recently that lack of space prevents us from doing more than listing them, planet by planet.

Mercury

From an analysis of the Doppler shift in radar beams reflected from the surface of Mercury, we now know that its period of rotation is 58.65 days, which is exactly $\frac{2}{3}$ of its period of revolution. This remarkable relationship can be understood if Mercury's equator is not circular but elliptical (a result of the sun's tidal action). The torque of the sun on this elliptical bulge would then cause the major axis of the equator of the planet to point toward the sun whenever Mercury is at perihelion (where the solar torque is largest). One can show that this

FIG. 12–19 *Taken only minutes after Mariner 10 made its closest approach to the planet Mercury on March 29, 1974, this is one of the highest resolution pictures obtained during the mission. Craters as small as 150 meters (500 feet) across can be seen.* (*From the National Aeronautics and Space Administration*)

will happen if Mercury's period of rotation equals $\frac{2}{3}$ of its period of revolution.

From data obtained from the Mariner 10 Mercury probe in 1974, we know that the temperature of the daylight surface of Mercury reaches 700°K at noon. The temperature of the dark surface drops to 100°K at midnight.

No evidence of any atmosphere on Mercury was found by Mariner 10, but a very thin layer of hydrogen gas was detected clinging to its surface (about 10^{16} hydrogen atoms above each square centimeter). This may be the result of the solar wind (streams of protons from the sun) which floods Mercury. The protons are probably trapped on Mercury's surface by the very weak magnetic field (hundreds of times weaker than the earth's magnetic field) detected there by Mariner 10.

Mariner 10, which passed Mercury at a distance of 9 500 km from its surface, televised back to the earth about 2 000 photographs (see Figures 12.18 to 12.22). These show clearly that the surface of this planet, with its hundreds of craters and flat, lavalike areas (maria), looks very much like the surface of the moon.

FIG. 12–20 *A field of bright rays created by ejecta from a crater, radiating to the north (top) from off camera at lower right, is seen in this view of Mercury taken September 21, 1974, by Mariner 10. Source of the rays is a large new crater to the south, near Mercury's south pole. Mariner 10 was about 48 000 km (30 000 miles) from Mercury when the picture was taken at 2:01 p.m. PDT just three minutes after the spacecraft was closest to the planet. Largest crater in this picture is 100 km (62 miles) in diameter.* (*From the National Aeronautics and Space Administration*)

Venus

The analysis of the Doppler shift in radar beams reflected from the surface of Venus shows that this planet has a retrograde rotation (from east to west) with a period of 243 days.

From Soviet space probes (for example Venera 4, June 11, 1967, and Venera 5, October 18, 1967, which dropped instruments on the surface of Venus by means of parachutes) and American probes, we now know that there is water vapor on Venus and that about 95 per cent of Venus' atmosphere is CO_2. This, in conjunction with the "greenhouse effect," accounts for Venus' very high surface temperature (about 750°K, as recorded by instruments from Veneras 6 and 7). The atmospheric pressure on Venus' surface is about 100 times as great as on the earth's surface.

Mariner 10 passed within 6 000 km of Venus' surface in 1974. The thousands of photographs it televised back to the earth show the following:

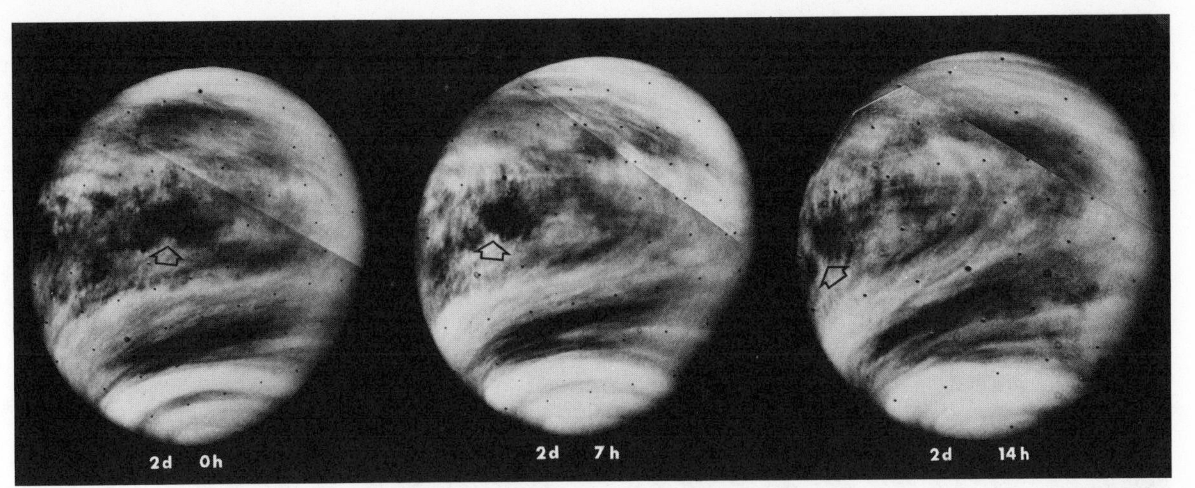

FIG. 12–21 *A series of photomosaics of Venus was taken at 7-hour intervals two days after Mariner 10 flew past the planet (from left, 10 a.m., 5 p.m., and 12 midnight, PDT, February 7, 1974). The pictures, taken through ultraviolet filters, show the rapid rotation of light and dark markings at the top of Venus' thick cloud deck. Size of the feature indicated by arrows is about 1 000 km (620 miles). (From the National Aeronautics and Space Administration)*

FIG. 12–22 *This view of Venus was taken from 720 000 km (450 000 miles) by Mariner 10 on February 6, 1974—one day after the spacecraft flew past Venus en route to Mercury. (From the National Aeronautics and Space Administration)*

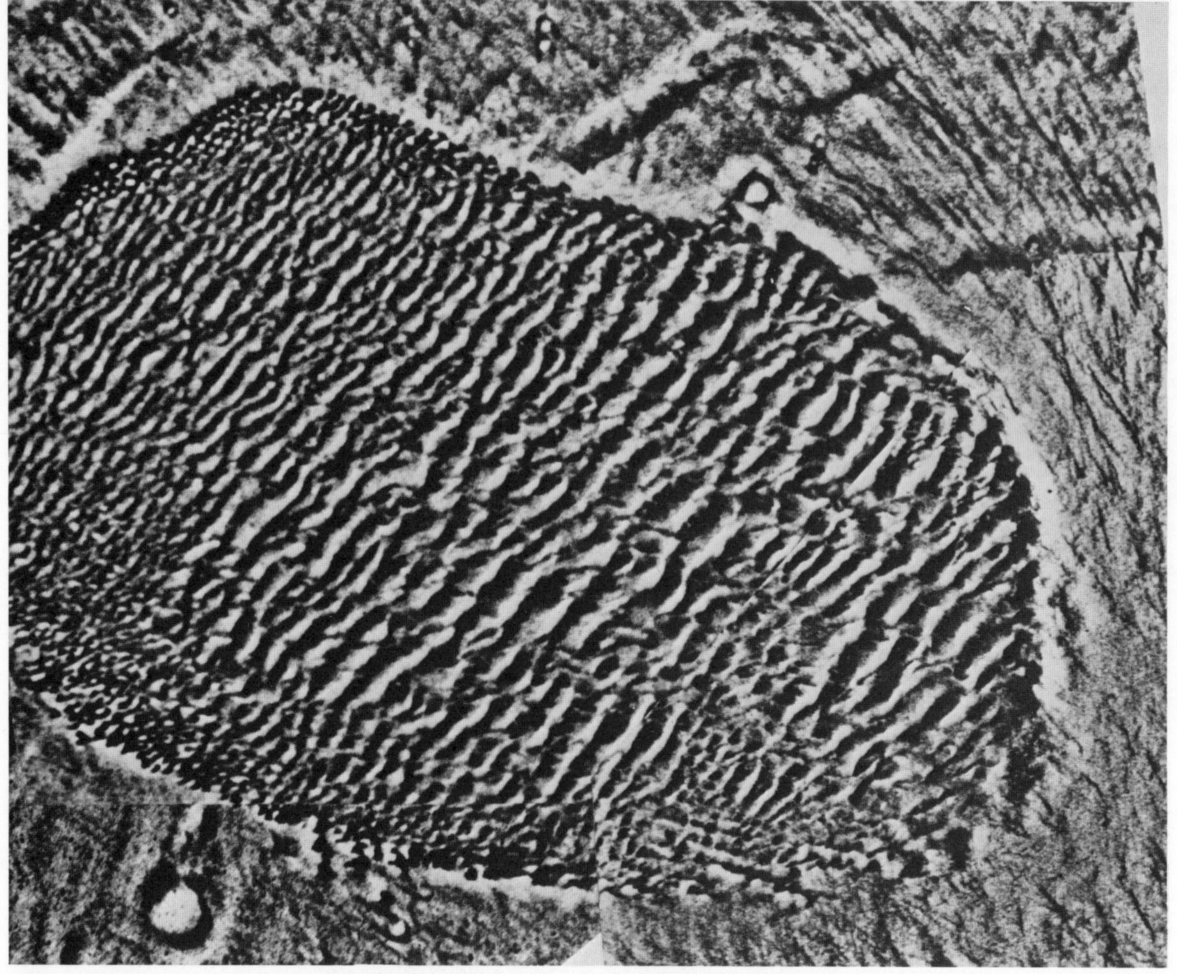

FIG. 12–23 *Mars as viewed by Mariner 9. This high-resolution photomosaic shows an elliptical dune field about 130 by 65 km. The dune field consists of series of subparallel ridges, 1 to 2 km apart, that closely resemble terrestrial transverse dunes. (From the National Aeronautics and Space Administration)*

1. There is a stratified hazy layer about 6 km thick above the clouds of Venus.
2. Dark and light bands alternate in the clouds and the upper atmosphere.
3. There are very rapid motions in the upper atmosphere, due to high-velocity winds (about 200 km/sec).
4. The temperature first drops with increasing height above the surface but then increases to a value of 600°K above the clouds.
5. Venus, like the earth, has an ionosphere.
6. A very weak magnetic field surrounds Venus.
7. The cloud cover consists of sulfuric acid droplets.

Radar echoes show an irregular surface with high peaks (up to 3 km high) and flat mare-like surfaces. The recent soft landings of recording instruments on Venus achieved by the Soviets (Venera 9 and 10) have transmitted direct photographs of the surface of Venus to the earth for the first time. One set of photographs shows rather sharp angular rocks, whereas a second set shows mountainous regions with smooth rounded rocks. The presence of sharp rocks means either that these rocks were recently ejected by volcanic action or that there is very little erosion on the surface of Venus. Temperatures of the order of 900°F and atmospheric pressures equal to about 90 earth atmospheres were measured. Contrary to what was expected, very low wind velocities (about 2 mi/hr) were detected at the surface. A high concentration of atmospheric carbon dioxide (about 97 per cent) was measured, in agreement with previous observations.

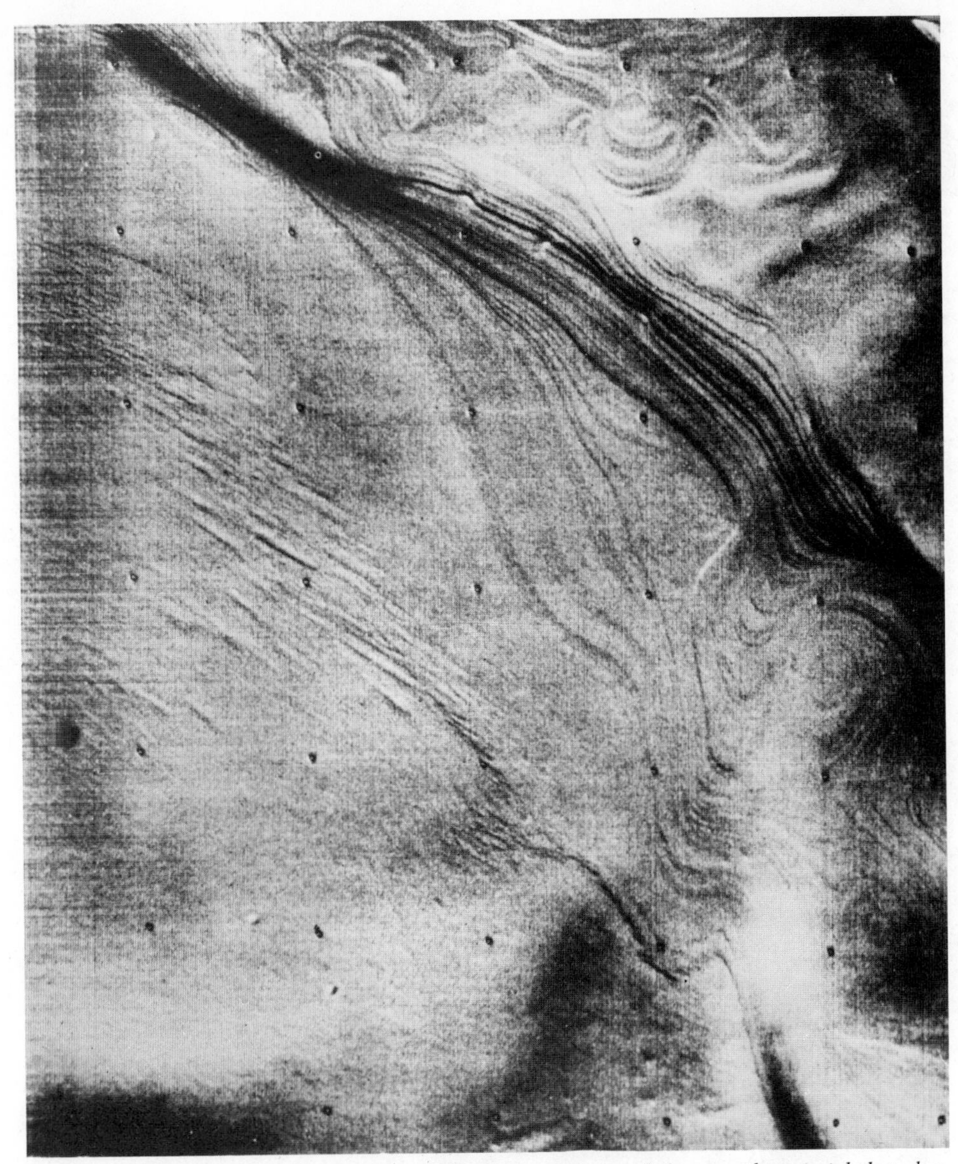

FIG. 12–24 *Mariner 9 photo of Mars. The laminar nature of the smooth material that characterizes much of the south polar region is revealed by these delicate patterns. The dark lines are ledges in the terrain, rather than dark deposits on a flat surface. (From the National Aeronautics and Space Administration)*

Mars

The findings listed below are based on data obtained from Mariners 6 and 7, which flew by Mars, and Mariner 9, which went into orbit around Mars in 1971. Mariners 6 and 7 transmitted more than 200 pictures of Mars, and Mariner 9 transmitted more than 7 200 photographs. Surface features as small as 100 meters across have been isolated on these photographic plates.

1. Mariner probes confirm the existence of clouds at great heights, as described in the main body of the text.
2. Great dust storms occur often and are practically opaque from heights of 11 km down to the surface.
3. The surface is similar to the moon's surface, with large numbers of craters of various sizes.
4. There are extensive smooth areas showing relatively recent lava deposits (within the last few hundred million years).

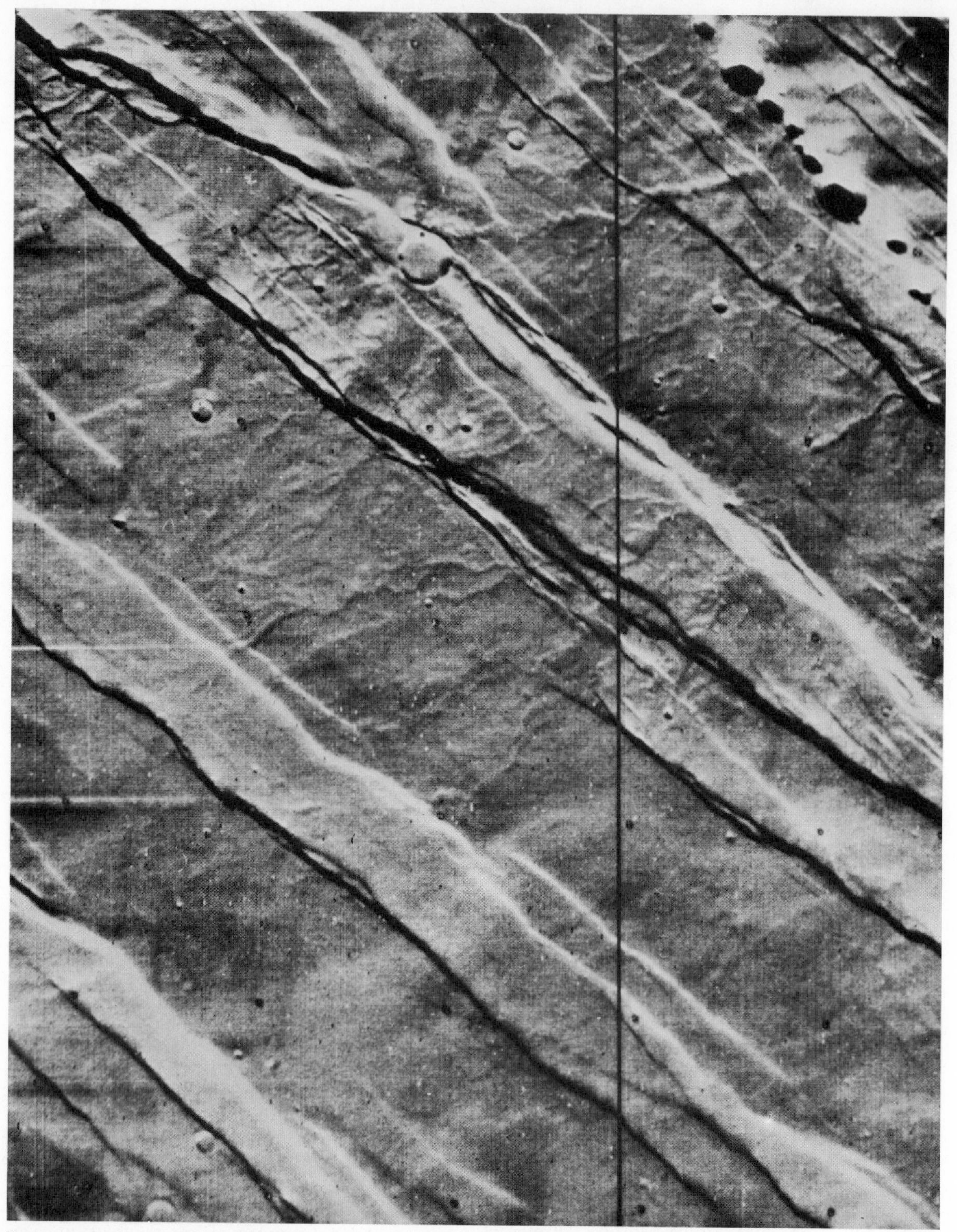

FIG. 12–25 *A detailed, narrow-angle view of part of the Martian graben system (about 60 by 80 km), as seen by Mariner 9. Multiple graben interrupt sinuous channels. Many fresh, raised-rim craters are younger than the broken surface. A linear crater chain is present at upper right. (From the National Aeronautics and Space Administration)*

FIG. 12–26 *Mosaic of two high-resolution scans of Mars by Viking 2, looking northeast to the horizon (3 km). The largest rock, at center, is about 60 cm long and 30 cm high. What appears to be a small channel winds from upper left to lower right. Horizon slopes owing to 8° tilt of the spacecraft.*

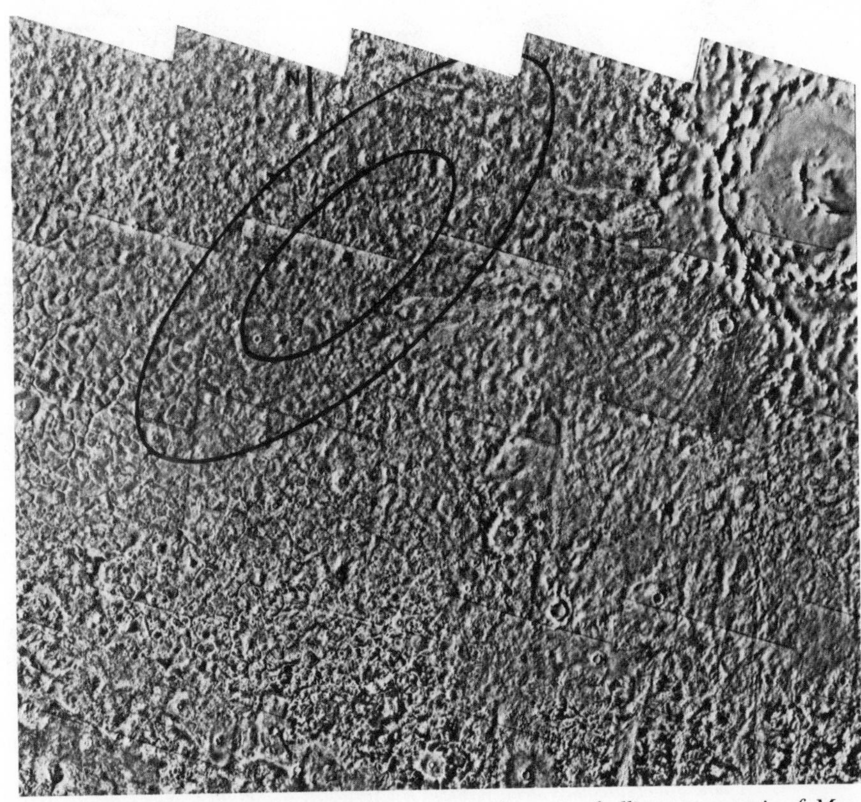

FIG. 12–27 *Targeted landing site for Viking 2 is centered in pair of ellipses on mosaic of Mars photos taken by Viking Orbiter 2. The landing site is on a mantle of young dunes laid down by Martian winds. The dunes cover the rocks thrown out of the large impact crater (Crater Mie) at right. Mie is 100 km in diameter and is similar to Copernicus on the moon.*

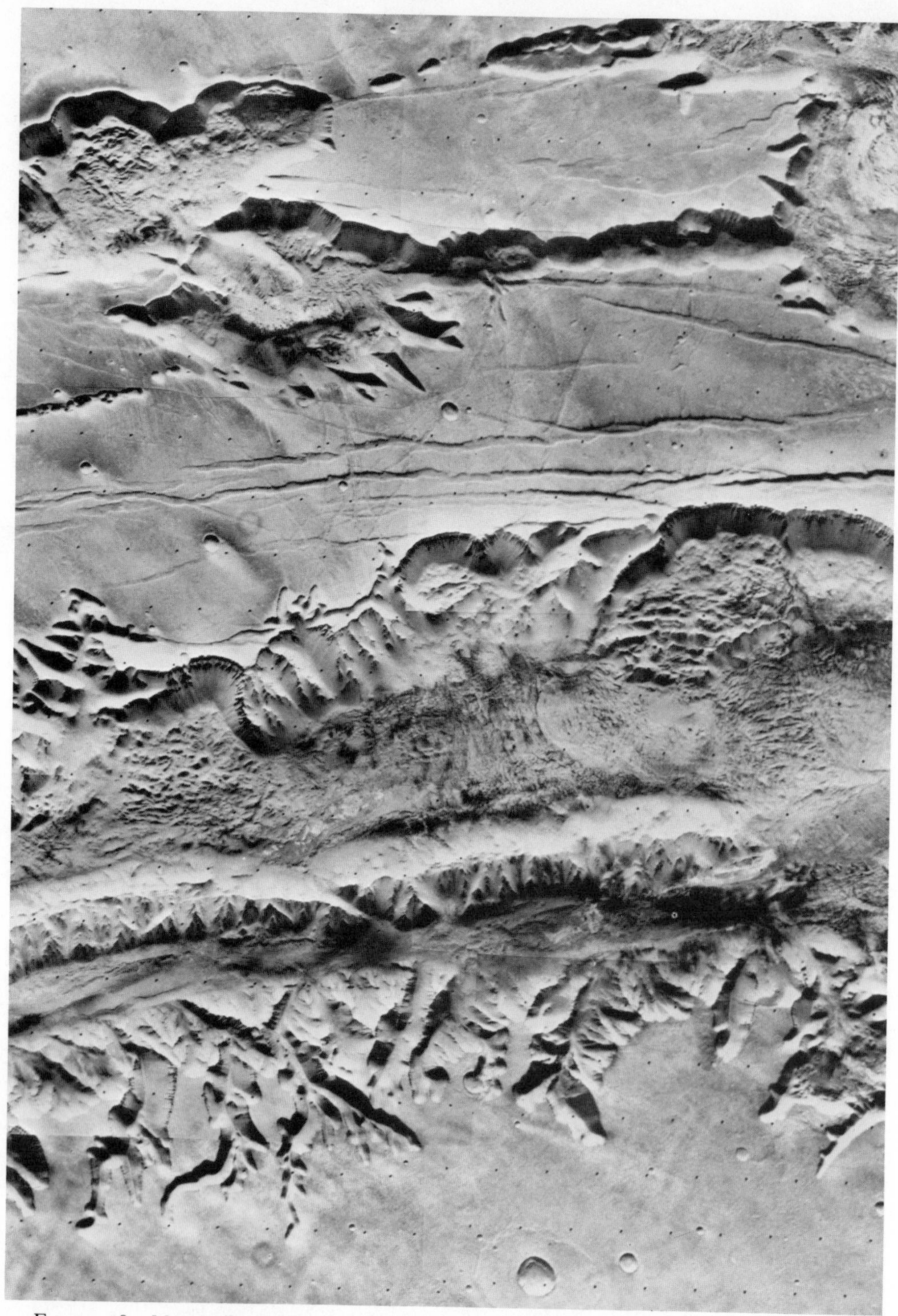

FIG. 12–28 *Mosaic showing part of the western end of the huge Martian equatorial canyon. A volcanic plateau is deeply dissected to form a complex series of interconnected depressions. Bedrock appears to be exposed along the crests of spurs jutting out into adjoining depressions. These features may result from a combination of wind excavation and mass wasting–slow downhill movement of debris during alternate freezing and thawing of ground ice.*

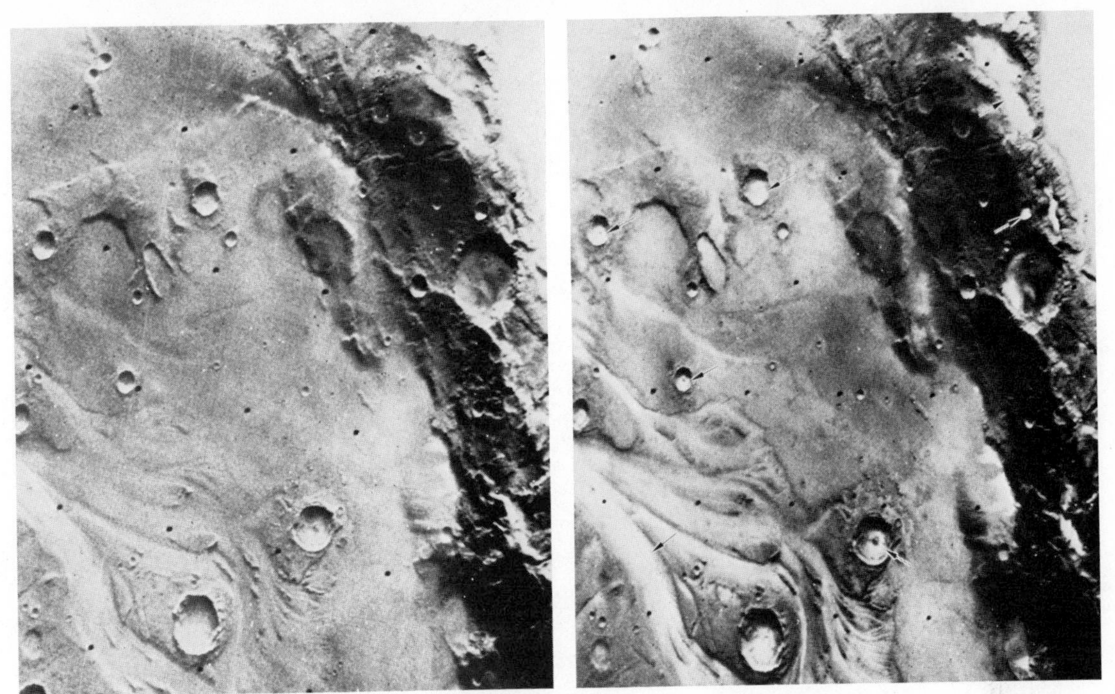

FIG. 12–29 *Early morning fog develops in low spots on Mars, such as crater and channel bottoms (see arrows, at right). Photo at left, taken by Viking Orbiter 1 shortly after Martian dawn, from 12 400 km; at right, 30 min later, from 9 800 km. Slight warming of the surface by the rising sun evidently has driven off a small amount of water vapor that has recondensed in the colder air just above the surface. Brightness measurements of fog patches indicate a film thickness of about 1 μ. These patches are the first direct, visible evidence showing where exchange of water between the Martian surface and atmosphere occurs.*

FIG. 12–30 *Mosaic of Arsia Mons, one of three large Tharsis volcanoes (among the youngest surface features on Mars). It stands about 19 km above surrounding terrain. The central caldera (collapse depression) is about 120 km across. Outside the caldera, many lava flows are visible as fine linear features. Vast amounts of lava appear to have flooded the surrounding plains.*

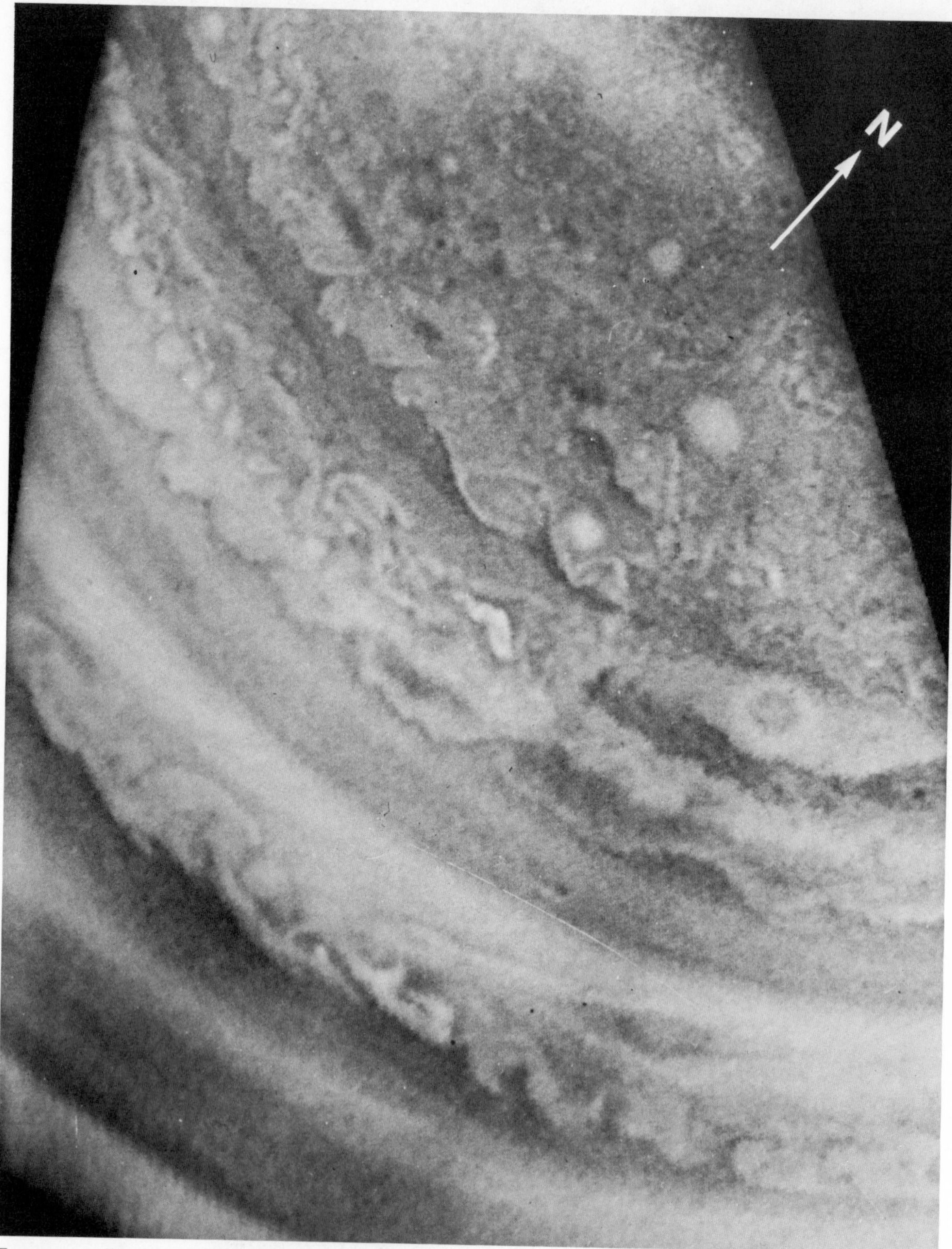

FIG. 12–31 *This enlarged portion of a picture of Jupiter, taken by Pioneer 11 in December 1974 from a distance of 600 000 km (373 000 miles), shows for the first time what appear to be hurricane-like storms in Jupiter's north polar region. The picture also gives the first detailed view of the breakup of Jupiter's alternating dark belts and bright zones as one goes toward the pole. (From the National Aeronautics and Space Administration)*

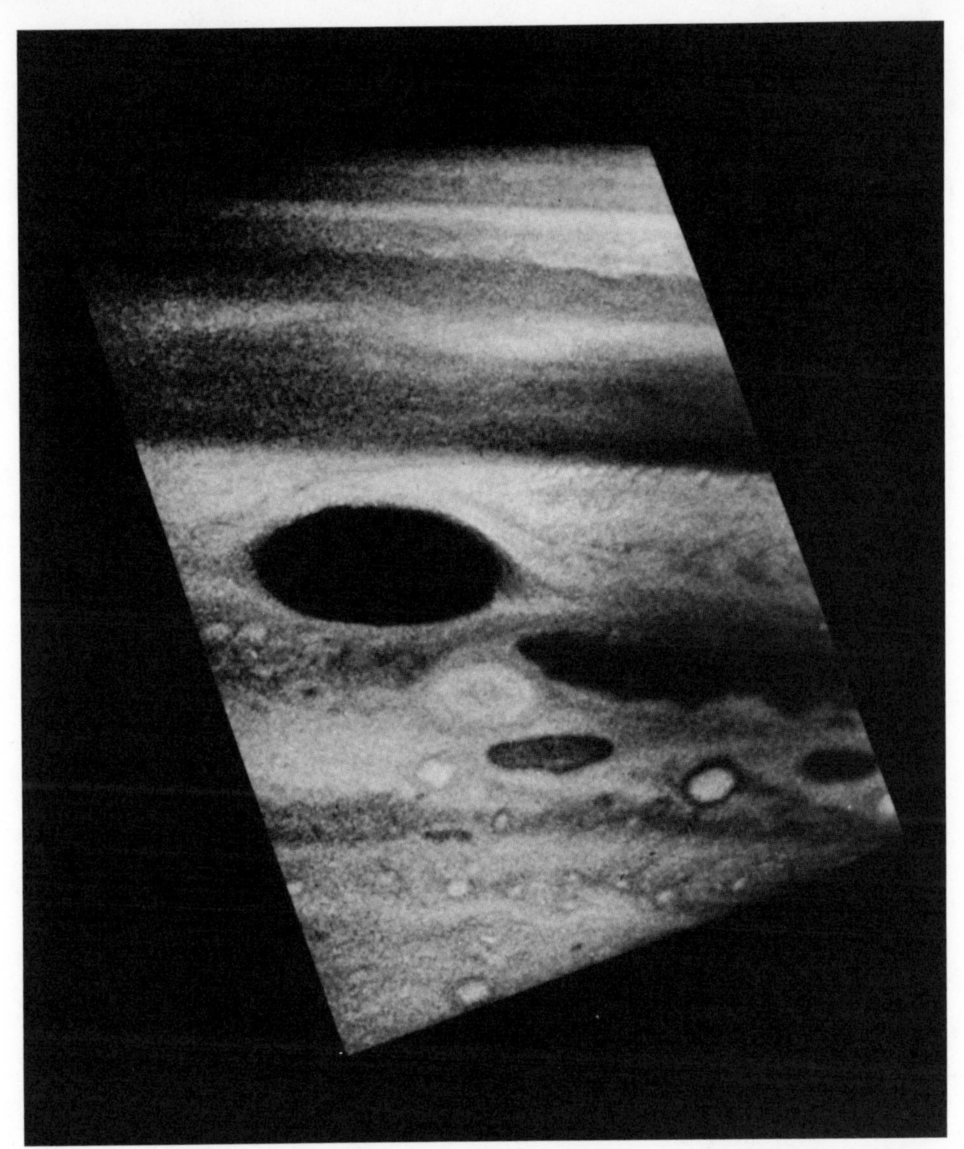

FIG. 12–32 *This view is the closest picture of Jupiter's Great Red Spot taken by Pioneer 11 (distance is 545 000 km or 338 000 miles). More details of the Great Red Spot and its surrounding region are visible here than have ever been seen before. The white oval below the right end of the Great Red Spot is one of three white ovals which are usually 120° apart around Jupiter. The oval's position relative to the Red Spot changes with time due to a different rate of flow of the cloud current which contains it. New details of the white oval in this picture, such as the circular "eye" in its center, strongly suggest rotational motion. Between the Red Spot and the white oval is a stream of brownish cloud material.* (*From the National Aeronautics and Space Administration*)

5. Congealed lava deposits and volcanic mountains are revealed in the Mariner photographs.
6. Various rills and branching channels (resembling the tributaries of dried water beds on the earth) are observed running into deep, wide rifts. This may mean that running water was present on Mars at some time in the past.
7. As measured by Mariner 7, the coldest regions are found at the south polar cap, where temperatures as low as −150°C occur. Measurements made by Mariners 6, 7, and 9 show a maximum temperature of 30°C at the equator. The maria were found to be about 10 to 15°C warmer than the surrounding areas.

The soft landing on Mars of the Viking 1 and 2 landers in 1976 has opened up a whole new era of Martian exploration. The excellent photographs sent back to earth from the Viking landers show windswept terrains of rocks, sand dunes, and cra-

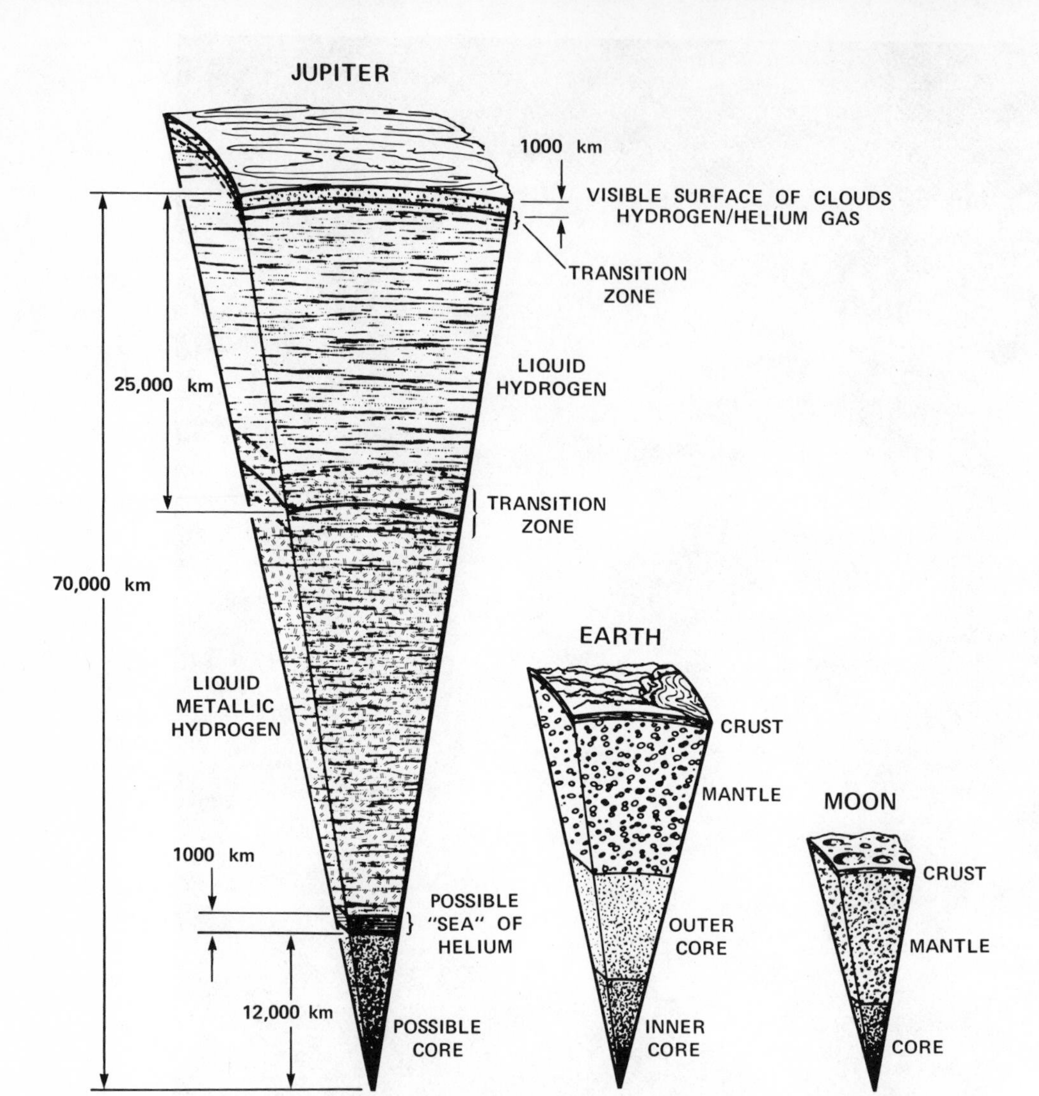

FIG. 12–33 *A graphic comparison of Jupiter's composition and those of the earth and the moon. (From the National Aeronautics and Space Administration)*

ters (see Figs. 12.23 to 12.30). The rocks are basalt-like and of volcanic origin. A remarkably high oxygen abundance (about 3 per cent) was found in the atmosphere (the abundance on earth is about 21 per cent), and a surprising amount of oxidation activity was found to occur in the soil.

Jupiter

Pioneer 10 was launched toward Jupiter on March 3, 1972 and passed within 130 000 km of the planet on December 3, 1973. Pioneer 11, which was launched on April 6, 1973, passed within 42 000 km of the cloudtops of Jupiter on December 3, 1974. Pioneer 11 is headed for an encounter with Saturn in 1979. Both spacecraft sent back vast quantities of data and color photographs of Jupiter. Each message (command and confirmation) between the earth and these vehicles, while in orbit around Jupiter, took about $1\frac{1}{2}$ hours to complete. Results of the Pioneer 11 mission to Jupiter are as follows:

1. Jupiter possesses a magnetosphere, magnetic fields, and radiation belts. The radius of the magnetosphere is about 80 times as large as Jupiter's radius.
2. Intense radiation belts of high-speed electrons and protons were found trapped by the magnetic fields at heights ranging from 40 000 miles to about 100 000 miles above Jupiter's surface. A flux of 10 million protons/sec/cm^2 was detected in these belts. These charged particles emit radio waves as they spiral around the magnetic lines of force.
3. Jupiter possesses a fairly intense magnetic field, which is about 20 times as strong at its

poles as the earth's magnetic field at the earth's magnetic poles. The interaction of Jupiter's magnetic field with the solar wind generates radio waves.

4. Infrared detectors on Pioneers 10 and 11 showed that Jupiter is radiating into space 2 to 2.5 times as much thermal energy as it receives from the sun per second. These detectors recorded a temperature of about 125°K.
5. As expected, free hydrogen and helium were detected in the atmosphere. A model atmosphere consisting of 85 per cent hydrogen and 15 per cent helium fits the observed temperature profile in the atmosphere quite well. The atmospheric pressure increases very rapidly with increasing depth and is so high a few hundred kilometers below the top of the atmosphere that hydrogen and other gases are liquefied.

CHAPTER 13: COMETS AND METEORS

Until quite recently, all observations of comets were made in earth-based observatories with visible light. Today we have cometary data obtained with radio telescopes on the earth and with space probes such as Mariner 10. The first comet studied in this way was Comet Kohoutek, which was first observed on March 7, 1973 and passed its perihelion (0.14 AU) on December 28, 1973. Radio waves from this comet showed the presence of hydrogen cyanide (HCN) and methyl cyanide (CH_3CN). Data from Mariner 10 showed that water was dissociating into atomic hydrogen H and the hydroxyl radical OH within 15 000 km of the nucleus. At distances beyond 45 000 km of the nucleus, the OH radical was dissociating into H and O.

This comet did not live up to expectations, and did not become a spectacular object in the heavens as predicted. However, it has provided a considerable amount of important scientific data.

13.22 Meteorites

The study of meteorites has aroused considerable interest recently because of the light they can shed on interstellar organic compounds. The chemical analysis of meteorites that have fallen on the earth during the last few years has demonstrated conclusively that amino acids are formed in interstellar space. Of the 20 amino acids that constitute living matter on earth, 19 have been identified on meteorites. The mixture of amino acids found on meteorites generally consists of equal quantities of *l*-type and *d*-type organic compounds, unlike the mixtures found on the earth, which consist of only *l*-type compounds. This finding confirms the work of radio and infrared spectroscopists, who have identified a large group of organic compounds in interstellar space.

CHAPTER 17: SPECTRAL CLASSIFICATION OF STARS

The H-R diagrams of open clusters are important in astronomy for two reasons, as discussed in Section 27.25.

1. These diagrams are the observational evidence for the theoretically derived fact (discussed in Chapter 23) that the more massive a star is when it contracts from a cloud of dust and gas, and begins its life on the main sequence, the higher up on the main sequence it lands.

2. The diagrams are the observational evidence for the theory of the evolution of stars; they show that the more massive a star is, the more rapidly it evolves (or ages) and the more quickly it leaves the main sequence. As can be seen from the H-R diagrams of the various open clusters, the blue stars of these clusters are off the main sequence (they have evolved away from it) whereas the yellow and red stars in the cluster are still on the main sequence.

In the composite H-R diagram of open clusters (Figure 17.9), we have assumed that the main sequences of all these clusters coincide. This would be true only if the chemical compositions of the dust clouds from which these clusters originated were the same. If, however, these open star clusters originated from clouds with different chemical compositions (different ratios of hydrogen content to helium content) the main sequences of the clusters would be paralled to each other but not coincident. The theory of stellar structure as discussed in Chapters 22 and 23 shows that the main sequence of a cluster is shifted upward (all the stars on the main sequence of the cluster are more luminous) if the helium content of the cloud is held constant while the hydrogen content is decreased (increase in the abundance of heavy elements). On the other hand, if the abundance of heavy elements in the cloud is held constant while the hydrogen content is decreased, the main sequence is shifted downward (all the stars on the main sequence are less luminous).

CHAPTER 20: THE ATMOSPHERE OF THE SUN

20.26 Solar Pulsations

Recent measurements of the sun's diameter by workers at the Santa Catalina Laboratory for Experimental Relativity by Astrometry (SCLERA) with an accuracy of 10 parts per million have demonstrated the existence of a periodic variation of the equatorial brightness of the sun. This discovery, and with it the discovery of a small oblateness of the sun (which is what one would expect from a sun rotating uniformly at its observed rate), makes it unnecessary to assume that the interior of the sun is rotating much more rapidly than the surface, as had been proposed previously by other astronomers. The observed variation of the equatorial brightness, as verified by astronomers at several observatories, indicates that the sun is pulsating in very complex modes with a period of 2 hr 40 min. These oscillations, when properly analyzed, can give astronomers a new insight into the internal structure of the sun.

CHAPTER 21: STELLAR RADII AND MASSES

21.6 The Application of the Interferometer to Measurement of Giant Stars

To reduce the effects of atmospheric fluctuations on stellar images, interferometers have recently been introduced that use electronic detectors instead of photographic plates. In the Narrabri interferometer, set up in Australia, two large mirrors (whose separation can be altered) focus light from a star on two different photoelectric detectors. When light strikes these detectors, output currents are generated which are sent into an electronic correlator. The electronic correlator compares the two currents, and the results of these comparisons are read off on the output indicator. The current from each detector fluctuates because the light which gives rise to this current fluctuates for various reasons. First of all, the atoms of a star emit light in a random fashion so that the light occurs in bursts of radiation from the surface of the star. Secondly, the radiation itself consists of a vast number of electromagnetic vibrations of different frequencies. The currents from the two photoelectric detectors are thus records of the fluctuations in the radiation of the two beams. If the mirrors are very close together, the fluctuations in the two currents are practically identical, so that there is a precise correlation as measured by the electronic correlator. But if the mirrors are far apart, the fluctuations are uncorrelated. The electronic correlator simply multiplies the fluctuations in the two currents, and the product is exhibited in the output indicator. If the product is $+1$, the two sets of fluctuations are perfectly correlated (they are exactly in phase); if the product is zero, the two sets of fluctuations are uncorrelated or completely different. This happens for a definite separation between the two mirrors. To measure the angular size of a star, one proceeds as follows: one points the two mirrors at the star and slowly separates them until the output indicator shows that the correlation of the two beams is zero. The angular diameter of the star is then equal to the wavelength of the light divided by the distance between the two mirrors. With this instrument, we can not only measure the angular diameter of a star but also find the shape of the star and determine how the light varies over the star's disk. In a sense we can see how the star would look if we could view it with a telescope whose mirror had a diameter equal to the maximum spacing between the two mirrors of this interferometer (600 feet for the Narrabri interferometer).

The Narrabri interferometer does not suffer from the disadvantages of the Michelson interferometer because the frequency of the fluctuations in the output currents is very small (about 10^8 cycles/sec) as compared to the frequency of light (about 6×10^{14} cycles/sec). Therefore the distances from the mirrors to the photodetectors can differ by as much as 30 cm if the mirrors are a few meters apart. This is easy to achieve. The irregularities in the atmosphere do not destroy the correlations of the two beams because these irregularities delay and alter the arrival of pulses at the two mirrors by times that are short compared to the period of fluctuations of these pulses.

In the Narrabri instrument the two mirrors are mounted on cars that are free to move on the rails of a circular track whose radius is 300 feet. The photoelectric detectors are mounted on a 30-foot-high tower at the center of the track. This instrument is intended to be used to measure the angular sizes of the 200 brightest stars in the sky.

In a recent development, electronic computer techniques have been employed to produce an enhanced photograph of the surface of Betelgeuse. This was done by first obtaining a large number of successive short-time-exposure images of Betelgeuse with a large optical telescope and then applying an electronic light intensifier to these images. In this way the intrinsic features of the light from the surface of Betelgeuse were enhanced and random fluctuations arising from atmospheric irregularities were minimized. The resulting photograph showed surface features on Betelgeuse similar to those observed on the sun's photosphere.

21.11 Mass-Luminosity Relationship

The reason that red supergiants such as Antares and Betelgeuse lie on the same mass-luminosity curve as main-sequence stars (as shown in Figure 21.5) is that such supergiants evolved from massive O and B main-sequence stars. Even though their internal structure and chemical compositions changed during this evolution away from the main sequence, the masses and luminosities of these O and B stars did not change much, so that when they became red supergiants they still obeyed the same mass-luminosity relationship.

CHAPTER 22: STELLAR INTERIORS

22.7 Convective Transport of Energy: Convective Equilibrium

Under certain conditions, convection may occur not only in the central core of a star but also in the outer envelop of a star. An analysis of the internal structures of main-sequence stars shows that the G, K, and M types have either no convective cores or very small ones. But these stars do have extensive outer convective envelopes. On the other hand, the O, B, A, and F main-sequence stars have well-defined convective cores but either very shallow or hardly any convective envelopes. This difference arises because energy is generated by the proton-proton chain in the lower main-sequence stars, whereas the carbon cycle dominates in the upper main-sequence stars.

22.11 Eliminating the Density ρ from the Equations

The four stellar interior equations given in this section in the main text apply only to regions of the star through which the energy is transported by radiation. If the region is in convective equilibrium, however, Equation (22.9) must be replaced by the equations

$$\Delta T = \left(1 - \frac{1}{\gamma}\right)\frac{T}{P}\Delta P = -\left(1 - \frac{1}{\gamma}\right)\frac{GH}{k}\frac{\mu M(r)}{r^2}\Delta r,$$

which follows directly from Equation (22.3) for an adiabatic process.

22.20 The Guillotine Factor and the Opacity Tables

The opacity of stellar material plays an important role in the structure of a star because it determines whether convection or radiation will transport energy. If the opacity is large enough, the stellar material will not allow radiation to get through and convection will have to transport energy. Convection will occur in a given shell at a distance r from the center of a star if the opacity of the material in that shell multiplied by the average rate at which energy is generated in the sphere of radius r exceeds the quantity

$$4\left(1 - \frac{1}{\gamma}\right)\bar{\kappa}\frac{L}{\mathscr{M}},$$

where $\bar{\kappa}$ is the average opacity for the whole star and L and $\mathscr{M}$ are the star's luminosity and mass, respectively. This tells us why convective cores exist in the upper main-sequence stars. The central temperatures in these stars are so high (of the order of 30 million degrees) that the amount of energy released per second is too great to be radiated out.

23.3 The Half-Life of the Neutron and β Radioactivity

In the modern picture of the neutron and proton, it is customary to consider these two particles as two different states of a single particle called a nucleon. The neutron, being slightly more massive than the proton, is considered to be an excited state of the nucleon, whereas the proton is the unexcited or ground state of the nucleon. The atomic weight A of a nucleus thus equals the number of nucleons in the nucleus. The de Broglie wavelength of a nucleon (neutron or proton) that is moving as fast as an electron is about 2 000 times smaller than that of the electron because the de Broglie wavelength of a particle varies inversely as its mass. It is owing to their small de Broglie wavelengths that nucleons can exist inside nuclei, which are very tiny compared to atoms.

23.6 Nuclear Forces and Mesons

Mesons as a group are particles whose spins, measured in units of $h/2\pi$, are 0, 1, 2, . . . , etc. (integral multiples of $h/2\pi$). The three π-mesons (or "pions") described in the text all have spin 0. Mesons are thus distinguished from neutrons, protons, electrons, and neutrinos, which have spins of $\frac{1}{2}$ a unit. The principal physical distinction between these two groups of particles, which is related to the difference in their spins, is that the particles with integral units of spin (the mesons) obey different statistical laws from those obeyed by the particles such as electrons and nucleons with spins equal to $\frac{1}{2}, \frac{3}{2}, \frac{5}{2}, \ldots$ units of spin. The particles in the latter group are governed by the Pauli exclusion principle (see Section 19.29) which assigns one and only one particle to a single energy state. The statistical laws obeyed by such particles were first discovered and stated by Enrico Fermi; therefore all these particles, as a group, are called fermions. The mesons and other such particles with integral multiples of spin (among which is the photon, which has 1 unit of spin) are not governed by the Pauli exclusion principle; they obey statistical laws that were first stated by Bose, and so these particles, as a group, are called bosons. Bosons (photon and mesons) are the quanta of the force fields that bind fermions to each other. Thus electrons are kept bound to the nucleus inside an atom by photons that are tossed back and forth between electron and nucleus and the nucleons in a nucleus are bound to each other by tossing pions back and forth.

The particles that we call μ mesons in the text (there are two kinds, μ^- with negative charge and μ^+ with positive charge) were discovered in cosmic rays before pions were, and it was then thought that these μ "mesons" were the bosons that cement nucleons together. This was later found to be a wrong assumption. We now know that the μ "mesons" are not the mesons that physicists were looking for. They are not bosons at all but rather fermions with spin $\frac{1}{2}$. For that reason we call them muons. The π^- and π^+ mesons decay into muons and neutrinos in about a millionth of a second, but the neutrinos in this case are not the same as those emitted by neutrons when neutrons change into protons. (When neutrons change into protons they actually emit anti-neutrinos and electrons.) There are thus electron-neutrinos ν_e and muon-neutrinos ν_μ. A muon decays spontaneously into an electron, an anti-electron neutrino $\bar{\nu}_e$, and a muon-neutrino ν_μ (the bar above the symbol for a particle stands for the anti-particle). The muons (which except for their mass are like electrons in all respects) are rather mysterious particles; no one knows what role they play in the scheme of things. Electrons, neutrinos, and muons as a group are called leptons, whereas nucleons and all heavy particles that decay into nucleons are called baryons. Baryons and mesons together are called hadrons or hyperons.

We note here an important neutrino characteristic that was discovered by T. D. Lee and C. N. Yang and led to the discovery that the universe when viewed in a mirror (the mirror-image universe) does not obey exactly the same laws as the real universe. The universe and its mirror image behave differently in any process involving neutrinos. The reason is that neutrinos are essentially left-handed particles (like left-handed screws) in the following sense: Whenever a neutrino is receding from an observer, the observer sees the neutrino spinning counterclockwise; when a neutrino is approaching an observer, it is always spinning clockwise. From this it follows that all neutrinos are moving at the speed of light and must therefore have a zero rest-mass. If this were not so, the same counterclockwise-spinning neutrino that was receding from one observer would appear to be approaching another observer who was moving in the same direction as the neutrino but with a greater speed than the neutrino and was thus overtaking it.

In radioactive decay of nuclei in which β-rays are emitted, anti-neutrinos are always emitted with electrons and neutrinos are always emitted with positrons.

23.16 The Proton-Proton Chain

The series of nuclear reactions (given in the box on page 394 of the main text) that we refer to as the proton-proton chain is only one of three competing sets that lead to the fusion of four protons into He^4. In the other two sets of reactions that go on simultaneously with the first set, steps one and two are the same as in the first set so that He^3 nuclei are still formed. But in these competing sets of reactions the He^3 nuclei combine with He^4 nuclei to form Be^7 (beryllium^7). The Be^7 then may either decay to Li^7 with the emission of an electron and an anti-neutrino (one competing set) or combine with a proton to form B^8 (boron^8), which then decays into Be^8 with the emission of a positron and a neutrino (the other competing set). In either case the final result is the formation of two He^4 nuclei. The important feature of these competing sets of reactions (particularly the one in which B^8 is formed) is that they lead to the emission of fairly large streams of neutrinos from the core of the sun; but the most careful attempts to detect these solar neutrinos have failed. The number of solar neutrinos that have actually been detected per year is smaller by at least a factor of 10 than that deduced from the theory. This has led to serious questions about the physical conditions at the core of the sun as deduced from the solar models that are computed from our basic stellar equations on page 379.

23.26 Surface and Center Conditions

From our discussion of the way the mathematical model of a star is constructed (using the basic equations on page 379), we see that a model cannot be constructed unless some definite information about the star is given; only then can we substitute numbers into the equations and begin our calculations. An analysis of this problem shows that if we specify the mass and the chemical constitution of the star (its hydrogen and helium content), the model of the star is completely determined. This relationship is known as the Vogt-Russell theorem. The physical basis of this theorem stems from the fact that a given mass (a cloud) of dust and gas (massive enough to contract gravitationally) having a definite chemical composition will contract down to one and only one final configuration (star). If we alter the mass of this initial cloud or its chemical composition or both, the final configuration will change. But since the mass and chemical composition of the cloud are the only things we can alter, we see that, once chosen, these two physical characteristics determine the star's structure and its future history.

23.29 A Chemically Homogeneous Model of the Sun

Even without solving the four interior stellar equations we can get a fair idea of the conditions in the deep interior of a chemically homogeneous star from general dynamic considerations. We can obtain an estimate of the mean temperature in a star from an important theorem in mechanics, known as the theorem of the virial, which states that a system of gravitationally interacting particles will be in dynamic equilibrium if the negative of the total potential energy of the system is twice the total kinetic energy of the particles. If M is the mass of the configuration (we assume the star to be a sphere) and R is its radius, then for equilibrium we must have

$$\frac{1}{2}\frac{GM^2}{R} = Nm\overline{v}^2,$$

where N is the total number of particles, m is the mass of a single particle, and $\overline{v}$ is the average speed of a particle. But from equation (16.7) we know that $m\overline{v}^2 = 3kT$. Hence, noting that $M = Nm$,

$$N3kT = \frac{1}{2}\frac{GMNm}{R},$$

or

$$T = \frac{m}{6k}\frac{GM}{R}.$$

If μ is the mean molecular weight of the star, we have $m = \mu H$, where H is the mass in grams of a hydrogen atom. Substituting this in the above equation we obtain

$$T = \frac{\mu H}{6k}\frac{GM}{R}.$$

This shows us how the average temperature in a star depends on its mass M, its radius R, and its chemical composition μ. Applying this to the sun, for which we assume μ to be 1, we find the average temperature to be about 4 million degrees.

From this figure we can obtain a lower bound on the central pressure. At the center of a star, the gas law tells us that the central pressure P_c, the central density ρ_c, and central temperature T_c are related by the equation

$$P_c = \frac{k\rho_c T_c}{\mu H}.$$

Since the central density is certainly larger than the mean density given by $M/\frac{4}{3}\pi R^3$ and the central temperature is certainly larger than the mean temperature deduced above, we must have

$$P_c > \frac{k}{\mu H}\frac{M}{\frac{4}{3}\pi R^3}\frac{\mu H}{6k}\frac{GM}{R}.$$

Hence

$$P_c > \frac{GM^2}{8\pi R^4}.$$

Applied to the sun, this tells us that the sun's central pressure is larger than 4.44×10^{14} dynes/cm^2 or 4.5×10^8 atmospheres.

To obtain an estimate of the central density of a homogeneous star, we must know its luminosity and the formula for its energy generation. Noting that the total luminosity L of a star must be less than $\epsilon_c M$, where M is the total mass of the star and ϵ_c is the rate of energy generation per gram at the center, we can estimate the central density from the energy formula for the proton-proton chain (23.1) or for the carbon cycle (23.2). For the sun we obtain an estimate of about 100 gm/cm^3 for the central density.

23.34 Thorough Mixing and Stellar Evolution

The path in the H-R diagram of Figure 23.8 showing the contraction of a star to the main sequence applies only if the star has no outer convective envelope. If the star does have a convective envelope, as in the case of stars like the sun, the path of its descent onto the main sequence in the H-R diagram is different from that shown. There is still a horizontal branch that differs in length, depending on the mass of the contracting star (the more massive the star, the longer the horizontal branch), but there is also a steep vertical path that takes the star from a low absolute magnitude (large luminosity) to the horizontal branch in a relatively short time.

The series of events that lead from a condensation in a cloud of gas and dust to a stellar configuration (a gas sphere) are then roughly as follows: Depending on the mass of the initial condensation, gravitational contraction occurs more or less rapidly (in anywhere from a few thousand years for massive condensations to a few hundred thousand years for less massive ones).

When contraction first occurs, much of the released gravitational potential energy escapes in the form of infrared radiation while the remainder is trapped in the condensation to increase its temperature. It is owing to the escape of gravitational energy that the condensation persists and continues to contract. Otherwise the particles in it would escape. Only if energy is radiated away, as just described, can stars be formed. As more and more material falls into the condensation, the pressure and temperature build up until the central core reaches a state of equilibrium with the outer material still falling in. As this continues, the gas and dust above the core become so opaque that the energy cannot escape by means of radiation; convection sets in, and the entire configuration reaches a state of convective equilibrium. Contraction continues but only very slowly, so that the structure is practically in a state of gravitational equilibrium, and only half the gravitational energy released by this slow contraction escapes. The other half remains within the configuration, raising the temperature and pressure gradually.

During this period in the pre–main-sequence stage of a star's life, its path in the H-R diagram is almost vertically downward (actually downward and slightly to the left, so that the temperature rises while the luminosity decreases; the configuration becomes smaller, hotter, and bluer). The vertical paths of all such condensations (regardless of their masses) are very nearly parallel to one another, and they all lie in a vertical band on the H-R diagram extending from 1600 degrees on the horizontal axis of the diagram to about 4000 degrees. The more massive a star is, the shorter is this vertical path (the less the time spent on this path). A star of 10 solar masses spends 10^3 years moving along this path, whereas a star of 1 solar mass spends over a million years on this path and a star of 0.1 solar mass is on this path for about 100 million years.

These vertical paths either end on the main sequence (for stars of masses smaller than 0.5 solar masses) or turn sharply to the left, becoming very nearly horizontal similar to the horizontal part of the path shown in Figure 23.8. This happens when most of the internal convection currents have disappeared (radiation prevails) and only the outer zone of the configuration is convective. The star then continues contracting slowly, moving along the horizontal path as it does so, with its surface temperature steadily rising until it reaches the main sequence (the zero-age line), when the thermonuclear phase of its life begins. For stars like the sun this happens when the central temperature reaches about 10 million degrees, which is high enough to trigger the proton-proton chain, and for upper main-sequence stars when the temperature reaches 20 million degrees, which is high enough to trigger the carbon cycle. The energy released by thermonuclear fusion prevents any further gravitational contraction; stars do not change in size while they are on the main sequence. When they reach the main sequence, the massive stars re-establish convective cores and lose their outer-zone convective currents. Stars more massive than the sun reach the main sequence in anywhere from ten thousand years to a few million years. Stars as massive as the sun reach the main sequence in about 30 million years, and those less massive than the sun may take 100 million years or more to become main-sequence stars.

23.40 Postgiant Evolution of Population II Stars

The comparison of theoretical evolutionary tracks of population II stars with observations has been difficult until now because little has been known from direct observation about the relative abundance of helium in these stars, and the helium content is important in the evolution of stars. Most of the stars in our Galaxy have very low atmospheric abundances of heavy elements (they are metal-poor) compared to the sun, but we know little about their helium abundances because their surface temperatures are too low to excite helium and therefore the absorption spectrum of helium cannot be observed. Owing to this fact, the abundances of helium in these metal-poor stars cannot be determined spectroscopically. Moreover, the masses of these stars are not well known, so we have two unknowns to contend with.

The H-R diagram of these population II stars in our Galaxy shows a very sparsely populated (it is practically absent) upper main sequence (above an effective temperature of 6300°K). This means that such stars are no longer being formed in the Galaxy. Those that are still present must be of relatively low mass (they have hardly evolved away from the main sequence) and are among the oldest objects in the Galaxy.

We may therefore assume that the abundances of elements found spectroscopically in the atmospheres of these stars are about the same as in the primordial gas from which the Galaxy and these old stars were fromed some 10^{10} years ago. This again leads to the conclusion that the relatively large abundances of heavy elements in the atmospheres of the young metal-rich (population I) stars like the sun must have been built up by nuclear transformations in the old first-generation (population II) stars and then ejected into interstellar space by some explosive process. Because recent evidence strongly supports the idea that all the helium now

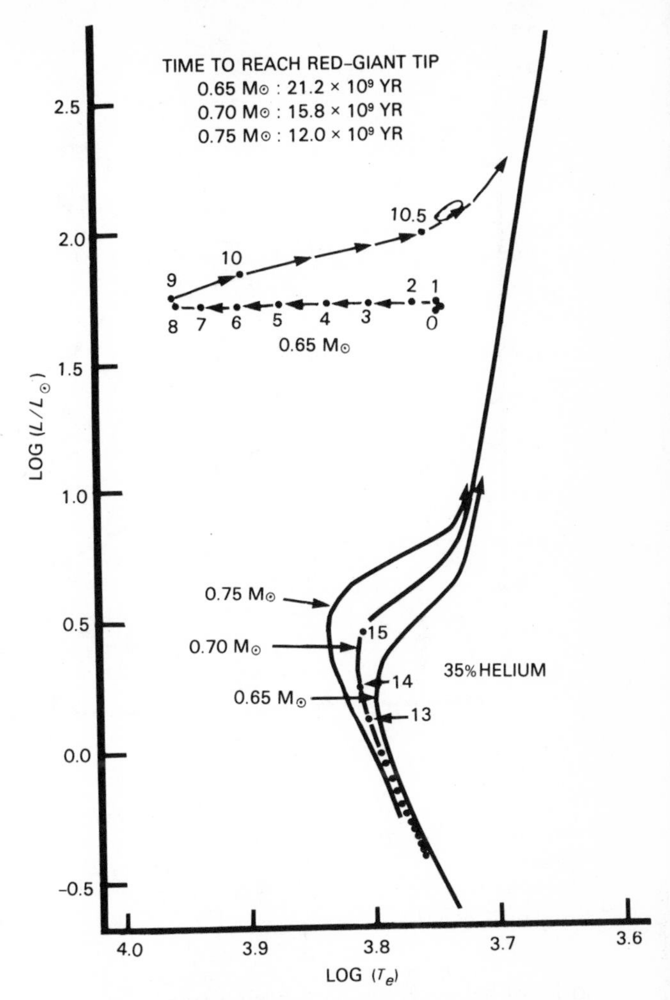

FIG. 23–12 *Evolutionary paths for metal-poor stars with a high initial abundance of helium. Helium is assumed to compose, initially, 35 per cent of each star's mass when it first reaches the main sequence. Intervals between consecutive circles along the 0.7 $M_\odot$ model track correspond to evolutionary times of 10^9 years. Intervals along the 0.65 $M_\odot$ dashed track correspond to 10^7 years.* (According to Iben and Faulkner)

present in population II stars was there initially and was not built up in these stars, we avoid the helium-abundance difficulty by the reasonable assumption that a considerable quantity of helium (in fact, as much as is required to account for the observations) was formed from the primordial neutrons and protons in the initial stages of the fireball, shortly after the "big bang," from which our universe originated. It is possible to find a set of values for the density and the temperature in this initial highly condensed fireball (a very hot, dense gas of neutrons and protons) such that large quan-

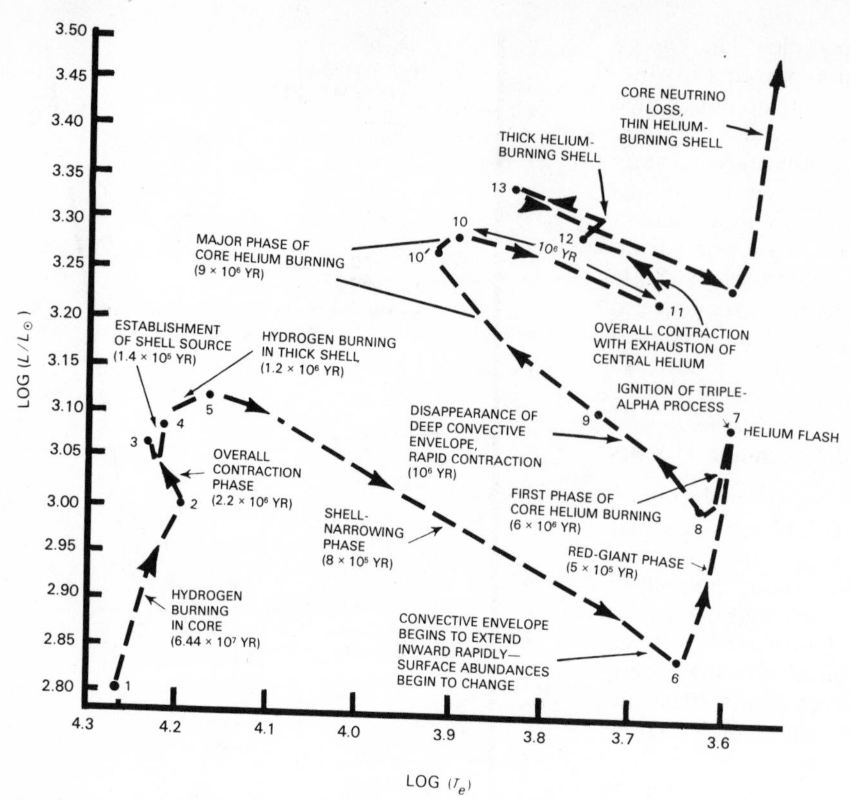

FIG. 23–13 *The path of a metal-rich 5 $M_\odot$ star in the H-R diagram. Luminosity is in solar units, $L_\odot = 3.86 \times 10^{33}$ erg/sec, and surface temperature T_e is in degrees Kelvin. Traversal times between labeled points are given in years. (According to Iben)*

tities of helium could have been formed while very small quantities of heavy elements (enough to account for the present very low metal abundances in population II stars) were formed simultaneously. Such temperatures and densities for the early fireball are in agreement with the observed temperature and density of matter in our universe if the expansion of the universe is taken into account.

If one now assumes that the initial helium abundance in metal-poor stars was the same as in the primordial material in the universe, one can construct evolutionary tracks for metal-poor stars with such helium abundances and compare them with the H-R diagrams of globular clusters, which consist of population II stars. I. Iben and J. Faulkner have done this for initial helium abundances of 10 per cent and 35 per cent and for a range of masses. They have shown that initial helium abundance in metal-poor stars was the same as in the give evolutionary tracks that best fit all the observed data, as shown in Fig. 23.12.

23.44 The Evolution of Population I Stars High Up on the Main Sequence

Although in a general way the evolutionary tracks of population I stars (metal-rich stars) in the H-R diagram are similar to those of population II stars, in the sense that these tracks lead from the main sequence to the red giant region of the diagram, important differences in the detailed features of the tracks are quite sensitive to the mass and chemistry of the evolving population I star. During the last few years considerable progress has been made in tracing in detail the evolution of such stars through their helium-burning phase (red-giant phase). Here we follow the work of I. Iben and give only the highlights of the various evolutionary stages as deduced from mathematical models. In Figure 23.13 the evolutionary path of a 5 $M_\odot$ star according to Iben is shown in the H-R diagram. In this stellar model the initial (main-sequence) hydrogen, helium, carbon, nitrogen, and oxygen concentrations (fractions of a gram) were taken as 0.71, 0.27, 0.0036, 0.0012, and 0.011. The labeled points on the curve are those at which the star undergoes sharp changes in its behavior (either its total behavior or its behavior in certain interior zones). The time it takes the star to pass from one of these points to the next on its evolutionary path is given in years in parentheses.

In the initial or "core-hydrogen-burning" phase (points 1 and 2) the star remains close to the main sequence as hydrogen is fused into helium by the carbon-nitrogen cycle. The convection in the core carries the helium out so that the hydrogen content throughout a fairly large central region drops steadily. The increase in the mean molecular weight (owing to the formation of helium) causes the pressure to drop and therefore overall contraction occurs. This causes the core temperature to increase quickly; hydrogen is then burned more rapidly in the carbon cycle and the luminosity rises until all the hydrogen in the core is exhausted. This ends the core-hydrogen-burning phase.

The shell-hydrogen-burning phase now begins, with the carbon-nitrogen cycle operating in a thick shell surrounding the pure helium core, which is now almost isothermal (point 5). The temperature throughout the core now drops because there is little or no nuclear energy generated in the core but there is a steady and fairly large flow of heat from the core to the shell, which is at a lower temperature. The core must then contract to replenish this outflowing energy from its store of gravitational potential energy. Along the track between points 2 and 5, the contraction is quite rapid as the core continues to cool and as the released gravitational energy flows into the hydrogen-burning shell. This rapid core contraction and cooling continues until the temperature is the same throughout the core (isothermal core) and equal to the temperature in the shell. At this point the core is still contracting, but extremely slowly, with very little release of gravitational energy. The luminosity of the star is due almost entirely to the carbon-nitrogen cycle in the thick shell. Shortly before this equilibrium stage is reached, the rapid contraction of the core causes some of the hydrogen in the shell to fall toward the hotter core and the hotter part of the shell. This hydrogen is ignited somewhat explosively, causing the entire star to expand a bit and to cool off slightly, lowering its luminosity as shown at point 3 in Figure 23.13.

A number of things now happen rather quickly. The mass of the pure helium isothermal core increases as the hydrogen-burning shell expands outward and adds layer upon layer of helium to the core until the mass of the core reaches about 10 per cent of the star's mass (the Schönberg-Chandrasekhar limit). The core now begins to contract rapidly and to heat up, with the result that as hydrogen in the shell falls toward the denser, hotter core, it is ignited explosively and the envelope of the star above the shell expands swiftly. During this period the energy-producing mass in the shell decreases quite rapidly as the hydrogen-burning shell contracts, so that the total rate of energy production decreases. Owing to these two processes (expansion of the envelope and contraction of the shell) the star becomes larger, cooler (redder), and less luminous. It thus moves to the right and downward in the H-R diagram. The drop in temperature in the expanding outer envelope causes the radiative opacity to increase to a point where convection sets in, playing the dominant role in the transport of energy through the outer envelope.

This downward-and-to-the-right motion in the H-R diagram halts suddenly and is replaced by a sharp upward turn (increasing luminosity, point 6) as the convective envelope of the star expands and the opacity of the envelope begins to drop owing to the decrease in the number of negative hydrogen ions (see Section 19.15). The nuclear energy produced in the shell is thus able to pass more readily through the envelope and contribute directly to the luminosity. While all this is happening in the envelope, the helium core contracts until the central temperature is high enough to trigger the triple alpha process and the helium flash occurs (point 7), as discussed in the text. This is the core-helium-burning phase of the star's evolution. The events that follow the helium flash are described briefly at various points in Figure 23.12 itself, and we need not go into them in detail. We merely note that as the helium burns, a carbon core develops, and this core is then surrounded by a shell of burning helium.

The description we have given above of the events in the post–main-sequence life of a 5 $M_\odot$ star applies with minor variations to stars of larger and smaller masses. In Figure 23.14 the evolutionary H-R tracks of 10 stars, with masses ranging from 0.25 $M_\odot$ to 10 $M_\odot$, are shown; in Table 23.2 the times spent by these stars in moving across the specified intervals in their evolutionary tracks are listed. We see from this analysis how important the mass is in setting the evolutionary time scale, the time it takes a star to go from the main sequence to the giant branch of the H-R diagram.

Although the theory of the evolution of popula-

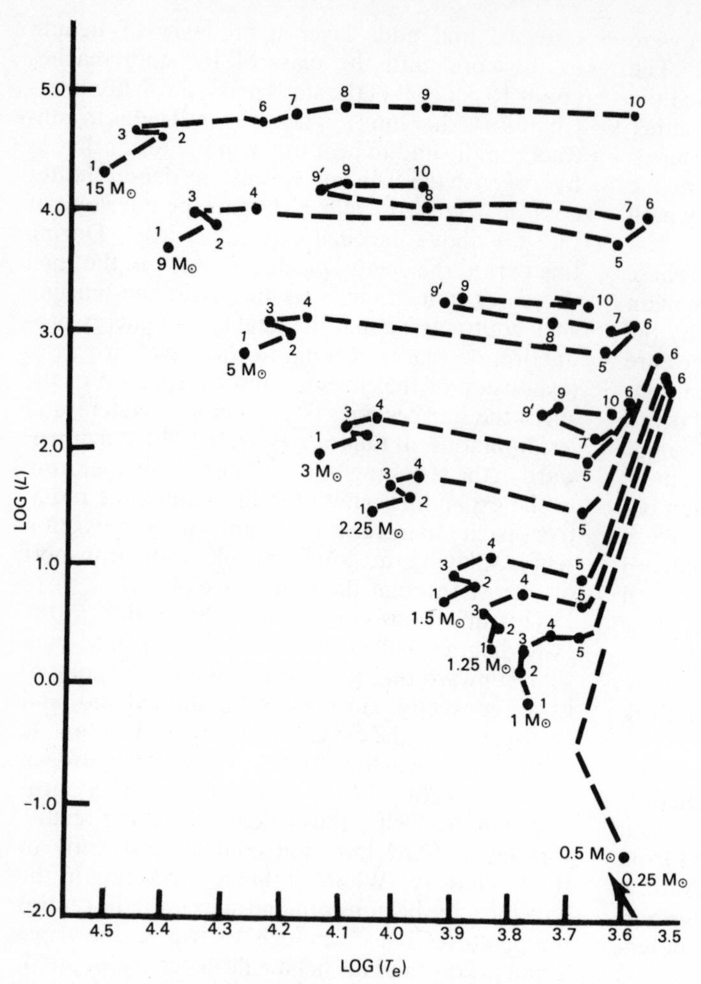

FIG. 23–14 *Paths in the H-R diagram for metal-rich stars of mass* $(M/M_\odot)$ *= 15, 9, 5, 3, 2.25, 1.5, 1.25, 1, 0.5, 0.25. Units of luminosity and surface temperature are the same as in Fig. 23–13. (According to Iben)*

TABLE 23.2

Time (in Years) Spent by Stars of Various Masses in Traversing the Intervals between Successive Labeled Points of Fig. 23–14. The Number in Parentheses beside Each Entry Indicates the Power of 10 to Which the Entry Is to Be Raised. (According to I. Iben)

Mass	Interval ($i-j$)								
($M_\odot$)	(1–2)	(2–3)	(3–4)	(4–5)	(5–6)	(6–7)	(7–8)	(8–9)	(9–10)
15	1.010(7)	2.270(5)		7.55 (4)		7.17(5)	6.20(5)	1.9 (5)	3.5 (4)
9	2.114(7)	6.053(5)	9.113(4)	1.477(5)	6.552(4)	4.90(5)	9.50(4)	3.28(6)	1.55(5)
5	6.547(7)	2.173(6)	1.372(6)	7.532(5)	4.857(5)	6.05(6)	1.14(6)	8.90(6)	9.30(5)
3	2.212(8)	1.042(7)	1.033(7)	4.505(6)	4.238(6)	2.51(7)	4.08(7)		6.00(6)
2.25	4.802(8)	1.647(7)	3.696(7)	1.310(7)	3.829(7)				
1.5	1.553(9)	8.10 (7)	3.490(8)	1.049(8)	≥2 (8)				
1.25	2.803(9)	1.824(8)	1.045(9)	1.463(8)	≥4 (8)				
1.0	7(9)	2(9)	1.20 (9)	1.57 (8)	≥6 (8)				

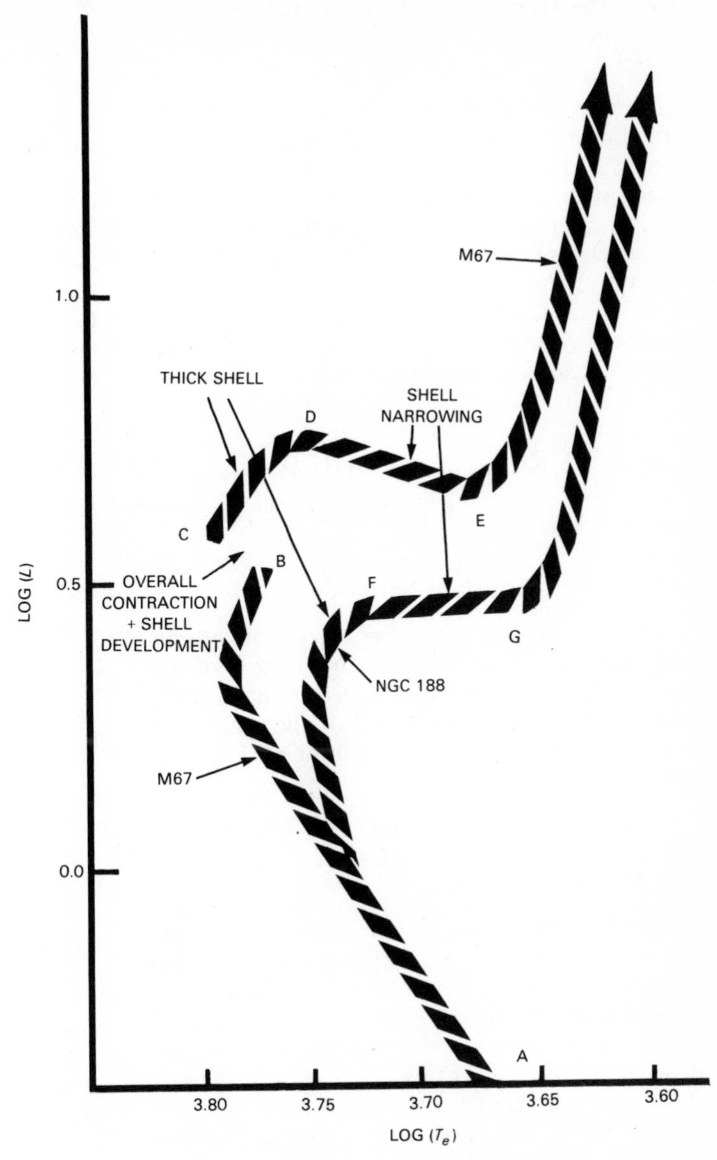

FIG. 23–15 *Schematic cluster loci for stars in the clusters M 67 and NGC 188. (According to Iben)*

tion I stars as described above is fairly well established, we cannot accept it as correct until we find that it agrees with observations. Such agreement has been found by comparing the theoretical evolutionary tracks with the present H-R diagrams of open clusters. As we have noted in our discussion of these H-R diagrams in Chapter 17, stars in such clusters not on the main sequence have evolved away from the main sequence at a rate that was determined by the masses of these stars when they were formed and began their thermonuclear lives on the main sequence. Since all the stars in a given cluster are the same age, we should be able to reproduce a schematic picture of the H-R diagram of a cluster if we know its age. To do this we need only draw a line through each point in Figure 23.14 (one point on each curve for a star of a given mass) that corresponds to the age of the cluster, using the numbers given in Table 23.2 to determine these points. This has been done by Iben for the open clusters M 67 and NGC 188. In Figure 23.15 the schematic forms of the H-R diagram of these two clusters are shown.

23.45 The Triple Helium Reaction and the Formation of C^{12}

There is direct observational evidence for the formation of heavy elements as described in this section, at least through C^{12}, in the existence of carbon stars. These objects are the most common form of later-type peculiar stars that make up a few per cent of the G, K, and M giants. The composition of these carbon stars, which show large overabundances of elements with isotopes that must have been synthesized by neutron and proton capture by nuclei formed during helium burning, are good evidence for the thermonuclear processes described in the preceding sections.

Additional evidence for evolution through the helium-burning phase can be found in a small group of hydrogen-deficient carbon stars. In these stars the hydrogen may be deficient by factors ranging from 100 to 100 000, and carbon is found to be overabundant by factors of 3 to 10. In addition these stars are rich in helium. They probably started on the main sequence as 1 $M_\odot$ stars and then, after their helium-burning phase began, lost their hydrogen-rich envelopes. They were thus left with very little or no hydrogen and with their helium-burning core exposed.

CHAPTER 24: VARIABLE STARS AND STARS WITH UNUSUAL CHARACTERISTICS

24.2 Classes of Intrinsic Variables

A number of new categories of variable objects that were not known when the first edition of this book was published must be added to the intrinsic variables listed in Section 24.2 of the main text. We list them here in the chronological order of their discovery.

1. Quasistellar objects (QSO's) or quasars.
2. Pulsars or neutron stars.
3. Variable X-ray stars.

Strictly speaking, the quasars which we discuss in Section 30.50 should not be listed among variable stars because they are not really starlike objects, even though they appear starlike on photographic plates. As we shall see, the preponderance of evidence is that they are galaxy-type objects. Nevertheless we list them here because some of them appear to be pulsating with remarkably short periods. Pulsars and variable X-ray sources are stars that have undergone drastic changes in their structures after they passed through their red-giant helium-burning phase and exhausted all their nuclear fuel.

24.10–24.11 RV Tauri and RR Lyrae Variables

About 100 RV Tauri stars have been found in the Galaxy and about 4 500 RR Lyrae variables are known.

24.22 The Origin of Pulsations

The relationship between the variations in temperature and the state of ionization of the hydrogen and helium in the outer zones of a pulsating star accounts not only for its pulsations but also for the phase lag between the light curve and the velocity curve. The outer ionization zones behave like a delayed driving mechanism for the pulsations. By undergoing ionization, they absorb a certain fraction of the radiation released by thermonuclear reactions in the deep interior when the star is most highly compressed (highest internal temperature), and then they release this ionization energy a quarter of a cycle later when the star is expanding most rapidly. Some of this released energy keeps the star expanding and some of it contributes to the total energy radiated by the star, thus giving the star its maximum luminosity when its expansion is most rapid.

Since ionization zones are present in most stars, one may ask why most stars are not variables. The reason is that ionization zones can act as pulsation-driving mechanisms only if a very narrow range of internal conditions is fulfilled. The theory of stellar evolution (as described in Chapter 23) has been applied to variable stars, and we now know that conditions for pulsations are fulfilled only when the evolving star lies in a very narrow, almost vertical strip (the instability strip) in the H-R diagram. This strip, which has a width of about 0.02 along the horizontal axis or about 50°C in the surface temperature, lies near point 9 in Figure 23.13, somewhat to the left of the red-giant tip (the tip where the helium flash occurs for the 5 $M_\odot$ star) and well above the main sequence. It is so narrow that a massive evolving star spends very little time in it (a few thousand years at most) when it is approaching the red-giant tip. There is thus very little chance of finding a star undergoing pulsations before it enters the helium-burning phase of its life. But after the star starts burning helium, if it is massive enough its evolutionary track will return to the instability strip and the star will remain there for a few hundred thousand years, and that is when it can be detected as a Cepheid. Thus the theory of stellar evolution tells us that Cepheid variables are massive stars in their helium-burning phases. This is confirmed by observations. Because the evolutionary path of a massive star may pass through the instability strip in the H-R diagram more than once, a given star, if massive enough, can become a variable (for intervals ranging from a few thousand to a few hundred thousand years) at several points during its life.

24.33 Spectra of Novae

If we accept Kraft's hypothesis that all ordinary novae are members of very close binaries, one of which is a hot white dwarf (the star that is to become the nova) and the other a star that is just about to enter the red-giant stage, we can explain the nova outburst as stemming from the accretion of hydrogen-rich material onto the hot surface of the white dwarf. To account for the existence of a binary system with the two stars at such different stages of evolution, we need merely assume that when the binary was formed one of the stars (the one that is presently a white dwarf) was a star of a few solar masses which, owing to its large mass, evolved rapidly to its present white-dwarf stage, whereas the red companion was a star of about one solar mass that evolved slowly and is just now approaching its red-giant stage. The white-dwarf member of the binary must have lost most of its mass before it could become a white dwarf (see Section 24.44), and it probably lost the mass when its outer zones expanded as the star became a red giant. This loss of mass could have happened in various ways, but most probably occurred as the result of a transference of mass to its companion. Such a process occurs in a region whose surface passes through the Lagrangian points of the binary system. These are equilibrium points at which the gravitational field arising from the two components of the binary is such that particles situated at these points move so as to maintain fixed distances from the centers of the two binary stars (see Section 12.35). If the surface of the expanding companion extends beyond the Lagrangian points, matter rushes from this expanding surface to that of the less massive star. In this way the rapidly evolving star loses enough mass to become a white dwarf.

The stage is now set for this white dwarf to become a nova as the surface of its red companion in turn expands beyond the Lagrangian surface and hydrogen-rich material passes from it to the white dwarf. One can show that, owing to the intense gravitational field on the surface of the white dwarf, the proton-proton reaction will be triggered explosively when a large enough quantity of the hydrogen-rich material from the expanding red star hits the surface of the white dwarf. As more and more of this material falls onto the white dwarf, it compresses the underlying hydrogen, raising its temperature to a point where hydrogen begins to burn. This occurs explosively, creating a nova; this hydrogen-accretion phenomenon may be repeated over and over, accounting for periodic novae.

24.39 Type I and Type II Supernovae

P. Morrison and L. Sartori have pointed out that the interpretation of the light curves of type I supernovae in terms of the decay of Cf^{254} is not warranted; there is no reason to think that if great quantities of Cf^{254} were produced during the nova outburst, the radioactive energy released by the Cf^{254} would be transformed to visible light with the same time dependence.

Their explanation of these supernova light curves is as follows. They assume that the bulk of direct energy released is a sudden burst of radiation that lasts for only a few days and is concentrated in the ultraviolet region and beyond. As the supernova expands, the output shifts toward lower frequencies so that the visible power increases even while the total emission is decreasing. This direct emission contributes to the observed luminosity only during the day of initial rise and fall. After maximum luminosity, the direct emission is assumed to diminish so rapidly that it soon becomes negligible. The light observed after about 15 days is then the result of fluorescence; we receive the light from the visible lines excited by the initial burst of ultraviolet radiation, as it is absorbed by the interstellar gas that surrounds the supernova. We thus observe fluorescence radiation emitted from regions of space progressively farther away from the explosion. The observed near-exponential decline in intensity with time is due to the exponential attenuation of the excited ultraviolet pulse as it moves outward in space. The ion HeII is the most important emitter of this fluorescent radiation. Its lines account for most of the observed spectral features. Morrison and Sartori explain most of the features of the light from Type I supernovae, and the light curves they obtain from their theory are in excellent agreement with the observed light curves.

24.50(a) Pulsars and Neutron Stars

When the theory of neutron stars was first developed by Oppenheimer and Volkoff, few astronomers believed that these objects would ever be discovered. We now know that the recently discovered "pulsars," which emit very small radio pulses with a periodicity (the time between the arrival of two successive pulses) that is extremely regular and constant compared to all the other observed properties of these objects, are neutron stars.

The first pulsar, now called CP 1919, was discovered in Vulpecula by A. Hewish, S. J. Bell, J. D. Pilkington, P. F. Scott, and T. A. Collins in the summer of 1967 with the aid of a new large radio telescope, operating at a frequency of 81.5 MHz (megahertz). When the signals were first received they appeared to be weak, sporadic interference phenomena and were dismissed as being terrestrial in origin, but their repeated occurrence at a fixed declination and right ascension showed that the source could not be terrestrial. The signals consist of a series of pulses, each lasting for about a third of a second, with an interpulse interval of 1.3372795 sec (a constancy accurate to one part in 10^7). Following the initial discovery by Hewish and his group, other pulsars, with properties quite similar to the Vulpecula pulsar but over a wide range of frequencies, were discovered by the Hewish group and by others. About 100 have since been discovered, with periods ranging from $\frac{1}{30}$ sec to about 3 sec.

Although the interpulse time intervals of pulsars are extremely constant, the amplitudes are not—these show a number of kinds of variation. First (taking the Vulpecula pulsar as an example), there is an amplitude variation from one pulse to the next, which is much greater for the lower frequencies than for the high frequencies. Second, there is an overall variation in the mean intensity of a pulse: when the source first becomes active, it emits a long series of intense pulses, but then the intensity diminishes for a period that is about three times longer than the duration of the intense pulses. Finally, there are variations that occur in a matter of milliseconds (msec, which means thousandths) within each individual pulse. A pulse from CP 1919 often consists of three subpulses, each of which has a duration of 12 msec. All three subpulses occur within a 37-msec interval, which is very sharply defined. The onsets of two successive 37-msec pulse groups are always separated by exactly 1.3373794 sec., as are also their completions. No radiation of a pulse nature is observed in this basic interval between the 37-msec pulse groups. Because of the short duration of this basic interval, the size of the source certainly cannot be greater than 11 000 km, and it may be no larger than 1200 km if we accept the evidence of the 12-msec subpulses. This value for the size of the source is obtained by multiplying the duration of a pulse by the speed of light. The reason for this procedure is that the waves of radiation of the pulse from the most distant point of a source and from the closest point start out together. Hence the distance between these two points divided by the speed of light cannot exceed the duration of the entire pulse.

An important feature of these pulses is that the various frequencies (or wavelengths) in the same pulse arrive at different times—the lower the frequency is (the longer the wavelength), the later is the time of its arrival. This delay is caused by the electromagnetic interactions of the radio pulses with the interstellar electrons along the line of sight from the earth to the pulsar. These electrons constitute a plasma that is set oscillating by the radio pulse from the pulsar so that the pulse itself is slowed down as it traverses the plasma. The theory shows that the delay in the time of arrival (compared to the time of arrival if there were no intervening electrons) of a particular frequency, as a result of plasma oscillations, varies inversely as the cube of the frequency. This relationship agrees with the observed delay in the pulsar radiation.

This phenomenon is essentially similar to the slowing down of light of different colors, by different amounts, when light enters a medium (such as glass). The electrons between us and a pulsar respond differently to the various frequencies of the pulsar radiation and thus slow down these frequencies by different amounts. Comparison of theory and observation of the delay in the various frequencies from CP 1919 shows that all frequencies between 40 and 430 MHz are emitted simultaneously from this pulsar with no greater discrepancy than 8 msec. The observations, which agree very well with theory, show that there is a 31.765-sec delay in arrival between a signal with a frequency of 430 MHz and one with a frequency of 40.12 MHz, with the higher-frequency signal arriving first. The theory also shows that an optical frequency signal will precede one of frequency 430 MHz by 0.279 sec.

We can use this theory to calculate the total number of electrons per cm^2 along the line of sight from the earth to the pulsar by comparing the measured delays of signals of different frequencies with the calculated delays for given numbers of electrons. The number of electrons per cm^2 between us and a pulsar is about 10^6 times the number of electrons per cm^2 found in the earth's atmosphere and about 10 times that in the solar corona at sunspot maximum. It is also about 1 per cent of the number found in the very bright regions of the galactic ionized hydrogen. These considerations rule out a planetary type of ionosphere, as well as an HII region, as the location of the plasma that delays the pulsar signals. An analysis of the formula for the propagation of radio waves through a plasma and the time delays shows that the electron plasma must be at least as large as 10^7 km and that the electrons causing the delay are part of the interstellar medium. If we assume this to be the case and assume the electron density of interstellar space to be 0.1 electron per cm^3, we obtain a distance of about 130 parsecs for this pulsar, showing that it lies in the Galaxy.

Remarkable scintillations have been observed recently in the signals from pulsars, which cannot be due to changes in the pulsars themselves. We now know that these random variations are due to changes in the interstellar medium; the pulsar scintillations are like the twinkling of a star because of changes in the earth's atmosphere. The pulsar scintillations arise from the motions of interstellar gas clouds.

24.50(b) Very Short Period Pulsars

Since the discovery of the first pulsar many others have been found, and among these are some with very short periods. These are interesting because they are probably very young pulsars and are therefore still emitting energy in the visible part of the spectrum as well as in the radio part. Two such pulsars are of special interest (the two with the very shortest periods) because they have been identified as the remnant cores of supernovae. The pulsars with the very shortest periods are believed to be the youngest because their periods are increasing more rapidly than those of the long-period pulsars. Although initially astronomers thought that the periods of pulsars were constant to an incredible degree, we now know that these periods do change. The first six pulsars that were discovered (which include the two supernovae remnants mentioned) have now been studied with sufficient accuracy and for a long enough time to show that their periods are increasing slowly. The greatest rate of increase is found for the very-short-period pulsars NP 0532 and PRS 0.0833-45. The first of these, which has a period of 0.033 sec (30 pulses per second) has now been identified as the hot central star in the Crab Nebula, and the second one as the residual core of the supernova in Vela.

The overwhelming evidence supports the theory that pulsars are rapidly rotating neutron stars and that the energy they emit as pulses of radiation comes from the kinetic energy of rotation of these neutron stars. It follows, then, that as they lose their rotational kinetic energy, their speed of rotation must slow down and the periods of their radio pulses must increase. Since the youngest pulsars have lost the least amount of energy up to now, they must be spinning fastest (shortest periods between pulses of radiation). They must also be losing energy at the greatest rate (because they have the greatest store of energy) and thus slowing down faster than the long-period pulsars. This is precisely what is found for NP 0532 and PSR 0.0833-45. That is why we believe that these are the youngest known pulsars.

24.50(c) Pulsars as Optical and X-Ray Objects

From the time that pulsars were first discovered, astronomers tried to identify them with optical objects. This goal was finally reached in 1969, when it was found that the hot white central star in the Crab Nebula blinks in its visible radiation at the same rate (30 times a second) as the radio pulses from pulsar NP 0532. This pulsar was already known to lie at or very close to the center of the Crab Nebula, so this was not too surprising. The central star in the Crab had been observed for many years, but its visible radiation always appeared to be quite steady when received through a telescope because it blinks so rapidly. A special technique (the equivalent of a stroboscope) had to be applied to show that the central star of the Crab sends out its visible radiation in pulses that arrive every 0.0331 sec (this is just the period of the radio pulsar in the Crab). The main optical pulse and interpulse last for 0.0014 and 0.0028 sec respectively, and their intensities are in the ratio of 3.7:1.

Quite recently the pulsar NP 0532 in the Crab was found to emit X-ray pulses as well as radio and optical ones. The remarkable thing about this pulsar is that it emits 100 times as much energy in the X-ray region as in the optical, and 100 times as much in the optical as in radio waves. The data show that the period of the optical pulses increased from 0.033097500 sec on February 16, 1969 to 0.033097845 on February 22, 1969.

For some time after the discovery of pulsars there was considerable conjecture and disagreement among astronomers concerning the nature of these objects and the ways in which they emit their radiant energy. From the very beginning it was clear that pulsars must be small, compact objects because of the short duration of the pulses emitted. As we have already noted, the size of the emitting area of the pulsar cannot exceed the duration of a single pulse multiplied by the speed of light. This means that pulsars cannot be larger than white dwarfs, and initially many astronomers suggested that pulsars were indeed pulsating white dwarfs. However, because the period of pulsation of a typical white dwarf, if it were pulsating, would be longer than 2 sec (the square of the period is proportional to the reciprocal of the density of the white dwarf; see the discussion of Cepheids in Section 24.20), such objects could not give the observed periods of most pulsars. Nor can white dwarfs rotate fast enough to account for the observations, if we accept rotation as the cause of the pulsar phenomenon.

We must therefore reject white dwarfs entirely and go to neutron stars (which we discussed in Section 24.50 in the main text) to explain pulsars. A typical neutron star, with a radius of about 10 km and a mass about equal to that of the sun, would pulsate much too fast (a period of milliseconds) to emit pulses with the observed periods. We are thus left with the rotation of neutron stars to account for pulsars, a mechanism that was first proposed by T. Gold and that now seems to account for many of the observed features of pulsars.

24.50(d) The Rotational Energy of Pulsars

As we have noted in our discussion of neutron stars, these objects result from the catastrophic collapse of the core of a giant star before the star becomes a supernova. Immediately after the supernova explosion, the surface of the residual core, a neutron star, is probably very hot and may also possess rotational, vibrational, and magnetic energy. The hot surface will initially emit very energetic photons (X-rays) and probably neutrinos and antineutrinos, resulting in a rapid cooling that will be accelerated if strong magnetic fields are present on the surface. After a thousand years (more or less), the surface of the neutron star will be too cool to contribute much thermal energy to the pulsar phenomena, and most of the radiated energy will come from the very rapid rotation that results from conservation of angular momentum. The angular momentum of a rotating sphere (our neutron star) of mass M and radius R is propotional to $MR^2\omega$, where ω is the angular speed of rotation. Since this quantity must be conserved as the core contracts (that is, as R gets smaller), ω must increase since M remains the same. Therefore the speed of rotation of the core rises rapidly as it collapses just before the supernova stage. But the energy of rotation is proportional to $MR^2\omega^2$ or to $J\omega$ where J is the angular momentum of the collapsing core. Since J remains constant as the core collapses and ω increases, the kinetic energy of rotation increases very rapidly.

Calculations show that this energy is enormous for a neutron star that is spinning some 10 or 30 times per second, and certainly enough to account for the energy emitted by pulsars. From the rate at which the period of the pulsar in the Crab is increasing, we deduce that it is losing energy at a rate of about 10^{38} ergs/sec. This energy that the pulsar pumps into the nebulosity of the Crab probably accounts for the total radiation emitted per second by the entire Crab in the X-ray, ultraviolet, visible, infrared, and radio parts of the spectrum. The initial store of rotational energy in a neutron star can easily be about 10^{52} ergs.

The mechanism by which the neutron star radiates away its rotational energy is not entirely clear, but magnetic fields probably play a very important role. It is not unrealistic to assume that the magnetic field strengths on the surfaces of pulsars are anywhere from 10^{10} to 10^{12} gauss. This is so because the intense collapse of the core of a supernova will compress even small magnetic fields enormously and increase their intensities a trillionfold. If the magnetic poles of a neutron star do not lie on the axis of rotation of the star (just as the earth's magnetic poles do not coincide with the earth's geographic poles), the intensity of the magnetic field as seen from the earth will vary as the star rotates (like a cone of light) and this will give rise to electromagnetic pulses of the right period. Moreover, the rotating magnetic field will greatly accelerate charged particles emitted by the neutron star (electrons and protons). These accelerated charges will then spiral around the magnetic lines of force and emit synchrotron radiation of the sort that has been observed from the Crab (in a cone of such radiation that co-rotates with the star). This model for the emission of radiation from a pulsar is essentially that of a collimated searchlight beam (like that from a lighthouse) that co-rotates with the star but whose axis is tilted with respect to the axis of rotation of the star.

Recently some sudden, large, but nonpermanent changes have been detected in the periods of certain pulsars. The periods of these pulsars were found to decrease quite suddenly and then to return slowly to their original values as though there had been a

TABLE 24.8

POSITIONS AND PERIODS OF SEVERAL PULSARS

Pulsar	Right ascension				Declination				Interpulse period (sec)
	h	m	s	s	°	′	″	″	
CP 0834	8	34	22	± 10	+6	07	00	± 180	1.2737642
CP 0950	9	50	28.95	± 0.7	+8	11	06	± 42	0.253065
CP 1133	11	33	36	± 3	+16	07	36	± 18	1.18740428
CP 1919	19	19	37	± 0.2	+21	47	02	± 10	1.33730109
"Blue Star"	19	19	36.88	± 0.1	+21	46	57.4	± 1.5	

sudden change in the dimensions of these pulsars followed by a readjustment. These phenomena, which have been referred to as "starquakes," have been explained as follows: It is believed that the surface of a neutron star is an exceedingly dense (of nuclear density) but very thin (a fraction of an inch) skin of iron that is under severe dynamical stress. Under such conditions a crack may develop in the surface of the pulsar, causing it to readjust itself as the surface of the earth does after an earthquake. This will cause the period of rotation to change suddenly, but then the period will return to its initial value when equilibrium is reestablished.

In the Table 24.8 we list the positions and periods of the first four pulsars discovered.

Most of the pulsars (about 100) that have been discovered lie close to the plane of the Galaxy but some have been found at high galactic latitudes. We believe that all known pulsars are galactic objects. In time, however, we undoubtedly will pick up pulsars in nearby galaxies.

24.51(a) Black Holes

In the section on hyperon stars in the text, we stopped short of discussing stars that collapse beyond the neutron and hyperon stages and become black holes; this is, in principle, the final stage of collapse of the core of a star if it is massive enough or if, for a given mass, the density is high enough. From general theoretical considerations, one can show that the core of a supernova can become a neutron star only if its mass lies in the range from about 0.40 to 1.6 solar masses. If the mass of the core is less than 0.05 solar masses, the gravitational force is not strong enough to keep most of the neutrons from breaking down into protons, electrons, and anti-neutrinos, and the core must become either a white dwarf or possibly a planet. If the mass of the core exceeds 1.6 solar masses, the core cannot remain in a gravitationally stable neutron state for any length of time but must go on contracting forever. Such configurations, which we may describe as "collapsing to a point" cannot be treated by classical (Newtonian) gravitational theory but must be analyzed within the framework of the general theory of relativity. When such an unlimited gravitational collapse occurs, the ultimate result is a black hole.

To see why we call such a configuration a black hole, we go back to our discussion of the behavior of clocks and rods in a gravitational field arising from a sphere of mass M and radius r (see Sections 11.25, 11.26). We saw that a rod placed vertically on the surface of such a sphere shrinks by the factor $\sqrt{1 - 2GM/c^2r}$, where G is the gravitational constant. We saw that clocks on the surface slow down by this same factor and that the speed of light along the vertical direction is decreased by the same factor. If the sphere contracts, so that r decreases steadily, the quantity $2GM/c^2r$, which is very, very small for ordinary stars, increases until it approaches 1. When that happens, vertically held rods shrink to almost zero length, clocks almost come to a halt, and the speed of light in the radial direction is almost zero. All this is the way things would appear to a very distant observer watching the sphere collapse. Light could escape from this collapsing surface only if it were emitted in an almost vertical direction (in a certain narrow cone around the vertical which becomes more and more narrow as r gets smaller). If it were emitted at an angle to the vertical larger than the angle of the cone, the intense gravitational field of the collapsing sphere would force the light back onto the surface.

Suppose now that the sphere contracts to such an extent that $1 - 2GM/c^2r = 0$, so that no light can escape from the surface even along the vertical direction. The radius of the sphere is then given by the formula $r = 2GM/c^2$, and we see that this is just the radius for which the speed of escape (see Section 6.24) from the surface of the sphere equals the speed of light. This value of r is called the Schwarzschild radius, and when a sphere collapses to this point, we say it enters the black hole stage or becomes a black hole. The reason we call it "black" is that no light can escape from it; an observer outside the Schwarzschild radius cannot see the sphere. Any light projected toward this collapsed sphere will not be reflected but will be trapped by the intense curvature of the space-time geometry surrounding the sphere. The Schwarzschild radius for a star of mass M is $2.94 \times M/M_{\odot}$ km where $M_{\odot}$ is 1 solar mass.

24.51(b) The Collapse to a Black Hole

An object of any mass can become a black hole if it is compressed to the Schwarzschild radius, but the gravitational force required for this to occur is so great that it will not happen to an ordinary object by itself; once an object has been compressed to the Schwarzschild radius, however, it will go on contracting endlessly.

To an observer standing on the surface of the collapsing sphere and falling in with it, the collapse to the Schwarzschild radius r_s occurs in a finite and relatively short time, depending on the mass M of the sphere and its initial radius R when it begins to collapse. This time is given by the formula

$$\frac{1}{3}\sqrt{\frac{2}{GM}}(R - r_s)^{3/2}$$

For a body like the sun, it is about 1 000 sec. To the distant observer the collapse to the Schwarzschild radius would in principle appear to take an infinite time, but actually the surface of a collapsing star would disappear quite suddenly as it approached the Schwarzschild radius. Fewer and fewer photons could leave the surface, and even these photons would be so drastically red-shifted by the gravitational field of the collapsing star that they would have practically no energy. The distant observer would first see the surface of the star collapse very rapidly and then slow down and appear to be frozen in time as it approached the Schwarzschild radius. Hence the Schwarzschild radius defines what is called the **event horizon** of the black hole, since no one from the outside can look beyond it into the black hole.

The reason the black hole is called a "hole" is that, inside the Schwarzschild radius, space and time are interchanged in the sense that only one direction of movement is possible—into the black hole. For r less than the Schwarzschild radius, the quantity $1 - 2GM/c^2r$ becomes negative; only decreasing values of r correspond to going from the past into the future, which is the natural state of events. To retrace one's steps and recede from the black hole through the event horizon would be as impossible as going from the future into the past. It is because of this situation that the black hole is truly a hole. Anything that falls through its event horizon can never come out again. This all stems from the fact that inside the event horizon the space-time geometry is such that the radial direction becomes time-like and intervals of time become space-like.

24.51(c) A Rotating Black Hole

Thus far we have been discussing a nonrotating black hole that is characterized by a single surface—the event horizon—that separates it from the outside. If, however, the collapsing sphere is rotating, there are other solutions of Einstein's general relativistic field equations that describe it. A rotating black hole is characterized by two surfaces, one of which is the event horizon as in the case of the nonrotating black hole, and the other of which is called the "static limit." The "static limit" lies outside the event horizon but touches it at the two poles of the rotating black hole. As an object approaches the rotating black hole, it is dragged around by the angular momentum of the black hole, revolving faster and faster as it approaches the static limit. When the object reaches the static limit, the dragging of the black hole is so powerful that no external force, however strong, can keep the object from being dragged around. One can stay in the same place on this surface, as viewed from a great distance, only by traveling at the speed of light in the opposite direction from the rotation of the black hole. That is why this surface is called the static limit: outside it, an observer in a rocket ship can apply the rocket motors to hold his ship fixed with respect to the distant stars; once he has penetrated the static limit, his ship must revolve around the black hole in the same direction in which the hole is rotating, no matter how powerfully he applies his rocket engines in the opposite direction.

The region between the static limit and the event horizon of the black hole is called the ergosphere. This is a region from which some of the rotational energy of the black hole can be extracted. This can be done by having a particle fall into the ergosphere from a great distance and then break into two parts in such a way that one of the parts falls through the event horizon into the black hole. The remaining part will then escape to a great distance, with more mass-energy than the initial undivided particle had.

This additional energy comes from the black hole.

The observational evidence for the existence of black holes is found in the emission of X-rays from very small but massive binary systems such as Cygnus X-1 which we describe below.

24.51(d) X-Ray Sources and Black Holes

With the launching of the UHURU space satellite on December 12, 1970, X-ray astronomy advanced from a rather hit-or-miss operation, involving high-altitude balloons and sounding rockets, to a precise new field of astronomy. The UHURU experiment collected far more X-ray data in its short life than had been obtained in the preceding decade by all other methods. These UHURU data showed the presence of a number of different kinds of X-ray sources which we list and describe below.

Supernova remnants, such as in the Crab Nebula, Cas A, Tycho SN, Vela X, Y, Z, Puppis A, and the Cygnus Loop. X-rays are emitted from supernova remnants as a natural result of the violent plasma vibrations following the initial explosion of the supernova.

Transient sources, such as were discovered by UHURU in Lupus. The few transient sources that have been discovered are sudden intense X-ray flare ups in regions where none had previously been observed. At their peak intensities, these transient sources are among the strongest in the sky, and they flare up and die down the way optical novae do. These sources, which increase in intensity by a factor of 1 000 or more, may disappear in a short time or last for a year or more.

Galactic X-ray sources. These are various types of celestial objects that can be identified with known classes of stars and stellar systems. The X-ray intensity or luminosity of these sources is of the order of 10^{37} ergs/sec. Among these sources are the following objects:

1. Sco X-1. A source of X-rays, radio waves, and optical radiation.
2. Cyg X-1 (probably a black hole). A source from which X-ray pulsations are emitted in intervals of about 0.1 sec, which means that the X-ray-emitting region in this variable source is extremely small (about as small as or smaller than a pulsar). Lying in the same region as this X-ray source is a variable radio source that is coincident with a 9th magnitude B0 I star that is revolving around a massive invisible companion every 5.6 days. If, in accordance with the mass-luminosity relation, the bright B0 star has a mass of about 20 $M_\odot$, the mass of the invisible X-ray companion must be about 13 $M_\odot$ and it must be smaller than 0.1 light-second. Since a 13 $M_\odot$ star is much too massive to be either a white dwarf or a neutron star, the invisible companion of the B0 star is probably a black hole. X-rays are emitted by this "black hole" as streams of ions flow from the bright B0 star to the dark companion.
3. Hercules X-1. A periodic X-ray source like Cyg X-1, with the additional property of being eclipsed periodically. The X-ray pulse period of 1.2 sec exhibits a variable Doppler shift with a period of 1.7 days; the eclipse period, in agreement with the Doppler period of the X-ray pulse, is 1.7 days. The X-ray source is coincident with the irregular optical variable HZ Hercules.
4. Can X-3. The first eclipsing, pulsing X-ray source that was observed. The pulse period is 4.8 sec and the eclipse period is 2.1 days. Here, too, the period of the X-ray pulses shows a variable Doppler shift that is in phase with the 2.1-day period of the eclipse. This X-ray source disappears in an erratic fashion from time to time for several days.
5. Cyg X-3. A variable X-ray source of great interest because of the intense flare that occurred in its radio component or companion. Although no significant change was observed in its X-ray luminosity, its radio luminosity increased suddenly in 1972 by a factor of 1 000.

Extra-galactic X-ray sources. A number of discrete X-ray sources in the Magellanic clouds were picked up by UHURU. X-ray sources have also been found in association with the quasar 3C273, the giant radio galaxy NGC 5128 and the Seyfert Galaxy NGC 4151. X-ray sources have also been discovered in association with rich clusters of galaxies such as the Virgo cluster.

Recently four X-ray sources have been identified with globular clusters. The importance of this discovery is that globular clusters are the oldest components of our Galaxy and presumably are free of gas and dust. It is therefore difficult to see how these objects can generate X-rays unless we suppose that accretion of some kind occurs on very massive but difficult to observe objects in these clusters.

24.52 Rapidly Rotating Stars

R. Kraft has carefully analyzed the question of the loss of spin (angular momentum) as a star evolves from its early contracting phase onto the main sequence, and he has concluded that this loss occurs through magnetic coupling with the stellar wind. Stars more massive than 1.20 $M_\odot$ begin their lives on the main sequence as F5 stars or of earlier spectral class. Such stars have no subphotosphere convective zones and hence no stellar winds. There is thus no coupling of magnetic lines of force with the surface and no magnetic slowing down of the rotation. But in main-sequence stars later than F5 (like the sun) the subsurface convective zones give rise to stellar winds that are coupled magnetically to the stellar surface. This causes the rotation of such stars to slow down to their presently observed rate of rotation in a few hundred million years. The magnetic lines of force carried out from the surface by the stellar wind act like rigid strands that resist the rotation.

CHAPTER 26: MULTIPLE STELLAR SYSTEMS

26.13 Using the Observed Motion to Detect Binaries

It should be noted that the binary stars studied by Van de Kamp and his colleagues at the Sproul Observatory lie within a radius of 10 parsecs. Since then the work on the observed perturbations in the motions of nearby stars has continued at the Sproul Observatory, and other nearby binaries or planetary systems have been discovered. In the study of six nearby red-dwarf stars (with parallaxes close to 0.1), the Sproul astronomers have found pronounced perturbations in the motion of CC 1299, which indicate a binary with a period of 40 years and an invisible companion with a mass of 0.14 $M_\odot$. Another of these red dwarfs, Cin 2347, also shows some perturbations corresponding to a period of 24 years (if it is a binary) and a mass of 0.02 $M_\odot$ for its dark companion. All these objects may, of course, be planetary systems.

26.30 Binaries in Which the Components are Very Close

The exchange of mass between the two components of a binary system is an important phenomenon and, as we have seen, has a very important bearing on the evolution of the two companion stars of the binary. Since many X-ray sources are associated with close binary stars (many of which are eclipsing variables), the exchange of mass between the components probably plays an important role in the generation of the X-rays emitted by such pairs.

CHAPTER 27: STAR CLUSTERS, STAR CLOUDS, AND THE MILKY WAY

27.9 The Milky Way at Intermediate Distances from the Sun

Our solar system appears to lie in a local concentration of stars known as Gould's belt, which contains many of the brightest stars in the southern sky and juts out from the lower edge of the Orion spiral arm (see Section 29.2) toward the center of the Galaxy. The extent of the Gould belt, which contains about 200 000 stars, is about 700 parsecs and its thickness is about 70 parsecs. Its absolute visual magnitude is -13 and its lifetime as deduced from its stellar motions (on the assumption that it expanded from a point source) is about 40 million years, which is in fair agreement with the evolutionary ages of the B stars in this belt. The sun is approximately 100 parsecs from the center of Gould's belt and 12 parsecs north of its plane. This belt of stars consists of luminous O and B stars, diffuse nebulae, extended dark nebulae, and clouds of neutral hydrogen.

27.24 The Relaxation Time of Clusters

We summarize below the important overall characteristics of open clusters.

Mean galactic latitude: 70 parsecs.
Total number in the Galaxy: 18 000.
Distribution in the galactic plane: 400 clusters per kiloparsec.
Distribution at a distance of 500 parsecs above or below the galactic plane: 4 clusters per kiloparsec.

If N is the number of stars in a cluster whose radius is R parsecs, then

$$\log N = 1.3 \log R + 2.0.$$

For a cluster to be stable, its density must be larger than 0.09 $\mathcal{M}_\odot$ per cubic parsec. The disruption time of a cluster of density ρ (expressed as $\mathcal{M}_\odot$ per cubic parsec) is $2 \times 10^8 \rho$ years.

27.28 The Time of Evolution and the Ages of Open Clusters

When Sandage did his important empirical work on the evolution and ages of open clusters, the theory of the internal structure of zero-age main-sequence stars was well established, but the theory of the evolution of stars was just beginning and the detailed evolutionary H-R tracks of stars were not known. It is therefore remarkable that the overall, gross picture of the evolutionary tracks of stars away from the main sequence as deduced by Sandage from the H-R diagram of open clusters agrees so well with the calculated evolutionary tracks given by Iben and others.

27.43 RR Lyrae Stars

Recently considerable work has been done on the theory of the evolution of stars along the horizontal branch of the H-R diagram of globular clusters (the branch along which the RR Lyrae variables lie). Combining theory and observation, workers in this field have deduced mass-luminosity relationships for RR Lyrae stars that are sensitive to the chemical compositions of these stars. These relationships place a restriction on the helium and heavy element abundances in RR Lyrae stars. Thus in the globular cluster M 3 the helium abundance Y cannot be less that 0.15 and the heavy-element abundance Z cannot be greater than .0003. For M 3 a mass-luminosity relationship exists of the form

$$\log L = 1.78 + 0.81 \log \mathcal{M},$$

where the luminosity L and the mass $\mathcal{M}$ are expressed in solar units.

An important parameter in the H-R diagram of globular clusters is the ratio of the luminosity of the horizontal-branch stars to the luminosity of stars that are just turning off the main sequence—the point where the effective temperature is a maximum in the main-sequence part of the H-R diagram of the clusters. Since the turn-off point in all globular clusters occurs at the same color in the H-R diagram as that at which the RR Lyrae gap in the horizontal branch begins, the difference between the apparent magnitude of this turn-off point and the apparent magnitude of the RR Lyrae variables in any given cluster gives us this ratio. If we know this ratio and also the luminosity of the stars at the turn-off point, we then have the luminosity of the horizontal-branch stars. For M 3 the logarithm of this ratio is 1.33. From all available data and from theoretical models, one must conclude that the horizontal-branch stars in globular clusters are in the core-helium-burning stage of their evolution, with the helium-burning core surrounded by a hydrogen-burning shell. The RR Lyrae variables lie in the region of the H-R diagram where the horizontal branch crosses the pulsation instability strip.

CHAPTER 28: INTERSTELLAR MEDIUM

28.1 Composition of the Interstellar Medium

A fourth category that must be added to the three listed in the main text is large organic molecules in interstellar space which emit a variety of spectral lines and bands (absorption and emission) in the infrared and radio parts of the spectrum. Very intensive work has been done in this field since 1969, and molecular spectroscopy in the infrared and radio regions of the spectrum has become a major astronomical tool. At present, 137 molecular spectral lines have been observed; they are emitted by some 35 terrestrially identified organic molecules, among which are hydrogen cyanide (HCN), formaldehyde (H_2CO), cyanoacetylene (HC_3N), and methyl alcohol (CH_3OH). Some of the organic molecules that have been identified from their spectral lines, such as CH_3CHO and CH_3C_2H, contain as many as seven atoms, and there is reason to believe that still larger molecules are present in the interstellar medium. Since both the formic acid molecule (HCOOH) and the methanimine molecule (CH_2NH) have been detected, the simple amino acid glycine (NH_2CH_2COOH) ought also be present in interstellar space because it is easily formed from the other two molecules. The presence of amino acids on meterorites confirms the synthesis of amino acids in interstellar space, which has important biological consequences. Of still greater biological significance is the recent work of F. Johnson, who has matched 16 laboratory spectral lines of the very large organic molecule $MgC_{46}H_{30}N_6$ (one of the porphyrins, an important constituent of blood) with the corresponding diffuse interstellar lines, within an experimental error of 2 Å.

The most widespread and abundant radio molecules (they emit in the radio wavelength region between 1.7 mm and 36 cm) are CO, OH, CS, HCN, H_2S, H_2O, H_2CO, and NH_3; most of these are found in the direction to the center of the Galaxy, which is the sole direction from which molecules with more than five atoms are observed. Great quantities of the molecules OH, H_2CO, and CO are found in dark, obscuring clouds. Following the molecule H_2, the most abundant is CO.

The first molecule to be detected with radio telescopes was the OH (hydroxyl) molecule, which presents some highly interesting and puzzling features. This molecule, which has an abundance of 10^{-7} compared to hydrogen, is identified by both its radio emission and absorption lines, all of which lie in the 18-cm microwave band. Most of the observed OH sources lie in areas near very hot stars (that is, in the HII regions where hydrogen is almost entirely ionized). These regions all lie close to the galactic equator.

Some remarkable and strange phenomena are associated with these OH sources. To begin with, the observed intensities of these lines differ from source to source and are in disagreement with the intensities as derived from the theory. The Doppler shifts in the absorption lines indicate that the OH molecules are moving toward the center of the galaxy at about 40 km/sec, in contrast to the atomic hydrogen, which appears to be moving away from the center with a velocity of 50 km/sec. There also appears to be a time variation in the intensity, which varies with random periods ranging from hours to weeks.

Another puzzling feature is the apparently very small size of the sources. Many of these features can be accounted for by assuming that some kind of maser-like action is responsible for the radio emission from the hydroxyl molecule (maser is an acronym for microwave amplified stimulated emission of radiation). In a maser, the excited molecules are stimulated to emit radiation by a beam of photons of the same frequency as those emitted by the molecules, so that great amplification of the beam occurs. For such a process, excited molecules (molecules in excited states) are required in large numbers. This is called an inversion of the population of molecular states because it is contrary to the normal situation, in which most molecules are in the ground state. The problem with the emission from the OH molecule is how to account for such an inversion in cold interstellar space, where it would seem that molecules should be in their ground state. This question still remains to be answered completely.

Associated with the discovery of the hydroxyl puzzle is the discovery of extraordinarily luminous infrared objects that seem to be generating energy in infrared regions far exceeding the luminosity of ordinary stars. It is not clear from the evidence whether these infrared objects are really starlike objects, in which case it would be extremely difficult to account for the vast amounts of energy they release, or whether they are huge clouds of dust that are collapsing catastrophically to form new stars. One noteworthy feature of these infrared objects is that they are closely associated in space with the hydroxyl regions.

28.12 Kinetic Temperatures of HI and HII Regions

The difference in temperature between the HII and HI regions is due entirely to the way energy is absorbed in these regions. In the HII regions the excess energy of the absorbed ultraviolet photons (the energy in excess of that needed to ionize the hydrogen atoms) is transferred to the ejected electrons, which then distribute this energy to other electrons by collisions. In this way the mean kinetic energy of the particles in the gas and hence the temperature of the gas increase. This goes on as long as there is neutral hydrogen available to be ionized, and the supply is ensured by the constant recombination of protons and electrons that occurs in the HII regions owing to the large numbers of free electrons. By contrast, cooling of the gas cloud occurs when the ejected electrons collide with and excite atoms such as OII, NII, or SII. Because the number of such cooling atoms is very small compared to the number of neutral H atoms, the heating mechanism is much more effective than the cooling mechanism, and HII regions are hot. In a stable state where the rate of cooling is exactly equal to the rate of heating, the temperature of the HII region is constant. If hydrogen ionization, hydrogen recombination, and cooling by atomic excitation are all properly taken into account, one finds that the temperature of the HII clouds should be about 10 000°K, which it is.

In the HI regions heating also takes place by ionization but not by the ionization of hydrogen. Heating is produced by the ionization of C, Si, S, and metals such as iron. But since few of these atoms are present, and since the number of free electrons is also small, the rate of recombination and hence the rate of ionization is very small. Therefore very little energy is absorbed by HI clouds. Moreover, since cooling in these clouds occurs by electron collisions with the atoms C, Si, Fe, etc., the cooling mechanism is about as effective as the heating mechanism and thus the temperature of the HI clouds is quite low.

28.15 The Emission of the 21-Centimeter Line

The reason that three-fourths of the neutral H atoms are in the upper sublevel (the electron and proton spins parallel) is that the total spin of this state is 1, so that three possible magnetic states are merged into this upper level; there are three possible values of the magnetic quantum number *m* for this state, namely $-1, 0, 1$ (see Section 19.26). In the lower level, with the electron and proton spins antiparallel, the total spin is 0, and only one magnetic state, $m = 0$, is associated with this level. Since the distribution of the neutral hydrogen atoms between the parallel (upper) and antiparallel (lower) spin states in the ratio of 3:1 is the result of thermal collision, the hydrogen gas is in thermal equilibrium and the 21-cm radiation is thermal radiation in the sense that it is excited by the kinetic energy of the atoms, which in turn is determined by the temperature of the gas. Hence if the gas cloud is thin, so that the 21-cm radiation can pass through easily, the intensity of the radiation is independent of the temperature and simply proportional to the number of atoms in the line of sight. We can then determine the number of neutral H atoms along the line of sight.

If the hydrogen cloud is so dense that it is practically opaque to the 21-cm radiation, the intensity of this radiation from the surface of the cloud is a maximum; its value is that of 21-cm black-body radiation from a black body at the temperature of the cloud. Consequently, we can determine the temperature of an opaque cloud from the intensity of its 21-cm radiation by using the Planck radiation formula with λ placed equal to 21 (see Section 16.30).

Although each neutral hydrogen atom when at rest absorbs and radiates in a very narrow wavelength interval on either side of 21 cm, the actual line from a fixed hydrogen cloud is less narrow because the random thermal motions of the atoms give rise to Doppler effects that depend on the atomic velocities. These random Doppler effects broaden the line. The width of the line then depends on the temperature. If the cloud of gas is moving as a whole or if there are currents within the cloud, the whole line will be shifted to one side or the other, or the shape will become more complicated.

28.17 The Interstellar Dust

The presence of dust in the plane of the Galaxy can be deduced from the following observations:

1. The apparent increase with distance in the size of open clusters (see Section 27.18).
2. The zone of avoidance: a strip 20° wide (10° on each side) along the galactic equator in which extragalactic nebulae are invisible.
3. The apparent bifurcation of the Milky Way owing to absorption of light by vast clouds of dust.
4. The reddening of distant stars as measured by the color excess (see Section 27.13). This shows an absorption of one stellar magnitude per thousand parsecs.

The dust is concentrated close to the galactic plane in a layer about 200 parsecs thick. It is mostly in the spiral arms, as demonstrated by the fact that distant galaxies and globular clusters can be observed between the spiral arms. As noted in Section 28.9, the dust is concentrated in discrete clouds.

28.18 The Structure of Dust Grains

As we have noted in the main text, the polarization of light from distant stars tells us that the grains of dust (which cause the polarization) in the dust clouds through which the light passes must be elongated, with their long axes lined up perpendicular to the spiral arms of the Galaxy. That the polarization is associated with dust clouds is evident from the fact that the more reddened the stars are (the greater the color index) by absorption, the greater is the polarization. That the long axes of the grains of dust are oriented perpendicular to the spiral arms is deduced from the fact that polarization is observed in the light from reddened stars only if the direction to these distant stars is perpendicular to the spiral arms. If one looks at stars along the spiral arms, no polarization is observed. This is so because each elongated grain of dust is rotating like a propeller around its own short axis, which is aligned along the spiral arm by the magnetic fields of the Galaxy, whose directions are also along the spiral arms. The dust grains, which are somewhat paramagnetic and are rotating, behave like tiny spinning magnets in the galactic magnetic fields (see Section 19.25) and become aligned as described. Such alignment is ensured if the strength of the galactic magnetic fields is of the order of 3×10^{-5} gauss; if the long axis of each grain is about three times the size of the short axis, the observed dependence of the reddening of the starlight on the polarization can be accounted for.

28.23 The Structure of Planetary Nebulae and Internal Motions

We noted in the text that a planetary nebula is an extended sphere of expanding gas with a very hot (temperatures range from 20 000°K to 100 000°K or more) and dense central star which, in some respects, is similar to a white dwarf. Although the light emitted by the bright gaseous envelope of a planetary nebula is dominated by the principal line of the Balmer series of hydrogen (the H_2 line at 6 563 Å; see Section 18.26), some 40 other bright lines have been identified. Recently infrared radiation has been detected from planetary nebulae in the wavelength range from 10^{-4} cm to 7×10^{-2} cm. This radiation is probably emitted by heated dust particles within or near the expanding gaseous envelope. Thermal radio waves with a continuous spectrum also have been observed from planetaries. This radiation probably arises from plasma vibrations in the expanding gaseous envelopes when these collide with interstellar gas clouds.

Attempts have been made, with some success, to fit planetary nebulae into stellar evolutionary sequences. According to these theories planetary nebulae originate from distended, irregular or long-period, red-giant variable stars that are burning helium in a shell around a carbon core. At this stage, instabilities within the helium-burning shell and in the extended envelope could lead to the escape of hydrogen and helium from the upper regions of the star's atmosphere, where the speed of escape is of the order of 20 to 30 km/sec. The combined action of intense radiation pressure and gas pressure resulting from a sudden increase in luminosity would be the mechanism for this ejection of gases from the star. As more and more of the envelope escapes, the remaining hot, dense core reaches a stage where it can remain in equilibrium as a white dwarf.

28.31(a) The Origin of the Speed of Relativistic Electrons

With the discovery that the central star of the Crab Nebula is a pulsar, (NP 0532 spinning at the rate of 30 times per second), the problem concerning the source of energy of the Crab Nebula was solved, as outlined in Section 24.50(a). The pulsar at the center of the Crab Nebula is slowing down and thus losing rotational energy at the rate of about 10^{38} ergs/sec (more than enough to account for the luminosity of the Crab) and transferring this energy to electrons by means of its intense magnetic fields, which are of the order of 10^{11} gauss. These relativistic electrons transfer enough energy to the expanding gaseous envelope of the Crab to give it its observed large luminosity.

CHAPTER 29: THE STRUCTURE AND ROTATION OF THE GALAXY

We give here a summary of the most recent data about the Galaxy.

The center of the Galaxy has been observed in the far infrared; the source of this infrared has a diameter less than 3′ and lies within 6′ of Sagittarius A. Whether this is a thermal or nonthermal source has not been determined. If it is a thermal source, it may consist of dust at a temperature of 20°K to 50°K. This result would be in conflict with the concept that the center of the Galaxy is dust-free. The intensity of the infrared radiation from the center would indicate a fairly large mass of dust if the infrared radiation is thermal.

The thickness of the Galaxy is 2 kiloparsecs.

The total mass of the Galaxy is $1.4 \times 10^{11} \mathscr{M}_\odot$.

The sun's distance from the galactic center is 10 ± 0.8 kiloparsecs.

The value of Oort's constant A (see Section 29.6) is 15.0 ± 0.8 (km/sec) per kiloparsec (kpsc).

The rotational velocity in the neighborhood of the sun is 250 km/sec. The escape velocity

from the galactic center is 700 km/sec;
from the neighborhood of the sun is 360 km/sec;
from the rim of the Galaxy is 240 km/sec.

The spacing between two neighboring spiral arms measured along a line from the center of the Galaxy to the sun is about 1.6 kpsc. Measured from the center of the Galaxy, the Sagittarius arm is 8.7 kpsc, the Orion arm is 10.4 kpsc, and the Perseus arm is 12.3 kpsc. The thickness of an arm is about 0.6 kpsc.

Additional evidence for large turbulences at the center of the Galaxy (as indicated by the hydrogen streaming there) may be found in the emission of gravitational waves from the center, which has been reported by J. Weber. The intensity of the gravitational waves reported by Weber is so large that, if he is correct, the gravitational energy released at the galactic center must correspond to the complete annihilation or transformation into gravitational radiation of one solar mass every minute, which means that the entire core of the Galaxy would collapse in less than a hundred million years. Thus far, other physicists have been unable to confirm Weber's observations; his data may not refer to gravitational waves at all. Nevertheless one would expect gravitational waves to be emitted from the core of the Galaxy if drastic gravitational events, such as the rapid collapse of stars, were occurring there. This is predicted by the general theory of relativity, but until Weber's data are confirmed we cannot say that intense gravitational waves are coming from the center of the Galaxy.

The general theory of relativity predicts that gravitational waves should be emitted by accelerated masses; in ordinary gravitational events such as the motions of binary stars around each other or the rotation of stars, the rate of emission of such waves is exceedingly small. Binary stars would have to revolve at speeds very close to the speed of light to emit any measurable quantity of gravitational radiation. This is also true of rotating bodies such as neutron stars. But when a star in its pre-supernova stage collapses violently, it can emit gravitational waves at the rate of 10^{50} ergs/sec. To account for Weber's observations, 400 stars like the sun would have to collapse violently every second at the center of the Galaxy.

29.1 The Spiral Structure of the Galactic Disk

A good deal more is understood now about the formation, stability, and permanence of the spiral arms of galaxies than just a few years ago. Spiral arms should be formed in a rotating system consisting of dust, gas, and stars if, like the Galaxy, the system is rotating differentially. Objects close to the center revolve rapidly, whereas those near the rim revolve slowly and lag behind those near the core. In this way trailing spiral arms are formed. The question that now arises is why these arms remain the way they are for billions of years instead of winding up and forming many closely spaced arms. Since the sun revolves around the core of the Galaxy every 250 million years, it must have gone around some 50 times since the Galaxy was formed, and the spiral arms ought to be wound around the Galaxy many times. Why is this not so? Why and how have the three spiral arms in the Galaxy maintained themselves as they are for some 10 billion years?

The stability and persistence of the spiral arms cannot be due to the magnetic fields in the Galaxy, because such fields are much too weak to confine the stars to the spiral arms. The difficulty presented by the winding up of the spiral arms can be eliminated by applying the concept of density waves. According to this picture, the spiral arms do not consist of the same stars and dust at all times, but instead are the regions where the density of matter at any moment is a maximum. The density maxima of the spiral arms and the density minima between them are the maxima and minima of a density wave that revolves around the Galaxy in a fixed pattern.

CHAPTER 30: GALAXIES

In the accompanying table we summarize the overall characteristics of galaxies as deduced from the most recent data. The total emission of radiation from galaxies is about 2.2×10^{-10} solar luminosities per cubic parsec of space. The average number of stars per galaxy is about 10^{11}. There are about 0.02 observable galaxies per cubic megaparsec.

	Spirals	Ellipticals	Irregulars
Mass (in solar units)	10^9 to 5×10^{11}	10^6 to 10^{13}	10^8 to 3×10^{10}
		mean mass = 9×10^{10}	
Diameters in kpsc	6.5 to 45	0.6 to 150	1.5 to 9
Luminosity (in solar units)	10^8 to 10^{10}	10^6 to 10^{11}	10^7 to 2×10^9
M_v	-15 to -21	-9 to -23	-13 to -18
		mean absolute magnitude = -20.3	
Spectral types	A to K	G to K	A to F
Interstellar matter	gas and dust	no dust, very little gas	gas and dust in large quantities

30.23 Radio Galaxies and Seyfert Galaxies

More than 100 extragalactic radio sources have been identified as individual galaxies. These can be divided into two groups:

1. Those like our Galaxy which are normal spirals and which, in addition to the 21-cm hy-

drogen line, emit a continuous spectrum of radio energy from their disks, their nuclei, and their coronas. The total radio energy emitted on the average by these galaxies is the order of 10^{38} ergs/sec, but for some galaxies it may go as high as 10^{41} ergs/sec. This is negligible compared to the total emission of these galaxies in the visible part of the spectrum.

2. The peculiar radio galaxies; these have very unusual structures and emit extremely large quantities of radio energy. The energy emitted in the radio part of the spectrum exceeds 10^{44} ergs/sec and is as large as or greater than that emitted in the visible part of the spectrum. The galaxy NGC 1068 emits 100 times as much radio energy as does a normal spiral, and the elliptical galaxy M 87 in the Virgo cluster [see Figure 30.6(a)] is 1 000 times as luminous in its radio spectrum as normal spirals.

The peculiar radio galaxies have complex structures and emit their radio energy in very unusual ways. In some cases the radio energy is emitted from a single concentrated source, whereas in others the emission occurs from a concentrated core and an extended halo. In still others the emission occurs from two extended sources; one on each side of the galaxy, which is generally a normal elliptical. Each of the two radio sources, which generally lie on a line passing through the center of the elliptical galaxy, is about one million light years from the center of the galaxy. If these radio sources originated in the core of the galaxy, the explosive events that gave rise to them must have been of incredible intensity to have projected the radio sources to such vast distances on either side of the core. The nature of such explosive events in galactic cores is not now known, but such possibilities as a chain reaction of supernovae, gravitational collapse of large quantities of matter, the annihilation of matter by antimatter, and large numbers of stellar collisions suggest themselves.

That violent events are occurring in the cores of certain galaxies is clearly evident in the case of the famous galaxies discovered in 1944 by Carl Seyfert. As one passes from the outer regions of Seyfert galaxies to their small luminous nuclei, there is a marked increase in brightness. The spectra of these dozen or more galaxies consist of strong, broad emission lines, which means that the cores of the galaxies contain concentrated regions of hot gases. The large widths of the spectral lines, which are caused by random Doppler effects, indicate that the hot gases are expanding rapidly, with speeds as high as thousands of kilometers per second. Seyfert galaxies are often strong radio emitters, and they all are intense sources of infrared radiation. The visual luminosities of the Seyfert galaxies are about the same as those of normal galaxies, but when the infrared radiation is added to the visible radiation the Seyfert galaxies are about 100 times as luminous as the normal ones.

Certain galaxies such as NGC 5128 (see Figure 30.16) and M 82, which has a very peculiar shape, are very strong radio emitters and show evidence of highly explosive events in their cores. These are not Seyfert galaxies but constitute an unusual group that have no special characteristics other than their explosiveness and unusual shapes. These galaxies are generally far more luminous than normal galaxies such as the Milky Way.

A group of very luminous galaxies known as N-type galaxies (where N stands for nucleus) have characteristics that lie between those of the Seyfert galaxies and those of the quasars. They have very small, bright nuclei (they look starlike on photographic plates) immersed in faint wispy, nebulous clouds.

Because Seyfert galaxies, N-type galaxies, and quasars have certain similar characteristics, various astronomers (and in particular S. Colgate) have suggested that the cores of Seyfert galaxies and N galaxies are members of a continuous sequence of objects that begin as ordinary galaxies and end up as quasistellar objects or quasars. According to this theory the Seyfert galaxies and N galaxies are intermediate stages in the evolution of galaxies that lie between ordinary galaxies and quasistellar objects. The regions outside the cores of nearby (that is, near enough to be observed) Seyfert galaxies look rather like the spiral arms of normal galaxies and their apparent sizes and speeds of recession at their observed distances are as expected. If we go to greater distances, however, only the nuclei of the Seyfert galaxies can be observed (N-type galaxies), and these objects then appear more like stars than like galaxies. This fact, coupled with their intense luminosities in the visible, the infrared, and the radio parts of the electromagnetic spectrum, suggests the possibility that Seyfert galaxies are related to the quasistellar objects discussed in Section 30.50.

30.48 The Problem of Stability of Clusters of Galaxies

The conclusions of Ambartsumian and some other astronomers, that clusters of galaxies such as the Virgo and the Corona Borealis clusters are not in equilibrium but are unstable and are disintegrating, are based on the assumption that almost all of the matter in these clusters is present in the form of the visible stars, dust, and gas in the individual galaxies. If this were so, then these clusters of galaxies would, indeed, be unstable because 25 times as much mass as is actually observed in the form of individual galaxies would be required to keep the Virgo cluster from disintegrating. Many astronomers now believe that this "missing matter" is present in some very unusual forms—possibly as mini black holes—in clusters of galaxies and that in time it will be observed. Thus far, however, nobody has detected enough intergalactic matter in any form to keep the galaxies in a cluster together. Without at least 25 times as much matter as is contained in all the visible galaxies of the Virgo cluster (some 2 500), the galaxies in this cluster could never have been drawn together or kept together even temporarily. Their motions within the cluster are much too rapid for them to have coalesced through their own gravitational attraction.

30.50 Quasistellar Objects (Quasars)

We give here a résumé of some of the latest data on quasars.

Up to the present time (1976), more than 300 quasars have been discovered. They all are at vast distances from us if their large red shifts mean that they are moving away from us at the speed of expansion of the universe, in accordance with Hubble's law of cosmological recession (see Section 31.7).

They all have the following characteristics in common:

1. They are starlike objects on photographic plates and emit radio waves.
2. Their radiation varies in intensity with periods that range from a day or less to a few years. In any case, the region from which the radiation of a quasar is emitted cannot be larger than the speed of light multiplied by the variation time or period of variation.
3. They emit an excess of ultraviolet radiation as compared to ordinary stars or galaxies.
4. They emit great quantities of infrared radiation that does not fit the usual infrared pattern of cool stars.
5. Broad emission lines are present in their spectra, and in some cases absorption lines are also present. The emission lines appear to arise in regions where the electron temperatures range from 10^4 to 10^5 °K. Not only the Balmer lines of hydrogen are present but also forbidden lines of ionized O, Ne, and Mg are found.
6. Both the emission and absorption lines in the spectrum of a quasar exhibit very large red shifts (no blue shifts have ever been found), but not of the same amount.
7. The z values of the red shifts of the absorption lines lie close to 1.95, where z is defined as $(\lambda - \lambda_0)/\lambda_0$ (here λ is the observed wavelength of an absorption line and λ_0 is its normal, unshifted value).
8. The absorption lines are much narrower than the emission lines, which are similar to those in the spectra of hot gaseous nebulae such as the planetary nebulae.
9. The red shifts range from a value of 0.16 for z, indicating a speed of recession of some 16 per cent of the speed of light, to a value of 3.53, indicating a speed of recession of 91 per cent of the speed of light. If Hubble's law applies to all quasars, the two quasars with red shifts of 3.4 (quasar OH471) and 3.53 (quasar OQ172) discovered early in 1973 represent the most luminous concentration of matter known. They are also the most distant objects known, revealing themselves the way they were some 10 billion years ago.
10. A comparison of the intensities of the radiation emitted in various parts of the spectrum by quasars with the intensities given by the Planck formula shows that the quasars are not radiating like black bodies. The quasar radiation is nonthermal. No single temperature can be assigned to a quasar to make its radiation agree with Planck's formula for such a temperature.

31.1 Olbers' Paradox

We call attention here to an important criticism of Olbers' paradox that was presented by E. R. Harrison, based on the conservation of energy. Olbers' paradox applies only if we assume that each star's luminosity has remained the same for a very long period of time (long enough to allow the radiation from the stars to fill the universe uniformly). Harrison has pointed out that this period is of the order of 10^{23} years or longer. But no star could have been radiating energy at its present rate for such a long period of time because there is not enough thermonuclear fuel available for this to happen. This means that Olbers' paradox could occur only if we deny the principle of the conservation of energy.

Harrison has pointed out that if stars have been radiating at their present rate for 10^{10} years there is no contradiction between the way the dark sky looks and the way that it ought to look according to the theory.

31.10 The Revised Hubble Constant

The best estimate of the Hubble constant at present, as given in a series of papers by Sandage and G. S. Tammann, based on their analysis of the angular sizes (apparent sizes) of the HII regions (the standard candles) in distant galaxies, is 55 km/sec/mpsc, with a probable error of ± 5 km/sec. This corresponds to a time span of $1/H = 18 \times 10^9$ years since the initial big bang. The figure 50 km/sec/mpsc is not the lowest estimate given by some astronomers, who propose 30 km/sec/mpsc for the Hubble constant. All that we can say with certainty is that the value of the Hubble constant lies between 30 and 75.

31.18 Time-Dependent Models of the Universe

It must be clearly understood that although all the distant galaxies and clusters of galaxies appear to be rushing away from us, we are not at the "center" of the universe. In fact there is no center in our three-dimensional space. The galaxies or clusters of galaxies are all rushing away from each other, so that no matter where we station ourselves in the universe the recession of the distant galaxies will appear exactly the same. An analogy will make this clear. Suppose the radius of the earth were increasing at a constant rate. We would then find all distances on the earth doubling when the earth's radius doubled; therefore all the cities in the United States would appear to be receding from New York at a rate proportional to their distances from New York. There would be no "center" on the surface of the earth from which the expansion was taking place; all observers on the earth, no matter where, would see the same thing—the distant cities would all appear to recede from them. If the earth's radius were to double every hour, a city such as Chicago, at a distance of 1 000 mi from New York, would appear to be receding at 1 000 mi/hr from an observer in New York, whereas San Francisco, at a distance of about 3 000 mi from New York, would be receding at a speed of 3 000 mi/hr from the New Yorker.

If we can speak of a "center of expansion" of the universe at all, it is a point on a line along a fourth dimension and thus perpendicular to our three-dimensional universe, as shown in Figure 31.6. The expansion of the universe increases the distances between galaxies or clusters of galaxies. It does not change the sizes of individual galaxies or the dimensions of electrons and protons. If electrons, protons, and atoms in general increased in size at the same rate as the universe does, owing to its expansion, we could never detect the expansion of the universe since the electrons and protons are the yardsticks we use to measure the expansion itself.

31.21 The Deceleration Parameter and the Observational Data

Equation (31.11) in the main text (which is used to measure the deceleration parameter) is called the magnitude-redshift relation to distinguish it from the distance-redshift relation (Equation 31.2 of the text), which is used to measure the Hubble constant H. The distance-redshift relation as given in Equation (31.2) is only a first approximation. When higher-order terms (higher powers of Hr) are taken into account, we obtain

$$cz = Hr + \frac{1}{2c}(1 + q)(Hr)^2$$
$$+ \text{ terms of the order of } (Hr)^3 \text{ and higher,}$$

where q is the deceleration parameter as defined in Section 31.20 and r is the distance from the earth to the distant galaxy or cluster of galaxies being considered. This distance-redshift relation can be understood from an elementary point of view if we note that the dependence of z on the product Hr

must contain higher powers of Hr if the universe originated in a "big bang." If the expansion of the universe were always the same (the same for all times in the past and the future), z would have the same value for all times. This is equivalent to saying that z would have the same value for all values of r, since different r values correspond to different times. We would then have

$$cz = Hr,$$

and the graph of z plotted against r would be strictly a straight line for values of r.

If, however, the universe originated in a big bang and has been slowing down since then (owing to the mutual gravitational pull of all particles of matter in the universe), higher powers of Hr must enter into the expression for z in its dependence on r. Taking into account only the next higher power of Hr, which is $(Hr)^2$, we have

$$cz = Hr + \frac{1}{2c}(1 + q)(Hr)^2.$$

The presence of this second term in the expression for z changes the curve in the graph of z from a straight line to a very shallow parabola, whose curvature depends on the value of q.

To obtain H from this distance-redshift relation, one plots the observed values of z for galaxies against their observed distances r. The slope of this z-r curve at the origin of the graph (the curve passes through the origin) equals the Hubble constant. The principal difficulty with this procedure is that the distance-redshift relation can be applied accurately only if many galaxies are used. To do so one must go out to fairly large values of r (large distances), but it is very difficult to measure such large distances accurately. The method of Sandage and Tammann, which we discussed in Section 31.10, is to use the largest and brightest clouds of ionized hydrogen (HII regions) in distant galaxies as standard candles, and to deduce the distance of these galaxies from the apparent brightness of their HII regions.

The principal difficulty involved in obtaining an accurate value of the deceleration parameter from Equation (31.11) stems from the requirement that this equation be applied to objects with the same absolute magnitude (the same luminosity L). One generally proceeds (as Sandage did) by assuming that the luminosity of the brightest galaxy in any regular cluster of galaxies is the same from cluster to cluster. There is some justification for this because such clusters contain mostly E-type (elliptical) galaxies, and the brightest such galaxies are remarkably similar from one cluster to another as shown by their spectra and by the almost perfectly straight line one gets when one plots their angular diameters versus their red shifts, or their apparent magnitudes versus their red shifts, or their apparent magnitudes versus their angular diameters. Nevertheless, this assumption about the brightest E galaxies in clusters of galaxies has a 25 per cent inaccuracy in it.

Other sources of inaccuracies in the determination of the deceleration parameter are: our lack of knowledge of the luminosity evolution of galaxies, the effect of intergalactic material on the apparent brightnesses of distant galaxies, and the amount of intergalactic material present in the form of massive particles of matter. The luminosity evolution of galaxies would tend to reduce the value of q, whereas the other two effects would increase it. Since very little is known about these three phenomena, we can only speculate about the correct value of q.

Taking only the present observations into account and assuming that there are no large quantities of hidden intergalactic mass, we find that q is of the order of 0.2, which means the universe is hyperbolic, open, and infinite. It will thus go on expanding forever, if the above conclusions are correct.

31.22 The Dependence of the Curvature of the Universe on the Density and Hubble's Constant

The significance of Equation (31.12) for the behavior of the universe can be understood in an elementary way if we introduce the concept of the speed of escape. If we consider matter in the three-dimensional shell expanding along the radius R at the rate $\dot{R}$, then we can consider Equation (31.12) or the equivalent Equation (31.7) as relating the rate of expansion $\dot{R}$ of the universe to the speed of escape to infinity of the matter in the universe. To see this, we multiply all the terms in Equation (31.7) by R^2 to obtain

$$\dot{R}^2 - \frac{2G\frac{4}{3}\pi R^3\rho}{R} = -kc^2.$$

Since ρ is the density of matter in the universe, the quantity $\frac{4}{3}\pi R^3\rho$ is just the total mass M of the universe, since $\frac{4}{3}\pi R^3$ is its volume. We thus obtain

$$\dot{R}^2 - \frac{2GM}{R} = -kc^2.$$

We recognize the quantity $2GM/R$ in Newtonian mechanics as the square of the parabolic speed of expansion or the square of the speed of escape from a sphere of mass M and radius R (see Section 6.24).

If we were dealing with Newtonian gravitational theory, the universe would stop expanding and begin to contract or go on expanding forever, depending on whether $\dot{R}^2$ were smaller or larger than the square of the Newtonian speed of escape $2GM/R$. In the general theory of relativity, however, an additional term appears $(-kc^2)$ which arises from Einstein's mass-energy relationship $E = Mc^2$. If $\dot{R}^2$ is smaller than the Newtonian speed of escape $2GM/R$, then, just as in Newtonian theory, the universe will stop expanding after a finite time and begin to contract. The constant k must then be a positive number (actually equal to 1), which means that the universe is finite, closed, and is governed by Riemannian geometry. If $\dot{R}^2$ exceeds the Newtonian speed of escape (the left-hand side of the above equation), the constant k must be negative (k is then -1). The universe is then an infinite open universe that will go on expanding forever.

31.28 Applying the Observational Evidence: The Pulsating Universe

At present the observational evidence as to the nature of the expansion of the universe is ambiguous and favors neither the closed ($k = +1$) or open ($k = -1$) model of the universe. A direct analysis of the red shifts of the distant galaxies as carried out by Sandage and others leads to a value of the deceleration parameter that lies between 0.65 and 1.03. This clearly favors the closed model of the universe as described in the main text. By contrast, an analysis of the known amount of matter in the universe leads to a density of matter that favors an open, infinitely expanding universe. To see what this means we note that Equation (31.13) in the text leads to a critical density $\rho_c = 1.7 \times 10^{27}$ (H^2/c^2) for the universe. If we use 55 km/mpsc for H and substitute the correct numerical value for c in this equation, we find that the critical density is about 10^{-29} gm/cm^3. The total amount of known matter gives a density that is only about one-third of this at most. From this point of view the universe is not closed unless a great deal of matter in some unusual and unobservable form such as mini black holes is present.

Some conclusion about the present density of matter in the universe can be drawn from the abundance of deuterium, which could only have been formed shortly after the big bang. The deuterium now present could not have been formed inside stars because deuterium cannot survive temperatures above 10 million degrees. Since the present abundance of deuterium depends on the initial density of protons and neutrons in the universe, which in turn determined the present density of matter, measuring the abundance of deuterium can give us some estimate of the present density (excluding mini black holes). We find from the deuterium abundance that the density of ordinary matter is about 10^{-30} gm/cm^3 and thus considerably less than the critical density.

31.29 Radio Sources as Evidence for Cosmological Models of the Universe

The recent observations on quasars and quasistellar objects in general also indicate that the steady-state theory is untenable. If one plots the number of quasistellar objects against their observed red shift z, one should obtain an overall constant distribution if the steady-state theory were correct. But D. W. Sciama and M. J. Rees have shown that the distribution is not the same out to all distances. As one goes to increasing values of z, the number of quasistellar objects per volume of space increases, which indicates that the distribution of matter in the universe millions of years ago was quite different from its present distribution.

31.30 The 2.7°K Black-Body Cosmic Radiation and the Big Bang Theory of the Universe

Perhaps the most powerful argument against the steady-state model of the universe is the recently discovered isotropic (same in all directions) 2.7°K background radiation. This radiation was first predicted by Gamow. It was discovered in 1965 by A. A. Penzias and R. W. Wilson (who had been tracking down the source of 75-cm background radiation) after it had been suggested by R. H. Dicke and his co-workers, who were unaware of Gamow's prediction. This radiation has been studied in great detail recently by P. Thaddeus and J. F. Clauser among others, to determine whether it is black-body radiation or not. If one plots the logarithm of the observed intensity of this microwave radiation against the frequency, the curve obtained agrees very well with Planck's black-body radiation curve for 2.7°K.

If we accept this deduction, we can draw some important conclusions about the origin of the universe, or at least about its cosmological state billions of years ago, because this radiation must have originated when the universe was formed. It can be shown from basic thermodynamical principles that the black-body nature of the radiation is not altered if one either slowly compresses or expands the region of space occupied by the radiation. In other words, the black-body radiation in an expanding universe remains black-body radiation but its temperature (i.e., its wavelength of maximum intensity) changes owing to the Doppler effect. That means that the black-body radiation that we observe now must have been black-body radiation at a much higher temperature when the universe was much younger and hence much more compact. Accepting this conclusion, we can now obtain a picture of the origin of the universe and of this black-body radiation by the following considerations.

Let us trace the universe back in time (that is, reverse the expansion we now observe) and see what we can deduce. When we do this, we see the universe getting smaller and smaller, with the background black-body radiation squeezed together and getting hotter and hotter (that is, bluer and bluer owing to the Doppler effect). This process can be traced back in time until the universe was concentrated into a very small region of extremely high density. If we may take this as the initial state of the universe, we then have the big bang theory of its origin, as first analyzed in detail by G. Gamow. According to Gamow's big bang theory the temperature of the universe about one second after the initial explosion (the big bang) exceeded 10^{10} °K, and so the universe was filled with very hot black-body radiation and neutrons. Gamow estimated that the expansion of the universe from this state up to the present time would result in a residual black-body radiation (owing to the continual red shift arising from the Doppler shift) having a temperature of about 6°K. This is in good agreement with the 2.7°K radiation now observed. Since the steady-state theory, with its eternal low density of matter, cannot possibly account for the origin of the background thermal cosmic radiation, we must reject this theory in favor of the big bang theory.

CHAPTER 32: COSMOGONY: THE ORIGIN OF THE UNIVERSE

The Alfvén theory of solar system formation, which is the most recent of the modern theories, also starts from a protosun but it takes into account electromagnetic forces as well as gravitational forces.

The initial gaseous cloud (which stretched over a large distance compared to the size of the solar system) is assumed to have been ionized so that it responded to magnetic forces as well as to the gravitational force from the central mass. If this cloud consisted of a mixture of different ions (for example, ionized H and He), a stratification of the cloud into layers of different chemical composition occurred as the temperature of the cloud dropped. The reason is that these ions became de-ionized at various times owing to their different ionization potentials. Thus the elements fell toward the central body at different times and at different rates. The first cloud to fall out this way was the A cloud, consisting mainly of helium together with some solid-particle impurities. This first cloud was followed by a B cloud containing mainly hydrogen, a C cloud containing mainly carbon, and a D cloud containing mainly silicon and iron. These various clouds stopped at different distances from the central body and condensed into planets. According to Alfvén, the type A cloud condensed into Mars and the moon (later captured by the earth). The B cloud and its impurities condensed into the earth, Venus and Mercury, and the C cloud condensed into the giant planets. The planet Pluto and the satellite Triton are pictured as having been formed from the D cloud.

In Alfvén's theory the various satellites and the asteroids were produced in the same way as the planets. Thus the Galilean satellites of Jupiter, Saturn's rings, and Saturn's inner satellites were formed from the inner C cloud, but the outer satellites of Saturn and those of Uranus condensed from the D cloud. In this theory the magnetic forces played an important role because they were responsible for the accumulation of the ionized plasma in the various regions where the secondary bodies were formed and also for the transfer of angular momentum from the central body to the plasma in such a way as to start the plasma revolving with approximately the Keplerian velocity.

32.9 The Importance of Turbulences

To what we have written in the main text about Urey's theory of the origin of the planets, we must add the following statements.

In the very earliest stage of the formation of the solar system, when the protosun and the disk of gas and dust were being formed (temperature less than 0°C), the gases were primarily H_2, various inert gases, and CH_4. The dust consisted of silicates, iron oxides and sulfides, solid H_2O, and solid ammonia, but little metallic iron.

In the preprotoplanet and early protoplanet stages, with the temperature at about 0°C, water vapor and ammonia were added to the gases, the dust was unchanged, but planetesimals were formed consisting of silicates, iron ores, ices, and chlorides.

When the temperature reached 2000°C on the protoplanets, iron oxides were reduced, gases and volatized silicates were lost, and large planetesimals began to accumulate and accrete. Small planetesimals were still accreting, and both kinds of planetesimals contained carbon.

A second low-temperature stage occurred (0°C), during which the planets lost most of their atmospheres. This is when the final accumulation of the earth occurred. The terrestrial planets were then nearly uniform mixtures of varying amounts of silicates and metallic iron-nickel. The iron-nickel core of the earth was probably formed at a much later stage—possibly during geologic time.

APPENDIXES

Appendix A

OPTICAL ASTRONOMICAL INSTRUMENTS

A.1 Disadvantages of the Unaided Eye

Until quite recently most of our knowledge about stellar bodies was obtained from the visible electromagnetic radiation emitted by these bodies, and although at present we are rapidly expanding our ability to extract information from the particles and the nonvisible radiation (cosmic rays, radio waves, infrared and ultraviolet radiation, X rays, and gamma rays) coming from stellar bodies, we still depend primarily on ordinary light for the vast amount of data that we must collect to further astronomical research. Since the amount of reliable information we can get about a source increases rapidly with the intensity of the signal (radiation) from the source that reaches us, it is to our advantage to enhance that signal as much as possible.

It is clear at once that the unaided eye is of little help in this because the aperture of the pupil, even at its greatest opening, is so small (about $\frac{1}{4}$ inch in diameter) that the retina of the eye receives only very faint signals from faint sources. Moreover, the retina has no way of accumulating or retaining the radiation energy it receives. To overcome these handicaps of the eye we replace its pupil and lens by a device that can collect much more light per unit time—in other words, by a large lens; and we replace the retina by a light-sensitive surface that can accumulate radiation over a long period of time and give us a permanent record of it, i.e., a photographic plate. The combination of a lens and a photographic plate is essentially a camera, but we can convert this astronomical camera into a telescope by removing the photographic plate and introducing an eyepiece with which the image formed by the front lens (the objective) can be examined.

A.2 The Index of Refraction of a Medium

To understand how a telescope works we must first consider how light behaves when it passes from empty space into a dense medium such as glass. We have seen that light (see Section 16.20) is a wave phenomenon which is propagated at a speed of 3×10^{10} cm/sec in empty space. When light enters a medium, however, the speed depends on the density of the medium and on the wavelength (that is, the color) of the light. The denser the medium is and the shorter the wavelength of the light (the bluer the light), the smaller, in general, is its speed in the medium. We may express this in the following way:

If c is the speed of light in a vacuum, and v is the speed of a beam of definite wavelength in a medium, then c and v are related by the equation

$$\boxed{n = c/v}, \qquad \text{(A.1)}$$

where n, called the **index of refraction** of this medium for the given wavelength, is larger, in general, for denser media and for shorter wavelengths. (For a detailed discussion of frequency and wavelength of light see Chapter 16, which deals with the nature of light.)

If the index of refraction of a medium depends on the wavelength of the light, the medium bends (refracts) different wavelengths by different amounts; we then say that the medium is **dispersive**. Of course, the indexes of refraction of all media depend to some extent on the wavelength of the light, but some media are much more dispersive than others, and this is of great importance in optical instrumentation.

TABLE A.1

INDICES OF REFRACTION OF OPTICAL GLASS

(From Bausch and Lomb)

For the Sodium *D* Line

Glass	n_D
Borosilicate Crown	1.517 0
Crown	1.523 0
Light Crown	1.572 5
Dense Crown	1.617 0
Light Flint	1.588 0
Flint	1.621 0
Dense Flint	1.720

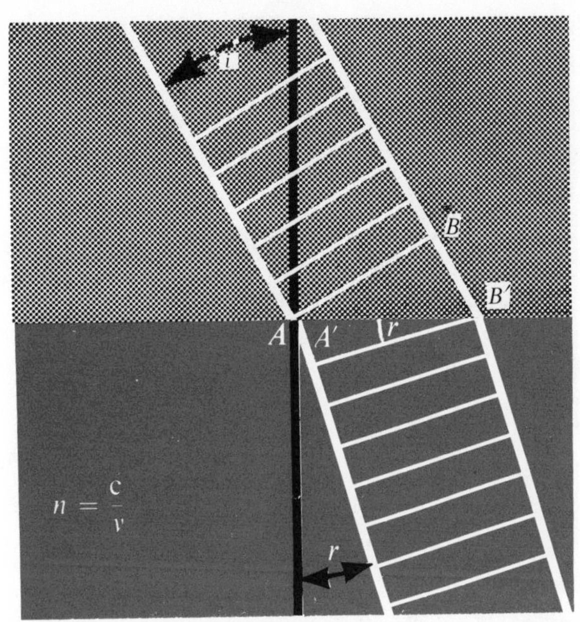

FIG. A–1 *A beam of light bends towards the normal when it goes from a less dense to a denser medium. By analyzing this figure we obtain Snell's law.*

A.3 Angles of Incidence and Refraction

We now consider a beam of monochromatic light (i.e., one definite wavelength) as it passes from empty space into a piece of glass having an index of refraction n for the given wavelength. This is shown in Figure A.1. If the interface between the vacuum and the glass is a plane, it is easy to see what happens as the light crosses this interface. To begin with, a small fraction of the light is reflected; this always happens when light passes an interface separating two media having different indices of refraction. We shall disregard this and consider the transmitted beam AB which enters the medium at an angle i to the normal of the surface (the **angle of incidence** of the beam).

Since the point A on the wavefront (a plane) strikes the glass surface before the point B does, the lower part of the wavefront is slowed down before the upper part is. As a result, the beam of light wheels around towards the normal. To understand this we note that by the time point B has reached the glass at B', the point A will only have gone as far as A' within the glass itself (because of the smaller speed of the light in the glass). Since AA' is smaller than BB', the beam turns towards the perpendicular (i.e., it is refracted by the glass) and the new wavefront in the glass is parallel to the line $A'B'$. The angle, r, which the perpendicular to this new wavefront in the glass makes with the normal to the glass surface is called the **angle of refraction** of this beam.

A.4 Snell's Law of Refraction

We see from the diagram that the angle of refraction, r, depends both on the angle of incidence, i, and on the index of refraction, n, of the glass medium. This dependence (i.e., the relationship between the quantities i, r, and n) can be derived very easily by noting that BB' is proportional to $\sin i$, whereas AA' is proportional to $\sin r$, and the proportionality factor (it is just the distance AB') is the same in both cases. In other words we have $\sin i/\sin r = BB'/AA'$. But the right-hand side of this equation is just the ratio c/v since BB' is the distance that the light has traveled in the vacuum during the time the light has traveled the distance AA' in the medium. But since c/v is just the index of refraction, we obtain the fundamental formula of geometrical optics for a beam of light going from a vacuum into a medium of index of refraction n,

$$\boxed{\frac{\sin i}{\sin r} = n}\,. \qquad \text{(A.2)}$$

This formula, **Snell's law**, can also be applied to

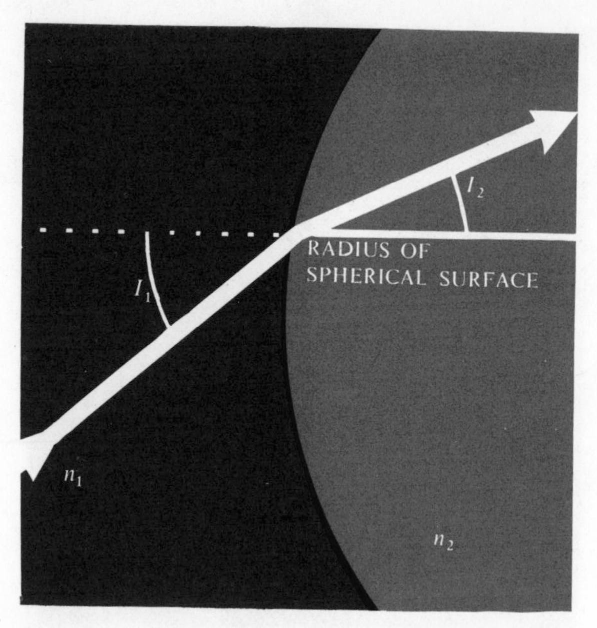

FIG. A–2 *Schematic diagram of a beam of light going from one medium of index of refraction, n_1, to another medium of index of refraction, n_2, across a spherical interface.*

the more general case of light going from any one medium into any other. If n_1 is the index of refraction of the first medium and n_2 that of the second medium, we have

$$\frac{\sin I_1}{\sin I_2} = \frac{n_2}{n_1},$$

where I_1 is the angle of incidence in the first medium and I_2 is the angle of incidence (or refraction, as the case may be) in the second medium. This is shown in Figure A.2. Note that we refer to both I_1 and I_2 as the angles of incidence since we can consider the beam as going in either direction.

With the aid of the above equation it is possible to determine how rays of light are bent as they pass through lenses, and therefore how lenses form images, on the assumption that light travels in a straight line in a homogeneous medium (a medium in which the index of refraction is the same at all points). That branch of optics which is based on this assumption is called **geometrical optics**. It neglects the wave structure of light and does not take into account the fact that light does not always travel in a straight line even in a vacuum. The branch of optics that deals with the wave structure of light and how this structure effects its propagation is called **physical optics.**

A.5 The Formation of Images by Lenses

Although we derived the fundamental Formula A.1 by considering light passing across a plane surface, this formula also applies to a beam moving from one medium to another across a curved surface—for example, a spherical surface. Because of this it is possible to apply this formula from surface to surface to find out exactly what happens to any beam of light that originates from any point on an object and then passes through a lens system such as shown in Figure A.3. In the diagram two rays are shown starting from the point A on the optic axis and then converging to the point A' on the optic axis behind the lens system. *The* ***optic axis*** *of a lens system is defined as the line that passes through the center of each lens surface and at the same time is perpendicular to each surface.*

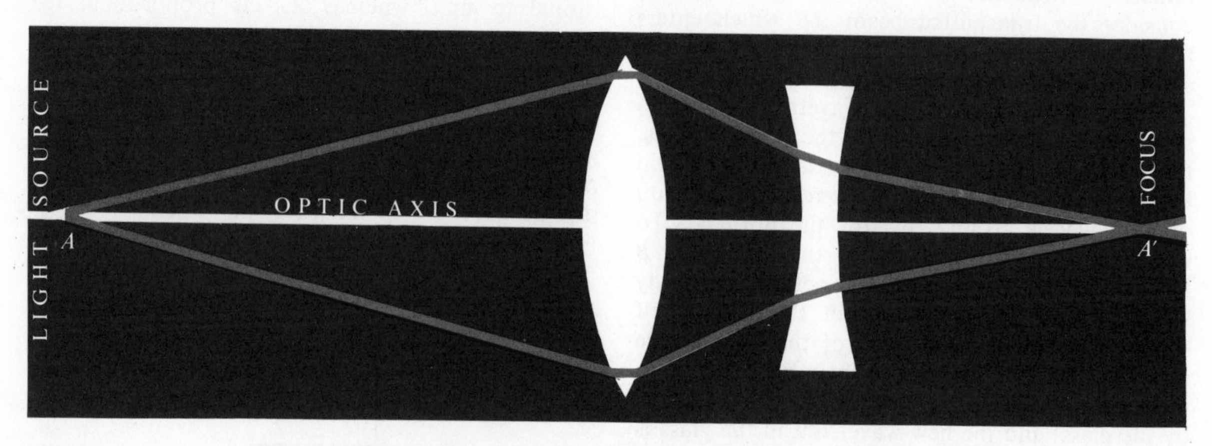

FIG. A–3 *Beams of light starting from the point A on the optic axis and passing through a lens system and out to a focus A′ on the optic axis.*

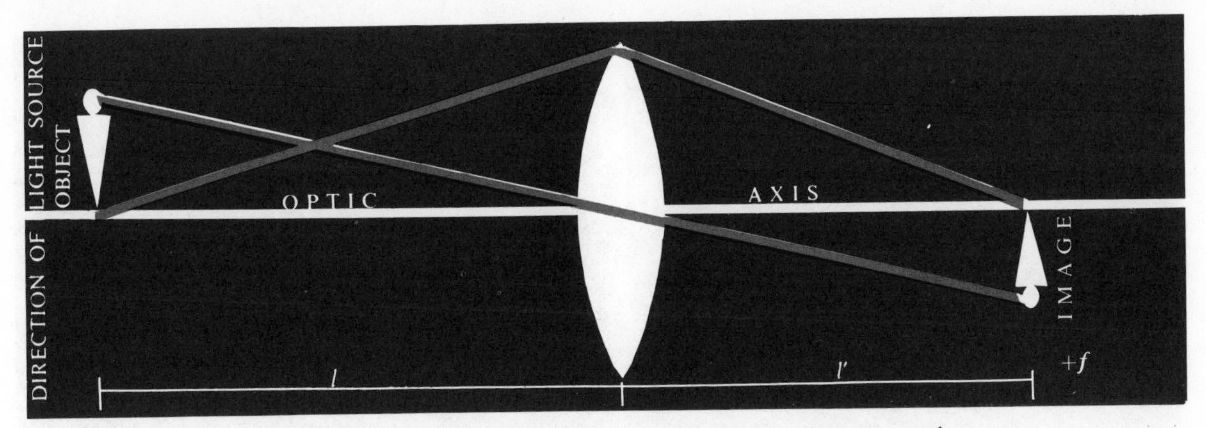

FIG. A–4 *(a) The formation of an image by a positive lens, such as a biconvex lens. The image is real and inverted.*

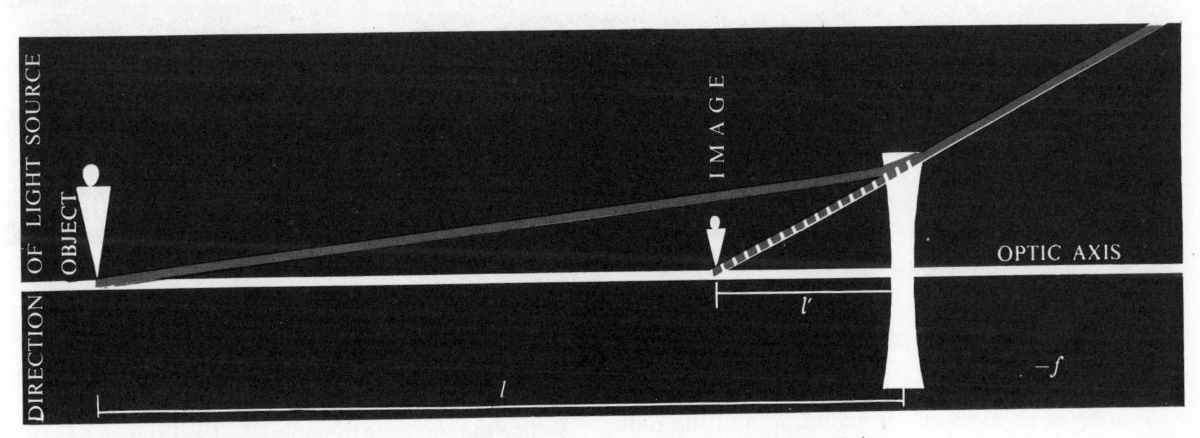

FIG. A–4 *(b) The formation of a virtual upright image by a negative (in this case a biconcave) lens.*

A.6 The Thin-Lens Formula

Although to find the position of an image accurately we must trace rays through the entire system using the exact trigonometric formula [Snell's law, Equation (A.2)] at each surface, we can get sufficiently precise image positions for most purposes when we are dealing with thin lenses by using the so-called **thin-lens formula**. If we apply Snell's law to a lens whose thickness compared to its diameter may be neglected, and if we consider only those rays for which i and r are so small that we may place $\sin i = i$, and $\sin r = r$ (in radian measure) at each surface, we obtain

$$\boxed{\frac{1}{l'} = \frac{1}{f} + \frac{1}{l}}, \qquad \text{(A.3)}$$

where l' is the distance of the image from the lens as measured along the optic axis, l is the distance of the object from the lens as measured along the optic axis, and f is the focal length of the thin lens. We may put this differently by stating that if a ray starts from a point on the optic axis at a distance l from a thin lens, it cuts the optic axis at a distance l' from the thin lens [Figures A.4(a) and (b)] if f is the focal length of this thin lens.

In this formula all three quantities l', f, and l may be either positive or negative. *l is negative if before entering the lens the ray (coming from the left and extended if necessary) cuts the optic axis to the left of the lens; it is positive otherwise.* If the ray (extended if necessary) after leaving the lens cuts the optic axis to the right of the lens, l' is positive; it is negative otherwise. The two cases for l', positive and negative, are shown in Figures A.4(a) and (b), respectively. If f is positive the lens is a **positive** or **collective lens**, and if f is negative the lens is a **negative** or **dispersive lens.**

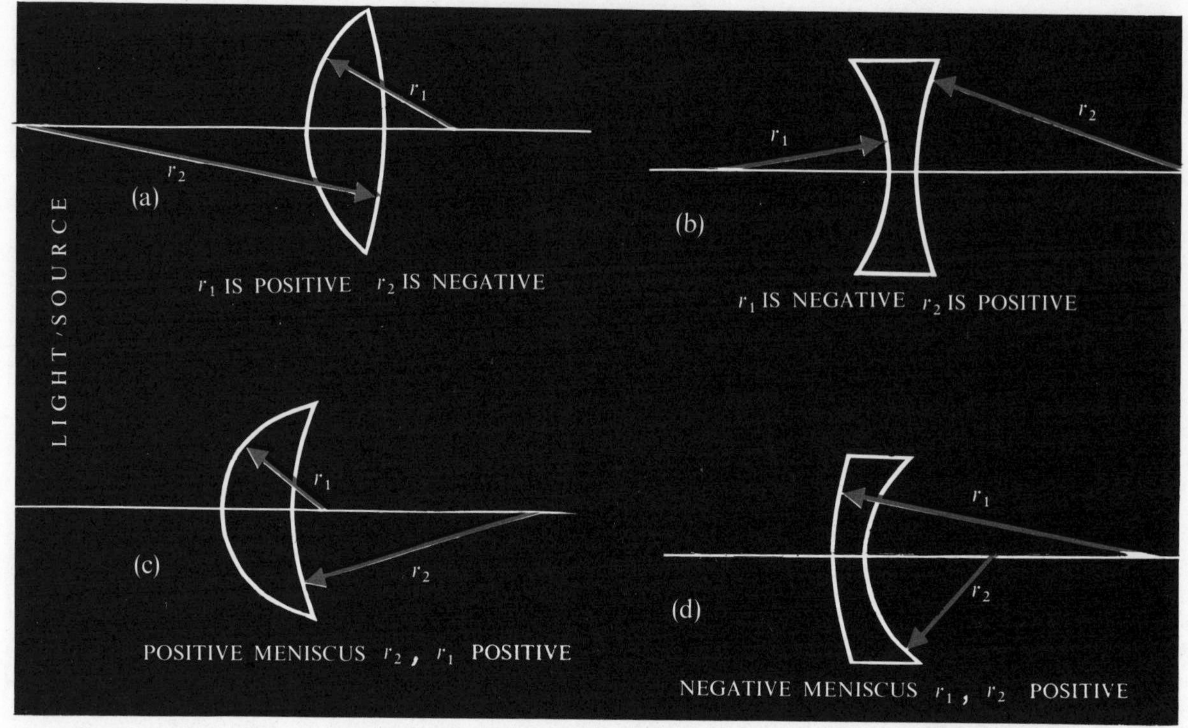

FIG. A–5 *Schematic drawing showing (a) positive, (b) negative, (c) and (d) meniscus lens surfaces.*

A.7 The Focal Length of a Thin Lens

The focal length, f, of a thin lens can be expressed in terms of its index of refraction, n, and the radii of curvature r_1 and r_2 of its two surfaces. (*Note:* the surface that is closer to the object is taken as the first surface.) We have

$$\frac{1}{f} = (n - 1)\left(\frac{1}{r_1} - \frac{1}{r_2}\right), \tag{A.4}$$

where the radius of a lens surface is to be taken positive if it is convex to the left (i.e., towards the advancing beam), and negative otherwise, as shown in Figures A.5(a) and (b), respectively. The radius of a flat surface is taken as infinite.

It follows from Equation (A.4) that the focal length of a biconvex lens [Figure A.5(a)] is positive, the focal length of a biconcave lens [Figure A.5(b)] is negative, and the focal length of a **meniscus lens** [Figure A.5(c) and (d)] is positive or negative depending upon whether the radius of the convex surface is smaller than or larger than the radius of the concave surface. In general, a lens that is thicker at the center than at the edge is positive, and one that is thicker at the edge than at the center is negative.

To see the physical significance of the focal length, we consider an object on the optic axis infinitely far away from a positive lens, so that $1/l = 0$. It follows then from (A.3) that $l' = f$. Hence the image is the point F at a distance f to the right of the lens. This point is called the **focal point** of the **thin lens**, as shown in Figure A.6(a). If rays come in parallel to the optic axis from the right and strike the lens, these rays converge at a point F to the left of the lens whose distance from the lens is also f. Thus, there is a focal point to the left and to the right of every lens. If the thin lens is negative, and the rays enter the lens from the left, the image is formed at the focal point, F, to the left of the lens, as shown in Figure A.6(b). In this case the rays do not converge to this focal point after they leave the lens but behave as though they were diverging from the focal point.

A.8 The Paths of Rays Passing through a Lens

Note that all the rays starting from a point on the optic axis infinitely far away are parallel to each other and to the optic axis when they reach the lens, and they are all brought to a focus at the point F. This is instructive since it tells us that any ray that is parallel to the optic axis is bent by the lens in

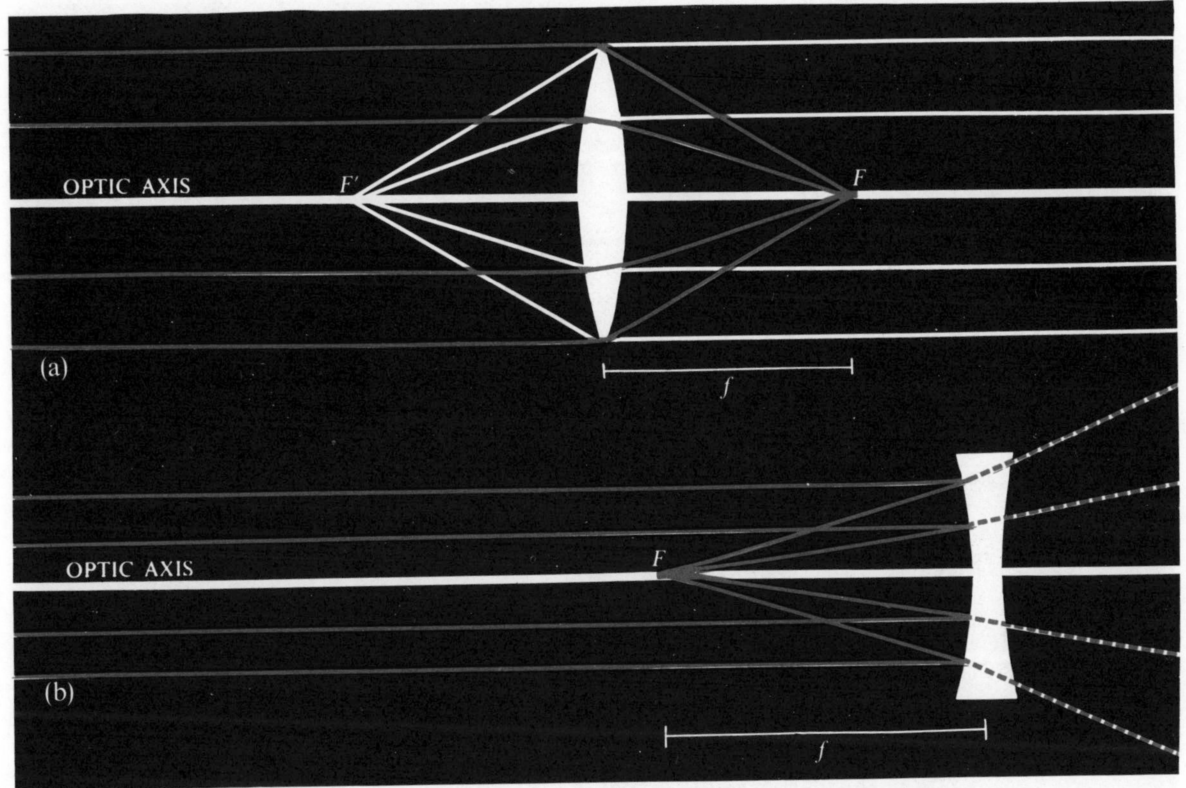

FIG. A–6 (*a*) *A ray coming from an infinite distance is brought to a real focus at the focal point F of a positive lens. Rays starting from the focal point of such a lens leave the lens parallel to the optic axis.* (*b*) *Parallel rays striking a negative lens diverge from the lens as though they originate from the focal point F.*

such a way that it passes through the focus after leaving the lens. This phenomenon enables us to obtain the image of an object graphically, if we use it together with the following thin-lens property. Since a ray that is parallel to the optic axis passes through the focal point of a lens after leaving the lens, any ray that first passes through the focal point of a lens and then strikes the lens must leave the lens parallel to the optic axis. The reason is that if the direction of a ray is reversed it retraces its original path. We introduce one more fact about thin lenses: *any ray that passes through the center of a thin lens is undeviated.*

A.9 Graphical Methods of Tracing Rays through Lenses and Obtaining Images

We can now find the image of an object by applying the following directions, as shown in Figure A.7. Starting from any point b on the object draw a ray parallel to the optic axis. This ray must pass through the focal point, F, after leaving the lens as described above. Now take a ray from b and aim it at the center of the lens. This ray continues undeviated as it passes through the lens. It therefore follows that the point b', where these two rays intersect, is the image of the point b. If we take a third ray and allow it to pass through F', it will leave the lens parallel to the optic axis and intersect the other two rays at point b'. This confirms that b' is the image of b. We thus obtain $a'b'$ as the image of the object ab.

The situation for a negative lens is also shown in Figure A.8. In this case the image we obtain is a so-called **virtual image** because the rays of light from b, after leaving the lens, do not converge to a point b' but appear to diverge from this point. Note that in this case the image is in front of the lens.

From the geometrical arrangement of the object and the image relative to the lens we see, by using the properties of similar triangles, that the ratio of the size of the image to the size of the object (which is defined as the **magnification**) is equal to the ratio of l' to l.

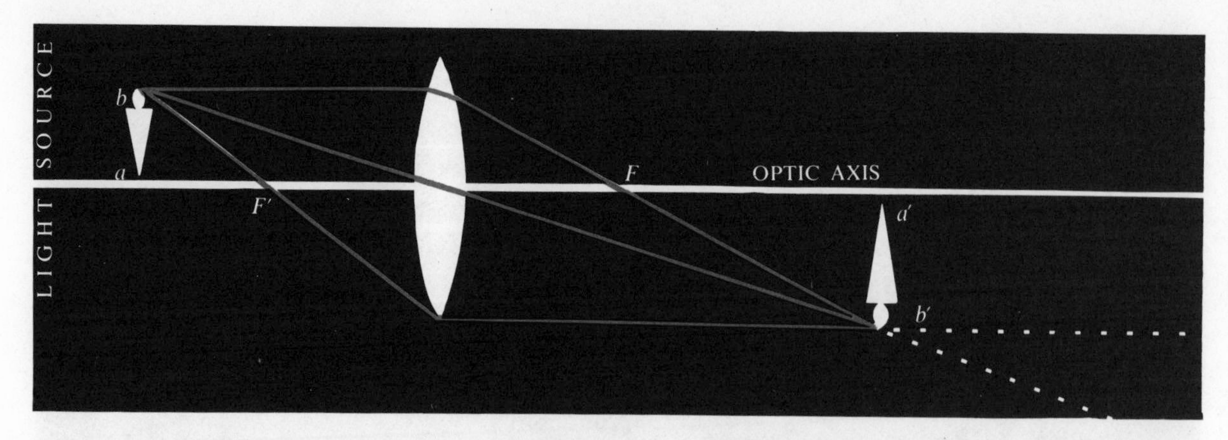

FIG. A–7 *The graphical method of tracing rays through a thin positive lens.*

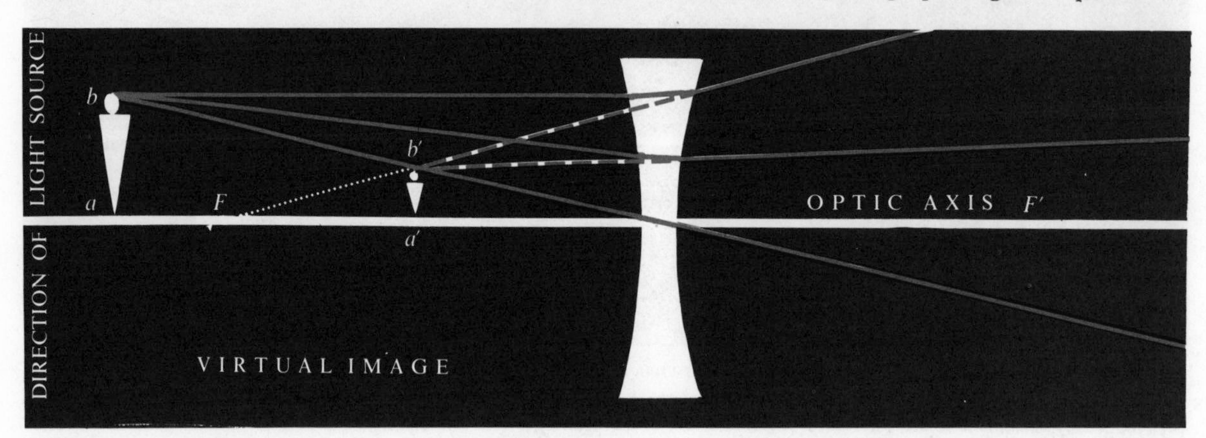

FIG. A–8 *The graphical method of tracing rays through a thin negative lens.*

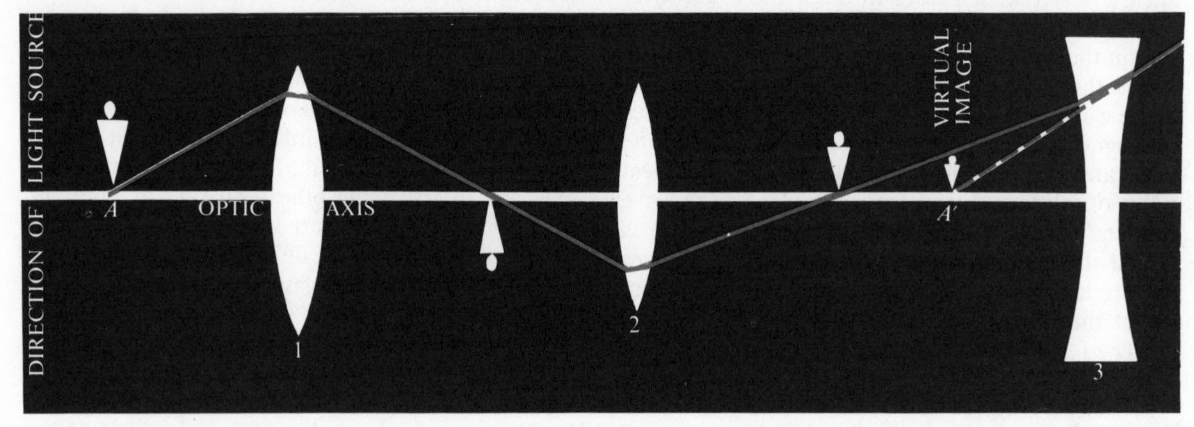

FIG. A–9 *The formation of an image by a compound lens system.*

If we are dealing with a system of multiple lenses, we can find the final image by treating one lens at a time. Thus we first find the image of the object formed by lens 1, and treat this image as the object for lens 2. We then find the image formed by lens 2 of the first image and treat it as the object of lens 3, etc., using the correct focal length for each lens in turn, as shown in Figure A.9. This procedure is very useful in tracing the path of a ray of light through any telescope.

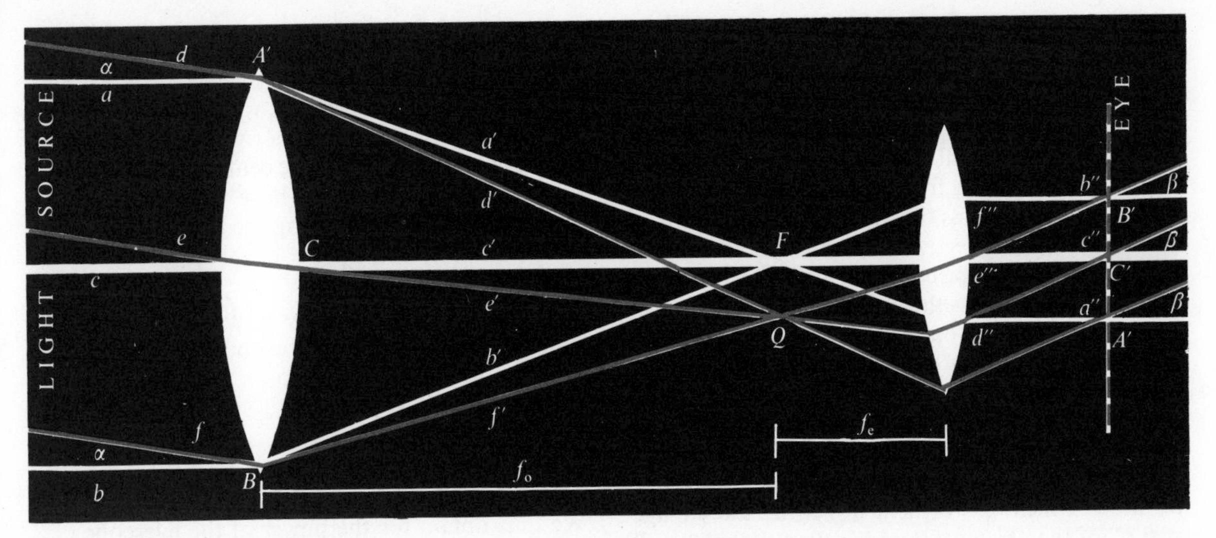

FIG. A–10 *A telescopic system consisting of an objective and an eyepiece. The focal plane of the eyepiece and the objective coincide at* F.

A.10 Refracting Telescopes

We may illustrate these ideas by considering the optical system of a refracting astronomical telescope, the largest of which is the 40-inch at Yerkes. (See Figure A.34(h)). This consists of a front lens system, the **objective**, which in general is a combination of two or more lenses, and a rear system of lenses called the **eyepiece**, which in general also contains more than one lens.

To simplify the discussion we take a single convex lens for the objective and also a single lens for the eyepiece as shown in Figure A.10. In this system the eyepiece and the objective are separated by a distance equal to the sum of the focal length f_o of the objective and the focal length f_e of the eyepiece. To see what happens we consider two rays, *a* and *b*, coming from an infinite point (let us say a star) on the optic axis. These rays are parallel to the optic axis and strike the objective at points *A* and *B*. It is clear that all points of the objective are pierced by parallel rays of this sort. After these rays have passed through the objective they all intersect at the focal point F, as shown in Figure A.10 (assuming that our simple lens formula applies to each ray no matter how high up above the optic axis it strikes the objective). This is only approximately correct and we shall see that the thin lens formula does not apply exactly to rays that strike near the edge of the lens. However, in order to discuss the basic properties of a telescope we shall make the assumption that the thin-lens formula applies to all rays.

Since the focal point F of the objective is also the focal point of the eyepiece, it follows that the rays a' and b' after passing through the eyepiece emerge parallel to the optic axis. If, then, the eye is placed on the optic axis so as to collect these rays, the lens of the eye forms on the retina an image of the original object.

A.11 The Angular Field of View of a Telescope

To determine where the eye must be placed to see the complete image, we consider rays coming from a point on the object that is not on the optic axis. Although these rays also form a parallel bundle (they come from infinity) this bundle is not parallel to the bundle defined by the rays *a*, *b*. This oblique bundle is shown by the rays *d*, *e*, *f*. Since these rays all come from the same infinite point, they converge after passing through the objective at the point *Q*, which is in the focal plane of the objective (i.e., the plane that contains F, and is perpendicular to the optic axis). *Q* is the image of the point on the object that is at an angular distance α from the optic axis. If this is the object point farthest from the optic axis that can be viewed with the telescope, then α is one-half the **angular field of view** of the telescope.

Let us now consider the rays d', e', f', after they strike and pass through the eyepiece. The central ray e' is now bent by the eyepiece, becoming the ray e'', and cutting the optic axis at a point C'. This point is clearly the image of the center of the objective formed by the eyepiece, and this is where the center of the pupil of the eye must be placed to receive this central ray of the oblique bundle. If the pupil of the eye is placed here and is wide enough, it receives all the other rays entering the objective and coming from object points within the field of view, α. Thus the ray d' which comes from point A on the objective is parallel to the ray e'' on leaving the eyepiece and enters the pupil at point A'; this point is the image formed by the eyepiece of the point A on the objective. In the same way the ray f' is bent by the eyepiece and leaves it parallel to the rays e'' and d''. It enters the pupil of the eye at point B' which is the image formed by the eyepiece of the point B on the objective.

If the pupil of the eye is not placed at this position which is coincident with the image $A'B'C'$ of the objective (and is called the **exit pupil** of the telescope just as the objective is called the **entrance pupil**) the eye cannot receive all of the rays and the best conditions for viewing the object are not fulfilled.

A.12 The Magnifying Power of the Telescope

Figure A.10 also instructs us about the magnifying power of the telescope, its light-gathering power, and so on. To find the magnifying power, we again consider the central ray e of the oblique bundle. If no telescope were present and the observer were looking at an infinitely distant object directly, this ray e would enter the pupil of his eye at an angle α with respect to rays coming from the center of the object [see Figure A.11(a), which is an exaggerated schematic drawing]. Thus, the half-size of the object would appear to be α to the naked eye.

With a telescope between the eye and the object, however, the situation is quite different. For now the eye receives the central ray e'' of the oblique bundle after it has passed through the objective and the eyepiece. It enters the eye at an angle β which, as shown in Figure A.11(b), is clearly greater than the angle α. In other words, *as seen through the telescope, the object appears to have the half-size β and appears to be inverted because the ray e'' enters the eye from the side of the optic axis that is opposite to the direction from which the ray e would enter the eye if no telescope were present.*

To find the magnification of a telescope all we need do, then, is obtain the ratio β to α, and we can do this from the geometry of the figure. We see that this ratio is equal to the ratio of the distance l of the objective from the eyepiece to the distance l' of the point C' from the eyepiece. We know that $l = f_O + f_e$, and hence we can find l' by applying the thin-lens formula (A.3) if we remember that C' is the image of C (the center of the objective) formed by the eyepiece whose focal length is f_e. We therefore have, on using the thin-lens formula,

$$\frac{1}{l'} = \frac{1}{f_e} + \frac{1}{l} = \frac{1}{f_e} - \frac{1}{f_O + f_e}.$$

Note that the minus sign must be used for l because C is to the left of the eyepiece. If we solve this equation for l', we have

$$l' = \frac{f_e(f_e + f_o)}{f_o}.$$

We thus obtain for the power of the telescope

$$\text{power} = \frac{\beta}{\alpha} = \frac{l}{l'} = \frac{(f_o + f_e)}{\dfrac{f_e(f_o + f_e)}{f_o}},$$

$$\boxed{\text{power} = \frac{f_o}{f_e}} \qquad \text{(A.5)}$$

Thus, the power of the telescope is equal to the ratio of the focal length of the objective, f_o, to the focal length of the eyepiece, f_e.

This is obvious because it is clear that the longer the telescope is (i.e., the longer the focal length of the objective is), the bigger is the image that is formed by the objective, and the smaller the focal length of the eyepiece is the closer we have to bring the eyepiece to the image formed by the objective to see this image, hence the larger this image appears. We can thus obtain as large a power with a telescope as we please by using an eyepiece having a small enough focal length. However, as we shall see, there is a limit to how large a magnification is practical for an objective of given diameter.

Since at present very little visual observing is done and most work is done with photographic film and photoelectric cells, an eyepiece is rarely used. The film or other recording device is placed at the focal plane of the objective and the entire telescope becomes a very large camera.

A.13 The Brightness of the Image: Light-Gathering Power

One great advantage of a telescope in viewing a point source image of light, such as a star, is that

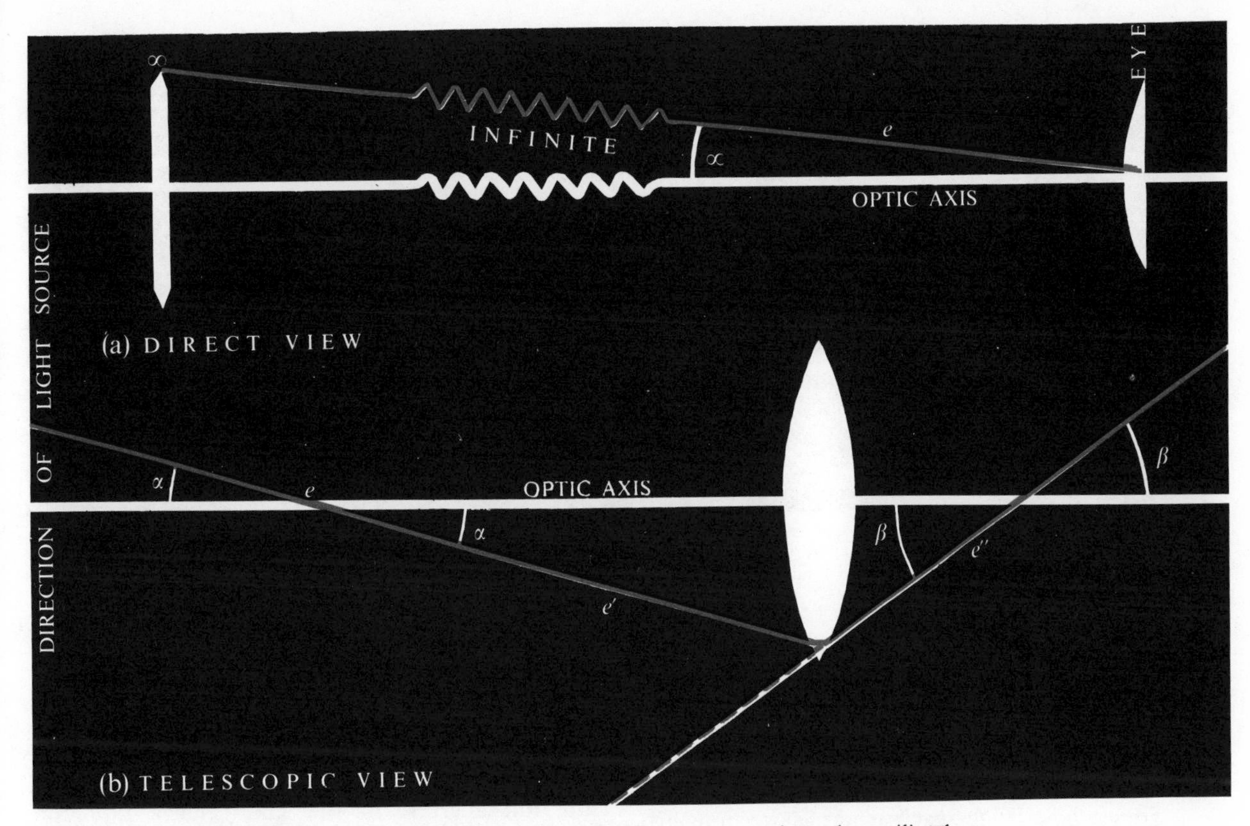

FIG. A–11 *(a) The apparent size of an object as seen by a naked eye is given as the angle α. (b) The apparent size of an object as seen through a telescope is given by the angle β. The magnification, or power, of the telescope is thus given as β/α.*

the telescope collects much more light per unit time than the pupil of the eye does. This light-gathering power therefore determines how bright the image of a point source is. To determine the light-gathering power of a telescope, consider the pupil of the eye placed behind the telescope looking at a point source of light on the optic axis, and compare it with the eye looking at the same point source without the telescope.

For a point source on the optic axis the eye placed behind the eyepiece receives the bundle of light $a''b''$ as shown in Figure A.10. This bundle contains all the light in the original parallel bundle ab, which is as wide as the objective AB. Without the telescope the eye receives a bundle that is only as wide as the pupil of the eye, $A'B'$. Since the total amount of light in a bundle depends upon the area of the bundle, it is clear from what we have just stated that the light-gathering power of a telescope depends upon the square of the diameter of the objective.

Since the pupil of the fully dilated eye is about $\frac{1}{4}''$, the amount of light that can enter the naked eye is proportional to $\frac{1}{16}$ square inches. However, the amount of light that can enter an objective of diameter d inches is proportional to d^2. Hence, we define the **light-gathering power** of a telescope as the ratio

$$\frac{d^2}{\frac{1}{16}}, \qquad \text{or } 16d^2,$$

which measures how many times as much light the objective can receive as the pupil of the eye. Thus the 200-inch telescope has a light-gathering power of 40 000 × 16 or 640 000. This expression of light-gathering power determines how bright a point source looks in telescopes of various sizes.

A.14 The Brightness of an Extended Image: The Speed of a Lens

If we are dealing with an extended object (so that we see it as a disk), the brightness of the image formed by the objective depends on the focal length as well as on the diameter of the objective. To see this let us suppose that we have two

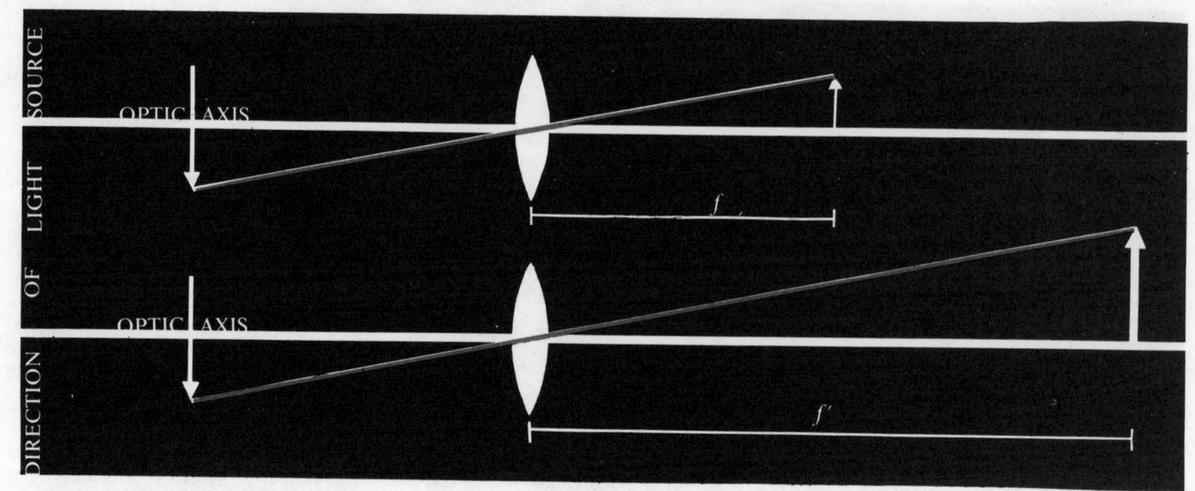

FIG. A–12 *The area covered by an image increases as the square of the focal length of the objective.*

objectives with the same diameter d, but with different focal lengths, f and f'. A ray from the edge of the object passing through the center of these two objectives hits the focal plane of each objective at different distances, as shown in Figure A.12, so that the two images formed by the two objectives are of different sizes.

To see how bright each image is we note first that the same amount of light, which is proportional to d^2, enters both objectives. However, the two images are not equally bright because they are not of the same size, so that this same amount of light is spread over different areas in both cases. Since the size of the image is proportional to the focal length, f, its area is proportional to f^2 and we see that the amount of light striking a unit area of the image (which is therefore a measure of the image's brightness) is directly proportional to d^2 and inversely proportional to f^2. Thus, the brightness of the image is measured by the ratio $(d/f)^2$. *For this reason the quantity f/d is referred to as the f-number, or the* ***speed of the objective.*** *The smaller this number is, the faster the lens is, and the brighter is the image formed by it.*

A.15 Flaws in Images Formed by Lenses

In our discussion of the properties of lenses we assumed that the light from a point source is focused to a point by a lens. But this does not occur in practice, for two reasons:

1. The thin-lens formula we used above applies only to rays of light that strike the lens very close to the optic axis, i.e., to rays of light that are contained in a thin bundle near the optic axis, as shown in Figure A.13.

This region is called the **paraxial region of the lens.** In practice it refers to a bundle of rays having a 2° spread, and this is generally the twentieth part of the full aperture of the lens. Hence Equation (A.3), introduced on page 635, is called the paraxial lens equation.

If we consider rays that strike the lens beyond the paraxial region, the formula does not apply. Thus, the rays striking the lens near the edge are brought to a focus at a point that differs from that determined by this formula. To find where such rays are brought to a focus by a lens system, we must trace the rays through the system using Snell's law of refraction at each optical surface. If rays of light coming from the same point on an object do not converge to the same point after leaving a lens we say that the lens has **aberrations.**

2. Rays of light from a point are not brought to a point focus by a lens (even if the lens had no aberrations of the kind mentioned above) because light does not move in straight lines, but has a wave structure which gives rise to a diffraction pattern when the light passes through a lens.

In other words, to describe precisely what happens to light when it passes through a lens system we must treat the light not by geometrical optics, as we did in the previous paragraphs, but by **wave** or **physical optics** which takes into account the wave properties of light.

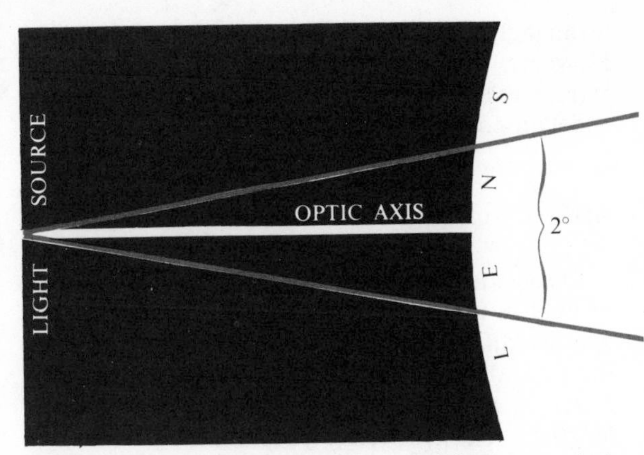

FIG. A–13 *The paraxial bundle of rays is defined as one which cuts out a 2° segment around the optic axis of the lens.*

We shall consider separately each of these causes of distortion of the image formed by a lens, and we shall consider lens aberrations first.

A.16 Aberrations of Lenses: Spherical Aberration

The aberrations arising from the lens structure itself are divided into two groups: those which are present in images of points on the optic axis; and those associated with images of points off the optic axis, **extra-axial image points**. There are just two kinds of aberration associated with the image of a point on the optic axis: **spherical aberration** and **chromatic aberration**.

We can easily explain spherical aberration for a simple convex lens by considering rays of light parallel to the optic axis striking the lens. Since Snell's law of refraction shows that the rays striking the lens further from the optic axis are bent more than those striking the lens in the paraxial region, the edge rays are brought to a focus, F, closer to the lens than the focus F' of the paraxial rays (see Figure A.14). The distance between the focal points of the paraxial and edge rays is called the *spherical aberration* of the lens. This aberration increases as the fourth power of the aperture (diameter) of the lens.

We can alter the spherical aberration of a simple thin lens without changing its focal length by changing r_1 and r_2 simultaneously in such a way that $1/r_1 - 1/r_2$ (the so-called **total curvature** of the lens) remains constant. This is called **bending the lens** as shown in Figure A.15. It can be shown that although the spherical aberration for a convex lens of given focal length can be minimized by the proper bending of the lens, it can never be reduced to zero but always remains positive. However, it is possible to eliminate spherical aberration by replacing the single positive lens by a combination of a negative and a positive lens of appropriate focal lengths.

Since the spherical aberration of the negative lens always remains negative but can be altered from a minimum negative value to an infinite negative value by proper bending, and the spherical aberration of the positive lens can be altered from a minimum positive value to an infinite positive value by bending, it is possible by bending both lenses independently of each other to find an arrangement in which the negative spherical aberration of the negative lens just compensates the positive spherical aberration of the positive lens. When one finds a system of two lenses of this sort, the system is said to be corrected for spherical aberration.

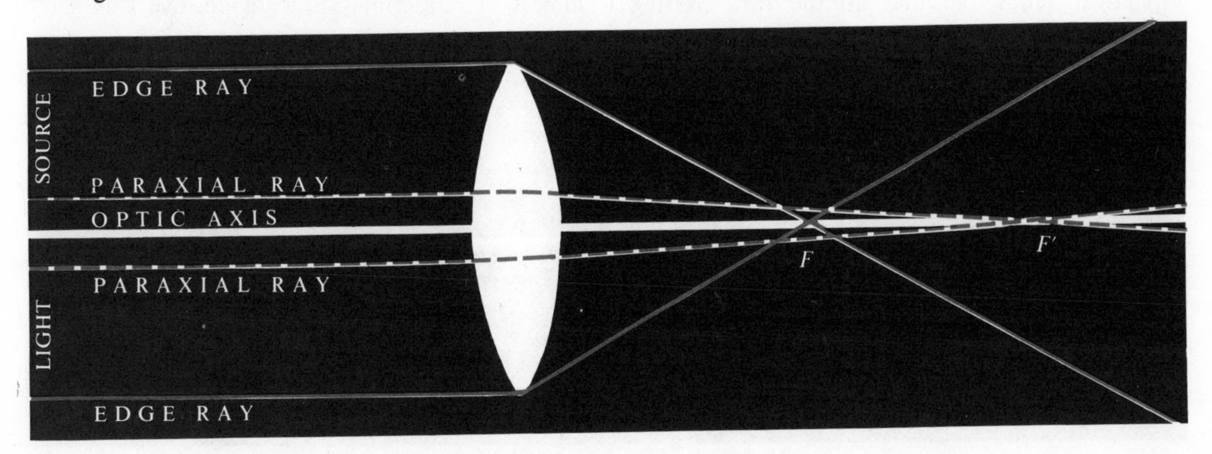

FIG. A–14 *Spherical aberration in a positive lens. The paraxial rays are brought to a focus F′ farther away from the lens than the focus F of the edge rays.*

FIG. A–15 *Bending a lens means changing its shape without altering its focal length.*

In such a lens system the edge rays are brought to the same focus as the paraxial rays. However, a ray striking the lens about three-quarters of the way up is, in general, not brought to this same focus and we say that the lens still has **zonal spherical aberration**. This can be corrected by combining three lenses of appropriate focal lengths and bending them independently. But even then not all the rays striking the lens at all heights above the optic axis are brought to the same focus. However, for most purposes we find that if the spherical aberration is corrected for the edge and the paraxial rays, the zonal spherical aberration is not too objectionable.

Aspherical Surfaces of Lenses. It is possible to correct the spherical aberration of a single lens if one or both of its surfaces are **aspherical** (Figure A.16). We can always design the contour of an aspherical surface so that all the rays passing through the lens are brought to the same focus. However, since such aspherical surfaces are extremely difficult to grind and polish they are not widely used.

A.17 Chromatic Aberration

Chromatic aberration arises because the speed of light in a medium such as glass depends on the wavelength of the light. We generally find over the optical range that the long wavelengths travel faster in a dense medium than the short wavelengths. This dispersive property of glass causes chromatic aberration of a simple lens (Figure A.17).

If we consider a beam of white light parallel to the optic axis, we see that the violet-blue rays are brought to a focus closer to the lens than the red rays. This difference in focus between the red rays and the violet rays is called **longitudinal chromatic aberration**; it is positive for a biconvex (positive) lens and negative for a biconcave (negative) lens. Whereas the spherical aberration of a lens depends only on the index of refraction of the lens (for a given focal length), the chromatic aberration depends both on the index of refraction of the glass and on its dispersion, which is measured by the difference between indices of refraction for red and blue light.

Glass that has a small dispersion is called **crown glass**, and glass that has a large dispersion is called **flint glass**. The dispersions depend on the chemical composition of the glass. By introducing appropriate amounts of various kinds of metals, such as barium and lead, into the glass melt it is possible to obtain glass with a wide range of dispersions and indices of refraction. Since a positive lens gives

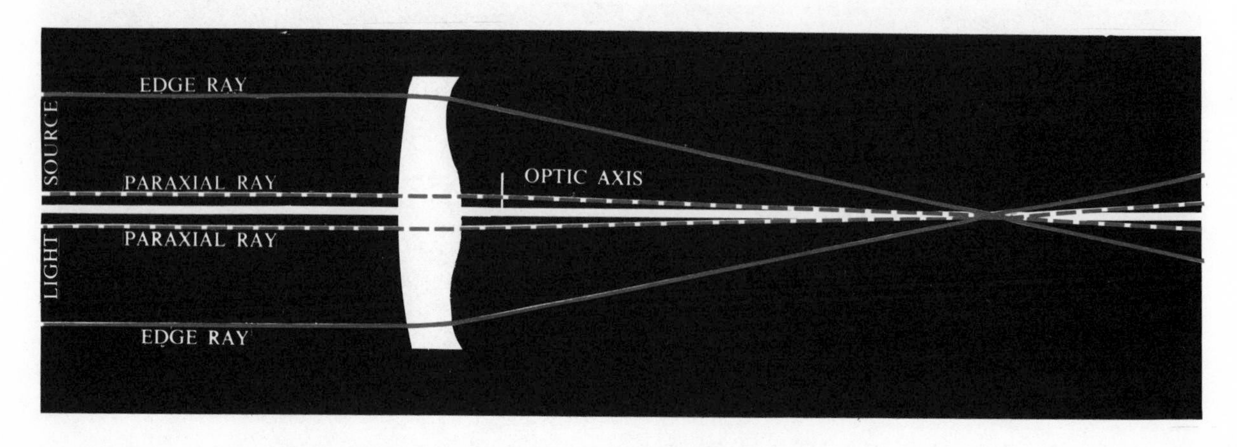

FIG. A–16 *The correction of spherical aberration by means of an aspherical surface.*

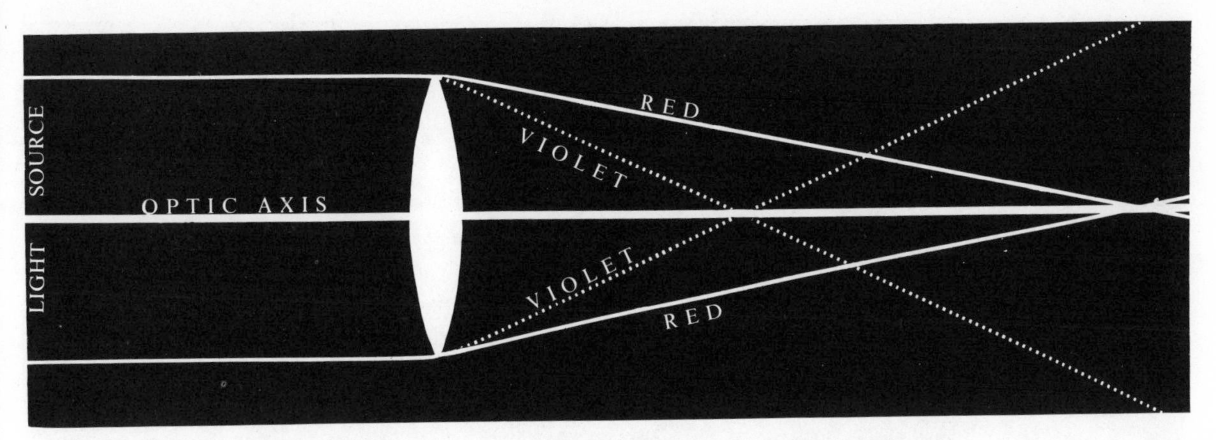

Fig. A–17 *Chromatic aberration. The red rays are brought to a more distant focus than the violet rays.*

positive chromatic aberration and a negative lens gives negative chromatic aberration, we can eliminate chromatic aberration for two distinct colors (i.e., bring, let us say, red light and blue light to the same focus) by an appropriate combination of positive and negative lenses of different dispersive powers. This is generally done by choosing crown glass for the positive lens and flint glass for the negative lens and cementing these lenses together as shown in Figure A.18.

A.18 Achromatic Objectives

Given two lenses with sufficiently different indices of refraction and sufficiently different dispersions, we can bend the two lenses and thus obtain a single system which is corrected for both chromatic and spherical aberration. This is so because with two lenses we have just two conditions to be fulfilled, namely the elimination of spherical and chromatic aberration. Such a lens is called an **achromat**.

Although in an achromat the two extreme colors are brought to the same focus, the intermediate colors such as yellow and green are not sharply focused. However, it is possible to bring three colors to a sharp focus by combining three lenses of different dispersions and different indices of refraction and bending them all independently. Such a system is called an **apochromatic system** if it is also corrected for spherical aberration. Apochromatic lenses are used for precise color photography.

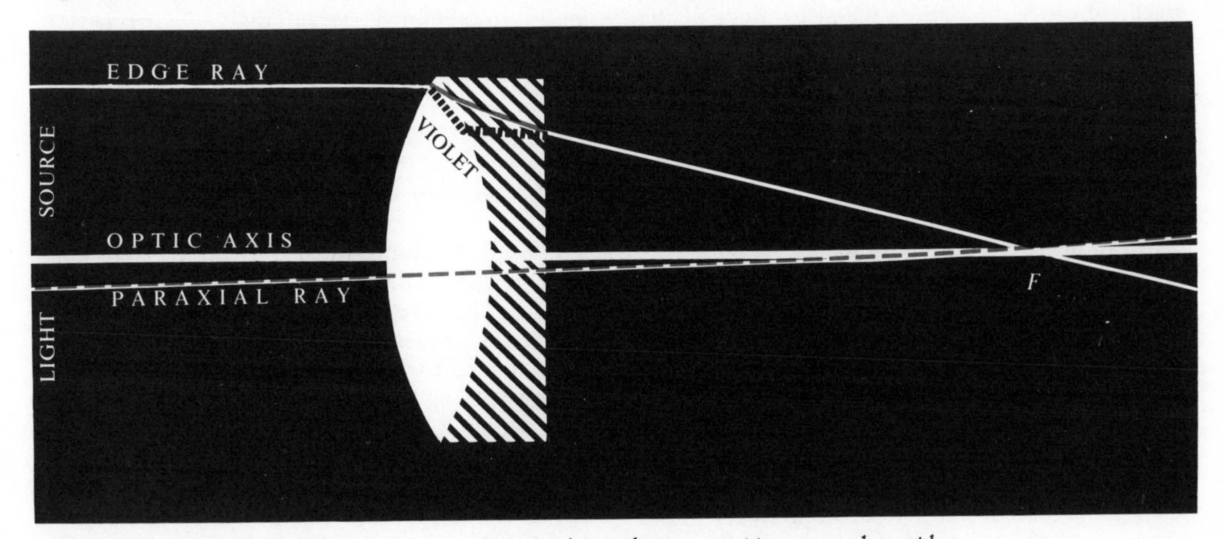

Fig. A–18 *Correcting spherical and chromatic aberration by combining a positive crown glass with a negative flint glass.*

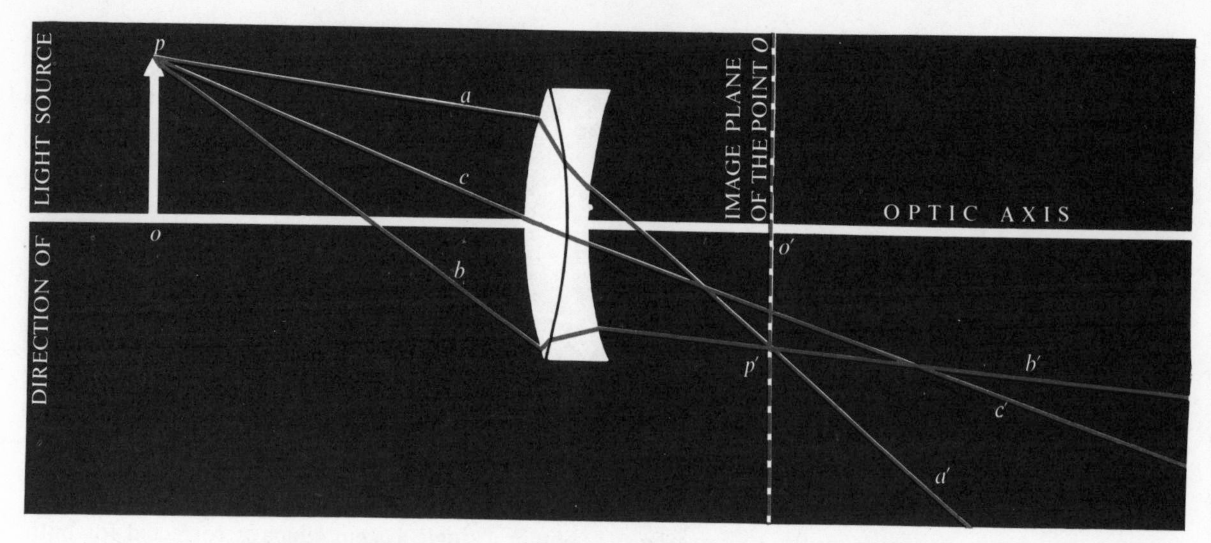

FIG. A–19 *Aberrations in a tangential (off-axis) bundle of rays originating from a point* not *on the optic axis.*

A.19 Aberrations for Off-Axis Points: Coma

Thus far we have been discussing the images of points on the optic axis, and we have seen that there are just two kinds of aberrations. However, if we are dealing with object points that do not lie on the optic axis, other aberrations arise which may be present even though spherical aberration has been eliminated for the axial image points. To understand the nature of these aberrations we consider a bundle of rays lying in the plane of the paper emanating from an off-axis point as shown in Figure A.19. Such a bundle is called a **tangential bundle of rays**. To discuss the aberrations of this bundle we must see what happens to all the rays in the bundle relative to the center ray which is called the **principal ray**, and which is aimed at the center of the first lens (the entrance pupil of the lens).

It is clear at once that not only is spherical aberration present in the bundle because the bundle is spread out over a surface of the lens from the top ray a to the bottom ray b, but there are also other aberrations present because the central or principal ray c is not an axis of symmetry of the lens system. Thus, as the top and bottom rays a and b pass from surface to surface of the lens system, they are bent differently relative to the central ray. After the bundle has left the last surface of the lens system, the rays a', b', and c' do not intersect a at a single point but form a small triangle. Thus the image of the point p is not sharp, but forms an asymmetrical comma-like structure. For this reason this asymmetrical aberration is called **tangential coma**. If all three rays a', b', and c' behind the lens intersect each other at the same point, coma is absent. It is possible by properly bending a system of lenses to eliminate this type of aberration.

If instead of taking rays that lie in the plane of the paper, we take a bundle that lies in a plane at right angles to the plane of the paper, we obtain another type of **comatic aberration** which is known as **sagittal coma**. This is generally equal to one-third of the tangential coma. Thus, if we eliminate tangential we automatically eliminate sagittal coma. It is possible by appropriately combining three lenses to obtain a system that is corrected for chromatic aberration, spherical aberration, and coma. Such a system is called an **aplanatic system**.

A.20 Curvature of Field and Astigmatism

Even if the above-listed aberrations were completely absent, other types of aberrations could still be present. We can see this by considering tangential bundles of rays coming from points at three different heights above the axis as shown in Figure A.20 in which the three principal rays are c_1, c_2, and c_3. If there is no coma in the lens system, then the rays in each bundle cut the principal ray of that bundle at the same point. However, in general, these points of convergence for each bundle do not lie in the image plane, but rather on the surface of a sphere. Only those

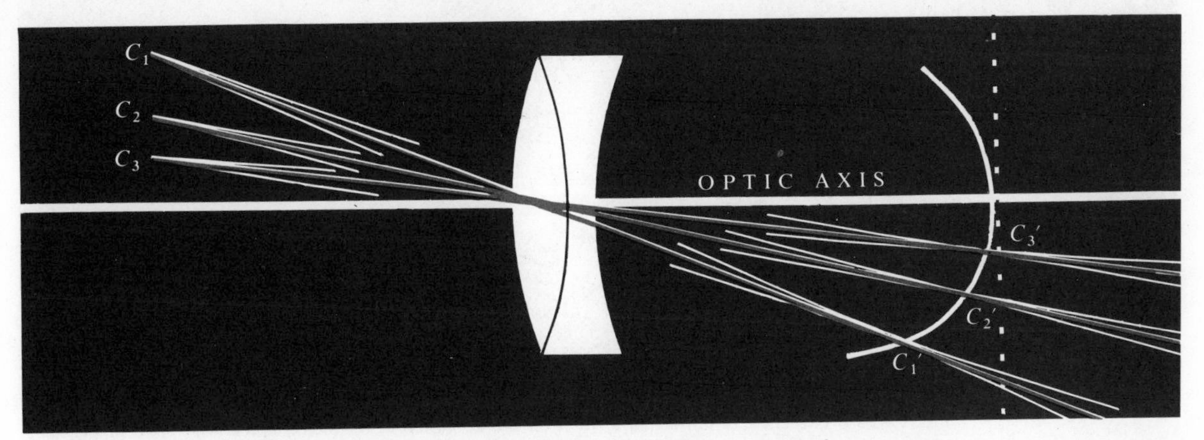

FIG. A–20 *Curvature of field arises when the rays from different points off the optic axis do not converge to points that lie on a plane but to points that lie on spherical surfaces.*

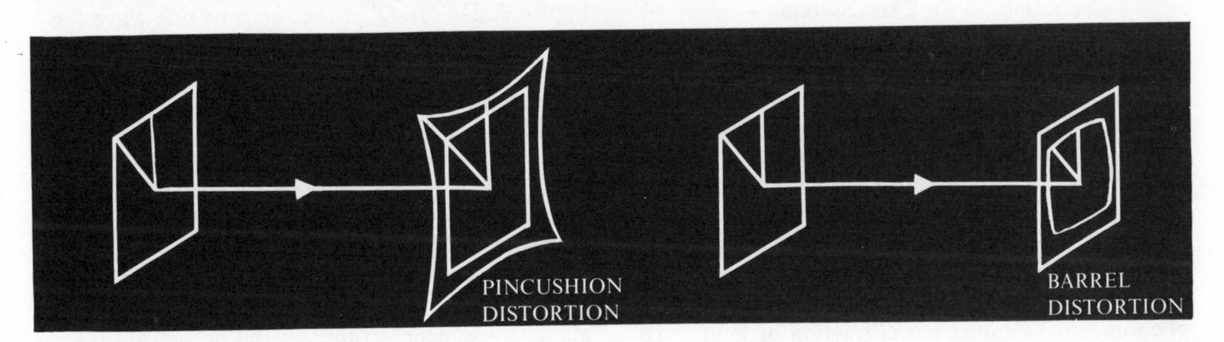

FIG. A–21 *(a) Pincushion and (b) barrel distortion in a lens.*

convergent points whose bundles of rays lie close to the optic axis lie close to the ideal image plane. This aberration is called **tangential curvature of field**.

In the same way we find that if we take sagittal bundles of rays, the convergent points do not lie on the image plane, and we have **sagittal curvature of field**. In general, the tangential and sagittal bundles coming from the same object point converge at different distances from the lens system (the convergent points lie on different spherical surfaces). This difference is called **astigmatism**. If the sagittal and tangential curvature of field are eliminated, astigmatism is eliminated.

A.21 Distortion and Transverse Chromatic Aberration

There are two further aberrations which we must mention to complete the roster. One of these is called **distortion** and arises because the magnification of the image formed by the lens system is not the same at all points of the image, but depends on the distance of the image point from the optic axis. Because of this, even though the image may be perfectly sharp, we find that straight lines at right angles to the optic axis but not intersecting it are imaged as curves, as shown in Figure A.21(a) and (b), where examples of pincushion and barrel distortion are illustrated. If the magnification increases with distance from the optic axis we have **pincushion distortion**. If the magnification decreases with distance from the optic axis we have **barrel distortion**.

Finally, there is another chromatic aberration called **transverse chromatic aberration** or **chromatic difference in magnification**. It arises because rays of light of different wavelength are magnified differently on passing through the lens, so that the image is bordered by a color fringe.

It is possible by combining a suitable number of lenses and bending them all independently to construct a lens system that is corrected for all the aberrations listed above. Such a lens system is referred to as an **anastigmat**.

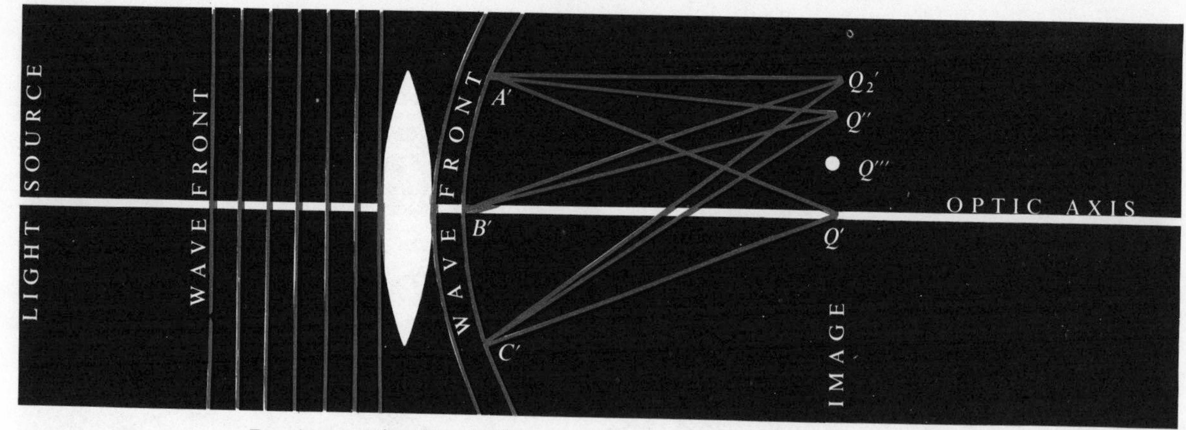

FIG. A–22 *The effect of diffraction of light on the formation of an image at the focal point of the lens.*

A.22 The Wave Structure of Light and the Aberrations of the Lens System

We must now consider the effect of the wave structure of light on the character of the image that is formed. To understand the situation we note that an image formed by a lens system is the result of the superposition in the region of the focus of the lens of waves coming from various parts of the lens system. The quality of the image depends on the distribution of the light as determined by the interference of these different waves. This phenomenon is called the **diffraction of light**. To see why it prevents the formation of a point image from a point source of light even if the lens system is completely free of aberration, we consider waves from a point entering a perfect lens system and passing through it as shown in Figure A.22.

The wavefronts of the light are expanding spherical surfaces which are bent by the lens system and, in general, are not spherical after they leave the system if the lens is not corrected for aberrations. *This departure from sphericity of the wavefront after it has left the lens system manifests itself as the various aberrations we have already discussed.* But if the lens system is corrected for all aberrations, the wavefront remains perfectly spherical after passing through the lens system, as shown in Figure A.22. In this case it is a converging spherical wave with its convex surface towards the lens.

A.23 The Diffraction of Light and Its Effect on Image Formation

From the wave theory of light we can show that each point on the wavefront $A'B'C'$ etc. emits waves in all directions to the right, as shown in Figure A.22. If Q' is the focus of this wave (i.e., $Q'B'$ is the radius of the wavefront), then this point Q' receives waves of radiation from every point on the wavefront, and these waves from these different points—all having the same distance to travel to reach Q'—arrive at Q' at the same time, i.e., **in phase.** Q' is therefore a point of strong concentration of light since the waves from all points on the wavefront reinforce each other there.

However, even if we move away from this central point, we still find some light striking the focal plane since each point of the wavefront sends waves in all directions. Thus, the point Q'' also receives light from all points on the wavefront, but the intensity of the light is diminished. *Because the waves from different points on the wavefront have different distances to travel, they do not reach this point at the same time so that they are* ***not*** *in phase and some destructive interference occurs.*

A.24 Formation of the Airy Spurious Disk

Thus, the light intensity falls off from a maximum at the center of the image Q' to a minimum value at some point Q''', for which the distance $A'Q'''$ is less than $B'Q'''$ by just half a wavelength, and $Q'B'$ is less than $C'Q'$ by half a wavelength. If this is so the contribution of light to the point Q''' from the upper half of the wavefront $A'B'$ is just canceled by the contribution of light to the point Q''' from the lower half $B'C'$ of the wavefront. This is so because the half-wavelength difference in the path length results in destructive interference. This holds for all points in the focal plane lying on the circumference of a circle having Q' as a center and having a radius equal to $Q'Q'''$. For any point lying closer to Q' than the point Q''', the contributions of light from points on the wavefront do not cancel out. We thus see that even though light originates from a point, the light in the image plane is distributed over a disk of radius $Q'Q'''$. This disk is called the **Airy spurious disk** (named after the Astronomer Royal G. B. Airy, who first calculated it in 1834).

If we move beyond the point Q''' we find a dark band which extends out to a point Q_2' such that the distance $B'Q_2'$ differs from $A'Q_2'$ by one wavelength, λ. Here we obtain a thin ring of light which is followed by another band of darkness, etc.

Thus, the image of a point source (i.e., of a star) is not itself a point but rather a central bright disk surrounded by a series of faint rings of light of diminishing intensity. Using the wave theory of light we can show that the angular diameter in radians of the spurious disk is given by

$$2\alpha = \frac{1.22\lambda}{A},$$

where A is the diameter (or aperture) of the lens system, and λ is the wavelength of the light expressed in the same units as A.

A.25 The Resolving Power of an Optical System

It is clear from this that we can reduce the diameter of the spurious disk and hence concentrate more and more light into something approximating a point by making the diameter of our lens system very large. If we express α in seconds of arc, the diameter of the lens in inches, and consider light whose wavelength is about that of the sodium D line, we can express the diameter of the spurious disk as

$$\boxed{2\alpha'' = \frac{5.7}{A}}. \qquad \text{(A.6)}$$

This quantity $2\alpha''$ defines what is called the **resolving power of the lens.**

To see what this means we consider a point source off the optic axis at an angular distance β from a point source on the optic axis. These two point sources give rise to two spurious disks which overlap if β is less than 2α. This follows since each point source gives rise to a spurious disk of radius α. *Thus,* if β is less than 2α the two point sources are indistinguishable, and *it follows therefore that we can see two point sources as distinct only if their angular separation is larger than* 2α. In the case of the 200-inch telescope, the diameter of the spurious disk is equal to 5.7/200, so that $\alpha = 5.7/400$. It follows from Equation (A.6) that in a 10-inch telescope the image of any body in the sky has a diameter that is at least 0.57 however small the actual image is. With a 100-inch refracting telescope this is reduced to 0.057 seconds.

The resolving power puts an upper limit on the magnification that can be used with any given telescope. It is clear that no advantage is gained by magnifying the diffraction pattern (i.e., the spurious disk) beyond a point where the central disk and its rings are clearly seen. For if we go beyond this point, the image is simply made more diffuse because the diffraction pattern itself is magnified. The maximum useful magnification corresponds to about 60 times the aperture measured in inches, because when this magnification is used the spurious disk is just large enough to be recognized by the normal eye as a disk.

A.26 Reflecting Telescopes: The Mirror Formula

Thus far we have been discussing optical systems consisting of transparent elements. However, it is possible to construct optical systems consisting of both reflecting and refracting elements. The formation of an image by a specular reflecting element (a mirror) is given by optical laws similar to those introduced above. Since reflecting elements have certain advantages over ordinary lenses we shall discuss in detail the **concave mirror**, the type of reflecting element most commonly used in astronomical telescopes.

If a point on the optic axis of a concave spherical mirror (see Figure A.23) at a distance l in front of the mirror is imaged at a distance l' from the center of the mirror, we have the simple formula

$$\boxed{-\frac{1}{l'} = \frac{1}{f} + \frac{1}{l}}, \qquad \text{(A.7)}$$

where f is the focal length of the mirror and is positive for a concave mirror (according to our usual sign convention described above). The signs of l and l' are chosen according to the same convention as in the case of lenses.

To see the significance of f for a concave mirror, we again consider how light from an infinite point on the optic axis (which is the radius passing through the center of the mirror) is reflected after it strikes the mirror. By using the law of reflection, namely that the angle of incidence of the incoming ray is equal to the angle of reflection, we find that the parallel rays on either side of the optic

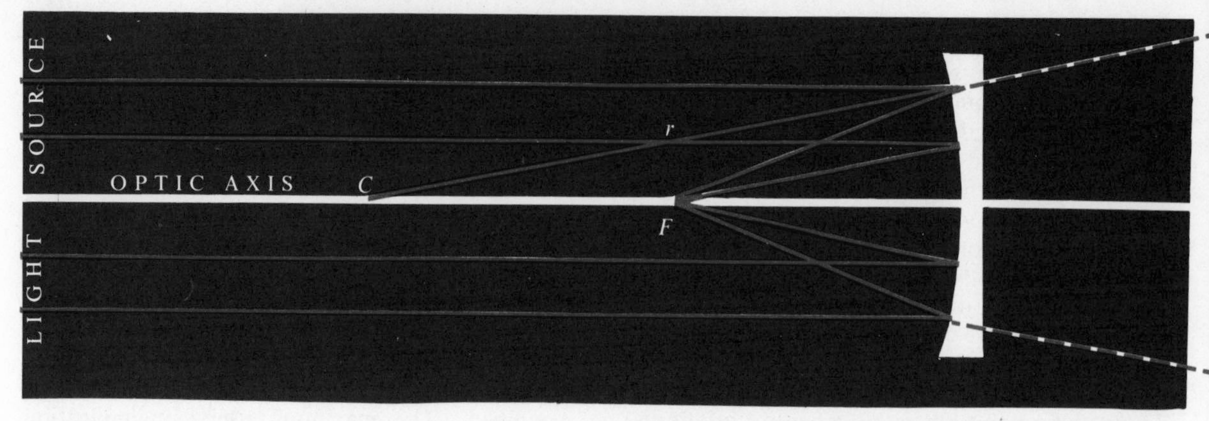

FIG. A–23 *The formation of an image by a concave mirror.*

axis (but not too far away from it) pass through the focal point F after reflection. We can show that for these rays the focal length, f, is equal to $-r/2$, where r is the radius of the concave surface. Thus, the focal point of a concave mirror is halfway between the center point o on the mirror, and the center of curvature c of the mirror.

A.27 Formation of Images by Mirrors

By using our simple formula (A.7), or by using a simple graphical construction, we can determine the position of the image, formed of any object by a concave mirror. To use the graphical construction we employ the following simple rules:

1. any ray parallel to the optic axis in the paraxial region of the mirror passes through the focal point on reflection;
2. any ray passing through the focal point and striking the mirror near the paraxial region is reflected parallel to the optic axis;
3. any ray passing through the center of curvature, C, is reflected right back upon itself because it moves along the radius of the mirror and hence strikes the mirror surface normally. This means that the image of any object placed at the position C coincides with the object.

In Figure A.24 the graphical method for constructing various images is illustrated.

A.28 Advantages of a Mirror over a Lens

There are certain geometrical and optical advantages in using a mirror rather than a system of lenses as the objective of a telescope. To begin with, there is no chromatic aberration whatever in a mirror so that one can obtain an achromatic system without having to combine a group of optical elements. In a reflector the objective consists of one polished surface. This means that it is possible to construct a very large reflector objective since one has only to grind and polish one surface.

It is not necessary to choose glass of any particular index of refraction because the light does not pass through the glass but is reflected at the surface. Moreover, as long as the polished surface is uniform, permitting the deposition of a uniform metallic reflecting film, it does not matter whether there are nonuniformities, such as bubbles, in the main bulk of the glass. These imperfections are not permissible in a lens.

Since the type of glass that is used in a mirror is not important, we can use glass such as Pyrex with a very small coefficient of thermal expansion to prevent daily variations in the dimensions of the mirror. Another advantage is that for a given focal length it is possible to obtain approximately twice as large a workable aperture for a mirror than for a lens because the focal length of a mirror is one-half the radius of curvature.

Finally, since the back of the mirror plays no role in determining the paths of the rays of light, the main body of the glass structure can be honeycombed as is the Mt. Palomar 200-inch mirror. This not only reduces the weight of the glass but also permits us to brace it by putting supports in its honeycomb sections, preventing sagging of the mirror. In this way the heavy weight of the glass can be supported over its whole back instead of just its edges.

A.29 Disadvantages of Reflecting Telescopes

There are certain disadvantages in reflecting telescopes which must be taken into account.

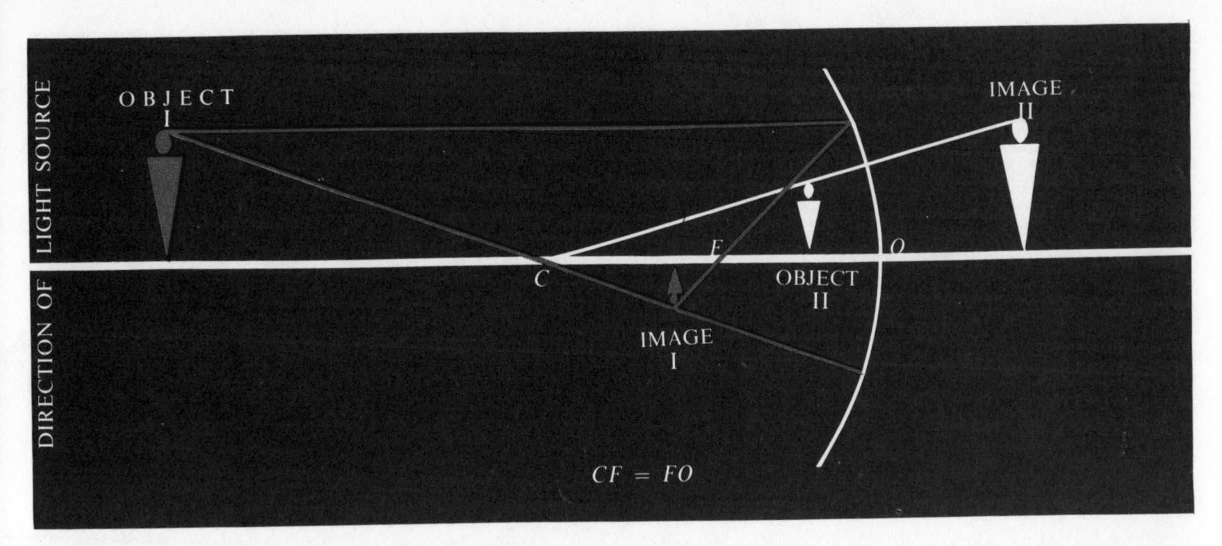

FIG. A–24 *The graphical method of constructing images in a spherical mirror.*

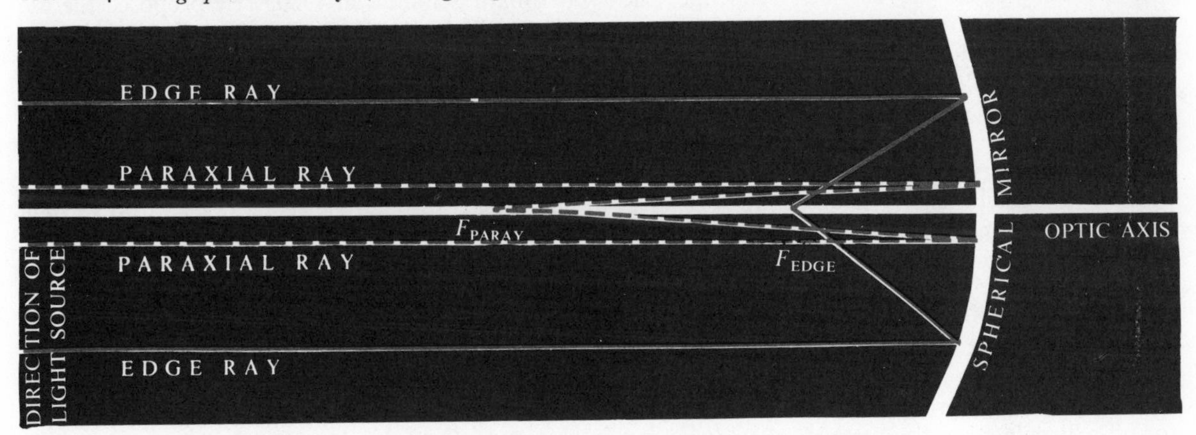

FIG. A–25 *The spherical aberration in the image formed by a spherical mirror.*

Except for the chromatic aberration, a spherical mirror possesses all the aberrations of spherical lenses discussed in previous sections. In particular, for objects on the optic axis, except at the point *C*, spherical aberration is present and the rays that strike the mirror high above the optic axis are focused closer to the center of the mirror than the paraxial rays are, as shown in Figure A.25. This aberration can be completely eliminated by **parabolizing** the surface of the mirror (i.e., by grinding the surface so that any plane containing the optic axis intersects the mirror in a parabola whose axis of symmetry is the optic axis). Most ordinary reflecting telescopes have parabolic mirrors as their objectives. Such a mirror is shown in Figure A.26.

Another disadvantage of a mirror is that the aluminized surface deteriorates constantly and the mirror must be periodically resurfaced. Furthermore, the entire surface of the mirror is not available to collect light since a central portion must be either cut out or blocked out to allow the light to be brought to an accessible focus. Finally, any slight irregularity on a reflective surface impairs the quality of the image much more seriously than the same irregularity on a lens surface.

A.30 Aberrations of a Parabolic Reflector

Although a parabolic mirror is completely free of spherical and chromatic aberration, it suffers from fairly severe coma, because it has only one axis of symmetry, the optic axis itself. If a bundle of light rays at an angle with respect to the optic axis is reflected from the mirror, the severe asymmetry of this bundle with respect to the surface of

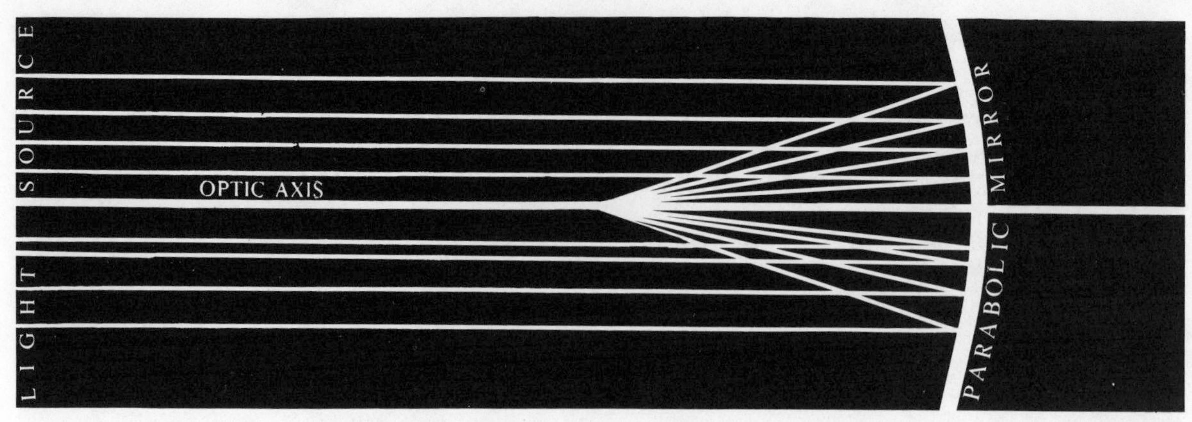

FIG. A–26 *All rays parallel to the optic axis are brought to the same focus at a focal point F.*

the mirror gives rise to considerable coma. If the angle a bundle makes with the optic axis is more than 40 minutes in the 200-inch telescope, the resulting coma is so severe that the image is too poor for observational work. For this reason large parabolic reflectors are used to photograph very small regions of the sky or individual objects such as galaxies and star clusters and nebulae. Because reflectors have large apertures, it is possible to obtain sharp images showing considerable detail and good resolution as long as the image lies close to the optic axis. See Figure A.26.

A.31 The Schmidt Telescope

Since parabolic mirrors give very poor imagery for off-axis points, another type of reflector is used to study wide regions of the sky. This objective is based on an optical design first introduced by Bernhard Schmidt, in 1930. He proposed the use of a spherical concave mirror (as the principal component of his objective) together with an aspherical lens appropriately placed in front of the mirror. The advantage of a concave spherical mirror for investigating objects at large angles from the optic axis is at once obvious if we note that any line passing through the center of curvature of such a mirror is an axis of symmetry as shown in Figure A.27. For this reason there is practically no coma present in the images of off-axis points.

Spherical aberration is still present, but this is eliminated by means of the **aspherical lens** (of which one surface is plano and the other appropriately contoured as shown in Figure A.27). If this aspherical lens (the **correcting plate**) is properly

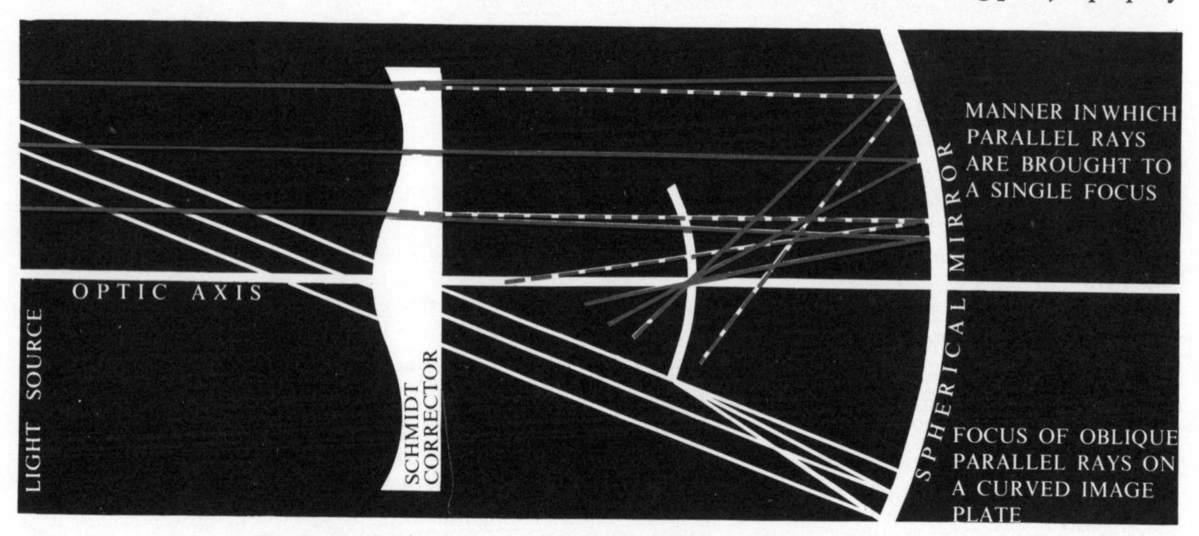

FIG. A–27 *A Schmidt system is free of off-axis aberrations because all lines through the center of curvature, C, of the mirror are axes of symmetry. Spherical aberration is eliminated by placing an aspherical correcting plate at the center of curve, C, of the mirror.*

contoured to eliminate spherical aberration, and is placed at the center of curvature of the mirror as shown, all but one of the other aberrations disappear. A small amount of chromatic aberration is introduced by the correcting plate, but this is negligible because the aspherical surface differs only slightly from a plane surface.

The only remaining serious aberration is curvature of field. Since the rays coming from an infinite object do not form an image of this object on a plane at right angles to the optic axis but rather on a spherical surface convex towards the mirror, curvature of field is present as shown in Figure A.27. This is not a serious drawback since the Schmidt camera is used only photographically and the photographic plate can always be molded to fit against the curved focal surface. Since the Schmidt camera has no coma, it can be used over fairly wide fields of view so that large regions of the sky can be photographed at the same time. Therefore this camera is very useful in obtaining quick surveys and in mapping large areas of the sky. The useful field for a Schmidt telescope is of the order of 12° on either side of the optic axis, but for astronomical purposes the field is generally limited to 25 square degrees. The second largest Schmidt telescope is the 48-inch at Mt. Palomar shown in Figure A.28. The largest is the 79-inch telescope at the Karl Schwarzschild Observatory in Jena, East Germany.

Such things as light-gathering power, magnification, resolving power, and so on are governed by the same laws for reflectors as they are for refractors.

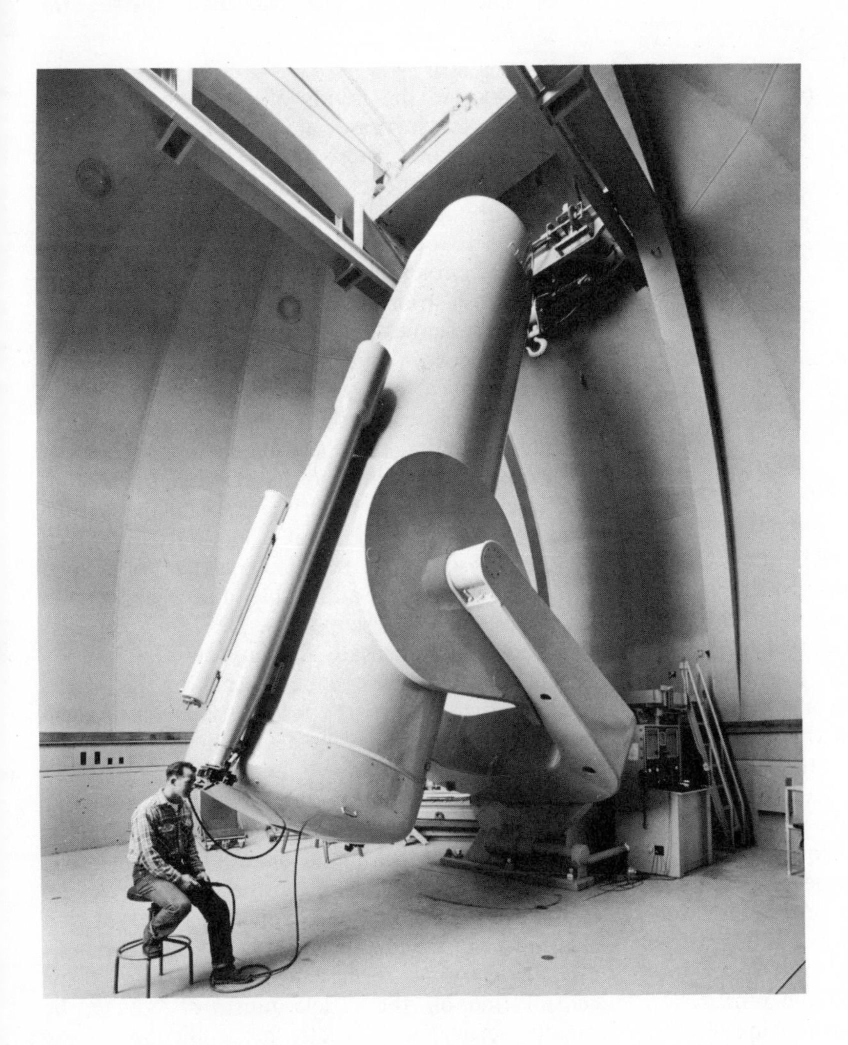

Fig. A–28 *The 48-inch Schmidt telescope, Palomar. (From Mount Wilson and Palomar Observatories)*

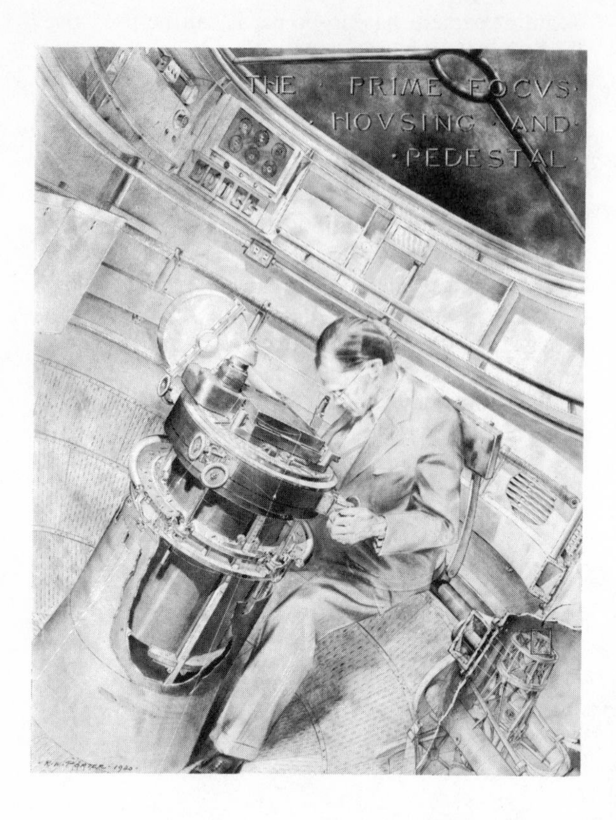

FIG. A–29 *200-inch Hale telescope, prime-focus housing, and pedestal. Drawing by Russell W. Porter. (From Mount Wilson and Palomar Observatories)*

A.32 Types of Mounts of Reflecting Telescopes

A reflecting telescope can be used with the focal point of its objective at various positions with respect to the tube of the telescope. Thus the focal point may be either directly in front of the mirror (the **prime focus**, see Figure A.29), directly behind the mirror (in which case the mirror is constructed with a hole at its center to allow the light to pass through it), or to the side of the tube. The various ways in which this can be done are diagrammed in Figure A.30. If the focal point is either behind the objective mirror or at the side of the tube, a **secondary mirror** must be used to reflect the converging rays of light in the desired direction. This mirror may be flat or convex, or may be a combination of a flat and a convex mirror.

If the primary focus is behind the mirror, the telescope is referred to as a **Cassegrainian reflector**. It is called a **Newtonian reflector** if the focus is at the side of the telescope. When the telescope is used as a Newtonian reflector the secondary mirror is flat because one does not have to change the focal length of the system, the reason being that the point of convergence of the rays is at the side of the telescope tube and hence not too far away. But if the telescope is used as a Cassegrainian reflector, the focal point must be shifted by a fairly large amount so that the entire focal length of the mirror must be increased. This is done by using a convex mirror for the secondary mirror. The advantage of the Cassegrainian mount is that an eyepiece can be placed at the lower end of the telescope so that viewing is somewhat easier.

A.33 The Transit; The Meridian Circle; the Altazimuth

Although in the minds of most people astronomical research is associated with the large optical telescopes such as those at Mt. Palomar and Mt. Wilson, very little astronomical work could be done without small, special-purpose telescopes that have to be designed and mounted with the greatest accuracy. Most important of these is the **transit instrument**, which consists of a small aperture telescope (diameter about 2 or 3 inches) mounted on two rigid vertical piers, and free to rotate about a horizontal axis that can be set exactly perpendicular to the plane of the observer's meridian. With this arrangement the transit telescope can be rotated so that its objective sweeps out a circle that is exactly in the plane of the observer's celestial meridian. The horizontal axis of the telescope is mounted on two V-shaped bearings which are adjustable and arranged so that the entire telescope can be lifted off and reversed to compensate for any residual errors. Special spirit levels can be placed on the pivots of the horizontal axis to test it for any deviation from the horizontal. A deviation can then be corrected by special adjustments.

The transit telescope is used in conjunction with an accurately designed eyepiece and a reticle placed exactly in the focal plane of the objective. This reticle has a series of parallel vertical lines on it, the central one of which coincides exactly with a small piece of the imaginary image of the observer's celestial meridian formed by the objective. This central line on the reticle must, of course, be exactly vertical and exactly perpendicular to the optic axis of the telescope, which in turn must be exactly perpendicular to the axis of rotation of the

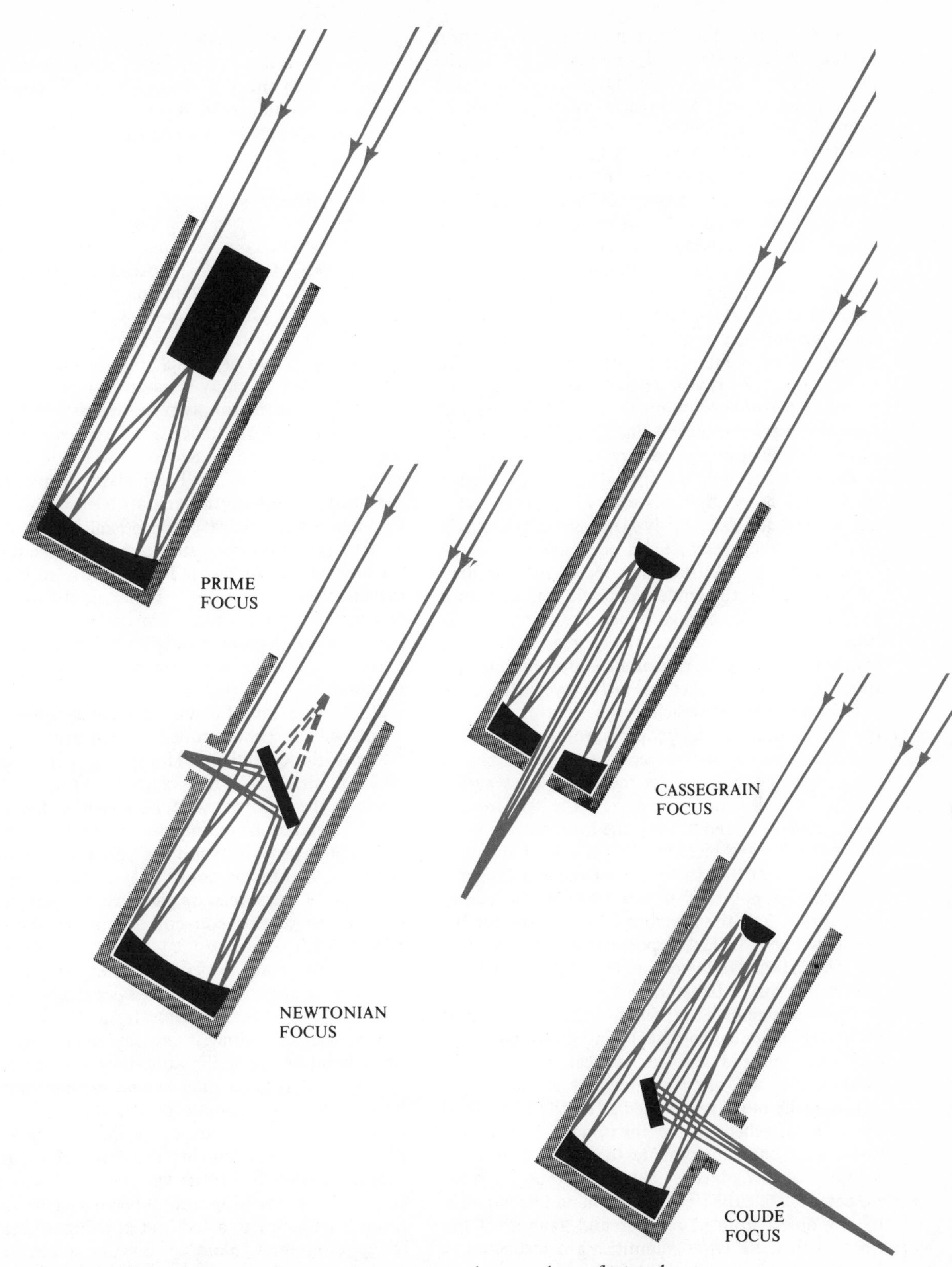

FIG. A–30 *Schematic diagram of various ways in which the eyepiece can be mounted in a reflecting telescope.*

telescope. Since the horizontal axis of rotation of the telescope is exactly perpendicular to the plane of the observer's celestial meridian, it must point in an exactly east-west direction so that its azimuth is precisely 90°.

The transit instrument is used together with a sidereal clock to determine the sidereal time of passage (that is, the **transit**) of a star across the observer's meridian. As we have seen in our discussion of celestial coordinate systems and sidereal time, this sidereal moment of passage is just the right ascension of the star. The precise moment of transit can be measured very accurately by observing the star as it moves successively across the different parallel vertical lines on the reticle. In this way a record can be made of the star's positions at different moments before and after it crosses the central line so that the time of transit can be accurately determined.

The **meridian circle** is a special case of the transit instrument, except that it is somewhat larger and it has a very accurate, fairly large, graduated circle attached to the telescope and perpendicular to the horizontal axis. As the telescope of the meridian circle is rotated, the graduated circle rotates with it so that the angle through which the telescope has been rotated can be accurately read off. This instrument is used to measure the declination of a star. The star is picked up in the eyepiece of the telescope before it reaches the central line on the reticle and the telescope is rotated until the horizontal line in the center of the reticle bisects the image of the star. As the star moves across the field of view in the eyepiece, the telescope is continuously rotated to keep the horizontal line on the reticle always bisecting the image of the star. At the moment that the center of the star image is right at the point of intersection of the center vertical line and the horizontal line on the reticle, the telescope is locked in position and the reading on the vertical graduated circle is taken. This is the declination of the star.

The **altazimuth** is a small telescope that is accurately mounted so that it can be rotated about a horizontal and also a vertical circle. Two graduated circles, one perpendicular to the vertical axis and the other perpendicular to the horizontal axis, are attached to the instrument. With this instrument one can measure the altitude and the azimuth of any object. This is very useful since we can calculate the right ascension and declination of any object from its altitude and azimuth if the time when these latter quantities are measured is known. It often happens that we must determine the celestial coordinates of a celestial object that does not transit our meridian at a convenient time. We can then use the altazimuth to determine its horizon coordinates at some moment and we can then calculate its right ascension and declination.

A.34 Spectrographs

With the telescope a number of accessory instruments may be used to take the greatest possible advantage of its light-gathering power. The most important of these is the **spectroscope**, which can be used in two different ways to give either the detailed spectrum of a single object, such as a star, or the spectra of many stars on a single plate. In the latter case one simply places a prism in front of the objective so that the light from each star (as explained in detail in Section 17.1) is spread out into a spectrum on the focal plane. In this case we speak of an **objective prism.**

When white light strikes a prism, it is bent as it enters the first polished surface and then bent again on leaving, as shown in Figure A.31, so that the dispersion is enhanced. This dispersion is increased still more when the prism is made of flint glass. Two or more prisms arranged in series can be used to obtain as great a dispersion as may be required.

If the spectrum of a single source is desired, one works with a spectroscope attached near the focal plane of the objective. The spectroscope consists of (1) a slit, *S*, which is placed right at the focal point of the objective; (2) a **collimating lens**, L_{coll}, whose distance from the slit is equal to its own focal length; (3) the **dispersing prism**, which breaks the light up into its colors; (4) a **converging lens**, L_{conv}; and (5) a **screen** or **photographic plate** at the focal plane of this converging lens, as shown in Figure A.32.

To understand the function of each optical component and the way this system operates we consider rays of light coming from the image of a star at the slit. Since these rays come from the focal point of the collimator, they are parallel to the optic axis when they leave this lens (the collimator lens is introduced to obtain a parallel beam) and when they strike the front polished face of the prism. On leaving the prism this parallel beam is dispersed into its various colors and then enters the converging (or collecting) lens which brings each color to a focus at a different point on the photographic plate or on the screen. This system of two lenses and a prism is just a device for

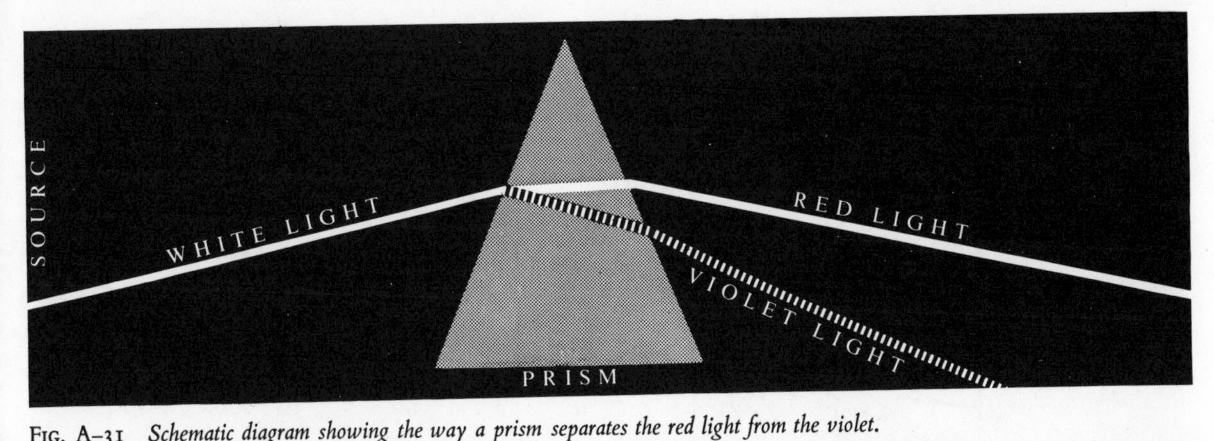

FIG. A–31 *Schematic diagram showing the way a prism separates the red light from the violet.*

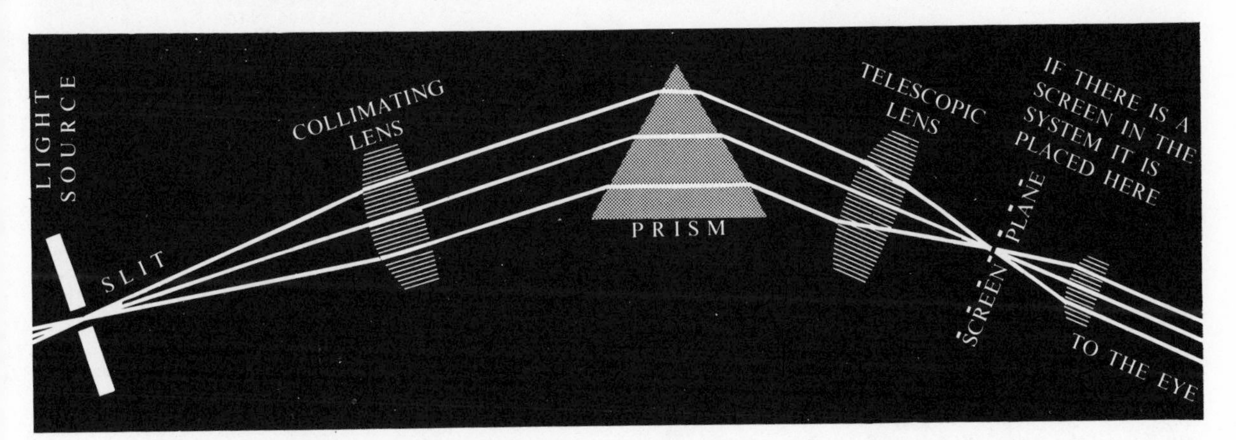

FIG. A–32 *Schematic diagram of the arrangement of lenses in a spectrograph.*

obtaining on the photographic plate a continuous array of images of the slit in different colors.

If spectral lines are present, each of these lines is an image of the slit; for this reason it is desirable to make the slit as narrow as possible in order to obtain sharp lines. This is particularly important when the lines are close together and one wants to separate them.

A.35 Diffraction Gratings

Instead of using a prism between the two lenses we can obtain a spectrum by using a **diffraction grating**. This consists of a transparent piece of glass on which a large number of very fine equidistant opaque lines are ruled (a good grating consists of at least 10 000 lines per inch). When a parallel beam of light strikes this **transmission grating**, as it is also called, the beam does not pass straight through but is bent (according to Huygens' principle—see Section 16.20) by an amount that depends on the wavelength of the light. The light is spread out into its component colors and a spectrum is obtained. It differs from the spectrum obtained by a prism in that the order of the colors is reversed. A diffraction grating can also be used in front of the objective of a telescope. If one wants to cover a large objective in this way, the grating may be a coarse one and may consist of thin parallel wires stretched in front of the objective.

It should be noted that the closer together the lines in a diffraction grating are, the more the spectrum is spread out and hence the greater is the resolving power of the grating.

A.36 The Michelson Interferometer

Another instrument that is used together with a telescope is the Michelson **interferometer** (which we discussed in detail in Section 21.5). To understand how this instrument is used to determine the diameters of stars, we consider a beam of light from

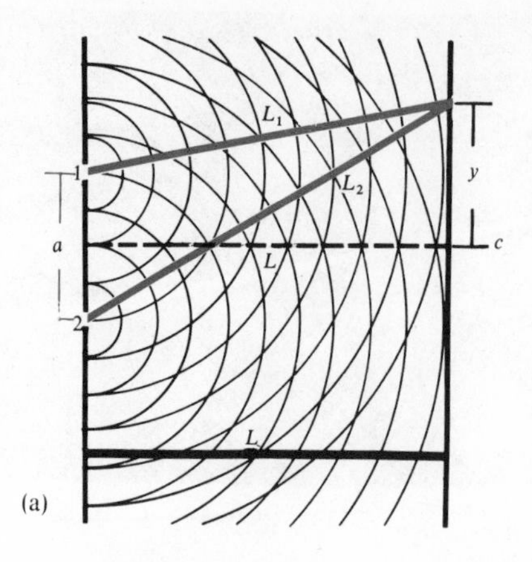

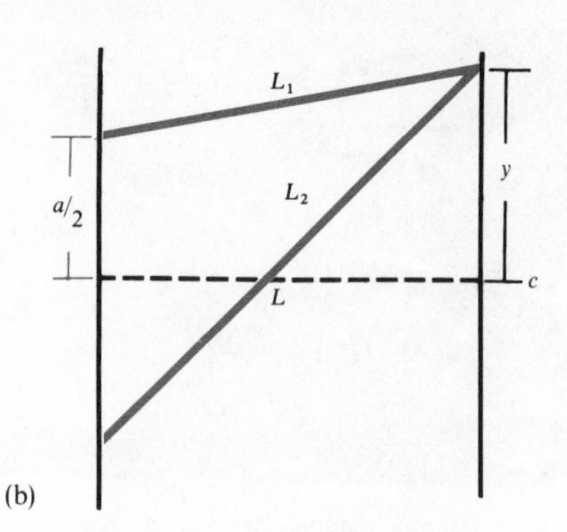

FIG. A–33 *(a) Interference of light passing through 2 slits in a screen. (b) The geometric analysis of the interference pattern.*

an infinite source striking a screen containing two parallel slits, as shown in Figure A.33.

According to Huygens' principle each slit becomes a source of secondary wavelets which spread out in all directions and interfere with or reinforce each other at various points. If, then, we have a second screen receiving these secondary wavelets of light, various points on this second screen are dark or bright depending upon whether the wavelets from the two slits interfere destructively or reinforce each other at these points. We thus get an interference pattern of light and dark bands.

We can easily calculate where the dark bands of the interference pattern lie on the second screen by means of simple geometrical arguments. Let the distance between the two slits be a, let the separation between the two screens be L, and let us consider a point on the second screen which is at a distance y from the center point c midway between the two slits. The two wavefronts from the upper and lower slits striking this point are out of phase because they have traveled different distances L_1 and L_2 to reach this point.

We have, using the properties of right triangles,

$$L_1^2 = L^2 + \left(\frac{a}{2} - y\right)^2$$

and

$$L_2^2 = L^2 + \left(\frac{a}{2} + y\right)^2.$$

If we now subtract these two expressions after performing the indicated algebraic operations we have

$$L_2^2 - L_1^2 = 2ay$$

or

$$(L_2 - L_1)(L_2 + L_1) = 2ay.$$

If we now consider the case for which $L \gg a$ so that L_1 and L_2 are practically equal to L, we have

$$2L(L_2 - L_1) = 2ay$$

or

$$L_2 - L_1 = \frac{ay}{L}.$$

Note we may place $L_1 = L_2$ when they are added to each other but not when they are to be subtracted from each other. If we now want y to be at the position of the first dark interference band (a bright band always lies at the position c) we must choose y so that the path difference $(L_2 - L_1)$ is just half a wavelength. We thus obtain

$$L_2 - L_1 = \frac{\lambda}{2} = \frac{ay}{L}$$

or

$$y = \frac{\lambda L}{2a}.$$

To use the Michelson interferometer to determine the radius of a star we place the screen with the two slits in front of the objective, and we place the second screen at the focal point of the objective so that $L = F$. We thus have the formula

$$y = \frac{\lambda F}{2a}.$$

The details of the procedure for using this type of interferometer to measure stellar diameters are given in Chapter 21.

Besides the interferometer one can attach to the telescope such devices as the **thermocouple** to measure the temperature of a source of radiation, **photoelectric elements**, **polarizing devices**, and others. The accompanying photographs (Figure A.34) show the 100-inch Mt. Wilson reflector, the 200-inch Mt. Palomar reflector, and the 150-foot solar tower telescope.

FIG. A–34 *(a) 200-inch Hale telescope dome exterior. Drawing by Russell W. Porter. (b) 200-inch Hale telescope looking east, showing the Cassegrain platform. Drawing by Russell W. Porter.*

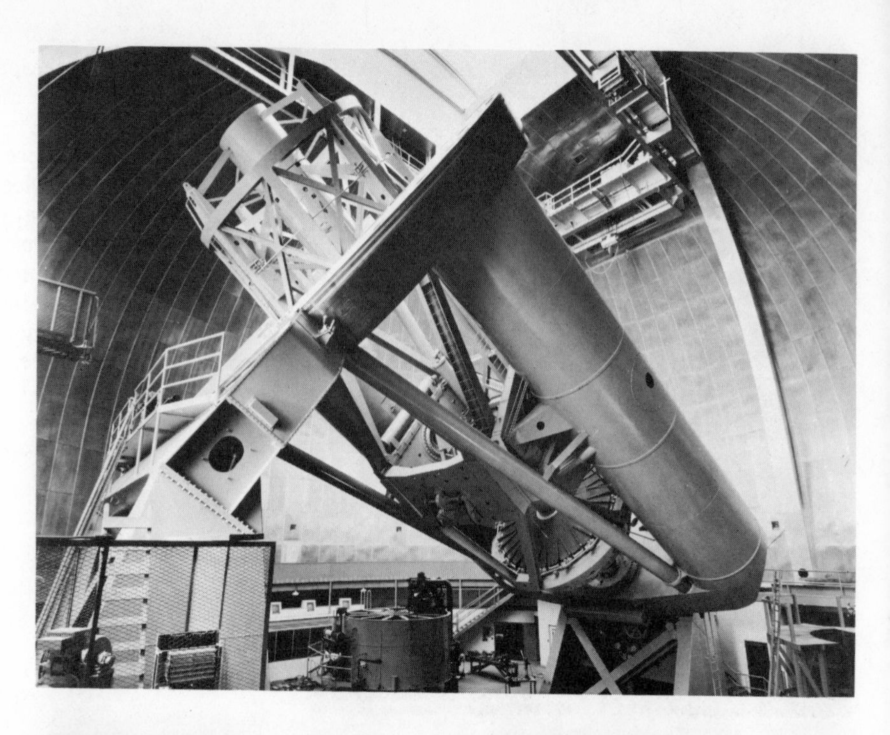

FIG. A–34 (*c*) 200-inch Hale telescope pointing north. (*d*) 200-inch Hale telescope showing observer in prime-focus cage and reflecting surface of 200-inch mirror.

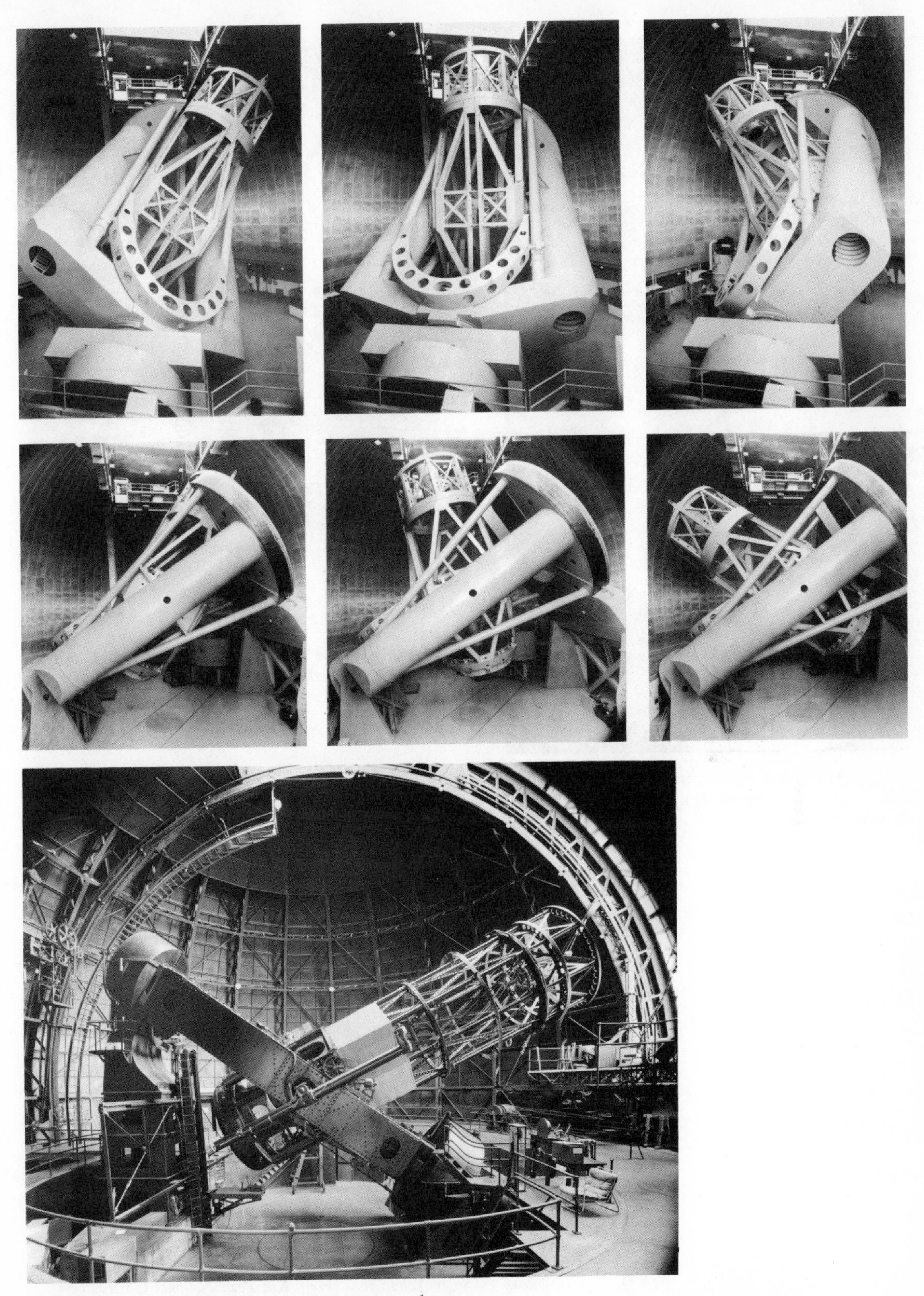

FIG. A–34 (*e*) *200-inch Hale telescope; six views showing various east-west and north-south positions of the telescope.* (*f*) *100-inch Hooker telescope.*

FIG. A–34 (*g*) *150-foot tower telescope, exterior view from northeast.* (*Fig. A34*(*a*) *through* (*g*) *from the Mount Wilson and Palomar Observatories*)

FIG. A–34 (*h*) *The 40-inch Yerkes refractor, rising floor at lowest position.* (*Yerkes Observatory photograph*)

FIG. A–34 (*i*) *Exterior view of Kitt Peak* (*Arizona*) *McMath Solar Telescope.* (*From Kitt Peak National Observatory*)

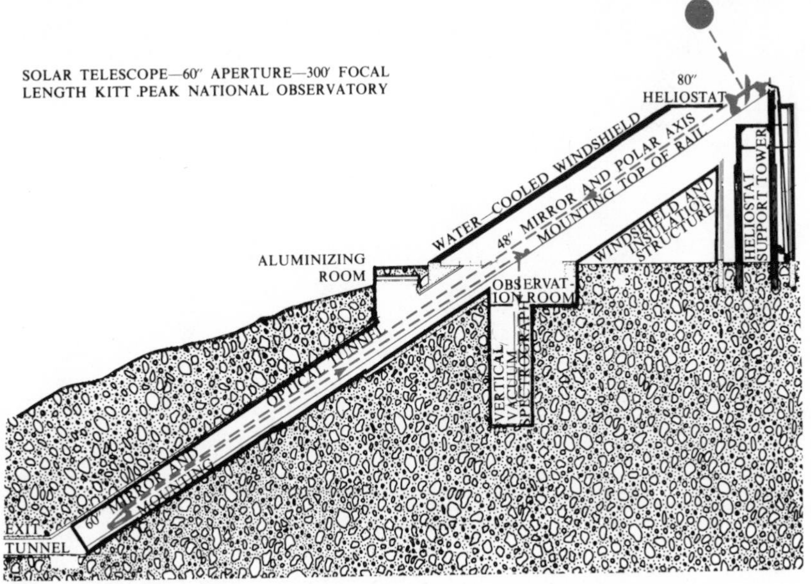

Fig. A–34 *(j) Close-up of heliostat of the Kitt Peak McMath Solar Telescope. (From Kitt Peak Observatory)*
(k) Schematic drawing of the Kitt Peak McMath Solar Telescope. (Adapted from a drawing of the Kitt Peak National Observatory)

EXERCISES

1. Compare the speed of light in glass with the speed of light in water. (Use $n = 1.333$ for the index of refraction of water.)

2. How does the speed of red light in ordinary glass compare with the speed of blue light? (Take the index of refraction of red light as 1.517, and the index of blue light as 1.534.)

3. (a) Calculate the focal length of a thin biconvex lens consisting of crown glass (index of refraction $n = 1.512$) if $r_1 = 15$ inches and $r_2 = -12$ inches. (b) Repeat (a) for a thin negative flint lens for which $n = 1.623$, $r_1 = -12$ inches, and $r_2 = -30$ inches.

4. If the two lenses in problem 3 were combined into a single thin doublet by cementing the back surface of the convex lens to the front surface of the negative lens, what would the focal length of this doublet be?

5. (a) How large is the image of the sun at the focal point of a telescope whose objective has a focal length of 15 ft? (b) How large would this image appear if viewed with a $\frac{1}{2}$-inch focal length eyepiece? (Assume that the image as viewed through the eyepiece is at the normal distance of 10 inches from the eye.)

6. Consider a telescope whose objective has a clear aperture of 36 inches and a focal length of 30 ft. Calculate the following properties of this telescope: (a) its light-gathering power; (b) the speed of its objective; (c) its resolving power; (d) its magnification when used with a $\frac{1}{4}$-inch eyepiece; (e) the maximum magnification that can be used with this telescope.

7. (a) When the 200-inch telescope is directed at a star of the twenty-first apparent magnitude, how much radiant energy enters the objective every second? (b) Assume the average wavelength of this light to be 6 000 Å. How many photons of this light enter the objective every second?

Appendix B

RADIO ASTRONOMY

B.1 The Nature of Radio Waves

In the main body of the text we discussed the information we can obtain about celestial bodies from the radio waves they emit. This phase of astronomy, which is called **radio astronomy**, has given us such diverse information as the surface temperature of Venus, the nature of the magnetic fields surrounding the planets, certain structural properties of gaseous nebulae such as the Crab Nebula, the spiral structure of our Galaxy, and the distribution of extragalactic stellar systems. Although we have outlined the way in which this information is obtained, we have not discussed the actual radio instrumentation involved nor the types of radio telescopes used. Before we can do this we must consider the nature of the radio waves themselves, and how the energy carried by these radio waves is transferred to the recording device used by the radio astronomer.

There is no qualitative difference between radio and visible radiation since they are both components of the electromagnetic spectrum. The radio wavelengths, however, are much longer than the wavelengths of visible radiation; radio astronomy deals with radiation having wavelengths that range from a fraction of a centimeter to many meters. The quantity that the astronomer measures is the **energy flux** in the electromagnetic field of the incident radio wave. This energy flux is given by the square of the electrical field intensity. In general, the radio wave is a mixture of all frequencies (or wavelengths) in the radio spectrum and therefore corresponds to the white light in the visible part of the spectrum. In studying visible radiation we must have an optical telescope to collect as much of the visible energy of all frequencies as we can, and, in addition, a spectroscope to break up this radiation into its separate wavelengths, So, too, in radio astronomy we must have a radio telescope to collect as much of the radio emission as we can, and a separate device to single out individual wavelengths and frequencies.

When a radio wave of a particular frequency ν reaches the radio telescope, it consists of an oscillating electric wave of intensity E given by

$$E = E_0 \cos 2\pi\nu t,$$

and a magnetic wave of intensity H given by

$$H = H_0 \cos 2\pi\nu t.$$

Here E_0 is the maximum amplitude of the electrical wave and H_0 the maximum amplitude of the magnetic wave, and t is the time as measured from some fixed moment. These two equations merely say that the radio wave consists of an electric field that oscillates ν times per second, and a magnetic field that does the same thing at right angles to the electric field. The amount of energy arriving per second at the radio telescope for this particular frequency ν is given by

$$\frac{c}{8\pi}(E_0^2 + H_0^2).$$

Since electromagnetic theory shows that $E_0 = H_0$ for waves coming from a very great distance, we may write for the energy flux

$$\mathrm{E}_{flux} = \frac{c}{4\pi} E_0^2.$$

But, in general, radio waves contain a whole range of frequencies. Hence the total energy flux consists of a sum of many such terms. The function of the radio telescopes is twofold:

1. to collect as much of this energy as possible and to record it, and
2. to analyze the energy into its various frequency (or wavelength) components.

The first of these tasks is achieved by a system consisting of an antenna, a receiver, and an output device that records the radio signals.

B.2 Radio Telescope Antennae

Radio telescope antennae, which correspond to the objectives of the optical telescopes, are of two kinds:

1. huge concave paraboloid dishes consisting of a metal mesh or solid metallic surface, and
2. arrays of individual dipoles, which are similar to outdoor television antennae.

In both of these types of radio telescopes the antenna collects the radio energy and transmits it to a receiver. In the case of the paraboloid dish the radio beam is reflected to a single dipole at the focal point of the parabola. (See Figure B.1.) The electrical energy is then conducted to a receiver where it is recorded by some mechanical, optical, or acoustical device (the optical device might resemble an oscilloscope or a television screen).

In the case of the array of individual dipoles each dipole conducts its own bit of electrical energy received from the radio beam to the receiver. The principal difference between the function of a single parabolic antenna and an array of dipoles is that the parabolic antenna simply collects all the energy, whereas the array of dipoles takes into account the difference in path lengths of various radio beams striking the dipoles and hence can be used for interferometer work in radio astronomy.

To understand the advantages of a radio telescope over an optical telescope and also to see what its limitations are, we must consider just how the antenna responds to the radiation it receives. We note first that although the total energy received per second by the antenna depends only on the maximum amplitude E_0 of the incoming electromagnetic wave, and hence is independent of the time, the actual instantaneous signal varies from moment to moment in a complicated way which depends upon the spectral composition of the incoming signal. This signal is a superposition of harmonics (different wavelengths) of continuously varying amplitudes and with each harmonic in a different phase. Therefore the output signal that is to be recorded varies with time in a complicated and unpredictable way. Since all the information about the radio source (e.g., some stellar body) contained in the radio wave is present in this output signal, we must be able to extract this information from the signal.

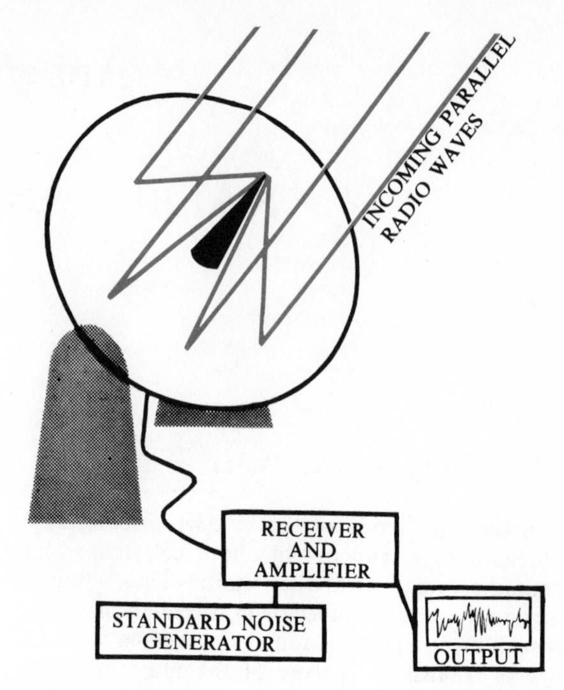

FIG. B–1 *A parabolic radio telescope focuses the parallel radio waves onto the dipole at the focal point of the parabolic disk.*

B.3 Radio Noise: Signal-to-Noise Ratio, Z

The problem is greatly complicated by the presence of so-called **noise**, which is essentially of a threefold nature:

1. general background radio noise arising from such things as electrical motors, spark plugs, terrestrial radio signals, lightning, and so on;
2. noise arising from the random motions of electrons that are always present in the receiver itself; this noise depends in part on the temperature of the input stage of the receiver.
3. thermal background noise emitted by all warm objects, such as the air, the ground, antennae, transmission lines, etc., that can radiate to the antennae.

Radio telescopes to be useful must first of all separate the actual information-carrying signal from noises of all kinds. Whether or not one can do this with a particular telescope depends upon the **signal-to-noise ratio**, Z. The larger this quantity is, the more efficient the radio telescope is and the greater the amount of information that can be obtained from the signal. If the useful power (energy per second) received by the antennae is S, and the power arising from the noise is S_n, the signal-to-noise ratio is given by

$$Z = \sqrt{S/S_n}.$$

Since there is no way of controlling S it is important for any given radio telescope to reduce the noise level S_n as much as possible. *The noise level sets a limit on the sensitivity of the radio telescope.*

It is clear that if the signal amplitude is much greater than the noise, the signal can be easily detected and measured. As the intensity of the incoming signal decreases, however, the signal becomes indistinguishable from the noise and one might arbitrarily define the limit of detectibility of a radio signal as one for which $Z = 1$. This at first hand seems reasonable since a signal whose amplitude is equal to or less than that of the noise would be hidden in the noise of the receiver itself. However, with modern correction techniques it is possible under certain conditions to detect signals whose intensities are only small fractions of the receiver noise. This is true if either the signal is fairly steady over a sufficiently long period of time, or if it has some unusual characteristic. In either case, many observations of the signal can be made with the radio apparatus, and the signal's stable characteristics, which distinguish it from the random fluctuations of the noise, can be enhanced electronically in the receiver and thus be detected.

What one actually observes in the radio telescope is, of course, the reaction of a recording device to the incident radio waves, just as in an optical telescope one deals with the effect of the light on either a photographic plate, the retina of the eye, or a photoelectric device. There is, however, one very important difference between optical and radio recording devices that one must take into account to understand the problem that arises with radio telescopes. When an optical signal of sufficient intensity strikes the eye, or any other light-sensitive device, there are very rapid fluctuations on the retina or any other surface resulting from the very rapid oscillations of the incident electromagnetic fields, if we assume for the moment that there is no intervening medium between the optical receiver (eye, etc.) and the source—that is, if the optical signal has no optical noise superimposed on it. Since the frequencies of the optical oscillations are then very large (because of the short wavelengths of light) the eye or any other optically sensitive device that is used smooths out all of these fluctuations during the time of observation and only a steady optical signal is obtained. In other words, the response time of the eye is so long that the rapid electromagnetic fluctuations are not detected. Note, however, that when one deals with optical devices in which light is chopped up (as in a motion picture projector or in a fluorescent light) or if the light passes through a fluctuating inhomogeneous medium, such as our atmosphere, optical noise is present, and an actual flicker that disturbs the measurements is observed. This, of course, places an important limitation on the optical instruments. We should also note that if the optical signal is very weak, statistical fluctuation becomes important because the individual photons arrive at a random rate, and we may not treat the signal as an electromagnetic wave.

In the case of the recording device in a radio telescope the fluctuations that occur in the signal are highly perceptible and important because of the much longer wavelengths and because of the relatively short response time of the receiver of the radio telescope.

B.4 Limitations on Sensitivity of Radio Telescopes

Since, as we have noted above, the information that can be abstracted from a radio signal by a telescope depends on the response of the telescope to the signal, we must analyze this response to both the internal noise of the receiver and to the signal if we are to determine the sensitivity of the telescope. If the internal noise were steady, even a very weak signal consisting of a series of pulses could be detected easily. But the noise itself consists of random fluctuations and this reduces the sensitivity of the receiver.

To see what is involved, let us suppose that the recording device attached to the receiver can detect n distinct pulses of current in the receiver in a time τ and let R be the average value of these readings given by the recorder. That is, let

$$R = \frac{R_i + R_2 + \cdots + R_n}{n} = \frac{\sum_{i=1}^{n} R_i}{n},$$

where R_i is the reading of the ith current pulse given by the recorder.

Now let ΔR be the root mean square deviation of any single reading from the average reading R.

Then one can show that ΔR is proportional to $R/\sqrt{n}$. In other words,

$$\frac{\Delta R}{R} = \frac{\alpha}{\sqrt{n}},$$

where α is a number, which according to statistical theory is about equal to 1 (and is determined by the construction of the receiver). Since n is the number of pulses that occur in the time interval of observation τ, we must have

$$n = \tau\,\Delta\nu,$$

where $\Delta\nu$ is what is called the **frequency bandwidth** of the noise signal admitted by the radio telescope. This is clear because $\Delta\nu$ represents the number of pulses that occur per second in the signal, so that in a time τ there are just $\tau\,\Delta\nu$ pulses. The bandwidth $\Delta\nu$ for a given frequency ν_0 may also be defined as the difference between the two frequencies on either side of ν_0 at which the output signal is half that of the input signal at ν_0. We thus have from the previous equation

$$\boxed{\frac{\Delta R}{R} = \frac{\alpha}{\sqrt{\tau\,\Delta\nu}}}\,. \tag{B.1}$$

This is a fundamental formula in radio astronomy and places a practical limit on the sensitivity of the telescope.

Although we have derived this formula for the fluctuations in the recorder arising from the noise in the receiver, it applies even if no noise is generated in the receiver itself, since the radio signal to be measured is itself of a random character and consists of a train of short pulses separated by long time intervals. Thus, this formula defines the uncertainty in the determination of the mean value of the signal.

Since it is desirable to make $\Delta R/R$ as small as possible in order to avoid violent fluctuations of the recording device (whether arising from noise or the randomness of the signals), one must either increase $\Delta\nu$ or τ or both. In the case of an optical device that works in a wavelength interval of about 1 angstrom ($1\text{Å} = 10^{-8}$ cm) we can show that the frequency bandwidth, $\Delta\nu$ of the optical signal, is 10^{11} cycles per second. This can be easily obtained from the wavelength interval $\Delta\lambda$ (in the above case 1 Å) by multiplying $\Delta\lambda$ by c/λ^2, where c is the speed of light. With $\Delta\nu$ so large as this, $\Delta R/R$ is very small (of the order of 10^{-5}) even if τ is only 0.1 seconds. In the case of a radio telescope, however, the situation is quite different because the frequency interval $\Delta\nu$ that can pass through the receiver is of the order of 10^6 to 10^7 cycles per second. If we now consider a time interval $\tau = 100$ seconds, we see that [using Equation (B.1)] $\Delta R/R$ is of the order of 10^{-4}.

To see how the formula for $\Delta R/R$ determines the smallest signal that can be detected by a given radio telescope, we first suppose that the antennae sweep over a part of the sky that is emitting practically no radiation. Then only the receiver noise is recorded and the fluctuations are given by $\alpha R/\sqrt{\tau\,\Delta\nu}$.

If, now, the telescope is turned to another part of the sky and receives a signal whose intensity is equal to or larger than this fluctuation, it is clear that this signal will be detected if its duration is long enough. For if the duration of the signal is $N\tau$ the noise fluctuation is given by $\alpha R/\sqrt{N\tau\,\Delta\nu}$, and this can be much smaller than the intensity of the incident signal if N is large enough.

B.5 The Internal Noise: The Sensitivity of a Radio Telescope

The internal noise arises from two sources:

1. the temperature, T, of the antenna itself, and
2. the thermal motions of the electrons in the various resistors in the receiver.

Let P_0 be the noise power (energy per unit time) generated within the antenna when no external signal falls on it. Since this is simply due to the temperature of the antenna we may write it (according to radio theory) as

$$P_0 = kT\,\Delta\nu, \tag{B.2}$$

where k is Boltzmann's constant and $\Delta\nu$ is the frequency range that can pass through the receiver. (Note that noises of all frequencies can arise, but clearly, the only noise that has a disturbing effect on the signal is the noise lying in a frequency range that the receiver is tuned to.) This formula has its origin in the equipartition theorem that we discussed in Section 16.10. Let P_i be the power generated by the noise in the receiver itself (exclusive of the antenna). We see then that the total noise that comes into the receiver is just $P_i + P_0$. However, the noise that the recording device receives is greater than this because of the amplification that occurs in the receiver. If the amplification factor is g (called the **gain**), the **recording or output device** receives an amount of power given by $g(P_0 + P_i)$ when no external signal is present. The deflection or reading, R, of the recording device when no signal is incident is therefore proportional to this quantity.

On the other hand, if the power of the signal striking the antenna is P, the amount of the deflection of the recording device ΔS arising from this signal alone is proportional to gP. We therefore have

$$\frac{\Delta S}{R} = \frac{gP}{g(P_0 + P_i)} = \frac{P}{(P_0 + P_i)},$$

$$\Delta S = \frac{RP}{(P_0 + P_i)}.$$

Since ΔS must be larger than or equal to ΔR for the signal to be easily detectable without long observing times or special techniques, we must have

$$\frac{RP}{(P_0 + P_i)} \geq \Delta R.$$

But from Equation B.1, $\Delta R = (\alpha R)/\sqrt{\tau\,\Delta\nu}$, hence

$$\frac{P}{(P_0 + P_i)} \geq \frac{\alpha}{\sqrt{\tau\,\Delta\nu}},$$

or

$$P \geq \frac{\alpha(P_0 + P_i)}{\sqrt{\tau\,\Delta\nu}}. \tag{B.3}$$

This gives us the minimum power in a signal of width $\Delta\nu$ that can be easily detected by a radio telescope tuned to this signal, which has the time τ (its **time constant**) in which to scan the source of the signal and which has a total antennae and internal noise of $(P_0 + P_i)$.

B.6 The Noise Factor, N, of a Radio Telescope

It is convenient to simplify this formula by introducing a quantity N that is called the **noise factor** of the radio telescope. We define this as the signal-to-noise ratio at the input of the receiver divided by the signal-to-noise ratio at the output. Since the only noise at the input of the receiver is the antenna noise, P, and the output noise of the receiver is $P_0 + P_i$, we have

$$N = \frac{P/P_0}{P/(P_0 + P_i)} = \frac{P_0 + P_i}{P_0} = \left(1 + \frac{P_i}{P_0}\right), \tag{B.4}$$

where P is the signal.

If we solve this for P_i we have

$$P_i = (N - 1)P_0 = (N - 1)kT\,\Delta\nu,$$

where we have introduced the temperature T of the antenna from Equation (B.2).

Since from Equation (B.4), $P_0 + P_i = NP_0 = NkT\,\Delta\nu$, we have on substituting this into Equation (B.3)

$$P \geq \alpha\frac{NkT\,\Delta\nu}{\sqrt{\tau\,\Delta\nu}},$$

If we now place $P = P_\nu\,\Delta\nu$, where P_ν is the power per frequency bandwidth at the frequency ν, we finally have

$$P_\nu \geq \alpha\frac{NkT}{\sqrt{\tau\,\Delta\nu}}. \tag{B.5}$$

We see then that the noise factor, N, determines how weak a signal we can detect. The right-hand side of the equation represents the weakest detectable signal. Hence if N were very small, we could detect very faint signals. That is, if the receiver were very quiet, the detection limits would be set by external noise alone.

Shklovsky points out that in good receivers, N ranges from 3 to 10 in the meter-to-decimeter wavelength range, and from 5 to 10 in the centimeter range. Since T is the surrounding temperature of the antenna, we may take it to be of the order of 300°K, and since $\alpha(\Delta\nu\tau)^{-1/2}$ is of the order of 10^{-4} for typical radio telescopes in a typical wavelength range, we have for $N = 5$,

$$P_\nu \geq 15 \times 10^{-2}k.$$

On substituting for k its value 1.4×10^{-16} ergs/deg, we have

$$P_\nu \geq 21 \times 10^{-17} \text{ ergs}.$$

Since the bandwidth $\Delta\nu$ of the frequency range for which our calculations were carried out is of the order of 10^6 cycles/sec, this means that we can detect a signal having the power $P = 2 \times 10^{-11}$ ergs/sec. We obtain this from P_ν by multiplying P_ν by the frequency range $\Delta\nu$ for a typical radio signal.

This is the amount of energy per second delivered to the entire paraboloid antenna. To get an idea of how sensitive our large modern telescopes are, we divide this by the area of the antenna. If we take a 100-foot paraboloid dish, these numbers indicate that such a telescope can respond to a radio signal amounting to 10^{-17} to 10^{-18} (ergs/cm^2) per sec. In the optical range in the spectrum this corresponds to a light source of the thirtieth apparent stellar magnitude, which is about seven magnitudes fainter than the faintest object that can be detected by the largest optical telescope.

FIG. B–2 *Nonthermal radio source in Cassiopeia, taken in red light. The faint wisps of gas in the upper central part of the photograph are the radio sources. (From Mount Wilson and Palomar Observatories)*

B.7 Thermal and Nonthermal Radio Sources

It is convenient in radio astronomy to express the brightness of a source in terms of an equivalent temperature, which plays a role similar to that of the effective temperature in dealing with stars. Although the effective temperature in optical astronomy was introduced from black-body considerations, it is a useful parameter to describe a star even though the star radiates only approximately like a black body. Actually the effective temperature is a simple numerical way of representing the total energy summed over all frequencies radiated by a star. In the same way we can introduce a temperature in the radio-wavelength regions, T_R, to describe the radio brightness of a source. We must, however, be very careful to distinguish between the radio energy emitted by a black body and that emitted by a nonthermal source.

If the radio energy we receive is emitted as a result of thermal motions of atoms and molecules in the source, which is at a temperature T, the source behaves exactly like a black body, and the temperature associated with this radio energy is a measure of the temperature of the source. One can then apply the usual laws of black-body radiation to the radio energy to obtain the temperature of the source. If, however, the radio energy emitted by a source is the result of nonthermal motions of charged particles (e.g., motions of electrons in a radio tube, the motions of charged particles in a lightning discharge or any other electrical discharge, alternating currents, etc.—see Figure B.2) then the radiant energy is nonthermal in nature and it can give us no clue as to the thermal temperature of the source. Nevertheless, we can still assign a kind of effective temperature to this radio energy to define its intensity. It is clear from this discussion that when we speak of the brightness temperature of a radio source to describe the intensity of the radio waves it emits, we must distinguish between thermal and nonthermal radio waves.

For thermal radiation, the brightness temperature is a measure of how hot the source is, whereas for nonthermal radiation this is not the case. We shall now see how the temperature of a source can be determined from a radio signal that is received by the antenna. To do this we assume that the source behaves like a black body of a particular temperature, T, and therefore emits energy in the radio part of the spectrum in accordance with Planck's law of black-body radiation (see Section 16.28). Since the wavelengths of the radio part of the spectrum are relatively long, Planck's formula for the intensity I_λ [see Equation (16.16)] in the radio-frequency range from ν to $\Delta\nu$ reduces to the following simple formula:

$$I_\lambda \, \Delta\nu = \frac{2kT}{\lambda^2} \Delta\nu, \qquad \text{(B.6)}$$

where we now refer to T as the brightness temperature of the source. This is also known as the Rayleigh-Jeans law since Rayleigh derived it in his unsuccessful attempt to explain black-body radiation.

We can now obtain the temperature of the source from the radio waves incident on the antenna of the radio telescope as follows: if our antenna were at the center of a sphere, the internal surface of which is a black body at temperature T_α, this surface would then emit radio waves to the antenna in accordance with the laws of black-body radiation, and the temperature of the antenna would be T_α also.

This is so because the laws of thermodynamics require that the antenna and the interior surface of the sphere be in equilibrium. In this case, of course, the antenna temperature would itself be an immediate measure of the black-body temperature of the sphere. We could then find the temperature

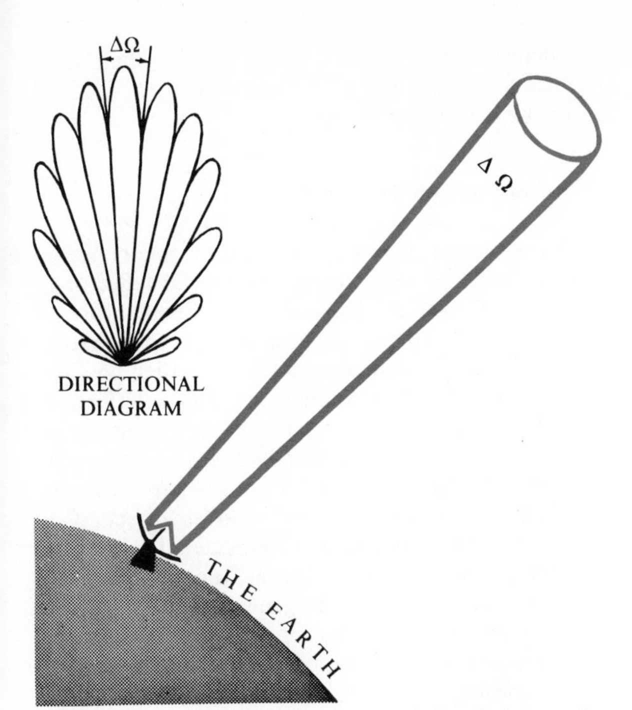

FIG. B–3 *A radio source subtends a solid angle* $\Delta\Omega$ *at the antenna.*

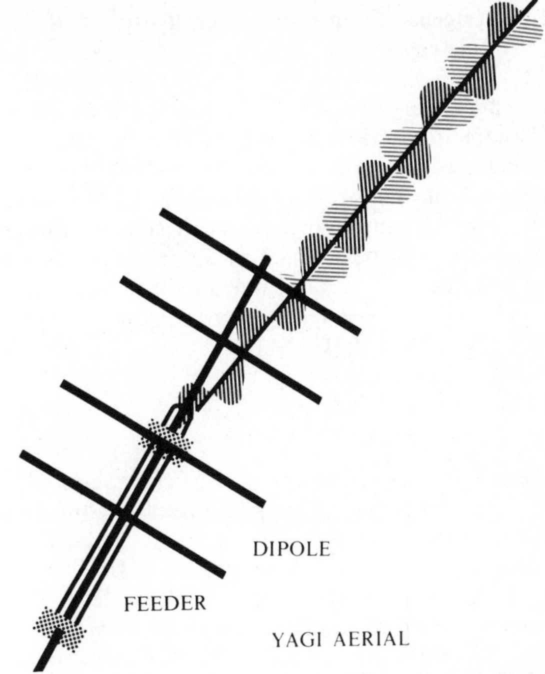

FIG. B–4 *Because of the polarization effect only one-half of the radio energy from a source is absorbed by the receiving antenna.*

of the surface of the sphere by noting that the power absorbed by the antenna in the frequency range $\Delta\nu$ is, according to Equation (B.2), $kT_\alpha\,\Delta\nu$. The temperature T_α may be referred to as the **antenna temperature**.

In general, however, we do not have this ideal situation, and the antenna temperature T_α is not equal to the brightness temperature of the source which we defined above. However, the two are related in a rather simple way. To get at this relationship we note that the intensity of the radio waves emitted by the source is not necessarily the same in all directions so that the brightness temperature, T_α, depends on the directions from which the source is viewed. In other words, *the intensity* I_ν *is dependent upon the direction of the source and the geometry of the antenna.*

If, then, we consider a small element of solid angle $\Delta\Omega$ (the element of solid angle subtended at the antenna by the source as shown in Figure B.3), and if we introduce A as the effective area of our antenna for this solid angle (essentially the area of the antenna sensitive to the radio waves coming from the source), the power absorbed by the antenna from the given direction is

$$P_{\text{absorbed}} = \tfrac{1}{2}AI_\nu\,\Delta\Omega\,\Delta\nu. \tag{B.7}$$

The factor $\frac{1}{2}$ appears here because only radio waves with their electric vector vibrating parallel to the antenna are completely absorbed (those vibrating perpendicular pass by), so that on the average only half the radio energy of the source is absorbed. This is called the **polarization effect** (see Figure B.4). From this equation and Equation (B.2) we see that the antenna temperature is given by

$$T_\alpha = \frac{AI_\nu\,\Delta\Omega}{2k}. \tag{B.8}$$

But if into Equation (B.8) we substitute the value of I_ν from (B.6) in terms of the temperature of the source, we have

$$T_\alpha = \frac{A\,\Delta\Omega T}{\lambda^2}. \tag{B.9}$$

This relates the antenna temperature to the brightness temperature, provided the solid angle subtended at the source by the antenna is very small. Note that the brightness temperature of a source is a measure of the thermal properties of the source only if the radio waves emitted by the source are thermal in nature, i.e., if the source itself behaves approximately like a black body. If this is not the case and the radio waves are nonthermal in nature, we may still refer to a brightness temperature of a source, but it is not the temperature that is measured by an ordinary thermometer.

B.8 Antenna Temperature—Sensitivity of Radio Telescopes

Just as we discussed the sensitivity of a radio telescope in terms of the minimum power that can be detected as given by Equation (B.5), we may speak of the temperature sensitivity of the radio telescope by considering the smallest temperature change that can be detected by the receiver. Since by definition $P_\nu = kT_a$, the smallest detectable change ΔT_a in the antenna is given by the formula (as derived from B.5)

$$\Delta T_a \geq \frac{\alpha NT}{\sqrt{\tau\,\Delta\nu}}, \tag{B.10}$$

where T on the right-hand side of the equation is the actual surrounding temperature of the antenna and is generally of the order of 300°K and α is of the order of 1. Since N ranges from 3 to 10 for good radio telescopes, the smallest detectable antenna temperature change can range from 0.06°K to 0.2°K for radio signals with frequency band $\Delta\nu$ equal to 10^6 cycles/sec and with a time constant τ of 100 seconds.

As an example of the application of the above formula, we note that in 1944 Reber, who was a pioneer in this field, used a 32-foot-diameter paraboloid reflector to detect radio frequencies of about 160 megacycles per second (which corresponds to a wavelength of 1.80 meters). The greatest intensity he obtained was 1.5×10^{-17} ergs/cm² per sec in the direction of the galactic center. From the Rayleigh-Jeans formula (B.6) we see that this corresponds to a brightness temperature of about 1200°K. The minimum intensity Reber was able to detect was about 21 times smaller than this, so that the temperature sensitivity ΔT_α of his antenna was about 60°K.

B.9 Interferometry in Radio Telescopes

Thus far we have been discussing telescopes in which only one antenna or paraboloid reflector is used. However, a good deal of additional information can be obtained from a radio source if a system of one or more antennae is used. It is then possible to arrange these antennae so that interference effects between different parts of the same radio wave occur. The way this happens is shown in Figure B.5, where just two antennae are indicated receiving a radio signal from a given direction in space. Both antennae are connected by leads to a central receiver so that when the signals reach this receiver they are, in general, out of phase. This difference in phase depends on the distance between the two receivers, d, and the angle θ that the wavefront of the radio waves makes with the line connecting the two antennae, as shown in Figure B.5. To obtain the phase difference between the parts A and B of the wavefront hitting the antennae, A_1 and A_2, we must calculate the path difference between A and B. From Figure B.5 we see that this is $d \sin \theta$. If λ is the wavelength of this wave, then the number of wavelengths contained in this path difference is just $(d \sin \theta)/\lambda$. By definition, then, the phase difference is just 2π times this or $(2\pi \sin \theta)/\lambda$.

Let us now consider the interference effects that may arise from such an arrangement of antennae. If the path difference between A and B is just one-

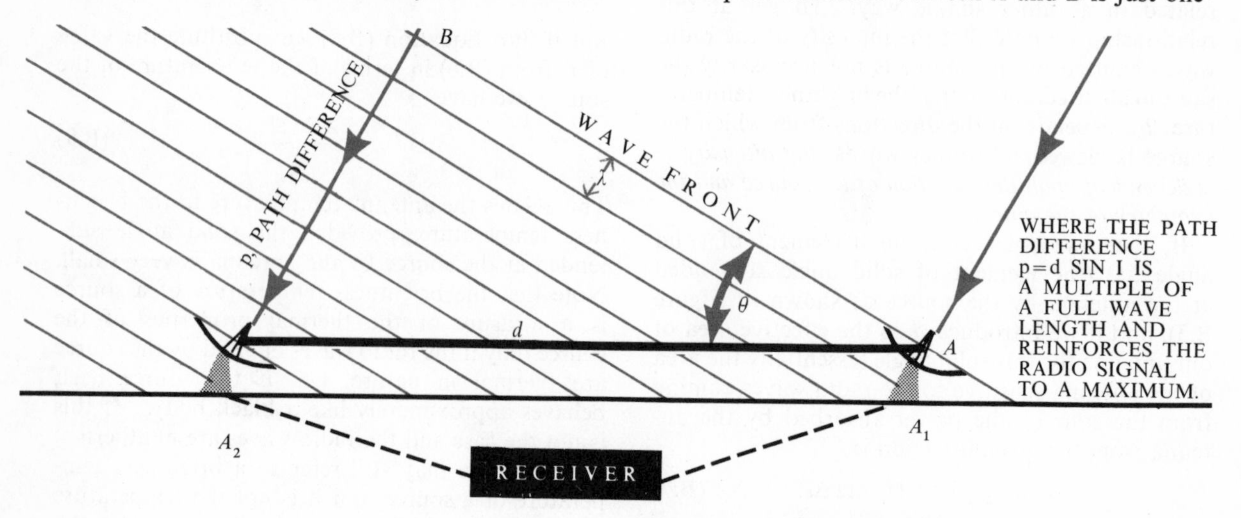

FIG. B–5 *A schematic diagram of a radio interferometer.*

half wavelength or any odd multiple of a wavelength, then the two waves interfere destructively and cancel each other out. On the other hand, if the path difference is a full wavelength or a multiple of a full wavelength, then the two parts of the beam reinforce each other and the radio signal is at its maximum. Thus if a point source of radio waves moves across the sky, so that the incoming radio beam sweeps across the two antennae at an angle θ, alternate maximum and minimum signals are recorded. We can find the right ascension of a source by this means if the antennae are lined up so that the point midway between them lies on the celestial meridian. The right ascension and the declination are then given at the moment that the central signal reaches its maximum.

If there are two distinct sources sending radio waves, each one has its own interference pattern, and we then obtain in the receiver a combined signal that is quite complicated. By analyzing this signal we can find the angular separation between the two point sources. By using a variation of this procedure which corresponds to the Michelson interferometer used in optical telescopes we can obtain the angular diameter of an extended radio source. This is of great importance in radio astronomy because radio telescopes have a low resolving power as compared with optical telescopes.

B.10 Radiation Gathering Power and Resolving Power of Radio Telescopes

The formula for the radiation-gathering power of a radio telescope is the same as for an optical telescope. It is given essentially by the cross-sectional area of the reflector (in this case called the **dish** or **dipole array**). Hence, just as in the case of an optical telescope, it is proportional to the square of the aperture (i.e., to the square of the diameter). Since radio telescopes can have diameters as high as 1 000 feet, the amount of radiant energy that can be gathered by such instruments is vastly larger than with optical telescopes.

FIG. B–6 (*a*) *The giant radio telescope at Jodrell Bank, Cheshire, England. The reflector bowl is about 250 feet in diameter. Designed by the radio-astronomy department of Manchester University. It is the world's largest fully steerable radio telescope. (British Official Photograph, courtesy of the Central Office of Information, London, England)*

FIG. B–6 (*b*) *Interior view of the Jodrell Bank radio telescope. The antenna system is shown, as well as the individual welded steel plates that form the surface of the bowl. The inside of the bowl is painted a matte white finish. (British Official Photograph, courtesy of the Central Office of Information, London, England)*

Fig. B–6 (*c*) *General view of the Jodrell Bank Experimental Station showing part of the mesh reflector of the 218-foot fixed radio telescope. In the background is the 250-foot steerable telescope together with various laboratories. (British Official Photograph, courtesy of the Central Office of Information, London)*

Fig. B–6 (*d*) *The bowl and aerial of the Jodrell Bank radio telescope as they appear at night. (British Official Photograph, courtesy of the Central Office of Information, London)*

Fig. B–6 (*e*) *The radio telescope at Jodrell Bank focused on the moon. Photograph taken from the roof of the Control Building. (British Official Photograph, courtesy of the Central Office of Information, London)*

Fig. B–6 (*f*) *The 60-foot radio telescope at the George R. Agassiz Station of Harvard Observatory.* (*From the Harvard College Observatory*)

Fig. B–6 (g) The 260-foot radio telescope of the Ohio State University. The radio waves, first received by the tiltable, flat reflector (top of photograph), are then sent to the paraboloidal section (bottom of photograph), and finally converge at the prime focus which is at the small radio-transparent plastic house (in front of rear flat reflector). This building houses the laboratory and the reference antenna. A curtain of fine wires hangs within the paraboloidal reflector and provides the reflecting surface. The three acres of ground between the two reflecting units are covered by a concrete floor with an aluminum-sheet surface. This ground plane provides an electrical image and guiding plane for the radio waves. It also reduces the antenna temperature. (From the Ohio State University Radio Observatory)

The formula for the resolving power (i.e., the size of the spurious disk in radians) for radio telescopes is the same as that for optical telescopes, namely:

$$R.P. = \frac{1.22\lambda}{a},$$

where λ is the wavelength of the radio beam and a is the aperture of the radio telescope. We see from this formula that radio telescopes as compared with optical telescopes have a much smaller resolving power. A simple comparison of the 4-foot Schmidt telescope with a 200-foot radio telescope shows this at once. If we consider a 6 000-Å optical beam and a 21-cm radio beam, we find that the spurious disk into which the radio telescope concentrates the 21-cm beam is almost 7 000 times larger in diameter than the spurious disk into which the 4-foot optical telescope concentrates the 6 000 Å beam. In other words, as long as we work with radio wavelengths of the order of 1 cm, even large radio telescopes cannot differentiate between two radio sources that are closer than some 10 minutes of angular measure. However, it is possible to overcome this disadvantage to some extent by working with the radio interferometers which we have already discussed.

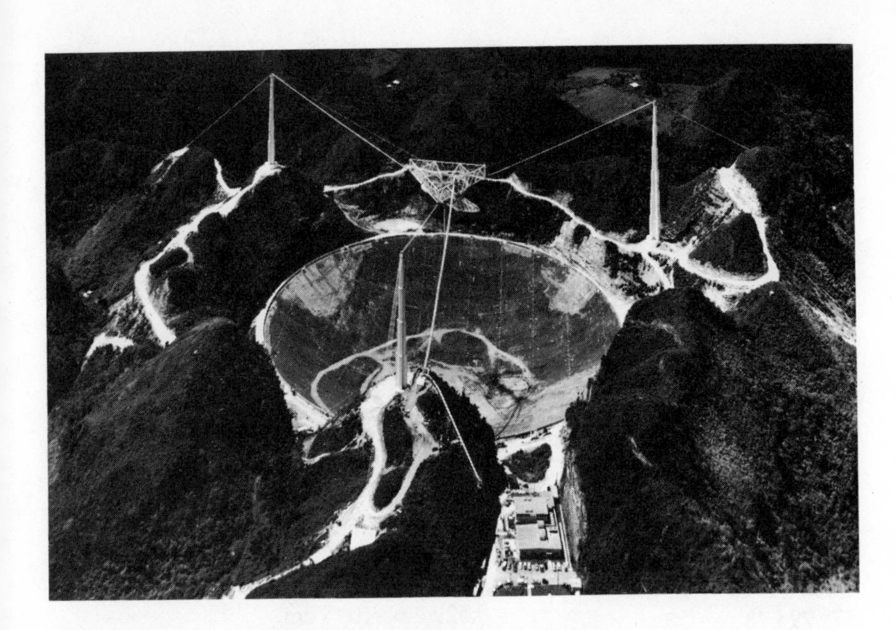

FIG. B–6 (*h*) *The Arecibo Ionospheric Observatory 1 000-foot radar-radio telescope, built and operated by Cornell University, is the largest of its type in the world. The ingenious steerable feed system (the needle-like piece pointing down towards the reflector) of this stationary instrument permits radio observations out to 20 degrees from the zenith. The base of the paraboloidal dish is a reshaped natural hollow in the earth. The steel-mesh lining provides the reflecting surface.* (*From Arecibo Ionospheric Observatory of Cornell University*)

Appendix C

CONSTANTS

Mathematical Constants

CONSTANT	NUMBER
π	3.141 59
e (base of natural logarithm)	2.718 28
$\log_{10} e$	0.434 29
1 radian	57°.295 78
	= 3 437′.746 77
	= 206 264″.806 25
1°	0.017 45 radians
1′	0.000 29 radians
1″	0.000 005 radians

Physical Constants

Velocity of light

$c = 2.997\,93 \times 10^{10}$ cm/sec

$\cong 3 \times 10^{10}$ cm/sec

$c^2 = 8.987\,58 \times 10^{20}$ cm²/sec²

Universal gravitational constant

$G = 6.668 \times 10^{-8}$ (dyn cm²)/gm²

Standard acceleration of gravity on earth

$g = 980.618$ cm/sec²

Planck constant

$2\pi\hbar = h = 6.625 \times 10^{-27}$ erg sec

Electron charge

$e = 4.803 \times 10^{-10}$ E.S.U. (electrostatic units)

Mass of electron

$m = 9.108 \times 10^{-28}$ gm

Mass of unit atomic weight (chemical scale)

$M_0 = 1.660\,26 \times 10^{-24}$ gm

C^{12} scale $= 1.660\,33 \times 10^{-24}$ gm

Boltzmann constant

$k = 1.380\,5 \times 10^{-16}$ erg/deg

Avogadro number (physical scale)

$N_0 = 6.024\,9 \times 10^{23}$ per mole

Standard atmosphere

$A_0 = 1\,013\,250$ dyn/cm²

$= 760$ mm of Hg

Atomic Constants

Rydberg constant for hydrogen

$R_H = 109\,677.58\ \text{cm}^{-1}$

1st Bohr Orbit radius

$\alpha_0 = h^2/4\pi^2 me^2$

$= 0.529\,2 \times 10^{-8}$ cm

Atomic unit of angular momentum

$\hbar = h/2\pi$

$= 1.054\,43 \times 10^{-27}$ g cm²/s

Rest mass energy of electron $= 8.186 \times 10^{-7}$ erg

Mass of hydrogen atom $= 1.673\,33 \times 10^{-24}$ gm

Mass ratio of proton to electron $= 1\,836.12$

Wavelength associated with 1 electron volt (1 ev)

$\lambda = 12\,379.7 \times 10^{-8}$ cm

Frequency associated with 1 ev

$10^8 e/hc = \nu_0 = 2.418\,1 \times 10^{14}\ \text{sec}^{-1}$

Energy of 1 electron volt

$E_0 = 1.602\,1 \times 10^{-12}$ erg

Radiation pressure constant

$a = 8\pi^5 k^4/15c^3h^3$

$= 7.564\,1 \times 10^{-15}$ erg/(cm³ deg⁴)

Stefan Boltzmann constant $ac/4$

$\sigma = 5.669\,2 \times 10^{-5}$ erg/cm² deg⁴ sec

Wien Displacement Law Constant

$C = 0.289\,780$ cm deg

1 calorie $= 4.18 \times 10^7$ erg

Astronomical Constants

The astronomical unit of distance

$$\text{A.U.} = 1.495\,985 \times \begin{cases} 10^8 \text{ km} \\ 10^{13} \text{ cm} \end{cases}$$

$= 9.295\,602 \times 10^7$ mi

Parsec = psc = 206 264.8 A.U. $= 3.085\,6 \times 10^{18}$cm

$= 1.92 \times 10^{13}$ mi = 3.261 5 light years

Megaparsec = Mpsc = 1 000 000 pc = 1 000 Kpsc

1 Light year
$= 9.460\,5 \times 10^{17}$ cm
$= 6.324 \times 10^{4}$ A. U.
$= 5.88 \times 10^{12}$ mi
Solar mass
$M_{\odot} = 1.989 \times 10^{33}$ gm
Solar radius
$R_{\odot} = 6.960 \times 10^{10}$ cm
Solar luminosity
$L_{\odot} = 3.90 \times 10^{33}$ erg/sec
Solar mean density
$\bar{\rho}_{\odot} = 1.41$ gm/cm^3
Earth mass
$M_{\oplus} = 5.977 \times 10^{27}$ gm
Earth mean density $\bar{\rho}_{\oplus} = 5.517$ g/cm^3
Earth equatorial radius $R_{\oplus e} = 6\,378.17$ km
Earth polar radius = 6 356.90 km
Earth mean radius = 6 371.23 km
Solar equatorial parallax = 8.″794 2
Moon's equatorial parallax = 3 422.″62
Constant of aberration = 20.″49
Constant of precession = 50.″256 4 per year
Mass ratios
$M_{\oplus}/M_{☾} = 81.33$
$M_{\odot}/M_{\oplus} = 332\,700$
$M_{\odot}/(M_{\oplus} + M_{☾}) = 328\,700$
Obliquity of ecliptic = 23°27′8″.26

Length of day (period of earth's rotation)
sidereal (referred to fixed stars)
$= 23^h56^m04^s.098$ mean solar time
mean solar day = $24^h03^m56^s.555$ sidereal time
Length of year
tropical (equinox to equinox)
= 365.242 198 78 mean solar days
$= 365^d5^h48^m45^s.98$
sidereal (fixed stars)
= 365.256 365 56 mean solar days
$= 365^d6^h9^m9^s.54$
= 31 588 149.5 sec
anomalistic (perihelion to perihelion)
= 365.259 641 34 mean solar days
$= 365^d6^h13^m53^s.01$
eclipse = 346.620 031 mean solar days
Julian calendar year = 365.25 mean solar days
Gregorian calendar year = 365.242 5 mean solar days
Length of the month
synodical (new moon to new moon)
= 29.530 588 days
$= 29^d12^h44^m2^s.8$
sidereal (fixed stars) = 27.321 661 days
$= 27^d7^h43^m11^s.5$
nodical (node to node) = 27.212 2
$= 27^d5^h5^m35^s.8$
Period of moon's node = 18.61 tropical years

Appendix D

CONVERSION FACTORS

To convert from units of	To units of	Multiply by a factor of
LENGTH		
kilometers km	meters m	1 000 or 10^3
	centimeters cm	100 000 or 10^5
	feet ft	3 280.83
	miles mi	0.621 37
meters	kilometers	0.001
	centimeters	100 or 10^2
	inches	39.37
	feet	3.280 8
	miles	6.21×10^{-4}
centimeters	kilometers	10^{-5}
	meters	0.01 or 10^{-2}
	inches	0.393 7
	feet	0.032 81
	miles	6.21×10^{-6}
inches	kilometers	2.45×10^{-5}
	meters	0.025 4
	centimeters	2.54
	feet	0.083 3
	miles	1.58×10^{-5}
feet	kilometers	3.05×10^{-4}
	meters	0.304 8
	inches	12
	miles	1.89×10^{-4}

To convert from units of	To units of	Multiply by a factor of
AREA		
square feet ft^2	square centimeters cm^2	929.03
cubic feet ft^3	cubic centimeters cm^3	28 316.9
solar volume $\frac{4}{3}\pi R_{\odot}^3$	cubic centimeters	1.412×10^{23}
cubic parsec	cubic centimeters	2.938×10^{55}
TIME		
hours	minutes	60
	seconds	3 600
days	seconds	86 400
tropical year	seconds	$3.155\,693 \times 10^7$
mean solar seconds	sidereal seconds	1.002 737 9
sidereal seconds	solar seconds	0.997 269 6
MASS		
kilograms	grams	1 000
pound (avoirdupois) U.S.	grams	453.592
	grains	7 000
pound(troy and apothecary)	grams	373.242
	grains	5 760

To convert from units of	To units of	Multiply by a factor of
LENGTH		
miles	kilometers	1.609 35
	meters	1 609.35
	centimeters	160 935
	inches	63 360
	feet	5 280
nautical mile	feet	6 076.1
	kilometers	1.852 0
degree	radians	0.017 453
minutes (angle)	radians	2.909×10^{-4}
seconds (angle)	radians	4.8×10^{-6}
angstrom unit Å	centimeters	10^{-8}
micron μ	centimeters	10^{-4}

To convert from units of	To units of	Multiply by a factor of
MASS		
grains	grams	0.064 799
pounds	poundal	32.174
ton	pounds	2 240
	kilograms	907.185
solar mass $M_{\odot}$	grams	1.989×10^{33}
atomic unit (electron) m	grams	9.108×10^{-28}
mass of unit atomic weight (chemical or C^{12} scale)	grams	$1.660\,3 \times 10^{-24}$

To convert from units of	To units of	Multiply by a factor of
ENERGY		
1 joule (M.K.S.) = 1 newton meter	ergs	10^7
1 erg = 1 dyne cm	joules	10^{-7}
erg/sec	horsepower	$1.341\,0 \times 10^{-10}$
calories	ergs	$4.185\,4 \times 10^7$
	joules (absolute)	4.185 4
kilowatt-hours (absolute)	ergs	$3\,600 \times 10^{13}$
	calories	$8.601\,3 \times 10^5$
foot-pounds	ergs	$1.355\,82 \times 10^7$
1 electron volt = E = ev = 10^{-6} Mev = 18^{-4} Gev, Bev = kMev	ergs	$1.602\,1 \times 10^{-12}$
energy of unit wave number	ergw	$1.986\,2 \times 10^{-16}$
mass energy of unit atomic weight	ergs	$1.491\,8 \times 10^{-3}$
	electron volts	$9.311\,4 \times 10^8$
energy associated with 1°K	ergs	$1.380\,5 \times 10^{-16}$
	electron volts	$0.861\,67 \times 10^{-4}$
POWER		
horse power (British and U.S.)	watts	746
watts	erg/sec	10^7
	joules/sec	10^7
solar luminosity	watts	3.90×10^{26}
ACCELERATION AND VELOCITY		
gravity (standard)	cm/sec²	980.665
	ft/sec²	32.174
meter per second (M.K.S.)	cm/sec	100
miles per hour	cm/sec	44.70
	ft/sec	1.466
light	cm/sec	$2.997\,93 \times 10^{10}$ $\cong 3 \times 10^{10}$
PRESSURE		
atmospheres (standard)	dyne/cm²	$1.013\,250 \times 10^6$
	mm Hg	760
	millibars	1 013.250
inches of mercury	dyne/cm²	$3.386\,38 \times 10^4$
PRESSURE		
pounds per square inch	dyne/cm²	$68\,947 \times 10^4$
	atmospheres	0.068 046
TEMPERATURE		
degrees Fahrenheit	Centigrade	subtract 32° and multiply by 5/9
degrees Centigrade	Fahrenheit	multiply by 9/5 and add 32°
degrees Centigrade	Kelvin	add 273°.155
temperature associated with 1 ev (E_0/k)	degrees Kelvin	11 605
ANGULAR MOMENTUM		
quantum unit ($\hbar = h/2\pi$)	g cm²/s	$1.054\,4 \times 10^{-27}$
ELECTRICAL		
Coulomb	electrostatic units E.S.U.	$2.997\,93 \times 10^9$
	electromagnetic units E.M.U.	0.10
electron charge e	E.S.U.	$4.802\,9 \times 10^{-10}$
	coulombs	$1.602\,1 \times 10^{-19}$

Appendix E

CONSTELLATIONS

Constellation	genitive ending	Meaning	Contractions	α	δ	Area
				h	°	$(^\circ)^2$
Andromeda	-dae	Chained maiden	And	1	40 N	722
Antlia	-liae	Air pump	Ant	10	35 S	239
Apus	-podis	Bird of paradise	Aps	16	75 S	206
Aquarius	-rii	Water bearer	Aqr	23	15 S	980
Aquila	-lae	Eagle	Aql	20	5 N	652
Ara	-rae	Altar	Ara	17	55 S	237
Aries	-ietis	Ram	Ari	3	20 N	441
Auriga	-gae	Charioteer	Aur	6	40 N	657
Boötes	-tis	Herdsman	Boo	15	30 N	907
Caelum	-aeli	Chisel	Cae	5	40 S	125
Camelopardus	-di	Giraffe	Cam	6	70 N	757
Cancer	-cri	Crab	Cnc	9	20 N	506
Canes Venatici	-num -corum	Hunting dogs	CVn	13	40 N	465
Canis Major	-is -ris	Great dog	CMa	7	20 S	380
Canis Minor	-is -ris	Small dog	CMi	8	5 N	183
Capricornus	-ni	Sea goat	Cap	21	20 S	414
Carina	-nae	Keel	Car	9	60 S	494
Cassiopeia	-peiae	Lady in chair	Cas	1	60 N	598
Centaurus	-ri	Centaur	Cen	13	50 S	1 060
Cepheus	-phei	King	Cep	22	70 N	588
Cetus	-ti	Whale	Cet	2	10 S	1 231
Chamaeleon	-ntis	Chamaeleon	Cha	11	80 S	132
Circinus	-ni	Compasses	Cir	15	60 S	93
Columba	-bae	Dove	Col	6	35 S	270
Coma Berenices	-mae -cis	Berenice's hair	Com	13	20 N	386
Corona Australis	-nae -lis	S crown	CrA	19	40 S	128
Corona Borealis	-nae -lis	N crown	CrB	16	30 N	179
Corvus	-vi	Crow	Crv	12	20 S	184
Crater	-eris	Cup	Crt	11	15 S	282
Crux	-ucis	S cross	Cru	12	60 S	68
Cygnus	-gni	Swan	Cyg	21	40 N	804
Delphinus	-ni	Dolphin	Del	21	10 N	189
Dorado	-dus	Swordfish	Dor	5	65 S	179
Draco	-onis	Dragon	Dra	17	65 N	1 083
Equuleus	-lei	Small horse	Equ	21	10 N	72
Eridanus	-ni	River Eridanus	Eri	3	20 S	1 138
Fornax	-acis	Furnace	For	3	30 S	398
Gemini	-norum	Heavenly twins	Gem	7	20 N	514
Grus	-ruis	Crane	Gru	22	45 S	366
Hercules	-lis	Kneeling giant	Her	17	30 N	1 225
Horologium	-gii	Clock	Hor	3	60 S	249
Hydra	-drae	Water monster	Hya	10	20 S	1 303
Hydrus	-dri	Sea-serpent	Hyi	2	75 S	243
Indus	-di	Indian	Ind	21	55 S	294
Lacerta	-tae	Lizard	Lac	22	45 N	201
Leo	-onis	Lion	Leo	11	15 N	947
Leo Minor	-onis -ris	Small lion	LMi	10	35 N	232
Lepus	-poris	Hare	Lep	6	20 S	290
Libra	-rae	Scales	Lib	15	15 S	538
Lupus	-pi	Wolf	Lup	15	45 S	334
Lynx	-ncis	Lyria	Lyn	8	45 N	545
Lyra	-rae	Lyre	Lyr	19	40 N	286
Mensa	-sae	Table (mountain)	Men	5	80 S	153

Constellation	genitive ending	Meaning	Contractions	α	δ	Area
				h	°	(°)²
Microscopium	-pii	Microscope	Mic	21	35 S	210
Moroceros	-rotis	Unicorn	Mon	7	5 S	482
Musca	-cae	Fly	Mus	12	70 S	138
Norma	-mae	Square	Nor	16	50 S	165
Octans	-ntis	Octant	Oct	22	85 S	291
Ophiuchus	-chi	Serpent bearer	Oph	17	0	948
Orion	-nis	Hunter	Ori	5	5 N	594
Pavo	-vonis	Peacock	Pav	20	65 S	378
Pegasus	-si	Winged horse	Peg	22	20 N	1 121
Perseus	-sei	Champion	Per	3	45 N	615
Phoenix	-nicis	Phoenix	Phe	1	50 S	469
Pictor	-ris	Painter's easel	Pic	6	55 S	247
Pisces	-cium	Fishes	Psc	1	15 N	889
Piscis Austrinus	-is -ni	S fish	PsA	22	30 S	245
Puppis	-ppis	Poop (stern)	Pup	8	40 S	673
Pyxis (= Malus)	-xidis	Compass	Pyx	9	30 S	221
Reticulum	-li	Net	Ret	4	60 S	114
Sagitta	-tae	Arrow	Sge	20	10 N	80
Sagittarius	-rii	Archer	Sgr	19	25 S	867
Scorpius	-pii	Scorpion	Sco	17	40 S	497
Sculptor	-ris	Sculptor	Scl	0	30 S	475
Scutum	-ti	Shield	Sct	19	10 S	109
Serpens (Caput and	-ntis	Serpent. Head	Ser	16	10 N	429
Cauda)		Tail		18	5 S	+208
Sextans	-ntis	Sextant	Sex	10	0	314
Taurus	-ri	Bull	Tau	4	15 N	797
Telescopium	-pii	Telescope	Tel	19	50 S	252
Triangulum	-li	Triangle	Tri	2	30 N	132
Triangulum Australe	-li -lis	S Triangle	TrA	16	65 S	110
Tucana	-nae	Toucan	Tuc	0	65 S	295
Ursa Major	-sea -ris	Great bear	UMa	11	50 N	1 280
Ursa Minor	-sea -ris	Small bear	UMi	15	70 N	256
Vela	-lorum	Sails	Vel	9	50 S	500
Virgo	-ginis	Virgin	Vir	13	0	1 294
Volans	-ntis	Flying fish	Vol	8	70 S	141
Vupecula	-lae	Small fox	Vul	20	25 N	268

THE GREEK ALPHABET

Alpha	Α	α	Iota	Ι	ι	Rho	Ρ	ρ
Beta	Β	β	Kappa	Κ	κ, ϰ	Sigma	Σ	σ, ς
Gamma	Γ	γ	Lambda	Λ	λ	Tau	Τ	τ
Delta	Δ	δ	Mu	Μ	μ	Upsilon	ϒ	υ
Epsilon	Ε	ϵ, ε	Nu	Ν	ν	Phi	Φ	ϕ, φ
Zeta	Ζ	ζ	Xi	Ξ	ξ	Chi	Χ	χ
Eta	Η	η	Omicron	Ο	ο	Psi	Ψ	ψ
Theta	Θ	θ, ϑ	Pi	Π	π, ϖ	Omega	Ω	ω

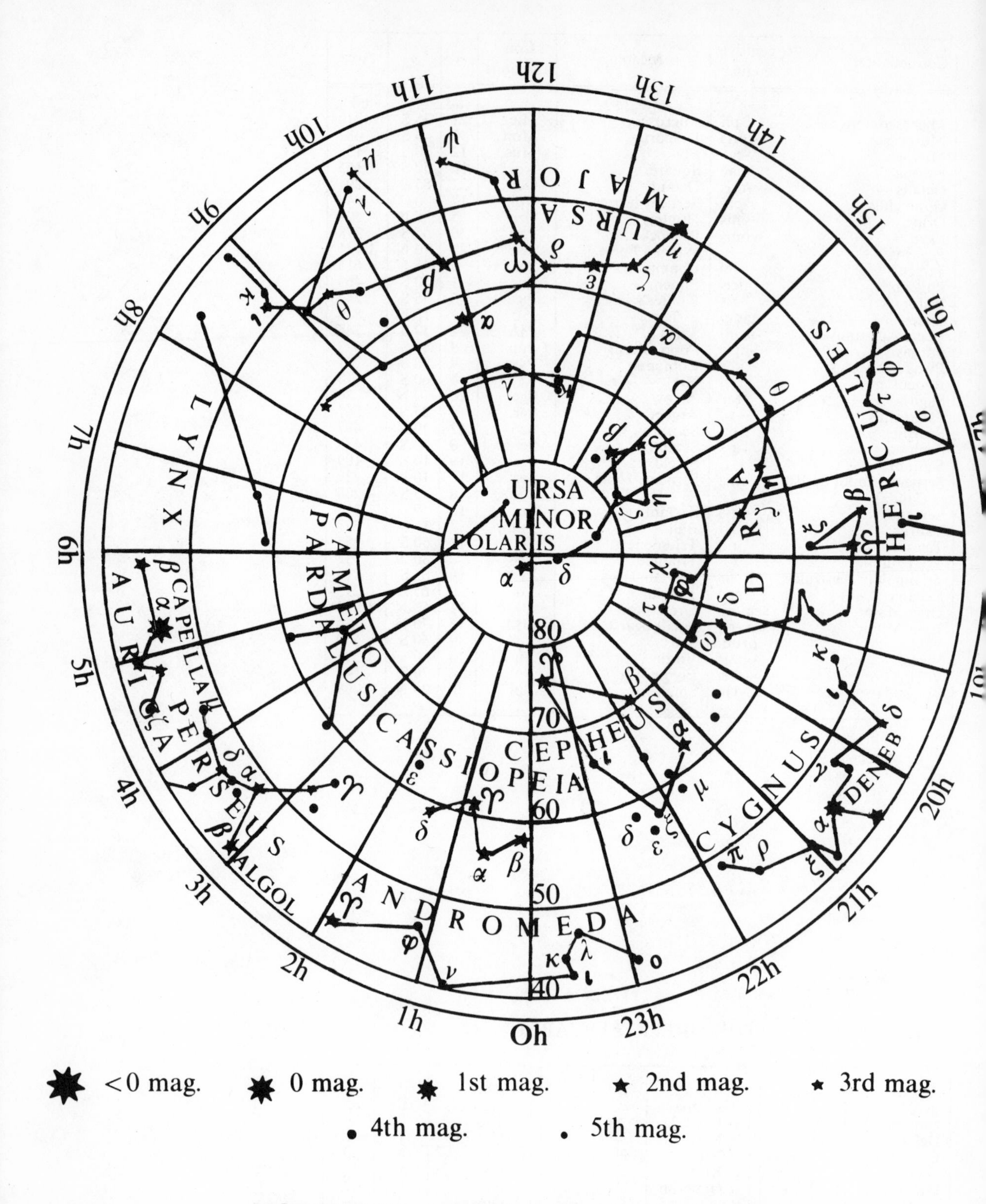

<0 mag. 0 mag. 1st mag. 2nd mag. 3rd mag.

4th mag. 5th mag.

NORTHERN CIRCUMPOLAR STAR CHART

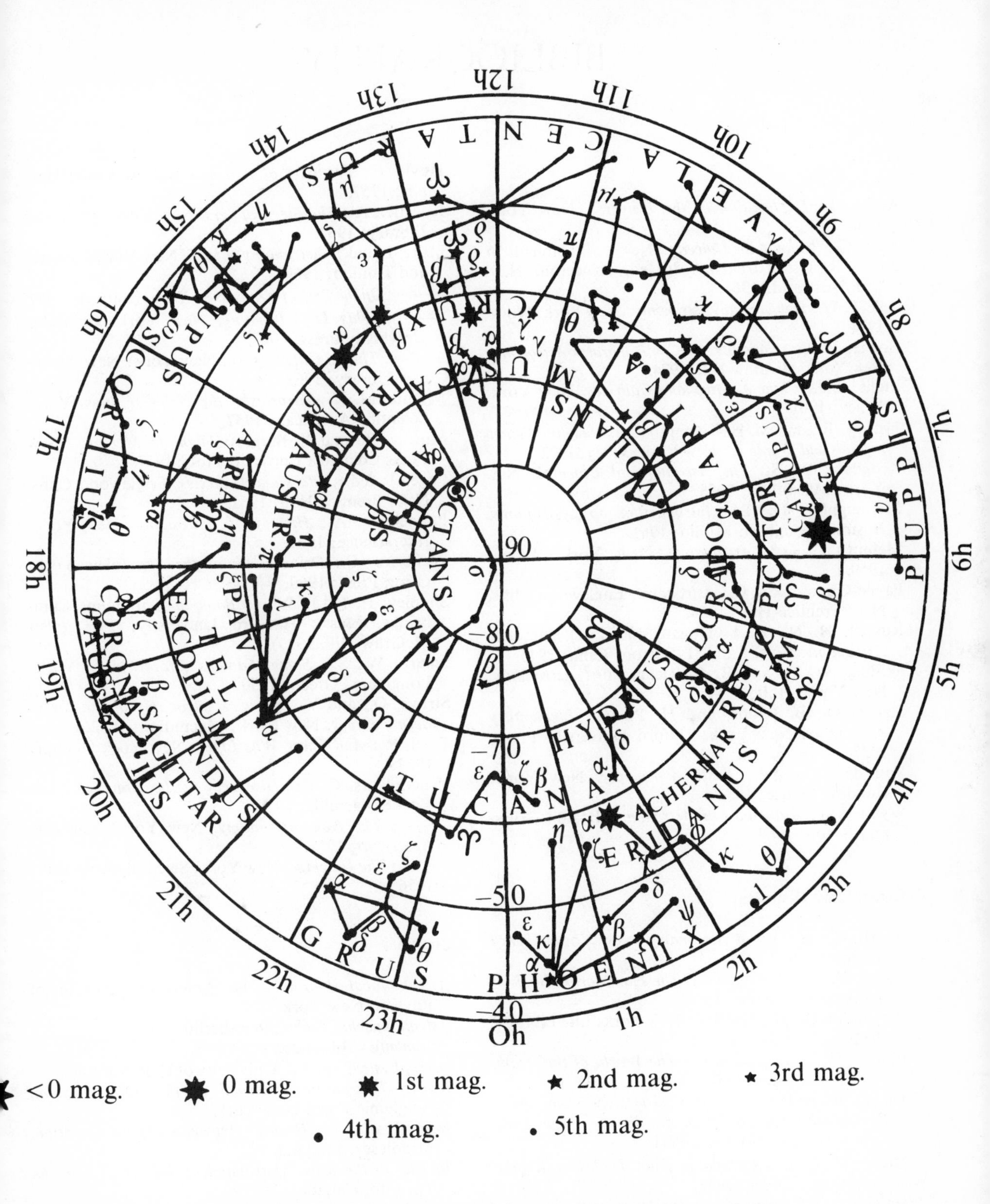

SOUTHERN CIRCUMPOLAR STAR CHART

BIBLIOGRAPHY

Textbooks

Abell, G.: *Exploration of the Universe*. New York: Holt; 1973.

Abell, G.: *Realm of the Universe*. New York: Holt; 1976.

Baker, R. H.: *Astronomy*. 8th ed. Princeton, N.J.: Van Nostrand; 1964.

Cole, F. W.: *Fundamental Astronomy*. New York: Wiley; 1974.

Hodge, P. W.: *Concepts of Contemporary Astronomy*. New York: McGraw-Hill; 1974.

Inglis, S. J.: *Planets, Stars and Galaxies*. New York: Wiley; 1961.

Jastrow, R., and M. H. Thompson: *Astronomy: Fundamentals and Frontiers*. New York: Wiley; 1972.

Krogdahl, W. S.: *The Astronomical Universe*. New York: Macmillan; 1958.

McLaughlin, D. B.: *Introduction to Astronomy*. Boston: Houghton Mifflin; 1961.

Mehlin, T. G.: *Astronomy*. New York: Wiley; 1959.

Payne-Goposchkin, C.: *Astronomy*. Englewood Cliffs, N.J.: Prentice-Hall; 1954.

Russell, H. N., R. S. Dugan, and J. Q. Stewart: *Astronomy*. Vols. I and II. Boston: Ginn; 1945.

Skilling, W. T., and R. S. Richardson: *Astronomy*. New York: Holt; 1954.

Struve, O., B. Lynds, and H. Pillans: *Elementary Astronomy*. New York: Oxford University Press; 1957.

Van de Kamp, P.: *Basic Astronomy*. New York: Random House; 1952.

Wyatt, S. P.: *Principles of Astronomy*. Boston: Allyn and Bacon; 1964.

General and Historical

Abetti, G.: *History of Astronomy*. New York: Henry Schuman; 1952.

Allen, G., and Unwin (eds.): *The Flammarion Book of Astronomy*. London; 1964.

Bergamini, D.: *The Universe*. New York: Life Library; 1962.

Boorse, H. L., and L. Motz: *The World of the Atom*. New York: Basic Books; 1966.

Davies, E., and B. C. Murray: *The View from Space: Photographic Exploration of the Planets*. New York: Columbia University Press, 1971.

Dreyer, J. L. E.: *Astronomy from Thales to Kepler*. New York: Dover; 1953.

Hawkins, G. S.: *Splendor in the Heavens*. New York: Harper; 1961.

Hoyle, F.: *Frontiers of Astronomy*. New York: Harper; 1957.

Koestler, A.: *The Sleepwalkers*. New York: Universal Library; 1963.

Motz, L. (ed.): *Astronomy A to Z*. New York: Grosset and Dunlap; 1964.

———: *On the Path of Venus*, New York, Pantheon, 1977.

———: *This Is Astronomy*. New York: Columbia University Press; 1963.

———: *This Is Outer Space*. New York: New American Library; 1963.

Munitz, M. K.: *Theories of the Universe*. New York: Scientific American; 1957.

Murray, B., and E. Burgess: *Flight to Mercury*. New York: Columbia University Press; 1977.

Page, T. (ed.): *Stars and Galaxies*. Englewood Cliffs, N.J.: Prentice-Hall; 1962.

Pannekoek, A.: *History of Astronomy*. New York: Interscience; 1961.

Pecker, J. C.: *Orion Book of the Sky*. New York: Orion Press; 1960.

Shapley, H. (ed.): *Source Book in Astronomy*. Cambridge, Mass.: Harvard University; 1960 (also McGraw-Hill, 1929).

Smart, W. S.: *Some Famous Stars*. New York: Longmans, Green; 1950.

Struve, O., and V. Zebergs: *Astronomy of the Twentieth Century*. New York: Macmillan; 1962.

Thiel, R.: *And There Was Light*. New York: Knopf; 1957.

Vaucouleurs, G. de: *Discovery of the Universe*. New York: Macmillan; 1957.

———: *The New Astronomy*. New York: Scientific American; 1955.

———: *The Universe*. New York: Scientific American; 1956.

Journals

Astronomical Journal, The American Institute of Physics, New York.

Astronomische Nachrichten, Berlin.

Astronomy. Milwaukee.

Astrophysical Journal, University of Chicago Press.

Icarus, International Journal of the Solar System, Academic Press, New York.

Journal of the British Astronomical Association, Middlesex, England.

Journal of the Royal Astronomical Society of Canada, Toronto, Canada.

Monthly Notices, Royal Astronomical Society, London, Burlington House.

NASA Activities. U.S. Government Printing Office.
The Observatory, Royal Greenwich Observatory, Herstmonceux Castle, Hailsham, Sussex.
Physics Today, The American Institute of Physics, New York.
Planetary and Space Science, Pergamon Press, New York.
Publications of the Astronomical Society of the Pacific, California Academy of Science.
Science, American Association for the Advancement of Science, Washington, D.C.
Sciences. New York Academy of Sciences.
Scientific American, New York.
Sky and Telescope, Cambridge, Mass., Sky Publishing Corp.
Die Sterne, Leipzig.

Yearbooks and Tables

Allen, C. W.: *Astrophysical Quantities*, London: University of London; 1963.
American Ephemeris and Nautical Almanac, Washington, D.C.
Berliner Astronomisches Jahrbuch.
The Nautical Almanac, London.

Aberration (See Spherical Astronomy)

Astrology (See General and Historical)

Astrometry (See Spherical Astronomy)

Astronautics

Baker, R. M., and M. W. Makemson: *Astrodynamics*. New York: Academic; 1960.
Dubridge, L. A.: *Introduction to Space*. New York: Columbia University Press; 1960.
Odishaw, H. (ed.): *Challenge of Space*. Chicago: University of Chicago Press; 1963.

Astrophysics

Aller, L.: *Astrophysics*. Vols. I and II. New York: Ronald; 1954.
Burbidge, G. and Burbidge, E., *Quasi-Stellar Objects*, San Francisco, W.H. Freeman 1967.
Goldberg, L., and L. Aller: *Atoms, Stars and Nebulae*. Philadelphia: Blakiston; 1943.
Handbuch der Physik, and *Encyclopedia of Physics*, S. Flugge (ed.). Vols. XLIX–LI. Berlin: Springer; 1962.
Hynek, J. A. (ed.): *Astrophysics*. New York: McGraw-Hill; 1951.
Menzel, D. H., P. L. Bhatnagar, and H. K. Sen: *Stellar Interiors*. New York: Wiley; 1963.
Motz, L., *The Universe: Its Beginning and Its End*, New York, Scribners, 1975.
Novotny, Eva, *Introduction to Stellar Atmospheres and Interiors*, New York, Oxford University Press, 1973.
Pecker, J. C., and E. Schatzman: *Astrophysique Generale*. Paris: Masson et Cie.; 1959.
Rosseland, S.: *Theoretical Astrophysics*. Oxford: Oxford University Press; 1936.
Siedentopf, H.: *Grundriss der Astrophysik*. Stuttgart; 1950.
Waldmeier, M.: *Einführung in die Astrophysik*. Basel; 1962.

Atomic Theory

D'Abro, A.: *The Rise of Modern Physics*. Vols. I and II. New York: Dover; 1950.

Celestial Mechanics

Brouwer, D., and G. Clemence: *Celestial Mechanics*. New York: Macmillan; 1961.
McCuskey, S. W.: *Introduction to Celestial Mechanics*. Cambridge, Mass.: Addison-Wesley; 1963.
Rand, J.: *Celestial Mechanics*. New York: Macmillan; 1962.
Ryabov, Y.: *Elementary Celestial Mechanics*. New York: Dover; 1961.
Smart, W. S.: *Celestial Mechanics*. New York: Wiley; 1953.
Sterne, T. E.: *An Introduction to Celestial Mechanics*. New York: Interscience Publishers; 1960.
Van de Kamp, P.: *Astromechanics*.

Chromosphere (See Solar Physics; Stellar Atmospheres)

Comets

Baldet, F., and G. De Obaldia: *Catalogue Général des Orbites des Comètes*. Paris; 1952.
———: *Liste Générale des Comètes*. Paris; 1950.
Bobrovnikoff, N. T.: "Comets," Chap. 7 in J. A. Hynek (ed.): *Astrophysics*. New York: McGraw-Hill; 1951.
Kuiper, G. P. (ed.): *The Moon, Meteorites and Comets*. Chicago: University of Chicago Press; 1962.
Olivier, C. P.: *Comets*. Baltimore: Williams and Wilkins; 1930.

Coordinates (See Spherical Astronomy)

Corona (See Solar Physics)

Cosmic Abundance of Chemical Elements (See also Astrophysics; Interstellar Matter; Stellar Atmospheres)

Strominger, D., J. M. Hollander, and G. T. Seaborg: *Reviews of Modern Physics*, XXX (1958), 585.
Suess, H. E., and H. C. Urey: *Reviews of Modern Physics*, XXVIII (1956), 53.

Cosmic Rays (See also Interstellar Matter)

Heisenberg, W.: *Cosmic Rays*. Berlin; 1951.
Morrison, P.: *Handbuch der Physik*. Vol. LI. Springer; 1961.

Cosmology and Cosmogony

Bondi, H.: *Cosmology*. Cambridge: Cambridge University Press; 1960.
———: *Rival Theories of Cosmology*. Oxford: Oxford University Press; 1960.
Gamow, G.: *The Creation of the Universe*. New York: Viking; 1952.
Hoyle, F.: *The Nature of the Universe*. Oxford: Oxford University Press; 1960.
Hubble, E.: *Observational Approach to Cosmology*. Oxford: Oxford University Press; 1937.
Jeans, J. J.: *Astronomy and Cosmology*. New York: Dover; 1961.
McVittie, G. C.: *General Relativity and Cosmology*. Urbana: University of Illinois; 1956.
———: *Facts and Theory in Cosmology*. New York: Macmillan; 1961.
Motz, L., *Astrophysics and Stellar Structure*, New York, Ginn, John Wiley, 1970.
Munitz, M. K.: *Space, Time and Creation*. Glencoe, Ill.: Free Press; 1957.
Öpik, E. J.: *The Oscillating Universe*. New York: New American Library; 1960.
Singh, J.: *Great Ideas and Theories in Modern Cosmology*. New York: Dover; 1961.
Whitrow, G. T.: *Structure and Evolution of the Universe*. London: Hutchinson; 1959.

Distance Determination (See Spherical Astronomy)

Double Stars

Aitken, R. G.: *New General Catalogue of Double Stars*. Washington, D.C.: Carnegie Institute; 1932.
———: *The Binary Stars*. New York: McGraw-Hill; 1935.
Binnendijk, L.: *Properties of Double Stars*. Philadelphia: University of Pennsylvania; 1960.
Kopal, Z.: *An Introduction to Eclipsing Variables*. Cambridge, Mass.: Harvard University; 1946.

Eclipses

Dyson, F., and R. Woolley: *Eclipses of Sun and Moon*. Oxford: Oxford University Press; 1937.
Mitchell, S. A.: *Eclipses of the Sun*. 5th ed. New York: Columbia University Press; 1951.
Oppolzer, Th. von: *Kanon der Finsternisse*. New York: Dover; 1963.

Epicycle Theory (See General and Historical)

Evolution of Stars (See also Astrophysics; Solar Physics)

Chandrasekhar, S.: *Stellar Structure*. New York: Dover; 1957.
Eddington, A. S.: *Internal Constitution of the Stars*. New York: Dover; 1959.
Johnson, Martin: *Astronomy of Stellar Energy and Decay*. London: Faber and Faber; 1949.
McCrae, W. H.: *Physics of Sun and Stars*. London: Hutchinson; 1950.
Schatzman, E.: *White Dwarfs*. New York: Interscience; 1958.
Schwarzschild, M.: *Structure and Evolution of Stars*. Princeton: Princeton University Press; 1958.
Struve, O.: *Stellar Evolution*. Princeton: Princeton University Press; 1958.

Expansion of the Universe (See also Cosmology; Extragalactic Systems)

Couderc, P.: *Expansion of the Universe*. New York: Macmillan; 1952.
Eddington, A. S.: *The Expanding Universe*. Cambridge: Cambridge University Press; 1933.
Humason, M. L., W. U. Mayall, and A. R. Sandage: "Red Shifts and Magnitudes of Extragalactic Nebulae." *Astronomical Journal*, LXI (1956), 97.

Extragalactic Systems

Burbidge, E. M., and G. Burbidge: "Stellar Populations." *Scientific American*, Vol. 199, No. 5 (1958).
Hubble, E. P.: *Realm of the Nebulae*. New York: Dover; 1958.
McVittie, G. C. (ed.): *Problems of Extragalactic Research*. New York: Macmillan; 1961.
Shapley, Harlow: *The Inner Meta-Galaxy*. New Haven, Conn.: Yale University Press; 1957.
———: *Galaxies*. Cambridge, Mass.: Harvard University Press; 1957.

Zwicky, F.: *Morphological Astronomy.* Berlin: Springer; 1957.

Formation of Stars (See also Evolution of Stars)

Goposchkin, C. E.: *Stars in the Making.* Cambridge, Mass.: Harvard University Press; 1952.

Galactic Nebulae (See also Astrophysics; Interstellar Matter)

Allen, L. H.: *Gaseous Nebulae.* New York: Wiley; 1956.
Dufay, J.: *Galactic Nebulae and Interstellar Matter.* New York: Philosophical Library; 1957.

Hertzsprung-Russell Diagram (See Astrophysics)

Interplanetary Matter

Parker, E. N.: *Interplanetary Dynamical Processes.* New York: Interscience; 1963.
Watson, F.: *Between the Planets.* Cambridge, Mass.: Harvard University Press; 1956.

Interstellar Matter (See also Astrophysics; Milky Way System)

Goldberg, L. (ed.): *Annual Review of Astronomy and Astrophysics.* Palo Alto, Cal.: Annual Reviews, Inc., 1963.
Pikelner, S.: *Interstellar Space.* New York: Philosophical Library; 1963.
Woltjer, L. (ed.): *Interstellar Matter.* New York: Benjamin; 1962.

Magnitudes (See Astrophysics; Photometry)

Meteors (See also Interplanetary Matter; Radio Astronomy)

Lovell, A. C. B.: *Meteor Astronomy.* Oxford: Oxford University Press; 1954.
Olivier, C. P.: *Meteors.* Baltimore: Williams and Wilkins; 1925.
Porter, J. C.: *Comets and Meteor Streams.* New York: Wiley; 1952.
Watson, F. G.: *Between the Planets.* Cambridge, Mass.: Harvard University; 1956.

Milky Way System

Bok, B. J., and P. F. Bok: *The Milky Way.* Cambridge, Mass.: Harvard University Press; 1957.
O'Connell, D. (ed.): *Stellar Populations.* New York: Interscience; 1958.

Minor Planets (See Planets and Satellites)

The Moon

Baldwin, R. B.: *The Face of the Moon.* Chicago: University of Chicago Press; 1949.
Kopal, Z.: *The Moon, Our Nearest Celestial Neighbor.* New York: Academic Press; 1960.
Kuiper, G. P. (ed.): *The Moon, Meteorites and Comets.* Chicago: University of Chicago Press; 1962.

The Moon's Orbit (See also Celestial Mechanics)

Brown, E. W.: *Lunar Theory.* New York: Dover; 1960.

Motions of the Stars

Campbell, W. W.: *Stellar Motions.* New Haven, Conn.: Yale University Press; 1913.
Chandrasekhar, S.: *Principles of Stellar Dynamics.* New York: Dover; 1960.
Eddington, A. S.: *Stellar Movements and Structure of the Universe.* New York: Macmillan; 1914.
Phalen, E. von der: *Stellar Statistik.* Leipzig; 1937.
Smart, W. S.: *Stellar Dynamics.* Cambridge: Cambridge University Press; 1938.

Novae and Supernovae (See Astrophysics; Evolution of Stars; Variable Stars)

Orbit Determination

Dubyago, A.: *Determination of Orbits.* New York: Macmillan; 1962.
Herget, P.: *Computation of Orbits.* Ann Arbor: University of Michigan; 1948.

Parallax (See Spherical Astronomy)

Photometry (See also Astrophysics)

Selwyn, E. W. H.: *Photography in Astronomy.* Rochester: Eastman Kodak; 1950.

Planetary Nebulae (See also Astrophysics; Interstellar Matter)

Wurm, K.: *The Planetary Nebulae*. Berlin; 1951.

Planets and Satellites

Alexander, F.: *The Planet Saturn*. New York: Macmillan; 1962.
Kuiper, G. P. (ed.): *The Atmosphere of the Earth and Planets*. Chicago: University of Chicago Press; 1952.
———: *The Earth as a Planet*. Chicago: University of Chicago Press; 1954.
———: *Planets and Satellites*. Chicago: University of Chicago Press; 1962.
Roth, G.: *The System of the Minor Planets*. London: Faber and Faber; 1963.
Sonder, W.: *The Planet Mercury*. London: Faber and Faber; 1963.
Vaucouleurs, G. de: *The Planet Mars*. New York: Macmillan; 1954.
Whipple, F. L.: *Earth, Moon, and Planets*. Cambridge, Mass.: Harvard University Press; 1963.

Positional Astronomy (See Spherical Astronomy)

Precession and Nutation (See Spherical Astronomy)

Radio Astronomy

Brown, R. H., and A. C. B. Lovell: *Exploration of Space by Radio*. New York: Wiley; 1958.
Davies, R. D., and H. P. Palmer: *Radio Studies of the Universe*. Princeton, N.J.: Van Nostrand; 1959
Lovell, A. C. B.: *Exploration of Outer Space*. New York: Harper; 1962.
———, and J. A. Clegg: *Radio Astronomy*. New York: Wiley; 1952.
Menzel, D. H.; *The Radio Noise Spectrum*. Cambridge, Mass.: Harvard University Press; 1960.
Palmer, H. P., R. Davies, and M. Lorge: *Radio Astronomy Today*. Manchester: Manchester University; 1963.
Pawsey, J. L., and R. N. Bracewell: *Radio Astronomy*. Oxford: Clarendon Press; 1955.
Piddington, J. H.: *Radio Astronomy*. New York: Harper; 1961.
Shklovsky, J. S.: *Cosmic Radio Waves*. Cambridge, Mass.: Harvard University Press; 1960.
Smith, F. G.: *Radio Astronomy*. London: Penguin; 1960.
Steinberg, J. L., and J. Lequeux: *Radio Astronomy*. New York: McGraw-Hill; 1963.
Wellman, P.: *Rädioastronomie*. Munich; 1957.

Refraction (See also Spherical Astronomy)

Dietze, G.: *Einführung in die Optik der Atmosphäre*. Leipzig; 1936.

Relativity

Bergmann, P.: *Introduction to the Theory of Relativity*. New York: Prentice-Hall; 1942.
Bohm, D.: *Relativity*. New York: Benjamin; 1965.
Bondi, H.: *Relativity and Common Sense*. New York: Doubleday; 1964.
Born, M.: *Einstein's Theory of Relativity*. New York: Dover; 1953.
D'Abro, A.: *The Evolution of Scientific Thought*. New York: Dover; 1950.
Møller, C.: *The Theory of Relativity*. Oxford: Clarendon Press; 1952.

Solar Activity (See Radio Astronomy; Solar Physics; Stellar Atmospheres)

Solar Physics (See also Astrophysics; Stellar Atmospheres)

Gamow, G.: *The Birth and Death of the Sun*. New York: Viking; 1940.
Kuiper, G. P. (ed.): *The Sun*. Chicago: University of Chicago Press; 1961.
Menzel, D. M.: *Our Sun*. Cambridge, Mass.: Harvard University Press; 1959.
Waldmeier, M.: *Sonnenforschung*. Leipzig; 1955.

Solar System (See also Comets; Meteors; Planets and Satellites)

Kuiper, G. P.: "Formation of the Planets." *Journal of the Royal Society of Canada*, Vol. L (1955).
Russell, H. N.: *The Solar System and Its Origin*. New York: Macmillan; 1935.
Sitter, W. de: *Kosmos*. Cambridge, Mass.: Harvard University Press; 1951.
Urey, H. C.: *The Planets, Their Origin and Development*. New Haven, Conn.: Yale University Press; 1952.

Spectroscopes (See Astrophysics; Telescopes)

Spectrum and Spectral Analysis (See Astrophysics)

Spherical Astronomy

Chauvenet, W.: *Spherical and Practical Astronomy*. New York: Dover; 1960.

Dick, J.: *Grundtalsachen der Sphärischen Astronomie*. Leipzig; 1956.
Prey, A.: *Einführung in die Sphärische Astronomie*. Vienna; 1949.
Schaub, W.: *Sphärische Astronomie*. Leipzig; 1950.
Smart, W. S.: *Spherical Astronomy*. Cambridge: Cambridge University Press; 1944.

Star Catalogues and Star Maps

Locations of Fundamental Stars (Heidelberg; annually).
Northcott, Ruth (ed.): *The Observer's Handbook*. Royal Astronomical Society of Canada, Toronto (annually).
Norton, A. P.: *A Star Atlas and Reference Handbook*. Harvard College Observatory, Cambridge, Mass.: Sky Publishing Co.
Rey, H. A.: *The Stars*. Boston; 1952.

Star Clusters (See Astrophysics; Milky Way System)

Stellar Atmospheres (See also Astrophysics)

Ambartsumian, V. A. (ed.): *Theoretical Astrophysics*. New York: Pergamon; 1958.
Chandrasekhar, S.: *Radiative Transfer*. New York: Dover; 1960.
Greenstein, J. (ed.): *Stellar Atmospheres*. Chicago: University of Chicago Press; 1960.
Unsöld, A.: *Physik der Sternatmosphären*. Berlin; 1955.
Woolley, R., and D. W. N. Stibbs: *The Outer Layers of a Star*. Oxford: Oxford University Press; 1953.

Stellar Statistics

Pahlen, E. von der: *Lehrbuch der Stellarstatistik*. Leipzig, 1937.
Trumpler, R. J., and H. F. Weaver: *Statistical Astronomy*. Berkeley, Cal.: University of California Press; 1953.

Telescopes

Dimitroff, G. F., and J. G. Baker: *Telescopes and Accessories*. Cambridge, Mass.: Harvard University Press; 1954.
Ingalls, A. G.: *Amateur Telescope Making*. New York: Scientific American; 1948.
King, H. C.: *The History of Telescopes*. London: Charles Griffin; 1955.
Kuiper, G. P., and B. M. Middlehurst (eds.): *Telescopes*. Chicago: University of Chicago Press; 1960.
Woodbury, D. O.: *The Glass Giant of Palomar*. New York: Dodd, Mead; 1963.

Three-Body Problem (See Celestial Mechanics)

Time and Time Measurement (See Spherical Astronomy)

Variable Stars (See also Astrophysics)

Campbell, L., and L. Jacchia: *The Story of Variable Stars*. Cambridge, Mass.: Harvard University Press; 1945.
Goposchkin, C. E.: *Galactic Novae*. New York: Interscience; 1957.
Merril, P. W.: *The Nature of Variable Stars*. New York; Macmillan; 1938.
Rosseland, S.: *Pulsation Theory of Variable Stars*. Oxford: Clarendon Press; 1949.

Zodiacal Light (See Interplanetary Matter)

INDEX OF NAMES

INDEX OF SUBJECTS